Handbook of HPLC
Second Edition

CHROMATOGRAPHIC SCIENCE SERIES

A Series of Textbooks and Reference Books

Editor: JACK CAZES

Handbook of HPLC

Second Edition

Edited by
Danilo Corradini
Consiglio Nazionale delle Ricerche
Rome, Italy

Consulting Editor
Terry M. Phillips
National Institutes of Health
Bethesda, Maryland, U.S.A.

CRC Press
Taylor & Francis Group
Boca Raton   London   New York

CRC Press is an imprint of the
Taylor & Francis Group, an **informa** business

CRC Press
Taylor & Francis Group
6000 Broken Sound Parkway NW, Suite 300
Boca Raton, FL 33487-2742

First issued in paperback 2020

© 2011 by Taylor and Francis Group, LLC
CRC Press is an imprint of Taylor & Francis Group, an Informa business

No claim to original U.S. Government works

ISBN-13: 978-0-367-57716-2 (pbk)
ISBN-13: 978-1-57444-554-1 (hbk)

Library of Congress Cataloging-in-Publication Data

Handbook of HPLC / editor, Danilo Corradini. -- 2nd ed.
    p. cm. -- (Chromatographic science series ; 101)
    Includes bibliographical references and index.
    ISBN 978-1-57444-554-1 (alk. paper)
    1. High performance liquid chromatography. I. Corradini, Danilo. II. Title. III. Series.

QD79.C454H36 2010
543'.84--dc22   2010005491

Visit the Taylor & Francis Web site at
http://www.taylorandfrancis.com

and the CRC Press Web site at
http://www.crcpress.com

To Daniela, Andrea, and Giorgio

Contents

PART I Fundamentals

PART II *Applications*

Preface

From the first separation performed by Tswett more than a century ago, a lot of changes have occurred in liquid chromatography, starting with the recognition of the importance of dimension and homogeneity of size and porosity of the chromatographic packing, which transformed classical liquid chromatography into an instrumental technique employing low permeability columns, mechanical pumps to force the mobile phase through them, and sophisticated detection systems. The transition from low to high efficient columns occurred in the early 1970s and the subsequent development of a wide range of stationary phases operating under different separation modes have progressively enlarged the area of application of HPLC. Another milestone has been the hyphenation of HPLC with spectroscopic techniques, which is employed for the identification and the structural elucidation of the separated compounds. Nowadays, HPLC is one of the most widespread analytical separation technique used for both scientific investigations and industrial and biomedical analysis.

The first edition of the *Handbook of HPLC* published in 1998 encompassed fundamental and practical aspects of HPLC organized in four distinct parts, which comprised the theoretical treatment of the chromatographic process, the description of the main separation modes of HPLC, an illustration of the instrumentation, and applications of HPLC in different areas of industry and applied research. This edition covers aspects of HPLC that have contributed to the further advancements of this separation technique in the last 12 years, avoiding theoretical and practical aspects that, due to the popularity of chromatography, are already part of the current background of practitioners at any level. Nevertheless, most of the latest innovative aspects of the majority of the subject matter covered by a specific chapter in the first edition are in any case discussed in this book. Hence, for example, although reversed-phase HPLC is not the subject of a specific chapter in this book, as it was in the first edition due to the large number of excellent publications on this topic, Chapters 2, 5, 9, 11, and 15 discuss several important aspects of RP-HPLC from different points of view. Similarly, chapters devoted to size-exclusion and normal-phase separation modes have not been proposed again in this edition, yet theoretical and practical aspects of these separation modes have been discussed in Chapter 16. Therefore, this edition encompasses aspects of HPLC that were not covered in the first edition and discusses them from a different perspective.

Emerging novel aspects of HPLC included in this edition comprise monolithic columns (Chapter 1), bonded stationary phases (Chapter 2), micro-HPLC (Chapter 3), two-dimensional comprehensive liquid chromatography (Chapter 4), gradient elution mode (Chapter 5), and capillary electromigration techniques (Chapter 6), which, in the first edition, were restricted to the description of a limited number of separation modes. Also not included in the previous edition are the chapters related to gradient elution mode (Chapter 5), LC–MS interfaces (Chapter 8), nonlinear chromatography (Chapter 10), displacement chromatography of peptides and proteins (Chapter 11), field-flow fractionation (Chapter 12), and retention models for ions (Chapter 15), which discuss the separation of either small or large ionic molecules. Part I (Fundamentals) includes chapters devoted to control and temperature effects (Chapter 9), affinity HPLC (Chapter 13), and ion chromatography (Chapter 14). Part II (Applications) is focused on four of the most significant areas in which HPLC is successfully employed, that is, chiral pharmaceutical (Chapter 17), environmental analysis (Chapter 18), food analysis (Chapter 19), and forensic sciences (Chapter 20). All chapters include extensive reference lists in addition to explanatory figures and summarizing tables.

We have written this edition with the purpose of reporting updated and detailed information on HPLC related to both conventional formats and more sophisticated novel approaches, which have been developed to satisfy the emerging needs in analytical separation science. Nowadays, analysts and scientists are dealing more often with samples of very limited amount and extremely complex composition, requiring the miniaturization of the analytical separation system or the enrichment of the trace components, or both. Also increasing is the need for molecular identification and structural elucidation of the separated compounds, necessitating the hyphenation of HPLC, or of a related separation method, with a suitable spectroscopic analytical system. The two examples mentioned above are explicative of the reasons that oriented us to write this book with particular attention to emerging novel aspects of HPLC and to expand the information on the recent development of the other two related separation methods based on the differential migration velocity of analytes in a liquid medium under the action of either an electric field (capillary electromigration techniques) or a gravitational field (field-flow fractionation).

We believe that this edition is suitable as a textbook for undergraduate college students having a general background in chromatography and for new practitioners interested in improving their knowledge on the current status and future trends of HPLC. Moreover, the book could be used as a valuable source of information for graduate students and scientists looking for solutions to complex separation problems, and for analysts and scientists currently using HPLC as either an analytical or a preparative scale tool.

Danilo Corradini
Terry M. Phillips

Editor

Danilo Corradini is research director at the Institute of Chemical Methodologies of the Italian National Research Council (CNR) and a member of the General Scientific Advisory Board of CNR. His involvement in separation science started in 1976 with his research work on chromatography and electrophoresis for his PhD studies in chemistry, which was carried out at Sapienza University of Rome, Italy, under the direction of Michael Lederer, founder and first editor of the *Journal of Chromatography*. In 1983–1984, he worked with Csaba Horváth, the pioneer of HPLC, at the Department of Chemical Engineering at Yale University, New Haven, Connecticut, where he initiated his first investigations on the HPLC of proteins and peptides, which he continued at the Institute of Chromatography of CNR after he returned to Italy.

Currently, Dr. Corradini is the head of the chromatography and capillary electrophoresis research unit of the Institute of Chemical Methodologies in Montelibretti, Rome. His research interests are focused on the theoretical and practical aspects of HPLC and capillary electromigration techniques for analytical scale separations of biopolymers, low molecular mass metabolites, and phytochemicals. His articles have been extensively published in international scientific journals; he has been chairman and invited speaker at national and international congresses and meetings, and serves on the editorial boards of several journals. He is the chairman of the Interdivisional Group of Separation Science of the Italian Chemical Society and a member of the International Advisory Board of the Mediterranean Separation Science Foundation Research and Training Center in Messina, Italy. In 2009, the Hungarian Separation Science Society assigned him the Csaba Horváth Memorial Award in recognition of his significant contribution to the development, understanding, and propagation of capillary electrophoresis throughout the world and cooperation in the development of separation science in Hungary.

Contributors

Dušan Berek
Laboratory of Liquid Chromatography
Polymer Institute
Slovak Academy of Sciences
Bratislava, Slovakia

Clemens P. Bisjak
Institute of Analytical Chemistry and
    Radiochemistry
University of Innsbruck
Innsbruck, Austria

Günther K. Bonn
Institute of Analytical Chemistry and
    Radiochemistry
University of Innsbruck
Innsbruck, Austria

Maria Concetta Bruzzoniti
Department of Analytical Chemistry
University of Torino
Torino, Italy

Francesco Cacciola
University of Messina
Messina, Italy

Achille Cappiello
Institute of Chemical Sciences
University of Urbino
Urbino, Italy

Alberto Cavazzini
Department of Chemistry
University of Ferrara
Ferrara, Italy

Catia Contado
Department of Chemistry
University of Ferrara
Ferrara, Italy

Lanfranco S. Conte
Department of Food Science
University of Udine
Udine, Italy

Danilo Corradini
Institute of Chemical Methodologies
National Research Council
Montelibretti, Rome, Italy

Francesco Dondi
Department of Chemistry
University of Ferrara
Ferrara, Italy

Giovanni Dugo
Dipartimento Farmaco-Chimico
University of Messina
Messina, Italy

Paola Dugo
Dipartimento Farmaco-Chimico
University of Messina
Messina, Italy

Heinz Engelhardt
Institute of Instrumental Analysis
University of Saarland
Saarbrücken, Germany

Giorgio Famiglini
Institute of Chemical Sciences
University of Urbino
Urbino, Italy

Attila Felinger
Department of Analytical and Environmental
    Chemistry
University of Pécs
Pécs, Hungary

Maria Carla Gennaro
Department of Environmental and Life
    Sciences
University of Piemonte Orientale Amedeo
    Avogadro
Alessandria, Italy

Valentina Gianotti
Department of Environmental and Life
    Sciences
University of Piemonte Orientale Amedeo
    Avogadro
Alessandria, Italy

Fabio Gosetti
Department of Environmental and Life
    Sciences
University of Piemonte Orientale Amedeo
    Avogadro
Alessandria, Italy

Andreas Greiderer
Institute of Analytical Chemistry and
    Radiochemistry
University of Innsbruck
Innsbruck, Austria

David S. Hage
Department of Chemistry
University of Nebraska–Lincoln
Lincoln, Nebraska

Ylva Hedeland
Department of Medicinal Chemistry
Uppsala University
Uppsala, Sweden

Nico Heigl
Institute of Analytical Chemistry and
    Radiochemistry
University of Innsbruck
Innsbruck, Austria

David E. Henderson
Department of Chemistry
Trinity College
Hartford, Connecticut

Christian W. Huck
Institute of Analytical Chemistry and
    Radiochemistry
University of Innsbruck
Innsbruck, Austria

Pavel Jandera
Department of Analytical Chemistry
Faculty of Chemical Technology
University of Pardubice
Pardubice, Czech Republic

Heather Kalish
Ultramicro Immunodiagnostics Section
Laboratory of Bioengineering and Physical
    Sciences
National Institute of Biomedical Imaging and
    Bioengineering
National Institutes of Health
Bethesda, Maryland

Tiina Kumm
Dipartimento Farmaco-Chimico
University of Messina
Messina, Italy

Luigi Mondello
Dipartimento Farmaco-Chimico
University of Messina
Messina, Italy

Sabrina Moret
Department of Food Science
University of Udine
Udine, Italy

Nicole Y. Morgan
Laboratory of Bioengineering and Physical
    Science
National Institute of Biomedical Imaging and
    Bioengineering
National Institutes of Health
Bethesda, Maryland

Pierangela Palma
Institute of Chemical Sciences
University of Urbino
Urbino, Italy

Luisa Pasti
Department of Chemistry
University of Ferrara
Ferrara, Italy

Curt Pettersson
Department of Medicinal Chemistry
Uppsala University
Uppsala, Sweden

Terry M. Phillips
Ultramicro Analytical Immunochemistry
  Resource
Division of Bioengineering and Physical
  Science
National Institutes of Health
Bethesda, Maryland

Stefano Polati
Department of Environmental and Life
  Sciences
University of Piemonte Orientale Amedeo
  Avogadro
Alessandria, Italy

Aldo Polettini
Unit of Legal Medicine
Department of Public Health & Community
  Medicine
University of Verona
Verona, Italy

Giorgia Purcaro
Department of Food Science
University of Udine
Udine, Italy

Corrado Sarzanini
Department of Analytical Chemistry
University of Torino
Torino, Italy

Paul D. Smith
Laboratory of Bioengineering and Physical
  Science
National Institute of Biomedical Imaging and
  Bioengineering
National Institutes of Health
Bethesda, Maryland

Jan Ståhlberg
Academy of Chromatography
Mariefred, Sweden

Lukas Trojer
Institute of Analytical Chemistry and
  Radiochemistry
University of Innsbruck
Innsbruck, Austria

Wolfgang Wieder
Institute of Analytical Chemistry and
  Radiochemistry
University of Innsbruck
Innsbruck, Austria

James A. Wilkins
Sensorin, Inc.
Burlingame, California

Part I

Fundamentals

1 Monolithic Stationary Phases in HPLC

Lukas Trojer, Andreas Greiderer, Clemens P. Bisjak, Wolfgang Wieder, Nico Heigl, Christian W. Huck, and Günther K. Bonn

CONTENTS

1.1   INTRODUCTION

1.1.1   BACKGROUND AND DEFINITION

Cross-linked polymer supports have been introduced in the 1940s as a result of rapidly growing research interest on solid phase synthesis, catalysis, and combinatorial chemistry. The polymerization of styrene in the presence of small amounts of divinylbenzene (DVB) as cross-linker resulted in polymer beads that showed distinctive swelling in good solvents without being totally dissolved [1]. However, the degree of swelling can easily be adjusted by varying the amount of DVB present during polymerization.

Based upon those observations, a series of resins were prepared by free radical cross-linking polymerization for various applications by using the suspension polymerization approach [2,3]. Even if these conventional (also referred to as *gel-type resins* or *homogeneous gels* [4]) polymers have been proven to be useful for many ion-exchange (IEX) applications, they are subject to a number of severe limitations.

The *molecular porosity* (porosity in the swollen state) of gel-like polymers, for example, is indirectly proportional to the amount of cross-linker. However, since the content of cross-linker is directly proportional to the chemical stability and degradation, the porosity has generally to be kept low. Moreover, gel-like resins do only exhibit a negligible molecular porosity in poor solvents [e.g., water or alcohols in the case of poly(styrene-*co*-divinylbenzene) (PS/DVB) supports], which severely restricts the use of such materials, predominately in the field of solid phase synthesis and combinatorial chemistry.

These problems were successfully solved in the late 1950s by the introduction of new polymerization techniques, enabling the synthesis of macroporous (macroreticular) cross-linked polymer resins. Their characteristics refer to the maintenance of their porous structure in the dry state and, therefore, in the presence of poor polymer solvents [5–9]. The synthesis of rigid macroporous polymer support is based on suspension polymerization in the presence of inert solvents (referred to as *diluents* or *porogens*), which are soluble in the monomer mixture, but possess poor ability to dissolve the evolving copolymer particles. The inert diluents thus act as pore-forming agents during the polymerization procedure, leaving a porous structure with sufficiently high mechanical stability after removal from the polymer network.

The development of those macroreticular polymer resins provided the basis for the area of rigid monolith research, since the manufacturing methods of monoliths and macroporous beads are essentially identical with regard to polymerization mixture components and composition. The only significant difference in their fabrication refers to the polymerization conditions. While beads and resins are prepared by suspension or precipitation polymerization in a vigorously stirred vessel, rigid monolithic structures evolve by bulk polymerization within an unstirred mold [10].

The structure of a typical rigid monolithic polymer is illustrated in Figure 1.1 (longitudinal section of a monolithic column). A monolithic stationary phase is defined as a *single piece of porous polymer* located inside the confines of a column, whereas the polymer network is crisscrossed by flow channels (also referred to as *gigapores* or *through-pores*), which enable a solvent flow through

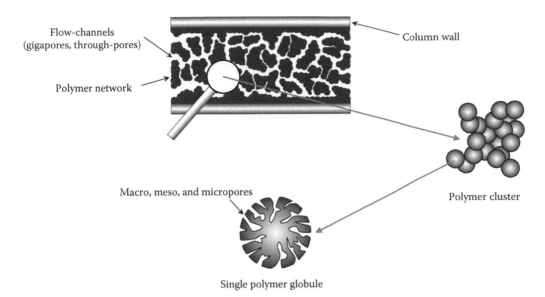

FIGURE 1.1 Schematic representation of the structure and the morphology of a typical monolithic polymer prepared in an HPLC column housing as an unstirred mold.

the entire monolith. The polymer network itself is structured by microglobules chemically linked to each other to yield globule clusters (Figure 1.1). The polymer globules, which present the smallest structural unit, are porous themselves, whereas the entire porosity can effectively be controlled by the polymerization mixture composition as well as by the polymerization conditions.

1.1.2 HISTORICAL ABSTRACT

In the 1950s, Robert Synge was the first to postulate polymer structures, which were similar to what is defined as monolith today [11,12]. However, the soft polymer materials available at that time (gel-type polymers) did not resist permanent pressure conditions.

This was confirmed in 1967 by Kubin et al., who were the first to accomplish polymerization directly in a glass column [13]. The resulting swollen poly(2-hydroxyethyl methacrylate-*co*-ethylene dimethacrylate) gel, which was prepared in the presence of 1% cross-linker, was strongly compressed after pressure application and consequently exhibited exceedingly low permeability (4.5 mL/h for a 25 × 220 mm column) and poor efficiency.

In the 1970s, several research groups came up with foam-filled columns for GC and HPLC [14–17]. These open pore polyurethane foam stationary phases, which were prepared via in situ polymerization, were shown to possess comparatively good column performance and separation efficiency. They could, however, not achieve general acceptance and broader application due to insufficient mechanical stability and strong swelling behavior.

Based upon the initial observations of Kubin et al., Hjérten introduced the concept of compressed gels in the late 1980s, which have also been referred to as continuous beds [18,19]. The copolymerization of acrylic acid and N,N'-methylene bisacrylamide resulted in highly swollen gels, which were deliberately compressed to a certain degree using a movable piston. Despite the high degree of compression, the gels exhibited good permeability and enabled efficient separation of proteins.

In the mid-1980s, Belenkii et al. comprehensively studied the separation of proteins on particle-packed columns of variational length and dimension under gradient conditions [20]. They concluded that a short distance of stationary phase is sufficient to enable protein separation with an acceptable resolution. With respect to that, Tennikova et al. came up with the concept of short monolithic separation beds, realized by copolymerization of glycidyl methacrylate as monomer and high amounts of ethylene dimethacrylate as cross-linker in the presence of different porogenic solvents in flat or tubular molds with diameters in the centimeter range [21–23]. The resulting highly cross-linked (and thus rigid) macroporous copolymer membranes or rods were then arranged in a pile or sliced into disks, respectively. Svec et al. extended the process of preparing highly cross-linked, rigid polymer disks or sheets in the presence of inert diluents to conventional HPLC column housings (8 mm I.D.), directly employed as polymerization molds [10,24]. The simple fabrication process, together with promising separation efficiency toward biomolecules, blazed for monoliths the trail to broader scientific interest.

Since 1992, a vast variety of rigid organic monolithic stationary phases with different chemistry, functionality, and column geometry has been reported for HPLC as well as CEC applications, as summarized by the number of excellent reviews [25–32,213]. The development and enhancement of monolithic stationary phases is still a rapidly growing area of research with scientific and industrial interest.

Almost at the same time (1996), Tanaka and co-workers [33] and Fields [34] expanded the research field of continuous polymer support by the introduction of (derivatized) inorganic silica rods, which were prepared by sol–gel process of silane precursors [tetramethoxysilane (TMOS) or dimethyloctadecylchlorosilane (ODS)] in the presence of poly(ethylene oxide) (PEO) as porogen.

A number of alternative names were introduced in literature to term the new types of stationary phases. Hjérten et al. used continuous polymer beds [18] to define the class of compressed polyacrylamide gels. Later on, the denotation stationary phases with reduced discontinuity rose for a single piece polymer support in order to express the diminishment of interparticulate voids, compared with particle-packed columns. The group of rigid macroporous polymers with cylindrical shape, being initiated by Svec in the early 1990s, have been referred to as *continuous polymer rods*, whereas this expression has been adapted to inorganic monoliths (porous silica rods) [33]. The term *monolith*, which probably is the most common expression for the new class of macroporous polymers, was first introduced to describe a single piece of derivatized cellulose sponge for the fractionation of proteins [35], but rapidly found general acceptance.

1.2 MONOLITHIC MATERIALS

As described in Section 1.1.2, modern monolith research was initiated by Tennikova and Svec, who prepared rigid, mechanically stable copolymers by a simple molding process in the presence of porogens, employing a high amount of cross-linking agent [10,21–23]. As almost all monolithic stationary phases are nowadays fabricated according to this basic concept, other (historical) approaches are not further considered.

The huge variety of different monolithic supports being introduced for HPLC applications can generally be divided into two main classes: monoliths based on organic precursors and monoliths based on inorganic precursors.

1.2.1 ORGANIC MONOLITHS

Organic monoliths are based on copolymerization of a monofunctional and a bifunctional (uncommonly trifunctional) organic precursor in the presence of a suitable initiator and a porogenic solvent. During the last 15 years, a vast number of different monomers and cross-linkers have been introduced and copolymerized using different polymerization techniques and initiators. A general survey of the tremendous amount of scientific contributions can be gained from numerous reviews [25–32].

Free radical copolymerization of a monovinyl compound and a divinyl cross-linker is by far the most commonly employed mode of polymerization for the preparation of organic monoliths.

- Styrene monoliths—thermal initiation
- (Meth)acrylate monoliths—thermal, photochemical, or chemical initiation
- Acrylamide monoliths—thermal, photochemical, or chemical initiation
- Norbornene monoliths—chemical initiation

In addition to thermally, photochemically, or chemically initiated free radical copolymerization of styrene, (meth)acrylate, or (meth)acrylamide building blocks, other polymerization techniques have been reported for the development of organic monolithic HPLC stationary phases. Ring-opening metathesis polymerization (ROMP) of norbon-2-ene and 1,4,4a,5,8,8a-hexahydro-1,4,5,8-*exo,endo*-dimethanonaphthalene (DMN-H6) initiated by a Grubbs-type catalyst resulted in hydrophobic monolithic rods, exhibiting chromatographic properties comparable to PS/DVB-based stationary phases [36–39]. Heat-induced polycondensation reactions of diamines (4-[(4-aminocyclohexyl) methyl]cyclohexlamine or *trans*-1,2-cyclohexanediamine) with an epoxy monomer (tris(2,3-epoxypropyl)isocyanurate) have recently proven to yield mechanically stable monolithic stationary phases applicable to HPLC analysis [40].

Table 1.1 gives a comprehensive, albeit fragmentary, summary of investigated organic monolithic polymer systems (based on all different kinds of styrene, acrylate, methacrylate, (meth)acrylamide building blocks, as well as mixtures thereof) together with their preparation conditions and utilization as stationary phase.

Despite the variety of polymerization principles and monomer systems available for the preparation of organic monolithic supports, the resulting polymer structure and morphology is surprisingly similar. Figure 1.2 illustrates the properties of a typical organic monolith in terms of surface morphology and porosity. Independent on the type of initiation of the free radical polymerization, the morphology of the resulting organic monolith is brush-like (Figure 1.2a) and reminds of the surface of cauliflower, being pervaded by cylindrical channels (flow channels). The porosity of a typical organic monolith is shown in Figure 1.2b, but it can be deduced that it is defined by a monomodal distribution of macropores.

1.2.1.1 Styrene-Based Monoliths

Most of the research on styrene-based column supports is focused on PS/DVB monoliths. The hydrophobic character of the monomers (styrene, divinylbenzene) results in a material, which can directly be applied to reversed-phase chromatography without further derivatization. After the introduction of rigid macroporous PS/DVB rods and their promising application to the separation of proteins in 1993 [41], the material has been employed for rapid separations of poly(styrene) standards [42] as well as for the separation of small molecules like alkylbenzenes, employing rods possessing a length of 1 m [43]. PS/DVB monoliths in capillary format—which have been commercialized by LC-Packings, A Dionex Company—have been demonstrated several times to exhibit high potential regarding efficient separation of proteins and peptides [44–46] as well as *ss*- and *ds*DNA [47–51].

Beside borosilicate and fused silica capillaries, PS/DVB monoliths have been fabricated within the confines of steel and PEEK tubings [52]. In order to increase the hydrophobic character of the supports, a Friedel-Crafts alkylation reaction was used for the attachment of C_{18}-moieties to the polymer surface. The derivatized material was demonstrated to be more retentive and to provide more efficient peptide separations compared with the original, nonderivatized monolith.

1.2.1.2 Acrylate- and Methacrylate-Based Monoliths

The main feature of (meth)acrylate-based support materials is the broad diversity of monomers that is commercially available and that can thus can be used for the fabrication of monoliths. The resulting (meth)acrylate monoliths consequently cover a wide spectrum of surface chemistries and properties. The scope of monomers includes hydrophobic, hydrophilic, ionizable, chiral, as well as reactive (meth)acrylate building blocks [53]—the most popular being mixtures of butyl methacrylate and ethylene dimethacrylate (BMA/EDMA) or glycidyl methacrylate and ethylene dimethacrylate (GMA/EDMA) as cross-linker.

The former polymer system represents a reversed-phase material, providing C_4-alkylchains, which has most frequently been employed for protein separation [53], whereas the latter carries reactive moieties that can easily be converted in order to yield the desired surface functionalities.

TABLE 1.1
Summary of Organic Monolithic Polymer Systems That Have Been Introduced in Literature Listed Together with Their Mode of Polymerization, Porogenic Solvent, and Their Key Application in HPLC and CEC Separation

Monomers	Initiator	Porogens	Application	References
		Styrene Supports		
S/DVB	Thermal, AIBN	1-Dodecanol	8 mm I.D., protein separation, separation of synthetic polymers	[24,134]
S/DVB	Thermal, AIBN	1-Dodecanol/toluene	8 mm I.D., protein peptides and small molecules	[135]
S/DVB	Thermal, AIBN	MeOH, EtOH, propanol/toluene, formamide	75 μm I.D., application to HPLC and CEC	[136]
S/DVB	Thermal, AIBN	1-Decanol/THF	200 μm I.D. capillary columns, IP-RP-HPLC of nucleic acids and RP-HPLC of proteins and peptides	[49,137]
4-Vinylbenzyl chloride/DVB	Thermal, AIBN	1-Decanol/toluene	8 mm I.D., derivatization and hydrophilization	[138]
MS/BVPE	Thermal, AIBN	THF, CH$_2$C$_{12}$, toluene/1-decanol	3 mm, 1 mm, and 200 μm I.D., IP-RP-HPLC of nucleic acids and RP-HPLC of proteins, peptides and small molecules	[139,140]
		Methacrylate Supports		
GMA/EDMA	Thermal, AIBN	Cyclohexanol/1-decanol	5 and 8 mm I.D., IEC of proteins, MIC of proteins by IMAC	[141–144]
GMA/EDMA	Thermal, AIBN	1-Propanol, 1,4-butanediol/water	250 μm I.D., IEC of metal cations	[145]
BMA, AMPS/EDMA	Thermal, AIBN	1-Propanol, 1,4-butanediol/water	100 and 150 μm I.D., CEC of small molecules (alkylbenzenes) and styrene oligomeres	[146–148]
BMA/EDMA	Photochemical, DAP	Cyclohexanol/1-dodecanol	200 μm I.D., HPLC separation of proteins	[149,150]
BMA/EDMA	Photochemical, AIBN	MeOH or mixtures MeOH with EtOH, THF, ACN, CHCl$_3$, or hexane	Capillary format and microchips, CEC application	[151]
BMA/EDMA	Thermal, AIBN	1-Propanol, 1,4-butanediol/water	100 μm I.D., HPLC separation of proteins, separation of small molecules	[150,152,153]
BMA/EDMA	Chemical, APS/TEMED	1-Propanol, 1,4-butanediol/water	320 μm I.D., HPLC of small molecules	[154]

TABLE 1.1 (continued)
Summary of Organic Monolithic Polymer Systems That Have Been Introduced in Literature Listed Together with Their Mode of Polymerization, Porogenic Solvent, and Their Key Application in HPLC and CEC Separation

Monomers	Initiator	Porogens	Application	References
HEMA/EDMA	Photochemical, AIBN or DAP	1-Dodecanol/ cyclohexanol or MeOH/hexane	Microfluidic devices, study of porosity, hydrodynamic properties and on-chip SPE	[155,156]
HEMA, MAH/ EDMA	Thermal, benzoyl peroxide	Toluene	Hydrophilic supports of AC	[157]
BMA, HEMA/ BDDMA GDMA	Thermal, AIBN	1-Propanol, 1,4-butanediol, or cyclohexanol/1-dodecanol	250 μm I.D., HIC of proteins	[158]
VAL, HEMA or acrylamide/EDMA	Photochemical, DAP	1-Decanol/cyclohexanol	100 μm I.D. capillaries of 50 μm I.D. Chips, online bioreactors	[159]
Acrylamide, VAL/EDMA	Thermal, AIBN	Tetradecanol or dodecanol/oleyl alcohol	1 mm I.D., high throughput reactors	[160]
GMA/TRIM	Photochemical, benzoin methyl ether	Isooctane/toluene	Study of porous properties	[107]
SPE/EDMA or TEGDMA	Photochemical, benzoin methyl ether	MeOH	2.6–2.7 mm I.D., IEC of proteins	[161,162]
SPE/EDMA	Thermal, AIBN	MeOH	100 μm I.D., HILIC of polar compounds	[163]
Acrylate Supports				
PEGMEA/PEGDA	Photochemical, DAP	Et$_2$O/MeOH or cyclohexanol/ dodecanol/hexane	75 μm I.D., HPLC of proteins and peptides	[164]
Butyl acrylate/ BDDA	Thermal, AIBN	EtOH/CH$_2$Cl$_2$/phosphate buffer, pH 6.8	75 and 100 μm I.D., CEC of small molecules	[165,166]
Butyl acrylate mixed with t-butyl or lauryl acrylate/ BDAA	Thermal, AIBN	EtOH/ACN/phosphate buffer, pH 6.8	75 and 100 μm I.D., CEC of small molecules	[166]
AMPS/PEDAS	Thermal, AIBN	Cyclohexanol/ethylene glycol/water	100 μm I.D., CEC of small molecules and amino acids	[167]
Ethyl, butyl, hexyl, lauryl acrylate/ BDDA	Photochemical, AIBN	EtOH/ACN/5 mM phosphate buffer, pH 6.8	100 μm I.D., CEC of small molecules, amino acids and peptides	[168,169]
HMAM, hexyl acrylate PDA	Chemical, APS/TEMED	Aqueous buffer	50 μm I.D., HPLC and CEC of small molecules	[170]
PA/1,2-phenylene diacrylate	Thermal, AIBN	2-Propanol/THF, CH$_2$Cl$_2$ or toluene	200 μm I.D., IP-RP-HPLC of nucleic acids and RP-HPLC of proteins	[171,172]

(continued)

TABLE 1.1 (continued)
Summary of Organic Monolithic Polymer Systems That Have Been Introduced in Literature Listed Together with Their Mode of Polymerization, Porogenic Solvent, and Their Key Application in HPLC and CEC Separation

Monomers	Initiator	Porogens	Application	References
(Meth)Acrylamide Supports				
Acrylamide/MBAA	Thermal, AIBN	DMSO/(C_1–C_{12})-alcohols	Hydrophilic supports of HPLC application	[106]
Acrylamide, butyl acrylamide/MBAA	Thermal, benzoyl peroxide	DMSO/(C_{12}–C_{18})-alcohols	8 mm I.D., HIC of proteins	[173]
MA, VSA, DMAA/PDA	Chemical, APS/TEMED	50 mM phosphate buffer, pH 7	75 µm I.D., CEC of small molecules	[174,175]
Acrylamide/MBAA acrylamide/AGE	Chemical, APS/TEMED	Water	10 mm I.D., supermacroporous monoliths for chromatography of bioparticles	[176–178]
IPA, MA/PDA	Chemical, APS/TEMED	50 mM (NH_4)$_2SO_4$	100 µm I.D., NP-HPLC of small molecules	[179]
Other Supports				
NBE/DMN-H6	Chemical, Grubbs type initiator	2-Propanol/toluene	5 mm, 3 mm, and 200 µm I.D., biopolymer chromatography and SEC of synthetic polymers	[39–42]
BACM, CHD/TEPIC	Thermal, –	Poly(ethylene) glycol 200 and 300	100 µm I.D., HPLC of small molecules, chiral separations	[40]
MS/BVBDMS	Thermal, AIBN	2-Propanol/THF, CH_2Cl_2 or toluene	200 µm I.D., IP-RP-HPLC of nucleic acids and RP-HPLC of peptides and proteins	[183]

S, styrene; DVB, divinylbenzene; AIBN, α,α′-azoisobutyronitrile; MS, methylstyrene; BVPE, 1,2-bis(p-vinylbenzyl)ethane; GMA, glycidyl methacrylate; PEGMEA, poly(ethylene glycol) methyl ether acrylate; PEGDA, poly(ethylene glycol) diacrylate; EDMA, ethylene dimethacrylate; BMA, butyl methacrylate; AMPS, 2-acrylamido-2-methylpropanesulfonic acid; DAP, 2,2-dimethoxy-2-phenylacetophenone; APS, ammonium peroxodisulfate; TEMED, N,N,N′,N′-tetramethylethylenediamine; HEMA, 2-hydroxyethyl methacrylate; MAH, N-methacryloyl-(L)-histidinemethylester; EGDMA, ethylene glycol dimethacrylate; BDDMA, 1,3-butanediol dimethacrylate; GDMA, glycerol dimethacrylate; VAL, 2-vinyl-4,4-dimethylazlactone; TRIM, trimethylolpropane trimethacrylate; SPE, N,N-dimethyl-N-methacryloxyethyl-N-(3-sulfopropyl) ammonium betain; TEGDMA, triethylene glycol dimethyacrylate; BDDA, butanediol diacrylate; PEDAS, pentaerythritol diacrylate monostearate; HMAM, N-(hydroxymethyl) acrylamide; PDA, piperazine diacrylamide; PA, phenyl acrylate; MBAA, N,N′-methylenebisacrylamide; MA, methacrylamide; VSA, vinylsulfonic acid; DMAA, N,N-dimethyl acrylamide; AGE, allyl glycidyl ether; IPA, isopropyl acrylamide; NBE, norbon-2-ene; DMN-H6, 1,4,4a,5,8,8a-hexahydro-1,4,5,8-exo,endo-dimethanonaphthalene; BACM, 4-[(4-aminocyclohexyl)methyl]cyclohexylamine; CHD, trans-1,2-cyclohexanediamine; TEPIC, tris(2,3-epoxypropyl) isocyanurate; BVBDMS, bis(p-vinylbenzyl)dimethylsilane.

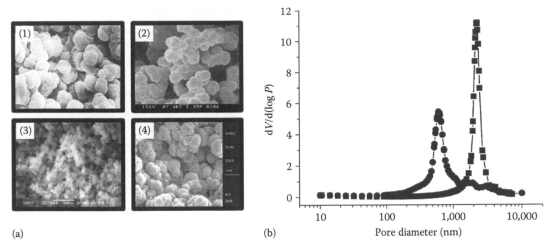

(a)   (b)

FIGURE 1.2 Morphology and porosity of a typical monolithic rod, prepared by copolymerization of organic precursors. (a) SEM micrographs of organic monoliths, being fabricated by different polymerization techniques: (1) thermally (AIBN) initiation, (2) photochemical (DAP) initiation, (3) chemical (APS/TEMED) initiation, and (4) norbornene monolith, Grubbs initiator. (Reprinted with permission from Wang, Q.C. et al., *Anal. Chem.*, 65, 2243, 1993. Copyright American Chemical Society; Lee, D. et al., *J. Chromatogr. A*, 1051, 53, 2004; Kornyšova, O. et al., *J. Chromatogr. A*, 1071, 171, 2005; Mayr, B. et al., *Anal. Chem.* 73, 4071, 2001. With permission from Elsevier.) (b) Differential pore size distribution curve of photochemically initiated poly(butyl methacrylate-*co*-ethylene dimethacrylate) monoliths, showing typical monomodal macroporosity. (Reprinted from Lee, D. et al., *J. Chromatogr. A*, 1051, 53, 2004. With permission from Elsevier.)

Several research groups, for instance, reported on the generation of weak anion exchanges by reacting the epoxy functionalities with diethylamine [54–57]. The resulting diethylaminoethyl monoliths—which are commercially available as CIM (Convective Interaction Media) in disk and column format—have frequently been used for protein and oligonucleotide separation [58] as well as for the purification of proteins and plasmid DNA [59–61].

On the other hand, cation-exchange monoliths based on GMA/EDMA monoliths have been realized by grafting with 2-acrylamido-2-methyl-1-propanesulfonic acid or by modification of epoxy groups using iminodiacetic acid [62,63].

In addition, the GMA/EDMA copolymer proved to serve as a basic unit for the fabrication of highly permeable bioreactors in capillary format. Trypsin immobilization after epoxide ring opening with diethylamine and attachment of glutaraldehyde is mentioned as the probably most prominent example [64]. The immobilization of trypsin was also carried out using another class of reactive monolithic methacrylate polymer, which is based on 2-vinyl-4,4-dimethylazlactone, acrylamide, and ethylene dimethacrylate [65]. In contrast to GMA/EDMA, trypsin can directly be immobilized onto this kind of monolith, as the 2-vinyl-4,4-dimethylazlactone moieties smoothly react with weak nucleophils even at room temperature.

To conclude, it seems to be obvious that (meth)acrylate monolithic supports that can be prepared by polymerization of a huge variety of chemically different monomers are very versatile due to their broad diversity of surface chemistries.

1.2.1.3 Acrylamide-Based Monoliths

The first hydrophilic monoliths based on acrylamide chemistry were based on copolymerization of acrylic acid and *N,N'*-methylene bis(acrylamide) in the presence of an aqueous buffer as porogen [66]. Shortly after, the first hydrophobic capillary support for hydrophobic interaction chromatography was fabricated by the substitution of acrylic acid by butyl methacrylate, whereas the monomer was copolymerized with N,N'-methylene bis(acrylamide) in a nonmodified (nonsilanized) capillary [67], which caused material compression with water as mobile phase. In addition, the monolith broke into several pieces and was therefore weak in performance. To solve the problem of support compression, fused silica capillaries, which served as mold for polymerization, were pretreated with a silanization agent in order to attach double bonds on the capillary inner wall. The modified capillary was then used for the preparation of an IEX material, based on methacrylamide, acrylic acid, and piperazine diacrylamide. As expected, the polymer was attached to the capillary inner wall and enabled the separation of proteins without being compressed, when solvent pressure was applied [68]. Since acrylamide-based monoliths generally represent polar support materials, they are predominately also used for separation in normal-phase mode. For that purpose, monolithic polymers were prepared by polymerization of mixtures containing piperazine diacrylamide as cross-linking agent and methacrylamide, N-isopropylacrylamide or 2-hydroxyethyl methacrylate, and vinylsulfonic acid as monomers.

1.2.1.4 Norbornene-Based Monoliths

In contrast to the most frequently employed free radical polymerization technique, Buchmeiser et al. introduced a novel class of monolithic polymer supports by employing ROMP [36,69]. This approach employed mixtures of norborn-2-ene and DMN-H6 that were copolymerized in the presence of appropriate porogenic solvents and a Grubbs-type ruthenium catalyst as initiator. The resulting hydrophobic polymers showed surprisingly similar morphological characteristics than that known for other organic polymer monoliths, prepared by thermally or photochemically initiated free radical polymerization (Figure 1.2a). ROMP-derived monolithic supports have been successfully applied to the separation of biopolymers in conventional column as well as in capillary format [38,70–72]. In addition, high-throughput screening of synthetic polymers can be accomplished [39].

1.2.1.5 Fabrication of Organic Monoliths

Because of the fact that organic polymers are known to suffer from swelling or shrinkage on changing the solvent [73,74], the inner wall of the column housings (fused silica capillaries or borosilicate columns) has—prior to polymerization—to be derivatized in order to provide chemical attachment of the monolith rod to the wall.

Even if this procedure does not influence (enhance) the swelling properties of the polymer itself, it prevents the packing from being squeezed out of the column housing on employing a weak solvent (e.g., water for hydrophobic polymers like PS/DVB) at high pressure. The most frequently employed method for inner wall derivatization relies on the condensation of surface silanol groups with bifunctional 3-(trimethoxysilyl)propyl (meth)acrylate according to the synthesis scheme, depicted in Figure 1.3a and b [75].

"In situ" (Latin for "in the place") polymerization means the fabrication of a polymer network directly in the finally desired shape and geometry. In the context of monolithic separation columns, the term in situ is referred to the polymerization in the confines of a HPLC column or a capillary as mold.

The preparation of a polymer monolithic column is relatively simple and straightforward compared with that of silica rod. Most frequently employed method of polymerization is the thermally initiated, free radical copolymerization. A mixture consisting of the monomer, a cross-linking agent, an initiator (often AIBN), and in the presence of at least one, usually two inert, porogenic solvents is put in a mold (typically a tube) or in a capillary column. Then the filled columns are sealed on both ends. The polymerization is started by heating the column in a bath at a temperature of 55°C–80°C or by UV light, depending on the initiator agent (see Figure 1.3c considering PS/DVB as example). After completion of the polymerization, the column is flushed with a suitable solvent to remove porogens and nonreacted residues.

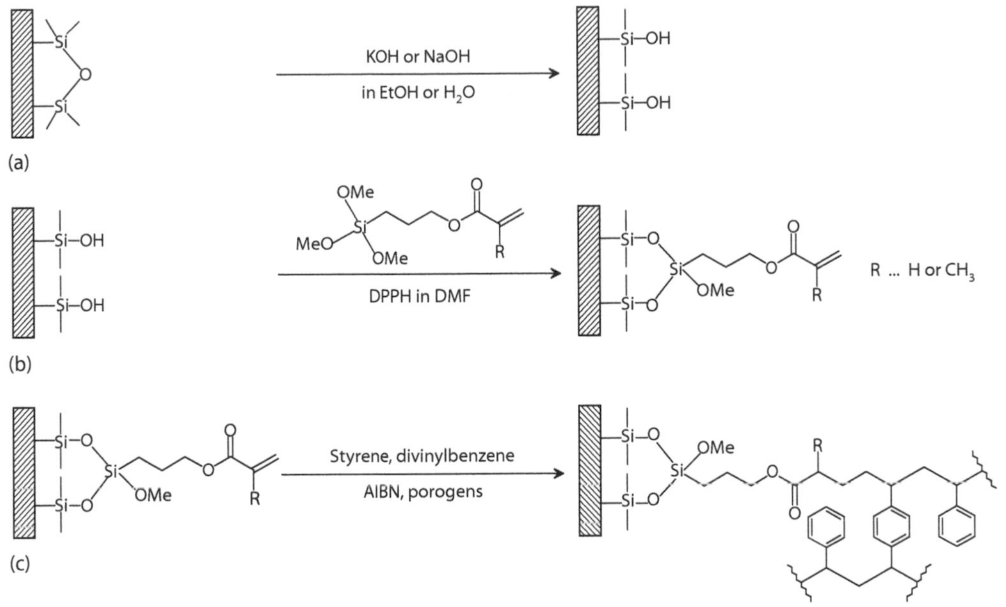

FIGURE 1.3 Schematic representation of the silanization procedure of borosilicate or fused silica capillary column inner walls. (a) Surface etching under alkaline conditions, (b) attachment of reactive groups by condensation with silanol, (c) chemical linkage of polymer (PS/DVB considered as example) by free radical polymerization.

1.2.2 Inorganic Monoliths

1.2.2.1 Silica-Based Monoliths

The basic studies dealing with the preparation of continuous porous silica materials date back to 1991 [76–78]. Two years later, Nakanishi and Saga applied for a patent describing the fabrication of monolithic silica rods for chromatographic application [79–81], whereas a second protocol for the preparation of continuous silica rods was independently filed by Merck KGaA in Germany [82].

First comprehensive investigations with respect to the properties of continuous porous silica rods were, however, carried out by Tanaka and Fields in 1996 [33,34,83], who reported on two different methods for the preparation of silica monoliths.

Fields used an approach similar to that of casting column end frits in fused silica tubings for particle-packed capillary HPLC columns. A monolithic reversed-phase column was fabricated by filling a fused silica capillary with a potassium silicate solution, followed by heating at 100°C and drying with helium at 120°C. Derivatization with hydrophobic end groups was accomplished flushing the column with a solution of ODS in dry toluene while heating at 70°C [34]. Unfortunately, the morphology of the silica material produced by this method was heterogeneous.

Silica monolith fabrication by a sol–gel approach was reported by Tanaka at the same time [33]. Following this protocol, the preparation of more uniform and homogeneous monolithic structures have been achieved, yielding continuous rods with 1–2 µm through-pore size, 5–25 µm mesopore size, and surface areas of 200–400 m²/g. Because of the capability of precisely controlling and varying the morphology and porous properties of the evolving inorganic monolithic structure, the sol–gel approach is nowadays most commonly applied for the preparation of silica-based monoliths.

The morphology of a typical inorganic monolith is fundamentally different from that of organic polymers (Figure 1.4a). The structure is rather sponge- than brush-like and is constructed by interconnected silica rods in the low micrometer size. This composition leads to a discrete distribution of flow channels, which can be deduced by comparison of macropore distribution of a typical organic

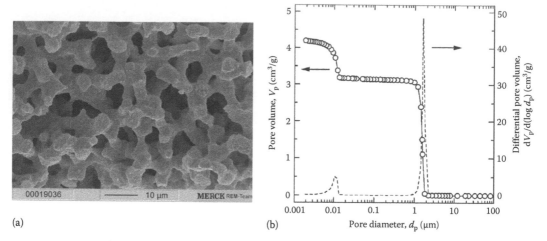

(a)   (b)

FIGURE 1.4 Morphology and porosity of a typical monolithic rod, prepared by copolymerization of silane precursors. (a) SEM micrograph of the fractured surface of a monolithic silica gel rod. (b) Pore size distribution of a representative monolithic silica rod. (Reprinted from Guiochon, G., *J. Chromatogr. A*, 1168, 101, 2007. Copyright 2007, with permission from Elsevier.)

(Figure 1.3b) and inorganic (Figure 1.4b) monolithic rod. The most important difference between the two types of monoliths, however, is directed to the distribution of mesopores. While organic monoliths do only possess a noteworthy amount of small pores (<50 nm), silica-based supports are characterized by a distinct bimodal pore size distribution of macropores (low micrometer range) and mesopores (5–30 nm, dependent on the conditions of polymerization) (Figure 1.4b).

Although inorganic, monolithic columns attracted considerable attention in the last 10 years, the preparation of silica-based monoliths does not yet offer the broad chemical variety of precursors and porogens for specific adjustment of separation compared with their organic counterpart. The preparation of silica monoliths uses the classical sol–gel process of hydrolysis and polycondensation of organosilicium compounds.

1.2.2.2 Fabrication of Silica-Based Monoliths

Monolithic silica columns can be prepared either in a mold (6–9 mm I.D. glass test tube) or in a fused-silica capillary. A considerable volume reduction of the silica monolith appears by the fabrication of a mold. The diameters of products are approximately 4.6–7 mm when a glass test tube of 6 and 9 mm I.D. was used. To cover the resulting silica rod, PTFE tubings or PEEK resins are frequently applied to produce a column for HPLC. The length of these columns is limited to about 15 cm or shorter. PEEK-covered monolithic silica columns, so-called chromoliths, are commercially available at 5–10 cm length and can withstand inlet pressure of up to 120 kg/cm² (Merck, Germany).

For capillary applications, the silica network structure must be grafted to the tube wall to fix the monolith in a fused silica capillary in order to prevent shrinkage of the skeletons. Smaller diameter tubes (50 μm I.D.) performed better than larger-sized tubes.

The preferentially employed approach for the fabrication of inorganic (silica) monolithic materials is acid-catalyzed sol–gel process, which comprises hydrolysis of alkoxysilanes as well as silanol condensation under release of alcohol or water [84–86], whereas the most commonly used alkoxysilane precursors are TMOS and tetraethoxysilane (TEOS). Beside these classical silanes, mixtures of polyethoxysiloxane, methyltriethoxysilane, aminopropyltriehtoxysilane, *N*-octyltriethoxysilane with TMOS and TEOS have been employed for monolith fabrication in various ratios [87]. Comparable to free radical polymerization of vinyl compounds (see Section 1.2.1.5), polycondensation reactions of silanes are exothermic, and the growing polymer species becomes insoluble and precipitates at a certain stage, being referred to as phase separation, whereas the solubility of the oligomers decreases with increasing degree of polymerization.

Even if Nakanishi and Soga initially employed polyacrylic acid as porogenic solvent [76–78], silica monoliths are nowadays generally fabricated in the presence of a defined porogen system including poly(ethylene glycol) and PEO [88]. The morphology of the macroporous monolithic silica network can effectively be controlled in wide ranges of volume fraction, pore connectivity, and average pore size (macropore formation) by modifying the solvent composition (water/alcohol) and the ratio of the porogen additive to monomers [88]. Other porogen additives have been reported. The addition of urea was reported by Ishizuka et al. [89]; Saito et al. investigated D-sorbitol [90]. However, poly(ethylene glycol) and PEO are by far the most routinely employed porogens.

In order to increase column stability (enhancement in stiffness and strength of the rod) and to increase the fraction of mesopores, the monolithic silica rods have to be subjected to an "aging" procedure after polymerization [76,91]. Aging in the presence of alkaline solutions has a strong influence on the size distribution of mesopores, whereas column treatment with acidic or neutral solution shows less or no effects regarding mesoporosity. While the pH of the aging solution mainly influences the average pore size of the support, increasing temperature during aging broadens the pore size distribution by controlling the formation rate of the pore network. That way, large average pore sizes have been obtained, but unfortunately being connected with in a considerable decrease of internal porosity of the gel. In order to yield mesopores in the range of 14–25 nm, frequently employed aging conditions are 0.01 M aqueous solution of ammonium hydroxide at 80°C–120°C [91].

The chemical procedure of attaching alkyl chains or other functional chemical group to bare silica monoliths essentially is the same as for conventional silica particles, whereas the rod is immersed in or flushed with an appropriate solution for the necessary period of time at a suitable temperature. For linking octadecyl groups on the silica surface, octadecyldimethyl-N,N-diethylaminosilane (ODS-DEA) is frequently used [92]. A remarkable approach for octadecylation of bare silica rods has recently been reported [93]. After derivatization of surface silanol groups with 3-methacryloxy-propyltrimethoxysilane, the silica capillary monoliths were grafted with octadecyl methacrylate to result in a inorganic poly(octadecyl methacrylate) (ODM) column. Comparison of the chromatographic characteristics of these ODM columns with ODS-DEA derivatized silica monoliths revealed that aromatic compounds with rigid and planar structures and low length-to-breadth ratios as well as acidic analytes seem to have more retention for polymer-coated stationary phase (ODM). That way, the polymer-coated octadecyl column enabled separation of some polycyclic aromatic hydrocarbons (PAHs), alkyl phthalates, steroids, and tocopherol isomers that could not be separated under the same conditions on ODS columns.

1.2.2.3 Metal Oxide and Carbon Monoliths

Metal oxides are inert materials that exhibit higher stability under strongly acidic, basic, or oxidizing solutions then conventional silica materials. They are even stable at elevated temperatures. All these advantageous properties attract scientific attention on metal oxide materials as new supports for enhanced HPLC application.

Monolithic columns consisting of various oxides, in particular aluminum, hafnium, and zirconium oxides [94,95], have recently been introduced. The authors report on the preparation of monolithic 50 μm capillary columns by in situ copolymerization of an aqueous solution of hafnium or zirconium chloride with propylene oxide in the presence of N-methylformamide as porogen. The polycondensation reaction was carried out in pretreated and sealed capillary tubes at 50°C. SEM pictures of fabricated hafnia and zirconia columns revealed microglobular, interconnected structures (one to a few micrometers in diameter), being criss-crossed by through-pores. NP chromatography of pyrazole and imidazole, however, exhibit exceedingly strong peak-tailing, which may indicate insufficient specific surface area and/or a heterogeneous surface of the stationary phase [95].

Randon et al. reported on an alternative approach for the preparation of zirconia monoliths [96]. The sol–gel process is initiated by hydrolysis of an ethanolic zirconium alkoxide solution, on addition of an aqueous solution of acetic acid at 30°C. Mixtures of polyethylene glycol and *n*-butanol have been employed as porogen. The resulting metal oxide monolithic rods were structures by porons in the range of 2 μm, resulting in average through-pore diameter of 6 μm.

Taguchi et al. [97] and Liang et al. [98,99] reported on the preparation of monolithic carbon columns, which exhibit a hierarchical, fully interconnected porosity. Silica particles (10 μm) have been suspended in an aqueous solution, containing ethanol, $FeCl_3$, resorcinol, and formaldehyde. After polymerization, the solid rod was dried, cured, and carbonized by raising temperature to 800°C and finally up to 1250°C. Finally, concentrated HF was used to remove silica and iron chloride. Even if carbon have been shown to possess a high specific surface area (up to 1115 m^2/g), their chromatographic efficiency is moderate (HETP of 72 μm).

1.2.3 CHROMATOGRAPHIC CHARACTERISTICS OF MONOLITHIC COLUMNS

Monolithic stationary phases have to be regarded as the first substantial further development of HPLC columns, as they present a single particle separation medium, made up of porous polymer. As a consequence of their macroporous structure, they feature a number of advantages over microparticulate columns in terms of separation characteristics, hydrodynamic properties, as well as their fabrication:

- Monolithic columns are comparatively easy to prepare. This is particularly true for capillary columns, which are known to be tedious to pack with particles. Furthermore, the reproducibility of microcolumn packing is low.
- Due to the fact that the polymer can chemically be attached to the column wall during polymerization, monolithic stationary phases do not necessitate frits to retain the column packing.
- Monolithic stationary phases do not possess interparticulate voids, which results in enhanced separation efficiency due to reduced band broadening.
- Due to the macroporous structure of monolithic stationary phases (flow channels), the solvent is forced to pass the entire polymer, leading to faster convective mass transfer (compared to diffusion), which provides for analyte transport into and out of the stagnant pore liquid, present in the case of microparticulate columns.
- Monolithic materials exhibit reduced flow resistance, which results in high permeability and consequently high speed of separation.

Even if the diminishment of interparticulate voids as well as the convective mass transfer are generally assumed to be the main reasons for the enhanced chromatographic properties of monolithic columns, the characteristics of organic and inorganic monolithic supports have to be separately discussed and evaluated, since they have been shown to complement one another regarding their applicability [29,100].

Organic monoliths have been proven to be efficient stationary phases for the separation of biomolecules, including proteins, peptides, oligonucleotides, as well as DNA fragments. This can be ascribed to their monomodal macropore size distribution (see Figure 1.2b), which satisfies all requirements for the resolution of high-molecular-weight compounds. Their chromatographic efficiency toward small molecules, however, has been shown to be extensively poor, due to missing or insufficiently available fraction of mesopores.

Inorganic silica monoliths, on the other hand, possess a bimodal pore size distribution of flow channels and mesopores (see Figure 1.4b), which substantiate their potential for the separation of low-molecular-weight compounds with high speed and resolution power. The analysis of biopolymers (especially biomolecules of high molecular weight, like proteins or DNA fragments), however, is limited due to the absence of macropores, being necessary for resolution of large analytes (average pore diameter: 50 to several 100 nm).

1.3   PORE FORMATION OF ORGANIC AND INORGANIC MONOLITHS

1.3.1   General Pore Formation Mechanism of Organic Monoliths

Due to the fact that thermally initiated free radical copolymerization is by far the most routinely employed method for fabrication of organic monolithic stationary phases, the pore formation mechanism is discussed for this particular kind of polymerization.

Other modes of copolymerization, like photochemically or chemically initiated free radical polymerization, ROMP, or polycondensation reactions, in the presence of inert diluents are, however, supposed to be comparable with respect to the formation of support porosity.

The polymerization mixture for the preparation of rigid, macroporous monolithic materials in an unstirred mold generally contains a monovinyl compound (monomer), a divinyl compound (crosslinker), an inert diluent (porogen), as well as an initiator. The mechanism of pore formation of such a mixture has been postulated by Seidl et al. [101], Guyot and Bartholin [102], and Kun and Kunin [103] and can be summarized as in the following text.

The thermal initiator, present in the polymerization mixture, decomposes at a certain temperature accompanied by disposal of radicals that initiate the polymerization reaction of monomer as well as cross-linking molecules in solution. After becoming insoluble in the employed polymerization mixture (strongly dependent on the nature of porogenic solvent and on the degree of cross-linking), the polymer nuclei precipitate.

This early stage of polymerization is referred to as *phase separation* or *gel point* and describes the transition from liquid to solid-like state. At this point in time, nonreacted monomers are thermodynamically better soluble in the swollen polymer nuclei than in the solvent, which causes the rate of further polymerization in the polymer globules to be larger than in the surrounding liquid (higher local monomer concentration in the swollen nuclei than in solution). The precipitated, insoluble nuclei thus increase in size as a result of polymerization in the polymer microspheres as well as of adsorption of polymer chains from the surrounding solution, whereas the high cross-linking character of the globuli prevents their mutual penetration and loss in individuality due to coalescence.

At a certain volume extension, the nuclei are subjected to chemical association (reaction of cross-linking agent) with other nuclei in their immediate vicinity in order to form polymer clusters (see Figure 1.1). These clusters still keep dispersed in the liquid porogen mixture, until their increase in size due to proceeding polymerization enables their mutual contact, thereby building a scaffolding structure that pervades the whole porogen mixture. Comparable to the polymer clusters, the development of the polymer scaffold is ascribed to cross-linking reactions that provide for chemical linkage among the clusters. Finally, the polymer skeleton is tightened by further capture and addition of polymer chains that still evolve in solution.

The resulting porosity of the monolithic polymer is thus defined as the space inside the polymer being occupied by porogens and—in case of uncomplete monomer conversion—nonreacted monomer as well as cross-linker. Consequently, the overall porosity is composed by three different contributions (listed in their chronological order of development during polymerization and in the order of increasing mean pore size):

- Free space inside the polymer microglobules that precipitate at early stages of polymerization as (monomer) swollen globules
- Free space inside the polymer clusters, arising after chemical linkage of microglobules in solution
- Space between the polymer clusters that build the scaffold by chemical linkage at late stages of polymerization

As being indicated above, the resulting overall porosity of the monolithic polymer can be influenced and controlled by the nature and composition of the porogenic solvent as well as the amount of cross-linker. Furthermore, a number of additional parameters have been described and discussed in literature in order to tailor and fine-tune the porous properties of organic monoliths.

1.3.2 CONTROL OF THE POROUS PROPERTIES

1.3.2.1 Influence of the Monomer to Cross-Linker Ratio

Increasing the amount of cross-linking agent (divinyl compound) at expense of monomer causes a decrease in pore size, which is accompanied by a distinct increase in surface area [101–104]. Even if this has been observed for macroporous beads prepared by suspension polymerization, the results can directly be transferred to the fabrication of rigid monolithic materials in an unstirred mold by thermally [105,106] as well as photochemically [107] initiated free radical copolymerization.

The experimentally elaborated effect of cross-linker on the porous properties of monolithic polymers is in accordance with the postulated mechanism of pore formation, presented in Section 1.3.1. The higher the amount of cross-linker, the higher the cross-linking degree of the dissolved polymer chains at early stages of the polymerization. This in turn causes an early occurrence of phase separation. Due to high cross-linking, the precipitated polymer globules exhibit a low degree of swelling with monomers, which keeps the rate of polymerization within the globules and thus the growth of the nuclei low.

On the other hand, the polymerization that occurs in the surrounding solvent (porogen mixture) is comparatively high. Furthermore, the high amount of good solvating monomer in solution causes the polymer chains to be subjected to a low probability of adsorption to the precipitated preglobules. As a result, the mean globule diameter of the polymer scaffold is reduced with increasing cross-linker content, leading to small interglobular voids and thus pore size.

1.3.2.2 Influence of the Porogenic Solvent

The formation of macroporous monolithic polymer supports is ascribed to a phase separation of small polymer nuclei due to their limit of solubility in the surrounding polymerization mixture (mixture of inert diluent and reactive monomers). The phase separation is thus a function of both the ability of the porogens to dissolve the growing nuclei as well as the degree of polymer cross-linking. At constant amount of cross-linking agent in the polymerization mixture, the point in time of phase separation is consequently only dependent on the choice and composition of the porogens.

Generally, the lower the dissolving properties of the porogenic solvent for a given evolving copolymer system, the larger the mean pore size of the polymer after complete monomer conversion [105]. Figure 1.5 illustrates two examples. Figure 1.5a shows the effect of the 1-dodecanol to cyclohexanol ratio on the pore size distribution of monolithic poly(glycidyl methacrylate-*co*-ethylene dimethacrylate). As cyclohexanol is—due to higher hydrophilicity—known to be a better solvent for this particular methacrylate system than 1-dodecanol, an increase in the fraction of the latter results in an increase in pore diameter. Figure 1.5b illustrates a similar study for monolithic PS/DVB. Regarding PS/DVB, 1-dodecanol is a poorer solvent than toluene, whose ability to dissolve styrene polymers is known to be excellent. Again, an increase in the solubility properties of the porogenic solvent (addition of toluene to 1-dodecanol) results in a tremendous decrease in pore size of the monolithic polymer.

Since the ability of the porogen or a porogen mixture to dissolve a certain polymer system can usually hardly be estimated without experimentation, the effect of porogens on the overall porosity of monolithic materials is widely empirical.

The fact that adding a better solvent to the mixture results in a shift of the distribution to smaller pore sizes has been explained by the mechanism of pore formation, postulated for macroporous resins in the late 1960s [101–103]. The addition of a poor solvent causes the phase separation to occur early, whereas the precipitated polymer nuclei are swollen with monomers, which present a better solvating agent than the porogen. Due to the high monomer concentration within the globuli,

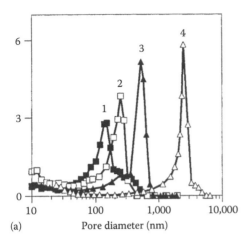
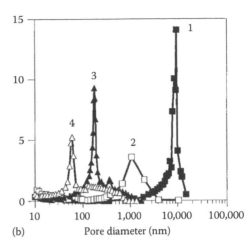

(a)   Pore diameter (nm)   (b)   Pore diameter (nm)

FIGURE 1.5 Influence of porogens on the porosity of poly(glycidyl methacrylate-*co*-ethylene dimethacrylate) and poly(styrene-*co*-divinylbenzene) monoliths. (a) Effect of 1-dodecanol in the porogenic solvent on differential pore size distribution curves of molded poly(glycidyl methacrylate-*co*-ethylene dimethacrylate). Conditions: polymerization time 24 h, temperature 70°C, polymerization mixture: glycidyl methacrylate 24%, ethylene dimethacrylate 16%, cyclohexanol and 1-dodecanol content in mixtures: 60% + 0% (1), 57% + 3% (2), 54% + 6% (3), and 45% + 15% (4). (b) Effect of toluene in the porogenic solvent on differential pore size distribution curves of molded poly(styrene-*co*-divinylbenzene) monoliths. Conditions: polymerization time 24 h, temperature 80°C, polymerization mixture: styrene 20%, divinylbenzene 20%, 1-dodecanol and toluene content in mixtures: 60% + 0% (1), 50% + 10% (2), 45% + 15% (3), and 40% + 20% (4). (Reprinted with permission from Viklund, C. et al., *Chem. Mater.*, 8, 744, 1996. Copyright 1996, American Chemical Society.)

the rate of polymerization there is higher than in the surrounding solution, which affects the nuclei rapidly to gain in size. In addition, polymer chains, growing in solution, are subjected to a high probability of adsorption to the chemically similar globules, which further increases their size.

The addition of a good solvent, on the other hand, causes the phase separation to occur at later stages of the polymerization, whereas the better porogenic solvent competes with the monomers in the solvation of the precipitated globules. As a consequence, the concentration gradient of monomers is not in that high gear; the growth of the nuclei is decelerated, while the polymerization in solution is promoted and the evolving polymer chains are subjected to a low probability for adsorption to the preglobules. As a result, the porous polymers, fabricated in the presence of good solvating solvents, exhibit smaller microglobules on average and thus a distinctive reduction in pore size.

1.3.2.3 Influence of the Polymerization Temperature

An increase in polymerization temperature decreases the mean pore size diameter, as it has been shown by bulk polymerization experiments with subsequent evaluation by mercury intrusion porosimetry (MIP) [108,109]. This is demonstrated in Figure 1.6a and b, where the overall porosity of poly(glycidyl methacrylate-*co*-methylene dimethacrylate) copolymers, resulting from different polymerization temperatures and polymerization techniques, is compared. The effect of the polymerization temperature is in accordance with the generally accepted mechanism of pore formation of thermally initiated polymerization in the presence of a precipitant (porogen) [101–103]. The higher the temperature, the faster the rate of initiator decomposition and the larger thus the number of free radicals available in solution. Consequently, the number of polymer chains and the number of precipitating globules at the point of phase separation is magnified. At constant monomer as well as cross-linker content, a larger number of microglobules necessarily results in smaller nuclei diameters, which in turn causes the interglobular voids as well as the voids between the chemically linked clusters to decrease.

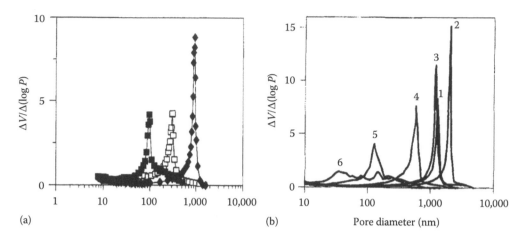

(a)

(b)   Pore diameter (nm)

FIGURE 1.6 Influence of the polymerization temperature on the porosity of poly(glycidyl methacrylate-
co-ethylene dimethacrylate) monoliths determined by MIP. (a) Differential pore size distribution curves of
the poly(glycidyl methacrylate-*co*-ethylene dimethacrylate) rods, prepared by 22 h polymerization at a temperature of 55°C (♦), 12 h at 70°C (■), and a temperature increased during the polymerization from 50°C to
70°C in steps by 5°C lasting 1 h each and kept at 70°C for another 4 h (□). (Reprinted with permission from
Svec, F. and Fréchet, J.M.J., *Chem. Mater.*, 7, 707, 1995. Copyright 1995, American Chemical Society.) (b)
Differential pore size distribution curves of the poly(glycidyl methacrylate-*co*-ethylene dimethacrylate) rods,
prepared by 22 h polymerization at a temperature of 55°C (3), 12 h at 70°C (1), and a temperature increased
during the polymerization from 50°C to 70°C in steps by 5°C lasting 1 h each and kept at 70°C for another 4 h
(2). (Reprinted with permission from Svec, F. and Fréchet, J.M.J., *Macromolecules*, 28, 7580, 1995. Copyright
1995, American Chemical Society.)

1.3.2.4   Influence of the Initiator

The choice of initiator is closely associated with the porosity of the resulting monolithic support,
provided that the decomposition rates of the initiators at a given temperature are different [109].
Substitution of AIBN by benzoyl peroxide, for example, causes a shift in the pore size distribution
to higher pores, which can be ascribed to the decomposition rate of benzoyl peroxide being four
times slower than that of AIBN [110]. The impact of the type of initiator is thus based on the same
explanation than the effect of the polymerization temperature (see Section 1.3.2.3). The higher the
decomposition rate, the higher the amount of polymer chains, evolving in solution, which results
in a large number of precipitated microglobules and finally small voids between them. In addition,
the initiator content acts on the same principle, as—at a given point in time—the number of free
radicals in solution is directly proportional to the original amount of thermal initiator used. The
higher the relative percentage of initiator, the smaller the mean pore size of the monolithic polymer
network after complete polymerization.

1.3.2.5   Influence of the Polymerization Time

The polymerization time as a polymerization parameter for adjustment of the porous properties of
thermally initiated copolymers has recently been characterized [111]. A polymerization mixture
comprising methylstyrene and 1,2-bis(*p*-vinylbenzyl)ethane as monomers was subjected to thermally initiated copolymerization for different times (0.75, 1.0, 1.5, 2, 6, 12, and 24 h) at 65°C. The
mixtures were polymerized in silanized 200 μm I.D. capillary columns as well as in glass vials for
ISEC and MIP/BET measurements, respectively.

   The results of the MIP analyses of the bulk polymers are illustrated in Figure 1.7. It could be demonstrated that the polymerization time is capable of influencing the shape of the pore distribution
itself, rather than shifting a narrow macropore distribution (and thus the pore-size maximum) along
the scale of pore diameter (see effect of the porogenic solvent in Section 1.3.2.2 and Figure 1.5). On

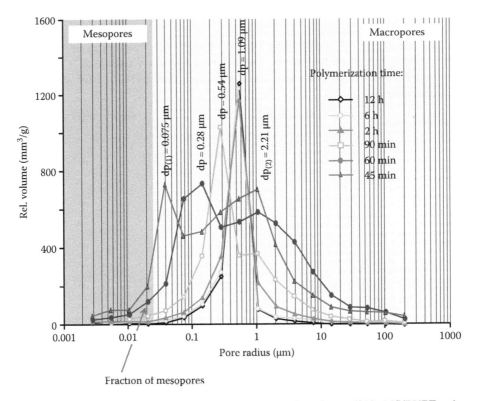

FIGURE 1.7 Influence of the polymerization time on the porosity of monolithic MS/BVPE polymer networks, determined by MIP. Reduction of the polymerization time converts a narrow monomodal pore distribution into a broad bimodal distribution, comprising mesopores.

a severe decrease in the polymerization time, a typical monomodal macropore distribution (being generated at a time >6 h) is stepwise converted into a comparatively broad bimodal distribution (see Figure 1.7, 60 and 45 min). At way, an initial pore maximum of 1.09 μm (12, 6, and 2 h) is systematically split up into two pore maxima of 0.28 and 2.21 μm in the case of a total polymerization time of 1 h, and 0.075 and 2.21 μm in the case of 45 min. As it can be derived from Figure 1.7, these addressed displacements and departments of the initial main pore maximum, being characteristic for long time free radical copolymerizations, are closely connected with a considerable increase in the fraction of small macropores (in the range of 50–200 nm) as well as in the fraction of mesopores (<50 nm), which in turn should be associated with an increase in specific surface area of the materials.

BET measurements (Table 1.2) prove the increase in mesopores, as decreasing the total polymerization time from 24 h to 45 min causes S_p to raise by a factor of 3, resulting in $S_p{\sim}80\,m^2/g$, which is comparable to silica particles with a mean pore diameter of 300 Å [112,113].

Even if MIP and BET are widely accepted regarding the characterization of HPLC stationary phases, they are only applicable to the samples in the dry state. In order to investigate the impact of polymerization time on the porous properties of "wet" monolithic columns, ISEC measurements of 200 μm I.D. poly(p-methylstyrene-co-1,2-bis(vinylphenyl)ethane) (MS/BVPE) capillary columns (prepared using a total polymerization time ranging from 45 min to 24 h) have been additionally evaluated (see Table 1.2 for a summary of determined ε values). On a stepwise decrease in the time down to 45 min, the total porosity (ε_t) is systematically increasing to about 30% in total (62.8% for 24 h and 97.2% for 45 min). This is caused by a simultaneous increase in the fraction of interparticulate porosity (ε_z) as well as the fraction of pores (ε_p). The ISEC measurements are in agreement with those of the MIP as well as BET analyses, as an increase in S_p should be reflected in an increase in ε_p and as the relative increase in the total porosity (caused by decreasing the polymerization time

TABLE 1.2

Influence of the Polymerization Time on the Porous Properties of Monolithic MS/BVPE Networks, Considering Capillary Columns (80 × 0.2 mm I.D.) for ISEC and Glass Vial Bulk Polymers for MIP and BET Measurements

Polymerization Time (min)	Porosity Data				Surface Area	
	ε_t (%)[a]	ε_t (%)[b]	ε_z (%)[b]	ε_p (%)[b]	S_p (m²/g)[a]	S_p (m²/g)[c]
45	89.9	97.12	71.39	25.73	75.5	77.2
60	81.4	91.12	68.18	22.94	50.3	52.4
90	78.4	89.03	67.49	21.54	49.5	48.2
120	76.0	82.21	60.99	21.22	33.3	43.7
360	67.9	76.55	55.61	20.94	25.1	30.0
720	67.0	70.20	49.76	20.44	23.8	27.2
1440	65.8	62.81	42.62	20.19	22.9	26.8

[a] Calculated from MIP data.
[b] Calculated from ISEC retention data.
[c] Calculated from BET.

from 24 h to 45 min) calculated from MIP as well as ISEC data is in the same order of magnitude (36% and 54% for MIP and ISEC, respectively) (see Table 1.2).

Figure 1.8 shows the influence of the polymerization time on the separation efficiency and resolution of MS/BVPE columns toward biomolecules (e.g., oligonucleotides) and small molecules (e.g., phenols).

1.4 CHARACTERIZATION OF MONOLITHS AND DETERMINATION OF THE POROUS PROPERTIES

Porosity is one of the most important properties of a stationary phase, since it severely influences the chromatographic column performance, the speed of separation, as well as the specific surface area and consequently loading capacity. Porosity refers to the degree and distribution of the pore space present in a material [114]. Open pores indicate cavities or channels, located on the surface of a particle, whereas closed pores are situated inside the material. The sum of those pores is defined as *intraparticular porosity*. *Interparticular porosity*, in contrast, is the sum of all void volume between the particles. According to their diameter, pores have been internationally (IUPAC) classified as follows [114]:

Micropores—pore diameter smaller than 2 nm
Mesopores—pore diameter bigger than 2 nm and smaller than 50 nm
Macropores—pore diameter bigger than 50 nm

For means of determination and quantification of the material porosity, different methods like mercury intrusion porosity, nitrogen gas adsorption, or inverse-size exclusion chromatography (ISEC) have been established and are nowadays routinely employed for that purpose. As an alternative to these well-known methods, a new approach based on near-infrared spectroscopy (NIR) for the characterization of monoliths is introduced in this chapter.

1.4.1 DETERMINATION OF THE POROUS PROPERTIES

1.4.1.1 Mercury Intrusion Porosimetry

MIP is a method for the direct determination of pore diameters (or a distribution of pore diameters) based on the volume of penetrating mercury as a not-wetting liquid at a certain pressure being applied.

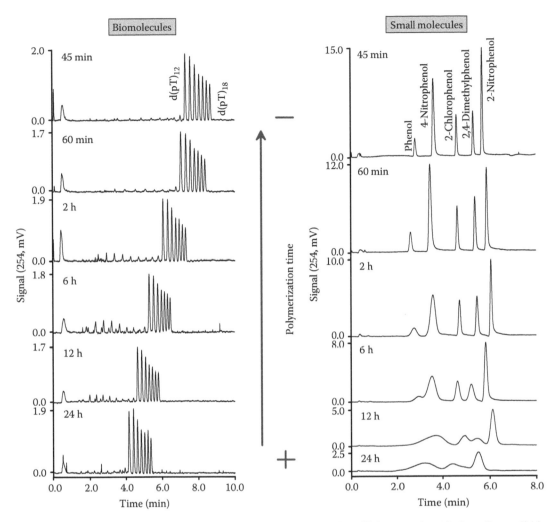

FIGURE 1.8 Influence of the polymerization time on the separation efficiency and resolution of monolithic MS/BVPE capillary columns (80×0.2 mm I.D.) toward biomolecules (considering oligonucleotides as example) and small molecules (considering phenols as example). Chromatographic conditions: oligonucleotides: 0%–20% B in 1 min and 20%–35% in 7 min, 7 μL/min, 60°C, UV 254 nm, inj.: 500 nL, 5 ng total; phenols: 0%–50% B in 5 min, 10 μL/min, 50°C, UV 254 nm, inj.: 500 nL, 10 ng each.

The principle of measurement is based on the fact that mercury does not wet most substances and thus, it will not penetrate pores by capillary action. Surface tension opposes the entrance of any liquid into pores, provided that the liquid exhibits a contact angle greater than 90° [115,116]. Therefore, external pressure is required to force the liquid (mercury in this case) into the pores of the material. The pressure that has to be applied to force a liquid into a given pore size is given by the Washburn equation,

$$p = -\frac{2\sigma\cos\Theta}{r} \tag{1.1}$$

where
  p is the applied pressure
  r is the pore radius
  σ is the surface tension
  Θ is the contact angle of the liquid

It has to be noted that this relation is only valid for pores, possessing cylindrical shape. From Equation 1.1, it gets apparent that under zero pressure, none of the nonwetting liquid will enter the pores of the immersed material. If now the pressure is raised to a certain level, the liquid will penetrate pores possessing radii greater than that calculated from Equation 1.1. Consequently, the higher the pressure that is applied, the smaller the pores that are penetrated by the liquid.

The experimental accomplishment of an MIP experiment can be summarized as follows: The porous sample is placed in a dilatometer. After evacuation, the dilatometer is filled with mercury, whereas it has to be taken care that no air bubbles remain. Finally, pressure is applied on the mercury column. Depending on the size of the pores, mercury is intruding the fraction of open pores at a given applied pressure. The change in volume, which is indicated on the scale of the dilatometer, is registered at each applied pressure, resulting in a graph that presents the cumulative intrusion volume as a function applied pressure. Since the pressure is indirectly proportional to the pore radius according to Equation 1.1, the size of pores can be plotted against the cumulative volume, which is described as the total volume of mercury, penetrating the porous material at a given pressure.

These raw data, provided by an MIP measurement, enable the calculation of a number of parameters that are necessary and helpful for the interpretation of a porous structure:

- The *volume pore size distribution*, which is defined as the pore volume per unit interval of the pore radius can be determined by building the first derivation of the cumulative volume by the pore radius.
- The *total pore volume* can directly be determined by the raw data, as it is equal to the cumulative volume at the highest pressure applied.
- The *specific surface* area is calculated as the area of the intrusion curve that results by plotting the cumulative volume versus the pore radius [117].
- The *mean pore diameter* is described by the pore diameter occurring with highest frequency and is the maximum of the volume pore size distribution curve and can be calculated from the total pore volume and the specific surface [118].

1.4.1.2   Nitrogen Adsorption

Nitrogen sorptiometry, also referred to as BET method (named after their inventors Brunauer [202], Brunauer and Emmet [203], and Teller and coworkers [204]), is an approach for the determination of the specific surface area of a (porous) support material based on the multilayer adsorption of nitrogen at the temperature of liquid nitrogen (77 K) according to following procedure:

Sample is placed in a U-shaped glass tube with defined weight, connected to the sorptiometer, and baked out under a constant carrier (He) gas flow to remove all adsorbed water. Afterward, the sample is cooled to RT and the volumetric flow rate of the carrier is determined. Adsorption gas (N_2) is added, the total volumetric flow is registered, and the sample is cooled to 77 K by means of immersing the U-shaped tube into liquid nitrogen. After removal of the nitrogen dewar, the desorption peak is registered by an appropriate detection unit. Finally, a defined volume of calibration gas (N_2) is injected and detected. This procedure is repeated for different adsorption gas flow rates.

The mole fraction of the carrier as well as of the adsorption gas can be calculated by their volumetric flow rates and enable the determination of the N_2 partial pressure at a certain air pressure. Desorption peak as well as calibration peak are integrated. The amount of injected calibration gas can be calculated according to the ideal gas equation. By comparison of the peak areas of the calibration and desorption signal, the adsorbed amount of N_2 can be determined. The measuring points (partial pressure of N_2 versus adsorbed amount of N_2) are then plotted according to the BET theory [204]. The resulting linear plot allows the calculation of the amount of N_2, being necessary for monolayer coverage (n_m), which further enables the calculation of the specific surface area by multiplication of n_m with the place, occupied by one adsorbed N_2 molecule at 77 K ($1.62 \times 10^{-20} \, m^2$) and the Loschmidt number.

1.4.1.3 Inverse Size-Exclusion Chromatography

ISEC, which was introduced by Halász and Martin in 1978 [119], represents a simple and fast method for the determination of the pore volume, the pore size distribution profile, and the specific surface area of porous solids. Generally, ISEC is based on the principle of SEC. SEC, also referred to as *gel permeation* or *gel filtration chromatography*, is a noninteractive chromatographic method that separates analytes according to their size by employing a stationary phase that exhibits a well-defined pore distribution.

As per definition, ISEC represents the inverse approach. Well-defined (monodisperse) polymer standards (e.g., PS standards) are employed for the determination of the porosity of a stationary phase, whereas principles, apparatus, and measurement method in ISEC are equal to that of HPLC.

In order to enable the calculation of relevant porosity parameters, a number of assumptions have to be defined:

1. The elution volume (V_e) of a solute—for a given porous structure of the stationary phase— is a function of the molecular size.
2. The solute does not adsorb at the surface of the support material.
3. A distribution equilibrium of the solute between the moving mobile phase and the stagnant liquid present in the pores has to exist.
4. The feasibility of the solute to stay in the moving mobile phase and in the stagnant pore liquid is proportional to the volume of moving mobile phase (interstitial volume) and to the volume of the mobile phase present in the pores (pore volume).
5. The eluted peaks can be described by a Gaussian distribution.
6. Operating parameters like temperature and flow rate have to be constant during measurement.

In order to satisfy all requirements, PS standards in "good solvents for polymers," like tetrahydrofuran (THF) or CH_2Cl_2, are used. Since PS is known to result in linear polymers that build random coils in solution, their molecular size is strictly weight dependent [119]. THF as well as CH_2Cl_2 can easily dissolve PS standards up to M_w of several million. Furthermore, these solvents prevent interactions between the polymer standards and the hydrophobic as well as hydrophilic support materials.

Determination of a pore size distribution profile requires a defined relationship between M_w and the PS diameter (ϕ). For that purpose, PS standards have been measured in SEC mode, on different silica materials with known porosity, employing CH_2Cl_2 and THF as mobile phase, resulting following correlation between M_w and ϕ (Å) for CH_2Cl_2 [119]:

$$M_w = 2.25\phi^{1.7} \tag{1.2}$$

For THF, the relation between M_w and ϕ [Å] has been determined to be [120]

$$M_w = 10.87\phi^{1.7} \tag{1.3}$$

By determining the retention data of polymer standards with known M_w and thus molecular diameter on a column with unknown porosity, a number of important parameters can be calculated:

- The interparticulate volume (V_z) is equivalent to the dead volume of the column and is defined by the retention volume of the largest PS standard. The corresponding porosity (ε_z) can be calculated by dividing V_z by the volume of the empty column.

- The pore volume (V_p) can be evaluated by subtracting V_z from the elution volume of the smallest PS standard (usually benzene or toluene), which is generally supposed to access all pores, being relevant for chromatography. The corresponding porosity (ε_p) is defined by dividing V_p by the volume of the empty column.
- The total column porosity (ε_t) is defined as the sum of ε_z and ε_p.
- Pore distribution curves can be obtained plotting the change in the sum of residuals, which can be calculated by experimental elution volumes of all standards against the mean diameter of the PS standards [119].

1.4.1.4   Comparison between MIP, BET, and ISEC

There is no recommendation of one of the introduced methods (MIP, BET, or ISEC) as the most accurate, reliable, and universally valid technique for the determination of the porous properties of a stationary phase. MIP, BET, and ISEC have rather to be regarded as three independent methodologies, those results complement one another to yield a precise estimation of the porosity of an investigated column packing. The most important characteristics, limitations, and methodological strengths of MIP, BET, and ISEC are intended to be discussed in this section.

The three techniques are characterized by severe differences in their range of measurement:

- Since determination of the pore diameter in case of MIP is proportional to the applied pressure on the mercury column, the lower limit of MIP is defined by the maximal pressure of the instrument. Typically, MIP instruments for routine analysis are constructed to work up to 2000–2500 bar, which corresponds to a pore diameter about 6 nm. The strength of MIP is focused on the macropore range, since it enables the precise determination of pores up to 500 µm.
- The lower limit of BET is set by the molecular diameter of the adsorption gas (N_2). Considering that the ratio of molecule diameter to pore diameter has to be lower then 0.2 for unrestricted excess [205], N_2 (assuming a molecular diameter of 3 Å) can penetrate all pores >1.5 nm. The upper limit of BET cannot be defined, as adsorption of N_2 even takes place at a plain (nonporous) surface.
- The measurement range of ISEC, finally, is defined by the PS standards, used for analysis. While benzene, toluene, or styrene can be employed as the lowest PS standard, the upper limit is characterized by the commercial availability of PS polymers with narrow distribution of M_r (~10,000,000 g/mol), which corresponds to a lowest and a highest pore diameter of 0.8 and 800 nm, respectively.

Figure 1.9 summarizes the measurement ranges for MIP, BET, as well as ISEC. In addition, the range of pore diameters, which are of relevance for chromatographic separation media, is depicted. It can be derived that none of the presented techniques is capable of providing information on all relevant pores. Even if the multiplicative distribution of analytes in the chromatographic process is limited to the fraction of intraparticular porosity (~3–100 nm) only, the determination of interparticulate pores (~0.1–10 µm) is of utmost significance for the evaluation of the quality of column packings and their hydrodynamic properties.

ISEC enables the investigation of the porous structure under chromatographic conditions, whereas MIP as well as BET determine pores in the dry state. Since it is known that particularly support materials based on organic polymers exhibit a certain degree of swelling/shrinkage on changing the solvent or drying (depending on their chemical properties and their degree of cross-linking) [24], ISEC is able to reveal the "true, pristine porosity" of a stationary phase.

Compared with MIP and BET, ISEC is, however, the less comprehensively studied and developed method. The differences in retention of the PS standards, which are proportional to the percentage of pores, present within a certain range, are minor. This makes great demand to the stability of the employed chromatographic system and the constancy of the applied flow rate. This is particularly

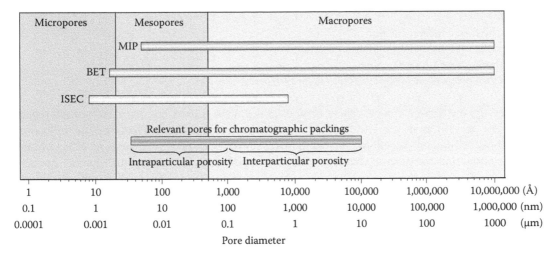

FIGURE 1.9 Schematic illustration of the measurement ranges of MIP, BET, and ISEC together with the pore range, being relevant for chromatographic packings.

true for the evaluation of narrow bore columns, where flow split devices have to be used. The theoretical applicability to capillary columns, however, has been reported in literature [140,172,206].

BET does not provide any information on the pore size distribution of the investigated medium. Consequently, nitrogen sorptiometry is restricted to fast and reliable determination of specific surface areas.

1.5 NEAR INFRARED SPECTROSCOPY

Absorbance signals seen in NIR consist of combination and overtone bands of hydrogen bonds such as C–H, N–H, O–H, and S–H, which are aroused by large force constants and small mass. NIR spectra thus cover precious information on chemical as well as physical properties of analyzed samples due to characteristic reflectance and absorbance patterns [121–123], which makes this analysis method applicable to the characterization of monolithic stationary phases.

In diffuse reflection spectroscopy, the spectrometer beam is reflected from, scattered by, or transmitted through the sample, whereas the diffusely scattered light is reflected back and directed to the detector. The other part of the electromagnetic radiation is absorbed or scattered by the sample [124,125]. Changes in band shapes or intensity as well as signal shifts can be affected by morphological and physicochemical properties of the sample or combinations thereof (e.g., chemical absorptions, particle size, refractive index, surface area, crystallinity, porosity, pore size, hardness, and packing density [126]). Therefore, NIR diffuse reflection spectra can be interpreted in dependence of various physical parameters [127].

In-line measurements are frequently used to perform kinetic studies to follow chemical reactions or to visualize emerging physical and chemical properties like quantities of analytes, particle, or pore size.

The absorption fraction of a particle is related to the volume of the particle. Thus, the larger the volume of a particle, the more of the incident light is absorbed. In contrast, reflectance is related to the particles surface area, being in turn dependent on material porosity. The absorption/remission function relates to the fraction of absorbed light, the fraction of remitted (or back scattered) light, and the fraction of light transmitted by a representative layer

$$A(R,T) = \frac{([1-R_S]^2 - T_S^2)}{R_S} = \frac{A_d(2 - A_d - 2R_d)}{R_d} \tag{1.4}$$

where
  T_S is the transmission fraction
  R_S denotes the measured remission fraction
  A is the absorbed light fraction
  Subscript d represents properties of a layer of thickness d

Generally, the surface of a particle or composite is not one mirror, but many, so for each reflecting surface at a different angle to the light, reflectance signal changes. Therefore, the reflections send light back in many directions; as a result diffuse reflectance occurs. There is no effect of sample absorption, thus the spectral character of the reflected light is the same as the spectral character of the light incident on the sample [128]. Here, absorption only occurs due to different path lengths traveled by the incident light due to scattering from the material. Therefore, changes in porosity and particle size can be interpreted by the collected reflectance spectra [129,130].

1.5.1 MECHANICAL STABILITY AND HYDRODYNAMIC PROPERTIES

Mechanical and chemical stability of novel stationary phases are basic requirements concerning their application. A lack in stability generally causes a loss in resolution and thus reduces column efficiency. In addition, the reproducibility of retention times, being important for qualitative analysis, may be affected. Evaluation of the mechanical stability of polymeric stationary phases is usually accomplished by the determination of the pressure drop across the column, when employing solvents of different polarity within a wide range of flow rates. A stationary phase can be considered as mechanically stable if a linear relationship between applied flow rate and resulting back pressure is obtained.

An important measure concerning column characterization in LC is the column permeability, which represents the capacity of the support to transport the mobile phase as consequence of a pressure drop occurring over the column. In other words, the permeability of a column determines the required pressure to achieve the desired flow rate. The linear flow velocity (u) across an empty cylindrical column is given by

$$u = \frac{F}{r^2\pi} \tag{1.5}$$

where
  F is the volumetric flow rate
  r is the radius of the column

Assuming a column filled with stationary phase, only a certain volume, namely the interparticulate or interstitial volume (ε_z), is accessible to the mobile phase by convection. Thus, the volumetric flow rate through a packed column can be formulated as

$$u = \frac{F}{\varepsilon_z r^2\pi} \tag{1.6}$$

Since chromatographic flow is laminar in nature, the specific permeability (B_0) of a packed column can be calculated using the Darcy equation [131,132]

$$B_0 = \frac{u\eta L}{\Delta p} \tag{1.7}$$

where

Δp is the pressure drop across the column

L is the column length

η is the viscosity of the mobile phase

Even if polymer-based chromatographic supports are generally known to be widely inert to chemical degradation, they suffer from swelling or shrinkage caused by solvent changes. An adequate measure regarding the swelling tendency of polymer-based packing materials is the swelling propensity (SP) factor, which was introduced by Nevejans and Verzele in 1985 [133]:

$$SP = \frac{p_{(solv)} - p_{(H_2O)}}{p_{(H_2O)}} \tag{1.8}$$

where p is defined as the ratio of back pressure to solvent viscosity ($p = P/\eta$). According to Equation 1.8, SP=0 for a nonswelling material. The higher the SP factor, the more a material swells and in contrast, shrinkage is indicated by a value below 0. SP can be determined by measuring back pressure versus flow rate curves for water and organic solvents. THF is frequently used for that purpose, as it is known to be an excellent solvent for organic polymers, causing extensive swelling of the support material. Additionally, acetonitrile (ACN) and methanol should be considered for the determination of the SP factor since they are frequently employed as mobile phase in chromatographic application.

1.5.2 REPRODUCIBILITY OF MONOLITHIC STATIONARY PHASES

The application of HPLC in routine environments, like pharmaceutical, food, or environmental analysis and particularly quality assurance, makes not only great demands on the robustness of HPLC hardware, comprising pumps, column thermostats, and detection units, but in addition to the column reproducibility. Column reproducibility can be investigated at different levels of complexity: Run-to-run reproducibility compares consecutive chromatographic runs, whereas long-term stability describes the column variance over several hundreds of injections. Column-to-column (batch-to-batch) reproducibility finally explores the match of independently fabricated chromatographic columns. Column characteristics that are routinely consulted for the determination of the robustness are retention, selectivity, column efficiency, and peak symmetry.

Monolithic stationary phases as a comparatively young species of HPLC column can only be accepted as a serious alternative to particle-packed columns if column robustness is in the same order of magnitude than their microparticulate counterparts.

Kele et al. reported on a comprehensive reproducibility study of retention times, retention factors, selectivity factor, peak efficiency, hydrophobic and steric selectivities, as well as peak asymmetry for 30 analytes on a set of 6 independently fabricated columns (Chromolith®, Merck) [207], whereas the (bath-to-batch) reproducibility of the columns was characterized by calculation of relative standard deviations (RSD). For most of the analytes, RSD of the absolute retention time as well as retention factors were smaller than 2% (~1.5% on average). In addition, reasonable %RSD have been found with respect to the reproducibility of the column efficiency. With few exceptions (7 of 30 analytes), column-to-column reproducibility has been determined to be >7%. Long-term stability of one selected column delivered considerably better results with deviations in retention time and column efficiency of ~0.2% and ~2%, respectively. The short-term repeatability data on monolithic silica column obtained in this study closely match to previously reported data of microparticulate HPLC column [208–212], which indicates that the fabrication of inorganic silica monoliths, comprising sol–gel polymerization, aging and monolith drying, cladding, as well as surface derivatization and end-capping procedures, is a highly reproducible approach.

1.5.3 PERMEABILITY OF MONOLITHIC STATIONARY PHASES

The absolute permeability (8×10^{-10} cm^2) of currently available silica monoliths (Chromolith, Merck) is similar to that of columns packed with 9 μm particles [213]. To increase efficiency, columns are usually packed with particles of 2, 3, and 5 μm, which possess absolute permeabilities of approximately 4×10^{-11}, 9×10^{-11}, and 2.5×10^{-10} cm^2, respectively. If we try to compare columns packed with 5 μm particles and monolithic columns with through-pores of an average size of 1.5 μm, we observe that the permeability of the monolithic columns is around three to four times larger than that of packed columns.

A study of the influence of the average sizes of the through-pores on the performance of a series of monolithic columns was carried out by Motokawa et al. [214]. This group found permeabilities between 8×10^{-10} and 1.3×10^{-8} cm^2.

The permeability of organic monoliths differ very strongly, depending mainly on the chemical nature of the monomers and porogens, the ratio of porogen to the initial solution, as well as the polymerization time and monomer conversion (see Section 1.3.2.5). Generally, the flow characteristics of organic monoliths are more than less comparable to that obtained for silica monoliths. Both offer significantly higher permeabilities than packed columns with similar chemistry and are highly applicable for high-speed separations.

For instance, Xie et al. prepared 50 and 150×4.6 mm PS/DVB columns to separate proteins and peptides [180]. The permeabilities of these columns are 27 times larger than those of a column packed with 10 μm particles of a similar polymer. Moravcova et al. characterized a number of polymeric narrow-bore (I.D. 320 μm) monolithic columns, using methacrylic monomers [215]. They applied polymer mixtures consisting of butyl methacrylate and ethylene glycol dimethacrylate as the monomers; mixtures of water, 1-propanol, and 1,4-butanediol as the porogens; and AIBN as the initiator of the polymerization. The permeabilities of these columns varied between 8.4×10^{-10} and 1.5×10^{-11} cm^2, depending on the composition of the monomers and the ratio of the porogen to the monomers. Since the decrease of the average sizes of the through-pores usually lead to enhanced column efficiencies, optimizing efficiency is practically limited by the column permeability and vice versa.

1.6 CHROMATOGRAPHIC APPLICATIONS OF ORGANIC AND INORGANIC MONOLITHS

A comprehensive survey of monolithic stationary phases that have been introduced in literature together with their key applications in separation science can be found in Table 1.1.

Styrene monoliths have been prepared by thermally (AIBN or benzoyl peroxide) initiated copolymerization of styrene and divinylbenzene to result mechanically stable, hydrophobic column supports for RPC as well as IP-RP-HPLC and CEC application [24,49,134–140].

Methacrylate monoliths have been fabricated by free radical polymerization of a number of different methacrylate monomers and cross-linkers [107,141–163], whose combination allowed the creation of monolithic columns with different chemical properties (RP [149–154], HIC [158], and HILIC [163]) and functionalities (IEX [141–153,161,162], IMAC [143], and bioreactors [159,160]). Unlike the fabrication of styrene monoliths, the copolymerization of methacrylate building blocks can be accomplished by thermal [141–148], photochemical [149–151,155,156], as well as chemical [154] initiation. In addition to HPLC, monolithic methacrylate supports have been subjected to numerous CEC applications [146–148,151]. Acrylate monoliths have been prepared by free radical polymerization of various acrylate monomers and cross-linkers [164–172]. Comparable to monolithic methacrylate supports, chemical [170], photochemical [164,169], as well as thermal [165–168,171,172] initiation techniques have been employed for fabrication. The application of acrylate polymer columns, however, is more focused on CEC than HPLC.

Acrylamide monoliths as well as methacrylamide monoliths [106,173–179] have been introduced as hydrophilic column support materials for CEC [175,176] of small molecules, as NPC [179] supports for the analysis of low-molecular-weight compounds, and as HIC supports for proteins [106,173]. They have been prepared by thermally (AIBN, benzoyl peroxide) [106,173] as well as chemically initiated (peroxodisulfate/N,N,N',N'-tetramethylethylenediamine) [174,176–179] free radical copolymerization.

Depending on the postpolymerization derivatization procedure, silica-based monolithic columns have been employed for NPC [95] (see also Merck KGaA; Darmstadt, Germany), RPC [189–193,196,197,200] (see also Merck KGaA; Darmstadt, Germany), IEX [194], and HILIC [84,194,198] application. Additionally, their use as efficient bioreactors has recently been reported [86,195].

Table 1.3 summarizes all commercialized monolithic columns that are currently available for HPLC separation. The steadily growing number of commercially available products, based on monolithic packings, express the potential of this kind of stationary phases in particular fields of chromatography and the increasing demand from customer side.

1.6.1 ANALYSIS OF BIOMOLECULES

1.6.1.1 Biopolymer Chromatography on Organic Monoliths

Due to their defined monomodal macropore distribution (see Section 1.2.1), monolithic stationary phases, based on polymerization of organic precursors, are predestined for efficient and swift separation of macromolecules, like proteins, peptides, or nucleic acids, as their open-pore structure of account for enhanced mass transfer due to convection rather than diffusion. In fact, most of the applications of organic monolith introduced and investigated in literature are directed to analysis of biomolecule chromatography [29].

Monolithic PS/DVB, as the most prominent example of a hydrophobic stationary phase, has been a subject of intense investigation over many years. In 1993, Wang et al. already reported on swift reversed-phase separation of proteins on highly permeable 8 mm I.D. monolithic PS/DVB rods (Figure 1.10a). Baseline separation of three proteins was achieved in less than 1 min, employing considerably high volumetric flow rates up to 25 mL/min [41]. Later on, Xie et al. reported on successful RP chromatography of peptides, using PS/DVB rod columns (Figure 1.10b) [180]. The applied 50 and 150×4.6 mm PS/DVB columns exhibited high permeabilities (up to 27 times lager than that of a column packed with 10 μm particles of a similar polymer) and external porosities (0.60 for the monolithic and 0.40 for the packed column). The fast mass transfer kinetics in polymeric columns is confirmed by the high dynamic binding capacity of the columns and the low influence of the flow rate on resolution.

Based on those promising results with respect to efficient separation of biopolymers, Huber worked on the miniaturization of PS/DVB monoliths by means of 200 μm I.D. fused silica capillaries [44–51]. The resulting capillary columns have been proven to be highly homogeneous, enabling high-resolution separation of the whole spectrum of biomolecules. Figure 1.11a and b exemplary illustrates examples for the application of those monoliths to ion-pair reversed-phase separation of homologous phosphorylated oligodeoxynucleotides and dsDNA fragments using tetra-alkylammonium acetate salts as ion-pairing additives. In addition, successful hyphenation of chromatography, employing monolithic PS/DVB columns, and electrospray ionization mass spectrometry have been demonstrated for various applications in genomics and proteomics by different groups [44,181].

Recently, novel monolithic "styrene-like" polymer systems have been introduced in literature. The copolymerization of MS and BVPE as cross-linking agent [139,140] yielded MS/BVPE capillary monoliths (200 μm I.D.) with enhanced stability and exceptionally low swelling in organic solvents. These polymer supports have been applied to efficient protein, peptide, as well as oligonucleotide separation and have successfully served as fast fractionation tool of highly complex mixtures in proteome research studies, prior to MALDI-MS/MS analysis [182]. Silane-based monoliths being synthesized by MS and bis(p-vinylbenzyl)dimethylsilane copolymerization in capillary format excellently performed with respect to protein, peptide, and oligonucleotide

TABLE 1.3
Summary of Commercially Available Monolithic HPLC Columns together with Their Preferred Field of Application

Support Material	Chemical Nature	Brand Name	Company	Dimension	Preferred Field of Application
Monolithic silica	C_{18} reversed phase	Chromolith® RP-18	Merck KgaA	100×4.6 mm 50×4.6 mm 25×4.6 mm	Swift RP analysis of small molecules
Monolithic silica	C_8 reversed phase	Chromolith® RP-8	Merck KgaA	100×4.6 mm	Analysis of small molecules and hydrophobic peptides
Monolithic silica	Underivatized	Chromolith® Si	Merck KgaA	100×4.6 mm	NP analysis of small molecules
Monolithic silica	C_{18} reversed phase	Chromolith® SemiPrep RP-18	Merck KgaA	100×10 mm	Semipreparative chromatography of small molecules
Monolithic silica	C_{18} reversed phase	Chromolith® Prep RP-18	Merck KgaA	100×25 mm	Preparative chromatography of small molecules
Monolithic silica	Underivatized	Chromolith® Prep Si	Merck KgaA	100×25 mm	Preparative chromatography of small molecules
Monolithic silica	C_{18} reversed phase	Chromolith® CapRod	Merck KgaA	150×0.1 mm	Proteome research, analysis of peptides and protein digests
Monolithic silica	C_{18} reversed phase	Onxy monolithic C_{18}	Phenomenex, Inc.	100×3 mm 100×4.6 mm 50×4.6 mm 25×4.6 mm	Swift RP analysis of small molecules
Monolithic silica	C_8 reversed phase	Onxy monolithic C_8	Phenomenex, Inc.	100×4.6 mm	Analysis of small molecules and hydrophobic peptides
Monolithic silica	Underivatized	Onxy monolithic Si	Phenomenex, Inc.	100×4.6 mm	NP analysis of small molecules
Monolithic silica	C_{18} reversed phase	Onxy monolithic C_{18}	Phenomenex, Inc.	100×10 mm	Semipreparative chromatography of small molecules

Material	Product name	Chemistry	Manufacturer	Dimensions	Application
Monolithic silica	Onxy monolithic C_{18}	C_{18} reversed phase	Phenomenex. Inc.	150×0.1 mm	Proteome research, analysis of peptides and protein digests
PS/DVB	PepSwift™	Reversed phase	Dionex Corporation	50×0.1 mm 50×0.2 mm 50×0.5 mm	Nano and cap-HPLC of peptides, protein digest and intact proteins
PS/DVB	ProSwift™ RP	Reversed phase	Dionex Corporation	50×4.6 mm	Swift separation of proteins
Polymethacrylate	ProSwift™ WAX	Weak anion exchanger (tertiary amine)	Dionex Corporation	50×4.6 mm 50×1 mm	Ion exchange chromatography of proteins
Polymethacrylate	ProSwift™ SAX	Strong anion exchanger (quaternary amine)	Dionex Corporation	50×4.6 mm 50×1 mm	Ion exchange chromatography of proteins
Polymethacrylate	ProSwift™ WCX	Weak cation exchanger (carboxylic acid)	Dionex Corporation	50×4.6 mm 50×1 mm	Ion exchange chromatography of proteins
Polymethacrylate	ProSwift™ SCX	Strong cation exchanger (sulfonic acid)	Dionex Corporation	50×4.6 mm 50×1 mm	Ion exchange chromatography of proteins
Organic polymer based	CIM®-dics	Broad variety of surface chemistry[a]	BIA separations	3×16 mm	Purification and fast fractionation of bioploymers and biological products
Organic polymer based	CIM®-tubes	Broad variety of surface chemistry[a]	BIA separations	105×45 mm (8 mL) 15×110 mm (80 mL) 65×150 mm (800 mL)	(Semi)Preparative chromatography of bioploymers, purification and bioconversion

[a] Available chemistry: quaternary amine, diethylamine, ethylenediamine, ethyl, butyl, hydroxyl, sulfonyl, carboxymethyl, epoxy, protein A, protein G, IDA, revered phase.

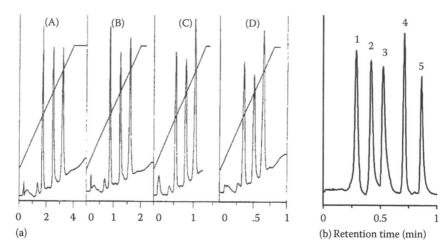

(a)

(b) Retention time (min)

FIGURE 1.10 Examples for the application of conventional monolithic PS/DVB columns for fast separation of proteins and peptides. (a) Separation of cytochrome *c*, myoglobin, and ovalbumin (order of elution) by reversed-phase chromatography on the continuous PS/DVB rod column at different flow rates separation conditions: column, 50×8 mm, mobile phase, linear gradient from 20% to 60% acetonitrile in water; flow rate, 5 mL/min (A), 10 mL/min (B), 15 mL/min (C), and 25 mL/min (D). (Reprinted with permission from Wang, Q.C. et al., *Anal. Chem.*, 65, 2243, 1993. Copyright 1993, American Chemical Society.) (b) Rapid separation of peptides at a flow-rate of 5 mL/min. Column: PS/DVB monolith (50×4.6 mm I.D.); mobile phase gradient: acetonitrile in 0.15% aqueous trifluoroacetic acid solution in 1.5 min. Peaks: bradykinin (1), leucine enkephalin (2), methionine enkephalin (3), physalaemin (4), and substance P (5). Detection: UV 214 nm. (Reprinted from Xie, S. et al., *J. Chromatogr. A*, 865, 169, 1999. Copyright 1999, with permission from Elsevier.)

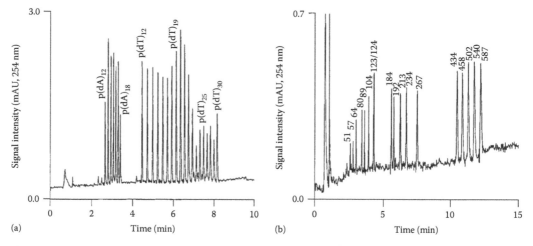

(a)   Time (min)   (b)   Time (min)

FIGURE 1.11 Application of monolithic PS/DVB capillary columns to the high-resolution separation of nucleic acids. (a) High-resolution capillary IP-RP-HPLC separation of phosphorylated oligodeoxynucleotide ladders in a monolithic capillary column. Column, continuous PS-DVB, 60×0.20 mm I.D.; mobile phase, (A) 100 mM TEAA, pH 6.97, (B) 100 mM TEAA, pH 6.97, 20% acetonitrile; linear gradient, 15%–45% B in 3.5 min, 45%–55% B in 2.5 min, 55%–65% B in 4.0 min; flow-rate, 2.5 μL/min; temperature, 50°C; detection, UV, 254 nm; sample, p(dA)$_{12-18}$, p(dT)$_{12-30}$, 40–98 fmol of each oligodeoxynucleotide. (b) High-resolution capillary IP-RP-HPLC separation of a mixture of double-stranded DNA fragments in a monolithic capillary column. Column, continuous PS-DVB, 60×0.20 mm I.D.; mobile phase, (A) 100 mM TEAA, pH 7.00, (B) 100 mM TEAA, pH 7.00, 20% acetonitrile; linear gradient, 35%–75% B in 3.0 min, 75%–95% B in 12.0 min; flow-rate, 2.2 μL/min; temperature, 50°C; detection, UV, 254 nm; sample, pBR322 DNA-Hae III digest, 1.81 fmol of each fragment. (Reprinted from Premstaller, A. et al., *Anal. Chem.*, 72, 4386, 2000. Copyright 2000, with permission from Elsevier.)

fractionation [183]. In addition, the silane group on the cross-linker can serve as functional group for further derivatization, either in situ (on column) as well as prior to polymerization. Marti prepared an acrylic acid-functionalized PS/DVB monolith, which has been utilized as a weak cation exchanger. Steep breakthrough curves have been observed for IgG; however, low specific surface area of the polymer support caused relatively low saturation capacities [184].

Even if a number of reversed-phase and ion-pair reversed-phase separations of biomolecules have been illustrated [mainly on poly(butyl methacrylate-co-ethylene dimethacrylate) or on poly(phenyl acrylate-co-1,4-phenylene diacrylate)] [171,172], monolithic (meth)acrylate columns are more directed to chromatographic applications, which separate biomolecules on the basis of weak hydrophobic (HIC) or (IEX) interactions. Figure 1.12 illustrates two examples for successful separation of oligothymidylic acids on diethylamino functionalized methacrylate monoliths by anion-exchange chromatography [55,185], whereas Figure 1.12a and b demonstrate the respective chromatograms obtained on a conventional IEX HPLC column (50×8 mm I.D.) and an IEX capillary column (80×0.2 mm I.D.), respectively. It can be concluded that miniaturized monolithic columns exhibit considerably better chromatographic performance than monoliths being prepared in conventional column housings, allowing fast and baseline separation of oligonucleotides within a couple of minutes. Similar observations have recently been published for reversed-phase and ion-pair reversed-phase separation of biomolecules on styrene monoliths [140].

Most applications of continuous hydrophilic gel columns are of biochemical nature. Hjerten and his group desorbed serum proteins from agarose gels by ethylene glycol, even in the presence of high salt concentrations, which generally weaken the hydrophilic interaction with the stationary phase. Aqueous solution of ethylene glycol and salts were used to suppress simultaneously hydrophobic interactions and electrostatic interactions, leading to the dissociation of antigens and antibodies bonds as well as bonds between membrane proteins. This method is suited to purify the human growth hormone and to solubilize membrane proteins [186].

Rezeli et al. synthesized antigen (e.g., a protein)-imprinted polymers by adding the protein to the monomer solution. After polymerization, the antigen is washed off by a solution applied to clean the

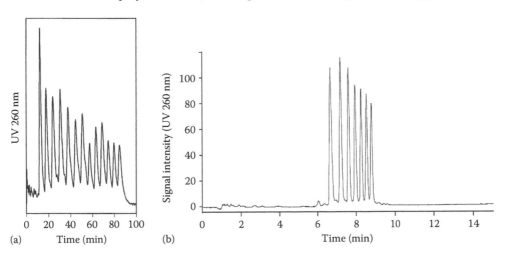

FIGURE 1.12 Anion-exchange separation of oligonucleotides diethylamino-functionalized methacrylate monoliths: (a) Conventional GMA/EDMA monolith (50×8 mm I.D.): separation of oligothymidylic acids [d(pT)$_{12-24}$], conditions: mobile phase gradient from 28% to 41% buffer B in A (A: 20% acetonitrile and 80% 20 mmol/L phosphate buffer, pH 7.0; B: 1 mol/L sodium chloride solution in A) in 90 min; flow rate 1 mL/min; UV detection at 260 nm. (Reprinted from Sykora, D. et al., *J. Chromatogr. A*, 852, 297, 1999. Copyright 1999, with permission from Elsevier.) (b) Capillary GMA/DVB monolith (80×0.2 mm I.D.): separation of oligothymidylic acids [d(pT)$_{12-18}$], conditions: mobile phase A: 20 mM KH$_2$PO$_4$, 20% ACN, pH 7.0; B: 1 M NaCl in A; gradient: 25%–55% B in 2 min, 55%–100% B in 7 min; flow rate: 2.3 µL/min; temperature: 25°C; detection: UV 260 nm.

column [187,188]. A prepared column that initially incorporates hemoglobin showed high selectivity toward hemoglobin leading to complete retaining, whereas cytochrome c and ribonuclease were only slightly retained and separated.

1.6.1.2 Biopolymer Chromatography on Inorganic Monoliths

In the field of biopolymer chromatography, monolithic columns, based on copolymerization of silane precursors, have predominately been investigated regarding separation of peptides and complex protein digests, whereas their application to analysis of high-molecules-weight analytes like proteins or dsDNA fragments is scarcely reported.

Minakuchi et al. investigated the separation of polypeptides on silica rod columns. They have been able to show that the separation of polypeptide mixtures could be carried out approximately three times faster on a silica rod columns than on conventional packed columns [189], which has been ascribed to extraordinarily low mass transfer resistance in monolithic column. Figure 1.13 demonstrates a comparison of a monolithic silica rod column (a) with conventional microparticulate columns with respect to peptide separation at a high linear flow velocity of 4 mm/s. The silica monolith delivered comparable results to that of a 1.5 μm nonporous silica column (Figure 1.13d) and succeeded all investigated porous 5 μm particle-packed columns with respect to resolution and efficiency. The authors concluded that the increase in band broadening with increasing mobile phase flow velocity due to mass transport restriction is less distinctive in case of monolithic rods than in their packed counterparts. A steep gradient can thus be applied for the separation to reduce analysis times at minor expense of resolution. On the basis of these investigations, monolithic silica columns have successfully applied to a number of applications in proteome research, including RP separations of peptides and proteins digests [190]. Kimura et al. worked on a 2D chromatographic approach, employing a 5-cm-long packed IEX column in the first dimension and a 2.5-cm-long silica rod column or a 10-cm-long narrow-bore silica column in the second dimension [191]. Investigation of BSA digests fragments revealed peak capacities around 700. Rieux et al. reported fast fractionation of peptides applying a 56-cm-long silica column (I.D. 50 μm) filled with a monolithic bed [192,193]. Observations of tryptic digests of cytochrome c resulted in high efficient and fast separation of peptides. Again, high mobile phase velocity and steep gradients were employed in order to decrease analysis time without loss of resolution.

In addition to classical reverse phase separation of peptides on octadecyl derivatized silica monoliths, sugars and peptides as well as proteins and nucleosides have been analyzed on a 20-cm-long silica-based poly(acrylic acid) column (I.D. 200 μm), employing HILIC and weak cation-exchange chromatography, respectively [194]. Furthermore, HILIC fractionation of polysaccharides delivered remarkable and promising results [84,194].

On-line hydrolysis of proteins catalyzed by trypsin or pepsin immobilized on monolithic silica beds was described by Kato et al. [86,195], whereas pepsin was encapsulated into the silica-gel matrix (75 μm capillary column), without loss in enzymatic activity [195].

1.6.2 ANALYSIS OF SMALL MOLECULES

1.6.2.1 Separation of Small Molecules on Organic Monoliths

In comparison to their silica counterparts, organic polymer monoliths generally exhibit lower efficiencies in the reversed-phase HPLC separation of small molecules. This reduced performance is primarily due to the lack of mesopores (see Section 1.3.2.5) and the presence of micropores in the polymer matrix, which cause slow internal diffusion. However, there are some promising approaches trying to accomplish the chromatography of small molecules on organic polymer monoliths.

Sinner and Buchmeiser prepared a new class of functionalized monolithic stationary phases (norbornene monoliths) by a ROMP process that tolerates a huge variety of functional monomers [69]. They used this approach in order to chemically attach β-cyclodextrin onto the monomer prior

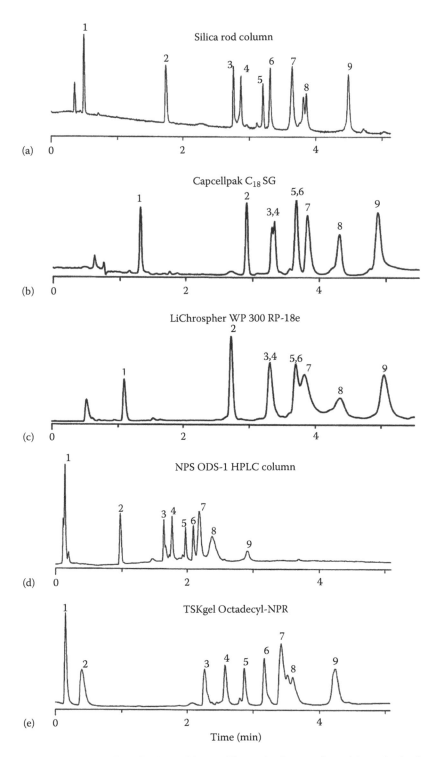

FIGURE 1.13 Gradient separation of polypeptides on silica rod column and particle-packed columns. Mobile phase velocity: 4 mm/s, gradient 5%–60% ACN in the presence of TFA, gradient time: 5 min, columns: (a) silica rod column, (b) Capcellpak C_{18} SG (5 μm), (c) LiChrospher WP 300 RP-18e (5 μm), (d) nonporous NPS-ODS-1 HPLC column (1.5 μm) (e) polymer-based TSKgel Octadecyl-NPR (2.5 μm). (Reprinted from Minakuchi, H. et al., *J. Chromatogr. A*, 828, 83, 1998. Copyright 1998, with permission from Elsevier.)

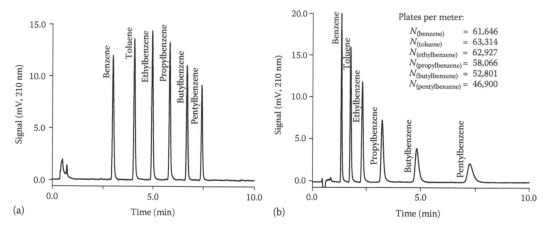

FIGURE 1.14 Reversed-phase separation of a homologous series of alkylbenzenes on a MS/BVPE capillary monolith (80×0.2 mm I.D.), being polymerized for 45 min. (a) Gradient separation; conditions: solvent A: water, solvent B: ACN, 30%–100% B in 10 min, 10 μL/min, separation at 25°C, 500 nL injection, UV 210 nm, 2.5 ng each alkylbenzene. (b) Isocratic separation; conditions: solvent A: water, solvent B: ACN, isocratic at 60% B, 7 μL/min, separation at 25°C, 500 nL injection, UV 210 nm, 2.5 ng each alkylbenzene.

to metathesis polymerization. The resulting monolithic rods have been successfully applied for enantioselective separation of proglumide in less than 3 min.

Hydrophobic monolithic methacrylate capillary columns have been introduced by copolymerization of butyl methacrylate and EDMA as cross-linking agent. The polymerization, however, was not thermally or photochemically but chemically initiated ammonium peroxodisulfate [154]. The resulting monolithic columns were applied to RP separation of small analytes like uracil, phenol, or alkylbenzenes. Reasonable results have been obtained under isocratic conditions, delivering typical values for theoretical plate height ranging between 40 and 50 μm.

Recently, polymerization time as a novel tool for control of pore size distribution was introduced [111]. By reducing the polymerization time, a broadening of pore size distribution together with the transition of a monomodal to a bimodal distribution curve has been observed, considering styrene-like MS/BVPE monolithic columns (see Section 1.3.2.5). That way, the fraction of mesopores could be increased, still keeping up sufficient macropores and flow channels in the micrometer range to apply high flow rates for fast separations. Figure 1.14 illustrates the performance of an MS/BVPE capillary monolith that has been polymerized for 45 min regarding the separation a homologous series of alkylbenzenes under gradient (a) as well as under isocratic conditions (b). Column efficiency in terms of plates per meter has been shown to range between 50,000 and 60,000, which is comparable with standard 5 μm particle-packed C_{18} columns.

1.6.2.2 Separation of Small Molecules on Inorganic Monoliths

Many contributions regarding silica monolithic columns were published by the group of Tanaka [93,189,196]. In their early work, they reported on the successful separation of alkyl benzenes, which are representative for the separation of many low-molecular-weight compounds, containing aromatic groups. Tanaka et al. also combined a conventional column in the first dimension with a silica rod column for the fractionation of aliphatic and aromatic hydrocarbons [197]. The successful separation of the 16 EPA priority pollutants PAHs was carried out by Nunez et al. [93] and is shown in Figure 1.15.

2D chromatography with silica rod columns for the separation of polar aromatic compounds was also employed by Ikegami et al. [197].

Inorganic ions (e.g., Li^+, Na^+, and K^+) were separated by a silica rod column under hydrophilic interaction mode (HILIC) [198]. The studies of Pack and Risley showed amazing high efficiencies at high mobile phase flow rates by applying HILIC conditions toward separation of inorganic ions.

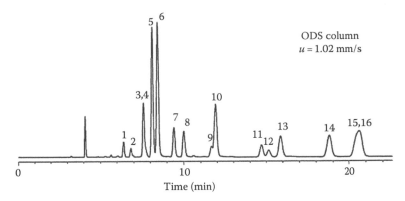

FIGURE 1.15 Separation of the 16 EPA priority pollutants PAHs with ODS column using an acetonitrile:water 70:30 (*v/v*) solution as mobile phase. Thiourea was used as t_M standard. Detection performed at 254 nm and 30°C. PAHs: 1, naphthalene; 2, acenaphtylene; 3, fluorene; 4, acenaphthene; 5, phenanthrene; 6, anthracene; 7, fluoranthene; 8, pyrene; 9, chrysene; 10, benz(*a*)anthracene; 11, benzo(*b*)fluoranthene; 12, benzo(*k*)fluoranthene; 13, benzo(*a*)pyrene; 14, dibenz(*a,h*)anthracene; 15, indeno(1,2,3-*cd*)pyrene; and 16, benzo(*g,h,i*)perylene). (Reprinted from Nunez, O. et al., *J. Chromatogr. A*, 1175, 7, 2007. Copyright 2007, with permission from Elsevier.)

Chankvetadze et al. prepared a 20-cm-long silica capillary column modified by in situ coating with amylase tris(3,5-dimethylphenylcarbamate) [199] for the fast separation of enantiomers. They showed the separation of 10 pairs of enantiomers. The monolith columns exhibit a slightly lower resolution but significant faster separation compared with a conventional 25 cm packed HPLC column.

The rapid separation of iridoid glycosides from extracts of medicinal plants on silica-based monolithic RPLC columns was investigated by Schmidt [200]. He studied the performances of these substances on packed and monolithic columns. While keeping the same degrees of precision and accuracy, and the same detection limit, higher efficiency at a higher flow rate could be achieved with the monolithic columns. Similar results were reported by McFadden et al. who investigated constituents of ecstasy tablets. He was able to separate these components in much less time by using silica-based monolithic columns.

Jia et al. reported of the fractionation of riboflavin and two of its derivatives applying a 25-cm-long ADS-silica monolithic column with 100 μm I.D. [201].

Merck KGaA (Darmstadt, Germany) designed a number of silica monolithic columns (Chromolith) with different dimensions (0.1, 3, 4.6, 10, and 25 mm I.D.; 10–15 cm in length), chemistries (Si, C_{18}, C_8), and applications (chromatography of biomolecules as well as chromatography of small molecules). They are commercially available and offer a broad range of chromatographic usage. Especially for the separation of small molecules like phenols, alkyl benzenes, alkaloids, sulfonamides, steroids, and so on, these types of monoliths show excellent results regarding resolution and separation time.

1.6.3 COMPARISON OF SILICA-BASED MONOLITHS AND ORGANIC MONOLITHS

The strict distinction between silica-based monolithic columns and polymeric columns in most papers and reviews is likely attributed to obvious differences with respect to preparation chemistry and consequently fabrication procedure. Because of the simplicity of the fabrication of organic monoliths, a huge number of monomer chemistry and a wide variety of different columns for a multitude of promising applications have been described in literature. Even the basic synthesis protocol of organic polymers seems to be straightforward, adjusting (fine tuning) the porous properties in order to yield the desired chromatographic characteristics toward the target analytes is critical and demands experience as well as know-how and scientific patience. Silica monoliths, on the other hand, are more difficult and challenging to prepare, in particular with respect to the sensitive "ageing" procedures, which have to be employed after the actual polymerization in order to obtain the desired mesoporous structure and pore-size distribution. Few groups reported on successful attempts for reproducing the preparation methods, initially introduced by Nakanishi et al. Up to now, literature on silica monoliths is mainly restricted to bare silica or C_{18} derivatized silica rods, lacking the huge range of different chemistries offered by their organic counterpart.

The second distinction between organic and inorganic (silica) monolith refer to the typical and preferred column dimension. Even if mechanically stable organic monolithic materials have been fabricated in almost every column dimension, ranging from nanocolumns (20 μm I.D.), over capillary columns to conventional column (2–8 mm I.D.) and even to the preparative format, their main focus with respect chromatographic application is put on capillary column with an I.D. >200 μm. This might be explained by the fact that free radical polymerization, which is a strongly exothermic reaction, creates a radial temperature gradient across the column, which is the more pronounced the larger the diameter of the column mold is [58]. This temperature gradient, in turn, influences the rate of polymerization and thus causes inhomogeneity of the resulting monolith porosity, which are closely associated with decreased column performance. In fact, a direct comparison of conventional HPLC columns (3 mm I.D.), based on a monolithic styrene network, has revealed a considerable loss in efficient than their capillary counterparts (200 μm I.D.) [140]. Silica monoliths, on the other hand, are mainly produced as conventional sized HPLC columns, even if recently Merck commercialized a silica monolith in a 100 μm fused silica capillary. However, narrow pore and capillary columns within a diameter range of 100–500 μm are scarcely reported in literature. This is attributed to the extensive shrinkage of the monolithic skeleton during the drying process, which demands for a cladding (leakproof fit into a column housing) procedure after polymerization. In case the silica rod is prepared in a sufficiently narrow (I.D. < 100 μm) silica tube, the gel remains glued to the wall. As the cladding process is from a technical point of view exceedingly difficult to realize for capillary columns up to 500 μm, miniaturized column between 100 and 500 μm I.D. cannot be prepared following the classical sol–gel approach. However, recently, alternative polymerization procedures have been introduced that allow the preparation of narrow pore and capillary silica monoliths without wall detachment [213].

Finally, silica and polymer monoliths significantly differ in their preferred field of application. While organic monoliths numerous times proved to enable highly efficient and swift resolution of large biomolecules, silica monoliths focus on the application of monolithic stationary phases by providing powerful and promising result for fast and high-resolution separation of low-molecular-weight compounds. The characteristic bimodal pore size distribution of silica monoliths establishes separation performance for small molecules, ascribed due to a high fraction of accessible mesopores, whereas the monomodal macropore distribution of typical organic monoliths comply with the requirements of macromolecule chromatography, mainly by reducing resistance to mass transfer by substituting analyte diffusion by solvent convention.

However, silica monoliths and organic polymers both exhibit very advantageous chromatographic characteristics: enhanced mass transfer characteristics, high reproducibility, and versatile surface chemistry, which make monolithic column attractive for a variety of forward-looking applications.

REFERENCES

1. H. Staudinger, E. Huseman. *Berichte der Deutschen Chemischen Gesellschaft* 68, 1618–1634, 1935.
2. G.F. D'Alelio. U.S. Patent 2,366,007, 1945.
3. K.W. Pepper. *J. Appl. Chem.* 1, 124–132, 1951.
4. I.M. Abrams. *J. Ind. Eng. Chem.* 48, 1469–1472, 1956.
5. O. Okay. *Prog. Polym. Sci.* 25, 711–779, 2000.
6. K. Dusek. *Chem. Prumysl.* 11, 439–443, 1961.
7. R. Kunin, E. Meitzner, N. Bortnick. *J. Am. Chem. Soc.* 84, 305–306, 1962.
8. J.R. Millar, D.G. Smith, W.E. Marr, T.R.E. Kressman. *J. Chem. Soc.* 2779–2784, 1963.
9. J. Malinsky, J. Rahm, F. Krska, J. Seidl. *Chem. Prumysl.* 13, 386, 1963.

10. F. Svec, F.M.F. Fréchet. *Anal. Chem.* 64, 820–822, 1992.
11. D.L. Mould, R.L.M. Synge. *Analyst* 77, 964–970, 1952.
12. D.L. Mould, R.L.M. Synge. *Biochem. J.* 58, 571–585, 1954.
13. M. Kubin, P. Spacek, R. Chromekec. *Coll. Czechosl. Chem. Commun.* 32, 3881–3887, 1967.
14. H. Schnecko, O. Bieber. *Chromatographia* 4, 109–112, 1971.
15. L.C. Hansen, R.E. Sievers. *J. Chromatogr.* 99, 123–133, 1974.
16. F.D. Hileman, R.E. Sievers, G.G. Hess, W.D. Ross. *Anal. Chem.* 45, 1126–1130, 1973.
17. T.R. Lynn, D.R. Rushneck, A.R. Cooper. *J. Chromatogr. Sci.* 12, 76–79, 1974.
18. S. Hjérten, J.L. Liao, R. Zhang. *J. Chromatogr.* 473, 273–275, 1989.
19. J.L. Liao, R. Zang, S. Hjérten. *J. Chromatogr.* 586, 21–26, 1991.
20. B.G. Belenkii, A.M. Podkladenko, O.I. Kurenbin, V.G. Maltsev, D.G. Nasledov, S.A. Trushin. *J. Chromatogr. A* 645, 1–15, 1993.
21. T.B. Tennikova, B.G. Belenkii, F. Svec. *J. Liquid Chromatogr.* 13, 63–70, 1990.
22. F. Svec, T.B. Tennikova. *J. Bioact. Compat. Polym.* 6, 393–405, 1991.
23. T.B. Tennikova, F. Svec. *J. Chromatogr.* 646, 279–288, 1993.
24. Q.C. Wang, F. Svec, J.M.J. Fréchet. *Anal. Chem.* 65, 2243–2248, 1993.
25. E.C. Peters, F. Svec, J.M.J. Fréchet. *Adv. Mater.* 11, 1169–1181, 1999.
26. F. Svec. *J. Sep. Sci.* 27, 1419–1430, 2004.
27. F. Svec, E.C. Peters, D. Sykora, J.M.J. Fréchet. *J. Chromatogr. A* 887, 3–29, 2000.
28. E. Klodzińska, D. Moravcova, P. Jandera, B. Buszewski. *J. Chromatogr. A* 1109, 51–59, 2006.
29. F. Svec, C.G. Huber. *Anal. Chem.* 78, A2101–A2107, 2006.
30. B.G. Belenkii. *Russ. J. Bioorg. Chem.* 32, 323–332, 2006.
31. F. Svec, J.M.J. Fréchet. *Ind. Eng. Chem. Res.* 38, 34–48, 1999.
32. K. Štulik, V. Pacakova, J. Suchankova, P. Coufal. *J. Chromatogr. B* 841, 79–87, 2006.
33. H. Minakuchi, N. Nakanishi, N. Soga, N. Ishizuka, N. Tanaka. *Anal. Chem.* 68, 3498–3501, 1996.
34. S.M. Fields. *Anal. Chem.* 68, 2709–2712, 1996.
35. J.F. Keenedy, G.O. Phillips, P.A. Williams. *Cellulosics: Materials for Selective Separations and Other Technologies.* Horwood, New York, 1993.
36. F. Sinner, M.R. Buchmeiser. *Macromolecules* 33, 5777–5786, 2000.
37. B. Mayr, R. Tessadri, E. Post, M.R. Buchmeiser. *Anal. Chem.* 73, 4071–4078, 2001.
38. B. Mayr, G. Hölzl, K. Eder, M.R. Buchmeiser, C.G. Huber. *Anal. Chem.* 74, 6080–6087, 2002.
39. S. Lubbad, M.R. Buchmeiser. *Macromol. Rapid Commun.* 23, 617–612, 2002.
40. K. Hosoya, N. Hira, K. Yamamoto, M. Nishimura, N. Tanaka. *Anal. Chem.* 78, 5729–5735, 2006.
41. Q.C. Wang, F. Svec, J.M.J. Fréchet. *Anal. Chem.* 65, 2243, 1993.
42. M. Petro, F. Svec, J.M.J. Fréchet. *J. Chromatogr. A* 752, 59, 1996.
43. Q.C. Wang, F. Svec, J.M.J. Fréchet. *J. Chromatogr. A* 669, 230, 1994.
44. A. Premstaller, H. Oberacher, W. Walcher, A.M. Timperio, L. Zolla, J.-P. Chervet, N. Cavusoglu, A. van Dorsselaer, C.G. Huber. *Anal. Chem.* 73, 2390, 2001.
45. W. Walcher, H. Oberacher, S. Troiani, G. Hölzl, P. Oefner, L. Zolla, C.G. Huber. *J. Chromatogr. B* 782, 111, 2002.
46. W. Walcher, H. Toll, A. Ingendoh, C.G. Huber. *J. Chromatogr. A* 1053, 107, 2004.
47. H. Oberacher, A. Krajete, W. Parson, C.G. Huber. *J. Chromatogr. A* 893, 23, 2000.
48. H. Oberacher, C.G. Huber. *Trend. Anal. Chem.* 21, 166, 2002.
49. A. Premstaller, H. Oberacher, C.G. Huber. *Anal. Chem.* 72, 4386–4393, 2000.
50. G. Hölzl, H. Oberacher, S. Pitsch, A. Stutz, C.G. Huber. *Anal. Chem.* 77, 673, 2005.
51. C.G. Huber, A. Krajete. *Anal. Chem.* 71, 3730, 1999.
52. X. Huang, S. Zhang, G.A. Schultz, J. Henion. *Anal. Chem.* 74, 2336, 2002.
53. F. Svec. *J. Sep. Sci.* 27, 747–766, 2004.
54. A. Podgornik, M. Barut, J. Jancar, A. Strancar. *J. Chromatogr. A* 848, 51, 1999.
55. D. Sykora, F. Svec, J.M.J. Fréchet. *J. Chromatogr. A* 852, 297, 1999.
56. F. Svec, J.M.J. Fréchet. *J. Chromatogr. A* 1995, 89, 1995.
57. E. Suarez, B. Paredes, F. Rubiera, M. Rendueles, M.A. Villa-Garcia, J.M. Diaz. *Sep. Purif. Technol.* 27, 1, 2002.
58. A. Podgornik, M. Barut, A. Strancar, D. Josic, T. Koloini. *Anal. Chem.* 72, 5693, 2000.
59. M. Barut, A. Podgornik, P. Brne, A. Strancar. *J. Sep. Sci.* 28, 1876, 2005.
60. J. Urthaler, R. Schlegl, A. Podgornik, A. Strancar, A. Jungbauer, R. Necina. *J. Chromatogr. A* 1065, 93, 2005.
61. M. Bencina, A. Podgornik, A. Strancar. *J. Sep. Sci.* 27, 801, 2004.

62. C. Viklund, F. Svec, J.M.J. Fréchet, K. Irgum. *Biotechnol. Prog.* 13, 597, 1997.
63. Q. Luo, H. Zou, X. Xiao, Z. Guo, L. Kong, X. Mao. *J. Chromatogr. A* 926, 255, 2001.
64. M. Petro, F. Svec, J.M.J. Fréchet. *Biotechnol. Bioeng.* 49, 355, 1996.
65. S. Xie, F. Svec, J.M.J. Fréchet. *Biotechnol. Bioeng.* 62, 30, 1999.
66. S. Hjertén, J.-L. Liao, R. Zhang. *J. Chromatogr.* 473, 273, 1989.
67. J.-L. Liao, S. Zhang, S. Hjertén. *J. Chromatogr.* 586, 21, 1991.
68. Y.-M. Li, J.-L. Liao, K. Nakazato, J. Mohammad, L. Terenius, S. Hjertén. *Anal. Biochem.* 223, 153, 1994.
69. F.M. Sinner, M.R. Buchmeiser. *Angew. Chem. Int. Ed.* 39, 1433, 2000.
70. C. Gatschelhofer, C. Magnes, T.R. Pieber, M.R. Buchmeiser, F.M. Sinner, *J. Chromatogr. A* 1090, 81, 2005.
71. S. Lubbad, B. Mayr, C.G. Huber, M.R. Buchmeiser, *J. Chromatogr. A* 959, 121, 2002.
72. B. Mayr, R. Tessadri, E. Post, M.R. Buchmeiser. *Anal. Chem.* 73, 4071, 2001.
73. L.A. Errede. *Polym. Preprints* 26, 77, 1985.
74. L.A. Errede. *J. Appl. Polym. Sci.* 31, 1746–1761, 1986.
75. J. Vidič, A. Podgornik, A. Štrancar. *J. Chromatogr. A* 1065, 51–58, 2005.
76. K. Nakanishi, N. Soga. *J. Am. Ceram. Soc.* 74, 2518, 1991.
77. K. Nakanishi, N. Soga. *J. Non-Cryst. Solids* 139, 1, 1992.
78. K. Nakanishi, N. Soga. *J. Non-Cryst. Solids* 139, 14, 1992.
79. K. Nakanishi, N. Soga, Inorganic porous column, Japan Patent 5-200,392, 1993.
80. K. Nakanishi, N. Soga, Production of inorganic porous body, Japan Patent 5-208,642, 1993.
81. K. Nakanishi, N. Soga, Inorganic porous material and process for making same, U.S. Patent 5,624,875, 1997.
82. K. Cabrera, G. Sättler, G. Wieland, Trennmittel (separator), European patent EP 0,686,258 b1, 1994.
83. N. Tanaka, H. Kobayashi, N. Ishizuka, H. Minakushi, K. Nakanishi, K. Hosoya, T. Itegami. *J. Chromatogr. A* 965, 35, 2002.
84. T. Ikegami, H. Kujita, K. Horie, K. Hosoya, N. Tanaka. *Anal. Bional. Chem.* 386, 578, 2006.
85. S. Wienkoop, M. Glinski, N. Tanaka, V. Tolstikof, O. Fiehn, W. Weckwerth. *Rapid Commun. Mass Spectrom.* 18, 643, 2004.
86. M. Kato, K. Inuzuka, K. Sakai-Kato, T. Toyo'oka. *Anal. Chem.* 77, 1813, 2005.
87. A.-M. Siouffi. *J. Chromatogr. A* 1000, 801, 2003.
88. A. Vegvari. *J. Chromatogr. A* 1079, 50, 2005.
89. N. Ishizuka, H. Minakuchi, K. Nakanishi, K. Hirao, N. Tanaka, *Colloids Surf. A* 187–188, 273, 2001.
90. H. Saito, K. Nakanishi, K. Hirao, H. Jinnai. *J. Chromatogr. A* 1119, 95, 2006.
91. K. Nakanishi, H. Minakuchi, N. Soga, N. Tanaka. *J. Sol–Gel Sci. Technol.* 13, 163, 1998.
92. C. Yang, T. Ikegami, T. Hara, N. Tanaka. *J. Chromatogr. A* 1130, 175, 2006.
93. O. Nunez, T. Ikegami, K. Miyamoto, N. Tanaka. *J. Chromatogr. A* 1175, 7–15, 2007.
94. L.A. Colon, D.C. Hoth, Group IV Metal Oxide Monolithic Columns, PCT Int. Appl. WO200509, 1972, 2005.
95. D.C. Hoth, J.G. Rivera, L.A. Colón. *J. Chromatogr. A* 1079, 392, 2005.
96. J. Randon, S. Huguet, A. Piram, G. Puy, C. Demesmay, J.-L. Rocca. *J. Chromatogr. A* 1109, 19, 2006.
97. A. Taguchi, J.-H. Smatt, M. Linden. *Adv. Mater.* 15, 1209, 2003.
98. C. Liang, S. Dai, G. Guiochon. *Anal. Chem.* 75, 4904, 2003.
99. C. Liang, Synthesis and applications of monolithic HPLC columns, PhD Thesis, University of Tennessee, Knoxville, TN, 2005.
100. E.C. Peters, C. Ericson. In: *Monolithic Materials, J. Chromatogr. Libr.*, Vol. 67. Elsevier, Amsterdam, the Netherlands, 2003.
101. J. Seidl, J. Malinsky, K. Dusek, W. Heitz. *Fortschritte der Hochpolymeren-Forschung*, 5, 113–213, 1967.
102. A. Guyot, M. Bartholin. *Prog. Polym. Sci.* 8, 277–331, 1982.
103. K.A. Kun, R. Kunin. *J. Polym. Sci.* 6, 2689–2701, 1968.
104. W.L. Sederel, G.J. DeJong. *J. Appl. Polym. Sci.* 17, 2835–2846, 1973.
105. C. Viklund, F. Svec, J.M.J. Fréchet, K. Irgum. *Chem. Mater.* 8, 744–750, 1996.
106. S. Xie, F. Svec, J.M.J. Fréchet. *J. Polym. Sci. A* 35, 1013–1021, 1997.
107. C. Viklund, E. Ponten, B. Glad, K. Irgum, P. Hörstedt, F. Svec. *Chem. Mater.* 9, 463–471, 1997.
108. F. Svec, J.M.J. Fréchet. *Chem. Mater.* 7, 707–715, 1995.
109. F. Svec, J.M.J. Fréchet. *Macromolecules* 28, 7580–7582, 1995.
110. J. Brandrup, E.H. Immergut. *Polymer Handbook*. Wiley, New York, 1989.

111. G. Bonn, S. Lubbad, L. Trojer. U.S. Patent 144,971 A1, 2007.
112. L. Trojer, G. Stecher, I. Feuerstein, G.K. Bonn. *Rapid Commun. Mass Spectrom.* 19, 3398–3404, 2005.
113. L. Trojer, G. Stecher, I. Feuerstein, S. Lubbad, G.K. Bonn. *J. Chromatogr. A* 1079, 197–207, 2005.
114. K.S.W. Sing, D.H. Everett, R.A.W. Haul, L. Moscou, R.A. Pierotti, J. Rouquerol, T. Siemieniewska. *Pure Appl. Chem.* 57, 603–619, 1985.
115. E.W. Washburn. *Proc. Natl. Acad. Sci. U.S.A.* 7, 115, 1921.
116. E.W. Washburn. *Phys. Rev.* 17, 273, 1921.
117. H.M. Rootare, C.F. Prenzlow. *J. Phys. Chem.* 71, 2733, 1967.
118. P.H. Emmett, T.W. DeWitt. *J. Am. Chem. Soc.* 65, 1253, 1943.
119. I. Halász, K. Martin. *Angew. Chem.* 90, 954, 1978.
120. M.E. Van Kreveld, N. Van den Hoed. *J. Chromatogr.* 83, 111–124, 1973.
121. H.W. Siesler., Y. Ozaki, S. Kawata, H.M. Heise, *Near-Infrared Spectroscopy—Principles, Instruments, Applications.* Wiley-VCH, Weinheim, Germany, 2002.
122. J. Workman, L. Weyer, *Practical Guide to Interpretive Near-infrared Spectroscopy.* Taylor & Francis, Boca Raton, FL, 2007.
123. D.A. Burns, E.W. Ciurczak, *Handbook of Near-Infrared Analysis.* CRC Press, Boca Raton, FL, 2001.
124. B.J. Berne, R. Pecora, *Dynamic Light Scattering.* Dover Publications, Inc. New York, 2000.
125. II.C. van de Hulst, *Light Scattering by Small Particles.* Dover Publications, Inc., New York, 1981.
126. D.J. Dahm, K.D. Dahm, K.H. Norris. *J. Near Infrared Spectrosc.* 10(1), 1–13, 2002.
127. C.W. Huck, R. Ohmacht, Z. Szabo, G.K. Bonn. *J. Near Infrared Spectrosc.* 14, 51–57, 2006.
128. D.J. Dahm, K.D. Dahm, *Interpreting Diffuse Reflectance and Transmittance.* IM Publications, Chichester, U.K., 2007.
129. N. Heigl, C.H. Petter, M. Rainer, M. Najam-ul-Haq, R.M. Vallant, R. Bakry, G.K. Bonn, C.W. Huck. *J. Near Infrared Spectrosc.* 15(5), 269–282, 2007.
130. C.W. Huck, N. Heigl, M. Najam-ul-Haq, M. Rainer, R.M. Vallant, G.K. Bonn. *Open Anal. Chem. J.* 1, 21–27, 2007.
131. C.A. Cramers, J.A. Rijks, C.P.M. Schutjes. *Chromatographia* 14, 439, 1981.
132. P.A. Bristow, J.H. Knox. *Chromatographia* 10, 279, 1977.
133. F. Nevejans, M. Verzele. *J. Chromatogr.* 350, 145, 1985.
134. M. Petro, F. Svec, I. Gitsov, J.M.J. Fréchet. *Anal. Chem.* 68, 315–321, 1996.
135. Q.C. Wang, F. Svec, J.M.J. Fréchet. *J. Chromatogr. A* 669, 230–235, 1994.
136. I. Gusev, X. Huang, Cs. Horvath. *J. Chromatogr. A* 855, 273–290, 1999.
137. W. Walcher, H. Oberacher, S. Troiani, G. Hölzl, P. Oefner, L. Zolla, C.G. Huber. *J. Chromatogr. B* 782, 111–125, 2002.
138. Q.C. Wang, F. Svec, J.M.J. Fréchet. *Anal. Chem.* 67, 670–674, 1995.
139. L. Trojer, S.H. Lubbad, C.P. Bisjak, G.K. Bonn. *J. Chromatogr. A* 1117, 56–66, 2006.
140. L. Trojer, S.H. Lubbad, C.P. Bisjak, W. Wieder, G.K. Bonn. *J. Chromatogr. A* 1146, 216–224, 2007.
141. F. Svec, J.M.J. Fréchet. *J. Chromatogr. A* 702, 89–95, 1995.
142. C. Viklund, F. Svec, J.M.J. Fréchet. *Biotechnol. Prog.* 13, 597–600, 1997.
143. D. Sykora, F. Svec, J.M.J. Fréchet. *J. Chromatogr. A* 852, 297–304, 1999.
144. Q. Luo, H. Zou, X. Xiao, Z. Guo, L. Kong, X. Mao. *J. Chromatogr. A* 926, 255–264, 2001.
145. Y. Ueki, T. Umemura, J. Li, T. Odake, K. Tsunoda. *Anal. Chem.* 76, 7007–7012, 2004.
146. E.C. Peters, M. Petro, F. Svec, J.M.J. Fréchet. *Anal. Chem.* 69, 3646–3649, 1997.
147. E.C. Peters, M. Petro, F. Svec, J.M.J. Fréchet. *Anal. Chem.* 70, 2296–2302, 1998.
148. G. Ping, L. Zhang, L. Zhang, W. Zhang, P. Schmitt-Kopplin, A. Kettrup, Y. Zhang. *J. Chromatogr. A* 1035, 265–270, 2004.
149. D. Lee, F. Svec, J.M.J. Fréchet. *J. Chromatogr. A* 1051, 53–60, 2004.
150. L. Geiser, S. Eeltnik, F. Svec, J.M.J. Fréchet. *J. Chromatogr. A* 1140, 140–146, 2007.
151. C. Yu, M. Xu, F. Svec, J.M.J. Fréchet. *J. Polym. Sci. A* 40, 755–769, 2002.
152. P. Coufal, M. Čihak, J. Suchankova, E. Tesařova, Z. Bosanova, K. Štulk. *J. Chromatogr. A* 946, 99–106, 2002.
153. D. Moravcova, P. Jandera, J. Urban, J. Planeta. *J. Sep. Sci.* 26, 1005–1016, 2003.
154. P. Holdšvendova, P. Coufal, J. Suchankova, E. Tesařova, Z. Bosakova. *J. Sep. Sci.* 26, 1623–1628, 2003.
155. T. Rohr, C. Yu, M.H. Davey, F. Svec, J.M.J. Fréchet. *Electrophoresis* 22, 3959–3967, 2001.
156. C. Yu, M.H. Davey, F. Svec, J.M.J. Fréchet. *Anal. Chem.* 73, 5088–5096, 2001.
157. L. Uzun, R. Say, A. Denizli. *React. Funct. Polym.* 64, 93–102, 2005.
158. P. Hemström, A. Nordborg, K. Irgum, F. Svec, J.M.J. Frechet. *J. Sep. Sci.* 29, 25–32, 2006.
159. D.S. Peterson, T. Rohr, F. Svec, J.M.J. Fréchet. *Anal. Chem.* 74, 4081–4088, 2002.

160. S. Xie, F. Svec, J.M.J. Fréchet. *Biotechnol. Bioeng.* 62, 30–35, 1999.
161. C. Viklund, A. Sjogren, K. Irgum. *Anal. Chem.* 73, 444–452, 2000.
162. C. Viklund, K. Irgum. *Macromolecules* 33, 2539–2544, 2000.
163. Z. Jiang, N.W. Smith, P.D. Ferguson, M.R. Taylor. *Anal. Chem.* 79, 1243–1250, 2007.
164. B. Gu, J.M. Armenta, M.L. Lee. *J. Chromatogr. A* 1079, 382–391, 2005.
165. L.J. Sondergeld, M.E. Bush, A. Bellinger, M.M. Bushey. *J. Chromatogr. A* 1004, 155–165, 2003.
166. B.L. Waguespack, S.A. Slade, A. Hodges, M.E. Bush, L.J. Sondergeld, M.M. Bushey. *J. Chromatogr. A* 1078, 171–180, 2005.
167. M. Bedair, Z. El Rassi. *Electrophoresis* 23, 2938–2948, 2002.
168. S.M. Ngola, Y. Fintschenko, W.-Y. Choi, T.J. Shepodd. *Anal. Chem.* 73, 849–856, 2001.
169. R. Shediac, S.M. Ngola, D.J. Throckmorton, D.S. Anex, T.J. Shepodd, A.K. Singh. *J. Chromatogr. A* 925, 251–263, 2001.
170. O. Kornyšova, A. Maruška, P.K. Owens, M. Erickson. *J. Chromatogr. A* 1071, 171–178, 2005.
171. C.P. Bisjak, S.H. Lubbad, L. Trojer, G.K. Bonn. *J. Chromatogr. A* 1147, 46–52, 2007.
172. C.P. Bisjak, L. Trojer, S.H. Lubbad, W. Wieder, G.K. Bonn. *J. Chromatogr. A* 1154, 269–276, 2007.
173. S. Xie, F. Svec, J.M.J. Fréchet. *J. Chromatogr. A* 775, 65–72, 1997.
174. D. Hoegger, R. Freitag. *J. Chromatogr. A* 914, 211–222, 2001.
175. D. Hoegger, R. Freitag. *Electrophoresis* 24, 2958–2972, 2003.
176. F.M. Plieva, J. Andersson, I.Y. Galaev, B. Mattiasson. *J. Sep. Sci.* 27, 828–836, 2004.
177. P. Arvisson, F.M. Plieva, V.I. Lozinsky, I.Y. Galaev, B. Mattiasson. *J. Chromatogr. A* 986, 275–290, 2003.
178. F.M. Plieva, I.N. Savina, S. Deraz, J. Andersson, I.Y. Galaev, B. Mattiasson. *J. Chromatogr. B* 807, 129–137, 2004.
179. A. Maruška, C. Ericson, A. Vegvari, S. Hjerten. *J. Chromatogr. A* 837, 25–33, 1999.
180. S. Xie, R.W. Allington, F. Svec, J.M.J. Fréchet. *J. Chromatogr. A* 865, 169, 1999.
181. G. Yue, Q. Luo, J. Zhang, S. Wu, B.L. Karger. *Anal. Chem.* 79, 938–946, 2007.
182. R.M. Vallant, Z. Szabo, L. Trojer, M. Najam-ul-Haq, M. Rainer, C.W. Huck, R. Bakry, G.K. Bonn. *J. Proteome Res.* 6, 44–53, 2006.
183. W. Wieder, S.H. Lubbad, L. Trojer, C. Bisjak, G.K. Bonn. *J. Chromatogr. A* 1191, 253–262, 2008.
184. N. Marti, PhD thesis, Institute for Chemical and Bioengineering, ETH, Zürich, Switzerland, 2007.
185. C.P. Bisjak, R. Bakry, C.W. Huck, G.K. Bonn. *Chromatographia* 62, 31–36, 2005.
186. S. Hjerten, H.J. Issaq. *A Century of Separation Science*, Marcel Dekker, Inc., New York, p. 421, 2001.
187. M. Rezeli, F. Kilar, S. Hjerten. *J. Chromatogr. A* 1109, 100, 2006.
188. J.-L. Liao, Y. Wang, S. Hjerten. *Chromatographia* 42, 259, 1996.
189. H. Minakuchi, K. Nakanishi, N. Soga, N. Tanaka. *J. Chromatogr. A* 828, 83, 1998.
190. B. Barroso, D. Lubda, R. Bishoff. *J. Proteome Res.* 2, 633, 2003.
191. H. Kimura, T. Tanigawa, H. Morisaka, T. Ikegami, K. Hosoya, N. Ishizuka, H. Minakuchi, K. Nakanishi, M. Ueda, K. Cabrera, N. Tanaka. *J. Sep. Sci.* 27, 897, 2004.
192. L. Rieux, H. Niederländer, E. Velpoorte, R. Bischoff. *J. Sep. Sci.* 28, 1628, 2005.
193. L. Rieux, D. Lubda, H.A.G. Niederländer, E. Verpoorte, R. Bischoff. *J. Chromatogr. A* 1120, 165, 2006.
194. T. Ikegami, K. Horie, J. Jaafar, K. Hosoya, N. Tanaka. *Biochem. Biophys. Methods* 70, 31, 2007.
195. M. Kato, K. Satai-Kato, H. Jin, K. Kubota, H. Miyano, T. Toyooka, M.T. Dulay, R.N. Zare. *Anal. Chem.* 76, 1896, 2004.
196. T. Ikegami, T. Hara, H. Kimura, H. Kobayashi, K. Hosoya, K. Cabrera, N. Tanaka. *J. Chromatogr. A* 1106, 112, 2006.
197. N. Tanaka, H. Kimura, D. Tokuda, K. Hosoya, T. Ikegami, N. Ishizuka, H. Minakuchi, K. Nakanishi, Y. Shintani, M. Furuno, K. Cabrera. *Anal. Chem.* 76, 1273, 2004.
198. B.W. Pack, D.S. Risley. *J. Chromatogr. A* 1073, 269, 2005.
199. B. Chankvetadze, C. Yamamoto, M. Kamigaito, N. Tanaka, K. Nakanishi, Y. Okamoto, *J. Chromatogr. A* 1110, 46, 2006.
200. A.H. Schmidt. *J. Chromatogr. A* 1073, 377, 2005.
201. L. Jia, N. Tanaka, S. Terabe. *J. Chromatogr. A* 1053, 71, 2004.
202. S. Brunauer. *Physical Adsorption*, Vol. 1. Princeton University Press, Princeton, NJ, 1945.
203. S. Brunauer, P.H. Emmet. *J. Am. Chem. Soc.* 57, 1754–1755, 1935.
204. S. Brunauer, P.H. Emmett, E. Teller. *J. Am. Chem. Soc.* 60, 309–319, 1938.
205. K.K. Unger, R. Janzen, G. Jilge. *Chromatographia* 24, 144–154, 1987.
206. H. Oberacher, A. Premstaller, C.G. Huber. *J. Chromatogr. A* 1030, 201–208, 2004.
207. M. Kele, G. Guiochon. *J. Chromatogr. A* 960, 19–49, 2002.

208. M. Kele, G. Guiochon. *J. Chromatogr. A* 830, 41–54, 1999.
209. M. Kele, G. Guiochon. *J. Chromatogr. A* 830, 55–79, 1999.
210. M. Kele, G. Guiochon. *J. Chromatogr. A* 855, 423–453, 1999.
211. M. Kele, G. Guiochon. *J. Chromatogr. A* 869, 181–209, 2000.
212. M. Kele, G. Guiochon. *J. Chromatogr. A* 913, 89–112, 2001.
213. G. Guiochon. *J. Chromatogr. A* 1168, 101–168, 2007.
214. M. Motokawa, H. Kobayashi, N. Ishizuka, H. Minakushi, K. Nakanishi, H. Jinnai, K. Hosoya, T. Itegami, N. Tanaka. *J. Chromatogr. A* 961, 53, 2002.
215. D. Moravcova, P. Jandera, J. Urban, J. Planeta. *J. Sep. Sci.* 27, 789, 2004.

2 Bonded Stationary Phases

Heinz Engelhardt

CONTENTS

2.1 INTRODUCTION

High-performance liquid chromatography (HPLC) has become the dominant analytical technique in pharmaceutical, chemical, and food industries, as well as in environmental laboratories, in clinical chemistry for therapeutic drug monitoring, and in bioanalysis [1]. The rise of HPLC to the most widely used instrumental analytical systems originates in part from the broad variety of selectivities introduced by the enormous number of stationary phases available and the easy adjustment of selectivity by changing the composition of the mobile phase. The classical separation systems based on pure silica or alumina—now called normal phase chromatography—with nonpolar mobile phases would not have provided the variety and the simplicity of separation methods, and the reproducibility, now state-of-the-art in HPLC. The availability of the so-called reversed phases (RPs) based on chemically modified silica, where adsorption is the highest from aqueous solutions, opened for HPLC direct access to aqueous, and hence, bioanalytical systems. These phases are the workhorses in HPLC. Their diversity allows us to select appropriate columns for a wide variety of applications ranging from separations of aromatic hydrocarbons, pharmaceuticals, and pesticides to applications in bioanalysis. All commonly used RP are based on silica. Their selectivity and efficiency depend on the physical and chemical properties of the base material silica and on the type and means of bonding of the alkyl groups for RP systems.

In literature there are many discussions on properties required for an optimal RP. According to Melander and Horvath [2], the ideal stationary phase for RP chromatography (RPC) should have the following properties:

- A high efficiency and allows for rapid analysis.
- Adequate stability and long life under widely varying operating conditions.
- The column should permit the modulation of retention behavior over a wide range of conditions. This means that the stationary phase is inert and does not exhibit specific interactions with certain functional groups of solutes with the concomitant advantage of rapid adsorption–desorption kinetics. Well-prepared hydrocarboneous-bonded phases should have properties that approach these requirements, which correspond to an ideal stationary phase.
- The column material must be available with different mean pore diameters to allow efficient separation of samples that fall in different molecular weight ranges.
- The homogeneity of the surface should be such that the free-energy changes and the adsorption–desorption rate constants associated with the chromatographic process on the molecular level fall within a narrow range.

The efficiency and the speed of analysis are related to the particle diameter of the stationary phase. Nowadays, chromatographic columns are packed with stationary phases with average particle diameters below 10 μm. The tendency is to use even smaller particles, down to less than 2 μm. The required theoretical plates for a given separation are then generated in much shorter columns, improving the speed of the analysis and the detection sensitivity. A discussion on the correlation of the particle diameter, the speed of the analysis, the column length and the required pressure drop is beyond the scope of this chapter and will be discussed elsewhere, e.g., in monographs on chromatographic theory [3].

Other properties postulated for an optimal stationary phase depend on the carrier material— almost exclusively silica—its physical and chemical properties, the type of bonded group (mainly octyl, C8 or octadecyl, C18), and the chemistry of the bonding reaction. There has been much improvement in our knowledge of the preparation of silica and the binding process in the course of RP development in the last 40 years approaching the desired goal. However, this is not a steady state and new and better, or at least different RPs, will become available. It is estimated that presently about more than 500 different RPs are on the market. Therefore, it will be beyond the scope of this chapter to summarize the commercially available RPs in a type of market overview. A discussion on the properties of RPs follows and an attempt is made to relate these properties to the type of silica carrier, the type of bonded organic moiety, and the means of binding reaction. Of course, this would not be possible without demonstrating and discussing the retention behavior of different classes of analytes in RPC.

The properties of RPs are determined by the

- Properties of the silica
  - Its specific surface area
  - Its pore size distribution
  - Its packing density
  - Its surface chemistry
- Chemical modification
  - The type of silane
  - Its reactivity

- The functional groups (alkyl; phenyl; fluoro; etc.)
- The course of reaction leading to
  - Different carbon content
  - Surface coverage
  - Residual silanols concentration
- End-capping or otherwise

2.2 THE BASE MATERIAL: SILICA

Different types of silica are used in the preparation of RPs. Silica are prepared by a condensation processes of soluble silicates like water glass, by destabilization of colloidal dispersions, or by polycondensation of ortho esters of silicic acid [4–6]. The formed hydrogels are aged, purified, and dehydrated to form xerogels. In the beginning (late 1960s) irregular silica prepared from sodium silicate was used (brand names, e.g., Lichrosorb, Bondapak, etc.). When these processes are performed in suspension between two immiscible liquids, spherical particles are obtained. The particle size can be adjusted by the stirring velocity. Spherical silica particles have also been prepared by emulsion polymerization in organic polymers. The organic part is removed by high temperatures, inducing a sintering process of initial nanospheres resulting in silica with much smaller porosities and higher packing densities (brand names, e.g., Hypersil, Spherisorb, Zorbax, etc.). The starting materials for these silica may contain heavy metals like iron, titanium, and alumina; consequently they have to be purified before use. As silica is a colloidal system, it changes its properties through weathering. To circumvent these problems, acid wash and rehydroxylation became popular in the 1980s. In the 1990s, very pure and metal-free silica were prepared, beginning with tetraethoxysilane (TEOS). These materials are optimal for RPC of basic solutes (brand names, e.g., Kromasil, Prontosil, Symmetry, etc.). By polycondensation of TEOS with alkylethoxysilanes, silica with an improved stability in alkaline solution can be obtained [7] (brand name, e.g., Xterra).

The physical properties of silica are determined by its specific surface area, pore volume, average pore diameter, porosity, and the particle diameter and shape [8]. The latter two are responsible for the efficiency, the physical stability and the pressure drop of the packed columns and do not contribute to retention and selectivity.

2.2.1 PHYSICAL PARAMETERS

Mostly fully porous particles are used in HPLC. The specific surface area and the pore diameter are interconnected in such a way that with an increasing surface area the average pore diameter decreases. Standard silica for HPLC have specific surface areas between 200 and $350 \, m^2/g$ and average pore diameters of 10–12 nm (100–120 Å). The surface area is exclusively within the pores. The outer surface of the spheres (geometrical surface) is negligible (<0.1%) for particles with diameters >3 μm. For special applications, like in protein analysis, wide pore silica has to be applied with pore diameters from 30 up to 100 nm and corresponding surface areas between 100 and $10 \, m^2/g$. As wide pore materials are prepared from standard 10 nm materials by a hydrothermal process, these materials still contain a portion of smaller pores. The average pore size is given by a distribution often spanning one decade.

The pore volume of the silicas is in the range of 0.7–1.0 mL/g. In connecting these physical parameters, it becomes obvious that the wall between the pores is extremely thin. Calculations demonstrate [9] that the wall thickness (spread to a two-sided sheet) would be around 2.5 nm, or more realistically, as silica is formed by the condensation of nanoparticles, the wall is composed of spheres with a diameter of 8 nm. The model of plain cylindrical pores is used for calculations only and does not reflect a real picture.

The specific pore volume is important in size-exclusion chromatography (SEC) [10] because the separation takes place there. In retentive chromatography, it is necessary to provide the surface area within the column for separation. As various silica differ in porosity and, hence, in their packing densities, for a comparison of their retentive properties, the surface area per unit column volume and not the specific surface area is important. The specific pore volume determines the packing density, and hence, the specific surface area per unit column volume. The latter value governs the strength of retention, i.e., proportional to the phase ratio. Silica (irregular) prepared in the classical way have packing densities of around 0.35–0.4 g/mL, whereas dense (spherical) materials have packing densities between 0.5 and 0.6 g/mL. Standard silica particles are of sufficient strength to withstand the rarely achieved pressure drops in HPLC of 2 MPa/cm. Breakage may occur only when the shear forces are high, e.g., at the beginning of column packing.

Non porous silica particles (NPS) with diameters of around 1.5 μm are used in high-speed separations [11]. Retention takes place here only at the geometrical surface, which amounts to 3 m²/g. By the preparation process, uniform spherical particles with narrow size distribution are obtained.

2.2.2 Measurement of Physical Parameters

The specific surface area is determined by the adsorption of liquid nitrogen at 77 K and by applying the BET equation. Under standardized conditions, the accuracy of the surface areas is around 5–10% [5]. By measuring the adsorption and the desorption curve over the whole range, the pore volume and the pore diameter distribution can be obtained by applying the Kelvin equation (cylindrical pores are assumed). However, when the pore diameter exceeds 15 nm, mercury intrusion has to be used. Here, the pore volume distribution is calculated from the volume of mercury intruded as a function of the pressure. The surface tension of mercury or nitrogen and their wetting angles with the surfaces (assuming a glass surface) are usually kept constant irrespective of whether plain silica or bonded phases are used. Consequently, data obtained for the RP by this means may not always reflect the reality. The data on the physical parameters are usually available from the manufacturer. With packed columns, the pore size distribution can also be obtained by inverse size-exclusion chromatography [12]. This technique gives a more realistic picture of the "real" pore size distribution within a chromatographic column.

From the standard silica with 10 nm pore diameter, polystyrene molecules with a molecular weight of 40,000 are totally excluded from the surface, while those with an MW < 10,000 have access to 50% of the surface area. For protein separations, silica with a pore diameter >30 nm is recommended. For the RP, the average pore diameter is always smaller than that of the parent silica (with RP, C18 the pore diameter is reduced by about 2 nm) [13].

2.2.3 Chemical Properties of Silica

Silica is an amorphous product with several types of silanol groups on its surface [4,5]. In Figure 2.1, the different types are shown. From the IR data, the presence of isolated silanol groups with a sharp absorption band at 3746 cm⁻¹ is evident. These isolated silanol groups are the more acidic ones (pK ~ 5) and can act as the cation exchanger. Their acidity depends on the surface impurities

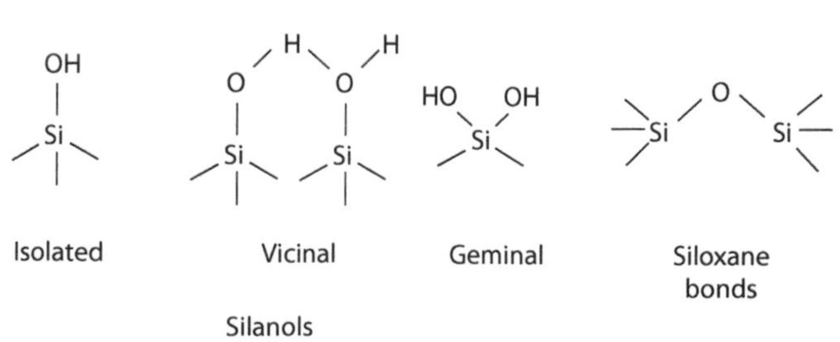

FIGURE 2.1 Type of silanol groups on the silica surface.

and is increased in the presence of alumina, titania, etc. surface ions. At pH values below 2.5, the silanols are no longer dissociated. Besides the isolated silanols, hydrogen-bridged silanol groups (vicinal silanols with a broad IR band at 3720–3730 cm^{-1}) (p$K \sim 8.5$) can be found. From NMR measurements also, the presence of geminal silanol groups (IR absorption at 3742 cm^{-1}) is made feasible. On all silanol groups water can be adsorbed physically (IR absorption at 3500–3400 cm^{-1}). By heating silica to temperatures below 400 K, only the physically adsorbed water is lost. In the temperature range between 400 and 900 K, the surface is dehydroxylated primarily by the reaction of vicinal silanol groups. This process is reversible. By boiling with water, the surface is rehydroxylated resulting in a more even distribution of silanols [14]. At temperatures above 900 K still more water is lost, however accompanied by irreversible structural changes (e.g., sintering). A totally dehydroxylated silica (heated to 1500 K) exhibits at the surface only siloxane groups (IR absorption 1250–1020 cm^{-1}), is hydrophobic, and is no longer wetted by water.

Due to differences in the preparation silica may contain different amounts of metal ions. Consequently, in aqueous suspension they show different pH values ranging from 3.9 (Zorbax) to 9.5 (Spherisorb). These differences are noticeable in NPC [15] and play an important role in RPC [16]. In Table 2.1, the metal content of different silica gels are summarized [16]. As can be clearly seen, the newer silicas based on TEOS are relatively free of metallic contaminations.

There are extensive and in part controversial discussions in literature about the overall concentration of surface silanols and the distribution between isolated and vicinal ones [17,18]. It is generally accepted that the overall concentration for fully hydroxylated silica is around 8 µmol/m^2 (or 4.6 groups/nm^2). The place requirement for a silanol group is 0.2 nm^2, which leads—assuming a planar surface—to an average distance of two silanol groups of approximately 0.4 nm. The relation between the isolated and the vicinal silanol groups is still under discussion. In earlier papers [19]—based on reactivity—an almost even distribution between isolated (reactive) and vicinal silanols is given (4.3 µmol/m^2 vs. 3.7 µmol/m^2), whereas newer papers [20], based on spectroscopic data, give a relationship of 1:4. The presence of geminal silanols is small compared to that of isolated and vicinal ones. The concentration of silanols depends on the treatment (history) of the silica, and varies through the aging processes. Before bonding, it is therefore advantageous to rehydroxylate the silica surface.

2.2.4 Bonding Technology

The surface silanols are usually reacted with chloro- or alkoxy silanes, differing in substitution (mono-, di-, and tri-chloro [or alkoxy] silanes) and alkyl chain lengths (C4–C30, dominating C18 and C8). The reaction can be performed in vacuum with the total exclusion of water or in an

TABLE 2.1
Metal Content of Old and TEOS Type Silicas Determined by ICP/AES after Dissolution of Silica

Column	Metal Content (ppm)								
	Na	Al	Fe	Ti	Zr	Ca	Mg	Total	Σ(−Na)
New									
SymC18	1	1	1					ca. 3	ca. 2
KroC18	25	<1	<0.2	2	<0.2	<1	<0.1	ca. 30	ca. 5
IntODS3	4	<1	6	<1				ca. 12	ca. 8
Old									
HypODS	2900	300	230	65		38	40	ca. 3573	673
ParODS1	23	80	445	235		216	79	ca. 1083	1060
LispRP18	2900	1100	445	246				ca. 4255	1780

appropriate solvent (mostly toluene) without or with a catalyst (usually an amine) to remove the hydrochloric acid. With alkoxysilanes, an acidic catalyst (*p*-toluene sulfonic acid) for propagation of the reaction is used [21]. The presence of a small amount of water (<500 ppm) improves the speed of reaction [21]. The reaction of the silica surface is unequivocal only with monofunctional silanes. When di- or tri-functional silanes are used, a higher coverage can be achieved, but on an average only 1.5 groups are able to react with the surface (for sterical reasons?) [22]. This leads to a better phase stability but introduces new silanol groups after hydrolysis. This is also true with trifunctional silanes when water is excluded from the reaction media. The longer the alkyl group of the silane is, the smaller the surface concentration becomes. The surface coverage is calculated from the carbon content measured by CHN combustion analysis. When bonding mono-chloro dimethyl octadecyl silanes, the surface concentration seldom exceeds 2.5 μmol/m^2. With tri-chloro octadecyl silanes, a surface coverage around 3.5 μmol/m^2 can be achieved. End-capping leads to an additional increase in the carbon content of about 1% [23], because "new" silanols are introduced. NMR techniques allow the characterization of the bonded phases for type of bonded groups, end-capping or not, and for remaining silanol groups. Regrettably quantitation is difficult to achieve here. A schematic scheme of an RP representing all groups present is shown in Figure 2.2.

In the presence of larger amounts of water and di- or tri-functional silanes, a polymerization of the silane may take place resulting in a higher coverage (up to 8 μmol/m^2). These stationary phases have been called polymeric phases [24], despite their polymeric nature being debatable. However, the selectivity of these phases for the separation of polynuclear aromatic hydrocarbons is superior compared to standard RP. NMR studies indicate that the geminal silanols react first, followed by the isolated ones [25]. Depending on the reaction conditions, the isolated silanols react to a certain extent, whereas the vicinal silanols are still present in the final product. The use of highly reactive silanes like organosilyltriflates [26], organosilylenolates [27], and *N*-(organosilyl)-*N*,*N*-dimethylamine [28] did not increase the bonding density at the silica surface. Also, with such silanes, the maximum coverage achievable is <4 μmol/m^2. This value corresponds to the concentration of isolated silanols. Therefore, at least half of the silanols remain unreacted on the surface after bonding. This applies to all bonded phases even when they are called "with maximal coverage" or "fully endcapped." These residual silanol groups are not accessible to additional bonding, but

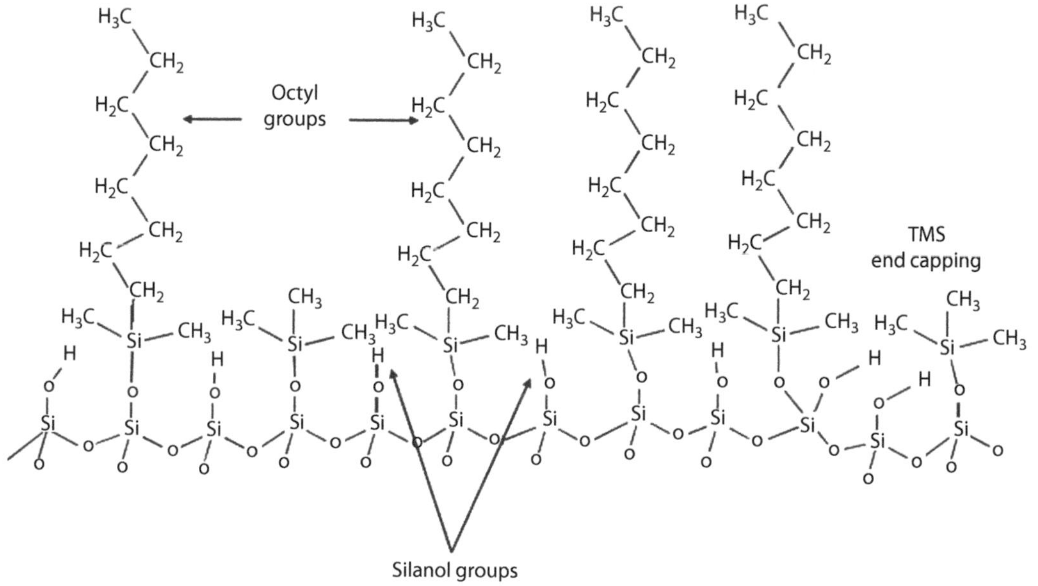

FIGURE 2.2 Surface of a bonded C8 phase with TMS end-capping. The presence of 50% unreacted silanols is indicated.

they may interact with analytes and contribute to the retention mechanism of the analytes. End-capping of bonded phases with more reactive small silanes (like hexamethyldisilazane) may help to reduce the influence of the residual silanols; however, it is not possible to make all the silanols react and exceed the concentration of bonded groups of $4\,\mu mol/m^2$. Even with the smallest silanization reagent, trimethylchlorosilane (TMS-Cl) only surface concentrations of $3.6–4.0\,\mu mol/m^2$ can be reached. This value corresponds to $2.2–2.45$ TMS groups/nm^2.

There has been a long discussion in literature about the reasons why not all surface silanols can react and be covered. The maximum bonding yield corresponds to the amount of isolated, more acidic silanols. The vicinal groups are not reactive. Steric reasons, however, give better arguments. When one calculates the portion of the surface which is covered by the projection of a TMS-group one gets a value around $0.43\,nm^2$ (corresponding to a coverage of $4.6\,\mu mol/m^2$). This number should be compared to the space requirement of $0.2\,nm^2$ for a silanol group. The average distance between two adjacent silanol groups is $0.4\,nm$ and that between two TMS groups is $6.5–7\,nm$ on the surface. Consequently around 50% of the surface silanols are still present but should be totally shielded by the TMS group. However, the assumption of a planar silica surface is far from reality. The mystery with surface silanols is, however, that it is easy to demonstrate their influence and to describe their contribution to analyte retention and separation selectivity, but it is extremely difficult to measure their concentration accurately. The quantization, e.g., with break through curves—a technique used with ion exchangers—for "naked" silica as well as for RP is possible, however, the concentrations measured are at least one order of magnitude too low [29].

As the not-reactive silanols still contribute to stationary phase selectivity and may have an unde-sired effect on the chromatographic retention (retention dependent on sample size) and on peak shape of strongly basic analytes, attempts have been made to create stationary phases, where the influence of the residual surface silanols is blocked chemically, either by end-capping with "hydro-philic" reagents or by introducing strong hydrogen bonds between polar groups (amide, carbamate, urea) in the alkyl chain and surface silanols [30,31]. RP with such polar-embedded groups close to the surface are advantageous in the separation of basic analytes, but exhibit additional polar selec-tivity, as will be shown later.

One disadvantage of all silica-based stationary phases is their instability against hydrolysis. At neutral pH and room temperature the saturation concentration of silicate in water amounts to 100 ppm. Solubility increases with surface area, decreasing particle diameter drastically with pH above 7.5. This leads also to a reduction of the carbon content. Hydrolysis can be recognized dur-ing the use of columns by a loss in efficiency and/or loss of retention. Bulky silanes [32], polymer coating [33], or polymeric encapsulation [34] have been used in the preparation of bonded phases to reduce hydrolytic instability, but most of the RPs in use are prepared in the classical way, by surface silanization. Figure 2.3 schematically shows these different types of stationary phases.

The nomenclature of the RP is not consequent. The RP most often used contains octyl (RP C8) or octadecyl (RP C18) groups. There is no differentiation even when two methyl groups are introduced additionally with the silane (as with monofunctional silanes) or only one (difunctional) or none (trifunctional silane). Some manufacturer use silanes with bulky side groups (e.g., isopropyl groups) to improve the hydrolytic stability of the bonded phases, but here also, only the longest alkyl group is used in nomenclature. RP C8 and RP C18 are the work horses in HPLC. Shorter chains (RP4) are used in protein separations, and special selectivity can be obtained with bonded phenyl, cyano, amino or fluoro groups.

2.2.5 BONDED PHASES WITH FUNCTIONAL GROUPS

Silanes with functional groups like amino, cyano, flour, phenyl, etc. are available commercially. The introduction of other functional groups in the bonded moiety can be accomplished in two ways: The silane with the functional group is prepared in situ. To prepare a silane with an amide group, tri-alkoxy-3-aminopropyl silane is reacted with an acid chloride (C8 or C18). The thus prepared amide

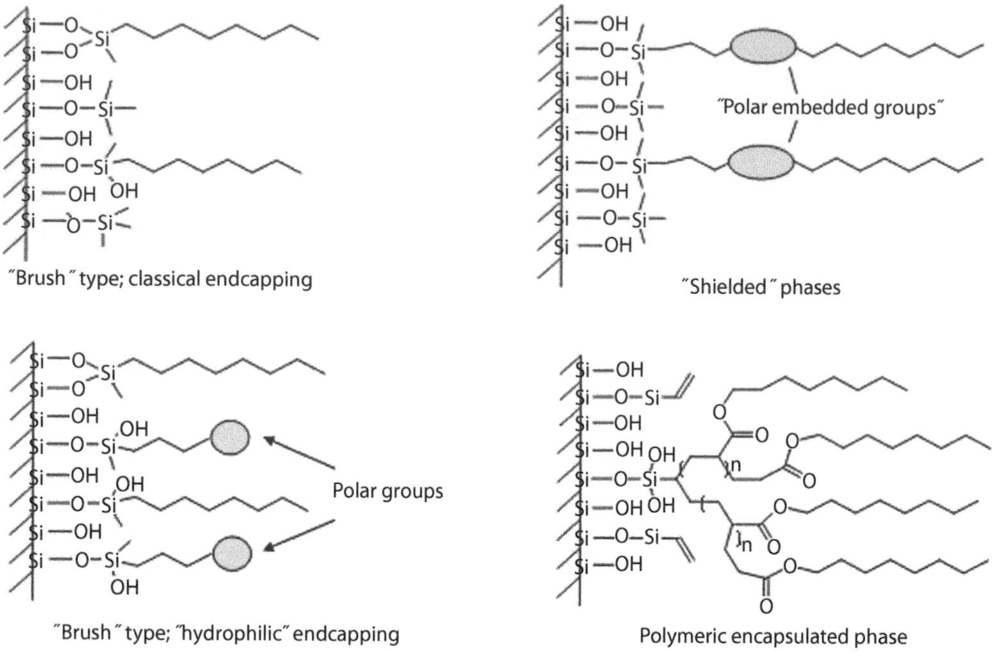

FIGURE 2.3　Schematic presentation of the different types of bonded phases.

silane is then (after purification) bonded to the silica (method A). The other way (method B) is to bind in a first step a silane with a reactive group, e.g., 3-aminopropyl-tri-alkoxysilane to the surface and then to react the amino-derivatized silica with the acid chloride. The direct way (method A) is the method to be used, leading to highly reproducible stationary phases. Method B, with a second coupling reaction at the surface, hardly yields reproducible stationary phases with an optimal and a uniform coverage. (Reaction at surfaces gives, in many cases, only 50% of the yield usually achieved with organic reactions in solution.) Similar findings have been made when 3-glycidoxy-propyl silane is used as an active coupling reagent at the surface.

2.2.6　RP with Shielding Groups (Polar-Embedded RP)

To reduce the influence of not-reacting silanols groups, the concept of polar-embedded groups in the alkyl chain has been developed. The not-derivatized silanols groups are able to form hydrogen bonds to polar groups like carbamate [29] or amide [30] groups in the bonded alkyl chain close to the surface. In most cases, these polar groups are separated with three methylene groups from the silica atom. The shielding effect with this type of embedded functionalities is demonstrated for the separation of strongly basic analytes (anti depressives) in Figure 2.4 with a classical RP column and an amide-embedded column prepared from the same silica with an identical carbon content. As can be seen, the peak shape of the basic analytes is significantly improved and the retention of all analytes is reduced. This can be attributed to a blocking of surface silanols via hydrogen bonds by the polar amide group and by the reduced hydrophobic selectivity of these polar-embedded stationary phases.

The polar groups are, on the other hand, responsible for an induced polar selectivity. Analytes able to form hydrogen bonds like phenols are retarded more strongly with polar-embedded stationary phases than with the corresponding classical RP of an identical carbon content. This is demonstrated in Figure 2.5 for the separation of polyphenolic compounds present in red wine. The retention time of the polyphenolic compound kaempferol with the shielded phase is more than three times longer than with the corresponding RP column of an identical carbon content. The polar

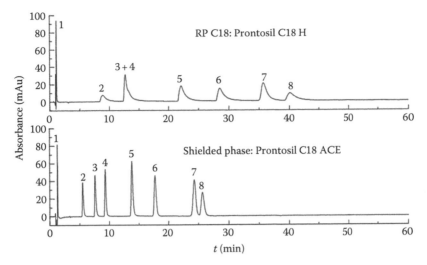

FIGURE 2.4 Separation of strong bases on a classical RP and on a shielded phase. Columns: Prontosil C18 H; Prontosil C18 ace EPS (150×4.0 mm). Mobile phase: MeOH/phosphate buffer 20 mM pH = 7, 65/35 v/v, 40°C. Analytes: 1, uracil; 2, protriptyline; 3, nortriptyline; 4, doxepine; 5, imipramine; 6, amitriptyline; 7, trimipramine; 8, clomipramine.

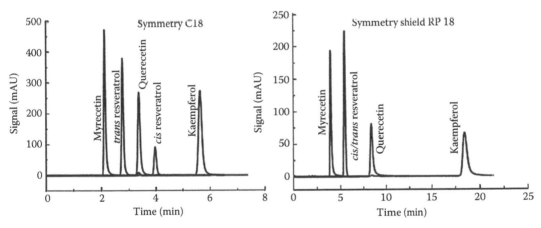

FIGURE 2.5 Separation of flavanoids on a classical RP and on a shielded phase. Columns: Symmetry C18; symmetry shield RP18 (150×4.6 mm). Mobile phase: ACN/10 mM PO_4^- buffer pH 3, 30/70 v/v, 40°C.

influence of the shielding groups gives rise to a special selectivity with many analytes thus extending the use of RPC.

2.2.7 RP with Other Functional Groups

Besides classical RPs with alkyl groups ranging from C4, frequently applied in peptide separations, to the work horses in HPLC C8 and C18, and up to C30 for the separation of long chain analytes, like carotinoids, many other organo silanes have been bonded to silica and used in HPLC.

Phases with phenyl groups are less retentive than C8 bonded phases. They show a pronounced selectivity and retention toward solutes with aromatic ring systems, attributed to π–π interactions. These π–π interactions can be improved when nitro substituted phenyl groups are bonded to silica. The strongest interactions have been reported with tetra nitro or tetra chloro phthalimide substituents.

Aminopropyl phases are also available commercially. They also show increased retention for aromatic hydrocarbons and can act as weak ion exchangers. Due to the basic functionality close to the surface they are not very stable against hydrolysis (self-catalyzed). As reactive groups, they can react with analytes, e.g., with aldehydes. The amino phases show good selectivity for the separation of sugars in acetonitrile–water mixtures at high acetonitrile (>80%) concentrations. The retention mechanism here is the partitioning of the sugars between a water-rich phase within the pores (caused by hydrate formation of the ionic groups) and the mobile phase that is higher in acetonitrile content [35].

Diolphases are prepared by reacting the silica surface with 3-glycidoxy-propyl silane and the opening of the epoxy ring. They are polar phases and show only weak retention under RPC conditions. They are used in SEC of polar polymers (peptides, proteins) in aqueous systems. When used in normal-phase chromatography, their retentive properties resemble that on plain silica without showing the strong influence of traces of water in the eluent. They are the best choice of bonded phases for reproducible normal-phase chromatography.

Cyano or nitrile phases are prepared by the reaction of silica with 3-cyanopropyl silane. They are also polar with weak retention in RPC. It is difficult to determine whether the polar interaction of this phase is caused by the dipole of the nitrile group or by residual silanols. There is a lack of significant studies of retention mechanisms with this phase.

Fluoro phases prepared with perfluoro-carbosilanes [36], which are increasingly gaining interest, show a significantly enhanced retention for fluorine-containing compounds through specific fluorine–fluorine interaction. Their retentive properties for hydrocarbons are very weak [37]. On the other hand, basic analytes are much more strongly retarded than classical or polar-embedded RP columns. This is shown in Figure 2.6 where the retention time for propranolol (peak 2) is increased from 4 to 6 min when compared to the RP and the fluoro phase where that of naphthalene (peak 5) and acenaphthene (peak 7) is reduced from 9 or 21 min to values around 3 min with the fluoro phase. These phases seem to also exhibit strong silanophilic interactions, because by end-capping with TMS, the retention of hydrocarbons increases tremendously and that of the strong base amitriptyline is reduced from 14 to 6 min as demonstrated in Figure 2.7. The high concentration of silanols groups

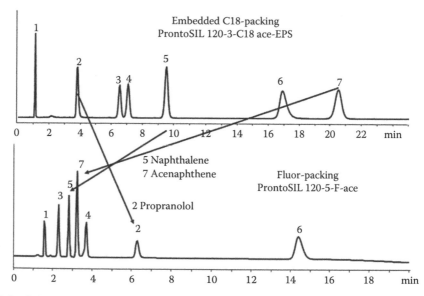

FIGURE 2.6 Selectivity of fluor phases. Columns: ProntoSIL 120-3-C18 ace-EPS; ProntoSIL 120-5-F-ace (150×4.0 mm). Mobile phase: MeOH/phosphate buffer 20 mM pH = 7 65/35 v/v. Analytes: 1, uracil; 2, propranolol; 3, butylparaben; 4, dipropylphthalate; 5, naphthalene; 6, acenaphthene; 7, amitriptyline.

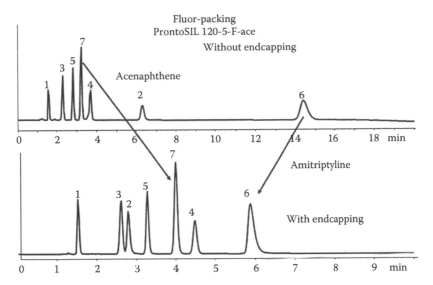

FIGURE 2.7 Selectivity of fluor phases; influence of end-capping. Conditions as in Figure 2.6.

makes these phases suitable for separations where normal-phase selectivity is required. Fluorous reversed phase silica gel (FRPSG) was introduced as an alternative to separate perfluoro-tagged compounds from non-tagged compounds by solid-phase extraction [38]. The fluorous-fluorous interactions could be employed for the non-covalent immobilization of catalysts and reagents [39]. After reaction, catalysts and reagents could be isolated by a filtration step.

2.2.8 ALTERNATIVE BONDING TECHNOLOGIES

To diminish the influence of surface silanols on retention attempts have been made to coat the surface with an insoluble polymeric layer. There are two different principles described in literature. One way is to anchor a silane with an olefin group, e.g., a vinyl silane, at the surface in the classical way. In a second step, the olefin silica is coated with a monomer and a radical starter by evaporating the solvent, and the polymerization reaction is started by increasing the temperature. Such polymeric encapsulated silica [33] based stationary phases can be prepared with a variety of functional groups and with improved hydrolytically stability. Polymethacrylates are also stable against hydrolysis; in alkaline solution the silica is dissolved below the polymeric layer, leading to a loss of efficiency (hole in the packing) at a constant retention. Improved efficiency is achieved when the polymer is allowed to grow in solution [40].

The other possibility is to coat the silica with a polymer of defined properties (molecular weight and distribution) and olefin groups, e.g., polybutadiene, and cross-linked either by radiation or with a radical starter dissolved in the polymer [32]. This method is preferentially used when other carriers like titania and zirconia have to be surface modified. Polyethylenimine has been cross-linked at the surface with pentaerythrolglycidether [41] to yield phases for protein and peptide chromatography. Polysiloxanes can be thermally bonded to the silica surface. Other technologies developed in coating fused silica capillaries in GC (polysiloxanes with SiH bonds) can also be applied to prepare RP for HPLC.

A totally different way has been used where the silica surface is reduced introducing SiH groups and reacting these groups with olefins. Not-reacted SiH groups may once again form silanols in the aqueous media. Consequently, these stationary phases contain a very high concentration of surface silanols and can be used in RP as well as in normal phase chromatography [42].

2.2.9   OTHER INORGANIC CARRIERS FOR BONDED PHASES

Hydrolytically stable inorganic carriers are alumina, titania, and zirconia. Their physical and chemical properties are quite different from that of silica. Their application as a base for bonded phases has been described in literature.

Alumina has been widely used in normal phase chromatography. Its surface exhibits amphoteric properties (available as basic, acidic, and neutral alumina) and can react as a cation or an anion exchanger. The surface area is in the usual range around $100\,m^2/g$, however, the pore structure is not very favorable as a relatively large portion of narrow pores (<2 nm) is present.

Like alumina zirconia is amphoteric but also exhibits additionally strong Lewis acidic sites at the surface, strongly interacting with analytes through a ligand exchange mechanism. The surface area of zirconia is around $20\,m^2/g$, but because of its 2.5 time larger packing density compared to silica, a sufficient active surface is within the column.

For these three materials, covalent bonding technologies cannot be used. With silanes, mixed anhydrides are formed lacking in hydrolytic stability. Coating with organic polymers [32] is the way to go. A bonded phase based on zirconia has been studied widely [43]. Method development strategies established with silica-based RP cannot be transferred to an RP bonded on zirconia. Selectivity is dependent, e.g., on the type of buffer used. Anions in the mobile phase influence retention. The kinetics of analyte interaction with the different active sites may lead to reduced efficiencies.

2.2.10   POLYMERS AS STATIONARY PHASES

The disadvantage of silica-based stationary phases is their hydrolytic instability at pH values >8. Many chromatographic separations, especially of strong basic solutes would benefit when the separations are possible at high pH values. Copolymers of styrene and divinyl benzene are stable over the entire pH range. All polymer-based packings available today exhibit inferior efficiencies when compared to well-designed silica-based packings in RPC. The reason for this seems to be the presence of micro-pores. Alkylation with octadecyl groups improves efficiency by blocking the micro-pores [44].

As carriers for ion exchangers they find application in HPLC. Cation exchangers are prepared by sulfonation, anion exchangers are prepared by nitration. Reduction to amines and alkylation, are achieved through quarternation and chloromethylation, respectively.

Polymeric stationary phases have many advantages when polymerized in capillaries (diameter < 0.5 mm) as "rods" in the presence of porogens to yield channels for mobile phase transport. They are frequently used in the analysis of peptides, proteins, and nucleic acids when directly coupled to electro spray mass spectrometry.

2.2.11   POROUS CARBON

Porous graphitized carbon, introduced in the 1980s [45], is fairly rigid and also stable over the whole pH range. For purely hydrophobic analytes, it shows similar retention characteristics as RP C18. However, with analytes with polar functionalities, already starting with aromatic rings, additional interactions come into play and the adjustment of the eluent composition for optimization of separations is not as straightforward as with standard RP columns [46]. Polar surface groups (from surface oxidation) and the π-electrons of the graphite structure may be responsible for these additional interactions. Graphitized carbon also shows retention in nonpolar eluents.

2.3   RETENTION WITH RP

RPC is the most common and important separation system in HPLC. The popularity of this technique is due to the versatility and convenience of the system. The separation principle is readily understood. Aqueous eluents (as mixtures with acetonitrile or methanol) can be used to accomplish almost all separation goals. Relative short time periods are required for an equilibration of the system after changing the eluent composition. With standard nonpolar RPC8 or RPC18 phases, the retention is proportional to the hydrophobicity of the analyte (size of hydrophobic area of the molecule), in a first approximation solute retention increases with its decreasing solubility in water. The attractive interactions between solutes and the stationary phase are based on van der Waals' forces (London dispersive forces). This "hydrophobic effect" plays a dominant role in the overall energetic of equilibrium distribution in RPC. As these van der Waals interactions also play a role between solute and eluent molecules, it is possible to adjust retention via changes in the eluent composition. Other contributions to retention like ionic interaction, hydrogen bonding, or other strong non-covalent interactions, e.g., formation of charge transfer complexes, are not expected to contribute to retention. However, as polar silanols groups are always present, such interactions certainly contribute to analyte retention and separation selectivity. The broad applicability of HPLC with the RP would never have been achieved if only pure hydrocarboneous stationary phases are available. The variability of RPC is caused by the wide differences of the RP columns on the market, differing in the basic properties of the carrier silica, the type of bonded silane, and last but not the least, on the differences of the coating levels and thus the presence of surface silanols at varying concentrations.

The theoretical treatment of the hydrophobic effect is limited to pure aqueous systems. To describe chromatographic separations in RPC Horvath and Melander developed the "solvophobic theory" [47]. In this theory, no special assumptions are made about the properties of solute and solvent, and besides hydrophobic interaction electrostatic and other specific interactions are included. The theory has been valuable to describe the retention of nonpolar [48], polar [49], and ionizable [50] solutes in RPC. The modulation of selectivity via secondary equilibria (variation of pH, ion pair formation [51]) can also be described. On the other hand, it is not a problem to find examples of dispersive interactions in literature, e.g., separation of carotinoids with a long chain (C30) RP gives a higher selectivity compared to standard RP C18; cyclohexanols are preferentially retarded on cyclohexyl-bonded phases compared to phases with linear-bonded alkyl groups.

A discussion on the theories of retention mechanisms and the type of forces generating retention and selectivity is not intended in this chapter. The interactions between solute, eluent components, and stationary phases with inhomogeneous surfaces are based on molecular recognition, where our understanding is still very limited. Therefore, an empirical description of retention and selectivity is preferred.

2.3.1 HYDROPHOBIC PROPERTIES OF RP

The retention of hydrocarbons is solely determined by the phase ratio (amount of "carbon" within the column), which is a function of the surface coverage with alkyl groups, the chain length of bonded silanes governing the carbon content, the packing density (porosity) of the silica gel, and the hydrophobicity—size of molecule, alkyl chain length—of the solute. The retention factor k increases linearly with the carbon content within the column, i.e., with the chain length of the bonded alkyl group when identical coverage with the same silica has been achieved. When looking carefully at the retention of a solute on the RP with different chain lengths it has been found that k reaches a plateau at critical chain lengths between 6 and 10 methylene groups [52]. Therefore, with RP C8 and RP C18 the retention of hydrocarbons is not a function of the alkyl chain length of the bonded phase. One reason for this may be that the C8 alkyl group has approximately the length of the toluene molecule. For analytes with longer chains, stationary phases with alkyl groups longer than C18 give better retention and selectivity. In these cases C30 alkyl groups are to be preferred.

Comparing commercially available RP C18 and RP C8 columns, it has been found that the differences in k-values for hydrocarbons (toluene and ethyl benzene) are greater within the group of RP C8 and RP C18 respectively than between RP C8 and RP C18 columns. From this it can be deduced that RP C8 and RP C18 can be exchanged without any problems only when neutral solutes are to be separated. With polar analytes, differences in residual silanols group concentration influence retention significantly. Consequently, the method transfer from RP C8 to RP C18 is much more complicated, especially when phases based on different silicas and/or from different manufacturers are used.

The logarithm of the retention factor k of the analyte increases linearly with the chain length of the sample within a homologous series. Every methylene group contributes the same increment to the retention factor. Other groups contribute smaller increments (CH of double bond) or reduce retention, e.g., polar groups like double bonds, hydroxyl, or carboxyl groups. The curves for different homologous series are parallel to each other. The differences in absolute retention with nonpolar components can be reduced by adjusting the mobile phase composition [23]. Approximately 10% more organic component has to be used with RP C18 to achieve the identical separation than with an RP C8 column. This can be done because the selectivity α for hydrophobic interaction (e.g., the α-value of ethyl benzene and toluene) is also independent of the alkyl chain length of the bonded silane when C8 or longer. The selectivity is then only a function of the eluent composition.

The hydrophobic RPs are wetted in organic solvents, and it is assumed and proved by NMR measurements, that the alkyl groups freely move around their anchor at the surface. The alkyl groups can interact with analytes. In water, of course, that cannot be the case; they form a more or less hydrocarboneous film at the surface. The eluent at which this happens has to contain more than 50% water, depending on the surface coverage and the pore structure. This normally does not affect separation, however, when one works with plain water as the mobile phase, the collapse of some of the RP may lead to a strong reduction of selectivity as only part of the alkyl groups are accessible for the analyte. It might also be noticeable in the gradient elution, when one starts from pure water. RPs specially designed for working with pure water are on the market.

In Figure 2.8, commercial RP (C8 and C18) columns are ordered according to increasing k values for toluene, corresponding to the phase ratio (carbon content). The selectivity α for the separation of ethyl benzene/toluene, however, depends on the accessibility of the alkyl groups within the pores. The hydrophobic selectivity is higher with wide pore material (pore diameter ≥12 nm) than with standard 10 nm phases, and lower with stationary phases with embedded polar groups. The highest selectivity is obtained with a 30 nm material, despite its low carbon content due to its smaller surface area.

2.3.2 RP WITH SHAPE SELECTIVITY

Stationary phases with a high density of bonded alkyl groups can differentiate between two molecules of identical size where one is planar and the other twisted out of plane. This shape selectivity has been described by Sander and Wise [53] for "polymeric" stationary phases, where in the preparation, water has been added on purpose and trichloro alkyl silanes have been used. The selectivity for the retention of tetrabenzonaphthalene (TBN) and benzo[a]pyrene (BaP) was taken as a measure to differentiate between "polymeric" and standard RP columns. With standard ("monomeric") RP columns, the twisted TBN elutes after the planar BaP, which on the other hand is more strongly retarded as TBN on "polymeric" stationary phases. In these cases the relative retention of TBN/BaP is smaller than 1, whereas with "monomeric" phases the value is >1.5. The separation of the standards on three different phases is shown in Figure 2.9. These stationary phases have superior selectivity for the separation of polyaromatic hydrocarbons in environmental analysis. Tanaka et al. [54] introduced the relative retention of triphenylene (planar) and o-terphenyl (twisted), which are more easily available, as tracers for shape selectivity. However, shape selectivity is not restricted to "polymeric" phases, "monomeric" ones can also exhibit shape selectivity when a high carbon content is achieved (e.g., with RP30) and silica with a pore diameter ≥15 nm is used [55]. Also, stationary phases with bonded cholestane moieties can exhibit shape selectivity.

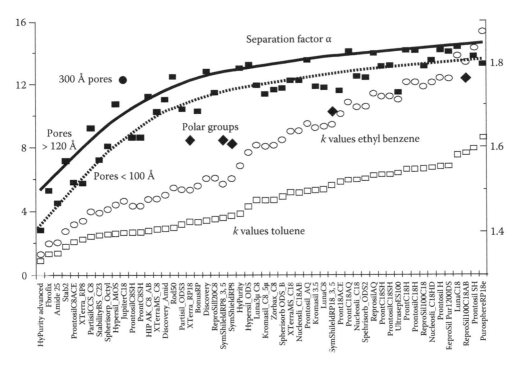

FIGURE 2.8 Hydrophobic retention and selectivity with RP columns. The stationary phases are ordered according to the increasing retention of toluene in methanol–water 50–50 v-v. Dashed line: Stationary phases with a silica pore diameter below 10 nm. Solid line: Stationary phases with a silica pore diameter >12 nm. (♦) Stationary phases with polar-embedded functional groups. ((●) Stationary phase based on a wide pore silica (30 nm)).

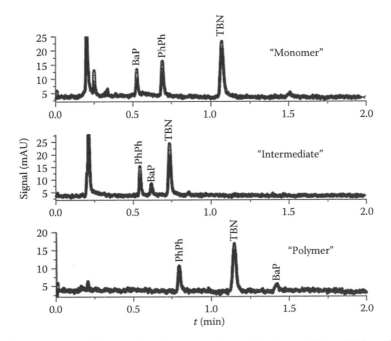

FIGURE 2.9 Demonstration of shape selectivity according to Sander and Wise. Differentiation between "monomeric" and "polymeric" RP columns. TBN: tetrabenzonaphthalene (twisted); BaP: benzo[*a*]pyrene (planar).

2.3.3 INFLUENCE OF ANALYTE STRUCTURE ON RETENTION

In general, in RP chromatography, analyte retention is inversely proportional to its solubility in water. The less soluble an analyte is, the higher is its retention. As already stated, log k increases within a homologous series linearly with the numbers of methylene groups. Branching reduces retention as does the introduction of double bonds: oleic acid (C18 with a double bond) co-elutes with palmitinic acid (C16 saturated). The position of the double bonds within the molecule also influences elution behavior as shown in Figure 2.10 for the separation of the fatty acid (ω3- and ω6-docosa pentaenic acid) of an identical chain length but of a different position of the double bond. The molecule with the longest "straight" chain length (ω6) is retarded the strongest.

Systematic rules for the influence of substituents can be derived as demonstrated for phenols in Figure 2.11. The introduction of chloro and methyl groups reduces water solubility and increases

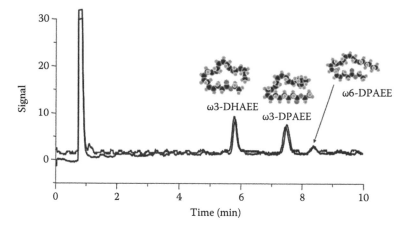

FIGURE 2.10 Demonstration of shape influence on solute retention. Separation of fatty acid esters of identical chain length but different position of double bonds. Column: Prontosil SH; 150×4 mm; mobile phase: acetonitrile–water 9–1, v–v. Samples: ω3-docosa hexaenic acid ethylester (DHAEE); ω3-docosa pentaenic acid ethylester (DPAEE); ω6-docosa pentaenic acid ethylester.

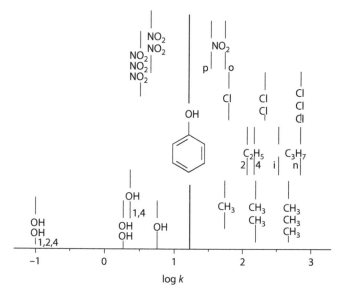

FIGURE 2.11 Influence of substitution on solute retention. Functional groups with phenol as base. Column: Lichrosorb RP C18; 250×4 mm; mobile phase: water.

retention. Also, here a logarithmic increase of k with the number of substituents is noticeable. Chlorine substituents increase retention more than methyl groups. One nitro group also increases retention. o-Nitro phenol is retarded stronger than p-nitro phenol, because of the internal hydrogen bond between the adjacent substituents. The introduction of a second nitro group increases the acidity, and hence the water solubility compared to un-substituted phenol, resulting in a significantly reduced retention. Hydroxyl groups also reduce retention as expected.

With analytes with dissociable substituents, the non-dissociated molecule is always more strongly retarded than the ionogenic one. Carboxylic acids are retarded more strongly at low pH values (pH<3), amines at higher pH values. Figure 2.12 shows the retention of the isomeric N,N-dimethyl aniline and N-ethyl aniline as a function of pH of the eluent. At low pH, the salts are present; the retention is small; the differences in molecular structure (hidden by the charge) do not contribute to retention. At high pH values the retention is much higher, but the identical hydrophobicity of the two bases also hinders separation. At intermediate pH values (5<pH<7), a separation of the two isomers is possible, probably caused by the small differences in the pK values. In this region, retention is strongly dependent on the pH of the mobile phase and the inflection point of the curves is close to the pK values of the analytes. For good reproducibility, the pH value has to be controlled well. As a consequence, carboxylic acids can best be separated at low pH values, where the hydrophobic increment is dominant. Problems with stability of RP columns are not to be expected. On the other hand, amines are better separated at pH values around 7. Higher pH values would be sometimes favorable, especially for strong bases, however, with most silica-based stationary phases problems with column stability start at pH values >7.5. In these cases, an increase of the ionic strength of the eluent reduces retention and helps to improve peak shapes, as demonstrated in Figure 2.13. The retention of the neutral analytes (phenol, toluene) is not affected while increasing the buffer strength from 1 to 20 mM, whereas the peak shape of the basic analytes (procaine, propranolol) is improved, and their retention is decreased, e.g., for propranolol for 6 min. Due to solubility problems with inorganic buffer components at a high organic eluent concentration, there are limitations of this optimization strategy. The addition of triethyl amine to the mobile phase also helps to reduce the influence of residual surface silanols on basic analyte retention. Selecting a different RP column with less accessible surface silanols may also help in solving such separation problems.

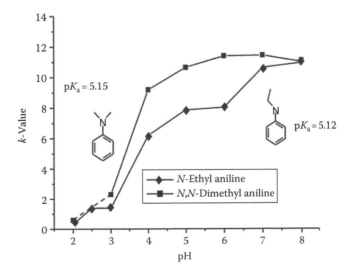

FIGURE 2.12 Influence of pH value on retention of basic analytes. Samples: N,N-dimethyl aniline; N-ethyl aniline. Column: symmetry shield C18; mobile phase: acetonitrile–20 mM phosphate buffers–water, 35–35–30, v–v–v.

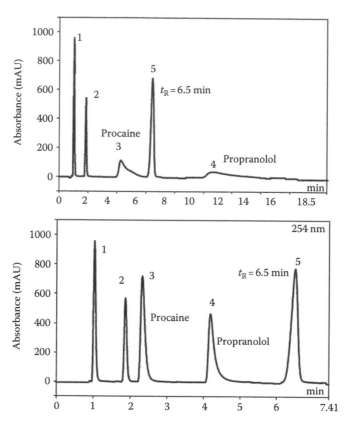

FIGURE 2.13 Influence of ionic strength on solute retention. Analytes: 1, uracil; 2, phenol; 3, procaine; 4, propranolol; 5, toluene. Column: Symmetry RP C8; 150×4 mm. Mobile phase: methanol–phosphate buffer pH 7; upper chromatogram: 1 mM; lower chromatogram: 20 mM.

2.3.4 INFLUENCE OF ELUENT COMPOSITION

Adsorption in RP is always strongest from water. The retention decreases with an increasing content of the organic modifier in the eluent. In principle, only methanol and acetonitrile are used in RP chromatography. The reasons for this are problems with the high viscosity of ethanol and isopropanol in aqueous mixtures, and as will be discussed later. The plots of log k vs. percentage of methanol or acetonitrile are more or less linear, as shown in Figure 2.14. In a first approximation, methanol and acetonitrile are interchangeable. About 10% more methanol is required to achieve iso-eluotropic conditions, i.e., to achieve an identical retention with the methanol–water and the acetonitrile–water mixture. The slopes of these plots are relatively steep, leading to the conclusion that small differences in the eluent composition do have a strong influence on analyte retention. A difference in the methanol content of 1% can change the retention factor from 3% to 10%. For exact and reproducible analytical methods, the preparation of eluent mixtures by weighing is highly recommended (exclusion of temperature influence on volume). HPLC instruments should be checked regularly for the proper functioning of the proportional valves and the exact preparation of mixtures. Slight malfunctioning of the valves and the volume contraction of eluents may influence reproducibility.

2.3.5 ACETONITRILE OR METHANOL

Differences between acetonitrile and methanol are in price, transparency, toxicity, and viscosity: methanol for HPLC is approximately by a factor of 3 cheaper than acetonitrile of identical purity;

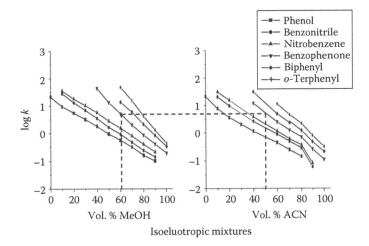

FIGURE 2.14 Influence of organic modulator concentration on solute retention. Comparison of methanol and acetonitrile; demonstration of isoeluotropic mixtures.

at low wavelength (<220 nm) methanol shows a higher transparency than acetonitrile; it is less toxic (MAK 250 mg/m³ vs. 34 mg/m³ for ACN). Significant differences exist in viscosity, especially in that of mixtures with water. Methanol–water mixtures show a viscosity maximum at about 40% methanol content, whereas the viscosity of acetonitrile–water mixtures decreases continuously with an increasing acetonitrile content. As the pressure drop over a column depends on eluent viscosity, in gradient elution with methanol–water mixtures the pressure drop increases first with the methanol content. This increase has to be considered when setting an upper pressure limit with HPLC instruments. The influence of viscosity on efficiency (changes in diffusion coefficient) can be neglected with modern high-efficiency columns (particle diameter ≤5 μm). Figure 2.15 shows the viscosity changes of acetonitrile–and alcohol–water mixtures. From this figure it can be clearly seen, that with ethanol or isopropanol as organic components, eluent viscosity becomes too high to be suitable for RP chromatography.

Selectivity differences between methanol and acetonitrile can also be observed. In general, basic analytes and those with polar groups (e.g., phenols, esters, etc.) are stronger retarded with acetonitrile as the organic modifier than with methanol when iso-eluotropic conditions for hydrocarbons are adjusted. This is demonstrated in Figure 2.16 for the separation of a standard mixture. The mobile phase had been adjusted in this way so that the retention of ethyl benzene is identical in both chromatograms: instead of a 55 Vol.% of methanol only a 42 Vol.% of acetonitrile had to be used. The retention of the amines, of phenol, and of the ester is higher with acetonitrile as the organic modifier. This may be caused by the presence of acidic surface silanols. At high acetonitrile concentrations (>60%) the retention of polar components (peptides, polyalcohol) can increase with an increasing acetonitrile content of the mobile phase: here the hydrophilic influence becomes dominant for solute retention [56]. With an RP of low silanols surface concentration or with methanol as the eluent component, this effect is no longer noticed.

Acetonitrile shows in mixtures with water, a better solubility for salts. It is therefore recommended in ion-pair chromatography [52]. Basic analytes also show better peak shapes in acetonitrile–buffer mixtures than with methanol. The proper selection, whether acetonitrile or methanol, should be used as the organic component in the mixture with a buffer, however, the type of RP column used: classical RP or a shielded RP column is also important. For demonstration with basic analytes, a standard mixture of anti depressives is used. Iso-eluotropic mixtures of methanol and acetonitrile are used. For standardization, the concentration of buffer components are also be kept constant. The analyte structures and the eluent mixtures are summarized in Table 2.2. As selectivity is worse in acidic eluents, a pH value of 7 has been used. Two phases

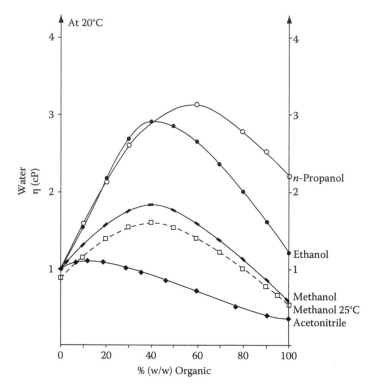

FIGURE 2.15   Viscosity of eluent mixtures.

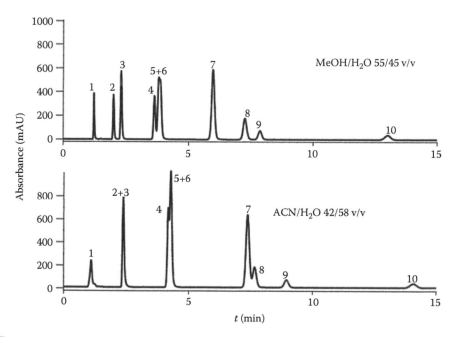

FIGURE 2.16   Influence of an organic modulator on selectivity. Comparison between acetonitrile and methanol. Isoelutropic conditions for ethyl benzene. Column: Symmetry C8, 150×4 mm. Mobile phase: methanol–water, 55–45, v–v. Acetonitrile–water, 42–58, v–v. Solutes: 1, thiourea; 2, aniline; 3, phenol; 4, o-toluidine; 5, m-toluidine; 6, p-toluidine; 7, N,N-dimethyl aniline; 8, benzoic acid ethylester; 9, toluene; 10, ethyl benzene.

TABLE 2.2
**Chemical Structure of Analytes and Composition
of the Eluent Employed for Testing the Capability
of Acetonitrile and Methanol in Separating
Basic Compounds**

Eluent Composition	Org. Modifier	Buffer pH 7	Water
Eluent 1: MeOH	65% MeOH	35% 20 mM	0%
Eluent 2: CAN	50% ACN	35% 20 mM	15%

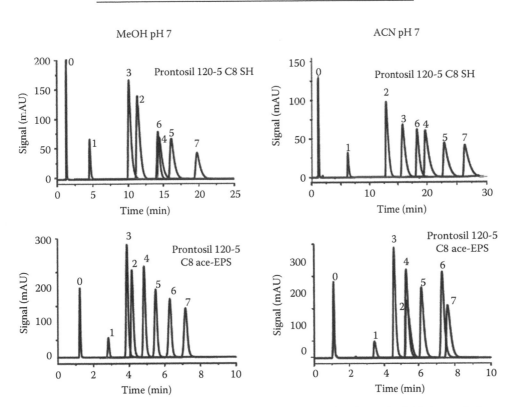

Isomers {
1: Propranolol ⟶
2: Nortriptyline
3: Doxepin
4: Imipramin
5: Amitriptyline ⟵ Difference
6: Trimipramin   1 CH₃ group
7: Clomipramin
}

FIGURE 2.17 Separation of strong bases (anti-depressive) on a classical RP and shield phases. Influence of organic modulator. Columns: Prontosil 120 C8 SH and Prontosil C8 ace-EPS (shield phase). Conditions and samples are summarized in Table 2.2.

based on identical silica and with identical carbon content have been used. In Figure 2.17, the four separations are compared. The upper row shows the separation on the classical RP column. The peak shape is better with acetonitrile (upper right chromatogram) as is the resolution of the components. The lower row shows the separation on the shield phases. Surprisingly here with methanol, the peak shape and the resolution is much better. Similar findings have been made with other brands of columns. It therefore follows that for basic analytes and classical RP columns

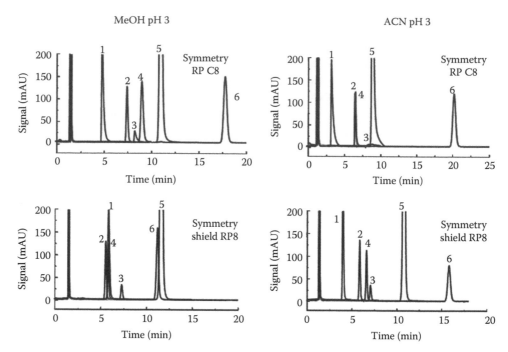

FIGURE 2.18 Separation of polar and acidic solutes on a classical RP and on shield phases. Columns: Symmetry C8 and symmetry shield C8. Mobile phases: 20% ACN, 10% H$_2$O, 70% 10 mM phosphate buffer pH 3 (v–v–v). 30% MeOH, 70% 10 mM phosphate buffer pH 3 (v–v). Samples: 1, 4-hydroxyisophthalic acid; 2, acetylic salicylic acid; 3, salicylic acid; 4, phenol; 5, methylparaben; 6, dimethylphthalate.

acetonitrile is the organic modifier to be chosen, whereas with shield phases, methanol is the organic modifier to be selected.

The opposite can be observed with acidic analytes as shown in Figure 2.18. Here, with classical RP and methanol at pH 3 the best selectivity and peak shapes were observed. On the other hand, with the shielded phases, acetonitrile gives the best separation. As already discussed, with shield phases the retention of nonpolar components (here peak 6: dimethyl phthalate) is reduced, whereas that of phenolic components (e.g., peak 5: *p*-hydroxy benzoic acid methyl ester) is increased.

The separation of polar analytes can also be optimized by applying ternary eluent mixtures. In this case, the retention is adjusted with binary mixtures with methanol–water for *k* in the desired range ($2 < k < 10$). An addition of 5%–20% of tetrahydrofuran and a simultaneous reduction of methanol content may help to improve the polar analytes resolution [57].

2.3.6 INFLUENCE OF TEMPERATURE

Increasing the temperature affects efficiency, retention, and selectivity. As viscosity decreases with increasing temperature and hence the diffusion coefficient increases the efficiency increases (increasing plate numbers) leading to narrower peaks. With increasing temperature, adsorption and hence retention decreases. As a rule of thumb: per degree centigrade the retention factor *k* is altered for about 3%. The retention of analytes varying in chemical nature may be differently affected by changes of temperature. This may cause changes in the elution sequence as shown in Figure 2.19. The retention of the hydrocarbons and of the esters is much more strongly reduced by increasing the temperature than that of the strong basic analytes. This indicates the contribution of different retention mechanisms on the various active sites at the surface of the stationary phase to analyte retention. From the van t'Hoff plots (ln *k* vs. T^{-1}) heats of adsorption can be determined and contributions to retention mechanisms discussed.

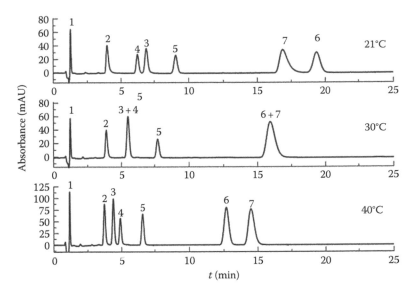

FIGURE 2.19 Influence of temperature on solute retention. Column: ProntoSIL C18ACE, 125×4 mm. Mobile phase: MeOH/20 mM phosphate buffer pH = 7, 65/35 v/v. Samples: 1, Uracil; 2, propranolol; 3, butylparaben; 4, dipropylphthalate; 5, naphthalene; 6, acenaphthene; 7, amitriptyline.

2.3.7 CHARACTERIZATION AND COMPARISON OF RP

Retention in RP chromatography is based on the interaction of the hydrophobic part of the analyte with the hydrophobic section of the stationary phase. This interaction can be modulated with the type and the concentration of the organic modifier in the mobile phase. The selectivity is mainly influenced by the interaction of the polar functional groups of the analyte with constituents of the mobile phase (buffer, salts, etc. in the aqueous part) and with the amount and activity of residual surface silanols, which are, of course, also modified by mobile phase constituents.

Characterization of stationary phases can be performed with bulk material, where all physical parameters of the bonded phase can be determined. The amount of bonded phase can be measured by the thermo gravimetric or the elemental analysis. Solid state NMR or other spectrometric methods allow to determine the type of bonded groups, way of bonding, end-capping or not, etc. However, the best method to characterize stationary phases, especially when only available in packed columns, is through the well-designed chromatographic test procedures. Such tests should enable the characterization of the packing materials for hydrophobic properties via the retention and the selectivity of hydrocarbons; to show the influence of residual silanols via peak shape or asymmetry of basic analytes. Other tests should allow the characterization of the stationary phases for shape selectivity or for the demonstration of the presence of metal ions on stationary phase surfaces. Due to the complexity of the problem several test procedures have been described in literature, mainly differing in types of test components and chromatographic conditions.

One of the earlier tests [58] contains as hydrophobic samples toluene and ethyl benzene, as a weakly acidic component phenol, and weakly basic analytes like aniline and the isomeric toluidines. Chemometric analysis showed the proper selection of the analytes for characterization, with the surprising result that N,N-dimethyl aniline is not a significant analyte in characterization for silanophilic interactions [59]. As the mobile phase, a mixture of 49 Vol.% methanol with 51 Vol.% water has been used. In the beginning, an unbuffered mobile phase has been used because silanophilic interactions can be blocked by buffer constituents. For better reproducibility and transferability, a 10 mM phosphate buffer of pH 7 is recommended. The comparison of RP columns for hydrophobic interaction by this test procedure is shown in Figure 2.8. The k value of toluene is directly related to the phase ratio, i.e., the amount of carbon within the column. The relative retention of ethyl benzene and toluene indicates the hydrophobic selectivity and the accessibility of the bonded groups. The three isomeric toluidines are of identical hydrophobicity, but differ in pK values. Consequently, they should not be separated when the influence of residual silanols is negligible. Because of their cancerogenic properties, they had to be replaced by p-ethyl aniline. The tailing of this peak is used as an indication of silanophilic activity, the presence of surface silanols.

A derivative of this test has been described by Neue [9]. Again a mixture of neutral hydrocarbons (naphthalene, acenaphthene), polar hydrophobic (dipropylphthalate), weakly acidic (butylparaben), and stronger basic analytes (propranolol, amitriptyline) has been used. A buffered mobile phase (65 Vol.% methanol, 35 Vol.% 20 mM phosphate buffer pH 7) is required. A typical chromatogram obtained with this test mixture is presented in Figure 2.19. The relative retentions of all analytes to acenaphthene are measured; that of the hydrophobic analytes describe the hydrophobicity of the packing. The tailing of the basic analytes and their relative retention to acenaphthene are correlated to the activity of silanols at a neutral pH value. Low silanophilic activity leads to a lower retention of amitriptyline compared to acenaphthene, measured at 25°C. As can be seen also in Figure 2.19, this test is very sensitive to changes of temperature so an exact control of the experimental conditions is required.

A combination of both tests, the asymmetry of p-ethyl aniline and the selectivity of butyl paraben and dipropylphthalate allows the characterization of a classical RP and the shielded phases also for the functionality of the shielding group. This is demonstrated in Figure 2.20. With a classical RP, the parabene is much more weakly retarded than the phthalate. With the increasing polarity of the shielding group (ester < carbamate < amide) the parabene is increasingly stronger retarded, thus permitting a recognition of the type of shielding group. At a stationary phase, with a free amino group (Bonus RP), the parabene is very strongly retarded. With this stationary phase, the acidic components in the mixture used in Figure 2.18 are not eluted with the conditions applied there. This stationary phase is only well suited for the separation of basic analytes. The figure also shows that all shielded phases have a low active influence of residual silanols, as the asymmetry of p-ethyl aniline is very small.

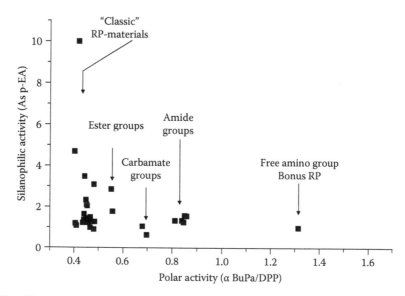

FIGURE 2.20 Characterization of RP columns for polar and silanophilic activity. Silanophilic activity: asymmetry of p-ethyl aniline. Conditions as in Figure 2.8. Polar activity: relative retention of butylparabene and dipropylphthalate. Conditions as in Figure 2.19.

A very elaborate test procedure has been proposed by Tanaka and coworkers [60]. Besides characterizing the stationary phases for their hydrophobic properties (retention value of pentyl benzene and selectivity pentyl benzene/butyl benzene) and shape selectivity (selectivity triphenylene as bent molecule and *o*-terphenyl as planar molecule), he tries to differentiate them by the test between the hydrogen-bonding capacity (relative retention of caffeine and phenol) and the "ion exchange capacity" of the bonded phase at pH 7.6 (most of residual silanols are here dissociated) and pH 2.7, where only very few and very acidic silanols should be contributing to retention. As test solutes, benzyl amine and phenol are used and the relative retention of these solutes is used to characterize the stationary phases. The hydrophobicity and the shape selectivity test is performed in methanol–water (80–20; v–v), the hydrogen-bonding test also in methanol–water (30–70; v–v); the "ion exchange capacity" test in a mixture of methanol with a 20 mM phosphate buffer (30–70; v–v) of pH 7.6 and 2.7, respectively. The whole test requires four different eluent mixtures and four different analyte mixtures.

The main feature in this test for "ion exchange capacity" is the retention of benzyl amine. As always found for strong bases, the retention can be also a function of its concentration (overloading), especially at low buffer concentrations. When comparing the retention of benzyl amine at different stationary phases as a function of pH, it can be seen in Figure 2.21 that the retention is already high or increases significantly with stationary phases prepared with "old" type silicas, e.g., with Hypersil and Ultrasep, due to the presence of many active residual surface silanols. With "new" metal free silicas prepared with TEOS a similar "ion exchange capacity" is obtained, and the change with pH is less pronounced. However, a direct correlation of these values with the "ion exchange capacity" of the RP is dangerous as the retention of amines also changes with pH as already demonstrated in Figure 2.12. At a low pH, the retention of the protonated amine is small.

With some stationary phases at low pH values (<4) benzyl amine as benzyl ammonium ion can be excluded by a Donnan potential from the pores, when positive charges are present at the surface. These could have stemmed from the manufacturing process or could have been introduced on purpose to shield amines from interacting with silanols. With an increasing pH, the Donnan exclusion decreases and at pH > 5 benzyl amine is retarded increasingly. An example of this effect with a "modern" RP with "low silanophilic" properties is demonstrated in Figure 2.22, where the elution peaks of benzyl amine are presented as a function of pH. With these stationary phases, basic analytes cannot be separated at low pH values.

The latest column test procedure has been introduced by NIST [61]. The test mixture contains toluene and ethyl benzene to describe hydrophobicity, amitriptyline as the basic analyte and

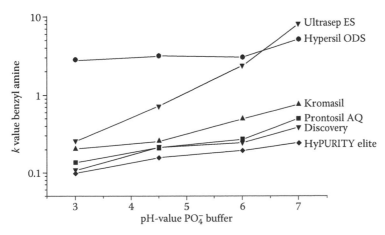

FIGURE 2.21 Retention of benzyl amine as a function of the pH value with different stationary phases. Eluent: MeOH/20 mM PO_4^- buffer 30–70; w–w. Sample: benzyl amine 0.5 g/1000 mL.

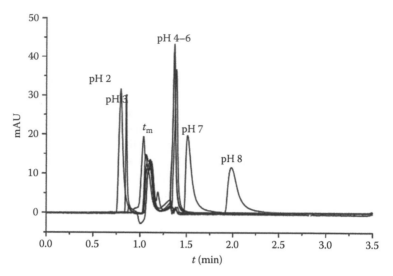

FIGURE 2.22 Influence of a pH value on retention of benzyl amine. Column: Symmetry C8. Mobile phase: acetonitrile–20 mM phosphate buffers–water; 35–35–30, v–v–v. Temperature: 40°C; benzylamine 125 ppm, injection 10 μL.

quinizarin as the chelating component for the detection of metal ions present at the surface. As the mobile phase, a mixture of methanol and a 20 mM phosphate buffer of pH 7 (80–20; v–v) has been recommended. (The initially used 5 mM phosphate buffer was too weak for use with "old" columns). The recommended concentration of amitriptyline is relatively high (28-fold of the concentration in the Neue test procedure of 100 ppm) so that peak tailing and retention time can be influenced by an overloading effect. A typical chromatogram is shown in Figure 2.23 that was achieved with an "old" type RP column. The overloading effect is also demonstrated by including the peaks of amitriptyline with lower concentrations than the recommended 2800 ppm. This column has a relatively high metal content; consequently, the quinizarin peak is very flat. Another problem with this test procedure is the co-elution of the quinizarin peak with amitriptyline with some RP C18 and with shield phases. A dual wavelength detector (quinizarin detection at 450 nm) can help to overcome this problem.

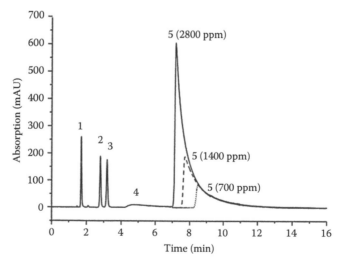

FIGURE 2.23 NIST test procedure. Column: Hypersil MOS2. Mobile phase: methanol–20 mM phosphate buffer pH 7, 80–20, v–v. Samples: 1, uracil; 2, toluene; 3, ethyl benzene; 4, quinizarin; 5, amitriptyline.

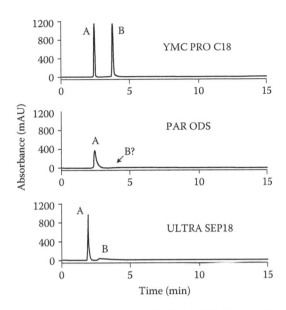

FIGURE 2.24 Metal test with RP columns. Columns: YMC-Pro C18; Partisil ODS; Ultrasep RP18. Samples: A, 4,4′-bipyridyl; B, 2,2′-bipyridyl. Mobile phase: methanol–water, 49–51, w–w; temperature: 40°C.

Surface contamination with metal ions can be determined by studying the elution behavior of acetyl acetone (2,4-pentanedione) [62], 1,5-dihydroxy anthraquinone [63], and by comparing the peak shapes of 2,2′-bipyridyl (chelating agent) and of 4,4′-bipyridyl (basic analyte, not chelating) to differentiate between silanophilic interaction and metal content [16]. Typical separations are depicted in Figure 2.24 for stationary phases with low residual silanols and a low metal content (YMC C18) as well as with high silanophilic interaction and a high metal content (Partisil ODS) and a column with a high metal content, but low residual silanols concentration (Ultrasep C18). The eluent composition has been methanol–water (49–51; v–v). When using a column for a longer period of time, the metal content increases continuously. Metal ions are accumulated on the column from frits, metallic tubing, etc. But even when using a total metal-free instrumentation (peek columns, frits, tubing, pump heads) metal ions are still concentrated at the surface originating from the mobile phase. By flushing the contaminated column with chelating agents like EDTA (ethylene diamine tetra acetic acid), the accumulated metal ions can be removed.

In comparing the various test procedures, there is always a good agreement found for hydrophobic retention and selectivity as well as for shape selectivity. However, the characterization of silanophilic interaction is still a matter of discussion. In part, the differences are due to the selection of the basic analyte. Therefore, the outcome of every test is different. It has been shown, that the peak asymmetry—used for detection of silanophilic interactions—does not correlate to the pK value of the basic test solute [64]. A closer look at these data leads to the assumption, that the differences are related to the structure of the basic solute, irrespective of whether a primary, secondary, or a tertiary amine is used. The presence of NH bonds seems to be more important in stationary-phase differentiation than the basicity expressed by the pK value. For comparable test procedures for silanophilic interactions further studies seem to be required.

All these test procedures cannot help in the selection of another column with identical properties when the used one fails and a replacement is not available or a different column has to be selected with similar selectivity. In these cases, the whole and often tedious procedure of method development with a new brand of column has to be started. Recently, on the basis of LFER (linear free enthalpy relationship), retentive parameters were determined by Snyder et al. to describe the properties of commercially available columns. The principles were presented in a series of papers [65,66].

In the development of this descriptive method, the retention of almost 100 components (ranging from neutral hydrocarbons, analytes with donor acidity, and basicity to strong acids and bases) had been studied with various mobile phases, unbuffered and buffered at pH 2.8 and pH 7. By empirical mathematical treatment, the following column parameter was evaluated:

H: relative column hydrophobicity
S^*: steric hindrance of bulky molecules to penetrate into the stationary phase
A: relative hydrogen-bonding acidity (related to the number and the accessibility of silanols)
B: relative hydrogen-bonding basicity
C: relative cation exchange capacity, differentiated for pH 2.8 and pH 7

As already discussed, retention is primarily determined by the H value (hydrophobicity), but its contribution to selectivity is low. Secondary interactions contribute to selectivity with increasing importance:

$$H < B < S^* < A \ll C$$

The practical application requires, however, several experiments to derive the data with analytes to be separated in one's own laboratory. H and S^* values of about 300 columns have been determined and are available with a corresponding software [67].

Regrettably, there is no way to use this technique for method development to solve pending analytical problems. Trial and error, based on fundamental knowledge, is still the way to go.

REFERENCES

1. F. Lottspeich, H. Zorbas, Eds., *Bioanalytik*, Spektrum-Verlag, Heidelberg, Germany, 1998.
2. W. R. Melander, Cs. Horvath, Reversed-phase chromatography, in Cs. Horvath (Ed.) *High-Performance Liquid Chromatography*, Vol. 2, Academic Press, New York, 1980.
3. J. C. Giddings, *Dynamics of Chromatography*, Dekker, New York, 1965.
4. R. K. Iler, *The Chemistry of Silica*, Wiley, New York, 1979.
5. K. K. Unger, *Porous Silica*, Elsevier, Amsterdam, the Netherlands, 1979.
6. K. K. Unger, Ed., *Packings and Stationary Phases in Chromatographic Techniques*, Dekker, New York, 1990.
7. U. D. Neue, E. Serowik, P. Iraneta, B. A. Alden, T. H. Walter, *J. Chromatogr. A*, 849, 87, 1999.
8. E. Papirer, Ed., *Adsorption on Silica Surfaces*, Dekker, New York, 2000.
9. U. D. Neue, *HPLC Columns: Theory, Technology and Practice*, Wiley-VCH, New York, 1997.
10. W. W. Yau, J. J. Kirkland, D. D. Bly, *Modern Size Exclusion Chromatography*, Wiley, New York, 1979.
11. H. Engelhardt, St. Lamotte, F.-Th. Hafner, *Am. Lab.*, 30, 40, 1998.
12. I. Halàsz, K. Martin, *Angew. Chem.* 90, 954, 1978.
13. I. Halàsz, W. Werner, *J. Chromatogr. Sci.*, 18, 277, 1980.
14. M. Mauß, H. Engelhardt, *J. Chromatogr.*, 371, 235, 1986.
15. H. Engelhardt, H. Müller, *J. Chromatogr.*, 218, 395, 1981.
16. H. Engelhardt, T. Lobert, *Anal. Chem.*, 71, 1885, 1999.
17. J. Nawrocki, B. Buszewski, *J. Chromatogr.*, 449, 1, 1988.
18. J. Köhler, D. B. Chae, R. D. Farlee, A. J. Vega, J. J. Kirkland, *J. Chromatogr.*, 352, 275, 1986.
19. A. V. Kiselev, Ya. I. Yashin, *Gas-Adsorption Chromatography*, Plenum Press, New York, 1969.
20. Y. Dong, S. V. Pappu, Z. Xu, *Anal. Chem.*, 70, 4730, 1998.
21. H. Engelhardt, P. Orth, *J. Liquid Chromatogr.*, 10, 1999, 1987.
22. H. Engelhardt, G. Ahr, *Chromatographia*, 14, 227–233, 1981.
23. H. Engelhardt, H. Schmidt, B. Dreyer, *Chromatographia*, 16, 11, 1982.
24. L. C. Sander, S. A. Wise, *J. Chromatogr.*, 656, 335, 1993.
25. R. W. Linton, M. L. Miller, G. E. Maciel, B. L. Hawkins, *Surf. Interf. Anal.*, 7, 196, 1985.
26. E. J. Corey, H. Cho, Ch. Rücker, D. H. Hua, *Tetrahedron Lett.*, 22, 3455, 1981.
27. G. Schomburg, A. Deege, J. Köhler, U. Bien-Vogelsang, *J. Chromatogr.*, 282, 27, 1983.
28. K. Szabo, N. Le Ha, Ph. Schneider, P. Aeltner, E. Kovats, *Helv. Chim. Acta*, 67, 2128, 1984.
29. H. Engelhardt, Ch. Blay, J. Saar, *Chromatographia*, 62, S19, 2005.

30. J. E. O'Gara, B. A. Alden, T. H. Walter, J. S. Petersen, C. Niederländer, U. D. Neue, *Anal. Chem.*, 67, 3809, 1995.
31. H. Engelhardt, R. Grüner, M. Scherer, *Chromatographia*, 53, 154, 2001.
32. J. J. Kirkland, J. L. Glaich, J. J. DeStefano, *J. Chromatogr.*, 635, 19, 1993.
33. U. Bien-Vogelsang, A. Deege, H. Figge, J. Köhler, G. Schomburg, *Chromatographia*, 19, 170, 1984.
34. H. Engelhardt, H. Löw, W. Eberhardt, M. Mauß, *Chromatographia*, 27, 535, 1989.
35. H. Engelhardt, P. Orth, *Chromatographia* 15, 91, 1982.
36. G. E. Berendsen, K. A. Pikaart, L. de Galan, *Anal. Chem.*, 52, 190, 1980.
37. H. Glatz, C. Blay, W. Bannwarth, H. Engelhardt, *Chromatographia*, 59, 567, 2004.
38. S. Barthelemy, S. Schneider, W. Bannwarth, *Tetrahedron Lett.*, 43, 80, 2002.
39. C. C. Tzschucke, C. Markert, H. Glatz, W. Bannwarth, *Angew. Chem.*, 114, 4678, 2002.
40. H. Engelhardt, M. A. Cuňat-Walter, *Chromatographia*, 40, 657, 1995.
41. A. J. Alpert, F. E. Regnier, *J. Chromatogr.*, 185, 375, 1979.
42. J. E. Sandoval, J. J. Pesek, *Anal. Chem.*, 63, 2634, 1991.
43. J. A. Blackwell, P. W. Carr, *Anal. Chem.*, 64, 853–863, 1992.
44. C. G. Huber, P. J. Oefner, G. K. Bonn, *Anal. Chem.*, 67, 575, 1995.
45. M. T. Gilbert, J. H. Knox, B. Kaur, *Chromatographia*, 16, 138, 1982.
46. M.-C. Hennion, V. Coquart, S. Guenu, C. Sella, *J. Chromatogr. A*, 712, 287, 1995.
47. Cs. Horvath, W. Melander, *J. Chromatogr. Sci.*, 15, 393, 1997.
48. I. Molnar, Cs. Horvath, *J. Chromatogr.*, 145, 371, 1978.
49. I. Molnar, Cs. Horvath, *J. Chromatogr.*, 142, 623, 1977.
50. Cs. Horvath, W. Melander, I. Molnar, *Anal. Chem.*, 49, 142, 1977.
51. B. L. Karger, J. N. LePage, N. Tanaka, Secondary chemical equilibria in HPLC, in Cs. Horvath (Ed.) *High-Performance Liquid Chromatography*, Vol. 1, Academic Press, New York, 1980.
52. G. E. Berendsen, L. de Galan, *J. Chromatogr.*, 196, 21, 1980.
53. L. C. Sander, S. A. Wise, *Anal. Chem.*, 56, 504, 1984.
54. N. Tanaka, Y. Tokuda, K. Iwaguchi, M. Arkari, *J. Chromatogr.*, 239, 761, 1982.
55. H. Engelhardt, M. Nikolov, M. Arangio, M. Scherer, *Chromatographia*, 48, 183, 1998.
56. K. E. Bij, Cs. Horvath, W. R. Melander, A. Nahum, *J. Chromatogr.*, 203, 65, 1981.
57. L. R. Snyder, J. L. Glaich, J. J. Kirkland, *Practical HPLC Method Development*, Wiley, New York, 1988.
58. H. Engelhardt, M. Jungheim, *Chromatographia*, 29, 59, 1990.
59. S.J. Schmitz, H. Zwanziger, H. Engelhardt, *J. Chromatogr.*, 544, 381, 1991.
60. K. Kimata, K. Iwaguchi, S. Onishi, K. Jinno, R. Eksteen, K. Hosoya, M. Araki, N. Tanaka, *J. Chromatogr. Sci.*, 27, 721, 1989.
61. L. C. Sander, S. A. Wise, *J. Sep. Sci.*, 26, 283, 2003.
62. M. Verzele, C. Dewaele, *Chromatographia*, 18, 84, 1984.
63. Y. Ohtsu, Y. Shioshima, T. Okumura, J. Koyama, K. Nakamura, O. Nakata, K. Kimata, N. Tanaka, *J. Chromatogr.*, 481, 147, 1989.
64. D. V. McCalley, *J. Chromatogr.*, 844, 23, 1999.
65. N. S. Wilson, M. D. Nelson, J. W. Dolan, L. R. Snyder, P. W. Carr, *J. Chromatogr. A*, 961, 171–193, 2002.
66. J. Gilroy, J. W. Dolan, L. R. Snyder, *J. Chromatogr. A*, 1026, 77–91, 2004.
67. L. R. Snyder, J. W. Dolan, in St. Kromidas (Ed.), *HPLC, Well Optimized*, Wiley, 2006.

3 Micro-HPLC

Heather Kalish and Terry M. Phillips

CONTENTS

3.1 INTRODUCTION

High-performance liquid chromatography (HPLC) has become a standard separation technique used in both academic and commercial analytical laboratories. However, there are several drawbacks to standard HPLC, including high solvent consumption, large sample quantity, and decreased detection sensitivity. Micro-HPLC (μHPLC) is a term that encompasses a broad range of sample volumes and column sizes (as shown in Table 3.1), but Saito and coworkers provided narrower definitions in their review based on the size of the columns.[1]

Micro-columns range in size from 0.5 to 1.0 mm internal diameter (i.d.); capillary columns range in size from 0.1 to 0.5 mm i.d., and nano-columns range in size from 0.01 to 0.1 mm i.d., as seen in Figure 3.1. Additionally, the size of the column tends to dictate the materials the columns are manufactured from.[1] Micro-columns are made from stainless steel tubing, while capillary and nano-columns are manufactured from mainly fused silica, glass-lined stainless steel, pressure-resistant plastic (polyetheretherketone—PEEK), or fused silica–lined PEEK (PEEKSil). Fused silica is advantageous to glass-lined stainless steel because it is flexible and inert, but it can allow for chemical bonding to the inner column wall.[1] Commercially available capillary and

TABLE 3.1
Units of Length and Volume Used in μHPLC

Symbol	Prefix	Use in Volume (L)	Use in Length (m)
c	Centi	10^{-2}	10^{-2}
m	Milli	10^{-3}	10^{-3}
μ	Micro	10^{-6}	10^{-6}
n	Nano	10^{-9}	10^{-9}
p	Pico	10^{-12}	10^{-12}
f	Femto	10^{-15}	10^{-15}

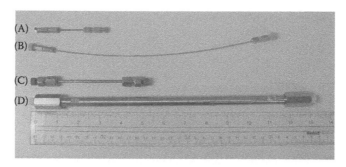

FIGURE 3.1 A comparison of columns used in routine and μHPLC. (A) A laboratory built PEEK nano-flow μHPLC column measuring 100 μm i.d. × 5 cm long, (B) A commercial C-18 PEEK capillary μHPLC column measuring 75 μm i.d. × 25 cm long, (C) A commercial C-8 stainless steel μHPLC column measuring 1 mm i.d. by 10 cm long, (D) a routine commercial reversed-phase stainless steel column measuring 4.6 mm i.d. × 25 cm long, and (E) A commercial HPLC size exclusion preparative column measuring 7.5 mm i.d. by 30 cm.

nano-columns are available from a number of companies including the following columns used in the author's laboratory: MicroTech Scientific (Vista, CA), Dionex/LC Packings (Sunnyvale, CA), Eksigent Technologies (Dublin, CA), Waters Corporation (Milford, MA), Agilent Technologies (Santa Clara, CA), and Shimadzu Scientific Instruments (Kyoto, Japan).

Horvath and coworkers first investigated miniaturization of the HPLC column in the late 1960s in a series of articles examining the separation of nucleotides.[2–4] The comparison of open tubular and stainless steel columns with i.d. of 0.5–1.0 mm packed with novel pellicular column materials indicated that the packed columns were superior for LC. Over the next decade, numerous research groups made significant advancements in the reduction of the column size, column construction materials, and packing supports. Ishii and coworkers continued to work with 0.5 mm i.d. columns, but investigated columns made of Teflon that were slurry packed with 30 μm particle diameter pellicular particles.[5–10] High speed, efficient separations were demonstrated by Scott and coworkers on 1.0 mm i.d. columns.[11–15] Novotny and coworkers further reduced the column i.d. to 50–200 μm, packed with 10–100 μm particles.[16,17] They concluded that with a 70 μm i.d. column packed with 30 μm particles, good efficiency could be obtained without excessive inlet pressure.[17] Over the next three decades, significant advancements have been made in the areas of column composition, detector interface, and hardware design, which are the subject of numerous review articles.[1,18–23]

3.2 μHPLC SYSTEMS

Today nearly all of the major HPLC companies offer a μHPLC system or at least the possibility to modify a standard instrument to accept micro-bore columns. In our laboratory, we routinely use the μHPLC systems Ultimate from Dionex/LC Packing (Figure 3.2), the Extreme Simple 4-D

(A)   (B)

FIGURE 3.2   (A) The Ultimate μHPLC system by Dionex/LC Packings, used in our laboratory for proteomics analysis. (B) A close-up of the injection port, columns, and switching valves on the Ultimate system.

(A)   (B)

FIGURE 3.3   (A) The Extreme Simple 4-D system by Micro-Tech Scientific, used in our laboratory for μHPLC work. The configuration shows the instrument setup as an eight-pump system (B) A close-up of the injection port, columns, switching valves, and four of the eight pumps on the Extreme Simple system.

system from Micro-Tech Scientific (Figure 3.3), and the 2-D system from Eksigent Technologies (Figure 3.4). Although the majority of these instruments are used for proteomics research, Sajonz et al.[24] used the Eksigent Express eight-channel μHPLC system to perform multiparallel, fast normal phase chiral separations, providing near "real-time" separations. Using a panel of test racemates, these investigators demonstrated rapid analyses, which were comparable to those obtained by conventional, but much slower HPLC procedures.

In addition to the commercially available systems, several authors have described laboratory-built systems using commercially available components from companies such as Upchurch Scientific (Oak Harbor, WA). One of the first reported laboratory-built micro-bore HPLC systems was described by Simpson and Brown,[25] which was a simple adaptation of a standard HPLC system to accept micro-bore columns built from guard columns. A complete system has been described based on dual microdialysis syringe pumps (CMA Microdialysis, Chelmsford, MA) or dual syringe pumps (Harvard Apparatus, Inc., Holliston, MA), a microinjection port, and a micro-column; the latter components being obtained from Upchurch scientific (Figure 3.5). This system was coupled with a laser-induced fluorescence (LIF) detector and used to measure neuropeptides in sub-microliter samples.[26] A further modification of this system was built to perform immunoaffinity isolations of biomedically important analytes from clinical samples.[27]

The advent of microfabrication greatly improved μHPLC design and will eventually provide the ultimate "lab-on-a-chip." Shintani et al.[28] built a multichanneled μHPLC for the separation of

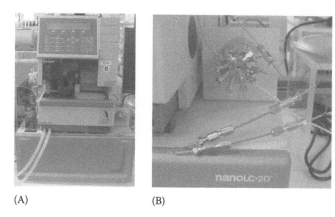

(A)   (B)

FIGURE 3.4   (A) The NanoLC 2-D system by Eksigent Technologies, used in our laboratory for μHPLC protein analysis. (B) A close-up of the injection port, columns, and switching valves on the NanoLC system.

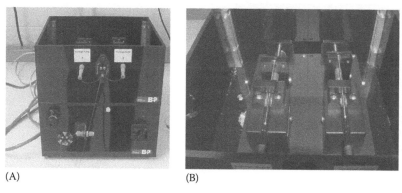

(A)   (B)

FIGURE 3.5   (A) A laboratory-built μHPLC system with dual syringe pumps, an electronic injector port, and an Upchurch nano-flow gradient mixer. (B) A picture of the two syringe pumps that comprise the gradient system. The injection valve, gradient system, and detector were controlled via a LabView interface.

multiple analytes within the same sample. This system employed an array of monolithic columns driven by a single HPLC pump and a chip-based microinjection device. Detection was achieved with a multichannel ultraviolet (UV) detector based on fiber optics. Further, Yin et al.[29] developed an entire μHPLC system on a microfabricated chip made from laminated polyimide layers. Following chromatographic separation on reversed-phase particles, the separated analytes were detected using an ion-trap mass spectrometer, a custom-built interface, and an integrated nanospray tip. A similar chip-based system has been described by Lazar and colleagues.[30] The μHPLC system was reported to compare well with a conventional HPLC in the fractionation of a protein tryptic digest.

3.3 ADVANTAGES OF μHPLC SYSTEMS

μHPLC has some significant advantages over traditional HPLC. The delivery of reliably small samples in μHPLC is often obtained using pre-columns, which concentrate large sample volumes. These pre-columns can accommodate large loading volumes that can dramatically reduce the over-all analysis time and provide protection of small-bore analytical columns from contaminants in the original sample.[1]

Biological samples, which are often available in limited amounts, can be separated and detected with a drastically improved signal-to-noise ratio, since the volumetric band-broadening (dilution) is much smaller on a capillary column.[1] Smaller columns also require less solvent which is an economical benefit, from the cost of both the solvent and the solvent disposal.[1] Finally, smaller columns promote the development of novel analytical columns. The reduced size makes investigating the correlation between the chemical structure of the packing material and its effects on the selectivity of the column more cost effective and allows researchers to use materials that are available in limited quantities.[1]

As the column diameter to particle diameter ratio decreases below 6, the core support region disappears and the support structure becomes dominated by the loosely packed wall region. The packing structure becomes more homogeneous, eliminating one cause of band broadening in HPLC.[31] Additionally, an analyte can diffuse easily across a column's cross section, which can average out any remaining cross-column differences in flow and retention.[32]

μHPLC systems have also enabled researchers to exploit temperature in liquid chromatographic separations with improved results and fewer problems. In narrow bore columns, thermal mismatches are reduced and can even be neglected due to fast heat transfer across the reduced column size.[33]

3.4 COLUMNS

In μHPLC, there are numerous types of columns used. The comparison and characterization of these columns are often discussed in terms of thermodynamic properties and kinetic characteristics. The retention factor, k, selectivity, α, and the peak asymmetry A_s are believed to be representative parameters for the thermodynamic properties, while the kinetic characteristics are often expressed in dimensionless magnitudes of reduced plate height, h, separation impedance, E, and flow resistance factor, ϕ.[23]

Three broad categories define the analytical columns used in μHPLC: Open tubular, semi-packed, and packed. Each of these three columns types is discussed briefly while making reference to the extensive review articles that cover each area.

3.4.1 OPEN TUBULAR

Open tubular columns are typically 10–20 μm in diameter and characterized by a stationary phase that is bound to the inner wall of the capillary rather than to the particles packed inside.[34] Kennedy and Jorgenson[34] effectively used open tubular columns to analyze single cells. The method was advantageous for the analysis of single cells because the columns require small samples, their resolving power equal or exceed conventional HPLC columns, and the detectors typically used with this method have good mass sensitivity. Several other research groups used open tubular columns in their research, many of which are referenced in a review by Saito et al.[1] However, for open tubular columns to be effective, Knox and Gilbert calculated that open tubular columns needed to have a diameter of 10 μm or less[35] and operating columns of 10 μm or less proved to be quite difficult, thus few reports exist on the use of this column in LC after the early 1990s.

3.4.2 SEMI-PACKED

Semi-packed columns, originally referred to as packed micro-capillaries, were developed by Tsuda and coworkers.[17,36] These columns are prepared by packing particles with diameters of 10–100 μm into glass tubes. The packed tubes are then drawn on a glass drawing machine so that the inner diameter is 50–200 μm and roughly two to three times the diameter of the particles.[37] Tsuda and coworkers establish that semi-packed columns are effective in significantly reducing the plate heights, but at a cost of decreased sample capacities[37] and significantly long analysis times.[38]

3.4.3 PACKED

Packed columns, being easier to prepare than semi-packed columns, offer a more feasible way to avoid many of the problems of open tubular columns while still retaining the advantages of micro-columns.[37] Packed columns used in µHPLC are either conventional fused silica capillaries packed with alkylated silica particles as the stationary phase or polymeric monolithic columns developed by in-situ polymerization. Numerous reviews on both packed and monolithic columns are found in the literature including the different methods used to pack capillary columns,[19] evaluation, and comparison of packed capillary columns with conventional-size columns,[20] the different organic polymers used to prepare monolithic columns,[22] silica gel-based monoliths prepared by the sol–gel method,[21] and a comparison of the efficiency of micro-particulate and monolithic capillary columns.[18]

Packed capillary columns have been used successfully in µHPLC for a number of reasons. It is easy to pack long columns, lengths of 1 m or greater, with 5 µm particles and still maintain low h values, therefore achieving over 100,000 theoretical plates.[37] Additionally, they are made from fused silica, which is superior to stainless steel used in conventional columns for several reasons, including increased wall smoothness, and good optical characteristics.[37] Development in packed capillary columns has focused on the reduction of the i.d., which leads to high separation efficiency. Karlsson and Novotny were successful in obtaining extremely high efficiencies with a column i.d. of 44 µm.[39] Kennedy and Jorgenson successfully reduced the column i.d. even further to 20–50 µm,[37] while Hsieh and Jorgenson further reduced the column i.d. to 12–33 µm.[32] An additional challenge in the preparation of capillary columns is the packing technique used to prepare the columns. Several methods, including gases, supercritical carbon dioxide, or liquids, have been used to help in transferring the packing material from an external reservoir to the column tubing. Lancas et al. have written an extensive review of these techniques and related work.[19]

3.5   STATIONARY PHASES

3.5.1   Micro-Particulate

The outcome of a chromatographic separation is also influenced by the stationary phase used to pack the column and is highly specific to the type of chromatography being carried out, whether it be normal or reversed phase, ion exchange, or affinity. The most common type of stationary phase for reversed-phase LC is nonpolar, hydrophobic organic species attached by siloxane bonds to the surface of a silica support according to Doyle and Dorsey, who have extensively reviewed the preparation and characterization of reversed-phase stationary phases.[40] According to Caude and Jardy, while materials such as alumina, zirconia, titania, and Florisil have been explored, the most common support for normal phase LC is bare silica.[41] For ion-exchange chromatography, the most popular supports are those that are based on poly(styrene) cross-linked with divinylbenzene,[42] while in high-performance affinity chromatography, the supports commonly used are modified silica or glass, azalacetone beads, and hydroxylated polystyrene media.[43]

3.5.2   Monolithic

Monolithic stationary phases are increasingly considered as a viable alternative for micro-particulate columns in HPLC.[18] The monolithic column bed has a uniformly porous integral structure thus eliminating the need for retaining frits and enhancing the column's mechanical stability during pressure changes.[44] Furthermore, a large number of readily available chemistries can be applied to functionalize surfaces, and column permeability can be easily adjusted by selecting the appropriate monomer, cross-linker, and porogen.[44] Four approaches have been utilized to prepare continuous beds[21]: (1) polymerization of an organic monomer with additives, (2) formation of a silica-based network using a sol–gel process, (3) fusing porous particulate packing material in a capillary by a sintering process, and (4) organic hybrid materials. Of the four techniques mentioned above, the first two have been the most widely utilized in µHPLC monolithic column preparation.

The first monolithic columns reported by Hjerten et al. were based on polyacrylamides.[45] This report was soon followed by Svec and Frechet, who reported the preparation of a novel, continuous bed column that incorporates both macroporosity and capacity.[46] They noted that these rod-shaped columns could be prepared using almost any monomer and offered a tempting alternative to particle-packed columns. Since then, thermally initiated free radical polymerization of acrylates, methacrylates, and dimethacrylates[46] and ring-opening metathesis polymerization of norbornane,[47] poly(styrene-co-divinylbenzene) prepared in nanospray needle,[48] and other monomers have been used to prepare monolithic column beds, and an extensive review article has been written detailing the numerous organic polymers used to prepare monolithic columns.[22] Polymer-based monolithic columns have clearly demonstrated their ability to afford excellent separations of peptides, proteins, oligonucleotides, and nucleic acids. Their ease of preparation, tolerance of high flow rates, and the rapid speed of separations that can be achieved at acceptable back pressures make this column format superior in many applications to particle-packed columns.[22] However, the efficiency of polymer-based monolithic columns remains low compared with modified silica-based monolithic columns, so further improvements are required.[22]

3.5.3 Monoliths Prepared by Porogen Alteration

Porogenic solvent plays a key role in determination of column morphology.[44] Altering the porogens in the polymerization of organic polymers to form monolithic column beds has been researched by Premstaller and Huber.[49,50] Premstaller et al. first experimented with using monolithic chromatographic beds for ion-pair, reversed-phase HPLC separation of single-stranded oligodeoxynucleotides and double-stranded DNA fragments.[49] In order to accomplish successful separations, the synthesis of the monolith had to be tuned such that its morphology resembled that of a chromatographic bed formed by nonporous particles. The use of decanol and tetrahydrofuran as porogens in the polymerization process resulted in monolithic column beds whose performance surpassed that of micro-particulate packed columns. Premstaller et al. further altered the morphology of monolithic column beds by using tetrahydrofuran/decanol in the polymerization process and produced reversed-phase monolithic columns whose separation efficiency of peptides and proteins ranging in molecular mass from a few hundred to more than 55,000 surpassed that of conventional particle-packed columns.[50]

3.5.4 Monoliths Prepared by Carbon Nanotube Incorporation

The alteration of monoliths by carbon nanotube (CNT) incorporation[51] or surface alkylation[52] has been investigated as another alternative to fine tune the separation efficiencies of organic polymer-based monolithic columns. Because of their curved surface, CNT are expected to show stronger binding affinity for hydrophobic molecules as compared with planar surfaces. Single wall carbon nanotubes (SWNT) consist of a graphene sheet rolled into a cylinder, with a typical diameter of 1 nm.[51] The challenge is maintaining their unique structure while obtaining a solubility that allows for their incorporation into the polymer. Since analyte retention after incorporation of SWNT was significantly enhanced without corresponding changes in column porosity, it is proposed that the specific structure, size, and charge of the characteristics of CNT all may play a role.[51] Analytes may be drawn onto the nanotube surface or channels between nanotubes due to surface tension and capillary effects and therefore exhibit longer retention times.[51] Huang and coworkers sought to increase the chromatographic resolution of peptides by octadecylating the surface of a poly(styrene-divinylbenzene) monolith.[52] Then by treating it with a solution containing a Friedel-Crafts catalyst, an alkyl halide, and an organic solvent, Huang et al. were able to alter the surface with C-18 alkyl groups. This surface alteration ultimately improved the reversed-phase LC separation of peptides over unmodified poly(styrene-divinylbenzene) monolithic columns.

3.5.5 Monoliths Prepared by Porogen Alteration and Surface Alkylation

A combination of surface alkylation and alteration of the porogens in the polymerization reaction was used by Li and coworkers to manufacture a surface alkylated poly(glycidyl methacrylate ethylene glycol dimethacrylate) monolithic column[44] that was copolymerized in situ with dodecanol and toluene as porogenic solvents and further functionalized by alkylation since alkyl groups are the most widely used retentive functions. The porosity of the polymeric monolith could be altered by numerous factors including using the appropriate monomer, cross-linker, porogens, the reaction temperature, or the concentration of initiator. Alkylation with linear octadecyl groups showed an appreciable improvement in the separation resolution for proteins over nonfunctionalized poly(glycidyl methacrylate ethylene glycol dimethacrylate) monolithic columns.

3.5.6 Monoliths Prepared by Photo-Initiated Polymerization

Lee and coworkers report on a photo-initiated polymerization for the preparation of poly(butyl methylacrylate-co-ethylene dimethylacrylate) within fused silica capillaries.[53] UV light-initiated polymerization is well suited to monolith formation in restricted spaces since polymer forms only in those areas that are exposed to irradiation. This leads to far greater control over the length and size of the monolithic column formed. The resulting columns are robust since their separation ability does not deteriorate with time or number of injections and their reproducibility is excellent.[53]

3.5.7 Monoliths Prepared by the Sol–Gel Method

The invention of monolithic silica-based columns can be regarded as a major technological change in column technology.[21] The preparation of a porous silica rod by the sol–gel method was reported by Nakanishi and Soga.[54,55] The porous silica rods were prepared by hydrolytic polymerization of tetramethoxysilane accompanied by phase separation in the presence of water-soluble organic polymers.[56] The morphology, determined by phase separation, is solidified by gel formation, resulting in a silica rod with a biporous structure that consists of micrometer-size through-pores and meso- or microporous silica skeletons.[56] Siouffi has published an extensive review that describes the sol–gel method of preparation in great detail and provides an extensive list of references.[21] Since the preparation of silica-based monoliths, research has been done on modifying them, in similar ways to organic polymer-based monoliths, to improve their separation performance. Minakuchi and coworkers derivatized silica-based monoliths to incorporate C_{18} alkyl groups on the surface.[56] The C_{18} silica rods showed much better performance at high flow rates than conventional columns packed with C_{18} particles; however, the separation efficiencies obtained were not impressive.[57] Luo and coworkers sought to improve the separation efficiencies of silica-based monolithic columns by preparing a 20 μm i.d. monolithic column.[57] These columns provided a high separation efficiency and sensitivity and show potential in analyzing small amounts of proteomic samples. Continuing to modify the preparation of these columns along similar lines may provide smaller i.d. columns as well.

3.6 GRADIENT ELUTION SYSTEMS

The delivery of accurate and reproducible gradients in μHPLC systems is one of the problems that many researchers are trying to solve. Simple gradients differ from each other in three respects, as seen in Figure 3.6: the shape of the gradient, the slope and curvature of the gradient, and the initial and final concentrations of the more efficient component B.[58] Several methods have been developed to deliver accurate and reproducible gradients including flow splitting, miniaturization of high-pressure gradient pumps, exponential gradient formation, preformed gradient loops and multiport switching valves, and high-temperature programming.

Splitting the solvent flow, delivered by the pumps, down to the required flow rate for nano-HPLC is achieved by inserting a variety of devices between the pumps and the injector. While Chervet

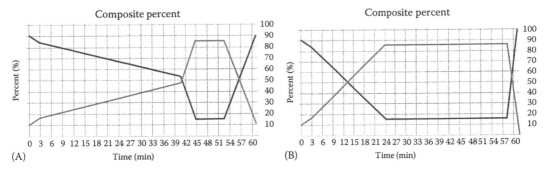

FIGURE 3.6 (A) An example of a gradual gradient on the Xtreme Simple by Microtech, in which solvent B is introduced gradually over a long period. (B) An example of a steep gradient on the Xtreme Simple by Microtech, in which solvent B is introduced quickly and used for the majority of the elution. In both chromatograms, the gradient ends with solvent A being restored as the major component.

and coworkers used a micro-flow processor from LC Packings[59,60] to generate the split flow and Alexander and coworkers used a modified stainless steel fused silica makeup adapter manufactured by Valco,[61] many groups simply use a stainless steel tee to split the flow pre-column.[62–64] However, since split flow gradient delivery requires flow rates higher than those commonly used, gradient distortions may occur due to column back pressure altering the split ratio, and mobile phase composition variation may be delayed due to reduced flow rates.[65]

Miniaturizing the overall pumping system generates solvent gradients at optimal flow rates for μHPLC without flow splitting. Zhou and coworkers developed a micro-flow pumping system capable of generating highly accurate and reproducible micro-flows.[66] The system uses dual pistons, with two independent drive systems to generate a continuous and constant flow. One piston was used to pre-compress the solvent in the piston chamber up to the pressure at the pump outlet, while the other piston performs its delivery stroke. Experimental results confirmed that the new micro-flow system could overcome mechanical limitations and imperfections of conventional piston and syringe pumps and provide conditions-independent flow rate accuracy and highly reproducible gradient.[66] Figeys and Aebersold further miniaturized the pumping system by constructing a microfabricated device consisting of three solvent reservoirs and channels etched in glass.[67] The solvent gradients and flows were generated by the computer-controlled differential electroosmotic pumping of aqueous and organic solvents and the shape, slope, and direction of the gradient could be controlled by voltages applied to the reservoirs and the surface chemistry of the transfer capillary. The newly constructed microfabricated microfluidic module was not only a success in the miniaturization of the overall pumping system but also achieved gradient elutions by producing exponential gradients.[67]

Exponential gradients deviate from linearity, either with concave or convex slopes and are advantageous because they provide a thorough mixing of solvents. The delivery of exponential gradients has been attained by the micro-fluidics device developed by Figeys and Aebersold, mentioned previously, by ultrahigh pressure LC and by a gradient elution system known as ηgrad. MacNair and coworkers developed an ultrahigh-pressure LC system and demonstrated its ability to deliver exponential gradients for protein analysis.[68,69] The development of the simple and low-cost device, "ηgrad," for the direct and reproducible delivery of exponential gradients was completed by Le Bihan and coworkers.[70] The device is composed of two reservoirs whose difference in solvent height generates a gravity-induced flow from reservoir B to reservoir A. The gradients that are produced can be modulated by using different connecting tubing between the two reservoirs, altering the starting height of the solvent in reservoir A, and altering the initial solvent compositions in both reservoirs. All three methods delivered exponential gradients capable of successfully separating proteins, but exponential gradients have several drawbacks including deviations from linearity, lack of flexibility, and the inability to reequilibrate columns for future analysis without reversing the solvents in the pumps and reservoirs.

Recently, several research groups have employed pre-formed gradient loops together with consequently switching valves, with more and more numerous ports, for the delivery of solvent gradients. In 1995, Davis and coworkers designed a solvent delivery system consisting of gradients stored in a length of narrow bore tubing (gradient loop) mounted on a standard high-pressure switching valve.[71] The advantages of this system included gradient operation without the use of split flows, pressure programmed flow control for rapid sample loading, recycling to initial conditions, and flow rates suitable for packed capillary columns. Natsume et al. expanded the idea of gradient loops to include 10-channel solvent reservoirs, connected between a 10-port manifold and a 10-position switching valve.[72] Their device, called the revolving nano-connection (ReNCon) system, generated a linear gradient by diffusion of the solvent boundaries during transfer to the column. A 14-port switching valve for the generation of smooth gradients was developed by Cappiello and coworkers.[65] Their multiport valve is equipped with 14 ports: 1 inlet, 1 outlet, and 12 ports, which support six loops, each containing a selected mixture of eluents. The first loop is filled with the weakest eluent, the last loop filled with the strongest, and an electric switch allows selection of a given loop at any given time. Basically, any gradient shape can be generated by the manipulation of two variables, eluent composition and switching time, but due to the limited number of loops, transitions between compositions can be sharp and stepwise. This is overcome by placing a mixing chamber between the multiport valve and injector. Finally, the asymptotic-trace 10-port-valve (AT10PV) nanoGR (GR being any desired gradient) generator (also known as the dual exchange gradient system, produced by Hitachi High-Technologies), as seen in Figure 3.7, developed by Ito and coworkers[73–75] consists of a conventional gradient pump with low-pressure gradient capability at micro-flow rates (micro-flow GR pump), an isocratic pump capable of delivering one solvent at nano-flow rates, a 10-port switching valve with 2 injection loops installed, a backpressure column or coil after the 10-port switching valve, and a

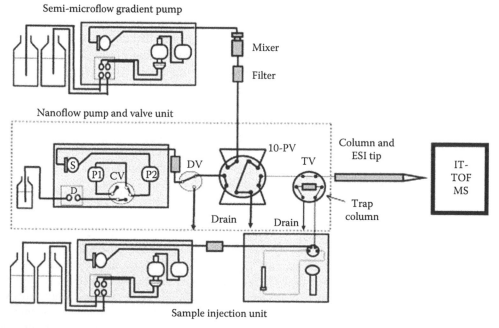

FIGURE 3.7 Schematic diagram of NanoFrontier (capillary HPLC/ESI-IT-TOF MS) system. The HPLC part consists of a conventional semi-micro-flow gradient pump, an isocratic nanoflow pump, a 10-port valve (10-PV) with two 1.0 μL injection loops, two 6-port values for drain (DV) and sample trapping (TV), and a sample injection unit (i.e., a conventional semi-micro-flow gradient pump and an auto-sampler). The AT10PV nanoGR generator consists of a semi-micro-flow gradient pump including a mixer and a filter, and a unit consisting of an isocratic nanoflow pump and valves indicated by the dotted line. (From Ito, S. et al., *J. Chromatogr. A*, 1090, 179, 2005. Copyright Elsevier 2005. With permission.)

controller to control the pumps and switching valve. Micro-flow GR pumps create an original gradient profile by mixing reservoir solvents A and B. The well-mixed solvents are delivered at a flow rate (e.g., 100 μL/min) into injection loop A or B. While A is being loaded, B is delivering solvent loaded in its loop to the capillary column at a nano-flow rate. The roles of A and B switch throughout the sample run and any nano-flow-gradient profile can be generated simply by shortening the switching period of the 10-port switching valve. All the systems mentioned above were able to effectively control the shapes of the gradients, producing accurate and reproducible gradients.

High-temperature generated gradients have been explored in μHPLC by Trones et al.[33,76,77] The dimensions of packed capillary columns enable a faster response to temperature changes and exhibit reduced temperature gradients, which contribute to band broadening, within the columns. However, the accuracy and reproducibility of the gradients are not impressive and the research is limited in this area.

3.7 DETECTORS

μHPLC systems have been designed with numerous detectors, which are ideally directly connected to the separation column. These detectors include UV, Raman, and infrared (IR) absorbance, fluorescence, electrochemical, nuclear magnetic resonance spectroscopy, evaporative light-scattering and electrospray ionization (ESI), and inductively coupled plasma mass spectrometry (ICP-MS). All of these detectors offer advantages and disadvantages, and some are more suitable to the small sample sizes that are collected at the end of a separation. While some of these detectors will be discussed briefly, an extensive review of their use has been published by Vissers et al.[23]

Raman and mid-IR detectors are of special interest due to the molecular specific fingerprint they provide.[78] In addition, their nondestructive character allows their use in sequence as well as coupling with more sensitive detection schemes.[78] However, direct application of Raman and IR spectroscopy is made difficult by the low concentration sensitivity of this technique.[78] Surowiec and co workers have worked to develop a flow-through micro-dispenser used as a solvent elimination interface between micro-bore HPLC and Raman/Fourier transform infrared (FTIR) spectroscopic detectors.[78] A dispensing frequency of 10 Hz was chosen to assure deposits with closer size and diameter to the diameter of the IR beam, which assured higher reproducibility and sensitivity. Improved signal reproducibility was obtained with IR than Raman spectroscopy due to a larger IR beam size measuring more of an average of the droplet than the Raman beam. While FTIR spectroscopy provides more reproducible measurements, Raman allows for the detection of minute amounts of sample.[78]

μHPLC has been successfully coupled to ICP-MS[79–81]; however, the use of gradient elution and high flow rates makes the technique unstable and therefore unusable in most applications. Trones and coworkers have coupled capillary HPLC with ICP-MS and replaced a liquid gradient elution with a temperature gradient elution to study organo-tin and organo-lead detection limits.[77] In contrast to other detection systems such as UV, temperature ramping had no effect on the ICP-MS, the limits of detection were much lower than those of conventional HPLC, and the repeatability of peak height and area were good.[77] Additional examples of analytes investigated by μHPLC-ICP-MS are found in Table 3.2.

TABLE 3.2
Examples of Analytes Investigated by μHPLC-ICP-MS

Technology Used	Analyte(s)
HPLC-ICP-MS	Selenomethionine, carboxymethylated selenocysteine[82]; Se-methylselenocysteine, γ-gluyamyl-Se-methylselenocysteine, selenosugar, trimethylselenonium, selenomethionine[83]; Asp-Tyr-SeMet-Gly-Ala-Ala-Lys peptide[84]; selenomethionyl calmodulin[85]; and tryptic digest of selenomethionyl calmodulin[86]

TABLE 3.3
Examples of Analytes Investigated by μHPLC–MS (Other Than ESI)

Technique	Analyte
HPLC-TOF-MS	Urinary metabolic profiles of samples from male and female Zucker rats[89]
HPLC-LTQ/FTMS	N-linked glycosylation structures in human plasma
HPLC-Ion Trap MS	Proteins expressing differences among isolates of *Meloidogyne* spp.[90]
HPLC-MALDI	Monosaccharide anhydride levoglucosan, galactosan, and mannosan in the PM10 fraction of ambient aerosols[91]; major histocompatibility complex (MHC)-associated peptides[92]; proteolytic digest of glycoproteins[93]

Adapting the evaporative light scattering device (ELSD) to μHPLC was investigated by Gaudin et al.[87] Quantitative analysis by ELSD is often hindered by nonlinearity; however, reduction of the flow rate, resulting in better homogeneity of droplet size distribution, has increased the linearity of the response with ELSD. Despite the predictable effect on droplet size in relation to the reduction of the inner diameter of the capillary inside the nebulizer, ELSD is relatively simple to adapt to micro/capillary LC.[87]

Liquid chromatography coupled to tandem mass spectrometry (LC–MS) is a powerful technique for the analysis of peptides and proteins.[88] While numerous methods for coupling MS to LC have been explored and used to analyze copious samples (as seen in Table 3.3), it is ESI that has transformed LC–MS into a routine procedure sensitive enough to analyze peptides and proteins.[88]

Miniaturizing the column i.d. is of great benefit to the sensitivity of ESI-MS, which behaves as a concentration-sensitive detection principle,[1] because the concentration of equally abundant components in the LC mobile phase is proportional to the square of the column internal diameter.[94] Column diameters from 150 to 15 μm with flow rates 20–200 nL improve detection limits of peptides 1–2 orders magnitude over microliter flow rates.[1] Several references referred to in other sections of this chapter discuss the use of LC-ESI MS to characterize separation products.[49,50,64,67,70,73,95] and a sample chromatogram from Ito and coworkers.[75] is seen in Figure 3.8. Table 3.4 provides additional and references that have used this technique.

A chip-based nanospray interface between an HPLC and the MS has been introduced by Advion Biosystems (Ithaca, NY). This instrument aligns a specialized pipette tip with a microfabricated nozzle, set in an arrayed pattern on a silicon wafer. The advantage of this interface is that each sample is sprayed through a new nozzle, thus virtually eliminating cross contamination.

3.8 APPLICATIONS OF μHPLC

3.8.1 PRE-CONCENTRATION

μHPLC coupled to MS is fast becoming the most widely used technique in proteomic research. However, the injection of large amounts of diluted sample is not possible without column overload, peak broadening, and poor separation performance.[94] Pre-concentration techniques are currently employed to reduce the sample volume, thus concentrating the analyte and remove any impurities, which may clog μHPLC analytical columns.

Pre-concentration by vacuum centrifugation is a common technique used in many versions of current in-gel digestion methods and μHPLC.[125] Throughout the process, the sample is exposed to many surfaces, which may contain a minimal number of active sites where losses can occur.[95] The amount of sample loss may be negligible for concentrated samples, but loses at lower peptide concentrations, such as the femtomole (fm = 10^{-15} M) level (see Table 3.1), increase markedly.[95,125]

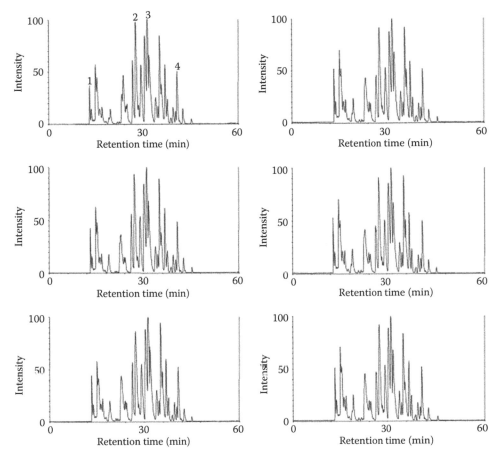

FIGURE 3.8 Total ion-intensity chromatograms (TIC) of BSA (100 fmol) peptides of six sequential runs at a flow rate of 50 nL/min. Solvents A: 0.1% formic acid in 2% acetonitrile and B: 0.1% formic acid in 98% acetonitrile. The composition of solvent B was linearly increased from 10% at 0.0 min to 40% at 70.0 min. Period of 10-port valve switching: 1 min. BSA peptide sample 1.0 μL (i.e., 100 fmol) was injected into a monolithic silica-ODS column (30 μm i.d., 150 mm in length) through a packed silica-ODS trapping column (particle size 5 μm, 150 μm i.d., 10 mm in length). The ESI voltage was 1.6 kV; curtain (nitrogen) gas was used at a flow rate of 1.0 L/min; and the scan mass range (m/z) was 300–2000. (From Ito, S. et al., *J. Chromatogr. A*, 1090, 181, 2005. Copyright Elsevier 2005. With permission.)

Electroextraction has been developed as a pre-concentration technique by Timperman and Aebersold.[95] Following in-gel digestion of the proteins, the resultant peptides are electrophoresed out of the gel and trapped on a micro-cartridge, containing a strong cation exchange material. The micro-cartridge is placed in-line with a capillary column and the peptides are transferred to this column by injecting 2 μL of elution buffer onto the micro-cartridge with a pressure vessel. This method gave modest improvements in sensitivity over the standard extraction method, which increased with decreasing sample size.

Murata and coworkers explored replacing the traditional trapping column with a membrane[126] because, at low flow rates, the void volume of trapping columns is often too large to allow rapid analysis.[127] The membrane unit consists of an in-line filter made of stainless steel used as the holder, with a stainless frit for the membrane unit and a membrane made of polytetrafluoroethylene fibers and a silica-based support modified with octadecyl groups. While the typical trapping column volume is 10 times higher than the membrane, they have the same trapping capacity and because back pressure of the membrane unit is much lower, samples can be loaded at 50 times higher flow rate,

TABLE 3.4
Examples of Analytes Investigated by μHPLC-ESI

Technique	Analyte(s)
HPLC-ESI	Flavonoids of lemon, grapefruit, bergamot, orange, and mandarin[96]; Bcl-2 antisense phosphothioate oligonucleotides G3139 and metabolites in plasma[97]; anthocyanins and derived components of red wines[98]; α-bisabolol in human blood[99]; anthocyanins in Sicilian wines[100]; ochratoxin A in grapes[101]; light-induced lateral migration of photosystem II antennae between appressed and nonappressed thylakoid membranes[102]; intact yeast proteins[103]; antiphosphotyrosine antibodies[104]; N-acylhomoserine lactones[105]
HPLC–MS-MS (ESI)	Integral plasma membrane proteins from a human lung cancer cell line (62 prenylated proteins and 45 Ras family proteins)[106]; urinary proteins[107]; AGP-derived glycoproteins[108]; major and minor populated isoforms of antithrombin[109]; regulatory lipids in breath condensate; Ser(150), Ser(418), and Ser(476) of human Grb10 zeta[110]; dissolved proteins in seawater samples[111]; human milk oligosaccharides derivatized with various esters of aminobenzoic acid[112]; sulphametoxazole, bezafibrate, metoprolol, carbamazepine, and bisoprolol in water samples[113]; 16 mycotoxins possibly related to "Sick Building Syndrome"[114]; 7-fluoro-4-nitrobenzoxadiazole (NBD-F) and 1-fluoro-2,4, dinitrobenzene (DNB-F) tagged amino acids[115]; six quinic acid derivatives isolated from *Baccharis usterii* Heering[116]; protease inhibitors and nonnucleoside reverse transcriptase inhibitors in dried blood spots from HIV/AIDS patients[117]; urinary 8-hydroxy-2'-deoxyguanosine[118]; site-specific horseradish peroxidase glycosylation[119]; "trypsinosome" from specific peptide characteristics[120]; soluble proteins in wine[121]
HPLC-ESI-Q-TOF-MS	Cereulide from *Bacillus cereus*[122]; human plasma proteome[103]; abiotic stress-tolerant (Mandolina) and an abiotic stress-susceptible (Jubilant) barley cultivar[123]; neuropeptides from isolated locust corpora cardiaca[124]

with as much as 50 μg sample, in spite of the small volume. One drawback observed by researchers was membrane durability that was impaired after 40 injections, but this could be improved by minor changes to the unit structure.

Licklider and co workers experimented with automating the sample introduction step in nano-scale LC–MS.[62] In order to achieve pre-concentration and desalting prior to sample analysis, they created a 2 cm vent after the head of the analytical column. Experimental results demonstrated 50 nanoliter (nL) elution peak volumes while retaining low-to subfemtomole detection levels. Additionally, implementing this pre-concentration technique requires minimal changes in current methods and equipment.

The most common form of pre-concentration involves the use of pre-columns.[94] Pre-columns combined with column switching techniques help avoid the problems mentioned in the introduction to this section. The type of pre-column used is dictated by experimental conditions. Large sample volumes can be eluted onto pre-columns using mobile phases with low elution powers, allowing for sample concentration and clean-up, prior to injection on the analytical separation column. However, the use of pre-columns for sample concentration can be plagued by void volumes, whose volumes can be several times greater than that of the pre-column, negating any initial sample pre-concentration and leading to poor separation performance.

One type of pre-analytical concentration that is particularly well suited to μHPLC is the incorporation of an affinity ligand pre-column. The addition of affinity and immunoaffinity ligands for pre-analytical concentration of samples has become popular in the analytical sciences.[128,129] The use of affinity pre-columns ranges from making sol–protein gel-derived monolithic columns containing the affinity ligand of interest[130] to the employment of true bioaffinity ligands and antibodies. Madera et al.[131] employed immobilized lectin micro-columns prior to reversed-phase analysis of glycoproteins in small samples of human serum. Likewise, Starkey et al.[132] used a combination of phenyl

TABLE 3.5
Examples of Analytes Investigated by Additional Detection Methods Than Those Mentioned in Section 3.7

Technique	Analyte(s)
HPLC	Clenbuterol in pork, beef, and hog liver[145]; Co(II) ion as 4-(2-thiazolylazo)resorcinol (TAR) or 5-methyl-4-(2-thiazolylazo)resorcinol (5MTAR)[146]; total phenols after nitrosation of USEPA classified 11 priority pollutant phenols[147]; polyprenol and dolichol[148]
HPLC-UV	Mebeverine hydrochloride in raw materials, bulk drugs and formulation[149]; meloxicam[150]; platinum in blood and urine samples after administration of cisplatin drug[151]; budesonide, a novel glucocorticoid prescribed for inflammatory bowel disease[152]; amikacin[153]; aromatic amines in water[154]; cetirizine in human plasma[155]; Fmoc and z-derivatives of natural and unnatural sulfur containing amino acids[156]; glycyl- and diastereomeric dipeptides and tripeptides[157]; gossypol in cotton[158]; West Nile and Sindbis virus PCR products[159]; resveratrol and resveratrol-glucosides in Sicilian wines[100]; glycyrrhizin and glycyrrhetic acid in licorice roots and candies[160]
HPLC-UV-electrochemical detection	Disodium-2,2'-dithio-bis-ethane sulfonate (BNP7787) intracellular conversion products[161]; honokiol and magnolol in fresh *Magnolia obovata*[162]
HPLC-UV-MS	Bioactive compounds from *Blumea gariepina*[163]; salicin, salicylic acid, tenoxicam, ketorolac, piroxicam, tolmetin, naproxen, flurbiprofen, diclofenac, and ibuprofen in pharmaceutical formulations and biological samples[164]
HPLC-UV-NMR	Bioactive compounds from *Blumea gariepina*[163]
HPLC-electrochemical detection	Baicalin and baicalein in rat plasma[165]; honokiol and magnolol in branches and leaves of *Magnolia obovata*[162]; quercetin in human plasma[166]
HPLC-chemiluminescence	Changes in catecholamines and 3-*O*-methyl metabolite concentrations in human plasma[167]
HPLC-fluorescence	Erythropoietin in pharmaceutical products[168]; 3,4-methylenedioxymethamphetamine, 3,4-methylenedioxyamphetamine, amphetamine, and methamphetamine in rat urine[169]
HPLC-diode array	Azoxystrobin, kresoxim-methyl, and trifloxystrobin fungicides[170]
HPLC-NMR	Shape constrained natural compounds (tocopherol homologues, vitamin E)[171]; protein kinase ZAP-70 tryptic fragment containing amino acids 485–496[172]; cartenoids from a spinach sample[173]; isoflavines in *Radix astragali*[174]

borate affinity pre-concentration followed by micro-column chromatography and atmospheric pressure photoionization MS to measure salsolinal and the major catecholamines in experimental animal brain tissue.

Antibody-based immunoaffinity pre-analytical concentrators have been widely used in routine chromatography (see Chapter 13) but are only just becoming introduced into μHPLC, although immunoaffinity pre-analytical concentrators have been described in capillary electrophoresis.[133,134] Most applications have employed the micro-columns as immunoaffinity chromatography with direct measurement of the captured analyte. Hodgson and colleagues[135] used an antibody-entrapped monolithic column to perform nanoflow immunoaffinity chromatography. Here the column was used as an immunoextraction concentrator with the released analyte being measured by LIF detection. However, as the authors rightly point out, such columns could easily be used as a pre-analytical concentrator.

3.8.2 MULTIDIMENSIONAL LIQUID CHROMATOGRAPHY

Holland and Jorgenson[136] reported separating amines using anion exchange and reversed-phase columns in 1995 and since then, there have been numerous reports of combining two LC columns (2D-LC) to achieve efficient sample separation.[137–143] In addition to the few references mentioned in this section, see Chapter 4 in this handbook on two-dimensional comprehensive liquid chromatography.

While 2D-LC can be accomplished using conventional LC columns, the field of proteomics has directed the 2D-LC field toward capillary columns.[1] Multidimensional chromatography appears to overcome problems associated with 2D gels, and the peak capacity of the LC system dramatically increases by combining two orthogonal separation techniques.[144] The rapid analysis of complex protein digests using a 2D-LC system was investigated by Mitulovic and coworkers.[144] A strong cation exchange column was used first to separate peptides based on their electric charge state and charge distribution, whereas a reverse-phase column was used second to separate proteins according to hydrophobicity. The system they used is fully automated, sample loss is low, and they demonstrated that the success of 2-D HPLC is determined by the nature of the loading solvent, the flow rate of the loading pump, and the sample loading time.[144]

3.8.3 OTHER APPLICATIONS

Many of the references cited throughout this chapter use specific analytes to validate the techniques they are researching. In addition to these references, there are numerous references that use additional techniques for specific analysis of reagents, proteins, environmental contaminants, etc. Table 3.5 is a summary of research cited from 2004 until early 2006, using μHPLC in conjunction with these methods of detection to study the specific analytes listed.

REFERENCES

1. Saito, Y., Jinno, K., and Greibrokk, T., Capillary columns in liquid chromatography: Between conventional columns and microchips, *Journal of Separation Science* 27(17–18), 1379–1390, 2004.
2. Horvath, C. and Lipsky, S. R., Rapid analysis of ribonucleosides and bases at picomole level using pellicular cation exchange resin in narrow bore columns, *Analytical Chemistry* 41(10), 1227–1234, 1969.
3. Horvath, C., Melander, W., Molnar, I., and Molnar, P., Enhancement of retention by ion-pair formation in liquid-chromatography with nonpolar stationary phases, *Analytical Chemistry* 49(14), 2295–2305, 1977.
4. Horvath, C. G., Preiss, B. A., and Lipsky, S. R., Fast liquid chromatography—An investigation of operating parameters and separation of nucleotides on pellicular ion exchangers, *Analytical Chemistry* 39(12), 1422–1428, 1967.
5. Ishii, D., Asai, K., Hibi, K., Jonokuchi, T., and Nagaya, M., Study of micro-high-performance liquid-chromatography. 1. Development of technique for miniaturization of high-performance liquid-chromatography, *Journal of Chromatography* 144(2), 157–168, 1977.
6. Ishii, D., Hibi, K., Asai, K., and Jonokuchi, T., Studies of micro high-performance liquid-chromatography. 2. Application to gel-permeation chromatography of techniques developed for micro high-performance liquid-chromatography, *Journal of Chromatography* 151(2), 147–154, 1978.
7. Ishii, D., Hibi, K., Asai, K., and Nagaya, M., Studies of micro high-performance liquid-chromatography. 3. Development of a micro-pre-column method for pretreatment of samples, *Journal of Chromatography* 152(2), 341–348, 1978.
8. Ishii, D., Hibi, K., Asai, K., Nagaya, M., Mochizuki, K., and Mochida, Y., Studies of micro high-performance liquid-chromatography. 4. Application of micro pre-column method to analysis of corticosteroids in serum, *Journal of Chromatography* 156(1), 173–180, 1978.
9. Ishii, D., Hirose, A., Hibi, K., and Iwasaki, Y., Studies on micro high-performance liquid chromatography. 5. Design of a microscale liquid chromatograph and its application to cation-exchange separation of alkali-metals, *Journal of Chromatography* 157(Sep), 43–50, 1978.
10. Ishii, D., Hirose, A., and Horiuchi, I., Studies on micro-high-performance liquid-chromatography. 6.Application of microscale liquid-chromatographic technique to anion-exchange separation of halide ions, *Journal of Radioanalytical Chemistry* 45(1), 7–14, 1978.
11. Kucera, P., Design and use of short microbore columns in liquid-chromatography, *Journal of Chromatography* 198(2), 93–109, 1980.
12. Reese, C. E. and Scott, R. P. W., Microbore columns—Design, construction, and operation, *Journal of Chromatographic Science* 18(9), 479–486, 1980.
13. Scott, R. P. W., Microbore columns in liquid-chromatography, *Journal of Chromatographic Science* 18(2), 49–54, 1980.

14. Scott, R. P. W. and Kucera, P., Mode of operation and performance-characteristics of microbore columns for use in liquid-chromatography, *Journal of Chromatography* 169(Feb), 51–72, 1979.

15. Scott, R. P. W., Kucera, P., and Munroe, M., Use of microbore columns for rapid liquid-chromatographic separations, *Journal of Chromatography* 186(Dec), 475–487, 1979.

16. Hirata, Y. and Novotny, M., Techniques of capillary liquid-chromatography, *Journal of Chromatography* 186(Dec), 521–528, 1979.

17. Tsuda, T. and Novotny, M., Packed microcapillary columns in high-performance liquid-chromatography, *Analytical Chemistry* 50(2), 271–275, 1978.

18. Eeltink, S., Decrop, W. M. C., Rozing, G. P., Schoenmakers, P. J., and Kok, W. T., Comparison of the efficiency of microparticulate and monolithic capillary columns, *Journal of Separation Science* 27(17–18), 1431–1440, 2004.

19. Lancas, F. M., Rodrigues, J. C., and Freitas, S. D., Preparation and use of packed capillary columns in chromatographic and related techniques, *Journal of Separation Science* 27(17–18), 1475–1482, 2004.

20. Pruss, A., Kempter, C., Gysler, J., and Jira, T., Evaluation of packed capillary liquid chromatography columns and comparison with conventional-size columns, *Journal of Chromatography A* 1030(1–2), 167–176, 2004.

21. Siouffi, A. M., Silica gel-based monoliths prepared by the sol-gel method: Facts and figures, *Journal of Chromatography A* 1000(1–2), 801–818, 2003.

22. Svec, F., Organic polymer monoliths as stationary phases for capillary HPLC, *Journal of Separation Science* 27(17–18), 1419–1430, 2004.

23. Vissers, J. P. C., Claessens, H. A., and Cramers, C. A., Microcolumn liquid chromatography: Instrumentation, detection and applications, *Journal of Chromatography A* 779(1–2), 1–28, 1997.

24. Sajonz, P., Gong, X., Leonard, W. R., Jr., Biba, M., and Welch, C. J., Multiparallel chiral method development screening using an 8-channel microfluidic HPLC system, *Chirality* 18(10), 803–813, 2006.

25. Simpson, R. C. and Brown, P. R., Development of a microbore high-performance liquid chromatographic system for biological applications, *Journal of Chromatography* 385, 41–54, 1987.

26. Phillips, T. M. S. P. D., Immunoaffinity analysis of substance P in complex biological fluids: Analysis of sub-microliter samples, *Journal of Liquid Chromatography and Related Technologies* 25, 2889–2900, 2002.

27. Peoples, M. C., Phillips, T. M., and Karnes, H. T., A capillary-based microfluidic instrument suitable for immunoaffinity chromatography, *Journal of Chromatography. B, Analytical Technologies in the Biomedical and Life Sciences* 843(2), 240–246, 2006.

28. Shintani, Y., Hirako, K., Motokawa, M., Iwano, T., Zhou, X., Takano, Y., Furuno, M., Minakuchi, H., and Ueda, M., Development of miniaturized multi-channel high-performance liquid chromatography for high-throughput analysis, *Journal of Chromatography A* 1073(1–2), 17–23, 2005.

29. Yin, H., Killeen, K., Brennen, R., Sobek, D., Werlich, M., and van de Goor, T., Microfluidic chip for peptide analysis with an integrated HPLC column, sample enrichment column, and nanoelectrospray tip, *Analytical Chemistry* 77(2), 527–533, 2005.

30. Lazar, I. M., Trisiripisal, P., and Sarvaiya, H. A., Microfluidic liquid chromatography system for proteomic applications and biomarker screening, *Analytical Chemistry* 78(15), 5513–5524, 2006.

31. Knox, J. H. and Parcher, J. F., Effect of column to particle diameter ratio on dispersion of unsorbed solutes in chromatography, *Analytical Chemistry* 41(12), 1599–1606, 1969.

32. Hsieh, S. C. and Jorgenson, J. W., Preparation and evaluation of slurry-packed liquid chromatography microcolumns with inner diameters from 12 to 33 μm, *Analytical Chemistry* 68(7), 1212–1217, 1996.

33. Greibrokk, T. and Andersen, T., High-temperature liquid chromatography, *Journal of Chromatography A* 1000(1–2), 743–755, 2003.

34. Kennedy, R. T. and Jorgenson, J. W., Quantitative-analysis of individual neurons by open tubular liquid-chromatography with voltammetric detection, *Analytical Chemistry* 61(5), 436–441, 1989.

35. Knox, J. H. and Gilbert, M. T., Kinetic optimization of straight open-tubular liquid-chromatography, *Journal of Chromatography* 186(Dec), 405–418, 1979.

36. Tsuda, T., Tanaka, I., and Nakagawa, G., Packed microcapillary liquid-chromatography with reduced Id columns, *Journal of Chromatography* 239(Apr), 507–513, 1982.

37. Kennedy, R. T. and Jorgenson, J. W., Preparation and evaluation of packed capillary liquid-chromatography columns with inner diameters from 20-Mu-M to 50-Mu-M, *Analytical Chemistry* 61(10), 1128–1135, 1989.

38. McGuffin, V. L. and Novotny, M., Optimization and evaluation of packed capillary columns for high-performance liquid-chromatography, *Journal of Chromatography* 255(Jan), 381–393, 1983.

39. Karlsson, K. E. and Novotny, M., Separation efficiency of slurry-packed liquid-chromatography microcolumns with very small inner diameters, *Analytical Chemistry* 60(17), 1662–1665, 1988.

40. Doyle, C. A. and Dorsey, J. G., Reversed-phase HPLC: Preparation and characterization of reversed phase stationary phases, in *Handbook of HPLC*, Katz, E. E., Eksteen, R., Schoenmakers, P., and Miller, N. (Eds.), Marcel Dekker, New York, 1998, pp. 293–323.

41. Caude, M. J. and Jardy, A., Normal-phase liquid chromatography, in *Handbook of HPLC*, Katz, E., Eksteen, R., Schoenmakers, P., and Miller, N. (Eds.), Marcel Dekker, New York, 1998, pp. 325–363.

42. Smith, R. E., HPLC of ions: Ion-exchange chromatography, in *Handbook of HPLC*, Katz, E., Eksteen, R., Schoenmakers, P., and Miller, N. (Eds.), Marcel Dekker, New York, 1998, pp. 365–411.

43. Hage, D. S., Affinity chromatography, in *Handbook of HPLC*, Katz, E., Eksteen, R., Schoenmakers, P., and Miller, N., Marcel Dekker, New York, 1998, pp. 483–498.

44. Li, Y., Zhang, J., Xiang, R., Yang, Y. H., and Horvath, C., Preparation and characterization of alkylated polymethacrylate monolithic columns for microHPLC of proteins, *Journal of Separation Science* 27(17–18), 1467–1474, 2004.

45. Hjerten, S., Liao, J. L., and Zhang, R., High-performance liquid-chromatography on continuous polymer beds, *Journal of Chromatography* 473(1), 273–275, 1989.

46. Svec, F. and Frechet, J. M. J., Continuous rods of macroporous polymer as high-performance liquid-chromatography separation media, *Analytical Chemistry* 64(7), 820–822, 1992.

47. Mayr, B., Holzl, G., Eder, K., and Buchmeiser, C. G., Hydrophobic, pellicular, monolithic capillary columns based on cross-linked polynorbornene for biopolymer separations, *Analytical Chemistry* 74(23), 6080–6087, 2002.

48. Moore, R. E., Licklider, L., Schumann, D., and Lee, T. D., A microscale electrospray interface incorporating a monolithic, poly(styrene-divinylbenzene) support for on-line liquid chromatography tandem mass spectrometry analysis of peptides and proteins, *Analytical Chemistry* 70(23), 4879–4884, 1998.

49. Premstaller, A., Oberacher, H., and Huber, C. G., High-performance liquid chromatography-electrospray ionization mass spectrometry of single- and double-stranded nucleic acids using monolithic capillary columns, *Analytical Chemistry* 72(18), 4386–4393, 2000.

50. Premstaller, A., Oberacher, H., Walcher, W., Timperio, A. M., Zolla, L., Chervet, J. P., Cavusoglu, N., van Dorsselaer, A., and Huber, C. G., High-performance liquid chromatography-electrospray ionization mass spectrometry using monolithic capillary columns for proteomic studies, *Analytical Chemistry* 73(11), 2390–2396, 2001.

51. Li, Y., Chen, Y., Xiang, R., Ciuparu, D., Pfefferle, L. D., Horwath, C., and Wilkins, J. A., Incorporation of single-wall carbon nanotubes into an organic polymer monolithic stationary phase for mu-HPLC and capillary electrochromatography, *Analytical Chemistry* 77(5), 1398–1406, 2005.

52. Huang, X. A., Zhang, S., Schultz, G. A., and Henion, J., Surface-alkylated polystyrene monolithic columns for peptide analysis in capillary liquid chromatography-electrospray ionization mass spectrometry, *Analytical Chemistry* 74(10), 2336–2344, 2002.

53. Lee, D., Svec, F., and Frechet, J. M. J., Photopolymerized monolithic capillary columns for rapid micro high-performance liquid chromatographic separation of proteins, *Journal of Chromatography A* 1051(1–2), 53–60, 2004.

54. Nakanishi, K. and Soga, N., Phase-separation in gelling silica organic polymer-solution—Systems containing poly(sodium styrenesulfonate), *Journal of the American Ceramic Society* 74(10), 2518–2530, 1991.

55. Nakanishi, K. and Soga, N., Phase-separation in silica sol-gel system containing polyacrylic-acid. 1. Gel formation behavior and effect of solvent composition, *Journal of Non-Crystalline Solids* 139(1), 1–13, 1992.

56. Minakuchi, H., Nakanishi, K., Soga, N., Ishizuka, N., and Tanaka, N., Octadecylsilylated porous silica rods as separation media for reversed-phase liquid chromatography, *Analytical Chemistry* 68(19), 3498–3501, 1996.

57. Luo, Q. Z., Shen, Y. F., Hixson, K. K., Zhao, R., Yang, F., Moore, R. J., Mottaz, H. M., and Smith, R. D., Preparation of 20-mu m-i.d. silica-based monolithic columns and their performance for proteomics analyses, *Analytical Chemistry* 77(15), 5028–5035, 2005.

58. Yan, C., Dadoo, R., Zare, R. N., Rakestraw, D. J., and Anex, D. S., Gradient elution in capillary electrochromatography, *Analytical Chemistry* 68(17), 2726–2730, 1996.

59. Chervet, J. P., Meijvogel, C. J., Ursem, M., and Salzmann, J. P., Recent advances in capillary liquid-chromatography—Delivery of highly reproducible microflows, *LC GC—Magazine of Separation Science* 10(2), 140–148, 1992.

60. Chervet, J. P., Ursem, M., and Salzmann, J. B., Instrumental requirements for nanoscale liquid chromatography, *Analytical Chemistry* 68(9), 1507–1512, 1996.

61. Alexander, J. N., Poli, J. B., and Markides, K. E., Evaluation of automated isocratic and gradient nano-liquid chromatography and capillary electrochromatography, *Analytical Chemistry* 71(13), 2398–2409, 1999.

62. Licklider, L. J., Thoreen, C. C., Peng, J. M., and Gygi, S. P., Automation of nanoscale microcapillary liquid chromatography-tandem mass spectrometry with a vented column, *Analytical Chemistry* 74(13), 3076–3083, 2002.

63. Martin, S. E., Shabanowitz, J., Hunt, D. F., and Marto, J. A., Subfemtomole MS and MS/MS peptide sequence analysis using nano-HPLC micro-ESI Fourier transform ion cyclotron resonance mass spectrometry, *Analytical Chemistry* 72(18), 4266–4274, 2000.

64. McCormack, A. L., Schieltz, D. M., Goode, B., Yang, S., Barnes, G., Drubin, D., and Yates, J. R., Direct analysis and identification of proteins in mixtures by LC/MS/MS and database searching at the low-femtomole level, *Analytical Chemistry* 69(4), 767–776, 1997.

65. Cappiello, A., Famiglini, G., Fiorucci, C., Mangani, F., Palma, P., and Siviero, A., Variable-gradient generator for micro- and nano-HPLC, *Analytical Chemistry* 75(5), 1173–1179, 2003.

66. Zhou, X., Furushima, N., Terashima, C., Tanaka, H., and Kurano, M., New micro-flow pumping system for liquid chromatography, *Journal of Chromatography A* 913(1–2), 165–171, 2001.

67. Figeys, D. and Aebersold, R., Nanoflow solvent gradient delivery from a microfabricated device for protein identifications by electrospray ionization mass spectrometry, *Analytical Chemistry* 70(18), 3721–3727, 1998.

68. MacNair, J. E., Lewis, K. C., and Jorgenson, J. W., Ultrahigh pressure reversed-phase liquid chromatography in packed capillary columns, *Analytical Chemistry* 69(6), 983–989, 1997.

69. MacNair, J. E., Patel, K. D., and Jorgenson, J. W., Ultrahigh pressure reversed-phase capillary liquid chromatography: Isocratic and gradient elution using columns packed with 1.0-mu m particles, *Analytical Chemistry* 71(3), 700–708, 1999.

70. Le Bihan, T., Pinto, D., and Figeys, D., Nanoflow gradient generator coupled with mu-LC-ESI-MS/MS for protein identification, *Analytical Chemistry* 73(6), 1307–1315, 2001.

71. Davis, M. T., Stahl, D. C., and Lee, T. D., Low-flow high-performance liquid-chromatography solvent delivery system designed for tandem capillary liquid-chromatography mass-spectrometry, *Journal of the American Society for Mass Spectrometry* 6(7), 571–577, 1995.

72. Natsume, T., Yamauchi, Y., Nakayama, H., Shinkawa, T., Yanagida, M., Takahashi, N., and Isobe, T., A direct nanoflow liquid chromatography—Tandem system for interaction proteomics, *Analytical Chemistry* 74(18), 4725–4733, 2002.

73. Ito, S., Yoshioka, S., Ogata, I., Takeda, A., Yamashita, E., and Deguchi, K., Nanoflow gradient generator for capillary high-performance liquid chromatography-nanoelectrospray mass spectrometry, *Journal of Chromatography A* 1051(1–2), 19–23, 2004.

74. Deguchi, K., Ito, S., Yoshioka, S., Ogata, I., and Takeda, A., Nanoflow gradient generator for capillary high-performance liquid chromatography, *Analytical Chemistry* 76(5), 1524–1528, 2004.

75. Ito, S., Yoshioka, S., Ogata, I., Yamashita, E., Nagai, S., Okumoto, T., Ishii, K., Ito, M., Kaji, H., Takao, K., and Deguchi, K., Capillary high-performance liquid chromatography/electrospray ion trap time-of-flight mass spectrometry using a novel nanoflow gradient generator, *Journal of Chromatography A* 1090(1–2), 178–183, 2005.

76. Trones, R., Andersen, T., Greibrokk, T., and Hegna, D. R., Hindered amine stabilizers investigated by the use of packed capillary temperature-programmed liquid chromatography. I. Poly((6-((1,1,3,3-tetramethylbutyl)-amino)-1,3,5-triazine-2,4-diyl)(2,2,6,6-tetramethyl-4-piperidyl)imino)-1,6-hexanediyl ((2,2,6,6-tetramethyl-4-piperidyl)imino), *Journal of Chromatography A* 874(1), 65–71, 2000.

77. Trones, R., Tangen, A., Lund, W., and Greibrokk, T., Packed capillary high-temperature liquid chromatography coupled to inductively coupled plasma mass spectrometry, *Journal of Chromatography A* 835(1–2), 105–112, 1999.

78. Surowiec, I., Baena, J. R., Frank, J., Laurell, T., Nilsson, J., Trojanowicz, M., and Lendl, B., Flow-through microdispenser for interfacing mu-HPLC to Raman and mid-IR spectroscopic detection, *Journal of Chromatography A* 1080(2), 132–139, 2005.

79. Branch, S., Ebdon, L., and Oneill, P., Determination of arsenic species in fish by directly coupled high-performance liquid chromatography-inductively coupled plasma-mass spectrometry, *Journal of Analytical Atomic Spectrometry* 9(1), 33–37, 1994.

80. Ebdon, L., Evans, E. H., Pretorius, W. G., and Rowland, S. J., Analysis of geoporphyrins by high-temperature gas-chromatography inductively-coupled plasma-mass spectrometry and high-performance liquid-chromatography inductively-coupled plasma-mass spectrometry, *Journal of Analytical Atomic Spectrometry* 9(9), 939–943, 1994.

81. Harrington, C. F., Eigendorf, G. K., and Cullen, W. R., The use of high-performance liquid chromatography for the speciation of organotin compounds, *Applied Organometallic Chemistry* 10(5), 339–362, 1996.

82. Encinar, J. R., Schaumloffel, D., Ogra, Y., and Lobinski, R., Determination of selenomethionine and selenocysteine in human serum using speciated isotope dilution-capillary HPLC-inductively coupled plasma collision cell mass spectrometry, *Analytical Chemistry* 76(22), 6635–6642, 2004.

83. Ogra, Y. and Suzuki, K. T., Speciation of selenocompounds by capillary HPLC coupled with ICP-MS using multi-mode gel filtration columns, *Journal of Analytical Atomic Spectrometry* 20(1), 35–39, 2005.

84. Polatajko, A., Encinar, J. R., Schaumloffel, D., and Szpunar, J., Quantification of a selenium-containing protein in yeast extract via an accurate determination of a tryptic peptide by species-specific isotope dilution capillary HPLC-ICP MS, *Chemia Analityczna* 50(1), 265–278, 2005.

85. Ballihaut, G., Tastet, L., Pecheyran, C., Bouyssiere, B., Donard, O., Grimaud, R., and Lobinski, R., Biosynthesis, purification and analysis of selenomethionyl calmodulin by gel electrophoresis-laser ablation-ICP-MS and capillary HPLC-ICP-MS peptide mapping following in-gel tryptic digestion, *Journal of Analytical Atomic Spectrometry* 20(6), 493–499, 2005.

86. Giusti, P., Schaumloffel, D., Encinar, J. R., and Szpunar, J., Interfacing reversed-phase nanoHPLC with ICP-MS and on-line isotope dilution analysis for the accurate quantification of selenium-containing peptides in protein tryptic digests, *Journal of Analytical Atomic Spectrometry* 20(10), 1101–1107, 2005.

87. Gaudin, K., Baillet, A., and Chaminade, P., Adaptation of an evaporative light-scattering detector to micro and capillary liquid chromatography and response assessment, *Journal of Chromatography A* 1051(1–2), 43–51, 2004.

88. Mann, M., Hendrickson, R. C., and Pandey, A., Analysis of proteins and proteomes by mass spectrometry, *Annual Review of Biochemistry* 70, 437–473, 2001.

89. Granger, J., Plumb, R., Castro-Perez, J., and Wilson, I. D., Metabonomic studies comparing capillary and conventional HPLC-oa-TOF MS for the analysis of urine from Zucker obese rats, *Chromatographia* 61(7–8), 375–380, 2005.

90. Calvo, E., Flores-Romero, P., Lopez, J. A., and Navas, A., Identification of proteins expressing differences among isolates of *Meloidogyne* spp. (Nematoda: Meloidogynidae) by nano-liquid chromatography coupled to ion-trap mass spectrometry, *Journal of Proteome Research* 4(3), 1017–1021, 2005.

91. Yttri, K. E., Dye, C., Slordal, L. H., and Braathen, O. A., Quantification of monosaccharide anhydrides by liquid chromatography combined with mass spectrometry: Application to aerosol samples from an urban and a suburban site influenced by small-scale wood burning, *Journal of the Air and Waste Management Association* 55(8), 1169–1177, 2005.

92. Hofmann, S., Gluckmann, M., Kausche, S., Schmidt, A., Corvey, C., Lichtenfels, R., Huber, C., Albrecht, C., Karas, M., and Herr, W., Rapid and sensitive identification of major histocompatibility complex class I-associated tumor peptides by nano-LC MALDI MS/MS, *Molecular and Cellular Proteomics* 4(12), 1888–1897, 2005.

93. Lochnit, G. and Geyer, R., An optimized protocol for nano-LC-MALDI-TOF-MS coupling for the analysis of proteolytic digests of glycoproteins, *Biomedical Chromatography* 18(10), 841–848, 2004.

94. Mitulovic, G., Smoluch, M., Chervet, J. P., Steinmacher, I., Kungl, A., and Mechtler, K., An improved method for tracking and reducing the void volume in nano HPLC-MS with micro trapping columns, *Analytical and Bioanalytical Chemistry* 376(7), 946–951, 2003.

95. Timperman, A. T. and Aebersold, R., Peptide electroextraction for direct coupling of in-gel digests with capillary LC-MS/MS for protein identification and sequencing, *Analytical Chemistry* 72(17), 4115–4121, 2000.

96. Dugo, P., Presti, M. L., Ohman, M., Fazio, A., Dugo, G., and Mondello, L., Determination of flavonoids in citrus juices by micro-HPLC-ESI/MS, *Journal of Separation Science* 28(11), 1149–1156, 2005.

97. Dai, G., Wei, X., Liu, Z., Liu, S., Marcucci, G., and Chan, K. K., Characterization and quantification of Bcl-2 antisense G3139 and metabolites in plasma and urine by ion-pair reversed phase HPLC coupled with electrospray ion-trap mass spectrometry, *Journal of Chromatography. B, Analytical Technologies in the Biomedical and Life Sciences* 825(2), 201–213, 2005.

98. Dugo, P., Favoino, O., Presti, M. L., Luppino, R., Dugo, G., and Mondello, L., Determination of anthocyanins and related components in red wines by micro- and capillary HPLC, *Journal of Separation Science* 27(17–18), 1458–1466, 2004.

99. Perbellini, L., Gottardo, R., Caprini, A., Bortolotti, F., Mariotto, S., and Tagliaro, F., Determination of alpha-bisabolol in human blood by micro-HPLC-ion trap MS and head space-GC-MS methods, *Journal of Chromatography. B, Analytical Technologies in the Biomedical and Life Sciences* 812(1–2), 373–377, 2004.

100. Dugo, G., Salvo, F., Dugo, P., La Torre, G. L., and Mondello, L., Antioxidants in Sicilian wines: Analytic and compositive aspects, *Drugs under Experimental and Clinical Research* 29(5–6), 189–202, 2003.

101. Timperio, A. M., Magro, P., Chilosi, G., and Zolla, L., Assay of ochratoxin A in grape by high-pressure liquid chromatography coupled on line with an ESI-mass spectrometry, *Journal of Chromatography. B, Analytical Technologies in the Biomedical and Life Sciences* 832(1), 127–133, 2006.

102. Timperio, A. M. and Zolla, L., Investigation of the lateral light-induced migration of photosystem II light-harvesting proteins by nano-high performance liquid chromatography electrospray ionization mass spectrometry, *Journal of Biological Chemistry* 280(32), 28858–28866, 2005.

103. Wang, H., Clouthier, S. G., Galchev, V., Misek, D. E., Duffner, U., Min, C. K., Zhao, R., Tra, J., Omenn, G. S., Ferrara, J. L. M., and Hanash, S. M., Intact-protein-based high-resolution three-dimensional quantitative analysis system for proteome profiling of biological fluids, *Molecular and Cellular Proteomics* 4(5), 618–625, 2005.

104. Ficarro, S. B., Salomon, A. R., Brill, L. M., Mason, D. E., Stettler-Gill, M., Brock, A., and Peters, E. C., Automated immobilized metal affinity chromatography/nano-liquid chromatography/electrospray ionization mass spectrometry platform for profiling protein phosphorylation sites, *Rapid Communications in Mass Spectrometry* 19(1), 57–71, 2005.

105. Frommberger, M., Schmitt-Kopplin, P., Ping, G., Frisch, H., Schmid, M., Zhang, Y., Hartmann, A., and Kettrup, A., A simple and robust set-up for on-column sample preconcentration–nano-liquid chromatography–electrospray ionization mass spectrometry for the analysis of N-acylhomoserine lactones, *Analytical and Bioanalytical Chemistry* 378(4), 1014–1020, 2004.

106. Zhao, Y. X., Zhang, W., Kho, Y. J., and Zhao, Y. M., Proteomic analysis of integral plasma membrane proteins, *Analytical Chemistry* 76(7), 1817–1823, 2004.

107. Hong, S. S. and Kwon, S. W., Profiling of urinary proteins by nano-high performance liquid chromatography/tandem mass spectrometry, *Journal of Liquid Chromatography and Related Technologies* 28(6), 805–822, 2005.

108. Imre, T., Schlosser, G., Pocsfalvi, G., Siciliano, R., Molnar-Szollosi, E., Kremmer, T., Malorni, A., and Vekey, K., Glycosylation site analysis of human alpha-1-acid glycoprotein (AGP) by capillary liquid chromatography-electrospray mass spectrometry, *Journal of Mass Spectrometry* 40(11), 1472–1483, 2005.

109. Plematl, A., Demelbauer, U. M., Josic, D., and Rizzi, A., Determination of the site-specific and isoform-specific glycosylation in human plasma-derived antithrombin by IEF and capillary HPLC-ESI-MS/MS, *Proteomics* 5(15), 4025–4033, 2005.

110. White, D. C., Geyer, R., Cantu, J., Jo, S. C., Peacock, A. D., Saxton, A. M., Mani, S., Jett, M., and Moss, O. R., Feasibility of assessment of regulatory lipids in breath condensate as potential presymptomatic harbingers of pulmonary pathobiology, *Journal of Microbiological Methods* 62(3), 293–302, 2005.

111. Langlais, P., Wang, C. H., Dong, L. Q., Carroll, C. A., Weintraub, S. T., and Liu, F., Phosphorylation of Grb10 by mitogen-activated protein kinase: Identification of Ser(150) and Ser(476) of human Grb10 zeta as major phosphorylation site, *Biochemistry* 44(24), 8890–8897, 2005.

112. Powell, M. J., Sutton, J. N., Del Castillo, C. E., and Timperman, A. I., Marine proteomics: Generation of sequence tags for dissolved proteins in seawater using tandem mass spectrometry, *Marine Chemistry* 95(3–4), 183–198, 2005.

113. Schmid, D., Behnke, B., Metzger, J., and Kuhn, R., Nano-HPLC-mass spectrometry and MEKC for the analysis of oligosaccharides from human milk, *Biomedical Chromatography* 16(2), 151–156, 2002.

114. Delmulle, B., De Saeger, S., Adams, A., De Kimpe, N., and Van Peteghem, C., Development of a liquid chromatography/tandem mass spectrometry method for the simultaneous determination of 16 mycotoxins on cellulose filters and in fungal cultures, *Rapid Communications in Mass Spectrometry* 20(5), 771–776, 2006.

115. Song, Y., Shenwu, M., Zhao, S., Hou, D., and Liu, Y. M., Enantiomeric separation of amino acids derivatized with 7-fluoro-4-nitrobenzoxadiazole by capillary liquid chromatography/tandem mass spectrometry, *Journal of Chromatography A* 1091(1–2), 102–109, 2005.

116. Simoes-Pires, C. A., Queiroz, E. F., Henriques, A. T., and Hostettmann, K., Isolation and on-line identification of antioxidant compounds from three Baccharis species by HPLC-UV-MS/MS with post-column derivatisation, *Phytochemical Analysis* 16(5), 307–314, 2005.

117. Koal, T., Burhenne, H., Romling, R., Svoboda, M., Resch, K., and Kaever, V., Quantification of antiretroviral drugs in dried blood spot samples by means of liquid chromatography/tandem mass spectrometry, *Rapid Communications in Mass Spectrometry* 19(21), 2995–3001, 2005.

118. Sabatini, L., Barbieri, A., Tosi, M., Roda, A., and Violante, F. S., A method for routine quantitation of urinary 8-hydroxy-2′-deoxyguanosine based on solid-phase extraction and micro-high-performance liquid chromatography/electrospray ionization tandem mass spectrometry, *Rapid Communications in Mass Spectrometry* 19(2), 147–152, 2005.

119. Wuhrer, M., Balog, C. I., Koeleman, C. A., Deelder, A. M., and Hokke, C. H., New features of site-specific horseradish peroxidase (HRP) glycosylation uncovered by nano-LC-MS with repeated ion-isolation/fragmentation cycles, *Biochimica et Biophysica Acta* 1723(1–3), 229–239, 2005.

120. Le Bihan, T., Robinson, M. D., Stewart, I. I., and Figeys, D., Definition and characterization of a "trypsinosome" from specific peptide characteristics by nano-HPLC-MS/MS and in silico analysis of complex protein mixtures, *Journal of Proteome Research* 3(6), 1138–1148, 2004.

121. Kwon, S. W., Profiling of soluble proteins in wine by nano-high-performance liquid chromatography/tandem mass spectrometry, *Journal of Agricultural and Food Chemistry* 52(24), 7258–7263, 2004.

122. Pitchayawasin, S., Isobe, M., Kuse, M., Franz, T., Agata, N., and Ohta, M., Molecular diversity of cereulide detected by means of nano-HPLC-ESI-Q-TOF-MS, *International Journal of Mass Spectrometry* 235(2), 123–129, 2004.

123. Sule, A., Vanrobaeys, F., Hajos, G., Van Beeumen, J., and Devreese, B., Proteomic analysis of small heat shock protein isoforms in barley shoots, *Phytochemistry* 65(12), 1853–1863, 2004.

124. Huybrechts, J., De Loof, A., and Schoofs, L., Melatonin-induced neuropeptide release from isolated locust corpora cardiaca, *Peptides* 26(1), 73–80, 2005.

125. Speicher, K. D., Kolbas, O., Harper, S., and Speicher, D. W., Systematic analysis of peptide recoveries from in-gel digestions for protein identification in proteome studies., *Journal of Biomolecular Techniques* 11(2), 74–86, 2000.

126. Murata, K., Mano, N., and Asakawa, N., Development of a large-volume, membrane-based injection system for gradient elution micro-liquid chromatography, *Journal of Chromatography A* 1106(1–2), 146–151, 2006.

127. Ishihama, Y., Proteomic LC-MS systems using nanoscale liquid chromatography with tandem mass spectrometry, *Journal of Chromatography A* 1067(1–2), 73–83, 2005.

128. Delaunay, N., Pichon, V., and Hennion, M. C., Immunoaffinity solid-phase extraction for the trace-analysis of low-molecular-mass analytes in complex sample matrices, *Journal of Chromatography. B, Biomedical Sciences and Applications* 745(1), 15–37, 2000.

129. Hage, D. S., Survey of recent advances in analytical applications of immunoaffinity chromatography, *Journal of Chromatography. B, Biomedical Sciences and Applications* 715(1), 3–28, 1998.

130. Hodgson, R. J., Chen, Y., Zhang, Z., Tleugabulova, D., Long, H., Zhao, X., Organ, M., Brook, M. A., and Brennan, J. D., Protein-doped monolithic silica columns for capillary liquid chromatography prepared by the sol-gel method: Applications to frontal affinity chromatography, *Analytical Chemistry* 76(10), 2780–2790, 2004.

131. Madera, M., Mechref, Y., Klouckova, I., and Novotny, M. V., Semiautomated high-sensitivity profiling of human blood serum glycoproteins through lectin preconcentration and multidimensional chromatography/tandem mass spectrometry, *Journal of Proteome Research* 5(9), 2348–2363, 2006.

132. Starkey, J. A., Mechref, Y., Muzikar, J., McBride, W. J., and Novotny, M. V., Determination of salsolinol and related catecholamines through on-line preconcentration and liquid chromatography/atmospheric pressure photoionization mass spectrometry, *Analytical Chemistry* 78(10), 3342–3347, 2006.

133. Guzman, N. A., Improved solid-phase microextraction device for use in on-line immunoaffinity capillary electrophoresis, *Electrophoresis* 24(21), 3718–3727, 2003.

134. Benavente, F., Vescina, M. C., Hernandez, E., Sanz-Nebot, V., Barbosa, J., and Guzman, N. A., Lowering the concentration limits of detection by on-line solid-phase extraction-capillary electrophoresis-electrospray mass spectrometry, *Journal of Chromatography A* 1140(1–2), 205–212, 2007.

135. Hodgson, R. J., Brook, M. A., and Brennan, J. D., Capillary-scale monolithic immunoaffinity columns for immunoextraction with in-line laser-induced fluorescence detection, *Analytical Chemistry* 77(14), 4404–4412, 2005.

136. Holland, H. A. and Jorgenson, J. W., Separation of nanoliter samples of biological amines by a comprehensive 2-dimensional microcolumn liquid-chromatography system, *Analytical Chemistry* 67(18), 3275–3283, 1995.

137. Masuda, J., Maynard, D. A., Nishimura, M., Uedac, T., Kowalak, J. A., and Markey, S. P., Fully automated micro- and nanoscale one- or two-dimensional high-performance liquid chromatography system for liquid chromatography-mass spectrometry compatible with non-volatile salts for ion exchange chromatography, *Journal of Chromatography A* 1063(1–2), 57–69, 2005.

138. Nagele, E., Vollmer, M., and Horth, P., Two-dimensional nano-liquid chromatography-mass spectrometry system for applications in proteomics, *Journal of Chromatography A* 1009(1–2), 197–205, 2003.

139. Shen, Y., Tolic, N., Masselon, C., Pasa-Tolic, L., Camp, D. G., Hixson, K. K., Zhao, R., Anderson, G. A., and Smith, R. D., Ultrasensitive proteomics using high-efficiency on-line micro-SPE-NanoLC-NanoESI MS and MS/MS, *Analytical Chemistry* 76(1), 144–154, 2004.

140. Shen, Y. F., Tolic, N., Zhao, R., Pasa-Tolic, L., Li, L. J., Berger, S. J., Harkewicz, R., Anderson, G. A., Belov, M. E., and Smith, R. D., High-throughput proteomics using high efficiency multiple-capillary liquid chromatography with on-line high-performance ESI FTICR mass spectrometry, *Analytical Chemistry* 73(13), 3011–3021, 2001.

141. Shen, Y. F., Zhao, R., Belov, M. E., Conrads, T. P., Anderson, G. A., Tang, K. Q., Pasa-Tolic, L., Veenstra, T. D., Lipton, M. S., Udseth, H. R., and Smith, R. D., Packed capillary reversed-phase liquid chromatography with high-performance electrospray ionization Fourier transform ion cyclotron resonance mass spectrometry for proteomics, *Analytical Chemistry* 73(8), 1766–1775, 2001.

142. Shen, Y. F., Zhao, R., Berger, S. J., Anderson, G. A., Rodriguez, N., and Smith, R. D., High-efficiency nanoscale liquid chromatography coupled on-line with mass spectrometry using nanoelectrospray ionization for proteomics, *Analytical Chemistry* 74(16), 4235–4249, 2002.

143. Wagner, Y., Sickmann, A., Meyer, H. E., and Daum, G., Multidimensional nano-HPLC for analysis of protein complexes, *Journal of the American Society for Mass Spectrometry* 14(9), 1003–1011, 2003.

144. Mitulovic, G., Stingl, C., Smoluch, M., Swart, R., Chervet, J. P., Steinmacher, I., Gerner, C., and Mechtler, K., Automated, on-line two-dimensional nano liquid chromatography tandem mass spectrometry for rapid analysis of complex protein digests, *Proteomics* 4(9), 2545–2557, 2004.

145. Chang, L. Y., Chou, S. S., and Hwang, D. F., High performance liquid chromatography to determine animal drug Clenbuterol in pork, beef and hog liver, *Journal of Food and Drug Analysis* 13(2), 163–167, 2005.

146. Chung, Y. and Chung, W., Determination of Co(II) ion as a 4-(2-thiazolylazo)resorcinol or 5-methyl-4-(2-thiazolylazo)resorcinol chelate by reversed-phase capillary high-performance liquid chromatography, *Bulletin of the Korean Chemical Society* 24(12), 1781–1784, 2003.

147. Chung, Y. S., Determination of total phenols in environmental waters by capillary-HPLC with USEPA classified eleven priority pollutant phenols after nitrosation and their visible spectrophotometric detection, *Bulletin of the Korean Chemical Society* 26(2), 297–302, 2005.

148. Bamba, T., Fukusaki, E., Minakuchi, H., Nakazawa, Y., and Kobayashi, A., Separation of polyprenol and dolichol by monolithic silica capillary column chromatography, *Journal of Lipid Research* 46(10), 2295–2298, 2005.

149. Arayne, M. S., Sultana, N., and Siddiqui, F. A., A new RP-HPLC method for analysis of mebeverine hydrochloride in raw materials and tablets, *Pakistan Journal of Pharmaceutical Sciences* 18(2), 11–14, 2005.

150. Arayne, M. S., Sultana, N., and Siddiqui, F. A., A new RP-HPLC method for analysis of meloxicam in tablets, *Pakistan Journal of Pharmaceutical Sciences* 18(1), 58–62, 2005.

151. Lanjwani, S. N., Zhu, R., Khuhawar, M. Y., and Ding, Z., High performance liquid chromatographic determination of platinum in blood and urine samples of cancer patients after administration of cisplatin drug using solvent extraction and N,N'-bis(salicylidene)-1,2-propanediamine as complexation reagent, *Journal of Pharmaceutical and Biomedical Analysis* 40(4), 833–839, 2006.

152. Gupta, M. and Bhargava, H. N., Development and validation of a high-performance liquid chromatographic method for the analysis of budesonide, *Journal of Pharmaceutical and Biomedical Analysis* 40(2), 423–428, 2006.

153. Ovalles, J. F., Brunetto Mdel, R., and Gallignani, M., A new method for the analysis of amikacin using 6-aminoquinolyl-N-hydroxysuccinimidyl carbamate (AQC) derivatization and high-performance liquid chromatography with UV-detection, *Journal of Pharmaceutical and Biomedical Analysis* 39(1–2), 294–298, 2005.

154. Yazdi, A. S. and Es'haghi, Z., Two-step hollow fiber-based, liquid-phase microextraction combined with high-performance liquid chromatography: A new approach to determination of aromatic amines in water, *Journal of Chromatography A* 1082(2), 136–142, 2005.

155. Kim, C. K., Yeon, K. J., Ban, E., Hyun, M. J., Kim, J. K., Kim, M. K., Jin, S. E., and Park, J. S., Narrowbore high performance liquid chromatographic method for the determination of cetirizine in human plasma using column switching, *Journal of Pharmaceutical and Biomedical Analysis* 37(3), 603–609, 2005.

156. Piccinini, A. M., Schmid, M. G., Pajpanova, T., Pancheva, S., Grueva, E., and Gubitz, G., Chiral separation of natural and unnatural amino acid derivatives by micro-HPLC on a Ristocetin A stationary phase, *Journal of Biochemical and Biophysical Methods* 61(1–2), 11–21, 2004.

157. Schmid, M. G., Holbling, M., Schnedlitz, N., and Gubitz, G., Enantioseparation of dipeptides and tripeptides by micro-HPLC comparing teicoplanin and teicoplanin aglycone as chiral selectors, *Journal of Biochemical and Biophysical Methods* 61(1–2), 1–10, 2004.

158. Meyer, R., Vorster, S., and Dubery, I. A., Identification and quantification of gossypol in cotton by using packed micro-tips columns in combination with HPLC, *Analytical and Bioanalytical Chemistry* 380(4), 719–724, 2004.

159. Hernandez, R., Nelson, S., Salm, J. R., Brown, D. T., and Alpert, A. J., Rapid preparative purification of West Nile and Sindbis virus PCR products utilizing a microbore anion-exchange column, *Journal of Virological Methods* 120(2), 141–149, 2004.

160. Fanali, S., Aturkil, Z., D'Orazio, G., Raggi, M. A., Quaglia, M. G., Sabbioni, C., and Rocco, A., Use of nano-liquid chromatography for the analysis of glycyrrhizin and glycyrrhetic acid in licorice roots and candies, *Journal of Separation Science* 28(9–10), 982–986, 2005.

161. Verschraagen, M., Boven, E., Torun, E., Hausheer, F. H., Bast, A., and van der Vijgh, W. J. F., Possible (enzymatic) routes and biological sites for metabolic reduction of BNP7787, a new protector against cisplatin-induced side-effects, *Biochemical Pharmacology* 68(3), 493–502, 2004.

162. Kotani, A., Kojima, S., Hakamata, H., Jin, D., and Kusu, F., Determination of honokiol and magnolol by micro HPLC with electrochemical detection and its application to the distribution analysis in branches and leaves of Magnolia obovata, *Chemical and Pharmaceutical Bulletin (Tokyo)* 53(3), 319–322, 2005.

163. Queiroz, E. F., Ioset, J. R., Ndjoko, K., Guntern, A., Foggin, C. M., and Hostettmann, K., On-line identification of the bioactive compounds from Blumea gariepina by HPLC-UV-MS and HPLC-UV-NMR, combined with HPLC-micro-fractionation, *Phytochemical Analysis* 16(3), 166–174, 2005.

164. Sultan, M., Stecher, G., Stoggl, W. M., Bakry, R., Zaborski, P., Huck, C. W., El Kousy, N. M., and Bonn, G. K., Sample pretreatment and determination of non steroidal anti-inflammatory drugs (NSAIDs) in pharmaceutical formulations and biological samples (blood, plasma, erythrocytes) by HPLC-UV-MS and micro-HPLC, *Current Medicinal Chemistry* 12(5), 573–588, 2005.

165. Kotani, A., Kojima, S., Hakamata, H., and Kusu, F., HPLC with electrochemical detection to examine the pharmacokinetics of baicalin and baicalein in rat plasma after oral administration of a Kampo medicine, *Analytical Biochemistry* 350(1), 99–104, 2006.

166. Jin, D., Hakamata, H., Takahashi, K., Kotani, A., and Kusu, F., Determination of quercetin in human plasma after ingestion of commercial canned green tea by semi-micro HPLC with electrochemical detection, *Biomedical Chromatography* 18(9), 662–666, 2004.

167. Tsunoda, M., Nagayama, M., Funatsu, T., Hosoda, S., and Imai, K., Catecholamine analysis with microcolumn LC-peroxyoxalate chemiluminescence reaction detection, *Clinica Chimica Acta* 366(1–2), 168–173, 2006.

168. Luykx, D. M., Dingemanse, P. J., Goerdayal, S. S., and Jongen, P. M., High-performance anion-exchange chromatography combined with intrinsic fluorescence detection to determine erythropoietin in pharmaceutical products, *Journal of Chromatography A* 1078(1–2), 113–119, 2005.

169. Wada, M., Nakamura, S., Tomita, M., Nakashima, M. N., and Nakashima, K., Determination of MDMA and MDA in rat urine by semi-micro column HPLC-fluorescence detection with DBD-F and their monitoring after MDMA administration to rat, *Luminescence* 20(3), 210–215, 2005.

170. De Melo Abreu, S., Correia, M., Herbert, P., Santos, L., and Alves, A., Screening of grapes and wine for azoxystrobin, kresoxim-methyl and trifloxystrobin fungicides by HPLC with diode array detection, *Food Additives and Contaminants* 22(6), 549–556, 2005.

171. Krucker, M., Lienau, A., Putzbach, K., Grynbaum, M. D., Schuger, P., and Albert, K., Hyphenation of capillary HPLC to microcoil H-1 NMR spectroscopy for the determination of tocopherol homologues, *Analytical Chemistry* 76(9), 2623–2628, 2004.

172. Hentschel, P., Krucker, M., Grynbaum, M. D., Putzbach, K., Bischoff, R., and Albert, K., Determination of regulatory phosphorylation sites in nanogram amounts of a synthetic fragment of ZAP-70 using microprobe NMR and on-line coupled capillary HPLC-NMR, *Magnetic Resonance in Chemistry* 43(9), 747–754, 2005.

173. Putzbach, K., Krucker, M., Grynbaum, M. D., Hentschel, P., Webb, A. G., and Albert, K., Hyphenation of capillary high-performance liquid chromatography to microcoil magnetic resonance spectroscopy—Determination of various carotenoids in a small-sized spinach sample, *Journal of Pharmaceutical and Biomedical Analysis* 38(5), 910–917, 2005.

174. Xiao, H. B., Krucker, M., Putzbach, K., and Albert, K., Capillary liquid chromatography-microcoil 1H nuclear magnetic resonance spectroscopy and liquid chromatography-ion trap mass spectrometry for on-line structure elucidation of isoflavones in Radix astragali, *Journal of Chromatography A* 1067(1–2), 135–143, 2005.

4 Two-Dimensional Comprehensive Liquid Chromatography

Luigi Mondello, Paola Dugo, Tiina Kumm,
Francesco Cacciola, and Giovanni Dugo

CONTENTS

4.1 INTRODUCTION

High-performance liquid chromatography (HPLC) represents a powerful method widely applied for the separation of real-world samples in several fields. Although this methodology may often provide sufficient resolving power for the separation of target components, many matrices present a complexity that greatly exceeds the separation capacity of any single chromatographic system. Therefore, despite a careful method optimization procedure, conventional HPLC may be inadequate when faced with truly complex samples. Moreover, peak overlapping may occur even in the case of relatively simple samples that contain components with similar properties. In such cases, the combination of more than one separation step is a convenient choice, generating a great increase in the resolving power of a chromatographic system, expanding separation space, enhancing peak capacity, and, thus, enabling the resolution of components even in highly complex matrices.

The most widespread among multidimensional HPLC techniques are two-dimensional (2D) methods, in which components migrate along two imaginary axes. Figure 4.1 illustrates such a process, where five fractions of the first-dimension separation are subjected to a second-dimension separation.

101

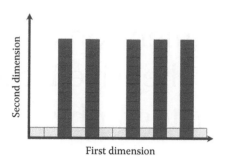

FIGURE 4.1   Schematic representation of a 2D separation.

If the separation mechanisms are independent, then the probability of component overlapping is quite small: components, unresolved in first separation according to one parameter, will probably be separated during the second separation. In the case where the two chromatographic separations are preceded by a solvent extraction step, the latter should be considered as a part of the multidimensional system, yielding a 3D system [1]. An additional dimension is also represented by a mass spectrometric detector and hence, even 4D systems are achievable.

Multidimensional HPLC analyses can be performed either off-line or on-line. In an off-line HPLC, the fractions eluted from the primary column are collected manually or by a fraction collector, concentrated if necessary and then reinjected into a second column. These techniques can be time consuming, operative intensive, and difficult to automate and reproduce. Furthermore, of greater importance in areas of quantitative trace analysis is the fact that off-line sample treatments can be susceptible to solute loss and contamination. However, the off-line approach is quite easy, because both analytical dimensions can be considered as two independent methods. This technique is most used when only parts of the first separation require a secondary separation. In an on-line HPLC system, the two columns are connected by means of a special interface, usually a switching valve, which allows the transfer of the fractions of the first column effluent onto the second column. There are two types of on-line multidimensional liquid chromatography: heart-cutting and comprehensive. While the former enables the bidimensional analysis of specific fractions of the initial sample, the second is a more powerful approach that extends the 2D advantage to the entire matrix. The heart-cutting techniques, extensively reviewed in previous papers [2,3], are not the argument of this chapter, devoted to the comprehensive approach for the characterization of complex matrices.

It has been stated that a comprehensive 2D separation should possess the following features [4,5]:

- All components in a sample mixture are subjected to two separations in which their displacement depends on different factors.
- Equal percentages (either 100% or lower) of all sample components pass through both columns and eventually reach the detector.
- Any two components separated in the first dimension must remain separated in the second dimension.
- The elution profiles from both dimensions are preserved.

The comprehensive HPLC is characterized by the following advantages if compared to the 1D mode:

1. Higher resolving power and peak capacity
2. Greater amount of information about the sample in a single analysis
3. Great potential for the identification of "unknowns" (formation of chemical class patterns on the 2D space plane, described later)

However, this technique presents a number of challenges:

1. More thought must be given to method and instrument design
2. Incompatibility of different mobile phases can occur: immiscibility, precipitation of buffer salts, incompatibility between the mobile phase from one column and the stationary phase of the other
3. Need for specific interfaces and software

4.2 HISTORY

The origin of MD liquid chromatography lies in planar chromatography. Initially, the major 2D separations were realized using thin-layer beds, originally in the form of paper chromatography (i.e., the partition between a liquid moving by capillary action across a strip of paper impregnated with a second liquid). The development of paper chromatography proceeded in parallel with the development of liquid–liquid partition chromatography on columns, and in 1944, Martin and coworkers [6] discussed the possibility of different eluents in different dimensions. In the following years, a variety of hyphenated techniques were developed, until the most important planar separation, based on the use of gel electrophoresis, was reported by O'Farrell in 1975 [7]: up to 1000 proteins from a bacterial culture were separated by using isoelectric focusing in one dimension and sodium dodecylsulfonate–polyacrylamide gel electrophoresis in the second. Most developments, in the past two decades, however, have involved coupled column systems which are much more amenable to automation and more readily permit quantitative measurements. Comprehensive 2D liquid chromatography was first introduced by Erni and Frei in 1978 [8], who analyzed a complex plant extract with a SEC column as the first dimension and a reversed-phase (RP) column as the second dimension, connecting the two columns with an eight-port valve. Next, the technique was evolved and improved by Bushey and Jorgenson in 1990 [9], who used a microbore cation exchange column for the first-dimension separation and a size exclusion column (SEC) for the second-dimension separation, using an eight-port valve for transferring the first column effluent into the second. The system was applied to the separation of protein standards and serum proteins.

In the last years, the use of comprehensive liquid chromatography has been greatly increased and it has been widely used to separate and characterize various complex samples, such as biomolecules [10–15], polymers [16,17], lipids [18–21], essential oils [22], acidic and phenolic compounds [23–28], pharmaceuticals and traditional medicines [29–31], etc. Comprehensive LC has been reviewed by several authors [32–37].

4.3 GENERAL PRINCIPLES

The complete separation of the compounds of interest must always be a primary goal in any chromatographic separation. The effectiveness of a separation may be measured by the peak capacity n_c, which shows the maximum number of peaks that fits side by side into the available separation space (with resolution equal to 1). It has been demonstrated that the frequency with which component peaks overlap depends upon the peak capacity [38]. High-resolution linear separation systems usually generate an n_c in the range of several hundreds, which is apparently enough for the separation of 100 peaks. However, to realize the maximum peak content, the peaks must be evenly spaced at their highest allowed density. In complex real-world samples, this occurs quite rarely, as the peaks tend to fall across the chromatogram in a random mode and co-elutions are observed. Therefore, the real number of components that may be isolated in most separation processes is much less than the peak capacity. Davis and Giddings [38] and Martin et al. [39] showed how peak capacity is only the maximum number of mixture constituents which a chromatographic system may resolve. Through the statistical method of component overlap (SMO) [38], it was estimated that no more than 37% of the peak capacity is used for peak resolution. This percentage corresponds to the number of visible peaks in a chromatogram.

Real-world samples are usually characterized by a variety of chemical groups and, consequentially, by random peak distribution, therefore requiring a high separating power. A practical method for enhancing the peak capacity, as mentioned before, can be achieved by using multidimensional separations. In MD systems, the peak capacity is the sum of the peak capacities of the 1D processes:

$$n_c(\max) = \sum_{i=i}^{k} n_{ci} \tag{4.1}$$

As mentioned above, the most common multidimensional separations are performed by using 2D systems. A considerable increase in peak capacity of the 2D system can be achieved if the whole sample is subjected on-line to two independent displacement processes (comprehensive MD separation) with peak capacities of n_z and n_y, respectively. If the two separations have different retention mechanisms (e.g., are orthogonal to each other), the maximum peak capacity n_{2D} of the system is approximately equal to the product of the peak capacities n_z and n_y [5]:

$$n_{2D} \cong n_z \times n_y \tag{4.2}$$

This is illustrated in Figure 4.2. The peak capacities of the two dimensions are shown as the number of adjacent Gaussian profiles that can be packed into the space along the respective separation coordinates. The separation plane is divided into rectangular boxes that represent the resolution units in the 2D plane. The total peak capacity n_{2D} is therefore approximately equal to the number of such boxes.

For example, peak capacities of 50 in both dimensions give a total peak capacity over 2000, which would require about 10 million theoretical plates in an 1D system. In reality, most of 2D systems have at least some retention correlation, and this decreases the optimum resolution and peak capacity of the system [40]. The same is caused by the additional broadening of the component zones in their migration along the second coordinate.

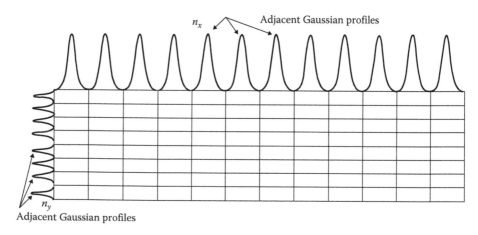

FIGURE 4.2   Peak capacity in a 2D system. (From Giddings, J.C., *J. High Resolut. Chromatogr. Chromatogr. Commun.*, 10, 319, 1987.)

The improvement in resolving power of MD chromatography is expressed also in terms of enhanced resolution. Giddings showed [1] how the contributors to the overall resolution, R_s, along the two axes, R_x and R_y, contribute to the final resolution according to

$$R_s \cong \left(R_x^2 + R_y^2 \right)^{1/2} \tag{4.3}$$

The isolation of a component from a neighboring one in a 2D system is much more probable than in a linear system, because the two displacements of the components are much less likely similar than in the case of a monodimensional separation.

Since the introduction of 2D chromatographic techniques, it has become clear that each separation step should ideally provide different selectivities (orthogonal system) as this maximizes the gain in peak capacity and hence the number of chromatographically resolvable components [8,40–42]. Through orthogonality, cross-information or synentropy existing between the two dimensions is minimized, resulting in maximum peak capacity and hence, high resolution. Minimizing synentropy maximizes the efficiency and the information content, generating a key for complex sample analysis [43–45].

The separation in both dimensions should be controlled by parameters that are relatively independent, otherwise the system has little effectiveness. With two independent migrations, we have two parameters characterizing each component peak emerging from a secondary column, the first identified by the position in the sequence of cuts from the first column and the second measured by the retention time in the secondary column. However, if the cut from the primary column is very broad (large Δt_r), then the component peaks emerging from the secondary column will have considerable uncertainty in the retention time for their migration through the primary column. The greater the "cut," the greater the degradation of the quality of information relative to each component's identity. Broader cuts also incorporate more components and increase the likelihood of interference in subsequent separation steps. Component separation will improve with decreasing Δt_r. Murphy et al. have stated that to obtain a high 2D resolution, each peak in the first dimension should be sampled at least three to four times [46].

The effectiveness of a 2D separation, where a first-dimension peak is unraveled in three compounds (*a*, *b*, and *c*), is illustrated in Figure 4.3, where the number of samplings over the first-dimension peak is 3.

For quantitation, the areas of all peaks relative to the same compound must be summed, as illustrated in Figure 4.4.

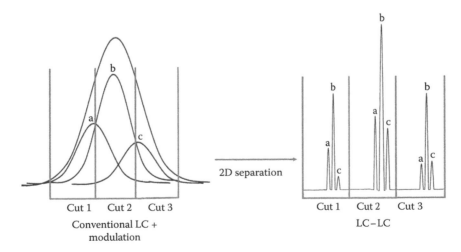

FIGURE 4.3 Effects of modulation on three co-eluting peaks.

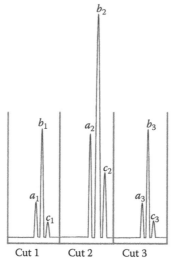

Area $a = a_1 + a_2 + a_3 = \Sigma$ Area a

Area $b = b_1 + b_2 + b_3 = \Sigma$ Area b

Area $c = c_1 + c_2 + c_3 = \Sigma$ Area c

FIGURE 4.4 Quantitative procedure in LCxLC.

4.4 INSTRUMENTATION

Various comprehensive HPLC systems have been developed and proven to be effective both for the separation of complex sample components and in the resolution of a number of practical problems. In fact, the very different selectivities of the various LC modes enable the analysis of complex mixtures with minimal sample preparation. However, comprehensive HPLC techniques are complicated by the operational aspects of transferring effectively from one operation step to another, by data acquisition and interpretation issues. Therefore, careful method optimization and several related practical aspects should be considered.

4.4.1 INTERFACE

A typical comprehensive 2D HPLC separation is attained through the connection of two HPLC systems by means of an interface (usually a high-pressure switching valve), which entraps specific quantities of first-dimension eluate, and directs them onto a secondary column. An ideal interface should retain the primary column eluate and reintroduce it as a sharp pulse when desired. Figures 4.5 through 4.7 show different types of multiport switching valves used in comprehensive LC approaches: an example of connecting two 6-port valves (Figure 4.5), an 8-port valve (Figure 4.5), and a 10-port valve (Figure 4.7). The valve is generally equipped with two sample loops (shown in the figures) or trapping columns.

In the previously described approaches, different HPLC pumps are used in the two dimensions. A different approach, where the flow from one single pump was splitted and introduced to one first- and two second-dimension columns, was used by Venkatramani and coworkers [45,47]. In this approach, a 12-port valve equipped with three loops [45] or with three guard columns [47] was used as an interface (Figure 4.8).

4.4.2 FIRST DIMENSION

Most of the frequently used comprehensive 2D LC systems employ a microbore HPLC column in the first dimension, operated at low flow rate, both under isocratic and gradient conditions. This enables the transfer of fractions of small volume via the multiport valve equipped with two identical

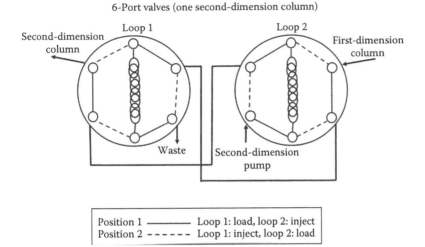

FIGURE 4.5 Example of coupling of two 6-port two-position switching valves. (Reprinted with permission from Tanaka, N. et al., *Anal. Chem.*, 76, 1273, 2004. Copyright 2004, American Chemical Society.)

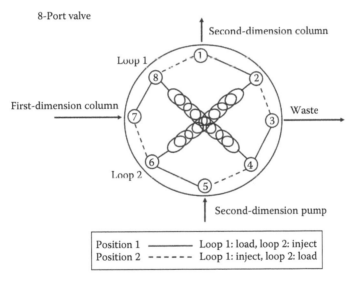

FIGURE 4.6 8-Port two-position switching valve.

sample loops, into the second-dimension column. The loop volume usually corresponds to the mobile-phase quantity per modulation time eluting from the first-dimension column.

Such a system has been used for the comprehensive 2D chromatography of proteins [9,14], synthetic polymers [16], oxygen heterocyclic fraction of cold-pressed citrus oils [22,29], carotenoids [39], triglycerides in fats and oils [18–21], pharmaceuticals [29], and acidic and phenolic compounds [27,28].

When a conventional column is used as a first-dimensional column, two different LCxLC configurations may be used, with either two trapping columns or fast secondary columns in parallel rather than storage loops. In the former setup, each fraction from the first dimension is trapped alternatively on one of the two trapping columns. At the same time, the compounds retained from the previous fraction on the other trapping column are back-flushed onto the analytical column for second-dimension analysis. In the latter setup, a fraction from the first-dimension column is trapped alternatively at the head of one of the two columns during the loading step in a one-column

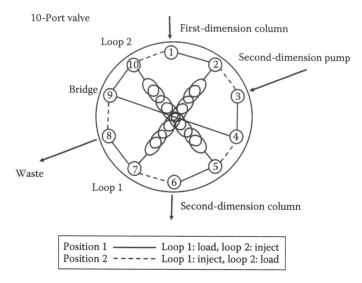

FIGURE 4.7   10-Port two-position switching valve.

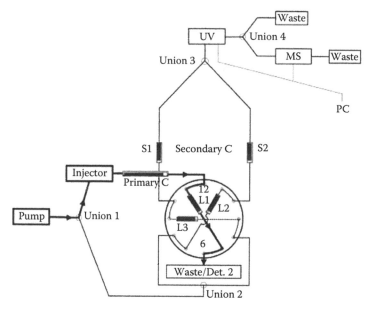

FIGURE 4.8 12-Port two-position switching valve. (Reprinted with permission from Venkatramani, C.J. and Zelechonok, Y., *Anal. Chem.*, 75, 3484, 2003. Copyright 2003, American Chemical Society.)

focusing mode, meanwhile in the other column the solutes, transferred in the previous cut from the first-dimension column, are separated.

These two approaches were investigated by several groups for the analysis of peptides and proteins [10,11,15,49–55] and for the analysis of phenolic antioxidants [24–26].

4.4.3   Second Dimension

The second-dimension separation of the transferred fraction must be completed before the injection of the successive fraction eluting from the first column and should be fast enough to permit that 3–4 fractions must be taken across the width of each first-dimension peak to avoid serious loss of information in the 2D separation due to under-sampling of first-dimension peaks, in agreement with Murphy et al. [46]. This consideration further emphasizes the need for very fast second-dimension separations in 2D LC. Moreover, short analysis times are particularly useful in the case of biological or other unstable samples, which risk to degrade or change with time. Second-dimension fast separation can be achieved in different ways. A common approach is to use short monolithic columns at high flow rates. The high permeability and good mass-transfer characteristics of monolithic stationary phases make them ideally suited for use in the second dimension because they enable to work at high flow rates without loss of resolution, thus reducing the second-dimension analysis time. Moreover, successive gradient cycles with a very brief equilibration time can be performed [19–22,24,26,56].

Another approach to improve the effective speed of the second-dimension separation is the use of an array of second-dimension columns, used in parallel [10,11,15,24,25,45,47,52–55,57], especially with 1.5 μm i.d. pellicular columns [10,15,54,55,58]. This approach is more complicated due to the fact that different columns are rarely identical, and it is critical to achieve precision of retention time in consecutive second-dimension separation containing the same analyte peak.

Another possibility to speed up the 2D analysis time is to use high temperatures, as described by Stoll et al. [13,59]: the decreased viscosity of the eluent at high temperature (100°C–120°C) allowed a much higher linear velocity through the column to allow faster gradient development without significant loss in efficiency. They demonstrated that excellent repeatability of retention times could be obtained using narrow-bore (2.1 mm i.d.), wide-pore (300 Å), and 3 μm carbon-coated zirconia columns at high temperature (100°C) with only one-column volume of flushing with the initial eluent. Regarding analyte thermal stability, with very short residence times (10–20 s) at high temperature, very little detectable degradation of either small, labile organic analytes or proteins occurred.

4.4.4 DETECTORS

Most of the traditional HPLC detectors can be applied to LCxLC analyses; the choice of the detectors used in comprehensive HPLC setup depends above all on the nature of the analyzed compounds and the LC mode used. Usually, only one detector is installed after the second-dimension column, while monitoring of the first-dimension separation can be performed during the optimization of the method. Detectors for microHPLC can be necessary if microbore columns are used. Operating the second dimension in fast mode results in narrow peaks, which require fast detectors that permit a high data acquisition rate to ensure a proper reconstruction of the second-dimension chromatograms.

PDA detectors can be operated at rapid data acquisition rates (up to 40 Hz) and are the most common used. Quadrupole MS systems are capable of supplying sufficient spectra for peak. For reliable component assignment, of course, TOF-MS systems possessing higher scan speed can be used.

Most of the detectors permit peak recognition but provide no structural information, which can be particularly important for identification of unknown compounds. From this point of view, the spectrometric detectors, specifically mass spectrometer and photodiode array detectors, add a third dimension to the multidimensional system and give additional information useful in components identification.

4.4.5 DATA ELABORATION AND SOFTWARE

LCxLC analyses produce great amounts of data, which contain retention information obtained from each separation. Such data is of high level of complexity and requires considerable data elaboration power and more sophisticated software which permit to collect all the information available and to compare the samples. Considering the overwhelming amount of data generated per run, the data handling may result in a real analytical problem, due to the fact that currently the area of dedicated software for LCxLC is poorly developed. In fact, dedicated software for LCxLC data processing is not yet commercially available as a complete package and therefore most of analysts use their own, homemade software.

With regard to comprehensive LC data elaboration, the acquired data is commonly elaborated with dedicated software that constructs a matrix with rows corresponding to the duration of the second-dimension analysis and data columns covering all successive second-dimension chromatograms. The result is a bidimensional contour plot, where each component is represented as an ellipse-shaped peak, defined by double-axis retention time coordinates. When creating a 3D chromatogram, a third axis by means of relative intensity is added. The colour and dimension of each peak is related to the quantity of each compound present in the sample. Figure 4.9 illustrates an example of data elaboration in comprehensive LC.

From the 3D plot, the individual first- or second-dimension separation can be deconvoluted, precluding the need for a 1D detector, which introduces band broadening into the system.

Different chemometric methods in a number of ways have been used to handle data generated by comprehensive analyses. The most relevant is the use of chemometric methods to reduce the dimensionality of the analyte and to allow comparison between different samples. This is usually done using principal component analysis (PCA), from which the entire multidimensional data set can be visualized as one point relating to the two principal components of the data. Points falling in particular regions of the plot can be categorized according to some standard training set, and differences between samples can be easily investigated.

Chemometric methods can greatly increase the number of analyzable peaks in MDLC; in particular, the generalized rank annihilation method (GRAM) can quantify overlapping peaks by deconvoluting the combined signal to those of each dimension. Standards with precise retention time are required, and there must be some resolution in both dimensions [60,61].

Peak integration and quantification can be performed by summing the areas of individual second-dimension peaks belonging to one analyte peak, which are integrated using conventional integration algorithms. It is, of course, necessary to confirm the identity of each peak prior to area summation.

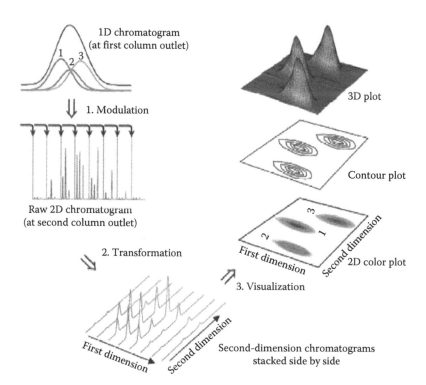

FIGURE 4.9 Data elaboration in comprehensive chromatography. (Reprinted from Dalluge, J. et al., *J. Chromatogr. A*, 1000, 69, 2003. Copyright 2003. With permission from Elsevier.)

4.5 METHOD DEVELOPMENT IN LCXLC

In the development and optimization of a comprehensive LCxLC method, many parameters have to be taken in account in order to accomplish successful separations. First of all, selectivity of the columns used in the two dimensions must be different to get maximum gain in peak capacity of the 2D system. For the experimental setup, column dimensions and stationary phases, particle sizes, mobile-phase compositions, flow rates, and second-dimension injection volumes should be carefully selected. The main challenges are related to the efficient coupling of columns and the preservation of mobile phase/column compatibility.

Most of the frequently used comprehensive HPLC are operated in a continuous mode, which means that the time of the second-dimension analysis corresponds to the transfer time of a fraction from the first into the second dimension. The total analysis time will be the product of the second-dimension analysis time and the total number of fractions injected onto the secondary column.

Comprehensive 2D HPLC can be also operated under stop-flow mode. In this case, after transferring a desired fraction volume onto the secondary column, the flow of the mobile phase in the first dimension is stopped and the fraction analyzed in the second dimension. When the separation is finished, the mobile-phase flow in the first dimension is switched on and the whole procedure is repeated again for the analysis of all the transferred fractions. The disadvantage of this procedure is the long analysis time, while the advantage can be that the second-dimension column can give higher plate numbers if compared to the continuous approach [23].

4.5.1 COUPLING WHEN PROBLEMS OF SOLVENT INCOMPATIBILITY OCCUR

LC techniques are characterized by a wider variety of separation mechanisms with truly different selectivities. As such, the number of theoretically achievable orthogonal combinations is high. It must be considered, though, that the combination of certain LC mode types can present difficulties if not impossibilities due to mobile-phase immiscibilities, precipitation of buffer salts, 1D mobile phase–2D stationary phase incompatibility. In comprehensive 2D LC, when selecting mobile phase for each dimension, compatibility should be considered. Generally, mobile phases used in RPxRP, RPxIEC, SECxRP, SECxNP are compatible. However, problems may occur if the mobile phases in the first and in the second dimension have great difference in viscosity. It is more difficult to combine NP and RP modes due to mobile-phase immiscibility.

If a large volume of an incompatible eluent is transferred from the first to the secondary column, broadened and distorted peaks can be obtained, and the solvent plug from the first chromatographic dimension can considerably alter the selectivity of the second dimension. If solvent incompatibility is unavoidable, the first-dimension eluate isolation and suitable solvent changeover can be performed. However, this can be complicated due to the need of extra equipment, increased maintenance, and thus longer sample treatment time. Literature reports two approaches used to overcome this problem in multidimensional heart-cutting LC. Sonnenfeld et al. [62] used a system in which the fraction of interest were transferred from the first (normal phase (NP)) column to a packed precolumn, and the NP eluent was removed by passage of on inert gas and vacuum. Once the solvent was removed, the precolumn was desorbed using a RP eluent and transferred to the second (RP) analytical column. Later on, Takeuchi et al. [63] used a microcolumn in the first dimension and a conventional-size column in the second dimension to interface NP and RP separations. Due to the reduced peak volume generated by the use of microcolumns, solvent removal was not required. This approach offers several advantages [3,58]:

- The small column i.d. helps to ensure a minimum of dilution and provides flow rates that are compatible with the sample volume for the secondary column.
- The dead volumes of the system are minimized.
- There is no need for a pre-concentration step at the head of the secondary column.

The low flow rate in the microbore column ensures sample volumes compatible with the secondary conventional column and permits the injection of a small volume onto the secondary column, making the transfer of incompatible solvents possible without peak shape deterioration or resolution losses [63]. The possible disadvantage could be the lower sample capacity of microbore LC columns. However, in LCxLC, a sensitivity enhancement can be obtained if the formation of compressed solute bands at the head of the secondary column is achieved during the transfer from the first to the second dimension. Moreover, a larger volume can be injected into the first-dimension microcolumn, used as a highly efficient pre-separation step, and a limited decrease in efficiency due to a large injection volume can be tolerated.

The combination of normal (silica) and reversed (C18) phase HPLC in a comprehensive 2D LC system was used for the first time for the analysis of alcohol ethoxylates [64]; the NP separation was run using aqueous solvents, so the mobile phases used in the two dimensions were miscible, resulting in the easy injection of the entire first-dimension effluent onto the second-dimension column.

The first fully comprehensive coupling of NP and RP, where the previously described difficulties related to solvents immiscibility were overcome, was developed by Dugo et al. and applied to the analyses of oxygen heterocyclic components of lemon essential oils [22]. Based on the configuration described in this work, other applications were developed for the analysis of carotenoids in citrus samples [48], citrus fruit extracts [29], pharmaceutical products [29], and triglycerides in fats and

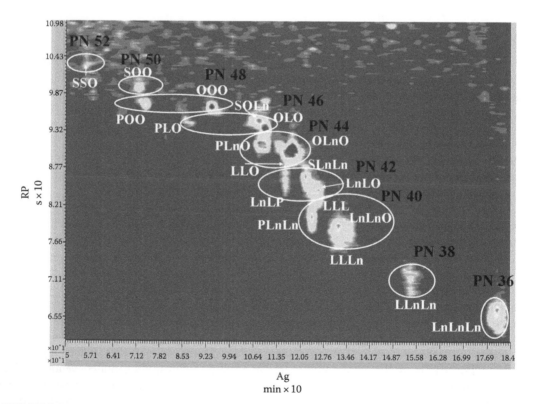

FIGURE 4.10 NPxRP LC contour plot for linseed oil TAGs. Comprehensive 2D silver ion–RPLC separation of the triglyceride fraction of linseed oil. First-dimension column: Ag-column (150×1 mm i.d., 5 μm dp); flow rate 11 μL/min; gradient elution using n-hex/ACN 99.3:0.7 (A) and n-hex/ACN 99.1:0.9 (B). Second-dimension column: monolithic C18 (100×4.6 mm i.d.); flow rate 4 mL/min; gradient elution using acetonitrile (A) and isopropanol (B); injection: 2 μL. Interface: 10-port valve equipped with two 20 μL sample loops. Detection was by using APCI-MS. S = stearic, O = oleic, P = palmitic, L = linoleic, Ln = linolenic. (Reprinted from Dugo, P. et al., *J. Chromatogr. A*, 1112, 269, 2006. Copyright 2006. With permission from Elsevier.)

oils [19–21]. Figure 4.10 shows an example of comprehensive NPxRP separation of a high number of triglycerides present in linseed oil.

Drawbacks that may occur working with incompatible solvents in the two dimensions may be system peaks which originate from the solvent incompatibility, hence reducing the available chromatographic separation space in the second dimension. When operating the second dimension at high flow rate, a solvent bump due to the poor mixing of the solvents in the sample loop was observed [29]. Both solvent peaks and "bump" are very reproducible over the run, so these factors can be accounted for if a blank run is performed and the 2D contour plot can be constructed after background subtraction.

4.5.2 Coupling without Problems of Solvent Incompatibility

In order to obtain 2D RPxRP separations with high degree of orthogonality, an accurate selection of RP columns, temperature, and mobile-phase composition are required. Adequate difference in selectivity can be obtained either by employing columns of analogous selectivity and different mobile phases or columns of different selectivity and analogous mobile phase. Ikegami et al. demonstrated that 2D HPLC systems employing equal C18 stationary phases with different organic modifiers in mobile phases for each dimension could produce large peak capacity for compounds with a large range of polarity [65]. Aromatic amines and non-amines were analyzed in comprehensive LC operating on both dimensions under comparable RP conditions [45]. In this case, the orthogonal separation was achieved by tuning the operating parameters, such as mobile-phase strength, temperature, and buffer strength, in conjunction with different secondary columns (ODS-AQ/ODS monolith, ODS/amino, ODS/cyano). On the other hand, various stationary phases, including polar-RP ether-linked phenyl phase with polar endcapping [66], PEG phase bonded on silica [23], or zirconium dioxide with deposited carbon layer [24,25], provide significant differences in selectivities with respect to C18 phases bonded on silica. Recently, 2D RPxRP separations by using various combination of silica-based columns in the two dimensions have been reported for the analysis of drugs and degradation products [67]. Figure 4.11 shows an example of comprehensive RPxRP separation of a mixture of phenolic and flavone standards, obtained using different stationary phases in the two dimensions.

Two-dimensional IECxRP setups by using an ion exchange column by means of salt gradient in the first dimension and an RP column in the second dimension have been extensively employed for

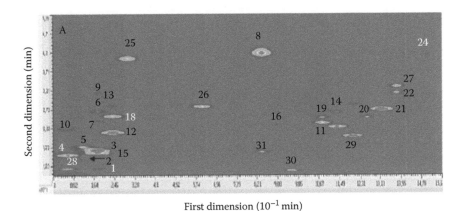

FIGURE 4.11 RPxRP LC separation of a mixture of phenolic antioxidant and flavonoid standards. First-dimension: phenyl column (50×3.9 mm i.d., 5 mm dp); flow rate 0.3 mL/min; gradient program using water and ACN; injection: 20 μL. Second-dimension: monolithic C18 column (100×4.6 mm i.d.); flow rate 2 mL/min; gradient program using water and ACN. Detection was by UV–vis at 280 nm. Interface: two C18 columns (30×4.6 mm i.d., 2.5 mm dp) mounted on a 10-ports valve. (Reprinted from Cacciola, F. et al., *J. Chromatogr. A*, 1149, 73, 2007. Copyright 2007. With permission from Elsevier.)

separation of proteins, peptide fragments, and other biopolymers. Combination of RP LC in the second dimension and pH gradients for protein separation according to their pI's in the first dimension seems to be a promising new 2D technique as potential alternative to 2D CE in proteomic research, because it is fast and can be easily automatized [49–51]. The resolving power of this technique is shown in Figure 4.12, which reports a comprehensive strong cation-exchange x RP chromatogram of a tryptic digest from bovine serum albumin (BSA).

Two-dimensional SECxLC has been applied mostly to synthetic polymers, (co)polymers, or polymer blends that are soluble in organic solvents, such as THF, dioxane, dichloromethane (DCM), etc., using SEC for separation according to the molar mass distribution in one dimension and NP LC [4,16,68,69], according to different functionalities (end-groups with different polarities "chemical composition" distribution) in the second dimension, under "near-critical" conditions, where compounds differing in the numbers of monomer units co-elute in a single peak [41,70]. In the NP dimension, usually a silica gel or cyanopropyl-modified silica column is used with THF–cyclohexane, DCM–heptane, DCM/ACN, DCM–methanol, trichloromethane–cyclohexane, etc., mixed mobile phases, either with isocratic or with gradient elution, whereas THF is most frequently used as the SEC mobile phase. Because conventional SEC columns are relatively long and the SEC analysis requires relatively long times, they are usually used in the first dimension, with an NP (or RP) column in the second dimension. In comprehensive chromatography, it is advantageous to use short SEC in the second dimension in combination with either isocratic [16,71] or gradient LC [72] in the first dimension to accomplish improved separation according to the chemical composition distribution of polymers. Figure 4.13 shows a NPLCxSEC chromatogram of a mixture polystyrene (PS) standards and polystyrene hydroxide (PS-OH).

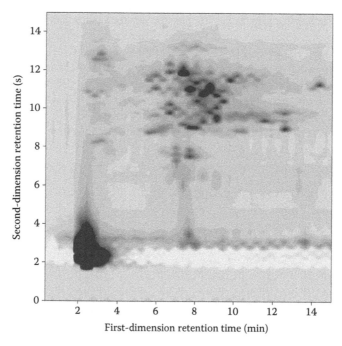

FIGURE 4.12 Two-dimensional chromatogram from the separation of tryptic peptides from BSA. First-dimension: strong cation-exchange column (50×2.1 mm i.d., 5 μm dp); flow rate 0.1 mL/min; salt gradient using 20 mM sodium dihydrogen phosphate, adjusted to pH 2.9 with phosphoric acid (A) and 500 mM sodium dihydrogen phosphate, adjusted to pH 2.9 with phosphoric acid (B); injection: 50 μL. Second-dimension (high-temperature LC): C18 column (50×2.1 mm i.d., 3.5 μm dp); flow rate 3 mL/min; gradient elution using 0.1% (v/v) trifluoroacetic acid (TFA) in water (A) and 0.1% (v/v) TFA in ACN (B). Detection was by UV–vis at 214 nm. Interface: 10-port valve equipped with two 35 μL sample loops. (Reprinted with permission from Stoll, D.R. and Carr, P.W., *J. Am. Chem. Soc.*, 127, 5034, 2005. Copyright 2005, American Chemical Society.)

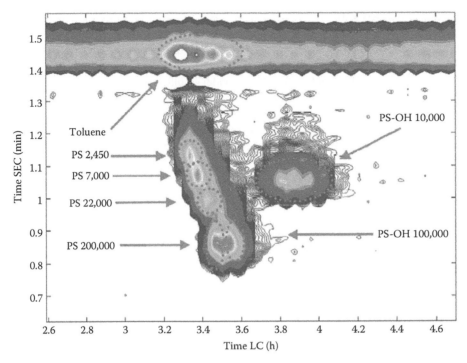

FIGURE 4.13 An LCxSEC chromatogram of a mixture of PS standards (M_w 2,450, 7,000, 22,000, and 200,000) and PS-OH (M_w 10,000 and 100,000) at critical conditions for PS. First-dimension: bare silica column (250×1.0 mm i.d., 3 μm dp), flow rate 4 μL/min; isocratic 42%/58% THF–hexane; injection 1 μL. Second-dimension: Mixed-C SEC column (75×4.6 mm i.d., 5 μm dp); flow rate 0.6 mL/min; isocratic 100% THF. Detection was by UV-absorbance detector at 254 nm. Interface: 10-way valve equipped with two sample loops. (Reprinted from van der Horst, A. and Schoenmakers, P.J., *J. Chromatogr. A.*, 1000, 693, 2003. Copyright 2003. With permission from Elsevier.)

Two-dimensional SECxRP can be also used for protein and peptide separation, and this approach has been used for separation of peptides in tryptic digests of ovalbumin and serum albumin [14].

4.5.3 PEAK FOCUSING AT THE HEAD OF THE SECONDARY COLUMN

As a rule, the sequence in which the columns are placed in a column-switching system has a marked effect on the results of a 2D separation. The final choice is dictated by the specific separation objectives. When subsequent fraction cuts have to be performed on the effluent from the first dimension, the column with a higher peak capacity should be placed into the first-dimension system and the flow rate should be matched to the fraction transfer switching period. The fractions transferred to the second-dimension column should be completely eluted before the subsequent fraction is transferred from the first to the second dimension. To increase the peak capacity in the first dimension, gradient elution is preferred to isocratic conditions.

Owing to the consecutive dilution of fractions transferred from the first dimension, detectability and sensitivity are of great concern when developing a 2D LC separation. This has to be considered and samples with the highest possible concentrations have to be injected into the 2D system. However, band broadening in the transferred fractions should be suppressed whenever possible, or reduced to a minimum. In comprehensive LC, band suppression effects can be achieved with a judicious selection of first- and second-dimension mobile and stationary phases. In particular, it is advantageous to select the chromatographic systems so that the mobile phase in the first dimension has lower elution strength than in the second dimension. In this way, the transferred fraction is concentrated (focused) at the head of the secondary column, obtaining the "modulation" effect [4,73]. Focusing is better achieved if the second dimension is operated under gradient elution, because even lower eluent strength can be used during the transfer of the fraction [59].

In NPxRP LC, mobile phase used in the first-dimension NP–LC separation is always stronger than the mobile phase at the head of the RP-secondary column [74,75]. However, it was demonstrated that the use of microcolumns in the first dimension operated at a high flow rate and a monolithic column in the second dimension operated at high flow rate in gradient mode can overcome problems of immiscibility of mobile phase. In particular, the use of a low percentage of strong solvent in the first step gradient of the 2D improves the chromatographic performance because of improved peak focusing.

REFERENCES

1. J.C. Giddings, *High Resolut. Chromatogr. Chromatogr. Commun.* 10, 319–323, 1987.
2. C. Corradini, Coupled-column liquid chromatography, in *Multidimensional Chromatography*, L. Mondello, C. Lewis, K.D. Bartle (Eds.) John Wiley & Sons, Chichester, U.K., 2001, pp. 109–129.
3. H.J. Cortes, *J. Chromatogr.* 626, 3–23, 1992.
4. P. Schoenmakers, P. Marriott, J. Beens, *LC–GC Eur.* 16, 335–338, 2003.
5. J.C. Giddings, *Anal. Chem.* 56, 1258A–1270A, 1984.
6. R. Consden, A.H. Gordon, A.J.P. Martin, *Biochem. J.* 38, 224–232, 1944.
7. P.H. O'Farrell, *J. Biol. Chem.* 250, 4007–4021, 1975.
8. F. Erni, R.W. Frei, *J. Chromatogr.* 149, 561–569, 1978.
9. M. Bushey, J.W. Jorgenson, *Anal. Chem.* 62, 161–167, 1990.
10. K. Wagner, K. Racaityte, K.K. Unger, T. Miliotis, L.E. Edholm, R. Bischoff, G. Marko-Varga, *J. Chromatogr. A* 893, 293–305, 2000.
11. G.J. Opiteck, J.W. Jorgenson, R.J. Anderegg, *Anal. Chem.* 69 2283–2291, 1997.
12. L.A. Holland, J.W. Jorgenson, *Anal. Chem.* 67, 3275–3283, 1995.
13. D.R. Stoll, P.W. Carr, *J. Am. Chem. Soc.* 127, 5034–5035, 2005.
14. G.J. Opiteck, K.C. Lewis, J.W. Jorgenson, R.J. Anderegg, *Anal. Chem.* 69, 1518–1524, 1997.
15. G.J. Opiteck, S.M. Ramirez, J.W. Jorgenson, M.A. Moseley III, *Anal. Biochem.* 258, 349–361, 1998.
16. A. van der Horst, P.J. Schoenmakers, *J. Chromatogr. A* 1000, 693–709, 2003.
17. K. Im, Y. Kim, T. Chang, K. Lee, N. Choi, *J. Chromatogr. A* 1103, 235–242, 2006.
18. H. Nakashima, Y. Hirata, K. Jinno, P. Sandra, A.J. Rackstraw (Eds.), *Proceedings of the 22nd International Symposium on Capillary Chromatography*, Gifu, Japan, Naxos Software Solutions, M. Schaefer, Scriesheim, Germany, 1999.
19. L. Mondello, P.Q. Tranchida, V. Stanek, P. Jandera, G. Dugo, P. Dugo, *J. Chromatogr. A* 1086, 91–98, 2005.
20. P. Dugo, T. Kumm, M.L. Crupi, A. Cotroneo, L. Mondello, *J. Chromatogr. A* 1112, 269–275, 2006.
21. P. Dugo, T. Kumm, B. Chiofalo, A. Cotroneo, L. Mondello, *J. Sep. Sci.* 29, 1146–1154, 2006.
22. P. Dugo, O. Favoino, R. Luppino, G. Dugo, L. Mondello, *Anal. Chem.* 76, 2525–2530, 2004.
23. E. Blahová, P. Jandera, F. Cacciola, L. Mondello, *J. Sep. Sci.* 29, 555–566, 2006.
24. F. Cacciola, P. Jandera, E. Blahová, L. Mondello, *J. Sep. Sci.* 29, 2500–2513, 2006.
25. F. Cacciola, P. Jandera, L. Mondello, *J. Sep. Sci.* 30, 462–474, 2007.
26. F. Cacciola, P. Jandera, Z. Hajdú, P. Česla, L. Mondello, *J. Chromatogr. A* 1149, 73, 2007.
27. J. Pól, B. Hohnová, M. Jussila, T. Hyötyläinen, *J. Chromatogr. A* 1130, 64–71, 2006.
28. M. Kivilompolo, T. Hyötyläinen, *J. Chromatogr. A* 1145, 155–164, 2007.
29. I. François, A. de Villiers, P. Sandra, *J. Sep. Sci.* 29, 492–498, 2006.
30. X. Chen, L. Kong, X. Su, H. Fu, J. Ni, R. Zhao, H. Zou, *J. Chromatogr. A* 1040, 169–178, 2004.
31. L. Hu, X. Chen, L. Kong, X. Su, M. Ye, H. Zou, *J. Chromatogr. A* 1092, 191–198, 2005.
32. Z. Liu, M. L. Lee, *J. Micro Sep.* 12, 241–254, 2000.
33. P.Q. Tranchida, P. Dugo, G. Dugo, L. Mondello, *J. Chromatogr. A* 1054, 3–16, 2004.
34. T. Stroink, M.C. Ortiz, A. Bult, H. Lingeman, G.J. de Long, W.J.M. Underberg, *J. Chromatogr. B* 817, 49–66, 2005.
35. S.P. Dixon, I.A.D. Perrett, *Biomed. Chromatogr.* 20, 508–529, 2006.
36. P. Jandera, *J. Sep. Sci.* 29, 1763–1783, 2006.

37. P.Q. Tranchida, P. Donato, P. Dugo, G. Dugo, L. Mondello, *Trends in Anal. Chem.* 26(3), 191–205, 2007.
38. J.M. Davis, J.C. Giddings, *Anal. Chem.* 55, 418–424, 1983.
39. M. Martin, D.P. Herman, G. Guiochon, *Anal. Chem.* 58, 2200–2207, 1983.
40. P.J. Slonecker, X. Li, T.H. Ridgway, J.G. Dorsey, *Anal. Chem.* 68, 682–689, 1996.
41. J.C. Giddings, *J. Chromatogr. A* 703, 3–15, 1995.
42. Z. Liu, D.G. Patterson Jr., M.L. Lee, *Anal. Chem.* 67, 3840–3485, 1995.
43. M.M. Bushey, J.W. Jorgenson, *Anal. Chem.* 62, 978–984, 1990.
44. C.J. Venkatramani, X. Jingzhen, J.B. Phillips, *Anal. Chem.* 68, 1486–1492, 1996.
45. C.J. Venkatramani, Y. Zelechonok, *Anal. Chem.* 75, 3484–3494, 2003.
46. R.E. Murphy, M.R. Schure, J.P. Foley, *Anal. Chem.* 70, 1585–1594, 1998.
47. C.J. Venkatramani, A. Patel, *J. Sep. Sci.* 29, 510–518, 2006.
48. P. Dugo, V. Škeříková, T. Kumm, A. Trozzi, P. Jandera, L. Mondello, *Anal. Chem.* 78, 7743–7750, 2006.
49. T. Greibrokk, M. Pepaj, E. Lundanes, T. Andersen, K. Novotna, *LC–GC Eur.* 18, 355–360, 2005.
50. M. Pepaj, S.R. Wilson, K. Novotna, E. Lundanes, T. Greibrokk, *J. Chromatogr. A* 1120, 132–141, 2006.
51. M. Pepaj, A. Holm, B. Fleckenstein, E. Lundanes, T. Greibrokk, *J. Sep. Sci.* 29, 519–529, 2006.
52. II. Liu, S.J. Berger, A.B. Chakraborty, R.S. Plumb, S.A. Cohen, *J. Chromatogr. A* 782, 267–289, 2002.
53. K. Wagner, T. Miliotis, G. Marko-Varga, R. Bischoff, K.K. Unger, *Anal. Chem.* 74, 809–820, 2002.
54. E. Machtejevas, H. John, K. Wagner, L. Standker, G. Marko-Varga, W. Forssmann, R. Bischoff, K.K. Unger, *J. Chromatogr. B* 803, 121–130, 2004.
55. K.K. Unger, K. Racaityte, K. Wagner, T. Miliotis, L.E. Edholm, R. Bischoff, G. Marko-Varga, *J. High Resolut. Chromatogr.* 23, 259–265, 2000.
56. N. Tanaka, H. Kobayashi, N. Ishizuka, H. Minatuchi, K. Nakanishi, K. Hosoya, T. Ikegami, *J. Chromatogr. A* 965, 35–49, 2002.
57. O.P. Haefliger, *Anal. Chem.* 75, 371–378, 2003.
58. A.P. Köhne, T. Welsch, *J. Chromatogr. A* 845, 463–469, 1999.
59. D.R. Stoll, J.D. Cohen, P.W. Carr, *J. Chromatogr. A* 1122, 123–137, 2006.
60. C.G. Fraga, B.J. Prazen, R.E. Synovec, *Anal. Chem.* 73, 5833–5840, 2001.
61. G.M. Gross, B.J. Prazen, R.E. Synovec, *Anal. Chim. Acta* 490, 197–210, 2003.
62. W.J. Sonnenfeld, W.H. Zoller, W.E. Maj, S.A. Wise, *Anal. Chem.* 54, 723–727, 1984.
63. T. Takeuchi, M. Asai, H. Haraguchi, D. Ishii, *J. Chromatogr.* 499, 549–556, 1990.
64. R.E. Murphy, M.R. Schure, J.P. Foley, *Anal. Chem.* 70, 4353–4360, 1998.
65. T. Ikegami, T. Hara, H. Kimura, H. Kobayashi, K. Hosoya, K. Cabrera, N. Tanaka, *J. Chromatogr. A* 1106, 112–117, 2006.
66. T.J. Whelan, M.J. Gray, P.J. Slonecker, R.A. Shalliker, M.A. Wilson, *J. Chromatogr. A* 1097, 148–156, 2005.
67. J. Pellet, P. Lukulay, Y. Mao, W. Bowen, R. Reed, M. Ma, R.C. Munger, J.W. Dolan, L. Wrisley, K. Medwig, N.P. Toltl, C.C. Chan, M. Skibic, K. Biswas, K.A. Wells, L.R. Snyder, *J. Chromatogr. A* 1101, 122–135, 2006.
68. T. Chang, *Adv. Polym. Sci.* 163, 1–60, 2003.
69. H. Pasch, *Adv. Polym. Sci.* 150, 1–66, 2000.
70. L.R. Snyder, J.W. Dolan, *Adv. Chromatogr.* 38, 115–187, 1998.
71. X. Jiang, A. van der Horst, V. Lima, P.J. Schoenmakers, *J. Chromatogr. A* 1076, 51–61, 2005.
72. S.J. Kok, T. Hankmeier, P.J. Schoenmakers, *J. Chromatogr. A* 1098, 104–110, 2005.
73. P. Marriott, R. Shellie, *TRAC—Trends Anal. Chem.* 21, 573–583, 2002.
74. N.E. Hoffman, S.-L. Pan, A.M. Rustum, *J. Chromatogr.* 465, 189–200, 1989.
75. P. Dugo, M. del Mar Ramírez Fernández, A. Cotroneo, G. Dugo, L. Mondello, *J. Chromatogr. Sci.* 44, 561–565, 2006.

5 Gradient Elution Mode

Pavel Jandera

CONTENTS

5.1 INTRODUCTION

In the simple isocratic HPLC elution mode, the chromatographic conditions are kept constant. Many complex samples contain compounds that differ widely in retention, so that HPLC in isocratic elution mode does not yield successful separation. If the isocratic conditions are adjusted for adequate retention of strongly retained solutes, some weakly retained components of complex samples may elute as poorly—if at all—separated bands close to the column hold-up time. On the other hand, if the operation conditions are adjusted so as to achieve satisfactory separation of weakly retained compounds, the elution of strongly retained sample components may take a very long time, their peaks may be broad, and their concentration in the effluent may be so low that the peak detection and integration becomes very difficult or even impossible. To solve this problem, the elution conditions can be changed according to a preset program during the elution.

The most commonly used programming technique in LC is solvent-gradient elution, where the composition of the mobile phase changes during the run by mixing two or more components. In solvent gradient elution, a weaker mobile phase is used in the initial part of the elution to provide adequate retention of weakly retained compounds, while the elution strength increases and consequently the retention of analytes gradually decreases during the gradient run, which results in improved resolution of weakly retained compounds and shorter retention times of the strongly retained ones. Further, decreasing retention is accompanied with narrower bandwidths and increased peak capacity with respect to isocratic separation. The original idea of solvent-gradient elution is attributed to A. Tiselius; it was used experimentally as early as in 1950 [1] and has been developed since by several groups (see, e.g., Refs. [2–4]). After commercial equipment had become available, gradient elution was gradually applied for practical solution of various LC problems.

The phenomena controlling gradient elution have often been misunderstood by many practicing chromatographers, who often regard it as subject to more experimental problems, less reproducible, slower, and more difficult to transfer from one instrument (laboratory) to another than isocratic elution, the reason being more complex equipment, more tedious method development, and more difficult interpretation of results. That is why some workers try to avoid gradient elution; however by doing so, they can miss undeniable benefits of this technique. Because of a higher number of experimental variables that should be taken into account, the effective use of gradient technique requires understanding how the gradient profile affects the separation. Some practitioners may be puzzled by observing that using a longer column or decreasing the flow rate of the mobile phase while keeping other gradient conditions unchanged may shorten the analysis time and decrease the resolution in gradient elution, contrary to the effects in isocratic HPLC. The theory of gradient can satisfactorily explain this behavior and provides tools to avoid unexpected results.

In the past decades, theoretical models describing gradient elution have been developed, so that the sample behavior in gradient elution is now well understood and can be explained and predicted. Various aspects of theoretical contributions and their impacts on the good practice of gradient elution have been presented and reviewed in several books and chapters in monographs [4–11]. This chapter intends to provide concise overview of the earlier work for improved understanding of the theoretical and practical aspects of gradient elution and to serve as the introduction for deeper study of this powerful technique for interested readers.

5.2   ADVANTAGES OF GRADIENT ELUTION AND CLASSIFICATION OF GRADIENTS

The main reasons for the use of gradient elution are

- Improving the resolution of samples with wide range of retention
- Increasing the number of peaks resolved within a fixed time of separation (increasing the peak capacity in single- or multi-dimensional LC)
- Separation of mixtures of compounds with high molecular weights, such as synthetic polymers, proteins, and other biopolymers whose retention changes markedly for small changes in the composition of mobile phases
- Providing economic and fast generic separation methods that can be applied with confidence in development and control laboratories to a large number of samples of variable composition to provide important information in short time to synthetic chemists, either for fast sample screening, or for generating impurity profiles
- Using initial "scouting" gradient experiments for efficient development of final gradient or isocratic methods
- Removing strongly retained interfering compounds in a separate sample pretreatment prior to analysis
- Suppression of tailing peaks, especially for samples containing basic compounds

Various criteria can be used to classify mobile phase gradients:

1. The number of mobile phase components whose concentrations change with time:
   a. Two-component (binary) gradients
   b. Multicomponent (ternary, quaternary, etc.) gradients
   c. So-called relay gradients, i.e., a special type of multicomponent gradients comprised of several subsequent steps with different solvents where the mobile phase changes from solvent **A** to solvent **B** in the first step, from solvent **B** to solvent **C** in the second step, etc.
2. The composition of the mobile phase in gradient LC can be changed according to a continuous gradient, the profile of which is characterized by three parameters affecting the elution behavior: (1) the initial concentration, (2) the steepness (slope), and (3) the shape (curvature) of the gradient. The gradient program can be also comprised of a few subsequent isocratic or gradient steps. Accordingly, the gradients can be classified as
   a. Linear gradients
   b. Nonlinear gradients with curved profile (concave or convex, rarely used, mostly substituted by segmented gradients)
   c. Step gradients, including several subsequent isocratic steps with increasing concentration of one or more strong eluting components of the mobile phase
   d. Segmented gradients, including several subsequent steps, usually linear with different slopes (ramps) or isocratic hold periods (most often at the end or at the beginning of elution, but also inserted between linear gradient steps)
   e. Reverse gradients with decreasing concentration of the strong mobile phase component are often used to restore the initial conditions before the next sample injection
3. The chromatographic mode and the type of the mobile phase:
   a. Reversed-phase solvent gradients—concentration(s) of one or more organic solvent(s) in water increase(s); nonpolar bonded columns
   b. Nonaqueous reversed-phase (NARP) solvent gradients—concentration(s) of one or more less polar organic solvent(s) in a more polar one increase(s)
   c. Ion-pair reversed-phase solvent gradients in mobile phases containing ion-pair reagents
   d. Ion-pair reagent concentration gradients in reversed-phase mode
   e. Reversed-phase pH gradients
   f. Reversed-phase decreasing ionic strength gradients (hydrophobic interaction chromatography, salting-out chromatography)
   g. Ionic strength gradients in ion-exchange LC (may be combined with solvent gradients)
   h. pH gradients in ion-exchange LC (may be combined with solvent gradients)
   i. Normal-phase solvent gradients – concentration(s) of one or more polar solvent(s) in a less polar one increase(s); nonaqueous mobile phases, polar adsorbent or polar bonded-phase columns
   j. Normal-phase hydrophilic interaction chromatography (HILIC) gradients – concentration of water (or of water and a more polar organic solvent) in a less polar solvent increase; aqueous mobile phases, polar adsorbent or bonded-phase columns
4. The objective of separation:
   a. Analytical separations (identification and determination of sample compounds), low sample load on the column, linear distribution isotherms
   b. Preparative separations (isolation or purification of compounds), high sample load on the column, usually nonlinear distribution isotherms

Solvent gradients are generally much more efficient to decrease the retention than programmed temperature. For example, the retention factors k of low-molecular-weight analytes in reversed-phase

LC decrease by a factor of 2–3 with a 10% increase in the concentration of organic solvent in aqueous-organic mobile phase, whereas an increase of temperature by 10°C usually leads to a decrease in k of nonionic compounds by 10%–20%. Further, many commercial HPLC columns do not tolerate increased temperatures of 60°C or higher [12]. Flow-rate programming during separation is not practical, as it is limited by maximum allowable operation pressure. Stationary phase programming is possible using off-line or on-line column switching in multidimensional LC, which is a very efficient tool for increasing the number of resolved peaks, but requires complicated method development, especially with respect to sample transfer between the individual dimensions.

Gradient-elution techniques can be combined with elevated temperature operation or temperature programs [13–17], flow-rate programming [18,19], column switching, and two-dimensional (2D) operation [20–24] to get full advantage of separation selectivity and to separate complex samples in as short a time as possible.

5.3 PRINCIPLES AND THEORY OF GRADIENT ELUTION

The theory of gradient-elution chromatography allows predicting the gradient-elution behavior of sample compounds and optimizing gradients in various reversed-phase, normal-phase, and ion-exchange systems, if the dependence of the retention data on the composition of the mobile phase is known. This information can be obtained from the isocratic retention data (or from two initial gradient experiments) [4–11]. Generally, the calculation of the entire band profile in gradient chromatography, either symmetrical for dilute samples in analytical chromatography (linear isotherm conditions), or nonsymmetrical for concentrated samples in preparative chromatography on overloaded columns (nonlinear isotherm conditions) is possible by numerical solution of the basic differential chromatographic mass balance equation (Equation 5.1), e.g., using finite difference method [25]:

$$\frac{d(c_m)}{d(t)} + f \cdot \frac{d(c_s)}{d(t)} + \frac{d(u \cdot c_m)}{d(l)} = D_a \cdot \frac{d^2(c_m)}{d(l)^2} \tag{5.1}$$

where
   c_m and c_s are the concentrations of the analyte in the mobile and in the stationary phase, respectively
   f is the phase ratio in the column
   u is the linear velocity of the mobile phase
   l is the column length
   t is the time elapsed from the start of elution
   D_a is the axial diffusion coefficient

However, the calculation requires different algorithms for each specific case. Hence, an explicit mathematical solution providing basing retention data (elution volumes or times, bandwidths, and resolution) is much more frequently used in analytical gradient chromatography, even at a cost of some simplifications [9–11].

5.3.1 PREDICTION OF RETENTION IN GRADIENT-ELUTION CHROMATOGRAPHY

The main difference from isocratic elution consists in continuous change (decrease) in the retention factors, k_i, in gradient chromatography with changing (increasing) concentration of the strong solvent **B** in the mobile phase. The change in mobile phase composition is first observed at the column inlet and propagates along the column until it arrives to the column outlet, delayed by the column

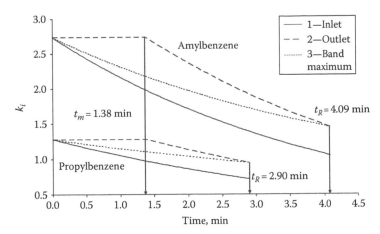

FIGURE 5.1 Instantaneous retention factors, k_i, of propylbenzene and amylbenzene in dependence on time elapsed since the start of the gradient, corresponding to the actual composition of the mobile phase: (1) at the column inlet (preset gradient program), (2) at the column outlet (delayed by the column hold-up volume), and (3) at the actual band maximum in the column. Column and separation conditions: Chromolith Performance RP-18e column, 80%–100% ACN in water in 9.3 min, $F_m = 1$ mL/min. No instrumental dwell volume, $V_D = 0$ mL. Parameters of Equation 5.7: propylbenzene: $a = 1.99$, $m = 2.68$, $R^2 = 0.9998$; amylbenzene: $a = 2.65$, $m = 3.16$, $R^2 = 0.9997$. (From Jandera, P., *J. Chromatogr. A*, 1126(1–2), 24, 2006. Copyright 2006. With permission from Elsevier.)

hold-up volume. Consequently, the retention factors of sample compounds continuously decrease as the sample bands move along the column, as shown in Figure 5.1, for two compounds differing in the retention. This means that unlike isocratic elution, the retention factors during gradient elution can be considered constant only within a very small (differential) time interval dt, in which the sample zone migrates along an infinitesimally small distance in the column, corresponding to the differential of the column hold-up volume, dV_m. The interval dt corresponds to a differential increase in the volume of the column effluent, dV, so that it contributes to the final net retention time, t'_R, and net retention volume, V'_R, by the increments $d(t'_R)$ and $d(V'_R)$, respectively:

$$d\left(t'_R\right) = k_i d\left(t_m\right); \quad d\left(V'_R\right) = k_i d\left(V_m\right)$$

(5.2a,b)

The differential equations (Equation 5.2a or b) can be solved by integration after introducing the actual dependence of k on the time, t (or on the volume of the eluate, V, which has passed through the column) from the start of the gradient until the elution of the band maximum. Freiling [26] and Drake [27] were the first to introduce this approach, which has been used later to derive equations allowing calculations of gradient retention data in various LC modes [2,4–7,28–30].

Assuming that the column hold-up volume does not change significantly with changing mobile phase in the gradient range, the change in $d(t'_R)$ during a gradient run is independent of the change in $d(t_m)$. Considering the change in the variables during gradient elution in the limits from 0 to t'_R for $d(t'_R)$ and from 0 to t_m for $d(t_m)$, Equation 5.2a can be integrated within these limits:

$$\int_0^{t_R - t_m} \frac{d\left(t'_R\right)}{k_i} = \int_0^{t_m} d(t_m) = t_m$$

(5.3)

Similar considerations apply for the integration of the volume elements in gradient elution. Setting the net (corrected) elution retention times (or volumes), $t'_R = t_R - t_m$, $V'_R = (V_R - V_m)$ as the upper integration limits in Equation 5.3 implies that the retention factor k_i relates to the actual position of the band maximum in the column, as shown by the middle plots (dot lines) in Figure 5.1, so that in the calculation only the volume of the mobile phase that has passed through the maximum of the analyte zone (or the corresponding time) is considered. The column hold-up volume (time) is not included in this solution, as the mobile phase contained in the column at the time of injection moves ahead of the injected sample and does not affect the retention during gradient elution. It should be noted that setting the uncorrected retention times or volumes as the integration variables instead of the net values necessitates other upper integration limits and may lead to difficulties in mathematical solution of Equation 5.3, so that the two approaches should not be confused [31,32].

Equation 5.3 can be solved by integration when the dependence of k_i on the time elapsed, t (or the volume of the eluate passed through the column, V), from the start of the gradient run is known. To allow general solution for various combinations of gradient profiles and HPLC modes, any dependence of k on V can be divided into two parts: (1) a dependence of k on the concentration of a strong eluting component in the mobile phase, φ controlled by the thermodynamics of the distribution process (the retention equation) and (2) the parameters of the gradient function describing the gradient profile, which is adjusted by the operator. Most frequent are linear gradients, but curved or segmented gradients can be also used, if they provide better resolution or shorter retention times.

Linear binary concentration gradients can be described by the gradient function [4–9,33]:

$$\varphi = A + B't = A + \frac{\Delta\varphi}{t_G}t = A + \frac{B'}{F_m}V = A + BV = A + \frac{\Delta\varphi}{V_G}V \qquad (5.4)$$

where
  A is the initial concentration φ of the strong solvent in the mobile phase at the start of the gradient (usually expressed in terms of volume fractions) and B or B' are the steepness (slope) of the gradient, i.e., the increase in φ in the time unit, or in the volume unit of the mobile phase, respectively
  V_G and t_G are the gradient volume and the gradient time, respectively, during which the concentration φ changes from A to the concentration $\varphi_G = A + \Delta\varphi$ at the end of the gradient
  $\Delta\varphi$ is the gradient range

A plethora of models for the description of retention in reversed-phase, ion-exchange, and normal-phase LC have been suggested during the past 40 years, resulting in various retention equations. For practical prediction and optimization of retention in gradient elution it is not very important whether the retention model is rigorously theoretical, semi- or fully empirical. What really matters is the goodness of the fit of experimental data at changing mobile phase composition and the number of model parameters. It should be kept in mind that the fit of the experimental data set to the model retention equation alone does not prove the validity of the underlying model. Generally, the more parameters are included into the retention equation, the better the model fits the data, regardless of the theoretical validity and the physical meaning of the parameters. Therefore, the model-based retention equations should be selected judiciously to obtain reliable parameters of a multiparameter retention equation. If the number of experimental data is too low, "over-fitting" of the equation with a high number of parameters may be due to a low number of degrees of freedom resulting to a close fit to the biased data subject to experimental errors rather than to better accuracy of the underlying theoretical model [34]. At a high enough number of experimental data, the random errors average and more reliable model parameters are obtained, but at the cost of increased experimental efforts; hence, two- or three-parameter retention equations are more useful in practice for calculations of the gradient retention data. Therefore, only simple equations yielding however reliable prediction of retention in reversed-phase, normal-phase, and ion-exchange gradient LC modes will be presented here. For more detailed discussion and comparison of other retention equations in various HPLC modes, the interested reader is referred to earlier literature (see, e.g., Refs. [4,9,10]).

In some cases, the solution is possible in explicit form allowing direct calculations of the retention data; however, for some combinations of gradient functions and retention equations iterative solution approach is necessary, which can be applied using standard calculation software. An overview of possible solutions of Equation 5.3 for various HPLC modes and gradient profiles was published earlier [4,33].

5.3.2 BANDWIDTHS AND RESOLUTION IN GRADIENT CHROMATOGRAPHY

The migration velocities of the bands along the column increase as the retention factors decrease during gradient elution. The consequence is the elution of various sample compounds with similar instantaneous retention factors, k_e, at the time the sample zones leave the column during a gradient run. The accelerated migration along the column causes significant reduction of the zones of later eluting compounds, so that the bandwidths in gradient-elution chromatography are approximately constant for both early and late eluting compounds, unlike isocratic conditions, where the bandwidths increase for more strongly retained compounds.

Bandwidths in gradient-elution chromatography can be determined exactly by numerical solution of Equation 5.1, which results in the calculation of the complete profile of the elution curve [35–37], or by numerical calculations of the accumulation of local diffusion coefficients in a series of discrete-time slices [38]; however, these approaches are tedious and hence rarely used for analytical gradient HPLC. Usually, a simplified approximate calculation procedure shown below provides results accurate enough for the practical gradient method development and optimization, by setting the bandwidths, w_g, in gradient elution equal to the isocratic bandwidths in the mobile phase at the column outlet at the elution time of the band maxima of the sample, which can be calculated as in isocratic elution using the corresponding instantaneous retention factor, k_e [4,6–11,28,29]:

$$w_g = \frac{4V_m\left(1+k_e\right)}{\sqrt{N}} \tag{5.5}$$

where
   N is the average number of theoretical plates determined under isocratic conditions in mobile phases within the gradient concentration range
   V_m is the hold-up volume of the column

It should be noted that the correct plate number cannot be determined directly from a gradient chromatogram, as the retention factors k are continuously changing during gradient elution. The retention factors k_e at the point of elution of peak maximum can be calculated using appropriate retention equation. Both the k_e and the bandwidths decrease as the steepness of the gradient increases [7,28]. The peaks in gradient-elution chromatography are generally narrower and higher in comparison to isocratic elution, which improves the detector response and the sensitivity of determination. However, the beneficial effect of gradient elution on increasing sensitivity is sometimes limited by increased baseline drift and noise, which can be suppressed by using high-purity "gradient grade" solvents and mobile phase additives.

One simplification adopted in derivation of Equation 5.5 consists in neglecting the effects of changing mobile phase composition on the diffusion coefficients, which affect the height equivalent of theoretical plate, H and $N=L/H$ [7]. The errors in the calculated gradient bandwidths caused by this simplification are estimated to be generally less than 1% for low molecular samples [39]. Neglecting additional band compression in gradient elution, which occurs as the trailing edge of the band moves faster in a mobile phase with higher elution strength whereas the leading edge migrates more slowly in a weaker mobile phase during gradient elution [5,7,11] is another possible source of errors in calculations using Equation 5.5, which may lead up to 10%–20% positive errors in the calculated bandwidths [40]. For practical method development, neglecting band compression may be used as a "safety tolerance" to compensate for additional band broadening effects caused by variation of the viscosity and diffusion coefficients during the gradient, or due to extra-column band broadening [11,39,40].

Combining the appropriate equations for the retention volumes of the solutes 1 and 2 with adjacent bands and Equation 5.5, we obtain Equation 5.6 for resolution in gradient LC [4,6,28], which can be used in the optimization of gradient separations:

$$R_s = \frac{V_{R(2)} - V_{R(1)}}{w_g} \tag{5.6}$$

where

$V_{R(1)}$ and $V_{R(2)}$ are the retention volumes of sample compounds 1 and 2, respectively
w_g is the average bandwidth calculated using Equation 5.5

5.4 GRADIENT ELUTION IN VARIOUS LC MODES

5.4.1 REVERSED-PHASE CHROMATOGRAPHY

Reversed-phase chromatography (RPC) is by far the most frequently used LC separation mode, as it is likely to provide satisfactory separation of a great variety of samples containing nonpolar, polar, and even ionic compounds according to the differences in hydrophobicity and (or) size of analytes [6,41]. To first approximation, the polar interactions in the mobile phase are the main factor controlling the retention. According to the solvophobic model, the transition of a solute molecule from the bulk polar mobile phase to the surface of the nonpolar stationary phase results from a decrease in the contact area of the solute with the mobile phase. Replacement of weaker interactions between a moderately polar solute and mobile phase by mutual interactions between strongly polar molecules of the mobile phase in the space originally occupied by a solute molecule results in overall energy decrease in the system, which is the driving (solvophobic) force of the retention in absence of strong (polar) interactions of the solute with the stationary phase [41]. In the real world, the interactions with the stationary phase contribute more or less to the retention, such as polar interactions of basic solutes with residual silanol groups remaining in silica-based bonded phases after chemical modification of the support. Further, organic solvents used as the components of the mobile phases in reversed-phase systems can be preferentially adsorbed by the stationary phase and modify its properties.

The sample retention increases as its polarity decreases and as the polarity of the mobile phase increases, i.e., as the concentration of the organic solvent **B** in aqueous-organic mobile phase decreases. The organic solvents used in gradient RPC include—in order of decreasing polarities and utility—acetonitrile, methanol, dioxane, tetrahydrofuran, and propanol. For successful separation of ionic, acidic, or basic substances, it is necessary to use additives to the mobile phase: buffers, neutral salts, weak acids, or surfactants as ion-pair reagents. Substances of very low polarity can be separated with a nonaqueous organic mobile phase by NARP chromatography [6].

By appropriate choice of the type (or combination) of the organic solvent(s), selective polar dipole–dipole, proton-donor, or proton-acceptor interactions can be either enhanced or suppressed and the selectivity of separation adjusted [42]. Over a limited concentration range of methanol–water and acetonitrile–water mobile phases useful for gradient elution, semiempirical retention equation (Equation 5.7), originally introduced in thin-layer chromatography by Soczewiński and Wachtmeister [43], is used most frequently as the basis for calculations of gradient-elution data [4–11,29,30]:

$$\log k = \log k_w - m\varphi = a - m\varphi \tag{5.7}$$

The constant a in Equation 5.7 increases as the polarity of a solute decreases and as its size increases and theoretically should be equal to the logarithm of the solute retention factor in pure water, k_w. However, the values of log k_w extrapolated to $\varphi=0$ do not describe accurately the real retention in water [44–46]. The constant m increases with decreasing polarity of the organic solvent **B** and with increasing size of the sample molecule [47,48].

Introduction of Equation 5.7 into the differential equation (Equation 5.3) and integration enables calculation of the elution volumes V_R in gradient RPC with linear gradients of organic solvent in water, described by Equation 5.4 [7–11,28,33]. Various forms of this solution were published, which all can be formally rearranged to Equation 5.8, assuming that the contribution of the instrumental dwell volume to the retention is small enough to be neglected:

$$V_R = \frac{1}{mB}\log\left\{ 2.31mB\left[V_m 10^{(a-mA)} \right] + 1 \right\} + V_m \tag{5.8}$$

Combining Equations 5.8 and 5.5, we obtain Equation 5.9 allowing prediction of bandwidths in RP gradient chromatography:

$$w_g = \frac{4V_m}{\sqrt{N}}\left[1 + \frac{1}{2.31mBV_m + 10^{(mA-a)}} \right] \tag{5.9}$$

5.4.2 Normal-Phase Chromatography

Normal-phase (straight-phase, adsorption) chromatography (NPC) is the oldest liquid chromatographic mode. The column packings include inorganic adsorbents (silica or, less often, alumina) and (cyanopropyl $-(CH_2)_3-CN$, diol$-(CH_2)_3-O-CH_2-CHOH-CH_2-OH$, or aminopropyl $-(CH_2)_3-NH_2$) polar-bonded phases on silica gel, zirconia dioxide, or other support [49]. The mobile phase is usually a mixture of two or more organic solvents of different polarities, a weak solvent **A** (a hydrocarbon—hexane or heptane) and a strong, more polar solvent **B**, which can be either non-localizing (e.g., dichloromethane), basic localizing (e.g., methyl-*tert*-butyl ether), or nonbasic localizing (alcohols, esters, nitriles, etc.). The selection of the polar solvent strongly affects the selectivity of separation [50–52].

The principal mechanism of adsorption chromatography is understood as the competition between the molecules of the solute and of the solvent for the localized adsorption centers on the adsorbent surface. The interactions in the mobile phase are assumed less significant and often are neglected in adsorption mechanism models. The stationary phase in NPC is more polar than the mobile phase and the sample retention is enhanced as the polarity of the stationary phase increases and as the polarity of the mobile phase decreases, opposite to RPC. The retention also increases with increasing number of adsorption sites in the column, i.e., on adsorbents with larger specific surface area. The type of the stationary phase affects much more strongly selective polar interactions in NPC than in RPC. Basic analytes are strongly retained on silica gel, whereas aminopropyl and diol-bonded phases prefer compounds with proton-acceptor or proton-donor functional groups (alcohols, esters, ethers, ketones, etc.), alumina favors interactions with π electrons and increased selectivity for separation of compounds with different numbers or spacing of double bonds than silica gel.

NPC is far less used than RPC, but it has several practical advantages: (1) because of lower viscosity, pressure drop across the column is lower than with aqueous-organic mobile phases used in RPC; (2) columns are usually more stable in organic than in aqueous-organic solvents; (3) columns packed with unmodified inorganic adsorbents are not subject to "bleeding," i.e., to gradual loss of the stationary phase, which decreases slowly the retention during the lifetime of a chemically bonded column; (4) some samples are more soluble or less likely to decompose in organic than in aqueous mobile phases; (5) because of fixed position of the adsorption sites, NPC is suitable for separation of various positional isomers or stereoisomers; (6) if sample pretreatment involves extraction into a nonpolar solvent, direct injection onto an NPC column is less likely to cause problems than the injection onto an RPC column; (7) gradient NPC is more suitable than RPC for the separation of synthetic polymers insoluble in water, but it has lower selectivity for the separation of molecules differing in the hydrocarbon part. Further, organic solvents are more expensive to purchase and dispose than water.

With some simplification, theoretical models of adsorption lead to Equation 5.10 describing the effects of increasing concentration of the stronger (more polar) solvent, φ, on decreasing retention factor, k in binary organic mobile phases [53]:

$$k = k_0 \varphi^{-m} \tag{5.10}$$

In the original adsorption models, φ is expressed as molar fraction, but it can be substituted by volume fraction without affecting significantly the accuracy of the prediction of retention. k_0 and m are experimental constants, k_0 being the retention factor in pure strong solvent.

Equation 5.10 applies in systems where the solute retention is very high in the pure nonpolar solvent. If this is not the case, another retention equation was derived [54–56].

$$k = (a + b\varphi)^{-m} \tag{5.11}$$

where a, b, and m are experimental constants depending on the solute and on the chromatographic system ($a = 1/(k_a)^m$, k_a is the retention factor in pure nonpolar solvent). Usually, Equation 5.11 only slightly improves the description of the experimental data with respect to Equation 5.10 [56].

In gradient-elution NPC, the solvent strength of the mobile phase gradually increases with increasing concentration of the polar solvent. In NP systems where Equation 5.10 applies under isocratic conditions, the elution volume V_R of a sample solute in NP LC with linear gradients described by Equation 5.4 can be calculated using [28]:

$$V_R = \frac{1}{B}\left[(m+1)B(k_0 V_m) + A^{(m+1)}\right]^{\frac{1}{(m+1)}} - \frac{A}{B} + V_m \tag{5.12}$$

Combining Equations 5.12 and 5.5, we obtain Equation 5.13 allowing prediction of bandwidths in RP gradient chromatography:

$$w_g = \frac{4V_m}{\sqrt{N}}\left\{1 + k_0\left[(m+1)Bk_0V_m + A^{(m+1)}\right]^{-\frac{m}{m+1}}\right\} \tag{5.13}$$

In NP systems where the retention is controlled by the three-parameter retention equation (Equation 5.11), the elution volumes in normal-phase gradient-elution chromatography can be calculated using [55–57]

$$V_R = \frac{1}{bB}\left\{(m+1)bBV_m + (a+Ab)^{(m+1)}\right\}^{\frac{1}{(m+1)}} - \frac{a+Ab}{bB} + V_m \tag{5.14}$$

and the bandwidths using

$$w_g = \frac{4V_m}{\sqrt{N}}\left\{1 + \left[(m+1)BbV_m + (a+Ab)^{(m+1)}\right]^{-\frac{m}{m+1}}\right\} \tag{5.15}$$

Chromatography on polar adsorbents suffers from a specific inconvenience—preferential adsorption of more polar solvents, especially water, which is often connected with long equilibration times if the separation conditions are changed such as in gradient NPC. Water is much more polar and hence is more strongly retained than any organic solvent on polar adsorbents. Adsorption of even trace amounts of water considerably decreases the adsorbent activity. Variations in the water contents in the mobile phase occurring in NP gradient LC by mixing continuously solvents with different trace concentrations of water impairs the reproducibility of retention, which can be considerably improved by using dehydrated solvents kept dry over activated molecular sieves and filtered just before the use and by accurate temperature control to ±0.1°C during the separation [8].

The uptake of water and strong polar solvents on the column in NP gradient chromatography may significantly change the properties of the stationary phase and the actual gradient profile, which shows a deficit of the adsorbed solvent, so that the actual elution strength of the mobile phase is lower than expected according to the preset gradient program, especially in the early part of the gradient. Consequently, the experimental elution volumes are higher than the values predicted by calculation. When the column adsorption capacity becomes saturated, the concentration of the strong solvent in the mobile phase can suddenly increase and can "sweep out" weakly retained sample compounds or impurities accumulated previously on the column top as an unexpected "ghost" peak at the polar solvent breakthrough time.

The solvent amount adsorbed on the column during gradient elution depends on the type of the stationary phase and of the polar solvent and is characterized by the adsorption isotherm of the solvent. It is far more significant in normal phase than in reversed-phase gradient LC, where it usually can be neglected. If the parameters of the adsorption isotherm of the strong solvent are known, the uptake of the solvent, the breakthrough time, and the deformation of the gradient profile can be predicted solving numerically the basic differential retention equation (Equation 5.1) [57]. Generally, the number of the adsorption sites controls the column saturation capacity for the polar solvent, whereas the steepness of the isotherm increases with increasing polarity of both the solvent and the stationary phase. Figure 5.2 shows several examples of calculated gradient profiles, which agree very well with the experiment. Generally, the effects of the preferential adsorption and the error

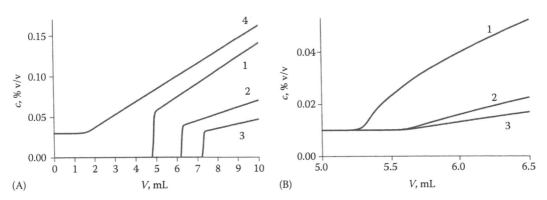

FIGURE 5.2   (A) Calculated gradient profile of 2-propanol in heptane on a Separon SGX silica gel column in normal-phase gradient-elution HPLC, simulated by numerical calculations using the experimental isotherm data and assuming $N = 5000$. Gradient dwell volume = 0.50 mL. Gradients: 0%–50% 2-propanol in 30 min (1), 0%–25% 2-propanol in 30 min (2), 0%–16.7% 2-propanol in 30 min (3), and 3%–50% 2-propanol in 30 min (4). c—concentration of 2-propanol in the eluate, V—volume of the eluate from the start of the gradient. (B) Calculated gradient profile of 2-propanol in dichloromethane on a Separon SGX silica gel column in normal-phase gradient-elution HPLC, simulated by numerical calculations using the experimental isotherm data and assuming $N = 5000$. Gradient dwell volume = 0.50 mL. Gradients: 1%–50% 2-propanol in 30 min(1), 1%–25% 2-propanol in 30 min (2), and 1%–16.7% 2-propanol in 30 min (3). c—concentration of 2-propanol in the eluate, V—volume of the eluate from the start of the gradient.

in the predicted retention data can be largely suppressed if the gradient starts at a higher than zero concentration of the strong solvent, preferably 3% or more, depending on the type of the adsorbent and of the polar solvent. More details can be found, e.g., in Refs. [9,57].

Aqueous-organic mobile phases are sometimes employed in normal-phase HILIC [58–60], employing more often with diol, aminopropyl, or specially designed HILIC (e.g., polyhydroxy-ethyl aspartamide) bonded phases than unmodified silica. The retention increases as the polarity of analytes increases and as the amount of water as the most polar component in the mobile phase decreases. This behavior is characteristic for NPC and opposite to RPC; Equation 5.10 can often be used to describe the effect of the concentration of water in the mobile phase on the retention in HILIC. Hence, the retention in gradient HILIC with increasing concentration of water (usually starting at 1%–10% water) [61] can be described by Equation 5.12. HILIC provides excellent separation selectivity for some strongly polar samples, such as some peptides and proteins, carbohydrates, oligonucleotides, etc., very weakly retained in RPC or very strongly retained in nonaqueous NPC [62].

Some columns may show U-turn dependence of retention on the increasing concentration of water in the mobile phase, with decreasing k in the low water concentration range (HILIC behavior) and increasing k in the high water concentration range (RP behavior).

5.4.3 GRADIENT SEPARATIONS OF IONIC COMPOUNDS

5.4.3.1 Reversed-Phase Chromatography

RPC separations of ionic samples usually require ionic additives to the mobile phase, which may cause problems in HPLC/MS operations. Completely ionized solutes are much less retained than the corresponding uncharged species and elute close to the column hold-up volume, often as asymmetric or even split peaks. Weak acids are eluted in order of decreasing K_a constants and weak bases in order of increasing K_a constants. By adjusting the pH, the degree of ionization and hence the separation selectivity of various weak acids or bases can be controlled. Weak bases are completely ionized at pH $< pK_a - 1.5$ and weak acids at pH $> pK_a + 1.5$, so that addition of a buffer to the mobile phase can often be used to suppress the ionization of acids at lower pH and of bases at higher pH and to eliminate undesirable chromatographic behavior of ionic species ($pK_a = -\log K_a$, K_a is the acidity, i.e., the dissociation constant) [5]. 10–50 mM phosphate buffers (pH 2.1–3.1 and 6.2–8.2) and acetate buffers (pH 3.8–5.8) have usually adequate buffer capacities for most HPLC separations. It should be noted that many bonded-phase silica-based columns are less stable outside the pH range 2–8 [5,8].

In gradient elution of weak acids or bases, gradients of organic solvent (acetonitrile, methanol, or tetrahydrofuran) in buffered aqueous-organic mobile phases are most frequently used. The solvent affects the retention in similar way as in RPC of nonionic compounds, except for some influence on the dissociation constants, but Equations 5.8 and 5.9 usually are accurate enough for calculations of gradient retention volumes and bandwidths, respectively.

pH gradients are less useful and more tricky to develop, but may be used for RPC separation of weak acids or bases, whose ionization and retention strongly depend on the pH of the mobile phase. The retention factor, k, of a non-dissociated acid or a base may be 10–20 times larger than that of their dissociated forms. Quasi-linear pH gradients can be accomplished by mixing universal buffers containing phosphoric, acetic, and boric acids in the part A and sodium hydroxide in the part B [63,64]. A reversed-phase column should be stable over a wide range of pH; bidentate bonded silica columns, hybrid silica–organic polymer matrix columns, or columns based on modified zirconium dioxide support are generally suitable for pH gradients. Kaliszan et al. [63,64] presented a model for reversed-phase pH gradients, enabling numerical calculation of the retention times of weak acids or weak bases in LC with linear pH gradients. The dependence of the dissociation constants on the concentration of organic solvent in the mobile phase may affect the accuracy of prediction [65,66].

RPC with mobile phases containing relatively high concentrations of salts (0.1–0.5 M) is occasionally employed for separations of organic acids or bases [67–69]. Equation 5.7 can often describe the effect of the concentration of organic solvents on the retention in these phase systems, so that Equations 5.8 and 5.9 can be principally used for calculation of the retention data in the LC with organic solvent gradients.

5.4.3.2 Ion-Pair Chromatography

Adjusting pH of the mobile phase to suppress the ionization fails with strong acid or basic compounds, which are ionized over a wide range of pH. Completely ionized substances can often be separated in ion-pair chromatography (IPC), where an ion-pair reagent with surface-active properties, containing a strongly acidic or strongly basic group and a bulky hydrocarbon part in the molecule, is added to aqueous-organic mobile phases. Salts of C_6–C_8 alkanesulfonic acids are used as ion-pairing reagents for basic analytes, whereas tetrabutylammonium or cetyl trimethylammonium salts for acidic samples, such as sulfonic acid dyes and intermediates [67,70]. Ion-pair additives greatly increase the retention and improve the peak symmetry through formation of neutral ionic associates with increased affinity to a nonpolar stationary phase. Aqueous-organic mobile phases in IPC usually contain a buffer enhancing the ionization of weak acids (higher pH) or bases (lower pH), opposite to the pH required for the separation of weak acids or bases in RP LC without ion-pairing reagents.

The retention in IPC can be controlled by the type or concentration of the ion-pair reagent, or by adjusting the concentration of organic solvent in the mobile phase. Like in RPC of nonionic solutes, the logarithms of retention factors decrease linearly with increasing concentration of the organic solvent in the mobile phase, so that Equation 5.7 can often be used to describe the effect of the organic solvent on the retention, Equation 5.8 to predict the elution volumes, and Equation 5.9 the bandwidths in IPC with organic solvent gradients [70]. The retention of ionic solutes generally increases with increasing number and size of alkyl substituents in alkanesulfonates or in tetraalkylammonium salts [70] and with increasing concentration of the ion-pair reagent in the mobile phase, until the column capacity for the reagent is saturated (in the range 10^{-3} to 10^{-1} mol/L in the mobile phase) [70]. Adequate ion-pair reagent concentrations in IPC are between 10^{-4} and 10^{-3} mol/L. In IPC, increasing concentration of the organic solvent during a solvent gradient affects the distribution equilibrium of the ion-pairing reagent between the stationary and the mobile phase [71,72], which may impair the accuracy of the predicted gradient data.

Major disadvantage of gradient IPC is slow column equilibration after changing the mobile phase. Complete washout of the adsorbed ion-pair reagent from the column may be difficult to achieve. Hence, it is not advisable to use a column in RPC without ion-pair reagents once it was run in the IPC mode.

5.4.3.3 Ion-Exchange Chromatography

Ion-exchange chromatography (IEC) is one of the oldest HPLC modes. Today, it is used mainly for separations of small inorganic ions, aliphatic carboxylic acids, or of ionic biopolymers such as oligonucleotides, nucleic acids, peptides, and proteins. Ion exchangers used in IEC contain functional ion-exchange groups carrying either a positive charge (quaternary or tertiary ammonium groups in anion exchangers) or a negative charge (sulfonic or carboxylic acid groups in cation exchangers) and retain ions with opposite charges by strong electrostatic interactions. The ion-exchange groups are covalently attached to a solid matrix, either organic such as, e.g., cross-linked styrene–divinylbenzene or ethyleneglycol–methacrylate copolymers, or inorganic, most frequently, silica gel support. The separation is based on stoichiometric ion exchange reaction between the solute ions and counterions of the same polarity. Hence, mobile phases must contain a salt or a buffer to achieve the elution.

The retention in IEC can be described by a stoichiometric model, which is similar to the competition/displacement model of NPC. With φ characterizing the molar concentration of electrolyte in the mobile phase and the parameter m as the stoichiometric coefficient of the ion exchange, Equation 5.10 can be formally used to describe the retention in many ion-exchange systems [4,33,54,73–78]. The exponent m characterizes the rate of decrease in log k with increasing counterion concentration, is close to 1 for the exchange of univalent ions, and is higher for multiply charged ionic solutes and lower for multiply charged counterions. The parameter k_0 is the retention factor in the mobile phase with 1 M electrolyte, which is proportional to the ion-exchange capacity of the column and increases with the affinity of the solute molecules to the matrix of the ion exchanger [55].

Because of the formal similarity of the competition models described by Equation 5.10, the elution volumes in IEC with linear salt concentration gradients can be calculated using Equation 5.12, like in NPC systems [4,8,9,53,76]. This approach was validated experimentally for ion-exchange separations of polyphosphates [79,80] oligonucleotides [76,81,82], peptides [83], and other compounds [55]. Organic solvents occasionally added to the mobile phase in IEC affect the sample solubility and the retention much like in RPC, by means of adjusting the solvophobic interactions with the ion-exchanger matrix. However, organic solvent gradients are rarely used in IEC.

pH affects the degree of dissociation and hence the selectivity of separation of weak acids and bases. The change in ionization is greatest when pH is close to pK_a. If the pK_a values are known, the change in ionization and in retention of the sample compounds resulting from a change in pH can be predicted. The use of pH gradients to accomplish the separation of solutes differing in the degree of ionization is more frequently applied in ion-exchange than in RPC, especially for the separation of biopolymers such as proteins, which elute roughly in order of their isoelectric points, pI.

5.4.4 GRADIENT SEPARATIONS OF LARGE MOLECULES

Size-exclusion chromatography (SEC) has been used for long years as the traditional method for the separation of macromolecular compounds and for the determination of the mass distribution of synthetic polymers. The interactions with the stationary phase can be also utilized for separations of large molecules in the "interaction chromatography" (IC) of polymers, including RPC or IEC generally used for biopolymers and NPC or NARP for synthetic polymers. To take full advantage of IC, wide-pore materials are preferred for separation of both biopolymers and synthetic polymers, because the narrow pores of the column packing material have limited accessibility for large molecules.

There is no sense in using gradient elution in the pure SEC mode, where the macromolecules are not retained on the column packing material and the mobile phase affects only the solubility. On the other hand, successful separation of polymer samples in interaction chromatography usually requires gradient elution because even a very small change in the mobile phase composition has much stronger effect on the retention of large molecules than on the retention of low molecular compounds. Hence it is very tedious—or almost impossible—to adjust the right isocratic conditions for separation of large molecules. Because of the "strange" differences between the behavior of small and large molecules, the mechanism of gradient-elution LC of polymers has been for a long time a matter of dispute. Various models were presented for gradient chromatography of polymers, such as the "precipitation–re-dissolution" mechanism, where gradient of increasing concentration of a "good" solvent (**B**) in a "poor" one (**A**) is based on the solubility of synthetic polymers and does not consider interactions of sample molecules in the stationary phase [84].

In mobile phases that are good sample solvents, the retention mechanism of high molecular compounds can be explained on the basis of "conventional theory" of RPC or NPC gradient elution applying for small molecules, considering the effect of increasing size of molecules on the values of the constants a, b, k_0, and m of the retention equations—Equation 5.7, 5.10, or 5.11.

In many separation systems, the contributions of the repeat structural units to the Gibbs free energy of retention, ΔG, are additive [85]. Hence, samples of homologues or oligomers with regular distribution of nonpolar repeat methylene groups, or another more or less polar structural units, such as methylene, oxyethylene, or other, show regular increase in retention at least over a limited interval of repeat monomer units [6,7,86–94]. Because the energy of retention, ΔG, is directly proportional to the logarithm of retention factor, k, the retention of a compound containing n repeat structural groups can be described by Equation 5.16 [44,89]:

$$\log k = \log \beta + n \log \alpha \tag{5.16}$$

The relative retention $\alpha = k_j/k_i$ is a measure of the separation selectivity for two compounds i and j with retention factors k_i and k_j, respectively, differing by one repeat structural unit, $\Delta n = 1$. β in Equation 5.16 is the end-group contribution to the retention factor. The conventional theory describes adequately the retention of oligomers and lower homopolymers and copolymers up to the molar masses 10,000–30,000 Da; for higher polymers the accuracy of the determination of retention model parameters is too low [95].

In reversed-phase LC, the retention in a series with repeat structural units depends on the volume fraction, φ, of the organic solvent in a binary aqueous-organic mobile phase and can be described combining Equations 5.7 and 5.16 as [44]:

$$\log k = a_0 - m_0 \varphi + n(a_1 - m_1 \varphi) \tag{5.17}$$

The constant a_1 means the contribution of the repeat units and a_0 that of the end group to a; m_1 characterizes the contribution of the repeat units and m_0 that of the end group to m in Equation 5.7 or 5.10.

Both the bandwidths and the differences in the elution volumes (times) of the compounds i and j with relative retention, $\alpha = k_j/k_i$ are approximately constant during a reversed-phase gradient run, except for very early-eluting compounds [4–10]. With some simplification, Equation 5.8 can be adapted to describe approximately the differences in gradient retention volumes, ΔV_R, of all but very early-eluting analytes:

$$\Delta V_R \cong \frac{(\Delta a - A \cdot \Delta m)}{mB} \cong \frac{\Delta n}{mB} \cdot \log \alpha_0 \tag{5.18}$$

where
   Δa and Δm are the differences between the constants a and m of Equation 5.7 for the compounds differing by Δn repeat structural units
   m is the average constant of Equation 5.7 for the sample solutes
   A is the initial organic solvent concentrations at the start of the gradient
   B is the gradient slope (steepness parameter), Equation 5.4

The retention, the band spacing in the chromatograms, and the number of well-resolved oligomers eluting before the end of the gradient can be conveniently characterized by the gradient oligomer capacity, $P(o)$, which is defined as the maximum number of peaks that can be separated with resolution $R_S = 1$ in between the peaks of the oligomers differing by one repeat unit:

$$P(o) = \frac{\sqrt{N}}{4} \frac{V_G}{V_m} \frac{a_1}{(m_0 + m_1 n) + 0.44 \dfrac{V_G}{V_m}} - 1 \tag{5.19}$$

The gradient oligomer capacity, $P(o)$, can be used to evaluate the column ability to resolve the individual oligomers in a certain polarity or molecular weight range according to the molar mass distribution [97]. $P(o)=0$ when the resolution between the oligomers differs by one repeat monomer unit $R_S=1$. $P(o)<0$ means that the oligomer resolution is insufficient, whereas $P(o)>1$ indicates some space between neighboring oligomer peaks available for separation of other compounds, e.g., isomers, or oligomers with different end groups. Generally, the gradient oligomer capacity increases with a higher amount of bonded carbon for samples with nonpolar repeat units, but $P(o)$ for oligomers with more or less polar repeat units, such as ethyleneoxide groups in polyethylene glycol samples, it is affected also by the chemistry of the bonded ligands, endcapping, and other stationary phase properties.

Although approximately constant selectivity for repeat oligomer groups results in almost regular oligomer peak distribution in an optimized gradient chromatogram, the type of the stationary phase may affect the range of oligomer distribution for best separation, as minor differences in resolution may be found at different parts of the chromatogram. Figure 5.3 compares oligomer capacity for oligo(ethylene glycols) on four columns providing different level of resolution for lower oligo(ethylene glycols) at the early and at the late parts of the chromatograms at standardized gradient conditions [96].

Equation 5.18 shows that the differences in the gradient-elution volumes decrease as the product $B.m$ increases. This means that for acceptable ΔV_R the macromolecules with large m generally elute within a narrow elution concentration interval in gradient-elution mode. The slope parameters m of the retention equations (Equation 5.7 or 5.10) for large polymers may be so high (even >100) that there is only a very small interval of φ available for the elution, often less than 1% organic solvent [11]. Outside this interval, the large molecules are either too strongly retained, or not retained at all. Consequently, sudden change from "full retention" to "no retention" (on–off: transition) may occur in gradient elution of large polymers.

In the synthetic polymer samples containing species with 50 or more monomer units, the individual oligomers usually cannot be resolved and the chromatogram shows a single broad peak,

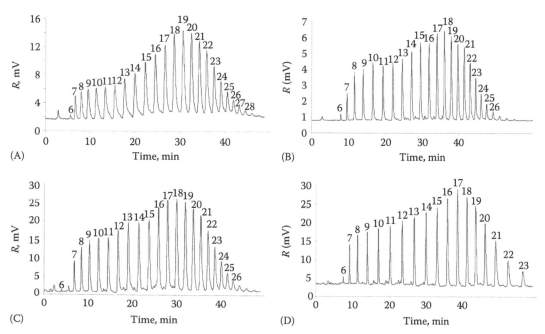

FIGURE 5.3 Gradient-elution separation of polyethylene glycol sample (PEG 1000) on C18 columns: (A) Zorbax RX C18, 250×4.6 mm i.d., (B) Alltima C18, 250×4.6 mm i.d., (C) Purospher RP-18e, 250×4.0 mm i.d., and (D) Zorbax SB-Aq, 250×4.6 mm i.d., all 5 μm particles. Gradients: Zorbax RX, Alltima, Zorbax SB-Aq: 30%–50% MeOH/40 min, 0.75 mL/min, Purospher RP: 30%–50% MeOH/30 min, 0.75 mL/min, 40°C. R—response of the ELS detector.

which, after calibration, may be used for the determination of molecular-weight distribution, much like the size-exclusion chromatograms. So far, the model for gradient elution of large molecules was simplified by neglecting the effect of the gradient on possible changes in diffusion coefficients in course of the elution, which is justified for small molecules and lower oligomers, however, may more significantly influence the bandwidths of polymers with molecular weights higher than 3000–5000 Da.

If a significant part of the pore volumes in the column packing material is not accessible to large molecules, the gradient retention volumes calculated using Equation 5.8, 5.12, or 5.14 with full column hold-up volumes may be subject to errors, which can be corrected for size-exclusion effects by using the size-exclusion volume, V_{SEC}, determined in a strong mobile phase where the adsorption is completely suppressed, instead of the column hold-up volume [9,10,47,87]. The correction for the accessibility of pores improves the agreement between the experimental and the calculated retention volumes of large proteins on columns with broad pore size distribution and significant proportion of small-diameter pores, but the correction for SEC effects has little influence on the accuracy of prediction of the retention of proteins on nonporous or superficially porous columns [97].

Gradient interaction chromatography can be also used for determination of chemical composition of samples containing blends of homopolymers or copolymers prepared from two or more different monomers [95]. For this purpose, "pseudo-critical" separation conditions are useful, under which the separation according to the molecular weight distribution or according to the distribution of a certain type of block copolymers are suppressed, so that the compounds co-eluting in one separation mode can be separated according to the second type of distribution [93,94]. More complete separation can be accomplished in 2D systems, where samples with multiple monomer units are separated according to the distribution of the number of each monomer units using different separation mechanisms in the first and in the second dimension [20,98].

Under "critical elution conditions," the SEC and interaction contributions may fully compensate each the other at a certain composition of the mobile phase for a specific polymer.

Similar behavior as with synthetic polymers was observed also in RP gradient chromatography of peptides. This may seem surprising as proteins and peptides are composed of various combinations of 20 different amino acids, but possibly the differences in the structures of the individual amino acids tend to more or less statistically level out in peptides and proteins containing tens or hundreds amino acids. The concentration of the organic solvent at the time of elution of peak maximum, φ_e, in the reversed-phase gradient HPLC generally increases for larger peptides or proteins. The values of the parameter m (Equation 5.7) are typically around 40 for peptides and even higher for proteins. The effect of increasing molecular weights of proteins, M_r, on increasing m can be predicted using [5]

$$m = 0.25 \cdot M_r^{0.5} \tag{5.20}$$

The experimental φ_e of peptides or proteins depend more on the type of the column than on the gradient program and are constant within 1% at various gradient times. The variance in the φ_e is lower as the size of molecules increases. The values of φ_e are slightly higher with totally porous particles than with columns packed with superficially porous, nonporous particles and monolithic columns [97].

Salt (ionic strength) gradients in IEC discussed in Section 5.4.3.3 are frequently used in the separation of complex peptides, proteins, and other biopolymer samples as a complementary technique to RP solvent gradient separations, often in a 2D setup [99,100]. The gradients usually start at a low salt (chloride, sulfate, etc.) concentration and typically run from 0.005 to 0.5 M. A buffer is used to control the pH; acetonitrile and methanol may be added to improve the resolution and urea to improve the solubility of proteins that are difficult to dissolve. Ion exchangers with not strongly hydrophobic matrices usually prevent protein denaturation in aqueous mobile phases.

pH gradients may be applied to accomplish ion-exchange separations of peptides and proteins, which elute roughly in order of their isoelectric points, *pI*. Proteins are focused in narrow bands during the separation and their band shape improves with respect to ionic-strength gradients. The focusing occurs because protein molecules carry multiple negative charges and are strongly retained on an anion exchanger at pH higher than *pI*, but as soon as the pH drops below the *pI* during gradient elution, the initially strongly retained protein gets positively charged and is released from the ion exchanger rapidly. The formation of linear pH gradients in IEC is more difficult than in RP LC, because the ion exchanger may show preferential adsorption for certain buffer components and consequently change the preset gradient [101]. For this reason, ion-exchange materials with small ion-exchange capacities are required in combination with buffer components that are not adsorbed on the ion-exchange column. Ethanol is added to the buffer to reduce the adsorption [102–104]. Here, the elution occurs at the time when the instantaneous pH at the column outlet reaches the *pI* of the solute. In combination with reversed-phase solvent gradient in second dimension, pH ion-exchange gradient in first dimension seems a promising emerging technique for 2D separation of proteins [104].

5.5 INSTRUMENTAL ASPECTS OF GRADIENT ELUTION

5.5.1 TYPES OF GRADIENT INSTRUMENTS

Two basic designs of gradient-elution HPLC instruments are commercially available: high-pressure and low-pressure systems, differing in whether the components of the mobile phase are blended at a high or at a low pressure. In high-pressure gradient systems, two pumps are used to meter the components **A** and **B** into the mixer at high pressure; the concentration proportions of the two components being controlled by changing the relative ratios of the flow rates of the two pumps while the total flow rate is kept constant. In a low-pressure gradient system, a single pump is used to deliver the blend of solvents premixed at a low-pressure side of the pump, most often by controlling the proportion of the opening and closing times of a set of two, three, or even four proportioning valves during a valve duty cycle time (e.g., 100 ms) [4,11]. In at least two commercial systems, exact metering auxiliary low-pressure pumps were used instead of proportioning valves to blend the solvents at the inlet of the high-pressure pump; unfortunately these highly accurate gradient liquid chromatographs are no longer available on the market. Other (hybrid) systems have proportioning valves mounted directly on the pump head and the solvents are proportioned and mixed within the pump head.

The solvents are blended according to a preset gradient program from the computer chromatographic workstation, which is also used for detector signal processing and data handling. The gradient program has a form of one or more subsequent linear segments, a few instrument types allow using curved gradient profiles or segments. The accuracy of the gradients is typically about ±1% or better, using more or less sophisticated electronic feedback system control.

It is not possible to recommend specific instrumentation as the best for all purposes, most modern gradient instruments offered by recognized manufacturers perform reliably and each type may have its pros and contras when used for specific applications. The instrumental difficulties in gradient elution may originate from improper mixing ratios of the gradient components caused by failure or poor design of the pumps or of the gradient mixer. In addition to traditional gradient system with the pressure limits of 30–40 MPa, high-pressure gradient systems operating up to 150 MPa are now available for rapid routine analyses.

Both high- and low-pressure gradient system designs have some general advantages and disadvantages of their own, which, however, may affect to different degree the performance of the specific instrumental models. High-pressure systems usually have smaller gradient dwell volume (the total volume of the gradient mixer and of the connecting tubing between the mixer and the column inlet, see Section 5.5.3) than the low-pressure systems, except for the newest ones, and can be more easily adapted for narrow-bore column applications (see Section 5.5.2.). Thorough degassing of the components of mobile phase (especially in RP applications) is more critical with low-pressure systems than with high-pressure systems. On the other hand, high-pressure systems require a high-pressure pump for each mobile phase component, therefore are less flexible to use for the operation with ternary or quaternary gradients, even though this is principally possible by using different solvent blends instead of pure solvents **A** and **B**. However, each pump can be used for independent isocratic operation, if necessary. Often, it is possible to add an additional pump into a high-pressure system for ternary gradient, but this is not very practical because of additional costs. Further, the accuracy of the gradient formation in the high-pressure systems is more affected by the compressibility of the mobile phase at high pressures than with the low-pressure systems. Compression of the solvents under high pressure causes lower-than-expected flow rate, which may be connected with some deviations from the preset gradient profile. Some manufacturers offer the possibility of electronic compressibility compensation in their systems, which the user can adjust according to the type of the blended solvents. The liquid mixtures have slightly smaller volume than the sum of the original pure liquids, which also affects to some extent the flow rate in gradient elution, but these changes are usually not very significant. Finally, the performance of the pump check valves and of the proportioning valves may impair when high-density or high-viscosity liquids are delivered as the components of a gradient.

The dispersion of the solvent **B** in the instrument parts between the gradient former and the column causes "rounding" of the gradient profile at the points where the gradient steepness changes. Rounding of a linear gradient is apparent as a higher-than-expected volume fraction of the solvent **B** at the start of the gradient and a lower than expected volume fraction of **B** at the gradient end and can reduce the retention times of the bands eluting near the start of the gradient and increase the retention times of bands eluting near the end of the gradient [43,105]. The rounding of the gradient depends on the mobile phase components and flow rate [4,7,11]. Last but not least, the volumes of the injecting valve, detector, and connecting tubing may significantly increase the retention times and contribute to band broadening in capillary LC [106].

5.5.2 GRADIENT ELUTION IN MICRO-HPLC

Micro-HPLC operation sets special demands on the gradient instrumentation. As the internal column diameter, d_c, decreases, lower flow rates should be used at comparable mean linear mobile phase velocities, $u = 0.2-0.3$ mm/s. At a constant operating pressure, the flow rate decreases proportionally to the second power of the column inner diameter, so that narrow-bore LC columns with 1 mm i.d. require flow rates in the range of 30–100 µL/min, micro-columns with i.d. 0.3–0.5 mm, flow rates in between 1 and 10 µL/min, and columns with 0.075–0.1 mm i.d. flow rates in the range of hundreds nL/min. Special miniaturized pump systems are required to deliver accurately mobile phase at very low flow rates in isocratic LC.

Simple gradient techniques employed in early micro-LC usually did not allow flexible variations of the gradient profiles, as different proportions of mobile phase components were sucked into a capillary tubing and stored there prior to the chromatographic run and then forwarded to the separation column by feeding the last solution from the pump [105,106]. Because of the check valve inertia, it is technically difficult to mix accurately at varying proportions solvents delivered at µL/min or lower flow rates by individual miniaturized pumps with small-volume piston chamber pumps. In contemporary micro- and capillary LC practice, conventional LC pumps with pre-column flow splitting are occasionally used to solve the problems with gradient delay caused by the dwell volume (see Section 5.5.3); more sophisticated instruments employ micro-pumps specially designed for operation at flow rates, ranging down to 50 nL/min. The pump can be equipped with a binary-to-quaternary low-pressure gradient former with high-speed micro-proportioning valves, or two high-pressure pumps can be used for high-pressure gradient formation, like in conventional gradient HPLC operation. High-pressure pre-column in-line filters or guard-columns contribute to the gradient dwell volume and, if used, their volume should be minimized in micro- and capillary LC.

Low flow rates in micro-LC systems can be achieved using conventional pumps and a flow-splitter, directing only a small part of the mixed mobile phase from the pump to flow through the column, while a larger part is diverted through a bypass restrictor capillary, whose inner diameter and length should be matched to the flow resistance of the LC column. The flow-splitter should be placed as close to the sample injector and column as possible, after the damper, mobile phase mixer, and high-pressure filter, to suppress the system dwell volume contributions to the retention times. Flow-splitting can seriously affect the accuracy of the flow rate and the reproducibility of the retention times in gradient micro-LC, as the mobile phase delivered by the pump is divided in between the column and the bypass path in the flow rate ratio controlled by the flow resistance in the two fluid paths, which depends on the mobile phase viscosity. The system with a flow-splitter operates at a constant pressure rather than at a constant flow rate and the split flow-rate ratio depends on the viscosity of the mobile phase in the column and in the bypass paths. During a gradient run, the viscosity and consequently the flow resistance change with time in a different way in the column and in the bypass resistor paths and consequently the split ratio does not remain constant. To suppress the dependence of the flow split ratio on the changes in solvent viscosity during a gradient run, an additional resistor may be used in the column fluid path, but at a cost of increased operation pressure. A more sophisticated approach correcting for this effect employs a feedback electronic control of the mobile phase flow in between the flow splitter and the column, based on measuring the migration time of a thermal pulse in the fluidic path. To achieve optimum gradient mixing, correct solvent compressibility values should be set for each mobile phase component mixed by the gradient pump [12].

5.5.3 GRADIENT DWELL VOLUME

The total volume of the gradient mixer and of the connecting tubing between the mixer and the column inlet, called the gradient dwell volume, V_D, which is usually between 0.5 and 3 mL with various commercial gradient instruments (and even larger if additional high-pressure pulse dampeners are used to suppress the pressure ripples as the source of signal noise for high-sensitivity analysis), may cause significant problems in the practice of gradient elution. The blended gradient needs some time to pass the dwell volume until it arrives to the top of the column, which causes "gradient delay" depending on the dwell volume and on the flow rate of the mobile phase. Even though the gradient delay may not decrease significantly the reproducibility of repeated runs, it increases the analysis time and may cause more or less significant deviations of the experimental retention times from the values predicted by theoretical modeling, if it is not appropriately accounted for. The instrumental dwell volume may differ significantly with various types of instruments, which may cause unexpected changes in the separation when a gradient method is transferred from one chromatograph to another. If some sample compounds injected at the start of the gradient run are very strongly retained, they do not move along the column in the mobile phase contained in the dwell volume for the whole gradient delay time. In this case, the dwell volume, V_D, just adds to the elution volumes predicted from Equation 5.8, 5.12, or 5.14. However, compounds less strongly retained in the starting mobile phase migrate a more or less significant distance along the column under isocratic conditions in the mobile phase contained in the dwell volume before they are affected by the gradient, depending on the size of the dwell volume and on the retention factors, k, of the analytes in the initial mobile phase [4,9,10,47]. (Some weakly retained sample compounds may even preelute from the column during the gradient delay period; their retention times should be calculated like in isocratic elution with the starting mobile phase.) Appropriate corrections should be adopted in Equations 5.8, 5.12, and 5.14 by introducing the terms with V_D to avoid errors in the calculated gradient-elution volumes, V_R, when modeling the gradient separation or when transferring gradient methods between instruments with different ratios of V_D to V_m. If V_D is taken into account, the retention volumes V_R in gradient-elution RPC can be calculated using Equation 5.21—for details of derivation see Refs. [4,9,10]:

$$V_R = \frac{1}{mB} \log\left\{ 2.31mB\left[V_m 10^{(a-mA)} - V_D \right] + 1 \right\} + V_m + V_D \qquad (5.21)$$

In gradient-elution NPC where Equation 5.10 applies, the elution volumes affected by the gradient dwell volume can be calculated using Equation 5.22 [57]:

$$V_R = \frac{1}{B}\left[(m+1) B\left(k_0 V_m - V_D A^m \right) + A^{(m+1)} \right]^{\frac{1}{(m+1)}} - \frac{A}{B} + V_m + V_D \qquad (5.22)$$

The value of the gradient dwell volume, which is necessary in the calculations, can be determined experimentally by evaluating the time delay of an experimental linear gradient baseline record with respect to the preset gradient profile. For this purpose, the experimental UV-detector baseline can be recorded when running a "blank" linear gradient of increasing concentration of a non-retained UV-absorbing compound (to avoid errors caused by the adsorption on the column). For example, pure acetonitrile or methanol can be used as the starting solvent (**A**) and 1% acetone dissolved in the solvent **A** as the solvent **B**, whose concentration increases during the gradient run. In this experiment, also the extent of gradient rounding caused by the gradient mixer at the initial and final stage of the gradient can be evaluated [105].

Figure 5.4A and B compares the uncorrected elution volumes, the elution volumes corrected by simple addition of V_D and the elution volumes calculated using Equation 5.21 in RP gradient elution on a conventional and on a micro-bore C18 column. The effect of the gradient dwell volume is more important for separations on short columns and especially on narrow-bore columns with i.d. ≤ 2 mm.

In micro- and capillary LC, the gradient dwell volume causes even more significant gradient delay than in operation on micro-bore columns, due to the low flow rates used. When conventional instrumentation is used in connection with micro-columns; the gradient delay may take even more time than the gradient run alone. Therefore, the connecting tubing, high-pressure in-line filters and especially additional mixers (often inserted between the pump and the sample injector to ensure good mixing of the gradient components) should be carefully minimized, as they contribute more significantly than in conventional LC to gradient dwell volume, V_D, between the gradient former and

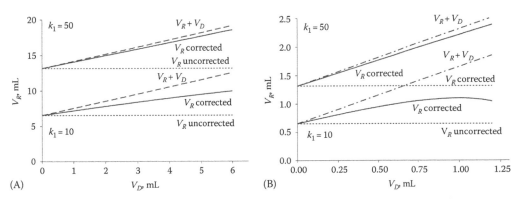

FIGURE 5.4 Effect of the gradient dwell volume, V_D, on the elution volume, V_R, in reversed-phase chromatography. Solute: neburon, retention equation (Equation 5.7) with parameters $a=4$, $m=4$. Linear gradients: 2.125% methanol/min: (a) from 57.5% to 100% methanol in water in 20 min ($k_1=50$); (b) from 75% to 100% methanol in water in 11.75 min ($k_1=10$). V_R *uncorrected*: calculated from Equation 5.8, V_R+V_D: V_D added to V_R *uncorrected*, V_R *corrected*: calculated from Equation 5.21. (A) A conventional analytical C18 column, hold-up volume $V_m=1$ mL; flow rate 1.0 mL/min. (B) A microbore analytical C18 column, hold-up volume $V_m=0.1$ mL; flow rate 0.1 mL/min.

the column, which delays the gradient elution by the time necessary to transport the gradient to the column and thus increases the time of analysis [12,42,105–108].

5.6 DEVELOPMENT, OPTIMIZATION, AND TRANSFER OF GRADIENT METHODS

The general strategies of the development and optimization of gradient methods can be focused (1) on tailor-made optimization of a gradient method for specific separation problem aiming at target analysis of known compounds and (2) on setting conditions for fast gradients or for separations of a large number of compounds in complex samples.

5.6.1 TARGET OPTIMIZATION OF BINARY GRADIENTS

Various strategies have been applied for the optimization of linear, step, nonlinear, and multi-segmented binary or ternary gradients and for simultaneous optimization of gradient elution and other operation variables. For appropriate characterization of the quality of separation over the whole chromatogram time range by a single value, various "elemental" or "sum" objective functions (such as chromatographic optimization function, COF, or chromatographic response function, CRF) [109] are used as optimization criteria for multiparameter optimization employing well-known statistical approaches such as Simplex or Grid-search methods [110]. Both approaches are time consuming, as they require a large number of experimental data. Further, the Simplex search method may "slide" into a region with a local rather than total optimum of the objective function in the retention surface, so that the result of optimization need not represent the real optimum combination of the operation parameters. However, these approaches are still used nowadays in automatized HPLC method optimization. Some recent Simplex applications were reported, including the PREOPT package for the optimization of binary step gradients [111]. Grid-search strategy can be employed for the optimization of multi-linear gradients with 2, 3, or 4 segments, within which Equation 5.7 is applied for calculations of the retention data [112].

In predictive single-parameter or multiparameter optimization methods, several initial experiments are performed to determine the constants of the equations controlling the retention (e.g., Equation 5.8 or 5.12). In single-parameter optimization, appropriate criteria describing the quality of separation for each pair of adjacent peaks in the chromatogram, usually the resolution, R_s (Equation 5.6), are plotted in dependence on the optimized parameter (gradient steepness, B, or the initial concentration of the solvent $\mathbf{B}$, A) as a "window diagram" [113], or—in multiparameter optimization—as an "overlapping resolution map"(ORM) [114]. In such plots, the areas are searched in which the resolution for all adjacent bands in the chromatogram is equal to or larger than the desired value (e.g., $R_S \geq 1.5$). The optimum gradient parameters are selected that either yield the maximum resolution for the "critical pair" of adjacent peaks most difficult to separate, or provide the desired resolution for all adjacent peaks in the chromatogram in the shortest run time [6,109,113]. The window diagrams or ORM maps offer detailed information about the separation of the individual sample compounds. Simulated optimum chromatograms may be constructed and observed, showing the optimized separation for all sample compounds, or—by simply moving a cursor on the monitor screen—the effect of the optimized parameter on the chromatogram.

Probably the best-known commercial optimization software for gradient elution based on these strategies is the DryLab G software, developed by Snyder et al. many years ago [6,115,116]. The retention data from two initial gradient runs are used to adjust the gradient steepness and the concentration range (initial and final %B) and if desired, other working parameters (pH, temperature) [117]. DryLab can be used also to optimize multi-segment or ternary gradients [118]. For the overview of various applications of the DryLab software, see Ref. [119].

As the gradient time and other parameters (temperature, pH, or initial concentration of solvent $\mathbf{B}$) often show synergistic effects on separation, simultaneous optimization of two or more parameters at a time can provide better results than their subsequent optimization. To vary band spacing in the chromatogram and to maximize the resolution, Snyder et al. [120–125] advocate simultaneous optimization of temperature, T, and gradient time, t_G, as a more efficient approach than the optimization of a single parameter, either T or t_G, alone.

Decreasing the gradient time (increasing steepness B in Equation 5.4), and increasing the initial concentration A of the strong solvent **B** result in lower elution volumes, much like increasing the concentration of the solvent **B** in isocratic LC. Traditionally, most effort in optimization of gradient elution is focused on the gradient steepness, B [119]. However, Equation 5.8 predicts strong effect of the initial concentration, A, on the retention, especially for weakly retained compounds with low $a = \log k_0$ in Equation 5.7. This is so because A appears in the exponent of the term $10^{(a-mA)}$, directly proportional to the gradient retention volume.

The effect of the initial concentration of acetonitrile, A, on the separation of homologous alkylbenzenes in acetonitrile–water mobile phases is illustrated in Figure 5.5A through C. To employ efficiently the synergistic effects of the two operation parameters in gradient elution, both the gradient range (the gradient steepness, B) and the initial gradient concentration, A, should be adjusted when developing a gradient HPLC method. To calculate the combined effects of A and B, Equation 5.8 can be used in reversed-phase systems and Equation 5.12 or 5.14 in normal-phase or in ion-exchange systems [9,10,113,126]. The optimization approach is based on the selection of a fixed gradient time t_G, or volume, V_G, and the final concentration, φ_G, of the solvent **B** (usually 100%, or the limit imposed by the solvent solubility). Under these conditions, the gradient steepness parameter B depends on the initial concentration, A, of the solvent **B**. After rearranging Equation 5.4 we obtain

$$B = \frac{\varphi_G - A}{V_G} \qquad (5.23)$$

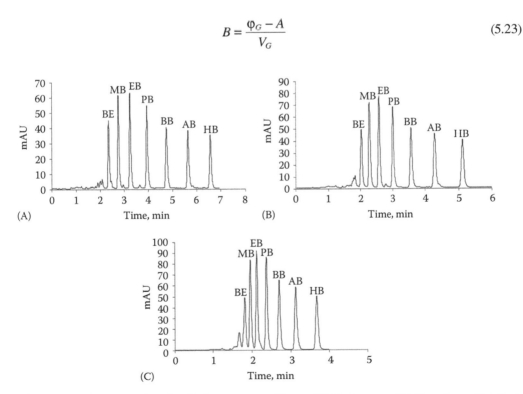

FIGURE 5.5 Gradient separation of n-alkylbenzenes on a Chromolith Performance RP-18e column. Gradient: (A) 60%–100%, (B) 70%–100%, and (C) 80%–100% ACN in water, gradient time 9.3 min. $F_m = 1$ mL/min, Instrumental dwell volume $V_D = 0.4$ mL. Temperature 40°C, UV detection at 254 nm. Solutes: benzene (BE), methylbenzene (MB), ethylbenzene (EB), propylbenzene (PB), butylbenzene (BB), amylbenzene (AB), and hexylbenzene (HB). (From Jandera, P., *J. Chromatogr. A*, 1126(1–2), 24, 2006. Copyright 2006. With permission from Elsevier.)

Based on the parameters of the adequate retention equation (e.g., Equation 5.7, 5.10, or 5.11) acquired in a few initial isocratic or gradient initial experiments [113,127], Equation 5.23 can be used for predictive calculations of the retention and of the resolution of the individual pairs of sample compounds. From these data, window diagrams are constructed as the plots of the retention volumes (times) or resolution *versus A*, from which optimum *A* and consequently *B* are determined. Even though the initial selection of V_G is not very critical for the results of the optimization, the calculations can be repeated for various preset values of V_G to find optimum conditions for the shortest time of separation [9,10,113,127]. The final gradient concentration can be set at a later stage, as the gradient can be stopped right after the elution of the most strongly retained compound. Programs for the optimization of one or two gradient parameters at a time using a common spreadsheet software such as Excel or Quattro table editors for stand-alone PCs are easy to write in either Windows or Macintosh format. The approach is general and can be used for optimization of reversed-phase, ion-exchange, or normal-phase [57] binary or ternary [113,127] gradients and even for simultaneous optimization of *A*, *B* and the shape (curvature) of nonlinear gradients [128]. An example of window-diagram optimization of the initial concentration of methanol in reversed-phase gradient separation of herbicides with fixed gradient time is shown in Figure 5.6. More details on this optimization approach and application examples can be found in Refs. [8–10,113] and the literature quoted therein.

Simultaneous optimization of the steepness of a salt gradient, column length, and mobile phase flow rate using iso-resolution plots was recently applied to cation-exchange chromatography of proteins on short columns [129,130].

As alternative to continuous gradients, step-gradient elution comprised of a sequence of isocratic steps can be used. The theory of step-gradient elution was developed more than two decades ago [4,33,131]. In each segment, the retention is controlled by appropriate retention equation describing the dependence of *k* on φ, e.g., Equation 5.7 in RP systems, or Equation 5.10 in NP or ion-exchange LC. The sequence of the gradient steps can be optimized by calculating the isocratic composition of the mobile phase necessary for resolution of a group of sample compounds in each step, taking into account the corresponding part of the column migrated by the compounds in the preceding steps [4,33]. The accuracy of prediction and the quality of optimization improve if the step changes are adjusted to the parts of the chromatograms where no peaks are expected, especially with detectors sensitive to changing composition of mobile phases [132].

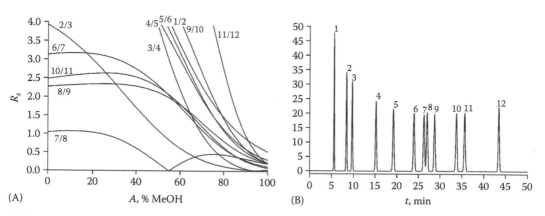

(A)   *A*, % MeOH
(B)   *t*, min

FIGURE 5.6   (A) The resolution window diagram for RP-gradient-elution separation of phenylurea herbicides on a Separon SGX C18, 7.5 μm, column (150×3.3 mm I.D.) in dependence on the initial concentration of methanol in water at the start of the gradient, *A*, with optimum gradient volume $V_G = 73$ ml; (B) The separation with optimized binary gradient from 24 to 100% methanol in water in 73 min. Flow-rate 1 ml/min, T=40°C. Column plate number *N*=5000; sample compounds: hydroxymetoxuron (1), desphenuron (2), phenuron (3), metoxuron (4), monuron (5), monolinuron (6), chlorotoluron (7), metobromuron (8), diuron (9), linuron (10), chlorobromuron (11), neburon (12).

Commercially available structure-based predictive software [133,134] (such as CHROMDREAM, CHROMSWORD, or ELUEX) for optimization of isocratic or gradient RPC incorporate some features of the "expert system," as the predictions employ the contributions to the retention by molecular structures for various sample compounds. However, the accuracy of the predictions of retention based on the structural parameters alone is too low for complete method development, which requires that the average error in resolution, R_S, should not exceed 0.2 units for samples containing 15 or more compounds. Hence, this approach is suitable rather for rough estimation of the range of operation variables than for final optimization of resolution. On the other hand, the accuracy of correlations is usually sufficient for using gradient experiments to predict hydrophobity of solutes including drugs and drug candidates using quantitative structure-retention relationships (QSRR) [135].

5.6.2 TERNARY GRADIENTS

Nowadays, multicomponent gradients are not very frequently used for method development and optimization due to increased complexity and lower accuracy of prediction with respect to binary gradients [136]. However, ternary [114,137] or quaternary [138] gradients may improve the resolution when the selectivity of binary gradients is not sufficient for the separation of a particular sample. In ternary gradients, the concentrations of two strong solvents **B** and **C** (such as acetonitrile and methanol in water in RPLC or propanol and dioxane in n-hexane in NPLC) usually change simultaneously, most often in a linear manner, so that the gradient of each of the two strong solvents is described by Equation 5.4 with different parameters A and B. In the "elution strength" (iso selective) ternary gradients the concentration ratio of the two strong solvents is kept constant to maintain constant selectivity, but the sum of the concentrations linearly increases [113], whereas in "selectivity" ternary gradients the sum of the concentrations of the two strong solvents is constant, but their concentration ratio changes [113]. If good separation of the early eluting compounds needs different selectivity than the separation of the late-eluting compounds, ternary gradients with simultaneous changes in both the sum and the ratio of the concentrations of the two solvents often can improve the separation. Figure 5.7 shows an example of the effects of ternary gradient on reversed-phase separation of phenols [139]. Whereas early-eluting compounds 2 and 3 are poorly separated using a binary gradient of methanol in water and strongly retained compounds 8 and 9 co-elute with a gradient of acetonitrile in water, the optimized ternary gradient of simultaneously increasing concentration of methanol and decreasing concentration of acetonitrile significantly improves the resolution of both the early- and the late-eluting compounds.

Ternary gradients employing simultaneous changes in pH and in the concentration of the organic solvent (methanol) in a binary aqueous-organic mobile phase, φ, run on an alkylsilica column were employed for simultaneous determination of the characteristics of drug candidates in a mixture, the acidity (pK_a) and the lipophilicity (φ_0, calculated from the parameters of Equation 5.7, log k_w/m, which is the concentration of the organic solvent at which $k = 1$ and is directly proportional to the partitioning coefficient between water and n-octanol, log P, a conventional measure of lipophilicity). The determination is based on fitting a theoretical model, assuming linear change of pK_a with the concentration of organic modifier, to the experimental gradient retention times [140].

So-called relay gradients represent a specific category of ternary gradients comprised of a train of two binary gradients, where the concentration of the solvent **B** in the solvent **A** is changed first until 100% **B**, followed then by the gradient of the solvent **C** in **B**. Such gradients are useful for reversed-phase separations of samples containing very hydrophobic compounds such as lipids, which are strongly retained in pure methanol or acetonitrile. The separation often improves if we combine an aqueous-organic and a nonaqueous gradient step, such as for the separation of mono-, di,- and triacyl glycerides using a gradient of acetonitrile (**B**) in water, followed by a gradient of 2-propanol (**C**) in acetonitrile, shown in Figure 5.8 [141].

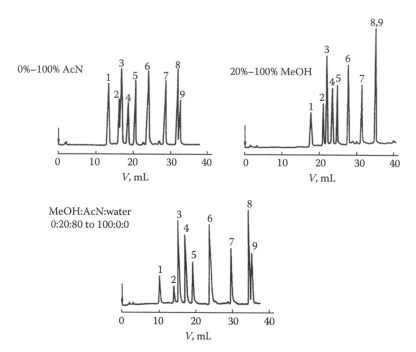

FIGURE 5.7 Effects of binary and ternary gradient elution with methanol and acetonitrile on separation selectivity in RP HPLC. Column: LiChrosorb RP-C18, 5 μm, 300×4.0 mm i.d.; flow rate 1 mL/min; UV detection, 254 nm. Sample: 4-cyanophenol (1), 2-methoxyphenol (2), 4-fluorophenol (3), 3-fluorophenol (4), 3-methylphenol (5), 4-chlorophenol (6), 4-iodophenol (7), 2-phenylphenol (8), and 3-*tert*-butylphenol (9).

Ternary "elution strength" and "selectivity" gradients can be optimized in a similar way as binary gradients [113,127,142,143], using "overlapping resolution maps" in solvent selectivity triangle [144–146], prism or tetrahedron [147–150], or window diagrams for simultaneous optimization of the initial composition of the ternary mobile phase at the start of the gradient and of the gradient time [113,128,142,143]. Another optimization approach is based on the substitution of linear gradients by several subsequent isocratic steps with various concentrations of methanol and acetonitrile in water [151].

5.6.3 Fast Gradients

Speed of separation is a crucial issue in improving the productivity of an analytical laboratory. Fast generic gradient methods are required for high throughput in food safety control, in environmental analysis, and especially in pharmaceutical laboratories throughout the whole drug analysis process, including drug discovery screening, raw material analysis, impurity profiling, pharmacokinetic studies, and final product stability tests. Clearly, the speed of separation increases at a higher flow rate of the mobile phase, traded for either decreased resolution or increased operation pressure, the latter being limited by the instrumentation. Monolithic columns allow approximately three times faster analyses than particulate packed columns at the same operating pressure, with separation efficiency comparable to the columns packed with 5 μm particles [152,153].

Fast gradient separations can be achieved, at a cost of decreased peak capacity, by using short packed columns and simultaneously decreasing the particle size, d_p, of the packing material to keep a constant ratio of the column length to the mean particle diameter, l/d_p, e.g., with a 3 cm column packed with a 3 μm material, or even a 1 cm, 1 μm column for very fast separation instead of a conventional 5 cm, 5 μm or a 10 cm, 10 μm column. At constant V_G/V_m and l/d_p ratios, the

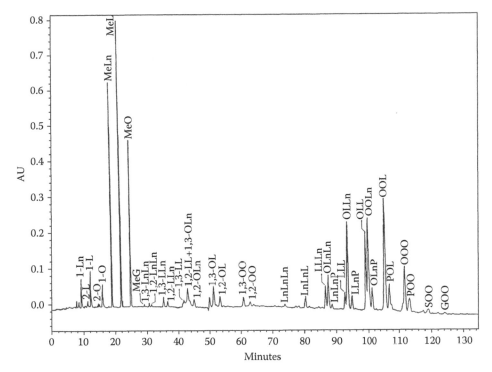

FIGURE 5.8 Relay gradient RP separation of partially transesterified rapeseed oil on two Nova-Pak C18 columns in series, 4 μm (3.9 × 150 mm i.d. each). Linear gradient from 70% acetonitrile to 100% acetonitrile in water in 20 min, followed by an isocratic hold-up step with 100% acetonitrile until 36 min and linear gradient from 100% acetonitrile to 60% 2-propanol in acetonitrile until 132 min. Flow rate 1 mL/min, detection UV, 205 nm. Peak notation—fatty acids in methyl esters (Me), mono-, di-, and triacylglycerols: Ln—linolenic, L—linoleic, O—oleic, P—palmitic, S—stearic, and G—gadoleic.

resolution only slightly impairs when decreasing the column length, but the time of separation decreases proportionally to the change in column length. In addition to short columns packed with small porous particles, nonporous [154] or superficially porous [155] particles offer decreased band broadening, at a cost of weaker retention and lower sample load capacity. Very fast gradients with short columns require a modified LC instrument with minimized extra-column volumes by using low-volume injectors and detector cells, fast autosamplers, detectors with short time constant, high signal sampling rates and flow splitting to decrease the effects of the gradient dwell volume. Over the last 2 or 3 years, several instrument manufacturers have introduced instrumental systems answering to these demands, operating at considerably higher pressures, up to 150 MPa. Figure 5.9 shows high-pressure fast-gradient RP separation of 14 phenolic acids and flavones on a column with superficially porous particles (Figure 5.9A) and on a column with small totally porous particles (Figure 5.9B).

Gradient separations can be accelerated at elevated temperature by reducing the viscosity of the mobile phase and hence the column operating pressure and by increasing the diffusion coefficients of the analytes, speeding up the mass transfer between the stationary and the mobile phase [156,157].

5.6.4 PEAK CAPACITY IN GRADIENT ELUTION

A convenient measure of the performance of a chromatographic system for the separation of complex samples is the peak capacity P, which is defined as the upper limit of the number of fully

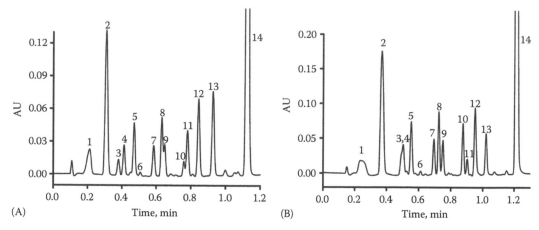

FIGURE 5.9 High-pressure fast-gradient separation of natural phenolic antioxidants. (A) Ascentis Express C18 column with superficially porous 2.7 μm C18 particles, 30×3.0 mm i.d. at 1.5 mL/min, 53 MPa. (B) Acquity BEH Phenyl column with totally porous 1.7 μm particles at 1.5 mL/min, 74 MPa. UV detection, 254 nm. Sample: gallic acid (1), protocatechine (2), esculine (3), chlorogenic acid (4), caffeine (5), epicatechine (6), vanilline (7), rutine (8), sinapic acid (9), hesperidine (10), 4-hydroxycoumarine (11), morine (12), quercetine (13), and 7-hydroxyflavone (14).

resolvable compounds under prescribed conditions in a single run (in a fixed time or volume of the eluate) [157,159]. Assuming that the contribution of the dwell volume to the retention can be neglected and that the bandwidths are approximately constant during the gradient run, the gradient peak capacity can be estimated in the time range given by the difference between the elution times of the last and of the first peaks in the chromatogram, $t_{R,Z}$ and $t_{R,1}$, respectively, using Equation 5.24 [160], or in the whole preset gradient time, $t_G, -t_m$, using Equation 5.25, assuming that the first peak elutes at the column hold-up volume and the last one at the end of the gradient [161]:

$$P = \frac{\sqrt{N}}{4}\left(\frac{t_{R,Z}}{t_{R,1}} - 1\right) \tag{5.24}$$

$$P = 1 + \frac{t_G - t_m}{w_g} = 1 + \frac{\sqrt{N}}{4}\left(\frac{t_G}{t_m} - 1\right)\frac{1}{(1+k_e)} \tag{5.25}$$

where

    N is the average column isocratic theoretical plate number

    k_e is the retention factor at the point of elution controlling the bandwidths in gradient elution—
        Equation 5.5

Introducing the average bandwidth, w_g, from Equation 5.5 and using the appropriate expressions for k_e at the point of elution into Equation 5.25, we obtain Equation 5.26 for the peak capacity in reversed-phase gradients [9]:

$$P = 1 + \sqrt{\frac{N}{4}}\frac{t_G}{t_m}\frac{1}{1 + \left[10^{(mA-a)} + 2.31m\frac{\Delta\varphi}{t_G}V_m\right]^{-1}} \tag{5.26}$$

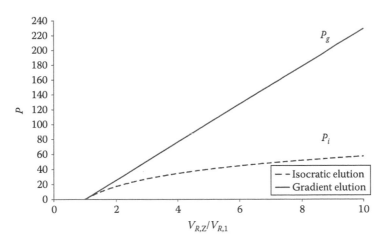

FIGURE 5.10 Comparison of isocratic (P_i, Equation 5.27) and gradient (P_g, Equation 5.24) peak capacity for a column with 10,000 theoretical plates (N).

Similar expressions for peak capacity in RP and IE gradient LC were derived using "average" retention factor under gradient conditions, $k_g = V_R'/Vm$ [162].

The gradient peak capacity is significantly higher than the isocratic peak capacity, P_i, within the same time limits, as illustrated in Figure 5.10.

$$P_i = \frac{\sqrt{N}}{4} \ln\left(\frac{V_{R,Z}}{V_{R,1}}\right) + 1 = \frac{\sqrt{N}}{4} \ln\left(\frac{k_z+1}{k_1+1}\right) + 1 \qquad (5.27)$$

Three factors contribute to the gradient capacity, P: (1) column efficiency (isocratic plate number, N), (2) gradient time normalized in the hold-up volume units, t_G/t_m, and (3) the retention factor at the point of elution, k_e. P increases with increasing N and t_G/t_m, and slightly decreases as k_e increases, because k_e depends to some extent on the parameters of the retention equation (Equation 5.7) (a, k_w, m) and somewhat decreases with steeper gradients. To first approximation, k_e are similar for all sample compounds eluting during the gradient run, so that the gradient peak capacity, P, is not much affected by the sample structure, provided that the diffusion coefficients are not very different and do not change very significantly during gradient elution.

Theoretical peak capacity, P, is much larger than the number of peaks that can be really separated in a single gradient run in structurally non-correlated samples with random distribution of retention. According to statistical treatment, a 90% probability that a compound of interest will be found as a pure peak in the chromatogram requires real peak capacity exceeding the number of sample compounds by a factor of 20, and that the peak capacity required to separate all sample compounds dramatically increases with the second power of the number of compounds [163–166].

5.6.5 GRADIENT ELUTION IN COMPREHENSIVE TWO-DIMENSIONAL LC × LC SEPARATIONS

Comprehensive 2D liquid chromatography is emerging as a new powerful technique for the separation of complex samples because of increased peak capacity, selectivity, and resolution in comparison to single-dimensional HPLC. 2D LC×LC systems essentially represent programming of stationary phases. "Comprehensive" LC×LC technique represents specific 2D mode, where all sample compounds eluting from the first dimension are subjected to separation in the second dimension [167]. The whole effluent from the first dimension is transferred into the second-dimension separation system in subsequent aliquot fractions collected in two alternating sampling loops or trapping columns in multiple repeated cycles [20,168–171]. The record of the detector at the outlet from the second-dimension column is transformed into a 2D chromatogram, which is usually represented as a contour plot with the separation time in the second dimension plotted *versus* the separation time in the first dimension.

The separation selectivities in the first dimension should largely differ from that in the second dimension. Best results are achieved in so-called orthogonal systems with non-correlated retention times in both dimensions [172,173]. Mobile phase, flow rate, and in some cases, temperature should be optimized in each dimension to increase the number of resolved compounds in a single run.

Gradient-elution operation is a useful means to suppress the band broadening in the second dimension and to increase the number of sample compounds separated in comprehensive LC×LC. Because of high peak capacity, gradient elution is widely used in the first dimension. As the separation time in the second dimension in comprehensive LC×LC is strictly limited by the fraction transfer cycle time from the first dimension, programmed elution in the second dimension may greatly improve the overall sample resolution. Very fast and steep gradients cover the whole mobile phase range providing adequate elution strength for all sample compounds in the individual fractions transferred to the second dimension.

Fast comprehensive LC×LC separation of bovine serum albumin tryptic digests with the total peak capacity of 1350 in 20 min was accomplished by using a fast (20 s) high-temperature binary gradient at 100°C on a short stable-bond C_{18} micro-column in the second dimension after first-dimension separation on a strong cation-exchange (SCX) micro-column [157]. However, the time necessary for re-equilibration of the column with the initial gradient conditions before the analysis of the next fraction diminishes the time really available for second-dimension separation. Hence, the equilibration time should be kept very short, so that complete equilibration is difficult to achieve. Further, successful application of this approach requires elimination of the gradient delay due to the dwell volume of the gradient liquid chromatograph (see Section 5.5.3).

Using two matching parallel gradients, one in the first and another in the second dimension, spanning over the whole separation time range in a comprehensive 2D system enable efficient utilization of the available time in the second dimension, as the re-equilibration of the second-dimension column to the initial conditions after the elution of every fraction is no more necessary [24]. The second-dimension column can be re-equilibrated at the same time as the first-dimension column, after the end of the whole 2D separation. Such parallel gradients may be useful for 2D systems with partially correlated retention, where the average elution times in the second dimension more or less increase for compounds more strongly retained in the first dimension. The parallel gradients enable to suppress the selectivity correlation between the two dimensions and to achieve more regular spacing of separated sample compounds in the available 2D retention plane [24].

5.6.6 Effects of Operating Conditions on Transfer of Gradient Methods

Practicing chromatographers often face the task of transferring once elaborated gradient methods between various instruments and columns with different geometry and (or) particle size. This is less straightforward than the transfer of isocratic HPLC methods, as changing column dimensions and flow rate affect the gradient volume (steepness) and consequently not only the elution times and the column efficiency (plate number), but also the selectivity of separation may change [8,9,90]. Hence, the gradient program should be adjusted to fit new column dimensions and (or) flow rate, which is not very difficult observing several rules following from the theory. Equation 5.8 for RP gradient-elution times or volumes can be rewritten as the dependence of the product of the net elution volume and of the gradient steepness parameter, $V_R' \cdot B$, on the product $V_m \cdot B = \Delta\varphi \cdot V_m/V_G$, as long as the product $V_m \cdot B$ and the gradient concentration range, $\Delta\varphi$, are kept constant [8–10,113]. The number of the column hold-up volumes necessary to elute a sample compound, V_R/V_m, is directly proportional

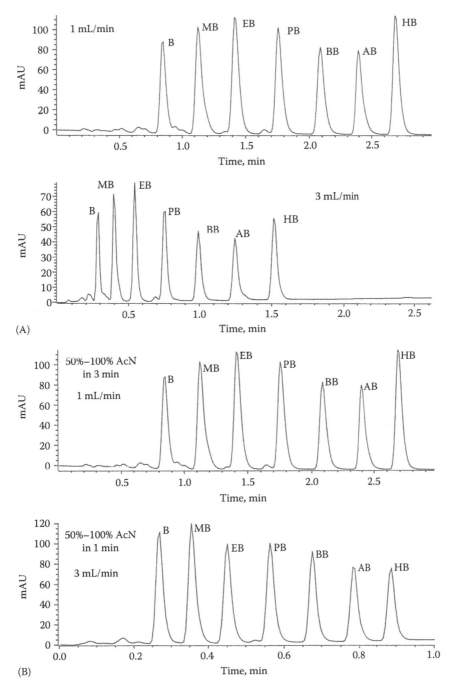

FIGURE 5.11 Gradient-elution reversed-phase separation of alkylbenzenes on a Purospher Star RP-18e, 3 μm, column (30×4 mm i.d.). (A) Non-adjusted linear gradients, 50%–100% acetonitrile in 3 min at 1 mL/min and at 3 mL/min. (B) Adjusted linear gradients, 50%–100% acetonitrile in 3 min at 1 mL/min and 50%–100% acetonitrile in 1 min at 3 mL/min. Conditions: 40°C, detection UV, 254 nm; sample: B—benzene, MB—toluene, EB—ethylbenzene, PB—propylbenzene, BB—butylbenzene, AB—amylbenzene, and HB—hexylbenzene.

to the ratio of the gradient volume, V_G, and the hold-up volume, V_m, and is independent of the flow rate and column dimensions:

$$\frac{V_m}{V_G} = \frac{V_m}{t_G F_m} = \frac{d_c^2 l}{t_G F_m} = \text{const} \tag{5.28}$$

This condition is similar to keeping a constant retention factor, k, at a constant mobile phase composition and temperature in isocratic LC. Hence, any change of column length, l, or diameter, d_c, at a constant gradient range $\Delta\varphi$ should be compensated by appropriate change in the gradient time, t_G, or flow rate of the mobile phase, F_m, to keep the ratio V_m/V_G constant [8–10,57]. In this case, a change in column length has the same impact on the column plate number, N, and time of separation as in isocratic chromatography. Figure 5.11 illustrates a less than proportional decrease in the elution times with increasing flow rate and non-adjusted gradient time (Figure 5.11A) in comparison to expected behavior with gradient adjusted to keep the product $t_G F_m$ constant (Figure 5.11B), in agreement with Equation 5.28).

If a column with different diameter, but the same length is used, the column hold-up volume changes in proportion to the second power of the column diameters. To keep the separation time constant, the flow rate of the mobile phase should be adjusted in the proportion of changing V_m, see Equation 5.28. Another possibility of compensation is by adjusting the gradient time to keep the ratio $V_m/(t_G \cdot F_m)$ constant [8,9].

The simple transfer rule expressed by Equation 5.28 does not take into account the differences between the instrumental gradient dwell volumes, V_D, in various commercial instruments, which may complicate the transfer of gradient methods. If two different instruments are used to run a gradient method on the same column and at the same flow rate, more or less significant differences in the retention times and unexpected changes in the band spacing and sample separation may occur, due to the migration of analytes injected at the start of the gradient run along the column under isocratic conditions in the mobile phase contained in the dwell volume at the start of the analysis, see Section 5.5.3. Experimental determination of the dwell volume, V_D, for the two instruments allows calculating the changes in the elution times using Equation 5.21 or 5.22, as appropriate.

The problem of different dwell volumes in the transfer of gradient methods can be solved to some extent either by late injection with respect to the start of the gradient for the systems with a larger dwell volume, or by adjusting the dwell volume by introducing an isocratic hold period at the beginning of the gradient so as to increase the actual dwell period of the equipment with a lower instrumental V_D [174]. Unfortunately some LC workstations do not allow using delayed injection. Furthermore, it is not always possible to merge a makeup initial hold-up time into the equilibration time between the subsequent runs in a sequence of analyses. Finally, a makeup gradient delay may increase significantly the time of analysis, especially in micro-column operations or with fast steep gradients on short columns.

The effect of the dwell volume on the retention times of analytes increases with decreasing retention factor k_i at the start of gradient elution and with increasing ratio V_D/V_m and becomes very significant in the instrumental setup with the dwell volume comparable to or larger than the column hold-up volume, which is more likely to occur in micro- or in capillary LC than in conventional analytical LC (see Figure 5.4) [12].

ACKNOWLEDGMENTS

Some results presented in this work are based on the research supported by the Ministry of Education, Youth and Sports of the Czech Republic under project MSM 0021627502 and by the Grant Agency of Czech Republic under project No. 203/07/0641.

REFERENCES

1. G.A. Howard, A.J.P. Martin, *Biochem. J.* 46, 532, 1950.
2. L.R. Snyder, *Chromatogr. Rev.* 7, 1, 1965.
3. C. Liteanu, S. Gocan, *Gradient Liquid Chromatography*. New York: Halsted Press, 1974.
4. P. Jandera, J. Churáček, *Gradient Elution in Column Liquid Chromatography*. Amsterdam, the Netherlands: Elsevier, 1985.
5. L.R. Snyder, Gradient elution. In: Cs. Horvath (Ed.), *High-Performance Liquid Chromatography. Advances and Perspectives*, Vol. 1. New York: Academic Press, 1980, p. 207.
6. L.R. Snyder, J.J. Kirkland, J.L. Glajch, *Practical HPLC Method Development*, 2nd edn. New York: Wiley-Interscience, 1997.
7. L. R. Snyder, J.W. Dolan, *Adv. Chromatogr.* 38, 115 1998.
8. P. Jandera, Comparison of various modes and phase systems for analytical HPLC. In: K. Valkó (Ed.), *Separation Methods in Drug Synthesis and Purification*. Amsterdam, the Netherlands: Elsevier, 2000, p. 1.
9. P. Jandera, *Adv. Chromatogr.* 43, 1, 2005.
10. P. Jandera, *J. Chromatogr. A* 1126, 195, 2006.
11. L.R. Snyder, J.W. Dolan, *High-Performance Gradient Elution. The Practical Application of the Linear-Solvent-Strength Model*. Hoboken, NJ: Wiley-Interscience, 2007.
12. P. Jandera, L.G. Blomberg, E. Lundanes, *J. Sep. Sci.* 27, 1402, 2004.
13. P.L. Zhu, L.R. Snyder, J.W. Dolan, N.M. Djordjevic, D.W. Hill, L.C. Sander, T.J. Waeghe, *J. Chromatogr. A* 756, 21, 1996.
14. P.L. Zhu, J.W. Dolan, L.R. Snyder, *J. Chromatogr. A* 756, 41, 1996.
15. P.L. Zhu, J.W. Dolan, L.R. Snyder, D.W. Hill, L.Van Heukelem, T.J. Waeghe, *J. Chromatogr. A* 756, 51, 1996.
16. P.L. Zhu, J.W. Dolan, L.R. Snyder, N.M. Djordjevic, D.W. Hill, J.-T. Lin, L.C. Sander, L. Van Heukelem, *J. Chromatogr. A* 756, 63, 1996.
17. C. Viseras, R. Cela, C.G. Barroso, J.A. Perez-Bustamante, *Anal. Chim. Acta* 196, 115 1987.
18. B. Bidlingmayer, K.K. Unger, N. von Doehren, *J. Chromatogr. A* 832, 11, 1999.
19. V. Lesins, E. Ruckenstein, *J. Chromatogr. A* 467, 1, 1989.
20. A. Van der Horst, P.J. Schoenmakers, *J. Chromatogr. A* 1000, 693, 2003.
21. E.L. Little, M.S. Jeansonne, J.P. Foley, *Anal. Chem.* 63, 33, 1991.
22. V. Wong, R.A. Shalliker, *J. Chromatogr. A* 1036, 15, 2004.
23. P. Dugo, O. Favoino, P.Q. Tranchida, G. Dugo, L. Mondello, *J. Chromatogr. A* 1041, 135, 2004.
24. F. Cacciola, P. Jandera, Z. Hajdú, P. Česla, L. Mondello, *J. Chromatogr. A* 1149, 73, 2007.
25. G. Guiochon, S. Golshan Shirazi, A.M. Katti, *Fundamentals of Preparative and Nonlinear Chromatography*. New York: Academic Press, 1994.
26. E.C. Freiling, *J. Am. Chem. Soc.* 77, 2067, 1955.
27. B. Drake, *Ark. Kemi* 8, 1, 1955.
28. P. Jandera, J. Churáček, *J. Chromatogr.* 91, 223, 1974.
29. L.R. Snyder, J.W. Dolan, J.R. Gant, *J. Chromatogr.* 165, 3, 1979.
30. P. Jandera, J. Churáček, L. Svoboda, *J. Chromatogr.* 174, 35, 1979.
31. P. Nikitas, A. Pappa-Louisi, *Anal. Chem.* 77, 5670, 2005.
32. M. Martin, *J. Liquid. Chromatogr. Relat. Technol.* 11, 1809, 1988.
33. P. Jandera, J. Churáček, *Adv. Chromatogr.* 19, 125, 1980.
34. G. Vivó-Truyols, J.R. Torres-Lapasió, M.C. García-Alvarez-Coque, *J. Chromatogr. A* 1018, 169, 2003.
35. M.Z. El Fallah, G. Guiochon, *Anal. Chem.* 60, 2244, 1988.
36. F.D. Antia, Cs. Horvath, *J. Chromatogr.* 484, 1, 1989.
37. P. Jandera, D. Komers, G. Guiochon, *J. Chromatogr. A* 760, 225, 1997.
38. H. Liang, Y. Liu, *J. Chromatogr. A* 1040, 19, 2004.
39. U.D. Neue, D.H. Marchand, L.R. Snyder, *J. Chromatogr. A* 1111, 40, 2006.
40. J.D. Stuart, D.D. Lisi, L.R. Snyder, *J. Chromatogr.* 485, 657, 1989.
41. W. R. Melander, Cs. Horváth Reversed-phase chromatography. In: Cs. Horváth (Ed.), *High Performance Liquid Chromatography, Advances and Perspectives*, Vol. 2. New York: Academic Press, 1980, p. 113.
42. K. Valkó, L.R. Snyder, J.L. Glajch, *J. Chromatogr.* 656, 501, 1993.
43. E. Soczewiński, C.A. Wachtmeister, *J. Chromatogr.* 7, 311, 1962.
44. P. Jandera, J. Kubát, *J. Chromatogr.* 500, 281, 1990.
45. P.J. Schoenmakers, H.A.H. Billiet, L. de Galan, *J. Chromatogr.* 282, 107, 1983.

46. R.M. McCormic, B.L. Karger, *Anal. Chem.* 54, 435, 1982.
47. M.A. Quarry, R.L. Grob, L.R. Snyder, *Anal. Chem.* 58, 907, 1986.
48. P. Jandera, *J. Chromatogr.* 314, 13, 1984.
49. K. Ballschmiter, M. Wössner, *Fresenius J. Anal. Chem.* 361, 743, 1998.
50. L.R. Snyder, J.L. Glajch, J.J. Kirkland, *J. Chromatogr.* 218, 299, 1981.
51. J.L. Glajch, J.J. Kirkland, L.R. Snyder, *J. Chromatogr.* 239, 268, 1982.
52. L.R. Snyder, *Principles of Adsorption Chromatography.* New York: Marcel Dekker, 1968.
53. P. Jandera, J. Churáček, *J. Chromatogr.* 91, 207, 1974.
54. P. Jandera, M. Janderová, J. Churáček, *J. Chromatogr.* 115, 9, 1975.
55. P. Jandera, M. Janderová, J. Churáček, *J. Chromatogr.* 148, 79, 1978.
56. P. Jandera, M. Kučerová, *J. Chromatogr. A* 759, 13, 1997.
57. P. Jandera, *J. Chromatogr. A* 965, 239, 2002.
58. J.J. Kirkland, C.H. Dilke, Jr., J.J. DeStefano, *J. Chromatogr.* 635, 19, 1993.
59. S.A. Churms, *J. Chromatogr.* 500, 555, 1990.
60. B.A. Olsen, *J. Chromatogr.* 913, 113, 2000.
61. T. Yoshida, *J. Biochem. Biophys. Methods* 60, 265, 2004.
62. A.J. Alpert, *J. Chromatogr.* 499, 177, 1990.
63. R. Kaliszan, P. Wiczling, M.J. Markuszewski, *Anal. Chem.* 76, 749, 2004.
64. R. Kaliszan, P. Wiczling, M.J. Markuszewski, *J. Chromatogr. A.* 1060, 165, 2004.
65. M. Rosés, E. Bosch, *J. Chromatogr. A* 982, 1, 2002.
66. S. Espinosa, E. Bosch, M. Rosés, *Anal. Chem.* 74, 3809, 2002.
67. P. Jandera, J. Churáček, B. Taraba, *J. Chromatogr.* 262, 121, 1983.
68. P. Jandera, J. Fischer, V. Staněk, M. Kučerová, P. Zvoníček, *J. Chromatogr. A* 738, 201, 1996.
69. H. Zou, Y. Zhang, L. Peichang, *J. Chromatogr.* 545, 59, 1991.
70. D. Vaněrková, P. Jandera, J. Hrabica, *J. Chromatogr. A* 1143, 112, 2007.
71. Y. Yokoyama, O. Ozaku, H. Sato, *J. Chromatogr. A* 739, 333, 1996.
72. M. Pathy, *J. Chromatogr.* 592, 143, 1992.
73. Y. Baba, N. Yoza, S. Ohashi, *J. Chromatogr.* 348, 27, 1985.
74. Y. Baba, *J. Chromatogr.* 485, 143, 1989.
75. W. Kopaciewicz, M.A. Rounds, J. Fausnaugh, F.E. Regnier, *J. Chromatogr.* 266, 3, 1983.
76. E.S. Parente, D.B. Wetlaufer, *J. Chromatogr.* 355, 29, 1986.
77. S.R. Souza, M.F.M. Tavares, L.R.F. deCarvalho, *J. Chromatogr. A* 796, 335, 1998.
78. J.E. Madden, N. Avdalovic, P.E. Jackson, P.R. Haddad, *J. Chromatogr. A* 837, 65, 1999.
79. Y. Baba, N. Yoza, S. Ohashi, *J. Chromatogr.* 350, 119, 1985.
80. Y. Baba, N. Yoza, S. Ohashi, *J. Chromatogr.* 350, 461, 1985.
81. Y. Baba, M. Fukuda, N. Yoza, *J. Chromatogr.* 458, 385, 1988.
82. Y. Baba, M.K. Ito, *J. Chromatogr.* 485, 647, 1989.
83. T. Sasagawa, Y. Sakamoto, T. Hirose, T. Yoshida, Y. Kobayashi, Y. Sato, *J. Chromatogr.* 485, 533, 1989.
84. G. Gloeckner, *Gradient HPLC of Copolymers and Chromatographic Cross-Fractionation*, Heidelberg, Germany: Springer, 1991.
85. A.J.P. Martin, *Biochem. Soc. Symp.* 3, 4, 1949.
86. T.H. Mourey, G.A. Smith, L.R. Snyder, *Anal. Chem.* 56, 1773, 1984.
87. M.A. Stadalius, H.S. Gold, L.R. Snyder, *J. Chromatogr.* 327, 27, 1985.
88. P. Jandera, *Chromatographia* 26, 417, 1988.
89. P. Jandera, *J. Chromatogr.* 449, 361, 1988.
90. C. Lochmüler, M.B. McGranaghan, *Anal. Chem.* 61, 2449, 1989.
91. C. Lochmüler, C. Jiang, M. Elomaa, *J. Chromatogr. Sci.* 33, 561, 1995.
92. P. Schoenmakers, F. Fitzpatrick, R. Grothey, *J. Chromatogr. A* 965, 93, 2002.
93. F. Fitzpatrick, R. Edam, P. Schoenmakers, *J. Chromatogr. A* 988, 53, 2003.
94. F. Fitzpatrick, B. Staal, P. Schoenmakers, *J. Chromatogr. A* 1065, 219, 2005.
95. L. Kolářová, P. Jandera, E.C. Vonk, H.A. Claessens, *Chromatographia* 59, 579, 2004.
96. P. Jandera, M. Halama, K. Novotná, *J. Chromatogr. A* 1030, 33, 2004.
97. J. Urban, P. Jandera, Z. Kučerová, M. A. van Straten, H. A. Claessens, *J. Chromatogr. A* 1167, 63, 2007.
98. P. Jandera, M. Halama, L. Koláčová, J. Fischer, P. Jandera, K. Novotná, *J. Chromatogr. A* 1087, 112, 2005.
99. D.A. Walters, M.P. Washburn, J.R. Yates III, *Anal. Chem.* 73, 5683, 2001.
100. M. Gilar, P. Olivová, A.E. Daly, J.C. Gebler, *Anal. Chem.* 77, 6426, 2005.
101. G. Wagner, F.E. Regnier, *Anal. Biochem.* 126, 37, 1982.

102. Y. Liu, D.J. Anderson, *J. Chromatogr. A* 762, 47, 1997.
103. L. Shan, D.J. Anderson, *Anal. Chem.* 74, 5641, 2002.
104. T. Andersen, M. Pepaj, R. Trones, E. Lundanes, T. Greibrokk, *J. Chromatogr. A* 1025, 217, 2004.
105. D.D. Lisi, J.D. Stuart, L.R. Snyder, *J. Chromatogr.* 555, 1, 1991.
106. D. Moravcová, P. Jandera, J. Urban, J. Planeta, *J. Sep. Sci.* 26, 1005, 2003.
107. T. Takeuchi,D. Ishii, *J. Chromatogr.* 218, 199, 1981.
108. K. Šlais, R.W. Frei, *Anal. Chem.* 59, 376, 1987.
109. P.J. Schoenmakers, *Optimisation of Chromatographic Selectivity*. Amsterdam, the Netherlands: Elsevier, 1986.
110. J.C. Berridge, *J. Chromatogr.* 244, 1, 1982.
111. R. Cela, C.G. Barroso, C. Viseras, J.A. Perez-Bustamante, *Anal. Chim. Acta* 191, 283, 1986.
112. V. Concha-Herrera, G. Vivo-Truyols, J.R. Torres-Lapasio, M.C. Garcia-Alvarez-Coque, *J. Chromatogr. A* 1063, 79, 2003.
113. P. Jandera, *J. Chromatogr.* 485, 113, 1989.
114. J.L. Glajch, J.J. Kirkland, K.M. Squire, J.M. Minor, *J. Chromatogr.* 199, 57, 1980.
115. L.R. Snyder, J.W. Dolan, D.C. Lommen, *J. Chromatogr.* 485, 65, 1989.
116. J.W. Dolan, D.C. Lommen, L.R. Snyder, *J. Chromatogr.* 485, 91, 1989.
117. J.W. Dolan, L.R. Snyder, D.L. Saunders, L. Van Heukelem, *J. Chromatogr. A* 803, 33, 1998.
118. T.H. Jupille, J.W. Dolan, L.R. Snyder, *Am. Lab.* 20(12), 20, 1988.
119. I. Molnár, *J. Chromatogr. A* 965, 175, 2002.
120. J.W. Dolan, L.R. Snyder, N.M. Djordjevic, D.W. Hill, T.J. Waeghe, *J. Chromatogr. A* 857, 1, 1999.
121. J.W. Dolan, L.R. Snyder, N.M. Djordjevic, D.W. Hill, T.J. Waeghe, *J. Chromatogr. A* 857, 21, 1999.
122. J.W. Dolan, L.R. Snyder, R.G. Wolcott, P. Haber, T. Baczek, R. Kaliszan, L.C. Sander, *J. Chromatogr. A* 857, 41, 1999.
123. R.G. Wolcott, J.W. Dolan, L.R. Snyder, *J. Chromatogr. A* 869, 3, 2000.
124. L.R. Snyder, J.W. Dolan, *J. Chromatogr. A* 892, 107, 2000.
125. J.W. Dolan, L.R. Snyder, T. Blanc, L. Van Heukelem, *J. Chromatogr. A* 897, 37, 2000.
126. P. Jandera, *J. Chromatogr. A* 797, 11, 1998.
127. P. Jandera, *J. Liquid Chromatogr.* 12, 117, 1989.
128. P. Jandera, *J. Chromatogr. A* 845, 133, 1999.
129. T. Ishihara, S. Yamamoto, *J. Chromatogr. A* 1069, 99, 2003.
130. S. Yamamoto, A. Kita, *J. Chromatogr. A* 1065, 45, 2003.
131. P. Jandera, J. Churáček, *J. Chromatogr.* 170, 1, 1979.
132. P. Nikitas, A. Pappa-Louisi, K. Papachristos, *J. Chromatogr. A* 1033, 283, 2004.
133. A. Drouen, J.W. Dolan, L.R. Snyder, A. Poile, P.J. Schoenmakers, *LC GC* 9, 714, 1991.
134. S.V. Galushko, A.A. Kamenchuk, *LC GC Int.* 8, 581, 1995.
135. R. Kaliszan, P. Haber, T. Baczek, D. Siluk, K. Valkó, *J. Chromatogr. A* 965, 117, 2002.
136. A.M. Siouffi, R. Phan-Tan Luu, *J. Chromatogr. A* 892, 75, 2000.
137. J.W. Weyland, C.H.P. Bruins, D.A. Doornbos, *J. Chromatogr. Sci.* 22, 31, 1984.
138. R. Phan-Tan-Luu, A. M. Siouffi, *J. Chromatogr.* 485, 433, 1989.
139. P. Jandera, J. Churáček, H. Colin, *J. Chromatogr.* 214, 35, 1981.
140. P. Wiczling, P. Kawczak, A. Nasal, R. Kaliszan, *Anal. Chem.* 76, 239, 2006.
141. M. Holčapek, P. Jandera, B. Prokeš, J. Fischer, *J. Chromatogr. A* 859, 13, 1999.
142. P. Jandera, B. Prokeš, *J. Liquid Chromatogr.* 14, 3125, 1991.
143. P. Jandera, L. Petránek, M. Kučerová, *J. Chromatogr. A* 791, 1, 1997.
144. J.J. Kirkland, J.L. Glajch, *J. Chromatogr.* 255, 27, 1983.
145. S.F.Y. Li, M.R. Khan, H.K. Lee, C.P. Ong, *J. Liquid Chromatogr.* 14, 3153, 1991.
146. J.W. Dolan, L.R. Snyder, *J. Chromatogr. Sci.* 28, 379, 1990.
147. J.L. Glajch, J.C. Gluckman, J.G. Charikofsky, J.M. Minor, J.J. Kirkland, *J. Chromatogr.* 318, 25, 1985.
148. J.L. Glajch, J.J. Kirkland, *Anal. Chem.* 54, 2593, 1982.
149. J.L. Glajch, J.J. Kirkland, *J. Chromatogr. A* 485, 51, 1989.
150. S. Heinisch, P. Riviere, J.L. Rocca, *Chromatographia* 32, 559, 1991.
151. J.A. Martinez-Pontevedra, L. Pensado, M. Casais, R. Cela, *Anal. Chim. Acta* 515, 127, 2004.
152. B. Bidlingmaier, K.K. Unger, N. von Doehren, *J. Chromatogr. A* 832, 11, 1999.
153. A.-M. Siouffi, *J. Chromatogr A* 1000, 801, 2003.
154. J.J. Kirkland, *Anal. Chem.* 64, 1239, 1992.
155. M. Hanson, K.K. Unger, *LC GC Int.* 9, 741, 1996.
156. U.D. Neue, J.R. Mazzeo, *J. Sep. Sci.* 24, 921, 2001.

157. D.R. Stoll, P.W. Carr, *J. Am. Chem. Soc.* 127, 5034 2005.
158. J.C. Giddings, *Anal. Chem.* 39, 1027, 1967.
159. J.C. Giddings, *Sep. Sci.* 4, 181, 1969.
160. Cs. Horváth, S.R. Lipsky, *Anal. Chem.* 39, 1993, 1967.
161. U.D. Neue, J.L. Carmody, Y.-F. Cheng, Z. Lu, C.H. Phoebe, T.E. Wheat, *Adv. Chromatogr.* 41, 93, 2001.
162. U.D. Neue, *J. Chromatogr. A* 1079, 153, 2005.
163. J.M. Davis, J.C. Giddings, *Anal. Chem.* 55, 418. 1983.
164. M. Martin, D.H. Herman, G. Guiochon, *Anal. Chem.* 58, 2200, 1986.
165. J.M. Davis, *Anal. Chem.* 63, 2141, 1991.
166. J.M. Davis, *Anal. Chem.* 65, 2014, 1993.
167. F. Erni, R.W. Frei, *J. Chromatogr.* 149, 561, 1978.
168. G.J. Opiteck, J.W. Jorgenson, R.J. Anderegg, *Anal. Chem.* 69, 2283, 1997.
169. G.J. Opiteck, K.C. Lewis, J.W. Jorgenson, R.J. Anderegg, *Anal. Chem.* 69, 1518, 1997.
170. S.T. Balke, R.D. Batel, *J. Polym. Sci. Polym. Lett. Ed.* 18, 453, 1980.
171. M. Bushey, J.W. Jorgenson, *Anal. Chem.* 62, 161, 1990.
172. J.C. Giddings, *Anal. Chem.* 56, 1270A, 1984.
173. J.C. Giddings, *J. High Resolut. Chromatogr. Commun.* 10, 319, 1987.
174. J.W. Dolan, L.R. Snyder, *J. Chromatogr. A* 799, 21, 1998.

6 Capillary Electromigration Techniques

Danilo Corradini

CONTENTS

6.1 INTRODUCTION

Liquid-phase separation techniques using columns of capillary size have recently gained large acceptance for qualitative and quantitative analyses as well as physicochemical characterizations. High throughput, small sample amount requirement, and high separation efficiency are the major advantages of these microscale separation techniques. Additional advantages include straightforward optimization of separation, use of small amounts of chemicals, reduced waste, and a wide range of possible detection methods, which include absorbance, fluorescence, electrochemical, nuclear magnetic resonance spectroscopy, and mass spectrometry.

A variety of microscale separation methods, performed in capillary format, employ a pool of techniques based on the differential migration velocities of analytes under the action of an electric field, which is referred to as capillary electromigration techniques. These separation techniques may depend on electrophoresis, the transport of charged species through a medium by an applied electric field, or may rely on electrically driven mobile phases to provide a true chromatographic separation system. Therefore, the electric field may either cause the separation mechanism or just promote the flow of a solution throughout the capillary tube, in which the separation takes place, or both.

The electrically driven flow of a liquid within a capillary tube, either open or packed or filled with a solid medium bearing a stationary phase, is caused by electroosmosis, which originates from the action of the electric field on the electric double layer formed at the plane of share between the inner surface of the capillary tube and the surrounding liquid (see Section 6.2). The flow of liquid caused by electroosmosis is termed electroosmotic flow (EOF) and displays a plug-like profile because the driving force is uniformly distributed along the capillary tube. Consequently, a uniform flow velocity vector across the capillary occurs. The flow velocity approaches zero only in the region of the capillary tube close to its inner surface. Therefore, no peak broadening is caused by sample transport carried out by the EOF. This is in contrast to the laminar or parabolic flow profile generated in a pressure-driven system where there is a strong pressure drop across the capillary caused by frictional forces at the liquid–solid boundary. A schematic representation of the flow profile due to electroosmosis in comparison to that obtained in the same capillary column in a pressure-driven system, such as a capillary HPLC, is displayed in Figure 6.1.

Separations in capillary electromigration techniques take place according to different separation modes in either open or packed or monolithic capillaries of typical inner diameter and length of 20–100 μm and 20–100 cm, respectively. The ends of the capillary are inserted into two distinct vessels (electrolyte compartments) containing the electrolyte solution, usually of same composition of that filled into the capillary, and the electrodes connected to a high-voltage power supply, typically up to 30 kV. The capillary format enables the application of high electric fields with minimal generation of Joule heat, which is efficiently dissipated by transfer through the tube wall as a result of the large surface-to-volume ratio of the capillary and, in open capillaries, ensures the absence of convective mixing of the separated zone in the electrolyte solution.

Samples are introduced into the capillary by either electrokinetic or hydrodynamic or hydrostatic means. Electrokinetic injection is preferentially employed with packed or monolithic capillaries whereas hydrostatic injection systems are limited to open capillary columns and are primarily used in homemade instruments. Optical detection directly through the capillary at the opposite end of sample injection is the most employed detection mode, using either a photodiode array or fluorescence or a laser-induced fluorescence (LIF) detector. Less common detection modes include conductivity [1], amperometric [2], chemiluminescence [3], and mass spectrometric [4] detection.

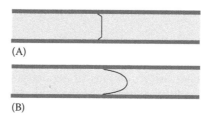

(A)

(B)

FIGURE 6.1 Schematic representation of the flow profile generated by (A) electroosmosis and (B) a mechanical pump.

On-line coupling of capillary electromigration techniques with nuclear magnetic resonance spectroscopy [5] and matrix-assisted laser desorption/ionization (MALDI) time-of-flight (TOF) mass spectrometry [6] has also been reported.

This chapter illustrates basic concepts, instrumental aspects, and modes of separation of electromigration techniques performed in capillary format. It should be noted that most of the fundamental and practical aspects of the electromigration techniques performed in capillary tubes also apply when the techniques are carried out in microchannels fabricated on plates of reduced dimensions, communally referred to as chips.

6.2 BASIC CONCEPTS OF ELECTROKINETIC PHENOMENA

The word electrokinetic implies the joint effects of motion and electrical phenomena. We are interested in the electrokinetic phenomena that originate the motion of a liquid within a capillary tube and the migration of charged species within the liquid that surrounds them. In the first case, the electrokinetic phenomenon is called electroosmosis whereas the motion of charged species within the solution where they are dissolved is called electrophoresis. This section provides a brief illustration of the basic principles of these electrokinetic phenomena, based on text books on physical chemistry [7–9] and specialized articles and books [10–12] to which a reader interested to study in deep the mentioned theoretical aspects should refer to.

6.2.1 ELECTROOSMOSIS

Electroosmosis refers to the movement of the liquid adjacent to a charged surface, in contact with a polar liquid, under the influence of an electric field applied parallel to the solid–liquid interface. The bulk fluid of liquid originated by this electrokinetic process is termed electroosmotic flow. It may be produced either in open or in packed or in monolithic capillary columns, as well as in planar electrophoretic systems employing a variety of supports, such as paper or hydrophilic polymers. The origin of electroosmosis is the electrical double layer generated at the plane of share between the surface of either the planar support or the inner wall of the capillary tube and the surrounding solution, as a consequence of the uneven distribution of ions within the solid/liquid interface.

6.2.1.1 Electric Double Layer

Usually a solid surface acquires a superficial charge when it is brought into contact with a polar liquid. The acquired charge may result from one or a combination of the following mechanisms: dissociation of ionizable groups on the surface, adsorption of ions from solution, by virtue of unequal dissolution of oppositely charged ions of which the surface is composed. This superficial charge causes a variation in the distribution of ions near the solid–liquid interface. Ions of opposite charge (counterions) are attracted toward the surface whereas ions of the same charge (co-ions) are repulsed away from the surface. This, in combination with the mixing tendency of thermal motion, leads to the generation of an electric double layer formed of the charged surface and a neutralizing excess of counterions over co-ions distributed in a diffuse manner in the polar liquid. Part of counterions are firmly held in the region of the double layer closer to the surface (the compact or Stern layer) and are believed to be less hydrated than those in the diffuse region of the double layer where ions are distributed according to the influence of electrical forces and random thermal motion. A plane (the Stern plane), located at about one ion radius from the surface separates these two regions of the electric double layer.

Certain counterions may be held in the compact region of the double layer by forces additional to those of purely electrostatic origin, resulting in their adsorption in the Stern layer. Specifically adsorbed ions are attracted to the surface by electrostatic and/or van der Waals forces strongly enough to overcome the thermal agitation. Usually, the specific adsorption of counterions predominates over co-ions' adsorption.

The variation of the electric potential in the electric double layer with the distance from the charged surface is depicted in Figure 6.2. The potential at the surface (ψ_0) linearly decreases in the Stern layer to the value of the zeta potential (ζ). This is the electric potential at the plane of shear between the Stern layer (and that part of the double layer occupied by the molecules of solvent associated with the adsorbed ions) and the diffuse part of the double layer. The zeta potential decays exponentially from ζ to zero with the distance from the plane of shear between the Stern layer and the diffuse part of the double layer. The location of the plane of shear a small distance further out from the surface than the Stern plane renders the zeta potential marginally smaller in magnitude than the potential at the Stern plane (ψ_δ). However, in order to simplify the mathematical models describing the electric double layer, it is customary to assume the identity of (ψ_δ) and ζ. The bulk experimental evidence indicates that errors introduced through this approximation are usually small.

According to the Gouy–Chapman–Stern–Grahame (GCSG) model of the electric double layer [12], the surface density of the charge in the Stern layer is related to the adsorption of the counterions, which is described by a Langmuir-type adsorption model, modified by the incorporation of a Boltzmann factor. Considering only the adsorption of counterions, the surface charge density σ_S of the Stern layer is related to the ion concentration C in the bulk solution by the following equation:

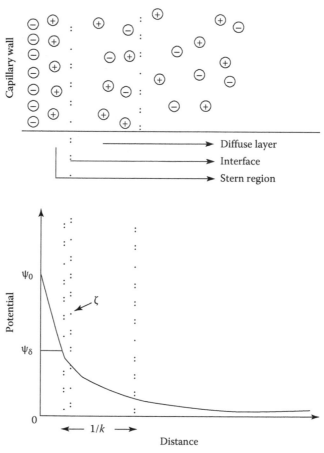

FIGURE 6.2   Graphical representation of the electric double layer at the solid–liquid interface within a capillary tube and diagram of the decay of the electric potential with distance from the capillary wall.

$$\sigma_S = \frac{zen_0 \dfrac{C}{V_m} \exp\left(\dfrac{ze\xi + \Phi}{kT}\right)}{1 + \dfrac{C}{V_m} \exp\left(\dfrac{ze\xi + \Phi}{kT}\right)} \tag{6.1}$$

where
  e is the elementary charge
  z is the valence of the ion
  k is the Boltzmann constant
  T is the temperature
  n_0 is the number of accessible sites
  V_m is the molar volume of the solvent
  Φ is the specific adsorption potential of counterions

The surface charge density of the diffuse part of the double layer is given by the Gouy–Chapman equation:

$$\sigma_G = (8\varepsilon kT c_0)\ \sinh\left(\frac{ze\xi}{2kT}\right) \tag{6.2}$$

where
  ε is the permittivity of the electrolyte solution
  c_0 is the bulk concentration of each ionic species in the electrolyte solution

At low potentials, Equation 6.2 reduces to

$$\sigma_G = \frac{\varepsilon\xi}{\kappa^{-1}} \tag{6.3}$$

where κ^{-1} is the reciprocal Debye–Huckel parameter, which is defined as the "thickness" of the electric double layer. This quantity has the dimension of length and is given by the following equation:

$$\kappa^{-1} = \left(\frac{\varepsilon kT}{2e^2 I}\right)^{1/2} \tag{6.4}$$

in which I is the ionic strength of the electrolyte solution.

Equation 6.3 is identical to the equation that relates the charge density, voltage difference, and distance of separation of a parallel-plate capacitor. This result indicates that a diffuse double layer at low potentials behaves like a parallel capacitor in which the separation distance between the plates is given by κ^{-1}. This explains why κ^{-1} is called the double layer thickness.

Equation 6.2 can be written in the form

$$\xi = \frac{\sigma_G \kappa^{-1}}{\varepsilon} \tag{6.5}$$

which indicates that the zeta potential can change due to variations in the density of the electric charge, in the permittivity of the electrolyte solution, and in the thickness of the electric double layer, which depends, according to Equation 6.4 on the ionic strength and consequently on the concentration and valence of the ions in solution. Sign and value of the zeta potential determine direction and velocity of the EOF, generated by applying an electric field.

6.2.1.2 Dependence of EOF on the Zeta Potential

The dependence of the velocity of the EOF (v_{eo}) on the zeta potential is expressed by the Helmholtz–von Smoluchowski equation [13]:

$$v_{eo} = -\frac{\varepsilon_0 \varepsilon_r \zeta}{\eta} E \qquad (6.6)$$

where
E is the applied electric field
ε_0 is the permittivity of vacuum
ε_r and η are the dielectric constant and the viscosity of the electrolyte solution, respectively

This expression assumes that the dielectric constant and viscosity of the electrolyte solution are the same in the electric double layer as in the bulk solution.

The Helmholtz–von Smoluchowski equation indicates that under constant composition of the electrolyte solution, the EOF depends on the magnitude of the zeta potential, which is determined by various factors influencing the formation of the electric double layer, discussed above. Each of these factors depends on several variables, such as pH, specific adsorption of ionic species in the compact region of the double layer, ionic strength, and temperature.

The specific adsorption of counterions at the interface between the surface and the electrolyte solution results in a drastic variation of the charge density in the Stern layer, which reduces the zeta potential and hence the EOF. If the charge density of the adsorbed counterions exceeds the charge density on the surface, the zeta potential changes sign and the direction of the EOF is reversed.

The ratio of the velocity of the EOF to the applied electric field, which expresses the velocity per unit field, is defined as electroosmotic coefficient or more properly, electroosmotic mobility (μ_{eo}) [13]:

$$\frac{v_{eo}}{E} = \mu_{eo} = -\frac{\varepsilon_0 \varepsilon_r \zeta}{\eta} \qquad (6.7)$$

Using the SI units, the velocity of the EOF is expressed in meters/second (m s^{-1}) and the electric field in volts/meter (V m^{-1}). Consequently, the electroosmotic mobility has the dimension of m^2 V^{-1} s^{-1}. Since electroosmotic and electrophoretic mobility are converse manifestations of the same underlying phenomena, the Helmholtz–von Smoluchowski equation applies to electroosmosis, as well as to electrophoresis (see below). In fact, it describes the motion of a solution in contact with a charged surface or the motion of ions relative to a solution, both under the action of an electric field, in the case of electroosmosis and electrophoresis, respectively.

6.2.1.3 Factors Influencing the EOF

According to Equation 6.6, the velocity of the EOF is directly proportional to the intensity of the applied electric field. However, in practice, nonlinear dependence of the EOF on the applied electric field is obtained as a result of Joule heat production, which causes the increase of the electrolyte temperature with consequent decrease of viscosity and variation of all other temperature-dependent parameters (protonic equilibrium, ion distribution in the double layer, etc.). The EOF can also be altered during a run by variations of the protonic concentration in the anodic and cathodic electrolyte solutions as a result of electrophoresis. This effect can be minimized by using electrolyte solutions with high buffering capacity, electrolyte reservoirs of relatively large volume, and by frequent replacement of the electrolyte in the electrode compartments with fresh solution.

Velocity and direction of the EOF also depend on the composition, pH, and ionic strength of the electrolyte solution. Both pH and ionic strength influence the protonic equilibrium of fixed charged groups on the surface and of ionogenic substances in the electrolyte solution, which affect the charge density in the electric double layer and consequently, the zeta potential. In addition, the ionic strength influences the thickness of the electric double layer (κ^{-1}). According to Equation 6.4, increasing the ionic strength causes a decrease in κ^{-1}, which is currently referred to as the compression of the double layer, with consequential reduction of the zeta potential. Hence, the practical effect in increasing the ionic strength is decreasing the EOF.

6.2.2 ELECTROPHORESIS

Section 6.2.1 has briefly examined the electrokinetic phenomena of electroosmosis, which refers to the motion of a liquid relative to a surface under the action of an electric field. This section examines the motion of ions in an applied electric field relative to the solution that surrounds them.

6.2.2.1 Electrophoretic Mobility

Under the action of an electric field, ions in solution migrate toward the electrode of opposite sign, i.e., positively charged ions migrate toward the cathode and negatively charged ions migrate toward the anode. The velocity v at which each ion migrates toward the electrode of opposite sign is proportional to the strength of the electric field E, which is expressed as the electric potential gradient in volts per unit length (in m) across the capillary tube:

$$v = \mu E \qquad (6.8)$$

The constant of proportionality in Equation 6.8 is called the electrophoretic mobility and expresses the velocity of the ion (in m s^{-1}) in the considered medium per unit electric field (in V m^{-1}):

$$\mu = \frac{v}{E} \qquad (6.9)$$

Therefore, the electrophoretic mobility is expressed in m^2 V^{-1} s^{-1}, and is a characteristic constant for any given couple "ion–medium."

The value of the electrophoretic mobility can be calculated considering the migration of an ion in an electrolyte solution at infinite dilution where no ionic interactions occur. Under the action of an electric field, the ion is accelerated by a force F_{el}, directed toward the appositively charged electrode, which is given by

$$F_{el} = qE \qquad (6.10)$$

where
   q is the electrical charge of the ion
   E is the electric field strength

This force is contrasted by an opposite force (F_v), due to the viscosity resistance of the solution, which increases as the ion velocity (v) increases:

$$F_v = fv \qquad (6.11)$$

The proportional constant f in Equation 6.11 is called the friction factor. When the two forces equal each other, the ion moves with a constant migration velocity, which is expressed by the following equation:

$$v = \frac{qE}{f} \tag{6.12}$$

Isolated ions can be assimilated to spherical particles for which, according to Stokes law, the friction factor is given by

$$f = 6\pi\eta r \tag{6.13}$$

where
η is the viscosity of the solution
r is the radius of the solvated ion, the so-called Stokes or hydrodynamic radius

Substitution of Equation 6.13 into Equation 6.12 yields

$$v = \frac{q}{6\pi\eta r} E = \mu^0 E \tag{6.14}$$

where μ^0 expresses the velocity of the ion per unit electric field.

6.2.2.2 Absolute and Effective Mobility

The constant of proportionality of Equation 6.14 is defined the absolute mobility or mobility at infinite dilution (μ^0):

$$\mu^0 = \frac{q}{6\pi\eta r} \tag{6.15}$$

Equation 6.15 is valid only for rigid, spherical ions. In addition, it originates from Equation 6.14, which is applicable only if the electric field at the ion is due to the applied electric field only, undisturbed by the effects of the other ions in solution. Consequently, Equation 6.15 ignores other forces originating in the counterion atmosphere, leaving the influence of the medium on the mobility unexplained.

In practice, electrophoresis is not performed at infinite dilution and other ions are present in the electrolyte solution. In addition, each ion is surrounded by oppositely charged counterions, forming the so-called ionic atmosphere, which tends to move in the opposite direction of the ion and, therefore, contrast the migration of the ion toward the electrode of opposite sign, lowering its electrophoretic mobility. Logically, the electrostatic interactions between ions of same sign and between each ion and its counterions are stronger with increasing electrolyte concentration with the result that the electrophoretic mobility at finite dilution is lower than that at infinite dilution.

In order to consider the influence of the ionic atmosphere on the electrophoretic mobility, the theoretical electrical charge of the ion q in Equation 6.14 is replaced by the smaller effective charge Q_{eff} and the hydrodynamic radius r by the effective radius R of the ion, which includes its ionic atmosphere:

$$\mu = \frac{Q_{eff}}{6\pi\eta R} \tag{6.16}$$

where μ is the effective mobility, which is always lower than the absolute mobility.

The effective mobility of ionogenic substances present in solution in different forms, which are in a fast dynamic equilibrium, is expressed as

$$\mu_A = \frac{1}{c_A} \sum_{i=1}^{n} C_i \mu_i = \sum_{i=1}^{n} x_i \mu_i \qquad (6.17)$$

where

μ_A is the effective mobility of the considered substance A

C_A is the analytical concentration

C_i, x_i, and μ_i, are concentration, molar fraction, and effective mobility of the component i, respectively

Under this condition, the ionogenic substance migrates as a single substance.

It has been pointed out above that electroosmotic and electrophoretic mobilities are converse manifestations of the same underlying phenomena; therefore the Helmholtz–von Smoluchowski equation based on the Debye–Huckel theory developed for electroosmosis applies to electrophoresis as well. In the case of electrophoresis, ξ is the potential at the plane of share between a single ion and its counterions and the surrounding solution.

6.3 INSTRUMENTATION FOR CAPILLARY ELECTROMIGRATION TECHNIQUES

6.3.1 SEPARATION UNIT

The instrumentation employed to perform capillary electromigration techniques is basically the same for all separations modes, either based on an electrophoretic or on a chromatographic separation mechanism. A schematic representation of a typical instrument for capillary electromigration techniques is reported in Figure 6.3. Commercially available instruments are equipped with a power supply allowing to apply up to 30 kV. Such limit depends on the possible corona discharges through the capillary and elsewhere in the instrument that may take place at values of applied voltage higher than 30 kV. Besides the applied voltage, the resulting driving current mainly depends on the composition of the electrolyte solution and is commonly maintained at values lower than 100–200 µA by regulating the applied voltage, which is usually set at a constant value. However, almost all commercially available instruments allow operation at constant current or constant power across the capillary, as well. Another desired feature of the power supply is the possibility to reverse the polarity, which allows to ground either electrode by a remote control.

The power supply is connected by a pair of isolated cables to the two electrodes that are immersed into the reservoirs containing the electrolyte solution and the ends of the capillary, either open

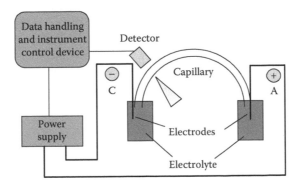

FIGURE 6.3 Scheme of a basic instrument for capillary electromigration separation techniques.

or containing a suitable chromatographic support (see Section 6.5), usually filled with the same electrolyte solution enclosed in the electrode compartments.

Tubular fused-silica capillaries, externally coated with a polyimide film to alleviate the fragility of this material, are generally employed. A small section of the polyimide coating is removed to generate a window in correspondence to the optical center of the detector to allow on-capillary detection. Capillaries made from organic synthetic material (e.g., Teflon or related materials, poly(etheretherketone) [PEEK], poly(vinylchloride), and polyethylene) have found limited applicability [14–16] due to the superior properties of fused-silica tubing, which include high thermal conductivity, high electrical resistance, good chemical inertness, good flexibility, high mechanical strength, and good UV-visible transparency in the zone where the polyimide external coating is removed. Also rarely employed are fused-silica capillaries of rectangular cross sections in spite of the significant improvements in absorbance detection sensitivity due to the flat walls that produce less optical distortion than circular capillary walls [17,18].

In most commercially available instruments, the capillary is mounted on a cartridge that is part of the cooling system devoted to maintain the capillary in an environment at constant temperature, either by circulating air or a cooling fluid through the cartridge, which also guarantees reliable alignment of the capillary detection windows in the optical center of the detector. An inert gas or air supply is used to rinse and fill the capillary with the electrolyte solution, to introduce the sample by hydrodynamic injection, and for the pressurization of packed and monolithic capillary column to prevent air bubble formation. In the case of rinsing and filling the capillary with solutions or introducing sample by hydrodynamic injection, only one electrolyte reservoir is pressurized, whereas the same pressure is applied on both reservoirs for the pressurization of capillaries filled or packed with a chromatographic support.

Other common sample injection techniques include siphoning and electrokinetic sampling, whereas the use of sampling devices, such as miniaturized sampling valve [19], rotary-type injectors [20], or electric sample splitter [21] have found very limited applications. Sample injection by siphoning is performed by introducing the appropriate end of the capillary into the sample vial and raising it up to a defined level above the electrolyte reservoir for a certain time. The resulting hydrostatic pressure generates a hydrodynamic flow that allows the introduction of a volume of the sample solution into the capillary proportional to the level and duration of raising the sample vial, pending on the fast and repeatable execution of all operations. In brief, the procedure comprises moving the capillary end from the electrolyte reservoir to the sample vial, raising the sample vial, returning it back to the reservoir level after the sampling time, and moving back the capillary end into the electrolyte reservoir. Therefore, the repeatability of these operations is low when they are done manually, as in the case of homemade apparatus. On the other hand, this injection mode allows introducing an aliquot of the sample representative of its composition.

Electrokinetic sampling is based on electrophoretic migration of charged components of sample and introduction of the sample solution into the capillary by the EOF. The main drawback of this sampling technique is the selective introduction of sample, which depends on the polarity of applied voltage, electrophoretic mobility of each sample component, and velocity of the EOF. For example, injecting at the anodic end in presence of cathodic EOF implies that positively charged analytes are selectively introduced into the capillary at a greater extent than negatively charged ones, which are introduced into the capillary only if their electrophoretic mobility is lower than the EOF, otherwise they do not enter the capillary at all. In such a case, the analytes introduced into the capillary tube are not representative of the sample composition.

In commercially available instruments, all vials containing either the sample solution or the solutions employed for separation and for rinsing the capillary between runs are held in an autosampler, which in most instruments can be thermostated at a desired temperature. The autosampler can also be programmed to carry out the collection of separated fractions for micropreparative applications. The operations performed by the autosampler and by the other equipments of the separation unit are controlled by suitable software loaded on a personal computer, also performing data acquisition and processing.

6.3.2 DETECTION

Typically, sample detection in electromigration techniques is performed by on-column detection, employing a small part of the capillary as the detection cell where a property of either the analyte, such as UV absorbance, or the solution, such as refractive index or conductivity, is monitored. This section briefly describes the major detection modalities employed in capillary electromigration techniques, which are accomplished using UV-visible absorbance, fluorescence spectroscopy, and electrochemical systems. The hyphenation of capillary electromigration techniques with spectroscopic techniques employed for identification and structural elucidation of the separated compounds is also described.

6.3.2.1 Absorbance

As in HPLC, the primarily used detector is based on the absorbance of UV or visible light, due to the great number of substances that absorb these radiations. The photometric on-column detection is performed through a "window" obtained by removing a small portion of the polyimide external coating at one end of the capillary tube. The UV or visible light is focused though the central bore (lumen) of the capillary tube by either fiber optics or high-quality lens, and is then detected by a suitable photosensor. The high transparency of fused silica allows detection at wavelength as low as 190 nm. The most common photometric detectors employed in electromigration techniques comprise fixed wavelength, variable wavelength, scanning monochromator, and photodiode array detectors. Similarly to the detectors used in HPLC, low pressure discharge lamps are employed as the sources of intensive line UV radiations, such as mercury (254 nm) or cadmium (229, 326 nm), whereas deuterium lamps, covering the range 190–700 nm, are employed in variable-wavelength and photodiode array detectors, where a tungsten–halogen lamp may be also employed to improve the performance in the visible region.

The main drawback of photometric on-column detection is the short distance the light travels through the capillary (the optical path length), which corresponds to the internal diameter of the capillary tube. This is because, according to the Beer's law, the sensitivity in photometric detection is proportional to the optical path length, which in the on-column detection is in the range of 20–100 µm, compared to the usual 1.0 cm path length of standard HPLC detectors. Moreover, the incident radiation enters the detection zone through a curve surface, which leads to light intensity loss. Even with detectors having noise levels as low as 10^{-5} absorption units (au), concentration detection limits are rarely lower than 10^{-6} M. However, since the injected sample volume is in the range of nanoliters (nL), for a 10.0 nL volume of injected sample at 10^{-6} M concentration, the corresponding injected sample mass is in the range of 10^{-14} mol. Therefore, at a relatively poor concentration, detection limit of 10^{-6} M corresponds an excellent mass detection limit of 10^{-14} mol.

Improvements in photometric detection obtained by increasing the optical path length include the use of capillaries having bubble shape at the detection zone [22], which is fabricated directly into the fused-silica tube, capillaries bent to a Z-shape configuration [23] or with modified surface to realize a multireflective absorption cell at the detection region of the capillary tube [24] (see Figure 6.4). Increasing the inner diameter of the capillary tube only at the detection window ("bubble" cell) offers the sensitivity of a wide inner diameter capillary and the low-current generation of a narrow one. The improvement in sensitivity is 3–5-fold over a capillary of same inner diameter, whereas resolution and peak shape are practically not affected. According to Beer's law, the sensitivity gains in a Z-shaped capillary of 50 µm inner diameter can exceed a factor 100 for an optical path length of 5 mm. However, the gain in sensitivity is at expenses of resolution that is lost for adjacent peaks. Resolution is also sacrificed using the multireflective absorption cell, which is made directly onto the capillary by silver coating a short portion of the capillary at the detection

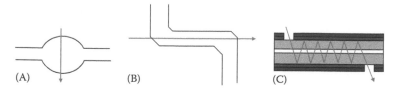

FIGURE 6.4 Schematic representation of capillaries with (A) bubble shape, (B) Z-shape, and (C) multireflective absorption detection zone configuration. The arrow indicates how the light beam travels through the capillary at the detection zone.

region in order to enlarge the volume of solution probed by the incident light, whose entrance is 1.5 mm apart from the exit port made in correspondence to the photosensor. Detection sensitivity improvement of 40 fold has been reported using laser illumination and multireflective absorption cell, in comparison to direct on-column detection [24].

A high-sensitivity detection cell that can be fully decoupled from the separation capillary, either open or packed has been introduced by Agilent [25]. The cell is constructed from silica parts fused together and is flanked by flat, clear windows in order to minimize stray light. The optical path length is 1.2 mm and a 20-fold increase in detection sensitivity is reported, however, resolution is lost when the interpeak volume is lower than the cell volume (1.2 nL).

Furthermore, as it has been mentioned above, the effects of limited optical path length can be alleviated using capillaries of rectangular cross sections and performing the detection through the long axis of the capillary with the additional advantage that the flat walls of rectangular capillaries produce less optical distortion and scatter compared to the walls of circular capillaries. However, higher cost and minor availability of rectangular capillaries over the circular ones have limited their use.

Besides the short optical path length, further sources of poor absorbance detection are the high background noise level due to low light intensity and challenging optical coupling resulting from the small size of the detection window, which may be improved using optical fibers to focus the incident radiation and to collect the light passed through the capillary. Other attempts to improve absorption detection include the use of light-emitting diodes (LEDs) as alternative light sources. LEDs are particularly attractive in absorbance detection due to their excellent output stability, reasonable monochromaticity, and the option of emission at a variety of wavelengths, covering UV, visible, and near-infrared region (280–1300 nm) [26]. The improvement in sensitivity in using a LED is primarily due to its low noise compared to traditional UV-visible light sources, its high intensity at nearly monochromatic output, and small size, which make easy the efficient coupling between light source, capillary, and photodiode detectors [27,28]. The reduced size of LEDs, their exceptional performance, long lifetime, and low cost have greatly contributed to increase the recent attention on these light sources [29,30], which are cool emitters and can be used to realize miniature absorbance detection systems for portable electrophoretic instrumentation [31]. Up to 10 time detection sensitivity improvement have been reported in absorbance detection with LEDs light sources, in comparison to traditional UV-visible light emitters [32]. However, also with LED light sources, the limits of detection are still in the range of 10^{-5} to 10^{-6} M [26].

6.3.2.2 Fluorescence and Laser-Induced Fluorescence

Fluorescence detection is widely employed in electromigration techniques for samples that naturally fluoresce or are chemically modified to produce molecules containing a fluorescent tag. Indirect detection incorporating a fluorescent probe into the electrolyte solution is also employed. One of the most common fluorophores used for this purpose is fluorescein, which is a water soluble, stable, and relatively cheap compound.

In a typical fluorescence detector, the excitation radiation emitted by the high-energy source passes through a suitable filter or monochromator and then is focused though the lumen of the detection window, which is made at the detection end of the capillary. The emitted fluorescence radiation is collected orthogonally to the excitation beam and passes a slit, to eliminate the scattered excitation radiation, and a second filter or monochromator, to select the emission wavelength, prior to be focused on the photosensor, which usually is either a photomultiplier or a photodiode. Typically, a mercury–xenon arc lamp is employed as the emitter of the excitation radiation.

The signal measured by the photosensor expresses the intensity of the emitted fluorescence (I_F), which depends on the intensity of the incident light (I_0) and sample concentration c, among other factors, such as the fluorescence (or quantum) yield (Φ), the molar absorption coefficient (ε), and the optical path length (d). At low concentration, the intensity of emitted fluorescence is given by the following equation [33]:

$$I_F = I_0 \varphi \, (2.3025 \varepsilon dc) \tag{6.18}$$

which indicates that at otherwise constant conditions the intensity of the emitted fluorescence directly augments with increasing the intensity of the incident light. Therefore, high-intensity monochromatic beams as lasers are preferred excitation sources due to their brightness and spatial beam properties, which allow the excitation energy to be efficiently focused in a very small area, such as the lumen of the on-column detection windows, whereas the high monochromaticity reduces stray light levels. These excitation light emitters are largely employed to realize LIF detectors, which usually employ microscope objectives or optic fibers to collect the fluorescence signal at the exit port of the detection window, as it has been firstly reported by Gassmann et al. [34].

LIF is to date the most sensitive commercially available detection system developed for electromigration techniques, having concentration detection limits ranging between 10^{-11} and 10^{-13} M and mass detection limits in the order of few molecules with ultrasensitive LIF detectors [35]. However, also the LIF detectors may suffer from various drawbacks, which include specular and diffuse scatter of the laser beam from the capillary, Raman scatter from the solvent, and luminescence from the optics and the background electrolyte solution. Several detector designs have been developed to minimize background radiation and scattering [36], including axial-beam illumination [37], two-beam line confocal detection geometry [38], noncircular cross-section capillary columns [39], and sheath-flow detection cells [40], which alleviate background drawbacks by focusing the sample into a narrow flow stream at the exit of the capillary column.

Among these, LIF with the use of a sheath-flow detection cell is the detector of choice for ultrasensitive detection in capillary electromigration techniques. A typical sheath-flow cuvette consists of a quartz flow cell with planar surfaces in which the capillary detection end is inserted. The sample stream exiting the capillary column is concentrically joined with a laminar flow of a sheath liquid of same composition of the solution employed for the separation in the capillary column, which flushes throughout the detection flow cell by a pump. The result is the formation of a stable narrow sample stream directed to the center of the flow cell where the excitation laser beam is focused using a microscope objective. Because sample and sheath solution have the same composition, no variations in refractive index and light scatter or reflection occur at their interface with the result of greatly reducing the background signal and improving detection sensitivity, which may lead to detection limits as low as one single molecule [41].

Gas lasers with powers ranging from a few to more than 10 W are the most commonly employed sources of the excitation beam in LIF detection. These include He–Cd laser (325 and 442 nm), He–Ne laser (543.6, 592.6, and 633 nm), KrF excimer laser (248 nm), Nd–YAG (yttrium–aluminum–garnet) laser (266 nm), and Ar+ ion laser (usually 488 and/or 514 nm) [35]. LEDs are very attractive alternatives to lasers as the excitation radiation source in fluorescence detection, due to their low cost, long lifetime, small size, high stability, and availability in a wide wavelength range [26]. Since their first use in capillary electrophoresis, proposed by Bruno et al. [42], LED-induced fluorescent (LEDIF) detection has started to be used with increasing frequency [43–45]. However, the divergent beam of the light emitted by LEDs require more complicated optical configuration than that employed with lasers, which are also superior in terms of monochromaticity. Therefore, using LEDs as the light source may necessitate to spectrally filter not only the fluorescence emitted beam but also the excitation light [46].

A more sophisticated mode of LIF detection is the multiphoton-excitation (MPE) fluorescence [47], which is based on the simultaneous absorption of more than one photon in the same quantum event and uses special lasers, such as femtosecond mode-locked laser [48] or continuous wave laser [49]. This mode of LIF detection allows mass detection limits at zeptomole level (1 zeptomole = 10^{-21} mol) due to exceptionally low detection background and extremely small detection volume, whereas detection sensitivity in concentration is comparable to that of traditional LIF detection modes. A further drawback is the poor suitability of MPE-fluorescence detection to the on-column detection configuration, which is frequently employed in conventional LIF detection.

6.3.2.3 Electrochemical

Electrochemical detection is based on the measurement of electrical properties of the solution transporting the analyte throughout the capillary column. The detection signal may originate either from a cumulative property of the solution, determined by its overall composition, such as conductivity, high-frequency impedance, permittivity (measurement of bulk property), or from a specific property of the solution related to the activity or concentration of the analyte of interest, such as the current resulting from the oxidation or reduction of an electroactive species (measurement of specific property). In principle, each of the main modes of electrochemical detection, which are based on conductometric, amperometric, voltametric, and potentiometric measurements, is applicable to capillary electromigration techniques [50]. However, technical limitations occurring in positioning the electrodes and in the isolation of the detection system from the high-voltage drop across the separation capillary are a challenge in adopting electrochemical detection in capillary electromigration techniques.

Conductivity detection is a universal detection mode in which the conductivity between two inert electrodes comprising the detector cell is measured. The different arrangements employed for the construction of these detectors include apparatus with a galvanic contact of the solution with the sensing electrodes (contact conductivity detection) [51] and detection systems without galvanic contact of the solution with the sensing electrodes (contactless conductivity detection) [1].

Contact conductivity detection can be performed in either on-column mode, realizing a direct or a solution-mediated contact of the analytes with the electrodes, or in end-column mode [1]. In the on-column mode, the sensing electrodes can be two platinum wires that are inserted into two holes made through the wall of the capillary tube, aligned with the help of a microscope and then sealed, as reported in one of the first papers describing the construction of an on-column conductivity detector [52]. The variations in the conductivity within the detection zone are measured by applying an alternating voltage between the sensing electrodes, in order to assure that the measurements are not affected by Faradaic reaction at the electrodes but solely by the concentration and mobility of the charged species in the solution.

In the end-column conductivity detection mode a sensing electrode is situated at the outlet of the separation capillary column and the conductivity measurement is made between this electrode and the ground electrode of the electromigration system used as the second sensing electrode. Advantages of the end-column detection in comparison to on-column mode include easier arrangement and minor difficulties in positioning the sensing electrode whereas detection sensitivity obtained with the two configurations is comparable (10^{-6} to 10^{-7} M) [53]. However, the dead volume between the separation column outlet and the sensing electrode is a source of band broadening that may greatly reduce efficiency and resolution.

Contactless conductivity detection mode, based on an alternating voltage capacitively coupled into the detection cell, is the practical and robust arrangement nowadays employed in commercially available detectors that has been independently developed in 1998 by Zemann et al. [54] and by Freacassi da Silva and do Lago [55]. This detection mode is based on two tubular electrodes,

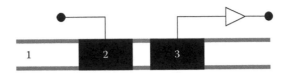

FIGURE 6.5 Schematic representation of contactless conductivity detection cell. (1) Capillary, (2) actuator electrode, and (3) pickup electrode.

consisting of either short metallic tubes [54] or a silver varnish [55], that are placed side by side around the separation capillary tube at a distance of few millimeters, which define the detection volume. Each electrode forms a capacitor with the electrolyte solution contained into the separation capillary whereas the electrolyte solution contained into the capillary forms a resistor. An alternating voltage is applied to one of the galvanically isolated electrodes and the resulting alternating current, which is affected by variations of conductivity into the detection volume, is measured at the second electrode. A schematic representation of a capacitively coupled contactless conductivity detector employing the axial arrangement of two tubular electrodes is reported in Figure 6.5.

Other electrode configurations, such as the radial arrangement consisting of four thin wires placed perpendicularly around the circumference of the separation capillary column, have found less application due to more complicated construction and restriction in space and diameter of the separation capillary [56]. Due to its low cost, robustness, minimal maintenance demands, possibility to be freely moved along the capillary [57], or combined with either UV-absorbance [58] or fluorescence [59] detection, the capacitively coupled contactless conductivity detector has recently gained wide acceptance not only for the determination of inorganic ions but also for biomolecules and organic ions, as it has been recently comprehensively reviewed by Kubáň and Hauser [1].

The amperometric detection modes are suited for monitoring analytes that can be either oxidized or reduced at an electrode surface (the sensing electrode) that is under the influence of an applied direct voltage, which is referred to as the working electrode. The potentiostatic circuitry is completed by the reference and counter electrodes, which, together with the working electrode, comprise the detection cell. Sample detection is carried out by controlling the potential applied to the working electrode, which is usually held at a constant value, and measuring the current resulting from the analyte oxidation or reduction at this potential. The current produced by either the oxidation or the reduction reaction is directly related to the analyte concentration.

Carbon is typically used to construct electrodes and alternative materials include transition or noble metals, such as copper, platinum, and gold. These electrodes are prone to fouling as the oxidation products of the analytes accumulate on the electrode surface. Such problems can be solved by using a pulsed amperometric detector (PAD) [60], in which the applied potential is accompanied by pulsed steps producing the electrode surface activation and electrode cleaning, or an integrated pulsed amperometric detector (IPAD) [61], which incorporates a scanning voltammetry in order to use a potential scan in the detection step along with the pulsed steps for surface activation and electrode cleaning.

The main challenge of coupling capillary electromigration techniques and amperometric detection is the performance of electrochemical measurements in the presence of a high-voltage electrical field. A possible approach to minimizing the interferences of the high-voltage electrical field with the detection circuit consists in making a fracture in the capillary near the end through which the electrophoretic current could be grounded. The fracture divides the capillary into two sections: a "separation capillary" before the fracture and a "detection capillary" beyond it. These two sections are maintained together by a fracture decoupler that in the first configuration described by Wallingford and Ewing was a porous glass capillary forming a joint for the two capillary pieces [62].

A variety of alternative materials can be employed in place of porous glass, including Nefion [63], which is the material of most frequent choice employed today to construct decoupler devices [2]. This arrangement is referred to as the off-column amperometric detection mode [64].

The fabrication and handling of decouplers are quite difficult and require time and manipulative skill. A more straightforward alternative arrangement that avoids the use of a decoupler device is the end-column detection mode, which was first suggested in 1991 by Huang et al. [65] and is widely employed nowadays [66]. In this detection mode, the working electrode is placed near to the outlet of the separation capillary tube and the voltage applied for separation is grounded in the detection reservoir *via* ground electrode. Separation columns of 25 μm internal diameter or smaller are usually employed in order to increase the electrical resistance within the capillary used for separation, with the consequence that the current associated with the electromigration separation system proportionally decreases and the electrical field falls more rapidly at the capillary exit. Typical distance of the working electrode from the outlet of the capillary column is in the order of tens of micrometers and the axial alignment between the capillary exit and the working electrode, which is crucial to achieve optimum detection sensitivity, is carried out by the help of an x,y,z-micropositioner and a microscope. Alignment by this process is quite complex and time consuming and, therefore, a variety of alternative approaches for positioning the working electrode have been developed, such as the use of integrated guiding systems [67] and capillary-electrode holders [68].

6.3.2.4 Other Detection Modes and Hyphenated Techniques

The three main detection techniques described above can also be applied in "indirect" detection mode, which implies to monitor a characteristic property of the background electrolyte solution, such as UV absorbance, fluorescence, or electrochemical property that is altered in the zone of the electrolyte solution containing the separated analytes. This detection mode is typically realized by incorporating into the electrolyte solution a ionic additive that can be easily revealed at low concentration by the selected detection mode, thus giving a highly background detection signal. For electroneutrality reasons, the migration of charged analytes within the electrolyte solution causes the displacement of the ionic additive from the zone where they are present, with consequent production of a negative peak in the background signal, which reveals their presence. Sensitivity is generally lower than that of the corresponding direct detection mode and has to be optimized by minimizing the concentration of the additive, whereas the ratio of background signal to background noise and the number of molecules of the ionic additive that are displayed by a molecule of the analyte (transfer ratio) should be as large as possible. An additional drawback of indirect detection is the linearity of detection response, which has generally a narrower range of that of the corresponding direct detection mode.

Other detection modes employed in capillary electromigration techniques include chemiluminescence [69–71], Raman spectroscopy [72,73], refractive index [74,75], photothermal absorbance [76,77], and radioisotope detection [78]. Some of these detection modes have found limited use due to their high specificity, which restricts the area of application and the analytes that can be detected, such as radioisotope and Raman-based detection that are specific for radionuclides and polarizable molecules, respectively. On the other hand, the limited use of more universal detection modes, such as refractive index, is either due to the complexity of coupling them to capillary electromigration techniques or to the possibility of detecting the analytes of interest with comparable sensitivity by one of the less problematic detection modes described above.

The hyphenation of capillary electromigration techniques to spectroscopic techniques which, besides the identification, allow the elucidation of the chemical structure of the separated analytes, such as mass spectrometry (MS) and nuclear magnetic resonance spectroscopy (NMR) has been widely pursued in recent years. Such approaches, combining the separation efficiency of capillary electromigration techniques and the information-rich detection capability of either MS or NMR, are emerging as essential diagnostic tools for the analysis of both low molecular weight and macromolecular compounds.

Mass spectrometry provides detailed information regarding molecular weights and structures from extremely small quantities of materials. Several types of ionization sources can be employed for the on-line hyphenation of capillary electromigration techniques with MS, which include electrospray ionization (ESI), atmospheric pressure chemical ionization (APCI), atmospheric pressure photochemical ionization (APPI), MALDI, inductively coupled plasma (ICP), and fast atom bombardment (FAB) [6,79–81]. At present, the most common approaches are based upon the atmospheric pressure ionization (API) interfaces APCI and ESI, whereas the earliest arrangements based on FAB interfaces, which are effected by several drawbacks, such as unstable electrical current, have been progressively abandoned. A major advantage of ESI is the softness of the ionization process by which ions in solution, also of high molecular mass such as proteins, can be transferred to the gas phase [82]. This characteristic results in natural compatibility with both biological samples and liquid-phase separation methods such as capillary electromigration techniques.

Another distinguished characteristic of ESI is the generation of a series of multiply charged intact ions for high molecular mass substances such as proteins. These ions are represented in the mass spectrum as a sequence of peaks, the ion of each peak differing by one charge from those of adjacent neighbors in the sequence. The molecular mass is obtained by computation of the measured mass-to-charge ratios for the multiple charged ions using a "deconvolution algorithm" that transforms the multiplicity of mass-to-charge ratio signals into one single peak on a real mass scale [83]. Obtaining multiple charged ions is advantageous as it allows the analysis of proteins up to 100–150 kDa using mass spectrometers with upper mass limit of 1500–4000 amu.

Several approaches are employed to connect the outlet of the separation capillary column to the MS ion source providing suitable arrangements to apply voltage to the capillary outlet and to obtain the flow rate adjustment between the separation capillary column (in the range of nL min^{-1}) and the MS ion source (in the range of μL min^{-1}), which is usually performed through an additional liquid (sheath liquid). Basically, three major types of interfaces are employed for coupling capillary electromigration systems with MS, which are identified as coaxial sheath-flow [84], sheathless [85], and liquid-junction interfaces [86,87].

In the sheathless interface, the electrical contact is obtained by coating with either a metal [85, 88–90] or a conductive polymer [91] the separation capillary outlet, which is shaped as sharp tip. Also employed are sheathless interfaces in which the electrical contact is established using a metal electrode or a conductive wire inserted into the outlet of the separation capillary [92]. A small gap between the separation capillary and the needle of the ionization device filled by a liquid is the approach made to establish the electrical contact in the liquid junction interface [86,87]. This arrangement is also realized by making porous through chemical etching the tip [93] or a small section of the wall [94] of the separation capillary at its outlet.

The most commonly used arrangement is the coaxial sheath interface, consisting of one or two concentric tubes surrounding the separation capillary through which are flushing the sheath liquid and the nebulizing gas [95,96]. Main advantages of this approach include the possibility of employing for the separation method solutions having limited compatibility with the ionization process and the formation of an electrical contact between the separation capillary and the electrode.

Capillary electromigration technique can also be successfully coupled to MALDI by several arrangements, either off-line [97] or on-line [6]. The off-line coupling is usually realized by collecting the effluent from the capillary column drop wise or by sample deposition from the outlet end of the separation capillary onto the MALDI target, either in the form of spots [98] or as a continuous steak [99], using either appropriated designed tips or sheath-flow-assisted deposition devices [100–103]. The matrix for MALDI is either deposited on the target plate, or mixed with the affluent from the column after the separation or added after sample deposition.

The on-line coupling of CE to MALDI reduces the sample handling steps and, therefore, is expected to be more effective for ultra-trace analysis. Several arrangements have been proposed for such approach including the direct introduction of the end of the separation capillary into the vacuum region of a time-of-flight (TOF) mass spectrometer and desorption of the separated samples by a laser light irradiating the capillary end [104]. A variety of vacuum deposition interfaces has been proposed too, which use either a rotating quartz wheel [105] or a disposable moving tape

[106,107], located in the vacuum of the mass spectrometer, onto which the effluent from the separation capillary is deposited after having been mixed with the matrix solution in a liquid junction located between the above separation capillary and the capillary tube used to deposit the sample–matrix mixture. Other proposed approaches employ a rotating ball interface with an open configuration that allows online sample and matrix contact deposition from the capillaries at atmospheric pressure [108,109].

The on-line approach shows high efficiency and sensitivity. However, off-line cleaning or replacement requirements of the interface result in limited operational time. On the other hand, problems have been reported with the off-line approach too. The essentially arbitrary collection of fractions (spots) clearly offsets the high resolution of the separation. Other problems may arise from the incomplete deposition of the analyte in the form of drops onto the target, which in turn results in a substantial fraction of the analyte being present in the next deposited drop.

Also promising is the hyphenation of capillary electromigration techniques with nuclear magnetic resonance spectroscopy (NMR), which is widely employed for identification and structural elucidation of organic compounds unattainable by other analytical methods. Direct on-line hyphenation can be realized by constructing a solenoidal geometry radio frequency (RF) NMR transmit/receive coil by wrapping a copper or gold wire around the fused-silica capillary at the end of the separation column, which is then located inside the magnet bore of the NMR instrument in position to ensure that the detection zone of the capillary and coil are perpendicular to the static magnetic field, as first reported by Sweedler and coworkers [110,111]. A saddle-type (Helmholtz) RF coil can also be used as the NMR probe [112]. The RF coil is the probe having the function of delivering the radio frequency energy needed to excite NMR active nuclei and of collecting signals from the sample. To gain the maximum NMR signal, a solenoidal RF coil must be aligned perpendicular to the static magnetic field whereas a saddle-type coil must be positioned with its axis parallel to the static magnetic field. Other specific requirements are limited to keeping magnetic objects far away from the magnet, usually several meters depending on the magnet field strength, except with shielded magnets, whereas nonmagnetic plastic electrolyte reservoirs and platinum electrodes can even been located within the magnet bore. Otherwise, the instrumentation employed to perform capillary electromigration techniques can be on-line hyphenated with the NMR spectrometer without major modifications.

Because capillary electromigration techniques employ limited volumes of solvents, their hyphenation with NMR makes economically practicable the use of deuterated solvents, avoiding the elaborate signal suppression techniques requested with the use of protonated solvents that lead to distortion of parts of the NMR spectra. The interpretation of NMR spectra can also be complicated by changes in the chemical shift caused by variations of temperature due to joule heating. Therefore, capillaries of smaller internal diameter are preferred for their higher capability at dissipating heat due to the greater surface-to-area ratio than larger capillaries. On the other hand, because of the low sensitivity of NMR as compared to other detection techniques, ranging between 10^{-9} and 10^{-11} mol [5], capillary of a relatively large internal diameter would be preferred. In practice, capillaries of 75–200 μm internal diameters are generally employed when electromigration separation techniques are hyphenated with NMR.

Using solenoidal RF coils the volume of detection is determined by the diameter of the capillary and the number of turns of the RF coil (determining the length of the probe), whereas the acquisition time of the NMR signal depends on the velocity at which each analyte passes the detection zone. Therefore, with capillary of 75 mm I.D. and solenoidal RF coil of 1 mm length the NMR detection cell volume is about 5 nL and the residence time of the analytes inside the detection cell is usually less than 60 s, which may adversely limit the number of possible NMR acquisitions and therefore the detection sensitivity. Bigger detection cells can be realized by enlarging the inner diameter of the zone of the fused-silica capillary at the end of the separation column where the RF coil is located [112]. Such arrangement results in increasing both detection volume and residence times at expenses of resolution of closely migrating/eluting analytes. Longer residence times have the effect of enabling prolonged NMR acquisition times and larger detection volumes, resulting in increased mass of the analyte exposed to the mass-sensitive RF coil with the consequence of enhancing sensitivity. However, the enlargement of the inner diameter of the capillary tube brings to a reduction of its thickness at the detection zone with the consequence that the coil comes into closer contact with the sample and magnetic susceptibility effects may occurs, resulting in degraded line shapes and loss of sensitivity. Moreover, a decrease in sensitivity is observed if the volume of the detection cell is larger than the volume occupied by the separated analyte in the capillary separation column due to an increase of the signal-to-noise ratio.

For a given diameter, solenoidal RF coils are 2–3 times more sensitive than saddle-type RF coils [113], which are more difficult to construct than RF coils with solenoidal geometry, especially at small dimensions. Another significant difference between the two RF coils is related to the influence of the magnetic field induced by the current resulting from the electric field applied across the ends of the separation capillary on the acquisition of the NMR spectra. With saddle-type RF coils, the separation capillary is aligned parallel to the NMR static magnetic field, which is practically not affected by the magnetic field induced by the current resulting from the electric field applied across the ends of the separation capillary, perpendicular to the current flow direction and, therefore, to the static magnetic field. On the other hand, using a solenoidal RF coil, positioned perpendicular to the static magnetic field in order to gain the maximum NMR signal, the current associated to the electromigration separation system induces a magnetic gradient along the static magnetic field, which causes the broadening of the NMR signals, resulting in a decrease of sensitivity and a loss of scalar coupling information that are hard to avoid using the shimming procedures generally employed to restore magnetic field homogeneity. Also of limited help are NMR post processing procedures appositively developed to extract information from NMR spectra distorted by inhomogeneous magnetic field resulting from the magnetic field induced by the current associated to the electromigration separation system [114].

The simplest approach to circumvent the detrimental effect of the magnetic field induced by the electromigration system consists in performing the acquisition of the NMR spectra during periodic interruptions of the applied electric field (stopped-flow method), which have also the beneficial effect of increasing the acquisition times and eliminating the observed degradation of NMR signals acquired with sample movement through the NMR probe [115]. Such procedure is also employed to perform signal acquisitions for two-dimensional NMR investigations [116]. However, the periodic interruption of the applied electric field may adversely affect the separation efficiency and reduce resolution due to the diffusion of the analytes over the stopped-flow periods, in addition to increasing the analysis time. These drawbacks can be avoided by splitting the flow from the separation capillary into two outlets, consisting of two capillaries each holding a RF coil, and alternating electromigration flow and NMR measurements between the two outlets in order to perform continuous electromigration separations with stopped-flow NMR signal acquisition [117].

Alternatively, the electric field can be solely applied along the portion of the capillary column, located outside the magnet bore of the NMR spectrometer, where the separation takes place and not across the whole capillary system. Such splitless arrangement can be realized by inserting between the separation and detection portion of the capillary a short stainless steel capillary tube, by using two zero dead volume unions, which is then electrically connected to the high-voltage power supply to close the electrical circuit of the electromigration separation system, excluding the capillary section holding the RF coil [118]. Besides its role for the identification and structural elucidation of compounds in complex mixtures, the hyphenation of capillary electromigration techniques with NMR is a powerful tool to monitor separation parameters [115] and to elucidate dynamic processes occurring during the separation, such as chiral recognition mechanisms [119] and Joule heating effects [120].

6.4 FACTORS INFLUENCING PERFORMANCE AND SEPARATION PARAMETERS

Besides electroosmosis, other phenomena that may affect the performance of any electromigration technique are the heat generated in the column by Joule effect, the untoward interactions of the analytes with the inner surface of fused-silica capillary tubes, sample diffusion, and electromigration dispersion. This section briefly describes the first two phenomena in Sections 6.4.1 and 6.4.2, respectively, whereas the motion of the analytes within the electrolyte solution due to chaotic Brownian movement (diffusion) and the variations in conductivity and electric field strength occurring in the sample zone during electrophoresis (electromigration dispersion) are discussed in Section 6.4.3 as factors contributing to the band broadening of charged analytes migrating under the influence of the applied electric field. Ssection 6.4.3 describes the separation parameters common to all electromigration techniques having electrophoresis as the dominant separation mechanism, whereas parameters related to a specific separation mode are introduced and discussed in each corresponding subsection.

6.4.1 JOULE HEATING

A capillary tube filled with an electrolyte solution behaves as a cylindrical electric conductor when an electric field is applied across its ends. Therefore, the passage of an electric current along the capillary generates Joule heat and the temperature of the electrolyte solution inside the capillary tends to increase if the produced heat is not efficiently dissipated through the capillary wall. In an open capillary tube of length L and cross-section A, filled with an electrolyte solution of conductivity σ, the amount of heat Q generated as a function of time depends on the applied electrical potential V and the current i passing through the capillary of resistance R, according to the following equation:

$$Q = i^2 Rt = \frac{V^2 t}{R} = V^2 \lambda t = \frac{V^2 \sigma A t}{L} \tag{6.19}$$

where λ is the conductance of the capillary that, like resistance R, depends upon the capillary dimensions as follows:

$$\lambda = \frac{\sigma A}{L} \tag{6.20}$$

The Joule heat generation in a packed capillary column is given by the following expression, similar to Equation 6.19:

$$Q = \frac{V^2 \sigma_\varepsilon A \varepsilon t}{L} \tag{6.21}$$

where ε is the column porosity, which for an open capillary is unity, and accounts for the reduced cross section in capillary tubes either packed or filled with a stationary phase, whereas σ_ε is an empirical equivalent conductivity parameter that accounts for the influence of the geometrical characteristics of the packed/filled capillary on conductivity and can be calculated from the measurements of current as a function of applied voltage. Therefore, as it can be verified experimentally, Joule heat increases with increasing applied voltage, capillary diameter, and conductivity of the electrolyte solution.

Heat dissipation takes place mainly by conduction through the capillary wall and by convection, radiation, and conduction in the medium surrounding the capillary tube, which can be either air or a cooling fluid, depending on the system employed to control the temperature of the capillary column. In the presence of low Joule heating and efficient heat dissipation, the current resulting from the applied electric field is expected to vary linearly with the applied potential drop, according to Ohm's law. Therefore, the experimental verification of linear increase of the electric current with increasing applied electric potential across the capillary tube is indicative of efficient heat dissipation through the capillary wall.

The generation of Joule heat is expected to produce temperature gradients inside the capillary tube that may affect EOF, migration velocity, band broadening of the analytes, and separation efficiency. As the temperature of the electrolyte solution increases, its viscosity and electrical permittivity decreases, whereas longitudinal diffusion of the analyte and the zeta potential at the solid/liquid interface increase. Therefore, poor repeatability of EOF and migration behavior of analytes may occur, as well as peak broadening of charged analytes due to the different electrophoretic mobility at the center and near the wall of the capillary owing to radial temperature differences inside the capillary tube. Further, peak broadening may occur as a consequence of increased longitudinal diffusion of the analytes.

The effects of Joule heating can be controlled by efficient heat removal using a thermostabilized heat exchanger in contact with the capillary tube and by the appropriate selection of the operational parameters, such as capillary size, composition of the electrolyte solution, and electric field strength. Therefore, Joule heating effects can be minimized by using capillary of small internal diameter, electrolyte solution of low conductivity, and by applying electric potentials which are within the linearity of Ohm's law. In fact, capillaries of smaller internal diameters are more efficient at dissipating heat due to the greater surface-to-volume ratio than larger capillaries and, in addition, less amount of Joule heat is generated in capillary of smaller cross section. Because Joule heating depends on the square of the current flushing through the capillary tube, the use of diluted solutions of low conductive electrolytes should be preferred whenever possible, recalling that in an open capillary tube the conductivity of an electrolyte solution is expressed as

$$\sigma = F^2 C \sum_j z_j^2 \nu_j \chi_j \tag{6.22}$$

where
 F is the Faraday constant
 C is the molar concentration of the electrolyte
 z_j, ν_j, and χ_j are valency, mobility, and number of moles of the jth ionic species per mole of electrolyte, respectively

However, it should be noted that using diluted electrolyte solutions imply lowering ionic strength and buffering capacity, leading to a variety of drawbacks that will be discussed later in this chapter.

Several models have been developed to describe the effects of Joule heating and to calculate the resulting temperature profile in open and packed/filled capillary columns [120–125]. The temperature inside the capillary tube assumes a parabolic profile with the maximum value at the center and drops according to a logarithmic curve in the capillary wall and in the medium surrounding the capillary tube. Accordingly, the temperature gradient between the center of the capillary and the surrounding medium is given by the following equation:

$$\Delta T = \frac{Q r_i^2}{2} \left[\frac{1}{\kappa_s} \ln\left(\frac{r_o}{r_i}\right) + \frac{1}{\kappa_w} \ln\left(\frac{r_c}{r_o}\right) + \frac{1}{r_c} \ln\left(\frac{1}{h}\right) \right] \tag{6.23}$$

where

r_i and r_o are the inner and outer radius of the fused-silica tube
r_c refers to the outer radius of the polyimide coating of the capillary tube
κ_s and κ_w are the thermal conductivities of the electrolyte solution and of the capillary wall, respectively
h is the thermal transfer rate from the capillary wall

Equation 6.23 evidences the importance of efficient heat removal from the outer capillary wall and is in agreement with the above observation on the advantageous use of capillary tubes of smaller inner radius. Capillaries with a larger outer radius reduce the effects of the thermal isolating properties of the external polyimide coating of the capillary tube, facilitating heat dissipation through the capillary wall.

6.4.2 Interactions with the Inner Wall of Fused-Silica Capillaries

Charged compounds bearing different functional groups may interact with a variety of active sites on the inner surface of fused-silica capillaries, which comprise inert siloxane bridges, hydrogen bonding sites, and different types of ionizable silanol groups (vicinal, geminal, and isolated), giving rise to peak broadening and asymmetry, irreproducible migration times, low mass recovery, and in some cases irreversible adsorption. The detrimental effects of these undesirable interactions are particularly challenging in analyzing biopolymers, owing to the general more complex molecular structure of these molecules, and basic compounds. One of the earliest and still more adopted strategy to preclude the interactions of interacting analytes with the wall of bare fused-silica capillaries is the chemical coating of the inner surface of the capillary tube with neutral hydrophilic moieties [126]. The chemical coating has the effect of deactivating the silanol groups by either converting them to inert hydrophilic moieties or shielding all the active interacting groups on the capillary wall. A variety of compounds, including alkylsilanes, carbohydrates, neutral polymers, acrylamide, and acrylamide derivatives can be covalently bonded to the silica capillary wall by using bifunctional reagents to anchor the coating to the wall [127]. Polyacrylamide (PA), poly(ethylene glycol) (PEG), poly(ethylene oxide) (PEO), and polyvinylpyrrolidone (PVP) can be successfully anchored onto the capillary surface treated with several different silanes including 3-(methacryloxy)-propyltrimethoxysilane, 3-glycidoxypropyltrimethoxysilane, trimethoxyallylsilane, and chlorodimethyloctylsilane. Alternatively, a polymer can be adsorbed onto the capillary wall and then cross-linked in situ. Other procedures are based on simultaneous coupling and cross-linking. The use of alternative materials to fused silica such as polytetrafluorethylene (Teflon) and poly(methyl methyacrylate) (PMMA) hollow fibers has found limited application.

The deactivation of the silanol groups can also be obtained by the dynamic coating of the inner wall by flushing the capillary tube with a solution containing a coating agent. A number of neutral or charged polymers with the property of being strongly adsorbed onto the capillary wall are employed for the dynamic coating of bare fused-silica capillaries [128]. Modified cellulose and other linear or branched neutral polymers may adsorb onto the capillary wall with the consequence of modifying the ξ–potential, increasing the local viscosity in the electric double layer and masking silanol groups and other active sites on the capillary surface. Hence, the dynamic adsorption of the coating agent results in lowering or suppressing the EOF and in reducing the interactions with the capillary wall.

Polymeric polyamines are also strongly adsorbed in the compact region of the electric double layer as a combination of multisite electrostatic and hydrophobic interactions. The adsorption results in masking the silanol groups and the other adsorption active sites on the capillary wall and in altering the EOF, which is lowered and in most cases reversed from cathodic to anodic. One of the most widely employed polyamine coating agents is polybrene (or hexadimetrine bromide), a linear hydrophobic polyquaternary amine polymer of the ionene type [129].

Alternative choices are polydimethyldiallylammonium chloride, another linear polyquaternary amine polymer, cationic amine surfactants, and polyethylenimine (PEI) [130]. Also interesting is the dynamic coating obtained with ethylenediamine-derivatized spherical polystyrene nanoparticles of 50–100 nm diameter, which can be successively converted to a more hydrophilic diol-coating by in situ derivatization of the free amino groups with 2,3-epoxy-1-propanol [131]. Particularly attractive is the use of aliphatic vicinal oligoamines, such as triethylenetetramine (TETA) and diethylenetriamine (DIEN), which are effective at masking the silanol adsorption sites of bare fused-silica capillaries for proteins and peptides, while being capable of controlling the protonic equilibrium in a wide pH range, when used as the BGE in combination with a polyprotic acid [132–134]. The pK_a values of the above aliphatic vicinal oligoamines are reported in Table 6.1 and an example of separation of basic proteins performed at pH 4.0 in a bare fused-silica capillary with the BGE consisting of 10 mM $N,N,N'N'$-tetramethyl-1,3-butandiamine incorporated into 40 mM DIEN-phosphate buffer, is reported in Figure 6.6. The efficient separation displayed in Figure 6.6 is the result of the capability of the two aliphatic oligoamines at masking the silanol active sites for proteins on the wall of bare fused-silica capillaries, also at a pH value at which the separation of basic proteins in uncoated capillaries is particularly problematic. This in consequence that at pH 4.0 the capillary wall and the basic proteins are oppositely charged, in view of the isoelectric points of the four proteins, which are ranging from 9.5 (cytochrome c) to 11.0 (lysozyme) and the ionization of the silanol groups, already significant at this pH value.

TABLE 6.1
Acronym, Structure, and pK_a Values of the Aliphatic Vicinal Oligoamines Diethylentriamine and Triethylenetetramine

Buffering Agent	Acronym	Structure	pK_a [133]
Diethylentriamine	DIEN	$H_2NC_2H_4NHC_2H_4NH_2$	4.23
			9.02
			9.84
Triethylentetramine	TETA	$H_2NC_2H_4NHC_2H_4NHC_2H_4NH_2$	3.25
			6.56
			9.08
			9.74

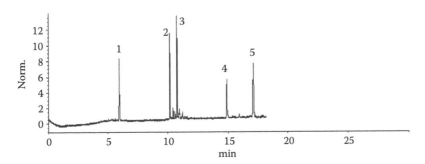

FIGURE 6.6 Separation of standard basic proteins in bare fused-silica capillary with BGE consisting of 10 mM $N,N,N'N'$-tetramethyl-1,3-butandiamine in 40 mM diethylenetriamine-phosphate buffer (pH 4.0). Capillary, bare fused silica 0.050 mm I.D., 0.375 mm O.D., total length 330 mm (245 mm to the detector); applied voltage 15 kV; cathodic detection at 214 nm; temperature of the capillary cartridge, 25°C; samples: (1) phenyltrimethylammonium iodide, (2) cytochrome c, (3) lysozyme, (4) ribonuclease A, (5) α-chymotrypsinogen A. (Personal collection).

6.4.3 SEPARATION PARAMETERS

The time required by a given analyte to migrate under the sole influence of the applied electric field across the capillary tube from the injection end of the capillary to the detection windows (migration distance) is defined as the "migration time" (t_m) and, similarly as the retention time in HPLC, is used for identification of sample components. It is given by

$$t_m = \frac{l}{v_{obs}}$$ (6.24)

where
  l is the migration distance
  v_{obs} is the observed migration velocity of the considered analyte

By analogy with chromatography, the record of the electrophoretic process is termed "electropherograms."

In the absence of EOF and separation mechanism other than electrophoresis, each analyte migrates with its own velocity which, according to Equation 6.8, is proportional to the strength of the electric field applied across the capillary tube. The constant of proportionality of the observed velocity of the charged analyte is defined as the "observed mobility" (μ_{obs}) and can be directly calculated by the migration time and the other experimental parameters, according to the following equation:

$$\mu_{obs} = \frac{v_{obs}}{E} = \frac{l}{t_m E} = \frac{lL}{t_m V}$$ (6.25)

where
  L is the total length of the capillary tube
  E is the strength of the applied electric field (V/L)

In the presence of EOF, the observed velocity is due to the contribution of electrophoretic and electroosmotic migration, which can be represented by vectors directed either in the same or in opposite direction, depending on the sign of the charge of the analytes and on the direction of EOF, which depends on the sign of the zeta potential at the plane of share between the immobilized and the diffuse region of the electric double layer at the interface between the capillary wall and the electrolyte solution. Consequently, v_{obs} is expressed as

$$v_{obs} = v_e + v_{eof}$$ (6.26)

where
  v_e is the electrophoretic velocity of the considered analyte in the absence of EOF
  v_{eof} is the velocity of the EOF, which is indicated with negative sign if the EOF is in the opposite direction to the migration of the analyte

The effective mobility, expressed by Equation 6.16, can be directly calculated from the observed mobility by measuring the electroosmotic mobility using a neutral marker, not interacting with the capillary wall, which moves at the velocity of the EOF. Accordingly, the effective mobility μ of cations in the presence of cathodic EOF is calculated from μ_{obs} by subtracting μ_{eof}:

$$\mu = \mu_{obs} - \mu_{eof}$$ (6.27)

Neutral substances commonly employed as neutral markers in measuring the EOF are methanol, acetone, mesityl oxide, and dimethylsulfoxide.

The main factor contributing to band broadening of analytes migrating under the sole influence of the applied electric field is longitudinal diffusion, considering negligible contributions due to convective motion, radial diffusion, and Joule heating, which are minimized by the use of a capillary tube. Therefore, under ideal conditions (primarily, small injection plug and absence of interactions of the analyte with the inner surface of the capillary), the variance of the migration band width (σ^2) is expressed similarly to the Einstein equation for diffusion as

$$\sigma^2 = 2Dt_m = 2D\frac{l}{v_{obs}} = 2D\frac{l}{\mu_{obs}E} = 2D\frac{l}{(\mu + \mu_{eof})E} \tag{6.28}$$

The variance of the migration band is related to the efficiency of the separation system, which by analogy with chromatography is expressed in terms of number of theoretical plates (N), according to the following equation [135]:

$$N = \frac{l^2}{\sigma^2} \tag{6.29}$$

Combining Equations 6.29 and 6.28 leads to

$$N = \frac{(\mu + \mu_{eof})El}{2D} = \frac{(\mu + \mu_{eof})V}{2D} \tag{6.30}$$

Since the number of theoretical plates is directly proportional to the strength of the applied electric field, the highest applied voltage possible is recommended to obtain high efficiency. This follows because on increasing the strength of the applied electric field the migration velocity of the analytes increases and, therefore, they have less time to diffuse. In addition, Equation 6.30 predicts better efficiency for larger charged molecules having small diffusion coefficients, which exhibit less dispersion than low molecular weight species.

The efficiency can also be estimated in terms of height equivalent to a theoretical plate, which expresses the differential increase of the band variance during the migration of a given analyte along the capillary and is given by

$$H = \frac{l}{N} = \frac{2D}{(\mu + \mu_{eof})E} \tag{6.31}$$

Both the number of theoretical plates and the height equivalent to a theoretical plate estimate the band broadening occurring in the separation system, either related to the whole capillary tube (N) or to the portion of the capillary occupied by the analyte (H). The higher is N (or lower is H) the narrower are the recoded peaks in the electropherogram.

For analytes migrating as symmetrical Gaussian peaks, the number of theoretical plates can be calculated directly from the electropherograms using the following equations:

$$N = 5.54\left(\frac{t_m}{2.355\sigma}\right)^2 = 5.54\left(\frac{t_m}{w_{1/2}}\right)^2 \tag{6.32}$$

or

$$N = 16 \left(\frac{t_\mathrm{m}}{4\sigma} \right)^2 = 16 \left(\frac{t_\mathrm{m}}{w_\mathrm{b}} \right)^2 \tag{6.33}$$

where $w_{1/2}$ and w_b are the peak width measured in time units at half-height and at the base of the detected peak, respectively. In practice, the efficiency determined by Equations 6.32 or 6.33 is generally lower than that calculated by Equation 6.31 which accounts only for peak broadening due to longitudinal diffusion whereas additional dispersive processes often contribute to the total variance of the system, according to the following equation:

$$\sigma^2 = \sigma_\mathrm{D}^2 + \sigma_\mathrm{E}^2 + \sigma_\mathrm{T}^2 + \sigma_\mathrm{I}^2 + \sigma_\mathrm{S}^2 \tag{6.34}$$

where the terms on the right side correspond to the contribution of band broadening due to diffusion (σ_D^2), electrodispersion (σ_E^2), Joule heat (σ_T^2), length of the sample injection plug (σ_I^2), and interactions of the sample with the inner surface of the capillary wall (σ_S^2).

Electrodispersion, also termed electrophoretic dispersion, describes the changes in conductivity and field strength occurring in the sample zone during the electrophoretic process. It conducts to band broadening and peak asymmetry or distortions, which can be controlled by matching sample and electrolyte solution conductivity and minimizing the concentration of the sample. Electrodispersion is almost negligible when the concentration of the sample in the migrating zone is sufficiently lower (generally at least 100 times) than that of the electrolyte solution [136]. Band broadening due to Joule heating is minimized by controlling the parameters influencing its generation (see Section 6.4.1) and by efficient heat removal from the outer capillary wall, whereas the broadening due to sample injection is simply controlled by minimizing the length of the sample plug introduced into the capillary. The untoward interactions of the analyte with the inner surface of the capillary may cause severe peak broadening and tailing, which require complex procedures to be minimized (see Section 6.4.2). If any of the above terms is predominant over the diffusion, only minimal improvements can be obtained by increasing the strength of the applied electric field.

The capability of separating two analytes is evaluated by the resolution (R_s), which can be simply calculated by dividing the distance between the peak maxima (Δx), expressing the difference between the migration distances of two adjacent peaks, to the average peak width at the baseline ($\overline{w}_\mathrm{b} = 4\overline{\sigma}$):

$$R_\mathrm{s} = \frac{\Delta x}{\overline{w}_\mathrm{b}} = \frac{\Delta x}{4\overline{\sigma}} \tag{6.35}$$

From the relationship existing between migration distance and electrophoretic mobility and between peak width and number of theoretical plates, the resolution can be expressed as

$$R_\mathrm{S} = \frac{\sqrt{N}}{4} \frac{\Delta v}{\overline{v}} = \frac{\sqrt{N}}{4} \frac{\Delta\mu}{\overline{\mu}_\mathrm{obs}} = \frac{\sqrt{N}}{4} \frac{\Delta\mu}{(\overline{\mu} + \mu_\mathrm{eof})} = \frac{\sqrt{N}}{4} \delta_\mathrm{m} \tag{6.36}$$

where Δv and $\overline{v}$ are the difference in electrophoretic velocity and the average electrophoretic velocity of the two analytes, respectively. The ratio $\Delta\mu/\overline{\mu}_\mathrm{obs}$ is the relative mobility difference of the two analytes being separated (δ_m), representing an operational value of selectivity that does not depend only on the properties of the two separated analytes, because it can be affected by the EOF. The number of theoretical plates requested to completely resolve adjacent peaks can be calculated by the following expression, which indicates the key role of the selectivity parameter δ_m [137]:

$$N = \frac{16 R_S^2}{\delta_m^2} \tag{6.37}$$

The ability to separate analytes on the bases of their electrophoretic mobility, in absence of secondary equilibrium in solution and interactions with the capillary wall, is expressed by the inherent selectivity, which is given by the ratio of the motilities of two analytes (1 and 2) migrating as adjacent peaks:

$$\alpha_{2,1} = \frac{\mu_2}{\mu_1} \tag{6.38}$$

The overall separation potential of an electromigration technique can be expressed by the peak capacity (n), which is defined as the maximum number of peaks that can be separated within a given separation time, usually coincident with the time interval between the first and last detected peak in the electropherogram, while retaining unit resolution for all adjacent peak pairs:

$$n = 1 + \frac{\sqrt{N}}{4} \ln t_\omega / \ln t_\alpha \tag{6.39}$$

where t_ω and t_α are the migration times of the last and first detected peak in the electropherogram, respectively [135]. Peak capacity and resolution increase with the number of theoretical plates N and, therefore, an optimization of operational parameters performed to obtain complete resolution and maximum number of separable peaks involves maximizing N.

6.5   SEPARATION MODES

Using the instrumentation described above, capillary electromigration techniques can be performed by a variety of modes, based on different separation mechanisms that can be selected by simply changing the operational conditions, specifically, composition of the electrolyte solution and/or capillary column. A fundamental aspect of each separation mode is the composition of the electrolyte solution, which may consist of either a continuous or a discontinuous electrolyte system. In continuous systems, the composition of the electrolyte solution is constant along the capillary tube, whereas in discontinuous systems it is varied along the migration path. In most capillary electromigration techniques, the same open capillary tube could theoretically be employed in a variety of separation modes, just varying the composition of the electrolyte solution, whereas other separation modes, such as capillary electrochromatography and capillary gel electrophoresis, require dedicated columns.

The names of the different separation modes of the capillary electromigration techniques described below are those recommended by the Analytical Chemistry Division of the International Union of Pure and Applied Chemistry (IUPAC) [138]. According to this recommendation, capillary electrophoresis is not used as a collective term for all capillary electromigration techniques performed in an open capillary tube. However, it should be noted that such term is generally accepted to identify any electromigration technique governed by separation mechanisms mainly based on electrophoretic principles, in analogy to the use of the term electrochromatography, which is referred to capillary electromigration techniques employing chromatographic columns, specifically, capillary tubes either filled, packed, or coated with a stationary phase [139]. Also, to consider is a number of borderline cases existing with respect to the naming of particular electromigration techniques governed by the simultaneous action of different separation mechanisms, including the cases where chromatographic and electrophoretic phenomena are superimposed.

The following subsections briefly describe the different separation modes of electromigration techniques. Most of the separation parameters utilized to evaluate performance and results of the different separation modes are common to the majority of them and have been discussed above, whereas terms and parameters related to a specific separation mode are introduced and discussed in each specific subsection.

6.5.1 Capillary Zone Electrophoresis

Capillary zone electrophoresis (CZE) is an electromigration separation mode that uses continuous electrolyte solution systems and constant electric field strength throughout the capillary length, either in aqueous or nonaqueous background electrolyte solutions. The separation mechanism is based on differences in the electrophoretic mobilities of charged species and, therefore, on differences in the charge-to-mass ratio (see Section 6.2.2). Under the influence of the electric field applied across the capillary tube, the analytes migrate with different velocities toward the corresponding electrode: the positively charged analytes toward the cathode and negatively charged analyte toward the anode. However, in the presence of sufficiently strong EOF, which is cathodic in bare fused-silica capillaries, cations, anions, and neutral species migrate in the direction of EOF, where sample detection takes place (see Figure 6.7). Consequently, the migration velocity of charged analytes is either the sum or the difference of electrophoretic velocity and EOF, whereas neutral molecules migrate with the velocity of EOF.

The concept of virtual migration distance has been introduced by Rathore and Horváth [140] to distinguish separative and non-separative components of the differential migration process occurring in CZE, as well as in HPLC and in other electromigration techniques. According to this concept, the length of the capillary is divided into two virtual parts, l_S and l_0. The virtual migration distance, l_S, represents the separation component of the overall migration process and it is considered as the virtual distance that the analytes move under the direct influence of the electric field. It is calculated as the product of the electrophoretic mobility of the analyte (μ) and its migration time (t_m):

$$l_S = \mu t_m \tag{6.40}$$

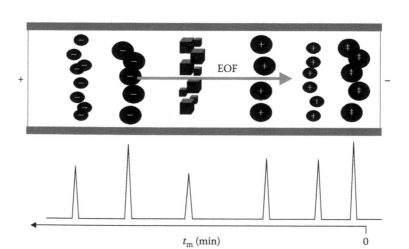

FIGURE 6.7 Representation of the migration behavior of cationic, anionic, and neutral analytes in CZE with sufficiently strong EOF to detect all analytes at the cathodic end of the capillary (panel A) and related electropherogram (panel B). Molecular masses proportional to the symbol size.

The virtual migration distance, l_0, arising from EOF is expressed as the product of the migration time of the analyte (t_m) and the electroosmotic mobility (μ_{eof}):

$$l_0 = \mu_{eof} t_m \tag{6.41}$$

Both l_s and l_0 can be expressed by either positive or negative numbers, depending on the direction of electrophoretic and electroosmotic mobilities and their algebraic addition equal to the real migration distance (L_m), which is the length of the capillary from the injection end to the detection window. Therefore, in co-electroosmotic mode of CZE, which means that the analytes migrate in the same direction of elecotoosmosis, both l_s and l_0 have the same sign, whereas in counter-electroosmotic mode of CZE, the two virtual migration distances have opposite signs.

The ratio of the two virtual lengths defines a parameter called the electrophoretic velocity factor k_e, which is analogous to the chromatographic retention factor and it is expressed as [140]:

$$k_e = \frac{l_s}{l_0} = \frac{\mu}{\mu_{eof}} = \frac{(L_m/t_m) - (L_m/t_{eof})}{L_m/t_{eof}} = \frac{t_{eof} - t_m}{t_m} \tag{6.42}$$

where t_{eof} is the migration time of the neutral tracer used to measure the EOF and the ratios L_m/t_m and L_m/t_{eof} express the velocity of the charged and neutral components of the sample, respectively. Therefore, the velocity factor k_e can also be considered as the dimensionless electrophoretic velocity of a charged analyte normalized to the velocity of the EOF. It is worth noting that since the analyte may migrate in the opposite direction to that of EOF, the velocity factor k_e can be negative.

The effective mobility relative to the total mobility identifies another dimensionless migration parameter termed electromigration factor (f_m), which is expressed as [141]:

$$f_m = \frac{\mu}{\mu + \mu_{eof}} = \frac{l_s}{(l_s + l_0)} = \frac{l_s}{L_m} \tag{6.43}$$

In terms of virtual migration distance, the electromigration factor is the dimensionless virtual distance of separative migration that measures the separative fraction of the total column length. Table 6.2 summarizes the fundamental migration parameters of CZE expressed according to the conventional formalism and to the concept of virtual migration distances.

In most applications, the electrolyte solution employed in CZE consists of a buffer in aqueous media. Whilst all buffers can maintain the pH of the electrolyte solution constant and can serve as background electrolytes, they are not equally meritorious in CZE. The chemical nature of the buffer system can be responsible for poor efficiency, asymmetric peaks, and other untoward phenomena arising from the interactions of its components with the sample. On the other hand, buffering agents and suitable additives incorporated into the BGE may be effective at preventing or minimizing the interactions of the analytes with the capillary wall, which are particularly challenging for proteins [142,143]. Most of the additives employed for this purpose act either as masking or competing agents for the silanol groups on the inner wall of the capillary, so that they are not accessible to the analytes. Others may function as strong ion-pairing or competing agents for the interacting moieties of the analytes exposed to the electrolyte solution, in order to subtract their availability to the interacting sites on the capillary wall. A schematic representation of the action of additives incorporated into the electrolyte solution on the electrophoretic behavior of analytes bearing multiple interaction sites is depicted in Figure 6.8.

In addition, the composition of the electrolyte solution can strongly influence sample solubility and detection, native conformation of biopolymers, molecular aggregation, electrophoretic mobility, and EOF, which can be altered as a consequence of the adsorption of the components of the BGE onto the capillary wall. Consequently, selecting the proper composition of the electrolyte solution

TABLE 6.2
Separation Parameters in CZE According to (A)
Conventional and (B) Virtual Migration Distance
Formalism

Parameter	A	B
Electrophoretic velocity factor, k_e	$k_e = \dfrac{t_{eof} - t_m}{t_m}$	$k_e = \dfrac{l_s}{l_0}$
Electromigration factor, f_m	$f_m = \dfrac{\mu}{\mu + \mu_{eof}}$	$f_m = \dfrac{l_s}{L_m}$
Selectivity, $\alpha_{2,1}$	$\alpha_{2,1} = \dfrac{\mu_2}{\mu_1}$	$\alpha_{2,1} = \dfrac{l_{s2} l_{01}}{l_{s1} l_{02}}$
Relative mobility difference, δ_m	$\delta_m = \dfrac{\Delta\mu}{(\bar{\mu} + \mu_{eof})}$	$\delta_m = \dfrac{\Delta l_s}{\bar{l_0}}$
Resolution, R_s	$R_s = \dfrac{\sqrt{N}}{4} \delta_m$	$R_s = \dfrac{\sqrt{N}}{4} \dfrac{\Delta l_s}{\bar{l_0}}$
Number of theoretical plates, N	$N = \dfrac{16 R_s^2}{\delta_m^2}$	$N = 16 R_s^2 \left(\dfrac{\bar{l_0}}{\Delta l_s} \right)^2$

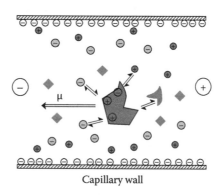

Capillary wall

FIGURE 6.8 Schematic representation of the action of additives incorporated into the electrolyte solution on the electrophoretic behavior of analytes bearing multiple interaction sites.

is of paramount importance in optimizing the separation of the analytes in CZE. The appropriate selection of the buffer requires evaluating the physicochemical properties of all components of the buffer system, including buffering capacity, conductivity, and compatibility with the detection system and with the sample.

Additives may also be incorporated into the electrolyte solution to enhance selectivity, which expresses the ability of the separation method to distinguish analytes from each other. Selectivity in CZE is based on differences in the electrophoretic mobility of the analytes, which depends on their effective charge-to-hydrodynamic radius ratio. This implies that selectivity is strongly affected by the pH of the electrolyte solution, which may influence sample ionization, and by any variation of physicochemical property of the electrolyte solution that influences the electrophoretic mobility (such as temperature, for example) [144] or interactions of the analytes with the components of the electrolyte solution which may affect their charge and/or hydrodynamic radius.

Several buffering agents and additives can influence selectivity by interacting specifically or to different extents with the components of the sample, also on the basis of their stereoisomerism. The large selection of compounds that can be employed as additive of the electrolyte solution for this purpose include zwitterions, anionic or cationic ion-pairing agents, complex-forming species, chiral selectors, organic solvents, surfactants, cyclodextrins, denaturing agents, and ionic liquids [145–151]. The related interactions may involve either chiral on non-chiral electrostatic or/and hydrophobic interactions, as well as hydrogen bonding, ion–ion, ion–dipole, dipole–dipole, and ion–dipole/ion-induced-dipole interactions. Moreover, several anions, such as phosphate, citrate, and borate, which are components of the buffer solutions employed as the BGE may act as ion-pairing agents influencing the electrophoretic mobility and, hence, selectivity [133,152].

Organic solvents may also be employed as an alternative to water in preparing the BGE used in CZE, which is referred to as nonaqueous capillary electrophoresis (NACE) [153–155]. The substitution of water by an organic solvent can be advantageous for increasing sample solubility, preserving the stability of analytes having the tendency to decompose in water or improving selectivity, which is expected to be affected by charge and solvation size of the analytes and by their interactions with the other components of the BGE, all influenced by the organic solvent to a different extent than water. In addition, the high volatility and low surface tension of many organic solvents make NACE suitable to be successfully hyphenated with mass spectrometry [156].

6.5.2 CAPILLARY AFFINITY ELECTROPHORESIS

Capillary affinity electrophoresis (CAE), also termed affinity capillary electrophoresis (ACE), is an electromigration separation mode for substances that participate in specific or biological-based molecular interactions, such as receptor–ligand interactions. CAE is widely employed for the characterization of biomolecules, for the analysis of specific interactions of a ligand (proteins, peptides, pharmaceutical active small molecules, etc.) with a receptor (typically a protein or a peptide), and for determining binding constants and binding stoichiometries. The wide array of interactions investigated by CAE include protein–drug, protein–DNA, peptide–carbohydrate, peptide–peptide, DNA–dye, carbohydrate–drug, and antigen–antibody [157–159].

According to the standard approach recalled in the IUPAC definition, CAE is performed incorporating one of the interacting partners into the BGE at varying concentrations, whereas the other specie participating in the specific interactions is injected into the capillary as the sample in mixture with a noninteracting standard and subjected to the electrophoretic run. In most applications, the receptor, which is very often a protein, is used as the sample, whereas the ligand, usually consisting of a less complex molecule, is incorporated into the BGE. When specific interactions between ligand and receptor determine the dynamic formation of a complex having charge-to-size ratio different from that of the receptor, a shift in the migration time of the receptor relative to the standard occurs, whose extension is a function of the ligand concentration. Therefore, the shift in the migration time of the receptor in response to the composition of the running BGE reflects the degree of interactions and, therefore, the binding constant, which can be calculated by linear or nonlinear regression analysis [160,161].

The most frequent deleterious effects that can spoil the estimation of the binding constants include the variations in migration times and losses of the receptor injected as the analyte, due to its interactions with the capillary wall, frequently occurring with uncoated capillaries, especially when the receptor is a protein. Coated capillaries and the experimental arrangement with the protein incorporated into the BGE and the ligand employed as the analyte are used to overcome such drawbacks. However, using UV detection, the high background absorption, due to the protein solution filled in the capillary, may annihilate the theoretical benefit of better peak shape, higher repeatability of migration time, and greater shift of the electrophoretic mobility expected using smaller and less complex molecules than proteins as the injected sample.

The loss in the detection sensitivity can be avoided by using the partial filling technique, in which only a small plug of the BGE containing the protein, or another UV-absorbing receptor, is introduced in the capillary filled with the BGE without the receptor. The ligand is injected as the sample and the electrophoretic run is performed under the conditions where the UV-absorbing receptor does not migrate toward the detection end of the capillary or migrates at a velocity sufficiently lower than the analyte, which, therefore, can be detected in the neat BGE without the interference of the UV-absorbing receptor [162]. These operational modes of CAE are currently referred to as "dynamic equilibrium capillary affinity electrophoresis" and besides being employed for studying molecular interactions and improving selectivity are widely utilized for the separation of enantiomeric molecules incorporating a chiral selector into the BGE [163].

If the interactions between the ligand and receptor are strong enough leading to the formation of a complex that can be considered irreversible in the capillary electrophoresis timescale, CAE is performed by incubating the interacting partners prior to be injected into the capillary containing the BGE without the interacting species. The separation of free ligand from complexed ligand is subsequently achieved upon application of an electrical field. The concentration of free ligand is determined from peak height or area measurement by the aid of external standards [164]. The third main format of CAE is realized by immobilizing the receptor (or the ligand) onto the capillary wall with the purpose of retaining or retarding the analytes having specific affinity for it, whereas the electrophoretic migration of all other noninteracting species are unaffected by the coating. Such approach is frequently applied using antibodies or antibody-related substances as the selective binding agent. This mode of CAE is also used for the enrichment, prior identification, and/or quantification of analytes present in diluted samples or at low concentration in complex matrices, performed by tandem on-line hyphenation of a packed or monolithic capillary, containing the immobilized selective interacting agent, with the separation capillary [165].

CAE employing antibodies or antibody-related substances is currently referred to as immunoaffinity capillary electrophoresis (IACE), and is emerging as a powerful tool for the identification and characterization of biomolecules found in low abundance in complex matrices that can be used as biomarkers, which are essential for pharmaceutical and clinical research [166]. Besides the heterogeneous mode utilizing immobilized antibodies as described above, IACE can be performed in homogeneous format where both the analyte and the antibody are in a liquid phase. Two different approaches are available: competitive and noncompetitive immunoassay. The noncompetitive immunoassay is performed by incubating the sample with a known excess of a labeled antibody prior to the separation by CE. The labeled antibodies that are bound to the analyte (the immunocomplex) are then separated from the nonbound labeled antibody on the basis of their different electrophoretic mobility. The quantification of the analyte is then performed on the basis of the peak area of the nonbonded antibody.

The competitive immunoassay is performed by incubating the sample with a known fixed quantity of a labeled substance competing with the analyte for the binding sites on the antibody (labeled analog) and a limiting amount of the antibodies that can bind to both the analyte and the labeled analog. The components of the incubating mixture are separated by CE and the amount of the analyte is determined on the basis of the relative amount of the labeled analog that is bound to the antibodies, or that remaining free in solution, in comparison to the amount that is obtained when performing the immunoassay with standard solutions containing known amounts of the analyte [167]. Other formats of CAE, developed to address particular applications, include Hummel–Dreyer method (HD), vacancy affinity capillary electrophoresis (VACE), vacancy peak (VP) method, frontal analysis capillary electrophoresis (FACE), and immunoaffinity capillary electrophoresis (IACE) [168].

6.5.3 CAPILLARY SIEVING ELECTROPHORESIS

The main separation mechanism in capillary sieving electrophoresis (CSE) is based on differences in size and shape of charged analytes migrating through a sieving matrix that is enclosed in the capillary tube. CSE is successfully employed for separating charged biological macromolecules having mass-to-charge ratios that do not vary with the molecular size. Typical examples of such compounds are the DNA restriction fragments, which have similar mass-to-charge ratios because each further nucleotide added to a DNA chain introduces an equivalent unit of charge and mass without affecting their ratio and, consequently, their electrophoretic mobility in free solution. Similarly, sodium dodecyl sulfate (SDS) protein complexes, which are generated by the binding of the anionic surfactant SDS to proteins in a constant ratio of 1.4 g of the surfactant per 1.0 g protein, have approximately unvarying mass-to-charge ratios resulting in comparable electrophoretic mobilities. Further examples of analytes having similar electrophoretic mobility in free solution and different molecular size and shape that can be separated by incorporating a sieving matrix in the capillary include oligonucleotides, RNA, double-stranded DNA, complex carbohydrates, and synthetic polyelectrolytes [169,170].

The sieving media may consist of permanent or chemical gels, physical gels, and polymer solutions of suitable viscosity [171–174]. Permanent cross-linked gels are prepared in the capillary by in situ polymerization processes based on adding a catalyst to the monomer solution, which is then pumped into the capillary tube, where the polymerization takes place. The gel is covalently bonded to the inner surface of the capillary to prevent its extrusion. Typical cross-linked sieving matrices are the polyacrylamide gels similar to those employed in slab gel electrophoresis, whose pore size can be adjusted by varying the cross-linker concentration in the reaction mixture and that can be easily anchored to the capillary wall, previously treated with 3-methacryloxypropyltrimethoxysilane.

Such arrangement of CSE is commonly referred to as capillary gel electrophoresis (CGE) and is comparable to conventional slab gel electrophoresis with the advantages of on-capillary detection, full instrumental automation, and the possibility of applying higher electric field, due to the better Joule heat dissipation in the narrow bore capillaries, with consequent shorter analysis time. On the other hand, with cross-linked gels, samples can be introduced in the capillary solely by electrokinetic injection, which is not recommended for sample solutions having high salt concentration. Further drawbacks include the introduction of air bubbles during filling the capillary, the possible shrinkage of the gel during polymerization, gel breaking during subsequent capillary manipulation, and gel degradation by hydrolysis. Clogging of the pores during electrophoresis is another factor which restricts the use of cross-linked capillaries to a limited number of runs.

Replaceable sieving matrices consist of either a physical gel or a polymer solution that can be easily renewed even after each run. Physical gels are formed within the capillary by dynamic noncovalent attractive interactions between the chains of a linear polymer, such as agarose, which is employed to form thermo-reversible gels for both traditional slab gel and capillary electrophoresis. This polysaccharide is not soluble in aqueous solutions at room temperature and goes into solution when heated. Therefore, agarose is dissolved in a warm aqueous electrolyte solution, which is fitted into the capillary. As the heated solution cools, agarose chains hydrogen bond to form double-helical structures, which in turn aggregate to form a three-dimensional network. The EOF should be suppressed to avoid gel extrusion, but the chemical bonding of the gel to the inside wall of the capillary is not required.

Nowadays, the sieving matrices most employed in CSE are polymer solutions that under suitable conditions may form a transient mesh or sieving matrix that provide the size-based separation of charged biopolymers. The polymer solutions can be formulated with linear acrylamide and N-substituted acrylamide polymers, cellulose derivatives, polyethylene oxide, and its copolymers or with a variety of polymers, such as polyvinylpyrrolidone (PVP), polyethylene oxide (PEO), and hydroxypropyl cellulose(HPC), which do not necessitate the preventive coating of the capillary wall due to their ability to dynamically coat the inner surface of the capillary, resulting in suppressed EOF and sample interactions with the capillary wall.

An interesting class of polymer matrices widely employed for CSE consists of temperature–dependent, viscosity-adjustable polymer solutions that are filled into the capillary at one temperature at low viscosities and are used in separation at another temperature at entanglement threshold concentrations and higher viscosities. Such viscosity-adjustable polymers are termed thermoresponsive and classified as thermoassociating, when viscosity increases and as thermothinning when viscosity decreases with increasing temperature, respectively [175]. Examples of thermoresponsive polymers are the triblock uncharged copolymers of the poly(ethylene oxide)–poly(propylene oxide)–poly(ethylene oxide) type, the block copolymer made of a hydrophilic polyacrylamide backbone grafted with poly-N-isopropylacrylamideside chains [176], and the block copolymers of N,N-dialkylyacrylamide with acrylamide, such as block copolymers of N,N-diethylacrylamide (DEA) with N,N-dimethylacrylamide (DMA) [177] and of linear acrylamide (LPA) with dihexylacrylamide (DHA) [178]. Most of these sieving matrices are successfully employed for DNA sequencing using automated multicapillary electrophoresis systems, consisting of an array of capillaries that can be simultaneously operated for increasing the throughput of DNA sequencing [179,180].

6.5.4 Capillary Isoelectric Focusing

In capillary isoelectric focusing (CIEF), amphoteric analytes such as peptides and proteins are separated according to their isoelectric points by the application of an electric field along a pH gradient formed in the capillary tube, which is filled with the electrophoresis medium containing the sample to be analyzed and a mixture of commercially available zwitterionic compounds, known as carrier ampholytes [181]. The anodic end of the capillary is placed into an acidic solution (anolyte), and the cathodic end in a basic solution (catholyte). The carrier ampholytes have closely spaced isoelectric point (pI) values encompassing a desired pH range, and upon applying the electric field across the capillary each ampholyte migrates toward the electrode of opposite charge generating a pH gradient.

The basic principle of CIEF is that under the action of an electric field and in the absence of EOF, an analyte migrates as long as it is charged and it stops migrating if it becomes neutral during its migration. Peptides and proteins are amphoteric analytes that are charged at pH values different from their isoelectric points, at which they are electrically neutral. Under the influence of an electric field in a pH gradient between the cathode and the anode, each negatively charged peptide or protein migrates toward the anode encountering progressively lower pH values with the result that its net negative charge and, consequently, its electrophoretic mobility are gradually reduced. When the considered analyte reaches the zone of the electrophoretic medium having pH value corresponding to its pI, it becomes neutral and stops migrating. Similarly, each positively charged peptide or protein migrates toward the cathode encountering increasing pH values, with consequent progressive reduction of its net positive charge and cathodic electrophoretic mobility, upon reaching the zone having pH value equal to its pI, where it stops migrating.

A steady-state condition is reached when all components of the sample have been focused in the zone of the capillary tube containing the electrophoretic medium at the pH value corresponding to their pI. The completion of the focusing procedure is monitored by a drop-off in current. Each sample component stops migrating in a self-sharpened band because if it diffuses out of the focused zone it acquires a charge with the consequence of being pulled back in the focused band where the analyte is electrically neutral. Therefore, a remarkable feature of isoelectric focusing (IEF) is the concentration of the separated analytes within a narrow band [182]. However, in order to avoid the precipitation of the focused analytes, additives, such as nonionic surfactants (e.g., Triton X-100 or Brij-35) and organic modifiers (e.g., glycerol, ethylene glycol) are often incorporated into the electrophoresis medium as dispersants to minimize aggregation.

The resolving power of IEF is measured in terms of minimum difference between the pI of two adjacent separated analytes, and is expressed by the following equation:

$$\Delta pI = 3\left[\frac{D(\mathrm{d}pH/\mathrm{d}x)}{E(-\mathrm{d}\mu/\mathrm{d}pH)}\right]^{1/2} \tag{6.44}$$

where

> D is the diffusion coefficient of the considered analyte
>
> dpH/dx is the rate of variation of pH with distance (pH gradient)
>
> E is the applied electric field
>
> $-$dμ/dpH is the variation of the electrophoretic mobility with pH in the vicinity of pI, which depends on the neat charge of the analyte near its pI

The operational parameters to be controlled are the applied electric field, with higher E leading to smaller ΔpI, and pH gradient, which should be as narrower as possible for higher resolution.

A peculiar aspect of performing isoelectric focusing in a capillary tube is that the anticonvective medium (usually a gel), requested in the classical version of the technique, is not used and relatively high electric field can be employed, due to the efficient Joule heat dissipation of small-diameter tubes. Other remarkable features include automation, the requirement of small amount of sample, high resolution, and fast analysis time. On the other hand, in classical gel-IEF, the separated zones are promptly visualized by staining, whereas CIEF performed on most capillary electrophoresis instruments (conventional CIEF) requires an additional step for visualizing the separation. These instruments use a single-point detection system located on one end of the capillary column. Therefore, after the IEF process is completed, a mobilization stage is required to move the focused analyte zones past the detection window. A schematic view of the separation mechanism operating in CIEF is reported in Figure 6.9.

Either two-step or one-step immobilization techniques are generally employed. Using the two-step techniques, the analytes are first focused in the capillary and subsequently displaced toward the detection window by either hydrodynamic or electrophoretic mobilization, as first reported by Hjertén and Zhu [183]. The hydrodynamic flow can be generated by either pressure or gravity, while the applied electric field is maintained to avoid the defocusing of bands by laminar flow. Otherwise, the focused zones can be moved in the direction of the detection window by changing the composition of the anolyte or catholyte solution by adding a salt in the proper reservoir and applying the electric field. This results in a variation of pH in the electrophoresis medium due to the migrations of anions and cations competing with the migration of hydroxyl and hydronium ions, respectively. Thus, as the pH is varied both ampholytes and analytes are charged and move

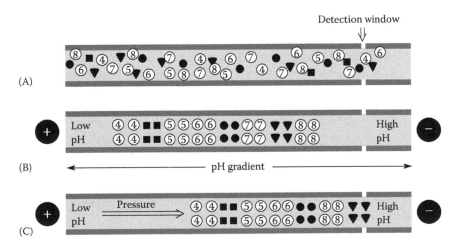

FIGURE 6.9 Schematic representation of CIEF with visualization of the separated analytes by simultaneous pressure–voltage mobilization. The analytes are filled into the capillary mixed with the carrier ampholytes (A). The application of the electric field generates a pH gradient and the analytes focus in narrow bands according to their pI (B). Once focused, the analytes are moved toward the detection window by hydrodynamic flow generated by applying pressure, while maintaining the applied electric field (C).

in the direction of the reservoir where the salt has been added. These approaches require the use of coated capillaries to suppress the EOF and to prevent troublesome interactions of the analytes with the inner surface of the capillary tube. A variety of covalent [183–186] and dynamic [187–189] coatings are employed for these purposes. Also popular is the use of capillaries covalently coated with a hydrophilic polymer in combination with a dynamic coating agent, such as zwitterions, polymers, and surfactants [190,191], which is incorporated into the electrophoresis medium.

The alternative approach is using a one-step technique, which allows performing the focusing process and the mobilization of the electrophoresis medium simultaneously. This is obtained by different methods, including the coincident application of electric field and pressure, the use of a catholyte (or anolyte) containing a salt, and performing CIEF with capillaries having a residual EOF, which can be controlled by incorporating an additive into the electrolyte medium [192,193]. One-step mobilization can also be performed with sulfonated polymer–coated capillaries, which provide sufficient and constant EOF for mobilization of the entire pH gradient to the capillary end with the detection windows [194].

Deformation of pH gradient, uneven resolution along the capillary column, and increased analysis times are the main drawbacks of the mobilization techniques used to move the train of focused analytes toward the detection window, which is requested when using classical instrumentation with detection at a fixed point. The ideal alternative is using real-time whole column imaging detectors (WCID), which allow for the simultaneous detection along the entire length of the stationary zones focused within the capillary column by IEF [195,196]. The different types of real-time WCID used in CIEF include refractive index gradient imaging detector, optical absorption imaging detector, and LIF imaging detector, which is the most sensitive detector among them [197–199]. The typical arrangement of a real-time WCI detector, operating in either optical absorption or fluorescence mode, consists of a laser or a LED light source whose beam is projected onto the whole separation length of the capillary column without the outer coating. Either the light intensity passing through the CIEF column or its fluorescence image emission is then focused onto a photodiode array or a charge-coupled device (CCD) camera. In an alternative arrangement, the entire column is transported past a single detection point, using either a modified UV-visible or fluorescence detector. However, in this case, a real-time detection mode is not performed, and the approach is more realistically defined capillary scanning detection. The hyphenation of CIEF with mass spectrometry (MS) is usually performed using electrospray ionization (ESI) or MALDI techniques. However, coupling CIEF with ESI–MS suffers more from signal suppression compared to MALDI-MS [200,201].

CIEF can also be performed in ampholyte-free mode using a tapered capillary column, which takes advantage of the Joule heat to form a pH gradient. Upon applying an electric field along a tapered capillary filled with tris-hydroxymethylaminomethane hydrochloric acid (Tris-HCl) buffer, the heat generated increases along the capillary axis as the inner capillary of the column decreases, forming a temperature gradient along the capillary, which is due to the sensitive variation of the pK_a value of this buffer with temperature [202,203]. Other CIEF modes comprise dynamic isoelectric focusing that employs capillaries having additional electrodes placed between those at the both ends of the capillary, which allow using multiple voltages within one capillary to manipulate the local pH gradient by the proper variation of the electric field [204]. Also attractive is the use of immobilized pH gradient in the form of a capillary monolithic column bearing ampholytes immobilized at different position in the column according to their pI [205].

6.5.5 Electrokinetic Chromatography

Electrokinetic chromatography (EKC) comprises a variety of electromigration techniques that use electrolyte solutions incorporating a separation carrier, which is called the pseudostationary phase, and are based on the distribution of the analytes between this phase and the surrounding solution.

The pseudostationary phases typically used in EKC can be either charged or neutral and comprise micelles, nanometer-sized droplets, nanoparticles, liposomes, etc. Their common characteristic is the capability of establishing a chemical equilibrium with the analytes dissolved in the BGE, which causes the selective variations of their migration velocity due to either electrophoresis or electroosmosis or both. Typically, charged pseudostationary phases migrate at lower velocity than EOF, which is cathodic with anionic and anodic with cationic pseudostationary phases. Therefore, they migrate in the same direction of EOF, but with a lesser velocity, whereas neutral pseudostationary phases migrate with the same direction and velocity of the EOF. Thus, the pseudostationary phase is typically transported through the electrolyte solution at a different rate than the analytes, whose velocity and separation is therefore influenced by their different affinity for the pseudostationary phase, in absence of which neutral analytes migrate unresolved with the same velocity of the EOF.

Therefore, EKC can be considered as a hybrid of electrophoresis and chromatography, separating neutral analytes exclusively on the basis of their differential partitioning between the electrolyte solution and the pseudostationary phase, whereas the charged analytes are separated on the basis of partitioning and electrophoresis. As mentioned above, pseudostationary phases can be either neutral or charged. It is evident that only charged analytes can be separated using neutral pseudostationary phases.

The first introduced and still more diffuse mode of EKC is micellar electrokinetic chromatography (MEKC), also referred to as micellar electrokinetic capillary chromatography (MECC), which employs micelles as the dispersed phase, formed by incorporating a surfactant into the electrolytes solution at a concentration higher than the critical micellar concentration (CMC) [206]. Composition and pH of the electrolyte solution are selected to generate a sufficiently high EOF to transport micelles and analytes toward the detection windows. Sodium dodecyl sulfate (SDS) is by far the most widely used surfactant, due to its relatively low cost, availability in highly purified forms, and low CMC in aqueous solution (8 mM), above which it forms almost spherical anionic micelles with the charged heads oriented toward the solution and the hydrophobic tails pointing inward the aggregate. SDS is incorporated into electrolyte solutions at neutral to alkaline pH values to ensure strong EOF with bare fused-silica capillaries. Under these conditions, the anionic micelles migrate in opposite direction to EOF and, therefore, are transported toward the cathode with a migration velocity slower than the stream of the bulk flow of BGE.

Neutral analytes not interacting at all with the micelles migrate at the same velocity of EOF, whereas the analytes being totally incorporated into the micelle migrate at the same velocity of the micelle. Neutral analytes participating in reversible interactions with the micelles migrate at the velocity of the micelles when they are associated with a micelle and at the velocity of EOF when they are in the bulk solution. Therefore, these analytes migrate at a velocity comprised between that of EOF and that of micelles, which is proportional to the time they spend associated with the micelles and, consequently, to the strength of their interaction with the dispersed phase. The partitioning of neutral analytes in and out of SDS micelles is mainly due to hydrophobic interactions. Therefore, the hydrophobic diazo dye Sudan III and methanol are generally employed as the markers to measure the migration velocity of micelles and EOF, respectively. The time interval between the migration time of methanol (t_0) and Sudan III (t_{psp}) specifies the migration time window within which neutral analytes can be separated by partitioning between the micellar and the aqueous phase (see Figure 6.10).

MEKC is also performed using cationic, nonionic, and zwitterionic surfactants. Widely employed are cationic surfactant consisting of a long chain tetralkylammonium salt, such as cetyltrimethylammonium bromide, which causes the reversal of the direction of the EOF, due to the adsorption of the organic cation on the capillary wall. Other interesting options include the use of mixed micelles resulting from the simultaneous incorporation into the BGE of ionic and nonionic or ionic and zwitterionic surfactants. Chiral surfactants, either natural as bile salts [207] or synthetic [208] are employed for enantiomer separations.

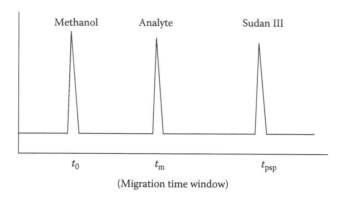

FIGURE 6.10 Schematic representation of the migration time window in MEKC.

The migration behavior of neutral analytes in MEKC can be described by parameters similar to those employed in liquid chromatography [209]. Hence, the ratio of the number of moles of the analyte in the micellar phase, the pseudostationary phase, (n_{psp}) to those in the surrounding solution (n_{lp}) defines the retention factor k:

$$k = \frac{n_{psp}}{n_{lp}} = \frac{t_m - t_0}{t_0\left(1 - (t_m/t_{psp})\right)} \tag{6.45}$$

where
t_m is the migration time of the analyte
t_0 and t_{psp} are the migration times of the markers employed to measure the EOF and the migration rate of the micelles, respectively

The retention factor is related to the distribution coefficient K of the analyte between the micellar phase and the aqueous phase, according to the following equation:

$$k = K\frac{V_{psp}}{V_{lp}} = K\frac{\overline{v}\left(C_{sf} - CMC\right)}{1 - \overline{v}\left(C_{sf} - CMC\right)} \tag{6.46}$$

where
V_{psp} and V_{lp} are the volumes of the micellar and aqueous (liquid) phase, respectively
C_{sf} is the concentration of the surfactant in the BGE
$\overline{v}$ is the partial specific volume of the micelle
CMC is the critical micellar concentration

As evidenced by the above equation, the distribution coefficient can be directly calculated from the retention factor and other easily measurable parameters. It is also worth noting that when the concentration of the micellar phase is sufficiently low, the denominator at the last term of the above equation can be approximated to be equal to unity and the retention factor is linearly proportional to the concentration of the surfactant into the BGE. Accordingly, Equation 6.46 is rewritten as

$$k \cong K\overline{v}\left(C_{sf} - CMC\right) \tag{6.47}$$

The degree of separation of two adjacent peaks is evaluated by the parameter resolution (see Section 6.4.3) that in EKC is expressed by the following equation:

$$R_S = \frac{\sqrt{N}}{4}\left(\frac{\alpha-1}{\alpha}\right)\left(\frac{\bar{k}}{1+\bar{k}}\right)\left(\frac{1-t_0/t_{psp}}{1+(t_0/t_{psp})\,\bar{k}}\right) \tag{6.48}$$

where
α is selectivity, given as the ratio of the retention factors of the two adjacent peaks
$\bar{k}$ is the mean retention factor

The above relationship is derived from Equations 6.36 and 6.45 and is similar to the equation correlating resolution to the separation conditions in liquid chromatography [210], except for the addition of the last term on the right-hand side, which accounts for the migration of the pseudostationary phase within the capillary column.

The equations reported above are related to MEKC and all other EKC separation modes performed with pseudostationary phase, analytes, and EOF moving in the same direction at different velocities. Such condition applies when the electroosmotic mobility is higher than the electrophoretic mobility of the pseudostationary phase migrating to the direction opposite to that of EOF.

EKC in the reversed direction mode is performed when analytes and pseudostationary phase move at different velocities in the same direction, which is opposite to that of EOF. In this case, retention factor and resolution are expressed by the following equations [211]:

$$k = \frac{n_{psp}}{n_{lp}} = \frac{t_m + t_0}{t_0((t_m/t_{psp})-1)} \tag{6.49}$$

and

$$R_S = \frac{\sqrt{N}}{4}\left(\frac{\alpha-1}{\alpha}\right)\left(\frac{\bar{k}}{1+\bar{k}}\right)\left(\frac{1+t_0/t_{psp}}{(t_0/t_{psp})\bar{k}-1}\right) \tag{6.50}$$

EKC is not restricted to the separation of neutral analytes, as it is widely employed for the simultaneous separations of charged and neutral analytes as well as of ionizable compounds having similar electrophoretic mobility. The separation of ionizable analytes by EKC is governed by differences in the partitioning between the pseudostationary phase and the surrounding electrolyte solution as well as electrophoretic mobility. For these analytes, the retention factor can be described by the following mathematical model:

$$k = \frac{t_m - t_{0psp}}{t_{0psp}(1-(t_m/t_{psp}))} \tag{6.51}$$

where t_{0psp} is the migration time of the analyte in the absence of the pseudostationary phase, under otherwise identical experimental conditions. However, t_{0psp} cannot be estimated without making several assumptions, such as negligible effect of the pseudostationary phase on the EOF, on viscosity and permittivity of BGE, and, in case of MEKC or microemulsion EKC (see below), absence of interactions of the analytes with surfactant monomers [212]. Also relevant is the influence of pH on the retention factor of an ionizable analyte, which can be calculated as the weighted average of the retention factor of its undissociated and dissociated forms:

$$k = \frac{k_{HA} + k_A \left(K_a / c \left[H^+ \right] \right)}{1 + \left(K_a / c \left[H^+ \right] \right)} \tag{6.52}$$

where k_{HA} and k_A are the retention factors of the undissociated (HA) and dissociated forms (A) of the analyte, whose dissociation constant and concentration are K_a and c, respectively, and [H^+] is the concentration of protons [213].

Selectivity in EKC can be modulated by changing the pH of BGE and using either a different pseudostationary phase or incorporating an organic solvent or a suitable additive into the electrolyte solution, such as ion-pairing or complex-forming agents, which are expected to influence the partitioning of the analytes between liquid and pseudostationary phases. A typical application of this approach is the separation of enantiomers, generally performed using electrolyte solutions that incorporate a chiral selector, such as cyclodextrins, crown ethers, or macrocyclic antibiotics. Also, organic solvents influence the analyte interactions with the pseudostationary phase, as well as their mobility and, in case of micelles, vesicles, and other aggregates, the physical structure, conformation, or self-association equilibrium of these pseudostationary phases. Another widely used additive is urea, which increases the solubility of very hydrophobic compounds in aqueous solutions, reducing their interaction with nonpolar pseudostationary phases.

Several limitations and shortcomings are associated with the use of micelles as the pseudostationary phase. Besides the irreversible incorporation of very hydrophobic compounds within the micelles, other analytes such as proteins may strongly interact with the free molecules of the surfactant in solution. Moreover, the significant influence of operational parameters such as temperature, pH, and composition of the BGE on the dynamic aggregation of the surfactant molecules may result in instable micelles and consequent irreproducibility.

One of the possible alternative to micelles are spherical dendrimers of diameter generally ranging between 5 and 10 nm. These are highly structured three-dimensional globular macromolecules composed of branched polymers covalently bonded to a central core [214]. Therefore, dendrimers are topologically similar to micelles, with the difference that the structure of micelles is dynamic whereas that of dendrimers is static. Thus, unlike micelles, dendrimers are stable under a variety of experimental conditions. In addition, dendrimers have a defined number of functional end groups that can be functionalized to produce pseudostationary phases with different properties. Other pseudostationary phases employed to address the limitations associated with the micellar phases mentioned above and to modulate selectivity include water-soluble linear polymers, polymeric surfactants, and gemini surfactant polymers.

Nanoparticles of diameter ranging from decades to hundreds of nanometers, bearing selected functionalities on their surface, such as ionizable groups or chiral selectors, are alternative choices as pseudostationary phases. Either polymer-based or silica nanoparticles are employed [215]. The nanoparticles are suspended in the electrolyte solution to form stable slurries, which are employed as the pseudostationary phase, according to two different operational modes utilizing either the partial or the total filling of the capillary with the slurry solution, as depicted in Figure 6.11. In the partial filling arrangement, a plug of the slurry solution is introduced into the capillary before the sample, and as the electric field is applied, the analytes migrate through the dispersed pseudostationary phase and reach the detection windows prior to the nanoparticle plug. In the total or full-filling arrangement the sample is injected in the capillary completely filled with the slurry solution and the analyte detection takes place in presence of the dispersed nanoparticles [216,217]. The advantage of the partial filling approach include the total absence of detection drawbacks due to light scattering or/and adsorption of the dispersed nanoparticles.

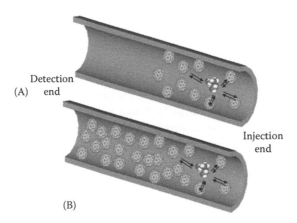

(B)

FIGURE 6.11 Schematic representation of (A) partial-filling approach and (B) full-filling approach of EKC with pseudostationary phases consisting of nanoparticles dispersed into the BGE.

As mentioned at the beginning of this section, pseudostationary phases also comprise nanometer-sized oil droplets. These carriers consist of droplets of a water-immiscible liquid dispersed in an aqueous electrolyte solution to form microemulsions. Heptane and octane are the widely used water-immiscible liquids used to form the microemulsion, although diethyl ether, amyl alcohol, cyclohexane, chloroform, methylene chloride, and several chiral compounds, such as (2R, 3R)-di-n-butyl tartrate, are alternative choices. The microemulsion is typically formulated by incorporating, in a suitable ratio, the water-immiscible liquid into the aqueous BGE with a surfactant (e.g., SDS) and a cosurfactant, generally a short-chain alkyl alcohol, such as 1-butanol. The surfactant is incorporated into the BGE to facilitate the formation of the droplets by lowering the surface tension, further lowered by the cosurfactant, which has the function of stabilizing the microemulsion. This EKC mode is termed microemulsion electrokinetic capillary chromatography (MEEKC) or microemulsion electrokinetic capillary chromatography (MEECC) [138]. It employs alkaline electrolyte solutions to generate high cathodic EOF, which transports the SDS-coated (or other anionic surfactant-coated) oil droplets to the cathode, in spite of their superficial negative charges. Neutral analytes are separated according to their partitioning between oil and aqueous phase whereas charged compounds may establish either attractive or repulsive electrostatic interactions with the negatively charged droplets, while attempting to migrate toward the anode against the EOF, which sweeps them through the detector windows at the cathodic end of the capillary. MEEKC offers more operational parameters than MEKC to modulate selectivity and to enlarge the separation windows, which include the concentration and choice of the water-immiscible liquid, surfactant, and cosurfactant. Also, efficiency may result superior in MEEKC than in MEKC due to faster mass transfer between the microemulsion and the surrounding aqueous phase.

The dispersed pseudostationary phase may also consist of liposomes. These are vesicles formed by phospholipids by a self-assembling process in aqueous solutions, which are composed of one or more lipid bilayer membranes that have entrapped a volume of the surrounding aqueous media during the self-assembling process [218]. Liposomes are classified on the basis of size and the number of lipid bilayers. Multilamellar vesicles (MLV) have more than one lipid bilayer and size up to 5.0 μm, whereas unilamellar vesicles consist of a single lipid bilayer and are distinguished in large unilamellar vesicles (LUV) and small unilamellar vesicles (SUV) if their size is in the range of 100–400 nm and 20–50 nm, respectively [219]. EKC using liposomes as the pseudostationary phase can be employed to separate neutral and charged low molecular size analytes [220] as well as proteins [221,222] and to study the interaction between liposomes and drugs [220,223].

6.5.6 CAPILLARY ELECTROCHROMATOGRAPHY

Capillary electrochromatography (CEC) can be considered as a variant of capillary HPLC in which the flow of the mobile phase through the column is propelled by electroosmosis instead of a mechanic pump. The capillary column can be either filled, packed, or coated with a stationary phase, which plays the dual role of providing sites for the required interactions with the analytes and charged groups for the generation of the EOF that ensures the movement of the mobile phase though the column. Stationary phases carrying cationic functional groups, such as amino or ammonium groups generate anodic EOF whereas stationary phases with anionic functionalities, such as sulfonic or acetic groups, generate cathodic EOF. The stationary phase may also carry zwitterionic groups and in this case exhibits either cathodic or anodic EOF, according to the pH of the mobile phase.

The capillary format of the column maximizes the surface area-to-volume ratio and, therefore, facilitates the rapid and efficient dissipation of the Joule heat, which can be of a marked extent when applying the elevated voltages requested to obtain high flow rates. As mentioned in Section 6.1, the flow of mobile phase propelled by EOF displays a flat plug-like profile, which does not contribute to peak broadening caused by the sample transport trough the column because the driving force is uniformly distributed along the capillary tube. This is in contrast to the laminar or parabolic flow profile generated in HPLC, where there is a strong pressure drop across the column caused by frictional forces at the liquid–solid boundary. The absence of column backpressure allows performing CEC also with capillary columns of low permeability, such as those packed with particles in the submicron size range [224], whereas the length is mainly limited by the value of the electric field requested to obtain the desired flow rate with commercially available power supplies, which usuallyoperate up to 30 kV.

Capillary columns packed with either silica-based or polymeric particles have been used since the earlier development of CEC [225–227] due to the availability and relatively easy preparation of a wide range of packing material of different diameter, pore size, and surface functionalities tailored for a variety of separation modes, such as in HPLC. Particles are usually packed into a fused-silica capillary tube by a variety of techniques, the most common of which is pressure packing of a liquid slurry using either water, organic solvents [228,229] or supercritical carbon dioxide as the carrier [230]. The generally accepted method employs a packing reservoir, such as a short 2.0 mm I.D. HPLC column, which is connected to the capillary tube to be packed to one end and to a high pressure solvent delivery pump to the other end. The slurry reservoir with the attached column is usually immersed in a ultrasonic bath to maintain the slurry homogeneity during packing, which is carried out by transporting the suspended packing material into the capillary column by pressure, typically between 35.0 and 70.0 MPa (≈5,000 to ≈10,000 psi). Alternative techniques include electrokinetic packing [231], procedures employing centripetal forces [232,233] or combination of high electric field and hydrodynamic flow [234] and packing techniques by gravity [235].

Any of these techniques require the fabrication of retaining frits within the capillary column, which have to posses high permeability to the mobile phase flow and, at the same time, must be mechanically strong to retain the packing material and to resist the pressure generated during the packing process and the flushing procedures requested to fill and condition the column with the mobile phase for its use. The frits are usually fabricated by sintering a plug of silica gel, usually wetted with either water or potassium silicate, introduced into the capillary tube. A heating filament or small coil, rather than the flame of a Bunsen burner used in the earlier methods, are generally employed for sintering the silica plug [236]. When the chromatographic support consists of silica-based material, the frits are generally fabricated by sintering the packing material, having the caution to minimize the degradation of the stationary phase bonded to the silica particles adjacent to the zone of the packing material heated to form the frit. Alternatively, the frits can be produced by polymerization of potassium silicate solution containing formamide, which is introduced into the capillary and polymerized by heating on a steam bath [237].

Sintered frits can be the source of several drawbacks, such as band broadening, bubble formation, EOF inhomogeneity, and insufficient reproducibility of the sintering method. An additional problem is the column fragility at the frit position due to the removal of the polyimide film during the thermal treatment, which may also cause alteration of the surface properties of the packing material when this is used to fabricate the frits. An alternative approach to frit formation is the use of capillaries with a restrictor or a fine tapered end, which retain the packing material inside the capillary column [238]. Tapered capillaries can be produced by drawing the fused-silica capillary through a high-temperature flame or using a laser-based micropipette puller. Such methods produce externally tapered capillaries having reduced outer and inner diameters, which are, therefore, very fragile. The alternative method, producing more robust columns, is carried out by melting the tip of the capillary in a high-temperature flame thereby sealing its end. Then the sealed end is carefully ground to produce an opening of the required diameter. Using this method, only the inner diameter of the capillary is reduced in dimension whereas the outer diameter is unvaried. Packed columns with either external or internal taper at the outlet of the capillary are suitable to be coupled with electrospray ionization and mass spectrometry (CEC-ESI-MS) [238,239].

A capillary restrictor can also be realized by joining two capillaries of same outer diameter (usually 375 μm) and different bore. The two capillaries, having uniform cuts at their ends, are joined by pushing them into a small length of PVC tube of 350 μm I.D. until their ends face each other [240]. Another possible approach is the use of fritless packed capillaries, which are fabricated without the frit at the detection end and with the tapered end at the injection side. These fritless capillaries can only be used with packing materials having electrophoretic mobility larger than EOF and directed to the opposite direction of it. For example, in the case of packing material consisting of negatively charged particles, upon applying the electric field the particles are attracted by the anode. If their electrophoretic mobility is large enough, it prevents the particles being flushed out from the column by the EOF, which transports the mobile phase and the analytes through the CEC column [241]. Besides the above limitations associated with the direction and magnitude of the electrophoretic mobility of the packed charged particles and the EOF, further problems arise from the poor stability of the packing bed when the electric field is not applied and from the fragility of the tapered column inlet.

The problems with packing and retaining particles in capillary tubes are eliminated by the use of columns containing in situ prepared monolithic separation media formed from either organic polymers or silica. Such monolithic columns are characterized by a bimodal pore structure consisting of large pores for flow the mobile phase through the stationary phase and diffusion pores for analyte-stationary phase interactions. The organic-based monolithic columns are generally made of a mixture of monomers, crosslinkers, and porogens by radical polymerization, which is conducted directly within the confine of a capillary. These monolithic columns have excellent pH stability. In addition, porous size and distribution, as well as EOF and retentive properties, can be easily tailored by tuning the composition of the reaction mixture that can be composed of a variety of ionizable monomers, neutral reactants, and porogenic solvents at different percentage [242,243]. The main drawbacks of organic polymer–based monolithic columns are their limited mechanical strength, in addition to the presence of micropores that may negatively affect efficiency and sample recovery. Also challenging is their tendency either to swell or to shrink when exposed to different organic solvents, which can be incorporated into the mobile phase.

Silica-based monolithic CEC column can be prepared by a multistep approach involving the preparation of a silica monolithic skeleton attached to the capillary wall, which is first treated with an alkaline solution to leaching the micropores and then chemically modified with different silylation reagents for binding the functionalities requested to operate the column at different separation modes [244,245]. Alternatively, hybrid organic–inorganic monolithic CEC columns can be prepared by hydrolytic polycondensation of siloxane and organosiloxane precursors, in presence of an acid catalyst and a water-soluble polymer acting as porogen, by one step, in situ, sol-gel process under mild reaction conditions. In such monolithic stationary phases, an organic moiety is covalently linked by a nonhydrolyzable Si–C bond to a siloxane specie, which hydrolyzes to produce a silica network [246]. The selection of proper sol-gel precursors allows the preparation of silica-based monoliths having ionizable groups and interaction sites that ensure the requested EOF and chromatographic behaviors [247]. Main advantages of silica-based monolithic columns are the high mechanical stability and the presence of surface silanol groups that provide reactive sides for the covalent bonding of an array of functional groups to generate stationary phases having different EOF characteristics and operating in a variety of separation modes.

Another alternative approach to packed capillaries are the so-called open-tubular columns consisting of a capillary tube whose inner surface is coated with a stationary which interacts with the analytes dissolved in the mobile phase transported by the EOF, as first reported by Tsuda et al. for the separation of aromatic compounds using an open silica capillary with octa-decylsilane-bonded inner surface [248]. Theoretically, any possible ligand suitable to prepare stationary phases employed in liquid chromatography can be used to make capillary columns for open-tubular capillary electrochromatography (OT-CEC). Also, wide is the array of methods employed for surface modification and immobilization of the stationary phase, which include adsorption [249], covalent bonding of monomers (with or without cross-linking) [250,251] chemical bonding after etching [250,252], grafting polymeric porous layer to the inner surface [253], chemical reactions by sol-gel processes [254], and coating the inner surface of the capillary with nanoparticles [255].

The most important advantages of open-tubular columns over packed capillary columns are common to those reported for monolithic columns, with additional merits arising from the possibility of preparing columns in the small internal diameter range of 20–25 µm, which may be requested in miniaturized analytical techniques. On the other hand, several drawbacks are typical of open-tubular capillary columns, such as the low sample capacity, low phase ratio, and short optical path length for on-column UV-visible and fluorescence detection methods. Therefore, most of the different approaches cited above and recent innovations in developing novel column for OT-CEC have been mainly aimed at increasing the surface area of the bonded stationary phase, besides improving selectivity for a given separation problem.

The separation of uncharged compounds in CEC occurs as in HPLC due to their partitioning between the stationary and the mobile phases, with the only difference that the movement of the mobile phase through the column is propelled by electroosmosis in CEC and by a mechanic pump in HPLC. Consequently, the retention of uncharged analytes can be described by the chromatographic retention factor, k, which is expressed by the well-known equation:

$$k = \frac{t_R - t_0}{t_0} \tag{6.53}$$

where t_R and t_0 are the retention (migration) times of the uncharged analyte and of an unretained uncharged tracer, respectively.

More complex is the separation of charged analytes in CEC, which is the result of the interplay of chromatographic and electrophoretic processes that is considered in the definition of the electrochromatographic retention factor, or overall retention factor, k_c, introduced by Rathore and Horváth [140]:

$$k_c = k + k k_e + k_e \tag{6.54}$$

where

k is the chromatographic retention factor

k_e is the electrophoretic velocity factor (see Equation 6.42) and the product $k k_e$ reflects the simultaneous occurrence of chromatographic and electrophoretic separation processes

It is worth noting that for uncharged compound k_e is zero and the electrochromatographic retention factor is equal to the retention factor defined by Equation 6.53.

The selectivity coefficient or separation factor, α, in CEC is given by the ratio of the electrochromatographic retention factors of the two analytes (1 and 2) migrating as adjacent peaks:

$$\alpha_{1,2} = \frac{k_{c,2}}{k_{c,1}} \tag{6.55}$$

where $k_{c,1}$ and $k_{c,2}$ are the electrochromatographic retention factors for the analytes 1 and 2, respectively, with $k_{c,2} > k_{c,1}$, such that $\alpha_{1,2} > 1$.

The separation factor influences the degree of separation of two adjacent peaks, which is also affected by the number of the theoretical plates, according to the following equation:

$$R_S = \frac{\sqrt{N}}{2} \left(\frac{\alpha_{1,2} - 1}{\alpha_{1,2} + 1} \right) \left(\frac{\bar{k}_c}{1 + \bar{k}_c} \right) \tag{6.56}$$

where $\bar{k}_c$ is the average value of the electrochromatographic retention factors of the two adjacent peaks.

The limits of the electrochromatographic retention factor defined by Equation 6.54 are that both k and k_e must be separately evaluated under specific experimental conditions used in CEC. These limitations can be avoided describing the retention of analytes in CEC by a peak locator that can be evaluated directly from the electrochromatogram as follows [256]:

$$k_{cc} = \frac{t_m - t_0}{t_0} \tag{6.57}$$

where t_m and t_0 are the migration times of the analyte and of a uncharged and chromatographically inert tracer, respectively. This peak locator is expressed in terms typical of retention factor in HPLC, as defined by Equation 6.53. However, in contrast to k in HPLC, k_{cc} may exhibit negative values for charged analytes eluting prior to the marker. It is worth noting that identical resolution values are obtained using either k_c or k_{cc}, whereas the values of selectivity calculated using k_c or k_{cc} differ from each other due to the different definition of the peak locators, as given in Equations 6.54 and 6.57.

A variant of CEC is pressurized capillary electrochromatography (pCEC), in which the mobile phase is propelled by both electroosmosis and pressurized flow generated by an HPLC pump [257,258]. The main advantage of pCEC is the possibility of regulating the flow rate independently by the applied voltage, which offers the opportunity to shortening the analysis time, in addition to minimizing the risk of bubble formation. Such advantages are obtained at the expense of column efficiency, which can be reduced as a consequence of the contribution to band broadening due to the parabolic flow profile produced by the mechanical pump [249]. Capillary electrochromatography is generally performed under isocratic elution mode due to technical difficulty to change the composition of the electrolyte solution at the inlet end of the CEC column, although examples of CEC under either step gradient or linear gradient elution mode have been reported, using appositively modified instruments [260–263].

6.5.7 CAPILLARY ISOTACHOPHORESIS

Capillary isotachophoresis (CITP) is an electromigration technique, which is performed using a discontinuous buffer system, formed by a leading electrolyte (LE) and a terminating electrolyte

(TE), containing a fast and a slow migrating component of like charge, respectively. The sample is introduced into the capillary filled with the leading electrolyte, which has higher electrophoretic mobility than any of the sample components to be separated. The terminating electrolyte, having electrophoretic mobility lower than any of the sample components, occupies the opposite reservoir. Therefore, the injected sample occupies a zone of the capillary column in between the leading electrolyte and the terminating electrolyte. In each single ITP analysis, either cations or anions can be separated [264].

Highly mobile zones have high conductivity, and as a result, have a lower voltage drop across the band. On the other hand, low mobile zones have low conductivity and consequently a higher voltage drop across the band occurs. Therefore, the electric field varies in each zone that moves as a band. Since the velocity of migration is the product of the electrophoretic mobility and the electric field and conductivity and voltage drop are inversely proportional, the individual band velocities are self-adjusting to a constant value. A consequence of the self-adjusting velocity of migration is that each band maintains very sharp boundaries with the neighboring faster and slower bands. Thus, if an ion diffuses into a neighboring zone, it will either speed up or slow down based on the field strength encountered and it returns to the original band. A schematic representation of the CITP separation process is depicted in Figure 6.12.

The main determining factor in CITP is the composition of the leading electrolyte and of the terminating electrolyte. To separate anions, for example, the LE must be an anion with electrophoretic mobility higher than that of each analyte, whereas the anion acting as TE must have an electrophoretic mobility lower than that of each analyte. When the electric potential is applied, the anionic components of sample, the LE and the TE start to migrate toward the anode. Since the leading anion has the highest electrophoretic mobility, it moves fastest, followed by the anion having the next fastest mobility, which moves faster than the anion having the next fastest mobility, and so on. The result is that the sample components are separated according to the order of their electrophoretic mobility into distinct zones, which are sandwiched between the leading and the terminating electrolyte, forming a front and a rear zone, respectively [265]. The separated zones are surrounded by sharp electrical field differences and their profiles can be negatively affected by the EOF, which is usually regarded as undesirable in CITP, although well-separated bands can be obtained under favorable conditions, such as EOF and isotachophoretic velocities directed in the same direction [266,267]. The EOF is generally suppressed, either using a covalently coated

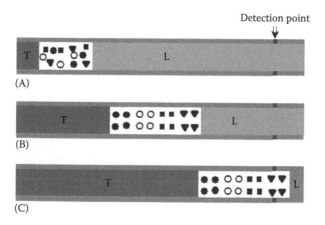

FIGURE 6.12 Schematic view of the CITP separation mechanism. The sample is introduced into the capillary between two electrolyte systems: a leading electrolyte (L), having electrophoretic mobility higher than any of the sample components to be separated and a terminating electrolyte (T), having electrophoretic mobility lower than any of the sample components (A). The sample components are separated according to the order of their individual mobility into distinct zones, which are sandwiched between T and L (B). The separated zones move with the same velocity toward the capillary end where they are detected as bands (C).

capillary or incorporating a suitable additive into the electrolyte solution, such as hydroxyethylcellulose, methylcellulose, and linear polyacrylamide, which increase the viscosity of the BGE and, therefore, minimize the EOF.

Crucial factors that must be considered in selecting the suitable composition of the electrolyte solution for CITP of either cations or anions include the choice of the pH range, which determines the sufficient ionization of the ample components. Another important factor is the range of electrophoretic mobility spanned by the LE and the TE, which determines the so-called mobility window of potential analytes that can be analyzed using a given LE–TE system [268]. Since adjacent bands are in contact with each other, a marker substance (spacer) having a mobility value that falls in between the mobilities of two peaks that need to be resolved is generally added to the sample in order to improve resolution. When UV-visible detection is employed, the spacer is a non-adsorbing substance in order to facilitate the analyte detection, which is otherwise traditionally performed by monitoring the variations in conductivity, although other detection methods are also possible.

Capillary isotachophoresis is usually performed in constant current mode, which implies the invariable ratio between concentration and electrophoretic mobility of ions. Therefore, bands that are less concentrated than the LE are sharpened, whereas those that are more concentrated than the LE are broadened to adapt their concentration to the requested constant value between concentration and electrophoretic mobility. The consequence of this unique property of CITP is that each sample component can be concentrated to an extent that depends on its initial concentration and the concentration of the leading electrolyte. Therefore, the opportune selection of composition and concentration of the leading electrolyte allows the enrichment of diluted analytes.

Sample enrichment techniques based on a transient isotachophoresis step performed in the capillary prior to start the selected electromigration method are widely employed. The transient isotachophoretic step is carried out by incorporating a high concentration of the leading and/or terminating electrolyte in either the sample or the BGE. According to this technique, relatively large volumes of samples introduced into the capillary can be concentrated in the isotachophoresis step by stacking the analytes in a very sharp zone [269–273]. Sample preconcentration can also be performed by coupling two capillary columns [274–276]. According to this approach, the sample migrates between a LE and a TE in the first capillary where it is concentrated by CITP, whereas in the second capillary, which is on-line connected with the exit of the first one, the analysis continues in another electromigration separation mode.

REFERENCES

1. P. Kubáň, P. C. Hauser, *Anal. Chim. Acta* 607, 15–29, 2008.
2. F.-M. Matysik, *Microchim. Acta* 160, 1–14, 2008.
3. X.-J. Huang, Z.-L. Fang, *Anal. Chim. Acta* 414, 1–14, 2000.
4. W. F. Smyth, *Electrophoresis* 27, 2051–2062, 2006.
5. D. A. Jayawickrama, J. V. Sweedler, *J. Chromatogr. A* 1000, 819–840, 2003.
6. C. W. Huck, R. Bakry, L. A. Huber, G. K. Bonn, *Electrophoresis* 27, 2063–2074, 2006.
7. P. C. Hiemenz, *Principles of Colloid and Surface Chemistry.* Marcel Dekker, Inc., New York, 1986.
8. A. W. Adamson, *Physical Chemistry of Surfaces.* John Wiley & Sons, Inc., New York, 1990.
9. R. J. Hunter, *Zeta Potential in Colloid Science.* Academic Press, London, U.K., 1981.
10. P. Jandik, G. Bonn, *Capillary Electrophoresis of Small Molecules and Ions.* VCH Publishers, Inc., New York, 1993.
11. J. P. Landers, *Handbook of Capillary Electrophoresis.* CRC Press, Boca Raton, FL, 1994.
12. D. C. Grahame, *Chem. Rev.* 41, 441–501, 1947.
13. M. von Smoluchowski, in I. Graetz (Ed.) *Handbuch der Elektrizitat und des Magnetismus*, Barth, Leipzig, Germany, 1921, pp. 366–428.
14. W. Schützner, E. Kenndler, *Anal. Chem.* 64, 1991–1995, 1992.
15. E. Sahlin, S. G. Weber, *J. Chromatogr. A*, 972, 283–287, 2002.
16. X. Huang, Cs. Horváth, *J. Chromatogr. A*, 788, 155–164, 1997.

17. T. Tsuda, J. V. Sweedler, R. N. Zare, *Anal. Chem.* 62, 2149–2152, 1990.
18. A. Cifuentes, M. A. Rodriguez, F. J. Garcia-Montelongo, *J. Chromatogr. A* 737, 243–253, 1996.
19. F. E. P. Mikkers, F. M. Everaerts, T. P. E. M. Vergheggen, *J. Chromatogr.* 169, 11–20, 1979.
20. T. Tsuda, T. Mizuno, J. Akiyama, *Anal. Chem.* 59, 799–800, 1987.
21. M. Deml, F. Foret, P. Boček, *J. Chromatogr.* 320, 159–165, 1985.
22. G. B. Gordon, U.S. Patent 5,061,361, October 29, 1991.
23. J. P. Chervet, M. Ursem, J. P. Salzmann, R. W. Vannoort, *J. High Resolut. Chromatogr. Chromatogr. Commun.* 12, 278–281, 1989.
24. T. Wang, J. H. Aiken, C. W. Huie, R. A. Hartwick, *Anal. Chem.* 63, 1372–1376, 1991.
25. Technical Note n. 5965–5984, Agilent Technologies, Santa Clara, CA, 1997.
26. D. Xiao, S. Zhao, H. Yuan, X. Yang, *Electrophoresis* 28, 233–242, 2007.
27. W. Tong, E. S. Yeung, *J. Chromatogr. A* 718, 177–185, 1995.
28. P. A. G. Butler, B. Mills, P. C. Hauser, *Analyst* 122, 949–953, 1997.
29. S. Casado-Terrones, S. Cortacero-Ramírez, *Anal. Bioanal. Chem.* 386, 1835–1847, 2006.
30. C. Johns, M. Macka, P. R. Haddad, *Electrophoresis* 25, 3145-31-52, 2004.
31. J. P. Hutchinson, C. J. Evenhuis, C. Johns, A. A. Kazarian, M. C. Breadmore, M. Macka, E. F. Hilder, R. M. Guijt, G. W. Dicinoski, P. Haddad, *Anal. Chem.* 79, 7005–7013, 2007.
32. M. C. Breadmore, R. D. Henderson, A. R. Fakhari, M. Macka, P. R. Haddad, *Electrophoresis* 28, 1252–1258, 2007.
33. J. S. Frits, G. H. Schenk, *Quantitative Analytical Chemistry*, 5th edn., Allin & Bacon, Inc., Boston, MA, 1987, pp. 369–421.
34. E. Gassmann, J. E. Kuo, R. N. Zare, *Science* 230, 813–814, 1985.
35. A. M. García-Campaña, M. Taverna, H. Fabre, *Electrophoresis* 28, 208–232, 2007.
36. M. E. Johnson, J. P. Landers, *Electrophoresis* 25, 3513–3527, 2004.
37. J. A. Taylor, E. S. Yeung, *Anal. Chem.* 64, 1741–1745, 1992.
38. P. G. Schiro, C. L. Kuyper, D. T. Chiu, *Electrophoresis* 28, 2430–2438, 2007.
39. T. Tsuda, J. W. Sweedler, R. N. Zare, *Anal. Chem.* 64, 967–972, 1992.
40. S. Wu, N. J. Dovichi, *J. Chromatogr.* 480, 141–155, 1989.
41. D. Y. Chen, N. J. Dovichi, *Anal. Chem.* 68, 690–696, 1996.
42. A. E. Bruno, F. Maystre, B. Krattiger, P. Nussbaum, E. Gassmann, *Trends Anal. Chem.* 13, 190–198, 1994.
43. S. Nagaraj, H. T. Karnes, *Biomed. Chromatogr.* 14, 234–242, 2000.
44. B. Yang, H. Tian, J. Xu, Y. Guan, *Talanta* 69, 996–1000, 2006.
45. J. Xu, S. Chen, Y. Xiong, B. Yang, Y. Guan, *Talanta* 75, 885–889, 2008.
46. E. P. Jong, C. A. Lucy, *Anal. Chim. Acta* 546, 37–45, 2005.
47. W. Du, S. Chen, Y. Xu, Z. Chen, Q. Luo, B.-F. Liu, *J. Sep. Sci.* 30, 906–915, 2007.
48. M. L. Gostkowski, J. B. McDoniel, J. Wie, T. E. Curey, J. B. Shear, *J. Am. Chem. Soc.* 120, 18–22, 1998.
49. J. E. Melanson, C. A. Lucy, C. A. Boulet, *Anal. Chem.* 73, 1809–1813, 2001.
50. K. Tóth, K. Štulik, W. Kutner, Z. Fehér, E. Lindner, *Pure Appl. Chem.* 76, 1119–1138, 2004.
51. A. J. Zemann, *Trends Anal. Chem.* 20, 346–354, 2001.
52. X. Huang, T. J. Pang, M. J. Gordon, R. N. Zare, *Anal. Chem.* 59, 2747–2749, 1987.
53. X. Huang, R. N. Zare, *Anal. Chem.* 63, 2193–2196, 1991.
54. A. J. Zemann, E. Schnell, D. Volgger, G. K. Bonn, *Anal. Chem.* 70, 563–567, 1998.
55. J. A. Freacassi da Silva, C. L. do Lago, *Anal. Chem.* 70, 4339–4343, 1998.
56. D. Kaniansky, V. Zelenskaè, M. Masaèr, F. Ivanyi, Sí. Gazdikova, *J. Chromatogr. A* 844, 349–359, 1999.
57. V. Unterholzner, M. Macka, P. R. Haddad, A. J. Zemann, *Analyst* 127, 715–718, 2002.
58. M. Novotný, F. Opekar, I. Jelínek, K. Štulík, *Anal. Chim. Acta* 525, 17–21, 2004.
59. F. Tan, B. C. Yang, Y. F. Guan, *Anal. Sci.* 21, 583–585, 2005.
60. P. L. Weber, T. Kornfelt, N. K. Klausen, S. M. Lunte, *Anal. Biochem.* 225, 135–142, 1995.
61. J. Wen, A. Baranski, R. Cassidy, *Anal. Chem.* 70, 2504–2509, 1998.
62. R. A. Wallingford, A. G. Ewing, *Anal. Chem.* 59, 1762–1766, 1897.
63. T. J. O'Shea, R. D. Greenhagen, S. M. Lunte, C. E. Lunte, M. R. Smyth, D. M. Radzik, N. Watanabe, *J. Chromatogr.* 593, 305–312, 1992.
64. R. Chen, H. Cheng, W. Wu, X. Ai, W. Huang, Z. Wang, J. Cheng, *Electrophoresis* 28, 3347–3361, 2007.
65. X. Huang, R. N. Zare, S. Sloss, A. G. Ewing, *Anal. Chem.* 63, 189–192, 1991.
66. F.-M. Matysik, *Electroanalysis* 12, 1349–1355, 2000.

67. M. Zhong, S. M. Lunte, *Anal. Chem.* 68, 2488–2493, 1996.
68. M. Chen, H. J. Hung, *Anal. Chem.* 67, 4010–4014, 1995.
69. M. Lowry, S. O. Fakayode, M. L. Geng, G. A. Baker, L. Wang, M. E. McCarroll, G. Patonay, I. M. Warner, *Anal. Chem.* 80, 4551–4574, 2008.
70. X. Huang, J. Ren, *Trends Anal. Chem.* 25, 155–166, 2006.
71. X.-B. Yin, E. Wang, *Anal. Chim. Acta* 533, 113–120, 2005.
72. E. V. Efremov, F. Ariese, C. Gooijer, *Anal Chim. Acta* 606, 119–134, 2008.
73. R. J. Dijkstra, F. Ariese, C. Gooijer, U. A. Th. Brinkman, *Trends Anal. Chem.* 24, 304–323, 2005.
74. H. Zhu, I. M. White, J. D. Suter, M. Zourob, X. Fan, *Anal. Chem.* 79, 930–937, 2007.
75. A. E. Bruno, B. Krattiger, F. Maystre, H. M. Widmer, *Anal. Chem.* 63, 2689–2697, 1991.
76. S. E. Johnston, K. E. Fadgen, J. W. Jorgenson, *Anal. Chem.* 78, 5309–5315, 2006.
77. M. Yu, N. J. Dovichi, *Anal. Chem.* 61, 37–40, 1989.
78. S. L. Pentoney Jr., R. N. Zare, J. F. Quint, *Anal. Chem.* 61, 1642–1647, 1989.
79. A. Gaspar, M. Englmann, A. Fekete, M. Harir, P. Schmitt-Kopplin, *Electrophoresis* 29, 66–79, 2008.
80. P. Schmitt-Kopplin, M. Englmann, *Electrophoresis* 26, 1209–1220, 2005.
81. C. W. Klampfl, *J. Chromatogr. A* 1044, 131–144, 2004.
82. M. Yamashita, J. B. Fenn, *J. Phys. Chem.* 88, 4451–4459, 1984.
83. M. Mann, C. K. Meng, J. B. Fenn, *Anal. Chem.* 61, 1702–1708, 1989.
84. R. D. Smith, J. A. Olivares, N. T. Nguyen, H. R. Udseth, *Anal. Chem.* 60, 436–441, 1988.
85. J. A. Olivares, N. T. Nguyen, C. R. Yonker, R. D. Smith, *Anal. Chem.* 59, 1230–1232, 1988.
86. E. D. Lee, W. Muck, J. D. Henion, *J. Chromatogr.* 458, 313–321, 1988.
87. E. D. Lee, W. Muck, J. D. Henion, T. R. Covey, *Biomed. Environ. Mass Spectrom.* 18, 844–850, 1989.
88. J. H. Whal, D. C. Gale, R. D. Smith, *J. Chromatogr. A* 659, 217–222, 1994.
89. M. S. Kriger, K. D. Cook, R. S. Ramsy, *Anal. Chem.* 67, 385–389, 1995.
90. L. Bendahl, H.S. Hansen, J. Olsen, *J. Rapid Commun. Mass Spectrom.* 16, 2333–2340, 2002.
91. E. P. Maziarz, S. A. Lorentz, T. P. White, T. D. Wood, *J. Am. Soc. Mass Spectrom.* 11, 659–663, 2000.
92. P. Cao, M. Moini, *J. Am. Soc. Mass Spectrom.* 9, 1081–1088, 1998.
93. J. P. Quirino, M. T. Dulay, B. D. Bennet, R. N. Zare, *Anal. Chem.* 73, 5539–5543, 2001.
94. J. T. Whitt, M. Moini, *Anal. Chem.* 75, 2188–2191, 2003.
95. R. Mol, G. J. de Jong, G. W. Somsen, *Electrophoresis* 26, 146–154, 2005.
96. C. Rentel, P. Gfrörer, E. Bayer, *Electrophoresis* 20, 2329–2336, 1999.
97. T. Johnson, J. Bergquist, R. Ekman, E. Nordhoff, M. Schürenberg, K.-D. Klöppel, M. Müller, H. Lehrach, J. Gobom, *Anal. Chem.* 73, 1670–1675, 2001.
98. K. L. Walker, R. V. Chiu, C. A. Monnig, C. L. Wilkins, *Anal. Chem.* 67, 4197–4204, 1995.
99. H. Y. Zhang, R. M. Caprioli, *J. Mass Spectrom.* 31, 1039–1046, 1996.
100. K. K. Murray, *Mass Spectrom. Rev.* 16, 283–299, 1997.
101. A. I. Gusev, *Fresenius' J. Anal. Chem.* 366, 691–700, 2000.
102. T. Rejtar, P. Hu, P. Juhasz, J. M. Campbell, M. L. Vestal, J. Preisler, B. L. Karger, *J. Proteome Res.* 1, 171–179, 2002.
103. M. Lechner, A. Seifner, A. M. Rizzi, *Electrophoresis* 29, 1974–1984, 2008.
104. S. Y. Chang, E. S. Yeung, *Anal. Chem.* 69, 2251–2257, 1997.
105. J. Preisler, F. Foret, B. L. Karger, *Anal. Chem.* 70, 5278–5287, 1998.
106. J. Preisler, P. Hu, T. Rejtar, B. L. Karger, *Anal. Chem.* 72, 4785–4795, 2000.
107. J. Preisler, P. Hu, T. Rejtar, E. Moskovets, B. L. Karger, *Anal. Chem.* 74, 17–25, 2002.
108. H. Orsnes, T. Graf, S. Bohatka, H. Degn, *Rapid Commun. Mass Spectrom.* 12, 11–14, 1998.
109. H. K. Musyimi, D. A. Narcisse, X. Zhang, W. Stryjewski, S. A. Soper, K. K. Murray, *Anal. Chem.* 76, 5968–5973, 2004.
110. N. Wu, T. L. Peck, A. G. Webb, R. L. Magin, J. V. Sweedler, *Anal. Chem.* 66, 3849–3857, 1994.
111. N. Wu, T. L. Peck, A. G. Webb, R. L. Magin, J. V. Sweedler, *J. Am. Chem. Soc.* 116, 7929–7930, 1994.
112. K. Pusecker, J. Schewitz, P. Gfrörer, L.-H. Tseng, K. Albert, E. Bayer, *Anal. Chem.* 70, 3280–3285, 1998.
113. D. I. Hoult, R. E. Richards, *J. Magn. Reson.* 24, 71–85, 1976.
114. Y. Li, M. E. Lacey, J. V. Sweedler, A. G. Webb, *J. Magn. Reson.* 162, 133–140, 2003.
115. D. L. Olson, M. E. Lacey, A. G. Webb, J. V. Sweedler, *Anal. Chem.* 71, 3070–3076, 1999.
116. J. Schewitz, K. Pusecker, L.-H. Tseng, K. Albert, E. Bayer, *Electrophoresis* 20, 3–8, 1999.
117. A. W. Wolters, D. A. Jayawickrama, A. G. Webb, J. V. Sweedler, *Anal. Chem.* 74, 5550–5555, 2002.
118. E. Rapp, A. Jakob, A. Bezerra Schefer, E. Bayer, K. Albert, *Anal. Bioanal. Chem.* 376, 1053–1061, 2003.
119. E. Bednarek, W. Bocian, K. Michalska, *J. Chromatogr. A* 1193, 164–171, 2008.
120. A. S. Rathore, *J. Chromatogr. A* 1037, 431–443, 2003.

121. C. J. Evenhuis, R. M. Guijt, M. Macka, P. J. Marriott, P. R. Haddad, *Anal. Chem.* 78, 2684–2693, 2006.
122. G. Chena, U. Tallarek, A. Seidel-Morgenstern, Y. Zhang, *J. Chromatogr. A* 1044, 287–294, 2004.
123. G. Y. Tang, C. Yang, C. J. Chai, H. Q. Gong, *Langmuir* 19, 10975–10984, 2003.
124. W. A. Gobie, C. F. Ivory, *J. Chromatogr.* 516, 191–210, 1990.
125. J. Knox, *Chromatographia* 26, 329–337, 1988.
126. S. Hjerten, *Chromatogr. Rev.* 9, 122–219, 1967.
127. E. A. S. Doherty, R. J. Meagher, M. N. Albarghouthi, A. E. Barron, *Electrophoresis* 24, 34–54, 2003.
128. C. A. Lucy, A. M. MacDonald, M. D. Gulcev, *J. Chromatogr. A* 1184, 81–105, 2008.
129. J. E. Wiktorowicz and J. C. Colburn, *Electrophoresis* 11, 769–773, 1990.
130. A. Eckhardt, I. Mikšík, Z. Deyl, J. Charvátová, *J. Chromatogr. A* 1051, 111–117, 2004.
131. Kleindiest, G., C. G. Huber, D. T. Gjerde, L. Yengoyan, G. K. Bonn, *Electrophoresis* 19, 262–269, 1998.
132. D. Corradini, L. Bevilacqua, I. Nicoletti, *Chromatographia* 62, 43–50, 2005.
133. D. Corradini, L. Sprecacenere, *Chromatographia* 58, 587–596, 2003.
134. D. Corradini, E. Cogliandro, L. D'Alessandro, I. Nicoletti, *J. Chromatogr. A* 1013, 221–232, 2003.
135. J. C. Giddings, *Sep. Sci.* 4, 181–189, 1969.
136. F. E. P. Mikkers, F. M. Everaerts, T. P. E. M. Verheggen, *J. Chromatogr.* 169, 1–10, 1979.
137. J. W. Jorgenson, K. D. Lukacs, *Anal. Chem.* 53, 1298–1302, 1981.
138. M.-L. Riekkola, J. Å. Jönsson, R. M. Smith, *Pure Appl. Chem.* 76, 443–451, 2004.
139. J. H. Knox, *J. Chromatogr. A* 680, 3–13, 1994.
140. A. S. Rathore, Cs. Horváth, *J. Chromatogr. A* 743, 231–246, 1996.
141. J. C. Reijenga, E. Kenndler, *J. Chromatogr. A* 659, 403–415, 1994.
142. D. Corradini, *J. Chromatogr. B* 699, 221–257, 1997.
143. J. Znaleziona, J. P. Radim Knob, V. Maier, J. Ševčik, *Chromatographia* 67, S5–S12, 2008.
144. J. C. Reijenga, L. G. Gagliardi, E. Kenndler, *J. Chromatogr. A* 1155, 142–145, 2007.
145. Z. Wang, J. Ouyang, W. R. G. Baeyens, *J. Chromatogr. B* 862, 1–14, 2008.
146. T. V. Popa, C. T. Mant, R. S. Hodges, *J. Chromatogr. A* 1111, 192–199, 2006.
147. P. Jáč, M. Polášek, M. Pospíšilová, *J. Pharm. Biomed. Anal.* 40, 805–814, 2006.
148. C. M. Shelton1, J. T. Koch, N. Desai, J. F. Wheeler, *J. Chromatogr. A* 792, 455–462, 1997.
149. Y. Yang, R. I. Boysen, J. I.-C. Chen, H. H. Keah, M. T. W. Hearn, *J. Chromatogr. A* 1009, 3–14, 2003.
150. A. Rizzi, *Electrophoresis* 22, 3079–3106, 2001.
151. M. López-Pastor, B. M. Simonet, B. Lendl, M. Valcárel, *Electrophoresis* 29, 94–107, 2008.
152. S. Mohabbati, S. Hjerten, D. Westerlund, *Anal. Bioanal. Chem.* 390, 667–678, 2008.
153. L. Geiger J.-L. Veuthey, *Electrophoresis* 28, 45–57, 2007.
154. S. P. Porras, E. Kenndler, *Electrophoresis* 26, 3203–3220, 2005.
155. H. Cottet, C. Simó, W. Vayaboury, A. Cifuentes, *J. Chromatogr. A* 1068, 59–73, 2005.
156. G. K. E. Scriba, *J. Chromatogr. A* 1159, 28–41, 2007.
157. A. C. Moser, D. S. Hage, *Electrophoresis* 29, 3279–3295, 2008.
158. C. Schou, N. H. H. Heegaard, *Electrophoresis* 27, 44–59, 2006.
159. Y. Tanaka, S. Terabe, *J. Chromatogr. B* 768, 81–92, 2006.
160. M. T. Bowser, D. D. Y. Chen, *J. Phys. Chem.* 103, 197–202, 1999.
161. K. L. Rundlett, D. W. Armstrong, *J. Chromatogr. A* 721, 173–186, 1996.
162. L. Valtcheva, J. Mohammad, G. Peterson, S. Hjerten, *J. Chromatogr.* 638, 263–267, 1993.
163. G. Gubitz, M. G. Scchmid, *Electrophoresis* 28, 114–126, 2007.
164. J. Østergaard, N. H. H. Heegaard, *Electrophoresis* 27, 2590–2608, 2006.
165. N. A. Guzman, T. Blanc, T. M. Phillips, *Electrophoresis* 29, 3259–3278, 2008.
166. L. K. Amundsen, H. Sirén, *Electrophoresis* 28, 99–113, 2007.
167. S. Lin, S.-M. Hsu, *Anal. Biochem.* 341, 1–15, 2005.
168. M. Gayton-Ely, T. J. Pappas, L. A. Holland, *Anal. Bioanal. Chem.* 382, 570–580, 2005.
169. A. Guttman, *LC-GC North Am.* 22, 896–904, 2004.
170. O. Grosche, J. Bohrisch, U. Wendler, W. Jaeger, H. Engelhardt, *J. Chromatogr. A* 894, 105–116, 2000.
171. S. Boulos, O. Cabrices, M. Blas, B. R. McCord, *Electophoresis* 29, 4695–4703, 2008.
172. I. Mikšík, P. Sedláková, K. Mikulíková, A. Eckhardt, T. Cserhati, T. Horváth, *Biomed. Chromatogr.* 20, 458–465, 2006.
173. V. Dolnik, W. A. Gurske, A. Padua, *Electrophoresis* 22, 707–719, 2001.
174. M. Bergman, H. Claessens, C. Cramers, *J. Microcol. Sep.* 10, 19–26, 1998.
175. F. Xu, Y. Baba, *Electrophoresis* 25, 2332–2345, 2004.
176. J. Sudor, V. Barbier, S. Thirot, D. Godfrin, D. Hourdet, M. Millequant, J. Blanchard, J. L. Viovy, *Electrophoresis* 22, 720–728, 2001.

177. C.-W. Kan, E. A. S. Doherty, B. A. Buchholz, A. E. Barron, *Electrophoresis* 25, 1007–1015, 2004.

178. T. Chiesl, W. Shi, A. E. Barron, *Anal. Chem.* 77, 772–779, 2005.

179. B. A. Buchholz, W. Shi, A. E. Barron, *Electrophoresis* 23, 1398–1409, 2002.

180. A. Tsupryk, M. Gorbovitski, E. A. Kabotyanski1, V. Gorfinkel, *Electrophoresis* 27, 2869–2879, 2006.

181. P. G. Righetti, C. Simò, R. Sebastiano, A. Citterio, *Electrophoresis* 28, 3799–2810, 2007.

182. P. G. Righetti, *Isoelectric Focusing: Theory, Methodology and Applications*, Elsevier, Amsterdam, the Netherlands, 1983.

183. S. Hjertén, M. D. Zhu, *J. Chromatogr.* 346, 265–270, 1985.

184. L. Gao, S. Liu, *Anal. Chem.* 76, 7179–7186, 2004.

185. M. Horká, J. Planeta, F. Růzička, K. Šlais, *Electrophoresis* 24, 1383–1390, 2003.

186. H. Tian, L. C. Brody, D. Mao, J. P. Landers, *Anal. Chem.* 72, 5483–5492, 2000.

187. K. K.-C. Yeung, K. K. Atwal, H. Zhang, *Analyst* 128, 566–570, 2003.

188. M. Poitevin, A. Morin, J.-M. Busnel, S. Descroix, M.-C. Hennion, G. Peltre, *J. Chromatogr. A* 1155, 230–236, 2007.

189. M. Horká, F. Růzička, *Anal. Bioanal. Chem.* 385, 840–846, 2006.

190. A. Palm, M. Zaragoza-Sundqvist, G. Marko-Varga, *J. Sep. Sci.* 27, 124–128, 2004.

191. P. G. Righetti, A. Bossi, C. Gelfi, *J. Capillary Electrophor.* 2, 47–59, 1997.

192. W. Thormann, J. Caslavska, R. A. Mosher, *J. Chromatogr. A* 1155, 154–163, 2007.

193. P. Lopez-Soto-Yarritu, J. C. Díez-Masa, A. Cifuentes, M. de Frutos, *J. Chromatogr. A* 968, 221–228, 2002.

194. J. R. Mazzeo, I. S. Krull, *Anal. Chem.* 63, 2852–2857, 1991.

195. J. Wu, J. Pawliszyn, *Anal. Chem.* 64, 224–227, 1992.

196. X.-Z. Wu, N. S.-K. Sze, J. Pawliszyn, *Electrophoresis* 22, 3968–3971, 2001.

197. X.-Z. Wu, T. Huang, Z. Liu, J. Pawliszyn, *Trends Anal. Chem.* 24, 369–382, 2005.

198. J. Wu, C. Tragas, A. Watson, J. Pawliszyn, *Anal. Chim. Acta* 383, 67–78, 1999.

199. X. Fang, C. Tragas, J. Wu, Q. Mao, J. Pawliszyn, *Electrophoresis* 19, 2290–2295, 1998.

200. H. Stutz, *Electrophoresis* 26, 1254–1290, 2005.

201. F. Foret, O. Muller, J. Thorne, W. Gotznger, B. L. Carger, *J. Chromatogr. A* 716, 157–166, 1995.

202. J. Pawliszyn, J. Wu, *J. Microcol. Sep.* 5, 397–401, 1993.

203. T. Huang, J. Pawliszyn, *Electrophoresis* 23, 3504–3510, 2002.

204. R. Montgomery, X. Jia, L. Tolley, *Anal. Chem.* 78, 6511–6518, 2006.

205. C. Yang, G. Zhu, L. Zhang, W. Zhang, Y. Zhang, *Electrophoresis* 25, 1729–1734, 2004.

206. S. Terabe, K. Otsuka, K. Ichikawa, A. Tsuchiya, T. Ando, *Anal. Chem.* 56, 111–113, 1984.

207. S. Terabe, M. Shibata, Y. Miyashita, *J. Chromatogr.* 480, 403–411, 1989.

208. A. Dobashi, T. Ono, S. Hara, J. Yamaguchi, *Anal. Chem.* 61, 1984–1986, 1989.

209. S. Terabe, K. Otsuka, T. Ando, *Anal. Chem.* 57, 834–841, 1985.

210. G. Guiochon. Optimization in liquid chromatography, in C. Horváth (Ed.) *High-Performance Liquid Chromatography*. Academic Press, New York, 1980, pp. 1–56.

211. P. Gareil, *Chromatographia* 30, 195–200, 1990.

212. H. Corstjens, H. A. H. Billiet, J. Frank, K. Ch. M. Luyben, *J. Chromatogr. A* 753, 121–131, 1996.

213. K. Otsuka, S. Terabe, T. Ando, *J. Chromatogr.* 348, 39–47, 1985.

214. P. G. H. M. Muijselaar, H. A. Classens, C. A. Cramers, J. F. G. A. Jansen, E. W. Meijer, E. M. M. de Brabander-van den Berg, S. van der Wal, *J. High Resolut. Chromatogr.* 18, 121–123, 1995.

215. C. Fujimoto, Y. Muranaka, *J. High Resolut. Chromatogr.* 20, 400–402, 1997.

216. P. Viberg, M. Jornten-Karlsson, P. Petersson, P. Spegel, S. Nilsson, *Anal. Chem.* 74, 4595–4601, 2002.

217. P. Spegel, L. Schweitz, S. Nilsson, *Anal. Chem.* 75, 6608–6613, 2003.

218. U. Sharma, S. Sharma, *Int. J. Pharm.* 154, 123–140, 1997.

219. M. N. Jones, *Adv. Colloid Interface Sci.* 54, 93–128, 1995.

220. Y. Zhang, R. Zhang, S. Hjerten, P. Lundahl, *Electrophoresis* 16, 1519–1523, 1995.

221. D. Corradini, G. Mancini, C. Bello, *Chromatographia* 60, S125–S132, 2004.

222. D. Corradini, G. Mancini, C. Bello, *J. Chromatogr. A* 1051, 103–110, 2004.

223. N. Griese, G. Blaschke, J. Boos, G. Hempel, *J. Chromatogr. A* 979, 379–388, 2002.

224. S. Luedtke, Th. Adam, N. von Doehren, K. K. Unger, *J. Chromatogr. A* 887, 339–346, 2000.

225. V. Preotorius, B. Hopkins, *J. Chromatogr.* 99, 23–30, 1974.

226. J. Jorgenson, K. D. Lukacs, *J. Chromatogr.* 218, 209–216, 1981.

227. S. Zhang, X. Huang, N. Yao, Cs. Horváth, *J. Chromatogr. A* 948, 193–201, 2002.

228. L. A. Colon, T. D. Maloney, A. M. Fermier, *J. Chromatogr. A* 887, 43–53, 2000.

229. F. Lynen, A. Buica, A. de Villiers, A. Crouch, P. Sandra, *J. Sep. Sci.* 28, 1539–1549, 2005.

230. M. M. Robson, S. Roulin, S. M. Shariff, M. W. Raynor, K. D. Bartle, A. A. Clifford, E. Myers, M. R. Euerby, C. M. Johnson, *Chromatographia* 43, 313–321, 1996.
231. C. Yan, U.S. Patent 5,453,163, September 26, 1995.
232. A. M. Fermier, L. A. Colon, *J. Microcol. Sep.* 10, 439–447, 1998.
233. T. D. Maloney, L. A. Colon, *Electrophoresis* 20, 2360–2365, 1999.
234. R. Stol, M. Mazereeuw, U. R. Tjaden, J. van der Greef, *J. Chromatogr. A* 873, 293–298, 2000.
235. K. J. Reynolds, L. A. Colon, *Analyst* 123, 1493–1495, 1998.
236. H. Rebscher, U. Pyell, *Chromatographia* 42, 171–176, 1996.
237. H. J. Cortes, C. D. Pfeiffer, B. E. Richter, T. S. Stevens, *J. High Resolut. Chromatogr. Chromatogr. Commun.* 10, 446–448, 1987.
238. G. A. Lord, D. B. Gordon, P. Myers, B. W. King, *J. Chromatogr. A* 768, 9–16, 1997.
239. G. Choudhary, Cs. Horváth, J. F. Banks, *J. Chromatogr. A* 828, 469–480, 1998.
240. M. E. R. Hows, D. Perrett, *Chromatographia* 43, 200–204, 1996.
241. M. Mayer, E. Rapp, C. Marck, G. J. M. Bruin, *Electrophoresis* 20, 43–49, 1999.
242. C. Fujimoto, *Anal. Chem.* 67, 2050–2053, 1995.
243. I. Gusev, X. Huang, Cs. Horváth, *J. Chromatogr. A* 855, 273–290, 1999.
244. N. Ishizuka, H. Minakuchi, K. Nakanishi, N. Soga, H. Nagayama, K. Hosoya, N. Tanaka, *Anal. Chem.* 72, 1275–1280, 2000.
245. D. Allen, Z. El Rassi, *Electrophoresis* 24, 408–420, 2003.
246. M. T. Dulay, J. P. Quirino, B. D. Bennett, M. Kato, R. N. Zare, *Anal. Chem.* 73, 3921–3926, 2001.
247. G. Ding, Z. Da, R. Yuan, J. J. Bao, *Electrophoresis* 27, 3363–3372, 2006.
248. T. Tsuda, K. Nomura, G. Nakagawa, *J. Chromatogr.* 248, 241–247, 1982.
249. Z. Liu, H. Zou, M. Ye, J. I. Ni, Y. Zhang, *Electrophoresis* 14, 2891–2897, 1999.
250. J. J. Pesek, M. T. Matyska, S. Cho, *J. Chromatogr. A* 845, 237–246, 1999.
251. J. J. Pesek, M. T. Matyska, S. Sentellas, M. T. Galceran, M. Chiari, G. Pirri, *Electrophoresis* 23, 2982–2989, 2002.
252. Z. J. Tan, V. T. Remcho, *Anal. Chem.* 69, 581–586, 1997.
253. X. Huang, J. Zhang, Cs. Horváth, *J. Chromatogr. A* 858, 91–101, 1999.
254. Y. Guo, L. A. Colon, *Anal. Chem.* 67, 2511–2516, 1995.
255. M. C. Breadmore, M. Macka, N. Avdalovic, P. R. Haddad, *Analyst* 125, 1235–1241, 2000.
256. A. S. Rathore, Cs. Horváth, *Electrophoresis* 23, 1211–1216, 2002.
257. T. Tsuda, *Anal. Chem.* 59, 521–523, 1987.
258. Z. Lin, Z. Xie, H. Lü, X. Lin, X. Wu, G. Chen, *Anal. Chem.* 78, 5322–5328, 2006.
259. P. Gfrörer, L.-H. Tseng, E. Rapp, K. Albert, E. Bayer, *Anal. Chem.* 73, 3234–3239, 2001.
260. C. Yan, R. Dadoo, R. N. Zare, *Anal. Chem.* 68, 2726–2730, 1996.
261. C. G. Huber, G. Choudhari, Cs. Horváth, *Anal. Chem.* 69, 4429–4436, 1997.
262. G. Choudhary, Cs. Horváth, J. F. Banks, *J. Chromatogr. A* 828, 469–480, 1998.
263. Q.-H. Ru, J. Yao, G.-A. Luo, Y.-X. Zhang, C. Yan, *J. Chromatogr. A* 894, 337–343, 2000.
264. P. Boček, M. Deml, P. Gebauer, V. Dolnik, *Analytical Isotachophoresis*, VCH, Weinheim, Germany, 1988.
265. A. J. P. Martin, F. M. Everaerts, *Anal. Chim. Acta* 38, 233–237, 1967.
266. J. C. Reijenga, G. V. A. Aben, Th. P. E. M. Verheggen, F. M. Everaerts, *J. Chromatogr.* 260, 241–254, 1983.
267. H. R. Udseth, J. A. Lao, R. D. Smith, *Anal. Chem.* 61, 228–232, 1989.
268. P. Gebauer, P. Boček, *J. Chromatogr.* 267, 49–65, 1983.
269. R. Aebersold, H. D. Morrison, *J. Chromatogr.* 516, 79–88, 1990.
270. M. Dankova, D. Kaniansky, S. Fanali, F. Ivanyi, *J. Chromatogr. A* 838, 31–43, 1999.
271. S. Fanali, C. Desiderio, E. Olvecka, D. Kaniansky, M. Vojtek, A. Ferancova, *J. High Resolut. Chromatogr.* 23, 531–538, 2000.
272. X. Fang, L. Yang, W. Wang, T. Song, C. S. Lee, D. L. DeVoe, B. M. Balgley, *Anal. Chem.* 79, 5785–5792, 2007.
273. P. Mikus, K. Marakova, J. Marak, A. Plankova, I. Valaskova, E. Havranek, *Electrophoresis* 29, 4561–4567, 2008.
274. F. Foret, E. Szoko, B. L. Karger, *J. Chromatogr.* 608, 3–12, 1992.
275. J. Safra, M. Pospisilova, J. Spilkova, *Chromatographia* 64, 37–43, 2006.
276. J. Safra, M. Pospisilova, J. Honegr, J. Spilkova, *J. Chromatogr. A* 1171, 124–132, 2007.

7 HPLC Detectors

Nicole Y. Morgan and Paul D. Smith

CONTENTS

7.1 INTRODUCTION

Since the development of HPLC as a separation technique, considerable effort has been spent on the design and improvement of suitable detectors. The detector is perhaps the second-most important component of an HPLC system, after the column that performs the actual separation: it would be pointless to perform any separation without some means of identifying the separated components. To this end, a number of analytical techniques have been employed to examine either samples taken from a fraction collector or the column effluent itself. Although many different physical principles have been examined for their potential as chromatography detectors, only four main types of detectors have obtained almost universal application, namely, ultraviolet (UV) absorbance, refractive index (RI), fluorescence, and conductivity detectors. Today, these detectors are used in about 80% of all separations. Newer varieties of detector such as the laser-induced fluorescence (LIF), electrochemical (EC), evaporative light scattering (ELS), and mass spectrometer (MS) detectors have been developed to meet the demands set by either specialized analyses or by miniaturization.

The use of a fixed wavelength UV detector for liquid chromatographic separations was first described by Horvath and Lipsky in 1966 [1], and is possibly the most popular HPLC detector in general use today. Although other detection techniques are more sensitive, the UV detector provides a simple and universal answer to the majority of HPLC applications [2]. Developed in 1982, the diode array UV detector measures the full absorption spectrum of each analyte peak, and was a significant advance on the single-wavelength detector [3]. This instrument was popular in the 1990s especially in toxicological or forensic analyses where additional information on separated analytes was required [4,5]. Diode array detectors have also been used to evaluate the homology of isolated peaks during pharmaceutical analysis [6]. Coupling diode array detection with mass spectrometry, as done by Nicoletti et al. for reversed-phase HPLC separations of stilbenes and other polyphenols, can provide still greater characterization of certain analytes [7].

The RI detector is another basic detector in general use in HPLC. Although the sensitivity is generally lower than that of most other LC detectors, RI detectors are especially useful for non-chromogenic compounds such as sugars [8], high molecular weight polymers [9], and some pharmaceuticals present in animal feeds [10]. RI has also been applied to microbore columns for the analysis of carbohydrates [11] and small molecules [12].

Evaporative light scattering is gaining popularity due to its ability to detect analytes on a nonselective basis. Basically, this detector works by nebulizing the column effluent, forming an aerosol that is further converted into a droplet cloud for detection by light scattering. This type of detector has been applied to studies of small molecule combinatorial libraries [13,14], carbohydrates [15], and lipids [16,17].

Although the majority of analytes do not possess natural fluorescence, the fluorescence detector has gained popularity due to its high sensitivity. The development of derivatization procedures used to label the separated analytes with a fluorescent compound has facilitated the broad application of fluorescence detection. These labeling reactions can be performed either pre- or post-separation, and a variety of these derivatization techniques have been recently reviewed by Fukushima et al. [18]. The usefulness of fluorescence detectors has recently been further demonstrated by the Wainer group, who developed a simple HPLC technique for the determination of all-*trans*-retinol and tocopherols in human plasma using variable wavelength fluorescence detection [19].

The development of microbore chromatography and nano-flow HPLC has greatly increased interest in detectors suitable for sub-microliter peak volumes. These include LIF, EC, and MS spectrometry detectors. Zare first described the application of lasers for detection in analytical chemistry in 1984, outlining the potential for the use of lasers in analytical science, especially their application to HPLC [20]. This was followed in 1988 by another review that focused more on HPLC applications and provided examples of the high sensitivity possible with LIF detection [21]. Another good review was written by Rahavendran and Karnes outlining the usefulness of LIF in pharmaceutical analysis [22].

The marriage of HPLC to mass spectrometry (MS), now developed into a mature instrumentation, continues to greatly impact many of the separation sciences, especially in pharmaceutical analysis where it has been used in new drug discovery [23,24] and in drug metabolite identification [25–27]. HPLC–MS has also made an impact on lipid research, providing a convenient approach to the analysis of phospholipids and fatty acids [28,29]. It has also greatly benefited the field of proteomics [30–34], especially analysis of protein structure and function.

When choosing an HPLC detector several important factors must be considered. First, consider the purpose of the separation: is it preparative or analytical, and what are the associated constraints on sample volume? Are there unknowns to be identified, which points toward a detection mode that provides as much information as possible, or do compounds only need to be quantified accurately? Is the desired detection mode sensitive to the analytes under study, and if not, is labeling practical? In addition, one has to take into account the concentration of the analyte within the samples to be examined, which determines the required sensitivity of the detector, as well as the complexity of the matrix that contains the analyte, which determines the required selectivity. Further complicating the decision, not all detectors are compatible with all solvents, or with commonly used techniques such as gradient elution. As a result, the type of chromatography to be used for the analysis and the physicochemical nature of the chromatographic packing also has to be considered. Unsurprisingly, this complexity generally leads to the need for several different detectors within the same analytical laboratory. For research purposes, the use of hybrid or multiple detectors on a single instrument is especially useful for providing more information about the analytes while conserving sample and measurement time.

7.2 DETECTOR CHARACTERISTICS

There are several major characteristics to consider when comparing different chromatography detectors. The main concerns are usually sensitivity, noise, and dynamic range, which together determine the minimum and the maximum detectable amount of analyte, but other features can also require careful matching to the desired application. The detection cell design, particularly the volume, should not contribute appreciably to peak broadening, but should provide good sensitivity within this constraint. A highly selective detector will respond to relatively few analytes; this can be extremely useful in identifying a few compounds of interest in a complex mixture, but in other applications could lead to missed peaks. A final consideration is whether preservation of the analyte peaks for further study is a priority. For example, an unknown compound may need identification, or an isolated protein may need to be tested for bioavailability.

7.2.1 SENSITIVITY

The sensitivity of a detector is defined simply as the change in the detector output as a function of the amount of analyte present. In a calibration curve, the sensitivity is the slope, and corresponds to the ability to distinguish between nearly identical amounts of analyte. This will depend greatly on the properties of the analyte, and potentially on the composition, temperature, and pH of the column effluent. When comparing different detectors, one of the most important factors is the limit of detection (LOD), which is simply the minimum amount of analyte that is distinguishable from zero. The LOD is determined by both the sensitivity and the noise of the detector, and is usually specified as the amount of analyte that yields a signal three times higher than the noise level. The value for the noise used is generally the root-mean-square (RMS) variation in the detector output at similar timescales to the signal peaks. However, there is considerable variety in the specification of the LOD between different manufacturers and/or researchers, and some care must be exerted in making comparisons between instruments. Some researchers will use a signal-to-noise ratio (SNR) of 2:1 rather than 3:1 for the LOD, or the specified noise may be peak-to-peak rather than RMS, or the bandwidth of the noise may be specified differently.

In order to obtain accurate peak areas for quantification, at least 10 data points should be recorded for each peak, giving a data-taking rate for the measurement. The measurement electronics or data post-processing should filter out any noise substantially faster than the data-taking rate; for a simple filter, the time constant should be about five times smaller than the interval between data points to avoid introducing artifacts. Very long timescale drift can be eliminated in post-analysis processing in order to obtain accurate peak areas for quantitation. For modern commercial detectors, the correction for baseline drift and peak integration for quantization are incorporated into the software.

7.2.2 DYNAMIC RANGE

The dynamic range of a detector is the range of sample concentrations over which the detector output changes, and is another important consideration when comparing detection modes. The detector response is generally not linear over the entire dynamic range, and so care is required when extrapolating a calibration curve to the extremes. Typically, the response flattens toward the upper end of the range, which means that the sensitivity of the detector is smaller in this region. Although the detector signal may be nonlinear in this region, it is almost always monotonic, and quantitation can still be achieved through careful calibration as long as the detector is not in saturation.

7.2.3 Detection Cell Volume

The detection cell must be designed with its volume small enough to prevent additional peak broadening in the detector; in practice, this means the detection cell volume should be at least 10 times smaller than the volume of the first, most narrow, chromatographic peak. For nano-flow HPLC systems, in which the peaks can be sub-microliter, detection cells with a volume on the order of 100 nL or less are appropriate. For conventional HPLC systems, in which the peak volume is tens of microliters, there is no benefit to having a detection volume smaller than a few microliters, and indeed most conventional HPLC flow cells are around 5–12 μL. It is best to ensure that the detection cell is well matched to the sample and peak volumes, as making the detection volume too small will unnecessarily decrease the sensitivity of the detector.

In comparing the sensitivity of different detectors, it is important to notice whether the LOD is stated in terms of concentration or mass. If desired, the detection cell volume can often be used for a rough conversion between these numbers for direct comparison between detectors. However, in the literature, sometimes the peak volume or the injected sample volume is given instead; if it is the latter, the sample dispersion during separation also needs to be taken into account when assessing the stated detector performance. In many modalities, including UV–Vis absorption and fluorescence, there is a loss in sensitivity, as expressed in concentration, as the detection cell volume decreases. However, the sensitivity expressed in mass remains constant or can even increase slightly, in part due to more efficient light collection or better background rejection. The particular application will determine whether a low LOD in concentration or mass is more significant. For trace analysis, a large detection volume with a very low concentration LOD is important, whereas a small detection volume and sensitive mass detection is preferable in microflow systems or where sample volume is limited.

7.2.4 Selectivity

A highly selective detector responds to relatively few compounds; whether this is advantageous depends on the particular application. Fluorescence, electrochemical, and radiochemical detectors are all extremely selective. These detection techniques can permit the identification and measurement of analytes within complex mixtures, especially when used in combination with chemical labeling of the compounds of interest or internal standards. Other detectors are more universal, which is an advantage if all analyte peaks need to be recorded, as when measuring unknowns. For a wide range of compounds that absorb in the UV range, especially certain pharmaceuticals, peptides, and proteins, UV absorption is considered a universal detection method. Other universal methods are evaporative light scattering, which generates a response for any compound that is less volatile than the eluting solvent, and RI detection.

There are several methods for improving detector response to a compound of interest. One such technique is chemical labeling, which will be discussed in detail in Section 7.3.3. Another possibility is to use indirect detection, in which the mobile phase is chosen or modified such that it gives rise to a strong detector signal by itself. When the analyte peaks reach the detector, they displace some of the mobile phase, and this gives rise to a change in signal [35]. In other cases, an additive to the mobile phase changes its physical properties in the presence of some chemical compounds of interest, leading to a signal at the detector.

7.2.5 Preservation of Sample

Another consideration when choosing a detector is whether it is important to preserve the separated analytes, either for use or for further analysis. Some methods, such as evaporative laser scattering detection and mass spectrometry, destroy the sample during the measurement. Other methods, such as fluorescence or radiochemical detection, may require chemical labeling of the analytes;

the labeling reaction inevitably alters some properties of the molecules, and these alterations could possibly interfere with later measurements or uses of the separated analytes.

7.3 TYPES OF DETECTORS

Over the years, many physical techniques, from nuclear magnetic resonance (NMR) to electrochemistry, have been applied to HPLC detection, and a number of them have become widely used. An analytical laboratory is virtually certain to need several different types of detectors, as no single modality will be suited to the range of analytes, sample volumes, and separation methods used. Hybrid detectors that combine two or more modes in a single instrument are becoming more popular, as they can extract more information from each sample run. Obtaining as much information as possible, as with the compound identification that can be performed by a mass spectrometer or a photodiode array detector, is especially important if the sample is destroyed in the measurement.

Although this section provides a brief description of most commonly used detectors for HPLC, most of the focus is on a few detection modes. Optical absorbance detectors remain the most widely used for HPLC, and are discussed in some detail. We also focus on fluorescence, conductivity, and electrochemical detection, as these methods were not widely used for HPLC in the past, but are especially well suited to micro- and nano-flow instruments because of their high sensitivity in small sample volumes. Mass spectrometry has also come into wide and routine use in the last decade, but as it is the subject of another chapter, it will not be further discussed here. Miniaturization has been particularly important for capillary and chip-based electrophoresis, which often employs sub-nanoliter detection volumes [36,37].

7.3.1 ABSORBANCE (UV–VIS)

The most commonly used detector for HPLC is still the UV–Vis absorbance detector. The amount of light transmitted through a solution of concentration c in a flow cell with path length ℓ is given by the Beer–Lambert law,

$$I_{tr} = I_0 \times 10^{-\varepsilon \ell c} \qquad (7.1)$$

where ε is the absorption coefficient or the molar extinction coefficient. The exponent, $A = \varepsilon \ell c$, is the absorbance of the solution, and is measured in absorbance units, or AU. This equation is only valid when no more than one absorbing molecule falls in the path of a single photon, which is to say that the solution is sufficiently dilute and the path length sufficiently short that the absorbers do not shadow each other. In practice, departures from the Beer–Lambert law are not measurable with most spectrometers provided the absorbance is less than 1.

The dependence of ε on the wavelength of light is known as the absorption spectrum, and varies so greatly between different compounds that it can often be used to identify them. For a particular compound, the absorbance at a single-wavelength scales linearly with the concentration; however, there can be significant variation with parameters such as solvent pH, solvent polarity, and temperature. When the absorbance is measured at a single fixed wavelength, it is typically in the UV, as most materials of interest have strong absorption in this region. For example, proteins and peptides are measured at a wavelength of 280 nm, or with the strong mercury lamp line at 254 nm.

The choice of solvent can have a major effect on detection, although many of the solvent properties will be determined by the separation requirements. Especially in the UV range, the solvent contribution to the absorbance can be substantial. Table 7.1 lists the cutoff wavelengths of several commonly used solvents and buffer components, where the cutoff wavelength is defined as the wavelength at which $A = 1$ for a 1 cm path length. It is important to realize that the absorbance of the solvent can have a significant effect well above this cutoff wavelength. In addition, the purity of the solvent is critical, as the UV absorbance can be dominated by the effect of a few

TABLE 7.1
Typical UV Cutoff Wavelengths for Some Commonly Used HPLC Solvents and Buffer Components, Taken Primarily from Manufacturer Product Specifications

Solvent	UV Cutoff (nm)
Acetonitrile	190
Dichloromethane	233
Ethyl acetate	256
n-Hexane	192
Methanol	205
2-Propanol	205
Water	<190
Trifluoroacetic acid, 0.1%	210
Potassium phosphate, pH 7.2, 0.1%	<200
Sodium citrate 10 mM	225
NaCl, 1 M	208

low-concentration impurities. HPLC grade solvents must be used, but even so, there can be variations in absorbance between different suppliers and different product lines. At a minimum, the absorbance specified by the chemical manufacturer or distributor should be checked; if the desired measurement is close to the cutoff wavelength, the absorbance spectrum of the solvent should be directly measured.

A typical configuration for an absorption detector is shown in Figure 7.1. The light source is usually a stable, high-intensity broadband lamp, such as a deuterium arc lamp, with a grating that disperses the light into its spectral components. The spectrum is projected onto a slit, selecting the wavelength of interest, and the spectral resolution of the measurement is determined by the width of the slit. A simpler configuration, for detection at a single fixed wavelength, uses an optical filter instead of the slit-grating combination to select the wavelength of light that impinges on the sample. The sensor element that detects the light is generally a photodiode, which is better suited

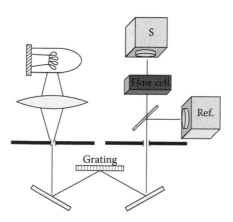

FIGURE 7.1 Schematic of a single-wavelength UV absorption detector. Broadband light is focused onto a slit and spectrally dispersed by a grating. A second slit selects the desired wavelength of light. A beamsplitter directs some of this light onto the reference photodiode R. The light transmitted through the sample flow cell is measured by the sample photodiode S.

for distinguishing between two nearly identical light levels than a photomultiplier tube. Most absorbance detectors also have a reference beam and an identical second sensor, which are used to correct for variations in the light source intensity. In the conceptually simplest configuration, the reference beam passes through a reference cell of identical construction to the sample cell, which also allows for a straightforward correction for light losses due to reflection, scattering, or absorbance from the cell itself, as well as solvent absorbance. However, for flow cells, especially when using gradient elution, this is not always practical. The alternative, shown in Figure 7.1, is to direct the reference beam immediately onto the second sensor. Although this second method does not correct for light losses associated with reflection and scattering at the sample cell, both methods enable compensation for lamp power fluctuations and drift, which are usually the dominant source of measurement noise.

As can be seen in Equation 7.1, the detection cell geometry has a large impact on the measurement sensitivity. The path length is directly proportional to the absorbance, so increasing the path length will increase the minimum detectable concentration. However, increase in the path length must be balanced with keeping the volume of the detection cell small enough that it is not a significant source of peak dispersion. As a result, most optical HPLC detectors use Z-cells, as seen in Figure 7.2, or cells with similar geometry. Similarly, for capillaries a simple laboratory U-cell can be made by bending the capillary. When designing these cells it is important to avoid creating any turbulence or dead volume, as these would prevent analyte peaks from washing smoothly out of the cell after measurement and lead to peak broadening. These cells do greatly improve the sensitivity of the measurement, but if the cross-sectional area is made too small other problems can arise. Scattering from the cell walls can lead to extra background noise, and the decrease in the total light transmission through the cell can lead to an increase in measurement noise [38].

Commercial detectors can have noise levels corresponding to a few millionths of an absorbance unit (AU), which corresponds to lower detection limits of 10–20 μAU. The upper LOD is usually around 2 AU, which gives a dynamic range of approximately five orders of magnitude. Some basic calculations are instructive. For tryptophan (MW 204), the amino acid with the strongest UV absorbance, the molar extinction coefficient in water is $5630 M^{-1} cm^{-1}$ at 280 nm [39]. For a cell with a 1 cm path length and a volume of 10 μL, a detection limit of 10 μAU corresponds to a concentration of 1.8 nM, or 360 pg/mL. Assuming that the total peak volume is 10 times the detection cell volume, this corresponds to a mass detection limit of 36 pg. However, most compounds and all proteins and peptides are not as strongly absorbing per unit mass as pure tryptophan, so limits of detection are routinely higher. The same detection limit with a 100 μm path length gives a minimum detectable concentration of 180 nM. However, in practice, the reduced light throughput with miniaturization would lead to a 10-fold increase in the detection limit, and so the expected minimum detectable concentration would be approximately 1.8 μM, for a 100 μm ×100 μm ×100 μm volume.

There are two main types of multiple wavelength detectors. In a scanning detector, different wavelengths of light are scanned through the sample by means of a rotating grating or mirror. A single sensor records the transmitted light at different wavelengths in series, and the detector

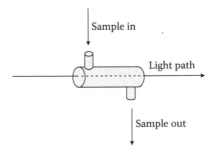

FIGURE 7.2 Schematic of a Z-cell, which is used to increase the optical path length for absorbance measurements. In order to avoid peak broadening, the corners must be designed to avoid dead volumes or flow disruptions.

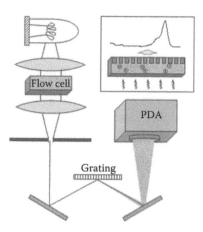

FIGURE 7.3 Schematic of a UV absorption photodiode array detector. Broadband light is collimated, passed through the sample flow cell, and focused onto a slit. A grating disperses the transmitted light onto a photodiode array, which continuously records the full absorption spectrum of the sample. Inset: schematic of photodiode array function. Spectrally separated light impinges on the photodiode array, creating electron–hole pairs. The negatively charged electrons are collected at the closest PIN junction, and when the junction array is read, the full absorption spectrum is obtained.

configuration can be identical to that shown in Figure 7.1. At present, this configuration is mainly used if the researcher wants to change the measurement wavelength between sample runs, or sometimes between well-separated peaks. A scanning configuration can be also be used to continuously record absorption spectra, but with reduced sensitivity and time resolution compared to a fixed wavelength measurement.

The second type of multiple wavelength detector uses a photodiode array as the sensor. With the increased availability of high-quality moderate-cost photodiode arrays, these are becoming increasingly common. In these detectors, the full spectrum from the lamp is focused onto the sample cell, as shown in Figure 7.3. The transmitted light passes through a slit and is dispersed by a grating onto the 1-D array of photodiodes, which records the absorbance at multiple wavelengths simultaneously. These detectors are capable of recording the full absorption spectrum at each time point with minimal loss of sensitivity or time resolution. Although they are more expensive than the single-wavelength detectors, the extra information can be extremely valuable. Even with measurements at only a few wavelengths, the relative absorbance is helpful in distinguishing between different compounds and in assessing peak purity. By measuring the absorbance at a range of wavelengths, a greater number of compounds can be measured, and the full absorption spectrum can be used to identify and characterize many compounds.

7.3.2 REFRACTIVE INDEX

The RI detector is a commonly used universal detector that measures a physical property of the bulk medium. RI is the physical parameter that characterizes the velocity of light in a medium and is used to describe the behavior of light as it passes between different materials. The deflection, or refraction, of light as it crosses the boundary between materials with different refractive indices is given by Snell's law:

$$n_1 \sin \theta_1 = n_2 \sin \theta_2 \qquad (7.2)$$

where
  n_1 and n_2 are the refractive indices of the two materials
  θ_1 and θ_2 describe the angle between the incident light and a line perpendicular to the interface

RI detectors measure this deflection, and are sensitive to all analytes that have a different RI than the mobile phase. There are two major limitations: First, RI detectors are very sensitive to changes in the temperature, pressure, and flow rate of the mobile phase, and so these measurement conditions must be kept stable in order to obtain low background levels. Second, RI detectors are incompatible with chromatographic separations using gradient elution. Furthermore, because RI detectors are nonselective, they must be used in conjunction with other detection methods if specificity is required. Nevertheless, they have found wide application in isocratic chromatographic analysis for analytes that do not have absorptive, fluorescent, or ionic properties, such as polymers and carbohydrates.

Several embodiments of the RI detector have been proposed. The most commonly used is the differential RI detector, which is commercially available from several manufacturers. This detector measures the deflection of light produced by the RI difference between the sample and reference compartments of an optical cell. This measurement is illustrated schematically in Figure 7.4. The deflection of the light as it passes through the cell depends on the analyte present in the cell, and the deviation from baseline is recorded by a position-sensitive detector. After calibration for a particular analyte, the detector signal can be processed to yield the concentration of analyte in the sample chamber. Typical limits of detection for commercial RI detectors are 10^{-9} refractive index units (RIU), which roughly corresponds to 10 ng/mL of analyte in a 10 µL flow cell. For aqueous media, the RI variation with temperature is approximately 10^{-4} RIU/°C, so reaching these LODs requires thermal stability to better than 10^{-4}°C.

RI detection for microbore HPLC needs to address the challenge of the small pathlength present in the sub-microliter sample volume of typical analytical cells. The same basic challenge is present in capillary electrophoresis (CE), and RI strategies developed for CE should be readily adaptable to microbore HPLC. A variety of schemes have been used, including forward scattering [40] and retroreflected interference techniques [41]. In the microinterference backscatter detector (MIBD), backscattered light from a helium neon laser incident on the capillary produces high contrast interference fringes that are incident on a CCD array detector [42]. Changes in RI within the capillary result in spatial displacement of the fringes, which can be monitored by computational algorithms to track the position of the minima/maxima or by using Fourier analysis methods [43]. Detection limits better than 10^{-7} RIU have been achieved with a MIBD. In measurements on a series of dyes with

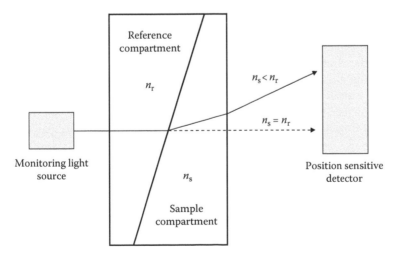

FIGURE 7.4 Schematic of a differential refractive index detection cell. The monitoring light passes through an optical cell comprised of reference and sample compartments, and is refracted at the boundary. The angle of deflection, which changes as the analyte peaks pass through the sample chamber, is recorded by a position-sensitive detector. The diagram shows the case for the sample having a lower refractive index than the reference.

high absorption coefficients in a 100 µm diameter capillary, MIBD was shown to be approximately a factor of 2 better than standard commercial UV–Vis absorption detectors [44].

7.3.3 FLUORESCENCE

There are several kinds of luminescence, or processes in which light is emitted from matter. The emitted light can arise from chemical reactions, as in chemiluminescence, or mechanical crushing or rubbing, as in triboluminescence. Fluorescence is a type of photoluminescence in which matter excited by light at one wavelength, or a range of wavelengths, emits light at a longer wavelength. In contrast to phosphorescence, another type of photoluminescence in which low-intensity light can be emitted for minutes or longer, fluorescence emission stops almost immediately after the excitation source is removed. As we discuss in Section 7.3.4, the shift in the wavelength of the emitted light, together with the rapid timescale for fluorescence emission, are what enable highly sensitive detection. Fluorescence detection is also extremely selective, as strong fluorescence from excitation with visible light is relatively uncommon in natural materials. Advances in low-cost solid-state lasers and compact high-sensitivity detectors such as charge-coupled devices (CCDs), photomultipliers (PMTs), and photodiodes have further facilitated the use of fluorescence detectors for a range of chromatographic applications.

A brief description of the physics of fluorescence follows, focusing on molecules for simplicity. A molecule has a number of molecular orbitals occupied by electrons; each of these orbitals has a number of sublevels corresponding to different vibrational states of the molecule. Although there are many electrons per molecule, for fluorescence processes we are only concerned with the electron in the highest energy occupied orbital. At room temperature, this electron is typically confined to its ground molecular orbital, although the vibrational and rotational levels can be accessible. When a photon of light is absorbed by a fluorescent molecule, the electron is excited from its ground state into a more energetic vacant state, as depicted in Figure 7.5. From there, the electron very rapidly (10^{-14} to 10^{-11} s) relaxes into the lowest energy excited state, with the extra energy transferred to the surrounding medium, typically as heat. When the electron subsequently falls from this lowest excited state into its ground state, a photon can be emitted. This emission is known as fluorescence, and it is isotropic, meaning that the emitted photon is equally likely to travel in any direction.

The energy-level diagram shown in Figure 7.5 is known as a Jablonski diagram. The lowest line in each grouping represents the lowest energy state of that orbital, and the thinner horizontal lines above represent higher-energy vibrational states of the orbital. Transitions that involve absorption or emission of a photon are represented as straight arrows, and non-radiative transitions are shown as wavy arrows. The absorption and fluorescent emission are represented as vertical lines, as they involve no change in angular momentum (or spin, a type of angular momentum). However, there are other possible transitions, known as intersystem crossings and shown as diagonal lines in the diagram, that do change the spin of the excited electron. The most common spin-changing transition is from the excited singlet state to a triplet state. This is much less probable than the singlet–singlet

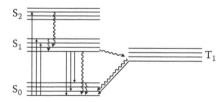

FIGURE 7.5 A Jablonski diagram. The solid horizontal lines represent molecular orbitals, with singlet states on the left and triplet states on the right. The arrows represent transitions between these levels, with straight lines for radiative transitions and wavy lines for non-radiative transitions. For the radiative transitions, absorption corresponds to the upward arrows at left, fluorescence corresponds to the downward arrows, and phosphorescence is represented by the diagonal arrow from T_1 to S_0.

transition that causes fluorescence emission; the rate can be well over 10 times smaller for a good fluorophore. Nonetheless, because the triplet state is relatively long-lived, the intersystem crossing can have a large impact on the quantum yield (defined below) of the molecule. Furthermore, molecules in the excited triplet state can transfer energy to nearby oxygen molecules, generating highly reactive singlet oxygen. The singlet oxygen and the reactive long-lived triplet state play major roles in photobleaching, a process in which the fluorophore undergoes an irreversible chemical change that destroys the fluorescence.

Looking at Figure 7.5, two other things are apparent. First, electrons can only be moved into the excited state by photons with particular energies that match the difference between the molecular orbitals. In practice, at room temperature, the absorption spectrum rarely falls to zero at higher energy, because the levels in the diagram are broadened by thermal fluctuations, and there are additional closely spaced electronic levels above those shown. Nevertheless, the level structure does lead to a peak in the absorption spectrum around that first transition. As a result, maximizing the fluorescence emission requires intense light at a particular wavelength or energy.

The second thing to notice from Figure 7.5 is that the energy of the emitted photon is generally less than that of the absorbed photon, as some energy is lost in the initial decay to the lowest level of the excited state. Since the energy of radiation is inversely proportional to the wavelength, this means that the emitted light will have a longer wavelength than the exciting light. This shift in wavelength is called the Stokes shift, and makes ultrasensitive detection possible by enabling efficient separation of the emitted fluorescence from the excitation light.

There are a few simple equations that describe these transitions. At steady state, the rate equation is

$$k_a = k_f + k_{nr} \tag{7.3}$$

where
k_a is the absorption rate
k_f is the rate of the fluorescence emission
k_{nr} is the sum of all other, non-radiative transitions

The fluorescence lifetime is given by $\tau_f = 1/(k_f + k_{nr})$, and is between 1 and 10 ns for most commonly used fluorophores. The probability of an absorbed photon of light yielding a photon of emitted light is known as the quantum yield, and is given by

$$\Phi = \frac{k_f}{k_f + k_{nr}} \tag{7.4}$$

The light absorbed by the sample can be obtained from the Beer–Lambert law (Equation 7.1):

$$I_{abs} = I_0 - I_{tr} = I_0(1 - 10^{-\varepsilon \ell c}) \tag{7.5}$$

and the fluorescence emission is then given by

$$I_{fl} = \Phi I_{abs} = \Phi I_0(1 - 10^{-\varepsilon \ell c}) \sim \Phi I_0 N \varepsilon \ell c. \tag{7.6}$$

The last approximation holds for $\varepsilon \ell c \ll 1$, with $N = \ln 10 = 2.3$. Equation 7.6 is only valid under two conditions. First, the solution must be dilute, so that absorption of the emitted light by other fluorophores in the sample is negligible. Second, there must not be significant depopulation of the ground state, as could arise from long fluorescence lifetimes, substantial population of the triplet state, or high incident light intensity. To move beyond this simple case, a quantitative treatment that uses molecular cross sections and includes photobleaching and intersystem crossing rates, as in Mathies et al. [45], is necessary.

The primary parameters that lead to strong fluorescence emission are a high molar extinction coefficient, which corresponds to a high probability that an incoming photon is absorbed; a short fluorescence lifetime, which corresponds to a fast decay from the excited state; and a high quantum yield. A low photobleaching rate is also desirable for stable and more sensitive measurements. For reference, fluorescein has $\varepsilon = 80,000\,M^{-1}\,cm^{-1}$ at 490 nm, a fluorescence quantum yield of 0.71, and a fluorescence lifetime of 4.5 ns [46]. The photobleaching quantum yield is roughly 3×10^{-5} in aerated aqueous solutions, meaning that on average, a single fluorescein molecule will emit roughly 24,000 photons before undergoing irreversible degradation.

Although there are some circumstances in which the analytes of interest have strong native fluorescence, particularly with UV excitation, fluorescence detection is usually combined with chemical labeling of the sample. This labeling can be performed before or after separation, by covalent linking to functional groups of interest. For example, many fluorophores are available with an attached thiocyanate or succinimide ester group, which will react with primary amines. This labeling can lead to some complications in quantitation: the fluorophores can undergo shifts in their emission or excitation spectra upon conjugation with other molecules, and the quantum yield may change considerably. Furthermore, the number of labels per analyte molecule can vary within a population, and the fluorescence does not increase linearly with the number of attached labels due to self-quenching effects. These effects have to be taken into account by careful calibration if quantification is desired. The use of two spectrally distinct labels can aid in rapid identification and quantification of analyte peaks, although a two-color fluorescence detector will be required [47].

There are several major components to any fluorescence detector: an excitation source, a filter or grating to separate the emitted light from the excitation, and a detector for measuring the emitted light. The following section discusses some considerations in choosing these components, with a focus on maximizing sensitivity. A typical arrangement is shown in Figure 7.6. There are many possible methods for coupling light into and out of the detection cell, some using lenses, others using optical fibers, and still others using direct coupling. In virtually all of the arrangements, the excitation and emission light paths are chosen to prevent direct passage of the excitation light into

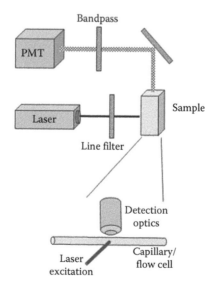

FIGURE 7.6 Schematic of a laser-induced fluorescence detector. A lamp with focusing optics and an appropriate band-pass filter could be used in place of the laser excitation when tightly collimated light is not required. The emitted fluorescence is detected by a PMT that can be operated in current mode or photon counting mode. Inset shows the mutually perpendicular arrangement of excitation, capillary, and detection optics.

the detector. The simplest arrangement, and the most widely used, is a perpendicular setup like that shown in the inset of Figure 7.6.

In the past, most chromatographic detectors have used a high-intensity broadband light source, such as a mercury or xenon arc lamp, as an excitation source. In this case, a narrow band-pass filter must be used before the sample, to select light near the excitation maximum of the fluorophore. The longer wavelength light must be eliminated to enable detection of the fluorescent emission, and the shorter wavelength light, especially in the UV, could lead to photodamage of the sample and/or the fluorophore. At present, with the advent of low-cost diode lasers, and the development of high-quality fluorophores with excitation maxima that match available laser wavelengths, the excitation source for a capillary electrophoresis fluorescence detector is almost always a laser. The collimated, monochromatic light is easily coupled into a small-volume detection cell with high intensity and easily rejected with a filter before the emission detector. However, for conventional HPLC systems, the detection volume is substantially larger; as a result, the collimation of the laser output is not a major advantage, and the intensity fluctuations of a laser are usually substantially larger than for a lamp source. For these systems, a broadband source might be preferable. When used with a tunable optical filter to select the excitation wavelength, a broadband source could also be used to measure a full excitation spectrum, at the expense of sensitivity and response time. A third possibility for an excitation source is a high-power light-emitting diode, or LED. These have broader emission spectra than lasers, and the light is not collimated. However, they are considerably less expensive than lasers, and are especially useful if miniaturization of the entire detector assembly is desired [48].

The primary function of an optical emission filter in a fluorescence detector is to reject stray excitation light. Even if a grating is used to obtain spectral information from the emitted light, a long pass filter should be used before the grating. If spectral information is not required, a band-pass filter, centered on the emission maximum of the fluorophore, can be used to further improve SNR by helping to discriminate against background fluorescence. Interference filters with greater than 80% transmission in the pass band and 10^6 rejection of the excitation light are available from several manufacturers. Several groups have found that optimal sensitivity can be achieved by using a notch or long pass filter followed by a band-pass filter [49].

In the past, most fluorescence detectors used photomultiplier tubes (PMTs) in current mode as the sensing elements, because of their high sensitivity at low light levels and reasonable dynamic range. More recently, photon counting modules, using avalanche photodiodes or PMTs as the sensing elements in concert with high-speed electronics, have been favored for ultrasensitive applications in which the detection volume is very small. For most HPLC applications, a PMT operated in current mode is still sufficient in the UV and visible range. If spectral information is desired, a high-sensitivity CCD array can be used with a grating.

For some laboratory-built systems, it is possible to detect on the order of 10 labeled molecules. In commercial systems, where the optical alignment from run to run has to be more robust, more typical limits for state-of-the-art instruments are a few hundred fluorophores, which for a detection volume of $(100\,\mu m)^3 = 1$ nL translates into a minimum detectable concentration of a few hundred femtomolar. Experimentally, there are several major factors that limit the sensitivity of detection [49]. For maximum sensitivity, the excitation light intensity should be sufficient to photobleach most of the fluorophores by the time they exit the detection volume. The collection optics are extremely important, and should be designed for spatial rejection of light originating from outside the detection volume as well as for efficient collection of as much of the isotropic fluorescence emission as feasible.

As long as the excitation source is sufficiently intense, and the detection electronics sufficiently sensitive, the measurement sensitivity of the instrument is typically limited by noise on the background signal. The background signal consists of several major components. Two of these, the scattering of the excitation from the detection cell walls, and Rayleigh scattering from the solvent in the detection cell, are greatly reduced by the optical emission filters; their contribution to the detected background then arises from the small fraction of excitation light that passes through the filters, and from fluorescence excited in the filter material itself. A third component of the background is

Raman scattering from the solvent; this can generally be rejected by carefully choosing the pass-band of the emission filter. The final component of the background is fluorescence arising from impurities in the detector cell walls or the solvent, or from sample components. Fluorescence from sample components is particularly a problem when analyzing labeled biological samples, in which the native fluorescence of proteins can interfere with detection of small amounts of labeled material. The effects of sample autofluorescence can be minimized by working at longer wavelengths; although many biologicals are fluorescent under UV illumination, there is relatively little fluorescence under illumination with red or near-infrared light. To eliminate scattering and fluorescence from the detection cell walls, some laboratory-built CE systems use a sheath-flow setup, in which the separation capillary empties into a larger tube filled with index matching fluid flowing in the same direction. This technique is extremely sensitive, but the sample is generally lost or very much diluted in the measurement, making further analysis impossible.

7.3.4 CONDUCTIVITY

Conductivity detectors can be used whenever the sample bands have different conductivity from the running buffer, although historically they have been used most extensively for the detection of small inorganic ions. They require less complicated instrumentation than optical detectors, and the sensors are more readily miniaturized, as there is no need for lamps or lenses. One major attraction of conductivity detectors is that there is no loss in sensitivity with miniaturization, which can be intuitively grasped from a simple model. The resistance of a volume of the column effluent is given by $R = \rho \ell / a$, where ℓ is the length of the volume measured, and a is the cross-sectional area. The resistivity, ρ depends on the composition of the effluent, and particularly on the concentration and mobilities of the ionic species. For a single ionic species, ρ is inversely proportional to the concentration. As the detection cell is made smaller, as long as the ratio of ℓ / a is kept constant, the resistance of the cell will be the same. In contrast, for absorbance measurements, $A = \varepsilon \ell c$, so the measured absorbance will decrease linearly with the light path length in the detection cell.

The prospect of ready miniaturization has led to considerable research interest in using conductivity detectors for capillary and chip-based electrophoresis [50,51]. Galloway et al. achieved a mass detection limit of 3.4 amol of alanine in a 425 pL injection volume using a two-electrode contact conductivity detector in a polymethyl methacrylate microfluidic (PMMA) device [52]. However, for larger detection volumes, the sensitivity of the technique limits its application primarily to the detection of small inorganic ions, which are difficult to detect with other methods. Another consideration with conductivity detection is the choice of the mobile phase; the highest sensitivity is achieved by using low conductivity solvent, which can create difficulties in the separation. To avoid this problem, and also to enable the use of conductivity detection with gradient elution, an ion suppressor can be used after the separation column and before the detector. For example, if the analyte of interest is anionic, an ion suppressor is chosen that exchanges the cations in the running buffer for hydronium ions, thereby decreasing the background conductivity and increasing the conductivity of the sample bands. A review by Haddad et al. discusses the variety of suppressor techniques and geometries currently in use [53]. Commercial detectors are generally able to detect less than 1 ppb F^- with 1 s or less integration times, which corresponds to a concentration of about 15 nM.

There are two main varieties of bulk conductivity detectors: contact and contactless. In a contact conductivity detector, the electrodes contact the column effluent directly. The electrodes are usually made of stainless steel, platinum, or gold in order to minimize electrochemical reactions, but they are still subject to fouling over time. In the absence of electrochemical reactions, there is no charge transfer between the solution and the electrodes, so the conductivity measurement is made with an oscillating or alternating voltage. The diagram for an equivalent circuit is shown in Figure 7.7. The current is given by

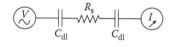

FIGURE 7.7 Simplified circuit diagram for a contact conductivity detector.

$$I = \frac{V}{Z} = \frac{V}{R_s + (1/i\pi f C_{dl})} \tag{7.7}$$

where

f is the frequency

R_s is the resistance of the solution

C_{dl} is the capacitance of the double layer formed when an ionic solution is placed next to a charged surface

For kilohertz measurement frequencies and above, the contribution of C_{dl} can generally be neglected. McWhorter and Soper employed a simple two-electrode contact conductivity detector with reverse-phase ion pair chromatography, and achieved a detection limit of 3.46 pg in a 100 nL volume (464 nM) for KCl [54]. Four-electrode setups, in which two leads are used to measure current and the other pair to measure voltage, can also be used in order to eliminate the effect of the impedance of the electrical leads on the measurement.

In a contactless conductivity detector, the electrodes are separated from the column effluent by insulating barriers, and are capacitively coupled to the sample. Because the electrodes are isolated, there is no possibility of fouling, so the contactless detection cells tend to be more robust. These cells can also be relatively simple in construction; one popular configuration in lab-built systems is the use of two open tubular electrodes that are placed directly over a polyimide-coated fused-silica capillary, shown in Figure 7.8, and first developed for use in CE [55,56]. A ground plane, placed around the capillary between the two electrodes, can be used to partially shield the direct capacitance between the electrodes. Kuban et al. used this detector configuration in reverse-phase LC separations of amino acids and drugs, achieving a 1 µM LOD in a capillary detection volume of approximately 30 nL (1 mm electrode spacing × 200 µm diameter) [57].

A simple equivalent circuit diagram for a two-electrode contactless conductivity measurement is shown in Figure 7.8. The impedance is given by

$$Z = \left[i2\pi f C_p + \frac{1}{R_s + (1/i\pi f C_e)} \right]^{-1} = \frac{R_s + (1/i\pi f C_e)}{i2\pi f R_s C_p + 2(C_p/C_e) + 1} \tag{7.8}$$

where

R_s is the solution resistance

C_p is the parasitic capacitance between the electrodes

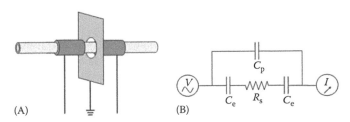

(A)　(B)

FIGURE 7.8 (A) Diagram for a laboratory-built detection cell for contactless conductivity measurements. The measurement electrodes can be wrapped around, threaded onto, or painted over a standard capillary. The use of a grounded shield in between the measurement electrodes greatly reduces stray capacitance. (B) Simplified circuit diagram for a contactless conductivity detector. C_e includes double layer capacitance C_{dl} as well as the capacitance across the capillary wall.

The capacitance between the electrode and solution, C_e, includes both the capacitance across the insulating barrier, such as the capillary wall, and the capacitance across the double layer. Without shielding between the electrodes, C_e dominates the response at low frequencies, and C_p dominates the response at high frequencies. Further complicating matters, the crossover between these regions depends on the relative size of R_s, C_p, and C_e, but as discussed above, the value of R_s depends on the effluent composition, and so changes over the course of the measurement. These difficulties can be easily avoided with reasonable ground-plane shielding between the electrodes, making C_p small enough that its contribution can be neglected except at very high frequencies. The impedance of the shielded configuration is then simply

$$Z = R_s + \frac{1}{i\pi f C_e} \tag{7.9}$$

As long as the measurement frequency $f \gg R_s/2\pi C_e$, the capacitive effects can be neglected, and the solution conductance $G = 1/R_s$ is measured directly.

7.3.5 ELECTROCHEMICAL

Electrochemical detectors can be among the most sensitive detectors used in separation science and are applicable to drug analysis, clinical chemistry, and neuroscience. The geometry is similar to that of a contact conductivity detector, in that the electrodes are in direct contact with the column effluent. The choice of electrode material, however, is very different, as the aim is to measure the result of electrochemical reactions at the electrodes rather than to minimize them. An electrochemical detector measures the electron flow that is generated at electrode surfaces during oxidation or reduction reactions. In cases where the reaction goes to completion, the total charge generated is proportional to the total mass of the reactant; this is known as coulometric detection. However, in detection for HPLC, typically the mobile phase will flow the reactant over the electrode, producing a current that varies with the concentration of the analyte being measured. This is amperometric detection and was until quite recently the most popular form of electrochemical detection.

Most electrochemical detectors require three electrodes: the working electrode, the auxiliary electrode, and the reference electrode. The working electrode is where the oxidative or reductive activity takes place, while the auxiliary and reference electrodes compensate for changes in the mobile phase conductivity. Figure 7.9 illustrates the most popular arrangements for the electrodes. In all cases, the electrodes are controlled by simple circuitry involving a regulated power supply, a potentiometer, and a series of amplifiers. The output from the electrodes is then fed into a recorder or more commonly a data acquisition system (Figure 7.10).

Electrodes can be constructed from a number of different materials but the major requirements are ruggedness and stability. One popular choice, carbon paste composed of graphite and a suitable dielectric substance, has several disadvantages, including solubility in a number of common solvents. Carbon in a vitreous or "glassy" form is a far better choice, especially when organic solvents are being employed in the separation. These electrodes are made by baking a carbon resin at elevated temperatures until it first becomes carbonized and then heating further until vitriation occurs. This process makes an ideal electrode; one which is mechanically strong, possesses good electrical properties, and can easily be regenerated and cleaned.

Electrochemical detection places a number of restrictions on the mobile phase used in the chromatographic separation but is well suited to reversed-phase separations. Even so, the mobile phase must be oxygen-free, which requires bubbling an inert gas such as helium through the mobile phase reservoir as well as the sample. Additionally, both the mobile phase and the sample must be free of metals to ensure baseline stability during the readings. Despite these constraints, electrochemical

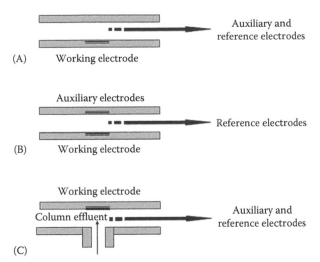

FIGURE 7.9 Three common electrode placements for an electrochemical detector. (A) The working electrode is integrated into the flow cell wall while the reference and auxiliary electrodes are placed downstream. (B) Both the working and auxiliary electrodes are integrated into the flow cell wall and the reference electrode is placed downstream. (C) The working electrode is integrated into the flow cell wall adjacent to the column effluent. As in example A, the reference and auxiliary electrodes are placed downstream.

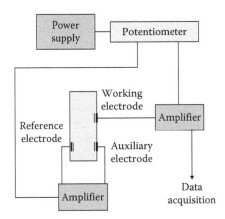

FIGURE 7.10 Block diagram of a basic electrochemical detector. All three electrodes are controlled by a regulated power supply coupled to a potentiometer and a series of amplifiers. The output from the electrodes is fed into a data acquisition system.

detectors have several advantages, in particular their suitability for small-bore chromatography, capillary electrochromatography, and chip-based systems [58]. This is due to the extremely low volume required to perform the analysis, which is often in the sub-microliter range.

For a more detailed overview of electrochemical detection, the interested reader is referred to Kissinger [59]. The range of applications suitable for electrochemical detection are discussed in an excellent review by Pacakova and Stulik [60]. Perhaps the most biomedically important application is the determination of catecholamines in biofluids [61–63]. Electrochemical detection has also lent itself to the detection of other medically important amines such as histamine [64] and neurotransmitters [65]. Additionally, electrochemical detection has been applied to the detection of certain antibiotics [66] and environmental carcinogens [67].

7.3.6 RADIOCHEMICAL

Radiochemical detectors used in HPLC come in two forms: those that use the addition of soluble chemical cocktails or scintillators, and those that use solid-phase scintillators. In either case, energy emitted from either β particles (tritium and carbon-14) or low-energy γ-ray sources (iodine-125) can be absorbed by the scintillators, which transform the energy into light in the range of 350–600 nm. This light can be detected by a photomultiplier, and its intensity is proportional to the energy of the radionucleotide emission. An overview of radiochemical detectors can be found in the book by Scott [68].

Detectors employed for β particle counting often use off-line detection, collecting fractions into a suitable scintillation cocktail, composed of organic solvent, a fluorophore, and a detergent, prior to measurement in a liquid scintillation counter. Alternatively, on-line detection can be achieved by adding a scintillation cocktail to the column effluent prior to detection in a flow cell surrounded by a pair of photomultipliers (Figure 7.11). Detection can also be achieved by passing the effluent through a solid scintillation core such as calcium fluoride or lithium glass, although this type of detector often suffers from low sensitivity and high background.

Gamma-ray counters are usually made from a cylinder of activated sodium iodide, which is enclosed in an aluminum shell, with one side of the cylinder lying in close proximity to a photomultiplier tube. The flow cell is composed of a "U" tube or coil (Figure 7.12) placed into the core of the cylinder. Radioactivity present in the column effluent is blocked from escaping by the dense sodium iodide, and the energy is converted into light, which is measured by the photomultiplier tube.

Unlike the majority of other HPLC detectors, radiochemical detection is based on time, and detection of low abundance peaks can be improved by adjusting the flow rate. Slower flow rates allow for more counting time and therefore increased accuracy and sensitivity. Even so, radiochemical detection often suffers from reduced sensitivity due to high background.

Radiometric monitoring of column effluents has been performed for some time and was in the past considered a highly sensitive means of analyzing trace materials. In 1982, Frey and Frey discussed the advantages and disadvantages of radiochemical detectors by comparing three "modern" detector systems [69]. Later, Wehmeyer et al. [70] reported the use of radiolabels for determining the metabolic pathways of vanilloids in rats using HPLC coupled with a on-line radiochemical detector, and recently de Bono et al. described a sensitive HPLC-radiochemical technique for studying nitric oxide synthesis in vascular tissue [71]. Radiochemical detection is generally considered

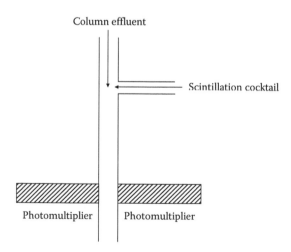

FIGURE 7.11 Schematic for radiochemical detection of analytes labeled with beta-particle emitters. Radio-labeled analytes in the column effluent are mixed with scintillation cocktail to produce light, which is detected and measured by the paired photomultiplier tubes and relayed to a data acquisition system.

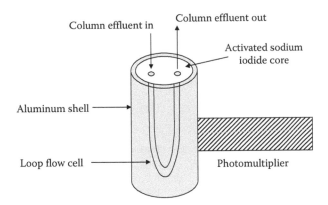

FIGURE 7.12   Schematic for radiochemical detection of gamma-emitting analytes. The analytes within each peak are measured by a photomultiplier tube placed in close proximity to or incorporated into the aluminum shell of the detector.

a good system for metabolic studies, particularly when coupled with UV detectors [72–74] or MS detectors [75,76] to obtain better characterization of the separated analytes.

7.3.7   EVAPORATIVE LIGHT SCATTERING

Evaporative light scattering detection (ELSD), which measures the particulate material that remains after the evaporation of the mobile phase, was first reported in 1966 and introduced commercially in the 1980s. It is considered a universal detection technique, in that the only requirement is that the analytes must be less volatile than all the components of the mobile phase. The ELSD signal is independent of the intrinsic characteristics of the analyte, including its absorptive, fluorescence, and electrochemical properties. ELSD is a mass detector and, in contrast to RI detection, is compatible with gradient elution and insensitive to temperature changes. However, it is also a destructive method, so no further analysis of the sample peaks is possible. This topic has recently been reviewed by Megoulas and Koupparis [77].

ELSD includes three distinct stages that need to be optimized in order to achieve low background, high sensitivity, and repeatability. These are the nebulization, evaporation, and optical detection stages, depicted schematically in Figure 7.13. First, the effluent of the chromatographic column is introduced into the nebulizer to create an aerosol. The nebulizer consists of a central orifice, through which the effluent exits, surrounded by an annulus through which an inert carrier gas flows (typically nitrogen, carbon dioxide, argon, or helium). The rapid flow of the carrier gas around the nebulizer orifice creates a region of low pressure that produces a narrow distribution of uniformly sized droplets. Many parameters that affect the diameter of the aerosol droplets are properties of the mobile phase, including the viscosity, density, and surface tension. As the mobile phase is typically chosen for optimal chromatographic separation, these parameters are not readily adjustable. The effluent flow rate through the chromatographic column is typically predetermined as well, chosen to optimize the analytical separation. As a result, the droplet diameter and distribution is controlled by the design of the nebulizer, in particular the size and shape of the orifice, and the flow rate of the carrier gas. Nebulizer orifices are designed to accommodate effluent flow rates from $\mu L\ min^{-1}$ to $mL\ min^{-1}$, with the carrier gas flow rate adjusted to produce the required droplet distribution. In general, lower carrier gas flow rates produce larger droplets, which carry more analyte and so lead to enhanced detection. However, these large droplets may not be efficiently evaporated in the second stage of the ELSD, giving rise to unwanted noise in the detector.

There are two basic designs for introducing the aerosol into the evaporation tube. A direct connection allows the complete distribution of droplets to reach the evaporation section, with the

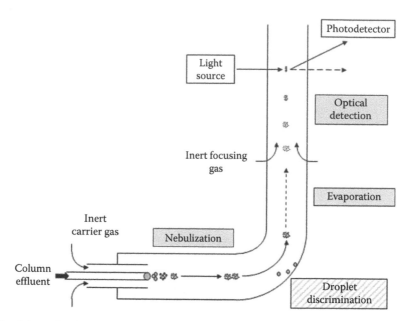

FIGURE 7.13 Schematic of an evaporative light scattering detector. The three stages are nebulization, in which column effluent is aerosolized; evaporation, in which the mobile phase is vaporized; and optical detection, in which the light scattering of the residual solute particles is recorded. Some detectors also include an obstacle in the flow path for droplet discrimination, which leads to a more homogenous distribution of droplet sizes.

advantage that all the analyte reaches the optical light scattering section. However, the droplet size distribution leads to a range in the mass of the analyte particles, and larger droplets may not be completely evaporated. A second design allows for the larger droplets to condense on the wall of the nebulization chamber by the introduction of an obstacle, such as a diaphragm or a bend, in the aerosol pathway. This design presents a more uniform size distribution of droplets for subsequent processing and analysis. The carrier gas flow transports the droplets into the evaporation stage, which is typically a tube heated between 30°C and 100°C. As the droplets move along this tube, the mobile phase is gradually vaporized, leaving behind solute particles of the analyte in solvent vapor. The use of lower evaporation temperatures, which reduces analyte evaporation and thermal decomposition, is especially important for measurement of higher volatility compounds. Using longer heated tubes, which increases the residency time in the evaporation section, and a carrier gas of higher thermal conductivity, such as helium, facilitates the use of lower temperatures. However, if the temperature is too low, the evaporation of the mobile phase will be incomplete, leading to elevated noise and background levels.

After evaporation, the solute particles enter the optical stage, where they interact with the probe beam; the scattered light is detected by a PMT or other photodetector. The scattered light falls into different scattering regimes depending on the relation of the wavelength of the interrogating light to the particle size, which is determined by operational conditions. Typically, particle diameters are of the order of the wavelength of light, leading to Mie scattering. Rayleigh scattering occurs if the particles are substantially smaller, and the interaction of the light with much larger particles is best described by reflection and refraction. The overlap of these differing scattering mechanisms for an inhomogeneous distribution of particle sizes leads to a nonlinear response, although the scattering is independent of the chemical composition of the particles. In some detectors, a second gas stream is introduced at the optical stage, focusing the particles into a smaller region in order to increase the signal strength, and also protecting the optical head from contamination. The background signal is purely from the contribution of the solvent vapor and can be essentially zero, provided that the solvent, modifiers, and carrier gas are of sufficient purity.

Some commonly used buffers, such as sodium and potassium phosphate, are incompatible with ELSD, but there are ready alternatives. For example, ammonium acetate has similar buffering properties to potassium phosphate, and ammonium carbonate, ammonium formate, pyridinium acetate, and pyridinium formate are options for different pH ranges. Typical mobile phase modifiers that do not meet the volatility criteria can be replaced by a wide variety of more volatile alternates. For example, phosphoric acid, commonly used as an acid modifier to control pH and ionization, can be replaced by trifluoroacetic acid; other acids that are sufficiently volatile for use with ELSD include, acetic, carbonic, and formic acids. Triethylamine, commonly used as a base modifier, is compatible with ELSD; other base modifiers that can be used are ethylamine, methylamine, and ammonium hydroxide [78].

Once the operational parameters, such as flow rate and evaporation temperature, have been optimized for a particular separation, ELSD is a highly reproducible detection scheme with low background signal and band broadening. Although ELSD is an order of magnitude less sensitive than other detection schemes, with an LOD of 1–10 ng of solute [79], it offers some real advantages. ELSD is a universal detector, allowing direct detection of compounds, such as carbohydrates, that are not amenable to UV absorption, fluorescence, or electrochemical detection without derivatization. In addition, ELSD permits the use of mobile phases with strong UV absorption, such as acetone or methyl ethyl ketone. As a result, ELSD has found application in a wide variety of separations, including the pharmaceutical, food products, and polymer industries.

7.3.8 CHEMILUMINESCENCE

Chemiluminescence is the emission of light as a result of a chemical reaction. To use this process for HPLC detection, the analyte of interest has to chemiluminesce when mixed with a particular reagent, cause chemiluminescence in other reagents, or cause quenching of a chemiluminescent reaction [80,81]. For on-line analyses, the reagents are introduced post-separation and mixed with the analytes immediately before the detector. The specificity of the chemical reaction can be helpful in selective detection: for example, the reaction between luminol and hydrogen peroxide can be used to identify phospholipid hydroperoxides in tissue extracts [82]. In other cases, pre-separation labeling of the analytes may be necessary in order for the chemiluminescence reaction to take place. Many fluorophores used in fluorescence detection can also be used in chemiluminescence detection, such as dansyl chloride-labeled amines, which can be excited using peroxyoxalate chemistry.

The detection hardware is similar to that of a fluorescence detector, in that efficient collection of the isotropically emitted light, and the ability to detect minimal amounts of light, are important. However, the major difference is that there is no need for excitation light, and no need to filter out scattered excitation light. The background in chemiluminescence detection can therefore be very low; the absence of scattered excitation light even provides an advantage over fluorescence detection. On the other hand, the emission of light will vary with time, starting when the reagents are mixed and peaking shortly afterward, with the detailed time dependence a function of the particular reaction chemistry. This places limits on the separation between the mixer and the detection zone, and can also constrain the flow rates.

7.3.9 OTHER METHODS

The chemiluminescent nitrogen detector is a specialized type of chemiluminescence detector, in which the separated analytes are combusted at temperatures in excess of 1000°C, causing the nitrogen atoms in the sample to form nitric oxide. The combustion products are then reacted with ozone, creating excited nitrogen dioxide, which then releases light [83]. Provided the analytes of interest contain nitrogen and their molecular formulae are known, this method provides reasonable single calibrant quantification [84].

NMR detectors use nuclear magnetic resonance spectroscopy to probe the relaxation times of nuclear spins in the eluting analytes, such ^1H or ^{31}P. The technique provides a wealth of structural information, but remains technically challenging, particularly for preserving chromatographic resolution and suppressing the solvent background signal [85,86].

ACKNOWLEDGMENTS

This work was supported by the Intramural Research Program of the National Institutes of Health, National Institute of Biomedical Imaging and Bioengineering. The authors would like to thank Terry M. Phillips for his guidance and for his valuable assistance, especially with the sections on electrochemical and radiochemical detectors.

REFERENCES

1. Horvath CG, Lipsky SR. Use of liquid ion exchange chromatography for separation of organic compounds. *Nature* 211, 748–749, 1966.
2. Rao RN, Nagaraju V. An overview of the recent trends in development of HPLC methods for determination of impurities in drugs. *Journal of Pharmaceutical and Biomedical Analysis* 33, 335–377, 2003.
3. Riordon JR. Diode array detectors for HPLC—High performance across the spectrum. *Analytical Chemistry* 72, 483A–487A, 2000.
4. Tracqui A, Kintz P, Mangin P. Systematic toxicological analysis using HPLC/Dad. *Journal of Forensic Sciences* 40, 254–262, 1995.
5. Lambert WE, VanBocxlaer JF, DeLeenheer AP. Potential of high-performance liquid chromatography with photodiode array detection in forensic toxicology. *Journal of Chromatography B* 689, 45–53, 1997.
6. Chan HK, Carr GP. Evaluation of a photodiode array detector for the verification of peak homogeneity in high-performance liquid-chromatography. *Journal of Pharmaceutical and Biomedical Analysis* 8, 271–277, 1990.
7. Nicoletti I, De Rossi A, Giovinazzo G, Corradini D. Identification and quantification of stilbenes in fruits of transgenic tomato plants (*Lycopersicon esculentum* Mill.) by reversed phase HPLC with photodiode array and mass spectrometry detection. *Journal of Agricultural and Food Chemistry* 55, 3304–3311, 2007.
8. Wang Q, Fang Y. Analysis of sugars in traditional Chinese drugs. *Journal of Chromatography. B, Analytical Technologies in the Biomedical and Life Sciences* 812, 309–324, 2004.
9. Liu M, Xie C, Xu W, Lu WY. Separation of polyethylene glycols and their amino-substituted derivatives by high-performance gel filtration chromatography at low ionic strength with refractive index detection. *Journal of Chromatography A* 1046, 121–126, 2004.
10. Golander Y, Schurrath U, Luch JR. Determination of pharmaceutical compounds in animal feeds using high-performance liquid-chromatography with refractive-index detection. *Journal of Pharmaceutical Sciences* 77, 902–905, 1988.
11. Hancock DO, Synovec RE. Microbore liquid-chromatography and refractive-index gradient detection of low-nanogram and low-ppm quantities of carbohydrates. *Journal of Chromatography* 464, 83–91, 1989.
12. Kenmore CK, Erskine SR, Bornhop DJ. Refractive-index detection by interferometric backscatter in packed-capillary high-performance liquid chromatography. *Journal of Chromatography A* 762, 219–225, 1997.
13. Kibbey CE. Quantitation of combinatorial libraries of small organic molecules by normal-phase HPLC with evaporative light-scattering detection. *Molecular Diversity* 1, 247–258, 1996.
14. Fang LL, Wan M, Pennacchio M, Pan JM. Evaluation of evaporative light-scattering detector for combinatorial library quantitation by reversed phase HPLC. *Journal of Combinatorial Chemistry* 2, 254–257, 2000.
15. Wei Y, Ding MY. Analysis of carbohydrates in drinks by high-performance liquid chromatography with a dynamically modified amino column and evaporative light scattering detection. *Journal of Chromatography A* 904, 113–117, 2000.
16. Stith BJ, Hall J, Ayres P, Waggoner L, Moore JD, Shaw WA. Quantification of major classes of Xenopus phospholipids by high performance liquid chromatography with evaporative light scattering detection. *Journal of Lipid Research* 41, 1448–1454, 2000.

17. Bravi E, Perretti G, Montanari L. Fatty acids by high-performance liquid chromatography and evaporative light-scattering detector. *Journal of Chromatography A* 1134, 210–214, 2006.

18. Fukushima T, Usui N, Santa T, Imai K. Recent progress in derivatization methods for LC and CE analysis. *Journal of Pharmaceutical and Biomedical Analysis* 30, 1655–1687, 2003.

19. Siluk D, Oliveira RV, Esther-Rodriguez-Rosas M et al. A validated liquid chromatography method for the simultaneous determination of vitamins A and E in human plasma. *Journal of Pharmaceutical and Biomedical Analysis*, 44, 1001–1007, 2007.

20. Zare RN. Laser chemical-analysis. *Science* 226, 298–303, 1984.

21. Belenkii BG. Use of laser detectors in capillary liquid-chromatography. *Journal of Chromatography. Biomedical Applications* 434, 337–361, 1988.

22. Rahavendran SV, Karnes HT. Solid-state diode laser-induced fluorescence detection in high-performance liquid-chromatography. *Pharmaceutical Research* 10, 328–334, 1993.

23. Jemal M, Xia YQ. LC-MS development strategies for quantitative bioanalysis. *Current Drug Metabolism* 7, 491–502, 2006.

24. Marzo A, Dal Bo L. Tandem mass spectrometry (LC-MS-MS): A predominant role in bioassays for pharmacokinetic studies. *Arzneimittelforschung Drug Research* 57, 122–128, 2007.

25. Korfmacher WA. Principles and applications of LC-MS in new drug discovery. *Drug Discovery Today* 10, 1357–1367, 2005.

26. Ma SG, Chowdhury SK, Alton KB. Application of mass spectrometry for metabolite identification. *Current Drug Metabolism* 7, 503–523, 2006.

27. Castro-Perez JM. Current and future trends in the application of HPLC-MS to metabolite-identification studies. *Drug Discovery Today* 12, 249–256, 2007.

28. Peterson BL, Cummings BS. A review of chromatographic methods for the assessment of phospholipids in biological samples. *Biomedical Chromatography* 20, 227–243, 2006.

29. Tan B, Bradshaw HB, Rimmerman N et al. Targeted lipidomics: Discovery of new fatty acyl amides. *AAPS Journal* 8, E461–E465, 2006.

30. Coon JJ, Syka JEP, Shabanowitz J, Hunt DF. Tandem mass spectrometry for peptide and protein sequence analysis. *Biotechniques* 38, 519–523, 2005.

31. Delahunty C, Yates JR. Protein identification using 2D-LC-MS/MS. *Methods* 35, 248–255, 2005.

32. Neverova I, Van Eyk JE. Role of chromatographic techniques in proteomic analysis. *Journal of Chromatography. B, Analytical Technologies in the Biomedical and Life Sciences* 815, 51–63, 2005.

33. Mitulovic G, Mechtler K. HPLC techniques for proteomics analysis—A short overview of latest developments. *Brief in Functional Genomics and Proteomics* 5, 249–260, 2006.

34. Carr SA, Annan RS, Huddleston MJ. Mapping posttranslational modifications of proteins by MS-based selective detection: Application to phosphoproteomics. In: Burlingame AL, ed., *Mass Spectrometry: Modified Proteins and Glycoconjugates, Vol. 405*, New York: Academic Press, 2005, 82–115.

35. Yeung ES. Indirect detection methods—Looking for what is not there. *Accounts of Chemical Research* 22, 125–130, 1989.

36. Landers JP. *Handbook of Capillary Electrophoresis*. Boca Raton, FL: CRC Press, 1997.

37. Swinney K, Bornhop DJ. Detection in capillary electrophoresis. *Electrophoresis* 21, 1239–1250, 2000.

38. Poppe H. The performance of some liquid-phase flow-through detectors. *Analytica Chimica Acta* 145, 17–26, 1983.

39. Pace CN, Vajdos F, Fee L, Grimsley G, Gray T. How to measure and predict the molar absorption-coefficient of a protein. *Protein Science* 4, 2411–2423, 1995.

40. Krattiger B, Bruin GJM, Bruno AE. Hologram-based refractive-index detector for capillary electrophoresis—Separation of metal-ions. *Analytical Chemistry* 66, 1–8, 1994.

41. Deng YZ, Li BC. On-column refractive-index detection based on retroreflected beam interference for capillary electrophoresis. *Applied Optics* 37, 998–1005, 1998.

42. Bornhop DJ. Microvolume index of refraction determinations by interferometric backscatter. *Applied Optics* 34, 3234–3239, 1995.

43. Markov D, Begari D, Bornhop DJ. Breaking the 10(−7) barrier for RI measurements in nanoliter volumes. *Analytical Chemistry* 74, 5438–5441, 2002.

44. Swinney K, Pennington J, Bornhop DJ. Universal detection in capillary electrophoresis with a micro-interferometric backscatter detector. *Analyst* 124, 221–225, 1999.

45. Mathies RA, Peck K, Stryer L. Optimization of high-sensitivity fluorescence detection. *Analytical Chemistry* 62, 1786–1791, 1990.

46. Tsien RY, Waggoner A. Fluorophores for confocal microscopy: Photophysics and photochemistry. In: Pawley JB, ed. *Handbook of Biological Confocal Microscopy*. New York: Plenum Press, 1995, pp. 267–279.

47. Morgan NY, Wellner E, Talbot T, Smith PD, Phillips TM. Development of a two-color laser fluorescence detector—On-line detection of internal standards and unknowns by capillary electrophoresis within the same sample. *Journal of Chromatography A* 1105, 213–219, 2006.

48. Xiao D, Zhao SL, Yuan HY, Yang XP. CE detector based on light-emitting diodes. *Electrophoresis* 28, 233–242, 2007.

49. Johnson ME, Landers JP. Fundamentals and practice for ultrasensitive laser-induced fluorescence detection in microanalytical systems. *Electrophoresis* 25, 3513–3527, 2004.

50. Guijt RM, Evenhuis CJ, Macka M, Haddad PR. Conductivity detection for conventional and miniaturised capillary electrophoresis systems. *Electrophoresis* 25, 4032–4057, 2004.

51. Sonlinova V, Kasicka V. Recent applications of conductivity detection in capillary and chip electrophoresis. *Journal of Separation Science* 29, 1743–1762, 2006.

52. Galloway M, Stryjewski W, Henry A et al. Contact conductivity detection in poly(methyl methacrylate)-based microfluidic devices for analysis of mono- and polyanionic molecules. *Analytical Chemistry* 74, 2407–2415, 2002.

53. Haddad PR, Jackson PE, Shaw MJ. Developments in suppressor technology for inorganic ion analysis by ion chromatography using conductivity detection. *Journal of Chromatography A* 1000, 725–742, 2003.

54. McWhorter S, Soper SA. Conductivity detection of polymerase chain reaction products separated by micro-reversed-phase liquid chromatography. *Journal of Chromatography A* 883, 1–9, 2000.

55. da Silva JAF, do Lago CL. An oscillometric detector for capillary electrophoresis. *Analytical Chemistry* 70, 4339–4343, 1998.

56. Zemann AJ, Schnell E, Volgger D, Bonn GK. Contactless conductivity detection for capillary electrophoresis. *Analytical Chemistry* 70, 563–567, 1998.

57. Kuban P, Abad-Villar EM, Hauser PC. Evaluation of contactless conductivity detection for the determination of UV absorbing and non-UV absorbing species in reversed-phase high-performance liquid chromatography. *Journal of Chromatography A* 1107, 159–164, 2006.

58. Ewing AG, Mesaros JM, Gavin PF. Electrochemical detection in microcolumn separations. *Analytical Chemistry* 66, A527–A537, 1994.

59. Kissinger P. Electrochemical detection. In: Cazes J, ed. *Encyclopedia of Chromatography*. New York: Marcel Dekker, 2001, pp. 276–279.

60. Pacakova V, Stulik K. Electrochemical detection in HPLC 2. Applications. *Nahrung-Food* 29, 651–664, 1985.

61. Hjemdahl P. Catecholamine measurements by high-performance liquid-chromatography. *American Journal of Physiology* 247, E13–E20, 1984.

62. Brandsteterova E, Kubalec P, Skacani I, Balazovjech I. HPLC-Ed determination of catecholamines and their metabolites in urine. *Neoplasma* 41, 205–211, 1994.

63. Nyyssonen K, Parviainen MT. Plasma-catecholamines—Laboratory aspects. *Critical Reviews in Clinical Laboratory Sciences* 27, 211–236, 1989.

64. Pihel K, Hsieh S, Jorgenson JW, Wightman RM. Electrochemical detection of histamine and 5-hydroxytryptamine at isolated mast-cells. *Analytical Chemistry* 67, 4514–4521, 1995.

65. Yi Z, Brown PR. Chromatographic methods for the analysis of basic neurotransmitters and their acidic metabolites. *Biomedical Chromatography* 5, 101–107, 1991.

66. LaCourse WR, Dasenbrock CO. High-performance liquid chromatography pulsed electrochemical detection for the analysis of antibiotics. *Advances in Chromatography* 38, 189–232, 1998.

67. Barek J, Cvacka J, Muck A, Quaiserova V, Zima J. Electrochemical methods for monitoring of environmental carcinogens. *Fresenius Journal of Analytical Chemistry* 369, 556–562, 2001.

68. Scott RPW. *Chromatographic Detectors: Design, Function, and Operation*. New York: Marcel Dekker, 1996.

69. Frey BM, Frey FJ. Three radioactivity detectors for liquid-chromatographic systems compared. *Clinical Chemistry* 28, 689–692, 1982.

70. Wehmeyer KR, Kasting GB, Powell JH, Kuhlenbeck DL, Underwood RA, Bowman LA. Application of liquid-chromatography with online radiochemical detection to metabolism studies on a novel class of analgesics. *Journal of Pharmaceutical and Biomedical Analysis* 8, 177–183, 1990.

71. de Bono JP, Warrick N, Bendall JK, Channon KM, Alp NJ. Radiochemical HPLC detection of arginine metabolism: Measurement of nitric oxide synthesis and arginase activity in vascular tissue. *Nitric Oxide-Biology and Chemistry* 16, 1–9, 2007.

72. Rajagopalan P, Gao ZL, Chu CK, Schinazi RF, McClure HM, Boudinot FD. High-performance liquid-chromatographic determination of (–)-beta-D-2,6-diaminopurine dioxolane and its metabolite, dioxolane guanosine, using ultraviolet and online radiochemical detection. *Journal of Chromatography. B, Biomedical Applications* 672, 119–124, 1995.

73. Visich JE, Byron PR. High-pressure liquid chromatographic assay for the determination of thyrotropin-releasing hormone and its common metabolites in a physiological salt solution circulated through the isolated perfused rat lung. *Journal of Pharmaceutical and Biomedical Analysis* 15, 105–110, 1996.

74. Kaivosaari S, Salonen JS, Mortensen J, Taskinen J. High-performance liquid chromatographic method combining radiochemical and ultraviolet detection for determination of low activities of uridine 5′-diphosphate-glucuronosyltransferase. *Analytical Biochemistry* 292, 178–187, 2001.

75. Nassar AEF, Bjorge SM, Lee DY. On-line liquid chromatography-accurate radioisotope counting coupled with a radioactivity detector and mass spectrometer for metabolite identification in drug discovery and development. *Analytical Chemistry* 75, 785–790, 2003.

76. Andrews CL, Li F, Yang E, Yu CP, Vouros P. Incorporation of a nanosplitter interface into an LC-MS-RD system to facilitate drug metabolism studies. *Journal of Mass Spectrometry* 41, 43–49, 2006.

77. Megoulas NC, Koupparis MA. Twenty years of evaporative light scattering detection. *Critical Reviews in Analytical Chemistry* 35, 301–316, 2005.

78. Chudy MR, Wilcox MJ. The use of volatile mobile phase modifiers for HPLC methods and evaporative light scattering detectors. Alltech Application Notes 2000.

79. Rissler K. High-performance liquid chromatography and detection of polyethers and their mono(carboxy) alkyl and -arylalkyl substituted derivatives. *Journal of Chromatography A* 742, 1–54, 1996.

80. Li FM, Zhang CH, Guo XJ, Feng WY. Chemiluminescence detection in HPLC and CE for pharmaceutical and biomedical analysis. *Biomedical Chromatography* 17, 96–105, 2003.

81. DeJong G, Kwakman P. Chemi-luminescence detection for high-performance liquid-chromatography of biomedical samples. *Journal of Chromatography. Biomedical Applications* 492, 319–343, 1989.

82. Miyazawa T, Fujimoto K, Suzuki T, Yasuda K. Determination of phospholipid hydroperoxides using luminol chemiluminescence high-performance liquid chromatography. *Oxygen Radicals in Biological Systems*, Pt. C 233, 324–332, 1994.

83. Fujinari EM, Courthaudon LO. Nitrogen-specific liquid-chromatography detector based on chemiluminescence—Application to the analysis of ammonium nitrogen in waste-water. *Journal of Chromatography* 592, 209–214, 1992.

84. Lane S, Boughtflower B, Mutton I et al. Toward single-calibrant quantification in HPLC. A comparison of three detection strategies: Evaporative light scattering, chemiluminescent nitrogen, and proton NMR. *Analytical Chemistry* 77, 4354–4365, 2005.

85. Lindon JC, Nicholson JK, Wilson ID. Directly coupled HPLC-NMR and HPLC-NMR-MS in pharmaceutical research and development. *Journal of Chromatography. B, Analytical Technologies in the Biomedical and Life Sciences* 748, 233–258, 2000.

86. Albert K. Online use of NMR detection in separation chemistry. *Journal of Chromatography A* 703, 123–147, 1995.

8 LC–MS Interfaces: State of the Art and Emerging Techniques

Achille Cappiello, Pierangela Palma, and Giorgio Famiglini

CONTENTS

8.1 INTRODUCTION

Mass spectrometer (MS) is considered the most powerful detector for liquid chromatography (LC), though only after the development of "soft" ionization techniques LC and MS were brought closer, and powerful instrumentations have permitted new challenging analytical applications. Although the coupling of LC and MS is a chemically and physically challenging operation and its practical application is still far from the immediacy and simplicity of gas chromatography (GC)-MS, it opens the door for the determination of large and thermally labile molecules in complex matrices.

GC-MS instrumentation is nowadays a well-established, relatively simple, and inexpensive instrumentation, while the LC–MS interfacing is still in progress, yet far from attaining its full growth. The strategies that LC–MS interfaces use to overcome the evident difficulties have led to the development of different approaches, each one with its own peculiarity and field of application.

Difficulty in LC–MS can be summarized as follows:

1. *Solvent restriction*: An LC mobile phase is a liquid of variable composition, and sample components are led to the detector together with a given solvent combination. Separation and ionization of analytes often depend on different and, sometimes, antagonist mobile-phase components.
2. *Sample restriction*: LC analyzes all samples that can be dissolved in solvents. The analytes may vary dramatically in weight, polarity, and stability, and this poses a huge variability in terms of response and system requirements on both interface and MS. In spite of this challenging premise, the demand for new LC–MS methods is increasingly strong in many fields of research.

Differently from GC-MS, the picture representing LC–MS is still composed into a puzzle with a number of techniques dedicated to solve a specific analytical task. The aim of this chapter is to give the reader a comprehensive and up-to-date view of the most advanced technical solutions keeping an eye on recent developments.

8.2 ELECTROSPRAY

Electrospray (ESI) is an atmospheric pressure ionization source in which the sample is ionized at an ambient pressure and then transferred into the MS. It was first developed by John Fenn in the late 1980s [1] and rapidly became one of the most widely used ionization techniques in mass spectrometry due to its high sensitivity and versatility. It is a soft ionization technique for analytes present in solution; therefore, it can easily be coupled with separation methods such as LC and capillary electrophoresis (CE). The development of ESI has a wide field of applications, from small polar molecules to high molecular weight compounds such as protein and nucleotides. In 2002, the Nobel Prize was awarded to John Fenn following his studies on electrospray, "for the development of soft desorption ionization methods for mass spectrometric analyses of biological macromolecules."

8.2.1 PRINCIPLES OF OPERATION

In ESI, ions and molecules of a solution are converted into ions in the gas phase by the vaporization of charged droplets of the solution [2–4]. Heat is applied to compensate for the heat of the vaporization of the solvent. The charged droplets are formed when a strong electrical field, of the order of 10^6 V/m, is applied to a liquid passing through a metal capillary tubing. The electrical field is achieved by applying a potential difference of 2–6 kV between the metal tubing and a counter-electrode placed at a distance that depends both on the flow rate and on the geometry of the ion source. This voltage induces the ions to move toward the surface of the liquid, lowering its surface tension due to the mutual repulsion of the ions. At a proper onset voltage, the solution forms a typical cone-shaped tip, the Taylor cone, from which a spray plume of charged droplets breaks out. A coaxial low gas flow confines the droplets in a limited space. The droplets contain both anions and cations, but one specie is in excess compared to the other, and therefore it is not neutralized. As the solvent evaporates, the droplets' size decreases and the ratio of the surface charge to the surface area increases. The point at which the charge repulsion completely offsets the surface tension is called the Rayleigh stability limit, and when it is reached, the droplet breaks out. It is defined by the equation:

$$q_{RY} = 8\pi(\varepsilon_0 \gamma R^3)^{1/2} \tag{8.1}$$

where

q is the droplet charge
R is the droplet radius
ε_0 is the permittivity of the vacuum
γ is the surface tension

As a consequence, the droplet breaks up into a stream of smaller droplets, each one continuing to shrink by evaporation until the Rayleigh stability limit is reached again. The process of droplet fission is repeated several times and it is called "uneven fission" or "droplet jet fission" [5,6].

A heated curtain gas or a heated capillary may cooperate in the evaporation of the solvent, causing the reduction of the droplet size.

Two models can explain the events that take place as the droplets dry. One was proposed by Dole and coworkers and elaborated by Röllgen and coworkers [7] and it is described as the "charge residue mechanism" (CRM). According to this theory, the ions detected in the MS are the charged species that remain after the complete evaporation of the solvent from the droplet. The "ion evaporation" model affirms that, as the droplet radius gets lower than approximately 10 nm, the emission of the solvated ions in the gas phase occurs directly from the droplet [8,9]. Neither of the two is fully accepted by the scientific community. It is likely that both mechanisms contribute to the generation of ions in the gas phase. They both take place at atmospheric pressure and room temperature, and this avoids thermal decomposition of the analytes and allows a more efficient desolvation of the droplets, compared to that under vacuum systems. In Figure 8.1, a schematic of the ionization process is described.

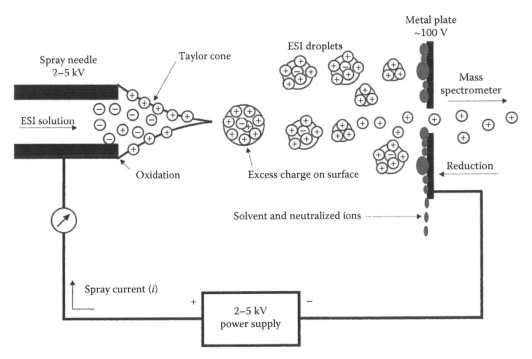

FIGURE 8.1 Schematic of the electrospray ionization process. The analyte solution is pumped through a needle to which a high voltage is applied. A Taylor cone with an excess of positive charge on its surface forms as a result of the electric field gradient between the ESI needle and the counter electrode. Charged droplets are formed from the tip of the Taylor cone, and these droplets evaporate as they move toward the entrance to the MS to produce free, charged analyte molecules that can be analyzed with their mass-to-charge ratio. (Reproduced from Cech, N.B. and Enke, C.G., *Mass Spectrom. Rev.*, 20, 362, 2001. Copyright 2001. Reprinted with permission from John Wiley & Sons.)

8.2.2 ION FORMATION

The ion formation may occur in the bulk solution before the electrospray process takes place or in the gas phase by protonation or salt adduct formation, or by an electrochemical redox reaction. Polar compounds already exist in solution as ions; therefore, the task of the electrospray is to separate them from their counterions. This is the case of many inorganic and organic species and all those compounds that show acidic or basic properties. Proteins, peptides, nucleotides, and many other bio- and pharmaceutical analytes are typical examples of substances that can be detected as protonated or deprotonated species.

Large molecules may have more than one protonation or deprotonation site, and they are therefore capable of forming multiple-charged ions, as illustrated in Figure 8.2. This is one of the most interesting aspects and it is peculiar of ESI to form ions such as $[M + nH]^{n+}$ and $[M - nH]^{n-}$. The degree of charging depends on the molecular mass. The higher the molecular mass, the higher the number of molecular ions that appear at different m/z values in the spectrum. This multitude of signals can be averaged in a deconvoluted mass spectrum, giving the true molecular mass of the analyte, as shown in Figure 8.3. The possibility of forming multiple-charged ions allows the detection of high molecular weight molecules with mass analyzers having a limited m/z scan range.

Polar analytes can also form adducts with various ions. A typical adduct ion that is formed in the positive ion mode is $[M + nNa]^{n+}$, but many other adduct ions with K^+, NH_4^+, Cl^-, acetate are frequently observed (Figure 8.4). Their formation normally occurs in the bulk solution before the charge separation process, in the electrospray droplets during evaporation, or in the gas phase. If the analyte is weakly basic or polar, salts bearing cations may be intentionally added to the sample to stimulate the formation of positive ions [10,11].

It is important to point out that the distribution of the charges, and therefore the final signal response, is strongly influenced by several factors that depend both on the nature of the solution (pH, solvent composition, additives...) and on instrumental conditions [12–25].

Noncovalently bound adducts can also be observed, involving enzyme and substrate, protein and ligand, protein and protein, antigen and antibody that can influence the biological activity of the cells [26]. The study and the detection of noncovalent complexes has the benefit of enormous advantages using ESI-MS because it requires a lower amount of material and a shorter time to provide structural information.

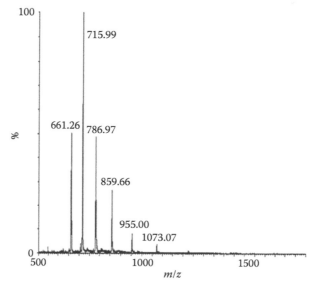

FIGURE 8.2 ESI-MS spectrum of ubiquitin, showing the multiple-charged ions.

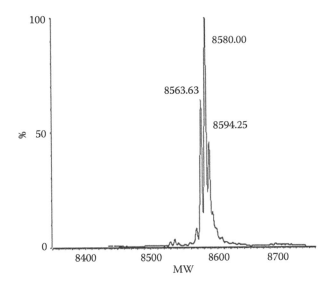

FIGURE 8.3   Deconvoluted spectrum of ubiquitin.

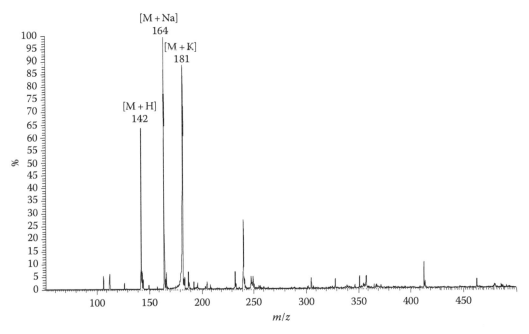

FIGURE 8.4   ESI-MS spectrum in FIA of metamidophos, showing adduct ions with Na^+ and K^+.

Electrolytic oxidation or reduction may occur upon contact with the electrospray solution, triggering the formation of ions that normally do not exist in the bulk. Derivatization can be used to induce ionization for those compounds that do not carry redox groups [27,28].

In ESI ion fragmentation is very poor or irrelevant; therefore, a mass spectrum only gives information on the molecular ion. The identification and the structural elucidation of unknown species can be performed only if two or more MS stages are available (MSMS). Using MSMS, fragmentation takes place inside the analyzer thanks to collisions with an inert gas (helium, argon), thus generating fragmentation spectra that allow qualitative and structural investigations of the analytes.

8.2.3 How Analyte Properties Play a Role in Selectivity and Interferences

It is of primary importance to understand how the chemical–physical properties of a given compound can influence its response in ESI-MS. In fact, ESI response can vary significantly among different analytes, depending not only on their concentration, but also on their properties [29–33].

The success of an analysis can depend on the suitability of a compound to undergo ESI conditions and on the presence of other species that can influence the sensitivity. In case of a weak response, it is possible to derivatize the analyte.

The electrospray response can be altered by the presence of coeluting compounds [34]. Although not fully described yet, this phenomenon, called matrix effects, is probably due to a sort of competition between an analyte and the coeluting, undetected matrix components.

The most important mechanistic aspects regarding ion suppression have been investigated in deeper detail than those regarding ion enhancement, which, in general, are less frequent. Ion suppression can occur both in the solution or in the gas phase [35].

In the bulk solution phase, a decrease in evaporation efficiency of the droplets can be related to the presence-interfering compounds in the matrix. For instance, a high concentration of volatiles can induce an increase in the boiling point of the solution, causing a decrease in the evaporation efficiency of the solvent. This occurrence limits the analyte in reaching the gas phase, and in some cases, endangers correct aerosol formation. Signal suppression can also result from a competition between the analyte and the interfering compounds for available charges. Another source of suppression is the solvent evaporation from the droplet surface that can induce analyte precipitation, both as pure compounds and as coprecipitates, especially when the matrix contains nonvolatile substances. The presence of ion-paring agents in the matrix can neutralize the charge of the analyte, thus causing signal suppression. Other studies demonstrate that interfering compounds bearing tensioactive properties can compete with the analyte for migration to the droplet surface, thus keeping the analyte from evaporating.

Ion suppression in the phase solution plays a predominant role in matrix effects; however, the analyte can also be neutralized in the gas phase through neutralization reactions, charge stripping, or charge transfer to another chemical species.

Other studies demonstrated that ESI response can be highly suppressed if the real samples do not undergo a pretreatment step, indicating the importance of this procedure in limiting matrix effects [36].

One hypothesis is that this phenomenon may be due to a competition between nonvolatile matrix components and analyte ions for access to the droplet surface for transfer to the gas phase. This competition can enhance or suppress the signal depending on the environment in which the ionization and the ion evaporation processes take place; therefore, the presence of a matrix can influence ion formation. Chemical–physical properties of the analytes, and in particular polarity, play a role on the degree of matrix effects that have a strong influence on the analytical precision of the method, and in particular, on the sensitivity and the limit of quantification [37].

As a consequence of ion suppression, the analytical method is affected in different ways [38]. First of all, it has an effect on the capability to detect analytes due to signal decrease. As a consequence, the true concentration of the analyte can be underestimated to the point of a "false negative." If signal suppression involves an internal standard only, one may be induced to overestimate analyte concentration to the point of a "false positive." Method precision response linearity can be affected because the degree of suppression can vary significantly in different samples. The presence of co-eluting compounds can modify the mass spectra, thus making the database search more complicated.

Endogen and esogen substances that co-elute with the analyte of interest can induce ion suppression. The first ones are present in the real sample and are injected together with the analyte, despite any purification step. They can be ionic species (i.e., salts), high polarity compounds, and other organic molecules that show affinity to the analyte. All the interfering compounds that are introduced in the sample during the sample preparation are considered esogen ion suppressors. They can be ion-paring agents, polymers, and phthalates from the labware, detergents, buffers, organic acids, or other materials released by the SPE stationary phases.

8.2.4 Interfacing ESI-MS with HPLC

Interfacing HPLC and electrospray has permitted remarkable advancements in biomedical analysis; however, the process of successfully coupling these two techniques has required considerable efforts to overcome the limitations imposed by their distinctive operating requirements.

8.2.4.1 Flow Rate Limitations

The first ESI design at the end of the 1980s proved to work properly as the HPLC interface with mobile phase flow rates between 1 and 10 μL/min. Meanwhile, the development of the HPLC instrumentation and columns was oriented in the mL/min flow rate mode. In addition, the nebulization process based only on the application of an electrical field does not produce a stable spray from aqueous mobile phases. A modified ESI source, called ionspray, was then introduced [39], in which the nebulization of a liquid solution is pneumatically assisted by a coaxial flow of nitrogen (sheath gas) that allows the formation of a stable aerosol at mobile-phase flow rates between 10 and 500 μL/min and the use of aqueous mobile phases. When working at higher flow rates (500–1000 μL/min), an additional nitrogen flow rate can be used (auxiliary gas) to assist the desolvation of the droplets. This modified source is called turboionspray.

Appropriate modifications of the capillary allow the nebulization at flow rates lower than 1 μL/min (nano and micro electrospray). As reported in Table 8.1, the electrospray can be successfully used at different flow rates, accomplishing hyphenation with LC and CE.

8.2.4.2 Mode of Operation

ESI analysis can be run either in the positive or in the negative ion mode. When dealing with complex mixtures, both positive and negative detection modes could be essential. In this case, the best and the most common way is to perform two separate analyses, even if it is possible to switch between modes during a single analysis. However, it should be taken into account that in the latter case, spray instability can cause loss of sensitivity. Negative ion ESI-MS analysis requires a more careful control over operative conditions to avoid corona discharge that can interfere with the normal operation and can be the source of background noise [40].

8.2.4.3 Mobile Phase Limitations

Solvent composition is dependent on the mode of operation (positive or negative), and the behavior of an analyte in ESI-MS can be strongly influenced by the properties of the mobile phase chosen. A proper electrospray is a perfect balance of flow rate, applied voltage, liquid surface, and mobile

---

TABLE 8.1
ESI Sources

Source	Flow Rate	Aerosol Generation
Nanoelectrospray	10–100 nL/min	Electrostatic
Microelectrospray	100 nL/min–4 μL/min	Electrostatic
Electrospray	1–10 μL/min	Electrostatic
Ionspray	10–500 μL/min	Pneumatic/electrostatic
Turboionspray	500 μL/min–1 mL/min	Pneumatic/electrostatic

phases normally used in the HPLC reversed phase mode, such as water, methanol, or acetonitrile, are most suitable to this use. Nonconductive mobile phases, such as those used in the HPLC normal-phase mode, are less appropriate for this technique. Only volatile buffers, such as acetic or formic acid, ammonium formate, and ammonium hydroxide, may be added to the mobile phase [41]. Acetic acid and ammonium hydroxide can also be used to adjust pH instead of HCl or NaCl that can introduce nonvolatile chlorine and sodium salts. Other modifiers, such as ion pairing agents and trifluoroacetic acid, can cause signal suppression and therefore should be avoided.

8.3 ATMOSPHERIC PRESSURE CHEMICAL IONIZATION

Atmospheric pressure chemical ionization (APCI) interface for LC–MS became a reality at the end of the 1980s [42]. Its sensitivity and robustness were immediately impressive in comparison with all the other interface mechanisms present on the market.

It was in the 1960s that Munson and Field presented a new ionization method for mass spectrometry [43]. They converted the so-called chemical ionization (CI) technique into a standard electron ionization (EI) source to obtain a less hard analyte fragmentation by the action of a reagent gas. In CI, the reagent gas is ionized under the impact of an electron beam. The charge is transferred to the analyte through a chemical process. The best result they obtained was to generate a more intense molecular-ion signal even with the compounds being more fragile. An additive, important new point with respect to EI was the possibility of acquiring mass spectra in both positive and negative polarities to have complementary data useful for structural elucidation. The mechanism involved in CI is an acid-base reaction in the gas phase that takes place by reactant ions and the neutral analyte molecules present in the ion source. In the early 1970s, the research group led by Horning developed an APCI source equipped either with a ^{63}Ni foil or a corona discharge needle as the primary source of electrons [44,45]. However, in those days no instrumentation was available to fit such a source. It was only at the end of the 1980s–beginning of the 1990s that a variety of instruments became available on the market [46]. It is not infrequent to have APCI and ESI interfaces installed on the same instrumentation.

8.3.1 PRINCIPLES OF OPERATION

The instrumentation is the same as the one used for ESI-MS, with a few exceptions. The eluate is either introduced directly or from a chromatographic column into a heated pneumatic nebulizer capable of accepting a flow rate ranging between 0.2 and 2 mL/min. The nebulizer converts the liquid into an aerosol with the assistance of a nitrogen high flow. Aerosol droplets are desolvated by passing through a heated quartz or a ceramic tube that acts as the desolvation/vaporization chamber. Solvent evaporation is induced by heating the nebulization zone. Different temperatures can be applied in different parts of the vaporization zone for a complete vaporization of the solvent and the analyte. A corona discharge needle for ions generation is placed in front of the nebulizer and kept at a voltage of 5–6 kV that generates a current of about 2–3 μA. The heated nebulizer is commonly positioned orthogonal to the ion-sampling orifice to avoid the contamination of the orifice itself by nonvolatiles that can be present in the liquid phase. This positioning is common to many ESI-MS systems as well. The heated nebulizer can generally accept LC flow rates as high as 1–2 mL/min. Many manufacturers produce combined ESI/APCI or APCI/APPI sources that can be operated independently or in combination, switching modes between scans. Ions are first lead to a pre-vacuum zone, then to the high-vacuum mass analyzer. The principles and apparatus that operate in an APCI interface are shown in Figure 8.5 [47,48].

The ionization in APCI is based on ion-molecule reactions activated by the electrons generated in the corona discharge and mediated by the solvent.

In the positive ion mode, the electrons from the corona discharge produce primary ions (mainly $N_2^{+\bullet}$, $O_2^{+\bullet}$, $H_2O^{+\bullet}$, and $NO^{+\bullet}$) through the EI of nitrogen and oxygen, as described below:

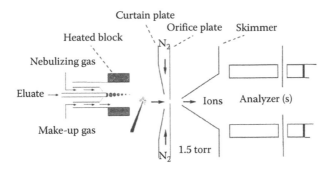

FIGURE 8.5 Schematic representation of an API source with a heated nebulizer interface for APCI. (Reproduced from Raffaelli, A., Atmospheric pressure chemical ionization (APCI), in Cappiello, A. (ed), *Advances in LC-MS Instrumentation*, vol. 72 (*Journal of Chromatography Library*), Elsevier, Amsterdam, the Netherlands, 2007, 11–25. Copyright 2007. With permission from Elsevier.)

$$N_2 + e^- \rightarrow N_2^{+\bullet} + 2e^- \tag{8.2}$$

or

$$O_2 + e^- \rightarrow O_2^{+\bullet} + 2e^- \tag{8.3}$$

In the presence of traces of water, a series of ion-molecule reactions lead to the formation of protonated water clusters $H_3O^+(H_2O)_n$. At atmospheric pressure, there is a significant interaction between reagent ions and analyte ions produced by the heated nebulizer and which will generate protonated analyte molecules $[M + H]^+$.

When APCI in used in combination with the normal phase LC, the nitrogen molecular ion will enter into a charge-transfer reaction with the organic solvent. Ion-molecule reactions lead to protonated solvent clusters that will react by proton transfer with the analyte molecules, forming $[M + H]^+$ ions.

In the negative ion mode, primary ions such as $O_2^{-\bullet}$, $O^{-\bullet}$, $NO_2^{-\bullet}$, $NO_3^{-\bullet}$, $O_3^{-\bullet}$, and $CO_3^{-\bullet}$ are produced and enter into a series of ion-molecule reactions with the solvent molecules that eventually lead to a reaction with the analyte molecules. When gas-phase acidity of the reagent gas is higher than the one of the analyte, a deprotonated $[M - H]^-$ specie is generated. Adduct ions such as $[M - HCOO]^-$ or $[M - CH_3COO]^-$ may also be formed. Some negative secondary processes may occur during the atmospheric pressure ionization: air components could contribute to plasma formation as well as solvent impurities and additives causing alteration in the APCI pathway. Ion suppression can occur especially when strong bases such as triethylamine or triethanolamine are used as mobile phase modifiers. Strong acidic additives could cause the formation of interfering adduct ions as well. Figure 8.6 illustrates the ionization in the positive and negative ion modes.

APCI is widely used nowadays in different application fields for low molecular weight analytes. Many of them can either be analyzed with ESI or APCI, and the choice of the method should take into account several aspects, such as the physical–chemical properties of the molecule, the mobile phase composition and the required flow rate, and possible matrix effects. Typical APCI applications are in pharmaceutical, environmental, and food safety analysis.

In the last decade, LC–MS system manufacturers have commercialized hybrid instruments that can operate in the ESI or the APCI mode with just a few simple modifications. Such instrumentation, based on Bajic's first prototype [49], is able to protonate compounds in the ESI mode together with those yielding characteristic APCI signature. Castoro [50] and Fischer et al. [51] first provided the utility of the dual atmospheric pressure ionization source.

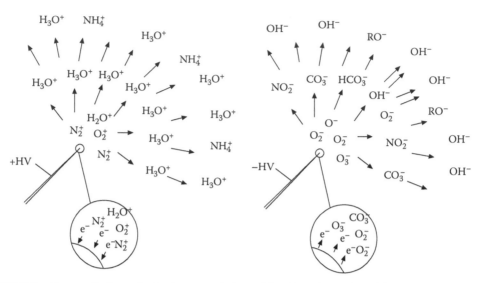

FIGURE 8.6 Corona discharge ionization cascade in positive (a) and negative (b) ion modes. (Reproduced from Raffaelli, A., Atmospheric pressure chemical ionization (APCI), in Cappiello, A. (ed), *Advances in LC-MS Instrumentation*, vol. 72 (*Journal of Chromatography Library*), Elsevier, Amsterdam, the Netherlands, 2007, 11–25. Copyright 2007. With permission from Elsevier.)

8.4 EMERGING TECHNIQUES

8.4.1 ATMOSPHERIC PRESSURE PHOTOIONIZATION

Atmospheric pressure photoionization (APPI) was recently introduced in the world of atmospheric pressure ionization techniques to analyze nonpolar molecules that are not efficiently ionized either by ESI or by APCI. Photoionization (PI) was already exploited some 30 years ago as a detection method for GC and LC, but only in recent times it has been used as an ionization method for mass spectrometry [52].

8.4.1.1 Principles of Operation

Two different geometries, in-line (see Ref. [39]) and orthogonal [53], that are based on the same operational principles, have been proposed.

The in-line source depicted in Figure 8.7 was designed by Bruins et al. to be mounted on a PE-Sciex triple quadrupole and it was derived from the standard heated nebulizer of their APCI source. The corona needle is replaced by a discharge lamp. Nitrogen is used as the nebulizing and the lamp gas, while air is used as the auxiliary gas. A dopant improves the efficiency of ionization and it is supplied through the auxiliary gas line and vaporized together with the solvent in the heated nebulizer.

The orthogonal geometry is based on the Agilent Technologies' APCI source, and it is illustrated in Figure 8.8: the heated nebulizer and the discharge lamp are respectively perpendicular and in-line with the MS inlet. This source design permits the achievement of significant ionization efficiency without the use of a dopant because of a greater lamp output, although overall sensitivity can benefit from the introduction of a dopant [54].

In both designs, the discharge lamp is filled with krypton and emits 10.6 eV photons, an energy lower than that of the major components of air and some of the commonly used solvents.

In direct APPI, the ionization process is activated by the photons emitted by the source and takes place if the ionization energies of the molecules are below 10 eV. This is the case of most analytes; however, the ionization energy of most HPLC solvents is higher. This implies that the sample must be vaporized prior to detection. The molecular ion is first generated by an impact with photons:

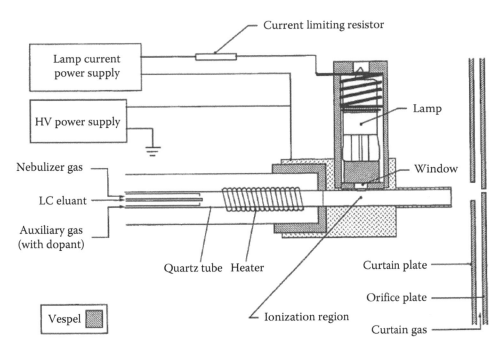

FIGURE 8.7 Schematic diagram of a complete APPI in-line ion source. (Reproduced from Robb, D.B. et al., *Anal. Chem.*, 72(15), 3653, 2000. With permission of American Chemical Society.)

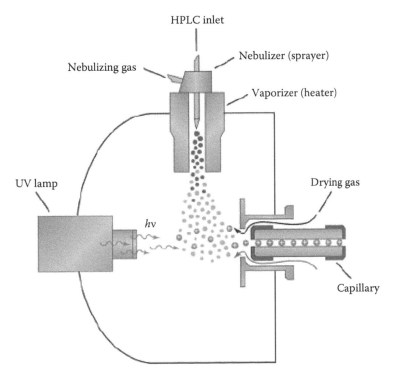

FIGURE 8.8 Schematic diagram of an orthogonal APPI source. (Courtesy of Agilent Technologies and Syagen Technology, Inc.)

$$M + h\nu \rightarrow M^{+\bullet} + e^{-} \quad \text{if IE(M)} < 10\,\text{eV} \tag{8.4}$$

Possible reactions between the molecular ion and other species in the gas phase include a charge exchange to a species with a lower ionization energy:

$$M^{+\bullet} + A \rightarrow M + A^{+\bullet} \quad \text{if IE(A)} < \text{IE(M)} \tag{8.5}$$

or hydrogen withdrawal from a protic solvent [55,56]:

$$M^{+\bullet} + S \rightarrow [M+H]^{+}[S-H]^{\bullet} \tag{8.6}$$

In case of a charge-exchange signal, lowering of the analyte is experienced. The formation of a protonated molecule does not affect the ionization efficiency and it is possible only if the analyte shows a certain proton affinity.

The low ionization efficiency can be boosted by adding a suitable substance, called dopant, at the same concentration level of the analyte [57] (see Ref. [49]). The dopant ionization energy must be lower than 10 eV, the energy of the photons, and is ionized producing a molecular ion:

$$D + h\nu \rightarrow D^{+\bullet} + e^{-} \tag{8.7}$$

The dopant molecular ion acts as an intermediate and reacts with the analyte through the charge exchange:

$$D^{+\bullet} + M \rightarrow D + M^{+\bullet} \quad \text{if IE(M)} < \text{IE(D)} \tag{8.8}$$

$$D^{+\bullet} + S \rightarrow [D-H]^{\bullet} + [S+H]^{+} \quad \text{if PA(S)} > \text{PA[D}-\text{H]}^{\bullet} \tag{8.9}$$

Mobile phase molecules can contribute to the formation of $[M+H]^{+}$ ions if the proton affinity of the analyte is higher to that of the solvent:

$$[S+H]^{+} + M \rightarrow S + [M+H]^{+} \quad \text{if PA(M)} > \text{PA(S)} \tag{8.10}$$

The mobile-phase composition can deeply influence sensitivity and its effects have been thoroughly investigated [58–60]. As a general rule, low water-content mobile phases produce an increment in sensitivity.

In the negative ion mode of the ionization process, an electron is released by the effect of the PI of the dopant:

$$D + h\nu \rightarrow D^{+\bullet} + e^{-} \quad \text{if IE(D) is} < 10\,\text{eV} \tag{8.11}$$

Electron affinitive species present in the atmospheric pressure source can capture such low energy electrons. In particular, oxygen can produce a superoxide ion:

$$M + e^{-} \rightarrow M^{-\bullet} \quad \text{if EA(m)} > 0\,\text{eV} \tag{8.12}$$

$$M + e^{-} \rightarrow [M-F]^{\bullet} + F^{-} \quad F = \text{fragment} \tag{8.13}$$

$$S - e^- \rightarrow S^{-\bullet} \tag{8.14}$$

$$O_2 - e^- \rightarrow O_2^{-\bullet} \tag{8.15}$$

O_2^- is a strong base; therefore, it is a proton acceptor from other species:

$$S + O_2^{-\bullet} \rightarrow [S-H]^- + HO_2^{\bullet} \quad \text{if } \Delta G_{acid}(S) < \Delta G_{acid}(HO_2^{\bullet}) \tag{8.16}$$

$$M + O_2^{-\bullet} \rightarrow [M-H]^- + HO_2^{\bullet} \quad \text{if } \Delta G_{acid}(M) < \Delta G_{acid}(HO_2^{\bullet}) \tag{8.17}$$

Proton transfer can also take place between the analyte and the solvent with a low gas-phase acidity:

$$M + [S-H]^- \rightarrow [M-H]^- + S \quad \text{if } \Delta G_{acid}(M) < \Delta G_{acid}(S) \tag{8.18}$$

Superoxide ion can also be responsible for the charge exchange with the solvent and the analytes:

$$M + O_2^{\bullet} \rightarrow M^{\bullet} + O_2 \quad \text{if } EA(M) > EA(O_2) = 0.451 \, eV \tag{8.19}$$

O_2 and O_2^- can react with some analytes through oxidation reactions:

$$M + O_2^{-\bullet} \rightarrow [M-X+O]^- + OX^{\bullet} \quad X = H, Cl, NO_2 \tag{8.20}$$

$$M^{-\bullet} + O_2 \rightarrow [M-X+O]^- + OX^{\bullet} \quad X = H, Cl, NO_2 \tag{8.21}$$

The mobile-phase composition must also be carefully evaluated in the negative ion mode because the response may vary significantly depending on the solvent's properties. APPI has been used in biological, environmental, and food analysis applications. Overcoming the limitations posed by APCI and ESI in the analysis of neutral compounds represents its point of strength and its potential. It can be combined with the other API sources to analyze compounds with very different properties, thus extending the range of applicabilities [61].

8.4.2 DIRECT-EI AND SUPERSONIC LC–MS

These are two approaches in the use of EI in coupling LC and MS.

Although there was an initial success of EI in coupling LC and mass spectrometry, it was rapidly outpaced by atmospheric-pressure ionization techniques (API), and the assumption that EI is only suited for GC-amenable compounds drove all of the research efforts into the direction of developing soft ionization-based interfaces. Nowadays, the situation is changing and interest is arising among researchers for the unique advantages of EI.

8.4.2.1 Principles of Operation

Electron Ionization, formerly called electron impact ionization, was developed by Dempster at the beginning of the last century and improved by Bleakney and Nier many years later [62,63].

Continuous, although less significant, improvements over the years have made it standard equipment for GC/MS. Analytes are ionized exclusively in the gas phase through interaction with high-energy electrons. The energy surplus, not directly involved in the ionization process, is responsible for the extensive fragmentation of the molecule, so the molecular ion is not always detected. However, because the fragmentation pattern is highly reproducible and characteristic of each molecular species, a direct comparison can be made with thousands of reference spectra for easier compound identification.

Electron energy is the key for achieving ionization and modulating fragmentation. Electrons are usually emitted by an electrically heated filament under high-vacuum conditions and accelerated by an electric field. Before being captured by the anode, electrons may intercept gas-phase molecules during their short journey in the ion source, producing an excited, positive radical ion, $M^{\bullet}+^{*}$. The ionization process is described by quantum mechanics because each moving electron is associated with an electromagnetic wave whose wavelength is given by h/mv, in which m is the mass of the electron, v is its velocity, and h is Planck's constant. The wavelengths are 2.7 Å for a kinetic energy of 20 eV and 1.4 Å for 70 eV. When the electron travels close to a gas-phase molecule, the electron's electromagnetic wave is distorted and becomes complex. The new wave can be described by the interference of a multiwave pattern of various frequencies. If one of the frequencies has an energy equal to a specific electronic transition in the molecule, the corresponding energy is absorbed. The transition usually exceeds the typical-ionization potential so that a thermal electron is expelled, thus leading to molecular ionization. Electron energies of 10–70 eV are sufficient to induce ionization in all organic species. Ionization does not occur at <≈10 eV, because this limit is the least amount of energy required for primary ionization and varies depending on the molecular structure. A compound with all bonds in its structure requires a higher energy for ionization because bonds are the strongest; π bonds and nonbonding electrons have a lower energy limit for primary ionization. The probability for a successful ionization event increases with the electron energy up to ~70 eV. Over this value, the impact section of the ionization process starts retreating, and electron–molecule interactions become weaker and unproductive. The secret of EI for generating classic, nicely matching spectra is in the very rapid sequence of events leading to fragmentation. For a 100-u molecular ion accelerated by a 1000 V potential toward the analyzer, the average time spent in the ion source depends on its speed and ranges between 10^{-6} and 10^{-7} s. The interaction between an electron and a gas-phase molecule is a very high-speed process. The electron can cover a distance of 1.88 Å across a molecule in 10^{-16} s, during which the interaction can be established and ionization may occur. The ionization process takes place under high-vacuum conditions, so any molecular collision is very unlikely. Under these conditions, the energy cannot be released by vibrational relaxation through the collision with other molecules. UV or visible-range radiation typically occurs in 10^{-8} s. If energy is distributed among the different parts of the molecule before fluorescence occurs, fragmentation is observed. Recombination is impossible because of the lack of collisions, so these unimolecular reactions depend only on the structure and the energy content within the isolated molecule (Figure 8.9). After 10^{-8} s, the whole process is complete, but the ions are still within the ion source. This occurs well before the ion leaves the source and explains why EI is reproducible from one spectrometer to another [64] with different instruments and different times of flight through changes in the accelerating potential (a few volts for a quadrupole, kilovolts for a sector instrument)—the spectrum profile is not substantially modified. Once all the instrumental parameters are established (i.e., ion source temperature, electron energy, etc.), EI spectra are the result of intramolecular reactions only and are not influenced by other complex analytical parameters used in other forms of ionization. EI spectra furnish data that are the most easily compared.

Its simple principle induces the ionization of any molecule present in the gas phase, regardless of its chemical background. The mobile phase does not play any role in the ionization process, and no molecule-ion interactions are normally observed. The typical EI spectrum is highly informative, and its reproducibility allows its comparison with thousands of spectra from commercially available sources (such as NIST or Wiley), generated over many years from a variety of instruments. In

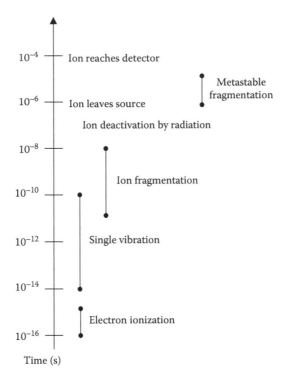

FIGURE 8.9 Time sequence during EI. (Reproduced from Cappiello, A. et al., *Anal. Chem.*, 75, 497A, 2003. With permission from American Chemical Society.)

addition, EI detection can benefit from a recent sophisticated algorithm developed by the National Institute of Standards and Technology (NIST) called AMDIS (Automated MS Deconvolution and Identification System), which extracts the analytes mass spectra in complex chromatographic mixtures with several overlapping peaks [65]. These aspects turn out to be essential in all cases where unknown peaks may be encountered or legal defensibility is a necessity; for example, in pharmaceutical development, drugs of abuse, or forensic applications.

Nowadays, there are two groups working constantly on the development of an efficient EI-based LC–MS interface: the group directed by Prof. Cappiello at the University of Urbino (Italy) who developed the Direct-EI interface, and the one directed by Prof. Amirav at Tel Aviv University (Israel) who developed the supersonic LC–MS. Each group is following a distinctive pathway to achieve this common goal.

8.4.2.2 Direct-EI Interface

The commercial availability of good, reliable nanoscale HPLC columns gave the impulse for the development of a very simple and straightforward LC-EI MS interface that was called "Direct-EI" [66–68]. Direct-EI is an almost no-interface device through which the HPLC effluent directly enters the EI source [69,70]. The Direct-EI project fully exploited this past experience and added fresh technology to avoid some of the weak points. The originality of this approach is to squeeze all the interfacing processes into the small volume of the EI ion source of the MS. A scheme of the interface is depicted in Figure 8.10.

The flow rate accepted by the interface ranges between 100 and 1000 nL/min for most applications. The interfacing mechanism is based on the formation of the aerosol in high-vacuum conditions, followed by a quick droplet desolvation and the final vaporization of the solute on a target surface prior to ionization. The process is fast and requires less than 8 mm of path length. At the core of the interface there are a nano-nebulizer and a treated surface. The nebulizer tip

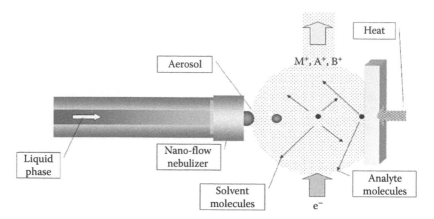

FIGURE 8.10 Schematic diagram of the Direct-EI interface. (From Cappiello, A. et al., *Anal. Chem.*, 79, 5364, 2007. With permission from American Chemical Society.)

protrudes a few millimeters into the ion source. Because of the reduced flow rate, droplets are quickly evaporated and converted into particles before reaching the surface. The high temperature of the ion source, 200°C–350°C, has a double function: it compensates for the latent heat of vaporization during the droplet desolvation and converts the solute into the gas phase upon contact with the hot target surface. The nebulizer is connected to the analytical nanocolumn by a convenient, small bore capillary tubing. The tubing is kept well insulated from the source heat to avoid premature mobile-phase vaporization and possible solute degradation or precipitation. Neither the electron path nor the electric fields are influenced by the interface intrusion into the ion volume, and high-quality mass spectra are thus produced. One obvious advantages is the extreme compactness and the simplicity without the need for additional complex devices between the liquid chromatograph and the MS. The lack of any particular transport mechanism involved in the interfacing process reduces sample losses, enhances sensitivity, and extends the range of possible applications. Direct-EI is even compatible with high concentrations of nonvolatile buffers in the mobile phase, opening the door to a wide range of challenging applications. The interface shows a superior performance in the analysis of small-medium molecular weight compounds especially when compared to its predecessors and a unique trait that excels particularly in the following aspects:

1. It delivers high quality, fully library-matchable mass spectra of most sub-1 kDa molecules amenable by HPLC.
2. It is a chemical ionization-free interface (unless operated intentionally) with accurate reproduction of the expected isotope ion abundances.
3. Response is never influenced by matrix components in the sample or in the mobile phase (nonvolatile salts are also well accepted).

Because its simplicity and the absence of signal suppression and other matrix effects, Direct-EI may offer a clear advantage over ESI in several small molecule applications. Many analytes (i.e., lindane) that give no signal with ESI, generate an intense and very informative spectrum (Figure 8.11).

When this spectrum was compared to a reference spectrum, a matching quality of 98% allows an appreciable identification in the real sample in multiresidue applications.

8.4.2.3 Supersonic LC–MS

Another promising approach to the use of EI coupled with LC is represented by the supersonic LC–MS. This interfacing mechanism was realized at Tel Aviv University by the group directed

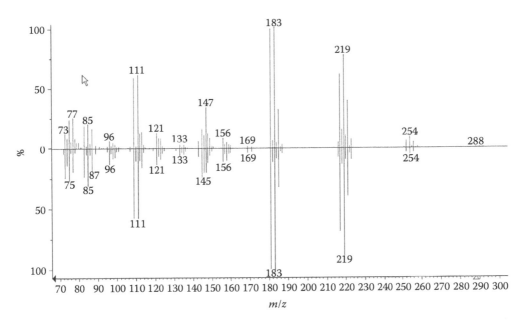

FIGURE 8.11 The Direct-EI mass spectrum of lindane is shown in the upper trace, and it is compared with the standard NIST EI library spectrum shown in the lower trace.

by Prof. Amirav [71,72]. Their studies started from the real consciousness that while standard 70 eV EI is a powerful ionization method for unknown sample identification, it is not ideal. About 30% of the NIST library compounds either have a weak (below 2% relative abundance) or no molecular ion. This problem is further amplified for LC related compounds that are larger and more thermally labile than standard GC-MS compounds. Furthermore, they are usually less volatile and hence they require higher EI ion source temperatures with the consequence of further intra-ion source degradation and weaker molecular ion production. Without a molecular ion, EI-based sample identification is not as trustworthy. Thus, the "ideal" ionization method should provide informative, library searchable EI fragments combined with enhanced molecular ions (relative to standard thermal EI) whose observation as the highest mass spectral peak should be trusted [73,74].

Their approach called LC-(Cold EI)-MS with a supersonic molecular beam (SMB) [75] is based on spray formation at high pressure, followed by full thermal sample particle vaporization prior to sample expansion as isolated molecules from a supersonic nozzle (Figure 8.12). This first step is similar to sample vaporization in APCI; however, instead of the next APCI step of corona discharge for inducing chemical ionization, the sample according to this method expands from the supersonic nozzle. Then, the supersonic-free jet is collimated and forms an SMB that contains vibrationally cold sample molecules. These molecules proceed axially along a fly through the Brink-type EI ion source for obtaining Cold EI mass spectra with an enhanced molecular ion. In the preliminary report, the problem of vaporization of intact thermally labile compounds was addressed through achieving fast sample vaporization followed by supersonic expansion cooling. The issue of the liquid solvent load on the vacuum pumps was addressed by differential pumping. Cluster formation was practically eliminated by using a relatively large diameter and a separately temperature-controlled nozzle. Vaporized solvent molecules served as the SMB carrier gas without adding another seeding gas.

The LC–MS with a supersonic molecular beam (LC-SMB-MS) apparatus, which is schematically shown in Figure 8.13, is based on a modified homemade GC-MS with an SMB system that was previously described [76]. The "heart and soul" of this system is the soft thermal vaporization nozzle (STVN) chamber. The STVN accepts the liquid flow from the LC or the flow injection liquid

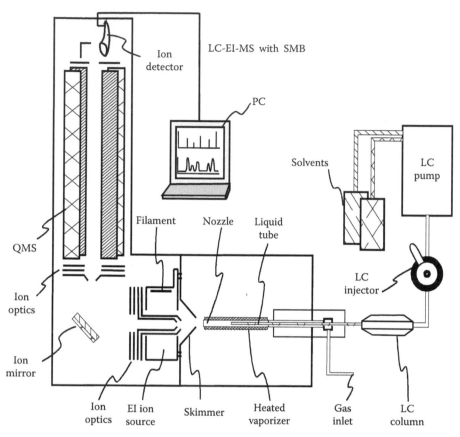

FIGURE 8.12 The supersonic LC-EI-MS apparatus. (Reproduced from Amirav, A. et al., *Rapid Commun. Mass Spectrom.*, 15, 811, 2001. Copyright 2001. With permission from Elsevier.)

transfer line tube and converts it into SMB of vibrationally cold undissociated sample compounds that are amenable for EI and mass analysis.

In Figure 8.13 [72], the STVN used in the third generation unit is schematically pictured. The liquid sample solution enters from the liquid transfer line tubing that is connected with a union to a heated liquid transfer line for obtaining a thermally assisted spray as in Thermospray [77,78] and a thermally assisted particle beam [79].

The supersonic LC–MS approach can be perceived as a synthesis of a few current and past approaches, but in fact, it is a combination of a few of these approaches plus its own unique ingredients. The initial step of spray formation relates to Thermospray or to the ThermaBeam Particle Beam system of Willoughby and coworkers, but unlike Thermospray this kind of spray is only thermally assisted, formed at non vacuum, at relatively high pressure and without the step of Thermospray ionization. Unlike the ThermaBeam and all other Particle Beam systems, this approach is characterized by complete sample particle vaporization prior to its supersonic expansion. It seems more akin to APCI in that it involves full sample vaporization at relatively high pressures, but unlike typical APCI, the spray is fully thermally assisted without using any gas for pneumatic nebulization. Furthermore, no high pressure CI is involved as Cold EI is a full in-vacuum ionization method. On the other hand, to a first approximation, the ability to handle thermally labile compounds is similar to that of APCI since both methods share similar high-pressure thermal sample vaporization. However, in contrast to APCI, the Cold EI can ionize all the sample compounds that were vaporized, regardless of their polarity, and no ion molecule reaction (collision-free molecular beam) can interfere in the obtained mass spectra.

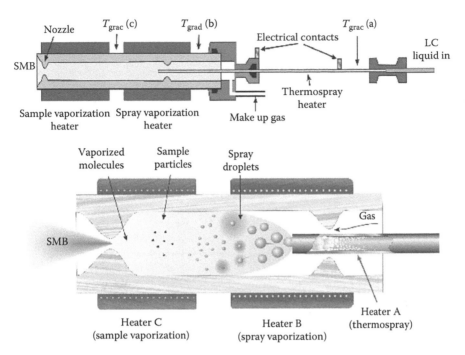

FIGURE 8.13 The STVN. (Reproduced from Amirav, A. et al., *Rapid Commun. Mass Spectrom.*, 15, 811, 2001. Copyright 2001. With permission from Elsevier.)

In Figure 8.14, the Cold EI mass spectrum of corticosterone in methanol solution is shown in the upper trace, and is compared with the standard NIST 98 EI library mass spectrum shown in the lower trace. Note the similarity of the library mass spectrum to that obtained with the SMB apparatus. All the major high mass ions of m/z 227, 251, 269, and 315 are with practically identical relative intensity and thus good library search results are enabled with the NIST library-matching factor of 829, and the reversed matching factor of 854% and 86.5% confidence level (probability) in corticosterone identification. In addition, the molecular ion at m/z 346 is now clearly observed while it is practically missing in the library (very small in the shown mass spectrum and absent in the other three replicate mass spectra).

8.4.3 Atmospheric Pressure Laser Ionization

Atmospheric pressure laser ionization (APLI) is a complementary technique to existing API methods. It is based on resonant or near-resonant two-photon ionization of aromatic ring systems. APLI utilizes resonantly enhanced multiphotonionization (REMPI) as the primary ion-production mechanism, nevertheless it occurs at atmospheric pressure [80].

8.4.3.1 Principles of Operation

The steps involved in the process can be illustrated in the following pathway [81–85]:

$$M + m\,h\nu \rightarrow M^* \quad m = 1, 2 \tag{8.22a}$$

$$M^* + n\,h\nu \rightarrow M^{+\bullet} + e^- \quad n = 1, 2 \tag{8.22b}$$

Reactions 8.22a and b are typical of REMPI ionization.

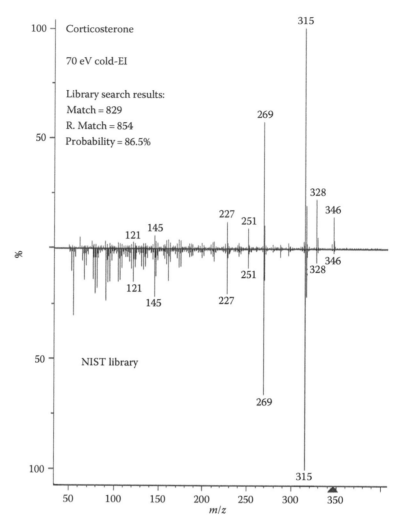

FIGURE 8.14 A comparison of the cold EI mass spectrum of corticosterone obtained with the supersonic LC-EI-MS system and its fitted NIST library mass spectrum, including the NIST library-matching factors and the probability of identification. Note the enhanced molecular ion ($m/z = 346$) exhibited. The NIST library-matching factors and the probability of identification are included. Corticosterone was flow injected using a 200 μL/min methanol solvent flow rate with a 1 ng/μL corticosterone sample concentration. (Reproduced from Amirav, A. et al., *Rapid Commun. Mass Spectrom.*, 15, 811, 2001. Copyright 2001. With permission from Elsevier.)

The precursor ion may fragment via ladder-switch mechanisms

$$M^{+\bullet} + x\,h\nu \rightarrow [F_1]^+ + F_2 \tag{8.23a}$$

$$[F_1]^+ + y\,h\nu \rightarrow [F_3]^+ + F_4 \tag{8.23b}$$

Rapid photodissociation at intermediate levels does not produce an adequate number of ions:

$$M^* \rightarrow D_1 + D_2 \tag{8.24}$$

While deactivation via intersystem, crossing appears to be less critical for higher molecular weight molecules:

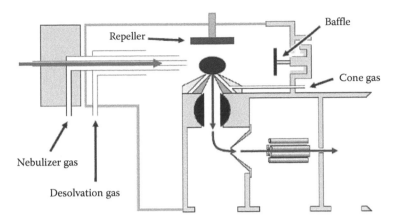

FIGURE 8.15 Schematic set-up of an APLI source based on the MicroMass Z-spray interface. The dark gray area indicates the laser beam position (from top). Light gray bold arrow: Liquid flow into the vaporization stage. Dark gray light arrows: Ion travel path. (Reproduced from Schmitz, O.J. and Benter, T., Atmospheric pressure laser ionization (APLI), in Cappiello, A. (ed), *Advances in LC-MS Instrumentation*, vol. 72 (*Journal of Chromatography Library*), Elsevier, Amsterdam, the Netherlands, 2007, 89–113. With permission.)

$$M^*(S_1) \rightarrow M^*(T_1) \tag{8.25}$$

The application of a second light source with higher energy photons can promote ionization. In APLI, as opposed to APPI, mobile-phase molecules of the HPLC stage or the solvent molecules when using direct syringe injection play a different role. Commonly used HPLC solvents, such as H_2O, CH_3CN, CH_3OH, are transparent to the UV wavelength range used, and therefore are not involved in the ionization processes.

Ionization takes place in a modified API source, as illustrated in Figure 8.15.

High-quality quartz windows are placed at the entrance and exit window of the source, where the laser beam position is simultaneously aligned.

If the MS operates with potentials close to ground at the sampling orifice, a repeller plate is required for optimum performance. A field gradient on the order of 50, ..., 200 V/cm, depending on the geometry, can be applied to the potential plate to improve the ion transmission. The laser-beam can be positioned in front of the nozzle; however, its positioning does not seem to be crucial. Pulsed fixed frequency lasers are satisfactory for the detection of most nonpolar aromatic compounds. They are very well ionized with APLI that to this extent can support other atmospheric pressure ionization techniques, as long as the analytes are soluble in a polar, aliphatic, or chlorinated organic solvent. If toluene is used as a solvent, it acts as a dopant in APPI, and can be deliberately added to the liquid flow for this purpose.

In summary, the distinctiveness of this soft ionization technique is that source optimization can be achieved through an independent adaptation of the ionization region, the electrical field gradient, and the gas flow. It is well suited for aromatic hydrocarbons that are selectively ionized, including nonpolar species, such as PAH, yielding to the lowest detection limits by orders of magnitude, as compared to all other API methods.

It is possible to successfully perform dopant-assisted DA-APLI upon addition of a dopant (e.g., toluene or anisole). DA-APLI currently shows the same performance as DA-APPI.

REFERENCES

1. J.B. Fenn, M. Mann, C.K. Meng, S.F. Wong, C.M. Whitehouse, *Science*, 246, 64, 1989.
2. N.B. Cech, C.G. Enke, *Mass Spectrom. Rev.*, 20, 362, 2001.

3. R.B. Cody, Electrospray ionization mass spectrometry: History, theory, and instrumentation. In: B.N. Pramanik, A.K. Ganguly, M.L. Gross (Eds.) *Applied Electrospray Mass Spectrometry*, Marcel Dekker, Inc., New York, 2002, pp. 1–104.

4. N.B. Cech, C.G. Enke, Electrospray ionization: How and when it works. In: W.M.A. Niessen, M.L. Gross, R.M. Caprioli (Eds.) *The Encyclopedia of Mass Spectrometry*, vol. 8, Elsevier, Amsterdam, the Netherlands, 2006, pp. 171–180.

5. P. Kebarle, L. Tang, *Anal. Chem.* 65, 972A, 1993.

6. P. Kebarle, Y. Ho, On the mechanism of electrospray mass spectrometry. In: R.B. Cole (Ed.) *Electrospray Ionization Mass Spectrometry*, Wiley, New York, 1997, p. 14.

7. G. Schmelzeisen-Redecker, L. Buttering, F.W. Röllgen, *Int. J. Mass Spectrom. Ion Proc.*, 90, 139, 1989.

8. J.V. Iribarne, B.A. Thomson, *Chem. Phys.*, 64, 2287, 1976.

9. B.A.Thomson, J.V. Iribarne, *J. Chem. Phys.*, 71, 4451, 1979.

10. R. Saf, C. Mirtl, K. Hummel, *Tetrahedron Lett.*, 35, 6653, 1994.

11. S.Z. Ackloo, R.W. Smith, J.K. Terlouw, B.E. McCarry, *Analyst*, 125, 591, 2000.

12. G. Wang, R.B. Cole, Solution, gas-phase, and instrumental parameter influences on charge-state distributions in electrospray ionization mass spectrometry. In: R.B. Cole (Ed.) *Electrospray Ionization Mass Spectrometry*, Wiley, New York, 1997, pp. 137–174.

13. J.A. Loo, C.D. Edmonds, H.R. Udseth, R.D. Smith, *Anal. Chem.*, 62, 693, 1990.

14. I.A. Kaltashov, C.C. Fenselau, *J. Am. Chem. Soc.*, 117, 9906, 1995.

15. I.A. Kaltashov, C.C. Fenselau, *Rapid Commun. Mass Spectrom.*, 10, 857, 1996.

16. E.R. Williams, *J. Mass Spectrom.*, 31, 831, 1996.

17. K. Vekey, *Adv. Mass Spectrom.*, 13, 537, 1995.

18. K. Vekey, *Mass. Spectrom. Rev.*, 14, 225, 1995.

19. S. Gronert, *J. Mass Spectrom.*, 34, 787, 1999.

20. G. Wang, R.B. Cole, *Anal. Chem.*, 66, 3702, 1994.

21. U.A. MIrza, B.T. Chait, *Anal. Chem.*, 66, 2898, 1994.

22. G. Wang, R.B. Cole, *Anal. Chem.*, 67, 2892, 1995.

23. G. Wang, R.B. Cole, *Anal. Chem.*, 70, 873, 1998.

24. R.H. Griffey, H. Sasmor, M.J. Greig, *J. Am. Soc. Mass Spectrom.*, 8, 155, 1997.

25. P.D. Schnier, D.S. Gross, E.R. Williams, *J. Am. Soc. Mass Spectrom.*, 6, 1086, 1995.

26. A.K. Ganguly, B.N. Pramanik, G. Chen, A. Tsarbopoulos, Detection of noncovalent complexes by electrospray ionization mass spectrometry. In: B.N. Pramanik, A.K. Ganguly, M.L. Gross (Eds.) *Applied Electrospray Mass Spectrometry*, Marcel Dekker, Inc., New York, 2002, pp. 361–430.

27. J.G. Van Berkel, S.A. Mc Luckey, G. Glish, *Anal. Chem.*, 64, 1586, 1992.

28. J.G. Van Berkel, J.M. Quirke, R.A. Tigani, A.S. Dilley, *Anal. Chem.*, 70, 1544, 1998.

29. Z.L. Cheng, K.W. Siu, R. Guevremont, S.S. Bergman, *J. Am. Soc. Mass Spectrom.*, 3, 281, 1992.

30. L. Tang, P. Kebarle, *Anal. Chem.*, 65, 972A, 1993.

31. N.B. Cech, C.G. Enke, *Anal. Chem.*, 72, 2717, 2000.

32. S. Zhou, K.D. Cook, *J. Am. Soc. Mass Spectrom.*, 12, 206, 2001.

33. N.B. Cech, J.R. Krone, C.G. Enke, *Anal. Chem.*, 73, 208, 2001.

34. L. Tang, P. Kebarle, *Anal. Chem.*, 65, 3654, 1993.

35. R.C. King, R. Bonfiglio, C. Fernandz-Metzler, C. Miller-Stein, T.V. Olah, *J. Am. Soc. Mass Spectrom.*, 11, 942–950, 2000.

36. D.L. Buhrman, P.I. Price, P.J. Rudewicz, *J. Am. Soc. Mass Spectrom.*, 7, 1099, 1996.

37. B.K. Matuszewski, M.L. Constanzer, C.M. Chavez-Eng, *Anal. Chem.*, 70, 882, 1998.

38. J.-P. Antignac, K. Wasch, F. Monteau, H. De Brabander, F. Andre, B. Le Bizec, *Anal. Chim. Acta*, 529, 129, 2005.

39. A.P. Bruins, T.R. Covey, J.D. Henion, *Anal. Chem.*, 59, 2642, 1987.

40. R.B. Cole, A.K. Harrata, *Rapid Commun. Mass Spectrom.*, 20, 287, 2002.

41. R.D. Voyksner, Combining liquid chromatography with electrospray mass spectrometry. In: R.B. Cole (Ed.) *Electrospray Ionization Mass Spectrometry*, Wiley, New York, 1997, pp. 323–341.

42. Perkin Elmer Sciex, *The API Book*, Perkin Elmer Sciex Instruments, Toronto, Canada, 1992 and references therein.

43. M.S.B. Munson, F.H. Field, *J. Am. Chem. Soc.*, 88, 2621, 1966.

44. E.C. Horning, D.I. Carroll, I. Dzidic, K.D. Haegele, M.G. Horning, R.N. Stillwell, *J. Chromatogr.*, 99, 13, 1994.

45. D.I. Carroll, I. Dzidic, E.C. Horning, R.N. Stillwell, *Appl. Spectrosc. Rev.*, 17, 337, 1981.

46. T.R. Covey, E.D. Lee, J.D. Henion, *Anal. Chem.*, 58, 2453, 1996.

47. A. Raffaelli, A. Saba, *Mass Spectrom. Rev.*, 22, 318, 2003.
48. A. Raffaelli, Atmospheric pressure chemical ionization (APCI). In: A. Cappiello (Ed.) *Advances in LC-MS Instrumentation*, vol. 72 (*Journal of Chromatograpy Library*), Elsevier, Amsterdam, the Netherlands, 2007, pp. 11–25.
49. S. Bajic, Electrospray and atmospheric pressure chemical ionization mass spectrometer and ion source, Patent 5756994, May 26, 1998.
50. J.A. Castoro, Investigation of a simultaneous ESI/APCI for LC/MS analysis, *Proceedings of the 50th ASMS Conference on Mass Spectrometry and Allied Topics*, Orlando, FL, June 2002.
51. S. Fischer, D.L. Gourley, R.L. Bertsch, Multimode ionization source, Patent 6646257, November 11, 2003.
52. D.B. Robb, T.R. Covey, A.P. Bruins, *Anal. Chem.*, 72, 3653, 2000.
53. J.A. Syage, K.A. Hanold, M.D. Evans, Y. Liu, Patent WO0197252, 2001.
54. K.A. Hanold, S.M. Fisher, P.H. Cormia, C. Miller, *Anal. Chem.*, 76, 2842, 2004.
55. J.A. Syage, M.D. Evans, K.A. Hanold, *Am. Lab.*, 32, 24–29, 2000.
56. J.A. Syage, *J. Am. Soc. Mass Spectrom.*, 15, 1521, 2004.
57. T.J. Kauppila, R. Kostiainen, A.P. Bruins, *Rapid Commun. Mass Spectrom.*, 18, 808, 2004.
58. J.P. Rauha, H. Vuorela, R. Kostiainen, *J. Mass Spectrom.*, 36, 1269, 2001.
59. T.J. Kauppila, T. Kuuranne, E.C. Meurer, M.N. Eberlin, T. Kotiaho, R. Kostiainen, *Anal. Chem.*, 74, 5470, 2002.
60. C. Yang, J. Henion, *J. Chromatogr. A*, 970, 155, 2002.
61. J.A. Syage, K.A. Hanold, T.C. Lynn, J.A. Horner, R.A. Thakur, *J. Chromatogr. A*, 1050, 137, 2004.
62. W.A. Bleakney, *Phys. Rev.*, 34, 157, 1929.
63. E.B. Winn, A.O. Nier, *Rev. Sci. Instrum.*, 20, 773, 1949.
64. E. De Hoffman, V. Stroobant, *Mass Spectrometry: Principles and Applications*, 2nd edn., Wiley, New York, 2001.
65. P. Ausloos, C.L. Clifton, S.G. Lias, A.I. Mikaya, S.E. Stein, D.V. Tchekhovskoi, O.D. Sparkman, V. Zaikin, D. Zhu, *J. Am. Soc. Mass Spectrom.*, 10, 287, 1999.
66. A. Cappiello, G. Famiglini, F. Mangani, P. Palma, *Mass Spectrom. Rev.*, 20, 88, 2001.
67. A. Cappiello, G. Famiglini, F. Mangani, P. Palma, *J. Am. Soc. Mass Spectrom.*, 13, 265, 2001.
68. A. Cappiello, G. Famiglini, P. Palma, *Anal. Chem.*, 75, 497, 2003.
69. A. Cappiello, G. Famiglini, P. Palma, A. Siviero, *Mass Spectrom. Rev.*, 24, 978, 2005.
70. A. Cappiello, P. Palma, Electron ionization in LC-MS: A technical overview of the direct-EI interface. In: A. Cappiello (Ed.) *Advances in LC-MS Instrumentation*, vol. 72 (*Journal of Chromatography Library*), Elsevier, Amsterdam, the Netherlands, 2007, pp. 27–44.
71. A. Amirav, A.B. Fialkov, A. Gordin, *Rev. Sci. Instrum.*, 73, 2872, 2002.
72. O. Granot, A. Amirav, *Int. J. Mass Spectrom.*, 244, 15, 2005.
73. M. Kochman, A. Gordin, P. Goldshlag, S.J. Lehotay, A. Amirav, *J. Chromatogr. A*, 974, 185, 2002.
74. A.B. Fialkov, A. Amirav, *Rapid Commun. Mass Spectrom.*, 17, 1326, 2003.
75. A. Amirav, Israel Patent Application Number 127217, November 1998; USA, Japan, and European Patent Applications, November 1999.
76. A. Amirav, A. Gordin, N. Tzanani, *Rapid Commun. Mass Spectrom.*, 15, 811, 2001.
77. C.R. Blakley, M.L. Vestal, *Anal. Chem.*, 55, 750, 1983.
78. W.M.A. Niessen, J. Van der Greef, *Liquid Chromatography-Mass Spectrometry Principles and Applications*, Marcel Dekker, Inc., New York, 1992, p. 157.
79. G.G. Jones, R.E. Pauls, R.C. Willoughby, *Anal. Chem.*, 63, 460, 1991.
80. O.J. Schmitz, T. Benter, Atmospheric pressure laser ionization (APLI). In: A. Cappiello (Ed.) *Advances in LC-MS Instrumentation*, vol. 72 (*Journal of Chromatography Library*), Elsevier, Amsterdam, the Netherlands, 2007, pp. 89–113.
81. M.B. Robin, *Appl. Opt.*, 19, 3941, 1980.
82. L. Goodman, J. Philis, Multiphoton absorption spectroscopy. In: D.L. Andrews (Ed.) *Applied Laser Spectroscopy: Techniques, Instrumentation, and Applications*, VCH Publishers, Inc., New York, 1992.
83. M.N.R Ashfold, J.D. Howe, *Ann. Rev. Phys. Chem.*, 45, 57–82, 1994.
84. J. Pfab, Laser-induced fluorescence and ionization spectroscopy of gas phase species. In: R.J.H. Clark, R.E. Hester (Eds.) *Spectroscopy in Environmental Science*, Wiley, New York, 1995.
85. U. Boesl, Multiphoton excitation in mass spectrometry. In: C. McNeil (Ed.) *Encyclopedia of Spectroscopy and Spectrometry*, Academic Press, New York, 1999.

9 Control and Effects of Temperature in Analytical HPLC

David E. Henderson

CONTENTS

9.1 INTRODUCTION

The effects of temperature as a variable in high-performance liquid chromatography (HPLC) are pervasive, influencing every aspect of the experiment. Temperature changes solvent viscosity and solute diffusion rates, which has led to a resurgence in interest in elevated temperature separations for a faster analysis. The increased backpressure required for sub-2 μm columns has been a major factor in this trend. Very large increases in speed, by an order of magnitude or more, have been made possible by the combination of ultrahigh-pressure liquid chromatography (UHPLC), narrow bore columns, and moderately elevated temperatures.

Changing the column temperature can produce a variety of additional effects. Temperature changes the balance between enthalpy and entropy effects on retention mechanisms. Changing the temperature changes the equilibrium constants of both solvent and solutes, and it changes the dielectric constant and thus the polarity of solvents. For simple systems, where all analytes are similar, the effects of temperature change are relatively predictable. For ionic, acidic, or basic compounds, the changes in chromatographic behavior can be unexpected and must usually be determined empirically. Temperature can also lead to changes in the conformation and degree of solvation of both solute and stationary phases. These can have profound and unpredictable effects.

Temperature was possibly the last variable in the liquid chromatographic separation to be fully exploited. The history of temperature effects is reviewed in detail by Dolan [1]. Researchers are learning to exploit temperature to their advantage and develop a theoretical understanding of its effects [2]. Zhu, Snyder, Dolan, and their coworkers [3–7] have produced a large body of work describing the effects of temperature and have included temperature as a variable in the DryLab® HPLC optimization software [8]. The mathematical basis for this is described in four articles, which provide the most comprehensive overview of temperature and gradient effects [3,9–11]. Their work covers a wide range of solute types and columns.

This chapter will describe both the theoretical understanding of temperature effects and the practical considerations needed in the laboratory. It will also provide a range of examples of how temperature has been applied to improve separation with emphasis on applications since the previous edition of the *Handbook of HPLC* [12]. It is not the intent to provide a complete review of all applications.

9.2 ELEVATED TEMPERATURE HPLC FOR HIGH-SPEED SEPARATION— EFFECTS ON VISCOSITY AND COLUMN EFFICIENCY

Much of the recent impetus for temperature control has focused on exploiting the effects of elevated temperature on viscosity and diffusion coefficients [2]. These lead to faster separations and also allow smaller particle diameters to be employed with conventional HPLC hardware. As the viscosity of solvents decreases, the column pressure drops. This can be exploited by using faster flow rates and smaller particle diameters. All of this leads to faster separations. In one experiment in this laboratory, a separation which required 8 min at room temperature was reduced to 2 min at 50°C without changing the column. Speed enhancements of as much as 50–100-fold have been reported [13] as shown in Figure 9.1.

A less obvious effect of elevated temperature is on plate height. The equation describing plate height (*H*) can be written as follows [14]:

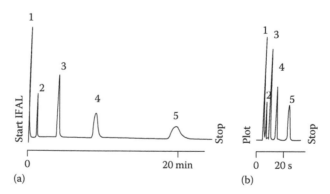

FIGURE 9.1 Chromatograms showing the effect of temperature on the separation of alkylphenones. Experimental conditions: mobile phase A, 30% ACN (v/v) and flow rate is 4 mL/min at 25°C; mobile phase B, 25% ACN (v/v) and flow rate is 15 mL/min at 150°C. Peaks: 1, acetophenone; 2, octanophenone; 3, decanophenone; 4, dodecanophenone; 5, tetradecanophenone. Column 50×4.6 mm packed with 2.5 μm polystyrene coated zirconia. (Reprinted from Yan, B. et al., *Anal. Chem.*, 72, 1253, 2000. Copyright 2000, American Chemical Society. With permission.)

$$H = \left[\frac{c_1 + c_2 f(T)}{u}\right] + \left[\frac{c_3 u}{f(T)}\right] + \left\{c_4 \left[\frac{u}{f(T)}\right]^{\frac{2}{3}}\right\} + \left[\frac{U}{g(T)}\right]$$ (9.1)

where
  $f(T)$ is the temperature-dependent diffusion rate of the solute
  $g(T)$ is the temperature-depended rate of desorption of the solute
  u is the linear velocity of the mobile phase
  the coefficients, c_i, are positive constants which depend on particle size
  $U = uk'/(1+k')^2$ where k' is the capacity factor of the solute on the column

The interplay of these factors is complicated, but the second, third, and fourth terms in the equation all lead to a reduced plate height as temperature increases. The first term, which increases plate height as temperature increases, is due to faster longitudinal diffusion in the column. The fact that this term decreases as linear velocity increases means that this term becomes minimal at high flow rates. Thus for most applications, a higher temperature produces a higher overall efficiency. Numerous studies have demonstrated that the minimum value of H shifts to higher linear velocities in the van Deemter plot [2]. The impact of the second through fourth terms on the plot is a reduction of the slope on the rising section of the van Deemter plot at higher flow rates, as shown in Figure 9.2.

This produces a very favorable situation in which the reduced viscosity makes higher flow rates practical and the increased rate constants $f(T)$ and $g(T)$ reduce the loss of efficiency for operating above the optimum flow rate. The impact of higher temperatures on gradient elution has also been reported to be consistent with these observations [15]. The one exception to this general rule may be the separation of macromolecules as reported by Antia and Horváth [14].

The other significant effect of elevated temperature is in the capacity factor, k'. In most cases, k' decreases as temperature increases, but there are enough exceptions to this general rule that a more detailed discussion is needed. This will be addressed in Section 9.3.

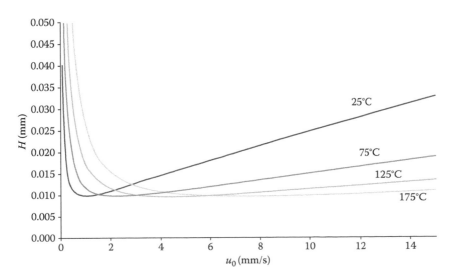

FIGURE 9.2 Theoretical plate-height curves for 5 µm particles, illustrating the effect of temperature on plate height and linear velocity. (Reprinted from Lestremau, F. et al., *J. Chromatogr.*, 1138, 120, 2007. Copyright 2007. With permission from Elsevier.)

9.3 COLUMN TEMPERATURE AND SOLUTE RETENTION

The retention of a solute in HPLC as a function of a temperature can, in theory, be described by the van't Hoff equation:

$$\ln k' = -\frac{\Delta H^\circ}{R}\frac{1}{T} + \frac{\Delta S^\circ}{R} + \ln \varphi \qquad (9.2)$$

where
  ΔH° and ΔS° are the enthalpy and the entropy of the interaction between the solute and the stationary phase
  φ is the phase ratio (the volume of the stationary phase divided by the void volume of the column)

Plots of $\ln k'$ vs. $1/T$ are expected to produce linear relationships. This assumes that the thermodynamics of the retention process are independent of temperature and that neither the solute nor the stationary phase changes with temperature. For a large proportion of systems and moderate temperature ranges, these assumptions are met. But there are many factors that can complicate this and lead to non-linear van't Hoff plots. Guillarme et al. [16] studied van't Hoff plots for various mobile phases over a very wide temperature range, from 30°C to 200°C. They found that methanol-water systems were more linear than acetonitrile-water systems and that the effects were not due to pressure. They also developed a rule of thumb that a 30°C–50°C increase in column temperature had the same effect on retention as a 10% increase in the organic component of the mobile phase.

As a general rule, one of the first experiments that should be done when attempting to use temperature as a variable is to generate van't Hoff plots for all of the solutes under study. It is advisable to obtain data for at least four temperatures over the desired range to assess linearity. An example of possible van't Hoff plots is shown in Figure 9.3.

The majority of solutes will show decreased values of $\ln k'$ as the temperature is increased. The slopes of the van't Hoff plots are often similar for compounds of the same functional group as shown by the three solid lines. The dashed lines show the irregular results often seen for solutes of different compound classes which can vary widely and have either a positive or a negative slope. All of these systems can be easily modeled with optimization software and require only two temperature points to define the system [4].

Problems arise for solutes that exhibit behavior shown in the dotted line. For these systems, simple models will not work. Reasons for non-linear van't Hoff plots will be discussed in Sections 9.3.2 and 9.3.3.

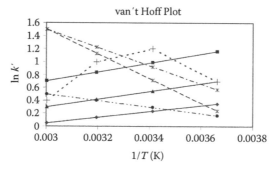

FIGURE 9.3 Example of a van't Hoff Plot showing regular (solid lines) and irregular (dashed lines) behavior.

9.3.1 SIMPLE SOLUTE BEHAVIOR

Non-ionic organic molecules with rigid conformations will almost always give linear van't Hoff plots. The retention times will normally decrease as the temperature is increased [11]. This provides a third advantage for operating at an elevated temperature. Not only does one achieve higher efficiency and faster flow rates, but the actual retention of the solutes is reduced. High-speed separations developed using this model are becoming increasingly popular, with analysis times for mixtures of 6–10 components at an elevated temperature falling below 60 s as shown in [17] (see Figure 9.4).

The application of high temperatures to increase the speed of HPLC separation extends to ion chromatography and to inorganic analysis. Le et al. [18,19] reported a 50% reduction in analysis time when a number of selenium and arsenic species including inorganic forms, organometallics, and compounds with amino acids and sugars were analyzed at 70°C.

In spite of the obvious advantages of elevated temperature, there are examples of cases where better separation is achieved at a reduced temperature, even for simple solutes. Craft et al. [20] recently demonstrated an improved separation of β and γ tocopherol at –20°C in THF/acetonitrile when compared to the ambient temperature separation of the compounds in Acetonitrile water. Bohm [21] reported the temperature dependence of the separation of a mixture of five xanthophylls and six carotenes on a C-30 column. The optimum temperature in this case was 23°C with a co-elution of some peaks at temperatures below 20°C and others above 35°C. In a study using a 300 Å pore C-18 column, Bohm [22] reported dramatic changes in the elution order over the temperature range –7°C to 35°C. On this column, the optimal separation was achieved at low temperatures

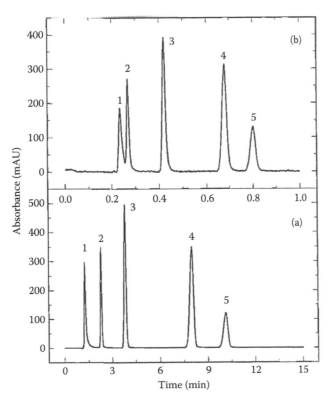

FIGURE 9.4 Chromatograms of a reversed-phase test mixture. Plot A is the chromatogram at 30°C and 1 mL/min, and plot B is the chromatogram at 100°C and 5 mL/min. Solutes: 1, uracil; 2, *p*-nitroaniline; 3, methyl benzoate; 4, phenetole; 5, toluene. 2.1% (w/w) poly-butadiene coated zirconia column, mobile phase 20% ACN, flow rate was 1.0 mL/min at 30°C and 5 mL/min at 100°C. detection 254 nm. (Reprinted from Li, J. et al., *Anal. Chem.*, 69, 3884, 1997. Copyright 1997, American Chemical Society. With permission.)

from −2°C to 9°C. Low-temperature operation has also been shown to be advantageous for the separation of lipids on silver-ion normal phase columns [23,24]. Unusual effects including decreasing retention at lower temperatures and non-linear van't Hoff plots were observed at temperatures from −40°C to 40°C.

9.3.2 Acids and Bases—Temperature Effects on p*K*

Acidic and basic compounds often show more complex behavior with non-linear van't Hoff plots and with an increased retention at high temperatures in some cases. This is due primarily to the impact of temperature on the various equilibrium constants at play in the solutions [25]. All equilibrium constants are temperature dependent. When the solute has multiple equilibrium forms, the retention depends on the fraction of the solute in each form, with the neutral form being more highly retained on the reverse phase HPLC column. The pK_w of water is also temperature sensitive, with the pH of a neutral solution shifting to a lower pH as the temperature increases.

If the HPLC mobile phase is operated close to the pK_a of any solute or if an acidic or basic buffer is used in the mobile phase, the effects of temperature on retention can be dramatic and unpredicted. This can often be exploited to achieve dramatic changes in the separation factor for specific solutes. Likewise, the most predictable behavior with temperature occurs when one operates with mobile phase pH values far from the p*K*'s of the analytes [10]. Retention of bases sometimes increase as temperature is increased, presumable due to a shift from the protonated to the unprotonated form as the temperature increases. As noted by Tran et al. [26], temperature had the greatest effect on the separation of acidic compounds in low-pH mobile phases and on basic compounds in high-pH mobile phases. McCalley [27] noted anomalous changes in retention for bases due to variations in their p*K*'s with temperature and also noted that lower flow rates were needed for optimal efficiency.

The effect of temperature on the acid base chemistry of the stationary phase can also play a role in separation. Free silanol groups on the stationary phase may exhibit changes in acid base chemistry with temperature [28]. Also, reverse phase columns with amine, amide, or acidic functional groups will be affected by the interaction of the temperature, the ionization state of the stationary phase, the mobile phase acidity, and the ionization state of the solute. Most non-linear van't Hoff plots can be rationalized in these terms, but it is difficult to predict *a priori* what the effects will be on a given system. Thus, it is important to characterize the system under study if a simple change in temperature produces unexpected effects.

9.3.3 Conformation and Solvation Changes with Temperature

Additional causes of non-ideal behavior are changes in the conformation of either the solute or the stationary phase. There have been several studies of the conformation and solvation of C-18 stationary phases as a function of temperature. ^{129}Xe NMR at various temperatures has allowed the exact nature of the solvation of C-18 reversed phase columns to be explored [29]. For low-polarity mobile phases, a stationary phase/solvent phase mixed phase was observed. The polarity and temperature dependence of this phase behavior can lead to non-linear van't Hoff plots. If the stationary phase undergoes a phase change, this will produce a discontinuity in the van't Hoff plot. A similar effect will be observed if the solute changes conformation or solvation. In these cases, the van't Hoff plot will often show two linear regions with different slopes. Macromolecules are the most likely to show this type of behavior due to the complexity of their available conformations. Chen et al. [30] found temperature effects on retention of D- and L-peptides. Helical peptides had a greater increase in retention than random peptides as the temperature increased with some reversals of the elution order as a function of temperature over the range 10°C–80°C. Pursch et al. [31] used the solid-state NMR of a C-30 stationary phase to understand the temperature-dependent behavior of retinol isomers. The NMR data showed a shift for the C-30 from a more rigid *trans* conformation to a more mobile *trans/gauche* conformation that began to occur around room temperature. At low temperatures, the most angled retinol had the longest retention but as the temperature increased, more linear isomers had the strongest interactions with the stationary phase. Separations of dansyl amino acids on a Human Serum Albumin (HAS) protein chiral column displayed an inverted U-shaped van't Hoff plot [32]. Differential Scanning Calorimetry studies allow this to be attributed to a phase transition in the HSA stationary phase between disordered and ordered states.

An extreme example of the effect of conformation is found in the analysis of taxans on C-8 and fluorinated reversed phase columns [33]. For these compounds, the van't Hoff plots on fluorinated stationary phases were generally linear while those on C-8 phases were extremely non-linear with several compounds showing increased retention at higher temperature. The surface excess isotherms for acetonitrile (ACN) showed that, as the ACN concentration increased, the water was expelled from the mobile phase and hydrogen bonded to the residual silanol groups. Because the retention of the compounds on the C-8 column required higher ACN concentrations, the effect was only noted on the C-8 column and not on the fluorinated phases. A related change in retention was observed in a separation of steroids using a β-cyclodextrin modified mobile phase [34]. The van't Hoff plots over the temperature range had a very pronounced peak with the most pronounced effect for these steroids with the greatest affinity for β-cyclodextrin.

Normal phase columns have also been shown to give non-linear van't Hoff plots in many cases. Bidlingmeyer and Henderson [35] found an improved separation of lipophilic amines at high temperature. They were unable to determine whether the lack of linearity was due to absorptive and electrostatic effects or changes in solvation of the silica as a function of temperature. They also noted a degradation of the silica support at elevated temperatures and found it necessary to use a pre-column to pre-saturate the mobile phase with silica.

9.3.4 Temperature-Responsive HPLC Stationary Phases

The impact of changing stationary-phase conformation has been exploited by several groups to create a new class of temperature-responsive stationary phases. These materials are made by bonding a functionalized poly-acrylamide polymer to aminopropyl silica or to polymer-based supports. The conformation of the poly-acrylamide polymer is a function of temperature. Various functionalities have been added to allow hydrophobic/hydrophilic interactions [36–40], ion exchange [41], and chiral separations [42]. The selectivity of the column can be programmed with temperature and pH rather than changing the solvent strength. Typical temperature ranges for these applications are from 5°C to 70°C. One advantage of these phases is that they can often be used with entirely aqueous mobile phases to minimize solvent effects on biological molecules. This interesting field was reviewed by Ayano and Kanazawa [43] and by Kanazawa and Matsushima [44]. Applications include biological molecules where organic solvents could lead to denaturation and environmentally sensitive applications where the elimination of organic waste is desirable, for example, the analysis of bisphenol-A reported by Yamamoto et al. [45]. They report that the aqueous mobile phase improved the background noise for UV and fluorescence detection. Applications of this technology have also included the separation of steroids [36,37] including commercial oral contraceptives [39], phenylthiohydantoin-amino acids [38,41], drugs [40], and polypeptides [37]. The polymer-based packings allow separate modifications of the internal and external surface, which can be useful for separating proteins in serum samples from drugs [40].

9.3.5 Chiral HPLC

There are numerous reports of the use of temperature to enhance chiral separations. In some cases, the optimal separation is achieved at elevated temperatures and in others at sub-ambient temperatures [46]. Tian et al. [47] noted that for the separation of chiral pesticides, most gave better separation factors at a low temperature. The exception was pyriproxyfen, which gave a larger separation factor at higher temperatures. Sun et al. [48] studied the chiral separation of clenbuterol and procatreol over the range −10°C to 30°C and found that a low temperature produced better separation. Peter et al. [49] found a better resolution of cyclic β-amino acids on a crown ether column at a low temperature. Schlauch and Frahm [50] found linear van't Hoff plots for cyclic α-amino acids on a Cu(II)-D-penicillamine chiral stationary phase. One of four enantiomers of 1-amino-2-cyclohexanecarboxylic acid (*cis*-1*R*,2*S*) exhibited a very different enthalpy term leading to several peak reversals over the 5°C–40°C range studied. The enthalpy of this enantiomer was also much more sensitive to acetonitrile concentration and pH than the other three.

There is no systematic way to predict the impact of temperature on chiral separations. It is advisable for anyone attempting to optimize a chiral separation to explore temperatures over the range available with their instrumentation to determine the best conditions.

9.3.6 ION EXCHANGE SEPARATIONS

Ion exchange HPLC provides yet another example of the types of behavior noted previously. Linear van't Hoff plots are observed for simple, mono-atomic ions in most cases, but complex equilibrium or multiple mechanisms of retention can lead to non-linear behavior. Hatsis and Lucy [51] studied the retention of common anions on Dionex anion exchange columns and found three different types of behavior. They observed significant changes in the temperature sensitivity of the retention with mobile phase type and concentration. Singly charged anions that are weakly retained showed a range of temperature effects, with some anions exhibiting increased retention at high temperatures and some the reverse. The slope of the van't Hoff plots of these ions sometimes changed from positive to negative with changes from the carbonate to the hydroxide mobile phase. Strongly retained singly charged anions such as iodide and perchlorate showed consistently decreasing retention as temperature increased. Multiply charged ions showed the opposite effect with a retention significantly increasing at an elevated temperature. Similarly, alkali and alkaline earth cations showed generally linear van't Hoff plots on silica-based cation-exchange columns as long as the cation exchanger did not have the capability to form chelates with the cation [52]. Separation of oligosaccharides was found to improve at sub-ambient temperatures and linear van't Hoff plots were obtained [53]. Temperature changes changed the elution order in some cases [54]. A series of aromatic alcohols gave a linear behavior on polymeric anion-exchange columns while aromatic carboxylic acids did not [55].

9.4 SUB-AMBIENT TEMPERATURE HPLC—SEPARATION OF LABILE/UNSTABLE SPECIES

The situation described in Equation 9.1 is reversed at a reduced temperature. The overall column efficiency decreases rather dramatically for most samples, but successful separations are still practical with the correct choice of parameters. The reduced longitudinal diffusion in the first term means that the optimal flow rate shifts to lower flow rates. The increased viscosity of the mobile phase requires lower flow rates as well. While at high temperatures one often operates the HPLC at flow rates many times the optimal value, in subambient work, it is best to sacrifice speed and work close to the optimal flow rate.

The use of low temperatures to separate labile molecules was first reviewed by Henderson and O'Connor [56]. All of the advantages of an elevated temperature on viscosity and efficiency are lost when one works at greatly reduced temperatures. The reason for working in this milieu is the possibility of separating species that are too unstable to be eluted at room temperature or to separate various forms of molecules that interconvert rapidly. Typically, the chromatogram of these species consists either of a single broad peak or two peaks separated by a long plateau. When the temperature is reduced sufficiently, distinct peaks are observed for each form. Alternatively, if the temperature is raised high enough to cause very rapid interconversion, then a single sharp peak will be observed. Henderson and coworkers [57] established this for *fac* and *mer* isomers of labile metal complexes of Cr(III), Co(III), Al(III), and Ga(III), operating at temperatures as low as −50°C.

Henderson and Horváth [58] demonstrated the separation of labile dipeptides at a reduced temperature, and Henderson and Mello [59] and Gustaffson et al. [60] extended this to larger peptides containing multiple proline residues. He et al. [61] saw a similar behavior for the racemization of oxazepam enantiomers during separation on a β-cyclodextrin derivatized chiral column. At temperatures below 13°C, racemization was slow enough that peaks for both enantiomers were observed. Kocijan et al. [62] noted the same behavior for angiotensin-converting enzyme inhibitors. Labile peptides were isolated after a low-temperature separation and their conformations confirmed by NMR [63]. The analysis of an unstable mesylate ester that cyclized during elution at room temperature was successful at −30°C on a diol column using a mobile phase of toluene, *n*-hexane, and ethyl acetate [64]. In some cases, even a low temperature is not sufficient to prevent undesired reactions. LoBrutto et al. [65] attempted to prevent the formation of gem diols during the separation of aldehydes. Even at −5°C they were not able to prevent this reaction and were forced to resort to on-column derivatization of the aldehydes.

9.5 VERY HIGH TEMPERATURE HPLC—DIELECTRIC CONSTANT EFFECTS

A few researchers are now doing HPLC at very high temperatures, up to 370°C [66]. At these extreme temperatures the changes in the dielectric constant of the solvent become a major factor. In the more typical range of temperatures available using commercial HPLC systems (typically 5°C–80°C), the temperature effects on the properties of the solvents used are normally ignored. The advantage of very high temperatures comes from the fact that water becomes increasingly non-polar as the temperature increases [67], reaching values comparable to pure room temperature acetonitrile and methanol at temperatures near 200°C [68] as shown in Figure 9.5.

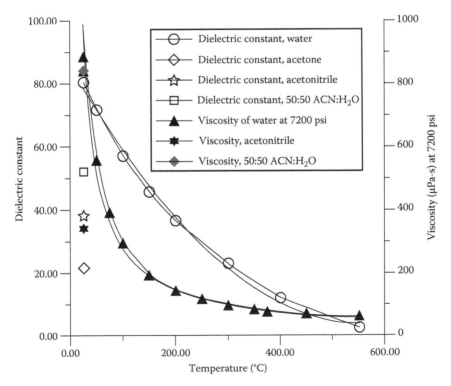

FIGURE 9.5 Dielectric constant and viscosity of water at 7200 psi along with the viscosity and the dielectric constant of both pure ACN and 50% ACN at room temperature. (Reprinted from Kephart, T.S. and Dasgupta, P.K., *Talanta*, 56, 977, 2002. Copyright 2002. With permission from Elsevier.)

This has allowed the reduction or the elimination of the use of an organic solvent from the reversed phase HPLC. The use of pure aqueous mobile phases has in turn allowed the use of the flame ionization detector for HPLC detection, either directly with narrow bore columns [66] or in the split mode with traditional columns [69]. Teutenberg et al. [70] compared the separation of four steroids at an ambient temperature using the traditional water-acetonitrile mobile phase to the separation in pure water at 185°C. Riddle and Guiochon [71] compared the separation of sterols from fruit juices on various high-temperature stationary phases at temperatures up to 150°C. The separation time was reduced from 17 min to 1.2 min. Edge et al. [72] found linear van't Hoff plots for nine drugs from 40°C to 180°C in purely aqueous mobile phases on several commercial columns. Column temperatures can be selected to provide a solvent strength equivalent to typical methanol-water [67] or acetonitrile-water mixtures [73]. This can allow temperature programming to replace gradient elution to increase solvent strength and the use of pure aqueous mobile phases or those containing only minimal amounts of organic solvents. One caution about operating at these high temperatures is that some compounds lack the thermal stability to elute without decomposition. A second important consideration is the stability of the column material at these temperatures, which is addressed in more detail in Section 9.8.2.

9.6 TEMPERATURE PROGRAMMING

While temperature programming has long been a staple of gas chromatography methods, it has not yet achieved much significance in HPLC methods. This may well be changing. Since increasing temperature frequently reduces retention, increasing temperature is an alternative to increasing the organic component of the mobile phase to elute strongly retained solutes. And in cases where retention is greater at elevated temperatures, the possibility of a reverse temperature program has also been explored [74]. Andersen et al. [75] obtained an excellent separation of polyethylene glycol oligomers starting at 80°C and programming at −1.5°C/min to 25°C. The reasons that temperature programming has not yet become popular in HPLC are practical rather than theoretical. Traditional analytical columns of a 4 mm and greater inside diameter do not equilibrate rapidly with temperature. This has prevented the routine application of temperature programming. Djordjevic et al. [76] have discussed instrumental considerations for temperature programming in detail. Some researchers have demonstrated successful programming even for typical analytical columns. Craft et al. [20] achieved a 60% reduction in analysis time for four retinol esters using a temperature program from 0°C–45°C over a 20 min period using 4.6 × 150 mm C-18 column.

Narrow bore columns of 2 mm i.d. and smaller have been shown to equilibrate much faster to changing temperature, and commercial temperature programming systems are becoming available for these columns. For columns 1 mm i.d. and less, the rates of temperature change are similar to those in GC, possibly up to 40°C/min or more. The value of temperature programming has been clearly demonstrated. Chen and Horváth [77] compared gradient elution and temperature programming at rates up to 30°C/min for alkyl benzenes and β-lactoglobulins and were able to obtain almost identical chromatograms in the two modes. These compounds displayed linear van't Hoff plots at a constant acetonitrile concentration and thus produced predictable behavior with temperature. They found that a 5°C temperature change had the same effect as a 1% increase in acetonitrile concentration. Molander et al. [78] demonstrated an excellent separation of three selective serotonin reuptake inhibitors using a temperature program from 35°C to 100°C. Houdiere et al. [79] combined temperature programming with flow programming to increase the separation speed on capillary columns as an alternative to gradient programming. Djordjevic et al. [80] use temperature programming with 2 mm i.d. columns to increase the speed of separation of medium and large oligonuceotides (25–60) by 75% using temperatures from 60°C to 80°C. Hayakawa et al. [81] used both a mobile-phase gradient and a two step temperature program to improve the separation of phenylthiohydantoin (PTH)-amino acids. They found that the separation of early eluting amino acids was better at a lower temperature while higher temperatures, up to 85°C was required to separate PTH-methionine from PTH-valine.

The fact that most commercial analytical HPLC systems have internal volumes in the connecting tubing and detectors, which are too large to allow an efficient use of narrow bore columns with very low volumes, continues to be an obstacle to the widespread application of either narrow bore columns or temperature programming. But the growth of high-speed separations is leading to design changes in instrumentation, which will overcome these obstacles.

9.7   INSTRUMENTATION FOR TEMPERATURE CONTROL

There are a number of ways to thermostat the HPLC column. Most manufacturers now offer column ovens. The majority of these are circulating air systems using either resistive heating for elevated temperature operation or Peltier effect systems, which can both heat and cool, usually from about 20°C below ambient temperature to a high value between 70°C and 100°C. Static air systems are also sold. It is also a simple matter to mount the column and even the injector valve in any commercial constant temperature bath for temperature control. Liquid systems have higher thermal conductivity and can remove heat from the column more rapidly than air systems. When using refractive index (RI) detection, the natural temperature cycling of the temperature bath can cause fluctuations of the baseline. These are reported to be the worst with water-THF and water-dioxolane mobile phases [82]. For the RI detector, circulating air systems produced more stable baselines. Static air systems without circulating fans have the lowest ability to remove heat from the column. Conductive systems are also used, with the column in physical contact with a heating block as noted in Table 9.1.

TABLE 9.1
Column Ovens by Selected Manufacturers

Manufacturer	Low Temp (°C)	High Temp (°C)	Temperature Programming	No. of Columns	Maximum Length (cm)
Agilent	**Ambient −10**	100	—	2 at independent temperatures	30
Cecil Instruments	Ambient +10	120	—	3	—
	Ambient −10	80			
Dionex	**5**	85	No	6	30
Eppendorf conductive	Ambient	150	No	2	
Jasco	**Ambient −15**	80	No	10	40
Hitachi	**Ambient −15**	60	—	3	25
Perkin Elmer	**Ambient −15**	90	Yes	6	30
Selerity Technologies	**Sub-zero**	200	Yes 30°C/min		25
Shimadzu	**Ambient −10**	85	Yes	—	—
Shimadzu conductive	**Ambient −15**	**Ambient +65**	No	2	25
Thermo-Fisher	**5**	90	Yes	—	25–30
Surveyor	Ambient +5	90			
Thermo-Fisher Acella	**4**	95		—	25–30
Waters Alliance	**4**	65	—	—	—
	Ambient +5	65			
Waters Acquity	**10**	90	—	4 2.1 mm	15

Notes: Conductive ovens are noted. Peltier effect systems in bold. Information from manufacturers' literature.

For column operation at very low temperatures, constant temperature baths using chillers and ethanol or ethylene glycol baths have been used [56,57,65]. The operation of these systems to temperatures as low as −70°C has been shown. At the opposite extreme, researchers working in the range from 100°C to 200°C have often used GC ovens for column heating.

The column temperature needs to be held constant to ±0.1°C. The equilibration rate is also an important consideration. Circulating air ovens with low thermal mass equilibrate relatively rapidly. Conductive systems typically have aluminum column holders, which clamp around the column and transfer energy from the heating and the cooling element. Table 9.1 shows some examples of typical commercial systems and their specifications. Model numbers are not shown because they change frequently. The information provided should give an indication of the present state of commercial systems and the variation in temperature ranges and size of ovens. These are considerations that should be made before making a purchase and will depend on the application. Many column ovens now come with optional switching valves that allow method development or routine analysis to have several columns available under software control. Some column ovens are now advertising temperature programming. Usually, this is relatively slow, 3°C/min–7°C/min. As temperature control becomes more common, the specifications of systems will evolve.

Elevated temperatures up to 100°C have also been used with integrated HPLC chips that include both a column and an electrochemical detector [83]. This system used a resistive heater and a mobile phase preheater and was used to separate o-phthaldialdehyde amino acid derivatives.

Several theoretical and empirical studies have been made of the impact of column ovens on separations [84]. The two most important factors that must be considered are the preheating of the mobile phase before it enters the column and the extra-column volume between the injector and the column. Work by Djordevic et al. [76] provides some of the most dramatic evidence for the impact of pre-heating the mobile phase. Their measurements of the axial temperature gradient in 4.6 mm columns at typical flow rates are shown in Table 9.2.

The effects of axial temperature gradients on the chromatographic peaks can produce serious asymmetry in peak shape. The need for significant column preheating is quite clear. The only way that Djordjevic et al. were able to achieve an accurate temperature program within the column was with capillary HPLC columns in a GC oven, where they were able to program at 10°C/min with good accuracy. A pre-heater of this length (4.5 m) would need to be placed in the flow system before the injector to avoid significant band broadening. The need for a pre-heater before the column can be a serious problem in systems using an autosampler. It requires the routing of the mobile phase from the pump to the column oven and then back to the injection valve. This problem can be solved with the use of a separate mobile phase pre-heater [69]. This increases the complexity of the instrumentation and control somewhat but allows the mobile phase temperature to be set independently

TABLE 9.2

Measured Axial Temperature Gradients for a 4.6 mm i.d. Column at Three Temperatures with and without a 4.5 m × 0.5 mm i.d. Preheater

Column Temperature (°C)	Axial Temperature Gradient (°C)			
	Without Preheater		With Preheater	
	1.0 mL/min	2.0 mL/min	1.0 mL/min	2.0 mL/min
70	6.50	10.80	2.32	2.67
90	9.53	16.03	3.35	3.73
110	12.83	20.87	4.60	5.00

Source: Adapted from Djordjevic, N.M. et al., *J. Microcolumn Sep.*, 11, 403, 1999.

of the column oven to ensure that the mobile phase enters the column with the correct temperature. Teutenberg et al. demonstrated the effectiveness of this approach with conventional 4.6 mm columns and a short heater (13 cm), which could also be used to cool the mobile phase for faster cycling. In some cases, the mobile phase can be heated to a different temperature to compensate for other temperature-related effects, especially on large diameter columns [84]. Alternatively, the injection valve can be placed in the oven itself as it is in the Thermo Accela instrument [85]. This minimizes the void volume between the valve and the column. At least one manufacturer has done this and others are certain to follow as the application of a high temperature increases in popularity. An additional complication is that the all volume between the pump and the column is significant for dead time in gradient elution. The problem of axial temperature gradients become less important as the column diameter is decreased. This is fortunate because narrower bore columns place greater constraints on extra-column volume, both to avoid peak broadening and to avoid long delays in the start of a mobile-phase gradient.

Direct resistive heating has also been employed for preheating the mobile phase [16] and as an alternative to column ovens for heating the column. The Selerity Caloratherm™ [86] is a commercial low-volume resistive preheater which can be placed on the column inlet to insure a proper mobile-phase temperature even at high flow rates on 4.6 mm i.d. columns. This approach solves the problem of where to place the injection valve and eliminates the problem of extra column volume as well. Having a temperature control system monitoring the mobile-phase temperature insures accurate results using this approach.

Another issue for high flow rates and small particles is the viscous heating of the mobile phase as it travels through the column. This produces an axial temperature gradient along the length of the column [87]. Still air-column ovens were found to produce larger axial gradients while water-bath thermostats produced larger radial temperature gradients. For large diameter columns, the radial temperature gradient [88,89] becomes increasingly pronounced. The radial gradient has a greater influence on the shape of the eluent profile and thus the overall efficiency of the separation than an axial gradient. One can achieve more efficient separations in large columns by adjusting the mobile-phase temperature to a value different from that of the column to minimize the impact of the radial gradient [81,90].

One problem with temperature programming in both GC and HPLC is the cooldown time after each run [70]. An interesting solution to this problem was the use of two thermostated baths that could be selectively circulated through the column jacket. This allowed the column temperature to be quickly cycled back to the initial temperature after a step-gradient elution.

9.8 LIMITATIONS

There are a number of limitations on the use of extremes of temperature in HPLC. Clicq et al. [91] note that instrumental issues become increasingly limiting as one goes to very high temperatures and flow rates. They suggest that most separations will occur below 90°C where there are less instrumental constraints. As detailed below, column bleed can limit the selection of columns. High-speed separations require a faster detector response than many systems allow and constrain extra column volume. This is especially true for narrow bore columns and sub-2 μm particles. In many cases, the additional speed gained above the temperature limits of commercial HPLC ovens will not be worth the additional expense and complexity required. For macromolecules, the effect of extreme pressure can also impact retention time as noted by Szabelski et al. [92].

9.8.1 SOLVENT TEMPERATURE LIMITS

As noted before, HPLC has been performed at temperatures as low as −50°C. The physical properties of various HPLC solvents are shown in Table 9.3. It is important to note that solvents cannot

TABLE 9.3
Physical Properties of HPLC Solvents

Solvent	Normal Freezing Point (°C)	Normal Boiling Point (°C)	Critical Temperature (°C)/Pressure (psi)
Water	0	100	374/3208 [95]
Acetonitrile	−45.7	81.6	274.7/701
Tetrahydrofuran	−65	66	267/753
Acetone	−95	56	235/682
n-Hexane	−95	68.7	234/436
Dichloromethane	−97	39.8	245/895
Methanol	−97.8	64.5	240/1142
Diethyl ether	−116	34.6	193.5/527

Source: Adapted from Chemical Hazards Response Information System, U.S. Coast Guard, http://www.chrismanual.com/findform.htm (accessed July 31, 2007).

be used below their freezing points, and for most solvents, freezing points increase as pressure increases. On the other hand, the high pressures in the HPLC system allow solvents to be used well above their normal boiling points. The critical temperature and pressure define the upper limit of operation in sub-critical conditions. Beyond this point, the separation is, by definition, super-critical fluid chromatography. Both sub-critical and super-critical operations require that a sufficient back-pressure is maintained for proper detector operation. This may require cooling the mobile phase below its boiling point before a spectroscopic detector. The use of water as a mobile phase at 370°C [66] shows that it is possible to work close to the critical point. As noted by Chester [93], there is no physical reason not to consider both the sub- and the super-critical operations as part of a continuum of the chromatographic process spanning the entire range from liquid to gas. However, working with binary or tertiary mobile phases in this region requires a knowledge of the detailed phase behavior to insure that the mobile phase is a single phase [94,95].

The minimum operating temperatures for various solvent mixtures used in the reversed phase HPLC are shown in Table 9.4. Values for acetonitrile were experimentally determined based on the temperature at which the system could no longer pump the mobile phase [56]. These values are approximate and will vary somewhat with pressure. Values are not shown for THF-water systems. While THF freezes at −65°C, work in our laboratory [56] has shown that water-THF mixtures separate and the water component freezes at the freezing point of water, making these mixtures unusable below 0°C [96]. No data is available for ternary mixtures, though the addition of another solvent may eliminate the separation of THF and water.

The freezing points are not the only difficulty in low-temperature work with reversed-phase systems. Solvent viscosity increases rapidly. Methanol-water mixtures are especially viscous due to hydrogen bonding and reach a maximum viscosity in the 40%–60% region that is several times higher than the viscosity of either pure solvent. Acetonitrile-water mixtures reach a maximum viscosity in the 10%–20% acetonitrile region [56]. In most cases where retention times increase as temperature decreases, higher acetonitrile concentrations are required. This tends to partially offset the viscosity disadvantage of reducing temperature.

9.8.2 COLUMN TEMPERATURE LIMITS

The interest in HPLC at high temperatures has led to studies of the limits of various commercial stationary phases. Claessens and van Straten [97] reviewed the thermal stability of stationary phases. Teutenberg et al. [98] and Marin et al. [67] studied column stability at temperatures up to

TABLE 9.4
Freezing Points (FP) of Reversed Phase Mixed Solvent Systems

Solvent % Water	FP Methanol	FP Acetonitrile
10	−6	
20	−13	−14
30	−21	
40	−33	−16
50	−46	
60	−62	−16
70	−86	
80	−107	−16
90	−118	−22

Source: Adapted from Landolt-Bornstein, *Zahlenwerte und Funktionen aus Physik, Chemie, Astronomie, Geophysik, und Technik*, II Band, 2 Teil, Bandteil b., 3-404, Springer-Verlag, Berlin-Gottingen-Heidelberg, 1962.

200°C by measuring column bleed under isothermal and temperature programming conditions and by measuring changes in k' for solutes as a function of time [99]. These studies show that silica-based phases are stable enough for use over the temperature range of most commercial HPLC systems up to limits of 80°C–100°C. Manufacturers' literature should be consulted before purchasing a column for such applications. However, a study of the column under the actual pH and mobile phase range to be used should be conducted to determine the actual stability of the column as part of any method-development process, especially as temperatures are raised toward the column limits.

At higher temperatures, zirconium dioxide and titanium dioxide supports gave much greater stability along with polymer-based supports [100,101] based on polystyrene-divinyl benzene (PS-DVB) such as PLRP-S noted in Table 9.5. PS-DVB supports have been reported to give a serious column bleed at 250°C [66]. Polybutadiene (PBD) modified zirconia columns have been used at temperatures up to 300°C and carbon-coated zirconia has been used at temperatures up to 370°C [66]. Applications have included the separation of steroids [73] and herbicides [102].The specific order of column bleed varied depending on the detection method as shown in Table 9.5.

The temperature stability of monolithic stationary phases based on alkyl methacrylate monomers in capillary HPLC has also been reported [103]. These columns allowed the separation time to be reduced by over 10-fold at temperatures up to 80°C. The upper-temperature limit for these columns was not reported.

TABLE 9.5
Column Bleed of HPLC Columns at 200°C in Water

Detector	Order of Column Bleed (High to Low)
Charged Aerosol	Luna C-18 > Thermo Hypercarb > ZirChromCarb > PLRP-S > TiO₂-Carb
UV 190 nm	Luna C-18 > PLRP-S_ZirChrom Carb > Thermo Hypercarb > TiO₂-Carb
UV 254 nm	Luna C-18 > PLRP-S > ZirChrom Carb > Thermo Hypercarb > TiO₂-Carb

Source: Adapted from Teutenberg, T. et al., *J. Chromatogr. A*, 1119, 197, 2006.

A final consideration at very high temperatures in fused silica capillaries is the solubility of the fused silica. Pre-saturation of the mobile phase using a silica precolumn has been recommended as a necessary step [66] and care must be taken to avoid the precipitation of silica as temperatures are reduced.

9.8.3 Detector Considerations

Commercial HPLC systems designed to operate in the range from 5°C to 90°C are designed to minimize temperature effects on detection, though detector cells are not typically thermostated. Spectroscopic absorption detectors are typically designed to minimize temperature effects both through the cell design and by using a small heat exchanger to insure constant temperature. At least one commercial system, the Agilent 1200SL, includes a post-column cooling section, which can be used on columns <100 mm length [85]. Refractive index (RI) detectors place the greatest constraint on temperature due to the large temperature coefficient of RI [104]. This is normally achieved using a large heat sink rather than with a thermostat, and thus RI detectors are not suitable for temperature programming. Conductivity detectors also require a constant temperature to avoid changes in calibration [105]. Other electrochemical detectors are less sensitive to temperature changes. Fluorescence often decreases as temperature increases due to collisional quenching. However, if the detector is operated at room temperature, the effects of temperature programming should be less than the variation in fluorescence due to the changing solvents in gradient elution [106].

Several things need to be considered when operating HPLC systems at extreme conditions. High-temperature systems often include a post-column cooling step prior to the detector or a restrictor after the column to prevent boiling in the detector. The detector must be capable of withstanding the pressure as noted by Yang [68], which may limit the flow rates if a post detector restrictor is used. This is needed to insure that the mobile phase does not boil in the detector as the pressure is reduced. Warming the eluent before a spectroscopic detector after a low-temperature separation is also useful to prevent moisture condensation on the optics. The use of pure water as a mobile phase also offers the possibility of using the flame ionization detector (FID). This is only practical at the flow rates of micro-bore columns as demonstrated by Yang et al. [69] for carbohydrates, amino acids, and other organic acids and bases or by splitting the flow.

9.8.4 Temperature Effects on Injection

The injection temperature can be a significant issue for thermally unstable samples or where samples are stored for hours in an autosampler prior to injection. For this reason, most manufacturers sell autosamplers with optional thermostated sample compartments. This can be done either by placing the sample tray in an air bath oven or by a conductive temperature control of the sample rack. The need to keep samples cool prior to injection when coupled with elevated temperature separation increases the complexity of the flow system required. For such application, a separate mobile phase pre-heater with a low volume placed between the injector and the column is a good choice. Alternatively, the injector valve would need to be mounted outside the autosampler or in the column oven to insure preheating of the mobile phase before the column.

For both capillary columns and packed columns, researchers have begun to explore what is common practice in GC, injecting samples at reduced temperature followed by temperature programming to allow larger sample loading. The theory of maximum injection volume indicates that the maximum amount of material that can be injected without band broadening is related to the retention time under the injection conditions [107]. This has been exploited by Rosales-Conrado et al. [108] to allow large injection volumes for trace analysis of herbicides in soil extracts and by Molander et al. [109] for separation of retinyl esters on capillary HPLC columns. Injections were made at a low temperature followed by a temperature program for elution. In a similar way, Holm et al. [110] developed a method for introducing a cold spot in a capillary HPLC column to allow the equivalent of cryofocusing in the GC injection. The column was mechanically moved to the higher-temperature region of the oven to begin elution. This allowed injections as large as 500 μL without degrading the peak shape.

9.9 CONCLUSIONS

The use of temperature as a variable can greatly enhance the range and speed of HPLC separations. Commercial instruments now allow temperature to be considered as a routine part of method development. The temperature range available in commercial systems is adequate for the majority of separation problems. However, researchers are exploring the limits of low-temperature and high-temperature sub-critical applications. These will play an increasingly important role in HPLC methods in the future as instrumentation for temperature control and columns stable at high temperatures become more readily available.

REFERENCES

1. J.W. Dolan, *J. Chromatogr. A* 965, 195–205, 2002.
2. F. Lestremau, A. de Villiers, F. Lynen, A. Cooper, R. Szucs, P. Sandra, *J. Chromatogr. A* 1138, 120–131, 2007.
3. P.L. Zhu, L.R. Snyder, J.W. Dolan, N.M. Djordjevic, D.W. Hill, L.C. Sanders, T.J. Waeghe, *J. Chromatogr. A* 756, 21–39, 1996.
4. J.W. Dolan, L.R. Snyder, N.M. Djordjevic, D.W. Hill, T.J. Waeghe, *J. Chromatogr. A* 857, 21–39, 1999.
5. J.W. Dolan, L.R. Snyder, T. Blanc, L. Van Heukelem, *J. Chromatogr. A* 897, 37–50, 2000.
6. J.W. Dolan, L.R. Snyder, T. Blane, *J. Chromatogr. A* 897, 51–63, 2000.
7. N.S. Wilson, M.D. Nelson, J.W. Dolan, L.R. Snyder, P.W. Carr, *J. Chromatogr. A* 961, 195–215, 2002.
8. Molnar-Instut, Berlin, Germany, http: //www.molnar-institut.com/cd/indexe.htm (accessed August 10, 2007).
9. P.L. Zhu, J.W. Dolan, L.R. Snyder, *J. Chromatogr. A* 756, 41–50, 1996.
10. P.L. Zhu, J.W. Dolan, L.R. Snyder, D.W. Hill, L. Van Heukelem, T.J. Wasghe, *J. Chromatogr. A* 756, 51–62, 1996.
11. P.L. Zhu, J.W. Dolan, L.R. Snyder, N.M Djordjevic, D.W. Hill, J.-T. Lin, L.C. Sanders, L. Van Heukelem, T.J. Wasghe, *J. Chromatogr. A* 756, 63–72, 1996.
12 J.K. Swadesh, Temperature control in analytical high-performance liquid chromatography. In E. Katz (Ed.), *Handbook of HPLC*, Marcel Dekker, New York, 1998, pp. 607–615.
13. B. Yan, J. Zhao, J.S. Brown, J. Blackwell, P.W. Carr, *Anal. Chem.* 72, 1253–1262, 2000.
14. F.D. Antia, Cs. Horváth, *J. Chromatogr.* 435, 1–15, 1988.
15. U.D. Neue, J.R. Mazzeo, *J. Sep. Sci.* 24, 921–929, 2001.
16. D. Guillarme, S. Heinisch, J.L. Rocca, *J. Chromatogr. A* 1052, 39–51, 2004.
17. J. Li, Y. Hu, P.W. Carr, *Anal. Chem.* 69, 3884–3888, 1997.
18. X.C. Le, X.-F. Li, V. Lai, M. Ma, M. Yalcin, J. Feldmann. *Spectrochim. Acta Part B* 53B, 899–909, 1998.
19. X.C. Le, M. Ma, N.A. Wong, *Anal. Chem.* 68, 4501–4506, 1996.
20. N.E. Craft, R.T. Tucker, J.E. Estes, S. Marin, *Am. Lab.* 37, 12–13, 2005.
21. V. Bohm, *J. Sep. Sci.* 24, 955–959, 2001.
22. V. Bohm, *Chromatographia* 50, 282–286, 1999.
23. R. Adlof, G. List, *J. Chromatogr. A* 1046, 109–113, 2004.
24. R. Adlof, *J. Chromatogr. A* 1148, 256–259, 2007.
25. C.B. Castells, L.G. Gagliardi, C. Rafols, M. Roses, E. Bosch, *J. Chromatogr. A* 1042, 23–36, 2004.
26. J.V. Tran, P. Molander, T. Greibrokk, E. Lundanes. *J. Sep. Sci.* 24, 930–940, 2001.
27. D.V. McCalley, *J. Chromatogr. A* 902, 311–321, 2000.
28. S.M.C. Buckenmaier, D.V. McCalley, M.R. Euerby, *J. Chromatogr. A* 1060, 117–126, 2004.
29. D. Chagolla, G. Ezedine, Y. Ba, *Micropor. Mesopor. Mater.* 64, 155–163, 2003.
30. Y. Chen, C.T. Mant, R.S. Hodges, *J. Chromatogr. A* 1043, 99–111, 2004.
31. M. Pursch, S. Stroschein, H. Händel, K. Albert, *Anal. Chem.* 68, 386–393, 1996.
32. E. Peyrin, Y.C. Guillaume, C. Guinchard, *Anal. Chem.* 69, 4979–4984, 1997.

33. R. Dolfinger, D.C. Locke, *Anal. Chem.* 75, 1355–1364, 2003.
34. P.K. Zarzycki, R. Smith, *J. Chromatogr. A* 912, 45–52, 2001.
35. B.A. Bildingmeyer, J. Henderson, *J. Chromatogr. A* 1060, 187–193, 2004.
36. H. Kanazawa, K. Yamamoto, Y. Matsushima, N. Takai, A. Kikuchi, Y. Sakurai, T. Okano, *Anal. Chem.* 68, 100–105, 1996.
37. H. Kanazawa, Y. Kashiwase, K. Yamamoto, Y. Matsushima, A. Kikuchi, Y. Sakurai, T. Okano, *Anal. Chem.* 69, 823–830, 1997.
38. H. Kanazawa, T. Sunamoto, Y. Matsushima, A. Kikuchi, T. Okano, *Anal. Chem.* 72, 5961–5966, 2000.
39. E. Ayano, Y. Okada, C. Sakamoto, H. Kanazawa, A. Kikuchi, T. Okano, *J. Chromatogr. A* 1119, 51–57, 2006.
40. K. Hosoya, K. Kimata, T. Araki, N. Tanaka, J.M.J. Fréchet, *Anal. Chem.* 67, 1907–1911, 1995.
41. C. Sakamoto, Y. Okada, H. Kanazawa, E. Ayano, T. Nishimura, M. Ando, A. Kikuchi, T. Okano, *J. Chromatogr. A* 1030, 247–253, 2004.
42. K. Kurata, T. Shimoyama, A. Dobashi, *J. Chromatogr. A* 1012, 47–56, 2003.
43. E. Ayano, H. Kanazawa, *J. Sep. Sci.* 29, 738–749, 2006.
44. H. Kanazawa, Y. Matsushima, *Yakugaku Zasshi* 117, 817–824, 1997.
45. K. Yamamoto, H. Kanazawa, Y. Matsushima, K. Oikawa, A. Kikuchi, T. Okano, *Environ. Sci. (Tokyo)* 7, 47–56, 2000.
46. R.J. Smith, D.R. Taylor, S.M. Wilkins, *J. Chromatogr. A* 697, 591–596, 1995.
47. Q. Tian, C. Lv, P. Wang, L. Ren, J. Qiu, L. Li,. Z. Zhou, *J. Sep. Sci.* 30, 310–321, 2007.
48. J. Sun, X. Tang, H. Song, Q. Tang, C. Fu, *Huaxi Yaoxue Zazhi* 21, 30–32, 2006.
49. A. Peter, G. Torok, F. Fulop, *J. Chromatogr. Sci.* 36, 311–317, 1998.
50. M. Schlauch, A.W. Frahm, *Anal. Chem.* 73, 262–266, 2001.
51. P. Hatsis, C.A. Lucy, *J. Chromatogr. A* 920, 3–11, 2001.
52. M.G. Kolpachnikova, N.A. Penner, P.N. Nesterenko, *J. Chromatogr. A* 826, 15–23, 1998.
53. C. Panagiotopoulos, R. Sempéré, R. Lafont, P. Kerhervé, *J. Chromatogr. A* 920, 13–22, 2001.
54. E. Landberg, A. Lundblat, P. Påhlsson, *J. Chromatogr. A* 814, 97–104, 1998.
55. H.K. Lee, N.E. Hoffman, *J. Chromatogr. Sci.*, 32, 97–101, 1994.
56. D.E. Henderson, D.J. O'Connor, *Adv. Chromatogr.* 23, 65–95, 1984.
57. D.E. Henderson, D.J. O'Conner, J.F. Kirby, C.P. Sears III, *J. Chromatogr. Sci.* 23, 477–483, 1986.
58. D.E. Henderson, Cs. Horváth, *J. Chromatogr.* 368, 203–213, 1986.
59. D.E. Henderson, J.A. Mello, *J. Chromatogr.* 499, 79–88, 1990.
60. S. Gustaffson, B.-M. Eriksson, I. Nilsson, *J. Chromatogr.* 506, 75–83, 1990.
61. H. He, Y. Liu, C. Sun, X. Wang, C. Pham-Huy, *J. Chromatogr. Sci.* 42, 62–66, 2004.
62. A. Kocijan, R. Grahek, D. Kocjan, L. Zupancic-Kralj, *J. Chromatogr. B* 755, 229–235, 2001.
63. A. Kálmán, F. Thunecke, R. Schmidt, P.W. Schiller, Cs. Horváth, *J. Chromatogr. A* 729, 155–171, 1996.
64. J.O. Egekeze, M.C. Danielski, N. Gringerg, G.B. Smith, D.R. Sidler, H.J. Pepall, G.R. Bicker, P.C. Tway, *Anal. Chem.* 67, 2292–2295, 1995.
65. R. LoBrutto, Y. Bereznitski, T.J. Novak, L. DiMichele, L. Pan, M. Journet, J. Kowal, N. Grinberg, *J. Chromatogr. A* 995, 67–78, 2003.
66. T.S. Kephart, P.K. Dasgupta, *Talanta* 56, 977–987, 2002.
67. S.J. Marin, B.A. Jones, W.D. Felix, D. Clark, *J. Chromatogr. A* 1030, 255–262, 2004.
68. Y. Yang, A.D. Jones, C.D. Eaton, *Anal. Chem.* 71, 3808–3813, 1999.
69. Y. Yang, A.D. Jones, J.A. Mathis, M.A. Francis, *J. Chromatogr. A* 942, 231. 2002.
70. T. Teutenberg, H.-J Goetze, J. Tuerk, J. Ploeger, T.K. Kiffmeyer, K.G. Schmidt, W. Kohorst, T. Rohe, H.-D. Jansen, H. Weber, *J. Chromatogr. A* 1114, 89–96, 2006.
71. L.A. Riddle, G. Guiochon, *J. Chromatogr. A* 1137, 173–179, 2006.
72. A.M. Edge, S. Shillingford, C. Smith, R. Payne, I.D. Wilson, *J. Chromatogr. A* 1132, 206–210, 2006.
73. S.M. Fields, C.Q. Ye, D.D. Zhang, B.R. Branch, X.J. Zhang, N. Okafo, *J. Chromatogr. A* 913, 197–204, 2001.
74. I.L. Skuland, T. Andersen, R. Trones, R.B. Eriksen, T. Greibrokk, *J. Chromatogr. A* 1011, 31–36, 2003.
75. T. Andersen, P. Molander, R. Trones, D.R. Hegna, T. Greibrokk, *J. Chromatogr. A* 918, 221–226, 2001.
76. N.M. Djordjevic, P.W.J. Fowler, F. Houdiere. *J. Microcolumn Sep.* 11, 403–413, 1999.
77. M.H. Chen, Cs. Horváth, *J. Chromatogr. A* 788, 51–61, 1997.
78. P. Molander, A. Thomassen, L. Kristoffersen, T. Greibrokk, E. Lundanes, *J. Chromatogr. A* 766, 77–87, 2002.
79. F. Houdiere, P.W.J. Fowler, N.M. Djordjevic, *Anal. Chem.* 69, 2589–2593, 1997.
80. N.M. Djordjevic, F. Houdiere, P. Fowler, F. Natt, *Anal. Chem.* 70, 1921–1925, 1998.

81. K. Hayakawa, M. Hirano, K. Masahiko, N. Katsumata, T. Tanaka, *J. Chromatogr. A* 846, 73–82 1999.
82. G. Openhaim, E. Grushka, *J. Chromatogr. A* 942, 63–71, 2002.
83. C.-Y. Shih, Y. Chen, J. Xie, Q. He, Y.-C. Tai, *J. Chromatogr. A* 1111, 272–278, 2006.
84. O. Dapremont, G.B. Cox, M. Martin, P. Hilaireau, H. Colin, *J. Chromatogr. A* 796, 81–99, 1998.
85. J.M. Cunliffe, S.B. Adams-Hall, T.D. Maloney, *J. Sep. Sci.* 30, 1214–1223. 2007.
86. Selerity Technologies, Salt Lake City, UT, www.selerity.com.
87. A. DeVilliers, H. Lauer, R. Szucs, S. Goodall, P. Sandra, *J. Chromatogr. A* 1113, 84–91, 2006.
88. A. Brandt, G. Mann, W. Arlt, *J. Chromatogr. A* 769, 109–117, 1997.
89. A. Brandt, G. Mann, W. Arlt, *J. Chromatogr. A* 796, 223–228, 1998.
90. C.B. Ching, Y.X. Wu, M. Lisso, G. Wozny, T. Laiblin, W. Arlt, *J. Chromatogr. A* 945, 117–131, 2002.
91. D. Clicq, S. Heinisch, J.L. Rocca, D. Cabooter, P. Gzil, G. Desmet, *J. Chromatogr. A* 1146, 193–201, 2007.
92. P. Szabelski, A. Cavazzini, K. Kaczmarski, X. Liu, J. Van Horn, G. Guichon, *J. Chromatogr. A* 950, 41–53, 2002.
93. T.L. Chester, *Microchem. J.* 61, 12–24, 1999.
94. Chemical Hazards Response Information System, U.S. Coast Guard, http: //www.chrismanual.com/find-form.htm (accessed July 31, 2007).
95. R.C. Reid, J.M. Prausnitz, B.E. Poling, *The Properties of Gases and Liquids,* 4th edn., McGraw-Hill, New York, 1987.
96. Landolt-Bornstein, *Zahlenwerte und Funktionen aus Physik, Chemie, Astronomie, Geophysik, und Technik,* II Band, 2 Teil, Bandteil b., 3–404, Springer-Verlag, Berlin/Gottingen/Heidelberg, Germany, 1962.
97. H.A. Claessens, M.A. van Straten, *J. Chromatogr. A* 1060, 23–41, 2004.
98. T. Teutenberg, J. Tuerk, M. Holzhauser, T.K. Kiffmeyer, *J. Chromatogr. A* 1119, 197–201, 2006.
99. J. Lin, P.W. Carr, *Anal. Chem.* 69, 837–43, 1997.
100. P. He, Y. Yang, *J. Chromatogr. A* 989, 55–63, 2003.
101. Y. Yang, T. Kondo, T.J. Kennedy, *J. Chromatogr. Sci.* 43, 518–521. 2005.
102. Y. Xiang, B. Yan, B. Yue, C.V. McNeff, P.W. Carr, M.L. Lee, *J. Chromatogr. A* 983, 83–89, 2003.
103. Y. Ucki, T. Umemura, Y. Iwashita, T. Odake, H. Haraguchi, K.-I. Tsunoda, *J. Chromatogr. A* 1106, 106–111, 2006.
104. M.N. Munk, Refractive index detectors. In T.M. Vickrey (Ed.), *Liquid Chromatography Detectors*, Vol. 23, Marcel Dekker, New York, 1983, pp. 165–204.
105. C.A. Pohl, E.L. Johnson, *J. Chromatogr. Sci.* 18, 442–452, 1980.
106. J.R. Lakowicz, *Principles of Fluorescence Spectroscopy*, Plenum, New York, 1983, Chapter 7.
107. L.W. Hollingshead, H.W. Habgood, W.E. Harris, *Can. J. Chem.* 43, 1560–1568, 1965.
108. N. Rosales-Conrado, M.E. Leon-Gonzalez, L.V. Perez-Arribas, L.M. Polo-Diez, *Anal. Chim. Acta* 470, 147–154, 2002.
109. P. Molander, S.J. Thommesen, I.A. Bruheim, R. Trones, Greibrokk, E. Lundanes, T.E. Gundersen. *J. High Resolution Chromatogr.* 22, 490–494, 1999.
110. A. Holm, P. Molander, E. Lundanes, T. Greibrokk, *J. Sep. Sci.* 26, 1147–1153, 2003.

10 Nonlinear Liquid Chromatography

Alberto Cavazzini and Attila Felinger

CONTENTS

10.1   INTRODUCTION

In analytical chromatography, the sample we analyze is usually rather dilute and allows the development of a rather straightforward method. Due to the minute concentrations we deal with in analytical chromatography, we face a linear behavior. The retention time of the analytes and the selectivity of a given separation can be forecast by simple rules that tremendously help us to develop efficient and fast separations. However, when we increase the sample size and a finite amount of sample is introduced in a chromatographic column, we leave the shelter of linear chromatography and have to cope with more complex peak shapes and phenomena.

When the amount of the sample is comparable to the adsorption capacity of the zone of the column the migrating molecules occupy, the analyte molecules compete for adsorption on the surface of the stationary phase. The molecules disturb the adsorption of other molecules, and that phenomenon is normally taken into account by nonlinear adsorption isotherms. The nonlinear adsorption isotherm arises from the fact that the equilibrium concentrations of the solute molecules in the stationary and the mobile phases are not directly proportional. The stationary phase has a finite adsorption capacity; lateral interactions may arise between molecules in the adsorbed layer, and those lead to nonlinear isotherms. If we work in the concentration range where the isotherms are nonlinear, we arrive to the field of nonlinear chromatography where thermodynamics controls the peak shapes. The retention time, selectivity, plate number, peak width, and peak shape are no longer constant but depend on the sample size and several other factors.

In preparative chromatography, we cannot avoid working in the nonlinear range of adsorption isotherms. The purification has to be economical and the productivity should be maximized, and therefore the separation columns will be heavily overloaded. In analytical chromatography column overload is undesirable since it will decrease the plate number and increase the asymmetry and width of the chromatographic peaks. In some instances, however, when the stationary phase has a rather limited capacity, one might accidentally overload the column. The typical areas for accidental column overload in analytical chromatography are capillary gas chromatography or the use of molecularly imprinted polymer (MIP) stationary phases in liquid chromatography.

The purpose of this chapter is to illustrate the fundamental phenomena occurring in nonlinear chromatography. Since nonlinear isotherms play a key role in the behaviors we observe in nonlinear chromatography, emphasis is put on the description of nonlinear isotherms and their determination. A complete treatment of the theory of nonlinear chromatography is discussed elsewhere [1]. Here, we furnish a brief summary of the fundamentals.

10.2   DIFFERENTIAL MASS BALANCE IN CHROMATOGRAPHY

The distribution of the solute between the mobile and the stationary phases is continuous. A differential equation that describes the travel of a zone along the column is composed. Then the band profile is calculated by the integration of the differential mass balance equation under proper initial and boundary conditions. Throughout this chapter, we assume that both the chemistry and the packing density of the stationary phase are radially homogeneous. Thus, the mobile and stationary phase concentrations as well as the flow velocities are radially uniform, and a one-dimensional mass balance equation can be considered.

As illustrated in Figure 10.1, an elementary slice of the column is regarded, and a mass balance is determined for that slice. The differential mass balance equation states that the rate of accumulation in a slice of Δz length is the sum of the accumulation by convection and the accumulation by diffusion.

The accumulation by convection is

$$u\varepsilon S[C(z - \Delta z, t) - C(z, t)] \tag{10.1}$$

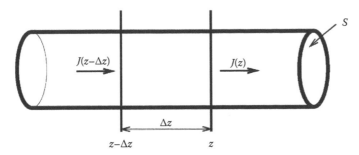

FIGURE 10.1   Differential volume element in a chromatographic column.

where
  u is the linear velocity of the mobile phase
  ε is the total porosity of the packing
  S is the cross section area of the column
  C is the concentration of the solute in the mobile phase
  t is the time

The accumulation by diffusion is given as

$$\varepsilon S\left[J\left(z-\Delta z,t\right)-J\left(z,t\right)\right] \tag{10.2}$$

where $J(z,t)$ is the flux of solute due to axial dispersion.
  The rate of accumulation in the slice is on the one hand the rate of change in the mobile phase

$$\varepsilon S \Delta z\frac{\partial C}{\partial t} \tag{10.3}$$

and, on the other hand, the rate of change in the stationary phase is

$$(1-\varepsilon)S\Delta z\frac{\partial q}{\partial t} \tag{10.4}$$

where q is the concentration of the solute in the stationary phase.
  The combination of the above equations results in

$$\varepsilon S\Delta z\frac{\partial C}{\partial t}+(1-\varepsilon)S\Delta z\frac{\partial q}{\partial t}=u\varepsilon S\left[C\left(z-\Delta z,t\right)-C\left(z,t\right)\right]+\varepsilon S\left[J\left(z-\Delta z,t\right)-J\left(z,t\right)\right] \tag{10.5}$$

which can be rearranged into

$$\frac{\partial C}{\partial t}+\frac{1-\varepsilon}{\varepsilon}\frac{\partial q}{\partial t}=u\frac{C\left(z-\Delta z,t\right)-C\left(z,t\right)}{\Delta z}+\frac{J\left(z-\Delta z,t\right)-J\left(z,t\right)}{\Delta z} \tag{10.6}$$

The flux due to the axial dispersion can be attributed to the concentration gradient and described by Fick's law

$$J(z,t)=-D_{\mathrm{a}}\frac{\partial C}{\partial z} \tag{10.7}$$

where D_a is an apparent dispersion coefficient (see further on).

The mass balance equation can be rewritten when Δz goes to zero as

$$\frac{\partial C}{\partial t} + \frac{1-\varepsilon}{\varepsilon}\frac{\partial q}{\partial t} + u\frac{\partial C}{\partial z} = D_a\frac{\partial^2 C}{\partial z^2} \qquad (10.8)$$

In the above derivation of the mass balance equation, we assume that the column is radially homogeneous, the compressibility of the mobile phase is negligible, the axial dispersion coefficient is constant, and the temperature is unchanged. Furthermore, no diffusion in the stationary phase is assumed.

When the adsorption isotherm is linear, the above mass balance has analytical solution for the cases of various initial and boundary conditions, and the proper functions that describe the peak shape can be obtained. The common feature of those solutions is that they all can well be approximated by a Gaussian function provided that the efficiency of the column is not extremely low.

The apparent dispersion coefficient in Equation 10.8 describes the zone spreading observed in linear chromatography. This phenomenon is mainly governed by axial dispersion in the mobile phase and by nonequilibrium effects (i.e., the consequence of a finite rate of mass transfer kinetics). The band spreading observed in preparative chromatography is far more extensive than it is in linear chromatography. It is predominantly caused by the consequences of the nonlinear thermodynamics, i.e., the concentration dependence of the velocity associated to each concentration. When the mass transfer kinetics is fast, the influence of the apparent axial dispersion is small or moderate and results in a mere correction to the band profile predicted by thermodynamics alone.

For the case of nonlinear adsorption isotherms, no analytical solutions exist; the mass balance equations must be integrated numerically to obtain the band profiles. Approximate analytical solutions are only possible for the cases where the solute concentration is low and accordingly, the deviation from linear isotherm is only minor. All the approximate analytical solutions utilize a parabolic adsorption isotherm $q = aC(1 - bC)$. This constraint prevents us from drawing general conclusions regarding most of the important consequences of nonlinearity.

10.3 EQUILIBRIUM MODEL

The equilibrium models of nonlinear chromatography assume that there always is an instantaneous equilibrium between the mobile phase and the stationary phase. That model is widely applied for the separation of small molecules, when mass transfer or diffusion in the stagnant pores of the mobile phase does not have a significant impact on the band profile.

The equilibrium models of chromatography are given by the mass balance equation given in Equation 10.8 and a proper isotherm equation, $q = f(C)$, should be used to relate the mobile phase and stationary phase concentrations.

The ideal model and the equilibrium–dispersive model are the two important subclasses of the equilibrium model. The ideal model completely ignores the contribution of kinetics and mobile phase processes to the band broadening. It assumes that thermodynamics is the only factor that influences the evolution of the peak shape. We obtain the mass balance equation of the ideal model if we write $D_a = 0$ in Equation 10.8, i.e., we assume that the number of theoretical plates is infinity. The ideal model has the advantage of supplying the thermodynamical limit of minimum band broadening under overloaded conditions.

The equilibrium–dispersive model is defined by Equation 10.8 and as with the ideal model, an isotherm equation should be used to relate the mobile phase and stationary phase concentrations.

The band profiles are obtained as the change of concentration against time at the column outlet. To better understand the effect of mobile phase dispersion, column length, mobile phase velocity, and other parameters, we introduce some normalization. We can rewrite Equation 10.8, using reduced time ($\tau = t/t_0$) and length variables ($x = z/L$), where t_0 is the holdup time of the column. Using these variables and substituting the apparent dispersion coefficient by

$$D_a = \frac{Hu}{2} = \frac{L^2}{2Nt_0}$$ (10.9)

the mass balance equation can be written as

$$\frac{\partial C}{\partial \tau} + \frac{1-\varepsilon}{\varepsilon}\frac{\partial q}{\partial \tau} + \frac{\partial C}{\partial x} = \frac{1}{2N}\frac{\partial^2 C}{\partial x^2}$$ (10.10)

The band profile obtained as a numerical solution of Equation 10.10 gives the concentration distribution as a function of the reduced time at the column end, i.e., at location $x = 1$, regardless of the column length. The band profile depends only on the column efficiency, the boundary conditions, the phase ratio, and the sample size (which is part of the boundary conditions). The mobile phase velocity has been eliminated from the mass balance equation and the apparent axial dispersion coefficient has been replaced by the plate number.

As long as the boundary and initial conditions remain unchanged, the band profiles on the reduced time and length scale depend only on the column efficiency. The conventional boundary and initial conditions for all modes of chromatography state that (1) the column is equilibrated with the mobile phase prior to the beginning of the separation (2) the sample is then injected as a rectangular pulse and (3) the separation proceeds as required by the specific mode selected. The amount of sample injected is determined by the volume and the concentration of the feed injected. As long as we avoid serious volume overload, the actual values of these two parameters are immaterial. Only their product, i.e., the amount injected, will influence the band profile.

Accordingly, two major parameters affect the band profiles in nonlinear chromatography: the column efficiency, and the amount of sample injected or loading factor. Parameter F (phase ratio) depends on the total porosity of the packing and cannot be changed in practice.

The degree of agreement between the ideal and the equilibrium–dispersive models depends on the value of the effective loading factor, a dimensionless number which is also known as the Shirazi number [1]:

$$m = NL_f \left(\frac{k}{k+1}\right)^2$$ (10.11)

where
   k is the retention factor at infinite dilution
   L_f is the loading factor, i.e., the ratio of the total amount of the components in the sample to the
      column saturation capacity

The analysis of Equation 10.10 indicated that the loading factor and the plate number are the essential parameters—besides the type of the isotherm—to control the band profile arising from column overload. The Shirazi number connects those two terms and can be regarded as a universal quantity that controls the band profile.

The band profiles calculated with the two models agree rather well provided that $m > 35$.

The Shirazi number combines parameters of fundamentally different origins. Therefore, it combines the effects of thermodynamics and kinetics on the sharpening of the bands. Figure 10.2 illustrates the effect of overload on the shape of the band at two different column efficiencies.

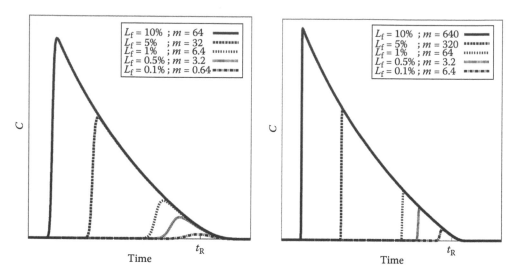

FIGURE 10.2   The effect of column overload on the shape of the band profiles. The number of theoretical plates are $N = 1,000$ and $N = 10,000$, respectively.

10.4   KINETIC MODELS

A number of kinetic models of various degree of complexity have been used in chromatography. In linear chromatography, all these models have an analytical solution in the Laplace domain. The Laplace-domain solution makes rather simple the calculation of the moments of chromatographic peaks; thus, the retention time, the peak width, its number of theoretical plates, the peak asymmetry, and other chromatographic parameters of interest can be calculated using algebraic expressions. The direct, analytical inverse Laplace transform of the solution of these models usually can only be calculated after substantial simplifications. Numerically, however, the peak profile can simply be calculated from the analytical solution in the Laplace domain.

The most detailed model is the general rate model [2]. This model has been studied by several authors [3]. The moments calculated from the general rate model allow the derivation of a most detailed plate height equation for both particulate and monolith columns [4].

The lumped pore model [5] is the result of a different simplification of the general rate model. It seems to be more detailed than the lumped kinetic model because, similarly to the general rate model, it takes into consideration the difference between the mobile phase percolating through the interparticle space (or the through-pores of a rod column) and the mobile phase stagnant in the stationary phase pores. However, it ignores the radial concentration gradient within a stationary phase particle.

The various kinetic models have been intensively used to model nonlinear separations. Their characteristics and behavior have been compared in various studies [6–8].

10.4.1   GENERAL RATE MODEL

The general rate model of chromatography is the most complex of all the models used in this field. In this model, it is assumed that the mobile phase percolates through the interstitial volume between stationary phase particles, diffusion takes place from this stream into the particles and inside the pores of the stationary phase particles, where the mobile phase is stagnant, and adsorption–desorption takes place between the stagnant mobile phase within the pores and the adsorbent surface.

Several mass balance equations are written for the kinetics of each step as the analyte is passing through the porous stationary phase. For the bulk mobile phase in the interstitial volume, the following differential mass balance equation is written

$$\frac{\partial C}{\partial t} + u_h \frac{\partial C}{\partial z} + \frac{3}{r_p} FN_0 = D_L \frac{\partial^2 C}{\partial z^2} \tag{10.12}$$

where
  C is the concentration of the analyte in the bulk mobile phase
  u_h is the interstitial velocity of the mobile phase
  r_p is the average radius of the stationary phase particles
  D_L is the axial dispersion coefficient that takes into account the mobile phase dispersion in the interstitial volume

D_L is governed by the molecular diffusion process and by eddy dispersion. F is defined as $F = (1 - \varepsilon_e)/\varepsilon_e$ where ε_e is the external porosity. N_0 is the mass flux of the analyte from the mobile phase to the external surface of the stationary phase which can be expressed as

$$N_0 = k_f \left[C - C_p(r_p) \right] = D_p \frac{\partial C_p}{\partial r} \bigg|_{r=r_p} \tag{10.13}$$

where
  C_p is the concentration of the analyte within the pores
  k_f is the external mass transfer coefficient
  r is the radial distance in the particle
  D_p is the pore diffusivity coefficient

For the diffusion of the analyte within the pores of the stationary phase particles, the following mass balance equation is written:

$$D_p \left(\frac{\partial^2 C_p}{\partial r^2} + \frac{2}{r} \frac{\partial C_p}{\partial r} \right) - N_p = \varepsilon_p \frac{\partial C_p}{\partial t} \tag{10.14}$$

where
  ε_p is the internal porosity of the stationary phase
  N_p is the mass flux of the analyte from the stagnant mobile phase in the mesopore space to the surface of the stationary phase:

$$N_p = (1 - \varepsilon_p) \frac{\partial q}{\partial t} = (1 - \varepsilon_p) k_a \left(C_p - \frac{q}{K_a} \right) \tag{10.15}$$

where
  k_a is the adsorption rate constant
  K_a is the equilibrium constant for adsorption

10.4.2 Lumped Pore Diffusion Model

The lumped pore model (often referred to as the POR model) was derived from the general rate model by ignoring two details of this model [5]. The first assumption made is that the adsorption–desorption process is very fast. The second assumption is that diffusion in the stagnant mobile phase is also very fast. This latter assumption leads to the consequence that there is no radial concentration gradient within a particle. Instead of the actual radial concentration profile across the porous particle, the model considers simply its average value.

Thus, two mass balance equations are written in the lumped pore diffusion model for the two different fractions of the mobile phase, the one that percolates through the network of macropores between the particles of the packing material and the one that is stagnant inside the pores of the particles:

$$\frac{\partial C}{\partial t} + u_h \frac{\partial C}{\partial z} + \frac{3}{r_p} F k_f'(C - C_p) = D_L \frac{\partial^2 C}{\partial z^2} \tag{10.16}$$

and

$$\varepsilon_p \frac{\partial C_p}{\partial t}(1 - \varepsilon_p)\frac{\partial q}{\partial t} = k_f' \frac{3}{r_p}(C - C_p) \tag{10.17}$$

where k_f' is the mass transfer rate constant. It can be related to the parameters of the more detailed general rate model as

$$\frac{1}{k_f'} = \frac{1}{k_f} + \frac{r_p}{5 D_p} \tag{10.18}$$

10.4.3 LUMPED KINETIC MODEL

The lumped kinetic model can be obtained with further simplifications from the lumped pore model. We now ignore the presence of the intraparticle pores in which the mobile phase is stagnant. Thus, $\varepsilon_p = 0$ and the external porosity ε_e becomes identical to the total bed porosity ε. The mobile phase velocity in this model is the linear mobile phase velocity rather than the interstitial velocity: $u = L/t_0$. There is now a single mass balance equation that is written in the same form as Equation 10.8.

In the lumped kinetic model, various kinetic equations may describe the relationship between the mobile phase and stationary phase concentrations. The transport-dispersive model, for instance, is a linear film driving force model in which a first-order kinetics is assumed in the following form:

$$\frac{\partial q}{\partial t} = k_f''\left(C - \frac{q}{K_a}\right) \tag{10.19}$$

Note that this kinetic equation is rather similar to Equation 10.15. The major difference between Equations 10.15 and 10.19 is that the general rate and the lumped pore models assume that adsorption takes place from the stagnant mobile phase within the pores, while the lumped kinetic model assumes that the mobile phase concentration is the same in the pores and between the particles.

The connection of the overall mass transfer coefficient of the lumped kinetic and the parameters of the general rate model is

$$\frac{1}{k_f''} = \frac{r_p}{3 k_f} + \frac{r_p^2}{15 D_p} \tag{10.20}$$

10.4.4 THOMAS KINETIC MODEL OF NONLINEAR CHROMATOGRAPHY

The solution of the simplest kinetic model for nonlinear chromatography the Thomas model [9] can be calculated analytically. The Thomas model entirely ignores the axial dispersion, i.e., $D_a = 0$ in the mass balance equation (Equation 10.8). For the finite rate of adsorption/desorption, the following second-order Langmuir kinetics is assumed

$$\frac{\partial q}{\partial t} = k_a(q_s - q)C - k_d q \tag{10.21}$$

where

k_a and k_d are the rate constants for adsorption and desorption, respectively

q_s is the specific saturation capacity of the stationary phase

For the case of a Dirac impulse injection, the solution of the Thomas model was given by Wade et al. [10]:

$$C = \frac{C_0}{\sqrt{\tau}} \frac{1 - e^{N_r L_f}}{N_r L_f} \frac{N_r I_1(N_r \sqrt{\tau}) e^{N_r(\tau+1)}}{1 - T(N_r, N_r \tau)(1 - e^{N_r L_f})} \tag{10.22}$$

where

$$\tau = \frac{t - t_0}{t_R - t_0} \tag{10.23}$$

is the dimensionless time, C_0 is the concentration of the solute injected, L_f is the loading factor, $N_r = k_a t_0$ is the number of mass transfer units, and

$$T(u, v) = e^{-v} \int_0^u e^{-t} I_0\left(\sqrt{2vt}\right) dt \tag{10.24}$$

is a Bessel function integral. $I_0(x)$ and $I_1(x)$ are the modified Bessel functions of the first kind. Since the Thomas model entirely ignores the dispersion processes occurring in the mobile phase and it attributes the band broadening solely to kinetic phenomena, it can be used to model chromatographic peaks when the kinetics of adsorption and desorption is extremely slow, e.g., in affinity chromatography.

10.5 BAND PROPAGATION IN NONLINEAR CHROMATOGRAPHY

When the concentration of the injected sample is not infinitesimally small but finite, the migration velocity of a given concentration zone, u_z, is not constant but is a function of the mobile phase concentration, C [1]:

$$u_z = \frac{u}{1 + F\dfrac{dq}{dC}} \tag{10.25}$$

where F is the phase ratio, $F = (1 - \varepsilon)/\varepsilon$.

Thus, the migration velocity depends on the mobile phase concentration C and on the local curvature of the isotherm, dq/dC. Therefore, the migration velocity of a given concentration zone is constant. Accordingly, for any Langmuir-type isotherm the rear side of the band spreads because the migration velocity of a concentration zone increases with increasing concentration. Furthermore, the front becomes steeper and steeper until a concentration discontinuity or shock develops. The concentration shock migrates at a velocity given by

$$u_z = \frac{u}{1 + F\dfrac{\Delta q}{\Delta C}} \tag{10.26}$$

where $\Delta q/\Delta C$ is the slope of the chord of the isotherm drawn to point (C, q) from the origin. A comparison of Equations 10.25 and 10.26 shows that—for a Langmuir-type isotherm—the migration velocity of a shock of concentration C is smaller than the migration velocity of concentration C on the rear part of the zone. Therefore, the shock continuously erodes and its velocity decreases.

10.6 BAND PROFILES IN MULTICOMPONENT SEPARATIONS

When the sample introduced into the column is composed of a number of components, the different components of the sample compete for adsorption on the surface of the stationary phase. That competition will affect the individual band profiles and the resulting band shape will depend on several factors, such as the isotherm, the separation factor, the loading factor, the relative concentration of the components, elution order, etc. [1,11].

When a major target compound and a minor impurity is to be separated, the minor component will have little or no effect at all on the band profile of the major component. The band profile of the minor impurity, however, is strongly affected by the presence of the major component.

Figure 10.3a presents a scenario where the relative composition of the mixture is 4:1, and the minor component is more strongly adsorbed than the major component. The solid lines give the individual band profiles of the components of the binary mixture. The dashed lines indicate the chromatograms we would obtain were the two compounds injected separately on the column. In the binary separation, the minor component exhibits a long plateau on the rear part of the elution profile. The second component is less retained in the presence of a large concentration of the major component than when it is alone. The reason for this is that during the separation process, in a significant part of the column, the two components travel together, and the presence of the major component prevents the minor component's adsorption. This phenomenon is known as the *tag-along* effect.

In contrast, in Figure 10.3c, the minor component is less retained than the major one. The component present in large excess in the sample displaces the other component out of the stationary phase, and the *displacement effect* of the first component, at the front of the second component band is observed. The first component elutes from the column sooner as if it were injected separately on the column.

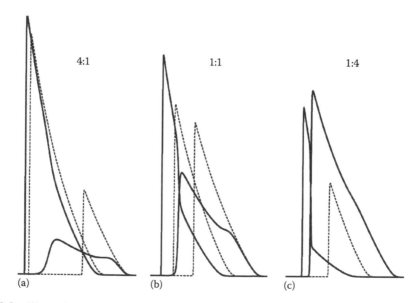

FIGURE 10.3 Illustration of the displacement and the tag-along effects.

Figure 10.3b shows that in intermediate cases, when the concentration of the two components is similar, both the displacement and the tag-along effects appear simultaneously and the band profiles of both components are strongly influenced by the presence of the other component.

10.7 ADSORPTION AT THE LIQUID–SOLID INTERFACE

Purification of liquids by solid adsorbents has been practiced from early times, and even today, with the tremendous expansion of chemical industry, the removal of undesired components from liquids is the main practical application of adsorption phenomena at solid–liquid interfaces. Both the range of substances to be purified by adsorption and the range of appropriate adsorbents has increased immensely, and thus the number of experimental investigations has also.

10.7.1 EARLY EXPERIMENTS

In the early experiments, when gas–solid adsorption was studied by some 200 years ago, charcoal was the most widely used adsorbent [12]. These investigations, generally limited to adsorption from dilute solutions, gave adsorption isotherms of the form shown in Figure 10.4. These isotherms could be fitted by the equation:

$$\frac{x}{m} = \alpha c^{1/n} \tag{10.27}$$

where

    x is the weight of adsorbate taken up by a weight m of solid
    c is the concentration of the solution at equilibrium
    α and n are constants

Equation 10.27 is generally known as Freundlich equation. Equation 10.27 with concentration replaced by pressure was also used to describe the adsorption isotherms of gases on solids, suggesting the incorrect idea that adsorption from solution by a solid could be paralleled with gas or vapor adsorption on the same adsorbents. Whereas in some cases the restriction to dilute solutions was imposed by the solubility of solids (e.g., benzoic acid in water or stearic acid in benzene) it was not imposed on the investigation of mixtures of completely miscible liquids, e.g., acetic acid in water.

Acetic acid and water are miscible in all proportions at room temperature. The isotherm for adsorption on charcoal, if plotted for the whole range of concentrations, has the form shown in Figure 10.5 (left), i.e., it passes through a maximum. In other cases (e.g., ethyl alcohol and benzene),

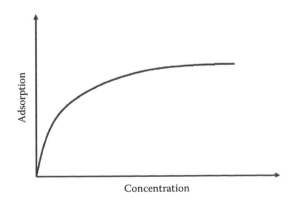

FIGURE 10.4 Adsorption isotherm from dilute solutions.

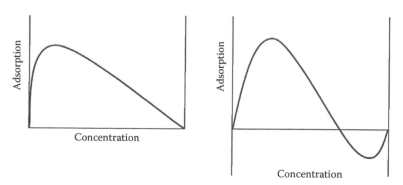

FIGURE 10.5   Isotherm of binary liquid mixtures for the whole range of concentrations.

the isotherm may have a so-called S-shape, as shown in Figure 10.5 (right), with both a maximum and a minimum.

The Freundlich equation proved to be applicable to the adsorption of liquids with only limited ranges of concentration. It was replaced by the Langmuir equation (see later on) and others which had a theoretical basis in the kinetic theory of gases. It is clear that neither the Freundlich nor the Langmuir equation can describe isotherms of the shape shown in Figure 10.5.

The data which are plotted as isotherms in the case of adsorption from liquid solutions on solid adsorbents are different in nature from those of gas (or vapor) adsorption on the same adsorbents. In fact, while the isotherm for adsorption of a single gas by a solid represents directly the quantity (weight or volume under standard conditions) of gas adsorbed per unit weight of the solid, the experimental measurement in adsorption from solution is the change in concentration of the solution which results from adsorption. The fact that a change in concentration is measured emphasizes that there are at least two components in the solution [13].

It was formerly assumed that the change in concentration was a measure of the extent to which one component ("the solute") had been adsorbed. The extent of adsorption was given by multiplying the change in concentration by the weight of solution used. The second component ("the solvent") was merely regarded as a "space in which the solute has play" [14]. Only for systems in which the solute has very limited solubility in the solvent, however, this concept may prove to be approximately valid. If two components are completely miscible, indeed, neither can be regarded as solvent or a solute over the whole range of concentrations. It thus becomes important to recognize that each component of the mixture may be adsorbed. Since there are not vacancies in the surface solution and the bulk solution, the number of molecules of a given component in the surface phase may increase only by displacing an equivalent number of molecules of other components from this phase to the bulk phase.

In light of these considerations, a qualitative interpretation of the isotherm courses shown in Figure 10.5 can be given. Over the first part of the concentration range, one component is preferentially adsorbed with respect to the other. This means that, at equilibrium, it is present in the adsorbed layer in greater proportion than in the bulk liquid. "Negative" adsorption of one component thus means preferential adsorption of the other component. For completely miscible liquids, the isotherm must fall to zero at each end of the concentration range, as no change in composition at these points (i.e., of the pure liquids) is possible.

10.7.2   EXCESS AND ABSOLUTE ISOTHERM

Two concepts of adsorption as applied to mixtures have to be recognized. The first is given by the experimental measurements described before, that is the determination of the change in concentration of the solution in question. Chromatographic methods suitable for measuring this change will be described later on. This kind of adsorption is referred to as preferential adsorption of a given component. Essentially, it corresponds to the expression "surface excess" in use to describe the same phenomena at liquid–vapor interface [15]. It is a measure of the extent to which the bulk liquid is impoverished with respect to one component because the surface is correspondingly enriched. By expressing the concentrations of the two components constituting the mixture as mole fractions, this change will be

$$\Delta x_1 = x_1^0 - x_1 \qquad (10.28)$$

where

    x_1^0 is the concentration in the original solution

    x_1 is the concentration after equilibration with the solid adsorbent of the preferentially adsorbed component 1

Δx_1 depends on the relative amounts of original solution and adsorbent used in the experiment. Letting these amounts be n^0 moles of solution and m grams of adsorbent, respectively, the preferential adsorption, which can be positive or negative, is defined as

$$\frac{n^0 \Delta x_1}{m} \qquad (10.29)$$

and it is a fact corroborated by experiment, that in the case of binary solutions, this specific adsorption is an unequivocal function of the equilibrium mole fraction x_1 only, in any given kind of system, at a given temperature (and pressure). The function

$$n_1^e = \frac{n^0 \Delta x_1}{m} \qquad (10.30)$$

represents the so-called reduced surface excess of component 1 [16,17] (also referred to as composite isotherm).

The second concept that has to be considered is that of absolute adsorption or adsorption of an individual component. This can be considered as the true adsorption isotherm for a given component that refers to the actual quantity of that component present in the adsorbed phase as opposed to its relative excess relative to the bulk liquid. It is a surface concentration. From a practical point of view, the main interest lies in resolving the composite isotherm into individual isotherms. To do this, the introduction of the concept of a Gibbs dividing surface is necessary. Figure 10.6 shows the concept of the surface phase model.

The composition profile is approximated by a step profile, with a uniform composition x_1^a in the surface layer ($0 < z < L$) and the composition of the bulk phase x_1 at $z > L$. It is assumed that the total amount of liquid can be divided into two parts with the first constituting the homogeneous bulk phase (mole numbers in it $n^b = n_1^b + n_2^b$) and the remainder standing under the influence of the forces emanating from the solid surface causing adsorption (mole numbers, referred to unit mass of adsorbent, $n^a = n_1^a + n_2^a$ the superscript a referring to adsorption) [17]. Simple mass balance considerations lead to the following expressions [12]:

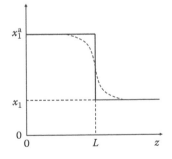

FIGURE 10.6 Surface phase model. The real mole fraction profile $x_1(z)$ is replaced by a step function (solid lines). L is the nominal thickness of the surface phase.

$$n^0 = n_1^{\text{b}} + n_2^{\text{b}} + n_1^{\text{a}}m + n_2^{\text{a}}m \tag{10.31}$$

Then, the concentration in the original solution x_1^0 and the concentration after equilibration with the solid adsorbent for the first and the second component, x_1 and x_2, respectively, are

$$x_1^0 = \frac{n_1^{\text{b}} + n_1^{\text{a}}m}{n^0} \tag{10.32}$$

$$x_1 = \frac{n_1^{\text{b}}}{n_1^{\text{b}} + n_2^{\text{b}}} \tag{10.33}$$

$$x_2 = \frac{n_2^{\text{b}}}{n_1^{\text{b}} + n_2^{\text{b}}} \tag{10.34}$$

Substitution of these relations into the expression of $\Delta x_1 (= x_1^0 - x_1)$ and simple algebraic transformations leads to the usual adsorption equation for the reduced surface excess

$$n_1^{\text{e}} = n_1^{\text{a}}x_2 - n_2^{\text{a}}x_1 = n_1^{\text{a}} - n^{\text{a}}x_1 \tag{10.35}$$

On division by S, the areal reduced surface excess, $\Gamma_1^{(n)}$, is defined as

$$\Gamma_1^{(n)} = \frac{n_1^{\text{e}}}{S} = \frac{n_1^{\text{a}} - n^{\text{a}}x_1}{S} = \frac{\left(x_1^{\text{a}} - x_1\right)n^{\text{a}}}{S} \tag{10.36}$$

A careful observation of Equations 10.35 and 10.36 prompts some caution. In principle, the isotherm is expressed in terms of specific amounts "adsorbed" n_1^{a} and n_2^{a}. These are however meaningless unless one is able to assess which part of the liquid is in the adsorption sphere (viz. the position of the Gibbs' dividing plane).

Since Equation 10.35 contains two unknowns, n_1^{a} and n_2^{a}, a second equation is needed before either term can be evaluated. There are both theoretical and experimental ways to provide this second equation. The complete discussion of these aspects, however, goes beyond the scope of this chapter. Interested readers are referred to specific literature [18–20].

In some exceptional cases, however, n_1^{e} can be given a definite meaning. This is so in the case of adsorption from dilute solutions [13,21]. Incidentally, these are the conditions usually encountered in preparative chromatography. Let us assume the compound numbered 1 to be the preferentially adsorbed. If its equilibrium concentration (x_1) is negligibly small, from Equation 10.36 one obtains

$$\Gamma_1^{(n)} = \frac{n_1^{\text{a}}}{S} \tag{10.37}$$

and thus the excess amount corresponds to the amount of component 1 in the surface phase. The maximum value of $\Gamma_1^{(n)}$ satisfying this condition is what is called the adsorption capacity of the adsorbent toward the liquid in question. It is important to underline that—even though it is probably true that at this point the adsorbent surface has reached a complete monolayer coverage by the preferentially adsorbed compound—neglecting the term $n^a x_1$ in Equation 10.36 does not assure that n_2^a has to be zero. So one cannot be sure that component 2 has been displaced completely from the interface.

At higher bulk concentrations, x_1, the term $n^a x_1$ becomes appreciable and thus $\Gamma_1^{(n)}$ is less than n_1^a /S. Furthermore, when x_1 is already high, it becomes increasingly difficult to accommodate more molecules of component 1 in the surface phase; hence $\Gamma_1^{(n)}$ passes through a maximum in this region. At even higher bulk concentrations, x_1, when the surface consists almost entirely of component $1(x_1^a \to 1)$, $\Gamma_1^{(n)}$ becomes nearly a linear decreasing function of x_1 with a slope $-n^a/S$. In some cases $\Gamma_1^{(n)}$ is positive in the region of low mole fraction x_1 and negative at high x_1. At the point where $\Gamma_1^{(n)}$ passes through zero, the compositions of the bulk and the surface phase are equal $(x_1^a = x_1)$. Such a point is called surface azeotrope. It is commonly observed when the surface consists of two different types of sites, which preferentially adsorb molecules of component 1 and 2, respectively.

10.8 ADSORPTION FROM DILUTE SOLUTIONS: ISOTHERM MODELS

In order to model preparative liquid/solid chromatographic processes one needs physically realistic model isotherm equations for the adsorption from dilute solutions. In this section a short survey of the adsorption isotherm models (for single-component systems) most often employed in nonlinear chromatography will be presented. The adsorption isotherm represents all the possible interactions, attractive as well as repulsive, between the solute molecules and the stationary phase [1]. There have been many attempts to classify adsorption isotherm depending on their shape. For the sake of simplicity, we will simply differentiate between isotherm models for homogeneous surfaces on which there are not significant adsorbate–adsorbate interactions (Langmuir, Jovanović); the adsorption isotherms for heterogeneous surface on which there are not significant adsorbate–adsorbate interactions (Bilangmuir, Langmuir-Freundlich, Tóth); the isotherm models for homogenous surfaces on which there are significant adsorbate–adsorbate interactions (S-shaped, quadratic, Fowler, Brunauer, Emmett, and Teller [BET]) and, finally, the isotherm models for heterogeneous surfaces on which adsorbate–adsorbate interactions take place to a significant extent (Martire, Bi-Moreau). In the following some of the most relevant models will be briefly described. The reader is addressed to Refs. [1,12,14,19,21,22] for a comprehensive treatment of adsorption isotherms.

10.8.1 LINEAR ISOTHERM

In analytical applications where the concentrations of analytes in mobile phase are very small the adsorption isotherm is usually considered to be linear, i.e., for each component involved in the separation the relationship between mobile (C) and stationary phase (q) concentrations is linear:

$$q = aC \tag{10.38}$$

The proportionality constant, a, is generally referred to as the Henry's constant of adsorption. Writing the mass balance equation (Equation 10.8) or the expression for the adsorption isotherm in terms of stationary and mobile phase concentrations corresponds to implicitly introduce into the model the mathematical boundary in the interfacial system–Gibbs dividing plane. Even though, as it was formerly mentioned, the definition of the position of the Gibbs surface is a fundamental issue for the physicochemical interpretation of the adsorption data, it is important to underline that for the construction and the solution of the mass balance equation the introduction of the Gibbs dividing plane is unnecessary, provided that excess adsorption concepts are used.

The assumption of linear chromatography reflects a condition in which there is no competition between analytes for the adsorption on the stationary phase. Molecules behave independently of each other as the "free" surface available for adsorption is much larger compared to molecular dimensions. As it was described before, even for systems composed by only one solute dissolved in a solvent (typical conditions in liquid chromatography), this is a very simplified description. In fact, it is the competitive nature of the adsorption that explains the origin of system peaks even in linear chromatography [1]. When some of the components of the mobile phase are adsorbed by the stationary phase, the chromatograms of even small samples are not predicted correctly by linear chromatography. Then the sample injection causes a perturbation of their equilibrium, which is not properly described by linear chromatography unless the equilibrium isotherms of these adsorbed components of the mobile phase are linear.

The assumption of linear chromatography fails in most preparative applications. At high concentrations, the molecules of the various components of the feed and the mobile phase compete for the adsorption on an adsorbent surface with finite capacity. The problem of relating the stationary phase concentration of a component to the mobile phase concentration of the entire component in mobile phase is complex. In most cases, however, it suffices to take in consideration only a few other species to calculate the concentration of one of the components in the stationary phase at equilibrium. In order to model nonlinear chromatography, one needs physically realistic model isotherm equations for the adsorption from dilute solutions.

10.8.2 LANGMUIR ISOTHERM

This isotherm assumes a homogeneous adsorption surface, i.e., the existence of only one type of adsorption sites for the adsorption of molecules. Even though this assumption is far from being realistic, this model is the one most frequently used to account for adsorption data in chromatography. Its mathematical formulation is

$$q = \frac{q_s bC}{1+bC} \tag{10.39}$$

where
q_s is the so-called saturation capacity (that is the maximum concentration that can be adsorbed on the stationary phase at a given temperature)
b is the adsorption constant

The slope of the isotherm at infinite dilution defines the Henry's constant of the adsorption, $a = q_s b$. In spite of its simplicity, experience shows that the Langmuir equation is an excellent approximation for single-component adsorption equilibrium in liquid–solid chromatography. Examples taken from literature [23–25] are given in Figures 10.7 through 10.10. Figures 10.7 and 10.8 show the adsorption isotherms measured for various small molecules on C_{18}-bonded silica. In particular, Figure 10.7 represents the adsorption isotherms of aromatic derivatives while adsorption isotherms of amino acids as a function of the chloride concentration in ion exchange chromatography are reported in Figure 10.8. Other examples include the adsorption equilibria of proteins represented in Figures 10.9 and 10.10.

10.8.3 BILANGMUIR ISOTHERM

This model is directly derived from the Langmuir isotherm. It assumes that the adsorbent surface consists of two different types of independent adsorption sites. Under this assumption, the adsorption energy distribution can be modeled by a bimodal discrete probability density function, where two spikes (delta-Dirac functions) are located at the average adsorption energy of the two kinds of sites, respectively. The equation of the Bilangmuir isotherm is

$$q = \frac{q_{s,1} b_1 C}{1+b_1 C} + \frac{q_{s,2} b_2 C}{1+b_2 C} \tag{10.40}$$

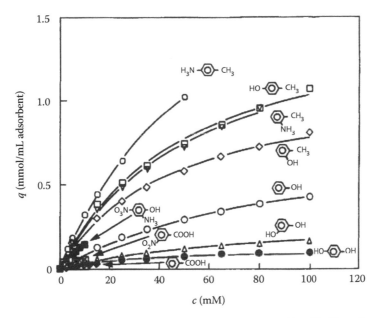

FIGURE 10.7 Single-solute isotherms on octadecyl-silica, as measured by FA. Experimental data points and the Langmuir correlations are shown by symbols and curves, respectively. Studied compounds were: p-toluidine, p-cresol, o-toluidine, o-cresol, phenol, 2-amino-4-nitrophenol, resorcinol, m-nitrobenzoic acid, hydroquinone, and benzoic acid. Conditions: column, 50.0×1.18 mm for phenol, resorcinol, and hydroquinone, 40.5×1.18 mm for the remaining compounds; mobile phase, 0.20 M sodium phosphate buffer (pH 6.3) for benzoic acids, water for the remaining compounds; flow rate, 70 μL/min; temperature, 25°C. (Reprinted from Jacobson, J. et al., *J. Chromatogr.*, 316, 53, 1984. With permission from Elsevier.)

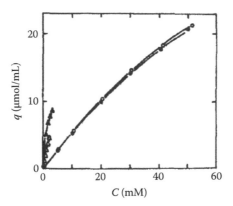

FIGURE 10.8 Langmuir adsorption isotherms of nucleosides on Spherisorb ODS-2 from 25 mM phosphate buffer (pH 7.0) at 25°C. (▲) Adenosine, (△) guanosine, (○) cytidine, (●) uridine. (Reprinted from Huang, J.X. and Horváth, Cs., *J. Chromatogr.*, 406, 275, 1987. With permission from Elsevier.)

where
  $q_{s,i}$ is the saturation capacity of ith site ($i = 1,2$)
  b_i is the adsorption constant on the ith site (whose Henry's constant is $a_i = b_i q_{s,i}$)

The other symbols have the same meaning as in Equations 10.38 and 10.39.

  Although the Bilangmuir isotherm is an ideal model (in the sense that real surfaces with exactly the characteristics described by the Bilangmuir model do not exist), it is often successful in describing the adsorption of enantiomers on chiral stationary phase. The reason is that, because of their

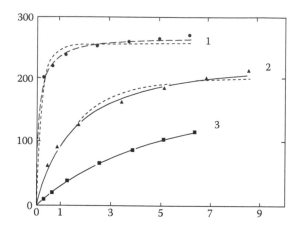

FIGURE 10.9 Experimental isotherms of Lysozyme in (1) 0.0 M, (2) 0.05 M, (3) 0.1 M $(NH_4)_2SO_4$ and 25 mM NaH_2PO_4. Best fit (- - -) Jovanovic and (—) Langmuir model. WCX-300H DuPont, 25°C. (Reproduced from Huang, J.-X. and Horváth, Cs., *J. Chromatogr.*, 406, 285, 1987. With permission from Elsevier.)

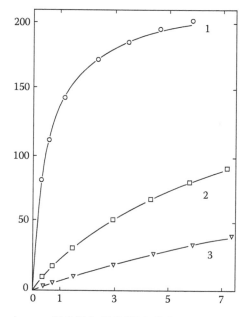

FIGURE 10.10 α-Chymotrypsinogen; (1) 0.0 M; (2) 0.05 M; (3) 0.1 M in the same buffer, on SCX-300. (Reproduced from Huang, J.-X. and Horváth, Cs., *J. Chromatogr.*, 406, 285, 1987. With permission from Elsevier.)

specific physicochemical characteristics, chiral systems fit most of the theoretical assumptions under which the Bilangmuir model holds. When the Bilangmuir isotherm is used to model the adsorption data of enantiomers, it is assumed that one type of site consists of the enantioselective sites (the stronger type) while the second type, the weaker one, accounts for all the other possible, nonselective interactions. Figure 10.11, for instance, shows the comparison between experimental data and the Bilangmuir isotherm for the adsorption of the enantiomers of *N*-benzoyl derivative of alanine adsorbed on bovine serum albumin immobilized on silica [26].

The adsorption isotherms of three basic drugs, the anxiolytic buspirone hydrochloride, the antidepressant doxepin hydrochloride, and the Ca^{2+} blocker diltiazem hydrochloride, were fitted to the Bilangmuir model, as shown in Figure 10.12 [27]. This finding suggested that the adsorption took place on two different types of sites and that there were two different adsorption mechanisms.

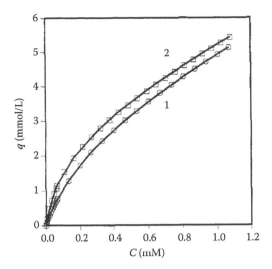

FIGURE 10.11 Adsorption isotherms of N-benzoyl-L-alanine (1) and N-benzoyl-D-alanine (2). Experimental data fitted to the Bilangmuir model. (Reproduced from Jacobson, S. et al., *J. Am. Chem. Soc.*, 112, 6492, 1990. With permission from American Chemical Society.)

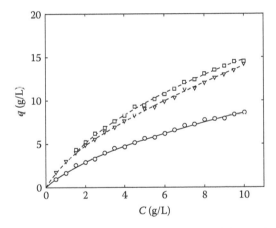

FIGURE 10.12 Adsorption data of buspirone (○), doxepin (∇), and diltiazem (□) and best Bilangmuir isotherms (continuous and dotted lines). Mobile phase is acetonitrile:buffer = 35:65; buffer is 0.1 M phosphate, pH 3.0, $T = 25°C$. (Reproduced from Quinones, I. et al., *J. Chromatogr. A*, 877, 1, 2000. With permission from Elsevier.)

Hydrophobic interactions took place on the low-energy sites (alkyl chains) and ion-exchange interactions on the high-energy sites (most likely, acidic silanols buried under the alkyl layer).

10.8.4 TÓTH ISOTHERM

As many other models, even the Tóth isotherm was originally derived for the study of gas–solid equilibria. However, it has been extended to the description of liquid–solid system. This isotherm assumes a continuous and wide adsorption energy distribution. The equation of the Tóth isotherm is

$$q = \frac{q_s bC}{(1 + (bC)^\nu)^{1/\nu}}$$ (10.41)

where ν is the heterogeneity parameter. This adsorption model has been used to account for the adsorption behavior of proteins, enantiomeric compounds, basic drugs, etc. This is demonstrated,

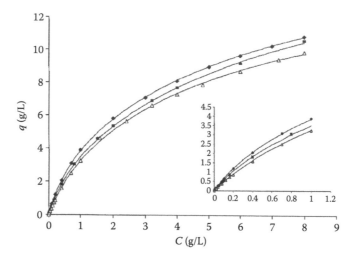

FIGURE 10.13 Experimental adsorption data (symbols) and the best Tóth isotherm (solid line) for porcine insulin, human insulin Lispro. (Reproduced from Liu, X. et al., *Biotechnol. Prog.*, 18, 796, 2002. With permission from John Wiley & Sons.)

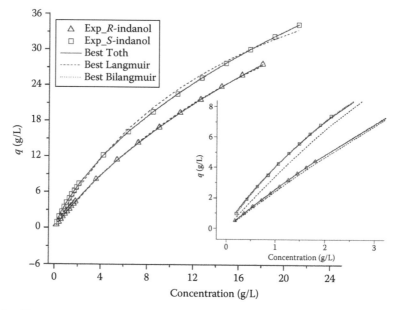

FIGURE 10.14 Single-component experimental isotherms for *R*- and *S*-1-indanol on cellulose tribenzoate fitted to three different models. The inset shows the low-concentration data. (Reproduced from Zhou, D., *Chem. Eng. Sci.*, 58, 3257, 2003. With permission from Elsevier.)

for instance, in Figure 10.13 where the adsorption isotherms for Lispro, human, and porcine insulin on C_{18} are plotted [28]. Figure 10.14 shows the comparison between empirical data and best Tóth, Langmuir, and Bilangmuir isotherms for the adsorption of the enantiomers of 1-indanol on cellulose tribenzoate from a solution of 2-propanol and *n*-hexane [29].

10.8.5 S-Shaped and Quadratic Isotherm

These isotherms are sometimes referred to as anti-Langmuir models because their initial curvature is convex down. The fact that the curvature of these isotherms at the origin and at low concentrations

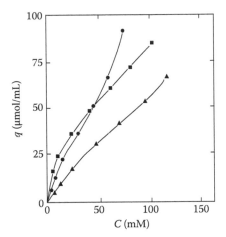

FIGURE 10.15 Adsorption isotherms of nucleotides on Spherisorb ODS-2 from 100 mM phosphate buffer (pH 7.0) at 25°C. (●) Guanosine 2′,3′-cyclic monophosphate, (■) adenosine 2′,3′-cyclic monophosphate and (▲) adenosine monophosphate. (Reprinted from Huang, J.X. and Horváth, Cs., *J. Chromatogr.*, 406, 275, 1987. With permission from Elsevier.)

is concave upward indicates that the amount adsorbed at equilibrium increases more rapidly than the concentration in the mobile phase. Such an effect results usually from strong adsorbate–adsorbate interactions, e.g., lateral interactions between hydrocarbon chains, stacking of nucleotides or large, planar, polycyclic aromatic compounds. Simple statistical thermodynamic models of adsorption suggest that the equilibrium isotherm should be written as the ratio of two polynomials of the same degree:

$$q = q_s \frac{b_1 C + 2b_2 C^2 + \cdots + n b_n C^n}{1 + b_1 C + b_2 C^2 + \cdots + b_n C^n} \tag{10.42}$$

the Langmuir isotherm is the first-order isotherm predicted by statistical thermodynamics. The second-order isotherm, obtained with $n = 2$, is called the quadratic isotherm. From Equation 10.42 it follows that the limit for infinite mobile phase concentration is $q = n q_s$. This is due to the model that considers that each site on the surface can accommodate n molecules and that there are q_s such sites on the surface. Figure 10.15 shows the adsorption isotherms for three nucleotides on Spherisorb ODS-2 [24].

10.8.6 Moreau Isotherm

The so-called Moreau isotherm is the simplest model for homogeneous adsorbent surface with lateral adsorbate–adsorbate interactions. The equation is

$$q = q_s \frac{bC + I b^2 C^2}{1 + 2bC + I b^2 C^2} \tag{10.43}$$

where q_s, b, and I are the monolayer saturation capacity, the equilibrium constant at infinite dilution, and the adsorbate–adsorbate interaction parameters, respectively.

10.8.7 Jovanović Isotherm

This model was derived to account for the adsorption of a gas onto a homogeneous surface, with no adsorbate–adsorbate interactions. It accounts for the fact that adsorption and desorption of solute molecules are not instantaneous but need a certain amount of time to happen. The equation of Jovanović isotherm is

$$q = q_s \left[1 - \exp(-bC) \right] \tag{10.44}$$

where the symbols have the same meaning as in former equations.

10.9 DETERMINATION OF ADSORPTION ISOTHERM BY MEANS OF CHROMATOGRAPHY

The empirical evaluation of adsorption isotherms represents the basis for the modeling and the optimization of separations in preparative or semi-preparative chromatography. It is still a tedious, time consuming, and expensive task. This serious obstacle has limited the role of computer optimization in method development in preparative chromatography. Frontal analysis (FA) is probably the easiest and certainly the most accurate method for measuring isotherms [1,30]. From FA data, the determination of the amount adsorbed as a function of the mobile phase concentration can easily be obtained. The perturbation on a plateau (PP) technique is also frequently used for determining adsorption data [31–34]. By contrast with FA, PP does not give the amount of material adsorbed on the stationary phase as a function of the mobile phase concentration. In the single-component case, the slope of the isotherm at any mobile phase concentration can be obtained from the retention time of perturbation peaks measured on the column equilibrated with a mobile phase containing that mobile phase concentration. The isotherm is derived from this set of data, through numerical integration. The most serious drawbacks of these two techniques (FA and PP) are the large amount of material required to equilibrate the column at a given mobile phase concentration and the time required to make the systematic determinations needed in a wide enough concentration range. The cost can be prohibitive when expensive compounds (pure enantiomers, polypeptides, proteins) are investigated. Different empirical strategies have been established.

Single-component adsorption isotherms can also be evaluated from the diffuse rear boundary recorded in FA or from the rear part of overloaded band profiles. Techniques such as FA by characteristic point (FACP) and elution by characteristic point (ECP) utilize this approach. Their major advantages over FA and PP are the smaller amount of material needed and the faster determination of an isotherm. However, both FACP and ECP are only suitable for single-component systems and can be used only when the efficiency of the chromatographic system used is at least several thousand theoretical plates. This is because both methods rely on the use of an equation of the ideal model, and hence include a model error that becomes important at lower efficiencies.

Numerical procedures can also be used for determining adsorption isotherms from overloaded profiles. The so-called inverse problem of chromatography consists of calculating the adsorption isotherm from the profiles of overloaded bands [35–37].

Among the different chromatographic approaches available for measuring adsorption isotherms, three fundamental techniques, the FA, the perturbation on the plateau (PP), and, finally, the inverse method (IM), will be discussed briefly.

10.9.1 FRONTAL ANALYSIS (FA)

In FA experiments, the chromatographic column has to be equilibrated with the solution of a known concentration of the studied compound. Successive abrupt step changes of increasing concentration are performed at the column inlet, and breakthrough curves are determined. From the breakthrough curves obtained in single-component FA experiments (Figure 10.16), the values of the stationary phase concentration in equilibrium with the inlet concentration can be obtained through the following equation:

$$q = \frac{(V_R - V_0)C}{V_{ads}} \qquad (10.45)$$

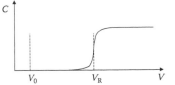

FIGURE 10.16 Determination of frontal analysis data. The breakthrough curve is the thick solid line.

where
 V_R is the retention volume of the self-sharpening shock
 V_0 is the column holdup volume (including the extra column volume)
 V_{ads} is the volume of the adsorbent material filling the column

Equation 10.45 assumes that the column was initially empty of solute and equilibrated with a stream of solution. A similar equation can be used if FA is carried out in the staircase mode.

10.9.2 PERTURBATION ON THE PLATEAU (PP)

The concept of *concentration velocity of a species* is of critical importance in nonlinear chromatography. It can be shown that a velocity is associated with each solute concentration and describes the rate of propagation of that concentration along the column. This concentration velocity gives or describes the propagation of a perturbation or disturbance of this concentration. This velocity is different from the velocity at which the actual molecules move along the column. The distinction is not commonly made in chromatography because both velocities are equal when the concentration is close to 0, i.e., under linear conditions. Neglecting all kinetic effects except the one due to convection (i.e., in ideal chromatography), Equation 10.8 becomes

$$\frac{\partial C}{\partial t} + F\frac{\partial q}{\partial t} + u\frac{\partial C}{\partial z} = 0 \qquad (10.46)$$

This equation states that the velocity, $u_{C'}$, at which a given concentration C' travels through the column is

$$u_{C'} = \frac{1}{1 + F(dq/dC)|_{C=C'}} \qquad (10.47)$$

where dq/dC is the slope of the isotherm at $C = C'$. Note that Equations 10.25 and 10.47 are identical. Equation 10.47 can be expressed in terms of the retention time of the concentration C', $t_R(C')$, as

$$t_R(C') = t_0\left(1 + F\left.\frac{dq}{dC}\right|_{C=C'}\right) \qquad (10.48)$$

where t_0 is the holdup time. By simply multiplying Equation 10.48 and the mobile phase flow rate, we obtain the retention volume of concentration C', $V_R(C')$:

$$V_R(C') = V_0\left(1 + F\left.\frac{dq}{dC}\right|_{C=C'}\right) \qquad (10.49)$$

Equations 10.47 through 10.49 constitute the basis for the method of isotherm determination by the perturbation method. The retention time of a small injection of a solute in a column equilibrated

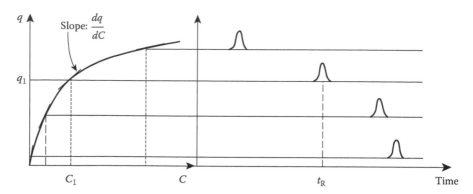

FIGURE 10.17 Relationship between the isotherm, its slope and the retention time of perturbation peak. (Reproduced from Tondeur, D. et al., *Chem. Eng. Sci.*, 51, 3781, 1996. With permission from Elsevier.)

with the pure mobile phase (typical condition of analytical chromatography), gives the retention time under linear conditions (i.e., at infinite dilution), through the following expression:

$$t_R = t_0(1 + aF) \tag{10.50}$$

where a is the initial slope of the isotherm. Similarly, the set of retention times of perturbation peaks measured in a column equilibrated with streams of the mobile phase of increasing concentrations allows the calculation of the slope of the isotherm at these different concentrations. This set of retention times constitutes the fundamental data set needed for determining the adsorption isotherm. The concepts on which the PP technique is based are represented in Figures 10.17 and 10.18.

10.9.3 Inverse Method (IM)

The IM of estimation of the best adsorption isotherm parameters is a promising alternative for measuring thermodynamic adsorption data. The algorithms employed are generally based on nonlinear least-squares methods. They calculate the best estimates of the isotherm parameters by minimizing the difference between one (or several) experimental overloaded profile and the corresponding calculated profile obtained by numerically integrating a mass balance equation (for instance, Equation 10.8), under the set of initial and boundary conditions describing the experiment performed. The starting point is the choice of an isotherm model, $q = f(C)$.

Although the examination of the shapes of experimental overloaded profiles (e.g., the fronting or tailing character of these profiles) does help in making this choice (e.g., upward or downward convex isotherm), this choice requires great attention. A good optimization program will almost always give a set of best values of the parameters: only careful statistical tests can justify the choice of one isotherm model rather than any other one. Numerous isotherm models can describe a convex upward isotherm. The "best" model is usually chosen according to its ability to predict well the experimental overloaded profiles when used to solve Equation 10.8. This turns out to be a more important criterion than its capacity to fit to the original set of experimental isotherm data. The IM involves the following steps:

- Selecting the isotherm model and initial estimations for the values of its parameters. For instance, in the case of the Langmuir isotherm (Equation 10.39), an estimation for the a parameter can be achieved by an injection made under linear conditions (through Equation 10.50). At this point, the least-squares estimation of b—starting from an initial guess—does not present any difficulty.
- An overloaded profile is calculated by integrating Equation 10.8 and using the initial guesses for the isotherm parameters.

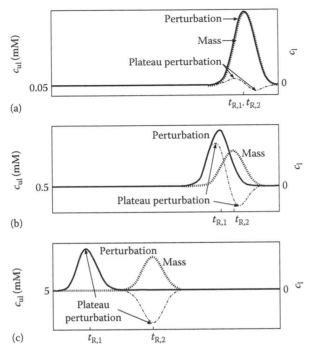

FIGURE 10.18 Illustration of the different types of possible peaks: (1) the perturbation peak, (2) the mass peak, and (3) the plateau perturbation peaks, on three concentration plateaus. A single Langmuir model was assumed with $a=2.0$ and $b=0.100$. (a) A linear plateau, $C=0.05$ mM. (b) A weakly nonlinear plateau, $C=0.5$ mM. (c) A clearly nonlinear plateau, $C=5$ mM. The chromatogram shows the result of an analytical injection of a mixture of labeled and unlabeled molecules on a concentration plateau of unlabeled molecules. The solid line shows the perturbation peak (left scale), the dashed-dotted line shows the plateau perturbation peaks (left scale), and the dotted line shows the mass peak (right scale). Here C_{ul} (mM) is the concentration of unlabeled molecules, C_l is the concentration of labeled molecules, and the x axis is time. The mean retention times, $t_{R,1}$ and $t_{R,2}$ are calculated according to Equation 10.47. (Reproduced from Samuelsson, J. et al., *Anal. Chem.*, 76, 953, 2004. With permission from American Chemical Society.)

- The sum of the squares of the differences between the empirical and the calculated band profiles is calculated.
- The isotherm parameters are changed to minimize the squares of the differences between calculated and experimental profiles.

The method appears to be particularly suitable for the study of expensive compounds and/or species available in very low amounts. Figure 10.19 shows the application of IM to the study of the adsorption equilibria of a polypeptide, nociceptin/orphanin FQ, on a C_{18} column [40]. Isotherms (Figure 10.20) were obtained by numerical procedure through the fitting of the overloaded band profiles obtained under different conditions. Adsorption data allowed for the prediction of the band profile under overloaded gradient conditions (Figure 10.21) with a minimal amount of compound consumption.

10.10 SCALE-UP AND OPTIMIZATION OF PREPARATIVE CHROMATOGRAPHY

The nonlinear nature of preparative chromatography complicates the separation process so much that the derivation of general conclusions regarding either the scaling up or the optimization is a rather difficult task. The optimization of preparative chromatography is further complicated by the fact that the choice of the objective functions in preparative chromatography is not as simple as in

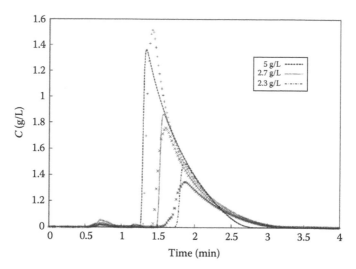

FIGURE 10.19 Comparison between empirical profiles (points) and peaks obtained via IM (lines) when a Langmuir adsorption isotherm is assumed. Mobile phase: ACN 16% (v/v) in aqueous TFA 0.1% (v/v) mixture; nociceptin-injected concentrations: 5.0, 2.7, and 1.3 g/L. (Reprinted from Marchetti, N. et al., *J. Chromatogr. A*, 1079, 162, 2005. With permission from Elsevier.)

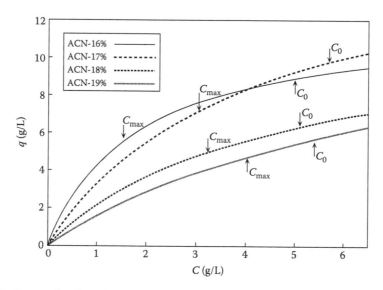

FIGURE 10.20 Langmuir adsorption isotherms under different mobile phase compositions. ACN-XX% indicates the CAN concentration (percent v/v) in aqueous TFA 0.1% (v/v) solutions. The vertical arrows represent, for any case, the maximum concentration recorded at the column outlet (C_{max}) and the maximum injected concentration (C_0). (Reprinted from Marchetti, N. et al., *J. Chromatogr. A*, 1079, 162, 2005. With permission from Elsevier.)

analytical chromatography. In industrial applications, the production cost would be the major factor to consider. However, many components of the production cost are beyond the scope of the separation process itself. Accordingly, a more straightforward approach is chosen and usually simply the production rate is maximized [1,41].

Although in several cases complex multicomponent mixtures are purified, the optimization problem can usually be reduced to the investigation of a binary mixture because there always exists a limiting impurity that elutes closest to the major component. The limiting impurity sets the constraint for the throughput, production rate, recovery yield, and other parameters.

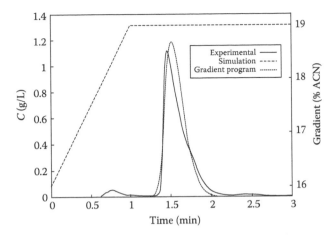

FIGURE 10.21 Comparison between experimental and simulated peaks in gradient elution (Langmuir adsorption isotherm). Injected concentration: 2.7 g/L; injected volume: 0.163 μL. Continuous line: experimental profile; dotted line: simulation. The shape of the gradient program is also represented. (Reprinted from Marchetti, N. et al., *J. Chromatogr. A*, 1079, 162, 2005. With permission from Elsevier.)

10.10.1 Objective Functions

The simplest is to maximize the productivity or the production rate itself. The production rate is the amount of purified compound produced per unit time and unit column cross/sectional area:

$$Pr = \frac{V_p C^0 Y}{\varepsilon S \Delta t_c} \tag{10.51}$$

where
V_p is the volume of sample fed into the column
C^0 is the concentration of the target compound in the sample
Y is the recovery yield
Δt_c is the cycle time, i.e., the time between two successive injections

The maximum production rate, however, often results in unacceptable recovery yields. Low recovery yield requires further processing by recycling the mixed fractions. The recovery yield at the maximum production rate strongly depends on the separation factor. In the cases of difficult separations, when the separation factor under linear conditions is around or lower than $\alpha = 1.1$, the recovery yield is not higher than 40%–60%. Even in the case of $\alpha = 1.8$, the recovery yield at the maximum production rate is only about 70%–80%. The situation is still less favorable in displacement chromatography, particularly if the component to be purified is more retained than the limiting impurity. In this case, from one side the impurity, whereas from the other side the displacer, contaminates the product.

When the production rate, Pr, and the recovery yield, Y, are simultaneously maximized via their product, $Pr \times Y$, the optimum conditions usually result in such a configuration in which the recovery yield is much higher with a small sacrifice in production rate [42].

The amount of solvent needed for the purification of a unit amount of target compound is conveniently described by the term "specific production." Thus, the minimum solvent consumption can be determined for a given purification. The amount of solvent pumped through the column during one cycle is proportional to the mobile phase flow rate and the cycle time. The amount of purified product made in one cycle is the product of the amount injected and the recovery yield. Thus, the specific production can be written as [43]

$$SP = \frac{V_p C^0 Y}{\Delta t_c F_v} = \frac{V_p C^0 Y}{\Delta t_c F_v} \tag{10.52}$$

Combining Equations 10.51 and 10.52 shows that

$$SP = \frac{t_p}{\Delta t_c} C^0 Y = \frac{Pr}{u} \tag{10.53}$$

where t_p is the injection time. This simple relationship between the production rate and specific production shows that, in order to operate chromatographic columns at minimum solvent consumption, we need to achieve a high production rate at a low mobile phase linear velocity.

10.10.2 PARAMETERS TO BE OPTIMIZED

The solvent consumption depends only on the column efficiency and the loading factor. The production rate depends on more parameters: it is also influenced by the linear velocity of the mobile phase. When all column design and operation parameters are optimized together for maximum production rate, the column should always be operated at the flow rate corresponding to the maximum inlet pressure allowed. Golshan–Shirazi and Guiochon showed that there is no independent optimum value for the column length, L, or for the particle diameter, d_p, but only the ratio d_p^2/L should be optimized. As long as d_p^2/L is unchanged, the production rate remains nearly constant [1,44].

Since it does not matter through which combination of the particle size and the column length the column efficiency is achieved, we can carry out a two-parameter optimization, the loading factor and the column efficiency (or the column length or the particle size, as most convenient). The loading factor is changed by setting the feed volume injected. As a second parameter, we can optimize the column length—which can conveniently be changed with axial compression preparative columns—keeping the value of the average particle diameter constant. From the values of the average particle size and column length, we can determine the maximum linear velocity allowed. Then, we can calculate the reduced plate height through a plate height equation, and consequently, the column efficiency.

The $Pr \times Y$ objective function can successfully be applied to the optimization of overloaded gradient elution chromatography. In Figures 10.22 and 10.23, we compare the chromatograms obtained under the optimum conditions given by the Pr and $Pr \times Y$ objective functions for the purification of the less and the more retained component of a binary mixture in overloaded isocratic separations. The chromatograms on the right-hand sides of Figures 10.22 and 10.23 were obtained for the optimum separation when $Pr \times Y$ is maximized instead of Pr. The optimum loading factor is significantly smaller for both components in the case in which $Pr \times Y$ is maximized and the recovery yield is significantly higher, whereas the production rate is only slightly smaller than in the case in which $Pr \times Y$ is maximized.

Furthermore, one should observe that the optimum conditions strongly depend on the elution order. If the target compound to be purified is less retained than the limiting impurity, a larger sample load is optimal than for the purification or the more retained compound. Therefore, the production rate is also larger if the target compound elutes first and then the limiting impurity. This is related to the band compressing effect of self-displacement or to the band spreading of the tag-along effect, as discussed earlier in this chapter.

Multiobjective optimization is an optimization strategy that overcomes the limits of a single-objective function to optimize preparative chromatography [45]. In the physical programming method of multiobjective optimization, one can specify desirable, tolerable, or undesirable ranges for each design parameter. Optimum experimental conditions are obtained, for instance, using bi-objective (production rate and recovery yield) and tri-objective (production rate, recovery yield,

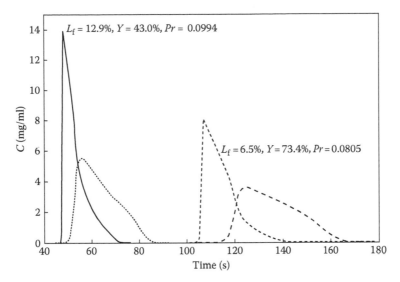

FIGURE 10.22 Optimum separations for isocratic overloaded elution for the purification of the less retained component. The production rate (left) and the product of the production rate and the recovery yield (right) were maximized, respectively, $\alpha = 1.2$; $k = 2$; $C_1^0 = C_2^0 = 100$ mg/mL. (Reproduced from Felinger, A. and Guiochon, G., *J. Chromatogr. A*, 752, 31, 1996. With permission from Elsevier.)

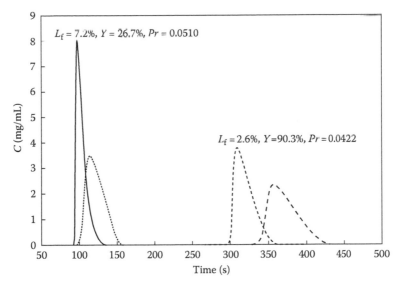

FIGURE 10.23 Same as Figure 10.22, except optimization was made for the more retained component. (Reproduced from Felinger, A. and Guiochon, G., *J. Chromatogr. A*, 752, 31, 1996. With permission from Elsevier.)

and product pool concentration) optimization. In Figure 10.24 a the Pareto frontiers* at 95% purity level and the concentration profiles of a ternary mixture are plotted using the bi-objective optimization. The Pareto frontier for the purification of the second component is below those for the other two components because it is surrounded with two impurities. In Figure 10.24b through d the optimum chromatograms for different scenarios are plotted for the three target compounds, respectively: reasonable recovery yield (Case 1) or reasonable production rate (Case 2) is the objective. Cases 11P

---

* A Pareto frontier is a limit for the solution along which the improvement in one objective can be achieved only at the cost of another objective.

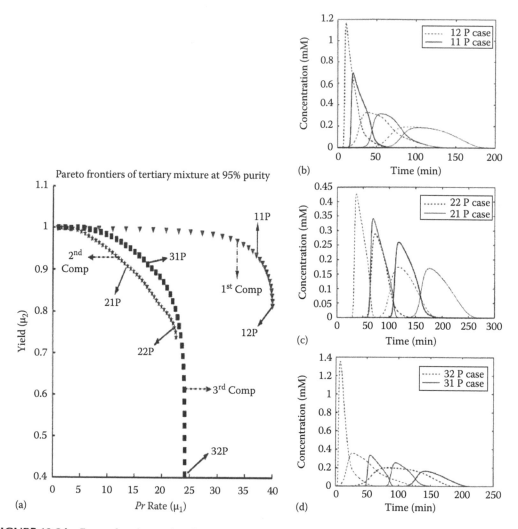

FIGURE 10.24 Pareto frontiers and outlet concentration profiles for a tertiary mixture (α-chymotrypsinogen A, ribonuclease A, and artificial component) on a 90 μm FF Sepharose stationary phase. The first component is the early-eluting ribonuclease A, the second component is the middle-eluting α-chymotrypsinogen A, and the third component is the later-eluting artificial component. The triangles, stars, and squares are the Pareto solutions for the optimization of the first, second, and later eluting components, respectively. Outlet concentration profiles for the tertiary mixture for the six cases indicated in (a) are shown in (b), (c), and (d) when the component of interest is the first, second, and the third component, respectively. Column diameter: 1.6 cm; length: 10.5 cm. Feed concentrations 0.5 mM of each component. The average separation factors are 1.55 between ribonuclease A and α-chymotrypsinogen A, and 1.35 between α-chymotrypsinogen A and an artificial-eluting component. (Reproduced from Nagrath, D. et al., *AIChE J.*, 51, 511, 2005. With permission from John Wiley & Sons.)

(high yield) and 12P (large production rate) are the optima for the production of the first component. The results of the bi-objective optimization correlate well with the results obtained with the $Pr \times Y$ objective function [45].

10.11 CONCLUSIONS

In this chapter, we provided a general overview of nonlinear chromatography. In nonlinear chromatography, the shape of the nonlinear isotherms and the molecular interactions leading to a nonlinear isotherm are of crucial importance. The migration velocity and the shape of concentration zones depend on a great number of operational parameters and on sample composition. Whether one wants to develop a preparative chromatographic purification method or one wants to better understand the nature of the interactions between solute molecules and the surface of the stationary phase, one has to leave the shelter of linear chromatography by overloading the chromatographic column. Then, the knowledge of the proper single-component or competitive isotherms is of utmost significance.

In preparative chromatography, the information provided by the isotherms and by the proper model of chromatography will help the scale-up and optimization of batch and continuous purifications with minimum use of solvents and sample. Although it is rather difficult to give straightforward instructions how to optimize a preparative separation, some general guidelines are provided in this chapter.

ACKNOWLEDGMENTS

This work has been supported by the Italian University and Scientific Research Ministry (CHEM-PROFARMA-NET, RBPR05NWWC_008), the University of Ferrara (F.A.R. 2007), grants OTKA 75717 and OTKA-NKTH NI 68863 from the Hungarian National Science Foundation, and the Hungarian–Italian Intergovernmental S&T Cooperation Program 2008–2010.

REFERENCES

1. G. Guiochon, A. Felinger, D.G. Shirazi, A.M. Katti, *Fundamentals of Preparative and Nonlinear Chromatography*, 2nd edn., Academic Press, Amsterdam, the Netherlands, 2006.
2. E. Kučera, *J. Chromatogr.*, 19, 237–248, 1965.
3. M. Suzuki, J.M. Smith, *Adv. Chromatogr.*, 13, 213–263, 1975.
4. K. Miyabe, G. Guiochon, *J. Phys. Chem. B*, 106, 8898–8909, 2002.
5. M. Morbidelli, A. Servida, G. Storti, S. Carra, *Ind. Eng. Chem. Fundam.*, 21, 123–131, 1982.
6. S. Golshan-Shirazi, G. Guiochon, *J. Chromatogr.*, 603, 1–11, 1992.
7. K. Kaczmarski, D. Antos, H. Sajonz, P. Sajonz, G. Guiochon, *J. Chromatogr. A*, 925, 1–17, 2001.
8. A. Felinger, G. Guiochon, *Chromatographia*, 60, S175–S180, 2004.
9. H.C. Thomas, *J. Am. Chem. Soc.*, 66, 1644–1665, 1944.
10. J.L. Wade, A.F. Bergold, P.W. Carr, *Anal. Chem.*, 59, 1286–1295, 1987.
11. H. Schmidt-Traub, *Preparative Chromatography of Fine Chemicals and Pharmaceutical Agents*, Wiley-VCH, Weinheim, Germany, 2005.
12. J.J. Kipling, *Adsorption from Solutions of Non-Electrolytes*, Academic Press, New York, 1965.
13. D.H. Everett (Ed.), in *Colloid Science*, Vol. 1, The Chemical Society, London, U.K., 1973, p. 49.
14. J. Tóth, *Adsorption. Theory, Modeling, and Analysis*, Marcel Dekker, New York, 2002.
15. G. Schay, in *Surface and Colloid Science*, Vol. 2, E. Matijević (Ed.), Wiley Interscience, New York, 1969, p. 155.
16. D.H. Everett, *Pure Appl. Chem.*, 31, 579–638, 1972.
17. D.H. Everett, *Pure Appl. Chem.* 58, 967–984, 1986.
18. M. Jaroniec, J. Goworek, A. Dabrowski, *Colloid Polym. Sci.*, 263, 771–777, 1985.
19. J. Oscik, *Adsorption*, Polish SciPubl and Harwood, Warsaw and Chichester, U.K., 1982.
20. A. Dabrowski, M. Jaroniec, J. Oscik, in *Surface and Colloid Science*, Vol. 14, E. Matijević (Ed.), Wiley Interscience, New York, 1987, p. 83.
21. G.H. Findenegg, in *Theoretical Advancement in Chromatography and Related Techniques*, F. Dondi, G. Guiochon (Eds.), Kluwer Academic Press, Dordrecht, the Netherlands, 1992, p. 227.
22. D.M. Ruthven, *Principles of Adsorption and Adsorption Processes*, J. Wiley & Sons, New York, 1984.
23. J. Jacobson, J. Frenz, Cs. Horváth, *J. Chromatogr.*, 316, 53–68, 1984.
24. J.X. Huang, Cs. Horváth, *J. Chromatogr.*, 406, 275–284, 1987.
25. J.-X. Huang, Cs. Horváth, *J. Chromatogr.*, 406, 285–294, 1987.
26. S. Jacobson, S. Golshan-Shirazi, G. Guiochon, *J. Am. Chem. Soc.*, 112, 6492–6498, 1990.
27. I. Quinones, A. Cavazzini, G. Guiochon, *J. Chromatogr. A*, 877, 1–11, 2000.
28. X. Liu, K. Kaczmarski, A. Cavazzini, P. Szabelski, D. Zhou, J. Van Horn, G. Guiochon, *Biotechnol. Prog.*, 18, 796–806, 2002.

29. D. Zhou, D.E. Cherrack, A. Cavazzini, K. Kaczmarski, G. Guiochon, *Chem. Eng. Sci.*, 58, 3257–3272, 2003.
30. H.L. Wang, J.L. Duda, C.J. Radke, *J. Colloid. Interface Sci.*, 66, 153–165, 1978.
31. J.H. Knox, R. Kaliszan, *J. Chromatogr.*, 349, 211–234, 1985.
32. J. Lindholm, P. Forssen, T. Fornstedt, *Anal. Chem.*, 76, 4856–4865, 2004.
33. Y.V. Kazakevich, R. LoBrutto, F. Chan, T. Patel, *J. Chromatogr. A*, 913, 75–87, 2001.
34. A. Cavazzini, G. Nadalini, V. Malanchin, V. Costa, F. Dondi, F. Gasparrini, *Anal. Chem.*, 79, 3802–3809, 2007.
35. A. Felinger, A. Cavazzini, G. Guiochon, *J. Chromatogr. A*, 986, 207–225, 2003.
36. R. Arnell, P. Forssen, T. Fornstedt, *J. Chromatogr. A*, 1099, 167–174, 2005.
37. A. Felinger, *J. Chromatogr. A*, 1126, 120–128, 2006.
38. D. Tondeur, H. Kabir, L.A. Luo, J. Granger, *Chem. Eng. Sci.*, 51, 3781–3799, 1996.
39. J. Samuelsson, P. Forssen, M. Stefansson, T. Fornstedt, *Anal. Chem.*, 76, 953–958, 2004.
40. N. Marchetti, F. Dondi, A. Felinger, R. Guerrini, S. Salvadori, A. Cavazzini, *J. Chromatogr. A*, 1079, 162–172, 2005.
41. A. Felinger, in *Scale-Up and Optimization in Preparative Chromatography*, A.S. Rathore, A. Velayudhan (Eds.), Chapter 3, Marcel Dekker, New York, 2002.
42. A. Felinger, G. Guiochon, *J. Chromatogr. A*, 752, 31–40, 1996.
43. A. Felinger, G. Guiochon, *AIChE J.*, 40, 594–605, 1994.
44. S. Golshan-Shirazi, G. Guiochon, *Anal. Chem.*, 61, 1368–1382, 1989.
45. D. Nagrath, B.W. Bequette, S.M. Cramer, A. Messac, *AIChE J.*, 51 511–525, 2005.

11 Displacement Chromatography in the Separation and Characterization of Proteins and Peptides

James A. Wilkins

CONTENTS

11.1 INTRODUCTION

Displacement chromatography is a technique that has long been recognized for its advantages in the production of high concentrations of pure materials and has therefore demonstrated its utility as a preparative method. On the other hand, the thermodynamic qualities of the method also make it highly useful for analytical purposes because of its unique ability to enrich low-concentration components in complex mixtures. Special emphasis will therefore be directed toward the enrichment of trace components and on studies in which displacement chromatography has been used in the separation/recovery and characterization of biomolecules with attention to the optimization of operating parameters that drive increased productivity and yield. With the advent and growth of the biotechnology industry and the steadily increasing demand for its products, methods for more efficient capture, purification, and characterization of therapeutic molecules are essential.

11.2 HISTORY

Displacement chromatography was named and introduced by Tiselius [1] who demonstrated its utility in both preparative and analytical applications. Tiselius explored separations of many different biological substances including amino acids and peptides [2,3]. The technique was also used

[4,5] to preparatively isolate rare earth metals for the Manhattan Project [6,7]. Much earlier, in the 1920s, a form of chromatography very similar to displacement had been used and was an established method for the analysis of petroleum and its products for many years [8].

It was not until the early 1980s however that Horváth began to investigate displacement with an eye toward its use in preparative chromatography of biomolecules. Interestingly, this coincided with the emergence of the then-nascent biotechnology industry. Since that time, various other investigators have sought to take advantage of the unique features of, and the advantages offered by, the displacement mode. Thus, by the beginning of the start of the new millennium, there was considerable new literature in the area dealing with the separation of biomolecules. Today, the technique offers a desirable option for both analytical and preparative applications, although the operational challenges of this nonlinear chromatographic mode continue to limit its broad application to important separation problems.

11.3   THEORY AND OPERATION

The theory of displacement chromatography has been ably reviewed by several authors including most recently by Guiochon [9]. It is not the intention of this review to address a detailed mathematical treatment of the subject; for this, the reader is referred to works by Helfferich [10], Frenz and Horváth [11], and by Katti and Guiochon [12]. Rather, the salient features of the method will be described. Displacement chromatography is a nonlinear chromatographic technique that relies on thermodynamic advantages offered by a situation in which analytes are loaded to a level where they compete with each other for binding to the stationary phase. This situation is in stark contrast to linear chromatography in which analytes are loaded at a small fraction of the total binding sites on the stationary phase. Often, the appropriate loading level in displacement chromatography is achieved by loading from 100 to 1000 times the amount typically loaded in an analytical separation. Under these conditions, and assuming that binding isotherms of the components all exhibit Langmuirian behavior, with "concave down" isotherms, analytes line up in zones on the stationary phase in the order of their relative affinities during the "feed" or loading stage of the separation. Unlike the conditions experienced in linear chromatography where an excess of stationary binding sites over analytes prevails, analytes in nonlinear chromatography compete with each other for stationary phase binding. Such a situation can be represented by an equation relating the stationary phase concentration q to the mobile phase concentration c of a given analyte; a and b are parameters in the model.

$$q = \frac{ac}{1+bc}$$

where both a and b are positive numbers

The multicomponent adsorption isotherms operative in displacement chromatography are described by the following equation that reflects the competitive nature of the process:

$$q_i = \frac{a_i c_i}{1 + \sum_{j=1}^{N} b_j c_j}$$

where
    q_i and c_i are the respective stationary and mobile phase concentrations of species i
    N is the number of sorbable components in the system
    a_i and b_j are parameters characteristic of species i or j

When $N = 1$, the model yields a concave down isotherm. The magnitudes of a_i and b_i for each species determine their relative affinities and therefore their ability to suppress each other's isotherms.

Following the loading stage, ideally the "zones" of analytes should be free of neighboring components and ready for displacement by an appropriate displacer. By definition, the displacer is a compound that has a higher affinity for stationary phase binding than any of the components in the analyte mixture. Following the loading ("feed") stage, the column is washed with the equilibration buffer until a stable baseline (e.g., UV absorbance measured at 214 nm) has been achieved. After the wash step, a solution containing the displacer at a selected concentration is directed to the column inlet. The concentration of the displacer and the flow conditions are chosen such that the bound analytes are displaced from the column in an "isotachic" (equal velocity) manner. Again, this contrasts with elution chromatography in which analytes traverse the column at different velocities in order to achieve separation. Of course, nonidealities are introduced into the system in the form of mass transport limitations of the stationary phase and by the formation of "shock layers" between high-concentration zones of analytes. The latter phenomenon has been addressed by Zhu and Guiochon [13] and by Natarajan and Cramer [14]. Both of the latter serve to limit the practical resolution attainable using this method. As shown in Figure 11.1, the concentration of the displacer determines the final concentrations of the analytes in the recovered fractions since the concentrations of the latter are defined by the intersection of the "operating line" with each component's isotherm. The operating line is defined as a line connecting the origin to the point on the isotherm of the displacer corresponding to its mobile phase concentration (Figure 11.1).

The advantages of the displacement mode derive from the above considerations and are manifested in the following ways:

1. Analytes with isotherms which intersect the operating line will be concentrated by the displacement process above their original concentrations in the load solution. This leads to interesting possibilities for trace component enrichment.
2. Purified zones of analytes can be obtained under favorable conditions in a far more efficient way than can be attained using overloaded elution chromatography.
3. Analyte loading levels can be of a similar or greater order of magnitude than those used in linear elution chromatography; therefore, the stationary phase utilization is more efficient, leading potentially to higher-throughput preparative separations.

A "textbook" example of a displacement separation, demonstrating the concentrating effect of the displacement regime is shown in Figure 11.2. In this case, Horváth et al. separated 4-hydroxy (30 mg), 2-hydroxy (35 mg), and 3,4 dihydroxy (45 mg)-phenyl acetic acids [15]. The separation was carried out using n-butanol as displacer. The figure demonstrates the almost-perfect square zones of the three analytes separated in the isotachic displacement train and also shows the concentrating effect of displacement on more highly retained analytes, since the final concentrations are based on the affinities of the components, not their initial load amounts.

An important advantage arising from the thermodynamics of the displacement mode is that low-abundance analytes are displaced from the stationary phase into the mobile phase at a concentration determined strictly by their isotherms and by the displacer concentration selected. This offers a particular advantage over the elution mode, which is normally dilutive for analytes and has encouraged several investigators to use displacement as an analytical technique for the detection of low-abundance substances in the presence of analytes at significantly higher concentrations (see below). This latter property has important implications for the field of proteomics where mixtures containing peptides of widely different concentrations (arising from biological systems) are routinely encountered, resulting, for example, in a dynamic range of human serum protein concentrations of at least 10^9 [16]. Low-abundance peptides present in such mixtures cannot be detected or analyzed even utilizing the most sensitive mass spectrometry instrumentation currently available.

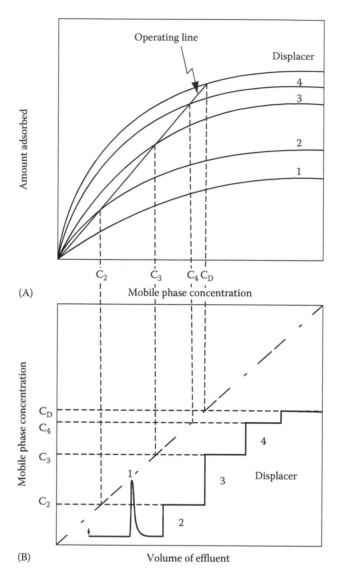

FIGURE 11.1 Graphical representation of the isotherms of the feed components and of the displacer as well as the operating line (panel A) and of the fully developed displacement train (panel B). Note that component 1 is eluted because its isotherm is not intersected by the operating line. Concentrations of the zones are determined by their intersection with the operating line and thus by the displacer concentration chosen. (Reprinted with permission from Elsevier from Frenz, J. and Horváth, Cs., High performance displacement chromatography, in Horváth, C. (ed), *High Performance Chromatography: Advances and Perspectives*, Academic Press, San Diego, CA, 212–314, 1988. Copyright.)

Chromatographic separation of these mixtures in the elution mode is incapable of resolving many thousands of peptides present in these mixtures, even when orthogonal, two-dimensional separations are performed. The investigator is left with little option for low-abundance peptide identification other than affinity approaches that target certain subclasses (e.g., phosphopeptides). While effective for certain applications, the latter allow for enrichment of only a small subset of low-abundance peptides. Because of its potential for broad applicability to the problem of low-abundance peptide enrichment, displacement chromatography remains a technique that offers great possibilities in this area.

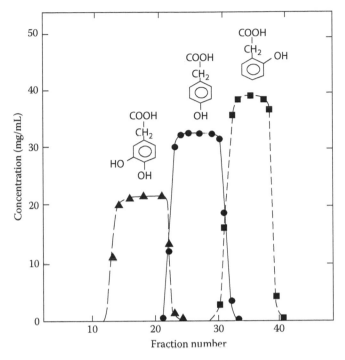

FIGURE 11.2 Separation of hydroxy-phenylacetic acids by displacement on a C18 reversed phase column (dimensions: 4.6×250 mm). Carrier 0.1 M phosphate; displacer 0.87 M aqueous *n*-butanol. Flow rate: 0.05 mL/ min; temperature: 25°C. (Reprinted with permission from Elsevier from Horváth, C. et al., *J. Chromatogr.*, 218, 365, 1981. Copyright.)

11.4 PROTEIN DISPLACEMENT CHROMATOGRAPHY

With the advent of the biotechnology industry and given the central role of chromatography in the production of therapeutic molecules, there has been tremendous interest in the development of technologies offering high throughput and high purity demanded by the application of these molecules in the treatment of human disease. As mentioned above, Csaba Horváth and his students and coworkers had a profound influence on the rediscovery and development of displacement chromatography as a preparative technique, reintroducing displacement chromatography in the context of the high resolution offered by separations performed on the HPLC platform using stationary phases with small particle size and high efficiency. This enabled resolution of components in displacement that was unattainable in Tiselius' day. Early on, it was realized that several factors including the nature of the displacers available and the operational difficulties of nonlinear chromatographic systems might limit the wide use of displacement chromatography in this application. However, Cramer and his colleagues have worked to address the above concerns through the innovation of new displacers, including small-molecule displacers [17]. These low-molecular-weight displacers were shown to allow "selective displacement" of desired components and enable the design of "tailor made" displacers for particular separations. The introduction of these new displacers has also been coupled to the use of the "steric mass action" (SMA) model which relates empirical protein adsorption data obtained in a series of elution experiments to the interactions of the protein with the stationary phase in order to predict analyte behavior in displacement separations [18,19]. The advantage of the SMA formalism is that it attempts to take into account the multiple charge interactions and steric effects experienced during ion exchange chromatography of proteins. In order to test the predictive power of the SMA-derived calculations, Gallant and Cramer [17] compared calculated and experimental results of the displacement of a cytochrome *c*/lysozyme mixture in a cation exchange displacement experiment. A low-molecular-weight displacer, neomycin, was used for the latter experiment.

SMA calculations were able to adequately describe the separation in a simulation. In the same paper, the authors investigated the effect of carrier salt and displacer concentration on the displacement of artificial components with either higher or lower affinities for the stationary phase than the "product" cytochrome c and showed the effects of different load levels and different displacer concentrations on the development of the displacement train in a simulated experiment.

In addition to antibiotics such as neomycin, the Cramer group has tested a number of other molecules as displacers in ion exchange chromatography such as protected amino acids [20], dendrimers [21], pentaerythritol- and phloroglucinol-based salts [22], and aminoglycoside-polyamines [23]. Unlike "traditional" displacement chromatography, selective displacement takes advantage of an induced salt gradient in ion exchange chromatography to elute less-strongly adsorbed components; as will be discussed later in this review, this can provide a significant advantage in targeting an analyte of interest.

Freitag [24] reviewed the role of displacement chromatography in biopolymer separations with a particular emphasis on protein separations. She points out the striking advantages sometimes demonstrated in the displacement mode especially in industrial-scale bioprocessing where economic and technical demands call for techniques capable of delivering proteins at high purity and concentration. In one study, the throughput and recovery of a CHO-cell-derived human blood factor (antithrombin III) were increased by an order of magnitude using displacement chromatographic separation on hydroxyl- and fluoroapatite stationary phases [25]. In other studies, Freitag and Vogt [26] compared particulate and monolithic anion exchange columns with polyacrylic acid as a displacer followed by Schmidt et al. [27] who investigated the utility of a polyelectrolyte (poly(diallyl-dimethylammonium chloride, PDADMAC)) as a displacer on both particle-based and monolithic cation exchange stationary phases. In the latter study, a model protein mixture containing ribonuclease A and alpha-chymotrypsinigen was used in the test. The authors sought to test predictions developed using SMA and dynamic affinity modeling on the two stationary phases in the two studies. As predicted, changes in displacer and protein concentration had a significant effect on the separation of the proteins tested. The authors ruled out the development of an induced salt gradient in an experiment where the displacer front did not meet the more highly retained component in the separation (chymotrypsinogen) and resolution was poor.

In the two studies cited above, although the SMA model was able to predict separation behavior on one type of column, this was not the case with the other column type, which yielded poor separations. Unfortunately, opposite results were obtained in the studies where SMA calculations were able to predict displacement behavior better for the monolithic anion exchange column [26]; the reverse was true for the cation exchange columns tested in the later study [27]. Trial-and-error optimization on the monolithic cation exchange column yielded a separation comparable to that shown in Figure 11.4b, but required a dramatic change in pH (from 7.2 to 5). Unfortunately, the authors made little comment in the later paper about the apparent discrepancy in these results.

Although much of the work in protein displacement has utilized ion exchange stationary phases, other media such as hydrophobic interaction [28,29] and metal affinity [30] have been tested. Recently, a protein separation was described [31] on a hydrophobic charge induction column. In hydrophobic charge induction, the stationary phase ligand is normally hydrophobic but differs from traditional hydrophobic interaction chromatography (HIC) ligands in that an ionizable group with a defined pK_a is attached to the hydrophobic ligand. In hydrophobic charge induction chromatography (HCIC), proteins can adsorb at relatively low salt concentrations when compared with HIC. Proteins can be eluted from HCIC by changing the pH of the mobile phase and thus the ionization of the charged groups [32]. For the displacement separation of lysozyme and alpha-chymotrypsinogen A benzethonium chloride, a water-soluble quaternary amine containing two aromatic ring structures was chosen (Figure 11.3).

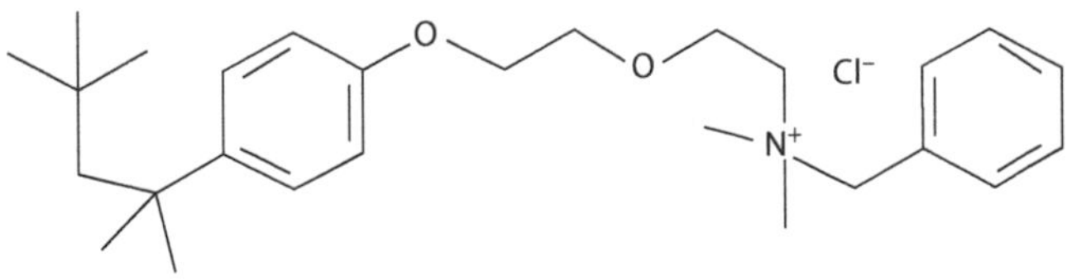

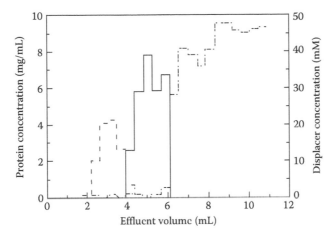

FIGURE 11.3 Structure of benzethonium chloride, the displacer used in the HCIC experiments. (Reprinted with permission from Elsevier from Gallant, S.R. and Cramer, S.M., *J. Chromatogr.*, 771, 9, 1997. Copyright.)

FIGURE 11.4 Displacement of lysozyme (dashed line; 5 mg loaded) and alpha-chymotrypsinogen (solid line; 10 mg loaded) by benzethonium chloride (alternating dots and dashes; 50 mM concentration) on a MEP-Hypercel HCIC column (dimensions: 5 × 100 mm) at a flow rate of 0.1 mL/min. (Reprinted with permission from Elsevier from Schmidt, B. et al., *J. Chromatogr.*, 1018, 155, 2003. Copyright.)

Based on capacity factor measurements, a pH of 5.0 was chosen for the separations. Adsorption isotherms were measured and found to exhibit almost ideal Langmuirian behavior for the proteins and for the displacer. In agreement with capacity factor measurements, a displacement performed at pH 5.0 produced an excellent separation of the two proteins (Figure 11.4).

Displacement at pH 7.0 was unsuccessful as was predicted by the capacity factor measurements. Interestingly, the capacity factor for lysozyme increases above pH 5.5 and exceeds that of chymotrypsinogen at pH 7 and so lysozyme was not displaced at the latter pH, while chymotrypsinogen exited the column partially mixed with the displacer. The authors point out that the pH effect may be used in HCIC to change the order of displacement or to selectively displace one component while leaving the other bound to the column. Column regeneration was easily achieved by lowering the pH of the mobile phase below the pK_a of the MEP-Hypercel resin (4.8). The easy regeneration of the stationary phase makes the HCIC resin particularly attractive since one of the problems inherent to displacement chromatography is the removal of tightly adsorbed displacer molecules.

11.4.1 DISPLACEMENT SEPARATIONS OF MORE COMPLEX PROTEIN MIXTURES

One of the promises of displacement chromatography in this context is to increase the overall throughput of column chromatographic separations by allowing high loading levels while delivering purified zones of the desired protein product. Several authors have tested this idea often with the focus on the separation of complex mixtures containing varying levels of therapeutic proteins of interest. Some examples of this type of separation are presented below.

Barnthouse et al. [33] demonstrated the separation of recombinant human-brain-derived neurotrophic factor (rHuBDNF) expressed in *Escherichia coli* in cation exchange chromatography using protamine as a displacer. In this case, it was necessary to separate BDNF from contaminating *E. coli* proteins as well as from BDNF variants which were very closely related to the desired product as shown in a figure from that paper. Linear elution of a BDNF sample separated by analytical reversed phase chromatography revealed the presence of variants and host cell proteins in the feedstock (Figure 11.5, panel a) while part b of the figure shows that separation of the same feedstock on cation exchange reveals very little resolution of the components as suggested by the single broad peak obtained.

Displacement separations were carried out using both small particle size (20 μm) and larger (50 μm) resins. As expected, the smaller particle resin outperformed the larger resin at equal displacer (protamine) concentrations. However, by lowering displacer concentration with the larger particle-size resin, the authors were able to attain comparable purities for the recombinant protein product. Interestingly, they also point out the advantage of higher loading levels in displacement, comparing purity and yield in a displacement separation at linear chromatography-type loads (2.1 mg/mL) to a separation in which the same column is loaded to 11 mg/mL. Purity and yield were increased in the case of higher loading versus lower loading with purity increasing to 92.8% in the higher load from 90.8% in the lower and yield increasing to 72.6% from 62.5%. This

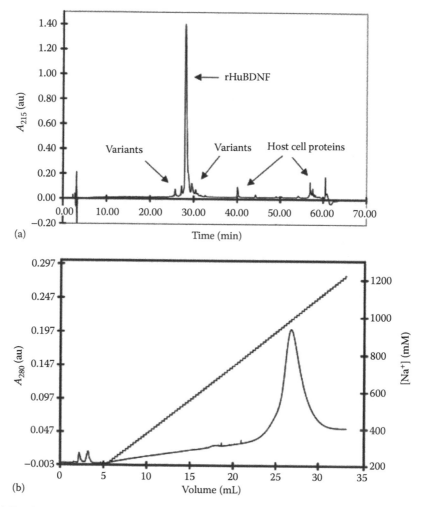

FIGURE 11.5 (Panel a) Reversed phase separation of rHuBDNF feedstock showing the positions of variants and host cell proteins. (Panel b) Ion exchange separation of the same feedstock on a 50 μm cation exchange resin. Loading was 20 μg of rHuBDNF. (Reprinted with permission from Elsevier from Zhao, G.F. and Sun, Y., *J. Chromatogr.*, 1165, 109, 2007. Copyright.)

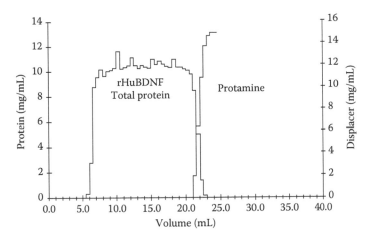

FIGURE 11.6 Displacement separation of rHuBDNF feedstock. Loading was 20 mg/mL on a 4.6×500 mm column of POROS HS/M resin. (Reprinted with permission from Elsevier from Zhao, G.F. and Sun, Y., *J. Chromatogr.*, 1165, 109, 2007. Copyright.)

result demonstrates the advantage of displacement where higher loading (up to a defined level) can lead to significant advantages in zone purity and in throughput in contrast to zone broadening and loss of resolution in elution separations performed at high stationary phase loading. An example displacement separation from Barnthouse et al. is shown in Figure 11.6.

The displacement separation yielded 72.9% of the product at 93.6% purity. These numbers compared favorably to a gradient separation performed on the same feedstock, the latter performed at eightfold lower loading. Comparable yields and purity were obtained using a column packed with 50 μm particles by lowering the displacer concentration sixfold to 0.5 mM from 3 mM used in the case of the 20 μm resin. The authors concluded that the reduced displacer concentration was required due to mass transfer limitations imposed using the larger particle-size resin. By lowering the displacer concentration, the maximal product concentration was lowered, but there is a corresponding decrease in overlapping product/impurity zones. In addition, the lower displacer concentration allows more equilibration time for zones to form. The productivity of the displacement process using the 50 μm resin (suitable for scaleup into manufacturing) was shown to significantly improve the production rate, yielding a 4.5-fold benefit over the then-current gradient process [33].

Shukla et al. [34] investigated the purification of a bacterially derived antigenic vaccine using selective displacement chromatography with low-molecular-weight displacers. The idea behind "selective displacement" is that a displacer is chosen such that the protein of interest is displaced while most of the impurities are either eluted by an induced salt gradient or exit the column mixed with the displacer zone as described in the paper. The challenge in this case was to purify the vaccine away from host cell proteins and to compare the selective displacement technique to conventional elution chromatographic approaches.

The feedstock containing the antigenic vaccine protein (AVP) is obviously complex as suggested by the analytical separations shown in the paper. *p*-Toluenesulfonic acid (PTS) was chosen as the displacer for AVP on an anion exchange resin based upon two criteria. First, the behavior of the AVP on size exclusion chromatography indicated that other displacers including sucrose octasulfate (SOS) cause aggregation of the protein while PTS did not. Second, dynamic affinity plots constructed for PTS, AVP, and the "high affinity impurity" identified in the analytical anion exchange runs indicated that PTS would be a good choice for selectively displacing AVP. A dynamic affinity plot was constructed using steric mass action parameters obtained from a k' (capacity factor) versus salt concentration plot showing predicted behavior of the components. The dynamic affinity plot predicted that the AVP protein would be selectively displaced while the high-affinity impurity would not. Based on an "operating regime plot" constructed using information obtained from the

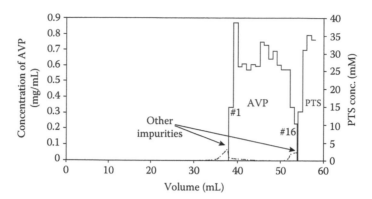

FIGURE 11.7 Selective displacement of AVP on DEAE Sepharose Fast Flow. Column dimensions were 10 × 290 mm; load was 6.5 mL AVP feedstock/mL resin. The flow rate was 1 mL/min. (Reprinted with permission from Elsevier from Barnthouse, K.A. et al., *J. Biotechnol.*, 66, 125, 1998. Copyright.)

above, the investigators selected a PTS concentration of 35 mM to perform the experiment. The results (see Figure 11.7) demonstrated that the selective displacement was largely successful with minor impurities displaced in front and behind the major product zone. The authors commented that, as predicted, weakly bound impurities eluted ahead of AVP while the high-affinity impurities were retained after displacer breakthrough.

The purity of the pooled fractions from the run was assessed by size exclusion chromatography to be 92.7% after pooling selected fractions. The authors did not comment however on the yield of the process or its suitability for further scaleup into manufacturing.

11.5 PEPTIDE DISPLACEMENT CHROMATOGRAPHY

Peptide displacement has received a great deal of attention and as mentioned above has been used in recent studies as an analytical tool. Because of their generally good retention behavior in reversed phase elution chromatography, peptides provide ideal bioanalytes for the study of the displacement process itself and for testing its strengths and weaknesses. Some early studies in the Horváth laboratory focused on peptides. Viscomi et al. [35] studied the behavior of a melanotropin (MSH) mixture derived from a pituitary extract in reversed phase chromatography first in the elution mode as shown

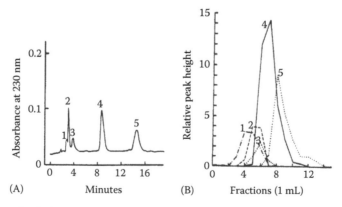

FIGURE 11.8 Isocratic elution separations of 20 μg (panel A) and 10 mg (panel B) of a melanotropin mixture. Column: 4.6 × 250 mm ODS; Eluent: 0.25% formic acid, 0.5% triethylamine, 24% acetonitrile in water. Components: 1, unknown; 2, MSH sulfoxide; 3, acetyl-MSH sulfoxide; 4, MSH; 5, acetyl-MSH. (Reprinted with permission from Elsevier from Barnthouse, K.A. et al., *J. Biotechnol.*, 66, 125, 1998. Copyright.)

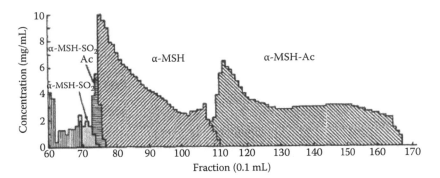

FIGURE 11.9 Displacement separation of a crude MSH mixture with the same column used as in Figure 11.8. Carrier: 0.25% formic acid, 0.5% triethylamine, 19% acetonitrile; displacer, 50 mM BDMA in 0.25% formic acid, 0.5% triethylamine, 21% acetonitrile. Flow rate: 0.1 mL/min; feed: 35 mg MSH. (Reprinted with permission from Shukla, A.A. et al., *Biotechnol. Prog.*, 14, 92, 1998. Copyright 1998, American Chemical Society.)

in Figure 11.8. This figure demonstrates the problems inherent in overloaded elution separations. On the left side of Figure 11.12, 20 μg of the crude MSH mixture was loaded on the 4.6×250 mm C18 column. In this isocratic elution, the mixture is separated into several nonoverlapping peaks identified in the figure caption. In the right side of the figure is shown the separation of 10 mg of the same mixture. As expected, peak broadening and tailing in the overloaded condition causes severe overlap and poor separation of all of the components.

A separation was then carried out in the displacement mode using the same column that yielded starkly different results as shown in Figure 11.9. In this experiment, the authors loaded 35 mg of the MSH mixture onto the column and displaced the column with a hydroorganic mixture of the detergent benzyldimethyldodecylammonium bromide (BMDA). As shown in the figure, the components were well separated in the displacement mode yielding large zones of purified products at a 3.5-fold higher loading than in the overloaded elution case.

Of particular interest in the case of the displacement separation of the melanotropins, acetonitrile was mixed with both the carrier and the displacer. The purpose was to suppress the sigmoidal isotherm of MSH which was then found to underlie the Langmuirian isotherm of the displacer in the area of operational interest. The idea of isotherm suppression through use of acetonitrile/displacer mixtures surfaced again later in the Horváth laboratory (see below).

In another study, Kalghatgi et al. [36] looked at the purification of mellittin, a 26-amino-acid peptide derived from bee venom, using micropellicular C18 stationary phases. Rapid HPLC analysis (1 min acetonitrile gradient at 80°C) of a commercially derived melittin sample revealed a complex mixture (Figure 11.14).

A sample similar to that shown in Figure 11.10 was used as the feed in displacement chromatography separations [36]. For these experiments, the authors employed a longer (105 mm) column of the same diameter as the analytical column. Displacement chromatography on 10 mg of a melittin sample was carried out at two temperatures to demonstrate the advantage of higher temperature on mass transfer kinetics in the separation (Figure 11.11).

Several interesting observations were made about the results of the experiment. The importance of increased temperature is clearly shown as the separation of the components took only 15 min at 40°C, whereas the same separation took 50 min to

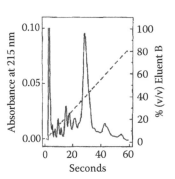

FIGURE 11.10 HPLC analysis of melittin on a micropellicular C18 silica column (dimensions 30×4.6 cm). Linear gradient of acetonitrile from 0% to 80% in 1 min. Temperature: 80°C. Flow rate: 3 mL/min. Load: 5 μg of a commercially derived melittin sample. (Reprinted with permission from Shukla, A.A. et al., *Biotechnol. Prog.*, 14, 92, 1998. Copyright 1998, American Chemical Society.)

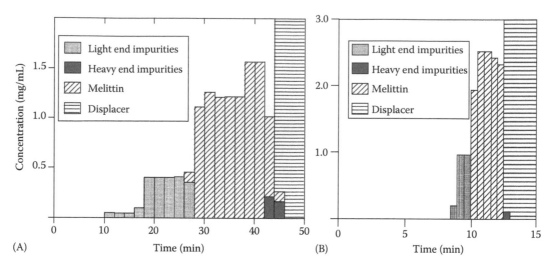

FIGURE 11.11   Separation of melittin mixture "P14" by displacement chromatography at two different temperatures: panel A, 23°C; panel B, 40°C. Column: micropellicular silica C18. 105×4.6 mm; carrier, 0.1% TFA in water; displacer, 25 mM benzyldimethylhexadecyl ammonium chloride in 10% (v/v) aqueous acetonitrile containing 0.1% TFA (v/v). Flow rate: 0.2 mL/min (panel A) and 0.6 mL/min (panel B). Fractions (0.5 mL each) were analyzed as in Figure 11.10. (Reprinted with permission from Shukla, A.A. et al., *Biotechnol. Prog.*, 14, 92, 1998. Copyright 1998, American Chemical Society.)

perform at 23°. Increasing the temperature benefited the separation because of increased mass transfer and also enabled much higher enrichment of the desired purified melittin product as narrower, higher concentrations of the products were obtained in the isotachic displacement train.

Ramanan and Velayudhan [37] used a binary mixture of *n*-formyl-Met-Phe (designated "P") and *n*-formyl-Met-Trp (designated "T") to examine operating parameters in reversed phase displacement chromatography. The authors looked at parameters such as choice of displacer, displacer concentration, mobile phase organic level, and flow rate to determine their effects on overall column productivity. Separations were performed on a conventional (3.9×150 mm) C18 column and fractions were collected for analysis. Based on their observations in the displacement separation of the two peptides, the authors point out the relative advantages, disadvantages, and compromises inherent in selection of the above parameters. Interestingly, by adjusting the parameters appropriately they were able to achieve displacement separations on the same time scale as that normally required for elution separations. This potentially overcomes one of the major criticisms of the displacement mode and will be discussed further below.

11.6   TRACE COMPONENT ENRICHMENT

The concept of using displacement chromatography for trace component enrichment has been discussed by several authors [38–40]. In Ramsey et al., the authors describe the use of displacement chromatography for trace enrichment and study the problem from the theoretical and experimental point of view. Their study was informed by earlier studies using overloaded elution separations in which significant enrichments of trace components had occurred [41,42]. Figure 11.12 shows a calculated displacement chromatogram from Ramsey et al. [38] in which a hypothetical three-component mixture containing a main component loaded at 20 mg and 2 minor components loaded at 0.1 mg each are separated.

The band of the second component in the displacement separation has an apparent efficiency of 160,000 theoretical plates while the simulated column only has 3,600 theoretical plates. The "squeeze effect" on component 2 in the displacement separation is important resulting in a band that is 7 times

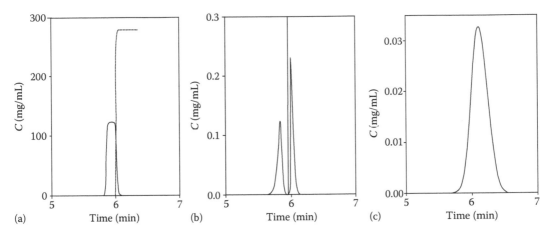

FIGURE 11.12 Hypothetical displacement separation of a three-component mixture containing two trace components loaded at 0.1 mg each and a major component loaded at 20 mg. Panel a, main component band and displacer front; panel b, peaks of the impurities; panel c, peak of the second component under analytical (linear elution) conditions. (Reprinted with permission from Elsevier from Viscomi, G.C. et al., *J. Chromatogr.*, 440, 157, 1988. Copyright.)

higher and 7 times narrower than a corresponding linear elution separation (compare panels b and c). As expected, the concentration effect on component 1 is less significant (given that its isotherm underlies that of component 2, this is not surprising). Even so, its efficiency is more than tenfold higher than that of the column at 56,000 plates and its shape is improved over a corresponding linear elution with 4 times greater height and 4 times narrower shape. The authors present several other interesting scenarios in which the effects of increasing the amount of the main component and increasing the column efficiency are studied. Increased column efficiency has a profound effect on trace enrichment. At a column efficiency of 10,000 plates, the peak efficiency (and corresponding enrichment) of the trace components increases dramatically. Displacement peak efficiency for component 2 goes from 160,000 plates in the less-efficient column to 1.4 million plates in the column with 10,000 plate column. There is a corresponding increase in efficiency for component 1 in the more efficient column.

The characterization of minor impurities present in purified recombinant protein samples has represented a difficult challenge to analytical protein chemists. One of the main tools for characterization of these complex products is peptide mapping, a process in which the protein is digested with an appropriate proteolytic enzyme, followed by analysis of the fragments by analytical HPLC. Today, these analyses are often carried out using online electrospray ionization (ESI) mass spectrometry which gives detailed information about the sequence of the peptides and hopefully reveals minor posttranslational modifications such as deamidation, oxidation, or proteolysis present in the sample including information about glycans attached to the peptides. Depending on the level present, these posttranslationally modified species may be difficult to detect using standard chromatographic approaches (i.e., linear elution chromatography). The reason for this difficulty arises from the structural similarity shared by many modified peptides and therefore their co-elution or very close elution during peptide mapping. This problem is compounded by the low abundance of the modified forms which may be difficult or impossible to detect using standard elution techniques. Although the mass spectrometer is a very sensitive instrument, capable of detecting peptides at attomole levels, the presence of high levels of other components can mask low-level ones due to ion suppression. Displacement chromatography can potentially address the problem of trace component enrichment because of its ability to concentrate components within the displacement train. This concept is illustrated schematically in Figure 11.13.

Frenz et al. applied the concept of trace enrichment to the tryptic digest of recombinant human growth hormone (rhGH). One of the innovations they introduced [39] was to directly monitor the

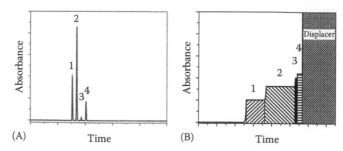

FIGURE 11.13 Separation of a hypothetical mixture containing trace component 3 by elution (panel A) or displacement (panel B) chromatography. Enrichment of 3 and 4 relative to 1 and 2 in B is shown. (Reprinted with permission from Elsevier from Viscomi, G.C. et al., *J. Chromatogr.*, 440, 157, 1988. Copyright.)

column effluent using online fast atom bombardment (FAB) mass spectrometry. This was an important innovation because one of the major difficulties in displacement of complex mixtures is that of identification. UV absorbance gives limited information about the identity of species present in the tightly packed displacement train. As will be shown below, with the advent of online ESI/MS, identification is instantaneous. In addition, the online MS detection allows targeted tandem mass spectrometry (ms/ms) and therefore sequence analysis on selected ions within the displacement experiment. Because the major peptides in the tryptic digest of rhGH were well known, it presented an excellent model system with which to explore the idea of trace enrichment. Using a 320 μm capillary C18 column, the authors showed separation of all of the major peptides in the tryptic digest of rhGH first by elution chromatography loading a 25 pmol sample of the digest. A capillary column of the same dimensions (0.32 × 150 mm) was then used for a displacement separation in which a 40-fold greater amount (10 nmol) of digest was fed onto the column. Cetyltrimethylammonium bromide (CTAB) was used as a displacer in an analysis requiring about 80 min. Analysis of the data indicated the presence of all the known tryptic peptides along with a number of peptides that were associated with nonspecific proteolysis and with autolysis of trypsin itself. In a later paper, the same authors again analyzed the tryptic digest of rhGH, this time with off-line analysis using ESI/MS. In this case, a larger column (4.6 × 150 mm) was used again using CTAB as the displacer. The separation time was substantially longer (>6 h). Among the peptides identified in the study were a deamidated version of fragment T13 and some missed cleavage-related peptides. Fraction collection also enabled large amounts of low-level variants to be isolated and further purified by elution chromatography. Overall the authors concluded that variants constituting less than 1% of the total digest could be isolated and identified.

11.6.1 ELUTION MODIFIED DISPLACEMENT CHROMATOGRAPHY

Building on the ideas first introduced by Viscomi et al. [35] and by Ramanan and Velayudhan [37], Wilkins et al. [43] studied the selective enrichment of trace peptides using carrier and displacer solutions containing targeted amounts of the mobile phase organic modifier (acetonitrile). The rationale for this approach is as follows. If a particular zone in an elution separation contains a trace component(s) of interest, that zone can be targeted for enrichment while other areas can be either eluted or allowed to exit the column mixed with the displacement front. The concept is illustrated in Figure 11.14.

The advantages offered by the method are as follow:

1. Shorter run times are achieved.
2. A particular area of the separation is targeted, allowing a wider window of separation, while components of less interest are either eluted or mixed with displacer.
3. Components are better resolved than in conventional displacement.

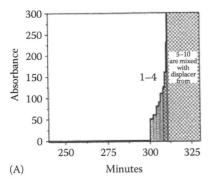
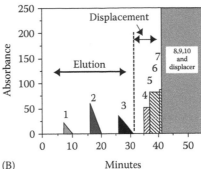

(A)   Minutes   (B)   Minutes

FIGURE 11.14 Concept of EMDC. Panel A, in conventional displacement, separation times are long (on the order of hours) and components form highly concentrated zones near the displacer front, some mixed with the displacer. Panel B, in elution modified displacement, acetonitrile is added to the carrier and to the displacer solution in a targeted way to effect elution of some components and displacement of others. Other components are mixed with the displacer. Run times are <1 h. (Reprinted with permission from Elsevier from Kalghatgi, K. et al., *J. Chromatogr.*, 604, 47, 1992. Copyright.)

The method was tested using a tryptic digest of rhGH that had been stored for more than a decade under refrigerated conditions and therefore contained increased amounts of low-level, modified variants. Conditions were chosen so that a "window" containing a defined set of components from the tryptic peptide map was targeted for enrichment. In Figure 11.15, the concept is illustrated for one such "window" in the tryptic map. The figure demonstrates the enrichment of low-level components in the tryptic digest. The experiment highlighted one of the advantages of the elution modified regime. Components are enriched, but also more resolved than in conventional displacement.

In a subsequent publication, the same authors reported moving the technique to an online, capillary chromatographic format linked to ESI/MS [44]. Low levels of a marker peptide (kemptide) were mixed with an rhGH tryptic and the mixtures were separated by elution and by elution modified displacement chromatography (EMDC) using the methodology described above. Although the trace peptide could not be detected in elution chromatography when mixed with the rhGH digest at a 1:300 (w/w) ratio, it was easily detected in EMDC. The authors went on to show that they could detect kemptide when mixed with the rhGH digest at levels down to 5 fmol, which represented a molar ratio of kemptide/rhGH of 1:300,000. Thus, EMDC allowed detection of peptides down to this level in high levels of a complex mixture.

The authors point out that enrichment methods such as displacement chromatography and EMDC are important in the area of proteomics where investigators are routinely confronted with separation problems involving mixtures with high dynamic concentration range. For example, the range of protein concentration in human serum is at least 12 orders of magnitude with albumin at the mg/mL concentration range and interleukin 12 present in ng/mL concentration. This presents a daunting problem in proteomics that will only be addressed by the introduction of new and more powerful separation technologies.

In line with this thinking, EMDC was shown to enable detection of low levels (20 fmol) of rhGH digest when it was mixed with a tryptic digest of whole human serum [44]. Figure 11.16 shows the EMDC total ion chromatogram (TIC), showing the positions of the rhGH tryptic peptides detected during the online separation.

This separation demonstrates the possibilities of the EMDC method when combined with mass spectrometry. It potentially provides a powerful tool for the proteomics area although to this point, it has not seen widespread adoption.

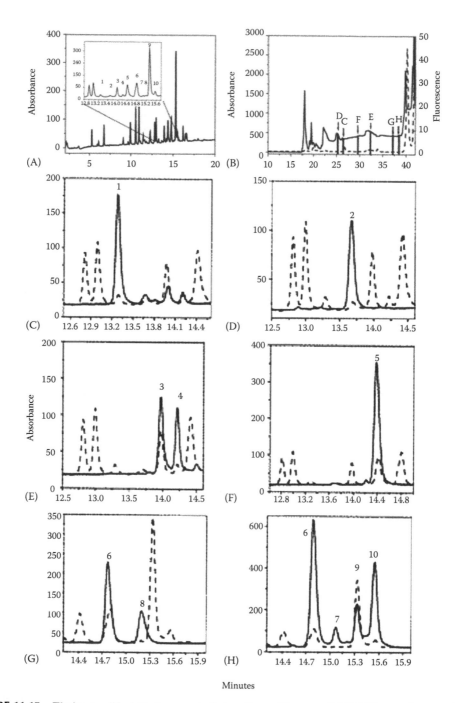

FIGURE 11.15 Elution modified displacement of a tryptic peptide map of rhGH. The carrier contained 10% (v/v) and the displacer (denzyldimethyldodecylammonium bromide) contained 20% acetonitrile. The elution and displacement chromatograms are shown in panel A and B, respectively. The "window" selected by using 20% acetonitrile is shown in the inset to panel A. Panels C through H show analysis by reversed phase elution chromatography of the C–H zones from the displacement chromatogram (solid traces; panel B). The dashed traces in panels C through H are extracted from the original elution chromatogram (inset, panel A). (Reprinted with permission from Elsevier from Kalghatgi, K. et al., *J. Chromatogr.*, 604, 47, 1992. Copyright.)

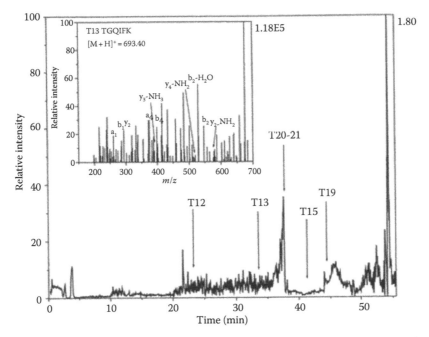

FIGURE 11.16 Total ion chromatogram; online detection using ESI/MS of capillary separation of human serum tryptic digest spiked with 20 fmol of rhGH digest. Zone containing the indicated tryptic fragments of rhGH were identified by data dependent ms/ms. The inset shows the ms/ms of tryptic peptide T13 showing the characteristic fragment ions. (Reprinted with permission from Ramsey, R. et al., *Anal. Chem.*, 62, 2557, 1990. Copyright 1990, American Chemical Society.)

11.7 LOOKING FORWARD: PROSPECTS FOR DISPLACEMENT CHROMATOGRAPHY

As outlined in this chapter, displacement chromatography has demonstrated its utility for both preparative and analytical separations of proteins and peptides. However, in spite of its promise in this area, it has received relatively little attention and use in recent years. The reasons for this are related primarily to the vagaries-associated nonlinear processes in general, but also, specifically in the case of protein separations, to the complexity and seemingly infinite variety of protein structures found in the proteome, leading to complex interaction behavior between protein analytes and stationary phases. This behavior in turn complicates protein separations and scaleup efforts. Although serious efforts such as Steric Mass Action formalism have been made to model these interactions, the reality is that protein feedstocks, even those derived from recombinant sources, contain thousands of host-cell-derived proteins of different types in addition to product-related impurities and other impurities such as DNA and endotoxins that must be separated away from the protein of interest. The chromatographic model most in use in the industrial setting remains overloaded elution chromatography which actually takes advantage of at least one of the characteristics of displacement separations, i.e., because the stationary phase is often loaded to near its capacity, analytes compete with each other for stationary phase interaction, leading to sharper zones and exclusion of less tightly bound impurities by the overwhelmingly higher concentration of the recombinant protein of interest. It is the latter characteristic that likely often drives separation of closely eluting variants from the product in manufacturing-scale elution separations. However, because of tailing associated with the overloaded elution condition, yields suffer in many cases.

Another factor that has historically limited the use of displacement chromatography, especially in the industrial setting, has been the lack of commercial availability of suitable displacers. Up until very recently, there was no dedicated commercial source for displacers. Sachem, Inc. has recently introduced a line of low-molecular-weight displacers for ion exchange and reversed phase displacement separations that can be removed from proteins using ultrafiltration. Hopefully, these materials will enable broader use of displacement chromatography in the future, especially for industrial purification of therapeutic and other proteins and peptides of interest.

As outlined above displacement chromatography may find its most important uses in the analytical area. The ability to enrich trace levels of components is ideally suited to the proteomics where more powerful tools are desperately needed to address the vast concentration ranges present in order to identify trace components. The technique also offers a way to isolate large quantities of protein variants which is important for the identification and characterization of minor product-related impurities commonly associated with therapeutic proteins.

REFERENCES

1. A. Tiselius *Ark. Kem. Miner. Geol.* 16A, 1, 1943.
2. R.L.M. Synge and A. Tiselius. *Acta Chem. Scand.* 1, 749–762, 1947.
3. R.L.M. Synge and A. Tiselius. *Acta Chem. Scand.* 3, 231–246, 1949.
4. F.H. Spedding, E.I. Fulmer, B. Ayers, T.A. Butler, J. Powell, A.D. Tevbaugh, and R. Thompson. *J. Am. Chem. Soc.* 70, 1671–1672, 1948.
5. F.H. Spedding *Discuss. Faraday Soc.* 7, 214, 1949.
6. F.H. Spedding, E.I. Fulmer, T.A. Butler, and J. Powell. *J. Am. Chem. Soc.* 72, 2349–2354, 1950.
7. F.H. Spedding, E.I. Fulmer, T.A. Butler, and J. Powell. *J. Am. Chem. Soc.* 72, 2354–2361, 1950.
8. B.A. Mair, A.L. Gaboriault, and R.D. Rossini. *Ind. Eng. Chem.* 39, 1072–1081, 1947.
9. G. Guiochon *J. Chromatogr.* 1079, 7–23, 2005.
10. F. Helfferich and G. Klein. *Multicomponent Chromatography—Theory of Interference*, Dekker, New York, 1970.
11. J. Frenz and Cs. Horváth. High performance displacement chromatography, In: C. Horváth, (Ed.) *High Performance Chromatography: Advances and Perspectives*, Academic Press, San Diego, CA, 1988, pp. 212–314.
12. A.M. Katti and G.A. Guiochon. *Adv. Chromatogr.* 31, 1–118, 1992.
13. J. Zhu and G. Guiochon. *J. Chromatogr.* 659, 15–25, 1994.
14. V. Natarajan and S.M. Cramer. *AIChE J.* 45, 27–37, 1999.
15. C. Horváth, A. Nahum, and J.H. Frenz. *J. Chromatogr.* 218, 365–393, 1981.
16. R. Pieper, C.L. Gatlin, A.J. Makusky, P.S. Russo, C.R. Schatz, S.S. Miller, Q. Su et al. *Proteomics* 3, 1345–1364, 2003.
17. S.R. Gallant and S.M. Cramer. *J. Chromatogr.* 771, 9–22, 1997.
18. C.A. Brooks and S.M. Cramer. *AIChE J.* 38, 1969–1978, 1992.
19. S.R. Gallant, A. Kundu, and S.M. Cramer. *J. Chromatogr.* 702, 125–142, 1995.
20. A. Kundu, S. Vunnum, G. Jayaraman, and S.M. Cramer. *Biotechnol. Bioeng.* 48, 452–460, 1995.
21. G. Jayaraman, Y.F. Li, J.A. Moore, and S.M. Cramer. *J. Chromatogr.* 702, 143–155, 1995.
22. A.A. Shukla, S.S. Bae, J.A. Moore, K.A. Barnthouse, and S.M. Cramer. *Ind. Eng. Chem. Res.* 37, 4090–4098, 1998.
23. K. Rege, S.H. Hu, J.A. Moore, J.S. Dordick, and S.M. Cramer. *J. Am. Chem. Soc.* 126, 12306–12315, 2004.
24. R. Freitag *Nat. Biotechnol.* 17, 300–302, 1999.
25. C. Kasper, J. Breier, S. Vogt, and R. Freitag. *Bioseparation* 6, 247–262, 1996.
26. R. Freitag, and S. Vogt. *J. Biotechnol.* 78, 69–82, 2000.
27. B. Schmidt, C. Wandrey, and R. Freitag. *J. Chromatogr.* 1018, 155–167, 2003.
28. K.M. Sunasara, F. Xia, R.S. Gronke, and S.M. Cramer. *Biotechnol. Bioeng.* 82, 330–339, 2003.
29. A.A. Shukla, K.M. Sunasara, R.G. Rupp, and S.M. Cramer. *Biotechnol. Bioeng.* 68, 672–680, 2000.
30. Y.J. Kim and S.M. Cramer. *J. Chromatogr.* 686, 193–203, 1994.
31. G.F. Zhao and Y. Sun. *J. Chromatogr.* 1165, 109–115, 2007.
32. S.C. Burton and D.R.K. Harding. *J. Chromatogr.* 814, 71–81, 1998.
33. K.A. Barnthouse, W. Trompeter, R. Jones, P. Inampudi, R. Rupp, and S.M. Cramer. *J. Biotechnol.* 66, 125–136, 1998.
34. A.A. Shukla, R.L. Hopfer, D.N. Chakravarti, E. Bortell, and S.M. Cramer. *Biotechnol. Prog.* 14, 92–101, 1998.

35. G. Viscomi, S. Lande, and C. Horváth. *J. Chromatogr.* 440, 157–164, 1988.
36. K. Kalghatgi, I. Fellegvari, and C. Horváth. *J. Chromatogr.* 604, 47–53, 1992.
37. S. Ramanan and A. Velayudhan. *J. Chromatogr.* 830, 91–104, 1999.
38. R. Ramsey, A.M. Katti, and G. Guiochon. *Anal. Chem.* 62, 2557–2565, 1990.
39. J. Frenz, J. Bourell, and W.S. Hancock. *J. Chromatogr.* 512, 299–314, 1990.
40. J. Frenz, C.P. Quan, W.S. Hancock, and J. Bourell. *J. Chromatogr.* 557, 289–305, 1991.
41. S. Jacobson, S. Golshanshirazi, A.M. Katti, M. Czok, Z.D. Ma, and G. Guiochon. *J. Chromatogr.* 484, 103–124, 1989.
42. A.M. Katti, R. Ramscy, and G. Guiochon. *J. Chromatogr.* 477, 119–130, 1989.
43. J.A. Wilkins, R. Xiang, and C. Horváth. *Anal. Chem.* 74, 3933–3941, 2002.
44. R. Xiang, C. Horváth, and J.A. Wilkins. *Anal. Chem.* 75, 1819–1827, 2003.

12 Field-Flow Fractionation

Luisa Pasti, Francesco Dondi, and Catia Contado

CONTENTS

12.1 INTRODUCTION

Field-flow fractionation, commonly designated as FFF, is a versatile family of separation techniques able to separate and characterize an enormous assortment of colloidal-supramolecular species in a wide range of dimensions/molecular weights. Giddings is considered the inventor of this technique since he contributed to the development of theory, different techniques, instrumentation, methodology, and applications [1], even if studies on the theoretical fundamentals of fractionation under force and flow fields had appeared before and/or independently [2].

FFF is a chromatographic elution technique and the instrumentation is very similar to that used in chromatography (Figure 12.1a). Like chromatography, typical FFF operations involve the injection of a narrow sample band into a stream flowing through a flow chamber where separation occurs. In chromatography, the flow chamber is called the column; in FFF, it is the channel. The FFF channel contains no packing; it is empty, that is, it is a one-phase system (see Figure 12.1b). Laminar flow in a thin, open channel is generally described by a parabolic flow profile. In parabolic flow, the flow stream velocity varies with the distance from the channel walls, that is, the velocity is zero at the walls, and gradually increases as one moves away from the walls, finally reaching the maximum at the center of the channel.

In FFF, retention and separation are achieved by a direct concert of the nonuniform flow profile in a liquid carrier with a force field applied at the right angle to the flow in the channel. Because

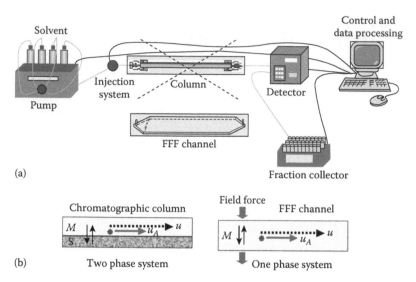

(a)

(b)

FIGURE 12.1   (a) FFF instrumental setup: comparison with a standard HPLC arrangement. (b) Separation principle for chromatography and FFF. M, mobile phase; S, stationary phase; u_A, velocity of the analyte A; and u, velocity of the mobile phase.

different species are held by the field in different regions of the channel and different regions have different velocities, the species migrate at different speeds. The differential migration of the species in FFF is thus achieved by the combined effect of the field and the flow velocity profile in a manner similar to the effect of the differential partition between the mobile and stationary phases in chromatography (see Figure 12.1b).

FFF can handle many different types of samples: polymers, various colloids, latexes, and biological components ranging from proteins to living cells. FFF can separate species according to different properties, for example, particle size, molar mass, density, magnetic susceptibility, dielectric permittivity, surface charge, and compressibility, depending on the type of applied field. Due to the theoretical tractability of the FFF separation process, various physicochemical properties can be determined from the fractogram (e.g., apparent mass, density, porosity, diffusion coefficient, and thermodiffusion coefficient) [3].

FFF is a still growing area with significant innovative advancements and promising developments. The field offers significant advantages and a broader range of applications in terms of molecular weight of the separated species than other chromatographic techniques (see Figure 12.2).

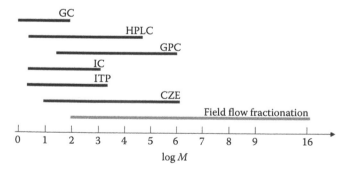

FIGURE 12.2   Working mass range of the FFF techniques compared with other common separation techniques. GC, gas chromatography; HPLC, high-performance liquid chromatography; GPC, gel permeation chromatography; IC, ionic chromatography; ITP, isotachophoretic separation; CZE, capillary zone electrophoresis.

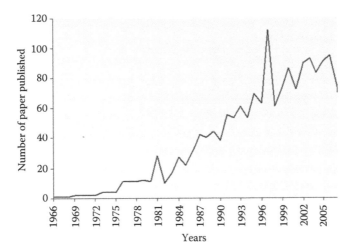

FIGURE 12.3 Number of published papers on FFF since its invention (1966). (From SciFinder 2007, CAS organization.)

The numerous variants in field type, operating mode, and channel configuration make FFF one of the most versatile families of separation techniques.

The elution character of the FFF techniques allows for it to be used in combination with other methods for further on-line or off-line characterization of the analytes (see Figure 12.1). FFF can be hyphenated with selective detection systems like mass spectrometry, multiangle laser scattering and can be combined with different separation techniques in multidimensional modes. In Figure 12.3, the trend in the number of published papers is reported.

12.2 PRINCIPLES OF SEPARATION

In FFF, separation is determined by the combined action of the nonuniform flow profile and transverse field effects. The classical configuration assumes the FFF channel as two infinite parallel plates (see Figure 12.4), of which the accumulation wall lies at $x=0$, where x is the cross-channel axis (directed upward from the accumulation wall). Inside the channel, the carrier fluid, assumed to have a constant viscosity, has a velocity profile $v(x)$ that takes the form

$$v(x) = 6\langle v\rangle\left[\frac{x}{w} - \left(\frac{x}{w}\right)^2\right] \tag{12.1}$$

where
$\langle v\rangle$ is the average cross-sectional velocity
w is the width of the channel

The interaction between particle and applied field determines the concentration profile distribution inside the channel. Theory assumes that the particles do not interact with each other, that is, the particle concentration is so low that they can be considered in a condition of infinite dilution and thus with an ideal behavior. Under these assumptions, the field–particle interaction has been classified into three major elution modes: Brownian (or normal), steric, and hyperlayer (or focusing). These elution modes, which are limiting cases, correspond to different mechanisms of separation and, theoretically, they can explain FFF retention on the basis of particle properties such as the

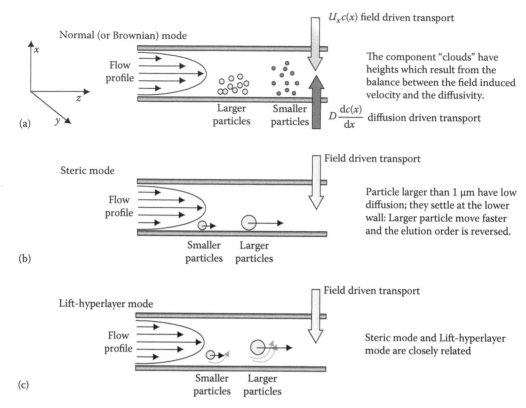

FIGURE 12.4 FFF particle equilibrium distribution in (a) Brownian (normal) mode, (b) steric mode, and (c) hyperlayer mode.

particle size or experimental variables such as the carrier flow rate. Experimentally, these mechanisms act together but with different weights. In the following, only the three major elution modes will be considered. Obviously, this oversimplification is valid only when one mechanism is strongly dominant over others; in practice, each real case is unique and the elution mechanism must be evaluated with the aid of appropriate literature.

As a general rule, species smaller than 1 μm are separated under Brownian mode, while larger species are separated in steric mode. The third mode—the hyperlayer—is achieved when the average flow rate is increased and secondary forces act significantly on the sample components. These limiting FFF modes are illustrated in Figure 12.4.

Brownian elution mode occurs when the field-induced velocity of the analyte $|U_x|$ in the separation channel is constant and comparable to its diffusive motion. Under these conditions, that is, when the field-induced flow and diffusive flow are equal (see Figure 12.4a), at the equilibrium, one has

$$U_x c(x) = D \frac{dc(x)}{dx} \qquad (12.2)$$

with
  D is the diffusion coefficient
  c is the concentration

The balancing of these opposing transport processes yields the equilibrium concentration profile of the analyte given by the well-known exponential relationship [4]

$$c(x) = c_0 \exp\left(\frac{xU_x}{D}\right) = c_0 \exp\left(-\frac{x|U_x|}{D}\right) = c_0 \exp\left(-\frac{x}{l}\right) \qquad (12.3)$$

where
  c_0 is the maximum concentration at the accumulation channel wall
  c is the concentration at distance x from the accumulation wall
  l is the mean thickness of the analyte layer

which can be expressed as the following ratio:

$$l = \frac{D}{|U_x|} = \frac{kT}{F} \qquad (12.4)$$

where
  T is the absolute temperature
  k is the Boltzmann constant
  F is the field force acting over the analyte

Note that, in Equation 12.3, U_x is a negative quantity since it is opposed to the direction x (see Figure 12.5) and this explains the negative sign in the second member of Equation 12.3 where the absolute quantity $|U_x|$ is introduced. Equation 12.4 relates U_x to F. Useful alternative expressions for F are obtained by combining Equation 12.4 with Einstein's equation

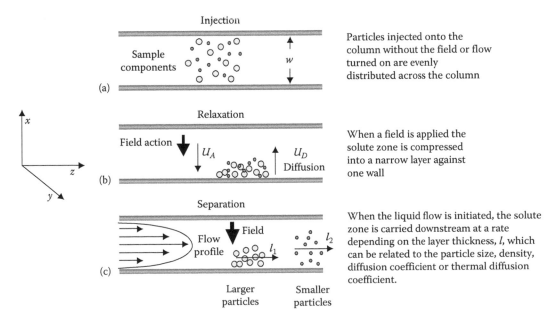

FIGURE 12.5 Brownian (or "normal") separation steps.

$$D = \frac{kT}{f} \qquad (12.5)$$

where f is the friction coefficient, giving

$$F = |U_x| \cdot f = kT \cdot \frac{|U_x|}{D} \qquad (12.6)$$

The mean thickness analyte layer (Equation 12.4) normalized vs. the channel width w (see Figure 12.5) defines the retention parameter λ

$$\lambda = \frac{l}{w} \qquad (12.7)$$

By combining Equations 12.4, 12.5, and 12.7, one obtains

$$\lambda = \frac{D}{U_x w} = \frac{kT}{Fw} \qquad (12.8)$$

Different species, belonging to the same sample, form exponential distributions or layers of different thickness l (see Figure 12.5c): the greater the thickness l, the higher the mean elevation above the accumulation wall and the further the penetration into the fast streamlines of the parabolic flow profile. The thickness is inversely proportional to the force exerted on the particle by the field (see Equation 12.8). Usually, this force increases with particle size and this defines the so-called "*normal*" mode of elution: smaller particles migrate faster and elute earlier than larger particles (see Figure 12.4a). This sequence is referred to as the "*normal elution order.*" The above-described equilibrium-Brownian mode will behave as normal mode. However, Brownian, equilibrium, and normal concepts are strictly interrelated.

Steric elution mode occurs when the particles are greater than 1 μm. Such large particles have negligible diffusion and they accumulate near the accumulation wall. The mean layer thickness is indeed directly proportional to D and inversely proportional to the field force F (see Equation 12.3). The condition is depicted in Figure 12.4b. The particles will reach the surface of the accumulation wall and stop. The particles of a given size will form a layer with the particle centers elevated by one radius above the wall: the greater the particle dimension, the deeper the penetration into the center of the parabolic flow profile, and hence, larger particles will be displaced more rapidly by the channel flow than smaller ones. This behavior is exactly the inverse of the normal elution mode and it is referred to as "*inverted elution order.*" The above-described mechanism is, however, an oversimplified model since the particles most likely do not come into contact with the surface of the accumulation wall since, in proximity of the wall, other forces appear—of hydrodynamic nature, that is, related to the flow—which "lift" the particles and exert opposition to the particle's close approach to the wall.

Lift-hyperlayer operation mode occurs when the lift forces are strong enough vs. the opposing field forces, or the latter are weak enough, to drive the particles away from the accumulation wall (see Figure 12.4c). Lift forces are usually significant for particles >1 μm. Under these conditions, particles will concentrate in thin layers due to the balancing of the two opposite forces: the applied field force and the lift force. In fact, the field forces work against the tendency of the particles to diffuse in the positive x direction, whereas the lift forces act against the tendency to diffuse in the negative x direction (see Figure 12.4c). Particles of different sizes will focus at different elevations, thus forming hyperlayers at different elevations. Larger particles are elevated at higher elevation distances vs. the accumulation wall and will thus move faster than smaller ones: the order of elution will thus be the same as that observed under steric mode: an inverted elution order. Sometimes the two processes are lumped together and referred to as steric-hyperlayer mode or as focusing mode. Finally, since lift forces depend on flow rate, steric-hyperlayer mode is achieved under high flow rate conditions and is thus found in the high-speed analysis of large particles.

12.2.1 SEPARATION PARAMETERS: RETENTION RATIO, PLATE HEIGHT, RESOLUTION, SELECTIVITY, AND FRACTIONATING POWER

FFF process can be theoretically described because (a) the flow regime inside the channel of well-defined geometry can be mathematically represented (see Equation 12.1), and (b) the tractability of the various force fields employed in the different techniques allows one to describe the analyte concentration profile (see Equation 12.3). The retention ratio expresses the retardation of an analyte zone caused by its interaction with the field, and it is given by the ratio of the average velocity of the analyte zone v_{zone} and the average velocity of the carrier liquid $\langle v \rangle$:

$$R = \frac{v_{zone}}{\langle v \rangle} \tag{12.9}$$

R can be expressed as a function of the retention time. In fact, by assuming L as the channel length, Equation 12.9 becomes

$$R = \frac{L/t_R}{L/t_0} = \frac{t_0}{t_R} \tag{12.10}$$

where t_0 and t_R are the void time and the retention time, respectively. Equation 12.10 defines the experimental retention ratio. The retention ratio can, in turn, be related to the λ parameter by

$$R = 6\lambda \left[\coth\left(\frac{1}{2\lambda}\right) - 2\lambda \right] \tag{12.11}$$

Equation 12.11 is obtained by substituting Equations 12.1 and 12.3 into Equation 12.9 and by performing the integration in order to calculate $\langle v \rangle$ [3]. Well-retained compounds have $\lambda < 0.1$ and, under these conditions, the retention ratio can be approximated as a linear function of λ:

$$R \cong 6\lambda \tag{12.12}$$

Equation 12.12 is accurate to within 5% when $\lambda < 0.02$ [3]. Consequently, under these conditions, by combining Equations 12.8 and 12.12, the retention time can be related to the force field as follows:

$$t_R \cong 6\frac{Fw}{kT}t_0 \tag{12.13}$$

It should be stressed that Equation 12.13 is approximated and it should be used cautiously to calculate physicochemical parameters from the retention data because of the associated error. However, it can be useful to obtain simplified, approximated relationships between retention and field–particle interaction as will be detailed below for the specific FFF techniques. The physicochemical parameters of the analytes can be calculated when Equation 12.11 is used in association with the pertinent λ expression. Table 12.1 reports the explicit expressions of λ for different subtechniques. Examples of

TABLE 12.1
Summary of the Applicable Physical Field in FFF from Which the FFF Subtechniques Originate

FFF Technique		Field	Drift Velocity	Retention Parameter	Force	Retention Parameter	Maximum Theoretical Selectivity ($\lambda < 0.01$)	Sample Component Parameters Characterized by Retention
Flow	Q_c	Volumetric cross-flow rate	$U_A = \dfrac{Q_c w}{V^0}$	$\lambda = \dfrac{DV^0}{Q_c w^2}$	$F = f\dfrac{Q_c w}{V^0}$	$\lambda = \dfrac{kTV^0}{fQ_c w^2}$	$S_d \approx 1$	D d_h stock diameter
Sedimentation	G	Gravitational acceleration	$U_A = sg$	$\lambda = \dfrac{D}{sgw}$	$F = m_{eff}g$	$\lambda = \dfrac{kT}{m_{eff}gw}$	$S_m \approx 1$	d_v volume diameter
	G	Centrifugal acceleration	$U_A = sG = \omega^2 r$	$\lambda = \dfrac{D}{sGw}$	$F = m_{eff}\omega^2 r$	$\lambda = \dfrac{kT}{m_{eff}\omega^2 rw}$	$S_d \approx 3$	ρ density
Thermal	$\dfrac{dT}{dx}$	Temperature gradient	$U_A = D_T\dfrac{dT}{dx}$	$\lambda = \dfrac{D}{D_T(dT/dx)w}$	$F = D\left(\dfrac{D_T}{D}+\gamma\right)\dfrac{dT}{dx}$	$\lambda = \dfrac{1}{((D_T/D)+\gamma)(dT/dx)w}$	$S_M \approx a_1$ (Equations 12.49 and 12.51)	D_T/D thermal factor (Soret coefficient)
Electrical	E	Electrical field	$U_A = \mu E$	$\lambda = \dfrac{D}{\mu Ew}$	$F = kT\dfrac{\mu E}{D}$	$\lambda = \dfrac{kT}{\mu Efw}$	$S_d \approx 1$	D μ mobility

D = mass diffumsion coefficient; D_T = thermal diffusion coefficient; f = friction coefficient; $G = \omega^2 r$ (centrifugal acceleration); k = Boltzmann constant; m_{eff} = particle effective mass; r = radius of centrifuge basket; s = sedimentation coefficient; T = absolute temperature; V^0 = geometric volume of the channel; w = channel thickness; γ = thermal expansion coefficient; μ = electrophoretic mobility; ω = angular rotation frequency.

physicochemical parameters that can be determined are molecular weight, hydrodynamic radius or equivalent particle diameter, density, porosity, and transport properties such as sedimentation coefficient, diffusion coefficient, thermal diffusion coefficient, electrophoretic mobility [3]. These same quantities are determined as distributions in the case of polydisperse materials.

As in chromatography, the most common variable to describe dispersion in FFF is the plate height (H), defined as the ratio between the spatial variance (σ_s^2) of the band and its mean position (s) as it migrates inside the separation medium of length (L):

$$H = \frac{\sigma_s^2}{s} = L\left(\frac{\sigma}{t_R}\right)^2 = \frac{L}{N} \tag{12.14}$$

where equivalent expressions are also reported. In Equation 12.14, σ is the peak standard deviation in time units and N the number of theoretical plates.

Thanks to the additivity of variance, the total plate height can be divided up into contributions due to independent factors. In FFF, this can be represented as

$$H = \sum_i H_i = H_d + H_n + H_r + H_p + H_e + H_s + H_o \tag{12.15}$$

where H_i are the various contributions to H corresponding to: axial diffusion (d), nonequilibrium (n), inadequate sample relaxation (r), sample polydispersity (p), channel triangular ends (e), injection of finite sample volume (s), and off-line analyte detection (o) (for a detailed analysis and explanations of these effects, see Ref. [3]).

In FFF, with a well-designed experimental setup, it can be demonstrated that the intrinsic sample polydispersity and nonequilibrium effect are the principal sources of peak dispersion [5]. The contribution to band broadening due to sample polydispersity is given by

$$H_p = LS^2 \frac{\sigma_a^2}{a^2} \tag{12.16}$$

where

a represent the analyte property responsible for fractionation (i.e., mass or diameter and σ_a^2 the variance of a)

S is the analyte property-based selectivity [3] (i.e., mass-based or diameter-based) of the system, as explained in the text below

As polydispersity is an inherent property of the sample being processed, it is not a system property and can be ignored when optimizing an instrument.

The nonequilibrium effect is due to the different velocities at which the components of the analyte are carried down the channel. The different velocities, in turn, originate from the laminar nature of the flow: since the constituents of the analyte are dispersed in these laminae, they undergo differential migration velocities in the axial direction. The expression of the nonequilibrium system dispersion takes the following form:

$$H_n = \frac{\chi w^2 \langle \upsilon \rangle}{D} \tag{12.17}$$

where χ is the nonequilibrium coefficient that has the following limits corresponding to no retention and high retention, respectively [3,6]

$$\chi = \frac{1}{105} \quad \lambda = \infty$$

$$\chi = 24\lambda^3 \quad \lambda \to 0 \tag{12.18}$$

The asymptotic plate height value, for $\lambda \to 0$, is given by

$$H_n = \frac{24\lambda^3 w^2 \langle v \rangle}{D} \tag{12.19}$$

From Equation 12.19, it can be seen that the maximum efficiency is achieved when retention is high (e.g., λ is low, see Equation 12.10) and the channel thickness is made as thin as possible.

As in chromatography, resolution (R_S) is a measure of separation between two components:

$$R_S = \frac{\Delta t_R}{2(\sigma_1 + \sigma_2)} = \frac{\Delta t_R}{4\bar{\sigma}} \tag{12.20}$$

where
Δt_R is the difference in the retention time of neighboring peaks
σ is the peak standard deviation

The subscripts refer to components 1 and 2 and $\bar{\sigma}$ indicates the average value of σ for the two components. Δt_R is due to a selective dispersion, the capability of the technique to separate two different components, whereas σ is an indicator of random dispersions.

Selectivity is defined as

$$S = \left| \frac{d(\log t_R)}{d \log a} \right| \tag{12.21a}$$

Generally, selectivity can be defined in terms of mass (m) or diameter (d), given, respectively, by

$$S_m = \left| \frac{d(\log t_R)}{d \log M} \right| \tag{12.21b}$$

$$S_d = \left| \frac{d(\log t_R)}{d \log d} \right| \tag{12.21c}$$

High selectivity makes FFF a powerful tool for separation and characterization because it means that small changes in particle size or molar mass cause significant changes in retention time.

FFF selectivities assume limiting values at high retention level: for sedimentation FFF (SdFFF), $S_d = 3$, whereas $S_m = 1$; for flow FFF (FlFFF), the limiting value of S_d is unitary. For ThFFF, the S_m of dissolved polymers depends on the polymer–solvent system with typical values in the 0.5–0.7 range. In ThFFF particle separation, a wide range of S_d values have been reported [3].

A more specific parameter that describes the separation capability is the fractionating power. Like selectivity, the fractionating power can be defined on the basis of the analyte property responsible for fractionation, that is, F_m or F_d for mass and diameter, respectively:

$$F_m = \frac{R_S}{\mathrm{d}m/m} \tag{12.22a}$$

$$F_d = \frac{R_S}{\mathrm{d}d/d} \tag{12.22b}$$

F_d expresses the resolution between particles whose diameters differ by the relative quantity $\mathrm{d}d/d$.

Fractionating power makes it possible to establish the relative increment in particle diameter or mass that can be separated with unit resolution. It can be demonstrated (by substituting Equations 12.14, 12.20, and 12.21b or c into Equation 12.22a or b, respectively) that F_d (or F_m) can also by expressed by

$$F_{d,(m)} = \left(\frac{1}{4}\right) N^{1/2} S_{d,(m)} \tag{12.23}$$

Equation 12.23 shows that the fractionating power depends on both random dispersion, which is inversely related to N, and selective dispersion, which is defined by S.

12.3 FFF TECHNIQUES

Different fields can be used in FFF, each defining a different range and area of applicability, depending both on the field type and on the nature of the analyte.

Figure 12.6 schematizes the application dimensional ranges of the different FFF techniques. Each field type requires a specific instrument design, and thus each field corresponds to an FFF technique. Although in some cases a channel utilizing simultaneously more than one field has been reported [7], because of strong instrumental and operative constrains, the multiple-field technique is not commercially available. A list of field types and corresponding techniques is reported in Table 12.1.

The most common, and commercially available, FFF variants are flow FFF (FlFFF), sedimentation FFF (SdFFF), and thermal FFF (ThFFF). These techniques are described in greater detail

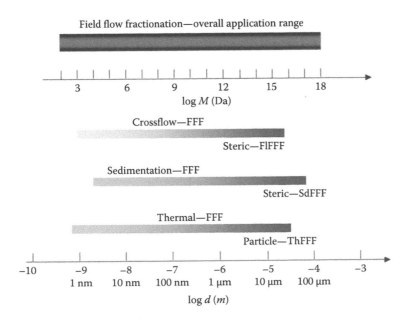

FIGURE 12.6 FFF application size and mass range.

below. Other fields such as electrical (ElFFF), force of gravity (GrFFF), magnetic (MgFFF), and a few less common ones have been applied as well [8,9].

12.3.1 Instrumentation: General Scheme

Like chromatography, the FFF instrument consists of a pump to deliver the carrier fluid, a separation medium (FFF channel), a detector responding to the eluting species, and a computer to control the operative parameters (e.g., field, flow) and to acquire the digitized fractogram (see Figure 12.1).

The most common FFF channels have a thin ribbon-like structure, of rectangular cross section, elongated along the flow axis with a length of tens of centimeters, 20–250 μm thick, with a breadth of few centimeters. Miniaturization of the separation channel was also realized. Other channel geometries have been used in different FFF subtechniques. Among them, one must recall (1) the trapezoidal cross section employed in asymmetrical FlFFF: it makes it possible to keep the axial-flow constant (since the cross-flow and the axial flow are generated by the same pump) and to reduce sample dilution; (2) the capillary fibers, that is, thin capillary tubes of round cross section that are employed in hollow fibers FlFFF (see Section 12.5); (3) the thin disc-shaped channel, where the carrier liquid flows radially from the center toward the circumference of the channel and a steady stream of the sample mixture is introduced continuously at a set distance from the carrier inlet, a technique which has been proposed for 2-D separation; and (4) the flat channels usually obtained by cutting and removing a piece of proper shape from a thin foil of plastic (i.e., the spacer). This spacer is sandwiched between two uniform walls of appropriate material.

The wall material must be able to transmit the field across the channel walls, and the wall surface must be uniform to guarantee unperturbed flow conditions. For instance, in FlFFF, the walls are permeable to fluid flow; those used in ThFFF must be thermally conductive. In Figure 12.7, sedimentation and ThFFF channels are sketched.

12.3.2 Detailed Instrumentation and Separation Operation Modes

12.3.2.1 Cross-Flow FFF

Cross-flow FFF or, as it was known in the past, FlFFF draws its name from the "field" type used to transport sample components across the channel thickness to the accumulation wall [3,8]. The

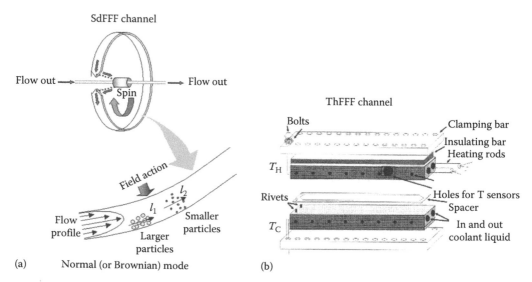

(a)   Normal (or Brownian) mode   (b)

FIGURE 12.7   (a) SdFFF and (b) ThFFF channel schemes: T_H, temperature of the hot wall and T_C, temperature of the cold wall.

second stream of carrier (cross-flow) is driven perpendicular to the axial flowstream and it acts equally—exerting equal force—on all the sample components (analytes), so that the entire sample is displaced regardless of mass, density, size, charge, etc. As a consequence, FlFFF is the most universally applicable FFF technique since the displacement by flow, acting as a field, is universal: separation is based only on the component size because a moving fluid is capable of displacing every suspended object, from molecules to particles, in its path. The application size range spans from M of 10^3–10^9 Da (macromolecules) to $100\,\mu m$ in hydrodynamic diameter (the diameter of a sphere with the same water density and particle velocity) for particulate samples [10]. The lower size limit is determined by the high pressures caused by the combination of low-molecular-weight cutoff membranes needed to retain the sample in the channel and the high cross-flow rates needed to counteract diffusion. The upper limit in particle diameter is generally about 20% of the channel thickness.

For materials smaller than ~1 µm, separation occurs in the normal mode [11]. According to this mechanism, the retention time (t_R) is inversely proportional to the diffusion coefficient D and proportional to the hydrodynamic diameter. The correspondence, for highly retained analytes (i.e., $R = 6\lambda$—Equation 12.12), is clear from

$$R = \frac{t_0}{t_R} \cong \frac{6kTV^0}{3\pi\eta w^2 d_h Q_c} = \frac{6DV^0}{w^2 Q_c} \qquad (12.24)$$

where
  k is the Boltzmann constant
  T is the absolute temperature during elution
  V^0 is the channel void volume
  η is the carrier viscosity
  w is the channel thickness
  Q_c is the volumetric flow rate of the cross-flow

The order of sample elution for the normal mode of operation is high diffusion coefficient (small particles or low M macromolecules) followed by decreasing D (large particles or high M macromolecules).

The alternative steric/hyperlayer mode can cover the 0.5–100 µm range [12]. In this operating mode, the larger particles elute first and this inversion in elution order is referred to as steric inversion. It generally occurs with diameters of around 1 µm when the Brownian motion of the molecules becomes too weak to oppose the field and all particles are initially forced onto the accumulation wall. The particles are also subjected to a lifting force from the channel flow along the membrane and reach an equilibrium position in the channel, at which the lift forces balance the cross-flow force (see Figure 12.4c). Larger particles experience greater lift and are therefore pushed further away from the membrane and consequently elute earlier than the smaller particles [13].

To generate this kind of field, that is, the cross-flow, the channel walls must be made semipermeable, and a membrane must act as an accumulation wall so that carrier liquid is allowed to pass but not the analyte. Cross-flow FFF can be broken down into symmetric, asymmetric, and cylindrical configurations (see Figure 12.8).

Symmetric FlFFF, which is the original configuration, utilizes two permeable walls with the cross-flow passing through both channel walls. Asymmetric FlFFF, which was introduced in 1987, has one permeable wall and one solid transparent wall. Despite more complicated theoretical equations, asymmetric FlFFF has become the more commonly used configuration because the transparent wall permits a visual performance check of the channel, higher efficiency has been observed [14], and more companies offer this design. A third configuration, that is not commercially available, uses a cylindrical tube such as a hollow fiber (HF) membrane [15] or a ceramic HF (Figure 12.8c).

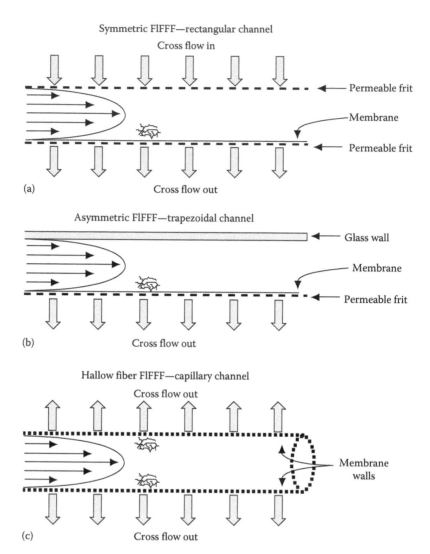

FIGURE 12.8 Cross-flow FFF variants. Side view of: (a) symmetric, (b) asymmetric, and (c) HF channels. Top view of the channels: (a) symmetric, (b) trapezoidal, and (c) capillary.

Recently, interest in the HF FlFFF channel has increased due to its simple channel construction and its low cost, both of which point toward the possibility of a disposable channel system. The idea of using HF membranes as cylindrical channels for FlFFF was reported as early as 1974. Compared with the classic FlFFF channel—having a rectangular cross section, in which separation is achieved through the application of cross-flow to the direction perpendicular to the migration—in HF FlFFF, flow separation is carried out by controlling radial flow through a porous membrane wall as the driving force of separation. Since the flow entering one end of the HF is divided into two parts (part of the flow leaves the channel wall through membrane pores and the remainder exits through the fiber outlet), in the HF FlFFF channel, the flow pattern resembles that found in asymmetrical FlFFF (see Figure 12.9).

For the two commercial FlFFF models, symmetric and asymmetric, it is possible to modulate the applied field strength during a run by programming the cross-flow rate; however, there are few examples of programmed analyses since, in FlFFF, the whole flowrate control is complex because of the presence of two flow axes, one for channel flow and the other for cross-flow. The back pressures on the channel and cross-flow outlets have to be adjusted, so that each pair of entering and

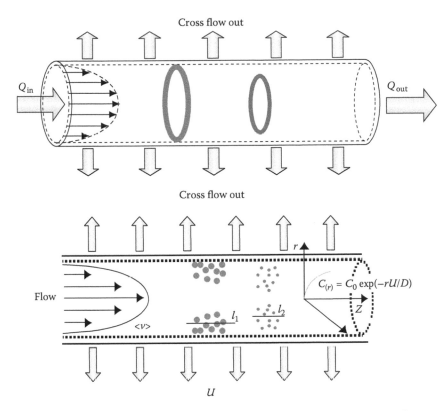

FIGURE 12.9 Schematic view of an HF FlFFF channel; Q_{in}, inlet flow rate; Q_{out}, outlet flow rate; l_1 and l_2, mean thickness of the analytes 1 and 2; U, field-induced flow; and $\langle v \rangle$, the average cross-velocity.

exiting flow rates are equal, that is, the inlet channel flow rate is equal to the outlet channel flow rate. For symmetric FlFFF, the program function applied thus far is

$$Q_c(t) = Q_{c,0} \left[\frac{(t_1 - t_a)}{(t - t_a)} \right]^p \tag{12.25}$$

where
    $Q_{c,0}$ is the initial cross-flow rate
    $Q_c(t)$ is the cross-flow rate at time t
    t_a is the asymptotic time
    p is the power

The requirements for these parameters are $t > t_a$, $t_1 > 0$, and $p > 0$. For normal mode FlFFF, the optimum condition, yielding uniform resolution, is obtained when $t_a = -pt_1$, and $p = 2$ [3] (see Figure 12.10).

When the channel has a trapezoidal shape [asymmetrical flow FFF (AsFlFFF)—see Figure 12.8], the cross-flow rate can be programmed to decrease linearly or exponentially. The commercially available power cross-flow program (Nova FFF® software) has the following expression [16]:

$$Q_c(t) = Q_{c,0} - \left(\frac{t - t_{start}}{t_{end} - t_{start}} \right)^p \left(Q_{c,end} - Q_{c,0} \right) \tag{12.26}$$

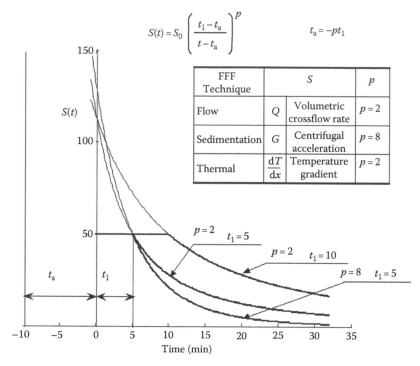

FIGURE 12.10 Example of programmed FFF decay.

where t_{start} and t_{end} are, respectively, the initial and the final time, that is, when the cross-flow becomes $Q_{\text{c,end}}$, the power p value is 0.50. The application of this power function is beneficial in comparison to a linear function, even if, on an old instrument—where the control software permitted decreasing the cross-flow rate according to linear functions—a series of linear decays have been applied in order to obtain a decreasing decay rate that closely approaches the exponential decay cross-flow function [17]

$$Q_{\text{c}}(t) = Q_{\text{c,start}}\,\text{e}\left[-\frac{\ln 2t(1/2)}{t}\right] \tag{12.27}$$

where
    $Q_{\text{c}}(t)$ is the cross-flow rate at time t
    $Q_{\text{c,start}}$ is the cross-flow rate at the start of the elution
    $t(1/2)$ is the half-time (1/2-time) constant determining how fast the function is decreasing

12.3.2.2 Sedimentation FFF

SdFFF has been a most successful technique for dealing with, and solving, the problems arising in particle and colloid science [18]. SdFFF is generally applicable in a size rage of 0.05–50 μm. SdFFF has the highest intrinsic resolving power of any of the current FFF techniques. In SdFFF, the multigravitational external field is generated by the rotation of the separation channel in a rotor basket, constituting of one of the most complex devices used in FFF separation. SdFFF fractionates according to the buoyant (effective) mass m_{eff}. Moreover, for SdFFF, there are two operating modes: the normal or Brownian mode that, as described above, is applicable to macromolecules and colloids of less than about 1–2 μm in size, and the steric-hyperlayer mode, applicable to micron-sized samples.

The explicit expression for the retention parameter λ (Equation 12.8) in SdFFF is given by [19]

$$\lambda = \frac{kT}{m_{\text{eff}} \omega^2 rw} \tag{12.28}$$

where ω is the angular velocity expressed in radiant $\omega = 2\pi(\text{rpm}/60)$, with rpm the number of revolutions per minute, r the radius of the centrifuge from rotation axis to the channel, and m_{eff} the effective mass of the particle suspended in the carrier given by

$$m_{\text{eff}} = m\frac{\rho_p - \rho_l}{\rho_p} = m\frac{\Delta\rho}{\rho_p} \tag{12.29}$$

where
  m is the particle mass
  ρ_p and ρ_l, respectively, the particle and liquid carrier densities and $\Delta\rho = (\rho_p - \rho_l)$

If $\Delta\rho$ is positive, particles accumulate at the outer wall of the SdFFF channel (which is the case described in Figure 12.7), whereas, if it is negative, they accumulate at the inner wall. If spherical particles are considered ($V_p = \pi d^3/6$), through Equation 12.29, Equation 12.28 becomes

$$\lambda = \frac{\rho_p kT}{m|\Delta\rho|\omega^2 rw} = \frac{6kT}{\pi d^3 |\rho_p - \rho_c|\omega^2 rw} \tag{12.30}$$

where d is the equivalent spherical particle diameter that can easily be calculated from the observed retention time t_R using the most simplified expression of retention (i.e., $R = 6\lambda$, Equation 12.12)

$$t_R = \frac{t_0 m|\Delta\rho|G}{6\rho_p kT} = \frac{t_0 \pi\eta|\Delta\rho|Gd^3}{36kT} \tag{12.31}$$

where $\omega^2 r$ is G. If the same particle type is modified by the surface adsorption of molecules of significant molecular weight (see Figure 12.11), thus creating an adsorbed film of mass m_c, because retention time depends on the effective particle mass, a retention time increment is observed (see Equation 12.31). In fact, the total effective mass of the coated particle $m_{\text{eff,p+c}}$ is given as [20]

$$m_{\text{eff,p+c}} = m_{\text{eff,p}} + m_{\text{eff,c}} \tag{12.32}$$

where $m_{\text{eff,p}}$ and $m_{\text{eff,c}}$ are the effective masses of the bare particle and the coating (adsorbed mass), respectively, expressed as

$$m_{\text{eff,p}} = m_p\left(\frac{\rho_p - \rho_l}{\rho_p}\right) \tag{12.33a}$$

$$m_{\text{eff,c}} = m_c\left(\frac{\rho_c - \rho_l}{\rho_c}\right) \tag{12.33b}$$

in which ρ_c is the density of the coating.

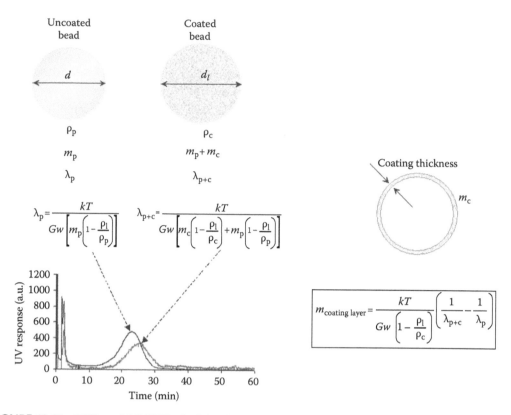

FIGURE 12.11 Differential SdFFF principles. The mass increase due to the adsorption of a molecular layer reflects a shift in retention time in the fractogram; the mass of the coating can be computed from the difference in λ values.

From Equations 12.28, 12.32, and 12.33, λ_p and λ_{p+c}, that is, the parameters experimentally determinable for the bare and coated particles, are

$$\lambda_p = \frac{kT}{Gw\left[m_p\left(1-(\rho_l/\rho_p)\right)\right]} \tag{12.34}$$

$$\lambda_{p+c} = \frac{kT}{Gw\left[m_c\left(1-(\rho_l/\rho_c)\right)+m_p\left(1-(\rho_l/\rho_p)\right)\right]} \tag{12.35}$$

By combining Equations 12.34 and 12.35, it is possible to calculate m_c:

$$m_c = \frac{kT}{Gw\left(1-(\rho_l/\rho_c)\right)}\left(\frac{1}{\lambda_{p+c}}-\frac{1}{\lambda_p}\right) \tag{12.36}$$

that is, by measuring the shift in the retention time t_R. The detection limits of the adsorbed mass $L(m_c)_D$ have been estimated to be of the order of 1×10^{-16} g [21].

For a sample of a homogeneous chemical composition, such as polymer lattices, Equation 12.30 can be rearranged in order to determine both particle diameter and sample density [22]:

$$\rho = \rho_p \pm \frac{6kT}{\pi w^2 dG\lambda} \tag{12.37}$$

This equation suggests that plotting carrier density vs. $1/G\lambda$ would yield a straight line with an intercept equal to the particle density, ρ_p, and with a slope related to the equivalent spherical diameter d.

The steric/hyperlayer mode developed into a powerful method for measuring the particle size distribution of particles in the 1–100 μm size range. However, in this separation mechanism, not only particle size, density, shape, and rigidity are involved but the channel geometry and flow rate characteristics also come into play. Although steric-hyperlayer FFF has the capability of producing fast separations, the lack of a precise model for lift forces (see Figure 12.4) initially hindered its implementation for the characterization of larger particles; however, the emergence of a calibration procedure that uses the principle of density compensation [23] has experimentally resolved this problem. The basic idea of the density compensation procedure is to drive different particles of identical size into the same equilibrium position in the channel, such as sample particles and latex standards used to calibrate the channel. This can be done by simply adjusting the field strength from that used in the standard run to a value that compensates for the different density of the sample. Thus, sample particles with a density that is different from that of the calibration standards are run at a field (rpm) that is defined by

$$\text{rpm} = \frac{\text{constant}}{\sqrt{\Delta\rho}} \tag{12.38}$$

where $\Delta\rho$ is the density difference between the sample and the carrier liquid. If $\Delta\rho$ is different for the sample compared to the standards, then the rotation rate (rpm) is adjusted according to Equation 12.38. As long as the sample run is carried out with the appropriately adjusted field, its particle size distribution can be obtained using calibration parameters determined from the standards. In the density compensation procedure, the accuracy of the sample density determines the accuracy of the resulting particle size distribution [23].

In SdFFF, the channel walls are built of polished stainless steel or polystyrene (PS) plates, depending on the applications, and are separated by a mylar spacer. These three pieces are coiled to fit the circumference of a centrifuge basket, which can spin at constant speeds (up to 3000 rpm—for commercially available instruments) or by following a field program decay. As in FlFFF, again in SdFFF, there are several forms of field programming, (see Figure 12.10) but the only one to yield uniform fractionating power over a broad diameter range is [13]

$$G(t) = G_0 \left[\frac{(t_1 - t_a)}{(t - t_a)} \right]^p \tag{12.39}$$

where G is the centrifugal field strength, which is keep constant at G_0 for a time t_1; the optimum condition, yielding uniform resolution, is obtained as in FlFFF with $t_a = -pt_1$ and $p = 8$. When power programming is used, the following relation exists between t_1 and d

$$t_1 - t_a \propto d^{1/3} \tag{12.40}$$

12.3.2.3 Thermal FFF

ThFFF is the FFF family technique that employs a temperature gradient as the applied field (see Figure 12.7). The presence of a thermal gradient in a fluid mixture induces a relative component matter flow known as thermal diffusion. Several terms are used to express the movement of material in response to a temperature gradient, including thermal diffusion, Soret or Ludwig–Soret effect (named after the discoverers of that effect in liquids [24]), and thermophoresis. Thermophoresis, the drifting of dispersed particles due to a thermal gradient, is the counterpart in macromolecular solutions or colloidal suspensions of the thermal diffusion (Soret effect), holding true in simple fluid mixtures.

A comprehensive physical understanding of thermal diffusion is still not fully developed and the microscopic picture is not yet complete, even though some new theories that improve the understanding of the phenomena have recently been proposed. Also, the direction of thermophoresis is unpredictable: in most cases, dispersed particles migrate toward the cold, displaying what we shall call "thermophobic" behavior, but examples of "thermophilic" motion (along the temperature gradient) have often been reported. A recent theory [25] developed for nanoparticles in water states that the Soret coefficient is given by the negative solvation entropy, divided by the thermal energy (kT), and the thermophilicity is attributed to increasing positive entropy of hydration. However, for colloids, the computation of solvation entropy is not a straightforward task, especially for complex analytical samples. A thermodiffusion theory based on hydrodynamic properties for mixtures of organic solvents has been also recently proposed [26]. The mathematical description of thermal diffusion is further complicated by the thermodynamic nonidealness of the experimentally studied systems [27]. Consequently, below, thermal diffusion is quantified, like other transport processes, by a phenomenological coefficient (diffusion coefficient) that defines the dependence of a mass or energy movement on a potential energy gradient. Thus, the thermal diffusion coefficient relates the thermophoretic velocity U_T induced in a material by the temperature gradient dT/dx applied along the cross section of the separation channel:

$$U_T = D_T \frac{dT}{dx} \tag{12.41}$$

where D_T is the thermal diffusion coefficient. Since concentration gradients relax via mass diffusion, the mass flow J can be written, as

$$J = -D \frac{dc}{dx} + U_T c \tag{12.42}$$

where
J (mol cm^{-2} s^{-1}) is the flux
D (cm^2 s^{-1}) is the diffusion coefficient
c is the concentration (Ref. [4]—Chapter 3, p. 45)

In the steady-state condition, that is, for $J=0$, by assuming that the thermal gradient is applied along the x direction, from Equations 12.41 and 12.42, one gets

$$\frac{dc}{dx} = -c S_T \frac{dT}{dx} \tag{12.43}$$

where

$$S_T = \frac{D_T}{D} \tag{12.44}$$

is the Soret coefficient.

Equation 12.6 in ThFFF becomes

$$F_T = kT S_T \frac{dT}{dx} \tag{12.45}$$

Equation 12.45 states that the ratio between the force (F_T) induced by the thermal gradient dT/dx and the thermal energy (kT) is proportional to the thermal gradient applied and the proportionality constant is the Soret coefficient. Equation 12.45 relates the Soret coefficient to F_T: thus, Equation 12.45 makes it possible to evaluate S_T once the force is known.

The ThFFF separation system is made up of a flat ribbon-like channel obtained by placing a trimming-spacer between two flat bars kept at different temperatures: T_H (at the upper wall) and T_C (at the lower wall), with $\Delta T = T_H - T_C$. The thickness of the spacer defines the channel thickness w. In the channel cross section, the thermal diffusion process pushes the analyte toward the so-called accumulation wall, usually the cold wall (thermophobic substances): the combination of the flow profile and the thermal diffusion produces the fractionation.

In ThFFF, the λ parameter (see Equation 12.8) is defined as

$$\lambda = \frac{1}{S_T(dT/dx)w} \tag{12.46}$$

Due to the channel symmetry, Equation 12.46 refers to the absolute values of S_T. In general, it is thus not possible to determine the sign of the Soret coefficient—for example, whether the thermophoresis is positive, with migration toward the hot wall, or negative in the opposite case—using ThFFF alone. Equation 12.46 holds true in an infinitesimal layer thickness, dx, located at value x and vertical coordinate ($0 \leq x \leq w$). The temperature does not change significantly within this dx layer located in the x position; moreover, here S_T is constant and referred to this temperature value.

The classical FFF retention equation (see Equation 12.11) does not apply to ThFFF since relevant physicochemical parameters—affecting both flow profile and analyte concentration distribution in the channel cross section—are temperature dependent and thus not constant in the channel cross-sectional area. Inside the channel, the flow of solvent carrier follows a distorted, parabolic flow profile because of the changing values of the carrier properties along the channel thickness (density, viscosity, and thermal conductivity). Under these conditions, the concentration profile differs from the exponential profile since the velocity profile is strongly distorted with respect to the parabolic profile. By taking into account these effects, the ThFFF retention equation (see Equation 12.11) becomes:

$$R = \frac{t_0}{t_R} = 6\lambda\left(v + (1 - 6\lambda v)\left(\coth\left(\frac{1}{2\lambda}\right) - 2\lambda\right)\right) \tag{12.47}$$

where v is a computable parameter [28], related to the changes in carrier viscosity η and carrier thermal conductivity κ in channel cross section and accounting for distortion of the ideal parabolic flow profile in the channel. Typical v values lie in the -0.1 to -0.15 range (e.g., the case of tetrahydrofuran having $\eta = 0.463$ cP, $\kappa = 1.66$ 10^4 erg cm^{-1} s^{-1} K^{-1}, (at $T = 298$ K) with $\Delta T = 20$–50 and $T_C = 300$–330 K [29]).

From Equations 12.46 and 12.47, one sees that ThFFF retention is related to ($1/S_T$), but not specifically to a conventional analyte property, such as molar mass M or particle diameter d. However, since $1/S_T$ depends on M or d, it also mediates the dependence between R and M or d. At constant T_C conditions, the relationship between λ, M, and ΔT can be experimentally exploited by using standards. For instance, in polymers mass characterization, monodisperse or polydisperse standards can be employed for a specific polymer–solvent system. Once the relationship λ vs. M, that is, the so-called calibration curve, is obtained, it is "universal," that is, valid for any ThFFF instrument [3]. A typical calibration function, which relates the instrumental response to analyte property (in this case M), is

$$\log(\lambda\Delta T) = a_0 + a_1 \log M \tag{12.48}$$

where a_0 and a_1 are suitable calibration constants. Under nonconstant T_C conditions, the dependence of the λ parameter on M and T_C can be modeled according to the following equation:

$$\log(\lambda\Delta T) = a_0 + a_1 \log M + a_2 \log\left(\frac{T_C}{298.15}\right) \quad (12.49)$$

The quantities a_i, $i = 0$–2 in Equation 12.49, are typical of the chosen calibration function, characteristic of the specific polymer–solvent system, and independent of the instrument.

When the simplified expression (see Equation 12.12) is employed and T_C is constant (see Equation 12.48), one has

$$t_R = \frac{t_0\Delta T}{6a_0 M^{a_1}} \quad (12.50)$$

Once the a_0 and a_1 coefficients are experimentally determined, they can be used to characterize an unknown monodisperse or polydisperse sample from the experimental determination of λ under given experimental conditions (see Equation 12.48).

For suspended particles, the calibration function of λ vs. d [30] is given by

$$\log(\lambda\Delta T) = a_0 + a_1 \log d \quad (12.51)$$

Different size dependencies of thermal diffusion in aqueous and organic carriers have been reported in the literature [31] and, to date, this phenomenon is still a research topic.

Generally, programming FFF consists of imposing a specific change in one of the "primary" variables governing the retention during the time interval: $0 \le t \le t_A$, where t_A is the analysis time which is usually greater than t_R. The term "primary" variables indicate the operative variables that affect retention such as the flow rate, the carrier composition, the applied force field (i.e., the thermal gradient), or a combination of these. Moreover, in ThFFF, the force field can also be programmed [3]. However, besides the force field (the primary variable here), which is intentionally changed by the set program, one must also carefully consider the other variables that undergo variation as a consequence of the primary variable variation, such as, for example, the carrier density and viscosity. The time dependence of ΔT is usually modeled according to the following functions:

$$\Delta T = \Delta T_0 \quad 0 \le t \le t_1 \quad (12.52a)$$

$$\Delta T(t) = \Delta T_0 \left(\frac{t_1 - t_a}{t - t_a}\right)^p \quad t \ge t_1 > t_a \quad (12.52b)$$

where t_1, t_a, and p are suitable parameters (see Figure 12.10), the subscript 0 refers to $t = 0$, that is, at the beginning of the program. Equations 12.52a and 12.52b describe the so-called power program. Normally, in ThFFF, p is chosen equal to 2. For a detailed study, see Ref. [32].

12.4 CHOICE OF THE MOST SUITABLE FFF TECHNIQUE: GENERAL CRITERIA

FFF can be used for the separation of a very broad size range (from sub-nanometers to millimeters) of analytes. The nature of the analyte, the dimensional range, and the nature of the solvent or dispersed medium are important factors in choosing the FFF technique. Concerning the analyte nature, it can be a dissolved macromolecule, a suspended colloid, or a particulate sample of biological, inorganic, or synthetic origin; the analyte size range may span from 500 Da to 100 μm. The solvent or dispersing medium can be categorized as aqueous or organic.

The choice of the FFF technique dictates which physicochemical parameters of the analyte govern its retention in the channel: FlFFF separates solely by size, SdFFF by both size and density, ThFFF by size and chemical composition, and ElFFF by mass and charge. The dependence of retention on factors other than size can be advantageous in some applications, and different information can be obtained by employing different techniques in combination or in sequence. On the other hand, the properties that can be characterized by FFF include analyte mass, density, volume, diffusion coefficient, charge, electrophoretic mobility, pI (isoelectric point), molecular weight, and particle diameter.

When different FFF techniques respond to the same physicochemical parameter such as the case of size, selection of the technique should also consider separation performance. Usually, SdFFF is preferred over FlFFF because retention time t_R depends on d (see Equations 12.21 and 12.22) and because band broadening is generally higher in FlFFF than in SdFFF. Consequently, for high polydisperse samples, FlFFF could outperform SdFFF. Moreover, FlFFF has a broader dimensional range of applicability (1 nm–100 μm) than SdFFF (20 nm–50 μm). SdFFF and FlFFF are generally applied to aqueous matrices. However, FlFFF can operate with organic solvents if the apparatus is equipped with an appropriate membrane. ThFFF is generally applied to macromolecules soluble in organic solvents. Nonetheless, for larger particles (gels and colloids), ThFFF can operate within an aqueous carrier [30]. ElFFF has been generally applied in water-based carriers.

Once the FFF field is chosen (i.e., the technique), one should select the proper operating mode that, as mentioned above, depends on particle size. Brownian (or normal) elution mode is employed in most applications in the lower range of particle dimensions. Sample retention R can be optimized by a proper choice of the field strength: the recommended R range lies between 0.25 and 0.05 because, at low retention, nonequilibrium effects can lead to significant deviation from standard retention as expressed by the theoretical FFF equations (see retention parameter in Table 12.1). When R is higher than 0.05, the solute zone is highly compressed inside the channel (see Figure 12.4). Consequently, particle–particle interactions as well as wall–particle interactions are enhanced. Nonequilibrium effects are generally reduced by increasing the relaxation time (see Figure 12.5 and the pertinent explanation in the text). Particle interactions also depend on the quantity and the concentration of injected samples: FFF techniques are prone to overloading effects, and injecting a minimum quantity of well-dispersed sample is always recommended. In general, a sample size of 1–50 μg is appropriate for most materials, the values close to the lower limit are being preferred when the detector has sufficient sensitivity.

The choice of carrier liquid is primarily based on the suspension stability for colloids or solvent goodness for macromolecule solutions. Moreover, surface-charged colloidal particles are also sensitive to ionic strength and addition of surfactants.

Finally, the liquid carrier velocity should be considered. The flow velocity is strictly related to the operating mode. In normal elution mode, both analysis time and resolution decrease as flow velocity decreases. In general, in normal mode, flow rates in the 0.1–0.5 mL min⁻¹ range are employed. Steric and hyperlayer modes operate at flow rate values higher than those typical of normal mode.

To sum up, the choice of operating conditions for a specific FFF application is made in a way that recalls the general criteria used in chromatography. An accurate search of literature addressed to similar samples that have been already analyzed by FFF techniques is very useful. A number of specific reviews have been published concerning, for example, environmental, pharmaceutical, and biological samples (see Section 12.5). As previously mentioned above, one of the most important factors is the stability of the considered colloidal system, for which a great deal of information can be obtained from specialized literature, such as colloid, polymer, and latex handbooks [33]. For example, the use of the proper surfactant (e.g., Fl-70) is common for SdFFF applications. Polymer analysis with ThFFF requires solvent types similar to those employed in size exclusion chromatography.

12.5   APPLICATIONS

12.5.1   CROSS-FLOW FFF

The applicability of FlFFF has been shown for the analysis of various water-soluble and biological macromolecules such as proteins, polysaccharides, ribosome, and tRNA in bacterial cells, starch, gelatin nanoparticle drug carrier systems, and pullulan. Several examples of pharmaceutical applications can be found in Ref. [34], where many biopolymeric applications of AsFlFFF are reported. The majority of the cases presented refer to protein analyses, although some examples of undissolved material and colloid separations are discussed. Applications of biological interest have seen the FlFFF used to analyze viruses, yeast cells, algae, human cells, and liposomes. An interesting link between biology and environmental problems is given by the use of FlFFF interfaced with ICP-MS.

Environmental applications of FlFFF have been carefully collected in a review by Gimbert et al. [35]. Separations of nanoparticles belong to the FlFFF tradition and this sector has recently found new, fully deserved impulse for microparticle separations. The FlFFF technique has been applied to analyze humic material and submicron Fe colloids. Coupled with ICP-MS, FlFFF has been applied to detect the major and trace element chemistry of aquatic colloids in groundwaters and to determine the trace element distribution in soil and compost-derived humic and colloidal fractions in municipal wastewater. Recently, the ICP-AES has also been proposed as a specific detector for FlFFF to analyze inorganic nanoparticles (Figure 12.12).

Among the many industrial applications, one can recall the analyses on carbon black, where FlFFF and SdFFF were used in synergy, and on carbon nanotube, for which a frit inlet AsFlFFF channel was used. Water-soluble polydisperse polymers were fractionated, with a very high selectivity, according to differences in the diffusion coefficient, yielding a diffusion coefficient spectrum which was then converted into a molecular weight (M) distribution curve based on the relationship between D and molecular weight [36].

From the technological point of view, the symmetrical FlFFF system has been technically bettered with a number of changes in order to improve separation performance: the frit inlet and frit outlet were proposed, respectively, to quickly relax sample components to their different equilibrium distances above the accumulation wall and for on-line concentration prior to detection. This last feature addresses the dilute nature of many biological samples and compensates for some of the dilution that occurs during the separation process.

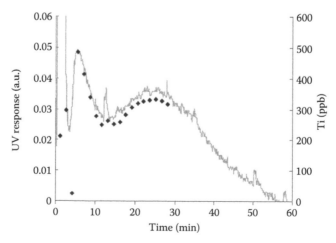

FIGURE 12.12   FlFFF fractogram achieved by injecting Degussa P-25 TiO$_2$ sample. On the right axis: the concentration (ppb=μg L^{-1}) of Ti determined by ICP-AES.

Diffusion coefficients and/or the hydrodynamic diameters and/or the MW of the sample components can be determined from a fractogram by using the theoretical FlFFF equations. The most commonly used FlFFF detector is the UV detector that suffices for many applications but M and M distributions require the use of calibration standards. In the absence of M calibration standards, the multiangle light scattering–differential refractometer (MALS–dRI) detector combination is used to measure the Ms and root mean square radii (rms) of the eluting samples. The MALS–dRI detectors offer a simplification of the FlFFF data analysis process; however, when analyzing large polymers (comparable with the size of the laser wavelength), MALS data analysis and interpretation have to be performed quite carefully [37].

Even if relatively new, HF FlFFF has been used to separate supramicrometer particles, proteins, water-soluble polymers, and synthetic organic-soluble polymers. Particle separation in HF FlFFF has recently been improved, reaching the level of efficiency normally achieved by conventional, rectangular FlFFF channels. With these channel-optimized HF FlFFF systems, separation speed and the resolution of nanosized particles have been increased. HF FlFFF has recently been examined as a means for off-line and on-line protein characterization by using the mass spectrometry (MS) through matrix-assisted laser desorption ionization time-of-flight mass spectrometry (MALDI-TOF MS) and electrospray ionization (ESI)-TOF MS, as specific detectors. On-line HF FlFFF and ESI-TOF MS analysis has demonstrated the viability of fractionating proteins by HF FlFFF followed by direct analysis of the protein ions in MS [38].

12.5.2 SEDIMENTATION FFF

The mass-sensitive SdFFF technique is a highly informative tool for the analysis of particles with sizes ranging from 20 nm to 1 μm, when used in the *normal* mode, or for particles up to roughly 50 μm, when used in the steric-hyperlayer mode. SdFFF has proved to be a very powerful technique for particle size analysis of submicrometer hydrosols, pharmaceutical colloids, inorganic, and organic species and it has proven to be an excellent tool for studying complex biological structures such as cells, viruses, bacteria, proteins, and nucleic acids. It has also demonstrated its strength and versatility in the study of various drug delivery vehicles, including vesicles, microcapsules, and emulsion.

The first field of application for SdFFF were latex beads, which were used either to test the channels or to produce separation results alternative to other separation techniques. PS nanoparticles used as model surfaces for bioanalytical work have been analyzed by SdFFF [39]. The appealing feature of SdFFF is its ability to characterize particle adlayers—by direct determination of the mass increase performed by observing the differences in retention between the bare and coated particles—with high precision and few error sources: the mass of the coating is determined advantageously on a "per particle" basis.

Inorganic particles such as the dispersion of carbon black, chromatographic silica, glass beads, silica beads, silica sol, silver sol, titania, and polydispersed zirconia colloidal suspensions have been characterized by SdFFF.

A number of environmental applications [3] have been performed in order to size characterize colloids collected in rivers (riverborne particles, SPM, and sediments), clay samples and ground limestone (from soils), coal particles, diesel soot particles (from combustion processes), or airborne particles in urban areas (from waste incinerators, vehicles, household-heating systems, and manufacturing). In many of these cases, not only the size but also the particle size distribution was important and thus, in conjunction with the traditional UV detector, specific detectors such as ETAAS, ICP-MS, ICP-AES were used [40] in order to obtain more detailed, more specific compositional information.

Pollen grains and starch granules, with major uses in the food, pharmaceutical, and cosmetic industries, have been separated by SdFFF, in which the results were complemented and compared with those achieved by using other FFF techniques (AsFlFFF and GrFFF) or flow cytometry.

Other pharmaceutical applications have seen the SdFFF applied successfully to monitor droplet size distributions in emulsions, together with their physical state or stability. Some examples are fluorocarbon emulsions, safflower oil emulsions, soybean oil emulsions, octane-in-water emulsions, and fat emulsions. SdFFF is also able to monitor changes in emulsion caused by aging or by the addition of electrolytes. SdFFF has been used to sort liposomes, as unilamellar vesicles or much larger multilamellar vesicles, the cubosom, and polylactate nanoparticles used as drug delivery systems [41].

Biological samples such as bacteria and viruses have been studied by SdFFF as well as FlFFF. These two techniques provided bacterial number, density, size, and mass distributions of bacterial cells of diverse shapes and sizes, molecular weights, sizes, densities, and diffusivities of viruses. SdFFF has been used to analyze protein particles, including casein derived from nonfat dry milk, albumin microspheres, and particles in cataractous lenses originating from the aggregation of lens proteins.

Hyperlayer SdFFF has proven to be the ideal elution mode for the separation of cells in different fields such as hematology, microorganism analysis, biochemistry/biotechnology, molecular biology, neurology, and cancer research. The first applications on cells were performed to characterize human cells, but more recently, this application field has had a great impulse. SdFFF has been used in monitoring cell apoptosis and cell differentiation [42]. Moreover, under strictly defined conditions and immunological cell characterization, SdFFF has provided a sterile, viable, useable, and purified immature neural cell fraction without inducing cell differentiation, an important result in stem cell preparation.

12.5.3 Thermal FFF

ThFFF is particularly suited for the separation of polymers that can be dissolved in an organic solvent, and it can be considered competitive with size exclusion chromatography (SEC). Water-soluble polymers can be separated as well by ThFFF. Retention of charged polymers can be improved by using mobile-phase additives or by employing proper aprotic solvents. ThFFF has been used to separate polymers with molecular weights ranging from 10^4 to more than 10^7. In fact, like SEC, ThFFF can separate a polymer mixture according to molecular weight (M), although in reverse order. In polymer characterization, the major limitation of ThFFF occurs in the separation of low-molecular-weight materials. This limit can be lowered by the proper use of solvent mixtures or by using special channels that are highly pressurized to prevent the solvent from boiling: this makes it possible to increase the thermal gradient utilized and, consequently, increase the retention. Moreover, because of different selectivity properties, ThFFF can also separate macromolecules of similar molecular weights but different compositions—such as in the case of PS (polystyrene) and polymethylmethacrylate of about 200,000 Da—as well as to obtain compositional information on copolymers [43]. The fact that the ThFFF separation system is made up of a flat, empty ribbon-like channel allows one to separate polymers and microgels without any of the adverse effects encountered in SEC [44]. ThFFF is also suitable for polyolefin analysis that usually requires high temperatures to achieve sample solubility. In fact, in ThFFF, the temperature limit is only dictated by the thermal stability of the channel spacer. Samples may contain components with a wide range of sizes or molecular weights that require differing field strengths for efficient separation. For these samples, the temperature gradient may be programmed with time (see Figure 12.13). In fact, programmed operating mode, achieved by varying the applied thermal field during the separation (i.e., ΔT), offers the following advantages: the mass-based fractionation power (i.e., mass selectivity) is kept uniform over the retention time axis; separation speed is increased; species detectability is kept uniform; and sample overloading is less critical [3].

Polymer analysis searches for M distribution. In both ThFFF and SEC, this is achieved by the calibration procedure that allows one to transform the retention time axis and the signal axis, respectively, of the fractogram or chromatogram into molar mass distribution [3]. In SEC, calibration has to be executed on each new column and repeatedly checked during its current employment

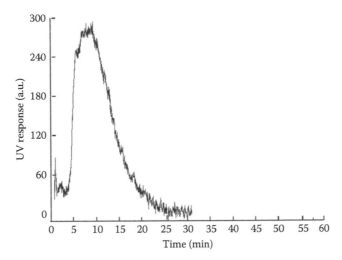

FIGURE 12.13 ThFFF fractogram of polypropylene $M_W = 1,470,000$ in decalin. $\Delta T = 51$ K, $T_C = 394$ K, flow rate $= 0.205$ mL min^{-1}.

through the use of a series of monodisperse standards, usually PS. This is because the porosity of the stationary phase determines the separation performance of the SEC column and this can differ from one column to the next and can even change during column use. However, column calibration performed using a given series of monodisperse standards can also be extended to other polymer types through Mark-Houwink constants.

Despite the above-mentioned features, ThFFF is not as widespread or popular as SEC. One possible reason is that calibration constants are only available for a very limited number of polymer–solvent systems. On the contrary, the SEC practitioner has available an extensive collection of Mark-Houwink constants for the different polymers, solvent types, and operating temperatures. In recent years, significant advances have been made in ThFFF calibration under isothermal conditions using a polydisperse standard as an alternative to a series of monodisperse standards. Recently, the calibration function was also made applicable to ThFFF field programming operation [32].

Calibration can be eliminated by coupling the SEC or ThFFF system to a proper detection system such as MALS detector or MALS and refraction index, which yields molar mass information without reference to standards [45]. The advantage of this coupling method is that compositional and dimensional information can be obtained regardless of whether the thermophoretic properties are understood. A continuous two-dimensional ThFFF channel for semipreparative separations of macromolecules has also been developed [46].

ThFFF is also suitable for studying the thermophoretic effect of particles (Figure 12.14), such as PS, modified PS, polybutadiene, copolymer latex particles, rubber particles, silica and modified silica particles, and metal particles (see Refs. [30,31,47] and referred literature). In these studies, both organic and aqueous carriers were employed. It was found that the thermophoretic effect depends on the (1) chemical nature of the particles, especially the chemical composition of the outer part, (2) particle size, and (3) nature and composition of the carrier (ionic strength and pH). By carefully selecting the carrier composition, very high size selectivity can be achieved. The use of ThFFF in determining the chemical composition of particles is, however, at the early stages. More systematic studies of particle thermodiffusion should be performed to further improve the applicability of ThFFF for particle composition analysis.

Recently, a miniaturized thermal apparatus, μ-ThFFF, was developed and applied to characterize the molar mass distribution of synthetic polymers in organic solvent as well to determine the particle size distribution of nanoparticles (PSs latex) in aqueous carrier. This μ-ThFFF proved to performed well in both macromolecule and particle analysis [48].

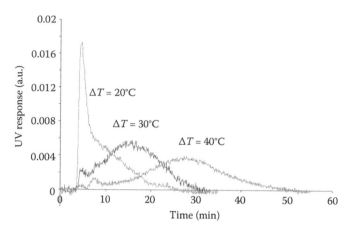

FIGURE 12.14 Elution profiles obtained at various thermal gradients; carrier: KNO_3 3 mM aqueous solution pH 9, flow rate: 0.304 mL min^{-1}, Si particle diameter = 300 nm.

12.6 PHYSICOCHEMICAL APPLICATIONS

The theoretical tractability of FFF makes it a tool for measuring physicochemical parameters [3]. Indeed, the physicochemical parameters derived from FFF measurements are essentially field-interaction parameters since FFF retention is associated with the action of a field. The parameters can be categorized as an intrinsic property of the analytes (e.g., mass, density, and molecular weight) or as solvent–analyte pair properties such as diffusion coefficient and phoretic mobility. In evaluating physicochemical parameters, it is crucial that retention measurements be as accurate as possible, that is, retention values should be in the range, in which secondary relaxation effects and nonideal sample behavior (overloading, particle-particle interactions) are negligible. Finally, an optimal apparatus setup—able to control and keep the variables governing retention stable—should be employed.

12.7 ADDITIONAL REMARKS

FFF is still a growing area of research and there are specific fields of application pushing toward innovations both in terms of instrumentation and methodology. For example, MgFFF has been developed specifically for stem cell research, and FlFFF (either HF FlFFF or AsFlFFF and traditional FlFFF) is driven overall by pharmaceutical–biological applications. In addition, as for others analytical techniques, the new tendency for FFF is toward miniaturization of the instruments. As examples we cite the "hollow fiber" channel for FlFFF [49], the microthermal unit developed by Janca [50], the microthermal–electrical unit by Gale and coworkers [51], or the SPLITT cells by Hoyos and coworkers [52]. In this regard, we should say that some important but very specialized topics were not reviewed here such as the SPLITT cells and separation channels similar to FFF that are useful for preparative aims, as they would require a detailed description which has already been reported in books and reviews [53,54].

Others FFF techniques, such as Acoustical FFF and Dielectric FFF are promising separation tools but they have not yet reached the commercial stage. Interested readers should address to specific literature [3].

LIST OF SYMBOLS

a analyte property responsible for fractionation
a_0, a_1, a_2 calibration constants
C, c concentration

c_0	concentration of accumulation wall
D	mass diffusion coefficient
D_T	thermal diffusion coefficient
d	derivative operator
d	particle diameter
d_h	hydrodynamic diameter
F	force on the analyte due to the applied field
F_d	diameter-based fractionating power
F_m	mass-based fractionating power
F_T	Force induced by the thermal gradient
f	friction coefficient
G	sedimentation field strength
G_0	initial sedimentation field strength in field programming
H	plate height
H_d	contribution to plate height from axial diffusion
H_e	contribution to plate height from channel triangular ends
H_i	ith contribution to plate height
H_{min}	minimum plate height
H_n	contribution to plate height from nonequilibrium dispersion
H_p	contribution to plate height from sample polydispersity
H_r	contribution to plate height from inadequate sample relaxation
H_s	contribution to plate height from injection of finite sample volume
H_o	contribution to plate height from offline analyte detection
k	Boltzmann's constant
J	mass flux
L	channel length
l	mean layer thickness of the analyte zone
M	molar mass
M_{ph}	mobile phase
m	mass of particle
m_c	mass of the coating
m_p	mass of particle
m_{eff}	effective (buoyant) mass
N	number of theoretical plates
p	exponent in power programming of the field
pI	isoelectric point
Q_c	cross-flow rate (volumetric) in flow FFF
$Q_{c,0}$	initial cross-flow (field strength)
$Q_{c,end}$	final cross-flow (field strength)
R	retention ratio
R_S	resolution
r	distance from rotor center or response factor of analyte (Equation 12.28)
S_d	diameter-based selectivity
S	analyte property-based selectivity
S_m	mass-based selectivity
S_{ph}	stationary phase
S_T	Soret coefficient, D_T/D
s	mean position inside the separation medium
T	absolute temperature
T_C	cold wall temperature in thermal FFF
T_H	hot wall temperature in thermal FFF

t_a	field programming parameter
t_A	analysis time
t_{end}	field programming final time
t_R	retention time
t_{start}	field programming initial time
t_0	void time
t_1	predecay time in field programming
u	velocity of the mobile phase–carrier
u_A	velocity of the analyte A
U_T	thermophoretic velocity
U_x	field-induced velocity of analyte
V^0	void volume or geometric volume of a channel
$\langle \upsilon \rangle$	cross-sectional average velocity of the carrier liquid
υ_{zone}	average velocity of analyte zone
w	channel thickness
x	cross-channel dimension in the FFF channel
$\Delta\rho$	density difference between the sample and carrier
ΔT	difference in temperature between hot and cold walls in ThFFF
ΔT_0	difference in temperature between hot and cold walls in ThFFF—initial condition in field programming
Δt_R	difference in retention time of neighboring peaks
χ	nonequilibrium coefficient
κ	thermal conductivity
λ	a dimensional retention parameter
λ_p	a dimensional retention parameter of a bare particle in SdFFF
λ_{p+c}	a dimensional retention parameter of a coated particle in SdFFF
σ	peak standard deviation
σ_a^2	analyte property variance
σ_s^2	band spatial variance
$\sigma_1\sigma_2$	standard deviation of peak 1 or 2
η	viscosity of carrier liquid
ρ_c	density of the coating layer
ρ_l	carrier density
ρ_p	particle density
ν	distortion flow profile parameter
ω	angular velocity expressed in radiant

ACKNOWLEDGMENTS

The authors gratefully acknowledge financial support for this work from the objective 2 (2000/06) Project LARA (FE120) and from Italian University and Scientific Research Ministry (FIRB RBNE03KN4S).

REFERENCES

1. J.C. Giddings, *Science*, 260, 1456–1465, 1993.
2. H.C. Berg, E.M. Purcell, *Proc. Natl. Acad. Sci.*, 58, 862–869, 1967.
3. M. Shimpf, K.D. Caldwell, J.C. Giddings (Eds.), *Field Flow Fractionation Handbook*, Wiley, New York, 2000.
4. J.C. Giddings, *Unified Separation Science*, Wiley, New York, 1991.
5. M.R. Shure, *J. Chromat. A*, 831, 89–104, 1999.
6. J.C. Giddings, Y.H. Yoon, K.D. Caldwell, M.N. Myers, M.E. Hovingh, *Sep. Sci.*, 10(4), 447–460, 1975.

7. X. Chen, K.-G. Wahlund, J.C. Giddings, *Anal. Chem.*, 60, 362–365, 1988.
8. K.D. Caldwell, L.F. Kesner, M.N. Myers, J.C. Giddings, *Science*, 176, 296–298, 1972.
9. J. Janča, *Field-Flow Fractionation*, Marcel Dekker Inc., New York, 1988.
10. S.K. Ratanathanawongs, I. Lee, J.C. Giddings, *ACS Symp. Ser.*, 472, 229–246, 1991.
11. J.C. Giddings, *Analyst*, 118, 1487–1494, 1993.
12. S.K. Ratanathanawongs, J.C. Giddings, *Anal. Chem.*, 64, 6–15, 1992.
13. P.S. Williams, J.C. Giddings, *Anal. Chem.*, 59, 2038–2044, 1987.
14. K.-G. Wahlund, J.C. Giddings, *Anal. Chem.*, 59, 1332–1339, 1987.
15. W.J. Lee, B.-R. Min, M.H. Moon, *Anal. Chem.*, 71, 3446–3452, 1999.
16. M. Leeman, M.T. Islam, W.G. Haseltine, *J. Chromatogr. A*, 1172, 194–203, 2007.
17. M. Leeman, K.-G. Wahlund, B. Wittgren, *J. Chromatogr. A*, 1134, 236–245, 2006.
18. J.C. Giddings, G. Karaiskakis, K.D. Caldwell, M.N. Myers, *J. Colloid Interfaces Sci.*, 92, 66–80, 1983.
19. J.C. Giddings, *Anal. Chem.*, 46, 1917–1923, 1974.
20. R. Beckett, *Langmuir*, 7, 2040–2047, 1991.
21. L. Bregola, C. Contado, M. Martin, L. Pasti, F. Dondi, *J. Sep. Sci.*, 30, 2760–2779, 2007.
22. J.C. Giddings, M.H. Moon, *Anal. Chem.*, 63, 2869–2877, 1991.
23. J.C. Giddings, M.H. Moon, P.S. Williams, M.N. Myers, *Anal. Chem.*, 63, 1366–1372, 1991.
24. H.J.V. Tyrrell, *Diffusion and Heat Flow in Liquids*, Butterworth & Co., London, U.K., 1961.
25. S. Duhr, D. Braun. *PNAS*, 103, 19678–19682, 2006.
26. S. Semenov, M.E. Schimpf, *Phys. Rev. E*, 72, 041202–041211, 2005.
27. E. Bringuier, A. Bourdon, *Physica A*, 385, 9–24, 2007.
28. M. Martin, *Advances in Chromatography*, Vol. 39. In: J.C. Brown, E. Grushka (Eds.), Marcel Dekker, New York, 1998, pp. 1–138.
29. W.-J. Cao, P.S. Williams, M.N. Myers, J.C. Giddings, *Anal. Chem.*, 71, 1597–1609, 1999.
30. S.J. Jeon, M.E. Schimpf, A. Nyborg, *Anal. Chem.*, 69, 3442–3450, 1999.
31. E.P.C. Mes, R. Tijssen, W.Th. Kok, *J. Chromatogr. A*, 907, 201–209, 2001.
32. L. Pasti, F. Bedani, C. Contado, I. Mingozzi, F. Dondi, *Anal. Chem.*, 76, 6665–6680, 2004.
33. R.J. Hunter, *Foundations of Colloid Science*, 2nd edn., Clarendon Press, Oxford, NY, 2001.
34. W. Fraunhofer, G. Winter, *Eur. J. Pharm. Biopharm.*, 58, 369–383, 2004.
35. L.J. Gimbert, K.N. Andrew, P.M. Haygarth, P.J. Worsfold, *Trends Anal. Chem.*, 22, 615–633, 2003.
36. M.-A. Benincasa, C.D. Fratte, *J. Chromatogr. A*, 1046, 175–184, 2004.
37. M. Andersson, B. Wittgren, K.-G. Whalund, *Anal. Chem.*, 75, 4279–4291, 2003.
38. D. Kang, M.H. Moon, *Anal. Chem.*, 77, 4207–4212, 2005.
39. M. Andersson, K. Fromell, E. Gullberg, P. Artursson, K.D. Caldwell, *Anal. Chem.*, 77, 5488–5493, 2005.
40. J.F. Ranville, D.J. Chittleborough, R. Beckett, *Soil Sci. Soc. Am. J.*, 69(4), 1173–1184, 2005.
41. A. Dalpiaz, C. Contado, E. Vighi, G. Tosi, E. Leo, Preparation and characterization of particulate drug delivery systems for brain targeting, in: M.N.V. Ravi Kumar (Ed.), *Handbook of Particulate Drug Delivery*, Vol. 2. American Scientific Publishers, Stevenson Ranch, CA, 2008.
42. D.Y. Leger, S. Battu, B. Liagre, P.J.P. Cardot, J.L. Beneytout, *J. Chromatogr. A*, 1157(1–2), 309–320, 2007.
43. M.E. Schimpf, L.M. Wheeler, P.F. Romeo, in: T. Provider (Ed.), *Chromatography of Polymers: Characterization by SEC and FFF*, Vol. 521 (ACS Symposium Series). American Chemical Society, Washington, DC, 1993.
44. P.M. Shiundu, E.E. Remsen, J.C. Giddings, *J. Appl. Polym. Sci.*, 60(10), 1695–1707, 1996.
45. S.K.R. Williams, D. Lee, *J. Sep. Sci.*, 29, 1720–1732, 2006.
46. P. Vastamäki, M. Jussila, M.-L. Riekkola, *Analyst*, 128, 1243–1248, 2003.
47. L. Pasti, S. Agnolet, F. Dondi, *Anal. Chem.*, 79, 5284–5296, 2007.
48. J. Janča, I.A. Ananieva, A.Y. Menshikova, T.G. Evseeva, *J. Chromatogr. B*, 800, 33–40, 2004.
49. P. Reschiglian, A. Zattoni, B. Roda, L. Cinque, D. Parisi, A. Roda, F. Dal Piaz, M.H. Moon, B.R. Min, *Anal. Chem.*, 77(1), 47–56, 2005.
50. J. Janca, *Microthermal Field-Flow Fractionation: Analysis of Synthetic, Natural, and Biological Macromolecules and Particles*, HNB Publishing, New York, 2008.
51. T. Edwards, B.K. Gale, A.B. Frazier, *Biomed. Microdevices*, 3(3), 211–218, 2001.
52. N. Callens, M. Hoyos, P. Kurowski, C.S. Iorio, *Anal. Chem.*, Web Release Date: May 31, 2008.
53. C. Contado, *Particle Size Separation—Split-Flow Thin Cell Separation, Encyclopedia of Separation Science*, Academic Press Ltd., London, U.K., 2000.
54. C.B. Fuh, *Anal. Chem.*, 72, 266A–271A, 2000.

13 Affinity Chromatography

David S. Hage

CONTENTS

13.1 INTRODUCTION

Affinity chromatography is a liquid chromatographic technique that uses a specific binding agent for the purification or analysis of sample components [1–6]. The retention of a solute or target substance in this method is based on selective and reversible interactions that occur in many biological systems, such as the binding of an enzyme with a substrate or a hormone with its receptor. These interactions are used in affinity chromatography by immobilizing one of a pair of interacting molecules onto a solid support and placing it into a column. The immobilized molecule is referred to as the *affinity ligand* and makes up the stationary phase of the affinity column [2,6].

The first known use of affinity chromatography was in 1910, when Starkenstein examined the binding of α-amylase to insoluble starch [7]. This was followed by many other reports using insoluble materials for the isolation of enzymes, antibodies, and other substances [2]. It was not until the late 1960s, however, that this method became common. There were several key developments that led to this change. One was the introduction of covalent coupling methods in 1936 by Landsteiner and van der Scheer [8] and the use of such methods with cellulose in 1951 by Campbell et al. [9]. The modern era of affinity chromatography began in the late 1960s following the creation of beaded agarose supports [10] and the discovery of the cyanogen bromide immobilization method [11]. These latter two methods were combined in 1968 by Cuatrecasas et al. to create immobilized nuclease inhibitor columns for purifying the enzymes staphylococcal nuclease, α-chymotrypsin,

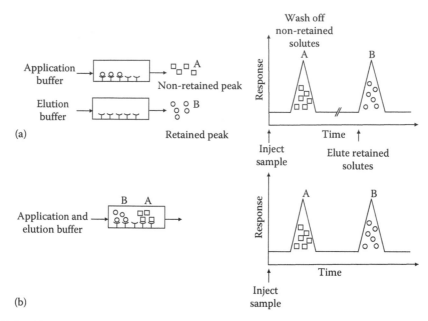

FIGURE 13.1 The (a) on/off elution mode of affinity chromatography that is used with high-affinity ligands and (b) the isocratic elution mode that is possible with low-affinity ligands. The circles represent the target analyte and the squares represent nonretained sample components. The "Y"-shaped objects are the immobilized ligands. (Adapted from Hage, D.S. et al., Application and elution conditions in affinity chromatography, in *Handbook of Affinity Chromatography*, 2nd edn., Hage, D.S. (ed.), CRC Press, Boca Raton, FL, 2005, Chap. 4.)

and carboxypeptidase A. It was also at this time that the term "affinity chromatography" was coined to describe this separation technique [12].

Figure 13.1a shows a typical scheme that is used to perform affinity chromatography. This particular scheme is known as the "on/off" or "step gradient" mode of affinity chromatography and is used for systems with relatively strong binding. In this scheme, a sample containing the compound of interest is injected onto the affinity column in the presence of a mobile phase that has the right pH, ionic strength, and solvent composition for solute–ligand binding. This solvent, which represents the weak mobile phase of an affinity column, is referred to as the *application buffer*. As the sample passes through the column under these conditions, compounds that are complementary to the affinity ligand will bind. Due to the high selectivity of this interaction, other solutes in the sample will tend to wash or elute from the column as a nonretained peak.

After all nonretained components have been washed from the column, the retained solutes are eluted by applying a solvent that displaces them from the column or that promotes dissociation of the solute–ligand complex. This solvent, which represents the strong mobile phase for the column, is known as the *elution buffer*. As the solutes of interest elute from the column, they are either measured directly or collected for later use. The application buffer is then reapplied to the system and the column is allowed to regenerate prior to the next sample injection [2,6].

Due to the strong and selective binding that characterizes many affinity ligands, solutes that are analyzed or purified with these ligands can often be separated with little or no interferences from other sample components. In many cases, the solute of interest can be isolated in only one or two steps, with purification yields of hundred- to several thousandfold being common [2,4–6]. In work with hormone receptors, purification yields approaching one millionfold have even been reported with affinity-based separations [5].

There are many types of affinity ligands that can be used in affinity chromatography. The wide range of ligands available for affinity chromatography makes this method a valuable tool for the purification and analysis of compounds present in complex samples. Areas in which affinity chromatography has been used include biochemistry, pharmaceutical sciences, clinical chemistry, and environmental testing [2]. Some specific applications of this technique are examined later in this chapter.

13.2 PRINCIPLES OF AFFINITY CHROMATOGRAPHY

13.2.1 THEORETICAL BASIS OF AFFINITY CHROMATOGRAPHY

A number of factors are important in determining the retention and elution of a target solute in an affinity column. These factors include the strength of the solute–ligand interaction, the amount of immobilized ligand present in the column, and the kinetics of solute–ligand association and dissociation. In the case where a solute (A) has single-site binding to an immobilized ligand (L), the following equations can be used to describe the interactions between the solute and the ligand in a column:

$$A+L \underset{k_d}{\overset{k_a}{\rightleftharpoons}} A-L \tag{13.1}$$

$$K_a = \frac{k_a}{k_d} \tag{13.2}$$

$$= \frac{\{A-L\}}{[A]\{L\}} \tag{13.3}$$

In these equations

K_a is the association equilibrium constant for the binding of A with L
A–L is the resulting solute–ligand complex
[A] is the mobile phase concentration of A at equilibrium
{L} or {A–L} represent the surface concentrations of the ligand or solute–ligand complex
k_a is the second-order association rate constant for solute–ligand binding
k_d is the first-order dissociation rate constant for the solute–ligand complex

At equilibrium, the retention of solute in the above system is given by the expressions shown below [13,14].

$$k = \frac{K_a m_L}{V_M} \tag{13.4}$$

$$= \left(\frac{t_R}{t_M}\right) - 1 \tag{13.5}$$

In these equations

k is the retention factor for the injected solute
t_R is the solute retention time
t_M is the void time of column
m_L represents the moles of active ligand in column
V_M is the column void volume

Equations 13.4 and 13.5 indicate that the retention factor of a solute, or its retention time, will depend on both the strength of solute–ligand binding (as represented by K_a) and the amount of ligand in the column (as represented by m_L or the ratio m_L/V_M) [14].

The strong retention of solutes seen on many types of affinity columns can be illustrated using Equations 13.4 and 13.5. As an example, for a porous silica-based column containing immobilized polyclonal antibodies, the association equilibrium constant would generally be in the range of 10^8–10^{10} M^{-1} [15] and the concentration of active immobilized ligand would be around 10 μM or greater [14]. Using these values with Equations 13.4 and 13.5 results in a retention factor of 1,000–100,000. For a 10 cm × 4.1 mm ID column operated at 1.0 mL/min, this would represent a mean elution time of about 1–90 days. The only way in which solutes could be eluted from this column on a reasonable timescale would be to change the system conditions in order to lower the effective equilibrium constant for solute–ligand binding. This is the purpose of the elution buffer used in Figure 13.1.

Although the "on/off" or "step gradient" elution shown in Figure 13.1a is a common practice in affinity chromatography, it is also possible to use isocratic elution or linear gradient elution methods. The scheme for isocratic elution is illustrated in Figure 13.1b. This technique can be employed if the value of the solute's association equilibrium constant (and retention factor) are small enough that elution can occur on the timescale of minutes, instead of days or hours. Fast association and dissociation kinetics are also necessary in order to allow a large number of solute–ligand interactions, thus producing good plate heights and narrow solute peaks. This method is called *weak-affinity chromatography* or *dynamic affinity chromatography*, and can be performed if the value of a solute's association equilibrium constant is less than or equal to 10^4 M^{-1}. Isocratic elution can also be used with affinity systems having association constants up to 10^6 M^{-1}, provided that the ligand concentration in the column (m_L/V_M) is sufficiently small [14,16–19]. Figure 13.2 shows an example of a chiral separation that was performed using isocratic elution and affinity chromatography.

13.2.2 Types of Affinity Ligands

The key factor in determining the success of any affinity separation is the type of ligand used within the column. A number of ligands that are commonly used in affinity chromatography are listed in Table 13.1. Many of these ligands are of biological origin, but a wide range of natural and synthetic molecules of nonbiological origin can also be used. Examples of specific applications for many of these ligands are presented later in this chapter.

All of the compounds listed in Table 13.1 can be placed into one of two categories: high-specificity ligands or general ligands [6]. The term *high-specificity ligand* refers to a ligand that will bind to only one or a few closely related molecules. This type of ligand is used in chromatographic systems in which the goal is to analyze or purify a specific solute. Typical high-specificity

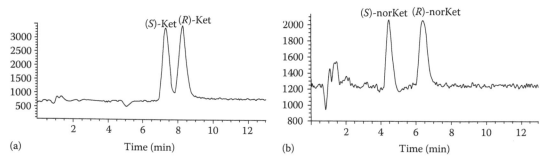

FIGURE 13.2 A chiral separation using an affinity column that contains the protein α_1-acid glycoprotein. This example shows the separation of the (R)- and (S)-enantiomers of (a) ketamine (Ket) and (b) norketamine (norKet) in human plasma using mass spectrometry for detection. (Reproduced with permission from Rosas, M.E.R. et al., *J. Chromatogr. B*, 794, 99, 108, 2003. Copyright 2003, Elsevier.)

TABLE 13.1
Common Ligands Used in Affinity Chromatography

Ligand	Retained Solutes
Biological ligands	
Antibodies	Antigens (drugs, hormones, peptides, proteins, viruses, and cell components)
Inhibitors, substrates, cofactors, and coenzymes	Enzymes
Lectins	Sugars, glycoproteins, and glycolipids
Nucleic acids	Complementary nucleic acids and DNA-/RNA-binding proteins
Protein A/protein G	Antibodies
Nonbiological ligands	
Boronates	Sugars, glycoproteins, and diol-containing compounds
Triazine dyes	Nucleotide-binding proteins, and enzymes
Metal chelates	Amino acids, peptides, proteins, and other agents that can bind to metal ions

ligands include antibodies (for binding antigens), substrates or inhibitors (for separating enzymes), and single-stranded nucleic acids (for the retention of a complementary sequence). As this list suggests, most high-specificity ligands tend to be biological compounds. These ligands also tend to have relatively large association constants for solutes and are generally eluted by a step gradient.

General, or *group-specific*, *ligands* are compounds that bind to a family or class of related molecules. These ligands are used in methods where the goal is to isolate a class of structurally similar solutes. General ligands can be of either biological or nonbiological origin. Examples given in Table 13.1 include proteins A and G, lectins, boronates, triazine dyes, and immobilized metal chelates. Many of these ligands have weaker binding for solutes than that seen with high-specificity ligands, but there are exceptions to this. For example, protein A has an equilibrium association of more than 10^8 M^{-1} for some types of antibodies [20,21]. Also, some molecules that are usually considered to be high-specificity ligands, such as antibodies, can be used to retain an entire class of solutes if they recognize a feature that is common to all of the desired analytes [22–24].

13.2.3 SUPPORT MATERIALS FOR AFFINITY CHROMATOGRAPHY

Another important factor to consider in affinity chromatography is the material used to hold the ligand within the column. Ideally, the support material should have low nonspecific binding for sample components and be easy to modify for ligand attachment. This material should also be stable under the flow rate, pressure, and solvent conditions to be used in the final application. In addition, the support should be readily available and simple to use in method development. The support should also possess a sufficient pore size and surface area to allow for the immobilization of a usable amount of ligand [25,26]. A wide variety of materials have been employed as supports in affinity chromatography. Table 13.2 lists some commercial matrices that are currently available. Suitable affinity supports can also be made in-house using underivatized agarose and silica or a variety of other materials [3,4,25,26].

Based on the type of support material used, affinity chromatography can be characterized as being either a low- or high-performance technique. In low-performance (or column) affinity chromatography, the support is usually a large diameter, nonrigid gel. Many of the carbohydrate-based supports and synthetic organic materials fall within this category. The low back pressure of these

TABLE 13.2
Commercial Supports Available for Affinity Chromatography[a]

Type of Material	Trade Name (Suppliers)
Agarose and cross-linked agarose	Affi-Gel (Bio-Rad Laboratories), Cellthru Big Beads (Sterogene), Mimetic Series (Prometic Biosciences/ACL), Sepharose Big Beads (Amersham/GE Healthcare), Sepharose FF (Amersham/GE Healthcare), Sepharose HP (Amersham/GE Healthcare)
Cellulose and modified cellulose	Cellufine (Chisso/Amicon/Millipore)
Polyacrylamide	Affi-Gel 601 (Bio-Rad Laboratories), Trisacryl (BioSepra/Pall)
Polymethacrylate	Affi-Prep (Bio-Rad Laboratories), Fractogel (Merck), SigmaChrom AF (Supelco), TSK-Gel (TosoHaas/Supelco)
Silica	Bakerbond (J.T. Baker) and ProteinPak (Waters)

[a]   Information in this table was obtained from Ref. [26]. and has been updated to indicate that Amersham is now part of GE Healthcare. Most of the listed supports are available in preactivated forms. Listed supports that are only available with specific ligands attached are Affi-Gel 601, Affi-Prep, the Mimetic series, Sepharose Big Beads, SigmaChrom AF, and Trisacryl.

supports means that they can be operated under gravity flow or with a peristaltic pump. This makes these gels relatively simple and inexpensive to use for affinity purifications. An additional benefit of working with many of these materials is their relatively broad range of pH stability. Disadvantages of the low-performance materials include their slow mass transfer properties and their limited stability at high flow rates and pressures. These factors limit the usefulness of these supports in analytical applications, where both rapid and efficient separations are often desired [6,26].

In high-performance affinity chromatography (*HPAC*), the support consists of small, rigid particles capable of withstanding the high flow rates and/or pressures that are characteristics of HPLC. Supports that can be used for this purpose include modified silica or glass, azalactone beads, and hydroxylated polystyrene media. (*Note*: Some types of cross-linked agarose may also fall in this category.) The mechanical stability and efficiency of these supports allow them to be used with standard HPLC equipment. Although the need for more sophisticated instrumentation does make HPAC more expensive to perform than low-performance affinity chromatography, the better speed and precision of this technique makes it the method of choice for analytical applications [6,26].

Most applications for affinity chromatography are based on porous, particulate supports, but there are several alternative types of supports that are gaining in popularity [26]. Nonporous beads can be useful in fast separations or in cases in which it is desirable to minimize stagnant mobile phase mass transfer effects [27]. Nonporous fibers provide high surface areas for ligand attachment along with low back pressures [28]. Membranes also can provide low back pressures, allowing their use in affinity separations that require large sample volumes or high flow rates [29,30]. Perfusion media and monolithic columns have been explored in affinity chromatography as another means for allowing both fast separations and low back pressures [31–34]. Supports designed for work with expanded-bed chromatography (see Figure 13.3) have also been employed to allow the efficient capture of target compounds from samples containing large amounts of debris or particulate matter [35].

13.2.4 Immobilization Methods for Affinity Ligands

A third item to consider in using affinity chromatography is the way in which the ligand is attached to the solid support, or the *immobilization method*. Several techniques are available for this, including both covalent and noncovalent coupling methods [25,36]. For a protein or peptide, this generally

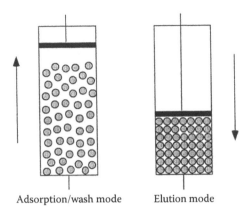

Adsorption/wash mode   Elution mode

FIGURE 13.3 Expanded bed chromatography. During the application of a sample, flow of the mobile phase is in an upward direction, allowing the support bed to expand and provide easy passage for debris and particulate matter. During the elution step, the flow of mobile phase is in a downward direction, allowing the support bed to become compact and helping to provide narrow peaks for the eluting solute. (Reproduced with permission from Gustavsson, P.-E. and Larsson, P.-O., Support materials for affinity chromatography, in *Handbook of Affinity Chromatography*, 2nd edn., Hage, D.S. (ed.), CRC Press, Boca Raton, FL, 2005, Chap. 2.)

involves covalently coupling the molecule through free amine, carboxylic acid, or sulfhydryl residues present in its structure. Immobilization of a ligand through other functional sites (e.g., aldehyde groups produced by carbohydrate oxidation) is also possible.

Table 13.3 summarizes various covalent immobilization methods that are used in affinity chromatography. Each of these methods involves at least two steps: (1) an activation step, in which the support is converted to a form that can be chemically attached to the ligand and (2) a coupling step, in which the affinity ligand is attached to the activated support. With some techniques, a third step, in which remaining activated groups are removed, is also required. The methods listed in Table 13.3 can be performed either in-house or can be used in the form of preactivated supports available from commercial suppliers (see list in Table 13.2) [25,36].

The choice of immobilization method is important since it can affect the actual or apparent activity of the final affinity column. If the correct procedure is not used, a decrease in ligand activity can result from multisite attachment, improper orientation, and/or steric hindrance (see Figure 13.4) [36]. *Multisite attachment* refers to the coupling of a ligand to the support through more than one functional group. Although multisite attachment does create a more stable linkage than single-site immobilization, this phenomenon can lead to distortion and denaturation of the ligand's active region. This effect can be minimized by coupling the ligand through functional groups that occur in only a few places in its structure or by using a support with a low density of activated sites.

Improper orientation can lead to a similar loss in ligand activity. This effect can be avoided by coupling the ligand through groups that are distant from its active region. An example would be the use of oxidized carbohydrate residues for the attachment of antibodies to a solid support [36]. If the actual location of the ligand's active site is not known or if no appropriate functional groups are available, it is sometimes possible to minimize this problem by empirically varying the procedure or conditions used for ligand attachment.

Steric hindrance refers to the loss of ligand activity due to the presence of a nearby support or neighboring ligand molecules. Steric hindrance produced by neighboring ligands can be minimized by using a low ligand coverage. Steric hindrance produced by the support can be reduced by adding a *spacer arm*, or tether, between the ligand and supporting material. The presence of a spacer arm is particularly important when using small ligands for the retention of large analytes. Examples of common spacer arms used in affinity chromatography include 6-aminocaproic acid, diaminodipropylamine, 1,6-diaminohexane, ethylenediamine, and succinic acid (anhydride) [25,36].

TABLE 13.3
Immobilization Methods for Affinity Chromatography

Group/Compound	Immobilization Technique
Amines	Azalactone method (for azalactone supports)
	Cyanogen bromide method
	N,N′-Carbonyl diimidazole method
	Divinylsulfone method
	Epoxy (bisoxirane) method
	Ethyldimethylaminopropyl carbodiimide method
	Fluoromethylpyridinium toluenesulfonate method
	N-hydroxysuccinimide ester method
	Schiff base (reductive amination) method
	Tresyl chloride/tosyl chloride method
Sulfhydryls	Azalactone method (for azalactone supports)
	Divinylsulfone method
	Epoxy (bisoxirane) method
	Iodoacetyl/bromoacetyl methods
	Maleimide method
	Pyridyl disulfide method
	TNB-thiol method
	Tresyl chloride/tosyl chloride method
Carboxylates	Ethyldimethylaminopropyl carbodiimide method
Hydroxyls	Cyanuric chloride method
	Divinylsulfone method
	Epoxy (bisoxirane) method
Aldehydes	Hydrazide method
	Schiff base (reductive amination) method
Nucleic acids	Carbodiimide method
	Carboxymethyl method (for cellulose)
	Cyanogen bromide method
	Cyanuric chloride method
	Diazonium (azo) method
	Epoxy (bisoxirane) method
	Hydrazide method
	Schiff base (reductive amination) method
	UV irradiation (for cellulose)

Source: Data compiled from Hermanson, G.T. et al., *Immobilized Affinity Ligand Techniques*, Academic Press, New York, 1992; Kim, H.S. and Hage, D.S., Immobilization methods for affinity chromatography, in Hage, D.S. (ed.), *Handbook of Affinity Chromatography*, 2nd edn., CRC Press, Boca Raton, FL, 2005, Chap. 3.

A variety of noncovalent methods can also be used for ligand immobilization. For instance, nonspecific adsorption can be utilized to attach ligands to some supports that have not been specifically functionalized for covalent attachment. This process generally involves forces such as coulombic interactions, hydrogen bonding, or hydrophobic interactions that can occur between the ligand and support [36–38]. This approach is relatively simple and inexpensive. One disadvantage is that this method is usually subject to problems with random orientation. Denaturation of the ligand through distortion of its structure during the adsorption process is possible as well. The activity of

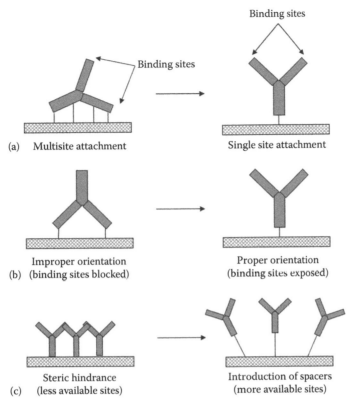

(a) Multisite attachment

Binding sites

Single site attachment

Improper orientation
(b) (binding sites blocked)

Proper orientation
(binding sites exposed)

Steric hindrance
(c) (less available sites)

Introduction of spacers
(more available sites)

FIGURE 13.4 The effects of (a) multisite attachment, (b) random or improper orientation, and (c) steric hindrance on an immobilized ligand. (Reproduced from Kim, H.S. and Hage, D.S., Immobilization methods for affinity chromatography, in *Handbook of Affinity Chromatography*, 2nd edn., Hage, D.S. (ed.), CRC Press, Boca Raton, FL, 2005, Chap. 3.)

the ligand should also be monitored over time, since leakage from the support may occur. However, it is often possible with nonspecific adsorption methods to later release the bound ligand and regenerate the support by applying a fresh batch of ligand to the system [36].

A more selective means for noncovalent immobilization is to use biospecific adsorption. This makes use of the binding between the ligand and a secondary ligand that has been previously attached to the support [36]. Many types of secondary ligands can be used. One common example is the use of immobilized avidin or streptavidin for the adsorption of biotin-containing ligands [39–42]. This technique has seen a great deal of recent interest in proteomics for the isolation of peptides with isotopic labels that also contain biotin tags [43,44]. Protein A and protein G are another common group of secondary ligands. These latter agents are bacterial proteins that are capable of binding to the constant regions of many immunoglobulins, making these secondary ligand useful for the production of immobilized antibody supports [23–25,36].

Yet another means for the noncovalent immobilization of an affinity ligand is to physically trap this in a support. This approach is known as entrapment or encapsulation and involves the use of a support with pores that are smaller than the ligand but larger than the desired target [36]. Entrapment has been performed by using sol–gels, where the support is prepared surrounding the ligand, or by using a cross-linking agent to prevent a ligand from leaving the support (see Figure 13.5) [45–47]. Liposomes have been used as a means for the entrapment of some small ligands [48–50]. In addition, freeze/thaw techniques have been used to entrap cells and proteoliposomes in porous supports [51–55].

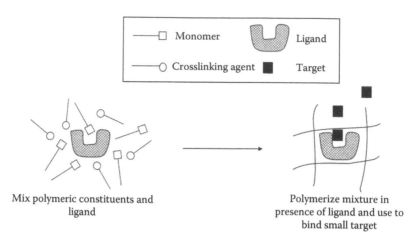

FIGURE 13.5 General approach for the entrapment of a ligand in a polymeric support.

13.2.5 APPLICATION CONDITIONS IN AFFINITY CHROMATOGRAPHY

The application buffer is another parameter that must be considered in the use of an affinity column [14]. Most application buffers in affinity chromatography are solvents that mimic the pH, ionic strength, and polarity experienced by the solute and ligand in their natural environment. Any cofactors or metal ions required for solute–ligand binding should also be present in this solvent. Under these conditions, the solute will probably have its highest association constant for the ligand and its highest degree of retention on the column.

The proper choice of an application buffer can help to minimize any nonspecific binding due to undesired sample components. For example, coulombic interactions between solutes and the support can often be decreased by altering the ionic strength and pH of the application buffer. In addition, surfactants and blocking agents (e.g., Triton X-100, Tween-20, bovine serum albumin, and gelatin) may be added to the buffer to prevent nonspecific retention of solutes on the support or affinity ligand.

The activity of the immobilized ligand should be considered in determining how much sample can be applied to the affinity column with each use. A rough indication of the maximum possible column-binding capacity can be made by assaying the total amount of ligand present. A better approach is to actually measure the ligand's activity. This can be accomplished by continuously applying a known concentration of solute to the affinity column (i.e., obtaining a breakthrough curve by performing frontal analysis) or by combining the immobilized ligand with a known excess of solute and measuring the amount of free solute that remains after binding has occurred. It is often important in affinity separations to have a sufficient amount of ligand not only to provide enough binding capacity for the analyte but also to provide multiple opportunities for this analyte to be captured as it passes through the column.

With some affinity systems, it is possible to see a large amount of nonretained solute during the application step even when the amount of injected sample is significantly less than known column-binding capacity [14,56]. This phenomenon, known as the *split-peak effect*, is caused by the presence of slow adsorption or slow mass transfer kinetics within the column. This effect has been reported with many types of affinity ligands, but tends to occur with high-performance supports because of the more rapid flow rates often used with these materials [14]. Ways in which this effect can be minimized include reducing the flow rate used for sample injection, increasing the column size, increasing the amount of immobilized ligand, and placing a more efficient support within the column. In some cases, changing to a different immobilization method may help by providing more active ligand or a ligand with more rapid binding kinetics [14,56].

13.2.6   Elution Conditions in Affinity Chromatography

The conditions used for the removal of a retained solute should also be considered in the design of an affinity separation. Just as the application conditions are selected to maximize specific solute–ligand interactions, the elution conditions are chosen to promote fast or gentle removal of solute from the column. The elution buffer used in affinity chromatography can be either a solvent that produces weak solute–ligand binding (i.e., a small association equilibrium constant) or a solvent that decreases the extent of this binding by using a competing agent that displaces solute from the column. These two approaches are known as *nonspecific elution* and *biospecific elution*, respectively [6,14].

Biospecific elution is the gentler of these two elution methods because it is carried out under essentially the same solvent conditions as used for sample application. This factor makes biospecific elution attractive for purification work, where a high recovery of active solute is desired. Biospecific elution may be performed either by adding an agent to the eluting solvent that competes with ligand for solute (i.e., normal role elution) or by adding an agent that competes with solute for ligand-binding sites (i.e., reversed role elution) (see Figure 13.6). In both cases, retained solutes are eventually eluted from the column by displacement and mass action.

The main advantage of biospecific elution is its ability to gently remove analyte from the column. The main disadvantages of this method include its typically slow elution times and its broad solute peaks. Another limitation is the need to remove the competing agent from the eluted solute. A further difficulty that may be encountered in analytical applications is the need to use a competing agent that does not produce a large background signal under the conditions used for analyte detection.

Many of these problems can be overcome by using nonspecific elution. This technique involves changing the column conditions in order to weaken the interactions between retained solutes and the immobilized ligand. This can be done by changing the pH, ionic strength, or polarity of the mobile phase. The addition of denaturing or chaotropic agents to the running buffer can also be used. This results in an alteration in structure for the solute or ligand, leading to a smaller association equilibrium constant and lower solute retention.

Nonspecific elution tends to be much faster than biospecific elution in removing analytes from affinity columns. This results in sharper peaks, which in turn produces lower limits of detection and shorter analysis times. For these reasons, nonspecific elution is commonly used in analytical applications of affinity chromatography. This elution method can also be used in purifying solutes, but there is a greater risk of solute denaturation than there is with biospecific elution. Also, care must be taken in nonspecific elution to avoid using conditions that are too harsh for the column. If this is not considered, it may result in long column regeneration times or an irreversible loss of ligand activity.

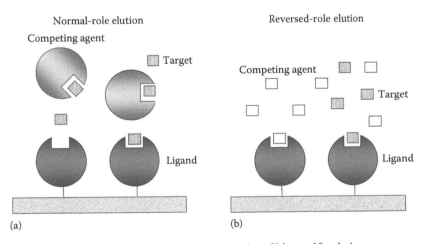

FIGURE 13.6   The (a) normal-role and (b) reversed-role modes of biospecific elution.

13.3 TYPES OF AFFINITY CHROMATOGRAPHY

13.3.1 Bioaffinity Chromatography

Bioaffinity chromatography, or *biospecific adsorption*, refers to affinity methods that use a biological molecule as the affinity ligand [21]. This was the first type of affinity chromatography developed and represents the most diverse category of this technique. The earliest known use of bioaffinity chromatography was in 1910, when Starkenstein used insoluble starch to purify the enzyme α-amylase [7]. However, it was not until the development of beaded agarose supports [10] and the cyanogen bromide immobilization method [11] that bioaffinity chromatography came into common use [12].

Table 13.1 lists a number of ligands that are used in bioaffinity applications. These, and many other biological compounds, can be attached to chromatographic supports using the materials and methods described previously in this chapter. Many commercial suppliers even provide these ligands already coupled to supports or packed in columns. Several hundred different types of immobilized biological molecules are presently available in this fashion [3–6,21,25].

The earliest applications of bioaffinity chromatography involved its use in enzyme purification (see Figure 13.7) [7]. Enzyme purification has continued to be a major application of this technique [57]. Some ligands that are employed for this purpose are enzyme inhibitors, coenzymes, substrates, and cofactors. Examples include methods that use nucleotide mono-, di-, and triphosphates for the

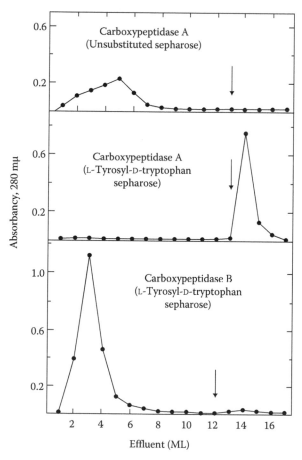

FIGURE 13.7 An early example of an enzyme purification performed by affinity chromatography. (Reproduced with permission from Cuatrecasas, P. et al., *Proc. Natl. Acad. Sci. U.S.A.*, 68, 636, 1968.)

purification of various kinases, the utilization of NAD for collecting dehydrogenases, the use of pyridoxal phosphate for the isolation of tyrosine and aspartate aminotransferases, and the utilization of RNA or DNA for the purification of polymerases and nucleases [5,57]. Another common example is the use of immobilized glutathione to isolate the enzyme glutathione-S-transferase or recombinant proteins that have been tagged with this enzyme [57]. The importance of bioaffinity chromatography in enzyme purification is illustrated by a number of reviews, which list hundreds of applications in this area [57,58].

Lectins are commonly used as general ligands in bioaffinity chromatography [21]. These are nonimmune system proteins that have the ability to recognize and bind certain types of carbohydrate residues [59]. Two lectins often used in affinity chromatography are concanavalin A, which binds to α-D-mannose and α-D-glucose residues, and wheat germ agglutinin, which binds to D-N-acetylglucosamines. Other lectins that can be employed are jackalin and lectins found in peas, peanuts, or soybeans. These ligands are used in the separation of many carbohydrate-containing compounds, such as polysaccharides, glycoproteins (e.g., immunoglobulins or cell membrane proteins), and glycolipids [3–6,21,25].

Another useful class of bioaffinity ligands that have already been discussed are bacterial cell wall proteins such as protein A from *Staphylococcus aureus* and protein G from group G streptococci [20,21,60,61]. As mentioned earlier, these ligands have the ability to bind to the constant region of many types of immunoglobulins. This makes them useful in antibody purification and as secondary ligands for antibody immobilization. Protein A and protein G have their strongest binding to immunoglobulins at or near neutral pH but readily dissociate from these solutes when placed into a lower pH buffer. These two ligands differ in their ability to bind to antibodies from different species and classes (see Table 13.4) [20,21,61]. However, a recombinant protein that blends the activities of these compounds, known as protein A/G, is now also available [21,62].

Nucleic acids and polynucleotides can act as either general or high-specificity ligands in bioaffinity chromatography. This area is known as *DNA affinity chromatography* [63]. When employed as high-specificity ligands, nucleic acids and polynucleotides can be used to purify DNA-/RNA-binding enzymes and proteins or to isolate nucleic acids that contain a sequence that is complementary to the ligand [63,64]. This approach is called *sequence-specific DNA affinity chromatography*. An immobilized nucleic acid can be utilized to purify solutes that share a common nucleotide sequence, giving a method called *nonspecific DNA affinity chromatography* [63]. Examples of the latter method include the use of immobilized oligo(dT) for the isolation of nucleic acids containing poly(A) sequences [63] and the use of calf thymus DNA or salmon sperm DNA fragments to purify DNA-binding proteins [65].

13.3.2 IMMUNOAFFINITY CHROMATOGRAPHY

In immunoaffinity chromatography (IAC), the ligand is an antibody or antibody-related reagent [23,24,66]. This technique represents a special subcategory of bioaffinity chromatography. Other methods sometimes included under the heading of IAC are those that use immobilized antigens for antibody purification [67,68]; however, depending on the nature of the antigen, the ligand used in this latter approach may or may not be of biological origin.

The high selectivity of antibody–antigen interactions and the ability to produce antibodies against a wide range of solutes have made IAC a popular tool for biological purification and analysis. Examples include methods developed for the isolation of antibodies, hormones, peptides, enzymes, recombinant proteins, receptors, viruses, and subcellular components (see Figure 13.8) [24,67–72]. The strong binding constants of many antibodies require that nonspecific elution be employed with most immunoaffinity columns [23,24,67]. Isocratic elution methods have also been reported with low-affinity antibody systems [16].

The first known application of IAC was reported by Campbell et al. in 1951, where an antigen immobilized to *p*-aminobenzyl cellulose was used for antibody purification [9]. Many current applications

TABLE 13.4

Binding of Various Types of Immunoglobulins to Protein A and Protein G[a]

	Type of Immunoglobulin	Protein A	Protein G
Cow	IgG	No binding	Weak
Dog	IgG	Weak	Weak
	IgM	Weak	No binding
Goat	IgG	Weak	Strong
Horse	IgG	Strong	Strong
Human	IgG_1	Strong	Strong
	IgG_2	Strong	Strong
	IgG_3	No binding	Strong
	IgG_4	Strong	Strong
	IgA	Weak	No binding
	IgM	Weak	No binding
	IgE	Weak	No binding
Murine	IgG_1	Weak	Moderate
	IgG_{2a}	Moderate	Moderate
	IgG_{2b}	Moderate	Moderate
	IgG_3	Weak	Moderate
	IgM	Moderate	Weak
	IgA	Moderate	Moderate
Rabbit	IgG	Strong	Strong
Rat	IgG	Weak	Moderate
Sheep	IgG	Strong	Strong

[a] This information was obtained from Ref. [69]. Immunoglobulins listed in this reference as showing no binding to protein A or protein G include chicken egg IgY, goat IgM, rabbit IgM, rat IgM, and sheep IgM.

are still based on the use of low-performance supports, particularly agarose [23,24]. Work has also been performed using derivatized silica, glass, and perfusion media with immobilized antibodies, giving a technique known as *high-performance immunoaffinity chromatography* [23,24,67].

The utilization of IAC in analytical methods has received increasing retention in recent years [23,24]. Of particular interest is the use of immobilized antibody columns in performing immunoassays, a technique known as a *chromatographic immunoassay* or *flow-injection immunoassay*. This approach has already been reported in a number of formats such as those involving simple analyte adsorption/desorption, sandwich immunoassays, competitive binding immunoassays, and multianalyte methods (see Figure 13.9) [23,24,73,74]. Typical advantages of these methods include decreased analysis times and improved precision versus manual immunoassays.

Although antibodies are usually thought of as being high-specificity ligands, they can be used to bind a general class of compounds that share a common structural feature. When used in this way, the resulting immunoaffinity column can serve either as a screening tool for the desired compound class or as a specific absorbent for the extraction and concentration of compounds before analysis by a second separation method (e.g., reversed-phase liquid chromatography). This approach is known as *immunoextraction* or *immunoaffinity extraction*. Several methods have been recently developed using this approach and related approaches for the quantitation of solutes in biotechnology, clinical chemistry, and pharmaceutical research [22–24,73]. A closely related technique known as *postcolumn immunodetection* uses immunoaffinity columns or chromatographic immunoassays to monitor the elution of specific analytes from a general chromatographic column [23,24,73–75].

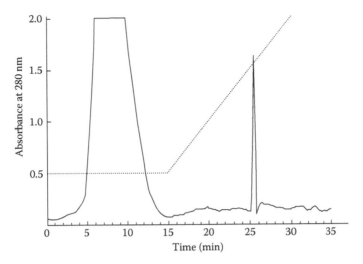

FIGURE 13.8 An example of a separation performed by immunoaffinity chromatography. This plot shows a chromatogram obtained by high-performance immunoaffinity chromatography for the isolation of complement component C3 from human serum. The first peak is due to nonretained sample components, and the second peak is due to the target eluted in the presence of a linear gradient of mobile phases containing 0–2.5 M sodium thiocyanate. (Reproduced from Hage, D.S. and Phillips, T.M., Immunoaffinity chromatography, in *Handbook of Affinity Chromatography*, 2nd edn., Hage, D.S. (ed.), CRC Press, Boca Raton, FL, 2005, Chap. 6.)

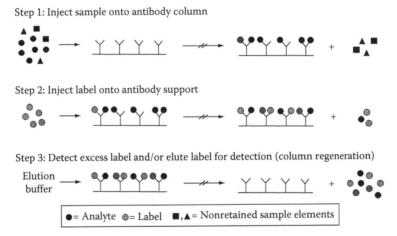

FIGURE 13.9 Scheme for a sequential injection, competitive binding immunoassay performed by high-performance immunoaffinity chromatography. In this method, the amount of injected analyte is measured by examining its effect on the amount of retained label (i.e., a labeled analogue of the analyte) that is later captured by the column during a separate injection. (Reproduced from Moser, A.M. and Hage, D.S., Chromatographic immunoassays, in *Handbook of Affinity Chromatography*, 2nd edn., Hage, D.S. (ed.), CRC Press, Boca Raton, FL, 2005, Chap. 29.)

13.3.3 DYE-LIGAND AND BIOMIMETIC AFFINITY CHROMATOGRAPHY

Another category of affinity chromatography is *dye-ligand affinity chromatography*. In this method, a synthetic dye is the immobilized ligand [76]. These ligands are often based on chlorotriazine polysulfonated aromatic dyes, which are commonly referred to as "triazine" dyes. An example of one such ligand is Cibacron Blue 3GA (see Figure 13.10). Other examples include Procion Red HE-3B,

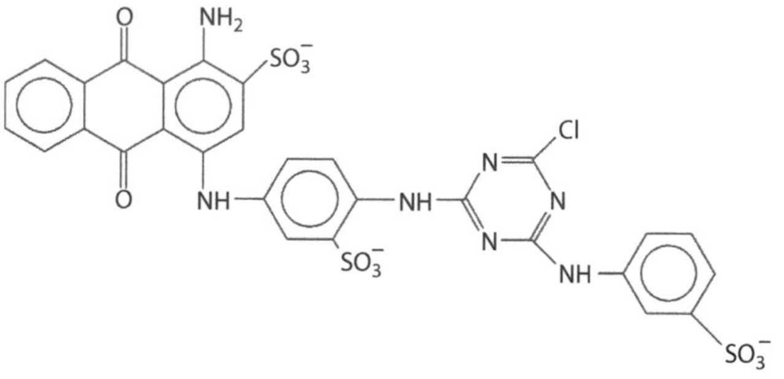

FIGURE 13.10 Structure of Cibacron Blue 3GA.

Procion Rubine MX-B, Procion Yellow H-A, and Turquoise MX-G. These dyes have found use over the last few decades in the purification of a broad range of proteins and enzymes, including albumin, decarboxylases, glycolytic enzymes, hydrolases, lyases, nucleases, oxidoreductases, synthetases, and transferases [77,78]. The first use of dye-ligand affinity chromatography was described by Staal et al. in 1971 [79]. Since that time, it has become an extremely popular tool for enzyme and protein purification, with hundreds of such compounds having been isolated by this technique [3–6,76–79].

The method of *biomimetic affinity chromatography* is similar to dye-ligand affinity chromatography in that a synthetic ligand is used for target isolation or analysis. However, computational methods or combinatorial chemistry are also now used to help identify and produce ligands for affinity columns [76]. Dyes have been explored in several studies as starting scaffolds for the creation of biomimetic ligands [76]. Other classes of molecules can also be employed as biomimetic ligands. Examples include large peptide libraries that have been produced by combinatorial chemistry [80–82], phage display technology [83], or ribosome display methods [84]. Other reports have used the SELEX method (i.e., systematic evolution of ligands by exponential enrichment) to create RNA- or DNA-based ligands known as *aptamers* [84–88].

13.3.4 IMMOBILIZED METAL-ION AFFINITY CHROMATOGRAPHY

In immobilized metal-ion affinity chromatography (IMAC), the ligand is a metal ion that is complexed with an immobilized chelating agent [89]. Other names for this technique are immobilized metal affinity chromatography, metal-chelate affinity chromatography, and metal-ion interaction chromatography. This method can be used to separate proteins and peptides that contained amino acids with electron donor groups, such as histidine. This type of affinity chromatography was first described by Porath et al. in 1975 [90]. Since that time, a large number of peptides and proteins have been purified by this method [89,91–94].

Selectivity in IMAC is determined by the chelating group and metal ion that make up the affinity ligand. Iminodiacetic acid is the most common chelating agent employed. Metal ions used with this agent and other chelating groups include Cu^{2+}, Zn^{2+}, Ni^{2+}, Co^{2+}, and Fe^{3+} [89]. The ability of histidine to bind Ni^{2+} ions has made Ni(II)-containing columns popular in the isolation of recombinant proteins with histidine tags (i.e., His-tag proteins) [89,95]. IMAC columns using Fe^{3+} or Ga^{3+} are popular for binding phosphorylated peptides [96].

13.3.5 ANALYTICAL AFFINITY CHROMATOGRAPHY

Besides its use in separating molecules, affinity chromatography can be employed as a tool for studying solute–ligand interactions. This particular application of affinity chromatography is called *analytical affinity chromatography* or *quantitative affinity chromatography*. Using this technique,

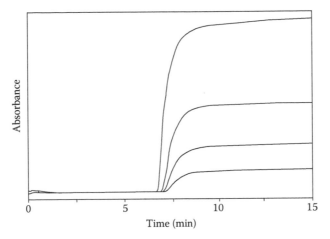

FIGURE 13.11 Typical frontal analysis curves obtained for the application of L-tryptophan to an affinity column containing immobilized human serum albumin. The concentrations of applied L-tryptophan (from left to right) were 100, 50, 25, and 12.5 μM. The flow rate was 0.25 mL/min and the column void time was 3.6 min. (Reproduced from Hage, D.S. and Chen, J., Quantitative affinity chromatography: Practical aspects, in *Handbook of Affinity Chromatography*, 2nd edn., Hage, D.S. (ed.), CRC Press, Boca Raton, FL, 2005, Chap. 22.)

information can be acquired regarding both the equilibrium and rate constants for biological interactions [13,51,97,98].

Information on equilibrium constants can be obtained by using the methods of *zonal elution* or *frontal analysis*. Zonal elution involves the injection of a small amount of solute onto an affinity column in the presence of a mobile phase that contains a known concentration of competing agent [97]. The equilibrium constants for binding of the ligand with the solute (and competing agent) can then be obtained by examining how the solute's retention changes with competing agent concentration. A typical zonal elution experiment is shown in Figure 13.7. This technique was first used to study biological interactions by Andrews et al. [99] and Dunn and Chaiken [100] in the mid-1970s. Since that time, this method has been used to examine a number of systems such as enzyme–inhibitor binding, protein–protein interactions, and drug–protein binding [13,51,97,98].

Frontal analysis, or frontal affinity chromatography, is performed by continuously applying a known concentration of solute to an affinity column at a fixed flow rate (see Figure 13.11). The moles of analyte required to reach the mean point of resulting breakthrough curve is then measured and used to determine the equilibrium constant for solute–ligand binding [97]. This method was used first by Kasai and Ishii in 1975 [101] to examine the interactions of trypsin with various peptide ligands. Frontal analysis has since been reported in many other applications [13,51,97,98]. One advantage of this approach over zonal elution is that it simultaneously provides information on both the equilibrium constants and number of active sites involved in analyte–ligand binding. The main disadvantage of this method is the need for a larger quantity of solute than is required by zonal elution.

Information on the kinetics of solute–ligand interactions can also be obtained using affinity chromatography. A number of methods have been developed for this, including techniques based on band-broadening measurements, the split-peak effect, and peak decay analysis [13,57,97,98]. These methods are generally more difficult to perform than equilibrium constant measurements but represent a powerful means of examining the rates of biological interactions. Systems that have been studied by these techniques include the binding of lectins with sugars, protein A with immunoglobulins, antibodies with antigens, and drugs with serum proteins [13,56,97,98,102–107].

13.3.6 MISCELLANEOUS METHODS

Besides the specific methods already discussed, a number of other ligands and techniques are often placed under the category of affinity chromatography. For example, boronic acid and its derivatives represent a class of general ligands that are usually considered to be affinity ligands. The resulting method is known as *boronate affinity chromatography*. At a pH of 8 or higher, boronate and related ligands have the ability to form covalent bonds with compounds that contain *cis*-diol groups in their structure. This makes these ligands useful for the purification and analysis of many compounds that contain sugar residues, such as polysaccharides, glycoproteins, ribonucleic acids, and catecholamines [108–110]. One important clinical application of immobilized boronate ligands is their use in the analysis of glycosylated hemoglobin in diabetic patients [111,112].

Other methods that are related to affinity chromatography include *hydrophobic interaction chromatography* and *thiophilic adsorption*. The former is based on the interactions of proteins, peptides, and nucleic acids with short nonpolar chains on a support. This was first described in 1972 [113,114] following work that examined the role of spacer arms on the nonspecific adsorption of affinity columns [114]. Thiophilic adsorption, also known as *covalent* or *chemisorption chromatography*, makes use of immobilized thiol groups for solute retention [115]. Applications of this method include the analysis of sulfhydryl-containing peptides or proteins and mercurated polynucleotides [116].

Another technique that is of growing interest is the use of molecularly printed polymers (MIPs) as affinity supports [117–119]. This type of polymer is typically prepared by using a template molecule that is present during the formation of a polymeric support. This template is similar or equivalent to the final target substance for which binding is desired. One way of forming this type of support is through a noncovalent imprinting method (see Figure 13.12), in which the target is combined with functional monomers that can interact with given groups on the target. These monomers are then polymerized in place about the target, with the target later being extracted from the final polymer. This leaves behind binding sites that now have the correct size and functional groups for binding the target from applied samples. A covalent imprinting method can also be used, in which the target itself has polymerizable side chains. The target is then later dissociated from these side chains and the polymer after MIP formation. MIPs have been used in several

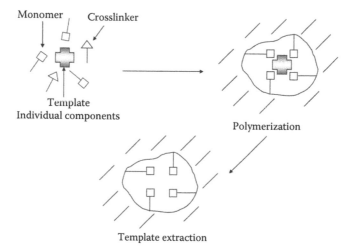

FIGURE 13.12 Preparation of a molecularly imprinted polymer. (Reproduced from Kim, H.S. and Hage, D.S., Immobilization methods for affinity chromatography, in *Handbook of Affinity Chromatography*, 2nd edn., Hage, D.S. (ed.), CRC Press, Boca Raton, FL, 2005, Chap. 3.)

studies for chiral separations. They have also been used in solid-phase extraction and membrane-based separations [117].

13.4 APPLICATIONS OF AFFINITY CHROMATOGRAPHY

13.4.1 PREPARATIVE APPLICATIONS

There are various ways in which affinity chromatography can be employed. One of the most important applications of this method is as a preparative tool for biological agents. Affinity chromatography has been utilized for this purpose ever since the first report of this method in 1910. It is now widely used for preparative applications in both large-scale industrial purification schemes and in small laboratory-scale separations [57,76–78,96,120,121]. There are several areas in which affinity chromatography has been used as a preparative tool. One good example is the use of this method for the isolation of enzymes [57,76–78]. Preparative affinity chromatography is also commonly used for the isolation of recombinant proteins, antibodies, and antigens [68,96,121].

As is discussed earlier in this chapter, there are many ligands and supports that can be used in affinity chromatography for preparative work. Agarose has been particularly popular as a support for this purpose because of its stability over a wide pH range. This support is also stable at high temperatures (allowing it to be autoclaved) and can withstand treatment with 0.5 M sodium hydroxide, a sanitizing agent that is often used to clean preparative columns after separations involving biological samples [26]. However, other supports can also be employed. These supports can be utilized in a standard packed bed format, but interest in formats such as expanded beds, membranes, perfusion media, and continuous beds is increasing for preparative work [26].

13.4.2 ANALYTICAL AND SEMIPREPARATIVE APPLICATIONS

The use of affinity chromatography for chemical analysis or in sample preparation is another important application of this technique. Semipreparative techniques based on this method are frequently used in biochemistry, molecular biology, and biomedical research for the isolation of specific compounds for analysis or study. This approach again includes targets such as enzymes, antibodies, and antigens [23,24,57,68]. Other targets that have been isolated by affinity chromatography for study include DNA or RNA and regulatory or signal-transducing proteins [122,123]. Numerous tags are also used with affinity columns for the isolation of recombinant proteins. Examples of such tags are histidine, calmodulin-binding peptide, glutathione-S-transferase, and maltose-binding protein [122].

Several applications of affinity chromatography in analytical chemistry have already been presented in this chapter. These applications have included the use of this method for the direct analysis of samples in HPAC [6,26], the development of chromatographic immunoassays [23,24,73,74], and the utilization of affinity chromatography as a tool to study biological interactions [13,51,97,98]. There are many fields in which analytical uses of affinity chromatography have been reported. Reviews on this topic have recently appeared for the fields of clinical chemistry [112], pharmaceutical analysis [112,124], biotechnology [96], environmental analysis [125], and molecular biology [122]. In addition, affinity chromatography has been combined with several other methods for chemical analysis. Examples are the use of affinity chromatography with capillary electrophoresis, mass spectrometry, and microanalytical systems [44,126,127].

REFERENCES

1. Ettre, L.S., Nomenclature for chromatography, *Pure Appl. Chem.*, 65, 819–872, 1993.
2. Hage, D.S. and Ruhn, P.F., An introduction to affinity chromatography, in *Handbook of Affinity Chromatography*, 2nd edn., Hage, D.S., Ed., CRC Press, Boca Raton, FL, 2005, Chap. 1.

3. Turkova, J., *Affinity Chromatography*, Elsevier, Amsterdam, the Netherlands, 1978.
4. Scouten, W.H., *Affinity Chromatography: Bioselective Adsorption on Inert Matrices*, Wiley, New York, 1981.
5. Parikh, I. and Cuatrecasas, P., Affinity chromatography, *Chem. Eng. News*, 63, 17–32, 1985.
6. Walters, R.R., Affinity chromatography, *Anal. Chem.*, 57, 1099A–1114A, 1985.
7. Starkenstein, E., Ferment action and the influence upon it of neutral salts, *Biochem. Z.*, 24, 210–218, 1910.
8. Landsteiner, K. and van der Scheer, J., Cross reactions of immune sera to azoproteins. *J. Exp. Med.*, 63, 325–339, 1936.
9. Campbell, D.H., Luescher, E., and Lerman, L.S., Immunologic adsorbents. I. Isolation of antibody by means of a cellulose-protein antigen. *Proc. Natl. Acad. Sci. U.S.A.*, 37, 575–578, 1951.
10. Hjerten, S., The preparation of agarose spheres for chromatography of molecules and particles. *Biochem. Biophys. Acta*, 79, 393–398, 1964.
11. Axen, R., Porath, J., and Ernback, S., Chemical coupling of peptides and proteins to polysaccharides by means of cyanogen halides. *Nature*, 214, 1302–1304, 1967.
12. Cuatrecasas, P., Wilchek, M., and Anfinsen, C.B., Selective enzyme purification by affinity chromatography. *Proc. Natl. Acad. Sci. U.S.A.*, 68, 636–643, 1968.
13. Chaiken, I.M., Ed., *Analytical Affinity Chromatography*, CRC Press, Boca Raton, FL, 1987.
14. Hage, D.S., Xuan, H., and Nelson, M.A., Application and elution in affinity chromatography, in *Handbook of Affinity Chromatography*, 2nd edn., Hage, D.S., Ed., CRC Press, Boca Raton, FL, 2005, Chap. 4.
15. Stites, D.P., Stobo, J.D., Fudenberg, H.H., and Wells, J.V., *Basic and Clinical Immunology*, 5th edn., Lange Medical Publications, Los Altos, CA, 1984.
16. Ohlson, S., Lundblad, A., and Zopf, D., Novel approach to affinity chromatography using "weak" monoclonal antibodies, *Anal. Biochem.*, 169, 204–208, 1988.
17. Wikstroem, M. and Ohlson, S., Computer simulation of weak affinity chromatography, *J. Chromatogr.*, 597, 83–92, 1992.
18. Ohlson, S., Bergstroem, M., Pahlsson, P., and Lundblad, A., Use of monoclonal antibodies for weak affinity chromatography, *J. Chromatogr. A*, 758, 199–208, 1997.
19. Strandh, M., Andersson, H.S., and Ohlson, S., Weak affinity chromatography, *Methods Mol. Biol.*, 147, 7–23, 2000.
20. Lindmark, R., Biriell, C., and Sjoquist, J., Quantitation of specific IgG antibodies in rabbits by a solid-phase radioimmunoassay with ^{125}I-protein A from *Staphylococcus aureus*, *Scand. J. Immunol.*, 14, 409–420, 1981.
21. Hage, D.S., Bian, M., Burks, R., Karle, E., Ohnmacht, C., and Wa, C., Bioaffinity chromatography, in *Handbook of Affinity Chromatography*, 2nd edn, Hage, D.S., Ed., CRC Press, Boca Raton, FL, 2005, Chap. 5.
22. de Frutos, M. and Regnier, F.E., Tandem chromatographic-immunological analyses, *Anal. Chem.*, 65, 17A–25A, 1993.
23. Hage, D.S., Survey of recent advances in analytical applications of immunoaffinity chromatography, *J. Chromatogr. B*, 715, 3–28, 1998.
24. Hage, D.S. and Phillips, T.M., Immunoaffinity chromatography, in *Handbook of Affinity Chromatography*, 2nd edn., Hage, D.S., Ed., CRC Press, Boca Raton, FL, 2005, Chap. 6.
25. Hermanson, G.T., Mallia, A.K., and Smith, P.K., *Immobilized Affinity Ligand Techniques*, Academic Press, New York, 1992.
26. Gustavsson, P.-E. and Larsson, P.-O., Support materials for affinity chromatography, in *Handbook of Affinity Chromatography*, 2nd edn., Hage, D.S., Ed., CRC Press, Boca Raton, FL, 2005, Chap. 2.
27. Lee, W.C., Protein separation using non-porous sorbents, *J. Chromatogr. B*, 699, 29–45, 1997.
28. Wikström, P. and Larsson, P.-O., Affinity fibre—A new support for rapid enzyme purification by high-performance liquid affinity chromatography, *J. Chromatogr.*, 388, 123–134, 1987.
29. Charcosset, C., Purification of proteins by membrane chromatography, *J. Chem. Technol. Biotechnol.*, 71, 95–110, 1998.
30. Roper, D.K. and Lightfoot, E.N., Separation of biomolecules using adsorptive membranes, *J. Chromatogr. A*, 702, 3–26, 1995.
31. Fulton, S.P., Meys, M., Varady, L., Jansen, R., and Afeyan, N.B., Antibody quantitation in seconds using affinity perfusion chromatography, *Biotechniques*, 11, 226–231, 1991.
32. Gustavsson, P.E., Mosbach, K., Nilsson, K., and Larsson, P.-O., Superporous agarose as an affinity chromatography support, *J. Chromatogr. A*, 776, 197–203, 1997.
33. Pålsson, E., Smeds, A.L., Petersson, A., and Larsson, P.-O., Faster isolation of recombinant factor VIII SQ, with a superporous agarose matrix, *J. Chromatogr. A*, 840, 39–50, 1999.

34. Mallik, R. and Hage, D.S., Affinity monolith chromatography, *J. Sep. Sci.*, 29, 1686–1704, 2006.
35. Voute, N. and Boschetti, E., Highly dense beaded sorbents suitable for fluidized bed applications, *Bioseparation*, 8, 115–120, 1999.
36. Kim, H.S. and Hage, D.S., Immobilization methods for affinity chromatography, in *Handbook of Affinity Chromatography*, 2nd edn., Hage, D.S., Ed., CRC Press, Boca Raton, FL, 2005, Chap. 3.
37. Chibata, I., *Immobilized Enzymes*, Wiley, New York, 1978.
38. Messing, R.A., Adsorption and inorganic bridge formations, *Methods Enzymol.*, 44, 148–169, 1976.
39. Schetters, H., Avidin and streptavidin in clinical diagnostics, *Biomol. Eng.*, 16, 73–78, 1999.
40. Hnatowich, D.J., Virzi, F., and Rusckowski, M., Investigations of avidin and biotin for imaging applications, *J. Nucl. Med.*, 28, 1294–1302, 1987.
41. La Rochelle, W.J. and Froehner, S.C., Immunochemical detection of proteins biotinylated on nitrocellulose replicas, *J. Immunol. Methods*, 92, 65–71, 1986.
42. Cole, S.R., Ashman, L.K., and Ey, P.L., Biotinylation: An alternative to radioiodination for the identification of cell surface antigens in immunoprecipitates, *Mol. Immunol.*, 24, 699–705, 1987.
43. Gygi, S., Rist, B., Gerber, S., Turecek, F., Gelb, M., and Aebersold, R., Quantitative analysis of complex protein mixtures using isotope-coded affinity tags, *Nat. Biotechnol.*, 17, 994–999, 1999.
44. Briscoe, C.J., Clarke, W., and Hage, D.S., Affinity mass spectrometry, in *Handbook of Affinity Chromatography*, 2nd edn., Hage, D.S., Ed., CRC Press, Boca Raton, FL, 2005, Chap. 27.
45. Bernfeld, P. and Wan, J., Antigens and enzymes made insoluble by entrapping them into lattices of synthetic polymers, *Science*, 142, 678–679, 1963.
46. O'Driscoll, K.F., Techniques of enzyme entrapment in gels, *Methods Enzymol.*, 44, 169–183, 1976.
47. Hodgson, R.J., Chen, Y., Zhang, Z., Tieugabulova, D., Long, H., Zhao, X., Organ, M., Brook, M.A., and Brennan, J.D., Protein-doped monolithic silica columns for capillary liquid chromatography prepared by the sol-gel method: Applications to frontal affinity chromatography, *Anal. Chem.*, 76, 2780–2790, 2004.
48. Cardile, V., Renis, M., Gentile, B., and Panico, A.M., Activity of liposome-entrapped immunomodulator oligopeptides on human epithelial thymic cells, *Pharm. Pharmacol. Commun.*, 6, 381–386, 2000.
49. Gregoriadis, G., McCormack, B., Obrenovic, M., Perrie, Y., and Saffie, Y., Liposomes as immunological adjuvants and vaccine carriers, *Methods Mol. Med.*, 42, 137–150, 2000.
50. Karau, C., Pongpaibul, Y., and Schmidt, P.C., Quantitative evaluation of human leukocyte interferon-entrapped in liposomes, *Drug Delivery*, 3, 59–62, 1996.
51. Winsor, D.J., Quantitative affinity chromatography: Recent theoretical developments, in *Handbook of Affinity Chromatography*, 2nd edn., Hage, D.S., Ed., CRC Press, Boca Raton, FL, 2005, Chap. 23.
52. Haneskog, L., Zeng, C.-M., Lundqvist, A., and Lundahl, P., Biomembrane affinity chromatographic analysis of inhibitor binding to the human red cell nucleoside transporter in immobilized cells, vesicles and proteoliposomes. *Biochim. Biophys. Acta*, 1371, 1–4, 1998.
53. Lu, L., Lundqvist, A., Zeng, C.-M., Lagerquist, C., and Lundahl, P., D-Glucose, forskolin and cytochalasin B affinities for the glucose transporter Glut 1. *J. Chromatogr. A*, 776, 81–86, 1997.
54. Lundqvist, A., Brekkan, E., Haneskog, L., Yang, Q., Miyaki, J., and Lundahl, P., Determination of transmembrane protein affinities for solutes by frontal chromatography, in *Quantitative Analysis of Biospecific Interactions*, Lundahl, P., Lundqvist, A., and Greijer, E., Eds., Harwood Academic, Amsterdam, the Netherlands, 1998, pp. 79–93.
55. Zeng, C.-M., Zhang, Y., Lu, L., Brekkan, E., Lundqvist, A., and Lundahl, P., Immobilization of human red cells in gel particles for chromatographic activity studies of the glucose transporter Glut 1, *Biochim. Biophys. Acta*, 1325, 91–98, 1997.
56. Hage, D.S., Walters, R.R., and Hethcote, H.W., Split-peak affinity chromatographic studies of the immobilization-dependent adsorption kinetics of protein A, *Anal. Chem.*, 58, 274–279, 1986.
57. Friedberg, F. and Rhoads, A.R., Affinity chromatography of enzymes, in *Handbook of Affinity Chromatography*, 2nd edn., Hage, D.S., Ed., CRC Press, Boca Raton, FL, 2005, Chap. 12.
58. Wilchek, M., Miron, T., and Kohn, J., Affinity chromatography, *Methods Enzymol.*, 104, 3–55, 1984.
59. Liener, I.E., Sharon, N., and Goldstein, I.J., *The Lectins: Properties, Functions and Applications in Biology and Medicine*, Academic Press, London, U.K., 1986.
60. Ey, P.L., Prowse, S.J., and Jenkin, C.R., Isolation of pure IgG, IgG$_{2a}$ and IgG$_{2b}$ immunoglobulins from mouse serum using protein A-Sepharose, *Immunochemistry*, 15, 429436, 1978.
61. Bjorck, L. and Kronvall, G., Purification and some properties of streptococcal protein G, a novel IgG-binding reagent, *J. Immunol.*, 133, 969–974, 1984.
62. Akerstrom, B. and Bjorck, L., A physiochemical study of protein G, a molecule with unique immunoglobulin G-binding properties, *J. Biol. Chem.*, 261, 10240–10247, 1986.

63. Moxley, R.A., Oak, S., Gadgil, H., and Jarrett, H.W., DNA affinity chromatography, in *Handbook of Affinity Chromatography*, 2nd edn., Hage, D.S., Ed., CRC Press, Boca Raton, FL, 2005, Chap. 7.

64. Potuzak, H. and Dean, P.D.G., Affinity chromatography on columns containing nucleic acids, *FEBS Lett.*, 88, 161–166, 1978.

65. Alberts, A.M. and Herrick, G., DNA-cellulose chromatography, *Methods Enzymol.*, 21, 198–217, 1971.

66. Calton, G.J., Immunosorbent separations, *Methods Enzymol.*, 104, 381–387, 1984.

67. Phillips, T.M., High performance immunoaffinity chromatography: An introduction, *LC Mag.*, 3, 962–972, 1985.

68. Phillips, T.M., Affinity chromatography in antibody and antigen purification, in *Handbook of Affinity Chromatography*, 2nd edn., Hage, D.S., Ed., CRC Press, Boca Raton, FL, 2005, Chap. 14.

69. Ehle, H. and Horn, A., Immunoaffinity chromatography of enzymes, *Bioseparation*, 1, 97–110, 1990.

70. Bailon, P. and Roy, S.K., Recovery of recombinant proteins by immunoaffinity chromatography, *ACS Symp. Ser.*, 427, 150–167, 1990.

71. Howell, K.E., Gruenberg, J., Ito, A., and Palade, G.E., Immuno-isolation of subcellular components, *Prog. Clin. Biol. Res.*, 270, 77–90, 1988.

72. Nakajima, M. and Yamaguchi, I., Purification of plant hormones by immunoaffinity chromatography, *Kagaku to Seibutsu*, 29, 270–275, 1991.

73. Hage, D.S. and Nelson, M.A., Chromatographic immunoassays, *Anal. Chem.*, 73, 198A–205A, 2001.

74. Moser, A.M. and Hage, D.S., Chromatographic immunoassays, in *Handbook of Affinity Chromatography*, 2nd edn., Hage, D.S., Ed., CRC Press, Boca Raton, FL, 2005, Chap. 29.

75. Irth, H., Oosterkamp, A.J., Tjaden, U.R., and van der Greef, J., Strategies for online coupling of immunoassays to HPLC, *Trends Anal. Chem.*, 14, 355–361, 1995.

76. Labrou, N.E., Mazitsos, K., and Clonis, Y.D., Dye-ligand and biomimetic affinity chromatography, in *Handbook of Affinity Chromatography*, 2nd edn., Hage, D.S., Ed., CRC Press, Boca Raton, FL, 2005, Chap. 9.

77. Labrou, N.E. and Clonis, Y.D., Immobilised synthetic dyes in affinity chromatography, in *Theory and Practice of Biochromatography*, Vijayalakshmi, M.A., Ed., Taylor & Francis, London, U.K., 2002, pp. 335–351.

78. Labrou, N.E. and Clonis, Y.D., The affinity technology in downstream processing, *J. Biotechnol.*, 36, 95–119, 1994.

79. Staal, G., Koster, J., Kamp, H., Van Milligen-Boersma, L., and Veeger, C., Human erythrocyte pyruvate kinase, its purification and some properties, *Biochem. Biophys. Acta*, 227, 86–92, 1971.

80. Huang, P.Y. and Carbonell, R.G., Affinity chromatographic screening of soluble combinatorial peptide libraries, *Biotechnol. Bioeng.*, 63, 633–641, 2000.

81. Lebl, M., Krchnak, V., Sepetov, N.F., Seligmann, B., Strop, P., Felder, S., and Lam, K.S., One-bead-one-structure combinatorial libraries, *Biopolymers*, 37, 177–198, 1995.

82. Kaufman, D.B., Hentsch, M.E., Baumbach, G.A., Buettner, J.A., Dadd, C.A., Huang, P.Y., Hammond, D.J., and Carbonell, R.G., Affinity purification of fibrinogen using a ligand from a peptide library, *Biotechnol. Bioeng.*, 77, 278–289, 2002.

83. Ehrlich, G.K. and Bailon, P., Identification of peptides that bind to the constant region of a humanized IgG$_1$ monoclonal antibody using phage display, *J. Mol. Recogn.*, 11, 121–125, 1998.

84. Pluckthun, A., Schaffitzel, C., Hanes, J., and Jermutus, L., In vitro selection and evolution of proteins, *Adv. Prot. Chem.*, 55, 367–403, 2001.

85. Tuerk, C. and Gold, L., Systematic evolution of ligands by exponential enrichment: RNA ligands to bacteriophage T4 DNA polymerase, *Science*, 249, 505–510, 1990.

86. Wilson, D.S. and Szostak, J.W., In vitro selection of functional nucleic acids, *Ann. Rev. Biochem.*, 68, 611–647, 1999.

87. Wen, J.D., Gray, C.W., and Gray, D.M., SELEX selection of high-affinity oligonucleotides for bacteriophage Ff gene 5 protein, *Biochemistry*, 40, 9300–9310, 2001.

88. Roming, T.S., Bell, C., and Drolet, D.W., Aptamer affinity chromatography: Combinatorial chemistry applied to protein purification, *J. Chromatogr. B*, 731, 275–284, 1999.

89. Todorova, D. and Vijayalakshmi, M.A., Immobilized metal-ion affinity chromatography, in *Handbook of Affinity Chromatography*, 2nd edn., Hage, D.S., Ed., CRC Press, Boca Raton, FL, 2005, Chap. 10.

90. Porath, J., Carlsson, J., Olsson, I., and Belfrage, B., Metal chelate affinity chromatography, a new approach to protein fraction, *Nature*, 258, 598–599, 1975.

91. Vijayalakshmi, M.A., Pseudobiospecific ligand affinity chromatography, *Trends Biotechnol.*, 7, 71–76, 1989.

92. Winzerling, J.J., Berna, P., and Porath, J., How to use immobilized metal ion affinity chromatography, *Methods: A Companion to Methods in Enzymology*, 4, 4–13, 1992.

93. Chaga, G.S., Twenty-five years of immobilized metal ion affinity chromatography: Past, present and future, *J. Biochem. Biophys. Methods*, 49, 313–334, 2001.

94. Mrabet, N.T. and Vijayalakshmi, M.A., Immobilized metal-ion affinity chromatography. From phenomenological hallmarks to structure-based molecular insights, in *Biochromatography-Theory and Practice*, Vijayalakshmi, M., Ed., Taylor & Francis, London, U.K., 2002, pp. 272–294.

95. Hochuli, E. and Piesecki, S., Interaction of hexahistidine fusion proteins with nitrilotriacetic acid-chelated Ni^{2+} ions, *Methods: A Companion to Methods in Enzymology*, 4, 68–72, 1992.

96. Jordan, N. and Krull, I.S., Affinity chromatography in biotechnology, in *Handbook of Affinity Chromatography*, 2nd edn, Hage, D.S., Ed., CRC Press, Boca Raton, FL, 2005, Chap. 18.

97. Hage, D.S. and Chen, J., Quantitative affinity chromatography: Practical aspects, in *Handbook of Affinity Chromatography*, 2nd edn., Hage, D.S., Ed., CRC Press, Boca Raton, FL, 2005, Chap. 22.

98. Hage, D.S. and Tweed, S., Recent advances in chromatographic and electrophoretic methods for the study of drug-protein interactions, *J. Chromatogr. B*, 699, 499–525, 1997.

99. Andrews, P., Kitchen, B.J., and Winzor, D., Use of affinity chromatography for the quantitative study of acceptor-ligand interactions: The lactose synthetase system, *Biochem. J.*, 135, 897–900, 1973.

100. Dunn, B.M. and Chaiken, I.M., Quantitative affinity chromatography. Determination of binding constants by elution with competitive inhibitors, *Proc. Natl. Acad. Sci. U.S.A.*, 71, 2382–2385, 1974.

101. Kasai, K. and Ishii, S., Quantitative analysis of affinity chromatography of trypsin. A new technique for investigation of protein-ligand interaction, *J. Biochem.*, 77, 261–264, 1975.

102. Chaiken, I.M., Quantitative uses of affinity chromatography, *Anal. Biochem.*, 97, 1–10, 1979.

103. Muller, A.J. and Carr, P.W., Chromatographic study of the thermodynamic and kinetic characteristics of silica-bound concanavalin A, *J. Chromatogr.*, 284, 33–51, 1984.

104. Walters, R.R., Practical approaches to the measurement of rate constants by affinity chromatography, in *Analytical Affinity Chromatography*, Chaiken, I.M., Ed., CRC Press, Boca Raton, FL, 1987, 1986, Chap. 3.

105. Loun, B. and Hage, D.S., Chiral separation mechanisms in protein-based HPLC columns. II. Kinetic studies of R- and S-warfarin binding to immobilized human serum albumin, *Anal. Chem.*, 68, 1218–1225, 1996.

106. Yang, J. and Hage, D.S., Effect of mobile phase composition on the binding kinetics of chiral solutes on a protein-based HPLC column: Interactions of D- and L-tryptophan with immobilized human serum albumin, *J. Chromatogr. A*, 766, 15–25, 1997.

107. Moore, R.M. and Walters, R.R., Peak-decay analysis of drug-protein dissociation rates, *J. Chromatogr.*, 384, 91–103, 1987.

108. Liu, X.-C. and Scouten, W.H., Boronate affinity chromatography, in *Handbook of Affinity Chromatography*, 2nd edn., Hage, D.S., Ed., CRC Press, Boca Raton, FL, 2005, Chap. 8.

109. Glad, M., Ohlson, S., Hansson, L., Mansson, M.-O., and Mosbach, K., High-performance liquid affinity chromatography of nucleosides, nucleotides and carbohydrates with boronic acid-substituted microparticulate silica, *J. Chromatogr.*, 200, 254–260, 1980.

110. Singhal, R.P. and DeSilva, S.S.M., Boronate affinity chromatography, *Adv. Chromatogr.*, 31, 293–335, 1992.

111. Mallia, A.K., Hermanson, G.T., Krohn, R.I., Fujimoto, E.K., and Smith, P.K., Preparation and use of a boronic acid affinity support for the separation and quantitation of glycosylated hemoglobins, *Anal. Lett.*, 14, 649–661, 1981.

112. Hage, D.S., Affinity chromatography: A review of clinical applications, *Clin. Chem.*, 45, 593–615, 1999.

113. Halperin, G., Tauber-Finkelstein, M., Schmeeda, H., and Shaltiel, S., Hydrophobic chromatography of cells: Adsorption and resolution of homologous series of alkylagaroses, *J. Chromatogr.*, 317, 103–118, 1984.

114. Shaltiel, S., Hydrophobic chromatography, *Methods Enzymol.*, 104, 69–96, 1984.

115. Brocklehurst, K., Carlsson, J., Kierstan, M.P.J, and Crook, E.M., Covalent chromatography. Preparation of fully active papain from dried papaya latex, *Biochem. J.*, 133, 573–584, 1973.

116. Mohr, P. and Pommerening, K., *Affinity Chromatography: Practical and Theoretical Aspects*, Marcel Dekker, New York, 1985.

117. Haupt, K., Molecularly imprinted polymers: Artificial receptors for affinity separations, in *Handbook of Affinity Chromatography*, 2nd edn., Hage, D.S., Ed., CRC Press, Boca Raton, FL, 2005, Chap. 30.

118. Sellergren, B., *Molecularly Imprinted Polymers—Man-Made Mimics of Antibodies and Their Applications in Analytical Chemistry*, Elsevier, Amsterdam, the Netherlands, 2001.

119. Komiyama, M., Takeuchi, T., Mukawa, T., and Asanuma, H., *Molecular Imprinting—From Fundamentals to Applications*, Wiley-VCH, Weinheim, Germany, 2002.

120. Subramanian, A., General considerations in preparative affinity chromatography, in *Handbook of Affinity Chromatography*, 2nd edn., Hage, D.S., Ed., CRC Press, Boca Raton, FL, 2005, Chap. 11.

121. Subramanian, A., Isolation of recombinant proteins by affinity chromatography, in *Handbook of Affinity Chromatography*, 2nd edn., Hage, D.S., Ed., CRC Press, Boca Raton, FL, 2005, Chap. 13.

122. Jurado, L.A., Oak, S., Gadgil, H., Moxley, R.A., and Jarrett, H.W., Affinity chromatography in molecular biology, in *Handbook of Affinity Chromatography*, 2nd edn., Hage, D.S., Ed., CRC Press, Boca Raton, FL, 2005, Chap. 20.

123. Rhoads, A.R. and Friedberg, F., Affinity chromatography of regulatory and signal-transducing proteins, in *Handbook of Affinity Chromatography*, 2nd edn., Hage, D.S., Ed., CRC Press, Boca Raton, FL, 2005, Chap. 15.

124. Wolfe, C.A.C., Clarke, W., and Hage, D.S., Affinity chromatography in clinical and pharmaceutical analysis, in *Handbook of Affinity Chromatography*, 2nd edn., Hage, D.S., Ed., CRC Press, Boca Raton, FL, 2005, Chap. 17.

125. Nelson, M.A. and Hage, D.S., Environmental analysis by affinity chromatography, in *Handbook of Affinity Chromatography*, 2nd edn., Hage, D.S., Ed., CRC Press, Boca Raton, FL, 2005, Chap. 19.

126. Heegaard, N.H.H. and Schou, C., Affinity ligands in capillary electrophoresis, in *Handbook of Affinity Chromatography*, 2nd edn., Hage, D.S., Ed., CRC Press, Boca Raton, FL, 2005, Chap. 26.

127. Phillips, T.M., Microanalytical methods based on affinity chromatography, in *Handbook of Affinity Chromatography*, 2nd edn., Hage, D.S., Ed., CRC Press, Boca Raton, FL, 2005, Chap. 28.

14 Ion Chromatography: Modes for Metal Ions Analysis

Corrado Sarzanini and Maria Concetta Bruzzoniti

CONTENTS

14.1 INTRODUCTION

Ion chromatography (IC), introduced by Small and coworkers [1], basically represents a continuous ion-exchange process where a water-based solution, "eluent" or "mobile phase," is fluxed through a column filled with an ion-exchange solid material, "stationary phase," that is characterized by the presence of fixed ionic sites within counterions on the surface, and eluted as "eluate" to the detector. Analytes are separated on the basis of their different relative affinities for the ionic sites of the stationary phase and the separation mechanism(s), hereafter detailed, is mainly an ion exchange occurring between the counterion on the stationary phase and ions (analytes or not) in the eluent. The most traditional and universal detector applied in ion chromatography (IC) was the conductometric detector. In respect to this kind of detector, two operating modes are defined, named nonsuppressed and suppressed ion chromatography. In the latter mode, a device "suppressor," located before the detector, enables the reduction or total removal of the contribute to the conducibility of the eluate due to the composition of the eluent, increasing the detection sensitivity. Spectrophotometric, amperometric, and spectrometric detection modes are also usually coupled to IC separations.

Ion-exchange principles and applications [2–5] have recently been reviewed by Lucy [6] and Fritz [7] and a wide number of books and reviews are available on ion chromatography [8–13] and on recent developments of new kind of suppressors [14–16] or different detection approaches [17].

It must be underlined that for IC, a pure ion-exchange was assumed as the dominant mechanism but, in many cases, secondary equilibria are involved. During the last few years, the term IC has also included other chromatographic techniques, e.g., ion-pair chromatography, chelation-ion chromatography and electrostatic-ion chromatography, which are ion chromatographic modes where secondary equilibria play a relevant role. In the following chapters, a survey of basic principles and models concerning ion exchange chromatography and related modes is made. Then, the evolution, mainly due to the development of new materials, will be discussed through some examples with particular regard to metal-species determination. The main detection modes in IC, principles, and instrumental evolution, are briefly described at the end for a better understanding.

14.2 ION CHROMATOGRAPHY

14.2.1 BASIC PRINCIPLES

The basic concept for ion chromatographic separation is the competition between the ions inside the eluent (analytes or not) and the counterions of fixed ionic sites on the stationary phase.

The simplest way to summarize reactions involved in IC separations is

$$x\left(A^y\right)_s + y\left(B^x\right)_m \leftrightarrow y\left(B^x\right)_s + x\left(A^y\right)_m$$

where

   s and m represent the stationary and mobile phase
   x and y represent the number of charges of competing B and A ions (sign omitted since the equation represents both anion and cation exchange process), respectively

The reaction is characterized by an equilibrium constant $K_{A,B}$, which referring to the competition between two analytes, is defined selectivity constant, and assuming that the acitivity coefficients are equal to the unit, can be written as

$$K_{A,B} = \frac{\left[B_s^x\right]^y \left[A_m^y\right]^x}{\left[A_s^y\right]^x \left[B_m^x\right]^y}$$

One other parameter characterizing the ion exchange is the concentration- [8] or weight-based [9] distribution coefficient $K_D = C_s/C_m$, where C_s and C_m represent the concentrations of analytes in the stationary and mobile phases, respectively.

From the chromatographic point of view, analytes are characterized by the retention factor, $k = (t_r - t_0)/t_0$, where t_r and t_0 are respectively, the retention time for the analyte and for a species which is not retained in the column. The retention factor is directly related to the partition coefficient or equilibrium constant through the ratio of the weight of the stationary phase w and the volume of the mobile phase V_m: $k = K_D [w/V_m]$ [9]. In addition, through the retention factors, the chromatographic separation factor $\alpha_{A,B}$ is defined ($\alpha_{A,B} = k_A/k_B$) for two analytes A and B. This is related to the weight of the stationary phase, the volume of the mobile phase, the charges of analytes, and the retention factor of the latter eluting component. In addition from this relation, it derives that for analytes with the same charge, the separation factor (log α) depends only on the selectivity coefficient and on the charge of solute ions.

The most important parameter to achieve a good separation is selectivity that has been recently defined as the relative ability of sample ions to form a cation-anion pair in the stationary phase. For example, for the separation of alkali metal ions on a surface sulfonated cation-exchange resin (simply based on a pure ion-exchange mechanism, see below), where the charge density of the analyte is assumed to be the main driving force, the retention order (selectivity) will be: Li^+, Na^+, NH_4^+, K^+, Rb^+, Cs^+. Selectivity, in ion chromatography, is effectively a function of the nature of the analyte (charge, radius, solvation), the electrostatic interaction between ions and ion-exchange groups and possible hydrophobic interactions, the solvation of both matrix and counterions, the physical and the chemical constitution of the stationary phase and the eluent composition [18]. All these aspects will be considered and have recently been discussed in depth [19].

14.2.2 Retention Models

The basic principles of thermodynamic aspects of chromatographic separations were described by Small [8]. Haddad et al. [9,20] derived a retention model for anions and cations, valid for eluents containing both a single competing ion and a complexing ligand. A model of the retention of cationic complexes was proposed [21] and a detailed retention model for the separation of anionic EDTA complexes was developed by Hajos et al. [22]. The role of complex-forming equilibria in the separation of metal cations was deeply discussed by Janoš [23]. Stahlberg [24] and Bartha and Stahlberg [25] used the theory of Gouy–Chapman to describe an electrostatic model for ion-exchange that is not stoichiometric, as the above mentioned models, but related to the electrostatic surface potential, the association constant of the eluent counterion, its concentration, and the surface concentration of fixed charges on the stationary phase. The retention of strong electrolytes on strong ion exchangers, assuming only the existence of a Stern layer, has also been treated [26], but more recently the same authors [27] showed that for the classical model, the parameters of the resulting system are not-self consistent and proposed a mixed mechanism based on an ion exchange with an incomplete Stern layer and an electrophoretic motion related to the movement of charges located in the diffuse Gouy–Chapman layer. A multiple species eluent/analyte retention model has been derived that also takes into account the protonation of sample anions [28] and results as a predictive tool for the retention behavior of anions. The model enabled the prediction of retention data for the separation of carboxylic acids as a function of pH and eluent composition [29]. Subsequently retention of the same analytes has been characterized according to analyte properties (alkyl chain, unsaturations, -OH substituents), stationary phase properties (ion-exchange site, hydrophilicity), and eluent modifiers (acetonitrile, methanol, and n-propanol) [30,31] underlying the effect on selectivity due to the nature of the eluent [31].

A comprehensive review of retention models for both cation- and anion-exchange separations [32] and a critical comparison of retention models for the separation of anions [33] have been published. Okada [34–37] elucidated on the electrostatic aspects of retention in some chromatographic models correlating the nature of ions and the separation selectivity [38] and investigated the effect of the local structures of ions at the ion-exchange resin/solution interface [39]. More recently, Horváth and Hajós [40] demonstrated that for latex based pellicular anion exchangers, the retention profiles of anions are related to the simultaneous ion-pair and ion-exchange competition in the analyte/eluent/ionic functional groups system. They developed a retention model based on the stoichiometric approach but introduced a new parameter: a fractional coefficient of the ion-exchange capacity due to the effect of the electrostatic potential as a structure-related parameter of the stationary phase confirming the predictive power of the model for monovalent and divalent anions. Considering both the complexation and the protonation equilibria, a theoretical retention model has been developed for a cryptand-based anion exchange column [41], which showed a good agreement between computed and experimental values for binding and exchange constants of five anions. Multilayer perceptron artificial networks have been investigated for retention modeling in isocratic ion chromatography [42–44] and more recently a radial basis function (RBF) artificial neural network gradient elution retention model has been proposed [45] and used to predict retention times for fluoride, chloride, nitrite, bromide, nitrate, sulfate, and phosphonate in relation to the starting time of the gradient elution and the slope of the linear gradient curve.

14.2.3 EVOLUTION

The stationary phases used in IC are both inorganic- and polymeric-based exchangers. Recently, with the development of new inorganic chemistries, the use of inorganic materials has received new interest. Stationary phases are divided into cation and anion exchangers. Eluents are usually water-based solutions (in some cases added with an organic modifier) containing acids, bases, or salts that furnish the counterions which compete with the analytes for the charged-fixed sites of the stationary phase. An analysis of transition and heavy metal ions can be performed by cation, anion, and the mixed-exchange mode. Since the selectivity coefficients for these ions are very similar, the separation is accomplished by adding a weak or strong chelating agent to the eluent so that secondary equilibria are induced enhancing the differences in the charge, the size, and the hydration of analytes and hence their selectivity for the stationary phase. Figure 14.1 shows a schematic illustration of the equilibria existing on a cation- or anion-exchanger between a solute cation, an added ligand, and an eluent cation.

14.2.4 CATION EXCHANGE ON POLYMER-BASED MATERIALS

The stationary phases used when IC was introduced [1] were polystyrene-divinylbenzene (PS-DVB) core particles, where the amount of DVB in the resin defined the optimum degree of cross-linking for their mechanical stability, usually between 2% and 5%. These particles, 20–25 μm in size, were surface-sulfonated and characterized by high capacity values. Due to the large differences in the selectivity of alkali cations towards the alkaline earth, monovalent cations were eluted with 2.5 mM HCl while a stronger divalent eluent component, m-phenylenediamine, was required for the elution of divalent cations (Figure 14.2). This was the main drawback of these materials since it was impossible to achieve the separation of both classes of cations by a simple isocratic elution in a reasonable period of time. The first improvement was realized by a new method of synthesis where the surface-sulfonated PS-DVB core particles were surrounded by a layer of anion-exchange latex particles and successively covered by a monolayer of sulfonated cation-exchange latex particles electrostatically

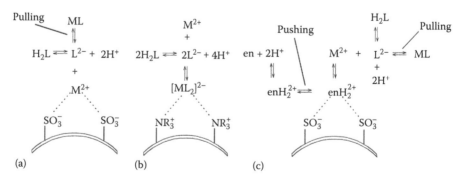

FIGURE 14.1 Secondary chemical equilibria in (a and c) cation- or (b) anion-exchange chromatography. Metal ion M^{2+} as such or as an anion complex: (a) cation exchange, the ligand (L^{2-}) exerts the pulling effect on the cation analyte (M^{2+}) through the complexation; (b) anion exchange, the deprotonated ligand (L^{2-}) competes (pushing effect) with the anion complex (ML_2^{2-}); (c) cation exchange, pulling and pushing effects of the ligand and of a competing cation on the retention of M^{2+}. (Reprinted from Sarzanini, C., High-performance liquid chromatography: Trace metal determination and speciation, in Brown, P.R. and Grushka, E. (eds.), *Advances in Chromatography*, Vol. 41, Marcel Dekker, New York, 2001. With permission.)

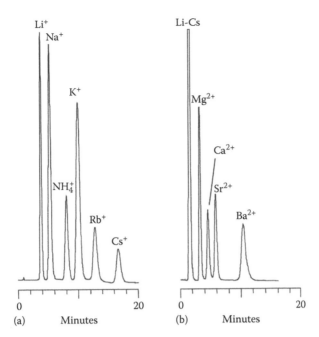

FIGURE 14.2 Separation of ammonium, alkali, and alkaline earth metals on IonPac CS1. Detection: suppressed conductivity. Eluent: (a): 5 mM HCl; (b): 2 mM HCl, 2 mM m-phenylenediamine dihydrochloride.

attached (IonPac CS3, Dionex Co., Sunnyvale, CA). The latex material enabled a higher mass transfer resulting in a significant improvement in peak efficiencies. Using a 2,3-diaminopropionic acid monochloride based eluent [46], this column enabled the simultaneous analysis of both alkali and alkaline earth elements in the presence of ammonium, but the analysis still required a long time without a complete baseline resolution. Kolla et al. [47] introduced a different kind of column containing carboxylate functionalities with a low selectivity for the hydronium ion. They used silica gel beads coated with a poly(butadiene-maleic acid) (PBMA), and with eluents containing slightly acidic complexing agents (e.g., tartaric acid) achieved the separation of mono and divalent cations plus ammonium in a single run in unsuppressed IC. The selectivity of carboxylic groups and the competition of the chelating agent in the eluent with the cation-exchange sites for divalent cations only, provided a good separation and a short analysis time. More recently, Fritz et al. [48] synthesized a similar weak acidic cation exchanger by polymerization of macroporous poly(ethylvinylbenzene-co-divinylbenzene) beads (5 μm) and maleic anhydride followed by the hydrolysis of the maleic anhydride groups. The column resulted solvent compatible and enabled the separation of alkali and alkaline earth metal ions in a single isocratic elution (3.5 mM sulfuric acid) with conductometric detection. Reversed stationary phases permanently coated with hydrophobic agents such as alkylsulfonates or alkylsulfates have also been used for the separation of transition and heavy metals [49,50]. The new philosophy for the stationary phases was to use columns with sulfonic groups at a decreased cation-exchange site density (IonPac CS10, Dionex) and/or columns with weaker functional groups, carboxylic (IonPac CS12, CS14), carboxylic, and phosphponic (IonPac CS12A) or carboxylic/phosphonic/crown ether (IonPac CS15). The use of a substrate based on ethylene-vinylbenzene (EVB) crosslinked with 55% divinylbenzene (DVB) covered by a 0.2%–5% crosslinked PS-DVB cation exchange latex, resulted in solvent-compatible materials. Using IonPac CS10, due to its decreased cation-exchange site density, the retention of monovalent and prevalently of divalent cations was reduced as was their interaction with the stationary phase [51]. An improvement in the separation was achieved by the IonPac CS12 [52] stationary phase at a higher selectivity for hydronium ion, which required only an acidic isocratic elution for the separation of alkali and alkaline earth plus ammonium in only 8 min. Fritz et al. [53] have recently developed an IC method for the determination of common cations and some transition metals by using a weak cation exchanger functionalized with carboxylic acid groups (250 mm × 4 mm I.D., IonPac SCS1, Dionex). Eleven cations (copper, lithium, sodium, ammonium, potassium, cobalt, nickel, magnesium, calcium, strontium, and zinc) were separated by isocratic elution (mobile phase: 2.5 mM methane sulfonic acid and 0.8 mM oxalic acid) and detected by non-suppressed conductivity (reversed mode) in beer and mineral waters with detection limits between 1.0 μg/L (Li+) and 48 μg/L (Ni+).

By moving from a microporous substrate bead to a highly cross-linked macroporous substrate bead with a large surface area and grafting on the surface a thin film, 5–10 nm in thickness, of polymer containing carboxylic and phosphonic groups, a new high efficiency stationary phase was produced [54]. This new stationary phase (IonPac CS12A), at high capacity (2800 μeq per column, 250 × 4 mm I.D.) and solvent compatible, resulted more hydrophilic enabling the separation of inorganic cations (I and II group), amines (methylamines, diamines, aliphatic amines), and quaternary ammonium ions by eluents based on sulfuric acid or sulfuric acid-acetonitrile (Figure 14.3). Since a particular feature of this column is its higher sensitivity to temperature, peak shapes and efficiencies are improved under elevated temperature; for the more hydrophobic organic analytes, where the separation mechanism was a combination of cation-exchange and reversed-phase interaction with the substrate, good resolutions were reached by working at an increased temperature (60°C) avoiding the use of the solvent. A detailed study on the effect of column temperature on retention of the above mentioned analytes for these kinds of columns has been made by Kulisa [55], which showed that for some cations complete and rapid separations were achieved by lowering the temperature down to 10°C rather than increasing it. To solve the problem arising from the analysis of samples of very different concentration ratios of analytes (e.g., ammonium-to-sodium) or for analytes with similar selectivity for carboxylated or sulfonated cation-exchangers (e.g., ethanolammonium and ammonium ions) a column-switching (dual column) technique has been proposed [56] speculating on the different selectivities of the stationary phases IonPac CS10 and IonPac CS12A. It should be remembered that in the column-switching mode, samples are eluted through two separate columns, each optimized for the chromatography of one group of cations. Separation is accomplished by switching the eluent flow paths from one column to the other. In the above-mentioned method [56] by an isocratic methanesulfonic-based elution, analytes flow through the first column

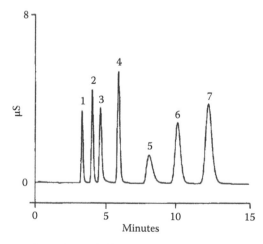

FIGURE 14.3 Isocratic separation of morpholine, alkali, and alkaline earth metals on IonPac CS12A column. Eluent: 10 mM sulfuric acid. Detection: suppressed conductivity. Peaks: 1, lithium (0.5 mg/L); 2, sodium (2 mg/L); 3, ammonium (2.5 mg/L); 4, potassium (5 mg/L); 5, morpholine (25 mg/L); 6, magnesium (2.5 mg/L); 7, calcium (5 mg/L). (From Rey, M.A. and Pohl, C.A., *J. Chromatogr. A*, 739, 87, 1996. Copyright 1996. With permission from Elsevier.)

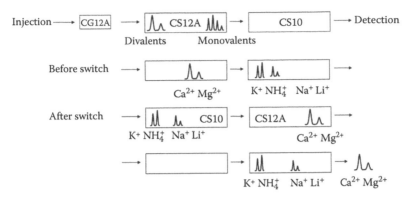

FIGURE 14.4 Step-by-step isocratic elution of mono- and divalent cations by column-switching. After elution of monovalent cations from CS12A to CS10, the valves are switched and the order of columns are reversed. (From Rey, M.A. et al., *J. Chromatogr. A*, 789, 149, 1997. Copyright 1997. With permission from Elsevier.)

(carboxylic-phosphonic) where divalent cations, which have higher selectivity are more retained, and monovalent cations enter the second column (sulfonated). At this point the valve is switched and during the elution of the divalent cation to the detector, the flux of monovalent cations is introduced again in the first column and after an improved separation reaches the detector (Figure 14.4). This method enabled a good separation of alkali and alkaline earth metal ions, with a reversed order of elution, and the detection at concentration ratios greater than 10,000: 1 for ammonium to sodium and *vice versa*. It must be underlined that the polymeric stationary phases used in IC and characterized by the presence of carboxylic or carboxylic/phosphonic functional groups are at the border line between IC and high pressure chelation ion chromatography. More recently, the IonPac CS12A column has also been used for the separation of a wide range of transition and heavy metal ions speculating on the chelating properties of the column [57]. The ion-exchange mechanism was hindered by working with nitric acid eluents at high ionic strength (e.g., 0.5 M potassium nitrate) and the resulting selectivity for cations was in agreement with the model stability constants of carboxylic groups.

The ability of macrocyclic ligands [58], such as crown ethers, to coordinate metal cations with different selectivities as a function of their cavity and metal ions diameter has been demonstrated [59] and macrocyclic ligands have been used as components of the stationary or mobile phases [60] for the separation of alkali metal cations, alkali earth metal cations, and ammonium. In this field, Lamb et al. [61] studied the performance of a chromatographic column based on tetradecyl-18-crown-6 macrocycle adsorbed onto a non-polar stationary phase (55% ETVB-DVB particles, $10\,\mu m$), coupled with suppressed detection. Conversely for the results obtained with water-methanol eluent systems, it has been shown that an acidic eluent avoids shifting retention times due to matrix counterions and that this kind of column allows greater sensitivity for Cs^+ [62,63]. Increase in the separation factors and a change in the elution order of alkali metal cations have also been obtained using $5\,\mu m$ macroporus cation-exchange resins (low capacity PS-DVB sulfonated resins) with aqueous-organic eluents [64]. This behavior was explained by two concomitant effects: (1) the lower dielectric of organic solvents favors the electrostatic attraction of cations to the sulfonic group of the resin increasing the retention time, (2) at a higher organic content cations are less solvated by water and more by the organic solvent molecules resulting in a larger radius [65] that inhibits the interaction of the cation with the resin reducing the retention time. The latter effect seems predominant and more pronounced for Li^+ and Na^+. The higher resolution obtained enabled the separation of ions that usually elute close together (Li^+/Na^+, K^+/NH_4^+) and both peak shape and resolution were improved by the addition of 18-crown-6 ether to the eluent. The selectivity due to the pure ion-exchange mechanism for alkali and alkaline earth cations was also modified by adding 18-crown-6 ether and acetonitrile to the mobile phase on a weak cation silica-based exchanger [66] and Ohta et al. [67]

described how the addition of the same macrocycle to the mobile phase improves the resolution between mono- and divalent cations on an unmodified silica gel column.

On the basis of these and similar results Rey et al. [68] developed a new chromatographic column (IonPac CS15) that consists of a polymeric macroporous substrate, 8.5 µm particle size with an average surface area of 450 m²/g, with carboxylic acid groups, phosphonate groups (like IonPac CS12), and 18-crown-6 ether groups permanently attached, grafted. This column resulted suitable for the separation of alkali and alkaline earth cations plus ammonium, giving a high resolution for the sodium-ammonium pair (eluent usually 5 mM sulfuric acid + 9% acetonitrile) at a sodium-to-ammonium concentration ratio of 4,000:1 or 10,000:1 respectively. This stationary phase also seems ideal for samples with a high concentration of potassium since, due to its high selectivity for 18-crown-6 ether, it is the last eluted. The resolution achieved by this column has been recently improved by using 18-crown-6 ether also in the eluent [69]. The addition of 18-crown-6 to the eluent (10.80 mM sulfuric acid) leads to the quantitation of Na^+ and NH_4^+ at relative concentrations of 60,000:1 and also improves the separation of amine cations from common metal ions. Saari-Nordhaus and Anderson, Jr. [70] proposed an alternative approach to enhance the selectivity for the above mentioned separations. They showed that different selectivities can be reached by coupling a conventional carboxylate exchanger (Alltech Universal Cation), consisting of silica coated with polybutadiene-maleic acid, with a crown ether packing produced by bonding an 18-crown-6 ether chlorosilane to silica. Experiments were done by packing the materials into a single or two different columns, with an methanesulfonic eluent and using conductivity detection. A new macroporous resin, 55% cross-linked EVB-DVB at a lower polymer particles diameter (5.5 µm) and a higher density of grafted carboxylic acid cation-exchange groups (capacity: 8400 µeq/column) was also developed [71] for different concentration ratios of adjacent peaks. Its performance was studied in depth also as a function of temperature [71,72] with methanesulfonic acid-based eluents and a suppressed conductivity detection. The resolution between the pairs sodium and ammonium, and calcium and strontium resulted significantly improved by the high capacity column which was also found to be tolerant to low pH values enabling the analysis of samples with a wide range of ionic strength, such as acidic soil extracts. It also seems remarkable that the separation of amines and alkanolamines from common cations, on this column, was reached by the use of a solvent-free (26 mM methanesulfonic acid) eluent. Among columns devoted to common cations and mainly to amines analysis, a macroporus cation-exchange column with a moderate capacity (1450 µeq/column, 250 mm × 4 mm I.D.) has been introduced [73]. The polymer particles (EVB-DVB 55% cross-linked, 7.0 µm) are first coated with a non-functional monomer, by a new graft technology, and weak carboxylic acid functional groups are successively grafted onto this coating. The non-functional monomer reduces the interaction of hydrophobic analytes with the substrate, optimizing the mass transfer characteristics, and due to very high selectivity for hydronium ions of exchange sites elution is reached by acidic elution also for polyvalent cations. Hydrophobic amines, diamines, and biogenic amines can be analyzed with a solvent-free eluent, dilute methanesulfonic acid, and suppressed conductivity detection. This column, in the suppressed conductometric detection mode, was successfully applied to the determination of biogenic amines in fish [74] and Saccani et al. [75] determined biogenic amines in fresh and processed meat by both suppressed conductivity and mass spectrometry detection. It must be underlined that tyramine, due to the loss of a proton upon suppression, cannot be detected by suppressed conductivity and was detected by UV spectrophotometry before the suppressor. Pastore et al. [76] used this column for biogenic amines quantitation in chocolate by coupling pulsed amperometric detection. The eluent was also in this case, aqueous methanesulfonic acid that required a post-column addition of a strong base for amperometric detection. More recently, a stationary phase devoted to the analysis of polar amines, namely Ion Pac CS18, was produced [77]. This column is more hydrophobic and has a lower capacity (290 µeq/column, 250 mm × 2 mm I.D.) than the previous one. It can be used with eluents of pH 0–7 and is compatible with 20% of organic solvents. This stationary phase is produced by a novel technology where a substrate (EVDB-DVB 55% cross-linked, 6.0 µm beads) is surface-modified before grafting onto it a functional layer obtained by alternating polymer graft with carboxylic acid groups. The good performance of this column has been widely demonstrated [77] and it was recently applied for the determination of biogenic amines in alcoholic beverages [78].

14.2.5 CATION EXCHANGE ON SILICA-BASED MATERIALS

Silica as a cation-exchange phase was used for alkali metals, alkaline earth metal ions, and ammonium separation with conductometric detection [79–81]. The drawback of these methods was the low sensitivity reached using weakly acidic or neutral eluents containing lithium ion as the competing cation, due to its low-limiting equivalent conductance. Ohta et al. [82] studied the properties of an unmodified silica gel in the acidic region and optimized the separation using 1.5 mM nitric acid and 0.5 mM 2,6-pyridinedicarboxylic acid eluent, and indirect conductometric detection. They showed that the cation efficiency is due to Al, present as an impurity in silica gels [83] and developed a highly selective separation and a sensitive determination of mono- and divalent cations on silica gel modified by coating with aluminum or zirconium [84,85]. Separations and determinations were achieved by using acidic eluents (nitric acid) containing 15-crown-5 (1,3,7,10,13-pentaoxacyclopentadecane) and indirect conductometric detection.

In recent years, there have been a number of reports dealing with preparation, properties, and applications of a new generation of stationary phases for liquid chromatography and capillary electrochromatography, namely monolithic columns [86–93]. Monolithic stationary phases, or continuous-bed stationary phases are based on modified silica gel or organic polymers [94]. Monolithic silica columns consist of a single piece, a rod, characterized by a network of macropores and mesopores (2 μm and 13 nm, respectively) that originate transport channels. In monolithic columns, the eluent is forced through these channels and, conversely to particle-based columns, the transport of the solute to the surface is prevalently due to a convective mass transfer [94] that is rapid. Additionally, monoliths at high flow rates show a limited back pressure and high efficiencies [95] and are suitable for fast analysis. The first application of methacrylate monoliths in IC was for the separation of proteins [96]. Minakuchi et al. [97] synthesized a uniform monolithic silica medium subsequently derivatizated to the C18 phase by an on-column reaction. This is remarkable in this contest since ion chromatographic separation of cations was later done mainly by reversed-phase silica based monoliths coated with anionic or amphoteric surfactants or by covalently modified bare silica columns [93]. Regarding rapid separation of cations using monolithic phases, Xu et al. [98] developed a method based on an octadecylsilyl (ODS) silica gel Chromolith column (5.0 cm × 4.6 mm I.D., Merck, Germany) modified with lithium dodecylsulfate (Li-DS) for the separation of hydronium from mono- and divalent cations. The eluent used 0.1 mM LiDS and 5 mM EDTA-2K enabled conductometric detection eliminating the effect of the coexisting cations NH_4^+ and K^+ since they have almost the same equivalent conductance and H^+ was eluted at 1 min. The drawback of the method was the poor peak shapes of calcium and the long retention time for magnesium. The method was later improved with the separation and the quantification of Ca^{2+} and Mg^{2+} also within 4 min by using a 2 mM ethylenediamine and 0.1 mM Li-DS solution as the eluent at pH 6.0 as shown in Figure 14.5 [99]. Fast separations of inorganic cations and anions were also obtained by Connoly et al. [100] with short reversed phase monolithic columns. Merck Chromolith RP-18e (2.5 cm × 4.6 mm) and Merck Chromolith Speed ROD RP-18e (5.0 cm × 4.6 mm) columns were permanently coated with double-chained surfactants, dioctyl sulfosuccinate (DOSS), and didodecyldimethylammonium bromide (DDAB). The columns were used individually or connected in parallel using a single combined eluent (2.5 mM phthalate, 1.5 mM ethylenediamine). This approach enabled the determination of five cations (copper, magnesium, calcium, strontium, and barium) in under 100 s with a flow rate at 8.0 mL/min and an indirect conductivity (Figure 14.6). Sugrue et al. [101] studied the organic solvent and buffer effects on the separation of alkaline earth and transition/heavy metal ions on a porous monolithic silica Performance SI (10 cm × 4.6 mm I.D., Merck, Germany). This study aimed to verify the hydrophilic interaction chromatography (HILIC)

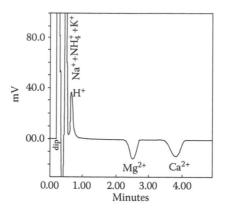

FIGURE 14.5 Analysis of rainwater spiked with $0.3\,mM$ H^+, Na^+, NH_4^+, K^+, Ca^{2+}, Mg^{2+}. Column: DS-coated monolithic stationary phase (Merck Chromolith, $50\times4.6\,mm$). Column temperature $35°C$. Eluent: $2\,mM$ ethylenediamine, $0.1\,mM$ Li-DS, pH 6. Flow rate: $4.0\,mL/min$. Loop volume: $100\,\mu L$. (From Xu, Q. et al., *J. Chromatogr. A*, 1026, 191, 2004. Copyright 2004. With permission from Elsevier.)

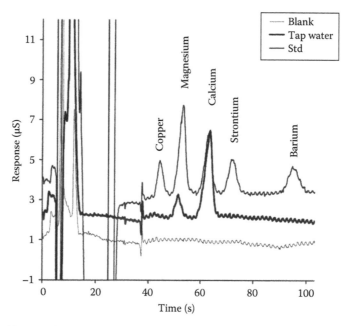

FIGURE 14.6 Rapid separation of inorganic cations on DOSS-coated Merck Chromolith SpeedROD RP-18e ($5.0\,cm\times4.6\,mm$). Eluent: $2.5\,mM$ phthalate/$1.5\,mM$ ethylenediamine, pH 4.5. Flow rate $8.0\,mL/min$. Column temperature $30°C$. detection: indirect conductivity. Loop volume: $50\,\mu L$. Concentrations: $25\,mg/L$ copper, $25\,mg/L$ magnesium, $50\,mg/L$ calcium, $50\,mg/L$ strontium, $100\,mg/L$ barium. (From Connoly, D. et al., *J. Sep. Sci.*, 27, 912, 2004. Copyright 2004, Wiley-VHC Verlag GmbH & Co. KgaA. Reproduced with permission.)

mode proposed by Pack et al. [102] for the separation of lithium, sodium, and potassium on a monolithic silica column. HILIC [103] refers to separations made with polar stationary phases and aqueous mobile phases at a high percentage of organic modifier and assumes that a stagnant water layer is originated on the surface of the stationary phase enhancing the weak interactions between the analyte and stationary phase (e.g., dipole-dipole, ion-dipole, hydrogen bonding) and the partition phenomenon. The authors [101] stated that the retention is due to an ion exchange enhanced by the organic nature of the eluent, and showed the separation of Cu^{2+}, Cd^{2+}, Ni^{2+}, Co^{2+}, and Mn^{2+} in less than 20 min by an 80% acetonitrile/$10.3\,mM$ ammonium acetate buffer. Among the new materials for the ion chromatographic separation of metal ions, more recently Nesterenko et al. [104] evaluated the use of microparticles of sintered ultradisperse detonation diamonds (SUDD) as a stationary phase on the basis of their content on the surface of both strong and weak carboxylic groups [105]. The authors verified that the working mechanism for the separation of alkaline earth and transition/heavy metal ions, is an ion-exchange, the selectivity is the same of a commercially available carboxylic cation exchanger, and finally showed for the first time a chromatographic separation of standard mixtures of Mg^{2+}, Mn^{2+}, Co^{2+}, Cd^{2+}, and Mg^{2+}, Ca^{2+}, Ba^{2+}, respectively on two columns ($100+50\,mm \times 4.0\,mm$ I.D.) and one column ($100\,mm \times 4.0\,mm$ I.D.) with $1.5\,mM$ HNO_3–$0.1\,M$ KNO_3, and $5\,mM$ HNO_3. Although the chromatographic peaks are tailing for the presence of pores in the SUDD inducing slow diffusion, this seems a very interesting and promising approach.

Permanently or dynamically coated reversed-phase monoliths with ion-exchangers suffer from variations in coating stability and are sensitive to temperature and changes in the composition of the eluent. Monolithic columns with chemically bonded ion-exchange groups could be a valid alternative to overcome these problems. Since at present this kind of stationary phases has been investigated but the separation mechanism is more a chelation than an ion exchange, they will be discussed in the appropriate chapter (chelation ion chromatography).

14.2.6 Ligand-Assisted Cation and Anion Exchange

Since the selectivity coefficients of transition and heavy metal ions (M^{n+}) for cation exchangers are similar, their separations using conventional ion-chromatographic columns (e.g., sulfonated cation exchangers) were performed with eluents containing one or more weak ligands (e.g., oxalic acid, citric acid, tartaric acid, 2-hydroxyisobutyric acid) that reduce the charge density of metal ions and change their selectivity through the formation of neutral or anionic complexes (MeL) [9]. In addition, the ligand avoids the formation of insoluble or hydroxo-species of metal ion when an acidic eluent is not suitable for separation. In cation exchange chromatography, metal ions, during the time spent in the eluent, are at equilibrium between their ionic free form, are available for the ion-exchange with the stationary phase, and their complex form originates with the ligand. For these separations, the chelating agent type, pH, and ionic strength play a relevant role. The type of ligand [106–108] influences the selectivity of the separation through the different stability constants with the same metal ions, and by modifying the pH, the stability constants of metal complexes are modified. This means, for example, that as the complex becomes more stable (higher pH values) the lower will be the time spent in the column. It must be underlined that the metal-ligand complex must be thermodynamically stable but kinetically labile. This means that the formation of the complex must be reversible under the chromatographic conditions used. Ionic strength means competing ions. The presence of monovalent cations in the eluent must be minimized since they do not modify the selectivity of the separation but their concentration in the eluent, which is greater than analytes, allows a loss of separation capability. For this reason, when a modification of pH is required, lithium hydroxide must be used in place of sodium hydroxide due to the lower affinity of Li^+ in respect to Na^+ for the cation exchangers.

Separation of metal ions can also be performed by anion exchange chromatography that involves their presence as negatively charged complexes. The ligand used must be characterized by a fast kinetic and must originate metal-complexes both thermodynamically and kinetically stable to avoid dissociation during the separation. Two main approaches are followed: (1) the complexation of analytes is made before the chromatographic separation, (2) the complexation is made, by adding the proper ligand to the eluent, during the injection of the sample. It must be stressed that in both cases metal-complexes must have a high stability constant to avoid decomposition during separation. In our opinion, anionic exchange offers many advantages compared to cation exchange. Different selectivity, reduced problems for metal-ion hydrolysis, and application to complex sample matrices could be reached by this approach. In addition, the nature of the stationary phase, cationic, induces electrostatic repulsion for the main ions of the matrix, e.g., alkali and alkaline earth metal ions which are uncomplexed. Another approach consists in using mixed-mode columns.

Many organic acids (e.g., carboxylic acids, α-hydroxyisobutyric acid, tartaric, citric, oxalic, pyridine-2,6-dicarboxylic acid, diethylenetriaminocyclohexanetetraacetic acid) have been evaluated for the simultaneous IC of anions, alkali, alkaline earth, and heavy metals [109,110]. Ethylenediaminetetraacetic acid (EDTA) played a fundamental role in this kind of determination [22,111–114]. Reiffenstuhl and Bonn [115] showed an interesting application of a multi-ligand eluent for the separation of mono- and divalent cations on a cation exchanger. Using a sulfonated porous 5 μm silica-based ion exchanger (TSK-IC cation SW, 50×4.6 mm I.D., Bio-Rad Labs.) coupled with 3.5 mM EDTA, 10.0 mM citric acid eluent at pH 2.8 they obtained the separation of 10 cations (Na, K, Cu, Ni, Co, Zn, Fe, Mn, Cd, and Ca) in about 15 min. Separation of metal-EDTA complexes (e.g., Cu^{2+}, Ni^{2+}, and Pb^{2+}) has also been studied with a cation-exchange polymer-based column (Dionex CS10, 250×4 mm I.D.) functionalized with sulfonic groups and possessing a medium hydrophobicity [116]. A detailed study on the behavior of metal-EDTA complexes at different pH values and ionic strength (eluent: 0.5 mM EDTA, pH adjusted with either $HClO_4$ or NaOH) provided data for a selective model taking into account both the ion-exchange and the adsorption phenomena of all differently charged or neutral species. On these considerations, an on-line preconcentration procedure was developed enabling the detection of lead at a 0.5 μg/L concentration. Bruzzoniti et al. [117] achieved the simultaneous separation of inorganic anions (BrO_3^-, SeO_3^{2-}, SeO_4^{2-}, $HAsO_4^{2-}$, WO_4^{2-}, MoO_4^{2-}, CrO_4^{2-}) and metal ions (Pb^{2+}, Cd^{2+}, Ni^{2+}, Cu^{2+}) precomplexed by EDTA in the presence of anions that commonly occur in real samples (Cl^-, NO_3^-, SO_4^{2-}) with a latex-based strong anion-exchange resin and a carbonate buffer as eluent. The separation was obtained with a column (Dionex AS9, 250 mm×4 mm I.D.), a 15 μm polystyrene-divinylbenzene substrate agglomerated with a completely aminated anion-exchange latex whose backbone is polyacrylate based, and a concentration gradient elution [HCO_3^-/CO_3^{2-}]. Detection was in the suppressed conductometric mode and detection limits ranged between 2 and 12 μg/L.

Besides the synthesized columns above mentioned, a new kind of stationary phase IonPac CS5A (9 μm bead diameter) has been obtained by agglomerating two layers of permeable ion-exchange latex particles [118]. The first sulfonic latex layer is attached to the surface of the structural polymer by a quaternary amine graft, the outer aminated latex layer is electrostatically bonded to the sulfonic layer. The unique feature of bearing both quaternary ammonium and sulfonate functional groups (mixed mode) means that simply changing the eluent can make major selectivity changes. The presence of both cation- and anion-exchange groups allows the separation of a broad range of metals using weak (e.g., oxalic acid) or strong (e.g., pyridine-2,6-dicarboxylic acid, PDCA) chelating agents. The mechanism could be considered in the first case, a cation exchange, and in the second the separation is achieved through the formation of anionic complexes of metal ions. An online chelation and preconcentration procedure coupled with an IonPac CS5 and a PDCA eluent enabled the determination of trace transition and rare-earth elements in seawater, and digested biological, botanical, and geological samples [119]. Many studies have been undertaken to optimize the analyte resolution for this kind of column [120–123] that allowed the simultaneous determination of heavy and transition metals in biochemical samples, nitrate/phosphate fertilizer solutions, water samples [124–126], as well as chromium speciation [127]. Lantanides have also been separated in the anion-exchange mechanism using the IonPac CS5, by coupling oxalate and diglycolate as ligands. This procedure enabled the determination of 11 elements (La, Ce. Pr, Nd, Sm, Eu, Gd, Tb, Dy, Ho, and Er) as impurities in a YbF_3 matrix [128]. A comparison of the chromatographic behavior of metal ions (Pb^{2+}, Cu^{2+}, Cd^{2+}, Co^{2+}, Zn^{2+} and Ni^{2+}) as a function of different oxalate-based eluents on this kind of column has been recently made by Lasheen et al. [129]. A tutorial review by Paull and Nesterenko [130] has recently been published on novel ion chromatographic stationary phases for the analysis of complex matrices.

14.3 ION-PAIR CHROMATOGRAPHY

Ion-pair chromatography (IPC), also called soap chromatography, dynamic ion-exchange chromatography, and ion-interaction chromatography, is a typical example of a chromatographic process based on secondary equilibria. In the commonest approach, this chromatographic mode uses HPLC stationary phases (usually reversed phase) with an aqueous mobile phase containing an ion-pairing reagent and an organic modifier. Ion-pairing reagents (IPR) are negatively or positively charged organic molecules with an inorganic co-ion. Among ion-pairing reagents, alkylammonium or alkane sulfonate salts are used for anionic or cationic analytes respectively [9].

The different names derive from the mechanisms proposed for the separation [131] and this means

- For ion-pair chromatography, it is assumed that during the elution, neutral ion-pairs are originated between analytes and the ion-pairing reagent, and retained by hydrophobic interactions [132,133].
- For dynamic ion-exchange, it is proposed that the ion-pairing reagent, due to the hydrophobic portion, is adsorbed on the surface of the stationary phase originating in ion-exchange sites available for ion-exchange between its counterions and analytes [49,134–136].
- For ion-interaction chromatography [131,137], it is assumed that molecules of the ion-pairing reagent are adsorbed on the surface of the stationary phase, and that the barrier of potential originated is immediately neutralized by a second layer of counterions (double-layer ion-pairing models). Analytes diffuse through this barrier originating pair of ions with the IPR and are separated by differences in their affinity for the pairing ion sites on the stationary phase.

The mechanism was quantitatively explained in terms of the Stern–Gouy–Chapman theory by assuming both ion-pair adsorption on the surface and dynamic ion exchange [135,138] or relating the retention factor k to the column phase ratio, Gibbs free energy of sorption, and to the difference of potential between the mobile phase and the surface of the stationary phase [139]. Retention and selectivity in ion pair chromatography are influenced by several experimental variables, including type, hydrophobicity, and concentration of the ion-pair reagent, the ionic strength, the concentration of the organic modifier, and the sorptive properties of the stationary phase [140]. Increasing the hydrophobicity and the concentration of the ion pair reagent leads to an increase in its concentration onto the stationary phase and to the higher retention of analytes [141]. The ion-paired complexes formation is affected by ionic strength modification due to the competition between eluent ions and analytes [142], but salting-in or salting-out effects could be originated [140]. Retention of analytes is reduced by increasing the concentration of the organic modifier since hydrophobic interactions between IPR, ion-paired species, and the stationary phase are reduced [143]. It has also been shown that the partitioning equilibrium of the pairing reagent depends on the nature of the co-ion [144,145] and the type of the eluent pairing ions [146]. This behavior is consistent with nonstoichiometric, double-layer ion-pairing models [147] and it seems that the mixed mode mechanism satisfies many of the ion-pair chromatographic separations [19]. A retention model [148], for singly and doubly charged analytes (anions), which can account for the main variables considered by stoichiometric [133,145,149–159] and thermodynamic methods, including electrostatic theories [25,139,142,160–166] was developed. The simultaneous effect of the ion-pairing reagent, the organic modifier, and the counterion present in the mobile phase was considered and the model concerns the retention process as the result of the contribution of adsorption and ion exchange equilibria between pairs of ions. This model was successfully applied for both prediction and interpretation of retention behavior for different structures and differently charged or uncharged analytes [167] and confirmed by comparison with a neural-network model [168].

IPC is usually performed in two ways: (1) by introducing the ion-pairing reagent inside the eluent and performing the separation of analytes with the proper eluent (dynamic system); (2) by preliminary modification of a reversed-phase column by an ion-pairing reagent, strongly (formally irreversibly) adsorbed, and by using this column as an ion-exchange column (permanent coating). Elution of cations is mainly achieved by complexation with a ligand (e.g., citric, tartaric, oxalic, and α-hydroxyisobutyric acids) and ion pairing of the negatively or positively charged complexes with IPR, so the conditional formation constants for the solute are of prime importance. Organic modifiers, and in some cases, the complexing agent, are also added to the eluent but since pH can alter the stoichiometry of the complexes and their overall charge, resolution is greatly dependent on the eluent acidity in addition to, as mentioned above, the nature and concentration of IPR. The great versatility of the system to manipulate selectivity is given by the wide number of parameters involved that govern the retention, but this becomes a drawback when the effect of each parameter has to be considered during the modeling of the ion-pairing mechanism. More attempts were probably made to apply this technique for the analysis of anions rather than cations for which some examples are given later.

Sirén [169] studied nitroso-naphthol metal complexes in depth, and developed a procedure where the metals are injected into a methanol-water eluent containing a quaternary ammonium bromide (e.g., cetyltrimethylammonium) and after the columns are mixed with a ligand solution (1-nitroso-2-naphthol-6-sulfonate) [170]. The separation of metal ions was found to be a function of the kinetics of formation of both complexes and ion-pairs and retention in the postcolumn mixer-reactor system. Azo ligands have also proven suitable for trace-metal ion separation and determination. A detailed study on the chromatographic behavior of 3-(-5-chloro-2-hydroxyphenilazo)-4,5-dihydroxy-naphthalene-2,7-disulfonic acid (Plasmocorinth B) metal complexes paired with different ion-pairing reagents [171] enabled the online preconcentration, separation, and determination of metal ions (Co, Cu, Fe, Ni, V) at μg/L levels in natural waters [172]. A 100 mL sample, 0.8 μM Plasmocorinth, and 0.32 mM tetrabutylammonium hydroxide, was loaded onto a LiChroCART 100 RP-18 (4×4 mm I.D.) (placed on the Rheodyne valve instead of on the loop) and after a washing step (ultrapure water), the retained analytes were eluted through a reversed phase column 10 μm LiChrospher 100 RP-18 (250×4 mm I.D.) with a methanol-water (50:50 v/v) mobile phase containing 10.0 mM acetic acid, 2.0 μM Plasmocorinth B, 1.6 mM TBA hydroxide, 12.5 mM sodium nitrate, and sodium hydroxide up to pH 6.3. Spectrophotometric detection (270 nm) of the eluted complexes enabled detection limits in the range 15–90 ng/L. Pyridylazosulfoaminophenol derivatives have also been used for the separation in IPC of metals as anionic chelates [173,174]. For these metal chelates, the retention behavior was explained as a function of the organic modifier employed in the water-based eluent (methanol or acetonitrile) [175] as a function of the volume fraction in water [176]. The strong effect that pH can play on the selectivity of IPC separations is clearly shown by a paper where the determination of the thermodynamic and conditional stability constants of Cu^{2+}, Ni^{2+}, Fe^{3+}, Al^{3+} with cyclo-tris-7-(1-azo-8-hydroxynaphthalene-3,6-disulfonic) acid (Calcion) enabled a specific separation of Al/Fe or Ni/Cu couples under the same conditions (column: 10 μm LiChrospher 100 RP-18, 250×4 mm I.D.; eluent: 53%–47% (v/v) methanol in water containing 40 mM buffer, 25 mM TBA, 25 mM NaCl, and 5.0×10^{-7} M Calcion) by working respectively at 4.5 or 7.5 pH values [177]. Transition metals were also separated in the presence of complexing agents on octadecyl-bonded silica permanently coated with sodium dodecylsulfate [50]. The separation mechanism shown in this case is similar to that of fixed-site exchangers where both the pushing effect of the eluting cation and the pulling effect of the complexing anion work, but the latter is dominant. Similar results were obtained by Cassidy and Sun [178] who compared the performance, in IPC, of an anion with a cation separation using cetylpyridinium chloride or n-octanesulfonate to modify a reversed phase. The two systems showed comparable efficiencies for the separation of transition metals (Mn, Co, Ni, Cu, Zn). Worthy of mention is a study on the performance of a stationary phase (5 μm silica-based Hypersil C18, HPLC Technology) modified with various alkanesulfonates (5–10 carbon atoms in the alkyl chains) that enabled the separation and determination of Fe(III), Cu, Pb, Zn, Ni, Co,

Cd, and Mn, in natural waters at µg/L levels [eluent: 5% (v/v) acetonitrile, 2 mM octanesulfonate, 50 mM tartaric acid, pH 3.4] with a post column reagent [PCR: 4-(2-pyridylazo)resorcinol (PAR)] and a spectrophotometric detection [179]. EDTA has also been considered in IPC and was coupled with tetraethylammonium (TEA) [180], tetrapropylammonium TPA [181], and tetrabutylammonium (TBA) [180,182] bromide ion-pairing agents for the separation and detection of metal species. The use of EDTA in the eluent enabled lower-detection limits and TBA was found to be the best-suited ion-pairing reagent [180,182]. Bedsworth and Sedlak [183] applied a similar IPC approach for identifying Cd(II), Co(II), Cu(II), Pb(II), and Zn(II) complexes of EDTA in municipal wastewaters and surface wastewaters. They used a 5 µm endcapped C18 column (250×4.6 mm I.D., Supelco Discovery) permanently coated with cetyltrimethylammonium bromide and a mobile phase 6 mM Na_2SO_4, and 1 mM 4-(2-hydroxyethyl)-1-piperazineethanesulfonic acid (HEPES) at pH 7. By a post-column reaction, where all the metal complexes are converted into $FeEDTA^-$, spectrophotometric detection furnished 20 µg/L detection limits. Ion-pair chromatography has also been coupled with flow injection techniques for metal ion determinations. Trojanowicz et al. [184] developed a method for the determination of Co(II), Ni(II), Cd(II), and Mn(II) in river water based on a preconcentration of the sample on a micro-column (20×5 mm I.D.) Cellex-P (cellulose matrix, phosphonic acid functional group: Bio-Rad Laboratories, Hercules, CA) followed by the recovery of the analytes with 0.1 M HNO_3 and their injection into a reversed-phase column 5 µm LiChrosphere 100 column preliminary coated with SDS. The separation was reached by a 100 mM tartrate solution containing 0.06 mM SDS, and 3% methanol at pH 3.5 in less than 20 min and spectrophotometric detection (PCR: PAR) furnished detection limits between 0.05 and 2.1 µg/L. Srijaranai et al. [185] coupled a reversed flow-injection procedure to an IPC system for the determination of Cr(VI), Ni, Co, and Cu. The metal chelates were obtained by an in-line reaction. 85 µL of PAR solution (18 mM) was injected into the sample stream and delivered within through a reaction coil (150 cm) to the injection port of the chromatographic system. A C_{18}-µBondapak column (300×3.9 mm I.D., Waters, Milford, MA, USA) with a mobile phase 37% acetonitrile, 6.2 mM TBA bromide, 3.0 mM acetate buffer at pH 6.0 were used for the separation. All the metals were well resolved in less than 14 min. Detection was spectrophotometric (530 – 440 nm). More recently, Threeprom et al. [186] developed an elegant method based on IPC for the simultaneous determination of Cr(III) and Cr(VI) and suitable for samples at high levels of salts and organic materials, e.g., wastewaters. The separation was achieved with a C-18 column (3 µm, 60×4.6 mm I.D., Nucleosil 100) dynamically coated with 5 mM octylammonium orthophosphate at pH 5.0 (20% CH_3OH, v/v) and samples analyzed after precomplexation of Cr(III) with potassium hydrogen phthalate to [Cr(III)-KHP]$^-$. The analysis time was less than 5 min and detection limits were 0.01 and 0.05 mg/L for Cr(III) and Cr(VI), respectively (Figure 14.7).

14.4 ELECTROSTATIC ION CHROMATOGRAPHY

Amphoteric surfactants, zwitterions, used as ion-pairing agents for the separation of nucleotides [187], as well as zwitterionic silica-based stationary phases, enabling the simultaneous separation of organic cations and anions, were described by Yu et al. [188,189]. Hu et al. [190], following the same approach, developed a new ion chromatographic mode defined electrostatic ion chromatography (EIC). They used a C_{18} reversed-phase silica column loaded with zwitterionic surfactants 3-[(3-cholamidopropyl)dimethylammonio]-1-propanesulfonate, 3-[(3-cholamidopropyl) dimethylammonio]-2-hydroxy-1-propanesulfonate, Zwittergent 3–14), and a pure water mobile phase. According to experimental results, they proposed that analytes and their counterions combine to make "ion-pairing-like" forms at some distance from the zwitterions' charges and are separated by the simultaneous electrostatic attraction and repulsion interactions with the zwitterionic charged stationary phase. A subsequent paper [191] showed that the same species of anion or cation appear in several single peaks in the chromatogram, but this problem can be solved simply by the addition of an electrolyte to the mobile phase or to the sample [192,193]. This problem may also be overcome

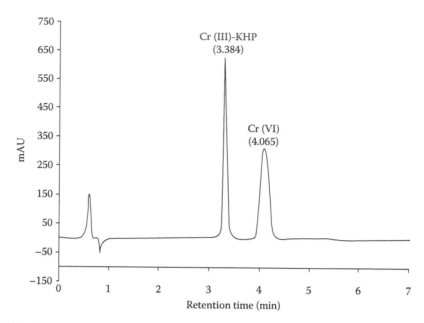

FIGURE 14.7 Determination of Cr(III) (5 mg/L) and Cr(VI) (10 mg/L). Column: C-18 (Nucleosil-100, 3 μm, 60 mm × 4.6 mm). Eluent: 5 mM octylammonium orthophosphate, pH 5.0, with 20% MeOH (v/v); Flow rate: 1.0 mL/min; detection at 200 nm. (From Threeprom, J. et al., *Talanta*, 71, 103, 2007. Copyright 2007. With permission from Elsevier.)

by varying the positions, spacing, or the strength of functional groups on the zwitterions [194]. Stationary phases are obtained by the adsorption of a zwitterionic surfactant onto a hydrophobic stationary phase [191,195,196] or by the chemical modification of the surface of a suitable substrate.

Jiang and Irgum [197], to overcome the problem arising from the limited stability of silica-based materials at pH <2 or >7, have synthesized a covalently bonded polymeric zwitterionic stationary phase. They activated 2-hydroxyethyl methacrylate-ethylene dimetacrylate copolymer beads with epichlorohydrin and successively coupled the epoxide groups on the surface with (2-methylamino) ethanesulfonic acid. They found that the retention of cations and anions, eluted with an aqueous solution of perchloric acid or perchlorate, depends on the individual ions in the analyte and not on the type of salt injected, conversely to the results obtained for ODS columns dynamically coated with zwitterionic surfactants and water as the eluent. Nesterenko and Haddad [198] published a comprehensive review on the classification and applications of zwitterionic ion-exchangers. The EIC mechanism has been deeply studied [199,200] and a serial study on the effect of salt concentration in the mobile phase on retention behavior of inorganic anions have demonstrated that ions in the mobile phase are capable of modifying the net potential of the zwitterionic stationary phase [200–202]. Cook et al. [200] developed a new EIC retention mechanism for anions. They defined the role of the ξ potential of zwitterionic surfaces in contact with electrolytes, and stated that two simultaneous effects, ion-exclusion and chaotropic interaction are responsible for the retention. Ion-exclusion is attributed to the terminal-charged groups of the bound zwitterions, which at the equilibrium with the eluent containing electrolyte(s) originate a charged layer acting as a Donnan membrane that in turn is a function of the nature of the mobile phase ions. Chaotropic interactions refer to the direct interaction of analytes with the inner charged groups of zwitterions, and their competiton with the mobile-phase anion. Since the mechanism does not involve electrostatic interactions, the name zwitterionic ion chromatography (ZIC) was suggested to be more appropriate than EIC. Many applications of zwitterionic ion chromatography are devoted to the analysis of anions and use sulfobetaine-type surfactant-coated columns, but this system was not efficient in the separation of cations. Hu et al. [203], after studies on the effect of the position of the zwitterion on the stationary phase, on its functional groups' strength, or on the use of a mixture of anionic and zwitterionic surfactants [194,204,205] for the separation of cations, proposed the use of a phosphocholine-type surfactant [*N*-tetradecylphosphocoline] (PC-ZS) for a new stationary phase. The peculiarity of this zwitterionic surfactant is that it has the negatively and positively charged functional groups reversed in respect to the sulfobetaine surfactants. The new stationary phase, obtained by flowing a 30 mM solution (water) of PC-ZS through an ODS-packed column (L-column, 250×4.6 mm I.D., Chemical Inspection and Testing Institute, Tokyo, Japan) enabled the separation of divalent cations (Mg, Ca, Ba, Mn, Cu, Zn) with water as the mobile phase. This study confirmed the separation mechanism based on coulombic effects and the formation of ion-pairs as previously proposed [190,191]. Cook et al. [206] studied the chromatographic behavior of cations on a C$_{18}$ column coated with *N*-tetradecylphosphocholine. The proposed mechanism for the separation is again an ion-exclusion effect and a predominant chaotropic interaction despite the lower chatropic nature of cations in respect to anions.

More recently, Sugrue et al. [207] prepared a zwitterionic-type stationary phase where lysine (2,6-diaminohexanoic acid) is covalently bonded to a bare silica monolith (10 cm×4.6 mm I.D., Performance SI, Merck) for the separation of both cations and anions. This column showed a variable capacity as a function of pH and enabled the separation of six anions in around 100 s, but a limited retention for alkali and alkaline earth cations. The separation of divalent transition and heavy metal ions was more successful and clearly in agreement with a complexation mechanism as usually shown by aminocarboxylic acid phases (Figure 14.8) [208]. Factors affecting selectivity in ZIC were also discussed in depth by Fritz [19] and, in general, ZIC seems particularly suitable for the analysis of samples of high ionic strength. As mentioned above, during the last few years, many attempts have been made at monolithic columns as shown in recent reviews [90,92,93], but up to now, little of the work has been devoted to the analysis of inorganic cations.

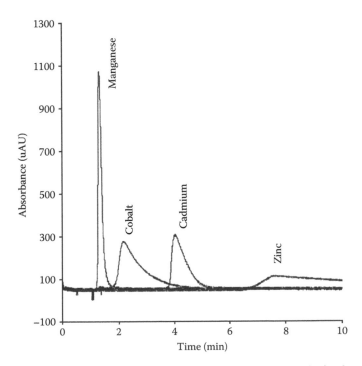

FIGURE 14.8 Overlay of Mn(II), Co(II), Cd(II), and Zn(II) chromatograms obtained on a lysine modified monolith. Eluent: 3 mM KCl, pH 4.5; Flow rate: 2 mL/min. Detection: post-column reaction with PAR, absorbance at 495 nm. (From Sugrue, E. et al., *J. Chromatogr. A*, 1075, 167, 2005. Copyright 2005. With permission from Elsevier.)

14.5 CHELATION ION CHROMATOGRAPHY

Chelation ion chromatography or, more recently, high-performance chelation ion chromatography (HPCIC), is mainly devoted to metal ions and trace-metal analysis for samples of high ionic strength where IC or IPC are unsuitable since the ion-exchange sites can become "swamped" with salt ions or the pairing mechanism is modified. The stationary phases are based on substrates (polymeric materials or silica) with both chemically bound chelating groups or coated with a ligand that is permanently or dynamically trapped on the surface. Permanent coating means that the column is prepared by fluxing a solution (e.g., water-methanol) containing the proper ligand that is adsorbed onto the substrate. The column is then washed (water at the proper pH) in order to remove the molecules of unretained ligands and finally to ensure the removal of any adsorbed metals, it is again washed with an acidic solution. Dynamic coating is similar to ion-pair chromatography where instead of an ion-pairing reagent, a chelating agent is added to the eluent. The retention mechanism involves both the formation of coordinate bonds and an ion-exchange process between metal ions and groups on the stationary phase. The thermodynamics and the kinetics of complex formation/dissociation are the key factors of the separation. Usually, to ensure that the coordination is the prevalent mechanism, eluents at high ionic strength (e.g., 0.5–1.0 M KNO_3) are used and the pH becomes the fundamental parameter to alter the separation.

Although chelating resins were widely used to preconcentrate and to separate elements and group elements in seawater [209], the first report on related HPCIC was made by Moyers and Fritz [210] in late 1977. The main principles of HPCIC were later formulated and studied by Timerbaev et al. [211] and Jones et al. [212].

14.5.1 PERMANENT COATING

Jones et al. [213] studied the separation of divalent (Mg^{2+}, Mn^{2+}, Zn^{2+}, Cu^{2+}) and trivalent (Al^{3+}, In^{3+}, Ga^{3+}) metal ions with a neutral polystyrene-based resin (Benson BPI-10, 10 μm particle size) permanently coated with dyestuffs mainly a hydrophobic chelating dye (Chrome Azurol S, CAS), which is characterized by chelating functions analogous to salicylic acid. A 0.2% methanol-water solution of CAS was pumped through the column ensuring the retention of the ligand on the surface. Then the column was washed with deionized water at pH 10.5 (NH_4OH) and 0.1 M HNO_3 to remove both an excess of the ligand and any trace metals polluting the surface. The authors showed that the separations are strongly affected by the pH of the eluent and little by ionic strength, and that this kind of stationary phase is suitable for the injection of a large volume sample (up to 7 mL), enabling by a pH gradient elution, both the preconcentration and the separation of analytes. The eluent was 1 M KNO_3 containing 50 mM lactic acid, and spectrophotometric detection was performed by a post-column reaction with calmagite or pyrocathecol violet respectively for divalent and trivalent analytes. Jones and coworkers [214] investigated this topic widely comparing the chelating ability of Xylenol Orange (XO) and CAS with small and large particle sized coated columns showing the weakest chelation for CAS as well as for large particle based columns. They showed that the efficiency for the separation of alkaline earth metal ions with XO column is similar to that obtained by a commercially available covalently bonded chelating column employing iminodiacetic acid functional groups (TOSOH TSK-GEL Chelate-5PW, 7.5 cm × 7.5 mm I.D., Tosoh Corp., Japan) [215]. Studies on the behavior of these stationary phases [214,216,217] allowed the preconcentration and the separation of trace metals in seawater [218]. A 10 μm particle size PS-DVB, 100 Å pore size, neutral PLPRS resin (Polymer Laboratories, Church Stretton, UK) was coated with Xylenol Orange and packed in a PEEK column (10 cm × 4.6 mm I.D.). This stationary phase resulted stable from pH 0.5 to 11.5 and enabled both the removal of Ca and Mg during the sample (20 mL) preconcentration at pH 6, and the separation and the determination of Zn, Pb, Ni, and Cu (μg/L levels) with a step-gradient pH elution. It seems interesting to remark that the retention order of metal ions, including earth metals, is reversed with respect to IC and so barium is eluted first with a sharp peak, followed by Sr, Mg, and Ca. This situation enabled the separation and the determination of barium, at a relatively low concentration (mg/L) in oil-well brine samples at 16,000 mg/L Ca concentration [219]. The chelating column (10 cm × 4.6 mm I.D., PEEK HPLC column), a 8.8 μm particle size, 12.0 nm pore size PS-DVB neutral hydrophobic resin (Dionex) impregnated with Methylthymol Blue (3,3′-bis[N,N-di(carboxymethyl)aminomethyl]thymolsulfonephtalein) (MTB) was coupled with 0.5 M KNO$_3$ + 0.5 M lactic acid or 1.0 M KNO$_3$ eluents and UV detection (PCR: PAR- ZnEDTA). Detection limits of 1 and 8 mg/L were reached respectively for Ba and Sr. Paull and Jones [220] studied, in depth, the parameters involved in the production of dye-impregnated columns evaluating 10 chelating dyes, mainly on triphenylmethane- or azo-based dyes, and a range of resins (4 PLRPS, 10 μm PS-DVB resins with pore size 100, 300, 1000, 4000 Å, Polymer Laboratories; 2 neutral hydrophobic macroporous PS-DVB based resins from Dionex, respectively 8.8 μm, 120 Å pore size and 7 μm, 300 Å pore size). The study was devoted to the preconcentration and the separation of alkaline earth, transition, and heavy metals at trace levels and deals with an application to Al determination in seawater.

The roughness of this kind of columns for sample at high ionic strength was also shown by Paull and Haddad [221]. They, with a Hamilton (Reno, NV) 5 μm PRP-1 reversed phase column (150 cm × 4.1 mm I.D.) coated with MTB analyzed samples at ionic strength up to 1.0 M NaCl and by large volume injection with a step-gradient elution (from 6 mM to 60 mM HNO$_3$) achieved the separation and the detection of UO_2^{2+} in a saline lake sample with a detection limit lower than 1 μg/L.

It must be underlined that both Xilenol Orange and MTB are characterized by the presence of two iminodiacetic functional groups (IDA) and metal ions are retained in respect to the conditional formation constants of their iminodiacetates, for separations based on these or similar ligands the selectivity is strongly affected by the control of the mobile phase pH. Since IDA is a weak acid, an ion-exchange mechanism at a higher pH or a low ionic strength can also be active, as discussed by Jones and Nesterenko [212], to ensure that chelation is the acting mechanism eluent at high ionic strength (e.g., 0.5–1.0 M KNO$_3$) are usually employed. Injecting the sample solution at a high pH, metal ions are completely retained as strong complexes on the top of the column and then by reducing the eluent pH by subsequent steps, the separation is achieved on the basis of the relative stability constants.

An aurin tricarboxylic acid (ATA) immobilized chelating polymer at a high efficiency was prepared by Jones et al. [222]. They, conversely to the traditional approach where the chromatographic support is coated by fluxing through it a buffered solution of the chelating agent, loaded the resin (7 μm PRP-1 PS-DVB resin, Hamilton) in a bath under ultrasonic stirring (10 h). The resin, after the optimization of a pH gradient elution (eluent 0.1 M KNO$_3$) resulted highly selective for the separation of Cd^{2+}, Pb^{2+}, and Cu^{2+} from alkaline earths and other transition metals. Another improvement was the enhancement of the spectrophotometric detection sensitivity reached by using the post-column reagent (0.1 mM PAR) buffered with di-sodium tetraborate instead of the conventional mixture ammonium phosphate/ammonium nitrate, which enabled detection limits of 1 μg/L for Cd^{2+} and 5 μg/L for Pb^{2+} and Cu^{2+} respectively.

14.5.2 DYNAMIC COATING

Since pre-coated columns were generally suitable for an isocratic separation of a limited number of metal ions and efficiencies resulted relatively poor when compared to covalently bonded chelating columns (see below), an alternative approach was investigated. As previously seen, to improve the selectivity for the separation of metal ions by cation-exchange chromatography (IC), weak ligands are usually added to the eluent, but since the stability constants can be modified through the pH also, strong chelating agents could be used at acidic pHs. This approach was followed for developing dynamic chelating chromatography.

Paull et al. [223,224] coupled an Hypercarb porous graphitic carbon reversed-phase column (10×4.6 mm I.D., Shandon HPLC, Runcorn, Cheshire, UK) to a 0.4 mM o-cresolphthalein complexone (3,3′-bis[N,N-bis(carboxymethyl)aminomethyl]-o-cresolphthalein) (OPC) based mobile phase. In this way, the partition of the ligand between the mobile phase and the surface of the stationary phase originated a dynamically coated stationary phase. Separation and direct determination of magnesium and calcium in saline samples with a direct spectrophotometric detection of calcium and magnesium chelates, originated at the working pH 10.5, were performed with a 58% methanol and 0.4 mM OPC [223]. The separation mechanism was a dynamic chelating ion-exchange since the contributes are due to chelation with OPC, both onto the stationary phase and in the eluent, and to an ion-exchange on carboxylic groups which at pH 10.5 are dissociated. Through a detailed study and optimization of the mobile-phase composition [224], the dynamic capacity of the column was increased and a more selective separation of magnesium and calcium was possible in samples containing 2.300 mg/L in excess of sodium with a higher sensitivity (detection limits: 0.05 mg/L and 0.10 mg/L for Mg and Ca respectively). Paull and coworkers' studies on polymeric [225–227] or silica-reversed stationary phases [227,228] coupled with the same chelating agent showed that different selectivities are reached by working in the permanent or the dynamic-coating mode for both alkaline earths [226,227] and heavy and transition metals (Figure 14.9) [225,227,228]. It must be stressed that the selectivities reached are a function of the nature of the ligand and the retention is mainly dependent upon the mobile phase pH and the ionic strength for permanent coating, but for the dynamically coated mode, it is also dependent on ligand concentration. In addition, including the ligand in the mobile phase, a higher dynamic capacity is obtained and peak efficiencies are improved. Dipicolinic acid dynamically sorbed to PS-DVB particles (5–10 μm) enabled the separation of high valence metals but with a poor resolution for the lower charged heavy and transition metals due to the low capacity of the column [229]. Shaw et al. [230] increased the dynamic loading of this kind of column by using 4-chlorodipicolinic acid, which through the chloride functional group strongly interacts with the π-electron donating resin support. They, with a PEEK (250×4.6 mm I.D.) column packed with a 7 μm PRP-1 PS-DVB resin (Hamilton) and a 1 M KNO_3, 0.25 mM chloropicolinic acid, and a 6.25 mM HNO_3 eluent (pH 2.2), obtained the separation of seven transition and heavy metal ions (Mn, Co, Ni, Zn, Cu, Pb, Cd) in less than 20 min.

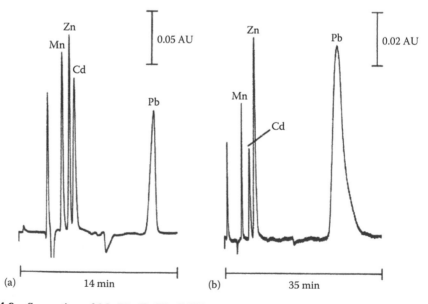

FIGURE 14.9   Separation of Mn(II), Zn(II), Cd(II), and Pb(II) using (a) a dynamically coated and (b) a pre-coated MTB ODS reversed-phase column. (From Paull, B. and Haddad, P.R., *Trends Anal. Chem.*, 18, 107, 1999. Copyright 1999. With permission from Elsevier.)

14.5.3 CHEMICALLY BOUND CHELATING GROUPS

Silica-based chelating stationary phases are also widely used and one of the most attached groups is iminodiacetic acid (IDA) [231,232]. Nesterenko et al. [233] studied the chromatographic behavior of some alkaline-earths, and heavy metals on a column (250×4 mm I.D.) packed with 6 μm particle size silica bonded with IDA (Diasorb IDA, Bio-Chemmack, Moscow, Russian Federation). They studied different eluent compositions and reached both on-column preconcentration and through a three step gradient elution (KCl-HNO₃, tartaric acid and picolinic acid) the separation of calcium, magnesium, manganese, cadmium, cobalt, zinc, nickel, and copper in high salinity samples. The same stationary phase was evaluated for the separation of 14 rare earth elements [234,235] that were isocratically eluted (eluent: 16 mM HNO_3, 0.5 M KNO_3) in about 65 min, with an enhancement of the selectivity by increasing the working temperature up to 65°C. A detailed study on the effect of the sorbent matrix and the porosity on the properties of the grafted IDA and on the conformational mobility and the grafting density of the bonded IDA groups in respect to the efficiency and selectivity of the separation of metal ions was done by Nesterenko and Shpigun [236]. More recently, the separation, in about 35 min, of 14 lanthanides and Y(III) on a column (150×4.0 mm I.D.), packed with 5 μm IDA-silica and 25 mM HNO_3, 0.75 M KNO_3 eluent at 75°C has been shown in Ref. [237].

Nesterenko et al. [238] synthesized an aminophosphonic silica-based cation exchanger (APAS). This stationary phase through the dissociation of phosphonic groups (in weak acidic or neutral solution) works as an ion exchanger but at acidic pHs (POH groups) or higher pHs (POH groups and the basic amino nitrogen), the main mechanism is chelation as discussed in depth in Refs. [239,240]. With a 5 μm particle size of this substrate packed in a PEEK column (50×4.6 mm I.D.), it has been shown that selectivity towards alkali and alkaline earth metal ions is modified by changing the pH and that the retention order for transition metal ions is correlated with the stability constants of their complexes with chloromethylphosphonic acid [238]. A selective procedure for the determination of Be^{2+} at trace levels in a sample containing in excess of 800 mg/L matrix metals (certified reference material GBW07311) was developed with the same column [241]. The eluent composition was 1 M KNO_3, 0.5 M HNO_3, and 0.08 M ascorbic acid, where the acidity enabled the complete elution of transition and heavy metals and ascorbic acid avoided interference from Fe^{3+} since it was reduced to Fe^{2+} and totally unretained. The analysis time was less than 6 min and a detection limit of 35 μg/L Be(II) was reached by a post-column reaction. The stationary phase resulted unmodified over 12 months. The separation of alkaline earth and metal ions (Ni, Zn, Fe(II), Cu, Pb, Cd, Mn) in less than 20 min, on an APA-silica column (3 μm, 150×4.0 mm I.D.) with a 50 mM HNO_3, 0.8 M KNO_3 eluent, has also been shown (Figure 14.10) [237]. Bashir and Paull [242], using an iminodiacetic acid-functionalized silica column (250×4 mm I.D., 8 μm IDA silica gel, BioChemMack, Moscow, Russia) developed a highly selective and sensitive method for the determination of beryllium (eluent 0.4 M KNO_3) also in high ionic strength samples, which is characterized by a detection limit of 4 μg Be(II)/L.

Sugrue et al. [243] covalently bonded iminodiacetic acid to a bare silica monolith (10 cm × 4.6 mm I.D., Performance SI, Merck). Selectivity for alkaline earth metal ions were similar to those reached with a commercially available IDA-modified silica gel column (250×4 mm I.D., BioChemMarck, Moscow, Russia) [244], but on the monolith, calcium and magnesium were totally base-line resolved (each at 10 mg/L concentration in a 2 M KCl sample) by a 1 M KCl eluent in only 40 s. Spectrophotometric detection was performed by post-column reaction with o-cresolphtalein complexone. In continuation of this study [245], they studied, in depth the selectivity of this kind of column towards alkali, alkaline earth, and transition metal ions. Alkali metals were generally co-eluted in a short time since IDA stationary phases show a poor selectivity for them, and their retention, due to simple ion exchange, is hindered by the usually high ionic strength of the eluent, and the separation of transition metals resulted strictly related to the pH values of the eluent. This stationary phase seems suitable for the determination of Mg, Ba, and Sr in the presence of a relatively high concentration of Ca (1.000 mg/L).

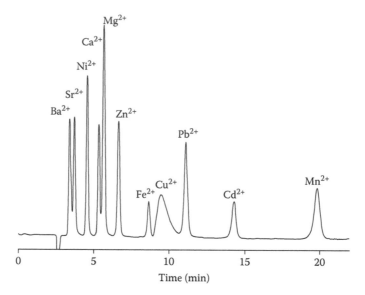

FIGURE 14.10 Separation of alkaline-earth and transition metal ions. Column: APA-silica, 150×4.0 mm, 3 μm. Eluent: 50 mM HNO_3–0.8 M KNO_3, detection at 510 nm with PCR reaction with PAR-ZnEDTA. (From Nesterenko, P.N. and Jones, P., *J. Sep. Sci.*, 30, 1773, 2007. Copyright 2007, Wiley-VCH Verlag GmbH & Co. KgaA. Reproduced with permission.)

Silica- or polymer-based materials with chelating agents grafted onto the surface, available on the market or laboratory made, or metallochromic ligands used in HPCIC have been extensively reviewed [130,212,227]. Recent developments in HPIC have been reported as well [91–93,237].

14.6   DETECTION

Detection in IC is strictly connected with the nature of eluents (composition, concentration), analytes and the sensitivity required. The ideal characteristics of a chromatographic detector are essentially the following: (1) high sensitivity, (2) low cell dead volume, (3) linear relationship between concentration and signal, (4) stable and low background noise, (5) high speed of response, and (6) no signal drift.

In IC, the election–detection mode is the one based on conductivity measurements of solutions in which the ionic load of the eluent is low, either due to the use of eluents of low specific conductivity, or due to the chemical suppression of the eluent conductivity achieved by proper devices (see further). Nevertheless, there are applications in which this kind of detection is not applicable, e.g., for species with low specific conductivity or for species (metals) that can precipitate during the classical detection with suppression. Among the techniques that can be used as an alternative to conductometric detection, spectrophotometry, amperometry, and spectroscopy (atomic absorption, AA, atomic emission, AE) or spectrometry (inductively coupled plasma-mass spectrometry, ICP-MS, and MS) are those most widely used. Hence, the wide number of techniques available, together with the improvement of stationary phase technology, makes it possible to widen the spectrum of substances analyzable by IC and to achieve extremely low detection limits.

14.6.1   CONDUCTIVITY DETECTION

Since all ions are electrically conducting, conductivity detection can be considered universal in response. Conductivity detection comprises non-suppressed and suppressed modes.

Non-suppressed conductivity detection furnishes a signal that is the sum of the conductance of the analyte ion, its co-ion, and the decrease in the eluent counterion that remains on the column owing to the stoichiometric replacement of the analyte ion. The change in conductance during the elution of the analyte is a function of the limiting ionic conductances (λ) of the analyte and eluent ions, and of analyte concentration.

For not completely dissociated analytes and/or eluents, whose degree of ionization depends on the pH of the separation conditions, the signal is proportional to the concentration of analytes and increases as the degree of ionization of the analyte in the eluent is increased.

Non-suppressed conductimetric detection can be applied in direct or indirect modes, according to the chemical properties of the eluent and analytes.

The direct mode is employed with eluents with significantly lower equivalent conductance than the analyte ion. Increase in sensitivity is obtained as the degree of the ionization of the eluent decreases, that is, with more weakly dissociated eluents, and non suppressed conductivity methods have been extensively developed using benzoate, phthalate [246], oxalate [53] or other partially ionized species as mobile phases. A key factor in the success of this technique is the use of an ion exchanger of low-exchange capacity, which in turn permits the use of a very dilute eluent.

On the contrary, eluents with high background conductance (e.g., NaOH) are amenable for indirect conductivity detection, where analyte elution is accompanied by a negative conductance change [247].

Suppressed conductivity detection is the most common mode of detection and differs from the previous approach for the use of an additional device, called *suppressor*, whose function is to reduce the background conductivity of the eluent prior to the conductivity cell and to increase the signal of the analyte.

Historically, the first suppressor was a packed-column in the H^+ or OH^- form. Introduced in 1975 [1], it actually represented the foundation on which the IC was built. To depict the reactions taking place during the suppression for an anion-exchange separation, the following equilibria must be considered:

$$\text{Resin} - H^+ + Na^+HCO_3^- \longleftrightarrow \text{Resin} - Na^+ + H_2CO_3$$

$$\text{Resin} - H^+ + Na^+ + A^- \longleftrightarrow \text{Resin} - Na^+ + H^+ + A^-$$

(analyte: A^-, eluent: $NaHCO_3$)

The net result is that, supplying H^+ ions, the detectability of the analyte is enhanced both for the decrease in eluent conductance and for the increase in that of the analyte by the replacement of Na^+ ion with H^+ (ion with a higher value of λ).

Parallelly, if OH^- ions can be added, the suppression is feasible also for cation-exchange separations.

The disadvantages of the packed-column suppressor (off-line regeneration, band broadening, and loss of efficiency in the IC separation) were overcome by the introduction of the hollow-fiber suppressor [248] where suppression is realized by passing the eluent inside a small internal diameter fiber and a suitable regenerant solution, providing H^+ or OH^- ions in the opposite direction outside the fiber itself. Since the low surface area means, in turn, low suppressor capacity and restricted eluent concentration ranges for separation, continuously regenerated flat-membrane suppressors (MicroMembrane Suppressor, MMS, Dionex) were introduced and marketed in 1991 [249,250]. These devices require a chemical regenerant (flowed from 3 to 10 times the eluent flow rate) that supplies the suppressing ions (H^+ or OH^-) and are designed in a sandwich layer configuration including a central chamber with ion-exchange membrane sheets as upper and lower surfaces where the eluent flows. The regenerant flows on the exterior sides of the membranes across and through the ion-exchange screens in a flow direction opposite that of the column effluent flow. The need to supply a chemical regenerant is overcome with the electrolytic membrane–based suppressor [251], Self-Regenerating Suppressor, SRS (Dionex), where the H^+ (OH^-) for suppression is generated by water electrolysis at a Pt anode (cathode) placed in the regenerant chambers between the regenerant screens and the outer hardware shell, in a flat membrane design. For anion analysis, H^+ ions generated at the anode migrate across the cation-exchange membrane to neutralize the basic eluent (e.g., NaOH) that goes to the detector cell. Na^+ cations from the eluent are driven toward the cathode and move across the cation-exchange membrane to pair with hydroxide ions generated by the cathode. The water for the electrolysis is derived either from an external reservoir (recycle mode) or from the suppressed eluent diverted back into the suppressor after detection. The current required to start the electrolysis is proportional to the eluent concentration that is to be suppressed. Continuous electrolytically regenerated packed bed suppressor, commercially known as the Atlas electrolytic suppressor, AES (Dionex), has been available since 2001 [252]. It includes cation-exchange monolith disks alternately sandwiched to impermeable flow-distributor disks (small holes allow the liquid to flow), between the ion-exchange membranes that separate the eluent chamber from the anode and cathode chambers. It should be mentioned that the continuous development in electrolytic devices [16] that utilize the electrolysis of water and charge-selective electromigration of ions through ion-exchange media have led to the release of new equipment also for on-line eluent generation for the production of high-purity electrolyte eluents (e.g., KOH, methanesulfonic acid) [253,254], continuously regenerated trap columns that remove ionic contaminants in the eluent [16], and to the concepts of ion reflux [255] or eluent recycling [256,257].

Other suppression devices using a packed-bed suppressor in disposable cartridges, Solid Phase Chemical Suppression [258], or solid phase electrochemical suppressor cells (Electrochemically Regenerated Ion Suppression, ERIS) [259,260] have also been marketed by Alltech. More recently, from the same corporation, a continuously electrochemically regenerated packed bed suppressor (DS-Plus suppressor) containing a degassing chamber placed after suppression but prior to the detection cell has been released [14]. The degassing unit has been conceived for the removal of the carbonic acid after the suppressor of a carbonate-hydrogencarbonate based eluent, whose partial ionization contributes to a conductive background. H_2CO_3 is removed diffusing the CO_2 produced after dissociation across a degassing membrane, thus decreasing the background conductivity and enhancing sensitivity. An overview in suppression technology for ions detection by conductivity is available from Ref. [15].

14.6.2 SPECTROPHOTOMETRIC DETECTION

Spectrophotometric detection can be divided into (1) direct; (2) indirect; (3) direct after post-column derivatization. The direct detection can be used for those analytes absorbing in the UV or Vis region (e.g., Br^-, BrO_3^-, NO_3^-, etc.). Some metal ions, such as Au, Pt, and Pd, can be directly detected as chlorocomplexes, while transitition, heavy, and alkaline earth metal ions can be complexed by suitable strong chromophore ligands either prior to separation and hence detected directly or else after separation and detected in the so-called post-column derivatization mode. The latter approach requires a device (pump or pressurized system) to deliver the post-column reagent (chromophore), and a reaction coil to ensure the completion of the derivatization reaction. The ideal post-column reagent (PCR) should be universal for as many metal ions as possible, and should form rapidly stable and high absorbing complexes with the metal ions. It is important that the post-column reagent can displace any eluent ligands that are complexed with the eluting metal ions. Popular reagents are 4-(2-pyridylazo)resorcinol (PAR) [124], 2,7-bis(arsenophenylazo)-1,8-dihydroxy-3,6-naphthalene-disulfonic acid (Arsenazo III) or 2-(5-bromo-2-pyridylazo)-5-diethylaminophenol (5-Br-PADAP) [121]. A survey on IC methods, including detection details, devoted to the determination of trace metals in environmental matrices has been presented by Shaw and Haddad [261].

Indirect detection exploits the absorbance of the eluent ion (e.g., NO_3^-, phthalate, benzoate, methylbenzylamine) that should absorb more than the analyte. The eluting analyte exhibits a negative peak. Nevertheless, this mode of detection has limited utilization since the detection limits are often inadequate.

14.6.3 Amperometric Detection

It is used in IC systems when the amperometric process confers selectivity to the determination of the analytes. The operative modes employed in the amperometric techniques for detection in flow systems include those at (1) constant potential, where the current is measured in continuous mode, (2) at pulsed potential with sampling of the current at defined periods of time (pulsed amperometry, PAD), or (3) at pulsed potential with integration of the current at defined periods of time (integrated pulsed amperometry, IPAD). Amperometric techniques are successfully employed for the determination of carbohydrates, catecholamines, phenols, cyanide, iodide, amines, etc., even if, for optimal detection, it is often required to change the mobile-phase conditions. This is the case of the detection of biogenic amines separated by cation-exchange in acidic eluent and detected by IPAD at the Au electrode after the post-column addition of a pH modifier (NaOH) [262].

14.6.4 MS Detection

Relatively new is the introduction of mass spectrometry in IC analysis. The combination of IC with MS has gained increased importance in the last few years due to the need to determine new contaminants at a high sensitivity and due to the fast development of interfaces that greatly simplified this combination.

Some combinations of IC with MS detection are available, including ICP-MS (element-specific detection), particle beam MS and MS with atmospheric pressure ionization (API) operated in either electrospray ionization (ESI) or atmospheric pressure chemical ionization (APCI) modes that give information on molecular ions or adducts.

In the ICP-MS technique, suitable for the detection of metal ions and metal complexes, the ions of the elements generated in the ICP after nebulization are extracted from the plasma into the MS part, organized in a multiple-stage differentially pumped interface. The first part of this interface is the sampling cone (cooled Ni or Pt) with an orifice (<1 mm). A rotary pump evacuates the expansion chamber and ensures a pressure below 3 mbar behind the sampler. The pressure difference creates a supersonic jet into the mass spectrometer. A second cone with a small orifice (skimmer) placed inside the supersonic jet preserves the required vacuum. Subsequently, an ion optic made of an extraction lens (to accelerate the positively charged ions), a photon stop or an off-axis mass analyzer and several extraction lenses focus the ion beam [17]. The mass analyzer is generally a quadrupole. The most critical units in the coupling of IC with ICP-MS are the nebulizer and the spray chamber; therefore compatible flow rates and eluent compositions and concentrations should be chosen taking into account factors like the volatility, the condensation at low temperatures (about 4°C) in the nebulization chamber, the formation of refractory species at the high temperatures of the plasma, particle dimensions after the desolvation step, nucleation and the growth of microcrystals inside the cones, and the formation of polyatomic species [263]. For these reasons, ammonium salts are preferred to sodium or potassium salts (e.g., $ArNa^+$ species interferes with the determination of Cu at m/z 63) and phosphate or borate and polyanions sulfur-derivative mobile phases are unadvisable. The powerful separation capabilities of IC when coupled to the high selectivity and sensitivity properties of the ICP-MS detection, reliably overcome the problems due to coelution on the chromatographic side and to spectroscopic interferences on the mass one.

In IC-MS systems, the core of the equipment is the interface. In fact, inside the interface evaporation of the liquid, ionization of neutral species to charged species and removal of a huge amount of mobile phase to keep the vacuum conditions required from the mass analyzer take place. Two main interfaces are used coupled to IC, namely electrospray ionization (ESI) and atmospheric pressure chemical ionization (APCI). In the ESI mode, ions are produced by evaporation of charged droplets obtained through spraying and an electrical field, whilst in the APCI mode the spray created by a pneumatic nebulizer is directed towards a heated region (400°C–550°C) in which desolvation and vaporization take place. The eluent vapors are ionized by the corona effect (the partial discharge around a conductor placed at a high potential), due to a metallic needle, which is at a potential of a few kilovolts, and react chemically with the analyte molecules in the gas phase.

General rules can be given to choose between ESI and APCI: ESI is preferred for compounds which are ionic or very polar or thermo labile, or with masses higher than 1000, whilst APCI is preferred for compounds of lower molecular mass that are not very polar [264]. The use of non-volatile buffers is usually avoided when performing ESI or APCI sources [265] and the analyst must make sure he uses eluents compatible with the stainless steel parts of the mass spectrometer and avoids inorganic salt buildup.

With the advent of mass spectrometry hyphenation in IC, continuously regenerated suppressors have not been surpassed; on the contrary, due to their efficiency in decreasing the electrolyte background noise and in minimizing the interference of the eluent in the ionization step [266], they greatly simplify the interface system to an easier and user-friendly device. Mass spectrometric detection has recently been included among those emerging detection techniques for ion chromatography [267] and considered as a current trend for ion analysis, including separation methods such as IC and capillary electrophoresis (CE) [268].

REFERENCES

1. H. Small, T.S. Stevens, W.C. Bauman. *Anal. Chem.* 47:1801, 1975.
2. F. Helfferich. *Ion Exchange*, Mac Graw Hill, New York, 1962.
3. J. Inczedy. *Analytical Applications of Ion Exchangers*, Pergamon Press, Oxford, NY, M. Williams, English translation ed., 1966.
4. W. Reiman III, H.F. Walton. *Ion Exchange in Analytical Chemistry*, Pergamon Press, New York, 1970.
5. H.F. Walton, R.D. Roklin. *Ion Exchange in Analytical Chemistry*, CRC Press, Boca Barton, FL, 1990.
6. C.A. Lucy. *J. Chromatogr. A* 1000:711, 2003.
7. J.S. Fritz. *J. Chromatogr. A* 1039:3, 2004.
8. H. Small. *Ion Chromatography*, Plenum, New York, 1989.
9. P.R. Haddad, P.E. Jackson. *Ion Chromatography: Principles and Applications*, Elsevier, New York, 1990; *J. Chromatogr. Libr.*, 46.
10. J. Weiss. *Ion Chromatography*, Wiley-VCH, Weinheim, Germany, 2001.
11. J.S. Fritz, D.T. Gjerde. *Ion Chromatography*, 3rd edn., Wiley-VCH, New York, 2000.
12. C. Sarzanini. *J. Chromatogr. A* 956:3, 2002.
13. C. Sarzanini, M.C. Bruzzoniti. *Anal. Chim. Acta* 540:45, 2005.
14. R. Saari-Nordhaus, J.M. Anderson Jr. *J. Chromatogr. A* 956:15, 2002.
15. P.R. Haddad, P.E. Jackson, M.J. Shaw. *J. Chromatogr. A* 1000:725, 2003.
16. Y. Liu, K. Srinivasan, C. Pohl, N. Avdalovic. *J. Biochem. Biophys. Methods* 60:205, 2004.
17. A. Seuber. *Trends Anal. Chem.* 20:274, 2001.
18. C.A. Pohl, J. R. Stillian, P.E. Jackson. *J. Chromatogr A* 789:29, 1997.
19. J. Fritz. *J. Chromatogr. A* 1085:8, 2005.
20. P.R. Haddad, R.C. Foley. *J. Chromatogr.* 500:301, 1990.
21. P. Janoš. *J. Chromatogr. A* 657:435, 1993.
22. P. Hajos, G. Revesz, C. Sarzanini, G. Sacchero, E. Mentasti. *J. Chromatogr.* 640:15, 1993.
23. P. Janoš. *J. Chromatogr. A* 699:1, 1995.
24. J. Stahlberg. *Anal. Chem.* 66:440, 1994.
25. A. Bartha, J. Stahlberg. *J. Chromatogr. A* 668:255, 1994.
26. G. Fóti, P. Hajós, E.sz. Kováts. *Talanta* 41:1073, 1994.
27. G. Fóti, G. Révész, P. Hajós, G. Pellaton, E.sz. Kováts. *Anal. Chem.* 68:2580, 1996.
28. P. Hajós, O. Horváth, V. Denke. *Anal. Chem.* 67:434, 1995.
29. M. C. Bruzzoniti, E. Mentasti, C. Sarzanini, P. Hajos. *J. Chromatogr. A* 770:13, 1997.
30. C. Sarzanini, M.C. Bruzzoniti, P. Hajos. *J. Chromatogr. A* 867:131, 2000.
31. M. C. Bruzzoniti, E. Mentasti, C. A. Pohl, J. M. Riviello, C. Sarzanini. *J. Chromatogr. A* 925:99, 2001.
32. P. Janoš. *J. Chromatogr. A* 789:3, 1997.
33. J.E. Madden, P.R. Haddad. *J. Chromatogr. A* 829:65, 1998.
34. T. Okada. *J. Phys. Chem. B* 101:7814, 1997.
35. T. Okada. *J. Phys. Chem. B* 102:3053, 1998.

36. T. Okada. *Anal. Chem.* 70:1692, 1998.
37. T. Okada, J.M. Patil. *Langmuir* 14:6241, 1998.
38. T. Okada. *J. Chromatogr. A* 850:3, 1999.
39. M. Harada, T. Okada. *J. Chromatogr. A* 1085:3, 2005.
40. K. Horváth, P. Hajós. *J. Chromatogr. A* 1104:75, 2006.
41. M.C. Bruzzoniti, P. Hajós, K. Horváth, C. Sarzanini. *Acta Chim. Slov.* 5:14, 2007.
42. J. Havel, E.M. Pena, A. Rojas-Hernandez, J.P. Doucet, A. Panaye. *J. Cromatogr. A* 793:317, 1998.
43. J. Havel, J.E. Madden, P.R. Haddad. *Chromatographia* 49:481, 1999.
44. J. Havel, M. Breadmore, M. Macka. P.R. Haddad. *J. Cromatogr. A* 850:345, 1999.
45. T. Bolanča, Š. Cerjan-Stefanović, M. Luša, H. Regelja, S. Lončarić. *Chemom. Intell. Lab. Syst.* 86:95, 2007.
46. R.E. Smith. *Ion Chromatography Applications*, CRC Press, Boca Raton, FL, 1988.
47. P. Kolla, J. Köhler, G. Schomburg. *Chromatographia* 23:465, 1987.
48. Y. Zhu, C. Yongxin, Y. Mingli, J.S. Fritz. *J. Chromatogr. A* 1085:18, 2005.
49. R.M. Cassidy, S. Elchuk. *Anal. Chem.* 54:1558, 1982.
50. P. Janos, M. Brol. *Fresenius J. Anal. Chem.* 344:545, 1992.
51. J. Weiss. GIT Spezial. *Chromatographie* 2:67, 1992.
52. D. Jensen, J. Weiss, M.A. Rey, C.A. Pohl. *J. Chromatogr. A* 640:65, 1993.
53. W. Zeng, Y. Chen, H. Cui, F. Wu, Y. Zu, J.S. Fritz. *J. Chromatogr. A* 1118:68, 2006.
54. M.A. Rey, C.A. Pohl. *J. Chromatogr. A* 739:87, 1996.
55. K. Kulisa. *Chem. Anal. (Warsaw)* 49:665, 2004.
56. M.A. Rey, J.M. Riviello, C.A. Pohl. *J. Chromatogr. A* 789:149, 1997.
57. M.J. Shaw, P.N. Nesterenko, G.W. Dicinoski, P.R. Haddad. *J. Chromatogr. A* 997:3, 2003.
58. R.M. Izatt, J.J. Christensen. *Synthesis of Macrocycles*, Wiley, New York, 1987.
59. S.A. Oehrle. *J. Chromatogr. A* 745:87, 1996.
60. J.D. Lamb, R.G. Smith. *J. Chromatogr.* 546:73, 1991.
61. B.R. Edwards, A.P. Giauque, J.D. Lamb. *J. Chromatogr. A* 706:69, 1995.
62. M. Takagi, H. Nakamura. *J. Coord. Chem.* 15:53, 1986.
63. K. Kimura, H. Hartino, E. Hayata, T.T. Shono. *Anal. Chem.* 58:2233, 1986.
64. P.J. Dumont, J.S Fritz. *J. Chromatogr.* 706:149, 1995.
65. W.R. Heumann. *Crit. Rev. Anal. Chem.* 2:425, 1971.
66. F. Steiner, C. Niederländer, H. Engelhardt. *Chromatographia* 43:117, 1996.
67. K. Ohta, K. Tanaka. *Anal. Chim. Acta* 381:265, 1999.
68. M.A. Rey, C.A. Pohl, J.J. Jagodzinski, E.Q. Kaiser, J.M. Riviello. *J. Chromatogr. A* 804:201, 1998.
69. D.A. Richens, D. Simpson, S. Peterson, A. McGinn, J.D. Lamb. *J. Chromatogr. A* 1016:155, 2003.
70. R. Saari-Nordhaus, J.M. Anderson Jr. *J. Chromtogr. A* 1039:123, 2004.
71. M.A. Rey. *J. Chromatog. A*, 920:61, 2001.
72. D.H. Thomas, M. Rey, P.E. Jackson. *J. Chromatog. A* 956:181, 2002.
73. M. Rey, C. Pohl. *J. Chromatogr. A* 997:199, 2003.
74. A.L. Cinquina, A. Calì, F. Longo, L. De Santis, A. Severoni, F. Abballe. *J. Chromatogr. A* 1032:73, 2004.
75. G. Saccani, E. Tanzi, P. Pastore, S. Cavalli, M. Rey. *J. Chromatogr A* 1082:43, 2005.
76. P. Pastore, G. Favaro, D. Badocco, A. Tapparo, S. Cavalli, G. Saccani. *J. Chromatogr A* 1098:111, 2005.
77. DIONEX, Application Note 182.
78. B.M. De Borba, J.S. Rohrer. *J. Chromatogr A* 1155:22, 2007.
79. R.L. Smith, D.J. Pietrzyk. *Anal. Chem.* 56:610, 1984.
80. D.M. Brown, D.J. Pietrzyk. *J. Chromatogr.* 466:291, 1989.
81. T. Iwachido, K. Ikeda, M. Zenki. *Anal. Sci.* 6:593, 1990.
82. K. Ohta, M. Sando, K. Tanaka, P.R. Haddad. *J. Chromatogr. A* 752:167, 1996.
83. K. Ohta, H. Morikawa, K. Tanaka, Y. Uryu, B. Paull, P.R. Haddad. *Anal. Chim. Acta* 359:255, 1998.
84. K. Ohta, H. Morikawa, K. Tanaka, P.R. Haddad. *J. Chromatogr. A* 804:171, 1998.
85. K. Ohta, K. Tanaka, P.R. Haddad. *Trends Anal. Chem.* 20:331, 2001.
86. H. Zou, X. Huang, M. Ye, Q. Luo. *J. Chromatogr. A* 954:5, 2002.
87. A. M. Siouffi. *J. Chromatogr. A* 1000:801, 2003.
88. C. Legido-Quigley, N.D. Marlin, V. Melin, A. Manz, N.W. Smith. *Electrophoresis* 24:917, 2003.
89. D. Moravcová, P. Jandera, J. Urban, J. Planeta. *J. Sep. Sci.* 27:789, 2004.
90. F. Svec. *J. Sep. Sci.* 27:747, 2004.
91. B. Paull, P.N. Nesterenko. *Trends Anal. Chem.* 24:295, 2005.
92. D. Schaller, E.F. Hilder, P.R. Haddad. *J. Sep. Sci.* 29:1705, 2006.

93. S.D. Chambers, K. M. Glenn, C.A. Lucy. *J. Sep. Sci.* 30:1628, 2007.
94. F. Svec, T.B. Tennikova, Z. Deyl., Eds. *Monolithic Materials: Preparation, Properties and Applications,* 1st edn., Elsevier, Amsterdam, the Netherlands, 2003.
95. M. Motokawa, M. Ohira, H. Minakuchi, K. Nakanishi, N. Tanaka. *J. Sep. Sci.* 29:2471, 2006.
96. F. Svec, J.M.J. Fréchet. *J. Chromatogr. A* 702:89, 1995.
97. H. Minakuchi, K. Nakanishi, N. Soga, N. Ishizuka, N. Tanaka. *Anal. Chem.* 68:3498, 1996.
98. Q. Xu, K. Tanaka, M. Mori, M.I.H. Helaleh, W. Hu, K. Hasebe, H. Toada. *J. Chromatogr. A* 997:183, 2003.
99. Q. Xu, M. Mori, K. Tanaka, M. Ikedo, W. Hu. *J. Chromatogr. A* 1026:191, 2004.
100. D. Connoly, D. Victory, B. Paull. *J. Sep. Sci.* 27:912, 2004.
101. E. Sugrue, P.N. Nesterenko, B. Paull. *Anal. Chim. Acta* 553:27, 2005.
102. B.W. Pack, D.R. Risley. *J. Chromatogr. A* 1073:269, 2005.
103. A.J. Alpert. *J. Chromatogr.* 499:177, 1990.
104. P.N. Nesterenko, O.N. Fedyanina, Y.V. Volgin, P. Jones. *J. Chromatogr. A* 1155:2, 2007.
105. G.A. Chiganova. *Kolloidnyi Zh.* 56:266, 1994.
106. G.J. Sevenic, J.S. Fritz. *Anal. Chem.* 55:12, 1983.
107. P.R. Haddad, P.W. Alexander, M. Trojanowicz. *J. Chromatogr.* 294:397, 1984.
108. P.R. Haddad, P.W. Alexander, M. Trojanowicz. *J. Chromatogr.* 324:319, 1985.
109. D. Yan, G. Schedt. *J. Chromatogr. A* 516:383, 1990.
110. D. Yan, G. Schedt. *Fresenius Z. Anal. Chem.* 338:149, 1990.
111. W. Zhou, W. Liu, D. An. *J. Chromatogr.* 589:358, 1992.
112. A.A. leGras. *Analyst* 118:1035, 1993.
113. W. Buchberger, P.R. Haddad, P.W. Alexander. *J. Chromatogr.* 558:181, 1991.
114. P. Hajos, G. Revsez, O. Horvath, J. Peer, C. Sarzanini. *J. Chromatogr Sci.* 34:291, 1996.
115. S. Reiffenstuhl, G. Bonn. *J. Chromatogr.* 482:289, 1989.
116. C. Sarzanini, G. Sacchero, E. Mentasti, P. Hajos. *J. Chromatogr. A* 706:141, 1995.
117. M.C. Bruzzoniti, E. Mentasti, C. Sarzanini. *Anal. Chim. Acta* 382:291, 1999.
118. P.E. Jakson, C.A. Pohl. *Trends Anal. Chem.* 16:393, 1997.
119. A. Sirirakis, H.M. Kingston, J.M. Riviello. *Anal. Chem.* 62:1185, 1990.
120. M. C. Bruzzoniti, E. Mentasti, C.Sarzanini. *Anal. Chim. Acta* 353:239, 1997.
121. N. Cardellicchio, S. Cavalli, P. Ragone, J. Riviello. *J. Chromatogr. A* 847:251, 1999.
122. X.J. Ding, S.F. Mou. *J. Chromatogr A* 920:101, 2001.
123. M.C. Bruzzoniti, N. Cardellicchio, S. Cavalli, C. Sarzanini. *Chromatographia* 55:231, 2002.
124. S. Motellier, H. Pitsch. *J. Chromatogr.* 739:119, 1996.
125. H. Lu. S. Mou, J.M. Riviello. *J. Chromatogr A* 857:343, 1999.
126. W. Al-Shawi, R. Dahl. *J. Chromatogr.* 391:35, 1999.
127. M. Derbyshire, A. Lamberty, P.H.E. Gardier. *Anal. Chem.* 71:4203, 1999.
128. M.C. Bruzzoniti E. Mentasti, C.Sarzanini, M. Braglia, G. Cocito. J. Kraus, *Anal. Chim. Acta* 322:49, 1996.
129. Y.F. Lasheen, A.F. Seliman, A.A. Abdel-Rassoul. *J. Chromatogr. A* 1136:202, 2006.
130. B. Paull, P.N. Nesterenko. *Analyst* 130:134, 2005.
131. B.A. Bidlingmeier. *J. Chromatogr. Sci.* 18:525, 1980.
132. C. Horvath, W. Melander, I. Molnar, P. Molnar. *J. Chromatogr.* 125:129, 1976.
133. C. Horvath, W. Melander, I. Molnar, P. Molnar. *Anal. Chem.* 49:2295, 1977.
134. E.P. Kroeff, D.J. Pietrzyk. *Anal. Chem.* 50:502, 1978.
135. A. Siriraks, C.A. Phol, M. Toofan. *J. Chromatogr.* 602:89, 1992.
136. J.M. Miller. *Chromatography Concepts and Contrasts*, John Wiley & Sons Inc., Hoboken, NJ, 2005.
137. B.A. Bidlingmeier, S.N. Deming, W.P. Price jr., B. Sachok, M. Petrusek. *J. Chromatogr.* 186:419, 1979.
138. F.F. Cantwell, S. Puon. *Anal. Chem.* 51:623, 1979.
139. J.J. Ståhlberg. *J. Chromatogr.* 356:231, 1986.
140. C.F. Poole, S.K. Poole. *Chromatography Today*, Elsevier Science Publishers, Amsterdam, the Netherlands, 1991, Chapter 4.
141. A. Bartha, H.A.H. Billet, L. De Galan, G. Vigh. *J. Chromatogr.* 291:91, 1984.
142. Y.K. Zhang, H.F. Zou, M.F. Hong, P.C. Lu. *Chromatographia* 32:538, 1991.
143. H.F. Zou, Y.K. Zhang, P.C. Lu. *J. Chromatogr.* 545:59, 1991.
144. Z. Iskandarani, D.J. Pietrzyk. *Anal. Chem.* 54:1065, 1982.
145. Z. Iskandarani, D.J. Pietrzyk. *Anal. Chem.* 54:2427, 1982.
146. J.M. Roberts. *Anal. Chem.* 74:4927, 2002.
147. S.G. Weber, L.L. Glarina, F.F. Cantwell, J.G. Chen. *J. Chromatogr. A* 656:549, 1993.

148. M.C. Bruzzoniti, E. Mentasti, G. Sacchero, C. Sarzanini. *J. Chromatogr A* 728:55, 1996.
149. J.C. Kraak, J.F.K. Huber. *J. Chromatogr.* 102:333, 1974.
150. K.G. Wahlund. *J. Chromatogr.* 115:411, 1975.
151. W.R. Melander, C. Horvath. *J. Chromatogr.* 201:211, 1980.
152. M. Johansson. *J. Liquid Chromatogr.* 4:1435, 1981.
153. E. Arvidsson, J. Crommen, G. Schill, D. Westerlund. *Chromatographia* 24:460, 1987.
154. E. Arvidsson, L. Hackzell, G. Schill, D. Westerlund. *Chromatographia* 25:430, 1988.
155. C. Petterson, G. Schill. *Chromatographia* 28:437, 1989.
156. X. Zhang, M. Wang, J. Cheng. *J. Liquid Chromatogr.* 16:1057, 1993.
157. S. Zappoli, C. Bottura. *Anal. Chem.* 66:3492, 1994.
158. Q. Xianren, W. Baeyens. *J. Chromatogr.* 456:267, 1988.
159. J.H. Knox, R.A. Hartwick. *J. Chromatogr.* 204:3, 1981.
160. J. Stranahan, S.N. Deming. *Anal. Chem.* 54:2251, 1982.
161. A. Bartha, G. Vigh, J. Stahlberg. *J. Chromatogr.* 506:85, 1990.
162. J. Ståhlberg, A. Furängen. *Chromatographia* 24:783, 1987.
163. J. Ståhlberg, I. Hägglund. *Anal. Chem.* 60:1958, 1988.
164. J. Ståhlberg, A. Bartha. *J. Chromatogr.* 456:253, 1988.
165. A. Bartha, J. Ståhlberg. *J. Chromatogr.* 535:181, 1990.
166. J. Narkiewicz-Michalek. *Chromatographia* 35:527, 1993.
167. C. Sarzanini, M.C. Bruzzoniti, G. Sacchero, E. Mentasti. *Anal. Chem.* 68:4494, 1996.
168. G. Sacchero, M.C. Bruzzoniti, C. Sarzanini, E. Mentasti, H.J. Metting, P.M.J. Coenegracht. *J. Chromatogr. A* 799:35, 1998.
169. H. Sirén. *Ann. Acad. Fennicae A II Chemia* 233:98, 1991.
170. H. Sirén, M.-L. Riekkola. *J. Chromatogr.* 590:263, 1992.
171. C. Sarzanini, G. Sacchero, M.Aceto, O. Abollino, E. Mentasti. *J. Chromatogr.* 640:179, 1993.
172. C. Sarzanini, G. Sacchero, M. Aceto, O. Abollino, E. Mentasti. *J. Chromatogr.* 640:127, 1993.
173. C. Ohtsuka, K. Matsuzawa, H. Wada, G. Nakagawa. *Anal. Chim. Acta* 252:181, 1991.
174. C. Ohtsuka, K. Matsuzawa, H. Wada, G. Nakagawa. *Anal. Chim. Acta* 256:1, 1992.
175. C. Ohtsuka, K. Matsuzawa, H. Wada, G. Nakagawa. *Anal. Chim. Acta* 294:69, 1994.
176. A. Alvarez-Zepeda, B.N. Barman, D.E. Martire. *Anal. Chem.* 64:1978, 1992.
177. C. Sarzanini, M.C. Bruzzoniti, O. Abollino, E. Mentasti. *J. Chromatogr. A* 847:233, 1999.
178. R.M. Cassidy, L. Sun. *J. Chromatogr. A* 654:105, 1993.
179. S. Zappoli, L. Morselli, F. Osti. *J. Chromatogr. A* 721:269, 1996.
180. M.L. Marina, P. Andrés, J.C. Diéz-Masa. *Chromatographia* 35:621, 1993.
181. S.R. Villaseñor. *Anal. Chem.* 63:1362, 1991.
182. G. Sacchero, O. Abollino, V. Porta, C. Sarzanini, E. Mentasti. *Chromatographia* 31:539, 1991.
183. W.W. Bedsworth, D.L. Sedlak. *J. Chromatogr. A* 905:157, 2001.
184. E. Poboży, R. Halko, M. Krasowski, T. Wierbicki, M. Trojanowicz. *Water Res.* 37:2019, 2003.
185. S. Srijaranai, S. Chanpaka, C. Kukusamude, K. Grudpan. *Talanta* 68:1720, 2007.
186. J. Threeprom, R. Meelapsom, W. Som-aum, J.-M Lin. *Talanta* 71:103, 2007.
187. J.H. Knox, J. Jurand. *J. Chromatogr.* 203:85, 1981.
188. L.W. Yu, T.R. Floyd, R.A. Hartwick. *J. Chromatogr. Sci.* 24:177, 1986.
189. L.W. Yu, R.A. Hartwick. *J. Chromatogr. Sci.* 27:176, 1989.
190. W. Hu, T. Takeuchi, H. Haraguchi. *Anal. Chem.* 65:2204, 1993.
191. W. Hu, H. Tao, H. Haraguchi. *Anal. Chem.* 66:2514, 1994.
192. W. Hu, P.R. Haddad. *Anal. Commun.* 35:317, 1998.
193. W. Hu, K. Hasebe, A. Iles, K. Tanaka. *Anal. Sci.* 17:1401, 2002.
194. W. Hu, P. Haddad. *Trends Anal. Chem.* 17:73, 1998.
195. P.N. Nesterenko, A.I. Elefterov, D.A. Tarasenko, O.A Shpigun. *J. Chromatogr. A* 706:59, 1995.
196. C. Viklund, K. Irgum. *Macromolecules* 33:2539, 2000.
197. W. Jiang, K. Irgum. *Anal. Chem.* 71:333, 1999.
198. P. Nesterenko, P.R. Haddad. *Anal. Sci.* 16:565, 2000.
199. W. Hu, P.R. Haddad, K. Tanaka, K. Hasebe. *Analyst* 125:241, 2000.
200. H. A. Cook, W. Hu, J.S. Fritz, P.R. Haddad. *Anal. Chem.* 73:3022, 2001.
201. W. Hu, P.R. Haddad, K. Tanaka, M. Mori, K. Tekura, K. Hasebe, M. Ohno, N. Kamo. *J. Chromatogr. A* 997:237, 2003.
202. W. Hu, P.R. Haddad, K. Tanaka, K. Hasebe. *Anal. Bioanal. Chem.* 375:259, 2003.
203. W. Hu, P.R. Haddad, K. Hasebe, K.Tanaka. *Anal. Commun.* 36:97, 1999.

204. W. Hu, H. Haraguchi. *Anal. Chem.* 66:765, 1994.
205. W. Hu, K. Hasebe, D.M. Reynolds, H. Haraguchi. *J. Liquid Chromatogr. Relat. Technol.* 20:1221, 1997.
206. H.A. Cook, G.W. Dicinoski, P.R. Haddad. *J. Chromatogr. A* 997:13, 2003.
207. E. Sugrue, P.N. Nesterenko, B. Paull. *J. Chromatogr. A* 1075:167, 2005.
208. P. Jones, M. Foulkes, B. Paull. *J. Chromatogr. A* 673:173, 1994.
209. J.P. Riley, D. Taylor. *Anal. Chim. Acta* 40:479, 1968.
210. E.M. Moyers, J.S. Fritz. *Anal. Chem.* 49:418, 1977.
211. A.R. Timerbaev, G.K. Bonn. *J. Chromatogr.* 640:195, 1993.
212. P. Jones, P.N. Nesterenko. *J. Chromatogr A* 789:413, 1997.
213. P. Jones, G. Schwedt. *J. Chromatogr.* 482:325, 1989.
214. O.J. Challenger, S.J. Hill, P. Jones, N.W. Barnett. *Anal. Proc.* 29:91, 1992.
215. J. Toei. *Fresenius Z. Anal. Chem.* 331:735, 1988.
216. P. Jones, O. J. Challenger, S. Hill, N.W. Barnett. *Analyst* 117:1447, 1992.
217. O.J. Challenger, S.J. Hill, P. Jones. *J. Chromatogr.* 639:197, 1993.
218. B. Paull, M. Foulkes, P. Jones. *Analyst* 119:937, 1994.
219. B. Paull, M. Foulkes, P. Jones. *Anal. Proc.* 31:209, 1994.
220. B. Paull, P. Jones. *Chromatographia* 42:528, 1996.
221. B. Paull, P.R. Haddad. *Anal Commun.* 35:13, 1998.
222. M.J. Shaw, J. Cowan, P. Jones. *Anal. Lett.* 36:423, 2003.
223. B. Paull, P.A. Fagan, P.R. Haddad. *Anal. Commun.* 33:193, 1996.
224. B. Paull, M. Macka, P.R. Haddad. *J. Chromatogr A* 789:329, 1997.
225. B. Paull, P. Nesterenko, M. Nurdin, P.R. Haddad. *Anal. Commun.* 35:17, 1998.
226. B. Paull, M. Clow, P.R. Haddad. *J. Chromatogr. A* 804:95, 1998.
227. B. Paull, P.R. Haddad. *Trends Anal. Chem.* 18:107, 1999.
228. B. Paull, P. Nesterenko, P.R. Haddad. *Anal. Chim. Acta* 375:117, 1998.
229. M.J. Shaw, S.J. Hill, P. Jones. *Anal. Chim. Acta* 401:65, 1999.
230. M.J. Shaw, S.J. Hill, P. Jones, P.N. Nesterenko. *Anal. Commun.* 36:399, 1999.
231. G. Bonn, S. Reiffenstuhl, P. Jandik. *J. Chromatogr.* 499:669, 1990.
232. A.I. Elefterov, S.N. Nosal, P.N. Nesterenko, O.A. Shpigun. *Analyst* 119:1329, 1994.
233. P.N. Nesterenko, P. Jones. *J. Chromatogr. A* 770:129, 1997.
234. P.N. Nesterenko, P. Jones. *Anal. Commun.* 34:7, 1997.
235. P.N. Nesterenko, P. Jones. *J. Chromatogr. A* 804:223, 1998.
236. P.N. Nesterenko, O.A. Shpigun. *Russ. J. Coord. Chem.* 28:726, 2002.
237. P.N. Nesterenko, P. Jones. *J. Sep. Sci.* 30:1773, 2007.
238. P.N. Nesterenko, O.S. Zhukova, O.A. Shpigun, P. Jones. *J. Chromatogr. A* 813:47, 1998.
239. M.S. Mahon, E.H. Abbott. *J. Coord. Chem.* 8:175, 1978.
240. S.K. Sahni, R.V. Bennekom, J. Reedijk. *Polyhedron* 4:1643, 1985.
241. M.J. Shaw, S.J. Hill, P. Jones, P.N. Nesterenko. *J. Chromatogr. A* 876:127, 2000.
242. W. Bashir, B. Paull. *J. Chromatogr. A* 910:301, 2001.
243. E. Sugrue, P. Nesterenko, B. Paull. *Analyst* 128:417, 2003.
244. W. Bashir, B. Paull. *J. Chromatogr. A* 907:191, 2001.
245. E. Sugrue, P. Nesterenko, B. Paull. *J. Sep. Sci.* 27:921, 2004.
246. S. C. Stefanovic, T. Bolanca, L. Curkovic. *J. Chromatogr. A* 918:325, 2001.
247. J.S. Fritz, D. Gjerde, R.M. Becker. *Anal. Chem.* 52:1519, 1980.
248. T.S. Stevens, J.C. Davis, H. Small. *Anal. Chem.* 53:1488, 1981.
249. J. Stillian. *Liquid Chromatogr.* 3:802, 1985.
250. C. Pohl, R. Slingsby, J. Stillian, R. Gajek. Modified membrane suppressor and method for use, U.S. Patent 4,999,098, 1991.
251. S. Rabin, J. Stillian, V. Barreto, K. Friedman, M. Toofan. *J. Chromatogr.* 640:97, 1993.
252. H. Small, Y. Liu, J. Riviello, N. Avdalovic, K. Srinivasan. Continuous electrolytically regenerated packed bed suppressor for ion chromatograph, U.S. Patent 6,325,976, 2001.
253. Y. Liu, N. Avdalovic, C. Pohl, R. Matt, H. Dhillon, R. Kiser. *Am. Lab.* 30:48C, 1998.
254. Y. Liu, H. Small, N. Avdalovic. Large capacity acid or base generation apparatus and method of use, U.S. Patent 6,225,129, 2001.
255. H. Small, J. Riviello. *Anal. Chem.* 70:2205, 1998.
256. H. Small, J. Riviello, Y. Liu, N. Avdalovic. Ion chromatographic method and apparatus using a combined suppressor and eluent generator, U.S. Patent 6,027,643, 2000.
257. H. Small, Y. Liu, N. Avdalovic. *Anal. Chem.* 70:3629, 1998.

258. R. Saari-Nordhaus, J.M. Anderson. *Am. Lab.* 26:28C, 1994.
259. R. Saari-Nordhaus, J.M. Anderson Jr. *J. Chromatogr. A* 782:75, 1997.
260. L.M. Nair, R. Saari-Nordhaus. *J. Chromatogr. A* 804:233, 1998.
261. M.J. Shaw, P.R. Haddad. *Environ. Int.* 30:403, 2004.
262. R. Draisci, L. Giannetti, P. Boria, L. Lucentini, L. Palleschi, S. Cavalli. *J. Chromatogr. A* 798:109, 1998.
263. R.G. Fernandez, J.I.G. Alonso, A. Sanz-Mendel. *J. Anal. At. Spectrom.* 16:1035, 2001.
264. E. de Hoffmann, J. Charette, V. Stroobant. *Mass Spectrometry, Principles and Applications*, Wiley, New York, 1996, pp. 21–33.
265. J. Mathew, J. Gandhi, J. Hedrick. *J. Chromatogr. A* 1085:54, 2005.
266. R. Roehl, R. Slingsby, N. Avdalovic, P.E. Jackson. *J. Chromatogr. A* 956:245, 2002.
267. W.W. Buchberger, P.R. Haddad. *J. Chromatogr. A* 789:67, 1997.
268. W.W. Buchberger. *J. Chromatogr. A* 884:3, 2000.

15 Retention Models for Ions in HPLC

Jan Ståhlberg

CONTENTS

15.1 INTRODUCTION

Ions constitute a common and important class of solutes in chromatography. Several different chromatographic techniques are used with many types of stationary and mobile phases. The purpose of this chapter to outline some general theoretical principles that are important for the understanding of the retention properties of ions. It is shown how the principles are applied to reversed-phase chromatography of ions, ion-pair chromatography, and ion exchange chromatography of small ions. Finally, the implications for ion exchange chromatography of proteins are discussed. For a more detailed review of the different theoretical aspects of the chromatography of ions, see Ref. [1].

The retention properties of an ion are determined by its physico-chemical behavior; it is, therefore, necessary to digress into the basic physics of charges. The first theoretical section begins with a short review of basic electrostatic concepts, and then we investigate their consequences are in chemical and chromatographic systems. Since charged stationary phases are often used in the chromatography of ions, it is important to discuss the physico-chemical properties of charged surfaces in contact with an electrolyte solution. Here, the Gouy–Chapman theory for the electrical double layer is briefly presented and applied to describe the retention of small ions in both ion exchange and ion-pair chromatography. The retention of small ions has been extensively studied experimentally for many years in ion exchange chromatography. Usually, the experimental data are fitted to an equation derived from a stoichiometric exchange of ions between the mobile phase and the stationary phase. This stoichiometric equation is originally derived from the Donnan description of the charged surface–electrolyte interface. Here, a new model is proposed. It is based on the contact value theorem for a charged surface in contact with an electrolyte.

In Section 15.6, the retention of proteins in ion exchange chromatography is discussed. The ions in the surrounding electrolyte form an electrical double layer around a charged macromolecule, e.g., a protein. The interaction between a protein and an oppositely charged surface can, therefore, be described as taking place between two overlapping double-layer systems.

The physical theories for electric charges have been incorporated in colloid and surface chemistry for many years. In the treatment presented here, these theories have been selected, adapted, and applied to describe the retention of ionic solutes. The approximations made in these models are well known and have limitations. Here, they are chosen from four requirements: they should have a meaningful physico-chemical interpretation, be easy to use, be easy to understand, and give practical and useful results. This implies that the presented models are useful starting points for describing and understanding the retention properties for the type of systems that are discussed here. When the experimental results deviate from the model, it may be possible to extend it within its framework. In other situations, empirically based functions may complement the model, or it may be necessary to resort to more sophisticated models.

15.2   RETENTION OF IONIC SOLUTES IN REVERSED-PHASE CHROMATOGRAPHY

Reversed-phase chromatography is often used to separate both neutral and ionic organic compounds. In this section, some important aspects for the understanding of the behavior of ionic compounds in reversed-phase chromatography are discussed. The important concepts introduced here are the electrical double layer and the electrostatic surface potential. It will be shown that they are essential for the understanding of the elution profile of ionic compounds. These concepts are further explored in the next section where theoretical models for ion-pair chromatography are discussed.

15.2.1   Coulomb's Law

From physics, we know that the interaction energy, $U(r)$, as a function of the distance, r, between two charges, Q_1 and Q_2, in vacuum is given by Coulomb's law:

$$U(r) = \frac{Q_1 \cdot Q_2}{4\pi \cdot \varepsilon_0 \cdot r} \tag{15.1}$$

where ε_0 is the relative permittivity of vacuum. If the two charges are instead situated in a medium with the dielectric constant ε, it can be shown that the interaction energy is inversely proportional to ε; i.e., we obtain

$$U(r) = \frac{Q_1 \cdot Q_2}{4\pi \cdot \varepsilon_0 \varepsilon \cdot r} \tag{15.2}$$

This equation has three important consequences for the retention theory of ions:

1. Compared to all other types of intermolecular interactions, the Coulomb interaction is a very strong interaction. E.g., for the interaction in vacuum between Na^+ and Cl^- at contact (2.76 Å), Coulomb's law gives that the interaction energy is approximately 500 kJ/mol, i.e., similar to the energy for a covalent bond.
2. Compared to all other types of intermolecular interactions, the Coulomb interaction is a very long ranged interaction. E.g., the interaction energy in water ($\varepsilon = 80$) between Na^+ and Cl^- at a distance 60–65 Å from each other is roughly the same as between two Ne atoms in contact.
3. Compared to all other intermolecular interactions, the Coulomb interaction is described by a simple law, i.e., Equation 15.2. A theory for Coulombic interaction, therefore, uses the concepts and laws that have been developed in classical electrostatics. However, it is worth pointing out that the dielectric constant is a macroscopic property and it is therefore, in principle, not correct to describe the solvent as a dielectric continuum on the molecular level. Nevertheless, experience has shown that it is in fact a useful approximation.

The first two points above have important consequences for the interaction between ions in chemical systems. In such systems, the interaction usually takes place in an electrolyte solution composed of a large number of ions. All the ions in the system are constantly in thermal motion and, due to the strength and long-range nature of the Coulomb interaction, the motion of a particular ion is affected by the continuous change in position of other ions or charged bodies in the system. The Coulomb interaction, therefore, is a many-body interaction, i.e., a particular ion is influenced by many other ions that are, on a molecular scale, quite far away. This is in contrast to the other types of intermolecular interactions where only the interaction between molecules in close contact is of significance.

The third point implies that it is possible to develop a physical theory for ionic interactions that is relatively simple and still useful. The most frequently used is the Poisson–Boltzmann (P-B) equation, which combines the Poisson equation from classical electrostatics with the Boltzmann distribution from statistical mechanics. This is a second-order nonlinear differential equation and its solution depends on the geometry and the boundary conditions.

In conclusion, it is easier to develop useful theories for the retention caused by the Coulombic interaction than for the other types of intermolecular interactions. Furthermore, the strength of the Coulombic interaction implies that it usually determines the physico-chemical properties of ions. Theories based on the solution of the P-B equation with appropriate boundary conditions are, therefore, a good starting point for the retention theories of ionic solutes.

15.2.2 BORN EQUATION

The Born equation is a simple equation that has interesting consequences for the understanding of the retention behavior of small organic ions in reversed-phase chromatography. It explains why the retention of the uncharged form of a solute has a retention factor 3–6 times higher than the corresponding charged form. E.g., if a carboxylic acid with pK_a around 4.5 has a retention factor of 10 at pH = 2, its retention factor at pH = 7 is probably around 2–3.

The Born equation describes the amount of work that is required to charge a body with radius R_1 from zero charge to charge Q_s in a medium of dielectric constant ε. The required work is called the solvation energy of the body in the medium. This is an important quantity because it has a strong influence on the dissolution of ions in a medium and how they partition between different solvents [2]. The Born equation shows that the solvation energy is higher in a medium with a low dielectric constant than in a medium with a high dielectric constant. The implication is that, when the two media are in contact, a charged molecule prefers to remain in the medium with a high ε.

The Born equation is derived as follows:

Consider a spherical body of radius R_1 that is charged from zero to Q_s in a medium with dielectric constant ε. At a certain stage of the charging process, the sphere has obtained the charge Q and the infinitesimally small charge dQ is added from infinity up to the sphere's surface at radius R_1. Putting $Q_1 = Q$ and $Q_2 = dQ$ in Equation 15.2, the energy required for this process is

$$dU = \frac{Q \cdot dQ}{4\pi \cdot \varepsilon_0 \varepsilon \cdot R_1} \tag{15.3}$$

The total energy required to bring the charge Q_s up to the charged surface is the integral of Equation 15.3 from $Q = 0$ to $Q = Q_s$, i.e.,

$$U = \int_0^{Q_s} \frac{Q \cdot dQ}{4\pi \cdot \varepsilon_0 \varepsilon \cdot R_1} = \frac{Q_s^2}{8\pi \cdot \varepsilon_0 \varepsilon \cdot R_1} \tag{15.4}$$

From this equation, the difference in the solvation energy for the sphere with radius R and charge Q_s between two solvents with dielectric constant ε_1 and ε_2 is found to be

$$\Delta U = \frac{Q_s^2}{8\pi \cdot \varepsilon_0 \cdot R_1} \left(\frac{1}{\varepsilon_1} - \frac{1}{\varepsilon_2} \right) \tag{15.5}$$

For an ion with valence z and radius R_1 given in nanometer, Equation 15.5 is approximated by [2]

$$\Delta U \approx \frac{69z^2}{R_1} \left(\frac{1}{\varepsilon_1} - \frac{1}{\varepsilon_2} \right) \quad \text{(kJ/mol)} \tag{15.6}$$

E.g., Equation 15.6 gives that 12 kJ/mol is required to transfer a monovalent ion with radius 0.5 nm from water ($\varepsilon_2 = 78$) to a solvent with $\varepsilon_1 = 10$.

In reversed-phase chromatography, the retentive layer is non-polar and therefore has a distinctly lower dielectric constant than the polar mobile phase. Furthermore, the retentive layer is thin at the same time as the surface area exposed to the liquid phase is extremely high. Taken together, these properties of a reversed-phase system in combination with the Born equation clearly indicate that the charged part of an ionic species will not enter the reversed-phase layer. For example, for an adsorbed carboxylic acid in its dissociated form this means that the charged carboxylic group remains in the polar mobile phase while the non-polar part of the molecule interacts with the non-polar retentive layer. The orientation of the charged group in the interface between the two phases imposes a restriction on the interaction between the non-polar part of the solute and the retentive layer. In the non-dissociated form, however, the non-charged carboxylic group does not impose this restriction and the retention factor is consequently higher by a factor of around 4.

15.2.3 pH Dependence of Retention for Protolytic Solutes

Because there is a large difference in retention between the charged and the neutral form of a solute, the retention factor depends on the pH of the mobile phase. The net retention factor k is the sum

$$k = f \cdot k_b + (1-f) \cdot k_a \tag{15.7}$$

where

k_b and k_a are the retention factors of the solute in its basic and acidic form, respectively

f is the fraction of the basic form of the solute

For an acid, f is a function of pH and the acid dissociation constant K_a:

$$f = \frac{K_a}{K_a + \left[H^+ \right]} \tag{15.8}$$

The combination of Equations 15.7 and 15.8 represents the net retention factor as a function of pH and results in a sigmoid curve with the same form as an acid-base titration curve.

The fact that the ionic form has a markedly lower retention factor than the neutral form for an acid or base makes it possible to regulate the retention by changing the pH of the mobile phase. Another practical consequence is that it is important to buffer the mobile phase properly. Furthermore, one should avoid using a mobile phase pH around the pK_a value, since small variations in the mobile phase pH result in comparatively large changes in the net retention factor.

15.2.4 Adsorption Isotherm of an Organic Ion in Reversed-Phase Chromatography

Consider a reversed-phase system with an organic ion (e.g., benzenesulfonate) as a solute and an inorganic salt (e.g. NaH_2PO_4) as a mobile phase electrolyte. Assume that the organic ion is more strongly adsorbed to the stationary phase than its electrolyte counter ion (i.e., Na^+ in this example). Thus, the organic ion is bound to the surface at a higher concentration than its counter ion and creates a charged layer at the mobile–stationary phase interface. The question is: What happens to the counter ions? These are attracted to the charged interface by Coulombic forces. However, the attraction is counteracted by the thermal motion (entropy), which strives to spread the counter ions uniformly in the mobile phase. The net result is the formation of an electrical double layer in which a balance is achieved between the electrostatic attraction, the entropic effect, and the way the counter ions shield each other (see Figure 15.1). The counter ions, therefore, form an ionic atmosphere, which is in rapid thermal motion close to the surface (i.e., of the order 1–2 nm). Viewed in this way, the arrangement of ions in the interface is similar to a plate capacitor with a very narrow gap between the two charged plates.

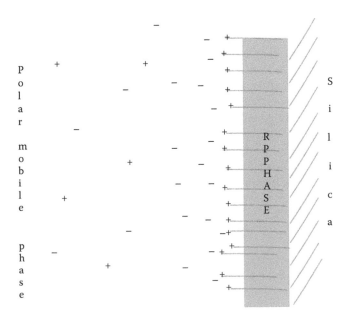

FIGURE 15.1 A schematic picture of the electrical double layer.

For our example, this means that the benzenesulfonate ions are adsorbed in a surface layer and that the sodium ions form an atmosphere of counter ions in which their concentration gradually decreases with the distance from the benzenesulfonate layer. A small separation between the negative and the positive charges has been created at the interface between the mobile and the stationary phases. As in a plate capacitor, this charge separation creates a difference in electrostatic potential between the charged stationary phase and the surrounding mobile phase. By convention, the surface has a negative electrostatic potential in this case.

The next step is to determine the electrical charge and potential distribution in this diffuse region. This is done by using relevant electrostatic and statistical mechanical theories. For a charged planar surface, this problem was solved by Gouy (in 1910) and Chapman (in 1913) by solving the Poisson–Boltzmann equation, the so called Gouy–Chapman (G-C) model.

For low surface concentrations of charged solutes, the G-C model shows that the electrostatic surface potential created by the adsorbed ions is proportional to the surface concentration of the ion according to the following equation [2,3]:

$$\Psi_0 = \frac{F \cdot n}{\kappa \cdot \varepsilon_0 \varepsilon} \ (\text{V}) \tag{15.9}$$

where
F is the Faraday constant
n is the surface concentration of the solute in mol/m^2
κ is defined as

$$\kappa = \left(\frac{F^2 \cdot \sum_i z_i^2 c_i}{\varepsilon_0 \varepsilon \cdot RT} \right)^{1/2} \ (\text{m}^{-1}) \approx 3.288 \cdot \sqrt{I} \ (\text{nm}^{-1}) \tag{15.10}$$

where c_i is the bulk concentration of an ion i (mol/m^3) and the summation is over all ions in the solution. The term $1/\kappa$ has the unit meter and is called the Debye length. It is a measure of the electrical double-layer thickness. It is easily calculated by the approximation in Equation 15.10 where I is the ionic strength of the solution in mol/dm^3 and the result is obtained as nm^{-1} [2]. For example, for a $0.1 \, \text{mol/dm}^3$ 1:1 electrolyte, the Debye length is $0.96 \, \text{nm}$.

In Equation 15.9, the bulk of the electrolyte is the reference point where the electrostatic potential is set to zero. A negative ion is electrostatically repelled from, and a positive ion attracted to, a surface with a negative surface potential, i.e., with negative ions adsorbed at the surface. The change in the adsorption free energy caused by electrostatic repulsion or attraction is the energy required to transfer a charge from the mobile phase, with electrostatic potential zero, to the surface with potential Ψ_0. This energy is known from physics to be the magnitude of the charge that is transferred times the difference in the electrostatic potential:

$$\Delta G_e^0 = zF\Psi_0 \ (\text{J/mol}) \tag{15.11}$$

where ΔG_e^0 is the electrostatic part of the standard free energy of adsorption. The total standard free energy of adsorption is now divided into two parts:

$$\Delta G_t^0 = \Delta G_n^0 + \Delta G_e^0 \tag{15.12}$$

where ΔG_n^0 is the standard free energy of the adsorption of the solute in the limit of zero surface concentration of solute, i.e., in the limit where Ψ_0 is zero.

From the previous discussion, it is clear that when an ionic solute adsorbs to the reversed-phase stationary phase, it will create an electrostatic surface potential that will repel ions with the same charge from the surface and that the magnitude of the repulsion is represented by Equation 15.11. With the help of Equations 15.11 and 15.12, the adsorption isotherm for an ionic solute can be derived in the following way:

Consider the equilibrium of an analyte A between the mobile and the stationary phase:

$$A^z(l) + S \leftrightarrow A^z S \qquad (15.13)$$

where
A^z(l) and A^{z}S represent the analyte in the mobile and on the stationary phase, respectively
S represents the surface

The thermodynamic condition of equilibrium requires that the electrochemical potentials for the right- and left-hand sides of the scheme in Equation 15.13 satisfy the condition [3]:

$$\mu_A + \mu_S = \mu_{AS} \qquad (15.14)$$

For the analyte in the mobile phase, we can write the electrochemical potential as

$$\mu_A = \mu_A^0 + RT \ln c_A \qquad (15.15)$$

To write down the electrochemical potential for the surface, we must consider that the ionic analyte changes the electrostatic surface potential, i.e.,

$$\mu_{AS} = \mu_{AS}^0 + RT \ln X_{AS} + zF\Psi_0 \qquad (15.16)$$

where X_{AS} is the fraction of the surface that is occupied by analyte ions. Similarly, for the unoccupied surface, the electrochemical potential is written as

$$\mu_S = \mu_S^0 + RT \ln X_S \qquad (15.17)$$

Where, it is implicitly assumed that $X_{AS} + X_S = 1$, and we set that

$$X_{AS} = \frac{n_A}{n_0} \qquad (15.18)$$

where n_0 is assumed to be the monolayer capacity of the analyte. Because of the low surface concentration of the analyte, we neglect for the moment a second complication— the entropy of mixing on the surface, i.e., we set $X_S = 1$. Inserting Equations 15.15 through 15.18 into Equation 15.14 and rearranging, we obtain

$$n_A = n_0 \cdot e^{\frac{\mu_{AS}^0 - (\mu_A^0 + \mu_S^0)}{RT}} \cdot e^{-\frac{F\Psi_0}{RT}} \cdot c_A \qquad (15.19)$$

Comparing Equations 15.19 and 15.12 we see that:

$$e^{-\frac{\mu^0_{AS}-(\mu^0_A+\mu^0_s)}{RT}} = e^{-\Delta G_n^{0/RT}} = K_A$$

where K_A represents the adsorption constant of A to the stationary phase surface. By inserting Equation 15.9 we finally obtain the adsorption isotherm for the solute A [4,5]:

$$n_A = n_0 \cdot K_A \cdot e^{-\frac{zF\Psi_0}{RT}} \cdot c_A = n_0 \cdot K_A \cdot e^{-\frac{z^2F^2n_A}{\kappa\varepsilon_0\varepsilon \cdot RT}} \cdot c_A \qquad (15.20)$$

where c_A is the mobile phase concentration of the solute. The last equality in the equation clearly demonstrates that, because of the repulsion, it becomes progressively more difficult to add a solute to the stationary phase surface the higher the surface concentration of the solute is.

The adsorption isotherm of a charged solute to a reversed-phase surface is therefore strongly non-linear. This is an important conclusion because, according to the basic chromatographic theory, a nonlinear adsorption isotherm for a solute results in a skewed peak of that solute. Unfortunately, Equation 15.13 does not have an algebraic solution in the form $n = f(c)$, so more elaborate methods are required to quantitatively evaluate the influence of the nonlinearity on the eluting peak profile. In cases where this has been done, an excellent agreement between the theory and the experiments is found. In Figure 15.2, an example is shown for varying amounts (10.6–106 nmol) of injected p-toluenesulfonate [5]. The increased tailing of the eluting peaks is evident as the amount injected gradually increases. The figure also demonstrates the good agreement between the theoretically predicted peak shape from Equation 15.20 (dotted line in the figure) and that obtained by experiments experimentally, especially for the higher amounts of injected analyte.

The origin of tailing peaks of basic compounds in reversed-phase chromatography has been an issue in the chromatographic literature for many years. From the foregoing discussion, it follows

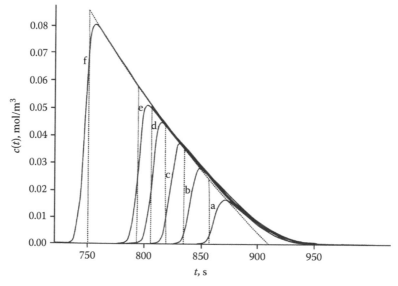

FIGURE 15.2 Experimentally measured (——) and theoretically calculated (·········) eluting peak profiles of different amounts of injected p-toluenesulfonate on a reversed-phase column. The amount injected is (a) 1.06×10^{-8} mol, (b) 2.12×10^{-8} mol, (c) 3.18×10^{-8} mol, (d) 4.24×10^{-8} mol (e) 5.30×10^{-8} mol, and (f) 10.6×10^{-8} mol. Mobile phase composition: sodium phosphate buffer at pH = 3 and ionic strength 25 mol/m³. (Reproduced with permission from Hägglund, I. and Ståhlberg, J., *J. Chromatogr. A*, 761, 3, 1997.)

that tailing is an inherent property for all charged solutes and can therefore not be completely avoided. In an interesting study, McCalley compared the column efficiency and the peak shape of neutral compounds, basic positively charged compounds, and a negatively charged compound [6] (Figure 15.3). It was found that the column efficiency as a function of the loading factor was very similar for the charged compounds and independent of its sign of charge. For neutral compounds, the column efficiency was markedly higher at low- and moderate-loading factors. Furthermore, the observed peak shapes were qualitatively similar for a positively and a negatively charged solute (Figure 15.4).

Since the negatively charged compound, naphthalenesulfonate, is not expected to interact with the solid support by specific interactions, the study by McCalley clearly indicates that, at least in this case, the tailing of the peaks has the electrostatic origin discussed here. Traditionally, the tailing of basic compounds has been attributed to interactions between the base and the solid support, e.g., silanol groups, acidic sites, etc. However, further quantitative studies in this area are needed to separate the electrostatic effect on peak asymmetry from other types of interactions.

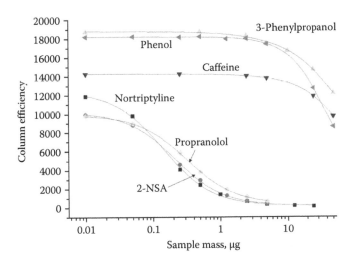

FIGURE 15.3 Plot of column efficiency against a sample mass for three neutral compounds (3-phenylpropanol, caffeine, phenol) and three charged compounds (propranolol, nortriptyline, 2-naphthalenesulfonate) on XTerra MS. For experimental details, see Ref. [6]. (Reproduced with permission from McCalley, D.V., *Anal. Chem.*, 78, 2532, 2006.)

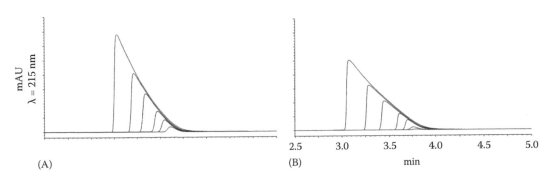

FIGURE 15.4 Overlaid chromatograms on XTerra MS for varying amounts of injected (A) propranolol (cationic) 0.05–5 µg and (B) 2-naphthalenesulfonate 0.05–5 µg. (Reproduced with permission from McCalley, D.V., *Anal. Chem.*, 78, 2532, 2006.)

15.3   ION-PAIR CHROMATOGRAPHY

Reversed-phase ion-pair chromatography is primarily used for the separation of mixtures of ionic and ionizable compounds. In this chromatographic mode, a pairing ion is added to the mobile phase in order to modulate the retention of the ionic solutes. The pairing ion is an organic ion such as alkylsulfonate, alkylsulfate, alkylamine, tetraalkylammonium ion, etc. Here, only a very brief description of the main ideas behind the electrostatic model for ion-pair chromatography is presented. For a complete discussion, the reader is referred to Ref. [7,8] and the references therein.

In the previous section, the physical mechanism behind the formation of the electrical double layer and the electrostatic surface potential is presented. From these discussions follow that the pairing ion adsorbs to the non-polar stationary phase so that the non-polar organic moiety interacts with the non-polar retentive layer and that the charged part remains in the polar mobile phase. In analogy with the arguments above, the pairing ion creates a charged layer on the stationary phase surface. Again, a difference in the electrostatic potential between the bulk of the mobile phase and the surface is created. Let us, for example, assume that sodium octylsulfonate is used as a pairing ion in a sodium phosphate buffered mobile phase. The reversed-phase column is equilibrated with the mobile phase, and during this process, the octylsulfonate ions are adsorbed at the stationary phase surface throughout the column.

The effect of the adsorbed octylsulfonate on the retention of a solute ion is straightforward. Positively charged solutes are attracted to the charged stationary phase and their retention increases. Negatively charged solutes are repelled from the surface and their retention decreases. The principle is the same irrespective of the sign of charge of the pairing ion, i.e., a positively charged pairing ion attracts negative solutes and repels positive solutes.

Basic chromatographic theory relates the retention factor of a solute to the equilibrium constant for adsorption, K_s, according to

$$k_s = \phi \cdot K_s = \phi \cdot e^{-\frac{\Delta G_{t,s}^0}{RT}} \tag{15.21}$$

where
$\Delta G_{t,s}^0$ is the standard free energy of adsorption of the solute to the stationary phase surface
ϕ is the column phase ratio

By using the previously introduced concepts, it is possible to develop a quantitative model for ion-pair chromatography. Thus, the standard free energy of adsorption of the solute is again divided into two contributions, the standard free energy in the limit of zero electrostatic surface potential, $\Delta G_{n,s}^0$, and the contribution from the electrostatic interaction with the charged surface, $\Delta G_{e,s}^0$ (see Equation 15.12). When no pairing ion is present in the mobile phase, the retention factor, k_{0s}, is determined by the magnitude of $\Delta G_{n,s}^0$ only, and we obtain the following relation for the retention factor in the presence of the pairing ion:

$$k_s = \phi \cdot K_s = \phi \cdot e^{-\frac{\Delta G_{t,s}^0}{RT}} = \phi \cdot e^{-\frac{\Delta G_{n,s}^0 + \Delta G_{te,s}^0}{RT}} = \phi \cdot e^{-\frac{\Delta G_{n,s}}{RT}} \cdot e^{-\frac{z_s F \Psi_0}{RT}} = k_{0,s} \cdot e^{-\frac{z_s F \Psi_0}{RT}} \tag{15.22}$$

where z_s is the charge of the solute. From this equation, it is clear that when the pairing ion and the solute are of opposite charge, the retention increases and vice versa. E.g., the combination of a negative pairing ion, i.e., negative Ψ_0, and a monocharged positive solute, $z_s = +1$, increases the retention of the solute. It is also interesting to note that the addition of a pairing ion can be used to decrease the retention of solutes that have the same sign of charge as the pairing ion.

Equation 15.22 shows that it is the electrostatic surface potential created by the adsorbed pairing ion that is the driving force behind the regulation of retention. The electrostatic surface potential

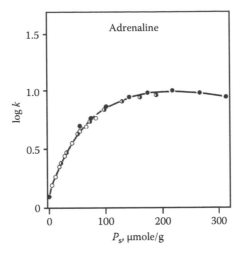

FIGURE 15.5 Retention factor for adrenaline as a function of stationary phase concentration n_A of (○) butyl- (◑) hexyl- and (●) octylsulfonate pairing ion at a constant ionic strength (Na⁺ = 175 mM). For experimental details, see Ref. [9]. (Reproduced with permission from Bartha, Á. et al., *J. Chromatogr.*, 303, 29, 1984.)

in its turn is determined by the surface concentration of the pairing ion according to e.g., Equation 15.9. This implies that the retention does not depend on the type of pairing ion, only on its surface concentration. An example is shown in Figure 15.5, where experimental data from Ref. [9] for the retention factor of adrenaline is plotted as a function of the surface concentration of the pairing ion on the stationary phase for three different pairing ions.

In Equation 15.9, a linear relation between Ψ_0 and the surface concentration of an ion, n, was used. This relation only holds for low Ψ_0 values ($\Psi_0 < 20\,mV$) and can safely be used for describing the retention of a solute in analytical applications. However, the pairing ion concentration on the surface will often be so high that Ψ_0 exceeds this value, and a more general relation must be used. For a planar surface, the G-C theory gives a more general relation between Ψ_0 and the surface concentration of a pairing ion, n_P:

$$\Psi_0 = \frac{2RT}{F} \ln \left(\frac{n_P z_P F}{\left(8\varepsilon_0 \varepsilon RT \sum_i c_i \right)^{1/2}} + \left(\frac{(n_P z_P F)^2}{8\varepsilon_0 \varepsilon RT \sum_i c_i} + 1 \right)^{1/2} \right) \quad (15.23)$$

Since chromatographic surfaces are not planar, this equation cannot be used to quantitatively describe the relation between n_P and Ψ_0 in chromatography. However, the relation tells us that the function is convex and qualitatively has the form found experimentally in Figure 15.5. Because of the highly irregular geometry of a reversed-phase surface, the exact form of the theoretical relation between Ψ_0 and n_P is almost impossible to obtain.

In practice, the experimenter controls the mobile phase concentration of the pairing ion and not its surface concentration. A practical and useful theory for ion-pair chromatography must therefore include a relation for the adsorption isotherm of the pairing ion. For low to intermediately high surface concentration of the pairing ion, the discussion leading to Equation 15.20 applies and we obtain the following expression for the adsorption isotherm of the pairing ion:

$$n_P = n_{0,P} \cdot \exp \left(-\frac{\Delta G_{t,P}^0}{RT} \right) c_P = n_{0,P} \cdot \exp \left(-\frac{\Delta G_{n,P}^0 + \Delta G_e^0}{RT} \right) c_P = n_{0,P} \cdot K_P \cdot e^{-\frac{z_P F \Psi_0 (n_P)}{RT}} \cdot c_P \quad (15.24)$$

where

n_P is the surface concentration of the pairing ion
c_P is its mobile phase concentration
K_P is its adsorption constant
n_{0P} is the monolayer capacity of the stationary phase surface for the pairing ion

In Equation 15.24, it is important to note that Ψ_0 is a function of n_P and the mobile phase ionic strength.

Up to now, we have only considered the ideal case of adsorption of the pairing ion, i.e., when the change in interaction with the surface is caused by varying the electrostatic surface potential only. This will be the case when the other ions in the mobile phase do not adsorb to the surface or interact with the pairing ion by specific interactions. There are good reasons to believe that e.g., Na^+ and $H_2PO_4^-$ are ions that in many cases will behave almost ideally. If $H_2PO_4^-$ is substituted by e.g., Br^- or another counter ion that may adsorb to the stationary phase and/or interact specifically with the pairing ion, the model above may easily be extended to a more general case.

Consider the adsorption of a pairing ion P^+ and its counterion ion C^- to the stationary phase surface. In analogy with Equation 15.13, we consider the following equilibria:

$$P^+(l) + S \leftrightarrow P^+S \tag{15.25}$$

$$C^-(l) + S \leftrightarrow C^-S \tag{15.26}$$

Let us also include the possibility of a specific interaction between the pairing ion and its counter ion, i.e., [4]

$$P^+(l) + C^-(l) + S \leftrightarrow PCS \tag{15.27}$$

In analogy with the previous discussion, the adsorption isotherms for both, the pairing ion and the counter ion are obtained from the equilibrium condition for the corresponding electrochemical potentials according to the scheme:

$$\mu_P + \mu_S = \mu_{PS} \quad \mu_C + \mu_S = \mu_{CS} \quad \mu_P + \mu_C + \mu_S = \mu_{PCS} \tag{15.28}$$

The electrochemical potentials for P and C in the liquid phase are (see Equation 15.15)

$$\mu_P = \mu_P^0 + RT \ln c_P \quad \mu_C = \mu_C^0 + RT \ln c_C \tag{15.29}$$

From Equations 15.16 and 15.17, the corresponding potentials for the adsorbed P, C, and S are obtained as

$$\mu_{PS} = \mu_{PS}^0 + RT \ln X_{PS} + F\Psi_0 \quad \mu_{CS} = \mu_{CS}^0 + RT \ln X_{CS} - F\Psi_0 \quad \mu_S = \mu_S^0 + RT \ln X_S \tag{15.30}$$

and for the neutral species PCS, the potential is

$$\mu_{PCS} = \mu_{PCS}^0 + RT \ln X_{PCS} \tag{15.31}$$

Furthermore, we assume that

$$X_S + X_{PS} + X_{CS} + X_{PCS} = 1 \tag{15.32}$$

where $X_i = n_i/n_0$ and n_i is the surface concentration of species i, and n_0 the monolayer capacity of all species. The adsorption isotherm for P^+ and C^-, respectively, is obtained by inserting Equations 15.29 through 15.31 into Equation 15.28, in combination with the condition in Equation 15.32. After some algebra, the following expressions are found:

$$n_P = \frac{n_0 \cdot \left(K_P \cdot e^{-\frac{F\Psi_0}{RT}} + K_{PC} \cdot c_C \right) \cdot c_P}{1 + K_P \cdot e^{-\frac{F\Psi_0}{RT}} \cdot c_P + K_C \cdot e^{-\frac{F\Psi_0}{RT}} \cdot c_C + K_{PC} \cdot c_C c_P} \tag{15.33}$$

and similarly for C

$$n_C = \frac{n_0 \cdot \left(K_C \cdot e^{-\frac{F\Psi_0}{RT}} + K_{PC} \cdot c_P \right) \cdot c_C}{1 + K_P \cdot e^{-\frac{F\Psi_0}{RT}} \cdot c_P + K_C \cdot e^{-\frac{F\Psi_0}{RT}} \cdot c_C + K_{PC} \cdot c_C c_P} \tag{15.34}$$

where K_P, K_C, and K_{PC} is the adsorption constant of P^+, C^-, and PC, respectively, to the stationary phase surface. It is important to note that the value for Ψ_0, in this case, is determined by the net charge on the surface, i.e., by the excess surface concentration of n_P, n_{P+}, which is determined by

$$n_{P+} = n_P - n_C - n_{PC} \tag{15.35}$$

Because of the highly irregular and ill-defined geometry of a reversed-phase surface, it is almost impossible to theoretically obtain the exact form of the relation between Ψ_0 and n_{P+} from the P–B equation. However, it is possible to obtain this relation from experimental retention data for charged solutes in combination with the adsorption data in analogy with the data in Figure 15.5. This can be done by using charged solutes as probes and apply Equation 15.22 to a set of retention data. For this purpose, it is advisable to use solutes of a different sign of charge. When e.g., a negatively and a positively charged solute give consistent values for Ψ_0 as a function of pairing ion concentration, it is highly probable that it is the surface potential that regulates the retention. Such data are usually consistent in polar solvents at low to immediate Ψ_0 values (0–40 mV) [10], but deviations may occur at higher values. Depending on the experimental conditions, the extension of the simple electrostatic ion-pair model, represented by Equation 15.22, can be performed by using the concepts presented in Equations 15.25 through 15.35.

A simple extension of the retention model discussed so far is to include the possibility for a specific interaction between the analyte and the pairing ion. By using a similar reasoning as for the electrochemical potential shown above, it is easily shown that the retention factor follows the relation

$$k_A = \frac{\phi \cdot n_0 \cdot \left(K_A \cdot e^{-\frac{z_A F\Psi_0}{RT}} + K_{AP} c_P \right)}{1 + K_P \cdot e^{-\frac{z_P F\Psi_0}{RT}} \cdot c_P} \tag{15.36}$$

where K_A and K_{AP} is the adsorption constant of the analyte and the AP ion pair to the stationary phase, respectively. Here, the denominator, as in previous equations, represents the steric effect caused by the limited surface area. For low surface concentrations, this effect is negligible and the denominator is unity. In this case, the retention factor can be expressed as

$$k_A = k_{0A} \cdot e^{-\frac{z_A F \Psi_0}{RT}} + k_{AP} \cdot c_P \tag{15.37}$$

where
  k_{0A} represents the retention factor in the absence of the pairing ion
  k_{AP} is the retention factor caused by the specific interaction between A and P

To evaluate this model, it is again necessary to determine Ψ_0 as a function of the pairing ion concentration in the mobile phase. However, because of the retention due to the specific interaction in this case, it is not possible to use the retention data for ions of charge opposite to the pairing ion to estimate Ψ_0. Instead, it would be possible to estimate Ψ_0 values from the decrease in retention for analytes of a charge similar to the pairing ion and from these data calculate the contribution of k_{AP} to retention. To the author's knowledge, no work along these lines has been performed.

In ion-pair chromatography, a great number of parameters influence the retention of a charged solute e.g., the type of solute, the type and the concentration of the pairing ion, the type and the concentration of the buffer, the mobile phase composition, etc. This makes ion-pair chromatography a versatile technique at the same time as it appears to be complicated and difficult to control. From the discussion above, it is clear that a few simple basic principles often can be used to understand the retention behavior when the experimental conditions are varied. In practical work, it may be desirable to make predictions of retentions from a limited set of retention data and without going into the more complicated theoretical models. For this purpose, an approximate equation was derived that considers most of the parameters in a simple and practically useful way. For the derivation of this simple version of the model and for a guide to its use and applicability, we refer to Ref. [8]. Here, we will only state the final equation and show one simple example of its use.

In practice, it is desirable to estimate what the effect on retention is when different experimental parameters vary. The parameters considered here are changes in (1) pairing ion charge and type, (2) pairing ion concentration, (3) organic modifier concentration, (4) analyte charge and type, and (5) mobile phase ionic strength. The following approximate equation can be derived from the more rigorous model:

$$\ln k_{cA} = \ln k_{0A} + \frac{z_A z_P}{1 + z_P^2} \cdot \ln\left(\left(\frac{n_0 \cdot K_P \cdot c_P}{\kappa}\right) + K\right) \tag{15.38}$$

where
  k_{cA} is the retention factor of the analyte in the presence of the concentration c_P of the pairing ion
  k_{0A} is its retention factor when $c_P = 0$
  z_A and z_P is the charge of the analyte and the pairing ion, respectively
  K_P is the adsorption constant for the paring ion to the stationary phase
  K is a constant
  κ is the inverse Debye length (defined in Equation 15.10)

Note that κ is proportional to the square root of the mobile phase ionic strength.

Because of the approximations made in the derivation of Equation 15.38, it cannot be used at zero or low concentrations of pairing ion (i.e., Ψ_0 values below 5 mV). In the example shown here, the equation is used to predict the retention as a function of pairing-ion concentration and organic modifier concentration in the mobile phase. In the equation, there are two constants, k_{0A} and K_P, that will change when the concentration of organic modifier changes.

From reversed-phase chromatography of uncharged analytes it is known that the retention factor as a function of the organic modifier concentration often follows the relation:

$$\ln k_{\phi A} = \ln k_{\phi=0,A} - S_A \cdot \phi \tag{15.39}$$

where

$k_{\phi A}$ is the retention factor when the organic modifier concentration is ϕ
$k_{\phi=0A}$ is the retention factor when $\phi = 0$
S_A is a constant that is characteristic of the analyte

This relation also holds well for the k_{0A} value of ionic analytes and also for K_P, each with its characteristic constant S_P. So, to make predictions according to Equation 15.38, the S value of the analyte and the pairing ion, respectively, must be determined in two chromatographic experiments with two different organic modifier concentrations in the mobile phase (note that the pairing ion must be present in the mobile phase in these experiments). From these two experiments, the two S values are determined and used in combination with Equation 15.38 to predict the combined effect of the modifier and the pairing-ion concentration on retention. In Figures 15.6 and 15.7, two examples are shown where octylsulfonate was used as the pairing ion and the analyte is positively and negatively charged, respectively. In the figures, the two filled points are used to calculate the S_A and S_P values, the full lines are the predicted retention factors and the open symbols are the experimentally measured values. It is seen that the predictions are reasonably good.

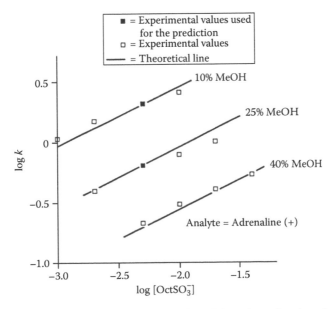

FIGURE 15.6 Retention factor for adrenaline as a function of the volume fraction of methanol (10%, 25%, and 40%) and the pairing ion (octylsulfonate) concentration. The theoretical line is obtained by combining Equations 15.38 and 15.39. Experimental data from Ref. [11]. (Reproduced with permission from Bartha, Á. et al., *J. Chromatogr.*, 506, 85, 1990.)

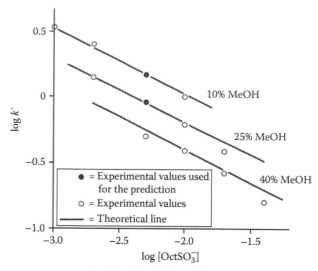

Analyte = Naphthalene sulfonate (−)

FIGURE 15.7 Retention factor for naphthalenesulfonate as a function of the volume fraction of methanol in the mobile phase (10%, 25%, and 40%) and the pairing ion (octylsulfonate) concentration. Mobile phase: phosphate buffer at pH = 2.1 and constant ionic strength (175 mM Na+). The theoretical line is obtained by combining Equations 15.38 and 15.39. Experimental data from Ref. [11]. (Reproduced with permission from Bartha, Á. et al., *J. Chromatogr.*, 506, 85, 1990.)

15.4 SUMMARY AND CONCLUDING REMARKS CONCERNING THE RETENTION OF IONIC SOLUTES IN REVERSED-PHASE CHROMATOGRAPHY

The model for ionic retention and ion-pair chromatography that are discussed in Sections 15.2 and 15.3 has been tested and applied to a number of different systems and works very well in most of the cases. From colloid and surface chemistry is known that the model has its limitations, and under certain chromatographic conditions, the presented model will not be valid. The limitations of the model when applied to reversed-phase chromatography of ions still need to be found. Some are self-evident, such as if the pairing-ion concentration is close or above the CMC or when the retention factor is very low so that the accumulation in the double layer is important in comparison to the adsorption, see Ref. [7] for a discussion concerning the accumulation in the double layer.

Other deviations and limitations are expected from the use of the Poisson–Boltzmann approach, two of which are

1. When ions of valence 2 or higher are used as an electrolyte in the mobile phase and if this ion is a counter ion to the analyte or pairing ion in ion-pair chromatography. E.g., if tetrabutylammonium ion is used as a pairing ion at pH = 9 in the presence of PO_4^{3-} in the electrolyte, it will probably not behave as ideally as at pH = 2 where phosphate will be in the $H_2PO_4^-$ form.
2. When the electrostatic surface potential becomes high or when the monolayer capacity of the surface is approached. So far the models agree with experiments up to 60–70 mV but the upper limit is probably below 100 mV.

More sophisticated models than the G-C model, also based on electrostatic concepts, have been used in both colloid and surface chemistry as well as in chromatography [12]. They divide the interface between the surface and the solution into different regions with different properties. These models can be applied to the interface between a molecularly smooth surface and an electrolyte solution, e.g., between liquid mercury and a solution. However, due to the irregularity and the softness of a chromatographic reversed phase–mobile phase interface, the use of such models is not physically motivated for the description of the reversed-phase chromatography of ions.

15.5 ION EXCHANGE CHROMATOGRAPHY OF SMALL IONS

In ion-pair chromatography, a charged surface is created by the adsorption of a charged pairing ion additive to the stationary phase surface. The most important parameters for regulating retention in ion-pair chromatography are the pairing-ion type and the concentration, and the organic modifier type and content. In ion exchange chromatography on the other hand, the charged groups are chemically bound to the stationary phase surface. Here, the most important retention-regulating parameter is the type and the concentration of the electrolyte in the mobile phase. The main difference between the two techniques is, from a theoretical point of view, that the concentration of the surface charges in ion exchange chromatography, n_S, is fixed. Another important difference is that the chromatographic conditions are usually such that the ratio between the counterion concentration in the double layer to its concentration in the eluent is usually much higher than its corresponding parameter in ion-pair chromatography. Altogether, this implies that we are in a region where there is a strong accumulation of counterions at the surface and a somewhat different and more general theoretical approach then used for ion-pair chromatography can be applied.

In this section the following nomenclature is used. The charges that are bound to the resin or the stationary phase are called the resin charges and are the reference point for the other ions. Ions in the electrolyte solutions with a sign of charge that is opposite to the resin charges are called counterions. Analogously, the charges that have the same sign of charge are called coions. Traditionally, the retention in ion exchange chromatography as a function of mobile phase electrolyte concentration is presented as [13–15]

$$z_B A_E + z_A B_R \leftrightarrow z_A B_E + z_B A_R \tag{15.40}$$

where
A is the analyte
B is the electrolyte counterion to the fixed resin charges
z_A and z_B is the charge of A and B, respectively
the subscripts "E" and "R" refer to the eluent and the resin phase, respectively

The equilibrium in Equation 15.40 leads to the following equation for the chromatographic selectivity coefficient [11,12]:

$$K_{AB} = \frac{[A]_R^{z_B} [B]_E^{z_A}}{[A]_E^{z_B} [B]_R^{z_A}} \tag{15.41}$$

where $[A]_R$ and $[A]_E$ denote the concentration of A in the resin phase and the eluent phase, respectively, and an analogous notation is used for ion B. Although Equation 15.41 is obtained from a pure stoichiometric approach from the equilibrium (Equation 15.40), it can also be derived from the Donnan model after some simplifying assumptions about the activity of ions in the resin phase. From Equation 15.41, a relation between the retention factor of A as a function of the concentration of B in the eluent phase is easily derived:

$$\log k_A = K - \frac{z_A}{z_B} \log c_{BE} \tag{15.42}$$

where

c_{BE} is the concentration of B in the eluent phase
K is called the selectivity constant

Historically, the introduction of the Donnan model was important. For a number of reasons, the model does not provide an adequate and consistent view of the properties of electrolyte solutions in contact with a charged surface. The main problem is that it assumes a constant electrostatic potential in the resin phase, an assumption which leads to inconsistencies in cases where the electrostatic potential varies within the system [16]. However, it is interesting that Equation 15.42 is found to hold well under many experimental conditions for monocharged and dicharged analytes. It is therefore important to find an alternative model, based on a more firm physical ground, and that at the same time agrees with the experimental results. It is the purpose of this section to present this. To the knowledge of the author, the model presented here has not previously been used to describe retention in ion exchange chromatography.

The main problem when developing a theory for ion exchange chromatography is that the geometry of the system is irregular. This implies that any model must be based either on simplifying the geometry or on theories that are not sensitive to the geometry. Several authors have suggested electrostatic models based on a planar surface i.e., the Stern–Gouy–Chapman theory or the Gouy–Chapman theory [17–22]. Although these theories are useful, they rely on assumptions concerning both the geometry as well as the properties of the ions, e.g., that they are point charges. A model is presented here, which is based on a theory that is more or less insensitive to the geometry and makes a minimum of assumptions regarding ionic properties. In the presentation, the G-C theory is used as a reference model to illustrate general ideas. It is therefore necessary to briefly examine the G-C theory in more detail.

15.5.1 Gouy–Chapman Theory

The G-C theory originates from a solution of the Poisson–Boltzmann equation for a charged planar surface in contact with a z:z electrolyte solution (i.e., the valence of the two ions is the same but of opposite charge). The electrolyte ions are assumed to be point charges and the surface charges are assumed to be evenly smeared out to give a homogenous charge density. For a derivation of the G-C equation, the reader is referred to standard textbooks in colloid and surface chemistry [3]. The resulting equations give a relation between the surface charge density, the electrolyte concentration, and the electrostatic surface potential, Equation 15.23, which at low surface potentials is approximated by Equation 15.9. Furthermore, it also gives a relation for the variation of the electrostatic potential with the distance x from the surface. It is found that

$$\Psi(x) = \frac{2RT}{zF} \ln\left(\frac{1 + \Gamma_0 \cdot e^{-\kappa x}}{1 - \Gamma_0 e^{-\kappa x}} \right) \tag{15.43}$$

where

$$\Gamma_0 = \frac{e^{\frac{zF\Psi_0}{2RT}} - 1}{e^{\frac{zF\Psi_0}{2RT}} + 1} \tag{15.44}$$

where z is the absolute value of the valence of the electrolyte ion, i.e., +1, +2, etc. The concentration of an ion in the double layer follows the Boltzmann equation:

$$c(x) = c_E \cdot e^{-\frac{zF\Psi(x)}{RT}} \tag{15.45}$$

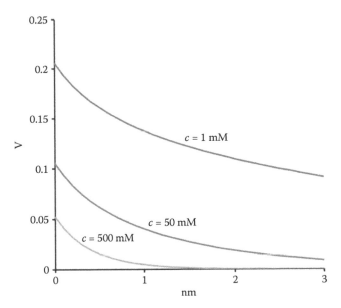

FIGURE 15.8 Electrostatic potential as a function of the distance from a surface according to the G-C theory for three different bulk concentrations of a 1:1 electrolyte 1, 50, and 500 mM, respectively. The surface charge density is kept constant at −0.1 C/m².

Equations 15.43 through 15.45 in combination with Equation 15.23 constitute the result of the G-C theory.

In Figure 15.8, $\Psi(x)$ is shown for a surface with the constant charge density −0.1 C/m² (i.e., approximately 1 μmol of negatively charged ions per m²) in contact with a 1:1 electrolyte solution with bulk concentrations 1, 50, and 500 mM, respectively. We see that as the electrolyte concentration increases, the Ψ_0 value decreases and the potential also decreases more rapidly with the distance from the surface. The interesting question is how the counter-ion concentration varies with x. Applying Equation 15.45 to the potential profiles in Figure 15.8 results in the concentration profiles for the counter ion shown in Figure 15.9.

Of special interest is that the concentration at the surface, $x=0$, is almost independent of the concentration in the bulk electrolyte. When the concentration increases by a factor of 500, from 1 to 500 mM, the concentration at the surface only increases by a factor of 1.3. This remarkable behavior is the result of the contact value theorem, which for the G-C model takes the form [3]:

$$\sum_i c_i(0) = \frac{F^2 n_P^2}{2\varepsilon_0 \varepsilon RT} + \sum_i c_{iE} \tag{15.46}$$

where
 c_{iE} is the bulk concentration of each ion i (in mol/m³ units)
 $c_i(0)$ is the concentration of ion i at the surface
 n_P is the surface charge density (in mol/m²)

By inspecting Figure 15.9 again, the equation is easily verified. For the surface charge density −0.1 C/m², the first term on the r.h.s. is around 2800 for water at room temperature and the surface concentration becomes that value plus the total ion concentration in the bulk water phase, i.e., $500+500$ mol/m³, for the 500 mM solution.

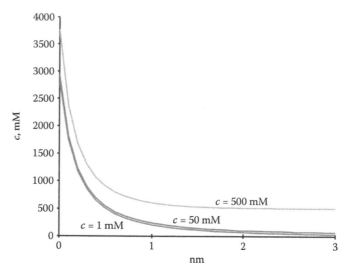

FIGURE 15.9 The counterion concentration as a function of the distance from a surface according to the G-C theory for three different bulk concentrations of a 1:1 electrolyte 1, 50, and 500 mM, respectively. The surface charge density is kept constant to −0.1 C/m².

15.5.2 Model for Ion Exchange Chromatography

There are two interesting and useful properties of Equation 15.46 that will be used here in order to develop a model for ion exchange chromatography. First, the validity of the equation is more or less general and independent of the model used to derive it. Its origin is from the condition of mechanical equilibrium at the surface where electrostatic Maxwell stress is balanced by osmotic forces. This condition must be met for all models and it is therefore used to test statistical mechanical models for the structure of ions close to a charged surface [23,24]. Second, the equation predicts that when the charge density is sufficiently high, the counterion concentration within a layer a few Å thick is so high that most of the surface charges are neutralized. Furthermore, the concentration of counter ions in the layer is almost constant and independent of its bulk concentration. In the following, this layer is called the contact layer.

To illustrate the principles, we simplify the ensuing discussion by considering an electrolyte consisting of two ions only, where B^{z_B} is the counterion to the fixed charges and C^{z_C} the coion. Keeping in mind that Equation 15.46 is of more general validity than the G-C model, it is now written

$$\sum_i c_i(0) - \sum_i c_{iE} = c_{BI} + c_{CI} - \left(c_{BE} + \frac{|z_B|}{|z_C|} c_{CE} \right) = \frac{F^2 n_P^2}{2\varepsilon_0 \varepsilon RT} \qquad (15.47)$$

In Equation 15.47, the contact layer is positioned at the surface, i.e., $x=0$, i.e., $c_i(0)=c_{i1}$. We now make the following assumptions: first, the electrostatic repulsion is so strong that the coion concentration in the contact layer is zero, i.e., $c_{CI}=0$, second, the accumulation of counterion is so high that c_{BI} is much higher than c_{BE} and Equation 15.47 is approximated to

$$c_{BI} = \frac{F^2 n_P^2}{2\varepsilon_0 \varepsilon RT} \qquad (15.48)$$

The ions in the contact layer are close enough to the surface that specific adsorption to the surface is possible. In analogy to Equation 15.20, this is described by

$$n_B = K_B \cdot c_{Bl} = K_B \cdot e^{-\frac{z_B F \Psi_l}{RT}} \cdot c_{BE} \tag{15.49}$$

where
n_B is the surface concentration of B
K_B its adsorption constant
Ψ_l the electrostatic potential of the contact layer

The adsorption of B changes the concentration of surface charges, n_P, according to

$$n_P = n_S - |z_B| n_B \tag{15.50}$$

where n_S is the concentration of the bound charges on the surface, i.e., n_P is the effective surface charge concentration.

Combining Equations 15.48 through 15.50 and solving the second-degree algebraic equation, Equation 15.51 is obtained

$$n_P = \frac{-1 \pm \sqrt{1 + 4|z_B| K_s K_B n_S}}{2|z_B| K_s K_B} \tag{15.51}$$

where

$$K_s = \frac{F^2}{2\varepsilon_0 \varepsilon RT} \tag{15.52}$$

is introduced to simplify the notation. The important conclusion from Equation 15.51 is that when n_P is so high that Equation 15.48 is valid, n_P is independent of the concentration of B in the eluent, c_{BE}.

In the discussion above, it is assumed that the concentration of c_{Bl} is determined by the electrostatic potential of the contact layer, Ψ_l, according to

$$c_{Bl} = c_{BE} \cdot e^{-\frac{z_B F \Psi_l}{RT}} \tag{15.53}$$

In the ideal case $\Psi_l = \Psi_0$, i.e., the surface potential obtained from the G-C theory. At first sight Equations 15.48 and 15.53 may seem to be contradictory. However, by inserting Equation 15.23 into Equation 15.53 under the condition that

$$\frac{(n_P z_P F)^2}{8\varepsilon_0 \varepsilon RT \sum_i c_i} \gg 1$$

which is identical to the condition used to formulate Equation 15.43, and using that $\sum_i c_i = c_{BE}$, Equation 15.53 is easily verified.

So far we have focused on the accumulation and the adsorption of the electrolyte counterion to the charged surface. Next, we will derive the retention factor of the analyte ion A^{z_A} as a function of the bulk concentration of B^{z_B}. We assume here that the concentration of A^{z_A}, c_A, is so small in comparison to c_B, that it does not influence the distribution B in the system. This condition is usually met in analytical applications of ion exchange chromatography.

The distribution of the analyte ion between the bulk solution and the contact layer is expressed by an equation analogous to Equation 15.53:

$$c_{AI} = c_{AE} \cdot e^{-\frac{z_A F \Psi_l}{RT}} \tag{15.54}$$

where c_{AI} and c_{AE} is the concentration of A^{z_A} in the contact layer and in the eluent, respectively. The analyte may also adsorb to the stationary phase surface, and in analogy with the treatment for the counterion, we assume that the adsorption is represented by

$$n_A = K_A \cdot c_{AI} \tag{15.55}$$

We now assume that the retention factor for A is represented by the ratio:

$$k_A = \frac{\text{Amount A in the contact layer} + \text{Amount A adsorbed to the surface}}{\text{Amount A in the eluent}}$$

which becomes

$$k_A = \frac{V_1}{V_E} \cdot \frac{c_{AI}}{c_{AE}} + \frac{A_s}{V_E} \cdot \frac{n_A}{c_{AE}} = \Phi_1 \frac{c_{AI}}{c_{AE}} + \Phi_s \frac{n_A}{c_{AE}} \tag{15.56}$$

where
  V_1, V_E, A_s are the hypothetical volume of the contact layer, the eluent volume, and surface area of the stationary phase, respectively,
  Φ_i is the corresponding phase ratios

Inserting Equations 15.54 and 15.55 into Equation 15.56 gives

$$k_A = \Phi_1 \frac{c_{A0} \cdot e^{-\frac{z_A F \Psi_l}{RT}}}{c_{A0}} + \Phi_s \frac{K_A c_{A0} \cdot e^{-\frac{z_A F \Psi_l}{RT}}}{c_{A0}} = e^{-\frac{z_A F \Psi_l}{RT}} (\Phi_1 + \Phi_s \cdot K_A) \tag{15.57}$$

From Equation 15.53 we find that

$$e^{-\frac{z_A F \Psi_l}{RT}} = \left(e^{-\frac{z_B F \Psi_l}{RT}} \right)^{\frac{z_A}{z_B}} = \left(\frac{c_{BI}}{c_{BE}} \right)^{\frac{z_A}{z_B}} = \left(\frac{K_s n_P^2}{c_{BE}} \right)^{\frac{z_A}{z_B}} \tag{15.58}$$

where $K_s n_P^2$ is a system constant according to Equation 15.51. Inserting Equation 15.58 into Equation 15.57 we finally obtain the retention factor of the analyte A^{z_A} as a function of the eluent concentration of the counter ion, c_{BE}:

$$k_A = \left(\frac{K_s n_P^2}{c_{BE}} \right)^{\frac{z_A}{z_B}} (\Phi_1 + \Phi_s \cdot K_A) = \frac{K_{IEX}}{(c_{BE})^{\frac{z_A}{z_B}}} \tag{15.59}$$

which is reformulated to the more frequently used form

$$\log k_A = K - \frac{z_A}{z_B} \log c_{BE} \qquad (15.60)$$

Consequently, we have derived an equation which is of the same form as the experimentally found relation in Equation 15.42.

15.5.3 Summary and Concluding Remarks Concerning Ion Exchange Chromatography of Small Ions

Here, a model for the retention of small ions in ion exchange chromatography is presented, called the contact layer model. The assumptions behind the model are based on the contact value theorem that is insensitive to both the geometry and to the specific ionic models. The critical assumption is most probably the approximation made when approximating Equation 15.47 by Equation 15.48, an approximation that seems to be satisfied in most experiments. It is evident that when the approximation is not valid, the retention of A will still decrease with an increasing c_{BE}, but because of the complex geometry, its mathematical form is difficult to model. Furthermore, under these circumstances, retention due to accumulation in the double layer becomes gradually more important, a contribution that is neglected here for distances beyond the contact layer.

The picture that emerges from this presentation is as follows: There is a layer close to the surface in which the eluent counter ions are strongly accumulated and are possibly also adsorbed to the surface by specific interactions. The analyte ions are distributed into this layer in proportion to their concentration in the eluent according to the relation

$$\frac{c_{AI}}{c_{AE}} = e^{-\frac{z_A F \Psi_I}{RT}} = \left(e^{-\frac{z_B F \Psi_I}{RT}} \right)^{\frac{z_A}{z_B}} = \left(\frac{c_{BI}}{c_{BE}} \right)^{\frac{z_A}{z_B}} \qquad (15.61)$$

that is easily obtained from Equations 15.53 and 15.54. This relation is of the same mathematical form as a stoichiometric exchange relation that has traditionally been used for many years:

$$\frac{c_{AR}}{c_{AE}} = K_E \left(\frac{C_R}{c_{BE}} \right)^{\frac{z_A}{z_B}} \qquad (15.62)$$

The arguments leading to Equation 15.62 can therefore be used also in the contact layer model with the important difference that it is c_{BI} that is constant because of the trace concentration of A, and not the resin charge concentration, C_R. In the contact layer model, the selectivity constant K_E is substituted by adsorption to the surface from the contact layer characterized by the constant K_A, see e.g., Equation 15.57.

15.6 ION EXCHANGE CHROMATOGRAPHY OF PROTEINS

Ion exchange chromatography is widely used for both analytical separation and preparative purification of proteins. The popularity of the technique is due to the easy methodology and to the preservation of biological activity of the protein after elution. The mobile phase is a buffer solution and the stationary phase usually consists of macroporous silica or polystyrene-divinylbenzene to which charged groups are chemically bound. The retention of the protein is modulated by varying the ionic strength of the mobile phase, usually by using gradient chromatography.

All proteins are polymers made from 20 different α-amino acids, each characterized by its side chain. Some of the amino acids have a basic and some have an acidic side chain. The protein therefore carries a net charge and its sign and magnitude depends on the pH value of the solution in which it is dissolved. At a certain pH value, the protein net charge is zero and this is the isoelectric point of that particular protein. The amino acid chain that constitutes the protein is folded in a certain geometrical form, the ternary structure. Here, we shall primarily discuss models for ion exchange chromatography of globular proteins in which the amino acid chain is organized into a compact structure. The globular proteins are an enormous and important class of proteins that perform most of the chemical work in a cell. Their size varies with the molecular weight and their radii are in the order of 2–5 nm, i.e., a factor of 10 or more bigger than the small ions in the surrounding electrolyte solution. Since the protein cannot be considered to be a point charge, another approach then previously used is needed to model the interaction between the charged surface and the charged protein. In Ref. [25], a general review concerning adsorption of proteins to surfaces is found.

In analogy with a charged surface, the protein will collect about itself a counter-ion atmosphere enriched in oppositely charged small ions. Consequently, the protein and the surrounding electrolyte also form an electrical double-layer system. The electrostatic interaction between a protein and an ion exchange stationary phase can therefore be modeled as an interaction between two double-layer systems of opposite charge. If the Poisson–Boltzmann equation is used to model the interaction, the main problem is to find a suitable geometric description of both the stationary phase and the protein. For a review concerning the different approaches for modeling the electrostatic interactions between a protein and a charge surface, see Ref. [26]. Here, we shall primarily discuss the simplest, and yet a practical, approach where the interaction is considered to be between two planar surfaces [27].

15.6.1 SLAB MODEL FOR ION EXCHANGE CHROMATOGRAPHY OF PROTEINS

Consider two planar oppositely charged surfaces in contact with an electrolyte solution. When the surfaces are far apart, an undisturbed electrical double layer exists close to each surface. Note that since the two surfaces have opposite signs of charge, the ions in the double layer of one of the surfaces have the opposite sign in relation to the other. We are interested in how the free energy for this system changes when the two surfaces approach each other. In the ensuing discussion, we will assume that the charge density on both surfaces is constant as they approach each other. By solving the linearized Poisson–Boltzmann equation for two oppositely charged surfaces, the following expression is obtained for the free energy per surface area ($\Delta G_{el}/A_P$) as a function of the distance (L) between the surfaces [27,28]:

$$\frac{\Delta G_{el}}{A_p} = \frac{1}{\kappa \cdot \varepsilon_0 \varepsilon} \cdot \left(\frac{(\sigma_p^2 + \sigma_s^2) \cdot e^{-\kappa L} + 2\sigma_p \sigma_s}{e^{\kappa L} - e^{-\kappa L}} \right) \tag{15.63}$$

where σ_p and σ_s are the surface charge densities (in C/m^2) of respective surfaces. The equation is illustrated in Figure 15.10 for varying ionic strengths in water and surface charge density 0.03 and $-0.16 \, C/m^2$, respectively.

A physical interpretation of Equation 15.63 shows that it is the increase in entropy that is the major contribution to the decrease in the free energy as the two surfaces approach each other. This can be understood as follows: When the surfaces are at an infinite distance from each other, there is an undisturbed electrical double layer associated with each surface. As they approach, the two double layers start to overlap and ions in the double-layer system are, loosely spoken, substituted by charge from the oppositely charged surface. Recalling that the formation of a double layer decreases the entropy of the ions forming the double layer (see discussion in Section 15.1), a release of ions from the two oppositely charged double layers leads to an increase in the entropy of the system. Furthermore, at a certain separation distance, L_m, the free energy goes through a minimum so that

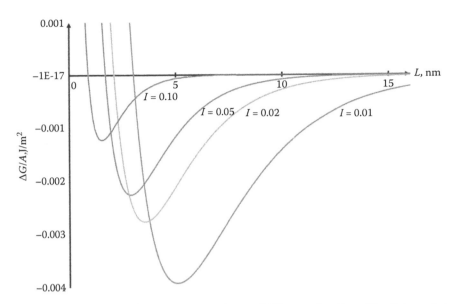

FIGURE 15.10 Plots of the Gibbs free energy per unit area, $\Delta G/A_p$, as a function of the distance between two oppositely charged planar surfaces, L, with the ionic strength as a parameter. The curves are calculated from Equation 15.63 with $\varepsilon = 80$, $\sigma_s = -0.16\,C/m^2$, and $\sigma_p = 0.03\,C/m^2$.

when the two surfaces come closer repulsion occurs. This also has an entropic origin and is caused by the difference in the charge density between the two surfaces. The reason is that electroneutrality needs to be maintained in the space between the two surfaces and that the difference in charge density implies that a certain number of electrolyte ions must remain in this space. As the two surfaces come closer than L_m, the volume available for these ions becomes so small that the entropy decrease for these ions offset the decrease in the electrostatic energy. There is, therefore, an equilibrium distance between the surfaces at which the free energy is a minimum.

By setting the derivative of the right-hand side of the equation with respect to L to zero, the distance for the minimum in free energy, L_m, is found to be

$$L_m = -\frac{1}{\kappa}\ln\left(\frac{-\sigma_p}{\sigma_s}\right) \quad \text{when } \sigma_p < \sigma_s \tag{15.64}$$

Inserting Equation 15.64 into Equation 15.63, the free energy at the distance for the minimum distance L_m is found to be

$$\frac{\Delta G_m}{A_p} = -\frac{\sigma_p^2}{\kappa\varepsilon_0\varepsilon} \quad \text{when } -\sigma_p < \sigma_s \tag{15.65}$$

Equation 15.65 shows that the interaction energy is determined by the surface carrying the lowest surface charge density and that it is independent of the surface charge density of the other surface. In practice, it is usually the protein that carries the lowest surface charge density so that the subscript p in the equations above refers to the protein.

Since the logarithm of the retention factor is proportional to the free energy of adsorption, we find for the slab model that

$$\ln k = \ln\Phi - \frac{\Delta G_{ads}^0}{RT} = \ln\Phi - \frac{\Delta G_m}{RT} = \ln\Phi + \frac{\sigma_p^2}{\kappa\varepsilon_0\varepsilon \cdot RT}\cdot A_p \tag{15.66}$$

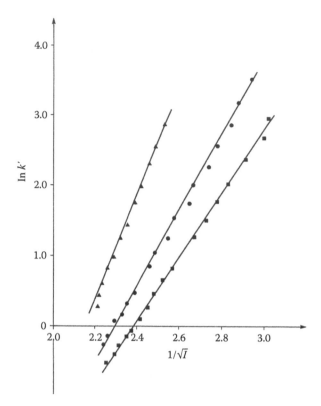

FIGURE 15.11 Plots of the logarithmic retention factor of ovalbumin as a function of the reciprocal square root of the mobile phase ionic strength, $1/\sqrt{I}$ at different pH values: 6.0 (■), 7.0 (●), and 8.0 (▲). Experimental conditions: Synchropak Q300 column, NaCl eluting salt under isocratic condition. Experimental data from Ref. [29]. (Reproduced with permission from Ståhlberg, J. et al., *Anal. Chem.*, 63, 1867, 1991.)

where Φ is the column phase ratio. Recalling that κ is proportional to the square root of the ionic strength of the eluent, $\ln k$ vs. $1/\sqrt{I}$ shall yield a linear plot with a slope proportional to the interacting surface area, A_p, and the square of the surface charge density of this area. Indeed, isocratic retention data usually give a straight line; an example is shown in Figure 15.11 where the retention factor for ovalbumin is plotted as a function of $1/\sqrt{I}$ for three different eluent pH values. Furthermore, the slope values in such plots are also reasonable when the interacting surface area is taken as half the protein area [28].

A complete description of the protein–stationary phase interaction involves many complications and a general model is extremely difficult to formulate. The slab model described above is very simple, yet it gives interesting physical insights and may be a useful starting point for more elaborate theories. Here, we shall only briefly discuss some of the challenges a more complete model meets. For more complete discussions see Refs. [1,24].

1. *Charge regulation.* When two charged surfaces approach each other, the electric field from the respective surface penetrates through the intervening solution and causes a drop in the absolute value of the electric potential on the other surface. For a protein, the effect is that the H^+ concentration at the surface changes and consequently the protein net charge changes. The net effect is that the interaction energy increases so that the energy calculated from Equation 15.65 is too low [30].
2. *The Poisson–Boltzmann equation.* The slab model is based on a solution of the linearized Poisson–Boltzmann equation that is valid only for low electrostatic surface potentials. As

discussed in (1) above, the potential decreases when the surfaces are close to each other, so this simplification is not as serious as it seems. The analogue of Equation 15.65 for the nonlinear Poisson–Boltzmann equation is [31,32]

$$\frac{\Delta G_m}{A_p} = -\frac{\sigma_p^2}{\kappa \varepsilon_0 \varepsilon} \cdot \frac{2}{\alpha^2} \cdot \left(1 - \sqrt{1+\alpha^2} + \alpha \cdot \ln\left(\alpha + \sqrt{1+\alpha^2}\right)\right) \qquad (15.67)$$

where

$$\alpha = \frac{F\sigma_p}{2\kappa \varepsilon_0 \varepsilon RT} \qquad (15.68)$$

Equation 15.67 gives a somewhat lower ΔG_m than Equation 15.66, but the difference is important for α-values above 1 only, e.g., when $\alpha = 10$, the ratio between ΔG_m(Equation 15.67)/ΔG_m(Equation 15.65) is around 0.4.

3. *The geometry.* It is clear that the geometry of the system is much simplified in the slab model. Another possibility is to model the protein as a sphere and the stationary phase as a planar surface. For such systems, numerical solutions of the Poisson–Boltzmann equations are required [33]. However, by using the Equation 15.67 in combination with a Derjaguin approximation, it is possible to find an approximate expression for the interaction energy at the point where it has a minimum. The following expression is obtained [31]:

$$\Delta G_m = -\frac{\sigma_p^2}{\kappa \varepsilon_0 \varepsilon} \cdot \frac{\pi R_p}{\kappa} \cdot \frac{10}{2 + |\alpha|} \qquad (15.69)$$

where
   R_p is the radius of the sphere (protein)
   α is given by Equation 15.68

However, because of the complicated geometry of the protein and the stationary phase interaction, it is impossible to find a general geometrical description.

4. *The van der Waals interaction.* As the protein comes close to the surface, the van der Waals interaction, between the two bodies may become important. In colloid and surface science, this interaction is included in the Hamaker constant and may also be incorporated in the interaction between a protein and a charged surface [34–36].

5. *The protein charge distribution.* This is obviously an important parameter but it requires a detailed knowledge about the protein structure complemented with titration data.

6. Changes in the protein ternary structure during adsorption. There is a difference in softness between different proteins and the ternary structure may change when it adsorbs.

15.6.2 Summary and Concluding Remarks Concerning Ion Exchange Chromatography of Proteins

Retention of proteins in ion exchange chromatography is mainly caused by electrostatic effects. Because both the protein and the surface have an electrical double layer associated to it, there is an increase in entropy when the two surfaces approach each other. This is due to a release of counter ions from the two double layers when they overlap. The model that is discussed here is based on a solution of the linearized Poisson–Boltzmann for two oppositely charged planar surfaces. We also show the result from a model where the protein is considered as a sphere and the stationary phase as a planar surface. This model is based on the Derjaguin approximation and the nonlinearized Poisson–Boltzmann equation. These simple models well describe the general relation between retention and the ionic strength of the mobile phase. Furthermore, they often give a semi-quantitative estimate of the protein net charge.

However, a complete understanding and a description of the adsorption of proteins to a charged surface requires more complicated tools than what is presented here. An interesting example of a study along these lines for ion-exchange chromatography is found in Ref. [37]. Some of the effects that must be included in a rigorous description are mentioned above. They require detailed information of the properties of the specific protein and surface. Since protein adsorption is very important in many technological and scientific areas, it is a field with high activity and much progress.

LIST OF SYMBOLS

c_i concentration of specie i in a solution, mol/dm³ or mol/m³
F Faraday constant, C/mol
f the fraction of the basic form for a protolyte
G Gibbs free energy, J/mol
K equilibrium constant
K_s defined by Equation 15.52
k retention factor
L distance, length, m
n_i surface concentration of specie i, mol/m²
n_0 the monolayer capacity of the stationary phase, mol/m²
Q electric charge, C
R the universal gas constant, J/(mol K)
R_i radius of a sphere, m
r radial distance, m
T temperature, K
U interaction energy, J
X_i the mole fraction of i, dimensionless
x a coordinate, distance, m
z ionic valence, dimensionless
ε dielectric constant, dimensionless
ε_0 permittivity of vacuum, F/m
Φ Column phase ratio, dimensionless
Γ_0 defined by Equation 15.44, dimensionless
κ the inverse Debye length, 1/m, defined by Equation 15.10
μ_i the (electro-) chemical potential of specie i, J/mol
Ψ electrostatic potential, V
Ψ_0 electrostatic potential at the plane of a surface, V

ACKNOWLEDGMENTS

I wish to thank David McCalley, University of Bristol, for discussions during the HPLC meeting in Ghent in June 2006 and for providing me with his figures, Figures 15.2 and 15.3 here. I also wish to thank my colleague Per Lövkvist for his valuable comments on the manuscript.

REFERENCES

1. J. Ståhlberg, *J. Chromatogr. A*, 855, 3–55, 1999.
2. J. Israelaichvili, *Intermolecular & Surface Forces*. London, U.K.: Academic Press, 1992.

3. D. F. Evans, H. Wennerström, *The Colloidal Domain*. New York: VCH Publishers Inc., 1994.
4. I. Hägglund, J. Ståhlberg, *Anal. Chem.*, 60, 1958–1964, 1988.
5. I. Hägglund, J. Ståhlberg, *J. Chromatogr. A*, 761, 3–11, 1997.
6. D. V. McCalley, *Anal. Chem.*, 78, 2532–2538, 2006.
7. Á. Bartha, J. Ståhlberg, *J. Chromatogr. A*, 668, 255, 1994.
8. J. G. Chen, S. G. Weber, L. L. Glavina, F. F. Cantwell, *J. Chromatogr. A*, 656, 549, 1993.
9. Á. Bartha, Gy. Vigh, H. A. H. Billiet, L. de Galan, *J. Chromatogr.*, 303, 29, 1984.
10. J. Ståhlberg, A. Furängen, *Chromatographia*, 24, 783–789, 1987.
11. Á. Bartha, Gy. Vigh, J. Ståhlberg, *J. Chromatogr.*, 506, 85, 1990.
12. F. F. Cantwell, S. Puon, *Anal. Chem.*, 51, 623, 1979.
13. C. E. Harland, *Ion Exchange: Theory and Practice*. Cambridge, U.K.: The Royal Society of Chemistry, 1994.
14. H. Small, *Ion Chromatography*. New York: Plenum Press, 1989.
15. P. R. Haddad, P. Jackson, *Ion Chromatography*. Amsterdam, the Netherlands, Elsevier Science Publisher B.V., 1990.
16. J. Th. G. Overbeek, *Prog. Biophys.*, 6, 57, 1956.
17. S. Afrashtehfar, F. F. Cantwell, *Anal. Chem.*, 54, 2422, 1982.
18. R. A. Fux, F. F. Cantwell, *Anal. Chem.*, 56, 1258, 1984.
19. T. Okada, *Anal. Chem.*, 70, 1692–1700, 1998.
20. T. Okada, *J. Chromatogr. A*, 850, 3–8, 1999.
21. T. Okada, *Anal. Chem.*, 73, 3051–3058, 2001.
22. J. Ståhlberg, *Anal. Chem.* 66, 440–449, 1994.
23. D. Henderson, L. Blum, J. L. Lebowitz, *J. Electroanal. Chem.*, 102, 315, 1979.
24. D. Boda, W. R. Fawcett, D. Henderson, S. Sokolwski, *J. Chem. Phys.*, 116, 7170–7176, 2002.
25. W. Norde, Driving forces for protein adsorption at solid surfaces, in *Biopolymers at Interfaces*, 2nd edn. M. Malmsten, Ed. New York: CRC Press, 2003.
26. C. M. Roth, A. M. Lenhoff, Quantitative modelling of protein adsorption, in *Biopolymers at Interfaces*, 2nd edn. M. Malmsten, Ed. New York: CRC Press, 2003.
27. V. A. Parsegian, D. Gingell, *Biophys. J.*, 12, 1192, 1972.
28. J. Ståhlberg, B. Jönsson, Cs. Horvath, *Anal. Chem.*, 63, 1867, 1991.
29. I. Mazsaroff, L. Varady, G. A. Mouchawar, F.E. Regnier, *J. Chromatogr.*, 499, 63, 1990.
30. J. Ståhlberg, B. Jönsson, *Anal. Chem.*, 68, 1536, 1996.
31. B. Jönsson, J. Ståhlberg, *Colloid Surf. B*, 14, 67–75, 1999.
32. H. Ohshima, *Colloid Polym. Sci.*, 253, 150, 1975.
33. J. Ståhlberg, U. Appelgren, B. Jönsson, *J. Colloid Interface Sci.*, 176, 397, 1995.
34. C. M. Roth, A. M. Lenhoff, *Langmuir*, 11, 3500–3509, 1995.
35. C. M. Roth, K. K. Unger, A. M. Lenhoff, *J. Chromatogr. A*, 726, 45–56, 1996.
36. J. Ståhlberg, B. Jönsson, Cs. Horvath, *Anal. Chem.*, 64, 3118, 1992.
37. Y. Yao, A.M. Lenhoff, *Anal. Chem.*, 77, 2157–2165, 2005.

16 Polymer HPLC

Dušan Berek

CONTENTS

16.1 INTRODUCTION

The high-molar-mass chemical species, macromolecules, represent the very important constituents of both organic and inorganic or natural and synthetic environments—from the living organisms, clays, and earth constituents to numerous synthetic objects such as the most rubbers, fibers, coatings, constructing materials, adhesives, etc. The separation and molecular characterization of macromolecules represents a demanding challenge for science, technology, and everyday life. For synthetic polymers, liquid chromatography is presently the most important tool for separation and molecular characterization. In this chapter, which is oriented to the everyday analytical practice, fundamentals of high-performance liquid chromatography of synthetic polymers (polymer HPLC) are presented. The differences between HPLC of small molecules and macromolecules are elucidated so that the experts from both fields can find the necessary information to make comparisons. The researchers in the area of polymer synthesis and application would acquire the basic knowledge about the potential and limitations of polymer HPLC: the former is often overestimated while the problems are neglected. Therefore, the present level of polymer HPLC is being critically assessed in this chapter and the necessity for further refinements and innovations is evidenced. Some anticipated future developments are also outlined. One of the objectives of this chapter is to help experimental workers in devising the optimum HPLC procedures for solving the particular separation/characterization tasks in polymer research and technology. This latter objective is accentuated by the unconventional organization of the text with numerous cross-quotations, which are intended to simplify the orientation of readers. In order to make the reading as easy as possible, examples of numerous practical applications are largely omitted; these can be found in the quoted references, especially in numerous monographs, handbooks, collections of papers, chapters in books, and review papers, for example [1–33]. It is amazing to find that many fundamental ideas and observations published in the first stages of the polymer HPLC development are overlooked or oblitered by some authors of new reviews and original studies. Honor is often not paid to the founders of certain methods, and numerous already published ideas are "re-discovered." Therefore, numerous references to the breakthrough articles and older books, reviews, and original papers are also included in this chapter. History of the particular procedures of polymer HPLC can be traced on the basis of those publications. It is also hoped that some forgotten ideas will be revived, and eventually brought to an application, taking advantage of a general progress in the theory, materials, and technology. The author apologizes to the pioneers not mentioned here: the limited scope of this chapter, short time available for the compilation, and difficult access to the literature—not the intention—are the reasons.

16.2 MACROMOLECULAR SUBSTANCES: THE POLYMERS

16.2.1 Basic Terms and Molecular Characteristics

In many aspects, the behavior of macromolecules substantially differs from the low-molar-mass substances. Therefore, it is necessary to briefly define the basic terms and explain the most important features of high-molar-mass systems related to liquid chromatography. To make the understanding as easy as possible, numerous simplifications have been adapted. This section is not intended for experts in polymer science and technology.

Many, but not all, macromolecules are created by the mutual chemical chain reactions of small molecules called *monomers* and the arising species contain repeated small units, *mers*. In that case they are designated *oligomers* or *polymers* depending on their molar mass. This means that all oligomers and polymers can be called macromolecular substances but not all macromolecular substances are of oligomeric or polymeric nature (lignin, humin substances, etc.). Properties of macromolecular systems depend on

- Molecular characteristics of constituents
- Mutual arrangement of macromolecular constituents
- Nature and amount of additives, which may be low- or high-molar-mass substances in liquid or solid state

The present chapter deals with molecular characteristics of synthetic polymers. Numerous natural polymers such as the most polysaccharides can be tentatively incorporated into this group of macromolecular substances because their behavior in many aspects resembles that of the synthetic polymers and also because they are often chemically modified to adjust their utility properties. The typical example is cellulose, the most abundant organic polymer on earth.

The molecular characteristics can be divided into the primary and the secondary ones.

The primary molecular characteristics of oligomers and polymers are

1. Molar mass
2. Chemical structure (composition)
3. Physical architecture

High molar mass of polymers is their most typical primary molecular characteristic. It is responsible for many of their valuable properties. Molar mass, M, is related to the number of *mers* entering their molecules and it is expressed in $g \cdot mol^{-1}$ or $kg \cdot mol^{-1}$. The alternative term is the unitless molecular weight. The inappropriate term "molecular mass" can be found in literature. In fact, it means the mass of a molecule and not of a mole. Molar mass of oligomers ranges from a few hundreds to a few thousands $g \cdot mol^{-1}$ while molar mass of polymers may reach several millions of $g \cdot mol^{-1}$, that is, thousands of $kg \cdot mol^{-1}$. The latter species are often termed *ultrahigh-molar-mass substances*. For the sake of simplicity, the oligomeric compounds will be included under the general term "polymers" except for the situation when specific properties of oligomers will be discussed. To distinguish macromolecules according to their size, the term *(high)polymers* will be applied in the latter case.

The term *chemical structure* of high polymers comprises several primary and secondary molecular characteristics. For example, the composition of copolymers or polymer blends, as well as the functionality of macromolecules, is included in the term chemical structure. Molecules of copolymers are built from two and more different monomers, the *comonomers*, to form the *statistical*, *block*, *graft*, *star*, *miktoarm*, etc. polymer species. In the linear copolymers or *terpolymers*, respectively, two or three different monomers are combined either in a fixed order or statistically. The example of former case represents the *alternating copolymers*. In the statistical copolymers or terpolymers the short blocks of identical monomeric units exhibiting different lengths are present, and the statistical copolymer exhibits certain *blockiness*. If different monomeric units are organized according to the random statistics they can be designated the *random copolymers*. When the blocks of identical *mers* are long enough to be themselves considered macromolecules of different nature, which are mutually connected with the chemical bonds into the linear structures, one speaks about di-, tri-, and multi-*block* copolymers. In the *graft* copolymers, the chemically different side chains, the *grafts*, are chemically attached to the main chain or backbone. The polymer chains of the same chemical nature oriented in different directions from a common center form the *star polymers* and when the chains in a star polymer exhibit different chemical nature, one speaks about the *miktoarm* polymers. A graft copolymer with one single side chain attached to the main chain is in fact the simplest form of a star polymer. The polymer blends exhibit as a rule differences in the molecular characteristics of their constituents. Most common are the blends of two and more chemically different polymers. For special purposes also, the polymers that differ only in their molar masses are blended. The nature, amount, and position of charged groups on macromolecules can be also considered their chemical structure. The charged macromolecules represent technologically important materials as the ionized groups grant them some specific properties. However, the charges carried by the polymer chains may cause additional problems in liquid chromatography. Therefore it is often attempted to suppress ionization processes in solutions of synthetic polymers by means of the appropriate low-molar-mass counterions. This is why the peculiarities of ionizable polymers will be mentioned in this chapter only as a certain warning.

All the above features of chemical structure of macromolecules can be found also in oligomers but the most important is the presence of the functional groups attached to their molecules. The functional groups to a large extent affect properties of oligomer molecules since their relative role is much more important than in (high)polymers. Chemical reactions of functional oligomers enable creation of the new kinds of important—often crosslinked—macromolecular systems. Molecules of oligomers can carry double bonds and in this case one speaks about the *macromonomers*. The double bonds render possible the mutual polymerization or copolymerization of macromonomers with other monomers. For example, macromonomers are employed in the synthesis of graft copolymers with controlled/known length of the side chains.

The physical architecture represents another group of the molecular characteristics of macromolecules. Due to the hindered rotation around single bond in polymer chains, the spacial orientation of side groups on the main chain plays an important role and may significantly affect properties of otherwise chemically identical macromolecules. The orientation of short side groups, for example, methyls or ethyls, decides the *stereoregularity* or *tacticity* of given polymer. In the *isotactic* polymer, the side groups are oriented in the same direction, in the *syndiotactic* species the orientation of side groups "oscilates," in the *heterotactic* macromolecules adjacent two side groups regularly alter their orientation, and in the *atactic* polymer, the side groups are oriented statistically. Stereoregularity of polymers influences not only their crystallability and thus their solubility—interaction with the mobile phase (Section 16.2.2)—but also their interaction with the column packing (Section 16.3), for example, their adsorption. Further important differences in the architecture of macromolecules of the same chemical composition represent, for example, the *cis* and *trans* structures for species containing double bonds in their main chain, as well as the linear, cyclic, long- and short-chain branched species, the abovementioned star-like polymers, etc., not to speak about the crosslinked systems with the "infinitive" molar mass. The architecture of macromolecules may affect their retention behavior within the liquid chromatographic columns (Section 16.3).

Each basic molecular characteristic may exhibit large interconnected variability, which is reflected in the secondary molecular characteristics. Differences in the secondary molecular characteristics may even appear within a group of polymers possessing the same overall chemical structure or architecture. For example, the particular side chains in the graft copolymers may have distinct compositions though the overall composition of macromolecules is equal. Similarly, various statistical copolymers may possess the same overall composition, while their blockiness or stereoregularity is different. It is evident that the properties of macromolecules may be extremely complex if the effects of two or even all three primary molecular characteristics are combined.

All molecular characteristics of synthetic polymers exhibit certain distribution so that they in fact represent multicomponent mixtures of distinct macromolecules. Except for the polymer blends, these distributions are as a rule continuous in nature. Distributions in the molecular characteristics of polymeric substances can be expressed in different ways. One can count or weight species of the same kind to express their relative concentration in system—their number or weight fraction—and to calculate the corresponding average (mean) values of molecular characteristics. Correspondingly, one speaks about *number average* or *weight average* of given molecular characteristic, for example, of molar mass. The difference between number and weight average values signalizes the width of the particular distribution. In the case of M, important and often used parameter is ratio of weight average molar mass $\bar{M}_w$ and number average molar mass $\bar{M}_n$ of given polymer, $(\bar{M}_w/\bar{M}_n)$, which is called *polydispersity* (or *polymolecularity*). Unfortunately, average values of polymer molecular

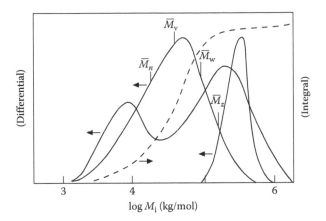

FIGURE 16.1 Schematic differential (——) and integral (- - -) representations of amount of polymer with certain molar mass present in the sample. The narrow, broad, and bimodal molar mass distributions are shown. The typical positions of weight, viscosity, number, and z-molar mass averages are depicted.

characteristics say little about the shape of distribution function, for example, about the molar mass distribution (MMD). For example, the $\bar{M}_w/\bar{M}_n$ values may be identical for both the (broad) unimodal and the multimodal distributed polymers. Specific and artificial averages were introduced to express certain behavior of macromolecules, for example, the *viscosity average* of molar mass $\bar{M}_v$ or the z and $z+1$ averages of molar mass, $\bar{M}_z$ and $\bar{M}_{z+1}$, respectively. The latter two averages are especially important for the broad molar-mass-distributed polymers. The meaning of above terms is evident from Figure 16.1.

The important feature of many polymers is simultaneous presence of distributions in two and several molecular characteristics. Polymers exhibiting multiple distributions are called the *complex polymers* or *complex polymer systems*. A detailed discussion of molecular characteristics of polymers and their average values and distributions can be found in numerous monographs and reviews, for example [34,35]. For the present purpose, it is important to repeat that all synthetic polymers and also polysaccharides are polydisperse in their nature. Only mother nature is able to produce macromolecules, for example many proteins, with uniform molar mass. The latter are often improperly called monodisperse(d) polymers.

There do exist various physicochemical methods for determination of the average values of molecular characteristics of polymers. Molar mass can be determined by, for example, the light scattering measurements, viscometry, osmometry, and ultracentrifugation for the (high)polymers and ebulliometry, cryometry, mass spectrometry, and vapor pressure osmometry for the oligomers. Recent progress in the mass spectrometry extended its application range to high molar mass area of hundred thousands of $g \cdot mol^{-1}$—the matrix-assisted laser desorption ionization (MALDI) and the electrospray mass spectrometries (Section 16.9.1). The chemical structure and architecture of macromolecules can be evaluated, for example, by the IR, UV, and NMR spectrometries.

In order to assess the distribution(s) of molecular characteristics, polymers must be separated. As to the molar mass, the classical measurement/separation methods employed, for example, the differences in the solubility, diffusion rate, or weight of macromolecules. Most of these techniques were partially (e.g., the conventional viscometry and light scattering) or almost fully (the Baker-Williams fractionation, osmometry, ultracentrifuge) substituted by the liquid chromatographic methods especially by size exclusion chromatography (SEC), while the light scattering and viscosity, together with NMR and the mass spectrometry measurements are utilized also in the detection of chromatographic column effluents (Section 16.9.1).

In this chapter, high-performance liquid chromatography of oligomers and (high) polymers (polymer HPLC) will be briefly presented. As mentioned in Section 16.1, there exist several monographs, chapter in books, and review papers on this subject, for example [1–33]. Most of them contain numerous examples of the HPLC separation and molecular characterization of particular macromolecular substances. Therefore, this chapter discusses almost exclusively the general principles of polymer HPLC and only few selected examples of practical applications will be mentioned for illustration.

16.2.2 MACROMOLECULES IN SOLUTION

In this section, attention will be paid again only to the oligomers and (high)polymers formed by the repeating units, *mers*. The basic principles and terms will be explained considering the linear polymer chains.

The dissolution of macromolecules is a prerequisite for the application of liquid chromatography for their separation and characterization. Compared to HPLC of small molecules, concentration of the polymer solutions injected into the analytical HPLC columns is higher and usually assumes $1\,mg \cdot mL^{-1}$ and more. This is mainly due to detection problems: the detectors used in polymer HPLC are much less sensitive (Section 16.9.1) than detectors for small molecules, which often carry the UV-absorbing chromophores. This means that samples subject to polymer HPLC must exhibit rather high solubility.

The well-known thermodynamic rule says that the two substances of different nature are miscible if the process brings about a gain in the value of the Gibbs function, ΔG, also called Gibbs energy or free enthalpy—that is, if $\Delta G > 0$. The Gibbs function is connected with further basic thermodynamic quantities enthalpy and entropy, by the relation

$$\Delta G = \Delta H - T\Delta S \tag{16.1}$$

where
H results from enthalpic interactions between molecules of solute and solvent
S is entropy of system
T is the temperature

The sign Δ indicates the change of particular quantity due to the mixing (dissolution) process.

The gain of the mixing entropy is high in case of small molecules. Therefore, the latter species may mutually dissolve even if the enthalpy of mixing is rather unfavorable. The situation is different for macromolecules, where the contribution of mixing entropy is much lower. Therefore, enthalpy of the process often decides the (in)solubility of macromolecules. In order to be dissolved, the attractive interactions between polymer segments and solvent molecules must be equal or stronger than the mutual interactions between segments themselves. This is generally the case of the thermodynamically good solvents. The crystallites must be destroyed by solvent molecules or by melting at elevated temperature to make crystalline polymers soluble. In solution, macromolecules can assume various conformations such as the rods, worms, globuli, gaussian statistical coils, etc., the last of which is the most suitable for polymer HPLC. The shape of polymer coils in solution is usually slightly ellipsoidal but is often approximated by the spherical one. The coil size for given (linear) macromolecule in solution depends on the solvent quality. In the thermodynamically good solvent, polymer coils are expanded. To estimate the solubility of a given polymer sample and to choose the appropriate solvent, the physical properties of latter should be considered. Knowledge of the solvent properties also helps to understand the solvent–packing interactions (Section 16.3.2). The most important is the polarity and hydrogen bonding ability of the solvent molecules. The solvent polarity is usually expressed by the Hildebrandt solubility parameters, σ, though the concept of σ was originally developed for mixtures of nonpolar substances [36]. The higher the σ value the more polar is the solvent: the σ values of alkanes are in the range of $14\,MPa^{0.5}$ while those for the highly polar organic solvents reach value $40\,MPa^{0.5}$. The solubility parameter of water is $47.9\,MPa^{0.5}$. It was proved many times that the solubility can be expected in the systems where polymer and solvent exhibit similar values of σ. This rule has, however, many exceptions. The values of solubility parameters for numerous polymers and liquids can be found in Ref. [37].

The polymer–solvent interaction is described by the famous Flory-Huggins equation [38]

$$\frac{\Delta G}{RT} = x_1 \ln \varphi_1 + x_2 \ln \varphi_2 + g\, x_1\, \varphi_2 \tag{16.2}$$

where
  R is the gas constant
  T is temperature [K]
  φ_1 and φ_2 are the volume fractions
  x_1 and x_2 are the mole fractions of solvent and polymer, respectively
  g is the polymer–solvent interaction parameter

Its related value was originally denoted as χ. Numerous χ values in terms of volume fractions are collected in Ref. [37]. Unfortunately the scatter in χ values found in the literature is large as they reflect also both the polymer source (e.g., narrow molar mass fractions or anionically prepared samples) and the method of measurement, for example, light scattering, osmometry, or inverse gas chromatography. The interaction parameters g (χ) for the polymer–good solvent systems assume values between 0 and 0.5 [37].

The alternative value, which describes the polymer–solvent interaction is the second virial coefficient, A_2 from the power series expressing the colligative properties of polymer solutions such as vapor pressure, conventional light scattering, osmotic pressure, etc. The second virial coefficient in $[mL \cdot mol^{-1}]$ assumes the small positive values for coiled macromolecules dissolved in the thermodynamically good solvents. Similar to χ, also the tabulated A_2 values for the same polymer–solvent systems are often rather different [37]. There exists a direct dependence between A_2 and χ values [37].

The thermodynamic quality of a solvent for a polymer can be also estimated from Kuhn-Mark-Houwink-Sakurada viscosity law (often called Mark-Houwink equation):

$$[\eta] = K_v\, M^a \tag{16.3}$$

where
  $[\eta]$ is the intrinsic viscosity (limiting viscosity number) of the linear homopolymer with molar mass M in the given solvent
  K_v and a are constants

The unitless exponent a characterizes polymer–solvent interactions of *linear flexible* coiled macromolecules. For good solvents, a ranges between 0.65 and 0.85. Intrinsic viscosities of macromolecules of the same chemical nature reflect their linearity/branching. Again, tabulated a values for the same polymer from different sources in the some solvent exhibit some scatter [37]. Equation 16.3 and a exponent are very popular in polymer HPLC. The product M $[\eta]$ represents the hydrodynamic volume of macromolecules in solution and it is the basis for the universal calibration dependence in SEC (Section 16.4.2). It is to be noted that Equation 16.3 loses its validity below molar mass about $10,000\,g \cdot mol^{-1}$. Equation 16.3 is generally invalid for branched polymers. Ratio of intrinsic viscosity for the linear and branched polymer of the same molar mass renders important information on polymer branching. It represents a basis for the assessment of polymer branching by SEC (Section 16.4.3).

The deterioration of the solvent quality, that is, the weakening of the attractive interactions between the polymer segments and solvent molecules, brings about the reduction in the coil size down to the state when the interaction between polymer segments and solvent molecules is the same as the mutual interaction between the polymer segments. This situation is called the theta state. Under theta conditions, the Flory-Huggins parameter χ assumes a value of 0.5, the virial coefficient A_2 is 0, and exponent a in the viscosity law is 0.5. Further deterioration of solvent quality leads to the collapse of coiled structure of macromolecules, to their aggregation and eventually to their precipitation, the phase separation.

The extent of the polymer–solvent interactions, the solvent quality can be affected by temperature and/or by the addition of further solvent components. In the latter case, one speaks about the mixed solvents (two- or multicomponent solvents). In most cases, polymer chains interact preferentially with one of the mixed solvent components so that its concentration increases in the domain of polymer coils. This phenomenon is called the preferential solvation. The preferential solvation gives rise to the system peaks or solvent peaks on chromatograms monitored by the nonspecific detectors such as differential refractometers [39,40] (Sections 16.4.5, 16.5.2, and 16.9.1). The preferential solvation is inevitably present also in most solutions of low molar mass substances in the multicomponent solvents but its role is likely less important than in case of polymers. The swelled polymer coils in solution exhibit rather large conformational entropy. Any change in the conformation of coiled macromolecules in solution is accompanied by important changes in this kind of entropy. Actually all processes taking place during chromatographic separation of macromolecules affect their conformation and, as a result, are accompanied by a large change, usually with a loss of conformational entropy of polymer species (Sections 16.3.1 and 16.4.1). This is why the conformational entropy of macromolecules plays a very important role in polymer HPLC. The ellipsoidal coils of macromolecules likely change their orientation in the chromatographic column and this process may cause variations in orientational entropy [41]. In conclusion, three kinds of entropy changes take place in the liquid chromatographic column, namely the mixing, orientation, and conformation, the most important is the last one. This shows further substantial difference between HPLC of small molecules and macromolecules.

Charged macromolecules exhibit particular behavior in solution. The repulsive intramolecular interactions of ions with the same charge cause large extension of polymer coils so that their effective size increases dramatically. This polyelectrolyte effect can be suppressed by addition of the low-molar-mass counterions to eluent. The preferential interactions of macromolecules with counterions, however, lead to similar effects as observed with uncharged polymer species dissolved in mixed solvents (Sections 16.4.5, 16.5.2, and 16.9.1). Rather strong ion–ion interactions may unexpectedly appear even with very low number of charged groups situated on the polymer chains in the cases when column packing also bears undesired charges (Sections 16.4.1 and 16.8.1). Some initially noncharged macromolecules (e.g., poly(ethylene oxide)s) selectively interact with the low-molar-mass ions present in solvent and become charged. This phenomenon leads to the pseudopolyelectrolyte effects. The role of ion nature in both the extent and the resulting effect of their interaction with macromolecules is so far not well understood. Large differences have been observed between ions of different nature already in the nineteenth century (Hofmeister series). Since recently, the interactions are intensively studied between proteins dissolved in water and various kinds of ions. It has been repeatedly confirmed that anions are more active than cations. The latest research has shown that the presence of ions extensively affects the water–air and water–macromolecule interfaces [42,43]. Similar phenomena may take place also in case of some synthetic polymers such as poly(N-isopropyl acryl amide) dissolved in organic solvents.

16.3 RETENTION MECHANISMS IN POLYMER HPLC

16.3.1 GENERAL CONSIDERATIONS

The most straightforward way for understanding the retention phenomena in polymer HPLC is afforded by thermodynamics. It is widely accepted that the retention of any kind of substance in any chromatographic column depends on its distribution between the stationary phase and the mobile phase:

$$V_R = V_0 + KV_s \tag{16.4}$$

where
V_R is the retention volume
V_0 is the interstitial volume of a column filled with the (particulate) packing
K is the distribution constant described as a ratio of solute concentration in the stationary and the mobile phase c_s/c_m
V_s is the volume of the stationary phase within column

Depending on the particular method of polymer HPLC, V_s is defined in different ways. It is the total volume of pores, V_p, in a porous packing but it can be also related to the total surface of packing (mostly to the surface situated within the pores) or to the effective volume of bonded phase. The volume of pores is relatively well defined in the case of many packings applied in polymer HPLC and plays an especially important role in the exclusion-based separations (Sections 16.3.3, 16.3.4, and 16.4.1). The exclusion processes, however, play an important role in the coupled techniques of polymer HPLC (Section 16.5). In the latter cases, the surface of packings and the effective volume of bonded phase are to be taken into account. In some theoretical approaches also, surface exclusion is considered.

Numerous theories attempted to define the quantitative relation between the distribution constant K and the molecular characteristic of polymer samples; so far such attempts have not been suc cessful. The uncertainty of V_s is one of the reasons why methods of polymer HPLC are considered nonabsolute.

To better understand the complexity of situation, it is useful to apply some thermodynamic considerations. The separation process is governed by the change of the Gibbs function due to transfer of solute molecules between the mobile and the stationary phase. One can write

$$V_R \sim K \frac{V_s}{V_m} \sim \exp\left(\frac{\Delta G}{RT}\right)\frac{V_s}{V_m} \tag{16.5}$$

and with respect to Equation 16.1,

$$V_R \sim \exp\left(\frac{\Delta H}{RT} - \frac{\Delta S}{R}\right)\frac{V_s}{V_m} \tag{16.6}$$

where in this case ΔG, ΔH, and ΔS denote the changes in thermodynamic quantities connected with the transfer of solute (macro)molecules between the mobile and the stationary phase, while V_s and V_m are the volumes of the stationary and the mobile phases, respectively. For a given chromatographic system the ratio V_s/V_m can be with a good approximation considered constant. In this way, the processes in the HPLC column can be divided into those driven by enthalpy and those controlled by entropy. If one managed to quantitatively express both ΔH and ΔS contributions to V_R in dependence on polymer molecular characteristics, the general quantitative theory of polymer HPLC would be created. Unfortunately, in spite of the numerous attempts, so far this was not the case. Neither the quantitative relation is known between the enthalpy and entropy changes during the chromatographic process nor the effect of molecular characteristics of separated polymers on ΔH and ΔS can be expressed. The mutual dependence between ΔH and ΔS is very complex because any change of enthalpic interactions within system affects conformational and possibly also mixing and orientation entropy of macromolecules and vice versa. Due to conformation barriers, only a fraction of polymer segments in a polymer coil (Section 16.2.2) can simultaneously interact with the column packing. In other words, behavior of polymer species within the HPLC system cannot be obtained by simple summing of the segmental interactions of macromolecules. In conclusion, there does not exist a theory, which would quantitatively describe and predict the elution behavior of species separated by polymer HPLC, so that it would be possible to directly assess their molecular characteristics from their retention volumes. Consequently, most polymer HPLC methods either need calibration or the fractions created are subject to further analyses and characterizations. Still, Equation 16.6 gives valuable hints for the design of chromatographic systems suitable for solving some complicated separation tasks. Usually the operators deliberately change experimental conditions to increase or to decrease the enthalpic interactions keeping exclusion (nearly) constant. The observed overall effects are evaluated in order to optimize the separation of given sample. Some general rules do exist in this respect, which facilitate the optimization procedures. These will be discussed in the following sections of this practically oriented chapter. For theoretical considerations on polymer HPLC we do refer the reader to the review papers [44–48].

In conclusion, large divergence does exist in the behavior of small and large (chain) molecules in the chromatographic systems. These are further augmented or suppressed by distinctions in the viscosity and mobility (diffusibility) of solutes with different sizes. It is necessary to consider these differences in order to devise the appropriate HPLC system for successful HPLC separation of macromolecules.

16.3.2 Role of Mobile Phase in Polymer HPLC

In Section 16.2.2, the role of the low-molar-mass liquids in dissolving/precipitation of macromolecules was discussed. It was shown that the same macromolecule can assume various conformations/sizes in solution, depending on the thermodynamic quality of solvent. The same considerations are to be applied also to liquid chromatographic systems where macromolecules interact with eluents. However, the third player has to be accounted for in polymer HPLC, namely the column packing and its interactions not only with the macromolecules of sample but also with the eluent molecules. In HPLC of small molecules, the latter interactions are denoted as the solvent strength, ε^0. The solvent strength concept and quantity ε^0 was originally introduced by Snyder [9] who considered interactions of bare alumina with liquids of different polarity. The term solvent strength reflects the extent of interaction between solvent (eluent) molecules and column packing (surface). Similar to HPLC of low-molecular-weight analytes, also in polymer HPLC, the mobile phase competes with the separated species for the active sites on the column packing surface. The retention of analytes is suppressed in a strong mobile phase. Very strong mobile phases may completely prevent enthalpic interaction of macromolecules with the column packing surface. Vice versa, the weak mobile phases allow interactions of analytes with the column packing and the retention volumes increase. The weak mobile phases, which interact with the column packing surface with a low intensity, promote even full retention of macromolecules within the column. Later, the solvent strength concept was extended also to the most common normal phase column packing, the bare silica gel. The corresponding ε^0 values are tabulated, for example, in Ref. [49] (see Table 16.1). As known, silica gels of diverse origin exhibit both different concentration and topology of the active silanol groups and consequently rather unlike surface properties. Therefore the values from Table 16.1 show the tendencies rather than the absolute quantities. Still, they are useful for orientation when selecting eluent components to control retention of analytes in polymer HPLC with bare silica based column packings of the same kind.

As mentioned, unlike small analyte molecules, macromolecules can be irreversibly retained on the packing surface in the given eluent and at given temperature (Section 16.6). In the case of interactions that lead to the adsorption of solute on the adsorbent surface (Sections 16.3.5 and 16.6) this situation is described by the high-affinity isotherms (Figure 16.2): up to a certain threshold concentration of given polymeric adsorbate, all its molecules are adsorbed and their concentration in

TABLE 16.1
Selected Properties of Some Solvents Used in Polymer HPLC

Solvent	Molar Mass (g·mol⁻¹)	Boiling Point (°C)	Refractive Index[a]	Density[b] (g·mL⁻¹)	Viscosity[c]	Elutropic Value (Silica)	UV Cutoff (nm)[d]
Acetone	58.08	56	1.3590	0.791	0.32	0.43	330
Acetonitrile	41.05	82	1.3440	0.786	0.37	0.50	190
2-Butanone	72.11	80	1.3790	0.805	0.40	0.39	330
Carbon tetrachloride	153.82	77	1.4595	1.594	0.97	0.14	263
Chlorobenzene	112.56	132	1.5240	1.107	0.80	Low	287
1-Chlorobutane	92.57	77–78	1.4024	0.886	0.35	Low	225
Chloroform	119.38	60.5–61.5	1.4460	1.492	0.57	0.31	245
Cyclohexane	84.16	80.7–81	1.4260	0.779	1.00	0.03	200
1,2-Dichloroethane	98.96	83	1.4438	1.256	0.79	Low	225
Dichloromethane	84.93	39.8–40	1.4240	1.325	0.44	0.32	235
N,N-Dimethylacetamide	87.12	164.5–166	1.4380	0.937	—	High	268
N,N-Dimethylformamide	73.10	153	1.4310	0.944	0.92	High	268
Ethylacetate	88.11	76.5–77.5	1.3720	0.902	0.45	0.45	260
Hexane	86.18	69	1.3750	0.659	0.33	0.00	200
Methyl alcohol	32.04	64.6	1.3290	0.791	0.60	0.73	205
1-Propanol	60.10	97	1.3840	0.804	2.256	High	210
2-Propanol	60.10	82.4	1.3770	0.785	2.30	0.63	210
Tetrahydrofuran	72.11	67	1.4070	0.886	0.55	0.35	215
Toluene	92.14	111	1.4960	0.867	0.59	0.22	285
Water	18.02	100	—	1.000	1.00	>0.73	—

Source: Gant, R., *Sigma-Aldrich Chromatography Products*, Sigma-Aldrich Corp., Milwaukee, WI, 1993.

[a] At 20°C.

[b] At 20°C ± 5°C relative to water at 4°C.

[c] In centipoise at 20°C.

[d] Wavelength at which absorbance is 1 Å for a good LC-grade solvent.

supernatant is negligible. The corresponding very weak mobile phase acts as adsorption promoting liquid or adsorli. The strong liquid, which promotes the complete desorption of the given analyte from the given adsorbent is called desorli. In conclusion, the strength of eluent decides about adsorption and desorption of macromolecules onto/from solid surfaces. At the same time the quality of eluent plays a much less important role in the adsorption control.

The behavior of the nonpolar bonded phases, as well as the column packings based on crosslinked organic polymers of low polarity, however, differs from that of polar column packings and the classical solvent strength concept should be reevaluated. This is especially important for the alkyl bonded phases (Section 16.8.1). In this case, surface and interface adsorption of polymer species (Section 16.3.5) plays a less important role and macromolecules are mainly retained by the enthalpic partition (absorption) (Section 16.3.6). In order to ensure this kind of retention of polymer species, the mobile phase must push them into the solvated bonded phase. Therefore the interactions of mobile phase with both the bonded phase and (especially) with the sample macromolecules—that is, the solvent quality—extensively controls retention of latter species within the alkyl bonded phases.

The bonded phases carrying polar groups such as –NH₂, –OH, –CN, etc., which are separated from the packing surface by a spacer (usually the n-propyl groups) behave in an intermediate manner between the solid surfaces and the nonpolar bonded phases though the effect of polar end groups usually dominates (Sections 16.3.5 and 16.3.6).

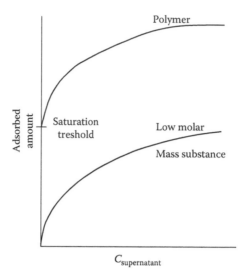

FIGURE 16.2 Typical adsorption isotherms, relations between amount of adsorbate attached to the surface of adsorbent and its concentration in supernatant. It shows the "high-affinity adsorption isotherm," which is typical for many polymers in contact with an active adsorbent in a weak solvent. Up to a certain extent of the surface saturation, the saturation threshold, *all* macromolecules are attached to the adsorbent surface.

The thermodynamic quality of mobile phase may be so poor that macromolecules precipitate within column or at least exhibit a tendency to phase separation. This is a specific feature of certain procedures of polymer HPLC (Sections 16.3.7, 16.5.3, and 16.5.6).

In conclusion, two different parameters of mobile phase are to be distinguished in polymer HPLC namely its strength toward column packing and its quality toward column packing and especially toward separated macromolecules.

16.3.3 PROCESSES IN SEPARATION COLUMNS

As mentioned, the quantitative description and prediction of all processes taking place within the separation column is so far not possible. Still, in order to design the appropriate separation system, it is necessary to understand some basic principles and tendencies. Their elucidation is attempted in this section. The result of retention processes is reflected in volume of mobile phase needed to elute particular macromolecule from the column, in its retention volume, V_R. For the sake of simplicity, let us consider the constant "exterior" condition within the column, that is, the eluent flow rate, temperature, and the pressure drop. The latter two parameters are dictated not only by the inherent hydrodynamic resistance of column influenced by the eluent viscosity, size, and shape of packing particles but also by the sample viscosity, which may be rather high in polymer HPLC. Further, only one molecular characteristic of a given polymer will be considered, namely its size in solution, which depends on its molar mass and on the solvent (eluent) quality. All other molecular characteristics are kept constant and/or the role of their variation is considered negligible. This is the case of the linear homopolymers with no side chains and with equal orientation of side groups. The eluent strength is varied and the polymer sample is dissolved and injected in eluent. Certainly, this all is only an approximation—a model system. The total volume of liquid within column, V_m depends on both the interstitial volume V_0 and the pore volume, V_p—that is, $V_m = V_0 + V_p$ (Section 16.3.1). V_m can be assessed from the retention volume of a deuterated analogue of eluent or of a nonretained low-molar-mass substance, etc. This value is called void volume in HPLC of low-molar-mass substances. The problems connected with its determination are discussed in numerous text books.

Let us first consider the situation when the polymer molecules do not exhibit any attractive or repulsive interaction with the porous column packing surface except for the effects caused by the imperviousness of the pore walls. This corresponds to $\Delta H = 0$ in Equation 16.6 and the sample retention volume is controlled by entropy of the process. The fictitious retention volume of eluent molecules corresponds to V_m because small molecules of eluent permeate all pores of the column packing. Under these conditions, the big molecules of polymers are partially or fully excluded from the packing pores and therefore they elute faster than the molecules of eluent. The polymer species outrun also their original solvent. This relative acceleration of macromolecules increases with their increasing size because the volume fraction of accessible pores decreases correspondingly. As a result, both the distribution constant K and the retention volume V_R of polymer molecules defined in Equation 16.4 decrease with their size in solution and with their molar mass. This situation is schematically represented in Figure 16.3a, which depicts the dependence of log M vs. V_R or log M [η] vs. V_R. The minimum value of V_R corresponds to $K = 0$ that is $V_R = V_0$. The maximum retention volume equals to V_m or, within a certain approximation, to the retention volume of the monomer forming the polymer under study. The shape of the particular log M vs. V_R dependence reflects the correlation between molar mass or the effective volume of polymer species in eluent—their hydrodynamic volume (Sections 16.2.2 and 16.4.2), and the packing pore both average size and distribution. The position of V_0 depends on the packing geometry (the arrangement of packing particles). The molar mass at which the acceleration of macromolecules compared to eluent molecules vanishes is called the excluded molar mass, M_{ex}. Under otherwise identical experimental conditions, value of M_{ex} reflects the size of the largest pores within the column packing. While the above exclusion processes play a rather inferior role in HPLC of low-molar-mass substances, they are the basis for the most important procedure of polymer HPLC, the size exclusion chromatography (SEC) (Section 16.4).

Let us now consider the case when the polymer molecules exhibit certain attractive interaction with the column packing, that is, $\Delta H \neq 0$ (Equation 16.6). This interaction decelerates but does not fully prevent elution of polymer species. Still, the extent of enthalpic interaction is large enough to overrun the effect of exclusion. In other words, exclusion of separated species does not dominate any more. As a rule, the extent of the attractive interactions of sample molecules with the column packing and the corresponding retention volumes rise exponentially with the increasing sample molar mass (Sections 16.3.5 and 16.3.6). The corresponding situation is depicted in Figure 16.3b. For comparison, the previous case (pure exclusion) (Figure 16.3a) is shown as a broken line (4). The K from Equation 16.4 is larger than unity in this case. The retention volumes of monomers are now higher than V_m. The courses of the log M vs. V_R dependences reflect the extent of the net attractive enthalpic interactions between macromolecules and column packing though the observed V_R values are affected also by the entropy changes. For large attractive interactions between the macromolecules and the column packing, which are only a little obstructed by the packing–solvent interactions, the retention volumes of polymer species rapidly increase with the raising log M [curve (1)]. Typically, already at relatively low molar mass (usually only few thousands g·mol^{-1}), the interactions between the packing and sample are so intensive that the macromolecules are fully retained within column packing and can be released only by changing the experimental conditions, for example, by increasing the eluent strength to decrease the net polymer–packing interaction. The fully retained molar mass (V_R is "infinitive") is marked with an empty circle in Figure 16.3b. The sample–packing attractive interactions depicted by the curves (2) and (3) are less intensive. The reduced sample recovery should be expected in many above systems already before the full retention appears. The eluted macromolecules may be not representative of the entire sample because the largest macromolecules exhibit the increased tendency to be fully retained.

A specific situation is depicted in Figure 16.3c, curve (1) where the basic feature of the exclusion mechanism still prevails but it is affected to a certain extent by the polymer–packing interactions. Again, the course of the log M vs. V_R dependence observed in the absence of enthalpic interactions is shown as a broken line (5). The interactions between column packing and macromolecules increase with the decreasing molar mass. This is typical for the oligomers carrying functional groups, which

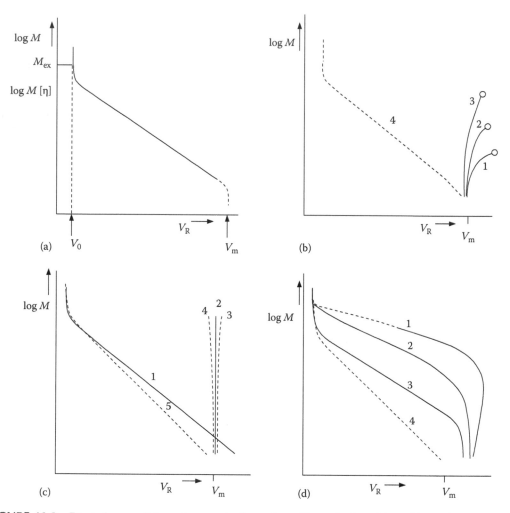

FIGURE 16.3 Dependences of the polymer retention volume V_R on the logarithm of its molar mass M or hydrodynamic volume $\log M\,[\eta]$ (Section 16.2.2). (a) Idealized dependence with a long linear part in absence of enthalpic interactions. V_0 is the interstitial volume in the column packed with porous particles, V_m is the total volume of liquid in the column and M_{ex} is the excluded molar mass. (b) $\log M$ vs. V_R dependences for the polymer HPLC systems, in which the enthalpic interaction between macromolecules and column packing exceed entropic (exclusion) effects (1–3). Fully retained polymer molar masses are marked with an empty circle. For comparison, the ideal SEC dependence (Figure 16.3a) is shown (4). (c) $\log M$ vs. V_R dependences for the polymer HPLC systems, in which the enthalpic interactions are present but the exclusion effects dominate (1), or in which the full (2) or partial (3,4) compensation of enthalpy and entropy appears. For comparison, the ideal SEC dependence (Figure 16.3a) is shown (5). (d) $\log M$ vs. V_R dependences for the polymer HPLC systems, in which the enthalpic interactions affect the exclusion based courses. This leads to the "enthalpy assisted SEC behavior" especially in the vicinity of M_{ex}. For comparison, the ideal SEC dependence (Figure 16.3a) is shown (4).

exhibit the attractive interaction with column packing. The effect of end group interaction decreases with the increasing molar mass of oligomer molecules, which may eventually behave as the non-functionalized species. This behavior can be utilized for enhancement of the SEC-like separation of oligomers [50,51]. A very specific case of the combined exclusion–interaction effects is shown in Figure 16.3c, curve (2). It represents the full mutual compensation of exclusion and interactions, which leads to the independence of the retention volumes of polymer molar masses. Immediately the question arises, how the $\log M$ vs. V_R curve (4) in Figure 16.3b (in absence of enthalpic interactions) and a dependence with the course as the curves (1–3) in Figure 16.3b can be overlaid to obtain the vertical course (2) or only slightly curved curves (3) or (4). It is, however, to be repeated that the course (2) of log M vs. V_R dependence in Figure 16.3b contains in itself also exclusion effects and that any interaction of macromolecules with the column packing affects in a complex way the exclusion behavior of macromolecules due to the changes in their conformational entropy. The situation shown in Figure 16.3c, courses (2–4), is very important for characterization of complex polymers. It allows eliminating or at least suppressing the effect of polymer molar mass so that a separation according to another molecular characteristic can be performed. The experimental conditions leading to this elution behavior are called the critical conditions of enthalpic interactions. They will be more in detail discussed in Section 16.5.2.

Figure 16.3d, curve (1) depicts the unusual situation when the retention volumes of macromolecules increase due to their interaction with the column packing only up to certain value of M and then again decrease [52,53]. This course can be tentatively explained by the permeation of certain segments of large (over M_{ex}) interacting macromolecules into the packing pores. The size of permeating segments is nearly constant but its relative role decreases with the increasing molar mass of polymer. The maximum effect of pore permeating and interacting polymer segments on V_R is situated in the vicinity of M_{ex}. Below M_{ex} the extent of enthalpic interactions between eluted macromolecules and column packing increases with polymer M but it remains constant over M_{ex}. In the case depicted by curve (2), the interactions (enthalpy), and the exclusion (entropy) nearly mutually compensate below M_{ex} (see above) but the effect of exclusion starts to prevail with further increase of sample molar mass over M_{ex}. As a result, the effect of the attractive enthalpic interactions on V_R passes though maximum and an increased (apparent) separation selectivity is observed (Section 16.5.5). The curve (3) in Figure 16.3d is further shifted in direction of log M vs. V_R dependence without presence of enthalpic interactions represented by the curve (4). The selectivity of separation is still higher than in the case of pure exclusion (Section 16.5.5). The shift of apparent M_{ex} to higher values with curves (1–3) can be due to the reduced sample recovery. The log M values in Figure 16.3d, curves (1–4) hold namely for injected polymer species but it is likely that only fractions with lower M are eluted.

In order to explain the dependences of log M vs. V_R depicted in Figure 16.3b through d and to understand particular processes in the polymer HPLC columns, the qualitative thermodynamic consideration, Equation 16.6 can be used.

16.3.4 Exclusion of Macromolecules and Related Processes

The effect of exclusion on the retention volumes of macromolecules was qualitatively explained above. The pioneering work of Casassa [54] has shown that the extent of pore exclusion of macromolecules is controlled by the changes in their (conformational) entropy. The principle is explained in a simplified form in Figure 16.4a through c. A zone of polymer solution with a nonzero concentration travels along a column packed with porous particles. Initially, the concentration of macromolecules within pores is zero (Figure 16.4a). The concentration gradient outside of pore ($c>0$) and within pore ($c=0$) "pulls" macromolecules into the pores.

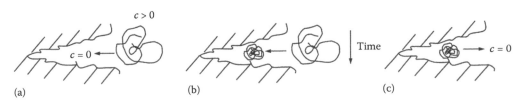

(a)   (b)   (c)

FIGURE 16.4 Three stages of the engagement of a macromolecule with a pore. Schematic representation of the entropy controlled exclusion process.

There is a tendency to equalize the chemical potentials in both volumes. Macromolecules entering the pore lose part of their conformational entropy.

When the penalty in entropy outbalances the pulling force, macromolecules stop their pore permeation; in other words the macromolecules cannot enter the pores, which would be large enough to accommodate their completely compressed size (Figure 16.4b). Next the polymer concentration outside the pores becomes zero and the macromolecules are pulled out of the pores to continue their elution (Figure 16.4c). Evidently, if the polymer solution contains the species of various sizes, these will permeate different pore volumes according to their size. The smallest species find the largest accessible pore volume. They are longer accommodated within the column packing pores so that their retention volume is higher than that of larger species. The smallest retention volume $V_R = V_0$ ($K = 0$) in Equation 16.4 exhibit macromolecules, which are

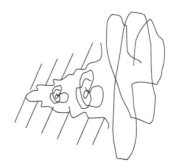

FIGURE 16.5 Schematic representation of a situation when the three macromolecules of different size are simultaneously in contact with a pore.

too large to enter any pore (Figure 16.5). This is the principle of separation applying size exclusion and the basis of SEC.

Surprisingly enough, the above processes are very fast and separation of macromolecules on this principle can be considered an equilibrium process. The precise measurements of retention volumes of polymers under conditions of their partial pore permeation in absence of enthalpic interaction did not reveal practically any effect of the eluent flow rate [55]. On the contrary, in a review, Aubert and Tirrell [66] have demonstrated that the SEC exclusion volumes can be flow rate dependent due to both anomalous and physical effects. The former are caused by

- Errors in estimation of the retention volumes from the apexes of skewed peaks (peak skewing depends on the flow rate)
- Degradation of polymers at the high flow rates (Section 16.4.5)
- Instrumental anomalies, the extent of which changes with the flow rate

The physical effects include the following:

- The possible change of the pore geometry in dependence of pressure and therefore the flow rate.
- The concentration dependence of retention volumes (Section 16.4.1).
- A possible (so far unjustified) enhancement of the polymer concentration within pores due to molecular migration phenomena. The enhanced pore concentration may result in greater partitioning into the very small pores and in the increased retention volumes.

It is to be noted that the long segments of the very large macromolecules may partially permeate the packing pores (Figure 16.6). This process is anticipated in the case of ultrahigh molar masses over 1000 and especially over 5000 kg · mol^{-1}. This kind of partial permeation likely contributes, together with the shearing on the outer surface of column packing particles, to the mechanical degradation of ultrahigh molar mass polymers in the SEC columns (Section 16.4.5).

In conclusion, the size exclusion of macromolecules in absence of any energetic interactions with the column packing is an

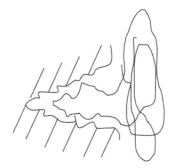

FIGURE 16.6 Schematic representation of partial permeation of a (large) macromolecule into a (narrow) pore.

entropy-controlled process. It is based on the distribution of solute molecules between two volumes of eluent—that is, between the interstitial volume and the pore volume within column. According to terminology of physical chemistry, such process is to be designated the *partition* and because it is controlled by entropy, it should be termed the entropic partition.

Skvortsov and Gorbunov [44,45] showed that it holds for the slit-like pores and for the (Gaussian-type) coiled macromolecules

$$K \sim 1 - 2\pi^{-0.5} \frac{\overline{R}}{\overline{d}} \quad \text{for } \overline{R} << \overline{d} \tag{16.7}$$

and

$$K \sim 8\pi^{-2} \exp\left[-\left(\frac{\pi\overline{R}^2}{2\overline{d}} \right)^2 \right] \quad \text{for } \overline{R} >> \overline{d} \tag{16.8}$$

where
$\overline{R}$ is the effective mean radius of macromolecules dissolved in the eluent
$\overline{d}$ is the effective mean pore radius

Equations 16.7 and 16.8 confirm the experimental observations that in SEC there does exist a distinct relation between effective sizes of pores and polymer coils.

As demonstrated by numerous experiments, temperature does not well influence the exclusion processes (compare Equation 16.6) in eluents, which are thermodynamically good solvents for polymers. In this case, temperature dependence of intrinsic viscosity [η] and, correspondingly, also of polymer hydrodynamic volume [η] M on temperature is not pronounced. The situation is changed in poor and even theta solvents (Section 16.2.2), where [η] extensively responds to temperature changes.

In the course of attempts to quantitatively and a priori describe separation of macromolecules by SEC, a theory was formulated, which considered the effect of porous structure of the column packing on the diffusion rate of separated macromolecules. The resulting theory of restricted diffusion by Ackers [56], however, rather contradicts the equilibrium conception of SEC.

A more realistic approach considers the contribution of the eluent flow patterns to the retention/ acceleration of macromolecules within the polymer HPLC columns. It is well known that the laminar flow of a liquid in a capillary is not fully homogeneous. The velocity of liquid is lower near the walls of capillary than in its middle part. The geometrical center of large macromolecule cannot approach the walls of capillary as can do the smaller species. Therefore, the average velocity of the larger macromolecules, which are accommodated mainly in the center of the capillary, is higher compared to the smaller species. This idea was for the first time presented by Pedersen [57] who has observed some size separation of polymers within a column packed with nonporous particles. Such system can be considered an array of capillaries. The idea was later elaborated by Small [58] and subsequently by numerous other authors, also applying single, long capillaries [59–62]. The corresponding separation method was termed the hydrodynamic chromatography (HDC). HDC is an especially valuable method for fractionation of solid particles and dispersions. This means that HDC can treat samples, for which SEC can be hardly used because of weak Brownian motion, which makes it unable to release the large particles of sample from the packing pores. The separation selectivity of HDC is rather low but the presence of flow inhomogeneity and the surface exclusion of macromolecules within columns likely augment the basic SEC separation mechanism based on the pore exclusion. HDC itself was partially substituted by the groups of methods termed field flow fractionation [63–65].

16.3.5 ADSORPTION

The adsorption is an enthalpy-driven process accompanied by the changes of entropy. The latter are as a rule especially large in case of polymers. According to the accepted terminology, adsorption is the distribution of solute molecules between the volume of solution and the adsorbent surface. The surface can be formed by a solid or by a liquid. Respectively, one speaks about the surface adsorption and the interface adsorption. The extent of adsorption depends on the affinity between the solute molecules and the surface as well as on the nature of solvent molecules. The solute–surface affinity depends on the nature and concentration of the interacting groups on both adsorbent (column packing) and adsorbate (separated macromolecules). Generally, polar–polar interactions are stronger than the interactions due to nonpolar London forces. Therefore the adsorption of polymers of the medium to high polarity on the polar surfaces is as a rule more intensive than the adsorption of the nonpolar macromolecules on the nonpolar surfaces. As mentioned, the extent of solute adsorption depends also on the interactions between the adsorbent and the solvent molecules, on the solvent strength (Section 16.3.2). If the solvent is very strong, it fully prevents adsorption of adsorbate on the given surface, it is a *desorli*. This is the situation welcomed in SEC, where the strong liquids/mobile phases and the nonadsorptive column packings are to be used. To be stressed again, if the eluent is a very weak solvent, it promotes the full adsorption of macromolecules, it is an *adsorli*. The latter extreme case is unwanted in most methods of polymer HPLC because if an adsorli is used as the eluent, polymer samples stay irreversibly retained within the column under given experimental conditions. It may be even difficult to quantitatively release the polymer species fully retained within porous column packing by subsequent application of a desorli (Section 16.5.3). The solvent strength of mobile phases can be adjusted by mixing appropriate liquids. The solvent strength of the mixture of liquids is usually considered additive [9], that is, the solvent strength of a mixture of liquids A and B, ε_{AB}^0, is

$$\varepsilon_{AB}^0 = \varepsilon_A^0 + \frac{\log N_B}{\alpha n_B} \tag{16.9}$$

where
ε_A^0 is the solvent strength of the weaker component A
N_B is the mole fraction of the stronger component B
n_B is the effective molecular area of an adsorbed molecule B

The adsorbent surface activity function, α, is defined as

$$\log K^0 = \log V_a + \alpha f(x,s) \tag{16.10}$$

where
K^0 is the sample adsorption distribution coefficient $(mL \cdot g^{-1})$
V_a is the volume of an adsorbed solvent monolayer per unit weight of adsorbent
$f(x,s)$ is a function describing properties of sample x and solvents

Equation 16.10 holds for mixed eluents A plus B with not very low concentration of stronger component B. The role of intermolecular interactions between eluent components A and B is not considered. The solvent strength of a binary eluent can be roughly estimated also from the relation

$$\varepsilon_{AB}^0 = \phi_A \cdot \varepsilon_A + \phi_B \cdot \varepsilon_B \tag{16.11}$$

although the difference between the estimated and the real values rapidly increases with the difference in polarity of the liquids A and B [9]. Equation 16.11 can be used for a coarse assessment

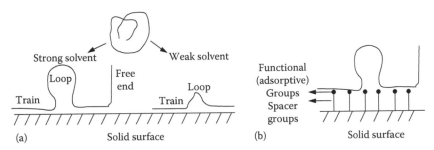

FIGURE 16.7 Schematic representation of the surface (a) and interface (b) adsorption of macromolecules. The role of solvent strength is depicted. Note large changes of macromolecular conformation which accompany the adsorption process.

of solvent strength of mixed eluents. In practice, however, the trial-and-error approach is widely used in polymer HPLC.

The process of polymer adsorption on the solid surface is schematically depicted in Figure 16.7a. The trains of adsorbed macromolecules are short, the loops are large, and the free ends are frequent if the solvent is strong. Under otherwise identical conditions—adsorbent, polymer, temperature—the trains of adsorbed polymer chain are longer when solvent is weaker. A typical example of the interfacial polymer adsorption represents its interaction with the bonded phase. The bonded phase can be considered a quasiliquid carrying either polar (e.g., –CN, –OH, –NH$_2$) or nonpolar (–CH$_3$) groups (Figure 16.7b). The interacting groups are separated from the solid surface by appropriately long (usually n-propyl) groups, spacers.

The important point is that the polymer coils change their conformation during the adsorption process and lose large part of their conformational entropy. As a result, the coiled (flexible) macromolecules may be less intensively adsorbed than the stiff (rigid) species of similar polarity [67]. It is accepted that the solvent quality toward polymer solutes plays less important though not negligible role in their adsorption processes. The exact role of the solvent quality in the polymer adsorption is not well understood. It is namely difficult to change the solvent quality keeping its strength constant. In any case, it is hardly possible to efficiently control the extent of adsorption and the retention volumes in polymers HPLC just by changing the eluent quality. On the other hand, the direct control of polymer adsorption is readily done by adjusting solvent/eluent strength. The extent of adsorption can be well affected also by temperature. Rising temperature as rule decreases adsorption of small molecules. This is not the general case for macromolecules. The conformational entropy of polymer species depends strongly on temperature. Consequently, the gain in enthalpy may be prevailed by the large losses of conformational entropy due to adsorption at increased temperature. This results in the possible though not frequent increase of adsorption with increasing temperature. Adsorption of samples within HPLC columns can be affected also by pressure. This effect may be direct [68] or indirect [69,70]. In the later case, the pressure affects the preferential sorption of mixed eluent components on the column packing and consequently also the extent of the polymer adsorption. In most adsorbate/adsorbent/solvent systems, both polymer adsorption and chromatographic retention increase with the molar mass of adsorbate. It was, however, shown that, at least in case of nonporous adsorbents, polymer species with lower molar mass may be adsorbed preferentially [71]. The above conclusions apply for many unmodified, bare polar column packings. Considering the role of changes in the conformational entropy in the adsorption processes may help explain some unexpected results in polymer HPLC.

It can be concluded that adsorption is an important tool for controlling retention volumes of samples in polymer HPLC. Eluent composition and temperature are the most feasible variables to affect adsorption in the given polymer–column packing systems. The thermodynamic quality of eluent plays less important role.

16.3.6 ENTHALPIC PARTITION (ABSORPTION)

Another enthalpy-driven process in polymer HPLC is the distribution of solute molecules between the volume of mobile phase and the volume of chromatographic stationary phase. The accepted terminology designates the processes of this kind the absorption or the partition, which is driven by enthalpy and therefore it is the enthalpic partition. In polymer HPLC, the only "nonbleeding" stationary phases with the defined volume are those chemically bonded on the solid surfaces. The physically immobilized stationary phases on the solid surfaces are rather instable. The so far attainable chemically homogeneous bonded phases with a large enough volume are mainly the "monomeric" or "polymeric" alkyl bonded phases on silica gel. Commercially available are –C-4, –C-8, –C-14, –C-18, –C-22, and –C-30 alkyl-bonded phase. The polymeric alkyl bonded phases are usually formed by the "layers" of alkyl groups (initially C-18), which are mutually crosslinked with the –Si–O–Si– bridges. The volume of the commercial poly(ethylene glycol) bonded phase is likely rather small. The poly(ethylene glycol) molecules seem to lie flat on the silica surface [72]. Synthetic and natural macromolecules immobilized on the surface or within the pores of solid particles constitute a challenge for polymer HPLC. Numerous different polymers were deposited on the silica gel surface [73] and subsequently crosslinked. Very attractive are the composite materials in which the pore volume is filled with a homogeneously crosslinked polymer network. The general problem with all the bonded phases is the relatively slow mass transfer and consequently the extensive broadening of chromatographic zones.

Enthalpic partition of macromolecules can result also from the preferential sorption of one mixed eluent component on the bare surface of column packing [74] because the composition of the eluent layer adjacent to the packing surface differs from the overall composition of mobile phase. This effect may be large in the case of homogeneously crosslinked polymer based soft column packings (Section 16.8.1) [75,76]. In this case the preferentially solvated polymer chains of the homogeneously crosslinked packing form a new phase. A particular *phase* with the non-negligible volume can be also formed by the solvated polymer chains, which protrude over walls of otherwise solid particles of heterogeneously crosslinked, polymer based column packings (Section 16.8.1) [77,78].

Surprisingly, in many papers, the terms adsorption and enthalpic partition are confused though the principal differences in their backgrounds are evident. The role of the mobile phase is also different when adsorption and enthalpic partition are compared. As explained, solvent quality toward macromolecules only marginally affects their adsorption on a given surface. The solvent strength is decisive for the extent of adsorption and therefore also for the adsorption-based retention of polymer species in the particular column packing–polymer system. The statements that can be often found in the literature such as "…add a nonsolvent to eluent to increase adsorption…" are completely wrong. The nonsolvent can be either an adsorli or a desorli. As explained in Section 16.3.2, the presence of a desorli strongly reduces the extent of adsorption irrespective of the eluent quality. On the contrary, solvent strength toward silica gel practically does not affect enthalpic partition of low-polarity polymers on the (alkyl) bonded phases. The driving force is the solvent quality. Eluent must be poor enough to push macromolecules into the volume of solvated alkyl bonded groups. The situation is schematically depicted in Figure 16.8.

The system is very complicated because the bonded groups are (preferentially) solvated by the eluent (components). Surprisingly enough, the nonpolar alkyl groups bonded to the silica surface can be solvated even with the molecules of polar solvents such as methanol. The extent of preferential solvation of the bonded groups by the mixed eluent components may be very large. Consequently, the driving force of the enthalpic partition due to mobile phase repulsion may be either reduced or augmented. For example, if the bonded alkyl groups are preferentially solvated with a good solvent for separated macromolecules the difference in the dissolving power between the mobile phase containing a nonsolvent on the one hand and the solvated bonded phase on the other hand may increase rather dramatically. Simultaneously, the macromolecules may be preferentially solvated by the

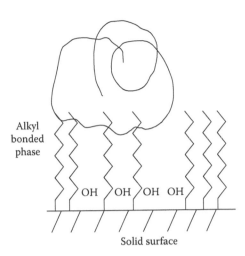

Alkyl
bonded
phase

OH   OH   OH   OH

Solid surface

FIGURE 16.8 Schematic representation of enthalpic partition (absorption) of a macromolecule between the mobile phase and the (solvated alkyl) *phase* bonded on the silica surface.

eluent component, which is not necessarily the same that preferentially solvates the bonded alkyl groups. As a result, even the polymers of medium polarity such as poly(methyl methacrylate) can be partitioned from certain eluents in favor of alkyl bonded phases [79]. Some liquids strongly support partition in favor of alkyl bonded phases of macromolecules for which they are still solvents   for example, dimethyl formamide for polystyrene and poly(n-butyl methacrylate) [80]. Enthalpic partition does not take place over the entire length of bonded alkyls: For example, polystyrenes are likely partitioned only within about six to seven –CH$_2$– groups of the alkyl bonded phases [81]. Therefore, the application of –C-18 alkyl bonded groups may turn inappropriate in some methods of polymer HPLC. The stretched –C-18 groups namely occupy about half of pore volume of silica gel with 10 nm (100 Å) pores and excessively decrease its pore size, both effective diameter and volume of pores [82].

The important feature of all available silica gel alkyl-bonded phases is the presence of rest silanols. In spite of the application of sophisticated end-capping procedures, about 50% of starting silanols remain unreacted in the best commercial –C-18 bonded phases. Unexpectedly, these silanols are accessible for macromolecules, which can thus became adsorbed on the alkyl bonded silica column packings [83]. A combination of two enthalpic retention mechanisms is usually unwanted in polymer HPLC because of difficulties with both control and interpretation of results. Therefore, as far as possible, mobile phase should be applied in order to suppress the adsorption phenomena when engaging the enthalpic partition retention mechanism in the separation of polymers of medium-to-high polarity. On the other hand, the controlled adsorption of polymer species on the alkyl bonded silica HPLC column packings allows application of macromolecular probes in the nontraditional testing of polymer silanophilic interactivity of alkyl bonded phases [83,84]. Of course, also the nonpolar interactivity of the alkyl-bonded phases can be evaluated on the basis of the enthalpic partition measurements with macromolecular probes [85]. It is important to remember the above facts when selecting the (mixed) mobile phase and the column packing for separation of the particular polymer sample.

In conclusion, the enthalpic partition processes in the columns for polymer HPLC substantially differ from the adsorption processes. Enthalpic partition can be employed for the separation of polymers of the low-to-medium polarity in combination with the alkyl bonded phases on silica gels. The extent of the enthalpic partition and consequently also of the polymer retention is controlled primarily by the thermodynamic quality of eluent toward separated species and by the extent of the bonded phase solvation.

16.3.7 PHASE SEPARATION

As shown in Section 16.2.2, the solubility of polymers is generally much lower than the solubility of low-molar-mass substances. When the solvent quality is deteriorated by adding a nonsolvent or by the change of temperature, polymer coils shrink, they begin to associate/aggregate and, eventually phase separation occurs. In the initial stage of the phase separation, the microdroplets of two immiscible liquids are usually formed—the concentrated and the diluted phase. In some cases, the "diluted phase" does not contain any polymer at all [86]. The existence of the microdroplets of phase separated polymer system within the HPLC column is undesirable. It should be noted that the decreasing temperature is the usual cause of the phase separation onset. The corresponding systems exhibit the upper critical solution temperature. In some cases, however, not the gain of enthalpy but the change of the conformational entropy due to increased temperature decides polymer solubility. Such systems exhibit the lower critical solution temperature, and the phase separation results from an increased temperature. Further, the important features of polymer solutions are the phenomena of the cosolvency and co-nonsolvency. In the former case, a mixture of the two nonsolvents for given polymer may became (even a very good) solvent [37,86] while the less common latter term describes the situation when a mixture of two solvents for given polymer becomes its nonsolvent [87,88].

Phase separation (precipitation) of a polymer strongly depends on all its molecular characteristics. On the one hand, this allows very efficient separations in polymer HPLC utilizing phase separation and re-dissolution processes [20]. On the other hand, due to complexity of phase separation phenomena, the resulting retention volumes of complex polymers may simultaneously depend on several molecular characteristics of separated macromolecules. This may complicate interpretation of the separation results. Both precipitation and redissolution of most polymers is a slow process. It may be affected by the presence of otherwise inactive surface of the column packing. Therefore, the applicability and quantitative control of the phase separation phenomena may be limited to some specific systems of polymer HPLC.

16.4 SIZE EXCLUSION CHROMATOGRAPHY

SEC is presently the most important method for separation and molecular characterization of synthetic polymers. The exclusion chromatography of lipophilic macromolecules is called also gel permeation chromatography (GPC) and for hydrophilic species, the term gel filtration chromatography (GFC) is often applied. It seems that the term gel permeation chromatography is "returning" probably because the abbreviation "SEC" means also Securities and Exchange Commission.

The separation of macromolecules on the insoluble, crosslinked porous media was attempted by several researchers just after the World War II. However, the first successful separation of proteins on the crosslinked dextran in aqueous eluents was reported by Porath and Flodin [89] as late as in 1959. Independently, Moore [90] separated the dissolved polystyrenes according to their molar masses on the porous crosslinked polystyrene beads, the raw material for the ion exchangers, in 1964. Above researchers are considered the founders of SEC.

SEC enjoys enormous popularity; practically all institutions involved in research, production, testing, and application of synthetic polymers are equipped at least with a simple SEC instrument. Numerous textbooks and review articles were devoted to this marvelous method [1–8,10–31]. SEC, however, exhibits also several weakpoints, which are mostly ignored by the method users. This section attempts to present a realistic picture of SEC, indicating not only its benefits but also its shortcomings, which deserve further research leading to innovations.

16.4.1 RETENTION MECHANISMS AND NONEXCLUSION PROCESSES IN SEC

The basic retention mechanisms of SEC were presented in Sections 16.3.3 and 16.3.4. These mainly include the steric exclusion of macromolecules from the pores of the column packing controlled by entropy and supported by the flow patterns. The separation procedures that employ exclusively these retention mechanisms are often denoted the "ideal" SEC. In other words, the absence of enthalpic interactions in the system ($\Delta H = 0$ in Equation 16.6) is a prerequisite of ideal SEC. Several undesirable processes, however, may influence the behavior of macromolecules in the SEC columns. These affect both the position and the width of chromatographic peaks and may lead to erroneous results. Moreover, additional peaks may appear on chromatograms. One deals with the "real" SEC. The most important contributors to the changeover from the ideal to the real SEC are following:

1. The secondary retention mechanisms, which result from the enthalpic interactions within the SEC column, cause adsorption, enthalpic partition, phase separation, incompatibility, and ionic effects. As shown in Section 16.3, both adsorption and enthalpic partition decelerate the elution of polymer species and increase the observed retention volumes of analytes. The processes that accompany the first stages of the phase separation—that is, the aggregation and association of macromolecules—increase the effective size of the separated species and decrease their retention volumes. The incompatibility of polymers is a consequence of the repulsive interactions between the macromolecules of different nature. The incompatibility of polymers is a general phenomenon of large technological importance: a great majority of mixtures of unlike macromolecules undergoes phase separation. The incompatibility between the column packing and the polymer sample decreases the effective volume of eluent in the packing pores, which is accessible for separated macromolecules and thus reduces their V_R values. The incompatibility may play a non-negligible role in the SEC systems where the injected polymer solutions are rather concentrated and where macromolecules that form column packing substantially differ in polarity from the separated species (Section 16.8.1). Ion interactions between sample molecules and column packing may either increase (ion inclusion) or decrease (ion exclusion) retention volumes. Most above enthalpic interactions respond to temperature changes.

2. The side processes bring about the changes of sizes of macromolecules in the course of their transport along the SEC column. These size variations may result from the de-coiling of macromolecules due to the flow, as well as from the degradation of polymer species by the mechanical shearing (Section 16.4.5), or by the oxidization—or just by interaction with the solvent molecules (e.g., poly(hydroxy butyrate)). The shear degradation can be reduced when applying low eluent flow rates and large (up to 20 µm) packing particles (Section 16.8.1). The opposite process is the aggregation/association of macromolecules within column, especially at the onset of the phase separation (compare ad 1). The continuous measurements of protein size changes due to association within columns were attempted already in the first stage of the SEC development [91–94]. Interesting results were obtained by the in-situ monitoring of light scattering and concentration evolution.

   Side processes may bring about either an increase or a decrease in measured retention volumes. Often, the presence of side processes in the SEC system is signalized by the increased dependence of polymer retention volumes on the operational parameters such as the sample concentration, flow rate, and temperature.

3. The parasitic processes include the axial and longitudinal diffusion in the interstitial volume, the local irregularities of flow due to mixing, and the viscosity effects. Most parasitic processes in the interstitial volume of the well packed SEC columns do not play an important role except for the viscosity effects, which may result in the extensive broadening of chromatographic peaks and in the appearance of multiple peaks. The injected polymer solutions usually exhibit much higher viscosity than the mobile phase. This difference produces sample zone perturbances in the form of "fingers." Therefore Moore [95] called the viscosity effect the "viscous fingering." The theory of viscous fingering in HPLC was elaborated by Martin and coworkers [96].

4. Osmotic effects play an important role mainly at high injected polymer concentrations. They may selectively affect the retention volumes of smaller polymer species contained in the polymer sample [97]. Osmotic effects within porous column packings form the basis of interesting preparative liquid chromatographic method developed by Teraoka and coworkers [98–101] and denoted the high osmotic pressure liquid chromatography.

5. The phenomenon of secondary exclusion is caused by the higher diffusion rate of the smaller polymer species compared with the larger ones. The smaller macromolecules may timely occupy the packing pores, which are otherwise accessible also for larger polymer species. Thus the elution rate of the large macromolecules is further accelerated. The term "secondary exclusion" was coined by Altgelt [102] and the theory of process was elaborated by Janča [103].

6. *The concentration effects.* This term describes variations in V_R of the polymer samples due to changes of injected concentration, c_i. The retention volumes as a rule increase with raising c_i in the area of practical concentrations [104–107]. This is mainly due to the crowding effects when the concentration of macromolecules is high enough so that they touch each other in solution and shrink due to their mutual repulsive interactions. The concentration effects in SEC are expressed by the slopes κ of the mostly linear dependences of V_R on c_i. κ values rise with the molar mass of samples up to the exclusion limit of the SEC column, M_{ex}. The concentration effects for the oligomers and also for the excluded (high) polymer species, are usually small or even negligible. κ values depend also on the thermodynamic quality of eluent [108] and the correlation was found between product A_2M and κ, where A_2 is the second virial coefficient of the particular polymer–solvent system (Section 16.2.2) and M is the polymer molar mass [109]. Concentration effects may slightly contribute to the reduction of the band broadening effects in SEC: the retention volumes for species with the higher molar masses are more reduced than those for the lower molar masses.

The concentration dependences of V_R may inverse their direction. Retention volumes may rapidly increase with decreasing c_i in the area of very low injected polymer concentrations, below the critical concentration, c^*, that is, when the crowding of macromolecules in solution starts to disappear.

The concentration dependences of V_R in SEC are likely affected also by the viscosity effects and by the secondary exclusion [103].

7. The preferential interactions of macromolecules with the mixed eluent components. The backgrounds of the preferential solvation of polymer chains in two-component solvents were elucidated in Section 16.2.2. In SEC, the consequences of preferential solvation are visible by the nonspecific detectors such as differential refractometers, which monitor the system peaks on chromatograms (Section 16.9.1). Evidently, the system peaks must be well separated from the sample peaks to avoid the mutual interference and erroneous results. To do so, the SEC column system must contain also narrow-pore packing material. It is well comprehensible that the system peaks may appear also with the eluents containing the dissolved solid additives, for example, the stabilizers and especially the (inorganic) salts added to the mobile phase in order to dissolve the sample or to suppress the undesired ion interactions. The inorganic salts are strongly hydrated with traces of water, which can be present in sample and also on the walls the container used for preparation of sample solution. As a result, additional system peaks may be observed, which belong to the hydrated salt molecules present in the injected sample. The typical example is SEC with the dimethyl formamide eluent that containing the salts as LiCl, LiBr, or KSCN [110].

8. *The SEC band broadening.* The band broadening phenomenon represents an important problem in all chromatographic methods. In HPLC of low-molar-mass substances, the band broadening limits the number of the base line separated substances, while in polymer HPLC it affects the calculated data of molar mass distribution, long-chain branching, etc. as these are directly assessed from the peak width. The extra column band broadening

is caused by the diffusion and the flow irregularities in the sample injectors, connecting capillaries, and detectors. The mixing processes within the column end-pieces can be also included into the extracolumn effects. However, the slow mass transfer in the packing pores was shown to be the most important source of the band broadening in SEC [111]. The measured peak width in SEC is composed from the two major contributions: the molar mass distribution of sample and the band broadening effect. The extent of band broadening often depends on the molar mass of sample. The strictly uniform (incorrectly called "monodisperse") synthetic homopolymers are not available. Therefore, it is difficult to determine the net band broadening effect and to introduce efficient corrections. Tung [112] proposed the famous method of reversed flow, which allows elimination of sample polydispersity when polymer species are returned back to the injector after they were first eluted in the direction of the column end. Band broadening was large with classical SEC columns and several methods of its correction, "BBC" were introduced [113–115]. BBC was abandoned when the SEC column technology was improved and the small column packing particles 10, 5, and even 3 μm in diameter were commercialized. Due to the small particle size and improved pore shape, the band broadening effects were strongly reduced in the modern SEC column packings. Still, the errors in the molar mass values due to band broadening were so far not completely eliminated. At present, the re-introduction of BBC is attempted in order to improve exactness of SEC measurements [116–120]. Some companies introduced simple BBC procedures in their software for processing the SEC chromatograms. It is hoped that a universal, efficient, and user-friendly BBC software will be soon available, possibly allowing also corrections for the concentration effects.

9. The possible flow rate effects on retention volume (Section 16.3.4).

This short overview illustrates the large complexity of the SEC processes and explains the absence of a quantitative theory, which would a priori express dependence between pore size distribution of the column packing—determined for example by mercury porosimetry—and distribution constant κ in Equation 16.4. Therefore SEC is not an absolute method. The SEC columns must be either calibrated or the molar mass of polymer species in the column effluent continuously monitored (Section 16.9.1).

16.4.2 CALIBRATION OF COLUMNS AND DATA PROCESSING IN SEC

The calibration of the narrow pore SEC columns intended for separation of lower oligomers is usually performed by the identification of one of the individual peaks, which appear on chromatogram of a polydisperse sample. This is done either by spiking the sample with a material of the same composition as the polydisperse sample but containing known number of *mers* in its molecule. Such material may be obtained from the polydisperse sample for example by the adsorption based HPLC (Section 16.5.1) or by molecular distillation. It is characterized by a conventional method, for example, by the vapor pressure osmometry. The peaks with the lower and higher retention volumes are then successively ascribed to the higher or lower numbers of mers in molecules, respectively. Alternatively, a baseline SEC-separated oligomer fraction is collected and subjected to an independent analysis. Evidently, this procedure is not suitable for the oligomers containing higher number of *mers* and for the (high)polymers because in both latter cases only an envelope-like chromatograms are obtained. Here, a series of the calibration materials, "standards" is successively injected into the SEC column and the resulting retention volumes are assessed from the positions of the polymer peak apexes or better from the peak moments. The calibrations materials are the narrow molar mass distribution species with known molar mass averages (Section 16.8.3). These produce narrow peaks, the apex of which can be precisely identified. The SEC calibration dependence of log M vs. V_R is constructed on the basis of data obtained (compare Figure 16.3a).

Soon after the birth of SEC, the researchers have revealed that the calibration dependences obtained with the same column did not coincide for the polymers of different nature and/or in the different eluents. In attempts to obtain one single calibration dependence for different polymers and different mobile phases, molar masses of polymers were substituted with the extended chain lengths, molar volumes, etc. The most successful procedure to create the universal SEC calibration dependence was that of Benoit with his coworkers [121,122]. They proposed plotting logarithm of the product of molar mass, M and corresponding intrinsic viscosity, $[\eta]$ (Equation 16.3) vs. V_R. The product $M. [\eta]$ is called hydrodynamic volume of the polymer coils (Section 16.3.4). The Benoit's concept turned out one of the most important breakthroughs in the SEC theory and practice. It is a basis for numerous different SEC applications [10–31] and serves also for assessment of both nonexclusion effects in SEC [74,123] and interactivities the silica gel alkyl bonded column packings intended for separation of both low- and high-molar-mass species [83–85]. The shapes of the SEC calibration dependences reflect—in absence of enthalpic interactions—the packing pore size distribution. They assume an S-shaped form, which can be expressed by the appropriate polynomials (Section 16.3.4). The polynomials of the third and fifth degrees show the best fit with the most experimental data. The column producers blend the column packing materials with different pore sizes so that large parts of the resulting SEC calibration dependence can be approximated with the straight lines (Figure 16.3a). It is then possible to write

$$V_R = A - B \log M \tag{16.12}$$

where A and B are the constants for the given column, eluent and polymer
or

$$V_R = D - C \log [\eta] M \tag{16.13}$$

where C and D are constants for the given column irrespectively of eluent and polymer nature. This is the expression of the linear part of the universal calibration dependence. Note that the V_R data are usually plotted on the abscissa, while for polynomials, the usual convency is applied, where the V_R values are plotted on the ordinate axis. The linear calibration dependences substantially simplify data processing in SEC though polynomials do not represent any problem for the advanced SEC software.

The calculation of molar mass averages and distributions for oligomers with at least partially discriminated peaks assumes determination of concentration of the particular *mers* in sample. The peak areas are considered either directly for the base line separated species or after a computer assisted deconvolution in case of partially resolved peaks. The detector response should be corrected for dependence of detector response on molar mass and functional groups of oligomers [124]. The principle of data processing for envelope-like chromatograms is evident from Figure 16.9. Computer divides the chromatogram into large number of slices, the fictitious fractions and calculates the concentration of polymer species from the slice heights or areas. The molar mass of each fraction is read from the calibration dependence. The entire computing normally takes seconds compared with hours needed for the hand-made calculation of each molar mass. From the point of view of chromatographers, each software represents an autonomously acting "black box." Still, it is important to properly set the base line and the peak limits (Section 16.4.5).

16.4.3 Applications of SEC

Applications of SEC are very broad and their detailed survey lies beyond the scope of this chapter. Numerous chapters in monographs as well as review papers are devoted to this matter [1–8,10–31]. The latest SEC applications can be found in the "R" designated reviews periodically published in

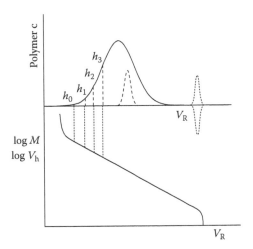

FIGURE 16.9 Schematic representation of principle of the molar mass calculation from the SEC chromatogram. The linear or polynomial calibration dependence log M vs. V_R or log M [η] vs. V_R is obtained with help of the narrow molar mass distribution calibration materials (Section 16.8.3). The heights h_i indicate concentration of each fraction i within sample ($c_i = h_i / \Sigma h_i$). The advanced computer software considers rather the areas of segments than their heights. The molar mass for each h_i or segment i is taken from the calibration dependence.

the journal *Analytical Chemistry*, for example [125,126]. The applications of SEC to copolymers are discussed, for example, in some journal articles [34,127].

Various criteria can be considered in the classification of the SEC applications. The most important are the analytical SEC procedures. The preparative applications, which encompass the purification of complex samples before their further treatment, draw rather wide attention. In this latter case, analytes are preseparated by SEC according to the size of their components and either macromolecular or low molecular fractions are subject to further analyses by other methods. The production oriented SEC did not find wide application in the area of synthetic polymers due to both the high price of organic solvents and the ecological considerations.

The analytical applications of SEC can be divided into the conventional and unconventional ones. The conventional SEC applications are based on the size separations of low-molar-mass substances, oligomers, and (high)polymers. In the case of oligomers and high(polymers), the objective is their molecular characterization, primarily in terms of their molar masses. As known, the separation selectivity of SEC is much lower compared to the interaction based HPLC. Therefore, SEC is only rarely applied to discrimination of complex mixtures of low-molar-mass substances aimed at their direct analysis. In this respect, SEC cannot compete with the enthalpic interaction–based HPLC methods. The exemption is the "finger print" identification of coal tars and crude oils [128–134]. The comparison of separation power of the enthalpic interaction dominated and the entropy controlled HPLC methods for oligomers does not close too favorably for SEC, either. Only the lowest members of oligomer samples—only exceptionally over $n=20$—can be at least in part mutually separated. The most important SEC applications concern (high)polymers. SEC allows determination of the molar mass averages and distributions, the long-chain branching parameters, the sizes of macromolecular coils in solution (in eluent) including their unperturbed dimensions, further the intrinsic viscosities (limiting viscosity members) and the constants in the Kuhn-Mark-Houwink-Sakurada viscosity law (Equation 16.3) of polymers in the eluent. The above data can be utilized for evaluation of various kinds of polyreactions leading to either building or degradation of macromolecules. Typically, these include the quantitative studies of polymerization and oxidization kinetics. Data attained by the fast or high-throughput SEC [11,135,278] can be applied for the direct control of the industrial polymerization processes and in the combinatorial polymer science.

The unconventional applications of SEC usually produce estimated values of various characteristics, which are valuable for further analyses. These embrace assessment of theta conditions for given polymer (mixed solvent–eluent composition and temperature; Section 16.2.2), second virial coefficients A_2 [109], coefficients of preferential solvation of macromolecules in mixed solvents (eluents) [40], as well as estimation of pore size distribution within porous bodies (inverse SEC) [136–140] and rates of diffusion of macromolecules within porous bodies. Some semiquantitative information on polymer samples can be obtained from the SEC results indirectly, for example, the assessment: of the polymer stereoregularity from the stability of macromolecular aggregates (PVC [140]), of the segment lengths in polymer crystallites after their controlled partial degradation [141], and of the enthalpic interactions between unlike polymers in solution (in eluent) [142], as well as between polymer and column packing [123,143].

In some cases, the selection of SEC for the particular application seems to be illogical. A typical example is the determination of the intrinsic viscosities of polymer samples by SEC. Here a simple and cheap glass viscometer is substituted by a rather expensive sophisticated SEC instrument. However, the advantage of SEC is evident when value of manpower, as well as speed and precision of determinations is taken into account.

Certain SEC applications solicit specific experimental conditions. The most common reason is the limited sample solubility. In this case, special solvents or increased temperature are inavoidable. A possibility to improve sample solubility and quality of eluent offer multicomponent solvents (Sections 16.2.2 and 16.8.2). The selectivity of polymer separation by SEC drops with the deteriorating eluent quality due to decreasing differences in the hydrodynamic volume of macromolecules with different molar masses. The system peaks appear on the chromatograms obtained with mixed eluents due to preferential solvation of sample molecules (Sections 16.3.2 and 16.3.3). The multicomponent eluents may create system peaks also as a result of the (preferential) sorption of their components within column packing [144,145]. The extent of preferential sorption is often sensitive toward pressure variations [69,70,146–149]. Even if the specific detectors are used, which do not "see" the eluent composition changes, it is necessary to discriminate the bulk sample solvent from the SEC separated macromolecules; otherwise the determined molecular characteristics can be affected. This is especially important if the analyzed polymer contains "a tail" of fractions possessing lower molar masses (Sections 16.4.4 and 16.4.5).

Numerous technologically important polymers, for example, many crystalline polyolefins are soluble only at the elevated temperature. Consequently, the high-temperature SEC instruments were designed, in which also dissolution and filtration of samples is performed at elevated temperature. Some SEC eluents such as hexafluoro isopropanol are very expensive, other ones are very aggressive—the extremes represent concentrated sulfuric acid [150] or hexamethyl phosphorotriamide [151] used as the mobile phase for some polyamides or cresols for polyesters [152]. Many biopolymers solicit the appropriate environment (eluent) and/or temperature of experiment below the ambient value to prevent their denaturation. Some polymers rapidly degrade in solution, for example, poly(hydroxy butyrate). These are to be separated as rapidly as possible and the data obtained should be corrected.

16.4.4 LIMITATIONS OF SEC

Numerous limitations of SEC result from the principles and the experimental arrangement of the method itself. Among them the necessity dominates to suppress both the enthalpic interactions and other nonexclusion processes within the column (Section 16.4.1). Consequently, the appropriate column packing and mobile phase should be chosen (Sections 16.8.1 and 16.8.2). Unfortunately this is often not the case in the common experimental practice (Section 16.4.5).

An important general feature of SEC is its limited separation selectivity. The separation itself proceeds in the pores of the column packing and its extent depends on both the effective pore diameter and the pore volume. The pore volume of a column is limited by the column size and by

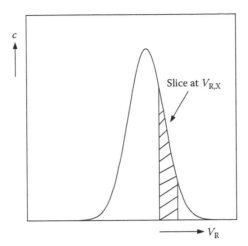

FIGURE 16.10 The SEC chromatogram of a statistical copolymer. Each slice contains macromolecules of nearly the same size but their molar masses may differ depending on their overall composition and architecture (blockines): hydrodynamic volume of macromolecules depends on all molecular characteristics of polymer species.

the mechanical instability of packings. The latter appears when the pore volume of otherwise solid column packing particles reaches about 85% of their total volume. The joining-up of several columns into a tandem is impractical as to time and eluent consumption. The problem can be also the rising pressure. The narrow pore size distribution SEC column packings afford increased separation selectivity but their molar mass separation range is reduced. For the sake of experimental simplicity, the universal "linear" SEC column packings are often applied. These provide simpler calibration dependence (Equations 16.12 or 16.13, Section 16.4.2) and allow separation of macromolecules over wide range of molar masses but their selectivity is further decreased (Section 16.8.1).

The significant intrinsic limitation of SEC is the dependence of retention volumes of polymer species on their molecular sizes in solution and thus only indirectly on their molar masses. As known (Sections 16.2.2 and 16.3.2), the size of macromolecules dissolved in certain solvent depends not only on their molar masses but also on their chemical structure and physical architecture. Consequently, the V_R values of polymer species directly reflect their molar masses only for linear homopolymers and this holds only in absence of side effects within SEC column (Sections 16.4.1 and 16.4.2). In other words, macromolecules of different molar masses, compositions and architectures may co-elute and in that case the molar mass values directly calculated from the SEC chromatograms would be wrong. This is schematically depicted in Figure 16.10. The problem of simultaneous effects of two or more molecular characteristics on the retention volumes of complex polymer systems is further amplified by the detection problems (Section 16.9.1): the detector response may not reflect the actual sample concentration. This is the reason why the molar masses of complex polymers directly determined by SEC are only semi-quantitative, reflecting the tendencies rather than the absolute values. To obtain the quantitative molar mass data of complex polymer systems, the coupled (Section 16.5) and two (or multi-) dimensional (Section 16.7) polymer HPLC techniques must be engaged.

16.4.5 Pitfalls of SEC

SEC is a very useful method. It is evident that it substantially influenced recent progress in both polymer science and technology. The SEC measurements are simple, fast, repeatable, and rather economic. The method, however, conceals several pitfalls. These are hidden in the measurements, in the data processing, and in the interpretation of results. Unfortunately, some SEC shortages are ignored by the method users and the results delivered by the computers are not evaluated from the point of view of their accuracy.

The recent round robin tests organized under auspices of IUPAC [153–155] have brought about alarming conclusions. Several commercial samples of oligomers and mostly linear (high)polymers were sent to participating laboratories. Intentionally, no measurement and data processing protocol was prescribed in most cases; everybody worked as he or she was used to. The standardization of the SEC measurements was attempted in the past but the resulting standard procedures differ in particular countries of EU, United States, and Japan. In fact, it was revealed [153] that especially the injected sample both volumes, v_i, and concentrations, c_i, differed quite substantially between participating laboratories. The dependence of the determined $\bar{M}_n$ and $\bar{M}_w$ values on v_i and c_i applied was quite evident. Further important observations were as follows [153]:

1. The participants reported high precision of results (intralaboratory repeatability of measurements). The $\bar{M}_w$ and $\bar{M}_n$ data obtained in particular laboratories scattered less than ±3%, often even in the range of ±1%. This is an excellent result, indeed. Unfortunately, the high repeatability of measurements may lead to a notion that the results are also exact.
2. The accuracy of results (the interlaboratory reproducibility of measurements) was poor or even dramatically low. The scatter in $\bar{M}_n$ and $\bar{M}_w$ values obtained for identical samples in the different participating laboratories reached several hundred of percent (2000% for poly(acrylic acid).
3. The majority of extreme data were received from the universities. It was evident that "the switch-on, inject, switch-off" approach was applied (likely by the students) without critical evaluation of data produced by the computers software. The least scattered M values were obtained in the industrial laboratories, in which evidently the skilled operators performed the measurements. Better data accuracy was obtained for polyamides [154] and for oligomeric polyepoxides [155] than for the unproblematic poly(dimethyl siloxane)s and even for the most "simple" polymer, polystyrene likely because only experts measured the latter "difficult" samples.
4. The important errors were produced by the drifts of base line and by the improper peak limits and base line settings. For example, the presence of a low-molar-mass tail in a polystyrene sample inflicted an error of 800% in its M_n value. In some cases, the low molecular tail was overlapped with the system peaks, while in other laboratories the software cut-off the low-molar-mass tail of chromatogram. It was likely considered just the base line perturbation by the computing system.

Neither the effect of the used instrumentation (the pumps, columns and detectors) nor the software applied for the data processing was evaluated from the point of view of accuracy of results. It is evident, that the worldwide standardization of both measurement and data processing are badly needed in SEC. At least the experimental conditions within the same laboratory should be kept constant. This may prove, however, difficult as the detectability of various samples differs substantially and the c_i/v_i is to be adjusted for each kind of sample.

Other important pitfalls lie again in the low selectivity of SEC, which does not allow identifying small amounts of the macromolecular admixtures that is the minor components of polymer blends. The bell-shaped chromatograms with a broad base and a slim upper part are often erroneously proclaimed to signalize the narrow molar mass distribution of sample. On the other hand, the accumulation peaks due to presence of macromolecules excluded from the packing pores (Section 16.8.1) are interpreted as the sign of sample bimodality. The "absolute" detectors may also contribute to erroneous conclusions concerning sample polydispersity (Section 16.8.1).

Evidently, any overlooked enthalpic interaction present in the system may substantially affect the calculated results because the measured retention volumes are shifted. The already mentioned system peaks caused by preferential solvation of polymer sample (Sections 16.2.2, 16.3.2, and 16.4.3), as well as peaks of air dissolved in sample solution may mask the sample fractions eluting at high retention volumes. Rather common is an increase of sample retention due to adsorption, enthalpic partition, and ion inclusion—all these phenomena are usually (improperly) denoted as "the adsorption." It has been also shown that the polar enthalpic interactivities of common polystyrene-*co*-divinylbenzene (PS-DVB) columns may be very large and depend on the column packing synthesis. The medium polarity poly(methyl methacrylate)s were to a different extent (or even fully) retained within some PS-DVB columns from a weak eluent, toluene [156] (Section 16.8.1). Large errors can be caused also by the sample fractions, which stay irreversibly retained within the SEC column from given eluent and at given temperature. It was shown that the pore volume and consequently the sample retention volumes of the non-retained macromolecules were substantially decreased as a result of the full retention of interactive macromolecules within the packing pores [157]. To correct the SEC data for this phenomenon, columns should be systematically recalibrated, regenerated, or exchanged. How many students do that?

In conclusion, it has to be remembered that even the outstanding, extremely useful, simple, robust, and repeatable SEC method may produce erroneous results if a desirable attemption is not paid to the conditions of measurement and to appropriate data processing.

16.5 COUPLED TECHNIQUES IN POLYMER HPLC

According to the accepted terminology, the term coupling denotes the combinations of the analytical procedures based on the same general physical principle, for example, the coupling of different retention mechanisms within the same HPLC column. The term hyphenation is reserved to the combinations of different physical principles, for example, in the detection. In polymer HPLC, especially when applying the porous column packing the entropy-controlled retention mechanism, exclusion, is always present. The addition of any enthalpy-driven retention mechanism leads to a coupled method of polymer HPLC. Practically all interaction or interactive procedures of polymer HPLC are in fact the coupled methods of polymer HPLC. The general aim of deliberate coupling the retention mechanisms within the same polymer HPLC column is as follows:

i. To substantially increase the separation selectivity according to certain molecular characteristic so that the effect of other characteristics can be—at least in the first approximation—neglected
ii. To suppress the effect of certain molecular characteristic on sample retention volume so that the resulting chromatogram (possibly) reflects mainly or even exclusively other characteristic(s). Usually, the effect of polymer molar mass on sample retention is partially or fully suppressed in order to determine the chemical structure and/or the physical architecture of sample. Ideally all but one molecular characteristics should be suppressed. To do so, however, a tandem of several columns would be needed. In the case of a two-component polymer system, the molar mass effect can be suppressed selectively for one component and the molar mass of the second component can be assessed by means of one single SEC measurement. A more general approach is represented by the two- and multi-dimensional procedures in polymer HPLC (Section 16.7), in which the coupled methods of polymer HPLC are included as an important (usually the first) separation steps.

16.5.1 ENTHALPY-CONTROLLED POLYMER HPLC

In this case, enthalpic interactions within the HPLC system exceed the exclusion effects (Figure 16.3b). The retention volumes of polymer species as a rule exponentially increase with their molar masses. The limitations of the resulting procedures were elucidated in Section 16.3: the retention of (high)polymers is usually so large that these do not elute from the column (Section 16.6). Therefore, the majority of enthalpy controlled HPLC procedures is applicable only to oligomers—up to molar mass of a few thousands $g \cdot mol^{-1}$. The selectivity of enthalpy-controlled HPLC separation of oligomers is much higher than in the case of SEC and, of course, the sequence of molar masses eluted from the column is reversed [158–160]. The so far mostly ignored problem of enthalpy controlled HPLC of oligomers may be the incomplete sample recovery. If the effect of enthalpy is reduced, problems with sample recovery are mitigated but at the same time the separation selectivity is reduced.

16.5.2 LIQUID CHROMATOGRAPHY UNDER CRITICAL CONDITIONS OF ENTHALPIC INTERACTIONS

The principle of the liquid chromatography under critical conditions (LC CC) was elucidated in Section 16.3.3. The mutual compensation of the exclusion—entropy and the interaction—enthalpy-based retention of macromolecules can be attained when applying in the controlled way the interactions that lead to either adsorption or enthalpic partition. The resulting methods are called LC at the critical adsorption point (LC CAP) or LC at the critical partition point (LC CPP), respectively. The term LC at the point of exclusion–adsorption transition (LC PEAT) was also proposed for the procedures employing compensation of exclusion and adsorption [161]. It is anticipated that also other kinds of enthalpic interactions, for example the ion interactions between column packing and macromolecules can be utilized for the exclusion–interaction compensation.

The basic experimental observations leading to establishment of LC CC were done in St. Petersburg by Belenkii and his coworkers [162–166]. Initially, the thin layer chromatography arrangement was employed and only later the LC CC idea was transferred into column liquid chromatography [163]. Other scientists from St. Petersburg, Skvortsov and Gorbunov [44,45], presented theoretical explanation of the unexpected experimental results when retention volumes became independent of polymer molar masses. Gorbunov and Skvortsov also introduced the term "chromatographic invisibility" for polymer chains eluted from porous packings irrespectively of their molar mass and proposed the important applications of LC CC: the assessment of molar masses of polymer blend components, which would coelute under SEC conditions [167], as well as characterization of the blocks in the block copolymers [168]. They have also anticipated that LC CC principle may be utilized in determination of functionality distribution in oligomers [169].

LC CC was independently applied to the characterization of oligomers by the group of Evreinov and coworkers in Moscow [170]. Experimental confirmation of characterization of polymer blends and block copolymers were again done in St. Petersburg [166]. Zimina continued her studies also in United Kingdom [172,173]. These researchers are considered founders of LC CC.

Numerous LC CC applications were performed by Pasch, Schulz, and Falkenhagen in Germany on (high)polymers and by Trathnigg in Austria on oligomers (for review see Ref. [23]). Presently, the LC CC method is rather widely used in the separation of complex polymer systems [26–28,31–33,173–194].

Unfortunately, the elegant and very useful idea of LC CC exhibits also some shortages. These were for the first time reviewed in [161] and include the following:

1. The extremely *high sensitivity of critical conditions* toward variations in the eluent composition and temperature, as well as toward the possible changes in the column interactivity. The latter are caused by the irreversible retention of some sample fractions. It may be very difficult to repeatedly prepare mixed eluents with exactly the same composition, as well as to prevent preferential evaporation from and absorption of humidity into eluent container. Philipsen et al. [188] proposed the temperature adjustments to cope with the steadily shifting critical conditions due to both the eluent composition and the column retentivity changes. To mitigate the above problems Pasch [23] proposed working near critical conditions, or in the critical range instead at the exact critical point.
2. *The possible temperature and pressure effects.* The friction between eluent molecules and column packing produces certain heat, which increases temperature within the column.

Both radial and axial temperature gradients may appear. As shown is Section 16.3.5, adsorption of polymers depends on temperature. Given the temperature and pressure dependence of the preferential sorption of the mixed eluent components within column packing [146–149], one can expect also considerable changes in the column interactivity with the temperature and pressure variations that may result in a possible gradual departure from the critical conditions.

3. The *excessive peak broadening*. Under critical conditions, the effect of molar mass distribution of sample disappears. Therefore, the peak width reflects only the chromatographic processes and these seem to be rather slow in the case of LC CC. The peak broadening in LC CC raises with the increasing polymer molar mass and with the decreasing packing pore size [180].

4. The *detection problems*. In LC CC, macromolecules elute together or in the vicinity of their initial bulk solvent. The latter exhibits changed composition compared to the mobile phase. This results from the phenomena of preferential solvation (Section 16.2.2) and complicates application of the nonspecific detectors (Section 16.9.1). The detection problem is also evident in the case of LC CC of block copolymers because the detector response as a rule depends or the nature and the relative length of both blocks.

5. The *limited sample recovery*. Parts of sample, preferably the fractions with the highest molar masses, may remain trapped within the LC CC columns. The irreversible sample retention increases not only with the increasing sample molar mass but also with the decreasing packing pore size [180]. The retained macromolecules affect interactive properties of the column (compare also Section 16.4.5). Unfortunately, the problem of sample recovery in LC CC of (high)polymers is mostly ignored though some new evidence was revealed [182].

LC CC proved especially useful in the characterization of functional oligomers [23,170,189–191]. Their main chain is eluted irrespectively of its molar mass and the observed variation of retention volumes is interpreted as the effect of nature and amount of the functional groups both average values and distributions. In some cases, even topology of chemically identical functional groups can be assessed. It is important to stress that the LC CC sample recovery is much less problematic with the oligomers than with the (high)polymers. Also the LC CC separation/characterization of polymer blends and block copolymers attracted rather wide attention [23,27,28,31–33,166,171–174,176,181,183]. One block is made chromatographically "invisible" and molar mass of the second block is assessed, usually applying the exclusion mode. The LC CC approach with another column packing and eluent that is by applying a different retention mechanism allows characterizing the second block. There are, however, some doubts on the effect of the "invisible" block on the retention of the SEC eluted block and vice versa [192,193]. Still, LC CC seems to produce at least semiquantitative valuable information on the block copolymers. LC CC enables also separation/characterization of one parent homopolymer present in a diblock copolymer [179,193].

The application of LC CC to the graft copolymers seems to be more problematic. The reasonable results were obtained only for those graft copolymers of polystyrene-*g*-poly(ethylene oxide), PS-*g*-PEO possessing short side chains of PEO [194].

The concept of entropy–enthalpy compensation resulting in the critical conditions of enthalpic interactions and the molar mass independent sample retention turned out useful also for the understanding several other coupled methods of polymer HPLC. It is accepted [195,196] that the polymer species tend to elute at the critical conditions also when either eluent strength or quality change within the HPLC system in the course of the HPLC experiment that is in the continuous and local gradient methods (Sections 16.5.3, 16.5.4, and 16.5.6). Irrespective of the problems and limitations of LC CC, its concept belongs to the important breakthroughs in polymer HPLC.

The identification of critical conditions for the given polymer-column packing system needs series of preliminary experiments. In the most practical LC CC experiments, the eluent strength or quality is adjusted tentatively to attain the molar mass independent elution at constant temperature.

The eluent strength is a parameter when adsorption retention mechanism is applied—LC CAP (Section 16.3.5), while the eluent quality is changed in the case of enthalpic partition—LC CPP (Section 16.3.6). A series of polymers under study—two different molar masses at minimum—are injected into eluents of different composition. The corresponding experiments are rather time and material consuming. Cools et al. [197] proposed an intelligent approach for identification of the critical conditions. The retention volumes are measured in the given column for the given polymer at different molar masses in eluents with different compositions. The intercept of the plots of distribution constants or retention volumes vs. eluent composition for particular molar masses indicates (the vicinity) of the critical conditions. The choice of eluent components depends on the retention mechanism. In LC CAP, a weak liquid (solvent) is mixed with a strong one while in LC CPP a poor solvent or nonsolvent is added to a good solvent (Section 16.3.2). It is advised to successively decrease the elution power of the mobile phase and not vice versa. Further, it is to be stressed that samples must be dissolved in particular eluent, otherwise different elution patterns are obtained, not LC CC, viz. liquid chromatography under limiting conditions of enthalpic interactions, (LC LC) (Section 16.5.6). A useful survey of published LC CC systems was compiled by Macko and Hunkeler [33]. In order to directly check the sample recovery, the second injection valve with a large loop can be mounted between the pump and the sample injector. The loop of this additional valve is filled with a liquid efficiently promoting polymer elution. Its injection reveals if any fraction of sample was retained within the column in the course of the LC CC experiment.

16.5.3 ELUENT GRADIENT POLYMER HPLC

Porath proposed to employ a gradient in the mobile phase composition for separation of macromolecules already in 1962 [198]. The experimental basis for eluent gradient polymer HPLC (EG-LC) was laid in the thin layer liquid chromatographic arrangement independently in Kyoto by the team led by Inagaki et al. [199] and in Skt. Petersburg (that time Leningrad) by the Belenkii's group [200]. Teramachi et al. [201] was the first to publish the successful polymer separations of polymers in the column arrangement applying adsorption retention mechanism. Glöckner [20] pioneered the column EG-LC polymer separations applying phase separation (precipitation–redissolution) retention mechanism. These authors can be considered founders of EG-LC of macromolecules.

Similar to other coupled methods of polymer HPLC, for example, LC CC (Section 16.5.2), the choice of the column packing and the mobile phase components for EG-LC depends on the retention mechanism to be used. Adsorption is preferred for polar polymers applying polar column packings, usually bare silica or silica bonded with the polar groups. The eluent strength controls polymer retention (Sections 16.3.2 and 16.3.5). The enthalpic partition is the retention mechanism of choice for the non polar polymers or polymers of low polarity. In this case, similar to the phase separation mechanism, mainly the solvent quality governs the extent of retention (Sections 16.2.2, 16.3.3, and 16.3.7). It is to be reminded that even the nonpolar polymers such as poly(butadiene) may adsorb on the surface of bare silica gel from the very weak mobile phases and vice versa, the polymers of medium polarity such as poly(methyl methacrylate) can be retained from their poor solvents (eluents) due to enthalpic partition within the nonpolar alkyl-bonded phases.

In EG-LC, polymer is injected into the mobile phase, which prevents its elution. It is an adsorli or a nonsolvent in case of either adsorption or enthalpic partition/phase separation retention mechanisms, respectively. Sample solvent must be weak enough or poor enough to avoid breakthrough of sample [202]: If the sample solvent efficiently prevents the full retention of sample, the latter travels along the column within the zone of its initial solvent. This is the situation, which appears in the isocratic liquid chromatography under limiting conditions of enthalpic interaction (Section 16.5.6). Subsequently, strength or quality of eluent is continuously or stepwise increased to allow polymer elution. The macromolecules of different chemical structure or architecture successively commence eluting to leave the EG-LC column in different retention volumes. The schematic representation of EG-LC separation of a two-component polymer blend by means of the continuous gradient is shown in Figure 16.11.

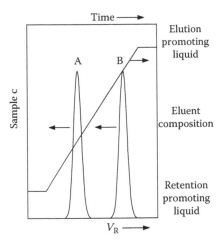

FIGURE 16.11 Schematic representation of eluent gradient polymer HPLC. Two polymer species A and B are separated. They exhibit different nature and different interactivity with the column packing (e.g., adsorptivity) or with the mobile phase (solubility). The linear gradient from the retention promoting mobile phase to the elution promoting mobile phase is applied. The focused peaks—one for each polymer composition/architecture—are formed in the appropriately chosen systems. Each peak contains species with different molar masses.

Various theories were developed to describe at least the elution tendencies. For example, Snyder et al. [203,204] and Shallikar et al. [205] considered the similarities between elution behavior of low and high-molar-mass sample molecules while Armstrong and coworkers [206–208] included also the specific features of macromolecular substances into their theoretical considerations. A simple kinetic model [209] explains the EG-LC separation mechanism of macromolecules on the basis of their higher mobility compared to the small molecules of mobile phase when eluted from the porous column packing (compare also Section 16.5.6). The role of difference in velocities of eluent and polymer in EG-LC was for the first time discussed by Glöckner for the phase separation retention mechanism [20]. Let us consider the adsorption retention mechanism. The injected sample is fully retained by adsorption near the column entrance. Next, the eluent strength is increased and the sample starts to desorb and to elute. In most cases, the smallest macromolecules are released first from the column packing so that a molar mass "gradient" is created in the mobile phase in the course of the initial stage of the desorption/elution process. The released macromolecules of certain nature move rapidly along the column to "catch" the eluent composition, which is weak enough to decelerate their continuing fast elution. It is likely the critical eluent composition [195,196]. The macromolecules of different molar masses successively accumulate on the "barrier" of eluent with the critical composition. From this moment, the polymer species of the same nature keep moving together, irrespectively of their molar mass. Macromolecules of another nature—composition or architecture—are released and eluted from the column packing at another eluent composition, with a different retention volume. The "barrier effect" of the eluent gradient is responsible for the peak focusing, which is typical for many EG-LC systems based on adsorption and enthalpic retention mechanisms. In the case of phase separation the situation is likely less favorable, mainly due to the slow dissolution/elution processes (Section 16.3.7).

EG-LC was successfully applied to many polymer separations, especially those aimed at the characterization of polymer blends and statistical copolymers [20,22,23,25–29,31,32,210–216]. Sato et al. [216] have demonstrated potential of EG-LC in the separation of chemically identical polymers according to their physical architecture.

The important shortage of some EG-LC methods is the reduced sample recovery, which results from the partial full retention of the sample within column packing [182]. This is rather comprehensible because the porous column packing may stay equilibrated with the retention promoting mobile phase during the EG-LC experiment. Surprisingly, however, a rather large part of sample may remain within EG-LC column even when the latter is eventually equilibrated with elution promoting liquid. The part of this retained sample is released when the blank gradient is repeated. As a result, the similar or even identical chromatogram can be generated without injection of any sample, just by the repeated running of the eluent gradient. The amount of released sample is small but not negligible [182]. This unexpected result needs further study.

In conclusion, eluent gradient polymer HPLC represents a useful tool for separation of complex polymer system. It belongs to the important constituents of several two-dimensional polymer HPLC procedures.

Interesting separations can be also obtained by the stepwise eluent gradient, which was for the first time described by Konáš and Kubín [217] and used also by Hutta and coworkers [218].

16.5.4 Temperature Gradient Polymer HPLC

In 1996, Chang with coworkers reported the high selectivity separations of some homopolymers according to their molar mass applying the temperature gradient within polymer HPLC columns [219]. This is considered the birth of the method, which is called temperature gradient interaction chromatography (TGIC). In TGIC, the column temperature is increased in (small) steps. It is likely that—similar to eluent gradient polymer HPLC—the fast elution of polymer species in the warmer eluent is decelerated by the slowly moving molecules of the cooler eluent. The cooler eluent is likely at the critical conditions, in this case at critical temperature. TGIC was used for the high selectivity analytical separations of homopolymers, and for the discrimination of polymer blend components including one parent homopolymer from the diblock copolymers [220–223]. In the latter case, one component of the mixture eluted unretained in the exclusion mode and the other one was subject to the temperature controlled elution. TGIC was applied also in the preparative fractionation of polystyrene to create the extremely narrow molar mass distribution fractions [224]. These fractions were used in the SEC band broadening studies (Section 16.4.1). Chang has likely managed to solve problems connected with the radial temperature gradients within TGIC columns. Still, the question of sample recovery may remain open. In any case, TGIC represents an interesting contribution to the coupled methods of polymer HPLC.

16.5.5 Enthalpy-Assisted SEC

The principle of enthalpy-assisted SEC (ENA SEC) is evident from Figure 16.3c and d (Section 16.3.3). The exclusion mechanism governs the order of elution that is the retention volumes decrease with the rising molar mass of sample. The presence of the controlled enthalpic interactions, however, raises the separation selectivity.

The simplest case of ENA SEC represents the separation of oligomers. The experimental conditions are selected so that only end groups (functional groups) on the oligomer molecules are retained due to enthalpic interactions. As the molar mass of oligomer rises, the effect of end groups decreases. This leads to the SEC-like elution order with increased selectivity (Figure 16.3c) [50,51]. The application of ENA SEC to the separation of (high)polymers is rare. It is evident that the attractive enthalpic interactions of eluted macromolecules with the column packing must be very weak—otherwise polymer species would be fully retained. The retention volumes of the interacting macromolecules increase compared with the noninteracting ones. The surface or bonded phase volume of the column packing available for interaction with the polymer species increases with the decreasing molar mass because of raising number of accessible pores. As a result, the increase of the retention volumes is more pronounced for smaller macromolecules. The log M vs. V_R dependence for interacting macromolecules is shifted toward higher retention volumes (Figure 16.3c) and separation selectivity increases. The typical example represents poly(methyl methacrylate) eluted from bare silica gel in tetrahydrofuran [225].

A rather new approach to ENA SEC of high polymers employs the critical conditions concept (Sections 16.3.3 and 16.5.2) [52,53]. The experimental conditions—eluent composition, column packing (and temperature)—can be identified, under which the retention volumes of the (high)polymers rapidly decrease with their increasing polymer molar mass. In other words, near the critical conditions, but on the exclusion side, the separation selectivity can be augmented for the macromolecules, which are large enough to be excluded from the column packing pores. Both adsorption [52] and enthalpic partition [53] retention mechanisms were applied in ENA SEC near the critical point. The important shortage of the ENA SEC methods, as with most coupled procedures of polymer HPLC, is the danger of reduced sample recovery.

16.5.6 LIQUID CHROMATOGRAPHY UNDER LIMITING CONDITIONS OF ENTHALPIC INTERACTIONS

The principle of the limiting conditions of enthalpic interactions (LC LC) has hardly any counterpart in the HPLC of low molecular substances. LC LC employs the combination of exclusion with enthalpic interactions in an unconventional way. A porous column packing is applied. The low-molar-mass substances permeate (practically) all pores of the column packing and therefore their elution rate is low. On the contrary, the polymer species are eluted rapidly from the same column provided they are fully or at least partially excluded from the packing pores. The appropriately chosen slowly eluting small molecules—usually one of the eluent components—promote the enthalpic interactions of macromolecules of certain nature within the LC LC column: adsorption, enthalpic partition, or phase separation. The zone of retention promoting small molecules does not allow fast elution macromolecules if situated in front of them. In this way, small molecules create a selective, slowly progressing barrier for interacting macromolecules. The barrier can be formed by the eluent itself while the sample is dissolved and injected in an elution promoting solvent: The sample molecules cannot leave the zone of their original solvent; otherwise they would be retained within the column (compare Section 16.5.3) [226,227]. Alternatively, the eluent promotes sample elution but the sample solvent creates the barrier [228]. Even more efficient approach is the injection of a narrow zone of retention promoting substance in front of sample solution [229]. This latter approach can be applied also if the barrier zone is a nonsolvent for the sample [230]. The interacting polymer is eluted slowly behind the barrier while the noninteracting macromolecules elute rapidly in the exclusion mode. In this way, the interacting macromolecules are rapidly and efficiently separated from the noninteracting ones. Employing the two above experimental arrangements and the three so far tested retention mechanisms, six LC LC procedures were established and successfully tested.

The first papers describing an LC LC separations of polymer blends due to the barrier effects were published in the mid-1990s [226,227] and the explanation of separation principle was originally presented in 1996 [231]. The review of both LC LC procedures and systems, as well as their first applications can be found in Ref. [28].

The principle of LC LC separation with a narrow zone of an adsorli injected before sample solution called liquid chromatography under limiting conditions of desorption (LC LCD) is evident from Figure 16.12. It is also possible to apply multiple barriers in order to separate three and more components, such as statistical copolymers of different compositions or parent homopolymers from diblock copolymers [232]. The basic features of the LC LC methods are the following:

1. The applicability of LC LC to the (ultra) high-molar-mass polymers. Contrary to liquid chromatography under critical conditions, LC LC likely exhibits no upper limit of the sample molar mass [234]. On the other hand, if the difference in elution rates of the barrier forming small molecules and the oligomeric species is too small, the LC LC separation becomes insufficient. As a result, LC CC and LC LC may be considered mutually complementary from the point of view of their applicability in the

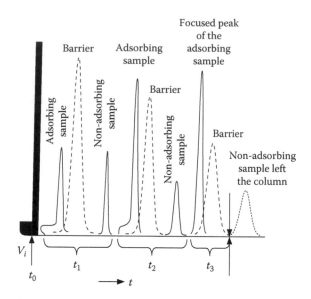

FIGURE 16.12 Schematic representation of the LC LCD procedure (adsorption retention mechanism with a narrow barrier of adsorli injected in front of sample). Large volume of sample is depicted. The sample contains two polymers exhibiting different adsorptivities. The fast (SEC-like) elution of adsorbing polymer is hindered by the barrier of adsorli while the non-adsorbing species are freely eluted in the exclusion mode. Four stages of the process are shown. The peak focusing is demonstrated.

    two-dimensional polymer HPLC (Section 16.7): LC CC preferably for oligomers and LC LC for (high)polymers.

2. The experimental simplicity: the common instruments can be utilized, preferably but not necessarily equipped with the column thermostat. An additional injecting valve can be easily mounted between the sample injector and the LC LC column in case of the LC LC procedures, which apply the zone of the barrier liquid injected in front of sample. Alternatively, an autosampler is to be used.

3. The unprecended robustness of most LC LC procedures. While variations in the eluent composition as small as 0.1 wt.% may produce large shifts in the LC CC retention volumes, a change in eluent composition as large as 20 wt.% is allowed in many LC LC systems. Similarly, also rather significant variations in temperature are often tolerated.

4. LC LC as rule produces narrow, focused polymer peaks because solutes accumulate at the edge of barrier. It is supposed that here again the critical conditions of adsorption and enthalpic partition represent the limits of the barrier composition, which hinders the fast progression of polymer species.

5. The peak focusing allows injections of the extremely large sample volumes, which may easily reach 10% of the total volume of the LC LC columns. This is an important advantage considering the two-dimensional liquid chromatography. For example, the LC LC column effluent can be directly forwarded into an online SEC column for further separation/characterization. The LC LC principle can be applied not only for polymer separations but also for reconcentration of polymer solutions, for example, of (diluted) effluents leaving other columns applied in polymer HPLC. This may be utilized in the multidimensional polymer HPLC.

6. The sample capacity of the optimized LC LC procedures is very high also in the terms of injected concentrations. This allows the discrimination and further direct characterization of minor (<1%) macromolecular admixtures in polymer blends, including parent homopolymers contained in block copolymers [232].

7. A very important advantage of LC LC procedures utilizing the small volume barriers lies in the high or even full sample recovery. Typical example is LC LCD. This is rather comprehensible because the LC LCD column packing is in a continuous equilibrium with the elution promoting mobile phase [235]. This equilibrium is likely only little perturbed in the course of the brief contact of the column packing with the retention promoting zone.

A unique application of LC LCD is the already mentioned separation of both parent homopolymers from the diblock copolymers [232], which is in fact a three-component polymer system. Therefore two barriers must be used. One block in the block copolymer and the corresponding homopolymer are noninteracting, that is, they do not exhibit any adsorption within a given column and a given mobile phase. The block copolymer contains the noninteracting block as a sort of a "diluent." This is why it is less adsorbed than the interactive homopolymer. The two barriers that possess a different blocking efficiency are introduced into the column behind the sample with appropriate time delays. In this way, the retention volumes of polymers, which are eluted behind the barriers can be properly adjusted. The first barrier is the more efficient one. It hinders the fast progression of the block copolymer. The less efficient barrier is introduced later, in front of sample solution. It does not decelerate the block copolymer but it still hinders the fast elution of the homopolymer. The action of two barriers is schematically evidenced in Figure 16.13. As demonstrated with the model systems of polystyrene-*block*-poly(methyl methacrylate) the LC LCD procedure could base line separate all three components within less than 2.5 min [232].

The selected latest LC LC studies are as follows: adsorption retention mechanism [233–236]; enthalpic partition retention mechanism [237]; and phase separation retention mechanism [229]. It is anticipated that the LC LC procedures will find numerous applications in the different areas of the polymer synthesis/characterization.

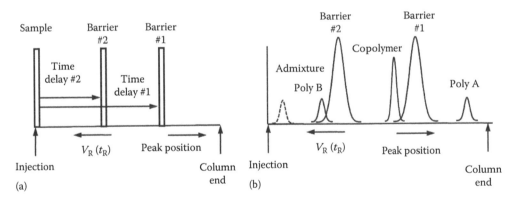

FIGURE 16.13 Schematic representation of separation of a block copolymer poly(A)-block-poly(B) from its parent homopolymers poly(A) and poly(B). The eluent promotes free SEC elution of all distinct constituents of mixture. The LC LCD procedure with two local barriers is applied. Poly(A) is not adsorptive and it is not retained within column by any component of mobile phase and barrier(s). At least one component of barrier(s) promotes adsorption of both the homopolymer poly(B) and the block copolymer that contains poly(B) blocks. (a) Situation in the moment of sample introduction: Barrier #1 has been injected as first. It is more efficient and decelerates elution of block copolymer. After certain time delay, barrier #2 has been introduced. It exhibits decreased blocking (adsorption promoting) efficacy. Barrier #2 allows the breakthrough and the SEC elution of block copolymer but it hinders fast elution of more adsorptive homopolymer poly(B). The time delay #1 between sample and barrier #1 determines retention volume of block copolymer while the time delay #2 between sample and barrier #2 controls retention volume of homopolymer poly(B). (b) Situation after about 20 percent of total elution time. The non retained polymer poly(X) elutes as first. It is followed with the block copolymer, later with the adsorptive homopolymer poly(B), and finally with the non retained low-molar-mass or oligomeric admixture. Notice that the peak position has an opposite sign compared to retention time t_R or retention volume V_R.

16.6 HPLC-LIKE METHODS: FULL RETENTION–ELUTION PROCEDURES

Compared to the low-molar-mass substances, it is much easier to attain the (selective) full retention of polymer sample component(s) within a HPLC column (Section 16.3). Subsequently, experimental conditions are abruptly changed and a fraction of sample—for example, a polymer of certain nature—is selectively and quantitatively released. The macromolecules of other nature are either unretained in the first step or not released in the second step. This is the basis for the full retention–elution separation procedures (FRE) [28]. FRE itself resembles the solid phase extraction (SPE) methods for preseparation/purification of low-molar-mass analytes. However, in the most SPE methods, the sample volume is limited because of the breakthrough phenomena, while even extremely large sample volumes can be processed in FRE of polymers. The only limit of the sample size is the packing capacity in term of its saturation. All three retention mechanisms discussed in this chapter (Section 16.3) can be utilized in FRE, however, both retention and elution processes must be very fast, otherwise the undesirable band broadening and even the zone splitting may occur. This is likely the case with the precipitation–redissolution and partially also with the enthalpic partition retention mechanisms. On the other hand, it was demonstrated that both attachment and detachment of macromolecules onto/from the nonporous surfaces due to adsorption processes are very fast and they can be easily controlled [238–240]. The resulting procedures are designated the full adsorption–desorption (FAD). The principle of FAD is demonstrated in Figure 16.2, where the difference is shown between rather common courses of adsorption isotherms for low- and high-molar-mass substances. Up to a threshold concentration of polymer, a saturation point, virtually all macromolecules are attached to the adsorbent surface and their concentration in supernatant is negligible. As mentioned in Section 16.2, this kind of dependence between the concentration of polymer in supernatant and the amount of adsorbed polymer is designated the high-affinity adsorption isotherm. Evidently, FAD works only with the sample concentrations below the saturation point. The complicated, slow exchange processes take place on the adsorbent surface above the saturation point. In this case, the adsorption is not complete and a fraction of the nonadsorbed polymer appears in the supernatant. The easiest way to make the polymer species quantitatively desorbed is the adjustment of the eluent strength either by the instanteous switching of the eluent composition or by the injecting of a defined zone of desorli into column. The desorbing power of eluent can be changed in several steps so that the adsorbed polymer sample is successively, in the well-controlled way, released from the FAD column. The FAD column can be easily on-line combined with the (noninteractive) SEC column (Figure 16.14). In this way, both the amount and the molar mass average and distribution of polymer released from the FAD column can be easily determined. Applying appropriate detector(s) (Section 16.9.1), also other molecular characteristics of the eluted macromolecules can be assessed. The FAD/SEC combination allowed separation and molecular characterization of the multicomponent polymer blends (so far up to six polymers of different nature) [241,242] including the minor macromolecular admixtures (<1%) present in the major component matrix [243]. The FAD procedure was also utilized for the separation of the noninteracting parent homopolymer from the block copolymer [244]. The solvent strengths that allowed the adsorptive attachment and the following release of macromolecules extensively depended on their molar mass. Therefore, the attempts to apply the FAD/SEC combination to separation of statistical [245] and block [246] copolymers were not fully successful: two molecular characteristics simultaneously affected elution of sample fractions. The FAD approach enabled the dynamic study of the polymer adsorption/ exchange/desorption processes [247], and the reconcentration of (very) diluted polymer solutions [248]. The reconcentration factor as high as 600 was easily attained. Consequently, the automated set of the FAD columns could be utilized in storing, reconcentration, and reinjection of fractions leaving the first dimension column in 2D-LC [249,250] (Section 16.7): The sample reinjection takes place in an appropriate and exactly known instant, which is especially important for the SEC "dimension" of the 2D-LC methods.

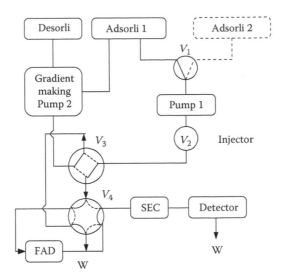

FIGURE 16.14 Scheme of the full retention/elution: SEC system. Adsorption/desorption mechanism serves as example. V_1, V_3, V_4 are the switching valves, V_2 is the sample injector (or another LC column). Instead of the gradient making device a set of containers with a series of mobile phases that differ in their strength and another switching valve can be used. The adsorli eluent transports the sample into the full adsorption/desorption (FAD) column, which is then successively flushed with eluents of the increasing strengths. The nonadsorbed or displaced fractions from the FAD column are transported into the SEC column, which has been previously equilibrated with the corresponding displacer.

16.7 TWO-DIMENSIONAL POLYMER HPLC

As explained in Sections 16.4 and 16.5, the comprehensive characterization of complex polymer systems is hardly possible by the SEC alone. SEC employs only one retention mechanism which simultaneously responds to all molecular characteristics of sample. Similarly, also the coupling of the different retention mechanisms within one single column only exceptionally allows fulfilling this task. Evidently several retention mechanisms should be applied in a tandem approach that is within at least two different on-line chromatographic systems. This is the basic idea of the two- and multidimensional polymer HPLC. In the present section, the principles of two-dimensional polymer HPLC, 2D polymer HPLC or (2D-LC) will be briefly elucidated. There are several reviews available [23–31,249,250] dealing with the 2D polymers. It is anticipated that also the three- and multidimensional polymer HPLC will be developed in future.

Numerous researchers attempted the two-dimensional separations of macromolecules. For example, Balke and Patel [251,252] employed two different SEC or SEC-like procedures for separation of statistical copolymers. It is likely that the first practically applicable 2D-LC separations of macromolecules were done by Kilz and coworkers [253,254] who pioneered modern two-dimensional polymer HPLC.

In any 2D HPLC, it is important to attain certain degree of both the complementarity and the orthogonality between the two separation "dimensions" [255–257]. The so far most universal approach to 2D polymer HPLC assumes the partial or possibly full suppression of the molar mass effect in the first dimension of the separation so that the complex polymer is separated mainly or even exclusively according to its chemical structure. Selected coupled methods of polymer HPLC are to be applied to this purpose. In the second dimension of separation—it is usually SEC—the fractions from the first dimension are further discriminated according to their molecular size. Exceptionally, SEC can be used as the first dimension to separate complex polymer system according to the molecular size. This approach is applicable when the size of polymer species does not depend or only little depends on their second molecular characteristic, as it is the case of the stereoregular polymers

[258,259]. The SEC fractions were further separated applying a coupled method of polymer HPLC, LC CC in the latter case. So far, the most important combinations are LC CC plus SEC and EG LC plus SEC. It is believed that also other combinations such as LC LC plus SEC will find applications in 2D polymer HPLC [250].

16.8 MATERIALS FOR POLYMER HPLC

16.8.1 COLUMN PACKINGS

The general requirements placed on the column packings for polymer HPLC are similar to qualifications of packings for HPLC of small molecules. These materials are subject to a very intensive research that results in large volume of literature data, many patents, and the known-how secrets of producers. This is comprehensible given the investment into and profit from the HPLC column production. The HPLC column packings must be mechanically resistant, and possess as regular as possible, preferentially strictly spherical particle shape. The spherical particles allow creation of stable packing beds with the minimized band broadening phenomena within interstitial volume (Section 16.4.1). Except for the few cases such as the FRE/FAD methods (Section 16.6), the totally porous particulate column packings are applied in polymer HPLC. The so far developed monolithic HPLC columns exhibit relatively large volumes of the flow-through pores (channels) and the low volume of mesopores (separation pores). This impairs applicability of monoliths in polymer HPLC [260]. The pore geometry must allow the fast mass transfer to keep the band broadening as low as possible. The interactivity of column packings must be well adjustable from the materials exhibiting minimum attractive/repulsive enthalpic interactions required in SEC (Sections 16.4.1 and 16.4.5) up to the highly interactive column packings designated for the coupled and HPLC-like methods (Sections 16.5 and 16.6, respectively).

The details on column packings for polymer HPLC can be found in numerous monographs and (review) papers. The entire monograph devoted to column packings for SEC was edited by Wu [25]. In this section, only general information will be presented.

As a rule, the particles of porous column packings for polymer HPLC are formed by the arrays of nonporous nanosized primary spherical particles, nodules. Such structure is often designated corpuscular or globular. The nodules arise as a result of the phase separation processes in the course of polymerization/crosslinking reactions. The phase separation is controlled by porogenes, usually the nonsolvents or poor solvents for polymer forming the nodules or by the appropriate diluents of the polymerization system. The porogenes are added to the polymerization system together with the crosslinking agent. Therefore the process is often called heterogeneous crosslinking. The pores in fact represent the free space among nodules. The larger nodules the larger are also the pores. The grape-like structure created by nodules contains numerous defects, which increase the pore volume and broadens the pore size distribution. The optimum pore volume within a particle is about 80%–85%, as the mechanical stability of material may rapidly decrease with the further increase in particle porosity. The corpuscular structure of column packing particles exhibits a rather broad pore size distribution. It seems that the sponge-like pore structure would be an optimum for the SEC column packings. The sponge-like structure is formed, for example, when the phase separation that accompanies the polymerization/crosslinking process in presence of porogenes is interrupted before spherical nodules are created. Unfortunately, the nonequilibrium phase separation in the polymerizing organic systems is difficult to control. The unfinished phase separation is also employed in creation of porous glasses. In the sponge-like structure, the pores exhibit narrow size distribution and the mechanical stability of particles is maintained up to a high pore volume.

In the SEC separations of synthetic polymers, the dominant position assume the poly(styrene-*co*-divinylbenzene) (PS/DVB) based materials. Divinylbenzene is the crosslinking agent for polystyrene chains. It allows creation of insoluble, mechanically stable three-dimensional structures. Chemically similar are the poly(divinyl benzene) (DVB) gels. It is likely that the walls of PS/DVB and DVB gels, as well as other column packings made of the organic polymers are not well defined. Free ends of macromolecules dip into the pore volume and they may form a solvated "liquid" phase [77,78], which is available for enthalpic partition. Surprisingly, the PS/DVB and DVB based SEC column packings of low polarity are used with the eluents of various strengths: from toluene to dimethyl formamide (Section 16.8.2) as well as for polymers differing in their polarity. Evidently, this is not an optimum approach because the nonsymmetricity of the chromatographic system may affect retention of macromolecules, for example, due to incompatibility effects (Section 16.4.1). As a result, the observed retention volumes may be shifted due to undesirable enthalpic interactions and the universal calibration (Section 16.4.2) is invalid. Kilz [261] proposed the logical concept of the *magic triangle*, according to which the polarity of the SEC column packing should match polarities of sample and eluent (Figure 16.15). Unfortunately, most SEC operators ignore problems connected with the non-symmetricity of the SEC systems. Apart from the rather high prices of the SEC columns, the possible reason is the convenience.

The SEC results can be impaired also by a rather high polar interactivity of some PS/DVB SEC columns [156]. The latter is likely caused by the presence of polar groups on the packing surface. The polar groups may result from the incorporated protective colloids applied in the course of packing synthesis such as derivatives of cellulose, poly(vinyl alcohol) or poly(*N*-vinyl pyrrolidone). The alternative sources of polar groups within the PS/DVB based SEC column packing are the reactions of secondary crosslinking [264]. Secondary crosslinking increases mechanical strength of porous particles and allows application of eluents exhibiting different polarities. The polar interactivity of the SEC columns is responsible for the shifts of the retention volumes of polymers with medium or high polarity in the low strength eluents. In the extreme cases, polymers may be fully retained within SEC columns, for example, poly(methyl methacrylate)s in toluene or poly(2-vinyl pyridine) s in tetrahydrofuran.

Besides the gels based on PS/DVB, several other materials are available in the form of heterogeneously crosslinked, hard porous particles suitable for SEC. These are for example based on

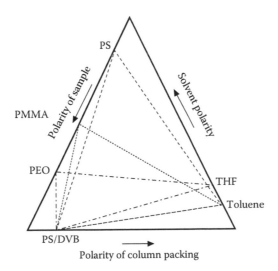

FIGURE 16.15 The principle of the "magic triangle" [261] that depicts the desirable symmetricity of the SEC system. The symmetric example (- - -) polystyrene-*co*-divinylbenzene (PS/DVB) column packing, polystyrene sample, toluene eluent. The nonsymmetric example: (· · ·) PS/DVB column, poly(methyl methacrylate) sample, toluene eluent. The highly nonsymmetric example (-----) PS/DVB column, poly(ethylene oxide) sample, THF eluent.

poly(hydroxyethyl methacrylate) also called ("hydroxylated polyacryl" and "poly(methacrylate)s") or poly(acryl amide). As a rule, the producers do not disclose the chemical composition of their nonPS/DVB SEC packings.

The silica gel–based column packings are the active materials of choice for polymer HPLC employing both exclusion and interaction retention mechanisms. These are either bare or bonded with various groups. C-18 alkyls and $–CH_2–CH_2–CH_2–NH_2$ groups are most popular for reversed-phase and normal-phase procedures of polymer HPLC employing the nonpolar and polar interactions, respectively.

The important parameter of porous packings for polymer HPLC is the pore size. In the broader sense, this term includes both pore diameter and volume. As the actual pore shape is unknown, the mean effective pore radius or diameter, $\bar{d}$ or $\bar{D}$, respectively, should be considered (Section 16.3.4, Equations 16.7 and 16.8, and Section 16.4.1). In SEC, $\bar{D}$ of pores must match the sizes of macromolecules dissolved in eluent. The role of effective pore diameter in the coupled polymers HPLC is so far not fully understood. It was shown [180] that too small $\bar{D}$ caused reduced sample recovery due to the flower-like conformation of macromolecules [235,262,263]. This is the situation similar to that depicted in Figure 16.6. Not only the concentration gradient but also the (strong) attractive interactions between polymer segments and column packing intensively pull macromolecules of large size into the pores. The direct contact between polymer chains and packing pores is created. The polymer in the pores forms a stem while the rest of macromolecule is the crown of a flower protruding over pores. The crown may prevent diffusion of the small molecules of displacer, for example, of desorli molecules into the pores. Macko et al. [265] has shown that the nonpolar polyolefins can reptate, even into the very narrow pores of zeolites to be strongly adsorbed; their release may be very difficult. On the other hand, the accessible surface and bonded phase volume of the narrow pores may be too small to warrant the necessary extent of the nonhindered enthalpic interactions. This may appear, when the attractive interactions between macromolecules and packing materials are too weak. In this case, the important role may be played by the interactions taking place in the pore orifices. This means that two different silica gels with the same mean effective pore diameters, for example, 6, 10, or 12 nm, with a similar both concentration and topography of free silanols and under otherwise identical experimental conditions may exhibit very unlike enthalpic interactivities. So far, the only way is the optimization of system by means of trial-and-error experiments. Many useful information on the silica-based HPLC column packings are collected in the monograph [266].

16.8.2 MOBILE PHASES

The dominating SEC mobile phase is tetrahydrofuran, THF. This solvent exhibits the following advantages:

- An appropriately low refractive index to warrant the use of refractometric and light scattering detectors (Section 16.9.1)
- Large dissolution power for many polymers from the non polar polydienes up to polar poly(vinyl chloride) or poly(2-vinyl pyridine)
- Good UV light transparency, down to about 215–230 nm wave length
- Relatively low toxicity
- Reasonable price

At the same time, THF forms

- Explosive peroxides and must be stabilized (it should be distilled under nitrogen, very carefully checking the actual boiling point)
- Charge-transfer complexes with oxygen, which absorb in the UV wavelength range
- Unwanted oligomeric substances

THF is highly flammable and its vapor pressure at ambient temperature is rather high. It may preferentially evaporate from various mixed mobile phases. The preferential evaporation brings about the base line drifts due the evaporation from the sample container and/or the pronounced system peaks due to the evaporation from the sample solution.

THF well dissolves oxygen from the air and the unwanted peaks are observed in the area of high retention volumes (Section 16.4.5). THF is highly hydroscopic and it readily absorbs large amounts of moisture. As a result, even the "well stored" THF eluents may contain the non-negligible amount of water, which may affect retention volumes of polymers both in the SEC and in coupled modes of polymer HPLC [28,267,268]. Azeotropic mixture of THF with water contains about 4.5 wt.% of water and its boiling point differs less than 3°C from the boiling point of dry THF at the atmospheric pressure.

Di- and tri-chlorobenzenes are used as the high-boiling-point eluents in SEC of polyolefins and the highly polar solvents such as poisonous and expensive hexafluoro isopropanol in SEC of polyesters and polyamides. Some unusual SEC eluents are mentioned in Section 16.4.3.

As a medium strength liquid (Table 16.1), THF is commonly used also in the coupled methods of polymer HPLC. It promotes desorption of medium polar polymers such as poly(acrylate)s and poly(methacrylate)s including poly(methyl methacrylate) from the nonmodified silica gel. Other strong(er) solvents widely used in the coupled polymer HPLC methods are acetonitrile that exhibits high UV transparency, and dimethyl formamide. The latter solvent readily decomposes into amine and formic acid and its strength may differ from batch to batch.

The popular weak solvents, which promote polymer adsorption are cyclohexane, dichloro methane, dichloroethane and toluene. Unfortunately, any standardization of the eluent purity/content of admixtures does not exist. This might be one of reasons why it is sometimes difficult to maintain long-term repeatability and interlaboratory reproducibility of measurements in polymer HPLC.

16.8.3 CALIBRATION MATERIALS

As explained in Sections 16.3.4, 6.4.1, and 16.4.2, SEC is a nonabsolute method, which needs calibration. The most popular calibration materials are narrow molar mass distribution polystyrenes (PS). Their molar mass averages are determined by the classical absolute methods—or by SEC applying either the absolute detection or the previously calibrated equipment. The latter approach may bring about the transfer and even the augmentation of errors. Therefore, it is recommended to apply exclusively the certified well-characterized materials for calibrations. These are often called "PS calibration standards" and are readily available from numerous companies in the molar mass range from about 600 to over 30,000,000 g·mol⁻¹. Their prices are reasonable and on average (much) lower than the cost of other narrow MMD polymers. Other available homopolymer calibration materials include various poly(acrylate)s and poly(methacrylate)s. They are, similar to PS, synthesized by anionic polymerization. Some calibration materials are prepared by the methods of preparative fractionation, for example, poly(isobutylene)s and poly(vinylchloride)s.

It is necessary to say that the molar masses of the calibration materials quoted by some suppliers may differ from the actual values. In other words, some commercial calibration materials may "jump out" from the calibration dependences because their molar mass determined by the producer is wrong.

It is to be stressed again (Section 16.4) that the molar masses of homopolymers different from PS, as well as the molar masses of copolymers determined by SEC directly from the polystyrene calibration dependences should be designated *polystyrene equivalent values*. The latter more or less differ from the actual values and should be used only for the assessments of tendencies.

All above homopolymers are used also for the identification of suitable conditions for the coupled polymer HPLC techniques. Typical examples are liquid chromatography under critical (LC CC) and limiting (LC LC) conditions, and eluent gradient liquid chromatography (EG LC). For the development of latter methods, several defined statistical and block copolymers are available.

These materials, however, as a rule exhibit rather broad chemical composition distribution. Block copolymers may contain important amounts of parent homopolymer(s) [232,244,269]. In any case, it is to be kept in mind that practically all calibration materials contain the end groups that differ in the chemical composition, size, and in the enthalpic interactivity from the *mers* forming the main chain. In some cases, also the entire physical architecture of the apparently identical calibration materials and analyzed polymers may differ substantially. The typical example is the difference in stereoregularity of poly(methyl and ethyl methacrylate)s: while the size of the isotactic macromolecules in solution is similar to their syndiotactic pendants of the same molar mass, their enthalpic interactivity and retention in LC CC may differ remarkably [258,259].

Numerous information on various polymers can be found for example in *Polymer Handbook* [37] and on the particular calibration materials in the monographs [23–25,30].

16.9 INSTRUMENTATION AND WORKING MODES IN POLYMER HPLC

The instruments for polymer HPLC except for the columns (Section 16.8.1) and for some detectors are in principle the same as for the HPLC of small molecules. Due to sensitivity of particular detectors to the pressure variations (Section 16.9.1) the pumping systems should be equipped with the efficient dampeners to suppress the rest pulsation of pressure and flow rate of mobile phase. In most methods of polymer HPLC, and especially in SEC, the retention volume of sample (fraction) is the parameter of the same importance as the sample concentration. The conventional volumeters— siphons, drop counters, heat pulse counters—do not exhibit necessary robustness and precision [270]. Therefore the timescale is utilized and the eluent flow rate has to be very constant even when rather viscous samples are introduced into column. The problems with the constant eluent flow rate may be caused by the poor resettability of some pumping systems. Therefore, it is advisable to carefully check the actual flow rate after each restarting of instrument and in the course of the long-time experiments. A continuous operation—24 h a day and 7 days a week—is advisable for the high-precision SEC measurements. THF or other eluent is continuously distilled and recycled.

An important challenge for pumping systems represent the frequent changes of mobile phase polarity and viscosity, especially in the course of the coupled polymer HPLC experiments. If possible, a separate SEC instrument should be under operation for each particular eluent.

16.9.1 DETECTORS

Sample detection in polymer HPLC is a large problem. Only few polymers bear chromophores that allow direct application of photometers, typical for HPLC of many low-molar-mass substances. The use of detectors that operate in the infrared region of spectrum is complicated by the absorptivity of eluents. Still, there is observed some revival in this area, especially in the SEC of polyolefins.

The most common SEC detectors are the devices continuously monitoring the refractive index of effluent, the RI detectors. These are nonspecific and monitor any change in eluent composition also due to the dissolved gases, humidity or preferential solvation (Sections 16.3.3, 16.4.1, and 16.4.5). Unfortunately, the RI detectors are sensitive also to temperature and pressure variations. This is why the RI detectors can be hardly used in the coupled methods of polymer HPLC that employ the continuous or local both eluent or temperature gradients. At the same time, the overall sensitivity of RI detectors to the sample concentration changes is rather limited. Therefore much larger volumes and concentrations of samples are to be applied in polymer HPLC compared to HPLC of low-molar-mass substances. Of the four types of RI detectors—deflection, reflection, Christiansen, and interferometers—the best known is the deflection type. The problems of the base line instability of some interferometric detectors were recently mitigated and their application becomes more common. Compared to deflexion and reflexion RI detectors, interferometers exhibit about one order of magnitude increased sensitivity. The monograph *Liquid Chromatography Detectors* [271] is a good starting point for discussion of principles of photometers and RI detectors as well as of other classical detectors.

Some specific properties of macromolecules are employed for the detection in polymer HPLC. It is primarily the large size of detected species, which is comparable with the wavelength of the visible light. As a result, the light beam interacting with the macromolecules is intensively scattered [272]. The extent of light scattering under otherwise constant experimental conditions depends on the molar mass of macromolecules.

$$\frac{R_\Theta}{K_{ls}c} = \bar{M}_w P(\Theta) - 2A_2 c M^2 P^2(\Theta) + \cdots \tag{16.14}$$

where
R_Θ is the excess of Rayleigh ratio [length^{-1}, as a rule cm^{-1}]
$P(\Theta)$ is the particle scattering function
A_2 is the second virial coefficient (Section 16.2.2) of polymer in eluent [mL$\cdot$mol^{-1}]
c is the polymer concentration in g$\cdot$mL^{-1}
M_w is it's the unitless weight-average molecular weight (Section 16.2.1)

K_{ls} is a constant expressed for the vertically polarized incident light as

$$K_{ls} - \frac{4\pi^2 n_0^2}{\lambda_0^4 N_A}\left(\frac{dn}{dc}\right) \tag{16.15}$$

where
λ_0 is the wavelength of the incident radiation in vacuum
n_0 is the refractive index of solvent at this wavelength
N_A is the Avogadro's number
dn/dc is the refractive index increment of given polymer in eluent (Section 16.2.2)

Numerous values of dn/dc are available in literature [37].
The excess Rayleigh ratio at the angle Θ is defined

$$R_\Theta = f\frac{I_\Theta - I_\Theta s}{I_0} \tag{16.16}$$

where
I_Θ is the intensity of light scattered at an angle Θ to the incident beam of the intensity I_0
$I_\Theta s$ is the intensity of light scattered by pure eluent
f is a constant for given instrument

The particle scattering function $P(\Theta)$ in Equation 16.14 equals R_Θ/R_0, where R_0 is the excess Rayleigh ratio at zero observation angle. Cantow et al. [273] pioneered the continuous light scattering detection in connection with the Baker–Williams polymer fractionation column. Ouano and Kaye [274,275] introduced light scattering detection of SEC column effluent employing laser light source. Wyatt [276] patented multiangle light scattering detector. The detailed theory and practice of the light scattering HPLC detection can be found for example in Refs. [276–279].

The flow-through light scattering detectors combined with the concentration detectors provide the values needed for the direct calculation of molar mass values for the given polymer in a given eluent. If the dn/dc and A_2 values are not available, they can be determined by the independent measurements. The modern flow-through light scattering detectors work either at a very low single angle (LALS detectors), where the angular scattering dependence can be neglected or at the 90° observation angle using right angle light scattering (RALS) detectors or simultaneously at up to 17 different angles using multiangle light scattering (MALS) instruments. The light source is as a rule a laser and the data are processed by the appropriate autonomous software. The variation of refractive index increment dn/dc with changing either polymer composition (e.g., in the case of copolymers) or eluent composition (e.g., in the case of eluent gradient polymer HPLC) brings important problems for the light scattering detectors. For example, following the proposal of Bushuk and Benoit [280], the light scattering measurements in three solvents should be performed in order to assess both molar mass and the compositional heterogeneity of copolymers. Consequently, the three SEC different mobile phases should be applied. It is important to note that not only the $\bar{M}_w$ but also the $\bar{M}_n$ values of polymer samples can be calculated from the light scattering detector response. Due to the band broadening in polymer HPLC, SEC in this case, the M_n values thus determined are, however, overestimated and the polydispersity indices (Section 16.2.1) are underestimated.

Another possibility for the direct determination of molar mass of macromolecules leaving the HPLC column offers the on-line continuous monitoring of effluent *viscosity*. The background for such measurements is the Kuhn-Mark-Houwink-Sakurada viscosity law (Equation 16.3), which does express the dependence of intrinsic viscosity of a polymer in solution [η] on its molar mass M (Section 16.2.2). Knowing the continuous changes of the [η] value of polymer in effluent the profile of molar mass changes can be assessed and averages values of sample molar mass can be calculated. This is the basis of the flow-through viscometric detectors, VISCO. First attempts to semicontinuously measure the viscosity of column effluent were likely done by Goedhard and Opschoor [281] and Mayerhoff [282]. The on-line continuous viscosity measurements were introduced by Ouano [283]. The intelligent "bridge arrangement" of VISCO detectors was proposed by Haney [284]. The pressure difference between the branch of the bridge in which pure eluent is transported and the effluent containing polymer is monitored (Figure 16.16). This arrangement allows compensation for the variations in the mobile flow rate. However, large changes in eluent viscosity inherently present in some methods such as eluent gradient polymer HPLC, EG LC (Section 16.5.3) or HPLC under limiting conditions of enthalpic interactions, LC LC (Section 16.5.6) render the application of viscometric detection practically impossible. Goldwasser [285] proposed a method of calculation the M_n values from the data obtained by the viscometric detector. Useful information on the VISCO detectors, their advantages, and limitations can be found for example in Refs. [31,278,279].

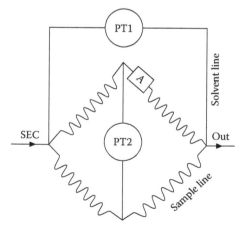

FIGURE 16.16 Schematic representation of the four capillary viscosity detector [284]. PT1 and PT2 are the pressure transducers. A is the hold-up reservoir. It decelerates the polymer solution so that the capillary which follows is still flushed with pure mobile phase when the sample solution has already entered the other capillaries.

Further important general characteristic of polymers, which is utilized in their detection is their nonvolatility. The evaporative light scattering (ELS) detectors continuously monitor intensity of light scattered by the (semi) dry polymer particles, which are formed by nebulization of the column effluent into microdroplets followed by the evaporation of eluent. The principle of ELS detectors was for the first time described by Smith [286] and the sophisticated versions of ELS detectors are presently available from several companies. The sensitivity of ELS detectors compared to RI detectors is generally about 20 times higher at a high baseline stability. In fact, the ELS detectors "do not see" eluent composition changes. This is why the ELS detectors are often used in coupled procedures of polymer HPLC. The process of nebulization, and the principles of light scattering on the particles are described in [287–291]. Unfortunately, the linearity of concentration response of ELS detectors is limited. Moreover, the size of the primary microdroplets depends on the eluent nature and on the polymer molecular characteristics—partially even on its molar mass [291–293]. This poses a limitation of applicability of ELS detectors for the quantitative measurements. Moreover, the quantitative ELS detection suffers from a rather poor day-to-day repeatability. An interesting analysis of the ELS detectors is presented in Ref. [291] together with a useful literature review.

A promising detection principle utilizes similar nebulization procedure as applied in ELS, namely the corona-charged aerosol detectors, CAD. In CAD, the aerosol particles interact with an ionized gas (usually nitrogen). The particles become charged and electrically detected [294]. It has been shown that the response of CAD does not depend on the nature of analyte. On the other hand, the size of the aerosol depends on the mobile phase composition and it has to be calibrated.

The non-volatility of polymers is also utilized in various interfaces. Column effluent is deposited on the appropriate moving support and eluent is continuously evaporated. The dry polymer is pyrolyzed and amount of carbon dioxide is monitored for example by means of the flame ionization detector. Various supports have been introduced: the endless wires or chains and also disks. The important problems include the low sample capacity of most kinds of supports and difficulties with their purification. The detector response can be extensively affected by the remnants of polymer sample, which was not completely pyrolyzed in the previous run. Boshoff et al. [295] have even proposed to deposit the column effluent on the moving chromatoplate and to apply the TLC detection principles. From the composition of pyrolytic gases, the polymer nature can be assigned. There is, however, a problem that the pyrolysis of a polymer blend may not pyrolyze in the same way as if the two individual polymers were pyrolyzed independently. Still the authors [297] claim to assess quantitatively even the compositional variation of some statistical copolymers from the constitution of their pyrolytic products. Schulz and King [296] proposed to discontinuously deposit the column effluent on the surface of vibrating piezoelectric crystal and to determine the amount of polymer after eluent evaporation monitoring the changes in vibration frequence. The modern approaches to the application of interfaces allow monitoring the selected properties of thus formed polymer film. The most attractive is the continuous monitoring of spectral properties of the polymer fractions deposited from the column effluent, primarily their absorption in the infra red domain of spectrum. In the latter case, the polymer film is created on the rotating germanium disk. As a rule, measurements are done at least at two appropriate wavelengths in order to correct for the film unevenness [298,299].

A very important tool for detection/characterization of macromolecules leaving the polymer HPLC became the mass spectrometry (MS) methods, the electrospray ionization, ESI and the matrix assisted laser desorption/ionization, MALDI mass spectrometries [300]. The modern MALDI instruments process the column effluent similar to the above described interfaces. However, the effluent, is continuously mixed with an appropriate matrix and deposited on the moving support. Next, the eluent is evaporated and the mixed film is irradiated by the pulses of the UV beam produced by the high-energy laser. The matrix absorbs the light pulses and transfers their energy to macromolecules, which "evaporate" in the ionized form. The stream of macromolecular ions is accelerated and detected in a time-of-flight mass instrument. In this way two different separation principles are hyphenated with the aim to suppress shortages of both methods: the limited selectivity of polymer

HPLC and the extensive complexity of ESI and MALDI spectra for multicomponent polymers with molar mass over about 10^4 g·mol^{-1}. Some applications of MALDI MS in polymer HPLC can be found for example in [300–303].

There were numerous attempts to apply other qualities proper to macromolecules in the SEC detection, such as turbidimetric titration [304], ebulliometry [305] and osmometry [306]. The precise flow-through osmometers would certainly bring enormous progress to polymer HPLC. Their production has been repeatably announced. However, it seems that the technical problems—for example with the preparation of appropriate semi-permeable membranes—connected with the osmometric detection were not yet solved at an acceptable level and the osmometric detectors so far did not appear in the market.

The recent improvements in the nuclear magnetic resonance (NMR) instrumentation allow its application as detectors in polymer HPLC [258,259,307]. Modern NMR techniques namely work with the decreased sample concentrations and enable the application of eluents acceptable for polymer HPLC.

The density detection in liquid chromatography was proposed by Fornstedt and Porath [308]. They have used a float in combination with electrobalance. The volume of measuring cell was rather large. Trathnigg [309] introduced flow-through microdensitometer as a detection tool in polymer HPLC. He used the oscillation method invented by Kratky and coworkers [310] and constructed the U-shaped microcell. The densitometer is a nonspecific detector. The volume of measuring cell is still relatively large (70 μL) and the overall sensitivity of detector is limited. Still, the densitometers can produce valuable data, especially for oligomers, which can be injected into the column at a relatively high concentration [311,312].

A useful approach to detection in polymer HPLC presents the on-line hyphenation of different measurement principles. For example, an RI detector combined with a UV photometer produces valuable additional information on the composition of some copolymers. Further progress was brought with the triple detection RI plus LALS plus VISCO [313], which is especially suitable for branched macromolecules and the tetra detection UV plus RI plus LALS plus VISCO, which enables characterization of some complex polymer systems, exhibiting a distribution not only in their molar mass and architecture, but also in their chemical composition such as long chain branched copolymers.

The volume of effluent among particular measuring cells has to be considered in the hyphenated detection systems. In spite of recent progress, the detection remains an important challenge in polymer HPLC, especially in case of complex polymer systems.

16.9.2 WORKING MODES

The dominating working mode of polymer HPLC is the straight elution of small volume of sample solution along the column. There were attempts to introduce the differential and vacancy procedures in SEC. In both cases, the eluent was a diluted solution of the polymer while either a proper sample or a reference polymer [314] or even pure eluent [315] was injected. A broader application of these proposals was hindered by both the high eluent viscosity and the large sample consumption. Moreover, the dependences of log M vs. V_R (Equation 16.12) obtained by the conventional and the differential/vacancy methods are mutually shifted. The deviation between both dependences was attributed to the nonequilibrium situation in the vacancy polymer HPLC [316].

In SEC, the recycling procedures were attempted. Column effluent was returned into the pump and transported back into the column [317,318]. The number of cycles was, however, limited because "the head easily caught the tail" that is the first fraction was mixed with the last one. The extensive band broadening took place within the older types of pumps. This last problem may be largely mitigated with the modern pumping systems, equipped with the low volume pump heads. Biesenberger et al. [319] proposed an interesting principle of the alternating recycling. Two parallel nearly identical columns A and B were used. By means of a switching valve the effluent from the column A was send to the column B and then again back to the column A. Sample did not return into the pump. However, rather large pressure strokes were produced during the recycling process because the column sequence was repeatedly altered. Consequently, the extensive variations of mixed eluent composition were observed due to column switching when the preferential sorption within column packing depended on pressure [69,70]. The local changes of eluent composition may appear also as a result of the column switching in 2D-LC, as far as this is accompanied with pressure (and temperature) variations. Grubisic-Gallot et al. [320] proposed a recycling method for the SEC band broadening assessment. The procedure was further elaborated in Ref. [321]. The general problems of recycling in the chromatography were discussed in Ref. [322].

16.10 CONCLUSIONS

High-performance liquid chromatography of synthetic polymers is a set of very useful experimental procedures allowing separation and molecular characterization of many kinds of macromolecules. All particular members of this group of methods and their mutual combinations necessitate further research. Even the oldest and likely the simplest method of polymer HPLC, namely SEC, which is often erroneously considered a mature procedure, deserves further intensive development. It is hoped that the basic information presented in this chapter will help understand not only the principles but also the challenges of polymer HPLC.

REFERENCES

1. M.J.R. Cantow, ed., *Polymer Fractionation*, Academic Press, New York, 1967.
2. H. Determann, *Gel Chromatography*, Springer, New York, 1968.
3. K.J. Bombaugh, in *The Practice of Gel Permeation Chromatography in Modern Practice of Liquid Chromatography*, J.J. Kirkland, ed., John Wiley & Sons, New York, 1971.
4. J. Cazes, ed., *Liquid Chromatography of Polymers and Related Materials*, vol. 8 (Chromatographic Science Series), Marcel Dekker, New York, 1977.
5. N.C. Billingham, *Molar Mass Measurements in Polymer Science*, Wiley, Halstedt, NY, 1977.
6. H.L. Tung, ed., *Fractionation of Synthetic Polymers*, Marcel Dekker, New York, 1977.
7. R. Epton, ed., *Chromatography of Synthetic and Biological Polymers*, vol. 1 (Column Packings, GPC, SFC and Gradient Elution), Ellis Horwood, Chichester, U.K., 1979.
8. W.W. Yau, J.J. Kirkland, D.D. Bly, *Introduction to Modern Liquid Chromatography*, Wiley-Interscience, New York, 1979.
9. L.R. Snyder, *Principles of Adsorption Chromatography*, Arnold, London, U.K./Marcel Dekker, New York, 1968; L.R. Snyder, J.J. Kirkland, *An Introduction to Modern Liquid Chromatography*, 2nd edn., Wiley-Interscience, New York, 1979.
10. T. Kremmer, L. Boross, *Gel Chromatography*, Akadémiai Kiadó, Budapest, Hungary, 1979.
11. W.W. Yau, J.J. Kirkland, D.D. Bly, *Modern Size-Exclusion Chromatography*, Wiley, New York, 1979.
12. J. Cazes, X. Delamare, eds., *Liquid Chromatography of Polymers and Related Materials, II*, vol. 13 (Chromatographic Sciences Series), Marcel Dekker, New York, 1980.
13. J. Cazes, ed., *Liquid Chromatography of Polymers and Related Materials, III*, vol. 19 (Chromatographic Science Series), Marcel Dekker, New York, 1981.
14. D. Berek, M. Kubín, K. Marcinka, M. Dressler, *Gel Chromatography*, Veda, Bratislava, 1983 (in Slovak); Państwowe Wydawnictwo Naukowe, Warszawa 1989 (in Polish).
15. J. Janča, ed., *Steric Exclusion Liquid Chromatography of Polymers*, Marcel Dekker, New York, 1984.
16. D. Berek, K. Marcinka, Gel chromatography, in *Separation Methods*, Z. Deyl, ed., Elsevier, Amsterdam, the Netherlands, 1984.
17. G. Glöckner, *Polymer Fractionation by Liquid Chromatography*, Elsevier, Amsterdam, the Netherlands, 1987.
18. B.J. Hunt, S.R. Helding, eds., *Size Exclusion Chromatography*, Blackie & Son, Glasgow, U.K., 1989.
19. H.G. Barth, J.W. Mays, eds., *Modern Methods of Polymer Characterization*, John Wiley & Sons, Inc., New York, 1991.
20. G. Glöckner, *Gradient HPLC of Copolymers and Chromatographic Cross-Fractionation*, Springer-Verlag, Berlin, Germany, 1992.

21. C.-s. Wu, ed., *Handbook of SEC*, Marcel Dekker, New York, 1995.
22. S. Mori, HPLC application to polymer analysis, in *Handbook of HPLC*, E. Katz, R. Eksteen, P. Schoenmakers, N. Miller, eds., Marcel Dekker, New York, 1998.
23. H. Pasch, B. Trathnigg, *HPLC of Polymers*, Springer, Berlin, Germany, 1998.
24. S. Mori, H.G. Barth, *Size Exclusion Chromatography*, Springer-Verlag, Berlin, New York, 1999.
25. C.-S. Wu, ed., *Column Handbook for Size Exclusion Chromatography*, Academic Press, San Diego, London, U.K., 1999.
26. P. Kilz, H. Pasch, Coupled liquid chromatographic techniques in molecular characterization, in *Encyclopedia of Analytical Chemistry*, vol. 9, R.A. Meyers, ed., Wiley, Chichester, U.K., 2000, p. 7495.
27. H. Pasch, *Adv. Polym. Sci.* 150, 1, 2000.
28. D. Berek, *Prog. Polym. Sci.* 25, 873, 2000.
29. J. Cazes, ed., *Encyclopedia of Chromatography*, Marcel Dekker, New York, 2001.
30. C.-S. Wu, ed., *Handbook of Size Exclusion Chromatography and Related Techniques*, vol. 91 (Chromatographic Sciences Series), Marcel Dekker, Inc., New York, 2004.
31. W. Radke, Chromatography of polymers, in *Macromolecular Engineering: Structure-Property Correlation and Characterization Techniques*, vol. 3, K. Matyjaszewski, Y. Gnanou, L. Leibler, eds., Wiley-VCH, Berlin, Germany, 2007; A.M. Striegel, J.J. Kirkland, W.W. Yau, D.D. Bly, *Modern Size-Exclusion Liquid Chromatography*, Wiley, Hoboken, New Jersey, 2009.
32. T. Chang, *Adv. Polym. Sci.* 163, 1, 2003.
33. T. Macko, D. Hunkeler, *Adv. Polym. Sci.* 163, 62, 2003.
34. L.H. Peebles, Jr., *Molecular Weight Distributions in Polymers*, John Wiley & Sons, New York, 1971.
35. G.R. Meira, J.R. Vega, Characterization of copolymers by size exclusion chromatography, in *Handbook of Size Exclusion Chromatography and Related Techniques*, vol. 91 (Chromatographic Sciences Series), C.-s. Wu, ed., Marcel Dekker, Inc., New York, 2004, p. 139.
36. J.M. Prausnitz, *Molecular Thermodynamics of Fluid-Phase Equilibria*, Prentice-Hall, Englewoods Cliffs, NJ, 1969.
37. J. Brandrup, E.H. Immergut, E.A. Gruelke, A. Abe, D.R. Bloch, eds., *Polymer Handbook*, Wiley, New York, 1999.
38. P.J. Flory, *Principles of Polymer Chemistry*, Cornell University Press, Ithaca, NY, 1953.
39. D. Berek, T. Bleha, Z. Pevná, *J. Chromatogr. Sci.* 14, 560, 1976.
40. D. Berek, T. Bleha, Z. Pevná, *J. Polym. Sci., Polym. Lett. Ed.* 14, 323, 1976; C.-H. Fischer, D. Berek, T. Macko, *Polym. Bull.* 33, 339, 1994.
41. P. Cifra, T. Bleha, *Macromol. Theory Simul.* 8, 603, 1999; *Polymer* 41, 1003, 2000; *Int. J. Polym. Anal. Charact.* 6, 509, 2001.
42. X. Chen, T. Yang, S. Kataoka, P.S. Cremer, *J. Am. Chem. Soc.* 129(12), 272, 2007.
43. J.D. Smith, R.J. Sayhally, P.L. Geissler, *J. Am. Chem. Soc.* 129(13), 847, 2007.
44. A.M. Skvortsov, A.A. Gorbunov, *J. Chromatogr.* 358, 77, 1986.
45. A.A. Gorbunov, A.M. Skvortsov, *Adv. Colliod Interface Sci.* 62, 31, 1995.
46. M.J.R. Cantow, J.F. Johnson, *J. Polym. Sci. A-1*, 5, 2835, 1967.
47. M. Le Page, R. Beau, J. de Vries, *J. Polym. Sci. C*, 21, 119, 1968.
48. J.C. Giddings, E. Kucera, C.P. Russel, M.N. Myers, *J. Phys. Chem.* 72, 4397, 1968.
49. R. Gant, *Sigma-Aldrich Chromatography Products*, Sigma-Aldrich Corp., Milwaukee, WI, 1993.
50. D. Berek, D. Bakoš, *J. Chromatogr.* 91, 237, 1974.
51. B. Trathnigg, A. Gorbunov, *J. Chromatogr. A*, 910, 207, 2001.
52. D. Berek, *Macromol. Symp.* 216, 145, 2004.
53. A. Russ, D. Berek, *J. Sep. Sci.* 30, 1852, 2007.
54. E.F. Casassa, *Macromolecules* 9, 182, 1976.
55. M.E. Kreveld, M.E. van den Hoed, *J. Chromatogr.* 149, 71, 1978.
56. G.K. Ackers, *Biochemistry* 3, 723, 1964.
57. K.O. Pedersen, *Ark. Biochem. Biophys.* (Suppl). 1, 157, 1962.
58. H. Small, *J. Colloid Interface Sci.* 48, 147, 1974.
59. E.A. di Marco, C.M. Guttman, *J. Chromatogr.* 55, 83, 1971.
60. G. Stegeman, A.C. van Asten, J.C. Kraak, H. Poppe, R. Tijssen, *Anal. Chem.* 66, 1147, 1994.
61. S.S. Huang, Molecular weight separation of macromolecules by hydrodynamic chromatography, in *Column Handbook for Size Exclusion Chromatography*, C.-s. Wu, ed., Academic Press, San Diego, CA, 1999, p. 597.

62. S.S. Huang, Size exclusion/hydrodynamic chromatography, in *Handbook of Size Exclusion Chromatography and Related Techniques*, vol. 91 (Chromatographic Sciences Series), C.-S. Wu, ed., Marcel Dekker, Inc., New York, 2004, p. 481.

63. J.C. Giddings, *Sep. Sci.* 1, 123, 1966.

64. J.C. Giddings, *J. Liquid Chromatogr.* 1, 1, 1978.

65. M. Martin, *Adv. Chromatogr.*, Eds. P.R. Brown, E. Grushka 39, 1, 1998.

66. J.H. Aubert, M. Tirrell, *J. Liquid Chromatogr.* 6(Suppl. 2), 219, 1983.

67. T. Sintes, K. Sumithra, E. Strade, *Macromolecules* 34, 1352, 2001.

68. V.L. McGuffin, C.E. Evans, S.H. Chen, *J. Microcol. Sep.* 5, 3, 1993.

69. D. Berek, M. Chalányová, T. Macko, *J. Chromatogr.* 286, 185, 1984.

70. T. Macko, M. Chalányová, D. Berek, *J. Liquid Chromatogr.* 9, 1123, 1986.

71. P.R. MacNair, K.C. Patel, R.D. Patel, *Stärke* 28, 267, 1976.

72. D. Berek, R. Mendichi, *J. Chromatogr. B* 800, 69, 2004.

73. M. Petro, D. Berek, *Chromatographia* 37, 549, 1993.

74. D. Bakoš, T. Bleha, A. Ozimá, D. Berek, *J. Appl. Polym. Sci.* 23, 2233, 1979.

75. R. Audebert, *Analysis* 4, 399, 1976; J. Lecourtier, R. Audebert, C. Quivoron, *J. Chromatogr.* 121, 173, 1976.

76. T. Bleha, D. Berek, *Chromatographia* 14, 163, 1981.

77. W. Heitz, *Z. Anal. Chem.* 277, 323, 1975.

78. W. Heitz, W. Kern, *Angew. Makromol. Chem.* 1, 150, 1967.

79. D. Berek, *Macromolecules* 37, 6096, 2004.

80. D. Berek, *J. Chromatogr. A* 1020, 219, 2003.

81. D. Berek, *Macromol. Chem. Phys.* 206, 1915, 2005.

82. I. Rustanov, T. Farcas, F.A. Ahmed, F. Chan, R. La Brutto, H.M. McNair, Y.V. Kazakcvich, *J. Chromatogr. A* 913, 41, 2001.

83. D. Berek, *J. Chromatogr. A* 950, 75, 2002.

84. D. Berek, J. Tarbajovská, *J. Chromatogr. A* 976, 27, 2002.

85. D. Berek, *J. Chromatogr.* 1020, 219, 2003.

86. R. Koningsveld, W.H. Stockmayer, E. Nies, *Polymer Phase Diagrams*, Oxford University Press, Oxford, NY, 2001.

87. B.A. Wolf, H.F. Bieringer, W. Breitenbach, *Polymer* 17, 605, 1976.

88. B.A. Wolf, M.M. Wilms, *Makromol. Chem.* 179, 2265, 1978.

89. J. Porath, P. Flodin, *Nature (London)* 183, 1657, 1959.

90. J.C. Moore, *J. Polym. Sci. A*, 2, 835, 1964.

91. G.K. Ackers, *Biochemistry* 3, 728, 1964.

92. E.E. Brumbaugh, G.K. Ackers, *Anal. Biochem.* 41, 543, 1971.

93. G.A. Gilbert, G.L. Kellet, *J. Biol. Chem.* 246, 6079, 1971.

94. P.A. Bakhurst, L.W. Nichol, A.G. Ogston, D.J. Winzor, *J. Biochem.* 147, 575, 1975.

95. J.C. Moore, *Sep. Sci.* 5, 723, 1970.

96. G. Rousseaux, A. De Wit, M. Martin, *J. Chromatogr. A* 1149, 254, 2007.

97. M. Schweiger, G. Langhammer, *Plaste Kautsch.* 24, 101, 1977.

98. M. Luo, I. Teraoka, *Macromolecules* 29, 4226, 1996.

99. I. Teraoka, M. Luo, *Trends Polym. Sci.* 5, 258, 1997.

100. D. Lee, I. Teraoka, *J. Chromatogr. A* 996, 71, 2003.

101. I. Teraoka, D. Lee, High osmotic pressure chromatography, in *Handbook of Size Exclusion Chromatography and Related Techniques*, vol. 91 (Chromatographic Sciences Series), C.-s. Wu, ed., Marcel Dekker, Inc., New York, 2004, p. 657.

102. K.H. Altgelt, *Makromol. Chem.* 88, 88, 1975.

103. J. Janča, *J. Chromatogr.* 134, 263, 1977; *Anal. Chem.* 51, 637, 1979; *J. Chromatogr.* 7, 1887, 1984.

104. M.J.R. Cantow, R.S. Porter, F.J. Johnson, *J. Polym. Sci. B* 4, 707, 1966.

105. A. Lambert, *Polymer* 10, 213, 1969.

106. Y. Kato, T. Hashimoto, *J. Appl. Polym. Sci.* 18, 1239, 1974.

107. L.-H. Shi, M.-L. Ye, W. Wang, Y.-K. Ding, *J. Liquid Chromatogr.* 7, 1851, 1984; O. Chiantore, M. Guaita, *J. Liquid Chromatogr.* 7, 1867, 1984.

108. D. Berek, D. Bakoš, L. Šoltés, T. Bleha, *J. Polym. Sci., Polym. Lett. Ed.* 12, 277, 1974.

109. T. Bleha, D. Bakoš, D. Berek, *Polymer* 18, 897, 1977; T. Bleha, J. Mlýnek, D. Berek, *Polymer* 21, 798, 1980.

110. P.L. Dubin, S. Koontz, K.L. Wright, *J. Polym. Sci. Polym. Chem. Ed.* 15, 2047, 1977; N.D. Hann, *J. Polym. Sci. Polym. Chem. Ed.* 15, 1331, 1977; D. Berek, T. Spychaj, J. Morong, Gel chromatography with mixed eluents dimethylformamide—Inorganic salt, in *Proceedings of the 6th Disc. Conference on Chromatography of Polymers and Polymers in Chromatography*, Prague, 1978, C25.

111. M.E. van Kreveld, N. Van den Hoed, *J. Chromatogr.* 149, 71, 1978.

112. L.H. Tung, *J. Appl. Polym. Sci.* 10, 1271, 1966.

113. A.E. Hamielec, W.H. Ray, *J. Appl. Polym. Sci.* 13, 1319, 1969.

114. L.H. Tung, J.R. Runyon, *J. Appl. Polym. Sci.* 13, 2397, 1969.

115. M. Potschka, *J. Chromatogr.* 648, 41, 1993.

116. J.P. Busnel, F. Foucault, L. Denis, W. Lee, T. Chang, *J. Chromatogr. A* 930, 61, 2001; J.L. Baumgarten, J.P. Busnel, G.R. Meira, *J. Liquid Chromatogr. Relat. Technol.* 25, 1967, 2002.

117. M. Netopilík, *J. Chromatogr. A* 1113, 95 and 162, 2005.

118. I. Schnöll-Bitai, *J. Chromatogr. A* 1084, 375, 2005.

119. M.M. Yossen, J.R. Vega, G.R. Meira, *J. Chromatogr. A* 1128, 171, 2006.

120. G. Meira, M. Netopilík, M. Potschka, I. Schnöll-Bitai, J. Vega, *Macromol. Symp.* 258, 186, 2007.

121. H. Benoit, Z. Grubisic, P. Rempp, D. Decker, *J. Chim. Phys.* 63, 1507, 1966.

122. Z. Grubisic, P. Rempp, H. Benoit, *J. Polym. Sci. B* 5, 753, 1967.

123. D. Berek, *Macromol. Symp.* 145, 49, 1999.

124. B. Trathnigg, M. Kollroser, *Int. J. Polym. Anal. Charact.* 1, 301, 1995.

125. H.G. Barth, B.E. Boyes, *Anal. Chem.* 64, R428, 1992.

126. H.G. Barth, B.E. Boyes, C. Jackson, *Anal. Chem.* 66, R595, 1994; 68, R445, 1996; *Anal. Chem.* 70, R251, 1998.

127. H.J.A. Philipsen, *J. Chromatogr. A* 1037, 329, 2004.

128. K.H. Altgelt, *Makromol. Chem.* 88, 75, 1965; *J. Appl. Polym. Sci.* 9, 3389, 1965.

129. H.H. Oelert, D.R. Latham, W.E. Haines, *Sep. Sci.* 5, 657, 1970.

130. K.H. Altgelt, T.H. Gouw, *Adv. Chromatogr.* 13, 71, 1975.

131. J. Aurenge, Z. Gallot, A.J. de Vries, H. Benoit, *J. Polym. Sci. Polym. Symp.* 52, 217, 1975.

132. A. Krishen, R.G. Tucker, *Anal. Chem.* 49, 898, 1977.

133. M. Minárik, Z. Šír, J. Čoupek, *Angew. Makromol. Chem.* 64, 147, 1977.

134. R.C. Davison, C.J. Glover, B.L. Burr, J.A. Bullin, Size exclusion chromatography of asphalts, in *Handbook of Size Exclusion Chromatography and Related Techniques*, vol. 91 (Chromatographic Sciences Series), C.-s. Wu, ed., Marcel Dekker, Inc., New York, 2004, p. 191.

135. P. Kilz, Methods and columns for high-speed size exclusion chromatography, in *Handbook of Size Exclusion Chromatography and Related Techniques*, vol. 91 (Chromatographic Sciences Series), C.-s. Wu, ed., Marcel Dekker, Inc., New York, 2004, p. 561.

136. J. Porath, *Pure Appl. Chem.* 6, 233, 1963.

137. R.N. Nikolov, W. Werner, I. Halász, *J. Chromatogr. Sci.* 18, 207, 1980.

138. K. Jeřábek, A. Revillon, E. Pucilli, *Chromatographia* 36, 259, 1993.

139. K. Jeřábek, *J. Appl. Polym. Sci.* 48, 745, 1993.

140. J. Lyngaae-Jorgenson, *J. Chromatogr. Sci.* 9, 331, 1971.

141. T. Williams, Y. Udagawa, A. Keller, I.M. Ward, *J. Polym. Sci. A-2* 8, 35, 1970.

142. D. Bakoš, D. Berek, T. Bleha, *Eur. Polym. J.* 12, 801, 1976.

143. T. Bleha, T. Spychaj, R. Vondra, D. Berek, *J. Polym. Sci. Polym. Phys. Ed.* 21, 1903, 1983.

144. D.J. Solms, T.W. Smuts, V. Pretorius, *J. Chromatogr. Sci.* 9, 600, 1971.

145. K. Šlais, M. Krejčí, *J. Chromatogr.* 91, 161, 1974.

146. T. Macko, D. Berek, *J. Chromatogr. Sci.* 25, 17, 1987.

147. D. Berek, T. Macko, *Pure Appl. Chem.* 61, 2041, 1989.

148. T. Macko, D. Berek, *J. Liquid Chromatogr. Relat. Technol.* 21, 2265, 1998.

149. T. Macko, D. Berek, *J. Chromatogr.* 592, 109, 1992.

150. M. Arpin, C. Strazielle, *Makromol. Chem.* 177, 293 and 581, 1976.

151. R. Panaris, G. Pallas, *J. Polym. Sci. C* 8, 441, 1970.

152. C. Dauwe, Size exclusion chromatography of polyamides, polyesters and fluoropolymers, in *Handbook of Size Exclusion Chromatography and Related Techniques*, vol. 91 (Chromatographic Sciences Series), C.-s. Wu, ed., Marcel Dekker, Inc., New York, 2004, p. 157.

153. D. Berek, Repeatability and apparent reproducibility of molar mass values for homopolymers determined by size exclusion chromatography, IUPAC Round Robin Test, General Assembly of IUPAC, Brisbane, Australia, 2001.

154. E.C. Robert, R. Bruessau, J. Dubois, B. Jacques, N. Meijerink, T.Q. Nguyen, D.E. Niehaus, W.A. Tobish, *Pure Appl. Chem.* 76, 2009, 2004.
155. Š. Podzimek, *Int. J. Polym. Anal. Charact.* 9, 305, 2005.
156. D. Berek, Interactive properties of polystyrene/divinylbenzene and divinylbenzene-based commercial size exclusion chromatography columns, in *Column Handbook for Size Exclusion Chromatography*, C.-s. Wu, ed., Academic Press, San Diego, CA, 1999, p. 445.
157. M. Šimeková, D. Berek, *J. Chromatogr. A* 1084, 167, 2005.
158. B. Trathnigg, M. Kollroser, *J. Chromatogr. A* 768, 223, 1997.
159. L. Kolářová, P. Jandera, E.C. Vonk, H.A. Claessens, *Chromatographia* 59, 579, 2004; P. Jandera, Gradient elution mode, in *Handbook of HPLC*, 2nd edn, D. Corradini, T.M. Phillips, eds., this book, Chapter 5.
160. P. Jandera, M. Holčapek, L. Kolářová, *Int. J. Polym. Anal. Charact.* 6, 261, 2001.
161. D. Berek, *Macromol. Symp.* 110, 33, 1996.
162. B.G. Belenkii, E.S. Gankina, M.B. Tennikov, L.Z. Vilenchik, *Dokl. Akad. Nauk SSSR* 231, 1147, 1976; *J. Chromatogr.* 147, 99, 1978.
163. T.M. Tennikov, P.P. Nefedov, M.A. Lazareva, S.Ya. Frenkel, *Vysokomol. Sojed. A* 19, 657, 1977.
164. B.G. Belenkii, E.S. Gankina, *J. Chromatogr.* 141, 13, 1977.
165. P.P. Nefedov, T.P. Zhmakina, *Vysokomol. Sojed. A* 23, 1797, 1981.
166. T.M. Zimina, E.E. Kever, E. Yu. Melenevskaya, V.N. Zgonnik, B.G. Belenkii, *Vysokomol. Sojedin. A* 33, 1349, 1991.
167. A.M. Skvortsov, B.G. Belenkii, E.S. Gankina, M.B. Tennikov, *Vyskomol. Sojedin. A* 20, 678, 1978.
168. A.M. Skvortsov, A.A. Gorbunov, *Vysokomol. Sojedin. A* 21, 339, 1979.
169. A.M. Skvortsov, A.A. Gorbunov, *Vysokomol. Sojedin. A* 22, 2641, 1980.
170. S.G. Entelis, V.V. Evreinov, A.V. Gorskov, *Adv. Polym. Sci.* 76, 127, 1987.
171. T.M. Zimina, A. Fell, J.Z. Castledine, *Polymer* 33, 4129, 1992.
172. T.M. Zimina, J.J. Kever, E.Y. Melenevskaya, A.F. Fell, *J. Chromatogr.* 593, 233, 1992.
173. H. Pasch, K. Rode, *Polymer* 39, 6377, 1998.
174. J. Falkenhagen, M. Much, W. Stauf, A.H.E. Müller, *Macromolecules* 33, 3687, 2000.
175. K. Baran, S. Laugier, H. Cramail, *Int. J. Polym. Anal. Charact.* 60, 123, 2000.
176. B. Lepoittevin, M.A. Dourges, M. Masure, P. Hemery, K. Baran, H. Cramail, *Macromolecules* 33, 8218, 2000.
177. K. Baran, S. Laugier, H. Cramail, *J. Chromatogr. B* 753, 139, 2001.
178. C.M. Guttman, E.A. DiMarzio, J.F. Douglas, *Macromolecules* 29, 5723, 1996.
179. M. Jančo, D. Berek, A. Önen, C.H. Fischer, Y. Yagci, W. Schnabel, *Polym. Bull.* 38, 681, 1997.
180. D. Berek, M. Jančo, G.R. Meira, *J. Polym. Sci. A Polym. Chem.* 36, 1363, 1998.
181. W. Lee, H. Lee, H.C. Lee, D. Cho, T. Chang, A.A. Gorbunov, J. Rooversi, *Macromolecules* 33, 8119, 2000.
182. D. Berek, A. Russ, *Chem. Pap.* 60, 249, 2006.
183. E. Beaudoin, P.E. Dufils, D. Gigmes, S. Marque, C. Petit, P. Tondo, D. Bertin, *Polymer* 47, 98, 2006.
184. E. Beaudoin, A. Favier, C. Galindo, A. Lapp, C. Petit, D. Gigmes, S. Marque, D. Bertin, *Eur. Polym. J.* 44, 514, 2008; A. Favier, C. Petit, E. Beaudoin, D. Bertin, *e-Polymers* 1, 009, 2009.
185. M. Jacquin, P. Muller, G. Lizarraga, C. Bauer, H. Cottet, O. Théodoly, *Macromolecules* 40, 2672, 2007.
186. X. Jiang, P.J. Schoenmakers, X. Lou, V. Lima, J.V.J. van Dongen, J. Brokken-Zijp, *J. Chromatogr. A* 1055, 123, 2004.
187. X. Jiang, V. Lima, P.J. Schoenmakers, *J. Chromatogr. A* 1018, 19, 2003.
188. H.J.A. Philipsen, B. Klumperman, A.L. German, *J. Chromatogr. A* 746, 211, 1996.
189. V.V. Guryanova, A. Pavlov, *J. Chromatogr.* 365, 197, 1986.
190. R.P. Krüger, H. Much, G. Schulz, *Macromol. Symp.* 110, 315, 1996.
191. B. Trathnigg, M. Kollroser, C. Rappel, *J. Chromatogr. A* 922, 193, 2001.
192. D. Berek, *Mater. Res. Innovat.* 4, 365, 2001; Round Table Discussion: Characterization of block copolymer with liquid chromatography under critical conditions of enthalpic interactions, in *International Conference on Coupled, Hyphenated and Multidimensional Liquid Chromatographic Procedures for Separation of Macromolecules*, Bratislava, Slovakia, 2001.
193. W. Lee, D. Cho, T. Chang, K.J. Haney, T.P. Lodge, *Macromolecules* 34, 2353, 2001.
194. R. Murgašová, I. Capek, E. Lathová, D. Berek, Š. Florián, *Eur. Polym. J.* 34, 659, 1998.
195. Y. Brun, *J. Liquid Chromatogr. Relat. Technol.* 22, 3027, 1999.
196. Y. Brun, *J. Liquid Chromatogr. Relat. Technol.* 22, 3067, 1999.
197. P.J.C.H. Cools, A.M. van Herk, A.L. German, W.J. Staal, *J. Liquid Chromatogr.* 17, 3133, 1994.

198. J. Porath, *Nature* (*London*) 196, 47, 1992.
199. H. Inagaki, H. Matsuda, F. Kamiyama, *Macromolecules* 1, 520, 1968.
200. B.G. Belenkii, E.S. Gankina, *Dokl. Akad. Nauk SSSR* 186, 573, 1969; *J. Chromatogr.* 53, 3, 1970.
201. S. Teramachi, A. Hasegawa, Y. Shima, M. Akatsuka, M. Nakajima, *Macromolecules* 12, 992, 1979.
202. X. Jiang, A. van den Horst, P.J. Schoenmakers, *J. Chromatogr. A* 982, 55, 2002.
203. M.A. Stadalius, H.S. Gold, L.R. Snyder, *J. Chromatogr.* 296, 31, 1984.
204. M.A. Quarry, M.A. Stadalins, T.H. Mourey, L.R. Snyder, *J. Chromatogr.* 358, 1, 1986.
205. R. Shallikar, P.E. Kavanagh, I.M. Russell, *Chromatographia* 39, 663, 1994.
206. D.W. Armstrong, K.H. Bui, *Anal. Chem.* 54, 706, 1982.
207. D.W. Armstrong, R.E. Boehm, *J. Chromatogr. Sci.* 22, 378, 1984.
208. R.E. Boehm, D.E. Martire, D.W. Armstrong, K.H. Bui, *Macromolecules* 16, 466, 1983; *Macromolecules* 17, 400, 1984; R.E. Boehm, D.E. Martire, *Anal. Chem.* 5, 471, 1989.
209. D. Berek, *Macromolecules* 32, 3671, 1999.
210. S. Mori, *J. Chromatogr.* 541, 375, 1991.
211. S. Teramachi, *Macromol. Symp.* 110, 217, 1996.
212. S. Mori, *Macromol. Symp.* 110, 87, 1996.
213. T. Mourey, *J. Chromatogr.* 357, 101, 1986.
214. H. Sato, K. Ogino, T. Darwint, I. Kiyokawa, *Macromol. Symp.* 110, 177, 1996.
215. T. Shunk, *J. Chromatogr. A*, 656, 591, 1993.
216. H. Sato, M. Sasaki, K. Ogino, *Polym. J.* 21, 965, 1989.
217. M. Konáš, M. Kubín, *J. Appl. Polym. Sci.* 47, 2245, 1993.
218. M. Hutta, R. Góra, *J. Chromatogr. A* 1012, 67, 2003; R. Góra, M. Hutta, *J. Chromatogr. A* 1084, 39, 2005; R. Góra, M. Hutta, M. Vrška, S. Katuščák, M. Jablonský, *J. Sep. Sci.* 29, 2179, 2006.
219. H.C. Lee, T. Chang, *Macromolecules* 29, 7294, 1996; *Polymer* 37, 5747, 1996.
220. T. Chang, H.C. Lee, W. Park, C. Ko, *Macromol. Chem. Phys.* 200, 2188, 1999.
221. J. Ryu, K. Im, W. Yu, J. Park, T. Chang, K. Lee, N. Choi, *Macromolecules* 37, 8805, 2004.
222. K. Im, Y. Kim, T. Chang, K. Lee, N. Cho, *J. Chromatogr. A* 1103, 235, 2006.
223. K. Im, H.-W. Park, Y. Kim, B. Chung, M. Ree, T. Chang, *Anal. Chem.* 79, 1067, 2007.
224. W. Lee, H. Lee, J. Cha, T. Chang, K. Hanley, T. Lodge, *Macromolecules* 33, 5111, 2000.
225. O. Chiantore, *J. Liquid Chromatogr.* 7, 1, 1984.
226. D. Hunkeler, T. Macko, D. Berek, in T. Provder, ed., *Chromatography of Polymers*, ACS Books, Washington, DC, 1993, p. C7; D. Hunkeler, M. Jančo, D. Berek, in T. Provder, ed., ACS Books, Washington, DC, 1996, Chapter 14.
227. A. Bartkowiak, D. Hunkeler, D. Berek, T. Spychaj, *J. Appl. Polym. Sci.* 69, 2549, 1998.
228. D. Berek, *Macromolecules* 31, 8517, 1998.
229. M. Šnauko, D. Berek, *J. Chromatogr. A* 1094, 42, 2005.
230. D. Berek, *Macromol. Chem. Phys.* 207, 893, 2006; *Chem. Pap.* 60, 71, 2006.
231. D. Berek, Coupled procedures in liquid chromatography of polymers, in *Proceedings of the Simposio Latino-Americano des Polímeros—SLAP*, Mar del Plata, 1996, p. 37.
232. D. Berek, *Macromol. Chem. Phys.* 209, 695, 2008; *Macromol. Chem. Phys.* 209, 2213, 2009; *Polymer* 51, 587, 2010.
233. M. Šnauko, D. Berek, *Macromol. Chem. Phys.* 206, 938, 2005.
234. M. Šnauko, D. Berek, *J. Sep. Sci.* 28, 2094, 2005.
235. M. Šnauko, D. Berek, *Chromatographia* 57, S-55, 2003.
236. D. Berek, *Macromol. Symp.* 231, 134, 2006.
237. D. Berek, I. Capek, R. Mendichi, S. Labátová, *Macromol. Chem. Phys.* 207, 2074, 2006.
238. M. Jančo, D. Berek, T. Prudskova, *Polymer* 36, 3295, 1995.
239. M. Jančo, T. Prudskova, D. Berek, *Int. J. Polym. Anal. Charact.* 3, 319, 1997.
240. S.H. Nguyen, D. Berek, *Colloids Surf. A: Physicochem. Eng. Aspects* 162, 75, 2000.
241. S.H. Nguyen, D. Berek, *Chromatographia* 48, 65, 1998.
242. S.H. Nguyen, D. Berek, *Colloid Polym. Sci.* 277, 318, 1999.
243. S.H. Nguyen, D. Berek, *Int. J. Polym. Anal. Charact.* 6, 229, 2001.
244. S. Park, I. Park, T. Chang, C.Y. Ryu, *J. Am. Chem. Soc.* 126, 8906, 2004.
245. D. Berek, S.H. Nguyen, J. Pavlinec, *J. Appl. Polym. Sci.* 75, 857, 2000.
246. D. Berek, S.H. Nguyen, G. Hild, *Eur. Polym. J.* 36, 1101, 2000.
247. D. Berek, S.H. Nguyen, *Macromolecules* 31, 8243, 1998.
248. S.H. Nguyen, D. Berek, *J. Polym. Sci., Part A: Polym. Chem.* 37, 267, 1999.
249. D. Berek, *Macromol. Symp.* 174, 413, 2001.

250. D. Berek, Two-dimensional liquid chromatography of synthetic macromolecules, in *Handbook of Size Exclusion Chromatography and Related Techniques*, vol. 91 (Chromatographic Sciences Series), C.-s. Wu, ed., Marcel Dekker, Inc., New York, 2004, p. 501; *Anal. Bioanal. Chem.* 396, 421, 2010.

251. S. Balke, R.D. Patel, *J. Polym. Sci. Polym. Lett. Ed.* 18, 453, 1980.

252. S. Balke, R.D. Patel, *Adv. Chem. Ser.* 203, 281, 1983.

253. P. Kilz, R.P. Krüger, H. Much, G. Schulz, *Adv. Chem.* 247, 223, 1995.

254. P. Kilz, H. Pasch, Coupled liquid chromatographic techniques in molecular characterization, in *Encyclopedia of Analytical Chemistry*, R.A. Mayers, ed., Wiley, Chichester, U.K., 2000, p. 7495.

255. E. Grushka, *Anal. Chem.* 42, 1142, 1970.

256. P. Jandera, J. Fischer, H. Lanovská, K. Novotná, P. Česla, L. Kolářová, *J. Chromatogr. A* 1119, 3, 2006.

257. T. Walerowicz, P. Jandera, K. Novotná, B. Buszewski, *J. Sep. Sci.* 29, 1155, 2006.

258. T. Kitayama, M. Jančo, K. Ute, R. Niimi, K. Hatada, D. Berek, *Anal. Chem.* 72, 1518, 2000.

259. M. Jančo, T. Hirano, T. Kitayama, K. Hatada, D. Berek, *Macromolecules* 33, 1710, 2000.

260. D. Berek, *Macromol. Symp.* 258, 198, 2007.

261. P. Kilz, Design, Properties and testing of polymer standards service size Exclusion chromatography (SEC) columns and optimization of SEC separations, in *Column Handbook for Size Exclusion Chromatography*, C.-s. Wu, ed., Academic Press, San Diego, CA, 1999, p. 267.

262. B.G. Belenkii, M.D. Valchikhina, I.A. Vakhtina, E.S. Gankina, O.G. Tarakanov, *J. Chromatogr.* 129, 115, 1976.

263. A.M. Skvortsov, L.T. Klushin, A.A. Gorbunov, *Macromolecules* 30, 1818, 1997.

264. B. Gawdzik, J. Osypjuk, *Chromatographia* 54, 323, 2001.

265. T. Macko, H. Pasch, R. Brüll, *J. Chromatogr. A* 1115, 81, 2006.

266. K.K. Unger, *Porous Silica*, Elsevier, Amsterdam, the Netherlands, 1979.

267. T. Spychaj, D. Lath, D. Berek, *Polymer* 20, 437, 1979.

268. T. Spychaj, D. Berek, *Polymer* 20, 1108, 1979.

269. C.Y. Ryu, J. Han, W. Kim, Large scale purification and fractionation of block copolymers using silica gels, *Proceedings of the IUPAC Congress on CD*, Turin, Italy, 2007.

270. T. Macko, O. Chiantore, D. Berek, *J. Chromatogr.* 364, 245, 1986.

271. R.P.W. Scott, *Liquid Chromatography Detectors*, Elsevier, Amsterdam, the Netherlands, 1977.

272. M. Huglin, *Light Scattering from Polymer Solutions*, Elsevier, Amsterdam, the Netherlands, 1987.

273. H.-J. Cantow, E. Siefert, R. Kuhn, *Chem. Ing. Technol.* 38, 1032, 1966.

274. A.C. Ouano, W. Kaye, *J. Polym. Sci. Polym. Chem. Ed.* 12, 1151, 1974.

275. A.O. Ouano, *J. Chromatogr.* 117, 303, 1976.

276. P.J. Wyatt, Light scattering and the solution properties of macromolecules, in *Handbook of Size Exclusion Chromatography and Related Techniques*, vol. 91 (Chromatographic Sciences Series), C.-s. Wu, ed., Marcel Dekker, Inc., New York, 2004, p. 623; *Anal. Chim. Acta* 272, 1, 1993.

277. Š. Podzimek, *Chromatographia* 33, 377, 1992; *J. Appl. Polym. Sci.* 54, 91, 1994; *J. Chromatogr. A* 677, 21, 1994; *Int. J. Polym. Anal. Charact.* 6, 533, 2001.

278. W.W. Yau, *Chemtracts-Macromol. Chem.* 1, 1, 1990.

279. C. Jackson, H.G. Barth, Molecular weight sensitive detectors for size exclusion chromatography, in *Handbook of Size Exclusion Chromatography and Related Techniques*, vol. 91 (Chromatographic Sciences Series), C.-s. Wu, ed., Marcel Dekker, Inc., New York, 2004, p. 99.

280. W. Bushuk, H. Benoit, *Can. J. Chem.* 36, 1616, 1958.

281. D. Goedhard, A. Opschoor, *J. Polym. Sci. A-2* 8, 1227, 1970.

282. G. Mayerhoff, *Makromol. Chem.* 118, 265, 1968; *Sep. Sci.* 6, 239, 1971.

283. A.C. Ouano, *J. Polym. Sci. A-1*, 10, 2169, 1972; *J. Polym. Sci. C* 43, 299, 1973.

284. M. Haney, *J. Appl. Polym. Sci.* 30, 3023 and 3037, 1985.

285. J.M. Goldwasser, *Proceedings of the International GPC Symposium*, Newton, MA, 1989, p. 150.

286. B.R. Smith, *Rubber Chem. Technol.* 49, 278, 1976.

287. S. Nukiyama, Y. Tanasawa, *Trans. Soc. Mech. Eng.* 4, 86, 1983.

288. J.M. Charlesworth, *Anal. Chem.* 50, 1414, 1978.

289. L.E. Oppenheimer, T.H. Mourey, *J. Chromatogr.* 298, 217, 1984.

290. T.H. Mourey, L.E. Oppenheimer, *Anal. Chem.* 56, 2427, 1984; *J. Chromatogr.* 323, 297, 1985.

291. S. Héron, M.-G. Maloumbi, M. Dreux, A. Tchapla, *LC.GC Eur.* 19(Dec.), 2, 2006.

292. R. Schulz, H. Engelhardt, *Chromatographia* 29, 517, 1990.

293. B. Trathnigg, M. Kollroser, D. Berek, S.H. Nguyen, D. Hunkeler, Quantitation in the analysis of oligomers by HPLC with ELSD, in *Chromatography of Polymers: Hyphenated and Multidimensional Techniques*, T. Provder, ed., ACS Symp. Ser., Washington, DC, 731, 178, 1999.

294. T. Górecki, F. Lynen, R. Szücs, P. Sandra, *Anal. Chem.* 78, 3186, 2006.
295. P.R. Boshoff, B.J. Hopkins, V. Pretorius, *J. Chromatogr.* 126, 35, 1976.
296. W.W. Schulz, W.H. King, Jr., *J. Chromatogr. Sci.* 12, 343, 1973.
297. E.R. Kaal, M. Kurano, M. Geissler, P. Schoenmakers, H.G. Janssen, *LC.GC* 20, 444, 2007.
298. L.M. Wheeler, J.N. Willis, *Appl. Spectrosc.* 47, 1128, 1993.
299. J.N. Willis, J.K. Dwyer, L.M. Wheeler, *Polym. Mater. Sci.* 69, 120, 1993.
300. R. Murgašová, D.M. Hercules, *Anal. Bioanal. Chem.* 373, 481, 2002.
301. K.-H. Spriestersbach, K. Rode, H. Pasch, *Macromol. Symp.* 193, 129, 2003.
302. M.S. Montaudo, C. Puglisi, F. Samperi, G. Montaudo, *Macromolecules* 31, 3839, 1998.
303. M.S. Montaudo, G. Montaudo, *Macromolecules* 32, 7015, 1999.
304. M. Hoffman, H. Urban, *Makromol. Chem.* 178, 2661 and 2683, 1977.
305. R.E. Poulsen, H.B. Jensen, *Anal. Chem.* 40, 376, 1969.
306. K. Lehmann, W. Köhler, W. Albrecht, *Macromolecules* 29, 3212, 1996.
307. V. Krämer, W. Hiller, H. Pasch, *Macromol. Chem. Phys.* 210, 1662, 2000.
308. N. Fornstedt, J. Porath, *J. Chromatogr.* 42, 376, 1969.
309. B. Trathnigg, *Monatsh. Chem.* 109, 467, 1967; *Angew. Makromol. Chem.* 89, 65 and 73, 1980.
310. H. Stabinger, H. Leopold, O. Kratky, *Monatsh. Chem.* 98, 436, 1967; O. Kratky, H. Leopold, H. Stabinger, *Z. Angew. Phys.* 1, 273, 1969.
311. B. Trathnigg, Ch. Jorde, *J. Liquid Chromatogr.* 17, 385, 1987.
312. B. Trathnigg, *Prog. Polym. Sci.* 20, 615, 1995.
313. M.A. Haney, *Int. Lab. News* 36(1), 6, 2006; 36(2), 12, 2006; 36(3), 20, 2006.
314. J.-Y. Chuang, J.F. Johnson, *J. Appl. Polym. Sci.* 17, 2123, 1973.
315. C.P. Malone, H.L. Suchan, W.W. Yau, *J. Polym. Sci. B* 17, 2123, 1973.
316. E.P. Otocka, M.Y. Hellman, *J. Polym. Sci. Polym. Lett. Ed.* 12, 439, 1974.
317. J. Porath, H. Bennich, *Arch. Biochem. Biophys. Suppl.* 29, 152, 1962.
318. K.J. Bombaugh, R.F. Levangie, *J. Chromatogr. Sci.* 8, 56, 1970; *Sep. Sci.* 5, 751, 1970.
319. J.A. Biesenberger, M. Tan, I. Duvdevani, *J. Appl. Polym. Sci.* 15, 1549, 1971.
320. Z. Grubisic-Gallot, L. Marais, H. Benoit, *J. Polym. Sci. Polym. Phys. Ed.* 14, 959, 1976.
321. D. Alba, G.R. Meira, *J. Liquid Chromatogr.* 9, 1141, 1986.
322. M. Martin, F. Verillon, C. Eon, G. Guiochon, *J. Chromatogr.* 125, 17, 1976.

Part II

Applications

17 HPLC in Chiral Pharmaceutical Analysis

Ylva Hedeland and Curt Pettersson

CONTENTS

17.1 INTRODUCTION

17.1.1 NEED FOR CHIRAL PHARMACEUTICAL ANALYSIS

At the beginning of the twentieth century, Abderhalden and Muller [1,2] reported the first observed difference in pharmacological activity for the two enantiomers of a drug. They found that (*S*)-adrenaline and (*R*)-adrenaline had different effects on the blood pressure of laboratory animals. Today, it is well known that the individual enantiomers of a drug often have different pharmacokinetic and pharmacological properties, as their target structures in the human body are chiral, as they also are in animals. There is a broad range of examples in which the enantiomers of drugs show differences in their bioavailability, distribution, receptor interaction, and metabolic and excretion behavior, and, thus, they should be considered to be two different compounds [3]. It has also been proven that the use of a single enantiomer may reduce the dose of a drug, simplify the dose–response relationship, and minimize the toxicity caused by the therapeutically less-active enantiomer [3]. The guidelines issued by authorities for the registration of new drugs state that a chiral impurity should be treated in the same way as any other impurity, and an enantioselective determination should be included in the specification of enantiomerically pure drugs [4,5]. The trend in drug discovery is toward single enantiomers. Fifty percent of the drugs approved by the U.S. Food and Drug Administration between 2000 and 2002 were single enantiomers, 6% racemates and 44% achiral [6]. That is a significant increase since 1983, when 37% of the new drugs that were registered worldwide were racemates and only 26% were single enantiomers (37% were achiral) [6]. The progress toward enantiomerically pure drugs makes the selective and rapid analysis of enantiomers an important issue in drug development, especially for chiral purity determinations of the lead compounds but also for studying the pharmacokinetic behavior, racemization, and stereoselective metabolism. Since pharmaceutical products often are produced as racemates, there is also a need for enantioselective purification of the bulk substances, that is, preparative separation techniques. However, the pioneering work in the field of asymmetric synthesis conducted by the Nobel laureates Knowles and coworkers [7–9] has been a major step forward toward minimizing this need.

The techniques that are used for chiral separations are high-performance liquid chromatography (HPLC), gas chromatography (GC), supercritical fluid chromatography (SFC), thin layer chromatography (TLC), capillary electrochromatography (CEC), and capillary electrophoresis (CE). This chapter covers chiral drug separations with HPLC conducted during the years over the 10 year period from 1996 to 2006. HPLC is the most frequently used separation technique for enantiomers today [10], but during the last two decades, CE has become of increasing interest [11–13]. For general reviews on chiral separations exemplifying the other above-mentioned techniques (i.e., those other than HPLC), see references by Zhang et al. [10] and Gubitz and Schmid [14].

17.1.2 SEPARATION OF ENANTIOMERS BY HPLC

Chiral separation methods using HPLC can be divided into two categories, depending on whether the chiral separation methods used are direct or indirect. The direct method is based on reversible diastereomeric complex formation by the addition of a chiral selector to the mobile phase (chiral mobile phase additive, CMPA) or by the immobilization of a chiral selector to the stationary phase (chiral stationary phase, CSP). The indirect method is based on covalent formation of diastereomers by reaction with an enantiomerically pure chiral reagent ("chiral derivatizing reagent") [15]. This technique facilitates the use of a more straightforward separation technique (i.e., achiral stationary phases and uncomplicated mobile phases) but requires a high-purity chiral derivatization reagent. The indirect separation method is less commonly used in pharmaceutical analysis with HPLC today [16–21].

For the majority of the direct chiral separations in HPLC, different kinds of CSPs are used. The CSP can consist of small chiral molecules or polymers [22] and is often immobilized on agarose [23], silica gel [23,24], or polymer particles [23]. During the two last decades, a number of new phases have been introduced. Most commonly used CSPs are listed in Table 17.1.

The mobile phases used for chiral separations on CSPs differ depending on the type of column and range from normal-phase systems containing high amounts of nonpolar solvent (e.g., hexane) to

TABLE 17.1
Chiral Stationary Phases

Families of Chiral Selectors	Example
Proteins [23]	α1-Acid glycoprotein, albumin and cellobiohydrolase I
Macrocyclic antibiotics [24]	Vancomycin and teicoplanin
Polysaccharides [22,25]	Cellulose tris(3,5-dimethylphenylcarbamate)
Oligosaccharides [26]	β-Cyclodextrin
Synthetic polymers [27]	Helical polymethacrylates
Molecularly imprinted polymers [28]	Methacrylic monomers cross-linked with ethyleneglycol dimethacrylate together with an enantiomeric template
Low-molecular-weight selectors/"Pirke-type" selectors [29]	O-(Tertbutylcarbamoyl)quinine and p-nitrobenzoyl leucine

Source: Reprinted from Welch, C.J. et al., *J. Liquid Chromatogr. Relat. Technol.*, 29, 2185, 2006. With permission.

reversed-phase systems with water-based buffers with or without low contents of organic modifier. For supplementary reviews on CSP and chiral chromatography, see Refs. [30–34].

When a CSP is applied, the separation mechanism is based on the differences in the interaction between the chiral selector in the stationary phase and the enantiomers of the solute. Depending on the nature of the selector and the type of the solute, the stereoselective interaction can be based on interactions of one or more different types such as inclusion complexation, $\pi-\pi$–interaction, dipole stacking, hydrogen bonding, electrostatic interaction, hydrophobic interaction, and steric interaction [35]. In order to obtain chiral discrimination between the enantiomers, a "three-point interaction" is required between at least one of the enantiomers and the CSP [36]. The interactions can be of attractive as well as repulsive nature (e.g., steric and electrostatic interactions).

When CMPAs are applied, an ordinary achiral stationary phase such as diol silica [37], porous graphitic carbon [38], or octadecylsilica (C18) [39] can be used. The mobile phases used with CMPA are, generally, based on organic solvents (i.e., in the normal-phase mode) [40]. The reversible formation of the diastereomeric complexes in the mobile phase can be based on one or more types of intermolecular interaction, for example, inclusion complexation, $\pi-\pi$–interactions, dipole stacking, hydrogen bonding, ion-pair formation, and ligand exchange [35]. The complexes can be classified into five different groups based on the structure of the selector, as shown in Table 17.2.

The separation can be based on one or more of three possible mechanisms as follows: (1) The two enantiomers of a solute have a tendency to form complexes with the selector in the mobile phase to different extents. The diastereomeric complexes formed and the free enantiomers have a different distribution to the achiral stationary phase. (2) The diastereomeric complexes formed have a different distribution to the achiral stationary phase. (3) The chiral selector adsorbs to the achiral stationary phase to form a chiral pseudostationary phase [49].

Since one or more of the interactions in these systems might originate from the stationary phase, only a "two-" or a "one-point" interaction between the solute and the selector is necessary for mechanisms (2) and (3) to occur [50]. However, some of the CMPAs used in HPLC [37,40,51,52] have also been used as chiral selectors in CE [53–56], which indicates that at least one of the separation mechanisms between the selector and enantiomers is selective complex formation in the mobile phase in these cases, since there is no stationary phase present in CE. A recent example by Yuan et al. [57] is presented in Figure 17.1. The authors introduced the use of (R)-N,N,N-trimethyl-2-aminobutanol-bis(trifluoromethane-sulfon)imidate as the chiral selector for enantioseparation in HPLC, CE, and GC. This chiral liquid serves simultaneously as a chiral selector and a co-solvent.

The use of CMPA is flexible and is convenient for exploring new chiral selectors. The stationary phases used for the CMPAs are less expensive than CSPs, whereas the additives are often quite expensive. Furthermore, the complex mobile phase often limits the choice of detection method (e.g., mass spectrometry [MS]) that could be used, which makes the CMPAs less commonly used than CSPs. Only a few applications have been published during the last 10 years [39,58–60]. A recent example with a chiral selector used as both the CSP and the CMPA is shown in Figure 17.2 [43]. For further reviews on the use of CMPA, see Refs. [35,40,49].

TABLE 17.2
Chiral Mobile Phase Additives

Families of Chiral Selectors	Example
Proteins	Albumin [41] and cellobiohydrolase I [42]
Macrocyclic antibiotics	Vancomycin [43]
Oligosaccharides	Cyclodextrins [44]
Metal ion–amino acid complex	Cu^{2+}-proline [45]
Low-molecular-weight selectors (counterions)	Camphersulfonic acid [37], quinine [46], dicyclohexyl tartrate [47], and N-benzoxycarbonylglycyl-L-proline [48]

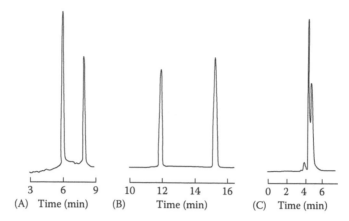

FIGURE 17.1 (*R*)-*N,N,N*-trimethyl-2-aminobutanol-bis(trifluoromethane-sulfon)imidate as the chiral additive. (A) CE BGE: Na-phosphate buffer pH 6.0 and 10 mM ionic liquid, analyte: propranolol. (B) HPLC, mobile phase H₂O: acetonitrile (AcN) (6:4) with 10 mM ionic liquid, analyte: 2,2′-diamino-1,1′-binafthalene. (C) GC capillary column coated with the chiral ionic liquid, analyte: citronella. (Reprinted with permission from *Anal. Lett.*, 39, 1447, 2006. Copyright 2006, Taylor & Francis.)

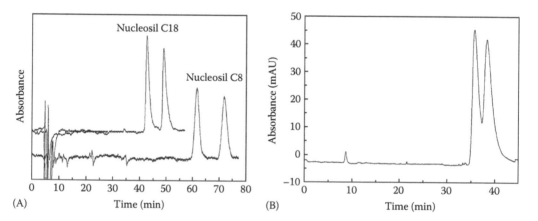

FIGURE 17.2 Chromatograms of (*RS*)-flurbiprofen with vancomycin as (A) CMPA and (B) CSP. Mobile phase: (A) 50% methanol (MeOH), 50% 0.1% triethylamine (TEA) (pH 5.0) with 4 mM vancomycin, and (B) 25% MeOH, 75% 0.1% TEA (pH 5.0). (From Kafkova, B. et al., *Chirality*, 18, 531, 2006. Reprinted with permission from Wiley-Liss Inc., a subsidiary of John Wiley & Sons Inc.)

17.2   METHOD DEVELOPMENT USING CSP OR CMPA

Many researchers have put a considerable amount of effort into studies of the chiral recognition mechanisms (using, e.g., NMR and molecular modeling), but yet the choice of chiral selector or chiral phase for a new compound is often based on trial and error. Different strategies for chiral method development have been presented by many of the retailers of chiral columns as a service for the customers. In addition to the information supplied by these retailers, another source of knowledge is Chirbase*, a database that contains more than 50,000 HPLC separations of more than 15,000 different chiral substances [61], which also can provide guidance to the analytical chemist.

Another practical guide for enantioseparation of pharmaceuticals with the most common types of CSPs has recently been published in a review by Thompson [34]. For polysaccharide and

---

* http://chirbase.u-3mrs.fr

macrocyclic glycopeptide columns, Anderson et al. [62] have developed an automated screening approach by introducing a column switching system and combinations of different CSPs and mobile phases. In this chapters, the authors used 55 enantiomeric pairs of drugs, intermediates and starting materials, and succeeded to enantioseparate 96% of them with their developed strategy. Yanik et al. [63] developed an automated method development strategy for the purification of chiral pharmaceutical candidates (i.e., preparative and semipreparative chromatography) using 10 different CSPs with different chemistry (ranging from the normal phase to the revered phase).

In a comparative study of 102 racemates of pharmaceutical interest with the three techniques, HPLC, CE, and SFC, HPLC was found to have the highest coverage of enantioseparation [64]. The authors state that this probably arose partly as a result of a higher variety of phases being commercially available for HPLC. The phases with the widest application ranges in that study were found to be Chiralpak AD and Chiralcel OD and OJ (i.e., polysaccharide phases).

As already mentioned, the enantiomers are considered to be two different compounds in enantiomerically pure products and a chiral impurity should be treated in the same way as the other impurities [4]. Thus, there are no specific guidelines for chiral bioanalytical or pharmaceutical drug analysis from the authorities apart from those relevant to achiral methods. The ICH guidelines for the validation of bioanalytical [65] and pharmaceutical methods for drug analysis [66,67] describe the general procedure for the validation of such methods and have been adopted by many of the authorities such as the U.S. Food and Drug Administration [4] and the European Pharmacopoeia [5]. These guidelines specify the requirements of the validation, and the bioanalytical guidelines include acceptance criteria for the method's accuracy (85%–115% of nominal value, except at the limit of quantification [LOQ], where 80%–120% is sufficient) and the precision (RSD ± 15%, except at the LOQ, where ±20% is sufficient). Since the guidelines for pharmaceutical products are lacking criteria for accuracy and precision [66,67], Carr and Wahlich [68] have suggested that an RSD of ±5% is satisfactory for the determinations of impurities in pharmaceuticals.

17.3 ENANTIOSELECTIVE DETERMINATIONS IN BULK MATERIAL AND PHARMACEUTICAL FORMULATIONS

The chiral purity determinations occur during the manufacturing of pharmaceutical formulations and bulk material, and it is not the only time that chiral separations are required [69]; it is also needed for studying the chiral inversion/racemization during storage [70] and for determining the absolute chiral configuration of new drug compounds [71].

The enantiomeric impurities above either 0.05% or 0.1% of the active substance in new products must be identified (the appropriate limit being dependent on whether the daily dose is >2 or <2 g) and reported to the authorities [4,5]. However, limitations arising from, for example, insufficient enantioresolution and/or low solubility sometimes make it difficult to obtain the required limits for the identification and for the qualification of the impurities. When trying to obtain baseline resolution between two peaks, it is often favorable to have the minor enantiomer as the first eluting one. The elution order can be altered by exchanging the chiral selector with its enantiomer or by changing to another CSP or CMPA (in the former case, the correspondence is straightforward, but in the latter, the outcome/result is harder to predict). However, it has also been shown that, in some cases, the temperature [72,73] or mobile phase [74–76] can alter the elution order on a particular CSP. Narayana et al. [77] developed a preparative chiral separation for linezolid on a Chiralpak AD column. The elution order was changed depending on the choice of mobile phase (Figure 17.3A through C). By altering the ratio of the components in the mobile phase and by exchanging 2-propanol (2-PrOH) for ethanol (EtOH), the undesired enantiomer (the (+)-enantiomer) was eluted as the first peak, which is favorable for the loading capacity for the purification. For further information about reversal of the elution order on polysaccharide, protein, and "Pirkle-type" CSPs, see the review by Okamoto [78].

The solubility of the sample in the mobile phase often limits the sample throughput in preparative chromatography and the limit of detection (LOD) in impurity determinations. Thus, for hydrophobic

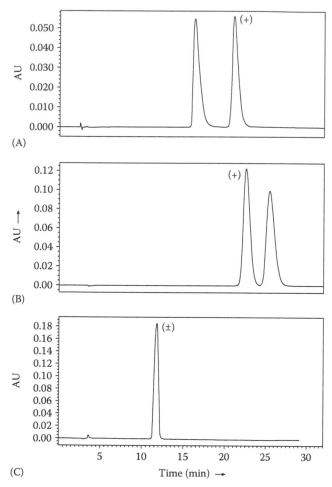

FIGURE 17.3 Separation of *rac*-linezolid on Chiralpak AD. Mobile phase: (A) hexane:2-PrOH:TFA (80:20:0.1 v/v/v), (B) hexane:EtOH:trifluoroaceticacid(TFA)(65:35:0.1 v/v/v), and (C) hexane:EtOH:2-PrOH:TFA (70:20:10:0.1 v/v/v/v). (Reprinted from Narayana, C.L. et al., *J. Pharm. Biomed. Anal.*, 32, 21, 2003. Copyright Elsevier science, 2003. With permission.)

compounds, the use of normal-phase chromatography or the "polar-organic mode" could be feasible. In the so-called "polar-organic mode" [79], polar-organic solvents (e.g., MeOH, EtOH) are used, often with the addition of acetonitrile (can) and a salt (e.g., ammonium acetate). The "polar-organic mode" is preferable to the normal phase when MS detection is used, since many of the frequently used normal-phase solvents require special care, for example hexane, which is flammable. For a *cis/trans* isomeric determination, Simms et al. [80] used a chiral (β-cyclodextrin) column in the "polar-organic mode" for an isomeric purity determination of a hydrophobic candidate drug in bulk. Their earlier efforts in the reversed-phase mode were unsuccessful and gave complex chromatograms, which probably originate from the precipitation of the analyte in the aqueous mobile phase, but in a mobile phase composed of AcN, MeOH, triethylamine (TEA), and acetic acid (HAc), the *cis* and the *trans* isomers of the drug candidate could be solved and separated from each other.

A majority of the chiral purity assays made with HPLC published during the last decade are based on separation on CSPs and subsequent UV detection (Table 17.3). The polysaccharide phases seem to be the dominating CSPs, but there is an even distribution in the methods that uses normal- and reversed-phase modes. A few of the methods utilize CMPA [39,59,60] or indirect separation by chemical derivatization [16,18]. However, it seems that the majority of the published papers

TABLE 17.3
Enantiomeric Determination in Bulk Material and Pharmaceutical Formulations

Substance	Column (Brand Name)	Mobile Phase (v/v)	LOD	Detection	Matrix	References
Anti-HIV nucleoside analogues	Chiralcel OD-RH	AcN:H$_2$O in different ratios	1.8 µg/mL, 0.4 µg/mL	UV, MS (SRM)	Bulk	[81]
Aminoglutethimide (deriv.[a] with dansyl chloride)	Chiralcel OD	EtOH:cyclohexane:MeOH (95:5:2)	20 ng/mL	F[b] λ_{ex} 395, λ_{em} 495 nm	Tablet	[17]
Aminoglutethimide (deriv. with fluorescamine)	Chiralcel OD-R	AcN:0.5% H$_3$PO$_4$ (85:15)	20.5 ng/mL	F. λ_{ex} 395, λ_{em} 495 nm	Tablet	[17]
Arginine	Chirobiotic T	MeOH:50 mM NaH$_2$PO$_4$ buffer pH 4.6 (2:8)	0.0025% 0.25 µg/mL	UV 214 nm	Capsule	[82]
Atropine	Chiral AGP	AcN:10 mM NH$_4$Ac buffer pH 6.2 (3:97)	<5%	UV, MS1 scan, 100–300 u	Bulk	[83]
Baclofen precursor (deriv. with FLEC[d])	Cellulose tris (3,5-di-methylphenylcarbamate) (made in-house)	Hexane:2-PrOH (80:20)	n.d.	UV/OR[e]	Bulk	[84]
Carnitine	LiChrospher 100 (C18)	AcN:TEA-PO$_4$ buffer pH 2.6 (27:73)	0.05% (LOQ)	F. λ_{ex} 260, λ_{em} 310 nm	Bulk	[18]
Citalopram	Shim-pack CN	0.1% TEAAc buffer (pH 4.0):AcN (90+10, v/v) and 12 mM β-CD	5.51 ng/mL	UV 240 nm	Tablet	[58]
Dexfenfluramine	Chiralcel OF and NH$_2$	n-Hexane:DEA[f] (99.9:0.1)	0.5%, 20 ng	UV 265 nm	Bulk, capsule	[85]
Donepezil	Chiralcel OD	n-Hexane:2-PrOH:TEA[g] (87:12.9:0.1)	20 ng/mL	UV 268 nm	Tablet	[86]
Doxazosin intermediate (deriv. with GITC[b])	YMC-Pack ODS (C18)	20 mM KH$_2$PO$_4$:MeOH (50:50)	0.5 mg/L	UV 250 nm	Bulk	[20]
Drug cand 30881[i]	Cyclobond I 2000	Gradient AcN:TEA:HAc[j] (100:0.3:0.2): AcN:MeOH:TEA:HAc (85:15:0.8:1.0)	n.d.	UV 230 nm, MS1 scan, 118–100u	Bulk	[80]
Drug candidate[k]	Chiralpak AD-RH or Chiralcel OD-RH	AcN: 0.01 M Ac buffer (pH 4.6) in different ratios	0.05%	UV 297 nm	Bulk	[87]

(continued)

TABLE 17.3 (continued)
Enantiomeric Determination in Bulk Material and Pharmaceutical Formulations

Substance	Column (Brand Name)	Mobile Phase (v/v)	LOD	Detection	Matrix	References
Drug candidate[l]	Chiralcel OD	EtOH:hexane:TFA[m] 23:77:0.1	1%–2%	UV 230 nm, OR, MS1 scan 100–1000 u	Bulk	[88]
Emtricitabine	Amylose tris [(S)-1-phenylethylcarbamate] (home packed)	AcN:MeOH (95:05) with 0.02% TEA and 0.02% HAc	0.06%	UV 280 nm	Bulk	[89]
Ketoprofen	Hypersil BDS C8	AcN:TEAA buffer (pH 5.2, 20 mM) (35:65) with 2.0 mM norvancomycin	0.20 ng	UV 290 nm	Capsule	[59]
Levamisole	Cyclobond I 2000 SN	AcN:0.5% TEAA buffer pH 5.0 (20:80)	0.02%	UV 254 nm, OR	Bulk, tablet	[90]
Levetiracetam	Chiralpak AD-H	Hexane:2-PrOH (90:10)	900 ng/mL	UV 210 nm	Bulk, formulation	[91]
Linezolid	Chiralpak AD	Hexane:2-PrOH:TFA (80:20:0.1)	123 ng/mL	UV 258 nm	Bulk, tablet	[77]
Mebeverine	Chiralcel OD	n-Hexane:2-PrOH:TEA (90:9.9:0.1)	0.05 μg/mL	UV 263 nm	Tablet	[92]
Methotrexate	Chirobiotic T	MeOH:HAc:TEA (100:0.2:0.1)	0.9 μg/mL	UV 303 nm	Tablet, inj. sol.	[93]
Methotrexate	Chiralcel OJ	AcN:H2O	n.d.	UV 303 nm	Bulk	[93]
Methotrexate	HSA	50 mM phosphate buffer pH 7.4:n-PrOH (99:1)	n.d.	UV 303 nm	Bulk	[93]
Metoprolol (deriv. with (−)-menthyl chloroformate)	Inertsil C8	73% MeOH in H2O	0.03%	F, λ_{ex} 276, λ_{em} 309 nm	Bulk	[16]
Metoprolol	Chiralcel OD	10 mM DEA, 28 mM H2O and 5 mM HAc in n-hexane:2-PrOH (85:15)	n.d.	UV 273 nm	Bulk	[94]
Nadolol	Chiralcel OD	Hexane:EtOH:DEA (80:20:0.4)	n.d.	UV 254 nm	Bulk	[95]
Naproxen	ODS[a]	AcN:H2O (20:80) with 20 mM methyl-β-CD and 50 mM NaAc pH 3	n.d.	UV 232 nm, amperiometric	Tablet	[39]
Ornidazole	Chiralcel OB-H	n-Hexane:MeOH:2-PrOH (95:4:1)	0.05% 0.2 μg/mL	UV 311 nm	Bulk, inj. solution	[96]
p-Chlorowarfarin	Cyclobond I RN	AcN:HAc:TEA (99.8:0.1:0.075) column temp. 0°C	n.d.	MS1 scan (150–500 u)	Bulk	[97]
Paroxetine	Chiralpak AD	n-Hexane:EtOH:DEA (94:6:0.5)	0.2%, 2 ng	UV 296 nm	Bulk, tablet	[98]
Pramipexole	Chiralpak AD	n-Hexane:EtOH:DEA (70:30:0.1)	300 ng/mL	UV 260 nm	Bulk	[99]

Propionyl carnitine	Chirobiotic TAG and Sperisorb S5 SCX	AcN:10mM NaH$_2$PO$_4$ pH 6.8 (65:35)	0.026%	UV 205 nm	Bulk	[100]
Propranolol	Chiralcel OD	Hexane:EtOH (75:25)	n.d.	UV 280 nm	Tablet	[101]
Suprofen	Cyclobond I RN	AcN:MeOH:HAc:TEA (95:5:0.2:0.2)	n.d.	MS1 scan (150–800u)	Bulk	[97]
Terbutaline	Sumichiral OA-4900	n-Hexane:ethylacetate:MeOH:TFA (24:25:2.5:0.1)	0.05%	UV 276 nm	Bulk	[102]
Tetramisole	Chiralcel OD	n-Hexane:2-PrOH:DEA (90:10:0.1)	1.6%–2.0% (LOQ)	UV 254 nm	Formulation	[103]
Thioridazine	ChiraDex	0.05M H$_3$PO$_4$ buffer (pH 6.5):AcN (50:50)	5 µg	UV 280 nm	Tablet	[104]
Thyroxine	Chiral AX QN-1	AcN:0.05M NH$_4$Ac buffer pH$_a$ 4.5	0.1 µg/mL	UV 240 nm	Tablet	[105]
Triiodothyronine	Chiral AX QN-1	AcN:0.05M NH$_4$Ac buffer pH$_a$ 4.5	0.5 µg/mL	UV 240 nm	Tablet	[105]
Tiaprofenic acid	Chiralcel OD or Chiralcel AD	n-Hexane:2-PrOH:TFA (98.5:1.5:0.1) or n-hexane:2-PrOH:TFA (94:6:0.1)	5 ng	UV 296 nm	Tablet, ampoule	[106]
Timolol	Chiralcel OD-H	Hexane:2-PrOH:DEA (965:35:1)	0.02% (0.27 µg/mL)	UV 297 nm	Bulk	[107]
Timolol intermediate	Chiralpak AS	n-Hexane:EtOH:FA:DEA (90:10:0.2:0.2)	3–7 ng/mL	MS (SRM 148→92)	Bulk	[108]
WCK771[o]	ODS[n]	AcN:K$_2$PO$_4$ buffer pH 7.3 (88:12) + 11.35 g/L β-cyclodextrin and TEA	0.015 µg/mL	UV 290 nm	Bulk	[60]

[a] Derivatization.
[b] Fluorescence.
[c] Ammoniumacetate.
[d] (+)-[1-(9-Fluorenyl)-ethyl]-chloroformate.
[e] Optic rotation.
[f] Diethylamine.
[g] Triethylamine.
[h] 2,3,4,6-Tetra-O-acetyl-β-D-glucopyranosyl isothiocyanate.
[i] Molecular mass 496, hydrophobic.
[j] Acetic acid.
[k] (R)-2-(4-Bromo-2-fluorobenzyl)-(1,2,3,4-tetrahydropyrrolo[1,2-a]pyrazine-4-spiro-3'-pyrrolidine)-1,2',3,5'-tetrone.
[l] (2S)-2-((2-Benzoylphenyl)amino)-3-(4-hydroxyphenyl)-propionic acid.
[m] Trifluoriacetic acid.
[n] Octadecylsilica (the brand name are not given in the reference).
[o] (S)-(−)-9-Fluoro-6,7-dihydro-8-(4-hydroxypiperidin-1-yl)-5-methyl-1-oxo-1H,5H-benzo[i,j]quinolizine-2-carboxylic acid l-arginine salt tetrahydrate.

concerning enantiomeric purity determination in bulk and in pharmaceutical formulations during the years 1996–2006 have used CE as the separation technique. Although the values of the LOD are presented in Table 17.3, it is difficult to compare the sensitivity since different definitions of the LOD are utilized in the papers.

Huang et al. [96] developed a method for the enantiomeric purity determination of (S)-ornidazole in raw material and injection solution that was used in an preclinical study. In this publication, a mobile phase of n-hexane, MeOH, and 2-PrOH (95:4:1) was used with a Chiralcel OB-H column. No chiral impurity (R)-ornidazole was detected above the LOD (0.05%) in either the raw material or the injection solution (see Figure 17.4D and E). The separation of the racemate is presented in Figure 17.4A, and the minor peak in Figure 17.4B corresponds to an enantiomeric impurity of 0.5%.

Marini et al. [107] validated fully a method, previously published by the same group [109], for the determination of (R)-timolol in (S)-timolol samples. They used a Chiralcel OD-H column and a mobile phase consisting of hexane:2-PrOH:diethylamine (965:35:1). The repeatability and intermediate precision of the method were good (RSD <1.3% and <2.0%, respectively) and the accuracy at the 15 µg/mL level (corresponding to a chiral impurity of 1%) was 5.4%. The selectivity of the method was studied by the injection of conceivable chiral and achiral impurities (Figure 17.5A), and by the injection of the blank solution (Figure 17.5B). A typical chromatogram of a standard solution at a chiral impurity of 0.2% is shown in Figure 17.5C. For chiral impurity determination of drugs, there is relatively little interference from the matrix, and the sensitivity obtained by UV detection is often sufficient. However, for those substances that are lacking chromophores and when mobile phases with a high UV cutoff are utilized, MS detection could be necessary to give the required sensitivity. Richards et al. [97] analyzed the enantiomers of p-chlorowarfarin on a β-cyclodextrin

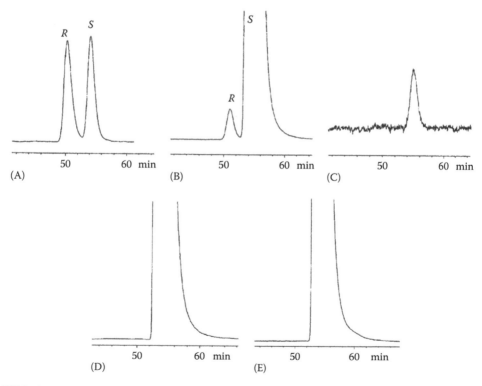

FIGURE 17.4 Chiral separation of ornidazole. (A) rac-ornidazole, (B) solution II, (C) 1 µg/mL (S)-ornidazole, (D) (S)-ornidazole raw material, and (E) (S)-ornidazole injection solution. Column: Chiralcel OB-H, mobile phase: n-hexane:MeOH:2-PrOH (95:4:1 v/v/v). (From Huang, J.Q. et al., Chirality, 18, 587, 2006. Reprinted with permission from Wiley-Liss Inc., a subsidiary of John Wiley & Sons Inc.)

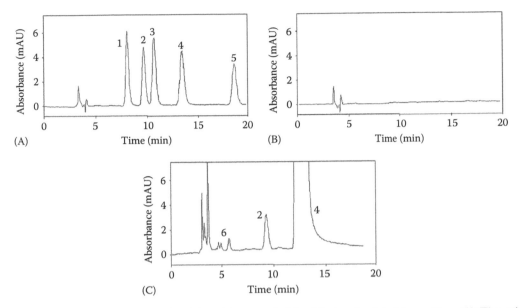

FIGURE 17.5 Separation of (*S*)-timolol and its conceivable chiral and achiral impurities. (A) The mixture solution containing the conceivable chiral and achiral impurities of (*S*)-timolol, (B) the dissolution solution, and (C) the standard solution at 0.2% enantiomeric impurity. Column: Chiralcel OD-H, mobile phase: hexane:2-PrOH:DEA (965:35:1 v/v/v). Peaks: (1) timolol dimer, (2) (*R*)-timolol, (3) isotimolol, (4) (*S*)-timolol, (5) dimorpholinothiadiazole, and (6) solvent front. Concentration of analytes: 5–10 µg/mL in (A) and 3 µg/mL (*R*)-timolol in (C). (Reprinted from Marini, R.D. et al., *Talanta*, 68, 1166, 2006. Copyright Elsevier, 2006. With permission.)

column using a mobile phase of AcN:HAc and TEA (99.8:0.1:0.075), which had a UV cutoff above the absorption maximum for *p*-chlorowarfarin. In this investigation, the molecular ion and its TEA adduct were monitored in the MS1 scan mode and an improved sensitivity was found compared with that obtained by UV detection. Furthermore, the typical isotopic pattern of chlorine-containing substances (100:22:35:7) was utilized for more selective identification of the *p*-chlorowarfarin.

Table 17.3 contains only few examples of where CMPA have been applied for pharmaceutical analysis. The majority of selectors that have been used are different types of cyclodextrines [39,60], but there is also a macrocyclic antibioticum in the list [59]. Guo et al. [59] applied norvancomycin as a CMPA for an assay of (*S*)-ketoprofen in capsules, the results are displayed in Figure 17.6B. They

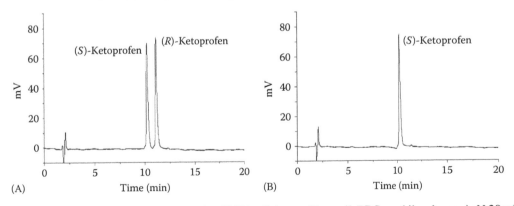

FIGURE 17.6 Norvancomycin used as the CMPA. Column: Hypersil BDS, mobile phase: AcN:20 mM TEA buffer (pH 5.2) containing 2.0 mM norvancomycin (35:65), (A) racemic solution of ketoprofen and (B) (*S*)-ketoprofen formulation. (Reprinted from Guo, Z.S. et al., *J. Pharm. Biomed. Anal.*, 41, 310, 2006. Copyright Elsevier, 2006. With permission.)

studied the influence of the CMPA and AcN concentration and also the pH in the mobile phase, and found a decreased retention time at higher pH and higher AcN concentrations. The optimized method showed good repeatability (an RSD of 2% or less), and the LOD was 0.20 ng for both the (R)-enantiomer and the (S)-enantiomer (UV detection at 290 nm).

Enantiomeric purity assays have also been performed without chromatographic separation being conducted prior to detection, for example, with circular dichroism (CD) and MS. Bertucci et al. [110] developed a chiral assay for pulegone, oxazepam, and warfarin by combining simultaneous UV, CD, and g factor detection on an achiral separation system with a Hypersil CN column and a mobile phase of hexane:2-PrOH (90:10). The precision (RSD%) of the method ranged from 0.6% to 2.6%, and the LOQs were between 0.1% and 1% (0.2–2.2 µg). For further information concerning the application of CD and polarometric detection for chiral detection, see the review by Bobbitt and Linder [111].

The determination of enantiomeric excess by MS detection without chromatographic or electrophoretic separation prior to detection has also grown in interest. The enantiomeric discrimination in the gas phase by MS can be obtained in different ways, for example, by the formation of diastereomeric complexes and from collision-induced dissociation in the MS/MS mode. For further information about this topic, see the review by Schug and Lindner [112].

17.4 APPLICATIONS IN BIOANALYSIS

Historically, most bioanalytical methods for chiral drug analysis were developed using HPLC with UV or fluorescence detection. The low efficiency often encountered when using CSPs results in high dilution of the sample, with a resultant decrease in the sensitivity compared with separations on achiral columns. During the last decade, MS detection has grown in interest. The high sensitivity and selectivity of MS make it a favorable alternative to UV detection, especially for complex matrices, where the orthogonal mass-to-charge separation power of the MS is complementary to the chromatographic separation. Thus, the use of chiral HPLC with MS/MS detection often offers complete resolution of the parent drug and its metabolites without the need for complex column switching systems or extensive sample clean-up procedures.

Today, the reversed-phase seems to be the dominant separation mode of the chiral pharmaceutical applications in biological matrices (compare Tables 17.4 and 17.5). The applications in which the normal-phase mode is used are limited almost exclusively to the polysaccharide phases, as can be seen in Table 17.4. The "polar-organic mode," discussed above, has also grown in interest during the last decade and is included in Table 17.5. The high compatibility for these types of mobile phases with MS has probably contributed to their increased use. The choice of the mobile phase composition in chiral separations is often more limited than in achiral ones. Exchanging from one buffer system to another might destroy the separation. Furthermore, the chiral phase materials are often more sensitive, which impose restrictions on the buffer pH, the kind of additive, and the total amount of organic solvent that can be used. For example, the protein-based columns, like α-1-glycoprotein, are used in water-based buffers (preferably phosphate or acetate) between pH 4 and 7 and with small amounts of organic additives (e.g., 2-PrOH or AcN). Up to 25% 2-PrOH could be added, but the highest enantioresolution is often found between 0.5% and 5% [113]. Even though protein columns are used for chiral application with MS detection, the MS ionization techniques that are most frequently used, namely, electrospray ionization (ESI) and atmospheric pressure chemical ionization (APCI), favor volatile organic mobile phases without detergents or inorganic buffers such as phosphate and borate [114,115]. Mobile phases with high aqueous content should be avoided if possible. Buffers containing, for example, ammonium acetate, ammonium formate, acetic, formic, or citric acid are preferable [114,116]. Strong volatile acids like trifluoroacetic acid are not usually compatible with MS detection since they cause strong ion pairing with basic analytes, making them neutral and resulting in the suppression of the analyte signal [114]. In "polar-organic" mode, however, the frequently used TEA and HAc have been successfully replaced by

TABLE 17.4
Chiral Separation in Biological Matrices Using the Normal-Phase Mode

Substance	Column	Mobile Phase (v/v)	Matrix	LOQ[a]	References
Azelastine and metab. (metabolites)	Chiralpak AD	n-Hexane:2-PrOH-DEA (95:5:0.6) or n-hexane-EtOH-DEA (97:3:9.0.7)	Rat plasma	nd	[124]
BMS-204352[b]	Chiralcel OD-H	2-PrOH and 0.1% FA[c] in hexane (10:90)	Plasma	0.10 ng/mL	[125]
Clevidipine	Chiralcel OD-H	"Normal phase"	Blood	0.5 nM	[126]
Donepezil	Chiralcel OD	n-Hexane:2-PrOH:TEA (87:12.9:0.1)	Rat plasma	20 ng/mL[d]	[86]
Doxazosin	Chiralpak AD	n-Hexane:2-PrOH:DEA (70:30:0.1)	Plasma	nd	[127]
Felodipine	Chiralcel OJ-R	2-PrOH:isohexan (11:89)	Plasma	0.25 nM	[128]
Fenozan B07	Chiralcel OG	Hexane:2-PrOH gradient (0%–30% 2-PrOH)	Dog plasma	2 ng/mL	[122]
Lercanidipine	Chiralpak AD	Hexane:EtOH:DEA (95:5:0.1)	Plasma	25 ng/mL	[129,130]
Lorazepam	Chiralcel OD and Chiralpak AS	n-Hexane-2-PrOH:EtOH (5:5:1)	Plasma	nd	[131]
Mebeverine	Chiralcel OD	n-Hexane:2-PrOH:TEA (90:9.9:0.1)	Rat plasma	0.1 μg/mL[d]	[92]
Metolachlor	Chiralpak AS H	n-Hexane:2-PrOH (90:10)	Surface and groundwater	0.10 ppb	[132]
Metrifonate	Chiralpak AS	2 mM NH₄Ac in n-heptane:EtOH (75:25)	Blood	5.0 μg/L	[133]
Metrifonate	Chiralpak AS	2 mM NH₄Ac in n-heptane:EtOH (75:25)	Brain	7.50 ng/g	[133]
MK-0767[e]	Kromasil CHI-DMB	Hexane:2-PrOH with 0.1 FA (81:19)	Plasma	1 ng/mL	[134]
Omeprazole	Chiralpak AD	EtOH:AcN:HAc (30:1:0.004)	Plasma	10 nM	[135]
Org 4428[f]	Chiralpak AD	n-Hexane:MeOH:EtOH (95:3:2)	Plasma	0.5 ng/mL	[136]
Oxybutynin and metab.	Chiralpak AD	n-Hexane:2-PrOH:DEA (90:10:0.1)	Plasma	nd	[127]
Sotalol	Chiralpak AD	EtOH:n-hexane:2-PrOH:DEA (30:63:7:0.17)	Plasma	nd	[127]
Sulfoxide drug cand.	Chiralpak AD	2-PrOH:hexane (80:20)	Plasma	5 ng/mL	[121]
Terazosin	Chiralpak AD	n-Hexane:2-PrOH with 0.05% DEA 65:35	Plasma	62.5 pg/mL	[137]
Tramadol	Chiralpak AD	Isohexane:EtOH:DEA (97:3:0.1)	Plasma	0.15 ng/mL	[123]
Verapamil and metab.	Chiralpak AD	n-Hexane:2-PrOH:DEA (92.5:7.5:0.1)	Plasma	nd	[127]

[a] All applications with the exception of the two marked with a UV detection were obtained using MS detection.

[b] (S)-2-((2-Benzoylphenyl)amino)-3-(4-hydroxyphenyl)-propionic acid.

[c] Formic acid.

[d] UV detection.

[e] A PPAR α/γ agonist.

[f] cis-1,3,4,13b-Tetrahydro-2,10-dimethyldibenz[2,3:6,7]oxepino[4,5-c]pyridin-4a(2H)-ol.

TABLE 17.5
Chiral Separations in Biological Matrices Using the Reversed-Phase and "Polar-Organic" Mode

Substance	Column	Mobile Phase (v/v %)	Matrix	LOQ	References
2-Hydroxyglutaric acid	Chirobiotic R	MeOH:5mM TEA buffer (pH 7.0) (9:1)	Urine	n.d.	[138]
3-Amino-2-fluoropropylphosphic acid	TBuCQN	0.2 M NH$_4$Ac (pH 5.0):MeOH (10:90)	Plasma	0.85 µM	[139]
Albuterol	Chirobiotic T	MeOH:HAc:28% NH$_4$ (1000:5:1)	Plasma	0.25 ng/mL	[140]
Albuterol	Chirobiotic T	MeOH with 0.02% FA and 0.1% NH$_4$ FAc	Dog plasma	2.5 nM	[141]
Amlodipine	Chiral AGP	10mM Ac buffer (pH 4.5):1-PrOH (99:1)	Plasma	0.1 ng/mL	[142]
Atenolol	Chirobiotic T	MeOH:AcN: 0.5 M NH$_4$Ac pH 4.5 (60:39:1)	Urine	400ng/mL	[143]
Atenolol	Chirobiotic V	MeOH:H$_2$O (90:10)+0.1% TEA adj. to pH 4.0 with HAc	Waste water	12–110ng/L (LOD)	[144]
Azelastine and metab.	Cyclobond I 2000	45mM NH$_4$Ac (pH 4.7):MeOH:AcN (70:21:9)	Rat plasma	125ng/mL	[124]
Baclofen	Crownpak CR(+)	10mM NH$_4$Ac (pH 6.8):MeOH (9:1)	Plasma/CSF	0.15 ng/mL	[145]
Benzodiazepine amides and metab.	PirkleDNBL	AcN:25 mM NH$_4$COOH (pH 7) (40:60)	Microsomes	nd	[146]
Biperiden	Cyclobond I 2000	AcN:MeOH:HAc:TEA (95:5:05:0.3)	Serum	1 ng/mL (LOD)	[147]
Bupivacaine	Chiral AGP	5mM NH$_4$Ac buffer (pH 7.0):2-PrOH (97:3)	Urine	nd	[148]
Carvedilol (derivatized with GITC)	Ace 3 C18	2mM NH$_4$ formate (pH 3):AcN (50:50)	Plasma	0.2 ng/mL	[21]
Chlorfeniramine and metab.	Cyclobond I 2000	DEAAc (pH 4.4):MeOH:AcN (85:7.5:7.5)	Plasma	0.25 ng/mL	[149]
CGS 26214[a]	Chiral AGP	n-PrOH:0.03% NH$_4$Ac (pH 7.0) (4:96)	Plasma	0.4ng/mL	[150]
Donepezil	Biooptick AV-1	10mM FA:MeOH (75:25)	Plasma	0.020ng/mL	[151]
Epibatidine	Chiral AGP	AcN:10mM NH$_4$COOH and 1mM HFBA (pH 7.4) (10:90)	Microsomes	nd	[152]
Fluoxetine	Chirobiotic V	MeOH with 0.075% NH$_4$TFAc	Plasma	2ng/mL	[153]
Fluvastatin	Chiralcel OD-R	AcN:MeOH:H$_2$O (24:36:40)	Plasma	5ng/mL	[154]
Ibuprofen	Chiralpak AD-RH	MeOH:0.1% H$_3$PO$_4$ (pH 2) (8:2)	Plasma	0.12 µg/mL	[155]
Ketamine and norketamine	Chiral AGP	2-PrOH:10mM NH$_4$Ac (pH 7.6) (6:94)	Plasma	1.0 ng/mL	[156]
Ketoprofen	Chirex3005	MeOH:30mM NH$_4$Ac (pH 3.5) (95:5)	Plasma	0.05 ng/mL	[157]
Keto-pyrrol analogues	Chiral AGP	5%–12% 2-PrOH in 0.1M NH$_4$Ac (pH 5–6.7)	Hepatocyte	nd	[158]
Lorazepam	Chiralcel OD-R	AcN:H$_2$O:HAc (80:20:0.1)	Plasma	0.2–1 ng/mL	[159]
			Urine	0.2–10ng/mL	[159]
Metabolite of FK778	Chiral AGP	20mM NH$_4$Ac (adj. to pH 6.5):2-PrOH (97:3)	Plasma	nd	[160]
Methadone	Chiral AGP	10mM NH$_4$Ac + 0.05% N-DMOA (pH 6.6): 2-PrOH (85:15)	Saliva	5ng/mL	[161]

Analyte	Column	Mobile phase	Matrix	LOD/LOQ	Reference
Methadone	Chiral AGP	2-PrOH:10 mM NH_4Ac (12:88;	Plasma	5.00 ng/mL	[162]
Methadone and metab.	Chiral AGP	AcN:5 mM NH_4Ac (pH 4.1) (2.5:97.5)	Serum	5 nM	[163,164]
Methadone and metab.	Chiral AGP	2-PrOH:2 mM NH_4COOH (pH 5.8) (gradient 8%–20% 2-PrOH)	Hair	nd	[165]
Methadone and metab.	Chiral AGP	20mM NH_4 formate (pH 5.7):MeOH (9:1)	Plasma	0.1 ng/mL	[166]
Methadone and metab.	Chiral AGP	AcN:10 mM NH_4Ac buffer (pH 7.0) (18:82)	Saliva	5 ng/mL, 0.5 ng/mL	[167]
Methadone and metab.	Chiral AGP	20mM HAc (pH 7.4):2-PrOH (93:7)	Hair	0.05 ng/mg	[168]
Methamphetamine	IAC[b]	50mM NH_4Ac pH gradient (7.0–3.5)	Urine	18 ng/mL (LOD)	[169]
Methamphetamine and amphetamine	ULTRON ES PhCD	10mM NH_4Ac (pH 5.0):MeOH:AcN (60:30:10)	Urine	20 ng/mL, 50 ng/mL	[170]
Methylphenidate	Chirobiotic V	0.05% NH_4TFAAc in MeOH	Plasma	87 pg/mL	[171]
Methylphenidate	Chirobiotic V	0.05% NH_4TFAAc in MeOH	Plasma (rat, rabbit and dog)	1.09 ng/mL	[172,173,174]
Methylphenidate	Chirobiotic V	0.05% NH_4TFAAc in MeOH	Fetal tissue	0.218 ng/mL	[175]
Metoprolol	Chirobiotic V	MeOH:H_2O (90:10) with 0.1% TEA, adj. to pH 4.0 with HAc	Wastewater	17–42 ng/L (LOD)	[144]
Montelukast[c]	Chiral AGP	AcN (20%–40%) in 1 mM NH_4Ac (pH 4.5)	Plasma, bile	n.d.	[176]
Nefopam and metab.	Chirobiotic V	0.1% NH_4F_3Ac in MeOH	Plasma/urine	0.875 ng/mL, 1 ng/mL	[177]
Omeprazole and metab.	Chiralpak AD-RH	AcN:20mM NH_4Ac (pH 4.65) (35:65)	Plasma	nd	[178]
Oxybutynin	Chiral AGP	AcN:10mM NH_4 format (90:10(sic!))	Plasma	0.5 ng/mL	[189]
Phenprocoumon	Chira-Grom-2	H_2O:AcN:FA (48:52:0.5)	Plasma	37.5 ng/mL	[180,181]
PGE-9509924[d] (sample derivatized with FLEC)	Xterra MS (C18)	MeOH:AcN:H_2O:FA (40:40:20:0.1)	Plasma (dog)	0.025 µg/mL	[19]
Pindolol	Chiral DRUG	AcN:H_2O (50:50) with 10mM NH_4Ac	Serum, urine	0.13 ng/mL (LOD)	[182]
Pindolol	Chirobiotic T	MeOH with 0.02% NH_4TFAAc	Plasma	250 pg/mL	[183]
Procyclidine	Cyclobond I	AcN:MeOH:HAc:TEA (95:5:0.5:0.3)	Serum	1 ng/mL (LOD)	[184]
Propafenone	AGP	10mM NH_4Ac (pH 5.96i:1-PrOH (100:9)	Plasma	20 ng/mL	[185]
Propranolol	Chirobiotic T	0.5 g NH_4TFA in 1 L MeOH	Rat plasma	2 ng/mL	[186]
Propranolol	Chirobiotic V	MeOH:H_2O (90:10) with 0.1% TEA, adj. to pH 4.0 with HAc	Wastewater	4.4–17 ng/L (LOD)	[144]
Rabeprazole and metab.	Chiral CD-Ph	0.5 M $NaClO_4$:AcN (6:4, v/v)	Plasma	5 ng/mL (UV-det)	[187]
R483[e]	Chiralcel OJ	AcN:MeOH:H_2O:FA (70:20:10:1)	Plasma	0.5 ng/mL	[188]
Salbutamol and metab.	Chirobiotic T	MeOH:HAc:NH_3 (1000:5:1)	Urine	1 ng/mL, 5 ng/mL	[189]
Salbutamol and metab.	Chirobiotic T	MeOH:HAc:NH_3 (1000:5:1)	Plasma	25 ng/mL	[189]

(continued)

TABLE 17.5 (continued)
Chiral Separations in Biological Matrices Using the Reversed-Phase and "Polar-Organic" Mode

Substance	Column	Mobile Phase (v/v %)	Matrix	LOQ	References
Sotalol	Chirobiotic T	MeOH:AcN:HAc:TEA (70:30:0.025:0.025)	Plasma	4 ng/mL	[190]
Thalidomide	Chirobiotic V	14% AcN in 20 NH_4-formate (adj. to pH 5.4)	Serum, tissue	0.05 µg/mL (UV det)	[191]
Terbutaline	Chirobiotic T	MeOH with 0.1% TFA	Plasma	1.0 ng/mL	[192]
Tetrahydro-*b*-carboline	Chiralcel OD	AcN:H_2O (gradient) both with 0.1% TFA	Blood	nd	[193]
Tetrahydro-*b*-carboline	Chiralcel OD	AcN:H_2O with 0.1% TFA	Urine	nd	[193]
Trihexyphenidyl	Cyclobond I 2000	AcN:MeOH:HAc:TEA (95:1:05:0.3)	Serum	1.3 ng/mL (LOD)	[147]
Trihexyphenidyl	Cyclobond I 2000	AcN:MeOH:HAc:TEA (95:5:0.5:0.3)	Serum	1 ng/mL (LOD)	[184]
Verapamil and norverapamil	Chiral AGP	AcN:0.02 M NH_4Ac buffer (pH 7.4) (15:85)	Plasma	50 pg/mL, 60 pg/mL	[194]
Warfarin	Cyclobond I	AcN:HAc:TEA (1000:3:2.5)	Plasma	1.0 ng/mL	[195]

a Synthetic thyromimetic agent with cholesterol lowering activity in laboratory animals.

b Immunoaffinity column made in-house.

c Drug candidate MK-476.

d 7-[3S-Aminopiperidinyl]-1-cyclopropyl-1,4-dihydro-8-methoxy-4-oxo-3-quinolinecarboxylic acid.

e 5-[[4-[3-5-Methyl-2-phenyl-oxazol-4-yl)ethoxy]benzo[2]thiophen-7-yl]methyl]-thiazolidine-2,4-dione.

ammonium trifluoroacetate for the separation of substances with amide or amine functionalities on glycopeptide columns [117]. Highly flammable or explosive normal-phase solvents, of which hexane is an example, should also be avoided [115]. Consequently, these mobile phases are often used with postcolumn addition of miscible solvents prior to detection to avoid explosion and to facilitate detection by promoting ion formation. Recently, Ding et al. [118] introduced a nonflammable fluorocarbon-ether as an alternative mobile phase solvent on vancomycin and teicoplanin CSPs in the normal-phase mode. They found comparable selectivity and higher or comparable sensitivity when ethoxynonafluorobutane was used instead of n-hexane or n-heptane.

CMPAs have rarely been used for chiral analysis in biological matrixes, and have, to the authors' knowledge, not been applied during the last decade. One of the reasons is probably the incompatibility between involatile CMPAs and MS detection. For recent reviews on chiral pharmaceutical analysis by LC/MS, see Refs. [114,116,119,120].

17.4.1 SEPARATIONS IN NORMAL-PHASE MODE

Only a limited number of biomedical applications have been published in the normal-phase mode, as can be seen in Table 17.4. However, the sensitivity of the methods seems to be comparable to the reversed-phase and polar-organic mode applications, although a detailed comparison is not feasible since the LOQ data are missing for the few substances that have been analyzed in both modes. The majority of the methods are based on MS detection, and APCI seems to be the predominant ionization mode for the applications in normal-phase mode.

One reason for the rare use of the normal-phase solvents in bioanalytical applications is probably the frequent use of MS detection, and another, the problems that have already been mentioned of combining flammable liquids with this detection technique. Miller-Stein and Fernandez-Metzler [121] used the normal-phase mode for a chiral assay of a sulfoxide drug candidate and its corresponding sulfide and sulfone metabolites in drug plasma. The separation was performed within 7 min and the structurally similar internal standard and sulfide metabolite was coeluted, but could be resolved by the MS. They used a Chiralpak AD column and a mobile phase of 80% 2-PrOH and 20% hexane with postcolumn addition of 75% 2-PrOH and 25% ammonium acetate. The makeup liquid was added to minimize the risk of explosion at the APCI interface. However, Maraschiello et al. [122] recently used 96% hexane and 4% 2-PrOH as the mobile phase for a chiral separation of a drug candidate without any reported drawbacks for the ionization in the APCI interface. Ceccato et al. [123], too, used a mobile phase containing isohexane, EtOH, and diethylamine (97:3:0.1) and claimed that the addition of a small amount of polar modifier made ignition, with a subsequent explosion, unlikely. They developed a method for simultaneous determination of the enantiomers of tramadol and its main metabolite in human plasma. These two substances and two additional metabolites were simultaneously enantioseparated by the LC/MS/MS system and the Chiralpak AD column (Figure 17.7) [123]. On the basis of the signal-to-noise ratio, the LOD for tramadol and its main metabolite was determined to be 0.05 and 0.09 ng/mL (with an LOQ of 0.3 and 0.33 ng/mL, respectively).

17.4.2 SEPARATIONS IN THE REVERSED-PHASE AND "POLAR-ORGANIC" MODES

The most frequently used CSPs for biological applications in the reversed-phase mode are based on macrocyclic antibiotics, proteins, or oligosaccharides, but some of the applications utilize phases based on polysaccharides, low-molecular-weight selectors, crown ethers, or columns based on immunoaffinity techniques (Table 17.5).

Plasma is the most frequently analyzed matrix, and enantioselective determination of a parent drug for pharmacokinetic and therapeutic monitoring the most frequent goal of the developed methods but assays with simultaneous determination of metabolism are also common (Table 17.5). The majority of the methods are based on MS detection, and ESI is the predominant ionization

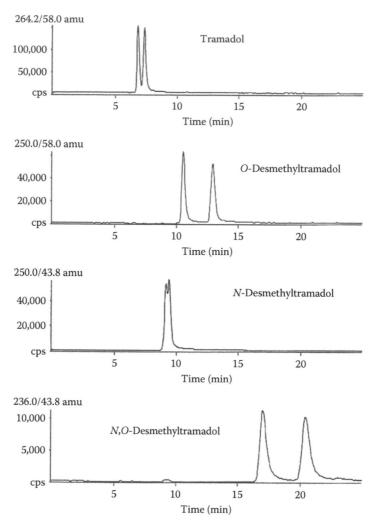

FIGURE 17.7 Chromatograms of tramadol and its three metabolites on a Chiralcel OD-H CSP. The chromatograms were obtained in the SRM mode. Mobile phase: isohexane:EtOH:DEA (97:3:0.1, v/v/v). (Reprinted from Ceccato, A. et al., *J. Chromatogr. B*, 748, 65, 2000. Copyright Elsevier, 2000. With permission.)

mode. ESI is known to be more sensitive to the matrix and the mobile phase composition than APCI [115,119]. The mobile phases used for the macrocyclic antibiotic and the oligosaccharide phases are based on polar-organic solvents with or without the addition of low amounts of water, resulting in a high compatibility with the ESI. But, despite the fact that the mobile phases used in combination with protein columns are based mainly on water-based buffers with only small amounts of organic solvent, these systems have been successfully applied in combination with MS detection [142,148,150–152,156,158,161–168,179,185,194]. The presence of ion suppression is often discussed and has been evaluated in different ways in the literature. One often used technique for the evaluation of ion suppression from the mobile phase and the matrix is to use continuous postcolumn infusion of the substance/internal standard and on-column injection of extracted blanks [119,162]. Ion suppression by the matrix has also been evaluated by comparing the peak areas of blank matrix (e.g., drug-free plasma) and pure solvent spiked with the substance and the internal standard [162]. Hedeland et al. [194] investigated the possibility of ion suppression originating from the co-elution of three of the analytes of interest [(S)-verapamil, d_3-(S)-verapamil, and (R)-norverapamil]. In this publication, the authors plotted the area quotient between d_3-(S)-verapamil and d_3-(R)-verapamil at

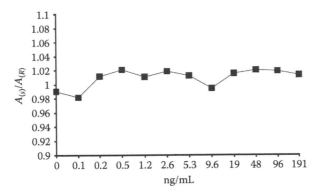

FIGURE 17.8 Peak ratio of *rac-d₃*-verapamil $[A_{(S)}/A_{(R)}]$ at a constant concentration (60 ng/mL) as a function of increasing *(R)*-norverapamil concentration. Column: Chiral AGP, mobile phase: AcN:0.02 M NH₄Ac buffer pH 7.4 (15:85). (Reprinted from Hedeland, M. et al., *J. Chromatogr. B*, 804, 303, 2004. Copyright Elsevier, 2004. With permission.)

increased concentrations of *(R)*-norverapamil in spiked plasma (Figure 17.8). The ratio varied only ±2% from its theoretical value of 1, thus no significant ion suppression from the co-elution was observed. The use of MS detection decreased the analysis time by providing detection selectivity between verapamil and its active metabolite norverapamil, both in reversed- [194] and normal-phase mode [127], without the use of a column switching system [196] or chemical derivatization [197], which it had been necessary to use in earlier investigations with UV and fluorescence detection.

However, the problems with ion suppression from the matrix are not exclusive to plasma and urine samples. Differences in the matrix composition from sample-to-sample might also influence the determination. Nikolai et al. [144] developed a method for enantioselective determination of four different β-blockers in raw and treated wastewater from two different treatment plants in Canada. The internal standard was (+)-levobunolol which the authors claim that is not likely to be present in the wastewater since it is used only in eye drops and not systemically. The enantiomers were separated on a vancomycin-based column in an LC/MS/MS system, with a mobile phase of 90% MeOH, and 10% H₂O containing 0.1% TEA adjusted to pH 4.0 with HAc. Atenolol and metoprolol were found in racemic composition in both the wastewater influent to the treatment plant and the wastewater effluent, whereas propranolol was not found in the influent and nonracemic in the effluent. Thus, the matrix effects were studied by spiking the influent and effluent water with racemic drugs, and the areas obtained for the enantiomeric pairs were compared. However, no difference in signal suppression was found between the enantiomers and the LOD for the analytes ranging from 4.4 to 17 ng/L in the effluent and 17–110 ng/L in the influent water.

Even though indirect separation of enantiomers by derivatization prior to separation often simplifies the chromatographic system and shortens the analysis time, it seems to be only rarely used in chiral HPLC separations today [16–21]. Yang et al. [21] enantioseparated carvedilol within 2.2 min on a C18 column using derivatization with 2,3,4,6-tetra-acetyl-β-glucopyranosyl isothiocyanate (Figure 17.9) (the origin of the figure is after a time of 1.4 min had elapsed). This procedure increased the sample throughput in comparison to the direct separation where a hydroxypropyl-β-cyclodextrin column was used in the normal-phase mode, for which the analysis time was more than 20 min.

The use of polysaccharide-based CSPs instead of protein-based CSPs often increases the peak efficiency and facilitates faster separations. Papini et al. [159] recently developed a method for the enantioseparation of lorazepam and on a Chiralpak OD-R column and an enzymatic hydrolysis was used to determine the amount of the glucoronide metabolite of lorazepam present. The separation was performed in 7 min with an LOQ of 1 and 10 ng/mL for lorazepam in plasma and urine, respectively. Another relatively fast separation for chiral analysis was published by Lausecker and Fischer [188]. They developed a method for determination of the drug candidate R483 within

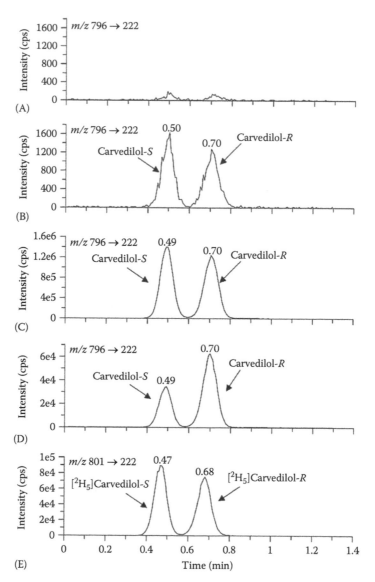

FIGURE 17.9 HPLC–MS/MS chromatograms of carvedilol. (A) A blank sample, (B) LLQ at 0.2 mg/mL, (C) HLQ at 200 mg/mL, (D) a clinical sample containing 4.3 ng/mL of (S)-carvedolol and 9.5 mg/mL of (R)-carvedilol, and (E) internal standard. The MS data acquisition started after 1.4 min has elapsed. Derivatization of the samples with 3,4,6-tetra-O-acetyl-β-glucopyranosyl isothiocyanate. Column Ace3 C18, mobile phase: AcN:2 mM NH₄FA buffer pH 3.0 (50:50 v/v). (Reprinted from Yang, E. et al., *Pharm. Biomed. Anal.*, 36, 609, 2004. Copyright Elsevier, 2004. With permission.)

4 min, and applied it on *in vivo* and *in vitro* assays in plasma. They used a Chiralcel OJ-R column with a mobile phase of AcN:MeOH:10 mM NH₄Ac:HAc (70:20:10:1 v/v/v/v) and an isotope labeled internal standard. The short run time meant that 200 samples could be run overnight.

The use of miniaturized systems might provide a feasible approach for speeding up the separations (given that smaller column dimensions, in terms of both the inner diameter and the length, decrease the dilution). However, miniaturization is not necessarily synonymous with fast separations, since problems often arise with dead volumes, caused by the connections. Nano-LC has been used with UV or MS detection for the analysis of atenolol in urine [143]. A homemade column with an internal diameter of 75 μm containing diol silica modified with teicoplanin was used as the CSP.

The mobile phase consisted of MeOH:AcN:0.5 M NH₄Ac (pH 4.5)(60:39:1 v/v/v). By exchanging the UV z-cell with MS detection, the LOD was decreased from 1.5 μg/mL to 50 ng/mL. Unfortunately, the analysis time increased from 9 to 19 min as a result of the increase in the distance from the end of the column to the detector.

17.5 FAST ANALYSIS, MINIATURIZATION, AND AUTOMATIZATION

The progress toward enantiomerically pure drugs makes the selective and rapid analysis of enantiomers an important issue, both for chiral purity determinations and for enantioselective bioanalysis. Chankvetadze et al. [198] performed enantioseparations within an analysis time of 1 min for each of two chiral compounds (1,2,2,2-tetraphenylethanol and 2,2′-dihydroxy-6,6′-dimethylbiphenyl) by using a homemade capillary column containing monolithic silica modified with amylose tris(3,5-dimethylphenylcarbamate) (Figure 17.10).

Often, the majority of the total analysis time is spent on sample pretreatment, with only a minor part being attributable to the chiral separation of the sample. Thus, automated on-line purification and sample enrichment should increase the throughput. Column switching has sometimes been utilized for on-line purification of biological samples; however, on-line systems are often more complicated when used in combination with CSPs, since the mobile phase composition that can be used with preserved enantioseparation tends to be limited. Katagi et al. [170] used on-line solid phase extraction (SPE) with an cationic exchange (SCX) column for trapping and purification of amphetamine (AP) and methamphetamine (MA). A previously optimized mobile phase composition for enantioseparation by the chiral β-phenylcyclodextrin column was utilized for elution from the SCX column with good recovery (94.1%–102.9%) of the analytes. The total analysis time of the method developed was approximately 30 min. This method was subsequently applied on urine samples to identify drug abuse of (S)-MA, since (R)-MA and (R)-AP could originate from Deprenyl, a drug prescribed for Parkinson's disease. Another interesting method for the determination of (S)-MA was developed by Lua et al. [169]. These researchers used an immunoaffinity column, fabricated in-house, with incorporating mouse monoclonal antibodies against (S)-MA. By the use of a pH gradient of ammonium acetate buffer in three steps (pH 7.0–6.6–3.5), (S)-MA could be enantioseparated from (R)-MA within 40 min, with an LOD of 18 ng/mL.

Motoyama et al. [182] applied on-line SPE for the enantioseparation of pindolol in urine and serum. The samples were injected into the system after filtration through a membrane filter. A cation exchanger with a mobile phase of 100 mM NH₄Ac (pH 5.0):AcN (90:10) was connected to a phenylcarbamate-β-cyclodextrin column. A mobile phase of AcN:H₂O (50:50) with 10 mM NH₄Ac

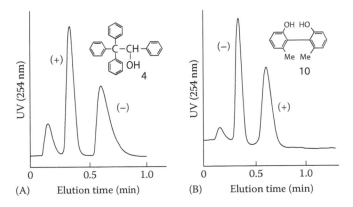

FIGURE 17.10 Fast enantioseparation of (A) 1,2,2,2-tetraphenylethanol and (B) 2,2′-dihydroxy-6,6′-dimethylbiphenyl. Stationary phase: monolithic silica modified with amylose tris(3,5-dimethylphenylcarbamate). Mobile phase: *n*-hexane:2-PrOH (9:1 v/v). The UV detection was conducted at 254 nm. (Reprinted from Chankvetadze, B. et al., *J. Chromatogr. A*, 1110, 46, 2006. Copyright Elsevier, 2006. With permission.)

was used first to elude pindolol from the precolumn and then for the separation on the chiral column. The system had a total cycle time of 16 min, including the preseparation and the postrun equilibration. Xia et al. [186] used a dual column on-line extraction system for the analysis of (S)- and (R)-propranolol in rat plasma, with a teicoplanin CSP and two Oasis HLBs* in parallel as the trapping column (the functional groups on the HLB stationary phase are divinylbenzene and N-vinylpyrrolidone). The samples were trapped by a mobile phase of 0.77% NH$_4$Ac in H$_2$O and eluted from the precolumn to the chiral column by 100% MeOH. The mobile phase was then changed to 0.0375% NH$_4$Ac in MeOH for the chiral separation. The sample throughput was high since the equilibration of the trapping column could be performed on one of the two parallel columns during the elution of the other [186,192] and one analysis was performed within 10 min [186].

Another way of increasing the sample throughput for the chromatographic run in the chiral separations is the use of staggered injection on parallel HPLC systems when "blank windows" are present in the chromatographic runs (i.e., sampling time that only contain registration of the base-line signal) [183].

As discussed above, the use of semiautomatic sample preparation procedures (liquid–liquid extraction or SPE) would speed up the total analysis time [153,157,172], but the choice of instrument also influences the total analysis time. Welch et al. [199] compared two commercially available multiparallel microfluidic HPLC systems, a 24-channel shared flow system and an 8-channel individual flow system (Table 17.6). They applied the two systems in their high-throughput screening in their pharmaceutical process research laboratory. They found that the shortest total analysis time was for a 96-well plate (taking around 35 min with analysis times of 2 min for each sample) and the highest peak symmetry was for the Eksigent system, which was more than 10 times faster than the Nanostream system. The lower throughput in the Nanostream system was primarily a result of the autosampler delay and because the equilibration of the column could not take place at the same time as an analysis. This void time accounted for 24 min of the total analysis time on each

TABLE 17.6
Comparison of between Two Different Microfluidic HPLC Systems

	Eksigent	Nanostream
Number of channels	8	24
Pumps	Individual channel control	Shared flow
	16 Pumps	2 Pumps
Columns	Microbore columns	Brio 24 column disposable cartridge
	Typically 0.3 mm ID	0.5 mm ID
	Many columns available	Few columns available
	Small particle/high pressure	Large particle/low pressure
	Compatible with NPLC	Not compatible with NPLC
Injection	Loop injection, 10–100 nL	Filled "pit" injection, 500 nL
Dwell volume	Small gradient dwell volume	Large gradient dwell volume
Autosampler	Dual LEAP, 18 microplates	8 or 12 head injector, single microplate
Detection	8 Individual detectors	Shared UV detector
	Diode array	Variable wavelength
	No fluorescence	Fluorescence detector available

Source: Reprinted from Welch, C.J. et al., J. *Liquid Chromatogr. Relat. Technol.*, 29, 2185, 2006.
Note: The Eksigent and the Nanostream systems are compared in the text, and as an aid to that discussion, some of the technical specifications are presented in the table.

* Hydrophilic–Lipophilic Balance sorbent (a reversed-phase sorbent).

well plate. The two above-mentioned automatic systems use microbore columns (0.3–0.5 mm ID) and injection volumes ranging from 10 to 500 nL. Even smaller systems using nanotechnology and microdevices (chip technology) can be used for high throughput separations; this issue has recently been reviewed by Eijkel and van den Berg [200].

17.6 CONCLUSIONS AND FUTURE TRENDS

The analysis time for chiral HPLC separations will probably remain relatively long until CSPs with higher efficiency than the present ones become available. But monolithic columns, columns with a smaller particle size (i.e., UPLC*), and miniaturized systems would increase the efficiency and speed up the enantioseparation of existing types of CSPs.

Many optimization strategies are available for the separation of different racemates, both from retailers and from research groups. However, the method development for new drug compounds will remain, at least part, "trial and error," until more general phases have been designed that include a more logical method development.

The trend in pharmaceutical analysis is toward utilizing more selective detectors, for example, MS. The use of MS detection minimizes the need for preseparation of the enantiomers of interest from endogenous substances and metabolites. However, some kind of purification is still necessary to minimize time-consuming cleaning of the MS interface and to increase the lifetime of the (often expensive) chiral columns. Since the purification and preconcentration steps are frequently more time-consuming than the analysis, especially in biological matrixes, automated systems that include these steps need to be developed. Often, the choice of mobile phase for the purification step is more limited than for achiral separations, and it is strongly dependent on which type of column is being used for the subsequent chiral separation.

APPENDIX 17.A

BRAND NAMES USED IN TABLES 17.3 THROUGH 17.5

Brand	Functional Group on Stationary Phase
Ace3 C18	Octadecylsilica
Biooptick AV-1	Avidin
ChiraDex	β-Cyclodextrin
Chiral AGP	α-1-Glycoprotein
Chiral AX QN-1	*tert*-Butyl carbamoylated quinine
Chiral CD-Ph	Phenylcarbamated β-cyclodextrin
Chiral DRUG	Phenylcarbamate β-CD
Chiralcel OB	Cellulose tribenzoate
Chiralcel OD	Cellulose tris(3,5-dimethylphenylcarbamate)
Chiralcel OF	Cellulose tris(4-chloro-phenylcarbamate)
Chiralcel OG	Cellulose tris(4-methylphenylcarbamate)
Chiralcel OJ	Cellulose tris(4-methyl-benzoate)
Chiraldex	β-Cyclodextrin
Chiralpak AD	Amylose tris(3,5-dimethylphenylcarbamate)
Chiralpak AS	Amylose tris((*S*)-α-methylbenzylcarbamate)
Chirex 3005	*R*-Naphtylglycine and 3,5-dinitrobenzoic acid
Chirobiotic R	Ristocetin A

(continued)

* Ultra Performance Liquid Chromotography.

Brand	Functional Group on Stationary Phase
Chirobiotic T	Teicoplanin
Chirobiotic TAG	Teicoplanin aglycone
Chirobiotic V	Vancomycin
Crownpak CR(+)	(S)-18-Crown-6 ether
Cyclobond I 2000 SN	Naphtylethylcarmamoylated-β-cyclodextrin
Cyclobond I 2000	β-Cyclodextrin
Cyclobond I RN	β-Cyclodextrin
Gira-Grom-2	A chiral polymer
HSA	Human serum albumin
Hypersil BDS	C8
IAC	Immunoaffinity column with monoclonal antibody against S-methamphetamine
Inertsil C8	C8
Kromasil CHI-DMB	0,0′-Bis (3,5-dimethylbenzoyl)-N,N′-diallyl-L-tartardiamide
LiChrospher 100	Octadecylsilica
ODS	Octadecylsilica (brand name are missing)
Pirkle DNBL	Dinitrobenzoyl leucine
Shim-pack CN	Cyanopropyl
Sumichiral OA-4900	Hydroxypropyl-β-cyclodextrin
TBuCQN	O-(tert-Butylcarbamoyl)-quinine
ULTRON ES PhCD	Phenylcarbamate β-CD
Xterra MS	Octadecylsilica

REFERENCES

1. AF Casy. *Stereochemistry and Biological Activity*. New York: A. Burger. 81 pp., 1970.
2. E Abderhalden, F Muller. *Z. Physiol. Chem.* 58:185, 1908.
3. J Caldwell. *J. Chromatogr. A* 719:3, 1996.
4. FDA. Federal Register 65:83041, 2000.
5. Council of Europe, European Pharmacopoeia 5th Ed., Edqm, Strasbourg, France, 2004.
6. H Caner, E Groner, L Levy, I Agranat. *Drug Disc. Today* 9:105, 2004.
7. WS Knowles, MJ Sabacky. *Chem. Commun.* 22:1445, 1968.
8. T Katsuki, KB Sharpless. *J. Am. Chem. Soc.* 102:5974, 1980.
9. A Miyashita, A Yasuda, H Takaya, K Toriumi, T Ito et al. *J. Am. Chem. Soc.* 102:7932, 1980.
10. Y Zhang, DR Wu, DB Wang-Iverson, AA Tymiak. *Drug Disc. Today* 10:571, 2005.
11. A Amini. *Electrophoresis* 22:3107, 2001.
12. TK Natishan. *J. Liquid Chromatogr. Relat. Technol.* 28:1115, 2005.
13. U Holzgrabe, D Brinz, S Kopec, C Weber, Y Bitar. *Electrophoresis* 27:2283, 2006.
14. G Gubitz, MG Schmid. *Biopharm. Drug Disp.* 22:291, 2001.
15. NR Srinivas, LN Igwemezie. *Biomed. Chromatogr.* 6:163, 1992.
16. KH Kim, PW Choi, SP Hong, HJ Kim. *Arch. Pharm. Res.* 22:614, 1999.
17. N Cesur, TI Apak, HY Aboul-Enein, S Ozkirimli. *J. Pharm. Biomed. Anal.* 28:487, 2002.
18. S Freimuller, H Altorfer. *J. Pharm. Biomed. Anal.* 30:209, 2002.
19. PH Zoutendam, JF Canty, MJ Martin, MK Dirr. *J. Pharm. Biomed. Anal.* 30:1, 2002.
20. ZQ Chen, Y Yu, LJ Li. *J. Sep. Sci.* 28:193, 2005.
21. E Yang, S Wang, J Kratz, MJ Cyronak. *J. Pharm. Biomed. Anal.* 36:609, 2004.
22. E Yashima. *J. Chromatogr. A* 906:105, 2001.
23. J Haginaka. *J. Chromatogr. A* 906:253, 2001.
24. TJ Ward, AB Farris. *J. Chromatogr. A* 906:73, 2001.
25. P Franco, A Senso, L Oliveros, C Minguillón. *J. Chromatogr. A* 906:155, 2001.

26. CR Mitchell, DW Armstrong. Cyclodextrin-based chiral stationary phases for liquid chromatography: A twenty-year overview. In G Gubitz, MG Schmid (Eds.) *Chiral Separations, Methods and Protocols*, Humana Press Inc., Totowa, NJ: Humana Press Inc. 61 pp., 2003.
27. T Nakano. *J. Chromatogr. A* 906:205, 2001.
28. B Sellergren. *J. Chromatogr. A* 906:227, 2001.
29. F Gasparrini, D Misiti, C Villani. *J. Chromatogr. A* 906:35, 2001.
30. J Kern, K Kirkland. *Chiral Separations*. New York: John Wiley & Sons. 537 pp., 1997.
31. TE Beesley, RPW Scott. *Chiral Chromatography*. New York: John Wiley & Sons, 1998.
32. NM Maier, P Franco, W Lindner. *J. Chromatogr. A* 906:3, 2001.
33. TE Beesley, JT Lee. *LC-GC Eur.* 16:33, 2003.
34. R Thompson. *J. Liquid Chromatogr. Relat. Technol.* 28:1215, 2005.
35. G Gubitz, MG Schmid. *Chiral Separations—Methods and Protocols*. Totowa, NJ: Humana Press Inc. 1 pp., 2003.
36. VA Davankov. *Chirality* 9:99, 1997.
37. C Pettersson, G Schill. *J. Chromatogr.* 204:179, 1981.
38. BJ Clark, JE Mama. *J. Pharm. Biomed. Anal.* 7:1883, 1989.
39. LO Healy, JP Murrihy, A Tan, D Cocker, M McEnery, JD Glennon. *J. Chromatogr. A* 924:459, 2001.
40. C Pettersson, E Heldin. *Kap 9 Ion-Pair Chromatography in Enantiomer Separations*. Weinheim, Germany: VHC Publishers. 279 pp., 1994.
41. C Pettersson, T Arvidsson, A-L Karlsson, I Marle. *J. Pharm. Biomed. Anal.* 4:221, 1986.
42. M Hedeland, R Isaksson, C Pettersson. *J. Chromatogr. A* 807:297, 1998.
43. B Kafkova, Z Bosakova, E Tesarova, P Coufal, A Messina, M Sinibaldi. *Chirality* 18:531, 2006.
44. J Debowski, D Sybilska, J Jurczak. *J. Chromatogr. A* 237:303, 1982.
45. PE Hare, E Gil-Av. *Science* 204:1226, 1979.
46. C Pettersson, K No. *J. Chromatogr. A* 282:1983.
47. E Heldin, NH Huynh, C Pettersson. *J. Chromatogr. A* 585:35, 1991.
48. C Pettersson, M Josefsson. *Chromatographia* 21:321, 1986.
49. C Pettersson. *TrAC* 7:209, 1988.
50. VR Davankov, VR Kurganov. *Chromatographia* 17:686, 1983.
51. C Pettersson. *J. Chromatogr.* 316:553, 1984.
52. C Pettersson, C Gioeli. *Chirality* 5:241, 1993.
53. AM Stalcup, KH Gahm. *J. Microcol. Sep.* 8:145, 1996.
54. I Bjørnsdottir, SH Hansen, S Terabe. *J. Chromatogr. A* 745:37, 1996.
55. Y Carlsson, M Hedeland, U Bondesson, C Pettersson. *J. Chromatogr. A* 922:303, 2001.
56. Y Hedeland, M Hedeland, U Bondesson, C Pettersson. *J. Chromatogr. A* 984:261, 2003.
57. LM Yuan, Y Han, Y Zhou, X Meng, ZY Li et al. *Anal. Lett.* 39:1439, 2006.
58. A El-Gindy, S Emara, MK Mesbah, GM Hadad. *J. AOAC Int.* 89:65, 2006.
59. ZS Guo, H Wang, YS Zhang. *J. Pharm. Biomed. Anal.* 41:310, 2006.
60. RD Yeole, AS Jadhav, KR Patil, VP Rane, ML Kubal et al. *J. Chromatogr. A* 1108:38, 2006.
61. P Piras, C Roussel, *J. Pharm. Biomed. Anal.* 46: 839, 2008.
62. ME Anderson, D Aslan, A Clarke, J Roeraade, G Hagman. *J. Chromatogr. A* 1005:83, 2003.
63. GW Yanik, RJ Bopp, MS Alper. *Chim. Oggi-Chem. Today* 21:10, 2003.
64. P Borman, B Boughtflower, K Cattanach, K Crane, K Freebairn et al. *Chirality* 15:S1, 2003.
65. VP Shah, KK Midha, JWA Findlay, HM Hill, JD Hulse et al. *Pharm. Res.* V17:1551, 2000.
66. Q2A Text on Validation of Analytical Procedures. Presented at *the International Conference on Harmonisation of Technical Requirements for Registration of Pharmaceuticals for Human Use (ICH)*, Geneva, Switzerland, 1995.
67. Q2B Validation of Analytical Procedures: Methodology. Presented at *the International Conference on Harmonisation of Technical Requirements for Registration of Pharmaceuticals for Human Use (ICH)*, Geneva, Switzerland, 1996.
68. GP Carr, JC Wahlich. *J. Pharm. Biomed. Anal.* 8:613, 1990.
69. S Husain, RN Rao. *Process Control Qual.* 10:41, 1997.
70. M Reist, B Testa, P-A Carrupt. In *Handbook of Experimental Pharmacology*, eds. M Eichelbaum, B Testa, A Somogyi. Berlin, Germany: Springer-Verlag. 91 pp., 2003.
71. RD Shah, LA Nafie. *Curr. Opin. Drug Disc.* 4:764, 2001.
72. A Karlsson, A Aspegren. *Chromatographia* 47:189, 1998.
73. K Fulde, AW Fraham. *J. Chromatogr. A* 858:33, 1999.
74. T Wang, YDW Chen. *J. Chromatogr. A* 855:411, 1999.

75. T Wang, YDW Chen, A Vailaya. *J. Chromatogr. A* 902:345, 2000.
76. TL Xiao, B Zhang, JT Lee, F Hui, DW Armstrong. *J. Liquid Chromatogr.* 24:2673 2001.
77. CL Narayana, T Suresh, SM Rao, PK Dubey, JM Babu. *J. Pharm. Biomed. Anal.* 32:21, 2003.
78. M Okamoto. *J. Pharm. Biomed. Anal.* 27:401, 2002.
79. DW Armstrong, S Chen, C Chang, S Chang. *J. Liquid Chromatogr.* 15:545, 1992.
80. PJ Simms, CT Jeffries, XM Zhao, YJ Huang, T Arrhenius. *J. Chromatogr. A* 1052:69, 2004.
81. N Mesplet, Y Saito, P Morin, LA Agrofoglio. *J. Chromatogr. A* 983:115, 2003.
82. HY Aboul-Enein, MM Hefnawy, H Hoenen. *J. Liquid Chromatogr. Relat. Technol.* 27:1681, 2004.
83. D Breton, D Buret, P Clair, A Lafosse. *J. Chromatogr. A* 1088:104, 2005.
84. V de Veredas, MJS Carpes, CRD Correia, CC Santana. *J. Chromatogr. A* 1119:156, 2006.
85. R Ferretti, B Gallinella, F La Torre, A Lusi. *J. Chromatogr. A* 731:340, 1996.
86. MA Radwan, HH Abdine, BT Al-Quadeb, HY Aboul-Enein, K Nakashima. *J. Chromatogr. B* 830:114, 2006.
87. M Kazusaki, H Kawabata, H Matsukura. *J. Liquid Chromatogr. Relat. Technol.* 23:2819, 2000.
88. CA Goss, DG Morgan, KL Harbol, TJ Holmes, J Cook. *J. Chromatogr. A* 878:35, 2000.
89. QB Cass, CSF Watanabe, JA Rabi, PQ Bottari, MR Costa et al. *J. Pharm. Biomed. Anal.* 33:581, 2003.
90. M Dolezalova, M Tkaczykova. *J. Pharm. Biomed. Anal.* 25:407, 2001.
91. BM Rao, R Ravi, BSS Reddy, S Chand, KP Kumar et al. *J. Pharm. Biomed. Anal.* 35:1017, 2004.
92. MA Radwan, HH Abdine, HY Aboul-Enein. *Biomed. Chromatogr.* 20:211, 2006.
93. DA El-Hady, NA El-Maali, R Gotti, C Bertucci, F Mancini, V Andrisano. *J. Pharm. Biomed. Anal.* 37:919, 2005.
94. S Svensson, J Vessman, A Karlsson. *J. Chromatogr. A* 839:23, 1999.
95. HY AboulEnein, LI AbouBasha. *J. Liquid Chromatogr. Relat. Technol.* 19:383, 1996.
96. JQ Huang, GY Cao, X Hu, CH Sun, JR Zhang. *Chirality* 18:587, 2006.
97. DS Richards, SM Davidson, RM Holt. *J. Chromatogr. A* 746:9, 1996.
98. R Ferretti, B Gallinella, F La Torre, L Turchetto. *J. Chromatogr. B* 710:157, 1998.
99. DB Pathare, AS Jadhav, MS Shingare. *J. Pharm. Biomed. Anal.* 41:1152, 2006.
100. I D'Acquarica, F Gasparrini, B Giannoli, E Badaloni, B Galletti et al. *J. Chromatogr. A* 1061:167, 2004.
101. M Santoro, HS Cho, ERM Kedor-Hackmann. *Drug Dev. Ind. Pharm.* 27:693, 2001.
102. KH Kim, HJ Kim, JH Kim, JH Lee, SC Lee. *Chromatographia* 53:334, 2001.
103. B Chankvetadze, N Burjanadze, M Santi, G Massolini, G Blaschke. *J. Sep. Sci.* 25:733, 2002.
104. R Bhushan, D Gupta. *J. Chromatogr. B* 837:133, 2006.
105. H Gika, M Lämmerhofer, I Papadoyannis, W Lindner. *J. Chromatogr. B* 800:193, 2004.
106. R Ferretti, B Gallinella, F La Torre, C Villani. *J. Chromatogr. A* 704:217, 1995.
107. RD Marini, P Chiap, B Boulanger, S Rudaz, E Rozet et al. *Talanta* 68:1166, 2006.
108. B Toussaint, F Streel, A Ceccato, P Hubert, J Crommen. *J. Chromatogr. A* 896:201, 2000.
109. RD Marini, P Chiap, B Boulanger, W Dewe, P Hubert, J Crommen. *J. Sep. Sci.* 26:809, 2003.
110. C Bertucci, V Andrisano, V Cavrini, E Castiglioni. *Chirality* 12:84, 2000.
111. DR Bobbitt, SW Linder. *Trac-Trend. Anal. Chem.* 20:111, 2001.
112. KA Schug, W Lindner. *J. Sep. Sci.* 28:1932, 2005.
113. Chromtech. *Chiral Application Handbook.* http://www.chromtech.se/pdf/handbook.pdf
114. CK Lim, G Lord. *Biol. Pharmacol. Bull.* 25:547, 2002.
115. R Bakhtiar, L Ramos, FLS Tse. *Chirality* 13:63, 2001.
116. M Hamdan. *Process Control Qual.* 10:113, 1997.
117. MJ Desai, DW Armstrong. *J. Chromatogr. A* 1035:203, 2004.
118. J Ding, M Desai, DW Armstrong. *J. Chromatogr. A* 1076:34, 2005.
119. JW Chen, WA Korfmacher, Y Hsieh. *J. Chromatogr. B* 820:1, 2005.
120. GL Erny, A Cifuentes. *J. Pharm. Biomed. Anal.* 40:509, 2006.
121. C Miller-Stein, C Fernandez-Metzler. *J. Chromatogr. A* 964:161, 2002.
122. C Maraschiello, J Vilageliu, I Dorronsoro, A Martinez, P Floriano, E Gomez-Acebo. *Chirality* 18:297, 2006.
123. A Ceccato, F Vanderbist, JY Pabst, B Streel. *J. Chromatogr. B* 748:65, 2000.
124. U Heinemann, G Blaschke, N Knebel. *J. Chromatogr. B* 793:389, 2003.
125. JN Zeng, H Palme, N Srinivas, M Arnold. *J. Chromatogr. B* 811:109, 2004.
126. H Ericsson, J Schwieler, B Lindmark, P Löfdahl, T Thulin, C-G Regårdh. *Chirality* 13:130, 2001.
127. T AlebicKolbah, AP Zavitsanos. *J. Chromatogr. A* 759:65, 1997.
128. B Lindmark, M Ahnoff, B-A Persson. *J. Pharm. Biomed. Anal.* 27:489, 2002.
129. VAP Jabor, EB Coelho, DR Ifa, PS Bonato, NAGd Santos, VL Lanchote. *J. Chromatogr. B* 796:429, 2003.

130. VAP Jabor, EB Coelho, VL Lanchote. *J. Chromatogr. B* 813:343, 2004.
131. H Kanazawa, Y Konishi, Y Matsushima, T Takahashi. *J. Chromatogr. A* 797:227, 1998.
132. AK Kabler, SM Chen. *J. Agric. Food Chem.* 54:6153, 2006.
133. D Zimmer, V Muschalek, C Muller. *Rapid Commun. Mass Spectrom.* 14:1425, 2000.
134. KX Yan, H Song, M-W Lo. *J. Chromatogr. B* 813:95, 2004.
135. H Stenhoff, A Blomqvist, PO Lagerstrom. *J. Chromatogr. B* 734:191, 1999.
136. JE Paanakker, J de Jong, J Thio, HJM van Hal. *J. Pharm. Biomed. Anal.* 16:981, 1998.
137. AP Zavitsanos, T Alebic-Kolbah. *J. Chromatogr. A* 794:45, 1998.
138. MS Rashed, M AlAmoudi, HY Aboul-Enein. *Biomed. Chromatogr.* 14:317, 2000.
139. C Fakt, BM Jacobson, S Leandersson, BM Olsson, BA Persson. *Anal. Chim. Acta* 492:261, 2003.
140. GA Jacobson, FV Chong, NW Davies. *J. Pharm. Biomed. Anal.* 31:1237, 2003.
141. ST Wu, JS Xing, A Apedo, DB Wang-Iverson, TV Olah et al. *Rapid Commun. Mass Spectrom.* 18:2531, 2004.
142. B Streel, C Laine, C Zimmer, R Sibenaler, A Ceccato. *J. Biochem. Biophys. Methods* 54:357, 2002.
143. G D'Orazio, S Fanali. *J. Pharm. Biomed. Anal.* 40:539, 2006.
144. LN Nikolai, EL McClure, SL MacLeod, CS Wong. *J. Chromatogr. A* 1131:103, 2006.
145. R Goda, N Murayama, Y Fujimaki, K Sudo. *J. Chromatogr. B* 801:257, 2004.
146. AP Watt, L Hitzel, D Morrison, KL Locker. *J. Chromatogr. A* 896:217, 2000.
147. V Capka, Y Xu, YH Chen. *J. Pharm. Biomed. Anal.* 21:507, 1999.
148. R Ledger. *J. Biochem. Biophys. Methods* 57:105, 2003.
149. KM Fried, AE Young, SU Yasuda, IW Wainer. *J. Pharm. Biomed. Anal.* 27:479, 2002.
150. TK Majumdar, LL Martin, D Melamed, FLS Tse. *J. Pharm. Biomed. Anal.* 23:745, 2000.
151. K Matsui, Y Oda, H Nakata, T Yoshimura. *J. Chromatogr. B* 729:147, 1999.
152. AP Watt, L Hitzel, D Morrison, KL Locker. *J. Chromatogr. A* 896:229, 2000.
153. ZZ Shen, S Wang, R Bakhtiar. *Rapid Commun. Mass Spectrom.* 16:332, 2002.
154. G Di Pietro, EB Coelho, TM Geleilete, MP Marques, VL Lanchote. *J. Chromatogr. B* 832:256, 2006.
155. PS Bonato, M Del Lama, R de Carvalho. *J. Chromatogr. B* 796:413, 2003.
156. MER Rosas, S Patel, IW Wainer. *J. Chromatogr. B* 794:99, 2003.
157. TH Eichhold, RE Bailey, SL Tanguay, SH Hoke. *J. Mass Spectrom.* 35:504, 2000.
158. SA Wood, AH Parton, RJ Simmonds, D Stevenson. *Chirality* 8:264, 1996.
159. O Papini, C Bertucci, SP da Cunha, NAG dos Santos, VL Lanchote. *J. Pharm. Biomed. Anal.* 40:389, 2006.
160. YL Chen, S Akhtar, H Murai, M Kobayashi. *Rapid Commun. Mass Spectrom.* 19:2681, 2005.
161. D Ortelli, S Rudaz, AF Chevalley, A Mino, JJ Deglon et al. *J. Chromatogr. A* 871:163, 2000.
162. HR Liang, RL Foltz, M Meng, P Bennett. *J. Chromatogr. B* 806:191, 2004.
163. ML Etter, S George, K Graybiel, J Eichhorst, DC Lehotay. *Clin Biochem* 38:1095, 2005.
164. DC Lehotay, S George, ML Etter, K Graybiel, JC Eichhorst et al. *Clin. Biochem.* 38:1088, 2005.
165. P Kintz, HP Eser, A Tracqui, M Moeller, V Cirimele, P Mangin. *J. Forensic Sci.* 42:291, 1997.
166. D Whittington, P Sheffels, ED Kharasch. *J. Chromatogr. B* 809:313, 2004.
167. MER Rosas, KL Preston, DH Epstein, ET Moolchan, IW Wainer. *J. Chromatogr. B* 796:355, 2003.
168. T Kelly, P Doble, M Dawson. *J. Chromatogr. B* 814:315, 2005.
169. AC Lua, Y Sutono, TY Chou. *Anal. Chim. Acta* 576:50, 2006.
170. M Katagi, M Nishikawa, M Tatsuno, T Miyazawa, H Tsuchihashi et al. *Jpn. J. Toxicol. Environ. Health* 44:107, 1998.
171. L Ramos, R Bakhtiar, T Majumdar, M Hayes, F Tse. *Rapid Commun. Mass Spectrom.* 13:2054, 1999.
172. R Bakhtiar, L Ramos, FLS Tse. *Anal. Chim. Acta* 469:261, 2002.
173. R Bakhtiar, L Ramos, FLS Tse. *Biomed. Chromatogr.* 18:45, 2004.
174. R Bakhtiar, FLS Tse. *Biomed. Chromatogr.* 18:275, 2004.
175. R Bakhtiar, L Ramos, FLS Tse. *Rapid Commun. Mass Spectrom.* 16:81, 2002.
176. SK Balani, X Xu, V Pratha, MA Koss, RD Amin et al. *Drug Metab. Dispos.* 25:1282, 1997.
177. J Chawla, M-E Le Guern, C Alquier, TF Kalhorn, RH Levy. *Ther. Drug Monit.* 25:203, 2003.
178. H Kanazawa, A Okada, M Higaki, H Yokota, F Mashige, K Nakahara. *J. Pharm. Biomed. Anal.* 30:1817, 2003.
179. RH Zobrist, B Schmid, A Feick, D Quan, SW Sanders. *Pharm. Res.* V18:1029, 2001.
180. B Kammerer, R Kahlich, M Ufer, S Laufer, CH Gleiter. *Rapid Commun. Mass Spectrom.* 18:458, 2004.
181. B Kammerer, R Kahlich, M Ufer, S Laufer, CH Gleiter. *Anal. Biochem.* 339:297, 2005.
182. A Motoyama, A Suzuki, O Shirota, R Namba. *J. Pharm. Biomed. Anal.* 28:97, 2002.
183. HP Wang, ZZ Shen. *Rapid Commun. Mass Spectrom.* 20:291, 2006.
184. V Capka, Y Xu. *J. Chromatogr. B* 762:181, 2001.

185. DF Zhong, XY Chen. *J. Chromatogr. B* 721:67, 1999.
186. YQ Xia, R Bakhtiar, RB Franklin. *J. Chromatogr. B* 788:317, 2003.
187. M Miura, H Tada, S Satoh, T Habuchi, T Suzuki. *J. Pharm. Biomed. Anal.* 41:565, 2006.
188. B Lausecker, G Fischer. *J. Chromatogr. B* 835:40, 2006.
189. KB Joyce, AE Jones, RJ Scott, RA Biddlecombe, S Pleasance. *Rapid Commun. Mass Spectrom.* 12:1899, 1998.
190. E Badaloni, I D'Acquarica, F Gasparrini, S Lalli, D Misiti et al. *J. Chromatogr. B* 796:45, 2003.
191. SF Murphy-Poulton, F Boyle, XQ Gu, LE Mather. *J. Chromatogr. B* 831:48, 2006.
192. YQ Xia, DQ Liu, R Bakhtiar. *Chirality* 14:742, 2002.
193. G Bringmann, M Munchbach, D Feineis, K Messer, S Diem et al. *J. Chromatogr. B* 767:321, 2002.
194. M Hedeland, E Fredriksson, H Lennernas, U Bondesson. *J. Chromatogr. B* 804:303, 2004.
195. W Naidong, PR Ring, C Midtlien, X Jiang. *J. Pharm. Biomed. Anal.* 25:219, 2001.
196. E Brandsteterova, IW Wainer. *J Chromatogr B* 732:395, 1999.
197. G Stagni, WR Gillespie. *J. Chromatogr. B* 667:349, 1995.
198. B Chankvetadze, C Yamamoto, M Kamigaito, N Tanaka, K Nakanishi, Y Okamoto. *J. Chromatogr. A* 1110:46, 2006.
199. CJ Welch, P Sajonz, M Biba, J Gouker, J Fairchild. *J. Liquid Chromatogr. Relat. Technol.* 29:2185, 2006.
200. JCT Eiijkel, A van den Berg. *Electrophoresis* 27:677, 2006.

18 HPLC in Environmental Analysis

Valentina Gianotti, Stefano Polati, Fabio Gosetti,
and Maria Carla Gennaro

CONTENTS

18.1 INTRODUCTION: GENERAL AND RECENT ASPECTS OF ENVIRONMENTAL ANALYSIS

Analytical methods based on high-performance liquid chromatography, hyphenated with different detection systems are the methods of election to study the multiform aspects of environmental analysis.

Due to the continuous progress of studies concerning the effects brought by the different anthropological activities to humans and environment, environmental analysis is undergoing continuous evolution. Always increasing is the number of the species to be controlled and of the aspects to be considered to safeguard health. Environmental chemical analysis is concerned with the identification and the determination of different micropollutants. To reach this aim, all the processes and the effects that favor diffusion in the environment and the bioavailability of the pollutants and of their potential metabolites must also be considered.

The natural biogeochemical cycles of the elements are heavily affected by the presence of anthropogenic species, which can undergo with the native ones different chemical and physical reactions as a function of the experimental conditions of pH, pE, temperature, sunlight, etc. The possible interactions are very complex, and mathematical models have been developed to help the study [1,2].

Relevant importance must be given to metal speciation in particular, as it concerns toxic metals. The different chemical species, under which a metal can be present and that are present characterized by different toxicity, affect the geoavailability, the bioavailability, and the diffusion of metals in the environment. So, for instance, mercury, that can be present under different inorganic oxidation forms and aggregation status as well as under the organometal species of different polarities and hydrophilicities, undergoes continuous reactions and interchanges in the hydrosphere, the atmosphere, the geosphere, and the biosphere.

Metal speciation in surface waters and in soil is advantageously studied by HPLC/mass spectrometry (MS) methods, interfaced with inductively coupled plasma (ICP)-MS technique [3], as described in detail in Section 18.2. The level of diffusion and the mobility of inorganic and organic pollutants in soil and surface waters are of paramount importance in determining their effective toxicity. If, for instance, a pollutant is so strictly bound to soil moieties to be considered "recalcitrant," its potential toxicity is lower than that of a pollutant, maybe less toxic, that is easily leached by meteoric waters and made so able to reach surface and also ground waters. The strength through which a pollutant is bound to the soil, together with the properties of the pollutant itself and of the matrix in which it is contained, addresses the choice of the extraction process to be used before analysis. It is worth underlining that the choice of the pre-treatment process of the sample, as well as the sampling technique that must guarantee the representativity of the sample collected, play determinant roles on the results. The method of extraction is usually chosen on the basis of the recovery yield of the analyte, that in turn is generally evaluated through suitable HPLC methods. Information and references concerning the extraction and the sample pre-treatment are given in detail in the following text for each class of pollutants considered.

Studies devoted to evaluate sorption properties of soil components (generally clays) towards chemicals characterized by different properties are performed for model systems. HPLC/UV techniques are employed to measure the amount of the chemicals still present after a suitable contact time in solutions containing a prefixed amount of clay and known concentrations of the chemical [4–7]. The selective and the irreversible ability of clays to retain pollutants can be advantageously used in remediation strategies, for example, as cheap surrogates to activated carbon. In addition, an evaluation of the sorption strength could suggest the effective pollution of a soil, not as a function of the total amount of the pollutants present, but only of that fraction that can be naturally leached by meteoric waters.

A relatively recent topic studies the mechanism of natural degradation undergone in the environment by different chemicals, as for example, by the so-called persistent organic pollutants (POPs), largely used in industry as intermediates or in agriculture as pesticides, that can reach very high concentrations in the environment. If water soluble, they can diffuse and reach large distances from the immission source. The presence and the amount of these chemicals in the environment have not been sufficiently considered in the past due to their relatively low toxicity. But, due to their high persistence in the environment, these species can often undergo degradation processes, induced by sunlight irradiation, hydrolysis reactions, or the action of microorganisms. Often, the degradation reactions do not lead to complete mineralization but to the formation of new organic species potentially more toxic than their precursors [8–10]. So, for instance, phenolic derivatives, more toxic than the precursor compounds, have been found in sewage treatments of non-ionic polyethoxylate surfactants [11]. In industrial wastes, aromatic amines (in particular chloroanilines) and chlorophenols are also widely present; methods for their identification and determination are given in Sections 18.3, 18.4, and 18.8.

Also, many pesticides can undergo degradation processes with the formation of species that often exhibit different chemical, physical, and toxicological properties [12]. It must be remembered that the general term of pesticide includes chemicals containing very different functional groups, such as charged ions (as quaternary ammonium cations), weak bases (triazines, anilines), polyethoxylated polymers, hydrophilic macrocycles, etc. The identification of the degradation products can be usefully studied by HPLC/MS/MS and in particular by ion trap analyzers in MS^n mode. In the identification process, the tandem MS characterization of the precursor species can be very useful, since in many cases the MS/MS fragmentation mimics the degradation pathway well.

HPLC–MS/MS methods are also essential for the identification of potentially toxic substances that form in the decolorization processes of dyes. The first interest for the presence of dyes in the environment and the first HPLC methods developed for their control was addressed to dyes characterized by potential carcinogenic activity such as those derived from benzidine, o-toluidine, and o-dianisidine [13,14]. The use of these dyes is on the decrease; aromatic sulfonate dyes are mainly used in textile and food industries. The dyes are generally not toxic but they could form toxic species in the decolorization processes that they undergo before the final disposal. To decolor industrial wastes, several advanced oxidative procedures (AOP) are applied, which consist of the addition of highly reactive chemical species, as for instance hydroxyl radicals, able to induce the degradation of stable molecules. But the loss of color does not necessarily correspond to the complete mineralization, because the treatments performed before disposal can lead to the formation of uncolored aromatic organic species, potentially toxic. A very hard goal is the identification of these species. To this end, HPLC/MS studies are sometimes supported by NMR and ion-chromatography measurements. Correlations between the structure of the dye and the rate of the degradation reaction have been proposed: so, for example, the presence of a conjugated system or of a higher number of aromatic rings in the molecular structure of the dye has been observed to increase the stability toward photo-catalytic treatments [15]. A detailed review of the papers present in literature dealing with dye decolorization and with degradation pathways proposed for benzene- and naphthalene-sulfonates is given in Section 18.5.

More recent interest in environmental analysis is represented by the so-called emerging contaminants, a general denomination that includes a great number of chemical compounds, as pharmaceuticals, drugs, steroid, sex hormones, and endocrine disrupting compounds (EDCs) that with ever increasing frequency are found in surface waters. In particular, carbamazepine, pirimidone, non-steroidal anti-inflammatory drugs (as naproxen and diclofenac), antibiotics (from human/veterinary medicine and from aquaculture), and growth promoters (as spiramycin and carbadox) are often present [12,16–18]. The diffusion of antibiotic residues in the environment represents a relevant threat to public health, since they induce resistance in bacterial strains and cause cross-resistances to the more frequently used antibiotics [19]. In surface waters are now also found pharmaceutical products such as tetracyclines, antipyretics, chloramphenicols, and sulfonamides, as well as antipsychotic and anxiolitics drugs (haloperidol, piperidine-derivatives), and cocaine [20]. The last is found worldwide in ever-increasing amounts in surface waters, in particular in urban areas. Its determination not only provides estimates of the consumption level of the drug, but also permits a detection of local variations in its diffusion. Some drugs present in wastes from intensive farming and fertilization agents can end up in sediments and soil. Methods for the determination of drugs in waters, soils, and sediments are described in Section 18.6.

The "emerging pollutants" also include perfluorooctanoic acid (PFOA) and perfluorooctanesulfonate (PFOS), used in the production of water-resistant coatings and polybrominated diphenylether (PBVDE) that find application in furniture, textiles, plastics, and paint industries. Due to their large industrial use, these chemicals are present profusely in the environment and also in the blood of people working in the fields [21].

Section 18.7 reports the methods for the determination of other emerging pollutants, whose presence in waters is due to the discharge of hygiene and cosmesis products, such as sunscreen, UV filters, alkylphenol, and ethoxylate surfactants.

Section 18.8 deals with the identification and the determination of disinfection products in waters and of the by-products formed during chlorination processes.

Among the toxic species present in waters, algal toxins must also be considered: HPLC methods for their determination are discussed in Section 18.9.

HPLC methods are also usefully employed in the analysis of the atmosphere; recent methods and applications are summarized in Section 18.10.

18.2 HPLC METHODS FOR METAL SPECIATION

More than 60 different organo-metal species have been found in the ecosystems of marine and terrestrial origins. The major elements are As, Bi, Hg, Pb, Se, and Sn, and the organic species are alkylated forms, fatty acids, acids, and sugar derivates, likely generated in the environment by both microbial and human metabolisms.

In the analysis, the most critical point necessary to obtain correct information on metal speciation consists in maintaining the native speciation and concentration along all the processes of extraction and the pre-treatment of the sample. This requirement is often seen as a challenge.

As suggested in a review concerning arsenic speciation, mild extraction methods are suggested to avoid possible chemical alterations [22] but, in any case, the validation of the method, performed on spiked samples, is always necessary. The pre-treatment procedures are generally based on traditional approaches such as solid–liquid extraction (SLE), liquid–liquid extraction (LLE), or solid-phase extraction (SPE). Sequential extractions are often used, methanol-water and acetonitrile-water mixtures being the most used solvents. Solid-phase micro extraction (SPME) methods have been recently employed in the determination of organometal species of lead, arsenic, mercury, tin, and selenium [23,24]. Ionic organometal compounds are derivatized to increase their stability and to improve chromatographic sensitivity: the crucial aspects are the choice of the derivatizing agent, the mode of SPME extraction (as for example, headspace or direct-immersion), and the desorption step before chromatographic analysis. For example, the use of alkyl- and arylborates [23], usefully employed in organometal determination, is not suitable for the simultaneous determination of organotin species. Tetrahydroborate and sodium tetraethylborate can be employed for lead speciation, but tetraethylborate is not suitable, because in its presence both ethyl-lead and lead inorganic species are transformed into tetraethyl lead. To derivatize organo-metals in aqueous samples tetra-n-propyl-, tetraammonium-, and tetrabutyl-borate are also used [23]. For most of the lead-containing species, after the derivatization reaction, a step of extraction and separation of the inorganic species is required, based on complexation reactions. Other derivatization methods involve hydride generation and the use of the Grignard reagent [25].

In a study devoted to arsenic speciation in soil and sediments, the use of different extraction agents has been compared [26]. The major problems encountered in arsenic speciation are the low recovery (20%–70%) of some organoarsenic species and the possible oxidation of As^{III} species to As^V ones: the reaction is prevented by the addition of a mixture of ascorbic and phosphoric acids. The extraction efficiency of inorganic forms of arsenic from drinking water is improved by the addition of ethylenediaminetetraacetate that, through the formation of complexes with Fe^{III} ions, prevents the precipitation of arsenic insoluble species [27]. Studies of speciation of arsenic in waters require particular attention due to the different toxicity of its species [3]. The liquid chromatographic techniques most commonly used are ion-pair chromatography (IP HPLC), reversed-phase chromatography (RP HPLC), ion-exchange (IE), or size-exclusion chromatography (SEC). In RP HPLC, C-18 stationary phases are generally used but a porous graphite carbon stationary phase has also been employed [28]. Two reviews report the most effective methods for arsenic speciation in ground waters, wastewaters, and sediments [29,30]. An anion exchange LC/ICP-MS method for the determination of arsenite, arsenate, dimethylarsinic acid, and monomethylarsonic acid in drinking waters has been developed and validated [31], and an LC/ICP-MS technique for measuring diphenylarsinic acid in ground waters has been developed [32].

In mercury speciation studies, pressurized liquid extraction (PLE), microwave-assisted extraction (MAE), and supercritical fluid extraction (SFE) are employed [33]. In particular, methyl-mercury is extracted by the "Westoo" method [33,34], which consists in a leaching process with hydrochloric acid, the extraction of the metal chloride into benzene or toluene, the addition of ammonium hydroxide that converts the metal species to hydroxide and the saturation with sodium sulfate. Most of the HPLC methods reported in literature for the determination of organomercury compounds (mainly monomethylmercury, monoethylmercury, and monophenylmercury) are based on reversed phase separation (RP), sometimes with the addition of chelating agents, ion pair (IP) reagents, or pH buffers to the mobile phase [33]. Detection is based on photometry, atomic absorption spectroscopy, MS, and the plasma technique.

For the determination of organotin compounds (tributyltin, triphenyltin, triethyltin, and tetra-ethyltin) a MAE is proposed before the normal phase (NP) HPLC/UV analysis [35]. In organotin and arsenic speciation studies, hydride generation is the most popular derivatization method, combined with atomic absorption and fluorescence spectroscopy or ICP techniques [25,36]. Both atmospheric pressure chemical ionization (APCI)-MS and electrospray ionization ESI-MS are employed in the determination of butyltin, phenyltin, triphenyltin, and tributyltin in waters and sediments [37]. A micro LC/ESI-ion trap MS method has been recently chosen as the official EPA (Environmental Protection Agency) method (8323) [38]: it permits the determination of mono-, di-, and tri- butyltin, and mono-, di-, and tri-phenyltin at concentration levels of a sub-nanogram per liter and has been successfully applied in the analysis of freshwaters and fish [39]. Tributyltin in waters has been also quantified through an automated sensitive SPME LC/ESI-MS method [40].

In the determination of organometal compounds, SPME HPLC/ICP-MS and SPME HPLC/ESI-MS methods are the most effective alternatives to the more diffused SPME GC/MS (or ICP-MS) method. The direct connection of the HPLC/MS and ICP-MS systems through a nebulizer spray chamber arrangement has recently been preferred. The HPLC/ICP-MS method is the best alternative for the determination of metalloids at trace levels. HPLC ion chromatography (IC) is also used: the buffer systems are phosphate, carbonate, phthalate, formate salts, and tetramethylammonium hydroxide solutions [22,29]. For the resolution of the metal species in their mixtures, the ESI-MS, as well as the MS/MS technique, offers useful structural information [29]. In the speciation of tetram-ethyl- and tetraethyl- lead, SPME HPLC/ESI-MS methods have been successfully employed [23]. Metal-inorganic anion complexes (i.e., aluminum fluoride complexes) and metals (Se, As, Hg, and Pb) covalently bound to organic molecules have been at trace levels in soils, through SEC coupled to ICP-MS and ESI-MS detection [41]. A crucial aspect related to the development of analytical methods based on MS is represented by the availability of organometal standards. In order to obtain accurate and matrix-independent results, isotopic dilution mass spectrometry (IDMS) is the optimal calibration technique [42].

On-line anion exchange LC/ICP-MS methods for selenium and chromium speciation have been published [30]. In studies of selenium speciation in environmental samples LC/ICP-MS and LC/ESI-MS methods have been used for the determination of both inorganic and organic selenium species [43]. For the determination of methyl selenide, strong anion exchange and RP chromatography have been employed [43], while for the determination of seleno-aminoacids, IP RP chromatography with on-line detection based on ICP-MS has been successfully employed [44].

18.3   HPLC METHODS FOR THE DETERMINATION OF AROMATIC AMINES

Aromatic amines can be introduced into the environment from different anthropogenic activities. They are often present in untreated wastewaters from dye industries [45–48]; in particular, sul-fonated amines have been found in municipal biological treatment plants and sediments of rivers, formed likely through the cleavage of the azo bonds, induced by anaerobic reduction [49].

Aromatic amines are generally extracted from the environmental samples by LLE [50] or SPE [51]. While the former method is time-consuming and requires large quantities of environmen-tally unfriendly solvents, in the latter, lower volumes of solvent are required, but larger volumes of sample. The SPME technique integrates into a single step the processes of sampling, extraction, and concentration and is often coupled with HPLC [52–54]: the SPME HPLC interface enables the mobile phase to contact the SPME fiber, to remove the adsorbed analytes, and to deliver them to the chromatographic column. Amines can also be removed from the fiber in a moving stream of the mobile phase (dynamic desorption) or, when the amines are more strongly adsorbed, the fiber can be soaked in the mobile phase or in a suitable solvent before injection into the column (static desorption).

HPLC methods can be utilized for the pre-concentration of aromatic amines from polluted waters on silica gel or octadecyl silica (ODS) columns [55]. The determination is then performed by RP HPLC using ODS packings as the stationary phases and a mixture of methanol, isopropanol, and water as the mobile phase [55]. RP HPLC with diode array detector (DAD) methods coupled on-line with a continuous sequential anaerobic/aerobic reactor system have been employed in wastewaters treatments [56]. A continuous monitoring of the possible presence of aromatic amines in azo dyes wastes is based on inducing in the waste, the reaction of a reduction of the dye, followed by HPLC/UV or HPLC/MS analysis [57–59]. The reducing agent solutions are sodium dithionite or tin(II) chloride in an aqueous acidic medium at 70°C, followed by SPE [58,59], LLE [60,61], or SFE [60–62].

Particular attention must be devoted to the presence in the environment of chloroanilines. They are widely diffused in the environment because they are released in soils and waters through degradation reactions of phenylureic herbicides and antibacterial agents (as chlorhexidine and triclocarban) present in hygiene products and because they are largely employed as intermediate agents in the synthesis of pigments. Their extraction from soils and waters is carried out through LLE, Soxhlet extraction, PLE [63], and liquid phase microextraction (LPME) [64]. An HPLC/UV method with a previous LPME allows LOD values always lower than $1.0\,\mu g\,L^{-1}$ [64]. 4-chloroaniline and 3,4-dichloroaniline have been determined, together with their precursor phenylurea pesticides, in drinking, ground, and surface waters through SPE preconcentration and HPLC analysis with ESI detection [65]. IE chromatography and HPLC with UV or fluorescence detection are widely used. Due to different toxicities of chloroaniline congeners, multiresidue separation methods are highly required. The simultaneous separation of mono- and di-chloroanilines was achieved through an ion-interaction reagent (IIR) RP HPLC/UV method [66] with detection limits always lower than $21\,\mu g\,L^{-1}$, while the simultaneous separation of 13 congeners (mono-, di-, and tri-chloroanilines) with LOD values lower than $60\,\mu g\,L^{-1}$ has been obtained through the development of an HPLC method, whose conditions have been optimized through an experimental factorial design [67].

18.4   HPLC METHODS FOR THE DETERMINATION OF PESTICIDES, THEIR DEGRADATION PRODUCTS, AND PHENOLS

LC techniques are widely diffused for the determination of hydrophilic but not volatile and thermally unstable pesticides. Since the European Community Directive [68] indicates $0.1\,\mu g\,L^{-1}$ as the concentration threshold level for a single pesticide in waters destined for human consumption, to quantify these concentration levels, suitable pre-concentration and extraction procedures must be generally performed prior to the HPLC determination. The extraction methods are based on LLE, MAE, on-line continuous flow liquid membrane extraction (CFLME), and mainly on SPE and SPME. Many SPE procedures are used: the packing materials are graphitized carbon, ODS, styrene-divinylbenzene co-polymers, or selective phases based on immunoaffinity. The extraction can be performed on- and off-line, manually, or in a semi-automated way.

Acidic herbicides, such as benzoate-, acetate- (2,4-D, 2,4,5-T, etc.), propanoate- (dichlorprop), butanoate- (2,4-DB, MCPB) derivatives, bentazone, dinoseb, dinoterb, and pentachlorophenols, are extracted by SPE and determined through gas or liquid chromatography. Silica bonded or graphitized carbon sorbents are generally employed, but also IE and IP stationary phases have been proposed to improve the cleanness of the extracts destined for HPLC analysis [69]. The SPME technique is receiving ever-increasing attention due to the advantages of low time consumption and low amounts of both the sample and the solvent required [70]. Polydimethylsiloxane, divinylbenzene, polyacrilate, carboxen, carbowax, and their mixtures are the principal materials employed in the fibers used in the extraction process of triazines (atrazine, prometryn, propazine, propham, simazine), carbamates (barbam, propham, chloropropham), organophosphorates, chlorinated and phenoxyacidic herbicides (dichlorprop, MCPA, 2,4-D, 2,4,5-T, 2,4-DB), phenylureas (linuron, isoproturon, monuron, neburon), and carbofurans from ground waters, river and sea waters, and soil leachates.

SPME methods are commonly coupled to LC methods, in particular HPLC/UV, HPLC/DAD, and HPLC/ESI-MS, which allow detection limits ranging from 0.5 to 50 µg L^{-1} for triazines and carbamates and lower than 1.0 µg L^{-1} for phenylureas and triazines [71] to be reached. MAE with a methanol/dichloromethane mixture has been successfully employed to extract sulfonylurea herbicides from soils [72].

A recent review reports the methods more frequently used for the extraction and for the LC tandem MS determination of many classes of pesticides as halogenated, phosphorated, carbamates, quaternary salts, N-heterocycles, phenyl- and sulfonylureas, amides, organometal species, surfactants, and phenols [12]. RP HPLC and IP RP HPLC are the most employed techniques. ODS, cyano, aminopropyl, phenyl, and styrene/divinyl benzene are the packings most often used of the stationary phases; the mobile phases are water/methanol (or acetonitrile) mixtures often added of a volatile buffer (ammonium acetate, ammonium formate, etc.) or of a volatile IP agent (i.e., tetrabutylammonium acetate (TBA). In the determination of phenols, stationary phases with porous graphitic carbon (PGC) packing are used, while in the determination of quaternary ammonium herbicides packings containing strong cation-exchange moiety are preferred [12]. As regards the detection technique, MS is largely employed, both in the APCI source, preferentially used in the determination of neutral and basic analytes, and in the ESI source, mainly used for cationic or anionic species. Triple quadrupole (TQ), quadrupole time of flight (Q-TOF), and ion trap (IT) are the mass analyzers frequently employed in LC hyphenated with tandem MS. This technique leads to a useful contribution to the identification process of the degradation products that form from pesticides due to the natural degradation processes and that often present more polar characteristics than the precursor molecule. The identification of the degradation products is often a challenge, and the possibility of successive fragmentation lends precious support.

SPE LC/MS/MS methods have been developed for the determination of polar pesticides (as propanyl, thiobencarb, diuron, simazine, bentazone etc.) and of their degradation products in waters with LOQ around 25 ng L^{-1} (Figure 18.1) [73], for the determination of chloroacetamide herbicides (LOD = 25 ng L^{-1}) [74], of sulfonylureas, bentazone, and phenoxyacids in river and drainage waters (LOD of the order of few ng L^{-1}) [9].

Sulfonylurea pesticides have been determined in soil and surface waters by HPLC/UV and HPLC/ESI-MS after LLE and SPE [75,76]. An HPLC/MS methodology with APCI and ESI sources has been developed for the determination of more than 75 pesticides belonging to different chemical classes and of their degradation products: phenylurea-, triazine-, bipyridylium herbicides, acetanilide metabolites, chlorophenoxyacidic herbicides, chlorophenols, and alkylsulfate surfactants [77]. The evaluation of the proton affinity of the species (protonation in the gas phase and polarity in solution) is very useful for the selection of the ion source more suitable in MS analysis, since pesticides may ionize in the positive ion (PI) or the negative ion mode (NI) or often in both. In general, the use of the APCI source in (PI) is more sensitive for neutral and basic pesticides, while ESI in the PI mode gives better sensitivities (LOD lower than 10 ng L^{-1}) for positively charged species. On the other hand, for acidic pesticides (as, for example, chloroacetanilide metabolites), the ESI interface gives greater (10–100 times) sensitivity than the APCI source [69]. In general, the use of the two stages in tandem MS analysis is a very helpful tool: in the product ion mode, the first stage pre-selects the ion while the second analyzes the fragments induced by the collision with inert gas. As a function of the problem to be solved, this dual analysis can be performed either in space or time. In the first case, where the two mass analyzers are in series, the two mass steps proceed simultaneously (TQ and Q-TOF). In the second case, the sequence of events takes place in the same space of the quadrupole IT but at different times [78]. Both the techniques are very helpful in the identification of the degradation products that can naturally form in the environment from pesticides. Among these, phenol derivatives have been identified as degradation products that form from phenoxyacidic pesticides [79]. Many congeners of phenols are also released into the environment from many different

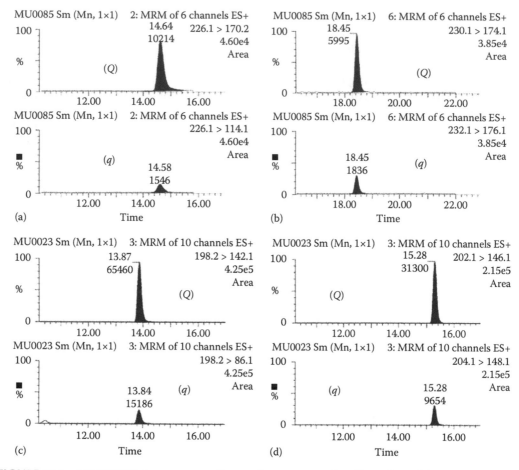

FIGURE 18.1 LC–MS/MS chromatograms for a surface water sample collected from Riu Verd (Massalaves, Valencia, Spain): (a) terbumeton 9 ng L^{-1}, (b) terbuthylazine 8 ng L^{-1}, (c) desethylterbumeton 86 ng L^{-1}, (d) desethylterbuthylazine 42 ng L^{-1}. Q, quantification transition; q, confirmation transition. (Reprinted from Marìn, J.M. et al., *J. Chromatogr. A*, 1133, 204, 2006. Copyright 2006. With permission from Elsevier.)

industrial wastes, for instance, from processes of synthesis of pharmaceuticals, dyes, pigments, and wood preservers. In the determination of phenols in waters, LLE is commonly employed, as well as SPE or sorption pre-concentration [76], that is generally more suitable for the extraction of polar species from aqueous matrices; PGC, ODS, and polymeric sorbent phases are generally used. Phenols are also extracted from water samples by LPME with back extraction (LPME/BE): HPLC/ UV analysis for chloro- and methyl-phenols allow detection limits ranging from 0.5 to 2.5 μg L^{-1} [80]. Liquid/liquid/liquid phase microextraction (LLLE) allow limits of quantification for mono-substituted phenols lower than 1 μg L^{-1} [81].

In particular, the priority pollutant phenols (PPP), identified by EPA since the 1970s are widespread water pollutants that must receive the greatest attention due to their recognized toxicity. For the separation of eleven PPP, an ion-interaction reagent (IIR) RP HPLC/UV method has been developed that allows limits of detection lower than 30 μg L^{-1} in river waters, after LLE in dichloromethane and a 500-fold pre-concentration [82]. Through on-line SPE followed by both UV and electrochemical detection [83], 16 priority phenols have been determined in water samples with the LOD value for chlorophenols lower than 1 ng L^{-1} [84]. LODs at ng L^{-1} levels were obtained for all the PPPs in samples of river water, employing a relatively small volume of sample through an on-line SPE HPLC/MS method with an APCI source.

18.5 HPLC METHODS FOR THE DETERMINATION OF ALIPHATIC AND AROMATIC SULFONATES, DYES, AND DEGRADATION PRODUCTS

Amino- and hydroxy-naphthalene sulfonates (ANS and HNS) are employed as intermediates in dye production while alkylated naphthalene sulfonates are more generally used as suspending and wetting agents, dispersants, and stabilizers. Condensation products, also referred to as sulfonated polyphenols, are used as synthetic tanning agents [85] and sulfonated naphthalene formaldehyde condensates as superplasticizers for cement [85]. Lignosulfonates find applications as dispersing and flotation agents and are employed to enhance oil recovery. The relatively less attention previously paid to the presence of aromatic sulfonates in aquatic environments with respect to linear alkyl benzene sulfonates (LAS) is due likely to their hydrophilicity. The aquatic toxicity of aromatic sulfonates appears to be relatively small, and the risk of bioaccumulation is low due to the pK_{ow} values that typically range below 2 and the consequent high solubility in waters [85]. LASs, in turn, are generally less soluble in waters, but their immission into the aquatic systems likely enhance, as a side effect, the solubility of hydrophobic xenobiotics already deposited within soils and sediments [85]. Contrary to LAS [85], aromatic sulfonates exhibit limited biodegradability, are potentially hazardous for surface and ground waters, and can heavily affect the quality of drinking water [85]. To investigate the fate of sulfonates in aquatic environments, sensitive analytical methods are required for the determination of alkyl and aromatic sulfonates and the identification of their degradation products [85].

IP LLE methods are generally employed for the extraction of LASs from river waters and the solvent sublation method of Wickbold is still used for their extraction from seawater [85]. SPE methods making use of C18 and C8 phases are largely employed [85]. The amount of sorbent is optimized as a function of the degree of pollution and the average composition of river waters [85]. The performances of C18 disks and C18 cartridges are compared [85].

The clean up of the extract containing LAS is performed by anion-exchange chromatography on C18 silica-based cartridges [85]. The use of deactivated charcoal is very efficient and it does not require a further clean-up step, but presents the risk of cartridge overloading [85]. LAS are preconcentrated on XAD-resins or on weak anion-exchangers [85]. Freeze-drying is also sporadically employed [85]. Several studies devoted to the identification of LASs in raw and treated sewage [86] and their biodegradation intermediates (as sulfophenylcarboxylates (SPC)) or by-products (as dialkyltetralin sulfonates (DATS)) have been carried out [87,88], based on the HPLC analysis (ESI source in NI mode) of sewage treatments [89], surface [90] and coastal waters [91]. The RP HPLC separation of LAS according to the alkyl chain length is straightforward [87,90], while the separation of SPC requires the presence of additives such as triethylamine (TEA) [90] or tetraethylammonium acetate (TEAA) [88]. When using TEAA, a suppressor agent must be introduced between the LC system and the mass spectrometer to remove the cationic additive that is not volatile [88]. Alternatively, SPC can be methylated prior to the LC/MS analysis [87]. An HPLC methodology with fluorescent detection has been developed that allows the determination of LAS homologues and isomers in urban wastewater and groundwater [92]; the stationary phase is a C8 column and the mobile phase is a mixture of methanol and 30.0 mM sodium dodecyl sulfate. The quantification of the total LAS is also achieved in flow injection analysis (FIA), through the direct infusion into the chromatographic system [92].

For the SPE of the more polar aromatic sulfonates, IPs with tetralkylammonium are previously formed [85,93,94]. The combination of the formation of IP with cetyltrimethylammonium and SPE is efficacious in the extraction of benzene and naphthalene sulfonates from an aqueous environment. Good recoveries are also obtained with octylammonium acetate as the IP agent, both when it is added to the aqueous sample prior to the extraction and when it saturates the solid phase to produce an anion-exchange column [95,96]. An alternative approach is based on the use of deactivated charcoal (Carbonpack B) [96] or of chemically modified polystyrene-divinylbenzene resins [85]. Often graphitized carbon black (GCB) is used for the clean-up process of the sample [96].

For the identification and the quantitation of aromatic sulfonates, LC is the technique most often used. The retention of isomeric non-substituted naphthalene sulfonic acids increases proportionally to the dipole moment, which can be explained by its effect on increasing the exposure of the naphthalene ring to hydrophobic interactions with non-polar stationary phases [97]. RP HPLC separations are obtained making use of mobile phases containing buffering systems [96] or strong electrolytes (as sodium nitrate, potassium phosphate, and sodium sulfate) to suppress ionic interactions with the stationary phase and so increase retention and separation selectivity [96]. It has been so possible to separate 1- and 2-naphthalenesulfonic acids [96] and more than 10 naphthalene congeners containing up to 4 sulfonic groups [96]. Retention times generally increase with the decrease of the number of sulfonic groups [96] in the molecule. The most often employed HPLC technique for the separation of polar aromatic sulfonates is IP HPLC [96] with UV [96,98], DAD [96,99], or fluorescence detection [92,96]. The retention depends, besides the nature of the ion-pairing reagent and its concentration, on the number of sulfonic groups present in the molecular structure, the kind of aromatic ring, and the nature, the number, and the position of the substituents [96,100]. IP HPLC is often coupled with off- or on-line [100,101] IP SPE, performed on C8 and C18 cartridges [96,100,101]. HPLC/ESI-MS methods in the selected ion monitoring (SIM) [102–104] or multiple-reaction monitoring mode (MRM) [105] have been used to separate mixtures of aromatic sulfonates in effluents [102,103,105], surface waters, and ground waters [104]. The addition of ammonium formate is used in the RP HPLC separation of benzene and stilbene sulfonic acids coupled to DAD and ESI-MS [106,107]. TBA is also used as an IP agent, but it must be removed on line before MS detection [108].

Benzene and naphthalene sulfonate moieties are present in the structures of many dyes that can be found in large amounts in wastewaters from textile and food industries. Even if wastes are decolored before the final discharge, not enough attention is nowadays devoted to the identification of possible uncolored degradation products, potentially toxic, that form during the decolorization process and are discharged into the aquatic systems. Besides sulfonate derivatives, aromatic amines have also been reported as possible degradation products of dyes [109].

SPE methods with different cartridge packings have been employed for the pre-concentration and clean up of sulfonated azo dyes from waters and soil extracts [110,111]. The extraction of solid samples has been carried out by sonication or Soxhlet extraction and the extracts treated like the water samples. C18 cartridges and columns [111] were used followed by the elution with aqueous organic solvents in the presence of TEA with recovery yields always greater than 65% [93,111]. Higher recoveries have been obtained by using C18 columns, pre-conditioned with an ammonium acetate buffer and eluted with methanol [111]. The use of styrene-divinylbenzene [93,112], as well as of cross-linked polymeric sorbents with sulfonate functions, was shown to be suitable in the SPE of the more polar compounds [111].

For the determination of sulfonated dyes and of their degradation products, RP IP chromatography with UV detection has been widely used [113]. RP [111,114,115] and polymeric material [116] are the stationary phase packings and the mobile phases are generally water/acetonitrile or water/methanol mixtures. When MS detection is employed, the mobile phases are buffered with volatile ammonium acetate or formate [113,117–123]. The IP agents most commonly used are TBA [124–126] and hexa-decyltrimethylammonium salts [96]. The resolution of isomers and congeners can be improved by inducing hydrophobic interactions (salting-out effects), favored by the addition of electrolytes, such as sodium sulfate [96,124], ammonium sulfate [98], and other salts to the mobile phase [96,124]: with UV detection LOD values around 100–250 μg L^{-1} are achieved [124]. LC/MS techniques offer better sensitivities in the analysis of industrial wastes and treated effluents (LODs range between 1 and 80 μg L^{-1}) [46].

The potentiality in the identification process [127] is increased by fragmentation experiments. A review reports the more recent analytical LC/MS methods suitable for the determination of sulfonated dyes in the effluents [45]. Atmospheric pressure ionization techniques are frequently used in the identification and the determination of sulfonated azo dyes [46,95,111,128,129] and their degradation products [86,130–135]. Detection is advantageously performed by ESI-MS in the NI mode. Since in MS detection, the presence of inorganic salts in the mobile phase leads both to signal suppression and to contamination of the ion source, more volatile additives should be used as, for example, ammonium acetate, that is very often used in the HPLC–MS determination of mono- and disulfonic acids [111,130,131,136]. It must be underlined, anyway, that under the same conditions, the species containing more than two sulfonate groups [46,104] are very poorly retained: this result possibly explains why the HPLC/MS technique has so far been used only for the separation of mono- and disulfonic dyes. To increase the retention of polysulfonated species, salts (acetate or formate) of mono-, di-, or trialkylamines are used as IP reagents [86,94,102,105,109,132,137,138]. Apart from the concentration of alkylammonium salts used as IP reagents, retention was also shown to depend on the number and the length of the amine alkyl chains [96,139]. TEA [98,102,137–140] and TEAA [141–143] are the most frequently used.

In MS amine, adducts can be detected at low cone voltages and their maximum number corresponds to the number of sulfonate groups minus one [144]. It is useful to underline that, since these adducts dissociate into sulfonic acid at elevated cone voltages, in the analysis of sulfonated dyes the addition of alkylamines to the eluent also helps to suppress the formation of multiply charged alkali cations [137,144,145] and leads to advantages in terms of sensitivity, clarity of spectra, and fragmentation behavior in collision-induced dissociation (CID). Nevertheless, the presence of alkylamine in the ESI source influences the ionization of sulfonates [144,146] that form clusters that in turn dissociate at elevated spray voltage. It has been also shown that an improved separation of naphthalene-, benzene-, and anthraquinonesulfonic acids (including two naphthalenetrisulfonic acids) is obtained when di-methylbutylammonium-, dihexylammonium-, or tributylammonium- acetate (TBAA) is used as ion-pairing additives, with respect to the use of TEAA. A compromise has been found between the optimization of the chromatographic performance (that improves with the concentration of TBAA) and the undesired occurrence in ESI of signal suppression (that increases for increasing concentrations of TBAA). For polysulfonated dyes, an enhancement of the ionization efficiency in the presence of diethylamine has been observed [144].

In the HPLC/MS analysis of azo dyes of low polarity, the particle beam interface with electron ionization can also provide useful structural information [96]. Soft ionization techniques such as thermospray, ESI, and APCI yield information on the molecular mass [86] and on the number of acidic groups for a wide polarity range of compounds [46,131]. For the structural characterization of ionic [46,111,131,145,147] and metal azo dye complexes [131,147], ESI is generally to be preferred to APCI. In order to obtain structural information, in connection with soft ionization techniques, the CID or the MS^n technique using an IT analyzer [86,147] is required. When using a TQ mass spectrometer, a characteristic fragment ion of m/z 80 (SO_3^-) is detected [111]. The induced cleavage of the azo bond can assist in confirming the dye structures [111], while in dyes bearing a carboxylate group, decarboxylation can be observed by the fragmentation [46] and CID technique [111].

The use of on-line HPLC monitoring of bioprocesses is critically discussed [148], mainly due to the long analysis time and the difficulty in sampling. In biological and medical monitoring [149], as well as in the control of wastewaters [150,151] and bioreactors [152–156], some approaches have been made to develop selective on-line HPLC methods. A microfiltration based on the continuous on-line sampling method based on HPLC/DAD has been developed: it monitors the presence of aminosulfonates and sulfonated azo dyes in anaerobic/aerobic bioreactors and an RP HPLC/ DAD system is combined with a quadrupole ion-trap hybrid MS system to monitor the biological treatment process of wastewaters containing azo dyes [152,154].

Chromatographic methods based on LC-phases with polar selectivity or TBA IP HPLC on C8 and C18 stationary phases combined with ion suppression and ESI hybrid MS spectrometry were also used in the monitoring of the biological treatment of a synthetic dye bath and in particular of the hydrolyzed azodye Reactive Black 5 [127]. Complementary information about dye degradation products was collected from MS analyses with different scan modes and DAD spectrophotometric analyses. The robustness of the analytical method was tested in the presence of different amounts of NaCl and Na_2SO_4, and it was shown that sodium sulfate concentrations greater than 0.5 M give high ionization suppression potential in ESI-MS/MS detection. The advantages offered by HPLC methods to monitor the aerobic wastewater treatment process, to characterize the biodegradability of specific chemicals, and to evaluate the selective adaptation of bacteria to a chemical pollutant must be underlined [127].

18.6   HPLC METHODS FOR THE DETERMINATION OF PHARMACEUTICALS, HORMONES, AND DRUGS

Due to their estrogenic activity, steroids have been included in preliminary lists of EDCs. These chemicals are even more diffusely found in waters, also due to sensitivities nowadays achieved in advanced LC/MS and LC/MS/MS instrumentations. The interfaces most widely used for the LC/MS determination of steroids, drugs, surfactants, and organic pollutants in an aquatic environment are ESI, which is particularly well suited for the analysis of polar compounds, and APCI, that is more effective in the analysis of medium- and low-polarity substances. LC/MS and LC/MS/MS have been mostly applied in the SIM mode and in the MRM mode.

Because of the complexity of the environmental matrices and because of the general trace concentration levels of micropollutants, a pre-concentration step is necessary to isolate the target compounds from the matrix and to achieve the LOD required.

The analytical methods present in literature for the determination of estrogens and progestogens in wastewaters have been recently reviewed [157]. While a filtration step (required if the samples contain suspended matter) and a concentration process (performed for evaporation or under a stream of nitrogen) do not lead to a significant loss of the analytes, the extraction process is to be considered as the critical step in sample pre-treatment. The SPE technique with sorbents of C18-bonded silica and GCB is usually employed for the extraction of estrogens and progestogens from waters [158,159] but successful examples of immunosorbent extraction are also reported [160]. Natural and synthetic estrogens are extracted from river sediments with acetone/methanol (1:1) mixtures and sonicated [161]. PLE is also employed, followed by a cleanup carried out by the SPE process on the C18 sorbent [162]. The methods recently published for the determination of estrogens and progestogens in solid environmental samples and in freshwater sediments are collected in two recent reviews [163,164]. An interesting approach is based on an LC–MS method with column switching and the use of restricted access material (RAM), a bifunctional sorbent that combines size-exclusion and RP retention mechanisms. The use of RAM-based pre-columns after the SPE-C18 cleanup process enables the LC/MS determination of steroid, sex hormones, and other pollutants in sediments at concentration levels of the order of ng g^{-1} [162].

The most frequently employed techniques for the quantification of sex hormones in the environment are immunoassay, GC/MS, and LC/MS [165,166]. LC–MS is the technique more diffusely employed, also because enzymatic hydrolysis and derivatization reaction are not required. The most used stationary phase is ODS and the mobile phases are generally water/methanol or water/acetonitrile mixtures. The LC/MS analysis of estrogens has been carried out with the ESI interface operating in NI, which offers greater sensitivities than in PI, as well as with APCI in the NI mode [167–171]. A review reports the most recent methods for the identification and the determination of many drugs in waters [172]. In the pre-treatment step, the aqueous samples are filtered and the pH adjusted as a function of the properties of the drug (e.g., at pH 2 for the acidic tetracyclines (TCs) and at pH 9 for the alkaline phenazone derivatives) and the samples are stored at about 4°C in dark bottles before analysis, to avoid degradation. Due to the general low concentration of pharmaceuticals in waters, an enrichment step is essential and is generally performed by SPE on RP sorbents; a tandem SPE and a mixed phase cation exchange (MPC) extraction is applied [173,174]. A lyophilization process has been also used, because it is efficient and fast, requires low amount of organic solvent, and can be easily automatizated and interfaced on-line [175,176]. The extraction process of TCs requires particular attention because silanol groups present in the sorbent or in glassware can easily bind the analytes in an irreversible way: the effect is avoided through the silanization (for instance with dimethyldichlorosilane [177]) of all the glassware or through the use of polytetrafluoroethylene containers; it is also suggested to wash off SPE sorbents with a diluted HCl of Na_2EDTA or other chelating agents [177].

SPME has been used to extract TCs from surface waters [178]. Extraction of drugs from solid samples is normally performed by sonication or by stirring the sample with water or polar organic solvents [163,179]. The cleanup of the extracts is carried out by SPE, LLE, gel permeation chromatography (GPC), and semipreparative LC. SPE is generally preferred, because it is fast, requires a low volume of organic solvent, and can be used on-line; an RP sorbent is generally used, but examples are found of the use of strong ion-exchange packings (Figure 18.2) [159,180]. In the LC methods employed in drug analysis, the most used stationary phase packing is ODSs and the mobile phases are water-methanol or water-acetonitrile mixtures at different pH values, modified by the addition of acetate [181–184] or formate [173,183–185]. In LC/MS and LC/MS/MS analyses, ESI ionization interface is preferentially used in both PI (Figure 18.3) and NI modes [172,177,182,186]: LODs lower than $1.10\,\mu g\,L^{-1}$ can be achieved. LC/MS/MS with an APCI source is employed for the determination of drugs and their degradation species produced through reactions of hydroxylation, reduction, or adducts formation [186].

A new drug now found in river waters is cocaine. A determination method for cocaine and its main urinary metabolite (benzoylecgonine, BE) in river waters and urban waste treatment plants is reported. A SPE is performed prior to the HPLC-MAS analysis. An RP column containing polar moieties is used and a mixture of acetonitrile/water (at pH 2.0 for formic acid) is the mobile phase flowing in gradient conditions. An MS detection is performed by a quadrupole mass spectrometer working with the ESI source in the PI mode, and data acquisition is performed by the MRM of selected fragmentation products. The identification is confirmed by ESI-MS/MS analysis. LOD values for cocaine and BE are 0.12 and $0.06\,ng\,L^{-1}$, respectively [20].

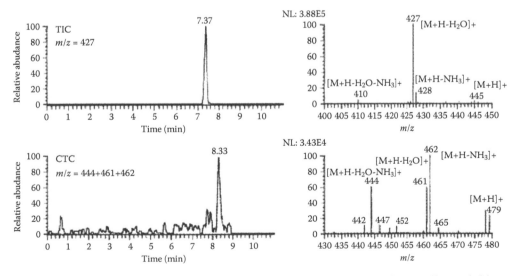

FIGURE 18.2 Ion chromatogram and the corresponding tandem mass spectra of tetracycline and chlortetracycline obtained for a sample of an agricultural soil amended with manure. (Reprinted from Lopez de Alda, M.J. and Barceló, D., *J. Chromatogr. A*, 1000, 503, 2003. Copyright 2003. With permission from Elsevier.)

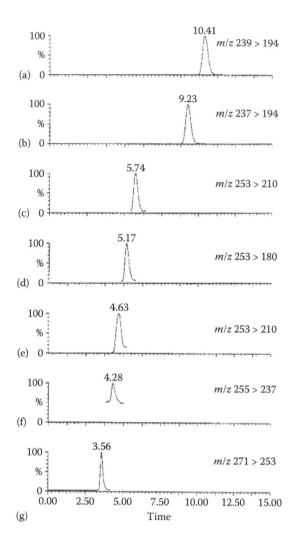

FIGURE 18.3 Time-scheduled SRM chromatograms of carbamazepine and its metabolites in an efflu-ent from the STP of Perborough: (a) 10,11-dihydroxycarbamazepine; (CBZ-DiH), internal standard; (b) carbamazepine (CBZ); (c) 3-hydroxycarbamazepine; (d)10,11-dihydro-10,11-epoxycarbamazepine; (e) 2-hydroxycarbamazepine; (f) 10,11-dihydro-10-hydrocarbamazepine; and (g) 10,11-dihydro-10,11-dihydroxy-carbamazepine. (Reprinted from Petrovic, M. et al., *J. Chromatogr. A*, 1067, 1, 2005. Copyright 2005. With permission from Elsevier.)

18.7 HPLC METHODS FOR THE DETERMINATION OF ALKYLPHENOL-ETHOXYLATE SURFACTANTS, SUNSCREEN, AND UV FILTERS

For the determination of alkylphenol-ethoxylate (APEOs) in water samples, a pre-concentration step is generally performed through SPE on packings of octadecyl bonded silica [187–190], strong anionic exchange (SAX) [191], and GCB [192]. Desorption is carried out with methanol, acetone, dichloromethane, or their mixtures. Selective elutions are also employed using solvents of different polarity, followed by FIA-MS analysis [193,194]. From solid environmental samples, alkylphenolic compounds are extracted by Soxhlet extraction, sonication, or extraction techniques based on high temperatures and pressures, such as PLE and MAE [195]. To clean up the extract from interferents that could suppress the analyte signal in LC–MS analysis, SPE is performed with different sorbents (C_{18}, CN, NH_2).

The general low volatility of APEOs also allows the LC determination of many oligomers: both NP and RP liquid chromatographic techniques have been used. In NP LC, APEOs have been separated according to the increasing number of ethylene oxide units, while RP LC allows the separation as a function of the properties of the hydrophobic moiety and is particularly suited for the separation of alkyl-homologues [159]. A method for the determination of alkylphenols in sediments and sewage samples is based on both SE and RP chromatographic techniques [187]. RP and NP HPLC methods hyphenated with an APCI-MS detector [196,197] and FIA [193,194] and RP HPLC methods coupled with ESI-MS detection [198–201] are proposed.

The need for methods for the identification and the determination of sunscreen and UV filters in waters is continuously increasing due to the large consumption of personal care products and the consequent higher amounts offloaded into waters and sediments. Amounts up to 10 ng L^{-1} have been found in coastal waters and in shower wastewaters [21]. Two kinds of UV filters can be distinguished: those based on the organic species that work by absorbing UV light and those based on the presence of TiO_2 and ZnO, that reflect and scatter UV light. Organic UV filters are contained in sunscreens, beauty creams, skin lotions, lipsticks, hair sprays, hair dyes, and shampoos: the most used are benzophenone-3 (BP-3), octyldimethyl-p-aminobenzoic acid (ODPABA), 4-methylbenzylidene camphor (4-MBC), ethylhexyl methoxycinnamate (EHMC), octylmethoxycinnamate (OMC), octocrylene (OC), butylmethoxydibenzoylmethane, terephtalylidene dicamphorsulfonic acid, ethylhexyltriazone, ethylhexyl salicylate, and 1-(4-ter-butylphenyl)-3-(4-methoxyphenyl)-1,3-propanedione (BMMP). Most of them are lipophylic species containing conjugated aromatic systems that absorb UV light in the wavelength ranges 280–315 nm (UV-B), 315–400 (UV-A), or both. Mixtures of UV filters, also in combination with inorganic micro pigments, are often employed [202,203]. Because of their widespread diffused use, these compounds can enter the aquatic environment indirectly from showering, cleansing, washing clothes, and wastewater treatment plants, as well as directly from swimming and sunbathing in lakes and rivers. For their determination in waters, extreme care must be taken during the sampling and the sample preparation, since these lipophylic chemicals are easily transferred to glassware and consumables and can heavily contribute to lead analytical blank problems [203]. The extraction from surface waters and wastewaters is generally performed by SPE on C_{18} disks and elution with ethyl acetate/dichloromethane mixture 1:1 v/v. RP HPLC is the most used technique for their determination, with C18 or C8 as the stationary phases. The mobile phases are mixtures of acetonitrile/water [204], tetrahydrofuran/water [205], acetonitrile/tetrahydrofuran/water [206], and methanol/tetrahydrofuran/water [207,208]. Due to the similar structure of some UV filters, the chromatographic resolution of mixtures often represents a problem, even when gradient elution and DAD are employed. The addition of sodium dodecyl sulfate to the mobile phase (acetonitrile/water 80/20 v/v) helps in improving HPLC separation: the method, based on UV detection, has been successfully applied in the analysis of coastal waters and shower wastewaters (average LOQ values around 10 ng L^{-1}) [209]. The use of HPLC/UV (313 nm) and HPLC/DAD methods, with the addition of a 65.4 mM solution of hydroxypropyl-β-cyclodextrin to the mobile phase, permits the identification (LOD values between 1.5 and 3.3 mg L^{-1}) and the quantification of the UV-absorbers benzophenone-4 (BP-4), BP-3, butyl methoxydibenzoylmethane, ODPABA, OMC, and octyl salicylate [211].

18.8 HPLC METHODS FOR THE DETERMINATION OF DISINFECTION BY-PRODUCTS

Disinfection by-products (DBPs) form an undesired species in the chlorine disinfection processes of waters (performed with chlorine, chlorine dioxide, and chloramines). The high priority DBPs include brominated, chlorinated, and iodinated species of halomethanes, brominated, and chlorinated forms of haloacetonitriles, haloketones, haloacids, and halonitromethanes, as well as analogues of 3-chloro-(4-dichloromethyl)-5-hydroxy-2(5H)-furanone. All the high priority DBPs included in the "Nation-wide DBP occurrence study" are listed in Table 18.1 together with other contaminants,

TABLE 18.1
**High-Priority DBPs Included in the Nationwide DBP Occurrence Study
Together with Other Contaminants Copresent[a]**

MX and MX Analogues

3-Chloro-4-(dichloromethyl)-5-hydroxy-2(5H)-furanone (MX)

3-Chloro-4-(dichloromethyl)-2(5H)-furanone (red-MX)

(E)-2-Chloro-3-(dichloromethyl)butenedioic acid (ox-MX)

(E)-2-Chloro-3-(dichloromethyl)-4-oxobutenoic acid (EMX)

2,3-Dichloro-4-oxobutenoic acid (mucochloric acid)

3-Chloro-4-(bromochloromethyl)-5-hydroxy-2(5H)-furanone (BMX-1)

3-Chloro-4-(dibromomethyl)-5-hydroxy-2(5H)-furanone (BMX-2)

3-Bromo-4-(dibromomethyl)-5-hydroxy-2(5H)-furanone (BMX-3)

(E)-2-Chloro-3-(bromochloromethyl)-4-oxobutenoic acid (BEMX-1)[a]

(E)-2-Chloro-3-(dibromomethyl)-4-oxobutenoic acid (BEMX-2)[a]

(E)-2-Bromo-3-(dibromomethyl)-4-oxobutenoic acid (BEMX-3)[a]

Haloacids

3,3-Dichloropropenoic acid

Halomethanes

Chloromethane

Bromomethane (methyl bromide)[a]

Dibromomethane

Bromochloromethane

Bromochloroiodomethane

Dichloroiodomethane

Dibromoiodomethane[a]

Chlorodiiodomethane[a]

Bromodiiodomethane[a]

Iodoform[a]

Chlorotribromomethane

Carbon tetrachloride

Halonitromethanes

Bromonitromethane

Chloronitromethane[a]

Dibromonitromethane

Dichloronitromethane[a]

Bromochloronitromethane[a]

Bromodichloronitromethane[a]

Dibromochloronitromethane[a]

Tribromonitromethane (bromopicrin)[a]

Haloacetonitriles

Bromoacetonitrile

Chloroacetonitrile

Tribromoacetonitrile

Bromodichloroacetonitrile

Dibromochloroacetonitrile

Haloketones

Chloropropanone

1,3-Dichloropropanone

1,1-Dibromopropanone

1,1,3-Trichloropropanone

1-Bromo-1,1-dichloropropanone

1,1,1,3-Tetrachloropropanone

1,1,3,3-Tetrachloropropanone

1,1,3,3-Tetrabromopropanone[a]

1,1,1,3,3-Pentachloropropanone

Hexachloropropanone

Haloaldehydes

Chloroacetaldehyde

Dichloroacetaldehyde

Bromochloroacetaldehyde[a]

Tribromoacetaldehyde[a]

Haloacetates

Bromochloromethyl acetate

Haloamides

Monochloroacetamide[a]

Monobromoacetamide[a]

Dichloroacetamide

Dibromoacetamide[a]

Trichloroacetamide[a]

TABLE 18.1 (continued)
High-Priority DBPs Included in the Nationwide DBP Occurrence Study
Together with Other Contaminants Copresent[a]

Nonhalogenated Aldehydes and Ketones

2-Hexenal	Methyl ethyl ketone (2-butanone)[a]
5-Keto-1-hexanal[a]	6-Hydroxy-2-hexanone[a]
Cyanoformaldehyde	Dimethylglyoxal (2,3-butanedione)

Volatile Organic Compounds (VOCs) and Miscellaneous DBPs

1,1,1,2-Tetrabromo-2-chloroethane	Methyl tert-butyl ether[a]
1,1,2,2-Tetrabromo-2-chloroethane[a]	Benzyl chloride

[a] DBP not originally prioritized but included due to similarity to other priority compounds.

---

often copresent in waters. Recent reviews report relevant information concerning epidemiology, toxicology, exposure to by-products [211], and formation [212], as well as the analytical methods used for their identification [212, 213]. In particular, a comprehensive study of a nationwide occurrence has been performed [214] that considers the occurrence of DBPs on the basis of a predicted adverse health effect. In this study, not available standards were synthesized, analytical methods developed, and the DBPs quantified in waters across the United States: the waters treated with the disinfectants more commonly used worldwide (chlorine, ozone, chlorine dioxide, chloramines), as well as source waters with a high bromide content were considered [215]. Continuously, new DBPs are being discovered, new water disinfectants are being proposed, and new chromatographic methods must be developed for water control. A chromatographic method hyphenated with tandem MS, based on a previous derivatization step with 2,4-dinitrophenylhydrazine (DNPH) [216] is particularly suitable for the determination of aldehydes, ketones, hydroxybenzaldehyde, and dicarbonyl DBPs in chlorinated drinking water, while an LC/APCI-MS method with a previous derivatization with 4-(dimethylamino)-6-(4-methoxy-1-naphthyl)-1,3,5-triazine-2-hydrazine has been proposed for the determination of highly polar carbonyl DBPs at concentration levels of a nanogram per liter without any pre-concentration or extraction step [217]. Good sensitivities are also achieved in UV and fluorescence spectroscopy. For measuring chlorinated and brominated haloalogenated acids (HAAs) in Barcelona drinking water, swimming pool water, and Portugal surface water, a very sensitive (detection limits in the range of low μg L^{-1}) SPE IP (TEA being the ion pairing reagent) LC/ESI-MS method has been developed [218]. The use of new derivatizing agents has also been explored for the identification of highly polar DBPs often present in treated drinking water. A derivatization step with DNPH, as well as with o-carboxymethylhydroxylamine (CMHA) prior to the LC tandem MS analysis was proposed [219]. The study evidenced that halogenated carbonyl DBPs can undergo in the presence of DNPH side reactions, in which chlorine atoms are substituted by DNPH. As a result, halogenated DBPs would be misidentified as non-halogenated, and identified as oxygenated DBPs. CMHA, in turn, was found to be a suitable alternative reagent because it does not participate in side reactions and in tandem MS analysis produces a greater number of structurally diagnostic ions than DNPH. Using DNPH and CMHA as derivatizing agents, highly polar aldehydes, ketones, hydroxybenzaldehyde, and dicarbonyl, DBPs have been identified in chlorinated drinking waters [219].

A new area of research concerning DBPs evidences the possible formation of a halogenated material at high molecular weight in the disinfection process: the study, based on the use of ESI-MS and matrix-assisted laser desorption ionization (MALDI)-MS, is receiving great interest but requires great effort, due to the complexity of the mass spectra obtained (known as "a peak at every mass"). The aim is to find out diagnostic ions able to select halogenated DBPs from the complex mixture of high molecular weight DBPs [220]. Tandem MS studies evidenced that some chlorine containing DBPs produce chloride ion fragments that can be used as a "fingerprint" for chlorinated DBPs. Fractions of chlorinated fulvic acids have been collected from the Suwannee river using SEC and have been analyzed by ESI-MS/MS. Hundreds of peaks have been observed in the different SEC fractions and it was possible to obtain preliminary MS/MS information in NI ESI-MS and ESI-MS/ MS. A distribution of ions from m/z 10–4000 has been observed, most of the ions being present in the region of m/z 100–500 [221].

Precursor ion scans of m/z 35 (chlorine) are useful to identify chlorine-containing DBPs, and product ion scans to confirm the presence of chlorine. Full scan product ion spectra are in general too complex to give definitive structural information but permit the conclusion that in the chlorination process, chlorinated DPBs at high molecular weight are preferentially formed. IC has been employed to monitor HAAs in drinking water [222]. A new HILIC (Hydrophilic Interaction LIquid Chromatography) LC/MS/MS method has also been developed to quantify dichloroacetic acid in drinking water without derivatization and with LOD values of around $5 \mu g L^{-1}$ [223]. This method presents remarkable advantages with respect to the methods that involve the use of methylating agents (e.g., diazomethane, H_2SO_4/methanol, BF_3/methanol) since recent studies showed that during the methylation process trichloroacetic acid could convert to dichloroacetic acid. The HILIC technique makes use of stationary phases containing a polar end group (e.g., an amino group) and retention is based on the affinity of the polar analyte for the charged end group of the stationary phase packing. The use of HILIC allows dichloroacetic acid to elute far away from the solvent front without the addition of IP agents to the mobile phase, that could suppress the ionization in MS detection. Nine HAAs in drinking waters were determined by a new ESI-FAIMS (high-field asymmetric waveform ion mobility mass spectrometry) method [224]. The FAIMS mode significantly reduces the chemical background in the ESI source and allows much lower detection limits (of the order of a submicrogram per liter), without requiring a sample derivatization or a previous chromatographic separation.

Dichloroacetamide present in chlorinated and chloraminated drinking waters was determined (at detection level of $23 \mu g L^{-1}$) through an LC/fluorimetric method and a post-column derivatization reaction with o-phthaldialdehyde in the presence of sulfite at pH 11.5, that leads to the formation of a highly fluorescent isoindole fluorophore [225].

18.9 HPLC METHODS FOR THE DETERMINATION OF ALGAL TOXINS

The increased incidence in surface waters of poisoning of shellfish and deaths of fishes, livestock and wildlife, as well as the occurrence of illness and death in humans has been correlated with the increasing presence of harmful algal blooms. The algal toxins that impact human health are generally neurotoxins or hepatotoxins, produced by dinoflagellates, diatoms, or cyanobacteria (blue-green algae). Dinoflagellates and diatom toxins impact humans mainly through the consumption of seafood while cyanobacterial toxins are directly assumed through drinking contaminated water. It has been evidenced that about one-third of source waters contain significant concentrations of cyanobacterial toxins and that trace levels ($<1.0 \mu g L^{-1}$) have also been detected in finished tap water. Different hypotheses are made to explain their presence in waters. The possibility that chlorine during the water treatment could destroy intact cells and release the intracellular toxins has been considered, as well as the possibility that the mechanical pumping leads to cell lysis and to toxin release. Also, the possible inability of chlorine to destroy microcystins and saxitoxins has been taken into consideration [21]. From a chemical point of view, most of these toxins are peptide-related species characterized by relatively high molecular weight and highly polar properties that hinder their presence in the environment, at least until the wide application of ESI- and APCI-MS techniques. The methods developed for the determination of algal toxins include enzyme-linked immunosorbent assays (ELISA), protein phosphatase inhibition assays,

LC/DAD, LC/ELISA, LC/MS, LC/MS/MS, MALDI-MS, and ESI-FAIMS-MS. These methods allow the achievement of detection limits as low as ng L^{-1}. The extraction process is very important, highest recoveries being obtained with hydrophilic-lipophylic balanced cartridges that are more efficient than C-18 and propylene cartridges [226]. In particular, ELISA, LC/DAD, and LC/MS/MS methods have been used to identify the presence of three cyanobacterial toxins in Australian drinking water plants [227], with quantification limits in the raw source waters of 8.0, 17.0, and 1.3 µg L^{-1} for microcystins, saxitoxins, and cylindrospermopsin, respectively [227]. A new LC/ESI-TOF MS method [228] allows detection limits lower than 1 µg L^{-1} for anatoxin-α, microcystin-LR, microcystin-RR, microcystin-YR, and nodularin, while an LC/ESI-MS/MS method allows detection limits of 2.6 ng L^{-1} [229] to be reached for microcystin-LR in surface waters. An LC/ESI-MS method allows the determination of 10 microcystins and nodularins [230] in an analysis time of 2.8 min for each sample and with detection limits of 5–10 pg/injection. The simultaneous determination of saxitoxin, anatoxin-A, domoic acid, nodularin, microcystins, okadaic acid, and dinophysistoxin-1 in a single chromatographic has been obtained by an LC/ESI-MS/MS method [231]. This method, associated to fraction collection, hydrolysis, and derivatization of free amino acids also allows the identification of new microcystins. LC/ESI-MS and LC/MS/MS methods have been used to identify new algal toxins, as for instance, a rare cyanobacterial toxin-homoanatoxin-α, found in some lake waters of Ireland [232]. Microcystins in lake waters of Algeria (concentrations ranging from 3 to 29 µg L^{-1}) have been determined by a MALDI-TOFMS method [233].

An international intercomparison exercise in the determination of microcystin, carried out by using the most common methods (LC/DAD, ELISA and LC/MS) indicated that LC/DAD is affected by lower precision [234], while the coupling of the LC technique with ELISA permit the achievement of high sensitivity and specificity in the determination of microcystins and nodularin [235] without the need of pre-concentration; the method meets the World Health Organization guidelines (1 µg L^{-1}). The combination of ELISA characterization and LC analysis with fluorescence, UV, and tandem MS detections, allowed the first identification of cylindrospermopsin, an algal toxin that caused the poisoning of up to 148 persons in Australia [236].

18.10 HPLC METHODS FOR THE ANALYSIS OF ATMOSPHERE

HPLC methods also find some advantageous applications in the analysis of the atmosphere.

The determination of pesticides belonging to different chemical classes and present at trace levels in the atmosphere (namely dinitroanilines, organochlorines, organophosphorous-, chlorophenoxy acids, triazines, substituted ureas, and carbamates) has been performed by employing an HPLC/UV method [237]. In particular, XAD-2 resins and glass fiber filters used in the sampling of atmosphere were undergone to Soxhlet extraction (diethyl ether/hexane mixture as the solvent) and the extract was analyzed by HPLC. The stationary phase is an NP packing and a mixture of *n*-hexane and methyl *tert*-butyl ether (MTBE) is the mobile phase flowing in gradient elution: LOD values range between 70 and 440 pg m^{-3}.

HPLC methods with fluorescence detection have also been developed for the determination of nitro-policyclic aromatic hydrocarbons (PAHs) (among which 9-nitroanthracene and 1-nitronaphthalene) [238] in atmosphere. Samples have been collected in a standard high-volume sampler with a Teflon-coated glass fiber filter, and the Soxhlet extraction was performed with dichloromethane as the solvent. RP HPLC/UV techniques are used for the determination of aldehydes, ketones, and carbonylic compounds after derivatization with DNPH [239].

Toxic nitrophenols (4-nitrophenol, 2-nitrophenol, and 2,4-dinitrophenol) present in air samples are sampled on silica gel or XAD-2 polymeric resin, extracted, and analyzed by the HPLC/DAD technique [240]. Nitrophenols are also determined in rain and snow precipitations through HPLC/UV analysis after SPE [241].

REFERENCES

1. C. Petit, R. Cabridenc, R.P.J. Swannell, R.S. Sokhi, *Environ. Int.* 21, 167–176, 1995.
2. R.A. Larson, E.J. Weber, *Reaction Mechanism in Environmental Organic Chemistry*, Lewis Publishers, Boca Raton, FL, 1994.
3. E. Terlecka, *Environ. Monit. Assess.* 107, 259–284, 2005.
4. G. Sheng, C.T. Johnston, B.J. Teppen, S.A. Boyd, *Clay Miner.* 50, 25–34, 2002.
5. S. Angioi, S. Polati, M. Roz, C. Rinaudo, V. Gianotti, M.C. Gennaro, *Environ. Pollut.* 134, 35–43, 2005.
6. S. Polati, S. Angioi, V. Gianotti, F. Gosetti, M.C. Gennaro, *J. Environ. Sci. Health B* 41, 333–344, 2006.
7. S. Polati, F. Gosetti, V. Gianotti, M.C. Gennaro, *J. Environ. Sci. Health B* 41, 765–779, 2006.
8. A. Laganà, G. Fago, L. Fasciani, A. Marino, M. Mosso, *Anal. Chim. Acta* 414, 79–94, 2000.
9. A. Laganà, A. Bacaloni, I. De Leva, A. Faberi, G. Fago, A. Marino, *Anal. Chim. Acta* 462, 187–198, 2002.
10. O. Pozo, E. Pitarch, J.V. Sancho, F. Hernandez, *J. Chromatogr. A* 923, 75–85, 2001.
11. S. Chiron, E. Sauvard, R. Jeannot, *Analusis* 28, 535–542, 2000.
12. C. Medana, P. Calza, C. Baiocchi, E. Pellizzetti, *Curr. Org. Chem.* 9, 859–873, 2005.
13. DHHS (NIOSH) Publication No. 78-148, Direct Blue 6, Direct Black 38 and Direct Brown 95 Benzidine Derived Dyes, Washington, DC, 1978.
14. DHHS (NIOSH) Publication No. 81-116, Health Hazard Alert—Benzidine-, o-Toluidine-, and o-Dianisidine-Based Dyes, NIOSH Alert, Washington, DC, 1980.
15. T. Reemtsma, *Trends Anal. Chem.* 20, 500–517, 2001.
16. D. Stevenson, *J. Chromatogr. A* 745, 39–48, 2000.
17. M. Anastassiades, E. Scherbaum, *Deut. Lebensm. Rundsch.* 96, 466–477, 2000.
18. I.H. Wang, R. Moorman, J. Burleson, *J. Chromatogr. A* 983, 145–152, 2003.
19. E.M. Thurman, I. Ferrer, R. Parry, *J. Chromatogr. A* 957, 3–9, 2002.
20. E. Zuccato, C. Chiabrando, S. Castiglioni, D. Calamari, R. Bagnati, S. Schiarea, R. Fanelli, *Environ. Health* 4, 14–20, 2005.
21. S.D. Richardson, T.A. Ternes, *Anal. Chem.* 77, 3807–3838, 2005.
22. C.B. Hymer, J.A Caruso, *J. Chromatogr. A* 1045, 1–13, 2004.
23. V. Kaur, A.K. Malik, N. Verma, *J. Sep. Sci.* 29, 333–345, 2006.
24. A.K. Malik, V. Kaur, N. Verma, *Talanta* 68, 842–849, 2006.
25. R. Morabito, P. Massanisso, P. Quevauviller, *Trends Anal. Chem.* 19, 113–119, 2000.
26. S. Karthikeyan, S. Hirata, *Anal. Lett.* 36, 2355–2366, 2003.
27. P.A. Gallagher, C.A. Schwegel, X.Y. Wei, J.T. Creed, *J. Environ. Monit.* 3, 371–276, 2001.
28. S. Mazan, G. Cretier, N. Gillon, J.M. Mermet, J.L. Rocca, *Anal. Chem.* 74, 265–274, 2002.
29. Z. Gong, X. Lu, M. Ma, C. Watt, C. Le, *Talanta* 58, 77–96, 2002.
30. Y. Martinez-Bravo, A.F. Roig-Navarro, F.J. Lopez, F. Hernandez, *J. Chromatogr. A* 926, 265–274, 2001.
31. J.A. Day, M. Montes-Bayon, A.P. Vonderheide, J.A. Caruso, *Anal. Bioanal. Chem.* 373, 664–668, 2002.
32. Y. Shibata, K. Tsuzuku, S. Komori, C. Umedzu, H. Imai, M. Morita, *Appl. Organomet. Chem.* 19, 276–281, 2005.
33. C.F. Harrington, *Trends Anal. Chem.* 19, 167–179, 2000.
34. G. Westoo, *Acta Chem. Scand.* 20, 2131–2137, 1966.
35. W. Wang, L. Ding, H. Zhang, J. Cheng, A. Yu, H. Zhang, L. Liu, Z. Liu, Y. Li, *J. Chromatogr. B* 843, 268–274, 2006.
36. D.Q. Hung, O. Nekrassova, R.G. Compton, *Talanta* 64, 269–277, 2004.
37. T.L. Jones-Lepp, G. Momplaisir, *Trends Anal. Chem.* 24, 590–595, 2005.
38. EPA Method No 8323: Determination of organotins by micro-liquid chromatography-electrospray ion trap mass spectrometry.
39. T.L. Jones-Lepp, K.E. Varner, D. Heggem, *Arch. Environ. Contam. Toxicol.* 46, 90–95, 2004.
40. J.C. Wu, Z. Mester, J.J. Pawliszyn, *Anal. At. Spectrom.* 16, 159–165, 2001.
41. R.N. Collins, *J. Chromatogr. A* 1059, 1–12, 2004.
42. A.V. Hirner, *Anal. Bioanal. Chem.* 385, 555–567, 2006.
43. M. Ocksenkuhn-Petropoulou, B. Michalke, S. Kavouras, P. Schramel, *Anal. Chim. Acta* 478, 219–227, 2003.
44. P.C. Uden, H.T. Boakye, C. Kahakachchi, J.F. Tyson, *J. Chromatogr. A* 1050, 85–93, 2004.
45. T. Reemtsma, *J. Chromatogr. A* 1000, 477–451, 2003.
46. C. Ráfols, D. Barceló, *J. Chromatogr. A* 777, 177–192, 1997.
47. J. Riu, I. Schönsee, D. Barceló, *Trends Anal. Chem.* 16, 405–419, 1997.

48. M. Bouzige, G. Machatalère, P. Legeay, V. Pichon, M.C. Hennion, *Waste Manage.* 19, 171–180, 1999.
49. H.M. Pinheiro, E. Tourald, O. Thomas, *Dye Pigments* 61, 121–139, 2004.
50. J.E. Bailey, *Anal. Chem.* 57, 189–196, 1985.
51. H. Bornick, T. Grischek, E. Worch, Fresen. *J. Anal. Chem.* 371, 607–613, 2001.
52. Y.C. Wu, S.D. Huang, *Anal. Chem.* 71, 310–318, 1999.
53. Z. Zhang, M.J. Yang, J. Pawliszyn, *Anal. Chem.* 66, 844A–853A, 1994.
54. J. Chen, J.B. Pawliszyn, *Anal. Chem.* 67, 2530–2533, 1995.
55. R.A.K. Rao, M. Ajmal, B.A. Siddiqui, S. Ahmad, *Environ. Monit. Assess.* 54, 289–299, 1999.
56. D.T. Sponza, M. Işik, *Process. Biochem.* 40, 2735–2744, 2005.
57. P. Sutthivaiyakit, S. Achatz, J. Lintelmann, T. Aungpraid, R. Chanwirat, S. Chumanee, A. Kettrup, *Anal. Biochem.* 381, 268–276, 2005.
58. M. Mayer, F. Mandel, P. Kesners, *GIT Chromatogr.* 1, 20–23, 1998.
59. A. Pielesz, I. Baranowska, A. Rybak, A. Wlochowiez, *Ecotoxicol. Environ. Saf.* 53, 42–47, 2002.
60. C.S. Eskilsson, R. Davidsson, L. Mathiasson, J. Chromatogr. A 955, 215–227, 2002.
61. M.C. Garrigos, F. Reche, M.L. Marlin, K. Perinas, A. Jimenez, *J. Chromatogr.A* 963, 427–433, 2002.
62. F. Planelles, E. Verdu, D. Campello, N. Grane, J.M. Santiago, *J. Soc. Leather Technol. Chem.* 82, 45–52, 1998.
63. S. Polati, M. Roz, S. Angioi, V. Gianotti, F. Gosetti, E. Marengo, C. Rinaudo, M.C. Gennaro, *Talanta* 68, 93–98, 2005.
64. J. Peng, J. Liu, G. Jiang, C. Thai, M. Huang, *J. Chromatogr. A* 1072, 3–6, 2005
65. A. Di Corcia, A. Costantino, C. Crescenzi, R. Samperi, *J. Chromatogr. A* 852, 465–474, 1999.
66. E. Marengo, M.C. Gennaro, V. Gianotti, E. Prenesti, *J. Chromatogr. A* 863, 1–11, 1999.
67. M.C. Gennaro, E. Marengo, V. Gianotti, S. Angioi, *J. Chromatogr. A* 945, 287–292, 2002.
68. Council Directive 80/778/EEC: Quality of water intended for human consumption.
69. M.J.M. Wells, L.Z. You, *J. Chromatogr. A* 885, 237–250, 2000.
70. J.S. Aulakh, A.K. Malik, V. Kaur, P. Schmitt-Kopplin, *Crit. Rev. Anal. Chem.* 35, 71–85, 2005.
71. L.J. Krutz, S.A. Senseman, A.S. Sciumbato, *J. Chromatogr. A* 999, 103–121, 2003.
72. N. Font, F. Hernandez, E.A. Hogendoorn, R.A. Baumann, P. Van Zoonen, *J. Chromatogr. A* 798, 179–186, 1998.
73. J.M. Marìn, J.V. Sancho, O.J. Pozo, F.J. Lopez, F. Hernandez, *J. Chromatogr. A* 1133, 204–214, 2006.
74. J.D. Vargo, Determination of chloroacetanilide and chloroacetamide herbicides and their polar degradation products in water by LC/MS/MS. In: *Liquid Chromatography/Mass Spectrometry, MS/MS and Time of Flight MS: Analysis of Emerging Contaminants,* I. Ferrer, E.M. Thurman (Eds.), American Chemical Society, Washington, DC, 2003.
75. S. Polati, M. Bottaro, P. Frascarolo, F. Gosetti, V. Gianotti, M.C. Gennaro, *Anal. Chim. Acta* 579, 146–151, 2006.
76. G.C. Galletti, A. Bonetti, G. Dinelli, *J. Chromatogr. A* 692, 27–37, 1995.
77. E.M. Thurman, I. Ferrer, D. Barcelò, *Anal. Chem.* 73, 5441–5449, 2001.
78. Y. Picò, C. Blasco, G. Font, *Mass Spec. Rev.* 23, 45–85, 2004.
79. S. Moret, M. Hidalgo, J.M. Sánchez, *Chromatographia* 63, 109–115, 2006.
80. L. Zhao, H.K. Lee, *J. Chromatogr. A* 931, 95–105, 2001.
81. A. Sarafraz-Yazdi, D. Beiknejad, Z. Es'haghi, *Chromatographia* 62, 49–54, 2005.
82. S. Angelino, M.C. Gennaro, *Anal. Chim. Acta* 346, 61–71, 1997.
83. D. Puig, D. Barcelò, *Anal. Chim. Acta* 311, 63–69, 1995.
84. R. Wissiack, E. Rosemberg, *J. Chromatogr. A* 963, 149–157, 2002.
85. T. Reemtsma, *J. Chromatogr. A* 733, 473–489, 1996.
86. T. Reemtsma, *J. Chromatogr. A* 919, 289–297, 2001.
87. A. Di Corcia, F. Casassa, C. Crescenzi, A. Marcomini, R. Samperi, *Environ. Sci. Technol.* 33, 4112–4118, 1999.
88. T.P. Knepper, M. Kruse, *Tenside Surf. Det.* 37, 41–47, 2000.
89. A. Di Corcia, L. Capuani, F. Casassa, A. Marcomini, R. Samperi, *Environ. Sci. Technol.* 33, 4119–4125, 1999.
90. E. Gonzalez-Mazo, M. Honing, D. Barcelò, A. Gomez-Parra, *Environ. Sci. Technol.* 31, 504–510, 1997.
91. J. Riu, E. Gonzalez-Mazo, A. Gomez-Parra, D. Barcelò, *Chromatographia* 50, 275–281, 1999.
92. M. del Olmo, A. Garballo, M. Nimer, I. Lopez, J.A. de Ferrer, J.L. Vilchez, *Chromatographia* 60, 157–164, 2004.
93. F. Gosetti, V. Gianotti, E. Mazzucco, S. Polati, M.C. Gennaro, *Dyes Pigments* 74, 424–432, 2007.
94. J.D. Vargo, *Anal. Chem.* 70, 2699–2703, 1998.

95. P. Jandera, *J. Liq. Chromatogr. Relat. Technol.* 30, 2349–2367, 2007.

96. S. Angelino, A. Bianco Prevot, M.C. Gennaro, E. Pramauro, *J. Chromatogr. A* 845, 257–271, 1999.

97. P. Jandera, S. Bunčeková, M. Halama, K. Novotná, M. Nepraš, *J. Chromatogr. A* 1059, 61–72, 2004.

98. R.N. Rao, S.N. Alvi, S. Husain, *J. High Resolut. Chromatogr.* 23, 329–332, 2000.

99. P. Jandera, J. Fisher, B. Prokeš, *Chromatographia* 54, 581–587, 2001.

100. F.T. Lange, M. Wenz, H.J. Brauch, *J. High Resolut. Chromatogr.* 18, 243–252, 1995.

101. S. Fichtner, F.Th. Langer, W. Schmidt, H.J. Brauch, *Fresenius J. Anal. Chem.* 353, 57–63, 1995.

102. M.C. Alonso, M. Castillo, D. Barceló, *Anal. Chem.* 71, 2586–2593, 1999.

103. M. Castillo, M.F. Alpendurada, D. Barceló, *J. Mass Spectrom.* 32, 1100–1110, 1997.

104. M.J.F. Suter, S. Riediker, W. Giger, *Anal. Chem.* 71, 897–904, 1999.

105. T. Storm, T. Reemtsma, M. Jekel, *J. Chromatogr. A* 854, 175–185, 1999.

106. R.N. Rao, N. Venkateswarku, S. Khalid, R. Narasimha, *Anal. Sci.* 19, 611–617, 2003.

107. V. Gianotti, F. Gosetti, S. Polati, M.C. Gennaro, *Chemosphere* 67, 1993–1999, 2007.

108. G. Socher, R. Nussbaumer, K. Rissler, E. Lankmayr, *Chromatographia* 54, 65–70, 2001.

109. F. Gosetti, V. Gianotti, S. Polati, M.C. Gennaro, *J. Chromatogr. A* 1090, 107–115, 2005.

110. M. Holčapek, K. Volná, D. Vaněrková, *Dyes Pigment* 75, 156–165, 2007.

111. I. Schönsee, J. Riu, D. Barceló, *Quim. Anal.* 16, S243–S249, 1997.

112. D. Puig, D. Barceló, *J. Chromatogr.* 733, 371–381, 1996.

113. D. Vaněrková, P. Jandera, J. Hrabica, *J. Chromatogr. A*, 1143, 112–120, 2007.

114. P.R.N. Carvalho, C.H. Collins, *Chromatographia* 45, 63–66, 1997.

115. M.M. Dávila-Jiménez, M.P. Elizalde-González, A. Gutiérrez-González, *J. Chromatogr. A* 889, 253–259, 2000.

116. S.C. Rastogi, V.J. Barwick, S.V. Carter, *Chromatographia* 45, 215–228, 1997.

117. A. Gnanamani, M. Bhaskar, R. Ganeshjeevan, R. Chandrasekar, G. Sekaran, S. Sadulla, G. Radhakrishnan, *Process Biochem.* 40, 3497–3504, 2005.

118. M. Karkmaz, E. Puzenat, C. Guillard, J.M. Herrmann, *Appl. Catal. B Environ.* 51, 183–194, 2004.

119. M. Hepel, J. Luo, *Electrochim. Acta* 47, 729–740, 2001.

120. X. Zhao, I.R. Hardin, *Dyes Pigments* 73, 322–325, 2007.

121. X. Zhao, Y. Lu, I. Hardin, *Biotechnol. Lett.* 27, 69–72, 2005.

122. J.S. Chang, C. Chou, Y.C., Lin, P.J., Lin, J.Y., Ho, T.L. Hu, *Water Res.* 35, 2841–2850, 2001.

123. D.P. Oliveira, P.A. Carneiro, C.M. Rech, M.V.B. Zanoni, L.D. Claxton, G.A. Umbuzeiro, *Environ. Sci. Technol.* 40, 6682–6689, 2006.

124. P. Jandera, J. Fischer, V. Staněk, M. Kučerová, P. Zvoníček, *J. Chromatogr. A* 738, 201–213, 1996.

125. V. Gianotti, S. Angioi, F. Gosetti, E. Marengo, M.C. Gennaro, *J. Liq. Chromatogr. Relat. Technol.* 28, 923–937, 2005.

126. L. Wojnárovits, T. Pálfi, E. Takács, S.S. Emmi, *Radiat. Phys. Chem.* 74, 239–246, 2005.

127. A. Plum, A. Rehorek, *J. Chromatogr. A* 1084, 119–113, 2005.

128. J. Yinon, L.D. Betowski, R.D. Voyksner, LC-MS techniques for the analysis of dyes. In: *Application of LC-MS in Environmental Chemistry*, D. Barceló (Ed.), Elsevier, Amsterdam, the Netherlands, 1996, pp. 187–218.

129. A. Preiss, U. Sänger, N. Karfich, K. Levsen, *Anal. Chem.* 72, 992–998, 2000.

130. J.E.B. Mccallum, S.A. Madison, S. Alkan, R.L. Depinto, R.U. Rojas Wahl, *Environ. Sci. Technol.* 34, 5157–5164, 2000.

131. M. Holčapek, P. Jandera, J. Prikryl, *Dyes Pigments* 43, 127–137, 1999.

132. T. Storm, C. Hartig, T. Reemtsma, M. Jekel, *Anal. Chem.* 73, 589–595, 2001.

133. W.F. Smyth, S. McClean, E. O'Kane, I. Banat, G. McMullan, *J. Chromatogr. A* 854, 259–274, 1999.

134. A. Sakalis, D. Vančrková, M. Holčapek, P. Jandera, A. Voulgaropoulos, *Chemosphere* 67, 1940–1948, 2007.

135. A.A. Bergwerff, P. Scherpenisse, *J. Chromatogr. B* 778, 351–359, 2003.

136. M. Chen, D. Moir, F.M. Benoit, C. Kubwabo, *J. Chromatogr. A* 825, 37–44, 1998.

137. F. Gosetti, V. Gianotti, S. Angioi, S. Polati, E. Marengo, M.C. Gennaro, *J. Chromatogr. A* 1054, 379–387, 2004.

138. F. Gosetti, V. Gianotti, M. Ravera, M.C. Gennaro, *J. Environ. Qual.* 34, 2328–2333, 2005.

139. E. Marengo, M.C. Gennaro, V. Gianotti, *Chemometrics Intell. Lab. Syst.* 53, 57–67, 2000.

140. P. Jandera, S. Bunčeková, J. Planeta, *J. Chromatogr. A* 871, 139–152, 2000.

141. R. Loos, M.C. Alonso, D. Barceló, *J. Chromatogr. A* 890, 225–237, 2000.

142. M.C. Alonso, D. Barceló, *Anal. Chim. Acta* 400, 211–231, 1999.

143. E. Pocurull, C. Aguilar, M.C. Alonso, D. Barceló, F. Borull, R.M. Marcé, *J. Chromatogr. A* 854, 187–195, 1999.
144. J.A. Ballantine, D.E. Games, P.S. Slater, *Rapid Commun. Mass Spectrom.* 11, 630–637, 1997.
145. J.A. Ballantine, D.E. Games, P.S. Slater, *Rapid Commun. Mass Spectrom.* 9, 1403–1410, 1995.
146. M. Holčapek, K. Volná, L. Kolárova, K. Lemr, M. Exner, A. Církva, *J. Mass Spectrom.* 39, 43–50, 2004.
147. K. Lemr, M. Holčapek, P. Jandera, A. Lyčka, *Rapid Commun. Mass Spectrom.* 14, 1881–1888, 2000.
148. L. Olssen, U. Schulze, J. Nielsen, *Trends Anal. Chem.* 17, 88–95, 1998.
149. W.M. Draper, *Anal. Chem.* 73, 2745–2760, 2001.
150. M. Bouzige, P. Legeay, V. Pichon, M.-C. Hennion, *J. Chromatogr. A* 846, 317–329, 1999.
151. Y. Coque, E. Tourand, O. Thomas, *Dyes Pigments* 54, 17–23, 2002.
152. A. Plum, G. Braun, A. Rehorek, *J. Chromatogr. A* 987, 395–402, 2003.
153. N. Daneshvar, M. Rabbani, N. Modirshahla, M.A. Behnajady, *J. Hazard. Mater. B* 118, 155–160, 2005.
154. A. Rehorek, K. Urbig, R. Meurer, C. Schäfer, A. Plum, G. Braun, *J. Chromatogr. A* 949, 263–286, 2002.
155. M. Işik, D.T. Sponza, *J. Hazard. Mater. B* 114, 29–39, 2004.
156. R. Brás, A.Gomes, M.I.A. Ferra, H.M. Pinheiro, I.C. Gonçalves, *J. Biotechnol.* 115, 57–66, 2005.
157. M.J. Lopez de Alda, D. Barcelò, *Fresenius J. Anal. Chem.* 371, 437–447, 2001.
158. M.J. Lopez de Alda, D. Barcelò, *J. Chromatogr. A* 938, 145–153, 2001.
159. M.J. Lopez de Alda, D. Barcelò, *J. Chromatogr. A* 1000, 503–526, 2003.
160. P.L. Ferguson, C.R. Iden, A.E. McElroy, B.J. Brownawell, *Anal. Chem.* 73, 3890–3895, 2001.
161. M.J. Lopez de Alda, A. Gil, E. Paz, D. Barcelò, *Analyst* 127, 1279–1282, 2002.
162. M. Petrovic, M. Solè, M.J. Lopez de Alda, D. Barcelò, *Environ. Toxicol. Chem.* 21, 2146–2156, 2002.
163. M.S. Diaz-Cruz, M.J.L. de Alda, D. Barcelò, *Trends Anal. Chem.* 22, 340–351, 2003.
164. M. Petrovic, E. Eljarrat, M.J. Lopez de Alda, D. Barcelò, *Trends Anal. Chem.* 20, 637–648, 2001.
165. S. Masunaga, T. Itazawa, T. Furuichi, D.L. Sunardi, K. Villeneuve, K. Kannan, J.P. Giesy, *J. Nakanishi Environ. Sci.* 7, 101–105, 2000.
166. K. Shimada, K. Mitamura, T. Higashi, *J. Chromatogr. A* 935, 141–172, 2001.
167. C. Baronti, R. Curini, G. D'Ascenzo, A. Di Corcia, A. Gentili, R. Samperi, *Environ. Sci. Technol.* 34, 5059–5066, 2000.
168. M.J. Lopez de Alda, D. Barcelò, *J. Chromatogr. A* 892, 391–406, 2000.
169. A. Laganà, A. Bacaloni, G. Fago, A. Marino, *Rapid Commun. Mass Spectrom.* 14, 401–407, 2000.
170. M. Seifert, G. Brenner-Weib, S. Haindl, M. Nusser, U. Obst, B. Hock, *Fresenius J. Anal. Chem.* 363, 767–770, 1999.
171. J. Rose, H. Holbech, C. Lindholst, U. Nurum, A. Povlsen, B. Korsgaard, P. Bjerregaard, *Comp. Biochem. Physiol. C* 131, 531–539, 2002.
172. M. Petrovic, M.D. Hernando, M. Silvia Diaz-Cruz, D. Barcelò, *J. Chromatogr. A* 1067, 1–14, 2005.
173. D.W. Kolpin, E.T. Furlong, M.T. Mayer, E. M. Thurman, S.D. Zaugg, L.B. Barber, H.D. Baston, *Environ. Sci. Technol.* 36, 1202–1211, 2002.
174. E.M. Golet, A.C. Alder, A. Hartmann, T.A. Ternes, W. Giger, *Anal. Chem.* 73, 3632–3638, 2001.
175. R. Hirsch, T. Ternes, K. Haberer, K.L. Kratz, *Sci. Total Environ.* 225, 109–118, 1999.
176. R. Hirsch, T. Ternes, K. Haberer, A. Mehlich, F. Ballwanz, K.L. Kratz, *J. Chromatogr. A* 815, 213–223, 1998.
177. M.E. Lindsey, M.T. Mayer, E.M. Thurman, *Anal. Chem.* 73, 4640–4646, 2001.
178. C.M. Lock, L.Chen, D.A. Volmer, *Rapid Commun. Mass Spectrom.* 13, 1744–1754, 1999.
189. M.D. Prat, D. Ramil, R. Compano, J.A. Hernandez-Arteseros, M. Granados, *Anal. Chim. Acta* 567, 229–235, 2006.
180. G. Brambilla, C. Civitareale, L. Migliore, *Quim. Anal.* 13, 573–577, 1994.
181. T. Ternes, M. Bonerz, T. Schmidt, *J. Chromatogr. A* 928, 175–285, 2001.
182. W. Ahrer, E. Scherwenk, W. Buchberger, *J. Chromatogr. A* 910, 69–78, 2001.
183. F. Sacher, F.T. Lange, H.J. Braunch, I. Blankenhorn, *J. Chromatogr. A* 938, 199–210, 2001.
184. G. Hamscher, S. Sczesny, H. Hoper, H. Nau, *Anal. Chem.* 74, 1509–1518, 2002.
185. J. Zhu, D.D. Snow, D.A. Cassada, S.J. Monson, R.F. Sapalding, *J. Chromatogr. A* 928, 177–186, 2001.
186. P. Calza, C. Medana, C. Baiocchi, E. Pellizzetti, *Curr. Anal. Chem.* 1, 267–287, 2005.
187. P.L. Ferguson, C.R. Iden, B.J. Brownawell, *Environ. Sci. Technol.* 35, 2428–2435, 2001.
188. H. Maki, H. Okamura, I. Aoyama, M. Fujita, *Environ. Toxicol Chem.* 17, 650–654, 1998.
189. M. Petrovic, A. Rodrigez Fernadez-Alba, F. Borrull, R.M. Marce, E. Gonzalez Mazo, D. Barcelò, *Environ. Toxicol. Chem.* 21, 37–46, 2001.
190. M.A. Blackburn, M.J. Waldok, *Water Res.* 29, 1623–1629, 1995.

191. J.A. Field, R.L. Reed, *Environ. Sci. Technol.* 30, 3544–3550, 1996.
192. C. Crescenzi, A. Di Corcia, R. Samperi, *Anal. Chem.* 67, 1797–1804, 1995.
193. H.Q. Li, F. Jiku, H.F. Schroeder, *J. Chromatogr. A* 889, 155–176, 2000.
194. H.F. Schroeder, K. Fytianos, *Chromatographia* 50, 583–595, 1999.
195. M. Petrovic, D. Barcelò, *Chromatographia* 56, 535–545, 2002.
196. S.D. Scullion, M.B. Clench, M. Cooke, A.E. Ashcroft, *J. Chromatogr. A* 733, 207–216, 1996.
197. P. Jandera, M. Holcapek, G. Teodoridis, *J. Chromatogr. A* 813, 299–311, 1998.
198. M. Castillo, F. Ventura, D. Barcelò, *Waste Manage.* 19, 101–115, 1999.
199. M. Castillo, M.C. Alonso, J. Riu, D. Barcelò, *Environ. Sci. Technol.* 33, 1300–1306, 1999.
200. M. Castillo, E. Martinez, A. Ginebreda, L. Tirapu, D. Barcelò, *Analyst* 125, 1733–1739, 2000.
201. M. Petrovic, D. Barcelò, *Anal. Chem.* 72, 4560–4567, 2000.
202. M.E. Balmer, H.R.Buser, M. D. Muller, T. Poiger, *Environ. Sci. Technol.* 39, 953–962, 2005.
203. T. Poiger, H.R. Buser, M.E. Balmer, M.E. Bergqvist, M.D. Muller, *Chemosphere* 55, 951–963, 2004.
204. L. Gagliardi, G. Cavazzutti, L. Montaranella, D. Tonelli, *J. Chromatogr.* 464, 428–433, 1989.
205. J.E. DiNunzio, R. R Gadde, *J. Chromatogr.* 519, 117–124, 1990.
206. K. Ikeda, S. Suzuki, Y.Watanabe, *J. Chromatogr.* 482, 240–245, 1989.
207. P. Schneider, A. Bringhen, H. Gonzenbach, *Drug Cosmet. Industry* 159, 32–38, 1996.
208. S. Scalia, *J. Chromatogr. A* 870, 199–205, 2000.
209. D.L. Giokas, V.A. Sakkas, T.A. Albanis, *J. Chromatogr. A* 1026, 289–293, 2004.
210. A. Chisvert, M.C. Pascual-Martì, A. Salvador, *J. Chromatogr. A* 921, 207–215, 2001.
211. S.D. Richardson, J. Simmons, G. Rice, *Environ. Sci. Technol.* 36, 198A–205A, 2002.
212. E.T. Urbansky, M.L. Magnuson, *Anal. Chem.* 74, 260A–267A, 2002.
213. S.D. Richardson, *J. Environ. Monit.* 4, 1–9, 2002.
214. H.S. Weinberg, S.W. Krasner, S.D. Richardson, A.D. Thruston, Jr., The Occurrence of Disinfection By-Products (DBPs) of Health Concern in Drinking Water: Results of a Nationwide DBP Occurrence Study, EPA/600/R02/068, U.S. Environmental Protection Agency, National Exposure Research Laboratory, Athens, GA, 2002.
215. S.D. Richardson, *Anal. Chem.* 75, 2831–2857, 2003.
216. C. Zwiener, L. Kronberg, *Fresenius J. Anal. Chem.* 371, 591–597, 2001.
217. S. Monarca, S.D. Richardson, D. Feretti, M. Grottolo, A.D. Thruston Jr., C. Zani, G. Navazio, P. Ragazzo, I. Zerbini, A. Alberti, *Environ. Toxicol. Chem.* 21, 309–318, 2001.
218. R. Loos, D. Barcelò, *J. Chromatogr. A*, 938, 45–55, 2001.
219. C. Zwiener, T. Glauner, F.H. Frimmel, Liquid chromatography/electrospray ionization tandem mass spectrometry and derivatization for the identification of polar carbonyl disinfection by-products. In: *Liquid Chromatography/Mass Spectrometry, MS/MS and Time of Flight MS: Analysis of Emerging Contaminants*, I. Ferrer, E.M. Thurman (Eds.), *ACS Symposium Series* 850, American Chemical Society, Washington, DC, 2003, pp. 356–375.
220. X. Zhang, R.A. Minear, Y. Guo, C.J. Hwang, S.E. Barrett, K. Ikeda, Y. Shimizu, S. Matsui, *Water Res.* 38, 3920–3930, 2004.
221. X. Zhang, R.A. Minear, S.E. Barrett, *Environ. Sci. Technol.* 39, 963–972, 2005.
222. B. Paull, L. Barron, *J. Chromatogr. A* 1064, 1–9, 2004.
223. A.M. Dixon, D.C. Delinsky, J.V. Bruckner, J.W. Fisher, M.G. Bartlett, *J. Liq. Chromatogr. Relat. Technol.* 27, 2343–2355, 2004.
224. W. Gabryelski, F.W. Wu, K.L. Froese, *Anal. Chem.* 75, 2478–2486, 2003.
225. Y.W. Choi, D.A. Reckhow, *Bull. Korean Chem. Soc.* 25, 900–906, 2004.
226. J. Rapala, K. Erkomaa, J. Kukkonen, K. Sivonen, K. Lahti, *Anal. Chim. Acta* 466, 213–231, 2002.
227. S.J. Hoeger, G. Shaw, B.C. Hitzfeld, D.R. Dietrich, *Toxicon* 43, 639–649, 2004.
228. M. Maizels, W.L. Budde, *Anal. Chem.* 76, 1342–1351, 2004.
229. L.F. Zhang, X.F. Ping, Z.G. Yang, *Talanta* 62, 193–200, 2004.
230. J. Meriluoto, K. Karlsson, L. Spoof, *Chromatographia* 59, 291–298, 2004.
231. J. Dahlmann, W.R. Budakowski, B. Luckas, *J. Chromatogr. A* 994, 45–57, 2003.
232. A. Furey, J. Crowley, A.N. Shuilleabhain, A.M. Skulberg, K.J. James, *Toxicon.* 41, 297–303, 2003.
233. A.B. Nasri, N. Bouaicha, J. Fastner, *Arch. Environ. Contam. Toxicol.* 46, 197–202, 2004.
234. J. Faster, G.A. Codd, J.S. Metcalf, P. Woitke, C. Wiedner, H. Utkilen, *Anal. Bioanal. Chem.* 373, 437–444, 2002.
235. A. Zeck, M.G. Weller, R. Niessner, *Anal. Chem.* 73, 5509–5517, 2001.
236. J. Burns, Cyanobacteria and Their Toxins in Florida Surface Waters and Drinking Waters Supplies, Report to the Florida Department of Health, Tallahassee, FL, 2003.

237. A. Sanusi, M. Millet, H. Wortham, P. Mirabel, *Analusis* 25, 302–308, 1997.
238. M. Dimashki, S. Harrad, R.M. Harrison, *Atmos. Environ.* 34, 2459–2469, 2000.
239. K. Muller, *Chemosphere* 35, 2093–2106, 1997.
240. R. Belloli, B. Barletta, E. Bolzacchini, S. Meinardi, M. Orlandi, B. Rindone, *J. Chromatogr. A* 846, 277–281, 1999.
241. W. Schussler, L. Nietshke, *Chemosphere* 42, 277–283, 2001.

.

19 HPLC in Food Analysis

Lanfranco S. Conte, Sabrina Moret, and Giorgia Purcaro

CONTENTS

19.1 INTRODUCTION

Among the several purposes of food analysis, the two main objectives are to verify authenticity and quality of food. The latter may be applied both in the evaluation of nutritional value (in this case, quantitative data are mandatory) and peculiar characteristics (age, applied technology, origin, etc.). In any case, useful information can be obtained as thorough knowledge of food composition is available.

Analytical chemistry applied to foods widely uses separation techniques in order to characterize foods on the basis of their major and minor constituent compounds. The minor compounds often highlight important differences rather than the major compounds. Chromatographic techniques have been widely applied; however, when only gas chromatography (GC) was available, the application of this technique was severely limited by the thermal stability of the analytes. Derivatization reactions are able to somewhat protect organic molecules from degradation, but generally these procedures are rather time consuming and, sometimes, artifacts could be formed.

High-performance liquid chromatography (HPLC) allowed resolving these problems, and the possibility to apply almost every kind of very sensitive, and sometimes selective, detector greatly improved its field of use.

Quantitative data can be obtained by means of the use of an internal or external standard, whose exact concentration can be checked by means of spectrophotometric detector, which measures specific absorbance or molar absorbance. Spiked authentic samples can be used to measure the limit of detection (LOD) and the limit of quantification (LOQ).

Nowadays, a large number of applications are available; therefore, in this chapter, a selection of more important and interesting ones will be presented. Particular attention will be paid to the official, reference or routine methods, as well as to selected experimental approaches. The number of applications in food analysis is increasing at a rapid pace and it seems rather difficult to be exhaustive in this field.

HPLC runs involve relatively large volumes of a solvent; however, by the use of dedicated interfaces, this is no more a problem, and bands eluted from HPLC can be transferred to a GC or to a mass spectrometry (MS). The interface is the hearth of the HPLC–MS hyphenation, and it must be able to eliminate solvent and to perform the ionization of the analytes in order to separate them on the basis of the ratio mass/charge.

Several interfaces had been developed, but at present, mainly electrospray ionization (ESI) and atmospheric pressure chemical ionization (APCI) are widely used. In the ESI interface, HPLC eluted band is sprayed and ionization is performed out of the vacuum area of the spectrometer, at atmosphere pressure, while in the APCI interface, the eluted form HPLC column is driven through an heated capillary tube to a heated probe, where solvent is evaporated as a result of an inert gas flow (usually nitrogen is used for this purpose). The HPLC–MS hyphenation, greatly improved the HPLC performances, as many compounds that could be degraded by GC run can reach the MS without any damage to their chemical structure. Phenolics, proteins, peptones, peptides can be characterized very efficiently in this way.

A number of criteria could be applied to organize this chapter, depending on the point of view by which foods are considered. In this chapter, application of HPLC to food analysis will be described considering homogeneous classes of food components: lipids, carbohydrates and related substances, proteins, peptides, amino acids, biogenic amines, phenolics, vitamins, and some selected contaminants.

The criteria are based on the fact that the HPLC determination changes little when different foods are analyzed, while sample preparation is usually the main problem, as different interfering substances can occur, depending on the chemical composition of different foods.

19.2 LIPIDS

Lipids are important components of the diet: fatty acids are the higher energetic source as they ensure 9 kcal/g. Furthermore, some peculiar fatty acids themselves and several components of the unsaponifiable fraction are biologically active molecules, as they can act as vitamins (tocopherols—vitamin E), provitamins (carotenes—vitamin A, cholecalciferol—vitamin D), vitamin-like (essential fatty acids), and hormones or hormone precursors (sterols—steroidal hormones).

Perhaps, depending on these important roles and on their not being homogeneous to other polar food compounds, so that it is easy to separate them, lipids have been thoroughly studied over a number of years. In 1775, Chevreul was able to isolate and characterize cholesterol and a few years later, the presence of glycerol in fats was demonstrated, and this paved the way for us to understand that fats are made (mainly) of triacylglycerols (TAGs).

GC has been widely applied to the study and characterization of food lipids, and from a historical point of view, it is worthwhile to remember that one of the earliest methods developed by James and Martin was the evaluation of fatty acid composition of butter. Nowadays, most of the instrumental analytical chemistry of lipids is performed by means of GC, often coupled to MS; this was helped by the improvement of the diffusion of capillary columns that started in the 1970s, depending on the commercial availability of silica-fused columns and a wide range of stationary phases. Capillary GC allows faster analysis and does not involve solvent consumption; however, few lipid molecules are suitable to be directly injected in a gas chromatograph, because of the high temperature of the injector and oven, so that they need long derivatization reactions. Furthermore, some molecules have significant molecular weights (e.g., TAGs) and this has greatly limited the possibility to get a separation that can give relevant information e.g., concerning the unsaturation level.

HPLC has been applied to lipid analysis mainly in consideration of the necessity to avoid high temperatures, so at the very beginning, its applications dealt with thermally unstable molecules (e.g., tocopherols, phenolics, oxidation products) and often it was used as an ancillary technique, as a preparative step prior to MS analysis. The limits were in the high volume of the HPLC band that strongly limited the possibility to transfer it to a GC or to a MS. Only in the last 20 years or somewhat less, this kind of hyphenation has become commercially available.

Application to high-molecular-weight compounds is dedicated mainly to TAGs and some related substances. In 1995, Ruiz-Gutierrez and Barron [1], in a review on the analysis methods of TAGs, stated that HPLC offers significant advantages over GC, even though in more recent years some improved applications of capillary GC have been obtained and published.

A limit in data quality obtained by usual reversed phase (RP)-HPLC analysis of TAGs depends on their different degree of unsaturation: this fact determines different spectroscopy characteristics, e.g., in the ultraviolet (UV) absorption, thus, UV detector that is widely used in HPLC is not suitable for this kind of analysis. Within a limited difference in molecular weight (MW), in fact, tristearine (SSS) has a molecular weight (MW) of approximately 891, trioleine (OOO), approximately MW 885, trilinoleine (LLL), approximately MW 879, and trilinolenine (LnLnLn) approximately MW 873, UV detector response would give important differences depending on the high absorption in UV region by polyunsaturated molecules. Despite the existence of refraction index differences between saturated and unsaturated molecules, they do not affect the accurate analysis of TAGs in HPLC, thus, refractive index (RI) detector was used for TAGs analysis, in official methods, too. The use of RI detector, however does not permit to change the composition of the elution mixture of the solvent, therefore only isocratic elution can be performed, consequently the separative possibilities are limited.

Another possibility is to use the evaporative light scattering (ELS) detector, which can be used independently of the composition or variations of the elution solvent mixture (presence of salt solutions is the only limit), however, it is not yet so diffused, even if its earlier applications date back to early 1980s.

HPLC analysis of TAGs was applied to olive oils and a limit for LLL was established at 0.5% of the total TAGs and enclosed in the European Commission (EC) Regulation on olive oils, as well as in the related international norms (International Olive Oil Trade Norm [2], Codex Alimentarius Standard [3]). The method performed the separation by RP-HPLC on a C18 Lichrosorb or Lichrosphere column, 25×0.46 cm, 5 μm of particle size, isocratic elution with acetone/acetonitrile (50/50 v/v) and RI detection. Figure 19.1 reproduced the HPLC traces annexed to the official method [4]: chromatogram A refers to 100% soybean oil, B to a mixture 50/50 soybean and olive, C 100% olive oil.

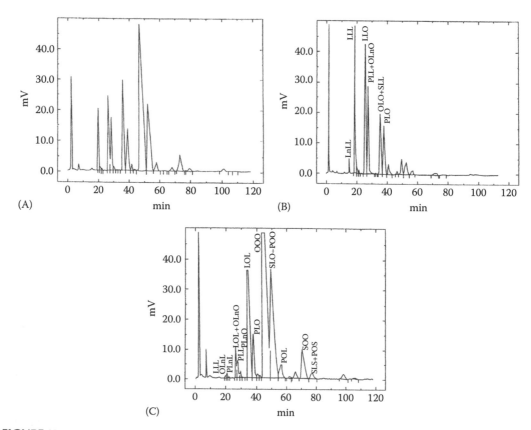

FIGURE 19.1 HPLC traces of TAG of (A) 100% soybean oil, (B) 50% soybean/50% olive oil, and (C) 100% olive oil. (From International Olive Oil Council, method COI/T.20/Doc. no. 20 /Rev. 1, 2001. With permission.)

The elution order depends on the increasing partition number (PN) defined as

$$PN = CN - 2DB$$

where
  CN is the carbon number of the molecule
  DB is the number of double bonds

The separation of TAGs with the same PN is not so efficient in the adopted chromatographic conditions, so it has been observed that some peak overlapping can take place, and that the phenomena affects the precision of LLL peak area.

In order to solve this problem, normative handling olive oils uses HPLC to obtain a percentage data of all TAG with PN = 42 (PN C18:2 = 18 − 4 = 14 LLL = 14 × 3 = 42). The term ECN42 was used for the first time by International Union of Pure and Applied Chemistry (IUPAC) in 1987 and has been adopted in further developed methods. The EC official method [5], starting from data of a simplified fatty acid composition, calculates the possible theoretical TAG with ECN 42, taking into account the rule of fatty acid distribution in TAG molecules, and then calculates the difference (as an absolute value) between the two data. The scientific basis to establish such a parameter is that TAG biosynthesis follows different pathways depending on the tissue where it is realized: fruit pulp (mother tissue) or seed (mother and father tissue). The approach was very successful and was the starting point for Arturo cert to study other significant ratios involving ECN42 and ECN44, which rises to the so called "global method" proposed within the framework of the International Olive Oil Council (IOOC) [6] to assess admixtures with seed oils with high oleic content. In this method, however, better resolution of some peaks (e.g., LOO/PLnP/PLO, SOL/POO) was obtained by using propionitrile as the eluent instead of the acetone/acetonitrile mixture, as proposed by Fiebig [7] and Moreda [8]. In 2006, Ollivier published a study in which a comparison between the two chromatograms was reported [9] (Figure 19.2).

At present (2007), IOOC is carrying out some ring tests in order to verify the accordance of the two methods of elution. Separation on the basis of unsaturation can be improved by modifying the stationary phase by Ag+ ions, as the dimension of Ag+ ion is suitable to interfere with π bonds. Christie [10–12] adopted this technique to carry out structural analysis of TAGs. TAGs were hydrolyzed by Grignard reaction, then the stereoisomers 1,2 and 1,3-diacylglyreols were transformed into diasteroisomers by derivatization with a chiral compound, and were separated on normal-phase (NP)-HPLC.

Recently, Mondello et al. [13] used silver ion HPLC to develop a comprehensive LC (LC×LC) system. The results obtained on rice oil sample led to the separation of a number of isomers that cannot be separated by the usual monodimensional HPLC. The use of APCI-MS as a detection system provided a more affordable method for the identification of different isomers.

When lipid oxidation takes place, among several possible radical reactions, termination reactions also can take place. If a termination reaction involves two oxidized TAGs, the possible product is a dimer TAG. The determination of polymerized TAGs was proposed by several authors [14–23] with different aims: to highlight the quality of oil from an oxidative point of view or to check for illegal heat treatment. A peculiar application deals with frying oils quality assessment. The method uses two or three gel permeation columns and a RI detector, and recently it has been adopted within the "German Official Methods of Analysis" [24,25] for fats and oils. Figure 19.3 reproduced the chromatograms obtained for an extra virgin olive oil and an oil with triglyceride polymers (TGPs).

Oxidation, as is well known, leads not only to the formation of heavy compounds but also of volatile compounds that are responsible for off-flavors. Usually they are evaluated by capillary GC, through several sampling techniques are available (purge and trap, head space solid phase microextraction (SPME)). Rovellini et al. [26] recently proposed the application of HPLC for identifying

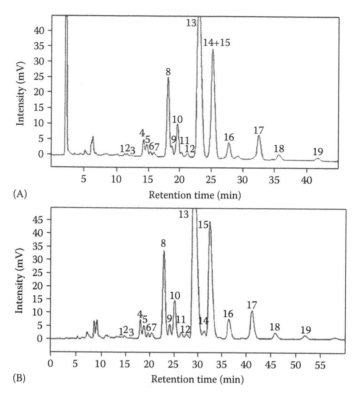

(A)

(B)

FIGURE 19.2 HPLC separation of TAG of a virgin olive oil sample. (A) Acetone /ACN elution, and (B) propionitrile elution. Peak identification: 1, LLL; 2, OLLn + PoLL; 3, PLLn; 4, OLL; 5, OOLn; 6, PLL; 7, POLn; 8, LOO + PLnP; 9, PoOO; 10, PLO + SLL; 11, PoOP; 12, PLP; 13, OOO; 14, SOL; 15, POO; 16, POP; 17, SOO; 18, POS; 19, POA. (From Ollivier, D. et al., *Food Chem.*, 97, 382, 2006. Copyright 2006. With permission from Elsevier.)

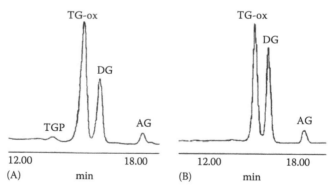

(A) min (B) min

FIGURE 19.3 HPLC (SEC) traces of an extra virgin olive oil: (A) sample containing TGP and (B) without TGP.

volatile compounds. Carbonyl compounds were derivatized as phenylhydrazones and analyzed by HPLC. Rovellini [27] also applied HPLC–MS for the evaluation of malonyldialdheyde, a well-known compound arising from oxidation, used mainly in the evaluation of meat lipid oxidation. Meat lipids are characterized by the presence of phospholipids, and this class of compounds has been analyzed by HPLC in meat and in dairy products [28–30], while Pacetti et al. [31,32] used HPLC–MS to elucidate the structure of phospholipds in egg yolk.

Not only fatty acids but every unsaturated lipid molecule can undergo oxidation. So sterol oxidation products have also been studied; in this case, HPLC is used both as a preparative step and for analytical purposes [33,34]. HPLC as a preparative step was proposed by several researchers to improve the speed of analysis in the case of sterols, stigmastadienes, and waxes [35,36]. Stigmastadiene, which is used as a marker in the refining process applied to vegetable oils, is determined by capillary GC; however, the International Organization for Standardization (ISO) method (ISO 15788-2:2003) [37] uses HPLC as a rapid screening technique.

Table 19.1 reports a summary of application of HPLC dealing with analysis of different classes of lipids in foods.

19.3 CARBOHYDRATES AND RELATED SUBSTANCES

19.3.1 ANALYSIS OF CARBOHYDRATES

The analysis of carbohydrates in foods is carried out with different aims: qualitative and quantitative composition of the sugar fraction depends on their natural source. For example, vegetables, fruits and derivates, and honeys can be characterized by means of sugars analysis. Furthermore, the concentration of sugars or their ratio to other compounds (e.g., organic acids) can be used as a measure of the ripening stage in fruits or can help in evaluating the shelf life of honeys, as well as their storage conditions; in the latter case, some sugar derivative molecules can highlight a possible damage caused by technology (heating).

Some sugar-related products, e.g., sugar alcohols, are used as sweeteners, and if used, they must be declared on the label; furthermore, Food and Drug Administration (FDA) requires sugar analysis when their concentration in foods is more than 1%. Another issue dealing with sugar analytical evaluation can depend on the need to highlight possible food adulterations: improving wine alcoholic content by adding extraneous fermentable sugars or adulteration of honeys by using hydrolyzed starch or dextrin.

Food industry needs to measure the sugar content e.g., when sugar is the main product obtained from raw vegetables (e.g., sugarcane, sugar rape and other sugar sources). Sugar analysis is important in wine making, and routine determination of sugar in grapes, musts and wines is undertaken to check for the potential amount of alcohols obtained by complete fermentation and to check for amount of residual sugars in sweet wines.

Polyols too are important in the wine industry, as glycerol is related to the sensory characteristic of smoothness, while inositol is used as a purity parameter for concentrated rectified musts [47]. Analysis of sugars is carried out in cider production [48].

The main compounds in honey are glucose and fructose, and their ratio can be used as a check for authenticity regarding their botanical origin [49], as shown for some honeys in Table 19.2.

In the presence of amino groups, reducing sugars can undergo a number of reactions named Maillard reaction, which, via Schiff bases and Amadori compounds, lead to the formation of different compounds, such as furosine, hydroxymethylfurfural (HMF), or "isomerized" sugars as happens in heated milk, where lactulose (made from fructose and galactose) rises from lactose (glucose and galactose). The compounds responsible for off-flavors development, such as Strecker aldehydes (isomethyl butanals and isovaleraldehydes) can arise from Maillard reactions, as well. Furosine is another compound that originates from amino acid glycosilation and it can be used as an aging marker, as well as a marker of heat treatments that are not in agreement with good practices of food production (e.g., the "mozzarella" cheese).

In the field of apiculture, legal limits were set for HMF content in honeys [49], as a tool to check both for applied technology (pasteurization) and for aging, while furosine is not an official parameter at present, but several researches were carried out on its application to royal jelly quality control [50,51].

Inulin (fructoligosaccharide) determination can be carried out by HPLC, too, and applications were developed for meat products.

TABLE 19.1

HPLC Applications for Lipid Analysis in Foods

Food/Topic	Sample Pre-Treatment	Column	Mobile Phase	Detector	References
COPs	SPE (silica)	Aquasil C18 (25×0.46cm, 5μm)	ACN/MeOH	APCI-MS	[34]
Beef meat/PHL	Folch extraction, SPE (amino phase)	Lichrosorb 60 (25×0.46cm, 10μm) Spherisorb Si (25×0.46cm, 10μm)	A: MeOH; B: NH$_4$OH	ELS	[28]
Coffee/TAGs	Solvent extraction, Ag+ TLC	Technosphere ODS (25×0.45cm, 5μm)	A: ACN; B: CH$_2$Cl$_2$/C$_2$H$_4$Cl$_2$	RI+ELS	[38]
Dairy products/PHL	Folch and Rose Gottlieb extraction (modified) + SPE (LC-SI, 1 g/6 mL)	Zorbax Rx-SIL (250×0.46cm, 5μm)	A: CHCl$_3$/MeOH/NH$_4$OH (80/19.5/0.5 v/v/v); B: CHCl$_3$/MeOH/NH$_4$OH/H$_2$O (60/34/0.5/5.5 v/v/v/v)	ELS	[30]
Egg yolk/COPs	Blight and Dyer extraction	Spherisorb S5 CN (25×0.46cm)	Hex/EtOH	ELS	[33]
Fish/PHL, TAGs	LLE Blight and Dyer method	Lichrosphere 100 Diol (25×0.4cm, 5μm)	PHL: A: Hex/IP/HAc (82/17/1 v/v/v); B: IP/H$_2$O/HAc (85/14/1 v/v/v)+TEA TAGs: A: Hex/HAc (99/1); B: Hex/IP/ HAc (84/15/1 v/v/v)	ELS	[39]
Oils/Stigmastadiens	LC (silica column)	RP-C18 (25×0.46cm, 5μm)	ACN/MTBE	UV (235nm)	[37]
Oils/TGP	SPE (Sep Pack Silica) (DGF CIII 3d only)	Two columns Phenogel (PS-DVB, 30×0.78cm, 5μm, 10Å)	THF	RI	[24,25]
Oils/TAGs	SPE Silica, Grignard reaction, naphtyl urethane derivatives, SPE C18	2 Silica column Hypersil (25×0.46cm, 5μm)	Hex/0.35% IP/2% H$_2$O	UV (280nm)	[40]
Virgin olive oil/ Wax and stigmastadienes	None	Supelcosil LC-Si (15×0.46cm, 5μm)	A: Hex; B: Et$_2$O	UV (217nm)	[36]
Virgin olive oil/ Volatile carbonyls	Derivatization as phenylhydrazones	C18 Spherisorb ODS-2 (25×0.40cm)	A: H$_2$O; B: ACN; C: MeOH	DAD (360nm)	[26]

Sample/analyte	Sample preparation	Column	Mobile phase	Detection	Reference
Virgin olive oil/TGP	IUPAC 2.507	Lichrosphere Si 60 (25×0.4cm, 10µm)	Hex/Et$_2$O (70/30 v/v)	DAD (230nm)	[41]
Virgin olive oil/ Triterphenic acids	SPE (SAX 50mm particle size, 70Å)	Spherisorb ODS2 (25×0.4cm, 3µm)	A: H$_2$O (0.2% H$_3$PO$_4$); B: MeOH; C: ACN	UV (210nm)	[42]
Virgin olive oil/ TAGs	None	RP-18 Supersphere 100 (25×0.4cm, 4µm)	A: (CH$_3$)$_2$CO/ACN B: Propionitrile	RI	[9]
Vegetable oils/ PGP	IUPAC 2.507	NARP Restek Ultra C18 (25×0.46cm, 5µm)	(CH$_3$)$_2$CO/ACN	UV (210nm)	[43]
Rice oil/TAGs	None	First dimension: Restek Ultra C18 (25×0.46cm, 5µm) Second dimension: chromspher 5 lipids (25×0.46cm, 5µm)	First dimension: (CH$_3$)$_2$CO/ACN Second dimension: Hex/ACN	APCI-MS	[44]
Rice oil/TAGs	None	Ag+ Hot Sep micro-bore Nucleosil 100-SSA (15×0.1cm, 5µm)	A: ACN; B: Hex	APCI-MS	[13]
Vegetable oils/TAGs	Filtration	Kromasil 100 C18 (25×0.46mm, 5µm)	A: (CH$_3$)$_2$CO; B: ACN	ELS	[45]
Vegetable oils/TAGs	None	First dimension: Hot Sep micro bore Nucleosil 100-5 SA (15×0.1cm, 5µm) Second dimension: Chromolith Performance RP18 (10×0.46cm)	First dimension: A: 0.7% Hex/ACN; B: 0.9% Hex/ACN Second dimension: A: IP; B: ACN	ELS, APCI-MS	[46]
Vegetable oils and chips/MDA	Reaction with DNPH	RP 18 Spherisorb ODS 2 (25×0.4cm, 5µm)	A: H$_2$O; B: MeOH; C: ACN	DAD (360 and 410nm) APCI-MS	[27]
Wheat flour/PHL	Liquid extraction after hydrolysis, SPE (silica)	Diol modified silica Lichrosphere 100 Diol (10×0.4cm, 5µm)	A: Hex/IP/HAc/TEA; B: IP/H$_2$O/HAc/TEA	ELS	[29]

TABLE 19.2
Sugar Composition of Different Honeys

Analytical Parameter	Citrus	Chestnut	Eucalyptus	Dandelion	Lime Tree	Thymus
HMF (mg/kg)	1.2–9.6	0.4–3.6	0.8–8.6	0–7.2	1.8–6.6	3.8–10.6
Fructose (%)	35.8–41.8	39.8–44.0	35.5–41.7	36.9–40.7	34.3–42.3	40.1–45.1
Glucose (%)	30.9–33.5	25.0–27.8	31.9–33.9	36.6–41.6	28.1–32.5	28.8–31.6
Σ Fruct. + gluc. (%)	67.4–74.6	65.5–70.9	68.2–74.8	74.2–81.6	63.5–73.9	69.7–75.9
Sucrose (%)	0–2.0	0–0.2	0.5–2.1	0–2.7	0–0.2	0–0.2
Maltose (%)	0.6–2.0	0.3–1.3	0.4–1.4	0.4–1.2	0.6–2.0	0.9–1.9
Isomaltose (%)	0.3–0.9	1.3–2.9	0.3–0.9	0–0.5	0.7–1.5	0.4–1
Trisaccarydes	Erlose	Melezitose	Erlose		Erlose + melezitose	
Gluc/water ratio	1.74–2.00	1.38–1.64	1.90–2.18	2.15–2.39	1.63–1.95	1.68–2.04

19.3.2 INSTRUMENTS AND METHODS

HPLC is known to be as much sensitive as an applied detection system. UV and spectrofluorometric detection permits very high sensitivity because of their specificity and good signal-to-noise ratio. However, it is not possible to detect sugars with UV or fluorescence without derivatization.

For a number of years, the only possibility had been to use RI detectors, which lead to severe limitations of sensitiveness and of separative performance, the latter because of the unsuitability of gradient elution. The first issue, usually, is not a very important problem because of high concentration of sugars occurring in foods.

A classic HPLC analysis of sugars is made by means of an isocratic elution with RI detection; hence the only "degree of freedom" is the choice of the column on which the level of separation depends. More recent issues that do not allow the use of isocratic elution are the use of an ELS detector or an electrochemical detector (ED).

As foods form complex matrices, usually a preparation step is required in order to remove possible interfering substances e.g., organic acids, phenolics, amino acids, if a simultaneous evaluation is not required.

Carbohydrates are usually separated by means of RP chromatography with bonded phases: amino groups bonded to silica or polymeric core are the most common choices [52–57]. In 2000, Wey and Ding [58] described a dynamic modification technique performed by applying a ethylenediamine layer on the silica core that had the advantage of making the stationary phase more stable. Ion exchange chromatography (IEC) was also applied in carbohydrates analysis by using cation or anion exchange resins [54,59–64]. Ion exclusion chromatography was applied in sugar separation [65].

Historically, RI detector was widely applied to carbohydrates analysis [52,55,59,66–71]. Other approaches involve the use of ED permitting a high sensitivity and selectivity even in the presence of complex matrices; further improvement in this method is the use of pulsed amperometric detector [69,70,72–78].

Although the concentration of sugars in foods is not usually a critical point, HPLC provides a suitable sensitiveness (LOD) even at low concentrations, as reported in Table 19.3 [79]. This can be useful for the evaluation of "minor" sugars, whose presence can be related to natural sources of food, as happens in honey.

Table 19.4 reports some applications concerning HPLC analysis of carbohydrates in food samples.

TABLE 19.3
Sensitivity of HPLC Analysis of Main Food Sugars and Related Substances

Analyte	LOD	Analyte	LOD
Glucose	1.26 µg/L	Cellobiose	1.5 µM
Fructose	0.14 mg/L	Galactose	0.14 mg/L
Sucrose	0.53 mg/L	Xylose	0.13 mg/L
Maltose	0.38 mg/L	Ribose	0.18 mg/L
Lactose	0.15 mg/L	Ramnose	0.36 mg/L
Raffinose	0.47 mg/L	Glycerol	4 mg/L
Arabinose	0.12 mg/L	Sorbitol	0.3 mg/L
Fucose	0.21 mg/L	Mannitol	0.13 mg/L
Mannose	0.23 mg/L	Maltitol, isomaltitol, lactitol, xylitol	0.3–1.1 mg/L

19.4 PROTEINS AND PEPTIDES

Food protein and peptides are a very heterogeneous group of compounds containing amino acids linked by an amide bond. The separation line between large peptides and small proteins is not well defined but usually it is assumed that peptides contain less than 100 residues (MW <10,000 Da) [104].

Proteins are important food components mainly due to their nutritional and functional value. Dietary proteins provide amino acids and nitrogen necessary for organisms. They also play a major role in determining the sensory and textural characteristics of food products. The functional properties are related to their ability to form viscoelastic networks, bind water, entrap flavors, emulsify fat and oil, and form stable foams [105].

On the other hand, peptides produced by protein hydrolysis have important sensory, biological, technological, and physiological properties. They are responsible for the sweet and bitter taste in foodstuffs and contribute to food flavors. Furthermore, some peptides (released during the gastrointestinal digestion, enzymatic hydrolysis, or food processing) exhibit antioxidant, toxicological, or antimicrobial activity. Milk proteins are the main source of biologically active peptides: opioid (casomorphins), antithrombotic, antihypertensive, immunomodulators, etc. [104].

19.4.1 HPLC ANALYSIS

HPLC analysis of food proteins and peptides can be performed for different purposes: to characterize food, to detect frauds, to assess the severity of thermal treatments, etc. To detect and/or quantify protein and peptide components in foods, a number of different analytical techniques (chromatography, electrophoresis, mass spectrometry, immunology) have been used, either alone or in combination. The main advantages of HPLC analysis lie in its high resolution power and versatility. In a single chromatographic run, it is possible to obtain both the composition and the amount of the protein fraction and analysis can be automated.

19.4.1.1 Sample Preparation

A sample preparation step aimed at isolating and concentrating the analytes from the matrix is often needed prior to HPLC. Solid samples are usually homogenized with a suitable solvent. Water, acid solutions, saline solutions, or buffers are usually used for peptide extraction from food, but hydrophobic peptides may require mixtures of chloroform or methylene chloride and methanol.

TABLE 19.4

HPLC Applications for Carbohydrate Analysis in Foods

Food/Topic	Sample Pre-Treatment	Column	Mobile Phase	Detector	Reference
Bakery	Extraction with H_2O, purification by Carrez reagent, centrifugation	Spherisorb S5 ODS2 (25 × 0.40 cm)	H_2O/ACN (95/5)	UV (284 nm)	[80]
Beet	Dilution with 0.2 M ammonium buffer, pH = 8.9, filtration through hydrophilic PVDF membrane	Waters 250 Ultrahydrogel (range 1–80 kDa)	0.2 M ammonium buffer (NH_4Cl/NH_3) (pH 8.9)/ H_2O (20/80 v/v)	UV/Vis, DAD (210–550 nm)	[81]
Beet	Dilution with H_2O	Zorbax ODS (25 × 0.4 cm)	H_2O	ELS	[82]
Beverages and foods	On-line clean up by dialysis	NH_2-Zorbax (25 × 0.46 cm) ION-300 (30 × 0.78 cm)	ACN/H_2O (75/25 v/v) 0.085N H_2SO_4	RI RI	[54] [83]
Chestnut	Sample freezing, sugar extraction with hydroalcoholic solution 80% (v/v), centrifugation	Amino (Kromasil 100 NH_2, 25 × 0.46 cm, 5 μm)	ACN/H_2O (80/20 v/v)	ELS	[83]
Dairy	Homogenization with H_2O, filtration through 0.45 μm filter,	Waters Sugar Pak (30 × 0.65 cm)	H_2O	RI	[70]
Dairy	Dilution with 0.3 N oxalic acid solution and deproteinization with TCA	Dionex PA1 ion exchange (25 × 0.4 cm)	1 M NaOH/H_2O/NaOAc	PAD	[84]
Dairy		Extrasyl ODS-2 S5 (25 × 0.40 cm, 5 μm)	NaOAc buffer 0.08 M	UV, DAD	
Extrusion cooked foods	Suspension in 0.02% $CaCl_2$, enzymatic hydrolysis, filtration	Spherisorb ODS2 column, (25 × 0.49 cm 5 μm)	Linear gradient: H_2O/MeOH	DAD (190–600 nm)	[85]
Fruit juices	SPE on SAX cartridge (3 mL/500 mg) elution with MeOH and NaOH 1 M	Aminex HPX 87H hydrogen form (30 × 0.78 cm)	H_3PO_4/ACN/MeOH	UV (210 nm), RI	[86]
Fruits	Extraction with H_2O	Polymer CHCA (30 × 0.65 cm) for sucrose, glucose, sorbitol, fructose	H_2O	RI	[87]
Cooked ham	Mincing, extraction with HCl 8 M at 110°C, filtration through 0.22 μm filter, purification on Sep Pack C18	C8 Furosine dedicated column (25 × 0.46 cm)	ACN/H_2O (80/20 v/v) A: HAc (0.4%) in H_2O; B: 0.3% KCl in solvent A	UV (280 nm)	[88]
HMF	Dilution with H_2O, filtration through 0.45 μm filter	Lichrosphere RP-18 (5 × 0.45 μm, 12.5 × 0.4 cm)	H_2O/HAc/MeOH	DAD (285 nm)	[89]
HMF	Dilution with H_2O, centrifugation at 14000 rpm,	Lichrocart purosphere cart (5.5 × 0.2 cm, 3 μm)	MeOH/H_2O	UV (280 nm)	[90]
HMF	Dilution with H_2O, deproteinization with TCA, centrifugation, filtration through 0.45 μm filter	C18 RP (Equisil ODS 5 μm, 25 × 0.46 cm)	NaOAc/MeOH (80/20 v/v) + HAc (pH 3.6)	UV (286 nm)	[91]
HMF alcoholic beverages	Sample dilution with ACN, filtration through 0.45 μm filter	C18 column (25 × 0.46 cm, 5 μm)	ACN/HAc/H_3PO_4	UV (280 nm)	[92]

Sample	Sample preparation	Column	Mobile phase	Detection	Reference
HMF Honey	Dilution with H$_2$O, purification by Carrez reactive, filtration, derivatization as DNPH-derivate	Supelcosil LP C18 (25×0.46cm, 5μm)	ACN/H$_2$O	UV (385 nm)	[93]
Honey	Dextrins separation by HCl, filtration	Spherisorb Amino (25×0.4cm, 3μm)	ACN/H$_2$O	RI	[94]
Honey	Solubilization with H$_2$O, purification by Dowex 50W (H+ form) and Amberlite IRA 68 (acetate form)	Tracer Kromasil (15×0.45cm, 5μm)	H$_2$O	RI	[95]
Honey	Solubilization with H$_2$O, treatment with activated carbon and celite	Bio Rad HPX-87C (30×0.78cm)	H$_2$O	PAD	[72]
Honey		Dionex Carbo Pac (25×0.4cm)	0.15M NaOH	PAD	[96]
Honey		Carbo Pac PA1 (25×0.4cm)	A: 1 M NaOH; B: 1 M NaOAc; C: H$_2$O	UV (285 nm)	[96]
Honey		RP C18 (25×0.46cm, 5μm)	H$_2$O/MeOH (40/10 v/v)	RI	
Honey		Aminopropylsylane (25×0.46cm, 5μm)	ACN/H$_2$O (80/20 v/v)		
Jams	Homogenization with H$_2$O	HMF:Nova-Pak1C18 column (15×0.39cm); Furosine: Spherisorb ODS2 (25×0.46cm)	Linear gradient: MeOH/H$_2$O; 5mM sodium heptane sulfonate with 20% ACN, 0.2% HCOOH	UV (283 nm); UV (280 nm)	[97]
Kiwi juice	Juice filtration through 0.45 μm filter	Aminex HPX 87C (30×0.78cm)	H$_2$O	RI	[98]
Milk	Acid hydrolysis, SPE (C18)	C8 (25×0.46cm) Altech furosine dedicated	A: H$_2$O/0.4% HAc B: H$_2$O/0.4% HAc+0.3% KCl	UV (280 nm)	[99]
Pasta	Hydrolysis with HCl 8 M at 110°C	C8 Furosine dedicated column, (25×0.46cm)	A: HAc (0.4%) in H$_2$O; B: 0.3% KCl in solvent A	UV (280 nm)	[100]
Polyalcohols	Sonication, extraction with H$_2$O, (chicory, toffees)/ purification by Carrez reactive (biscuits, roasted malt, chocolate, cream)	Carbopac PA 100–quaternary amine functionalized (25×0.40cm) or Carbopac MA1 (25×0.4cm, 8.5μm)	NaOH solution	PAD	[75]
Royal jelly	Purification with Carrez reactive	Pinnacle II Amino (25×0.32cm, 3μm)	ACN/H$_2$O	RI	[101]
Royal jelly	SPE (Sep Pak)	Alltech furosine- dedicated (25×0.46cm)		DAD (280 nm)	[50]
Royal jelly	SPE (Sep Pak)	Alltech Furosine dedicated column		DAD (280 nm)	[51]
Sourdough	Homogenization with H$_2$O/HClO 1 M, centrifugation, + KOH 2M, filtration through 0.45 μm filter	Ion Exclusion ORH-801 (30×0.65cm); C18 (15×0.45cm) for sucrose maltose separation only	0.001 N H$_2$SO$_4$	RI	[102]
Sugar alcohols	Homogenization with H$_2$O, centrifugation, vacuum filtration, purification by Carrez reactive, filtration; Homogenization with 30% EtOH, centrifugation, derivatization with PNBC at 50°C	Inertsil Ph 3 (25×0.45cm)	ACN/H$_2$O (67/33 v/v)	UV	[103]

Precipitation of proteins and large peptides or selective peptide and protein fractionation can be accomplished with a solution containing organic solvents (ethanol, methanol, acetone), acids (trichloroacetic, sulfosalicylic, phosphotungstic), by means of salting-out precipitation (ammonium sulfate) or by pH adjustment to the isoelectric point. After precipitation, a filtration or centrifugation step is usually carried out [104,106]. When the precipitated proteins have to be analyzed, they are solubilized by adding a reagent for reduction (β-mercaptoethanol, dithiothreitol), destruction of H-bounding (urea), or destabilization of hydrophobic interaction (SDS) [107].

Dialysis and ultrafiltration have been largely applied to isolate and fractionate food proteins and peptides. To isolate the protein fraction from wine and must samples, different authors used dialysis followed by lyophilization to concentrate the dialyzed samples [106,108,109]. Depending on the application, membrane of different material, filtration surface and cut-off, able to fractionate the molecules in function of their molecular size, can be used to remove either proteins and other macromolecules or amino acids and small peptides.

Classical low-pressure chromatographic methods such as size exclusion chromatography (SEC), IEC, and affinity chromatography (AC) are traditionally used to separate peptides from interfering substances (sugars, organic acids, salts, amino acids) [104,107]. More recently solid phase extraction (SPE) has been advantageously introduced for preparative isolation and fractioning of peptide from complex matrices. Several SPE sorbents of different polarity (from C2 to C18) can be employed for this purpose. SPE on C18 sorbents allows not only peptide fractioning, but also peptide enrichment and purification (from salts and proteins) [104]. Using IEC SPE (cationic and anionic cartridges in series) Bennett [110] fractionated a mixture of peptides in neutral, acidic, and basic pools. Depending on the complexity of the matrix, a combination of different sample preparation procedures can be necessary to achieve a suitable sample purification and enrichment, prior to HPLC analysis [104].

19.4.1.2 Chromatography

Peptides and proteins have a different number of active sites able to interact with the stationary phase and mobile phase, and consequently have a different chromatographic behavior. The smallest peptides exhibit properties similar to those of individual amino acids and therefore their chromatographic behavior depends on the character of their side chains and substituent groups (strength of the basic and acid ionizable groups, degree of hydrophilicity). Their retention is also strongly dependent on the pH. With an increasing number of amino acid residues, the importance of peptide primary structure increases. With more than about 15 amino acid residues, secondary (and progressively tertiary and quaternary) structure begins to play a role, and the conformation of the largest peptides can decisively affect their retention behavior [111].

19.4.1.2.1 Supports for Protein and Peptide Separation

In the last 30–40 years, many advances have taken place with respect to the chromatographic supports for protein separation. The first chromatographic supports employed for the analysis of proteins were polysaccharides (cellulose, agarose, and cross-linked dextrans), named "soft gels." They had the advantages of being cheap, easily derivatizable, and highly porous (high surface area), but they were not able to sustain high backpressures [112]. Later, inorganic supports of higher mechanical strength, such as silica, became available for protein analysis. As silica-based materials often exhibit secondary interactions (due to the presence of residual silanols on the sorbent) and have a limited stability at high pH values (>9), a number of other materials, such as inorganic oxides, organic resins of polymethacrylate, and polystyrene–divinylbenzene (PS-DVB), have been proposed as supports. Supports based on organic polymers are stable at a wider range of pH (1–12), have shorter run-to-run re-equilibration times and better cleaning ability. On the other hand, they usually exhibit lower separation efficiencies [111,113].

Both completely porous and pellicular supports (mainly of silica) consisting of spherical nonporous particles covered with a very thin layer of porous adsorbents are used in peptide separations. A pore size of 100 Å is commonly recommended for this purpose, but wider pore materials (300 Å) are preferred in separations of large peptides. Nonporous pellicular silica packing materials (with pores of 2–4 Å inaccessible for solutes), introduced for peptide separation, have the advantages to allow fast mass transport (as restricted pore diffusion is eliminated), high surface accessibility, and minimal void volumes (which decreases solvent consumption). Furthermore columns with pellicular sorbents require eluents with lower organic modifier contents. On the other hand, their small outer surface area (two orders of magnitude lower than that of porous packings), leads to low retention and mass loadability [111].

One of the most important causes of poor resolution in protein and large peptide analysis with conventional phases is slowness of mass transport and of sorption/desorption. When selecting a column for a protein separation, it is important to look for obtaining rapid separations in order to limit protein denaturation.

In order to facilitate the intraparticle mass transfer, different possibilities have been proposed over the years, such as the use of particle of smaller size (to reduce the length of the diffusion path) and the use of pellicular packings. The main drawbacks associated with these supports were increased column backpressure and limited loading capacity. Later, chromatographic membranes (consisting of hollow fibers), composite stationary phases (comprising rigid porous particles filled with a gel of high binding capacity), and "tentacular sorbents" were designed to accelerate the contact between proteins and interactive groups, without losses in mechanical strength and loading capacity.

The introduction of "large pore supports," which are able to increase particle permeability and hence mass transfer, represented a turning point in protein analysis [113]. Around 1990, a new HPLC technique, using nonporous supporting materials characterized by presenting two set of pores, was introduced by Afeyan et al. [114]. A brief description of this HPLC technique, which is gaining still more popularity in protein analysis, is discussed in the next section.

19.4.1.2.2 Chromatographic Modes

Several researchers have used liquid chromatography in its different modes, such as reversed phase chromatography (RPC), hydrophobic interaction chromatography (HIC), IEC, SEC, and AC, to analyze and fractionate food proteins and peptides.

Among these, RPC is by far the most widely used for separation of peptides and proteins. RPC separation of peptides is usually carried out with silica-based, alkyl-bonded phases. When large peptides or protein has to be separated, alkyls no longer than butyl are preferred, because they have a lower retention capacity and make possible the use of mobile phases with higher water content that do not cause important denaturing of the analytes, while C8 and C18 are the choice for small and less hydrophobic peptides [106,111]. Polymeric reversed phases of PS-DVB and composite materials (silica particles with a polymeric coating) operating in a wider pH range are also available [104,115].

When using RPC, the mobile phase usually consists of a mixture of water and an organic solvent such as acetonitrile, methanol, or isopropanol, the former being the most suitable due to its low UV absorbance, low viscosity, and high elution strength. In most cases, 0.1%–0.2% of an organic modifier such as trifluoroacetic acid (TFA), which acts as a counterion of any charged amino acid increasing peptide retention, is also present. Furthermore, the acid modifier also allows to reduce peptide interaction with the charged silanols of the silica support, which may decrease the chromatographic resolution. Even though different modifiers (formic acid, acetic acid, phosphoric acid, hydrochloric acid, heptafluorobutyric acid, ammonium acetate, and triethylammonium phosphate) have been proposed for peptide separation, TFA remains the most popular due to its strong ion-pairing capacity, UV transparency, and volatility. Peptide retention increases as the hydrophobicity of the ion pair-forming reagent increases and this effect is more marked in basic peptides. As demonstrated by Kálmán et al. [116], which separated β-casomorphin peptides on different silica-bonded phases, the composition of the mobile phase affects the separation more than the kind of bonded stationary phase used.

RPC of small peptides can also be used for peptide identification, based on their amino acid composition. Computer programs able to predict peptide retention time are available, thus simplifying peptide identification. These programs generally work well with peptide up to 20 amino acids, but fail with larger peptides for which the secondary structure may contribute significantly to the retention.

Compared to other HPLC techniques, RPC has a higher resolution power and allows protein analysis at low ionic strengths. On the other hand, it can be responsible for protein denaturation, loss of biological activity and interferences of hydrophobic contaminants [107].

HIC can be considered a further extension of RPC, which uses less hydrophobic stationary phases, allows obtaining large peptide and protein separation in milder conditions, with a mobile phase of high ionic strength, limiting analyte denaturation. HIC stationary phases, based on silica modified with ether or alkyl ligands (methyl, butyl, phenyl), are available.

Since peptides and proteins can exist either as an anion or a cation, IEC is widely used for their separation. Both silica- and polymeric-based columns operating in anion- or cation-exchange mode are available for peptide and protein separation [107]. IEC exhibits poorer separation efficiency compared to RPC and large peptides may be strongly sorbed. Nevertheless silica-based strong cation-exchange resins (SCX) are ideal for routine peptide [117]. Silica-based materials are unsuitable for anion-exchange chromatography of peptides, as the pH values required for deprotonation leads to silica degradation. IEC is especially important in the separations of strongly basic peptides, and when it is important to use mild eluent conditions in order to maintain the biological activity of peptides. A salt gradient (from pure buffer to a buffer and salt mixture) is usually employed as mobile phase [111].

SEC is a technique that enables the separation of analytes according to their molecular weight. The stationary phases used in this method consist of porous materials (silica or polymeric-based) with size-defined cavities. Compared with other chromatographic modes, SEC provides limited resolving power, but has the advantage of not promoting protein denaturation. In fact, mobile phases used in SEC normally consist of buffers at pH ranging from 6.5 to 8. In some cases, a small proportion of an organic modifier (acetonitrile or methanol) is also added [118]. Initially these packings were more suitable for the separation of proteins than that of peptides, however today, supports able to fractionate peptides with molecular weights <10,000 Da are available. SEC is advantageously used as the first step in the multidimensional separation of complex protein or peptide mixtures, prior to RPC. In ideal SEC, analyte retention time depends only on its molecular size. Due to a hydrophobic and ionic interaction occurring between the analyte and the phase surface, silica-based phases exhibit nonideal SEC behavior. A series of synthetic peptides has been recommended for monitoring the departure of SEC columns from ideal behavior [119]. As high ionic strength reduces ionic interactions but promotes the formation of hydrophobic interactions to minimize both ionic and hydrophobic interactions, the mobile phase should have an ionic strength between 0.2 and 0.5 M and low percentage of an organic modifier able to decrease hydrophobic interactions and increase the peptide solubility [120]. Mobile phase pH affects the net charge of the analyte and any nonspecific interactions between the solute and the sorbent [111]. Since in ideal SEC there is a linear relationship between the logarithm of the molecular weight and the elution volume of a protein, it can be used to determine the molecular weight of a protein [107]. A protocol for rapid HPLC determination of molecular weight of peptides by SEC has recently been proposed by Irvine [120].

AC is based on highly specific biological interactions that make use of very high selectivity and mild separation conditions limiting solute denaturation. Conventional AC is restricted peptides or proteins that are capable of specific binding to the immobilized affinity counterpart. Peptides are often used as stationary phase ligands [111]. It is possible to introduce a specific binding property giving unique binding properties to the target peptide into the peptide or protein of interest through genetic fusing of an affinity tail in to its N- or C-terminus [121]. Another mode of affinity chromatography, used for peptide purification, is immobilized metal-ion affinity chromatography (IMAC). This technique is based on the selective complexation of electron-donating groups, such as phosphate, cysteine, tryptophan, and histidine, with an appropriate immobilized metal ion [118].

Even though NP chromatography is not common in peptide and protein separations, examples of applications for the analysis of peptides produced by the cleavage of membrane proteins are reported in the literature [122].

19.4.1.2.3 Perfusion Chromatography

García et al. [113] reviewed perfusion chromatography for protein separation. The name perfusion chromatography derives from the particles used as stationary phase, which resemble the body's own system of arteries and capillaries transporting nutrients to cells [114]. As large pore supports, perfusion media are constituted by two sets of pores, but what changes is the size of these two set of pores. Stationary phase particles are designed to facilitate analyte access and present: *through-pores* (6000–8000 Å), which cross the stationary phase particle from side to side allowing molecule transport into the interior of the particle by convective flow and *diffusive pores* (800–1500 Å), interconnecting the *through-pore* network and enabling the transport by diffusion. In this structure, molecules cross the stationary phase particles by means of a combination of convective and diffusive transport, thus accelerating the mass transport of molecules. As a consequence, rapid separations (10–100 times faster than in conventional chromatography) are achieved, while maintaining good resolution and loading capacity. The reduction of the analysis times occurring in perfusion chromatography limits protein denaturing and improves recoveries [113].

Due to its mechanical and chemical stability, highly cross-linked PS-DVB is the support most used in perfusion chromatography [123]. Perfusion media for rapid protein separation are available in different chromatographic modes: RPC, IEC, HIC, and AC. The use of a perfusion matrix in mode different from the RPC requires a tightly cross-linked hydrophilic surface layer that masks the hydrophobicity of the matrix [124]. For this purpose, the matrix is coated with an hydroxylated polymer where hydroxylated groups are substituted with functional groups forming different chromatographic modes [125].

Most of the applications rely on RPC employing linear binary gradients (water and acetonitrile), which include an ion-pair agent such as TFA. Flow rates ranging from 1 to 10 mL/min and from 0.025 to 0.125 mL/min are used when using analytical columns and microbore capillary columns, respectively. Analysis times are generally between 0.2 and 15 min [113].

19.4.1.2.4 Chiral Separations

Three approaches can be employed to separate peptide stereoisomers and amino acid enantiomers: separations on chiral columns, separations on achiral stationary phases with mobile phases containing chiral selectors, and precolumn derivatization with chiral agents [111]. Cyclodextrins are most often used for the preparation of chiral columns and as chiral selectors in mobile phases. Macrocyclic antibiotics have also been used as chiral selectors [126]. Very recently, Ilsz et al. [127] reviewed HPLC separation of small peptides and amino acids on macrocyclic antibiotic-based chiral stationary phases.

Stationary phases with specific molecular recognition properties for D,L-enantiomers of peptides have been tailored using the molecular imprinting technique. A template molecule is added to suitable monomer(s), the system is polymerized, and the chiral template molecule is washed out [128].

19.4.1.2.5 Detection

Detection of peptides in HPLC can be achieved by measuring natural absorbance of peptide bonds at 200–220 nm. Unfortunately at these wavelengths a lot of food components and also the solvents used for analysis absorb, demanding an intensive sample pretreatment and clean-up [129]. Peptides with aromatic residues can be detected at 254 nm (phenylalanine, tyrosine, and tryptophan) or 280 nm (tyrosine and tryptophan). Taking advantage of the natural fluorescence shown by some amino acids (tyrosine and tryptophan), detection by fluorescence can also be used for peptides containing these amino acids [106].

In order to improve selectivity and sensitivity, for those peptides that do not contain natural chromophores or fluorophores, pre- or postcolumn derivatization is usually applied. Koller and Eckert [129] presented a comprehensive review of derivatization methods suitable for the chromatography of peptides. A derivatization step can be also introduced in order to reduce the hydrophilicity of the analytes to enable RP-HPLC or to label racemic compounds by a chiral reagent to separate them as their diasteromers [129].

The use of photodiode array detector (DAD) allows obtaining complete UV spectra of the HPLC peaks, thus facilitating the checking of peak purity. According to Dziuba et al. [130], DAD enables to detect the presence of phenylalanine, tyrosine, and tryptophan in proteins and peptides, as these three amino acids show characteristic UV spectra. These authors developed a strategy suitable for protein identification using the second and fourth derivatives of UV spectra (in the 270–300 wavelength range) obtained with DAD on-line with HPLC assembly. Application of UV spectra may allow obtaining univocal identification when different proteins or peptides with the same or similar retention times occur in the same mixtures.

Electrochemical detection is limited to peptides that contain tyrosine, thryptophan, methionine, cysteine, or cystine, but can be used to determine the location of disulfide bonds. On-line scattering detectors give direct information on molecular weight that is independent of elution position, and have been applied to the detection of protein and their aggregates. As the sensitivity of light scattering is proportional to molecular weight, very low sensitivity are achieved for small peptides [120].

The advent of "soft" ionization techniques, starting from 1981 with the introduction of fast atom bombardment (FAB), followed by ESI and matrix-assisted laser desorption/ionization (MALDI), has made analyses by MS possible for proteins and peptides [105]. MS was initially used as an off-line technique in protein and peptide determination, but the development of new interfaces and ionization systems has allowed HPLC–MS coupling. Nowadays on-line HPLC–MS is increasingly applied in detection and identification of peptides and proteins. In particular, ESI and MALDI, coupled with quadrupole or flight time analyzers, or with MS/MS tandem systems, can be used to obtain accurate molecular weight determination, peptide sequencing, peptide purity information, detection of protein modifications induced by food processing, identification of protein genetic variants, and peptide posttranslational modifications, characterization of enzymes and allergenic proteins, and information on protein interactions [105,106,131,132]. Several review papers on analysis of food peptides and proteins by chromatography and mass spectrometry have appeared in the last 10 years [104–106,132,133]. In 2001, Pastorello and Trambaioli [134] reviewed the analytical problems concerning isolation of food allergens.

19.4.1.3 Applications

Leonil et al. [105] and more recently Careri and Mangia [132] reviewed the application of HPLC–MS to the analysis of food proteins and derived peptides. The most studied matrices were milk and derived products, cereals, meat, and wine.

19.4.1.3.1 Animal Proteins

Due to the complexity of milk composition, identification and quantification of all milk proteins is a challenging and difficult task. Furthermore, the natural variation of milk composition and genetic polymorphism within milk proteins complicated the analysis. Bordin et al. [135] proposed an optimized method able to separate and quantify the seven major milk proteins (k-casein, α_{s2}-casein, α_{s1}-casein, β-casein, α-lactalbumin, β-lactoglobulin B, and β-lactoglobulin A) in one run, without classical preseparation of whey proteins from caseins. The identification of each protein was ascertained by using retention times, second derivative UV spectrophotometry (which reveals the fine structure of the spectra), and by calculating the peak area ratio at two different wavelengths ($A_{214\,nm}/A_{280\,nm}$), which is an index of the relative proportion of the total aromatic amino acid content of a

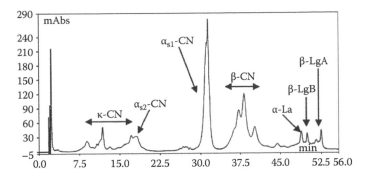

FIGURE 19.4 RP chromatographic profile of a CRM 063R skimmed milk powder sample. Vydac C4 150×2.1 mm column; linear gradient from 26.5% to 28.6% B in 7 min, then from 28.6% to 30.6% B in 10 min and from 30.6% to 36.1% B in 11 min, followed by an isocratic step at 36.1% B during 10 min and a final increase to 43.3% B in 18 min, where eluent A is composed of 10% 21 (v/v) acetonitrile and 0.1% TFA in ultra pure water and eluent B of 10% water and 0.1% TFA in acetonitrile; flow-rate 0.25 mL min at 40°C; UV detection at 214 nm. (From Bordin, G. et al., J. Chromatogr. A, 928, 63, 2001. Copyright 2001. With permission from Elsevier.)

given protein. The procedure was applied to various types of raw and commercially available bovine milk in order to assess fraudulent addition of nonmilk proteins to milk or mixing milks of different species. Figure 19.4 shows the chromatogram of the proteins of a skimmed bovine milk. Trujillo et al. [136,137] investigated the complexity of ovine and caprine milk proteins (from whole skimmed milk), mainly due to genetic polymorphism, posttranslational changes (phosphorylation and glycosylation), and the presence of multiple form of proteins by coupling HPLC and ESI-MS.

HPLC of milk proteins has also been largely used for the detections of frauds. Ferreira and Caçote [138] developed an analytical protocol that uses a PS-DVB column to better separate whey proteins from bovine, ovine, and caprine milks. Enne et al. [139] developed a simple RP-HPLC method to detect fraudulent addition of bovine milk during manufacturing of buffalo "Mozzarella" (in governing liquid as well as in raw milk and cheese). Analyses were based on the measurement of quantity ratios within β-lactoglobulin protein family. In order to achieve a reliable estimation of bovine milk addition, the use of a matrix-specific calibration curve was essential. To reduce dairy surpluses, EC regulation subsidizes the use of skimmed milk powder in compound feeding stuffs. The low price of some plant proteins (soybean, pea) make them attractive as potential adulterant for these milk powders. Luykx et al. [140] developed a sensitive method for identification of illegal plant protein addition at concentrations ranging from 1% to 5%. The method included a prefractionation step to enrich plant proteins (by using a borate buffer able to precipitate most of the plant protein), followed by RP perfusion chromatography and selective intrinsic fluorescence detection. After tryptic digestion, nano HPLC–MS/MS was applied to the enriched protein fraction from adulterated skimmed milk powders in order to identify soy and pea peptides. To determine protein content in infant formulas, Hong et al. [141] coupled gel electrophoresis followed by in-gel digestion with nano-HPLC-ESI-MS/MS. A total of 154 peptides, corresponding to 31 unique proteins, were identified in commercial infant formula profiled in this study. This approach represents a valid analytical tool to investigate adulteration in infant formulas.

HPLC analysis of milk proteins represents a useful tool to assess the severity of thermal treatments. Heating is known to induce structural changes in milk proteins (particularly β-lactoglobulin) depending on the duration and the severity of heat treatments. These modifications can be exploited to differentiate raw, pasteurized, and UHT milk, to assess the severity of heat treatments applied to milk (which is of relevance for processing control and regulatory purposes), and to assess allergenic potential of whey proteins (that is of paramount importance to safeguard the health of allergic consumers). Hau and Bovetto [142] used RP-HPLC coupled with ESI-MS to study the modification induced by industrial processing on commercial whey proteins. Henry et al. [143] used on-line HPLC-ESI-MS to study the covalent interactions between casein micelles and β-lactoglobulin from goat milk, and found that the occurrence of a covalent linkage between β-lactoglobulin and k-casein could account for the low heat stability of goat milk. Monaci and van Hengel [144] exploited the potential of HPLC-ESI-MS to investigate β-lactoglobulin modification as a consequence of thermal treatments in milk samples. They concluded that the degree of lactosylation in raw milk samples provides a potential marker to trace heat treatments. An overview of the most recent HPLC applications for milk protein analysis is given in Table 19.5.

Meat proteins comprise a water-soluble fraction (containing the muscle pigment myoglobin and enzymes), a salt-soluble fraction composed mainly of contractile proteins, and an insoluble fraction comprising connective tissue proteins and membrane proteins. As reviewed by Dierckx and Huyghebaert [107], HPLC analysis of meat proteins has been successfully applied to evaluate heat-induced changes in the protein profile, to detect adulterations (addition of protein of lower value, the replacement of meat from high-value species with meat from lower-value species, etc.), and for specie identification in noncooked products (also for fish sample).

19.4.1.3.2 Plant Proteins

Most of the applications of HPLC for protein analysis deal with the storage proteins in cereals (wheat, corn, rice, oat, barley) and beans (pea, soybeans). HPLC has proved useful for cultivar identification, protein separation, and characterization to detect adulterations (illegal addition of common wheat flour to durum wheat flour) [107]. Recently Losso et al. [146] have reported a rapid method for rice prolamin separation by perfusion chromatography on a RP POROS RH/2 column (UV detection at 230 nm), sodium dodecyl sulfate (SDS)-polyacrylamide gel electrophoresis (PAGE), and molecular size determination by MALDI-MS. DuPont et al. [147] used a combination of RP-HPLC and SDS-PAGE to determine the composition of wheat flour proteins previously fractionated by sequential extraction.

The increasing interest in nutritional and functional properties of soybean protein has promoted their use in the manufacturing of foods for human consumption. Soybean products (particularly infant formulas and soybean dairy-like) may also represent an interesting substitute for infants and people allergic to milk proteins. On the other hand, due to their technological properties and low cost, soybean proteins are increasingly employed as ingredients in milk, bakery, and meat products, in which their addition is forbidden or allowed up to a certain limit.

Saz and Marina [148] published a comprehensive review on HPLC methods and their developments to characterize soybean proteins and to analyze soybean proteins in meals. In the case of soybean derived products, a number of papers dealing with cultivar identification [149,150], quantification of soybean proteins [151–154], detection of adulteration with bovine milk proteins [151,155–158], and characterization of commercial soybean products on the basis of their chromatographic protein profile [159,160] have been published in the last years. Other studies deal with the analysis of soybean proteins added to meat [161–165], dairy [151,165–167], and bakery products [156,163,168,169]. The same research group developed perfusion RP-HPLC methods for very rapid separation of maize proteins (3.4 min) and characterization of commercial maize products using multivariate analysis [170], and for the characterization of European and North American inbred and hybrid maize lines [171].

Typical conditions for separating vegetable proteins are RP-HPLC with linear gradient of water and acetonitrile containing 0.05%–0.1% TFA) and UV detection at 254 or 280 nm. The introduction of perfusion chromatography (on PS-DVB beads) allowed to reduce the chromatographic run time from 20–70 to 2–5 min, depending on the application. Very recently Heras et al. [149] designed a perfusion IEC method for the separation of soybean proteins. Marina et al. [156] used a 150 μm i.d. capillary column packed with a C18 stationary phase (4 μm, 300 Å) for simultaneous separation of soybean proteins from cereal (wheat, rice, and corn) and milk proteins.

TABLE 19.5
HPLC Applications for Milk Protein Analysis

Application	Sample Pre-Treatment	Column	Mobile Phase	Detector	References
Detection of plant protein addition in skimmed milk powder Identification of plant protein	Borate buffer treatment (pH 8.3), vortex-mixing, centrifugation, pellet dissolved in 6 M urea and dithiothreitol Tryptic digestion of borate buffer precipitate	POROS R2/10 (RP) (50×4.6mm, 10μm), $T=60°C$	A: 0.1% TFA in H_2O; B: 0.1% TFA in ACN; flow rate: 2 mL/min	FLD ($\lambda_{ex}=280$ nm; $\lambda_{em}=340$ nm)	[140]
Quantification of α-lactoalbumin in human milk	Centrifugation at 5000rpm (for skimming), filtration, dilution	Chrompack P 300 RP (150×4.6 mm, 8μm, 300Å)	A: 0.1 TFA in H_2O: B: 0.1% TFA in 95% ACN-5% H_2O; flow rate: 1 mL/min	UV (280nm)	[145]
Protein determination in infant formulas	Dissolution in Phosphate buffer, gel electrophoresis, in gel tryptic digestion, petide extraction ($CH_3CN/H_2O/TFA$), cleaning on ZipTip C18	C18 capillary column (70mm×75μm)	A: 2% ACN/97.9% H_2O/0.1% HAc; B: 90% ACN/9.9%H_2O/0.1% HAc; flow rate: 0.2μL/min	ESI-MS/MS	[141]
Detection of β-lactoglobulin modification in milk samples as a consequence of heat treatments	Acidification (pH 4.6 with 5% HAc) to precipitate caseins, centrifugation	Discovery C5 (Supelco, 100×2.1mm, 3μm), $T=25°C$;	A: 0.1% formic acid in ACN/H_2O (90/10); B: 0.1% formic acid in ACN/H_2O (10/90);	DAD ESI-MS	[144]
Detection of β-lactoglobulins as a marker to detect fraudulent addition of bovine milk during manufacturing of buffalo Mozzarella	For cheese sample: Ultraturrax homogenization (two cycles) with H_2O, centrifugation (for skimming), acidification at pH 4.2–4.6 (for casein precipitation), centrifugation, filtration	Phenomenex C4 (250×4.6mm; 5μm, 300Å)	A: 0.1% TFA in H_2O; B: 0.1% TFA in ACN; flow rate: 1mL/min	UV (205nm)	[139]
Detection and quantification of bovine, ovine, and caprine mixtures in milk and cheese	For cheese sample: extraction with H_2O (sonication); for milk: centrifugation, acidification (pH 4.6) to precipitate caseins, centrifugation, filtration	Chrompack P 300 RP (PS-DVB) (150×4.6mm; 8μm, 300Å); Chrompack P RP (24×4.6mm) pre-column. $T=45°C$	A: 0.1% TFA in H_2O; B: 0.09% TFA in 80% ACN-20% H_2O; flow rate: 0.5mL/min	UV (215nm)	[138]
Identification and quantification of major bovine milk proteins in milk samples	Milk: Centrifugation at 4°C (for skimming)	Vydac C4 (150×2.1mm, 5μm, 300Å), $T=40°C$	A: 10% ACN and 0.1% TFA in H_2O; B: 10% H_2O and 0.1% TFA in CH_3CN; flow rate: 0.25mL/min	DAD	[135]
Ovine milk protein characterization	Centrifugation at 30°C (for skimming), acidification (pH 4.6) to precipitate caseins, purification of whey protein by GPC	Jones Chromatography C18 (250×4.6mm, 7μm, 300Å)	CH_3CN and H_2O in 0.1% TFA; $T=46°C$; flow rate: 1 mL/min	UV (214nm) ESI-MS	[136,137]

19.4.1.3.3 Food Peptides

Concerning food peptides, most of the HPLC applications deal with food characterization (based on peptide profile), peptide separation, and identification [172,173] and detection of frauds in milk-based products, meat, and protein hydrolysates [115].

Different authors used RP-HPLC and UV detection to monitor peptide formation during cheese ripening [174–178], providing valuable information about proteolysis. When large hydrophobic peptide need to be separated an IEC represents the best choice [179]. Nevertheless, the identification of these peptides is essential for the complete understanding of the proteolytic process. The peptides eluted from the LC column can be subjected to ESI-MS for molecular weight determination and MS/MS for amino acid sequence determination, which allow rapid peptide identification [172]. HPLC-ESI-MS and MS/MS techniques have been successfully used for peptide mass fingerprint purposes for sequence analysis of purified albumin from *Theobroma cacao* seeds [180,181].

Peptide composition can be used to characterize foods and protein hydrolysates by applying multivariate analysis to the peptide pattern [175,182]. Table 19.6 reports basic information on some of the most recent HPLC applications for peptide analysis.

19.5 AMINO ACIDS

As already known, amino acids are the building units of peptides and proteins and they play an important plastic, energetic, and regulatory role in all living organisms. Almost all foods contain amino acids, either in the bound (partially hydrolyzed or intact proteins) or in the free form. High amounts of free amino acids can be found in some fermented foodstuffs as a consequence of proteolytic processes.

Some free amino acids (and their analogs) exert important functions as neurotrasmitters, signal transducers, and free radical scavengers [190]. In addition to their nutritional value, free amino acids are also important nonvolatile taste-active components. Aspartic acid and glutamic acid contribute to the characteristic "umami" taste of food, while alanine and glycine confer a sweet taste [191,192]. A glutamic to aspartic acid ratio of 4:1 brings out the genuine tomato taste [193]. Free amino acids (particularly glutamic acid, aspartic acid, alanine, and glycine) were also reported to affect the taste of mushrooms [194,195], seaweeds [192], crustaceans, and molluscs [191].

Amino acid analysis has long been of importance for nutritional purposes. Recently, there has been an increasing interest in assessing the exact amino acid composition of new protein sources, infant formulas, nutraceutical-type products, or supplements targeted at people with unusual metabolic needs. There is also a regulatory necessity to have rapid methods available for amino acid determination to verify product consistency with that declared on the label [196].

On the other hand, free amino acid analysis represents a powerful tool to characterize different foods and beverages, monitor proteolysis, assess freshness, detect adulterations, and safeguard consumer health. The occurrence of some potentially toxic nonprotein amino acids (some of which are neuroexcitatory) in commercially available seedlings has been reported by different authors [197–199]. Due to the incapability of humans to utilize the D-isomers of amino acids (some of which are thought to be toxic), the enantiomeric separation of D- and L-form of amino acids is also an area of growing interest [196].

19.5.1 HPLC Analysis

Diverse analytical methods have been proposed for the analysis of amino acids, including GC, HPLC, and capillary electrophoresis. The preferred method at present is RP-HPLC with precolumn derivatization, which has the advantages of requiring short analysis time, simple instrumentation, and low cost [192].

Sample preparation for HPLC analysis of free amino acids usually involves an extraction step (for solid matrices), followed by purification to remove possible impurities. When the aim of the

TABLE 19.6
HPLC Applications for Food Peptide Analysis

Application	Sample Pre-Treatment	Column	Mobile Phase	Detector	References
Separation of water-soluble peptides in Emmental cheese	Extraction with H_2O (politron homogenizer), centrifugation, filtration	C18 (ET 250/8/4 Nucleosil 300-5); pre-column: C18 (KS 11/6/4 Nucleosil 300-5)	A: 0.1% TFA in H_2O; B: ACN/H_2O (60/40); flow rate: 1 mL/min	UV (210nm)	[174]
Peptide identification in Cheddar cheese	Ether extraction (to eliminate fat), extraction with 0.1M HCl, centrifugation	C18 (250×4.6mm)	A: 0.1% TFA in H_2O; B: ACN/H_2O (70/30); flow rate: 1 mL/min	DAD ESI-MS and MS/MS	[172]
Effect of milk heating on peptide formation during cheese ripening	Extraction with 0.1M trisodium citrate, centrifugation, acidification of supernatant at pH 4.6, centrifugation	C18 Nucleosil (250×4.6mm, 5μm, 100Å)	A: 0.1% TFA in H_2O; B: 10% H_2O and 0.1% TFA in ACN; flow rate: 1 mL/min	UV (214nm)	[175]
Identification of small peptides released during milk sterilization	SPE (C18) enrichment and purification	C18, Simmetry, phase separation (150×2.1mm), $T=40°C$.	A: 0.1% TFA in H_2O; B: 20% H_2O and 0.1% TFA in ACN; flow rate: 0.25 mL/min	ESI-MS MS/MS	[173]
Characterization of soft cheese proteolysis. Effect of cheese freezing on peptide formation during ripening	Extraction of water-soluble and water-insoluble peptides	C18, SynChropak RPP, (250×4.6mm, 300Å), $T=30°C$	A: 0.1 TFA in H_2O; B: 0.1% TFA in ACN	UV (220nm)	[177,178]
Peptide profiles of Teleme cheese (different ages) made with goat milk, ewe milk or mixtures	Extraction with H_2O	C18 (Nucleosil 250×4.0mm, 5μm, 300Å), pre-column (40×4.0mm) of the same material.	A: 0.1% TFA in H_2O; B: 0.085% TFA in ACN/H_2O (60/40); flow rate: 0.8 mL/min	UV (214nm)	[176]
Study on the role of alkaline phosphatase in cheese ripening	Extraction with H_2O	Ultrasphere ODS (250×4.6mm, 5μm, 80Å)	A: 0.1% TFA in H_2O; B: 0.1% TFA in ACN/H_2O (75/25); flow rate: 0.75 mL/min	UV (230)	[183]
Detection of hen's egg white lysozyme in milk and dairy products	Homogenization with NaCl solution (pH 6.0; 40°C), acidification (HCl), centrifugation, filtration	PLRP-S (250×4.6mm, 5μm, 300Å), $T=45°C$	A: 0.1% TFA in H_2O; B: 0.1% TFA in ACN; flow rate: 1 mL/min	UV (280nm) Fl ($\lambda_{ex}=280$nm; $\lambda_{em}=340$nm)	[184]

(continued)

TABLE 19.6 (continued)
HPLC Applications for Food Peptide Analysis

Application	Sample Pre-Treatment	Column	Mobile Phase	Detector	References
Peptide characterization of whey hydrolysates	Dilution with H_2O	Chrompack P-300-RP	A: 0.1% TFA in H_2O; B: 0.1% TFA in ACN/H_2O (80/20); flow rate:0.5 mL/min	UV (215nm)	[182]
Isolation and identification of peptides formed during beef storage	Extraction with H_2O (homogenization), protein precipitation (TCA 5%), centrifugation, filtration	Lichrosphere 100 RP-18 (250×4 mm, 5 μm); Nuclesil 100 RP-18 (250×10 mm, 7 μm)	A: 0.05% TFA in H_2O; B: 0.05% TFA in ACN/H_2O (80/20); flow rate:0.5 mL/min	UV (214 nm)	[185]
Peptide separation for assessment of pork meat quality	Homogenization with 0.01 M HCl (4°C), centrifugation, filtration, purification on C18 cartridge	Column 1: Sherisorb S10SCX (250×10 mm) Column 2: C18 (symmetry, 250×4.6 mm, 5 μm, 300 Å)	Mobile phase 1: A: 20% ACN in 6 mM HCl; B: 20% ACN in 6 mM containing 1 M NaCl; flow rate 3.6 mL/min Mobile phase 2: A: 0.1% TFA in H_2O; B: 0.085% TFA in ACN; flow rate: 0.9 mL/min	UV (214 nm)	[186,187]
Determination of peptides formed during processing of Iberian ham	Homogenization with 0.6 N PCA, centrifugation, filtration, pH adjustment (6.0), filtration	C18 (250×4.6 mm, 5 μm)	A: H_2O; B: 0.1% TFA in ACN	UV (214 nm)	[188]
Separation of peptides formed during the ripening of semi-dry fermented sausages	Preparation of 2% TCA- and 5% phosphotungstig-soluble extracts by homogenization with Ultra Turrax and centrifugation	C18 symmetry (250×4.6 mm, 5 μm, 300 Å)	A: 0.1% TFA in H_2O; B: 0.1% TFA in ACN; flow rate: 1.2 mL/min	UV (214 nm)	[189]

analysis is to determine the total amino acid composition of food proteins, a preliminary hydrolysis step has to be accomplished to liberate protein-bound amino acids.

19.5.1.1 Sample Preparation

19.5.1.1.1 Free Amino Acids

Water or diluted acids are usually employed to extract free amino acids from solid matrices. In many cases, when liquid samples have to be analyzed, filtration or centrifugation and sample dilution are the only steps required before injection.

Among food components, proteins and peptides are a frequent cause of loss of resolution and column damage. Different precipitation agents, such as picric acid, perchloric acid, sulfosalicylic acid, tungstic acid, ammonium sulfate or organic solvents, such as acetonitrile, have been used to eliminate proteins and large peptides. Ultrafiltration and SPE on disposable C18 cartridges (able to retain proteins as well as lipids) represent an alternative to precipitation agents.

Each of these methods has its advantages and disadvantages and can lead to incomplete recoveries of certain amino acids [196].

19.5.1.1.2 Total Amino Acid Composition

An excellent review on protein hydrolysis for amino acid composition analysis has been published by Fountoulakis and Lahm [190]. Hydrolysis can be performed by either chemical (under either acidic or basic conditions) or enzymatic means. The acidic hydrolysis itself can be carried out in a liquid or a gas-phase mode. The conventional acid hydrolysis uses 6 M HCl for 20–24 h at 110°C under vacuum [200]. In these conditions, asparagine and glutamine are completely hydrolyzed to aspartic acid and glutamic acid, respectively. Tryptophan is completely destroyed (particularly in the presence of high concentrations of carbohydrate), while cysteine and sometimes methionine are partially oxidized. Tyrosine, serine, and threonine are partially destroyed or hydrolyzed and correction factors have to be applied for precise quantification [190,201].

Hydrolysis is the major source of inaccuracy in an amino acid analysis. Since hydrolysis and destruction of the liberated residues proceed simultaneously, the aim of every hydrolysis method is the quantitative liberation of all amino acids with minimal losses. Some amino acid sequences (formed by combinations of valine, leucine, and isoleucine) are particularly resistant to acid treatment and would need more drastic conditions to be completely hydrolyzed. On the other hand, this would lead to higher losses of more sensitive amino acids. Several authors investigated on the effects of time and temperature on the completeness of hydrolysis [201–204]. In general, elevated temperatures and shorter hydrolysis times give similar or superior results to those of conventional hydrolysis at 110°C for 24 h [203]. To reduce losses during hydrolysis, several authors investigated the possibility to use different protective agents (such as phenol, thioglycolic acid, mercaptoethanol, indole, or tryptamine) [200], and alternative nonoxidizing organic acids such as 4 M methanesulfonic acid, 3 M p-toluenesulfonic acid, and 3 M mercaptoethanesulfonic acid [196]. To obtain acceptable recoveries of the more labile amino acids for complex food matrices, laborious and time-consuming sample clean-up is needed in order to remove the carbohydrates and fats prior to acid hydrolysis [190].

The introduction of microwave radiation energy (in specially designed pressurized instruments) for hydrolysis of protein samples allowed a reduction of the overall hydrolysis time from many hours to a few minutes [205–209]. Joergensen and Thestrup [210] hydrolyzed food proteins containing carbohydrates, lipids, nucleic acids, and minerals, with HCl in a microwave oven at 150°C for 10–30 min, obtaining results similar to those of conventional hydrolysis.

Alkaline hydrolysis (with NaOH, KOH or more seldom with Ba(OH)$_2$) is almost exclusively applied for the determination of tryptophan and phosphoamino acids. Serine, threonine, arginine, and cysteine are completely destroyed by alkaline hydrolysis, while other amino acids are racemized [190]. Since racemization also occurs during acid hydrolysis, when it is important to analyze the D-amino acid content of the sample, an enzymatic hydrolysis has to be applied. It has also the advantage of allowing a correct quantification of sensitive residues. The quantification of the D-amino acid forms in a polypeptide requires the use of specific hydrolysis methods involving multiple enzymatic digestions [211]. Csapó et al. [212] studied the effects of temperature on the racemization of amino acids during acid hydrolysis.

19.5.1.1.3 Particular Amino Acids and Di- or Tri-Peptides

Sometimes, for different reasons, it can be necessary to quantify only one of few sensitive amino acids. In this case, particular attention has to be paid to optimize recoveries of these amino acids.

The sulfur-containing side chains of cysteine and methionine are particularly prone to oxidation and the standard hydrochloric acid hydrolysis cause the partial conversion of these amino acids in a number of oxidation products. The classical strategy to overcome this problem is to drive the oxidation to completeness. Cysteine and methionine can be quantified as cysteic acid and methionine sulfone, after oxidation with performic acid [213] or 0.2% sodium azide [214], or after reaction with an alkylating agent able to convert them into acid-stable species [215–217]. Even though an additional sample preparation step is required to remove the excess of the alkylating reagent (prior to the hydrolysis or HPLC analysis), this method has the advantage of improving chromatographic retention of the analyte, as cysteic acid usually elutes near the void volume of the column.

If the major goal of the analysis is the determination of tryptophan only, alkaline hydrolysis is the most proper digestion method [218]. Alternatively when other amino acids have to be determined together with triptophan and a high accuracy is not required, tryptophan can be determined following the hydrolysis with HCl in the presence of various tryptophan protecting additives, or with methanesulfonic acid in the presence of 3-(2-aminoethyl)indole or mercaptoethanesulfonic acid [190]. Nevertheless, according to Kivi [196], attempts to modify the standard hydrolysis, by adding protective agents or using alternative organic acids, fail to give acceptable recoveries of tryptophan in food matrices. In 1997, Molnar-Perl [219] reviewed tryptophan analysis in peptide and proteins.

During standard acid hydrolysis, glutamine and asparagine are deaminated to corresponding carboxylic acids (aspartic acid and glutamic acid, respectively). When it is important to analyze correctly one of these amino acids, as in the case of glutamine content in special dietary supplements devoted to patients suffering from gut trauma, special strategies aimed at avoiding deamination have to be followed. The first strategy consists in converting the side chain of these amino acids into a moiety that is stable to acid hydrolysis, through reaction with bis(thrifluoroacetoxy)iodobenzene (BTI) [220]. The possibility to employ enzymatic digestion has also been investigated, although there are fewer applications, probably due to the lengthy and complex procedure and the possibility of a spurious contribution due to autodigestion [196,221].

Lysine is an essential amino acid with an ε-amino group on the side chain that can react with various food components. As known, reaction of the ε-amine can render lysine nutritionally unavailable reducing the nutritional value of food. While the determination of total lysine is straightforward (it is stable to acid hydrolysis), the determination of available lysine is difficult as lysine adducts are labile to the standard acid hydrolysis. A solution to this problem consists of derivatizing the ε-amino group with a chromophore such as 1-fluoro-2,4-dinitrobenzene (FDNB) to form a derivate which is stable to optimized hydrolysis conditions [222].

19.5.1.2 Chromatography

For many years, automated amino acid analyzer using IEC and postcolumn derivatization with ninhydrin (or less frequently with fluorescamine) has been the most popular technique for amino acid determination. Amino acids are separated in their free form by employing stepwise elution with sodium- or lithium-based buffers.

In recent years, the wide diffusion of precolumn derivation agents able to increase analyte hydrophobicity and hence its retention on an apolar phase allowed a gradual replacement of dedicated amino acid analyzer with more versatile and less expensive RP-HPLC systems.

C18 column from different suppliers (3–5 μm of particle size) are commonly used for amino acid determination, with mobile phases consisting of an acetate or phosphate buffer and an organic solvent such as acetonitrile, methanol, or tetrahydrofuran. To avoid peak tailing for the basic amino acids (in the presence of uncapped silanols), triethylamine is often added as a modifier. Depending on the complexity of the sample, complicated tertiary gradients can be necessary to separate all the researched amino acids [196].

Naulet et al. [223] compared three chromatographic techniques: IEC on an automatic amino acid analyzer, HPLC (with phenylisothiocyanate derivatization) and GC for the determination of the free amino acid content in 64 orange juices from different countries. The consistency of the different methods was estimated by considering the mean standard deviation for the set of amino acids observed. The best consistency was observed between IEC and HPLC.

19.5.1.3 Derivatization and Detection

With the exception of amino acids that have an aromatic structure, other amino acids absorb at very short wavelengths (around 210 nm), where many other food components, and often solvents, absorb. For this reason, a derivatization step is usually needed. Postcolumn derivatization has the advantage of simplifying sample preparation directed to eliminate components that may interfere with the derivatization step, but require expensive extra equipment (such as reagent pump, mixing manifold, heated reaction coil) that needs maintenance and can cause band broadening. Furthermore, only those reagents exhibiting sufficiently fast reaction kinetics can be conveniently used as postcolumn derivatizing agents. On the other side, precolumn derivatization has the big advantage of allowing faster RP separations of derivatized amino acids, but it requires the formation of stable derivatives [196].

Different derivatization agents, whose main properties are summarized in Table 19.7, have been proposed for pre- and/or postcolumn derivatization of the amino groups. Among these derivatizing agents the most utilized for amino acid derivatization are phenylisothiocyanate (PITC), o-phthalaldehyde (OPA), and 9-fluorenylmethylchloroformate (FMOC-Cl). Each of these reagents has specific advantages and limitations [224]. PITC derivatization methods are poorly sensitive compared with methods based on fluorometric detection but allow to obtain rapid HPLC separations. Figure 19.5 shows the HPLC trace of PITC derivatives of amino acids obtained from a red seaweed sample after acid hydrolysis. The method enabled simultaneous analysis of 16 amino acids in less than 30 min [192]. OPA does not react with secondary amino acids while FMOC-Cl reacts rapidly. The combined use of OPA/FMOC-Cl has been recommended to overcome these problems, with UV detection of the resulting derivatives at the optimum wavelengths for both primary and secondary amino acids to maximize sensitivity [192,225]. The reaction with OPA that leads to the formation of an isoindole derivative, occurs in the presence of a thiol, such as 2-mercaptoethanol, which act as a reductive coagent. A number of other thiols have been tested in combination with OPA: 3-mercapto-1-propanol, ethanothiol or 3-mercapto propionic acid (3-MPA), the latter being able to react with primary and secondary amino acids. The principal drawback of OPA is that OPA derivatives are not stable. For this reason it is best used as a postcolumn derivatizing agent and, when used for precolumn derivatization it necessitates well-controlled and possibly automated handling to ensure uniformity of the time elapsed from reaction to injection [196].

There are two major approaches to achieve enantiomeric separation of D- and L-amino acids. The first involves precolumn derivatization with a chiral reagent, followed by RP-HPLC [226], while the second involves direct separation of underivatized enantiomers on a chiral bonded phase [227]. Weiss et al. [209] determined D- and L-form of amino acids by applying derivatization with OPA and chiral N-isobutyryl-L-cysteine.

Depending on the derivatizing agent used, spectrophotometric or fluorometric detectors are usually employed. Electrochemical detection of underivatized amino acids is limited to amino acids possessing aromatic or sulfur-containing side-chain, even if derivatization procedures to attach electrochemical active moieties to the amino acids can be employed, as well as other approaches

TABLE 19.7
Most Common Derivatizing Agents for Amino Acid Analysis by HPLC

Derivatizing Reagent	Detection	Properties
Ninhydrin (post-column derivatization)	UV (550/570 nm; 440 nm for proline) Fluorescence	React (at 125°C–135°C) with both primary and secondary amino groups.
Fluorescamine (post- and pre-column derivatization)	Fluorescence (ex: 390 nm; Em: 475/500 nm)	Reacts rapidly (1–2 min at 20°C) at alkaline pH with primary amino groups without forming side-products. Derivatives are stable for 1 h.
Dansyl chloride (Dns-Cl) (pre-column derivatization)	UV (254 nm) Fluorescence (ex: 320/360 nm; em: 430/470 nm)	Reacts with both primary and secondary amino groups in alkaline medium (1 h at 40°C). Forms stable derivatives (days). It gives by-products reactions. Different strategies can be used to separate the reagent excess. Detection limits: picomoles (with fluorescence detection).
Dabsyl chloride (pre-column derivatization)	UV (420/450 nm)	Reacts with both primary and secondary amino groups in alkaline medium (10 min at 70°C). Used as alternative to Dns-Cl.
Phenylisothiocyanate (PITC) (Pre-column derivatization)	UV (254 nm)	Reacts with both primary and secondary amino groups. The reaction is completed in 10–20 min but a long time (more than 1 h) is needed to remove the excess of the reagent under vacuum. Detection limits: picomoles.
9-Fluorenylmethylchloroformate (FMOC-Cl) (post- and pre-column derivatization)	UV (260 nm) Fluorescence (ex: 262 nm; em: 313 nm)	Reacts rapidly (90 s at 20°C), under slightly alkaline conditions, with both primary and secondary amino groups. Forms stable derivatives. Derivatizing reagent excess and hydrolyzed by-products have to be removed from the reaction mixture. Detection limits: femtomoles.
o-Phthalaldehyde (OPA) (mostly used for post-column derivatization)	Fluorescence (ex: 330/340 nm; em: 450/475 nm)	Reacts rapidly (1 min at 20°C) with only primary amino group at alkaline pH (9–11) in the presence of a thiol, without forming interfering side products. Derivatives are not stable. Detection limits: fentomoles.
6-Aminiquinolyl-N-hydroxysuccinimidyl carbamate (AQC)	Fluorescence (ex: 250 nm, em:395 nm)	Reacts rapidly (in less than 1 min at 20°C) with both primary and secondary amines. Forms highly stable derivatives. It does not require intensive sample purification and extraction of derivative excess. Detection limits: picomoles.

based on the use of immobilized enzymes able to react with amino acids producing hydrogen peroxide, which is then quantified by amperometric detection [228].

Quantification of the separated amino acids is usually performed by using external calibration or the internal standard method. Due to the large differences in chemical structure exhibited by the various amino acids, there is not a single ideal standard for the overall amino acid profile. Nevertheless, a suitable internal standard must be stable to hydrolysis and offer chromatographic resolution. The most popular choices comprise norleucine, norvaline, and α-amino-n-butanoic acid (AABA) [196].

19.5.1.4 Applications

Recently, Peace and Gilani [229] reviewed chromatographic determination of amino acids. HPLC analysis of free amino acids can be a powerful tool in the control of food authenticity. Cotte et al. [230] demonstrated that HPLC analysis of free amino acids, followed by statistical processing of

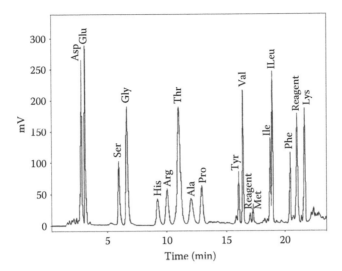

FIGURE 19.5 HPLC profile of PITC derivatives of amino acids from a red seaweed (UV 254 nm). (From Sánchez-Machado, D.I. et al., *Chromatographia*, 58, 159, 2003. With permission from Springer Science + Business Media.)

the data, enables to distinguish certain varieties of honey and to detect the addition of sugar syrup to rape and fir honeys. Giraudo et al. [231] used the free amino acid profile to assess authenticity of fresh and concentrated orange and lemon juices.

Free amino acid profile has also been exploited to characterize food products and for differentiation purposes. Naulet et al. [223] found that amino acids provide useful criteria for characterizing orange juices from different countries. Some authors [232,233] found significant differences in free amino acid composition of a limited number of almond cultivars from the same region and among almonds cultivated in different geographical areas. Serine and asparagine proved to be suitable parameters for differentiating cultivars, whereas arginine resulted more associated with localities. Other authors used amino acid profile to differentiate San Marzano tomato varieties [234] or to classify white wines according to grape variety, vintage, geographic origin and type of vinification [235]. According to Cornet and Bousset [236], free amino acids and some dipeptides can be used to differentiate muscles of the same species and to explain their differences in taste. The amino acid content of muscle is closely related to the metabolic type of fibers. Oxidative muscles contain more aspartic acid, glutamine and taurine, while glycolytic muscles contain high amount of β-alanine and carnosine.

A number of authors investigated the possibility to use free amino acid content as a marker of freshness for different vegetables [237–239]. Boggio et al. [240] found that free amino acid composition changes markedly during tomato fruit ripening. Particularly, an increase in the relative content of glutamate, and a negative correlation between the relative glutamate content and fruit shelf life were also found [241].

Storage conditions and preservation treatments also can affect the free amino acid content. Some authors [242] reported changes in the concentration of free amino acids of broccoli florets stored in air or in controlled atmospheres. Arginine concentration varied greatly during air storage, while γ-amino butyric acid, alanine and an unidentified amino acid accumulated in response to low O_2 and/or high CO_2 treatment.

Free amino acids and other nonprotein nitrogen compounds have been used as indicators of protein fragmentation during the ripening of cheese [176], meat and meat products, including dry-cured ham [243]. Martín et al. [188] investigated free amino acids formed during processing of Iberian ham, demonstrating that their determination could provide a maturation index for this product.

Free amino acids have been used as quality indices in various fish and crustacean species. Fish processing and storage conditions decisively influence the free amino acid profile [244]. According to Ruiz-Capillas and Moral [245], alanine and 1-methylhistidine can be used, in combination with the dipeptide anserine and tryptophan, as quality indices in both ice- and atmosphere-stored hake. In the case of molluscs and crustaceans, most authors recommend ornithine as a freshness index [191,244]. Ornithine and tryptophan demonstrated to be a suitable freshness index only for Norvey lobsters stored in ice, while the dipeptide anserine revealed to be a suitable freshness index for lobster stored either in ice or protective atmospheres [244].

Table 19.8 summarizes some of the works on HPLC determination of free amino acids, published in the last 10 years.

19.6 BIOGENIC AMINES

Biogenic amines are low-molecular-weight organic bases with aliphatic (putrescine, cadaverine, spermine, and spermidine), aromatic (tyramine, phenylethylamine), or heterocyclic (histamine, tryptamine) structures, which are mainly produced by microbial decarboxylation of amino acids (particularly hystidine, tyrosine, lysine, ornithine, and arginine), and transamination of aldehydes and ketones. Due to the high concentration of precursors (i.e., amino acids), and the presence of decarboxylating microflora, high amounts of biogenic amines can originate during proteolysis in fermented foods (such as cheese, sausages, wine, and beer) or as a consequence of food spoilage [257].

The interest in amine determination is mainly due to their effect on the human vascular and nervous systems. Ingestion of foods containing large amounts of biogenic amines (particularly tyramine and histamine), can cause rash, headache, nausea, hypo- or hyper-tension, cardiac palpitation, intracerebral hemorrhage, and anaphylactic shock, especially when the detoxifying enzymes monoamino oxidase (MAO), present in the gastrointestinal tract, are impaired as in the case of patients treated with MAO inhibitors (including painkillers and drugs used for the treatment of stress, depression, and Parkinson's disease) [258–261]. Tyramine intake exceeding 6 mg is considered dangerous for these patients [262].

Although no standards or guidelines regarding permissible concentrations of histamine or other biogenic amines have been established, the U.S. FDA has set a tolerance limit in fresh fish of 50 mg/kg, whereas European legislation [263] established a maximum histamine concentration of 100–200 mg/kg in fishery products. In general, concentrations greater than 100 mg/kg are not recommended, especially for consumers with a deficiency of the detoxification system. Some European countries have fixed a maximum allowed concentration of the biogenic amines in wine (that ranges between 2 and 8 mg/L) [264]. Recently, Wohrl et al. [265] concluded that 75 mg of pure histamine, a quantity that can be ingested in a single portion of affected cheese, can provoke immediate as well as delayed symptoms in 50% of healthy volunteers [266].

Polyamine, such as putrescine, cadaverine, spermidine, and spermine, can enhance the toxic effects of tyramine and histamine by competing for the detoxifying enzymes [267,268]. Furthermore they can act as precursors of carcinogenic nitroso-amines [269] and, according to Bardócz [270] and Eliassen et al. [271], a dietary regime high in polyamines can promote tumor growth. Polyamines, which are involved in a number of cell processes [272], are practically ubiquitous in all vegetables where they are present at a level of few mg/100 g of fresh weight [273–277]. Tyramine and other aromatic amines are less widespread in vegetables than polyamines, but they can reach particularly high concentrations in some vegetables, where they seem to have a defensive role against insects and herbivores. Prolonged ingestion of plants rich in biologically active amines (tyramine and related phenolic amines) by cattle, horses, sheep, and goats, has been often associated with neurological diseases [278–280].

There are two main reasons for determining biogenic amines in foods: (1) their potential toxicity and (2) the possibility to use them as food quality markers (quality control of raw materials, monitoring fermentation processes, process control, etc.) [281].

TABLE 19.8
HPLC Applications for Free Amino Acids Analysis in Foods

Sample	Sample Preparation	Column	Mobile Phase	Derivatization	Detector	References
Orange juices	Centrifugation	AA analyzer Cation exchange resin (Li$^+$) (200×4.6 mm)	Lithium citrate buffer	Ninhydrin (post-column)	UV-Vis (570 and 440 nm)	[223]
		Supershere RP18, (250×4.0 mm)	A: 0.14 M NaOAc/ 4% TEA (pH 6.4 with H$_3$PO$_4$); B: ACN/H$_2$O (60/40)	PITC	UV (254 nm)	
Soybean-fermented food	Extraction with EtOH (70%)	Alltima C18	A: 20 mM NH$_4$H$_2$PO$_4$ (pH 6.5 with NH4OH)/15% MeOH; B 90% ACN in H$_2$O	FMOC-Cl	FLD (λ_{ex} = 263 nm, λ_{em} = 313 nm)	[246]
Almond	Extraction with MeOH/H$_2$O (10/2 v/v) on defatted sample	Waters Accq (C18)	0.02 M Phosphate buffer (pH 6.4)/ 2% THF (adjusted to an ionic strength 0.08 with NaNO$_3$)/MeOH/ACN	OPA	FLD 1(λ_{ex} = 330 nm, λ_{em} = 450 nm), 2(λ_{ex} = 340 nm, λ_{em} = 425 nm)	[232,233]
Mushrooms	Extraction with 0.1 N HCl	Phenomenex, Prodigy 5 ODS2 (25×4.6 cm)	A: 50 mM NaOAc (pH 5.7)/B: H$_2$O; C: MeOH; flow rate: 1.2 mL/min	OPA	FLD (λ_{ex} = 340 nm, λ_{em} = 450 nm)	[194]
Broccoli	Extraction with boiling MeOH (70%) for group separation of neutral, acid and basic AA on micro-columns	Spherisorb S3 ODS2 (150×4.6 mm)	Phosphate buffer (0.2% NaNO$_3$)/ACN	OPA/2-mercaptoethanol	UV-Vis (340 nm)	[242,247]
Legumes and seedlings	Extraction with EtOH (70%)	C18	A: 0.1 M NH$_4$OAc; B: 0.1 M NH$_4$OAc/ ACN/MeOH 44/46/10 (pH 6.5).	PITC	UV (254 nm)	[198,199]
Green tea	Dilution	Dionex AminoPac PA10 (250×2 mm)	Ternary gradient eluent consisting of: water/0.25 M NaOH/1.0 M NaOAc	No	ED (equipped with a thin layer type amperometric cell)	[248]
Edible seaweeds	Sample hydrolysate (HCl 6N, 1% phenol 110°C×24 h in tubes closed under N$_2$)	Teknokroma, ODS2, (250×4 mm)	A: 0.14 M NH$_4$OAc buffer with 0.05% TEA (pH 6.4 with HAc); B: MeOH/ H$_2$O (60/40)	PITC	UV (254 nm)	[192]
Dry fermented sausages	Not reported	Sperisorb S5 ODS2 (250×4.6 mm)	Not reported	PITC	UV (254 nm)	[249]

(continued)

TABLE 19.8 (continued)
HPLC Applications for Free Amino Acids Analysis in Foods

Sample	Sample Preparation	Column	Mobile Phase	Derivatization	Detector	References
Dry-cured ham	Extraction with 0.1 M HCl	Nova-Pack C18 (150×4 mm, 4 μm) and RP guard column	A: 0.05 M NaOAc buffer (pH 6.85)/1% MeOH/ 0.8% THF; B: 0.05 M NaOAc buffer (pH 5.35)/65% MeOH; flow rate: 1.6 mL/min	OPA	FLD ($\lambda_{ex}=330\,nm$, $\lambda_{em}=445\,nm$)	[243]
Meat essences Mackerel hydrolysates	Extraction with TCA (7%)	Hitachi 2622 SC, (60×4.6 mm)	Lithium citrate buffer	Ninhydrin (post-column)	UV-VIS (570 and 440 nm)	[250,251]
Iberian ham	Extraction with sulfosalicylic acid (10%)	Supelcosil LC 18, (250×4.6 mm), T = 35°C	A: 0.03 M NaOAc with 0.05% TEA (pH 6.8 with HAc); B: MeOH/H$_2$O (90/10); flow rate: 1 mL/min	PITC	UV (254 nm)	[188]
Pork meat	Extraction with 0.01 N HCl, centrifugation, filtration through glass wool	Nova Pak C18 (300×3.9 mm), T = 52°C	A: 70 mM NaOAc (pH 6.55)/2.5% ACN; B: H$_2$O/ACN MeOH (40/45/15)	PITC	UV (254 nm)	[187]
Porcine muscles	Extraction with H$_2$O, TCA (0.5%)	2 ODS2 columns connected in series (Shandon HPLC, 120×4.6 mm, 3 μm), T = 54°C	A: 0.07 M NaOAc with 1.6% ACN and 0.4% MeOH (at pH 6.5 with phosphoric acid); B: 0.07 M NaOAc with 40% ACN and 10% MeOH (at pH 6.5 with phosphoric acid); flow rate: 0.8 mL/min	PITC	UV (254 nm)	[236]
Fermented shrimp waste	Extraction with borate buffer (sonication)	Hypersil ODS 5, (250×4.6 mm, 5 μm); T = 38°C	A: 30 mM ammonium phosphate (pH 6.5) in MeOH/H$_2$O (15/85); B: MeOH/H$_2$O (15/85); C: ACN/H$_2$O (90/10); flow rate: 1.2 mL/min	FMOC-Cl	FLD ($\lambda_{ex}=270$, $\lambda_{em}=316$)	[252]
Norway lobsters	Extraction with PCA (6%)	Lithium cationic exchange (Li$^+$, 150×3 mm)	Lithium citrate buffer	OPA (post-column)	FLD ($\lambda_{ex}=330\,nm$, $\lambda_{em}=465\,nm$)	[244]

Sample	Sample preparation	Column	Mobile phase	Derivatization	Detection	Ref.
Wine	Filtration	Adsorbosphere XL C18 (250×4.6mm, 5μm, 90Å)+Kromasil C18 (150×4.6mm, 5μm precolumn)	A: 6.804g NaOAc 3H$_2$O and 50mL THF in a 1L H$_2$O (pH 5.7); B: MeOH; flow rate: 2mL/min	OPA/2 mercaptoethanol	FLD (λ_{ex}=340nm, λ_{em}=450nm)	[235]
Musts and wine	Centrifugation and dilution (for musts) Filtration (for wine)	Lichrocart, Superspher 100, RP18 (250×4.6mm, 5μm), T=42°C	A: 20mM NaOAc buffer/0.018% TEA/0.3% THF/0.01% EDTA 4% (pH 7.2 with HAc); B: 20% NaOAc buffer (pH 6.0)/40% MeOH/40% ACN/0.018% TEA	OPA/3-MPA/FMOC-Cl	FLD (λ_{ex}/λ_{em} 340/450nm for primary amino acids; 237/340nm for secondary amino acids)	[225]
Cheese	Extraction with 0.1N HCl, centrifugation, supernatant added with TCA (40%), centrifugation	Waters PicoTag C (RP), T=46°C.	A: 70mM NaOAc (pH 6.55 with HAc)/2.5% ACN; B: 45% ACN, 40% H$_2$O/15% MeOH.	PITC	DAD (254nm)	[253,254]
Cheese	Extraction with H$_2$O, centrifugation, supernatant+sulfosalicylic acid, centrifugation	Ultrasphere ODS (250×4.6mm, 5μm), T=30°C.	A: THF/MeOH/0.05M NaOAc (pH 5.9) (1/19/80); B: MeOH/0.05M NaOAc (pH 5.9) (80/20); flow rate: 1.7mL/min	OPA	FLD (λ_{ex}=305/395, λ_{em}=430/370nm)	[255, 176]
Honey	Dilution with H$_2$O, filtration	Hypersyl ODS (200×2.1mm, 5μm) with pre-column (20×2.1mm)	A: 20mM NaOAc/0.018% TEA/0.3% THF (pH 7.2); B: 20% 10mM NaOAc (pH 7.2)/1% HAc/40% ACN/40% MeOH; flow rate 0.45–0.7mL/min	OPA/FMOC	FLD (for primary amino acids λ_{ex}/λ_{em} 340/450nm, for secondary amino acids λ_{ex}/λ_{em} 266/305nm)	[230]
Infant formulas (taurine, methionine, arginine, tyrosine)	Dilution with H$_2$O+MeOH (protein precipitation), centrifugation	Superspher 100 RP-18 (125×4mm) with pre-column (4×4mm)	A: 0.1M NaOAc (pH 5.8)/MeOH (80/20); B: 0.1M NaOAc (pH 5.8)/MeOH (20/80); flow rate: 1.0mL/min	OPA	FLD (λ_{ex}=230nm, λ_{em}=426nm)	[256]

19.6.1   HPLC Analysis

Several analytical techniques including capillary electrophoresis, thin layer chromatography (TLC), GC, IEC, and HPLC, have been proposed for the determination of biogenic amines in various foods. Among these, RP-HPLC is considered the most suitable one. HPLC methods used for amine determination usually involve two steps: amine extraction from the matrix and analytical determination. Depending on the complexity of food matrix and the selectivity of the final analytical determination, a further purification step may be necessary prior to the analytical determination. To ensure adequate sensitivity, a derivatization step is generally required before injection [282].

19.6.1.1   Sample Preparation

Most methods for amine extraction from solid matrices (cheese, meat, fish) involve homogenization/extraction in an acid medium (0.1 M HCl, 0.2–1.0 M HClO$_4$ or 5%–10% thrichloroacetic acid). If a further purification step is needed, the acid extract can be basified and extracted with an organic solvent (buthanol, buthanol/chloroform) able to extract selectively free amines, letting free amino acids in the aqueous layer. Since different amines show different optimum pH for extraction, in order to assure satisfactory recoveries and reproducibility, a strict control of this parameter is needed [282]. Direct derivatization of the acidic extract has been successfully used by a number of authors [261,280,283,284]. When using dansyl chloride (Dns-Cl) as derivatizing agent, the derivate so obtained can be injected directly or undergo liquid–liquid extraction (LLE) with diethyl ether [285]. When there are no problems of interference by the derivatized amino acids, direct injection has the advantage of giving quantitative recoveries of all the researched amines. When cheese samples with particularly high free amino acids are processed, an extraction step with diethyl ether (or toluene) may be useful in order to avoid interference problems for the first eluted peaks. In this way, only dansyl amines are extracted from the organic solvent, while dansyl amino acids remain within the aqueous layer [284]. This simple extraction step requires about 10 min, with minimal solvent consumption. A further improvement can be obtained by adding an appropriate amount of L-proline to the derivatized mixture, in order to neutralize the excess of Dns-Cl that otherwise can cause the appearance of an interfering peak [280,284,286]. Duflos et al. [287] used 0.2 M HClO$_4$ to extract biogenic amines from fish tissue and applied direct derivatization with Dns-Cl, followed by proline treatment and LLE with toluene, before injection into a RP-HPLC. They found that both fish specie and spoilage degree influence amine extraction, and demonstrated that, to achieve a quantification that is closest to reality, the use of the standard addition method should be preferred, rather than the method of direct determination using standard solution.

For liquid samples, such as wine and beer, both direct derivatization without any other sample pretreatment [288–294] and sample purification/preconcentration by LLE [295,296] or SPE [297,298], have been proposed by different authors. Preconcentration by LLE or SPE has the advantage of lower detection limits. Busto et al. [298] used two different commercially available SAX and C18 cartridges for sample clean-up (to remove interfering compounds such as amino acids and polyphenols), before on-column derivatization with OPA. Very recently, Soufleros et al. [299] used polyvinyl pyrrolidone in order to selectively eliminate phenolic compounds from wine sample before dansylation, followed by SPE on C18 and RP-HPLC. Some authors [300], evaluated amine extraction efficiency of different SPE sorbents. C18 silica cartridge was selected as optimal phase with recoveries nearly of 100%. The use of strong cationic exchange cartridges (SCX) makes the elution of the adsorbed compounds more difficult, and the addition of ion pairing agents is often required. According to Hernández-Orte [289], none of the sorbents tested is suitable for the extraction of all the compounds of interest in a single step. These results suggest that direct analysis of derivatized amines, avoiding extraction processes, should be preferred.

19.6.1.2 Chromatography

According to some recent literature, typical conditions for biogenic amine determination are: precolumn derivatization with Dnsl-Cl followed by separation on C8 or C18 column with gradient elution with mobile phases consisting of water or phosphate buffer and acetonitrile (or methanol) or postcolumn derivatization with OPA and gradient elution with mixtures of sodium acetate buffer and methanol (or acetonitrile). In the latter case, a counterion (such as hexanesulfonic or octansulfonic acid) is usually added in the mobile phase.

As an alternative to RP columns some authors used cation-exchange columns. Salazar et al. [301] used an Alkion column (4×150 mm) heated at 40°C to separate OPA derivate of biogenic amines extracted from complete feeds and animal tissues, while Saccani et al. [302] used an IonPac CS17 column and a gradient elution of methanesulfonic acid to separate biogenic amines in meat products.

19.6.1.3 Derivatization and Detection

Moret et al. [280] applied precolumn derivatization with Dns-Cl and OPA to the same sample (vegetables) extracts and compared the results. Figure 19.6 reports HPLC traces obtained for a spinach sample (Kromasil C18 column, 250×4.6 mm i.d., 5 μm particle size).

A good agreement between results obtained with the two procedures was found. Derivatization with Dns-Cl has the advantage of UV detection (254 nm) of both primary and secondary amines

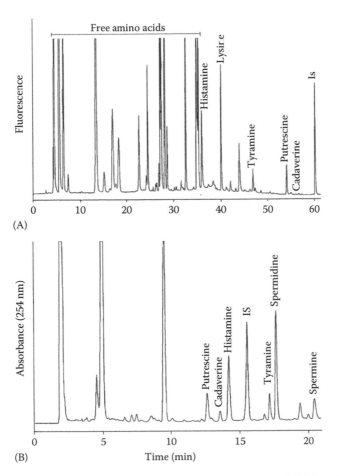

FIGURE 19.6   HPLC trace of a spinach sample derivatized with (A) OPA and (B) Dns-Cl. (From Moret, S. et al., *Food Chem.*, 89, 355, 2005. Copyright 2005. With permission from Elsevier.)

and rapid elution time (using simple gradient elution program consisting of water and acetonitrile or methanol). Both UV (254 nm) and fluorometric detection (excitation 330–340 nm; emission 450–470 nm) can be used [303]. The reaction of dansylation is time-consuming (typical conditions: 1 h at 40°C) but it can be accelerated by using microwave assisted derivatization [304]. On the other hand, OPA derivatization is very rapid (1–2 min) but gives unstable derivatives, so, when using precolumn derivatization, the injection has to be done after a fixed time (or using automated systems). It allows detection of a restricted number of amines (only primary amines react with OPA), requires fluorometric detection (at excitation/emission wavelengths of 340/455 nm), longer elution time, and mobile phases consisting of phosphate buffer and acetonitrile. Nevertheless, OPA derivatization represents a good choice when simultaneous determination of free amines and amino acids is of concern [280,284,305]. Krause et al. [306] developed a protocol for RP-HPLC of the dabsyl derivate of both free amino acids and amines in food samples where more than 40 compounds could be separated simultaneously. Herbert et al. [225] obtained simultaneous quantification of primary and secondary amino acids and biogenic amines in musts and wines using OPA/(3-MPA)/FMOC-Cl fluorescent derivatives, while Draisci et al. [307] used integrated pulsed amperometric detection for simultaneous determination of underivatized biogenic amines and their precursor amino acids.

Other derivatizing agents, such as fluorescamine, PITC, and FMOC-Cl, have been used as an alternative to Dns-Cl and OPA. PITC derivatives cannot be detected by fluorescence, and the method used for them lacks sensitivity. FMOC gives good sensitivity and stable derivatives, but produces multiple products from a single amine, so it is not suitable when dealing with complex samples [297]. According to Busto et al. [298], 6-aminoquinolyl-N-hydroxysuccinimidyl carbamate (AQC) reagent represents a good alternative for determining biogenic amines in wines. It gives stable derivatives of both primary and secondary amines, allowing direct injection of the derivatized sample, with higher sensitivity values than those reported for OPA. Hwang [308] optimized a sensitive derivatization procedure with benzoyl chloride for rapid RP-HPLC determination (less than 15 min) of 9 biogenic amine in fish samples. Both temperature and duration of the benzoylation reaction affected amine peak height, reproducibility, and the presence of interfering peaks. The optimal benzoylation conditions for biogenic amines was incubation at 30°C for 40 min. Kirschbaum et al. [309] tested 3,5-dinitrobenzoyl chloride for precolumn derivatization of biogenic amines and found that it reacts rapidly at room temperature and forms stable derivatives that are well resolved on a C18 column using a ternary gradient system. In 2006, García-Villard et al. [295] described a novel and sensitive HPLC method involving precolumn labeling of the analytes with 1,2-naphthoquinone-4-sulfonate (NQS) and LLE of derivatives with chloroform for analyte preconcentration and sample clean-up in wine samples. In 2007, Gosetti et al. [310] presented a new HPLC method hyphenated with mass spectrometry detection for the separation and determination of six amines in cheese.

Amine quantification is usually accomplished using the internal standard method. Different amines have been used for this purpose: 1,7-diaminoheptane [282], 1,3-diaminopropane [287], and benzylamine [311].

19.6.1.4 Applications

Similarly to free amino acids, biogenic amines may affect food sensory properties [312] and they have been proposed for food characterization. Soleas et al. [293] reported cultivar-related differences in biogenic amine concentration of Ontario wines, while Romero et al. [291] demonstrated that the amine generated during malolactic fermentation could be used as chemical descriptors to characterize table wine samples. Other authors [313,314] used biogenic amine content as an indicator of the degree of proteolysis and the typical character of some particular cheeses. The presence of biogenic amines at relatively high concentrations in meat, fish, and vegetables has been also associated with spoilage due to prolonged storage or food processing, and storage under inappropriate conditions [261,280,281,287,315]. Kumudavally et al. [316] demonstrated the suitability of measuring cadaverine for rapid assessment of bacterial quality of fresh and stale mutton. Moreover, biogenic amine content represents a parameter of hygienic quality in cheese-making [283,317,318] and has been proposed as an index of defective food manufacturing process related to poor sanitary condition [268].

Several authors reviewed occurrence and HPLC determination of biogenic amines in different foods [257,264,271,274,281,303,319,320].

19.7 PHENOLICS

Phenolic substances are widely distributed in the vegetal kingdom and because of this, they are components of a number of foods. Bearing in mind their important biological activity, food technology is generally capable of preserving them in processed foods. The biological activity of phenolic substances mainly depends on their antioxidant activity that has been demonstrated in both in vitro and in vivo. Another important characteristic of some phenolic substances is their capability to determine a number of sensory attributes to foods, of course, because of their molecular weight, they are involved in the taste but not in the odor. Nevertheless, they are very important for the pungency, bitterness, astringency, and other related issues.

The antioxidant activity of phenolic substances makes them easily prone to oxidation. Thus, the analytical techniques must be aimed at not applying pH and temperature variations that cause oxidation.

For a number of years, phenolic substances were dosed by colorimetric techniques, based on redox reactions usually known as "Folin Ciocalteau" methods, even if a number of adjustments were developed to fit different matrix characteristics. The Folin Cioalteau reagent is a mixture of phosphomolybdic and phosphotingstic acids, with molybdenum in the 6+ oxidation state and, when the reaction takes place, it is reduced to form a complex called molybdenum blue and tungsten blue. In this complex, the mean oxidation state is between 5 and 6 and the formed complex is blue so it can be read spectrophotometrically at 750 nm.

Colorimetric determination of the total phenolic substances in foods and beverages were carried out through a number of years; however, then the availability of separation techniques suggested the possibility to better highlight the composition of this very complex mixture. Different approaches were developed mainly by means of GC and GC-MS, with an obviously long procedure of extraction and derivatization (usually trimethylsilyl derivatives were used) that strongly limited the analytical approach to these classes of compounds, and were mainly limited to wine and somewhat less to olive oils.

19.7.1 PHENOLIC SUBSTANCES IN BEER

The major source of phenolic substances in beer is the malt, but the hops, contributes to the total amount as well [321]. Phenolic substances in beer are involved in the physical and chemical stability, as well as in the froth maintenance [322–324]. Despite their important role, the number of scientific papers dealing with phenolic evaluation in beer is not so high [322,325–327].

Montanari [327] applied two HPLC approaches, both using an octadecylsilane (ODS)-3 column and a gradient elution with a buffered mobile phase added with sodium docecyl sulfate, coupled to different detection techniques: amperometric electrochemical detection (AED) and DAD, the latter being in accordance with what Montedoro et al. [328] developed for olive oils. No statistical significant differences resulted between the two methods for gallic acid, 3,5,-dihydroxybenzoic acid, vanillic acid, 3-OH-benzoic acid, syringic acid, chlorogenic acid, m- and o-coumaric acid, while differences exist for protocatechuic acid, p-OH –benzoic acid homovanilic acid, caffeic acid, p-coumaric acid and ferulic acid: higher values were obtained for protocatechuic, p-coumaric and ferulic acids, and lower concentration were measured for p-OH –benzoic, caffeic and homovanilic acids.

19.7.2 PHENOLIC SUBSTANCES IN FRUIT JUICES

Coppola [329] suggested that phenolic acid profile could be used as an indicator of apple juice adulteration. Shui and Leong [330] reported the simultaneous determination in apple juice on an ODS column with methanol/sulfuric acid/water as the mobile phase within 80 min. However, (−)-epicatechin and p-coumaric acid may be eluted with a similar retention time, so that the spectrum must be recorded in order to identify single compounds. The method was validated and intraday and interday relative standard deviation (RSD%) data resulted within 0.3% and 1.6%, respectively. The paper also reports LOD data that range between 0.03 mg/L (syringic acid) and 0.18 mg/L (benzoic acid).

Cranberry juice was also studied by Chen et al. [331] who found that benzoic acid is the major phenolic acid in this kind of sample, and it is mainly present as a free acid, unlike flavonoids, which are mainly present in the esterified form.

19.7.3 PHENOLIC SUBSTANCES IN OLIVE OILS

The importance of the Mediterranean diet led to increased research activity on olive oils, which mainly focused, among other issues, on phenolic substances because of their biological activities and of the fact that virgin olive oils are the only vegetable oils that contain them, while tocopherols are widely distributed in such foods. Phenolic substances are responsible for the typical taste of olive oils, characterized by fruity, bitter, and pungent flavors.

Cantarelli [332] and Montedoro and Cantarelli [333] noted the presence of phenolic compounds in olive oils and they established a set of research priorities related to these compounds that are practically still carried out nowadays. Research issues deal with the development of analytical procedures to quantify phenolic compounds in oils, the relationship between these compounds and the characteristics of the olive fruit, as determined by variety, ripeness degree, and other sources of variation, the relationship between the profile, concentration, and extraction technology of phenolic compounds, the biological role in the antioxidant activity in vitro and in vivo [334].

Montedoro [335] and Vazquez Roncero [336] first observed the presence of benzoic acid and cinnamic acid and their derivatives. Montedoro et al. [328,337,338] described that phenolic substances are present in virgin olive oil mainly as secoiridoids in the dialdehydic form of elenolic acid bonded to the phenolic alcohols hydroxytyrosol (3,4-dihydroxyphenyl-ethanol, 3,4,-DHPEA-EDA) and tyrosol (p-hydroxyphenyl-ethanol p-HPEA-EDA), and as a form of oleuropein aglycone (3,4-DHPEA-EA); they also assigned them the molecular formula. Recently [339,340], oleuropein and ligstroside aglycones were also detected in virgin olive oil. Owen et al. [339,341] and Brenes et al. [342] isolated and characterized the lignans (+)-1-acetoxypinoresinol, (+)-pinoresinol, and (+)-1-hydroxypinoresinol. The peculiar characteristics of lignans is that even if they have an antioxidant activity similar to other phenolics substances as 3,4-DHPEA-EDA, they do not give bitter taste to oils. Bianco et al. [343] characterized more recently a new class of phenolic compounds named hydroxy-isochromans, among which 1-phenyl-6,7-dihydroxy-isochroman and 1-(3′-methoxy-4′-hydroxy)phenyl-6,7-dihydroxy-isochroman. The molecular formulas of olive oil phenolics are reported in Figure 19.7 [334].

Sample preparation is usually carried out by means of two different approaches: LLE or SPE, as recently reviewed by Servili et al. [344]. LLE is carried out by the partition of the nonpolar fraction (mainly TAGs) in n-alkane solvent (usually n-hexane) and of the polar fraction in a polar phase, usually methanol or methanol–water at different ratios [336,341,345–349]. SPE is carried out by means of C18 cartridges [343,350–354]. Some authors performed LLE extract purification through a C18 cartridge [343].

The phenolic compounds content of virgin olive oils is not a composition parameter enclosed among those listed by the normative (EC Regulations, International Olive Oil Trade standard, Codex Alimentarius Standard), probably because of the widespread distribution of values

Phenolic acids Benzoic acid derivatives Caffeic acid Gentisic acid Benzoic acid Vanllic acid Protocatechuic acid p-Hydroxybenzoic acid Syringic acid 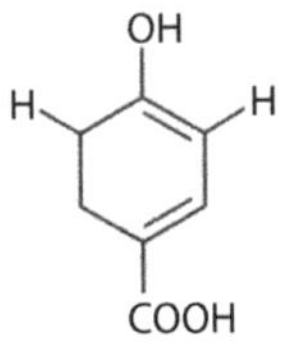	Hydroxy-isochromans .1-phenyl-6,7-dihydroxy-isochroman 1-(3′-methoxy-4′-hydroxyphenyl-6,7-dihydroxy-isochroman 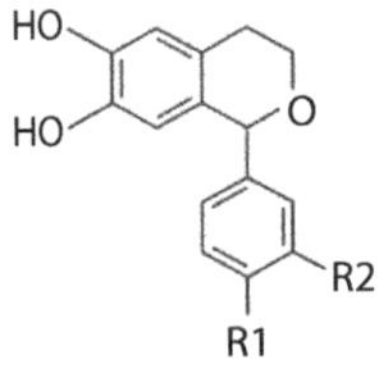
Cinnamic acid derivatives Caffeic acid p-Coumaric acid o-Coumaric acid Ferulic acid Cinnamic acid Sinapinic acid 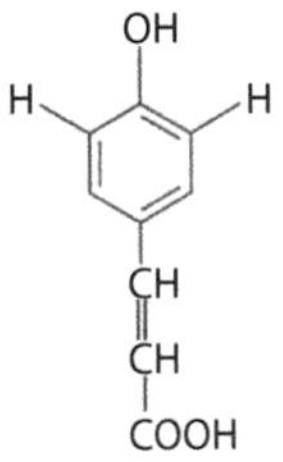	Flavonoids Flavones Apigenin Luteolin Flavanonol (+)-Taxifolin 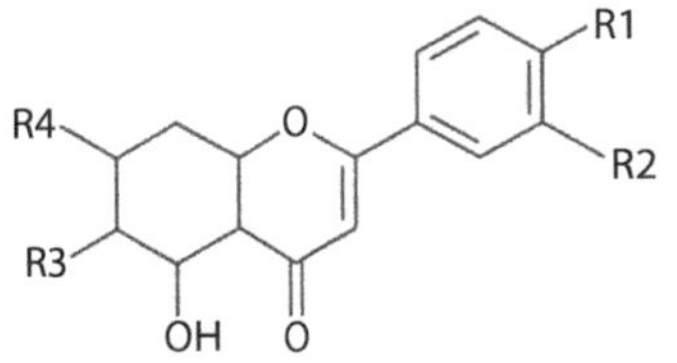
Other phenolic acids and derivatives 4-(Acetoxythyl)-1,2-dihydroxybenzene Dopac (3,4-dihydroxyphenylacetic acid) 4-hydroxyphenylacetic acid 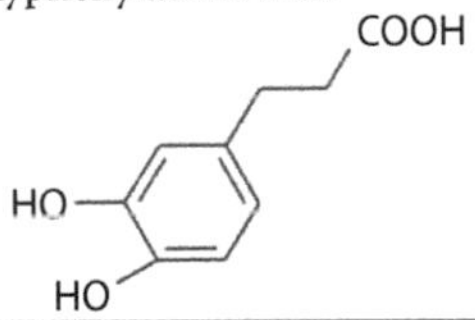	Lignans (+)-1-Acetoxypino resinol (+)-Pinoresinol (+)-Hydroxypinoresinol 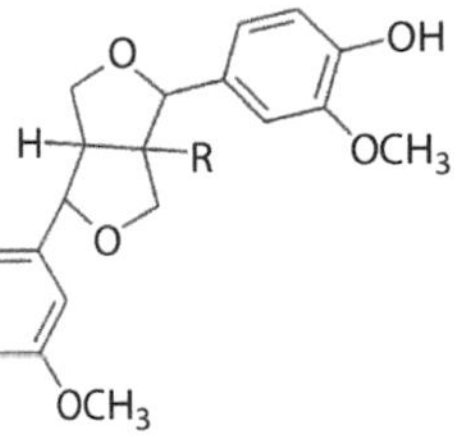
Phenolic alcohols 3,4-Dihydroxyphenyl-ethanol (3,4-DHPEA) p-Hydroxyphenyl-ethanol (p-HPEA) (3,4-Dihydroxyphenyl)ethanol-glucoside 2-(4-Hydroxyphenyl)ethylacetate	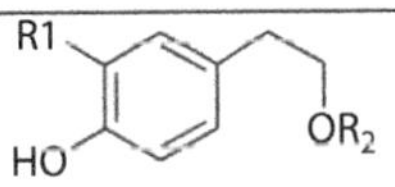
Secoiridoids Oleuropein Dialdehydic form of oleuropein aglycone Dialdehydic form of ligstroside form aglycone Ligstroside aglycone Oleuropein aglycone (3,4-DHPEA-EA) Dialdehydic form of decarboxymethyl elenolic acid linked to 3,4-DHPEA (3,4-DHPEA-EDA)	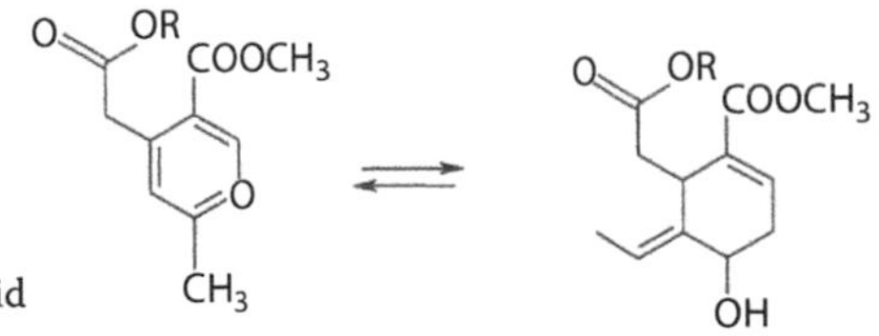
Dialdehydic form of decarboxymethyl elenolic acid linked to p-HPEA (p-HPEA-EDA)	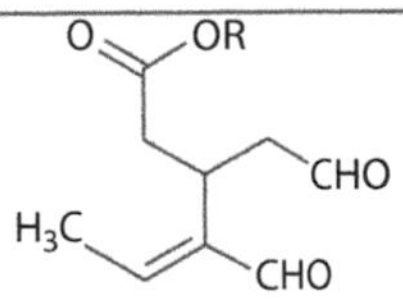

FIGURE 19.7 Molecular structures of olive oil phenolic substances. (Adapted from Carrasco-Pancorbo, A. et al., *J. Sep. Sci.*, 28, 837, 2005. With permission.)

(depending on cultivar, stage of ripening, technology applied for extraction, interaction among cultivars, the stage of ripeness, and environmental aspects), that make it hard to establish a limit.

The differences between oils obtained through different cultivars and in different areas is well highlighted by HPLC traces of Figure 19.8 [355]. Picual and Arbequina are typical Spanish products and Saggianese is typical of Tuscany in Italy.

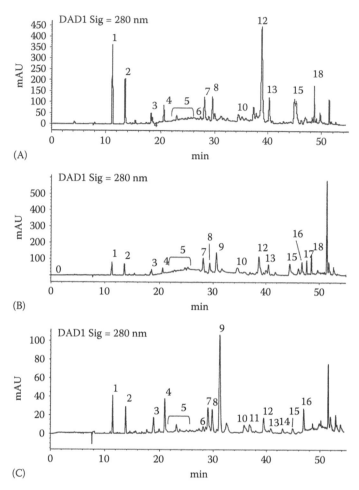

FIGURE 19.8 HPLC traces of phenolic substances of three different extra virgin olive oils: (A) Picual from Granada, Andalucia, Spain, (B) Arbequina from Malaga, Andalucia, Spain, and (C) Saggianese from Tuscany, Italy. (From Oliveras-Lopez, M.J. et al., *Talanta*, 73, 726, 2007. Copyright 2007. With permission from Elsevier.)

Perhaps due to lack of regulation, no official, harmonized or internationally validated method is available at present. Despite lack of official standards, many protocols of Protected Denomination of Origin list limits for "total polyphenols" and this data is often used in trade.

In 2006, International Olive Oil decided to validate a method for phenolic substance evaluation and the decision was to validate both a colorimetric and a HPLC method. The HPLC method is the one already validated by the Italian Technical Committee for Fats and Oils, based on the Cortesi's method. Phenolic substances are extracted by means of methanol/water 80/20 and injected on a RP C18 Spherisorb ODS-2 type column, 25×0.46 cm, 5 μm particle size, and the detection is carried out at 280 nm. Syringic acid is used as the internal standard. 27 peaks are separated and identified, enclosing some oxidized forms of aglycones.

In 2007 IOOC organized a ring test involving about 20 laboratories in order to standardize the method. The results showed a mean RSD of about 20%. In the following sections, analytical approaches will be discussed for different kinds of food.

19.7.4 PHENOLIC SUBSTANCES IN PLANTS: MISCELLANEOUS

As stated in the introduction, phenolic substances are widely distributed in vegetal kingdom and are present in a number of foods and food products. Sometimes the phenolic content is interesting from a pharmaceutical point of view, mainly when borderline characteristics are considered, as happens for "nutraceuticals" and "functional foods."

Zgorka and Kawka [356] applied RP-HPLC both with UV detection and fluorometric detection in assessing the phenolic concentration in *Eleutherocossus senticosus* root preparation whose properties are considered similar to Ginseng. Amaral et al. [357] fully characterized the phenolic profile of walnut leaves extract, which is used as a folk medicine. Juglone (5-hydroxy-1,4-naphtoquinone), which is characteristic of the fresh leaves, was found in very low concentration, due to polymerization phenomena. Therefore, this compound is unsuitable for quality control of this drug.

More usual vegetables were studied by Nuutila et al. [358] who characterized onions and spinach for the phenolic composition, with and without previous hydrolysis. The authors performed the simultaneous determination of phenolic acids, flavonols, flavones glycosides, and cathechins. Figure 19.9 reproduces the separation obtained for the standard mixture in this study.

Tea is known to be a good source of phenolic substances and its earliest HPLC characterization was published in 1976 [359]; The introduction of DAD as a detection technique for HPLC greatly improved the possibility of fully characterizing it. In 2004, Yao et al. [360] published a study on tea cultivated in Australia: six phenolic acids were identified, including *p*-coumarylquinic, *p*-coumaric, and 3-(*p*-hydroxyphenyl) propionic acid. Previously, Bonoli et al. [361] carried out a comparison of HPLC with capillary electrophoresys and concluded that the latter shows advantages in terms of the time of analysis with respect to HPLC, even if in 2004, Pelillo et al. [362] were able to develop a fast analysis by means of HPLC.

A number of phenolic substances were identified by Sterbova et al. [363] in different plants by a BDS C18 columns with 3 μm particle size, and a comparison was carried out on different extraction methods, by the comparison of microwave-assisted extraction (MAE) to the traditional approach.

Caffeic, *p*-coumaric, ferulic acids concentrations were measured in tomato extracts, after HPLC separation on RP C18 column, with formic acid and methanol as the two solvents used for gradient elution [364].

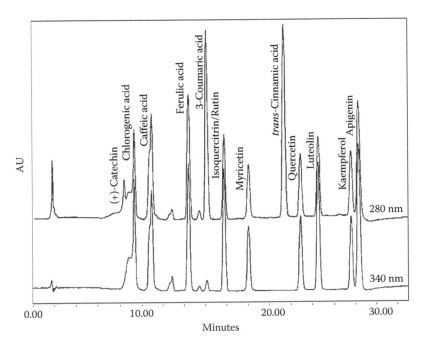

FIGURE 19.9 HPLC chromatogram of standards of phenolic acids, flavonols, flavones, and glycosides isolated in onions and spinach. (From Nuutila, A.M. et al., *Food Chem.*, 76, 519, 2002. Copyright 2002. With permission from Elsevier.)

Besides aromatic plant belonging to *Labiatae* family [365] and other plants [366], rice was studied because of its peculiar content of derivatives of ferulic acid [367]. In this case, hydroxybenzoic acid was detected at 280 nm, while hydroxycinnamic acid and its derivatives were detected at 325 nm.

19.7.5 PHENOLIC SUBSTANCES IN WINE

The evaluation of phenolic substances in wine is a historical topic, and colorimetric methods were developed at first for quality control of wine. Nowadays, it is still performed, and a parameter named "optical density" is still measured as an index of color, while other indices have been developed for a rapid evaluation of phenolic compound content. Moreover, the form of some phenolic compunds (mono- or diglucoside) is used to distinguish between wines obtained by the fermentation of the juice of *Vitis vinifera* or *Vitis rupestris*.

As nowadays wine making has factory dimensions, the need for a rapid method is important, so a flow injection technique for liquid–solid extraction coupled to an HPLC was developed [368]. The target analytes were extracted from wine in a continuous way by means of a minicolumn packed with C18, the elution was carried out by means of acid water (pH 2) and acetonirile, then the solvent was evaporated by nitrogen stream on-line to a HPLC injector. Malvidin-3-glucoside, cyaniding-3-glucosyde, and peonidin-3-glucoside antocyanins were determined in a more rapid, accurate, and sensitive way than with the traditional methods.

Selectivity and sensitivity were increased by Rodriguez-Delgado et al. [369] using a fluorometric detector, and 17 phenolic substances were separated and their concentration was measured in about 35 min of chromatographic run.

Five phenolic antioxidants were dosed in Greek red wines by Sakkidi et al. [370]. *trans*-resveratrol, a stilbene compound typical of red wines was recently thoroughly studied in connection with its biological activity of lowering serum lipids and inhibiting placket aggregation, so methods dedicated to the analytical evaluation of this compounds were developed. Malovana et al. [371] were able to develop a method that allows the simultaneous determination of *trans*-resveratrol and other phenolic substances in Canary Islands red wine. By using a three-stage linear gradient, *trans*-resveratrol was eluted in 25 min and all phenolic compounds were eluted in 35 min. Celotti et al. [372] also developed a rapid method for *trans*-resveratrol evaluation in Italian Red wines Recioto and Amarone by HPLC with UV detection. Similar approaches were carried out by other researchers in wines from different geographical areas.

Phenolic compounds in Sicilian wines were directly detected by La Torre et al. [373] using an HPLC with a DAD coupled on-line with a MS system equipped with ESI source operating in the negative-ion mode and a quadruple mass analyzer. The structure was elucidated by recording MS spectra at different voltages, in addition to the molecular mass information. The method allowed both the identification and determination of 24 phenolic compounds in 22 different commercial Sicilian red wines by direct injection without any prior purification of the sample. Figure 19.10 reproduced an HPLC trace obtained in this work.

Phenolic and antioxidant substances have usually studied in red wines, however, recently, interest has increased in the study of bioactive phenolics in white wines; Frega et al. [374] isolated and measured concentration of ethyl caffeoate in "Verdicchio" white wine by HPLC-tandem-mass spectrometry (HPLC-ESI-MS/MS) and they also determined its effects on hepatic stellate cells and intracellular peroxidation. The results were interesting in the light of other studies demonstrating the relationship between reactive oxygen species, chronic liver injury, and hepatic fibrosis.

The development of HPLC technique led to an increase in the number of scientific papers dealing with phenolic evaluation in foods and, in the meantime, it also improved the number and type of foods in which phenolic substances were evaluated. Despite the high number of scientific papers, HPLC analysis of phenolic substances, except for wine and tocopherols in fats and oils, has never became an official method of analysis, so the number of scientific papers has ever more increased,

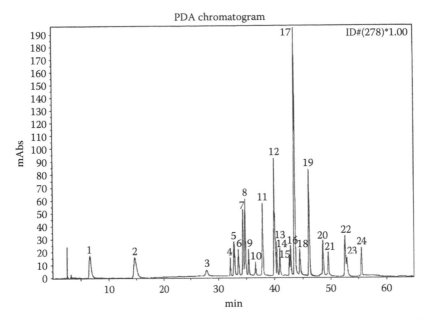

FIGURE 19.10 HPLC–PDA trace of a standard mixture of phenolic compounds. Detection at 278 nm. Peak identification: 1, gallic acid; 2, protocatechuic acid; 3, tyrosol; 4, vanillic acid; 5, procyanidin B1; 6, (+)-catechin; 7, caffeic acid; 8, syringic acid; 9, procyanidin B2; 10, (–)-epicatechin; 11, ethylgallate; 12, ferulic acid; 13, rutin; 14, isoquercitrin; 15, kaempferol-3-O-glucoside; 16, isorhamnetin-3-O-glucoside; 17, p-coumaric acid; 18, myricetin; 19, trans-resveratrol; 20, quercetin; 21, cis-resveratrol; 22, kaempferol; 23, isorhamnetin; 24, rhamnetin. (From La Torre, G.L. et al., *Food Chem.*, 94, 640, 2006. Copyright 2006. With permission from Elsevier.)

not only depending on new scientific knowledge and instrumental development, but also on the lack of harmonized and standardized methods. In Table 19.9, a list of several HPLC approaches to phenolic substances analysis is reported; however, it is not exhaustive as the number of papers dealing with this subject is very high. The purpose is just to give an idea of the scientific activities in this field. The reader will note that the kind of columns used for phenolic analysis in foods does not vary so much, as well as the detection methods. New approaches have been developed in more recent years based on capillary electrophoresis techniques. Another common issue is the use of a more acidic mobile phase (formic acid, phosphoric acid, acetic acid), which is used with the aim to avoid the ionization of phenolic compounds because of their weak acidity that could lead to a poorer separation.

19.8 VITAMINS

A vitamin has been defined [399] as a biologically active organic compound that is essential for the organism's normal metabolic and physiologic functions. Vitamins are not produced by human body itself, thereby they have to be obtained from the diet. As micronutrients, in contrast to nonessential chemical substances, they present both a minimum and a maximum level of intake beyond which arise risks of deficiency conditions or adverse effects [400].

According to the directive 90/496/EEC on nutrition labeling of foodstuffs [401], the vitamin levels in food have to be reported as relative daily allowance (RDA%). This information is mandatory on the label if any claim about their content [402] or addition of vitamins [403] is made. The European Union Regulation 1925/2006 stipulates the vitamin formulations that may be added to foods [403].

Considering the importance of vitamins in human health and the legislation regarding them, the analysis of these compounds appears to be very important. Indeed, vitamin assays in food

TABLE 19.9
HPLC Applications for Phenolic Analysis in Foods

Food/Topic	Sample Pre-Treatment	Column	Mobile Phase	Detector	References
Beer	Dilution with mobile phase and filtration	Inertsil ODS-3 (25×0.46cm, 5μm)	A: MeOH/H$_2$O (50/50 v/v)/34.7 μM SDS; B: MeOH/H$_2$O (50/50 v/v)/173 μM SDS	ED	[327]
Fruit juices	Centrifugation at 5000rpm and filtration through 0.45 μm filter	Shim Pack VP-ODS (25×0.46cm)	A: aqueous H$_2$SO$_4$; B:MeOH	UV (215 nm)	[330]
Fruit juices	SPE (Sep-Pak C18) and filtration Hydrolysis, SPE (Sep-Pak C18) and filtration	Eclipse XDR-C18 RP (15×0.46cm, 5μm)	A: H$_2$O/HAc (97/3 v/v); B:MeOH	UV (280 and 360 nm)	[331]
Olive oil	LLE	Lichrosphere 100 RP18 (25×0.4cm, 5μm)	A: H$_2$O/HAc; B: MeOH	UV (280, 285 nm)	[375]
Olive oil	Adjusting at pH 3.0, extraction with EtAc, evaporation to dryness and redissolution in MeOH	C18 RP (Symetry 10×0.46cm, 3.5μm)	A: H$_2$O/0.01% HAc; B: ACN	UV (280 nm) and MS for identification	[348]
Olive oil	SPE (C8 500 mg, phenolic elution with ACN)	ODS-1, ODS-2, C8, C1, PHENYL (Spherisorb, 25×0.46cm, 3μm)	10^{-3} M H$_2$SO$_4$/ACN	UV (225 nm), DAD	[351]
Olive oil	LLE with H$_2$O/MeOH	Apex octadecyl 104 C18 (25×0.4cm, 5μm)	A: 2% HAc; B: MeOH; C: ACN; D: IP	UV (280nm)	[349]
Olive oil	LLE with EtOH/H$_2$O (pH 2.5 by HCOOH) 70/30 (v/v)	Luna RP 18 (25×0.46cm, 5μm) Lichrosorb PC18 (25×0.46cm, 5μm)	A: H$_2$O (pH 3.2 with HCOOH); B: ACN	DAD APCI Electrospray-MS	[355]
Olive oil	SPE (C18)	Ultrasphere ODS C18 (25×0.2cm, 5μm)	A: H$_2$O; B: MeOH	UV (225 nm)	[376]
Olive oil		Spherisorb (25×0.46cm, 10μm)			[377]
Olive oil	LLE with MeOH/H$_2$O (80/20 v/v)	Spherisorb ODS2 (25×0.46cm)	A: 0.5%HAc; B: ACN	UV, ESI-MS (positive ion mode)	[378]
Olive oil	LLE with MeOH	μ Bondapak C18 (15×0.46cm)	A: MeOH/2%HAc; B: MeOH	DAD (280nm)	[379]
Olive oil	Aqueous Na$_2$SO$_3$/EtOH SPE	Lichrosorb RP18 (25×0.46cm, 5μm)	A: H$_2$O (pH 3.2 with H$_3$PO$_4$); B: ACN	MS DAD	[380]

Sample	Sample preparation	Column	Mobile phase	Detection	Ref.
Olive oil	LLE	Hibar RP18 (25×0.46cm, 7µm) Spherisorb ODS C18 (25×0.46cm, 5µm)	A: 2%HAc; B: ACN; C: MeOH	UV	[346]
Olive oil	LLE with MeOH/H_2O	Erbasil C18 (15×0.46cm)	A: 2% HAc (pH 3.2); B: MeOH	UV	[328,353]
Olive oil	LLE with MeOH/H_2O	RP18 nPecosphere (8,3×0.45cm, 3µm)	A: 2% HAc; B: MeOH/HAc 2%	DAD	[381]
Olive oil	LLE with MeOH	RP 18 Latex (25×0.4cm, 5µm)	A: 2% HAc (pH 3.2); B: MeOH	UV, ESI-MS (positive and negative ion mode)	[339,341]
Olive oil	LLE with buffer, SPE (Phenyl)	Nucleosil ODS (25×0.21cm. 5µm)	H_2O/ACN (17/83 v/v) with 0.08% HAc	UV, FLD MS, MS/MS	[382]
Olive oil	LLE and SPE (diol phase)	Lichrosphere 100 RP 18 (25×0.4cm, 5µm)	A: 3%HAc; B ACN/ MeOH (50/50 v/v)	UV, DAD	[383]
Olive oil	LLE with MeOH/H_2O and SPE (C18)	Nucleosil ODS (25×0.21cm, 5µm)	A: H_2O/HCOOH; B: MeOH/HAc	MS, MS/MS (APCI-MS) (positive and negative ion mode)	[343]
Olive oil	LLE and SPE (diol phase)	Spherisorb S3 ODS 2 (25×0.46cm, 5µm)	A: H_2O/HAc (95/5 v/v); B: MeOH; C: ACN	DAD (240, 280, 335nm)	[384]
Plant	ISO 3103, 1980 method	Hypersil ODS S5 (25×0.46cm, 5µm)	A: ACN/HAc 2%; B: ACN	UV (280, 310, 340, 380, 450, 510nm)	[360]
Plant (fennel)	Extraction with 20% MeOH	Chromolith Performance RP-C18 (10×0.46cm)	A: ACN/HCOOH 5%; B: ACN/0.1% HCOOH	UV (330nm)	[366]
Plant (Rosemary)	Extraction with MeOH/H_2O	Chromolith Performance RP-18e (10×0.46cm)	A: ACN (65.1%)/H_2O (34.9%)/H_3PO_4 (0.02%); B: ACN (22%)/H_2O (78%)/H_3PO_4 (0.25%)	UV (230, 270, 280, 330nm)	[385]
Plants	Extraction with H_2O/MeOH (on dried sample) refluxed at 90°C filtration	Spherisorb ODS 2 (25×0.46cm. 5µm)	A: H_2O+HAc 1%; B: H_2O+HAc 6% C: H_2O/ACN (65/30, v/v) +5% HAc	UV (280nm)	[365]
Plants	MAE+SPE (RP-105)	Hypersil BDS C18 (10×0.46cm, 3µm)	A: H_2O/HAc 3%; B: MeOH	UV (280 and 320nm)	[363]
Plants	Extraction with MeOH+SPE (C18, 500 mg+quaternary amine, 500mg)	Hypersil ODS (20×0.46cm, 5µm) Simmetry C18 (25×0.46cm, 5µm)	MeOH/H_2O/HAc (23/77:1, v/v/v) MeOH/0.01 M H_3PO_4 (23:77 v/v)	UV (254 nm) FLD	[356]

(continued)

TABLE 19.9 (continued)
HPLC Applications for Phenolic Analysis in Foods

Food/Topic	Sample Pre-Treatment	Column	Mobile Phase	Detector	References
Plants (onion and spinach)	Hydrolysis HCl, Reflux 80°C,	Symmetry C18 (15×0.39 cm, 5 μm)	A: MeOH; B: H_2O/ 0.1 M TFA	UV (280, 340 nm)	[358]
Rice	Extraction with 70% EtOH/ H_2O+SPE (Sep Pak C18, 1 g)	Cosmosil 5C18-MS-II-RP (15×0.46 cm)	A: ACN; B: 0.025% TFA in H_2O	UV (280, 325 nm)	[367]
Tea	Extraction with H_2O/ACN (1:1)	Wakosil-II 5C18 HG (15×0.30 cm)	H_2O/MeOH/ H_3PO_4 (85/15/0.1)	DAD	[386]
Tomato	Freeze-drying, then saponification or liquid extraction with MeOH/ H_2O (80/20 v/v)	RP Luna C18 (15×0.46 cm, 5 μm)	A: HCOOH 0.1%; B: MeOH	UV (270, 310, 325 nm)	[364]
Walnut	Extraction with: (1) MeOH (2) $CHCl_3$ (3) H_2O (pH 2) SPE (Isolute C18)	LichrocartvRP 18 (25×0.4 cm, 5 μm)	TFA 0.1% in H_2O/ MeOH	DAD (280, 320, 350 nm) MS	[357]
Wine		UltraSep ES RP18 MLD II, (2.5×0.4 cm, 5 μm)	A: ACN/phosphate buffer (5/95 v/v); B: ACN/phosphate buffer (50/50)	UV-Vis (518 nm)	[387]
Wine	C18 (Sep-Pak)	Inertsil ODS-3 (15×0.3 cm, 3 μm)	A: 50 mM NaOAC buffer (pH 5)/ MeOH (80:20 v/v); B: 50 mM NaOAc buffer (pH5)/ MeOH/ACN (40:40:20 v/v/v)		[388]
Wine	LLE with Et_2O	ODS Hypersil RP (10×0.21 cm, 5 μm)	A: H_2O/PCA 0.2%; B: MeOH	UV (280 nm)	[389]
Wine	Wine filtration through 0.45 μm filter	Discovery C18 (15×0.21 cm, 5 μm)	A: HCOOH in H_2O; B: HCOOH in ACN	ESI-MS (negative ion mode)	[373]
Wine	Hydrolysis in 50% M (v/v) containing HCl	C18 Symmetry column (25×0.46 cm, 5 μm)	A: H_2O/5%HAc; B: ACN	DAD	[390]

Sample	Sample preparation	Column	Mobile phase	Detection	Reference
Wine	LLE with EtAc	Partisphere C18 (12.5×0.46cm. 5μm)	H2O/HAc/ACN	UV	[372]
Wine	Dilution with H_2O, filtration	Nucleosil 100 C18 (25×0.4cm, 5μm)	A: H_2O/HAc; B: MeOH	FLD	[391]
Wine	Filtration	Lichrospher 100 RP 18	A: H_2O/ HCOOH/ACN (87/10/3 v/v/v); B: H_2O/ HCOOH/ACN (40/10/50 v/v/v)	UV-Vis (518 nm)	[392]
Wine	None	Hypersil ODS (25×0.4cm, 5μm)	A: HAc; B: MeOH; C: H_2O	UV (280, 306, 360 nm)	[393]
Wine	Dilution and filtration on 0.20 μm PTFE membrane	Cromo lith RP-18 (10×0.46cm)	A: MeOH/H_2O (2.5/97.5 v/v, pH 3 with H_3PO_4); B: MeOH/H_2O (50/50 v/v, pH 3 with H_3PO_4)	DAD (200–405 nm)	[394]
Wine	LLE with Et_2O SPE (Sep-PAk C18)	Nova Pak C18 (15×0.39cm, 4μm)	A: MeOH/Hac/H_2O (10/2/88); B: MeOH/HAc/H_2O (90/2/8)	UV (280 nm)	[371]
Wine	SPE (ODS, 100 mg/SAX Oasis 200 mg)	Hypersil ODS (20×0.46cm, 5μm)	A: HAc/H_2O (2/98); B: HAc/ACN/ H_2O (2/20/78)	UV (254, 280, 340 nm)	[395]
Wine	Isotachophoretic separation of analytes, by a 90 mm×0.80 mm i.d. capillary tube made of FEP	Discovery RP Amid C16 (25×0.30cm, 5μm)	A: MeOH; B: ACN; C: 0.085% H_3PO_4	UV (280 nm)	[396]
Wine	Preparative HPLC on Spherisorb S5 ODS2 column (25×0.20cm, mobile phase HCOOH 4.5% in H_2O/ACN	Chrompack, Chromspher C18 (25×0.46cm, 5μm)	A: HCOOH 4.5% in H_2O; B: ACN	DAD (200–700 nm) ESI-MS	[374,397]
Wine	Adjustment at pH 2 LLE with Et_2O	Nova-PAk C18 (15×0.39cm, 4μm)	A: MeOH/HAc/H_2O (10/2/88); B: MeOH/HAc/H_2O (90/2/8)	UV (280 nm) FLD ($\lambda_{ex}=278$ nm, $\lambda_{em}=360$ nm; $\lambda_{ex}=330$ nm, $\lambda_{em}=374$ nm)	[369]
Grape and wine	LEE with MeOH/H_2O/HCOOH Filtration through 0.45 μm filter	ACE 5 C18 (25×0.46cm)	A: $NH_4H_2PO_4$ 50mM (pH 2.6); B: $NH_4H_2PO_4$ 50 mM (pH 2.6)/ACN (20/80 v/v); C: H_3PO_4 200mM (pH 1.5)	DAD (280, 320, 360, 520 nm) FLD ($\lambda_{ex}=280$, $\lambda_{em}=320$ nm)	[398]

are carried out to verify the correspondence with the label declaration, to assess quality of foods supplements, for acquiring data for food composition tables, for shelf life studies following vitamins contents during processing, packaging and storage, and so on.

Vitamins are generally labile compounds and many of them are susceptible to oxidation and breakdown. Since the mid-1970s, the most applied method for vitamins analysis has been HPLC, because this technique does not need hard derivatization and its nondestructive nature allows the use of HPLC both as a preparative purification method as well as for quantification.

Many reviews [404–406] and book chapters [407–409] have been published on vitamins analysis by HPLC. The aim of this chapter is not to be an exhaustive review of vitamin analyses, but just an overview of the possible applications of HPLC in this field. The attention has been focused on studies done in the last 10 years.

19.8.1 Fat-Soluble Vitamins

19.8.1.1 Vitamin A and Provitamin A (β-Carotene)

This class includes both vitamin A and the provitamin A carotenoids. All the compounds related to all-*trans*-retinol (Figure 19.11) are known as vitamin A. These compounds, together with their metabolites and synthetic derivatives, exhibiting the same properties are called retinoids. Vitamin A is found in animal products as retinyl esters (mainly palmitate).

Carotenoids are isoprenoid compounds that are biosynthesized only by plants and microorganisms. Some carotenoids (α- and β-carotene, β-cryptoxanthine) can be cleaved into vitamin A (retinol) by an enzyme in the small intestine. Carotenoids have been reported to present some effects in the prevention of cardiovascular diseases [410] and in the prevention of some kind of cancers [411]. Furthermore, antioxidant activity has been widely reported [411–414] but a switch to pro-oxidant activity can occur as a function of oxygen concentration [415,416].

The activity of vitamin A is related to vision process, tissue differentiation, growth, reproduction, and the immune system. A deficiency of this micronutrient mainly leads to visual problems, impaired immune function, and growth retardation in children. Hypervitaminosis could lead to hepatotoxicity, affect bone metabolism, disrupt lipid metabolism, and teratogenicity [417]. The isomerization of β-carotene, due to technological processes in foods, leads to a reduction of the vitamin A activity; it is therefore important to analyze it.

The vitamin formulations that may be added to foods are retinol, retinyl acetate, retinyl palmitate, and β-carotene [403]. Many reviews [418] and book chapters [419,420] report the development of retinoids analysis over the years. The introduction of HPLC allowed for a significant advancement in this field, allowing the separation of retinoids isomers. This group of compounds presents a high sensitivity to light, heat, and oxygen. Furthermore the extraction is complicated due to some properties like insolubility in water and strong protein binding.

19.8.1.1.1 Sample Preparation

19.8.1.1.1.1 Retinoids The challenge in fat-soluble vitamins analysis is to separate them from the lipid fraction that contains interferents. Alkaline hydrolysis, followed by LLE, is widely applied to remove triglycerides. This technique converts the vitamin A ester to all-*trans*-retinol. A milder process, which does not hydrolyze vitamin A ester, is alcoholysis carried out with methanolic KOH solution under specific conditions that favor alcoholysis rather than saponification. A more accurate explanation of this technique is reported in the book *Food Analysis by HPLC* [409]. For some kind of matrices a simple liquid extraction can be sufficient with [421–423] or without [424,425] the purification

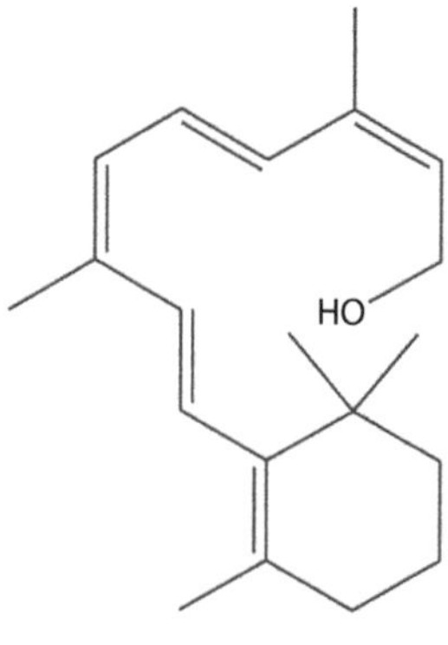

FIGURE   19.11   Structure of retinol.

step. The clean-up was carried out mainly using an open column chromatographic purification, but more recently has been substituted by SPE cartridges, as C18 [422], C2 [423], and Chromabond XTR® [426].

19.8.1.1.1.2 Carotenoids A large number of solvents have been used for extraction of carotenoids from vegetables matrices, such as acetone, tetrahydrofuran, *n*-hexane, pentane, ethanol, methanol, chloroform [427–431], or solvent mixtures such as dichloromethane/methanol, tetrahydrofuran/methanol, *n*-hexane/acetone, or toluene or ethyl acetate [424,432–435]. SPE has been used as an additional purification step by some authors [422,426]. Supercritical fluid extraction (SFE) has been widely used, as an alternative method, also adding CO_2 modifiers (such as methanol, ethanol, *n*-hexane, water, methylene chloride) to increase extraction efficiency [436–438]. In addition, saponification can be carried out, but a loss of the total carotenoid content has been observed and, furthermore, direct solvent extraction has been proved to be a valid alternative [439].

For more comprehensive information, Rodrìguez-Bernaldo and Costa [440] have recently made an exhaustive review about carotenoid analysis in vegetable samples. Carotenoids can easily isomerize and be oxidized, leading to artifacts in the analysis. Therefore, the extraction must be carried out quickly, avoiding exposure to light, oxygen, and high temperatures.

19.8.1.1.2 Chromatography
Both NP (428, 441, 442) and RP (427, 436, 437) chromatography has been applied to analysis of vitamin A and provitamin A. A Lichrosorb Si60 column (250 mm × 4.6 mm) has been used for NP chromatography using hexane and hexane/2-propanol as mobile phase [441]. More widespread are C8, C18, and C30 column with acetonitrile, methanol, water, and mixtures of these as the most used mobile phase [424,427,428,443]. Chávez-Servín et al. [442] reported the use of a short narrow-bore column (50 mm × 2 mm id), which enables less solvent consumption and higher mass sensitivity.

19.8.1.1.3 Detection
Detection can be performed using UV, UV-Vis, DAD, fluorometric detector (FLD) or ED. Gatti et al. [444] report an on-line photochemical derivatization to enhance the selectivity. A photo reactor system was inserted on-line between the column and the FLD. The eluate was irradiated at 254 and 366 nm inducing alterations that modified spectral properties of the analytes. Recently an increasing number of studies have reported the use of LC–MS for the identification of both retinoids and carotenoids [426,433,445–447].

19.8.1.1.4 Applications
A list of selected applications of HPLC for vitamin A analysis is reported in Table 19.10.

19.8.1.2 Vitamin E
Vitamin E is the term used to describe eight naturally occurring components related to fat-soluble tocochromanols, four tocopherols and four tocotrienols, which exhibit antioxidant activity and are nutritionally essential. The basic structure is a chromanol ring with an isoprenoid side chain, which can be saturated (tocopherol) or unsaturated (tocotrienol). Both occur in α-, β-, γ-, and δ-form, according to the number of methyl group present on the chromanol ring. α-Tocopherol (three methyl groups) is the most active homologue, followed by β-, γ-, and δ-tocopherol (Figure 19.12).

The main action of vitamin E in human tissue is to prevent oxidation of polyunsaturated fatty acids (PUFA), thereby protecting lipid and phospholipids in membranes. Vitamin E interacts synergically with other nutrients, such as vitamin C, selenium, and zinc, which are also involved in the oxidation pathway. The recommended intake is strongly related to the quantity of PUFA consumption. Some studies [454–456] on animal models and epidemiological trials in human suggest

TABLE 19.10
HPLC Applications for Vitamin A and Provitamin A Analysis in Foods

Application	Sample Pre-Treatment	Column	Mobile Phase	Detector	Reference
Determination of retinol palmitate in fortified cereal products	SFE	C8 (Altex C8, 150mm×3.9mm)	ACN/2-IP/25 mM aqueous NaClO$_4$ (45/45/10, v/v/v); flow rate: 2.0mL/min	ED	[425]
Determination of vitamins A and E in milk powder	SFE with 5% of MeOH in CO$_2$; saponification	Lichrosphere RP-18 (250mm×4mm, 5μm)	MeOH/H$_2$O (96/4 v/v); flow rate: 1.0mL/min	UV-Vis (325nm)	[448]
Determination of retinol in flour and milk	Alkaline hydrolysis (NaOH 50% w/v) LLE (Et$_2$O/petroleum ether) (1/1 v/v)	Lichrosorb RP 18 (125mm×4.5mm); T=30°C	ACN/H$_2$O (80/20 v/v); flow rate: 0.8mL/min	DAD (325nm)	[449]
Determination of carotene in carrot juice	Extraction with (CH$_3$)$_2$CO/Hex (1/1 v/v)/emulsion removed by NaCl (10% wt:v); Hex washed with H$_2$O	C30 (250mm×4.6mm i.d.)	A: MeOH/MTBE/H$_2$O (81/15/4 v/v/v); B: MTBE/MeOH/ H$_2$O (90/6/4 v/v/v); flow rate: 1mL/min	UV-Vis (452nm)	[424]
Determination of vitamin A and β-carotene in emulsified nutritional supplements	SPE C18 eluted with EtOH	Inerstil ODS 80A (150mm×0.46cm i.d.)	Vit.A: EtOH/MeOH (1/1 v/v); flow rate: 0.6mL/min; carotene: ACN/EtOH (70/30 v/v); flow rate: 1.6mL/min	VitA: FLD (λ_{ex}=350nm, λ_{em}=480nm)	[422]
	SPE C2 eluted with EtOH, using MSG as dissolving agent	Inerstil ODS 80A (150mm×0.3cm i.d.)	Vit.A: EtOH/MeOH (1/1 v/v); flow rate: 0.4mL/min	Carotene: UV-Vis (450nm)	[423]
Determination of provitamin A in fruit	Extraction with THF/MeOH (1/1 v/v); centrifugation; filtration	Lichrosphere C18 (150mm×4.6mm)	A: ACN/0.05% TEA/0.1%BHT; B: MeOH/EtAc (1/1 v/v);	UV-Vis, DAD (300–600nm)	[450]
Determination of provitamin A in pandanus fruit	Homogenization with (CH$_3$)$_2$CO plus MgCO$_3$ and Na$_2$SO$_4$; extraction with (CH$_3$)$_2$CO	Novopack C18 (300mm×39mm)	MeOH/THF (9/10 v/v); flow rate: 0.5mL/min	UV (463nm)	[451]

Determination of vitamin A in animal livers	Extraction with CHCl$_3$/MeOH (2/1 v/v); filtration	AQUASIL C18 (150 mm × 4 mm)	ACN/MeOH (85/15 v/v)	DAD (320 nm)	[443]
Determination of vitamin A, provitamin A and vitamin E in animal and plant samples	Extraction with 2-propanol/ CH$_2$Cl$_2$ (2/1 v/v); centrifugation or filtration	Microsorb-MV column (100 mm × 4.6 mm)	A: MeOH/ H$_2$O (3/1 v/v) with 10 nM NH$_4$OAc; B: MeOH/ CH$_2$Cl$_2$ (4/1 v/v); flow rate: 0.8 mL/min	DAD	[421]
Determination of β-carotene in beverages	Direct injection	Spherisorb 12% ODS 5 μm (15 cm × 4 mm i.d.)	THF/ H$_2$O (75/25 v/v); flow rate: 0.8 mL/min	UV-Vis (455 nm)	[452]
Determination of carotenes and tocopherols in ATBC drinks	Liquid extraction with (CH$_3$)$_2$CO/Hex (1/1 v/v)	C30 (250 mm × 4.6 mm i.d.)	A: MeOH/ MTBE/ H$_2$O (81/15/4 v/v); B: MTBE/MeOH/H$_2$O (90/6/4 v/v); flow rate: 1 mL/min	UV-Vis (453 nm, all-*trans*-β-carotene; 445 nm all-*trans*-α-carotene	[453]
Determination of vitamin A and E in infant milk-based formulae	Extraction with EtOH and Hex; filtration	Pinnacle II short narrow-bore column (50 mm × 2 mm i.d.)	0.5% of EtAc in Hex; flow rate: 0.4 mL/min	DAD (326 nm)	[442]
Determination of carotenoids in pumpkins	Homogenization with MgCO$_3$ and THF/MeOH, filtration; re-extraction with petroleum benzine	201 TP54 (250 mm × 4.6 mm, 5 μm)	ACN/MeOH/ CH$_2$Cl$_2$ (76/20/4 v/v/v) with 0.1% BHT and TEA; flow rate: 1.0 mL/min	APCI-MS	[433]
Determination of carotenoids in vegetable juice	Extraction with MTBE	ProntoSil silica gel (250 mm × 4.6 mm, 3 μm) modified with triacontyltrichlorosilane	MeOH/ MTBE; flow rate: 1 mL/min	APCI-MS	[445]
Determination of vitamin A, D3 and E in fortified infant formula	Saponification; SPE (Chromabond XTR®)	Silica-based column, Nucleosil 100-5 (250 mm × 4.6 mm i.d.)	Hex/dioxin/2-IP (96.7/3/0.3 v/v/v); flow rate: 1.45 mL/min	APCI-MS	[426]
Determination of carotenoid esters in mandarin essential oil	Filtration	YMC 30 (250 mm × 4.6 mm, 5 μm)	MeOH/MBTE (95/5 v/v); flow rate: 1.0 mL/min	UV-Vis, MS	[447]

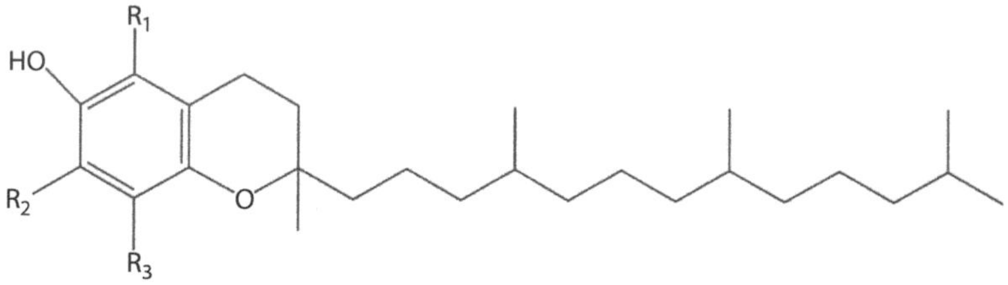

FIGURE 19.12   Structure of tocopherol. α: R_1, R_2, R_3= methyl, β: R_1, R_3=methyl; γ: R_2, R_3=methyl; and δ: R_3=methyl.

a probable effect of cancer prevention. Vitamin E deficiency leads to a progressive necrosis of the nervous system and muscle, while an overdose could decrease blood coagulation [417].

The main sources of these compounds are vegetable oils, unprocessed cereal grains, and nuts. The vitamin formulations that may be added to foods are D-α-tocopherol, DL-α-tocopherol, D-α-tocopheryl acetate, DL-α-tocopheryl acetate, and D-α-tocopheryl acid succinate [403].

19.8.1.2.1   Sample Preparation

The main problem in the vitamin E analysis is that it is easily oxidized, thereby an antioxidant, such as butyl hydroxy toluene (BHT) or ascorbic acid, is added to prevent degradation during the extraction step. The traditional method for extraction of tocopherols and tocotrienols in foods is solvent extraction (like soxhlet) and saponification with KOH [457,458]. Some authors have recently proved that saponification is not necessary [459–462], nevertheless, it has been widely applied until the present day.

Several methods apply LLE followed by a direct injection [453,463,464] or by a simple purification step, like SPE [465,466]. Bonvehi [466] compares saponification to SPE method using a Shandon Hypersep Silica Sep-Pak (500 mg), and the results show a 20% higher recovery using the latter method.

Some authors [467,468] use novel extraction techniques, such as pressurized liquid extraction (PLE) and SFE obtaining recovery higher than with saponification.

19.8.1.2.2   Chromatography

HPLC is preferred to GC for vitamin E analysis because it does not require derivatization and the preparation step is easier. Generally RP columns are used less, despite their better stability, because most of the traditional ODS are not able to separate the β-, γ-isomer of tocopherols and tocotrienols [469,470], but several authors have achieved a better resolution using new stationary phase, such as pentafluorophenylsilica [471], octadecylpolyvinyl alcohol [472,473], and long-chain alkylsilica [474,475]. The RP column allows the use of ED, which provides the highest sensitivity; this detector needs the presence of electrolytes on the eluents, which are generally insoluble in NP eluents. Since most foods do not contains all the eight vitamers or it is not always necessary to separate all of them, RP chromatography has been used in any case.

Despite the low reproducibility and short life time of the NP columns, they are widely used owing to their better selectivity and resolution power, which enable the separation of β- and γ-isomers easily. The most used stationary phases are silica, aminopropyl- or diol-bonded [476]. A more accurate description of the column used can be found in a review about tocol-derivatives analysis by Abidi [477]. Kamal-Eldin et al. [478] compare the performance of new silica-type columns, six different silica columns, three amino columns, and one diol column. The new generation column results are much more repeatable and therefore suitable for vitamers analysis.

More recently, some studies have reported the use of supercritical fluid chromatography (SFC) [479,480]. Coupling SFC with SFE, sample extraction, preconcentration, and quantification can be performed in a single step. The mobile phase, carbon dioxide, can be modified by adding different quantities of modifiers, such as methanol, ethanol, propanol, methylene, etc. Using this technique, the β- and γ-isomers are resolved even in an ODS column [481].

19.8.1.2.3   Detection

The most sensitive and specific detector is ED, but the insolubility of the electrolytes in NP eluents preclude its wide application. A FLD is generally preferred due to its higher selectivity and specificity than a UV method, but the latter may be preferred in simultaneous analysis of other lipophilic compounds such as TAGs, sterols, and vitamins A, D, and K [448,482,483]. MS and tandem MS have been recently introduced as detector for vitamin E.

19.8.1.2.4   Applications

A list of selected applications of HPLC for vitamin E analysis is reported in Table 19.11.

19.8.1.3   Vitamin K

Vitamin K is the name of a series of fat-soluble compounds, 2-methyl-1,4-naphto-quinone, with a lipophilic side chain at position 3. The most important compounds are vitamin K_1 (phylloquinone) (Figure 19.13) with a phytil group in position 3, isolated from green plants, and vitamin K_2 (menaquinones) with an unsaturated multiprenyl group in 3, synthesized by bacteria, including the intestinal ones. The phylloquinone concentration in most foods is very low (<10 μg/100 g). The main uptake is obtained from a few leafy green vegetables and four vegetable oils (soybean, cottonseed, canola, and olive) that contain high amounts of the vitamin [495,496].

The absorption of vitamins K_2, which are found mainly in cheese, curd cheese, and natto, is much higher and may be almost complete. Thus the nutritional importance of menaquinones is often underestimated. The vitamin K activity is related to the activation of specific proteins involved in blood clotting and bone metabolism. Clinical vitamin deficiency due to dietary inadequacy is rare or nonexistent in healthy adults, thanks to the widespread distribution of the vitamin K in foodstuffs and the microbiological flora of the gut, which synthesizes menaquinones. Only infants up to 6 months are at risk of bleeding due to a vitamin K deficiency. No data on negative effects of an overdose of vitamin K are found [417].

Among the vitamin K vitamers, only phylloquinone is accounted for routine food analysis. Furthermore infant formulas, both milk-based and soy protein-based, are supplemented with a synthetic preparation of phylloquinone (the only form admitted) [403], which usually contain about 10% of the biologically inactive *cis*-isomer [497].

The development of the HPLC technique in the 1970s has given the possibility to measure with a good confidence a low level of vitamin K in foods, in concentrations of about 1 μg/100g.

19.8.1.3.1   Sample Preparation

Vitamin K vitamers are sensible to UV radiation, alkali, acids and reducing agents, but they are reasonably stable to oxidation and heating. During the analysis, it is important to protect the sample from direct light.

Vitamin K is traditionally extracted with solvents such as acetone or ethanol, followed by a sample clean-up, such as enzymatic hydrolysis [498,499], column chromatography [308], TLC [500,501], SPE [502,503], or a combination of them. Gao and Ackman [504] couple two different methods, an SPE purification, previously reported by Ferland and Sadowski [505], preceded by a modified enzymatic digestion from the original procedure of Bueno and Villalobos [506]. They observe that SPE step is not enough to purify the sample. Jacob and Elmadfa [507] report an extraction from food with different solvents (according to the type of food), the evaporation of these and re-dissolving in hexane to perform a simple LLE as the clean-up step using methanol/water. SFE has been successfully applied for phylloquinone extraction from powdered infant formulas [497].

TABLE 19.11
HPLC Applications for Vitamin E Analysis in Foods

Application	Sample Pre-Treatment	Column	Mobile Phase	Detector	Reference
Determination of tocopherols in malt sprouts	SFE (250 bar, 80°C, CO_2 flow rate = 1 mL/min) using ODS as adsorbing material	Zorbax (250 mm × 4.6 mm, 5 μm)	MeOH/H_2O (98/2 v/v); flow rate 2.0 mL/min; room temperature	FLD (λ_{em} = 328 nm; λ_{ex} = 303 nm)	[484]
Determination of tocopherol in *Rosmarinus officinalis*	Probe sonication; extraction with $(CH_3)_2CO$; centrifugation; filtration	C18 Nucleosil (250 mm × 4.6 mm, 5 μm)	A: H_2O; B:MeOH/ACN (30/70 v/v) with 0.1% HAc; flow rate: 2 mL/min	DAD	[485]
Determination of tocopherols and tocotrienols in biscuits	Extraction with petroleum ether; SPE with Shandon Hypersep silica Sep pak (500 mg)	μBondpak-C18 (300 mm × 3.9 mm i.d., 10 μm)	MeOH/ H_2O: flow rate 1.0 mL/min	FLD (λ_{ex} = 296 nm; λ_{em} = 330 nm)	[466]
Determination of vitamin E isomers in seeds and nuts	PLE with ACN (50°C, 1600 psi, 2 × 5 min)	Synergi 4 μ Hydro-RP 80A column (250 mm × 4.6 mm i.d.)	A: 2.5 mM HAc/NaOAc in MeOH; B: H_2O (99.9/1 v/v)	ED	[467]
Determination of tocopherol and tocotrienol in rice bran	Extraction with MeOH, centrifugation, filtration	Nva-Pak C18 (150 mm × 3.9 mm, 4 μm)	ACN/MeOH/IP/1% HAc in H_2O (45/45/5/5 v/v/v/v)	UV-Vis (292–298 nm)	[486]
Investigation of vitamin E content in olives	Homogenization with H_2O and $(CH_3)_2CO$, extraction with Et_2O	Spherisorb W silica column (250 mm × 4.6 mm, 5 μm)	2-IP/Hex (0.5/99.5 v/v)	FLD (λ_{em} = 290 nm; λ_{ex} = 330 nm)	[487]
Determination of tocopherol and tocotrienol in olive oils	Dilution	Inertsil 5 SI (250 mm × 3 mm)	1,4-Dioxane/Hex (3.5/96.5 v/v)	DAD, FLD (λ_{em} = 330 nm; λ_{ex} = 290 nm); UV-Vis (295 nm); ELS	[488]
Determination of vitamin E in buckwheat	Extraction with MeOH, centrifugation	Phenomenex Luna C 18 (150 mm × 2 mm, 3 μm)	ACN with 0.15% TFA; flow rate 0.25 mL/min	DAD (220 nm)	[489]
Determination of tocopherols in vegetable oils	Saponification with KOH, filtration, LLE	Phenomenex Luna C 18 (150 mm × 2 mm, 3 μm)	ACN/H_2O (95/5 v/v); flow rate: 0.4 mL/min	DAD (295 nm); FLD (λ_{ex} = 290 nm; λ_{em} = 325 nm)	[490]
Determination of tocopherols and tocotrienols in rice bran	Extraction with 2-IP, centrifugation	Inertsil Sil 100A-5 (250 mm × 4.6 mm)	Hex/1,4-dioxane/2-IP (100/40/5 v/v); flow rate: 1.0 mL/min	FLD (λ_{ex} = 296 nm; λ_{em} = 326 nm)	[491]
Determination of tocopherols in tomato	SFE, saponification, extraction with Hex	Nucleosil 100 (240 mm × 4.6 mm, 10 μm)	Hex/EtOH (99.6/0.4 v/v); flow rate: 1.2 mL/min	FLD (λ_{ex} = 295 nm; λ_{em} = 330 nm)	[492]
Determination of vitamin E different matrices	Extraction with Hex	Lichrosphere 5160 (250 mm × 4.6 mm)	Hex/dioxane (96/4 v/v); flow rate: 2.0 mL/min	FLD (λ_{ex} = 292 nm; λ_{em} = 335 nm)	[493]
Determination of all-rac-α-tocopherol in virgin olive oil	LLE with 50% EtOH aqueous and Hex/EtAc (4/1 v/v), centrifugation	Inertsil ODS-2 (250 mm × 4.6 mm, 5 μm)	ACN/MeOH (95/5 v/v), flow rate: 2.0 mL/min	DAD (λ_{ex} = 295 nm; λ_{em} = 325 nm)	[494]

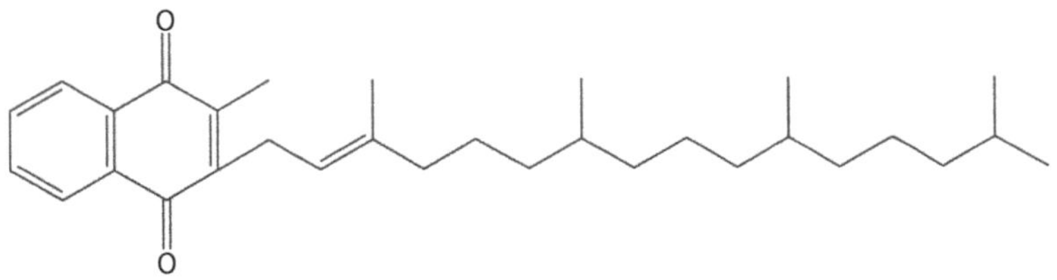

FIGURE 19.13 Structure of phylloquinone.

19.8.1.3.2 Chromatography

NP chromatography enables the separation of the *cis*-isomer from the active *trans*-isomer, nevertheless, it has been applied mainly as preparative chromatography using silica columns. RP chromatography instead is widely used as the separation step, even if the resolution between *cis*- and *trans*-isomer is not achieved. Menaquinone-4 (MK-4) [504,508–510] or 2′,3′-dihydrophylloquinone [517] are frequently used as an internal standard.

When fluorescent detection is used, the mobile phase contains zinc chloride as the reduction agent for vitamin K derivatization. The most used mobile phases are methanol and dichloromethane or water.

19.8.1.3.3 Detection

Vitamin K can be measured using UV, ED, FLD, and mass MS detector. FLD is the most widespread technique, but it needs previous reduction to the hydroquinone form to increase selectivity and sensitivity. This derivatization can be carried out "in-line" by heating the solvent containing the reduction agent [511], by photochemical [512,513], electrochemical [514], or solid phase procedures, using platinum [515] or zinc [503]. Solid-phase procedures usually use dry zinc powder as a stationary phase of a reduction column after the analytical column and before the FLD. Photochemical reactions proceed via free radical residues; therefore the reaction rates are very fast. Pérez-Ruiz et al. [513] use the capability of phylloquinone to extract hydrogen from SDS by irradiation with UV. The photoreactor was constructed from a polytetrafluoroethylene (PTFE) capillary tube with irradiation at 300 nm, while SDS was added as micellar medium into methanol mobile phase. The stability of the photoreducted vitamin K is higher than that obtained using chemical, electrochemical or other photochemical reduction system. Ahmed et al. [516] use the peroxyoxalate chemiluminescence (PO-CL), generated by UV irradiation of vitamin K, to increase the sensitivity. Indeed, PO-CL is based on the reaction between hydrogen peroxide and aryloxalate, which produces strong luminescence in the presence of a fluorophore. When vitamin K is subjected to UV irradiation, it generates hydrogen peroxide and a fluorophore, thereby mixing this solution with aryoxalate. The PO-CL reaction occurs and helps to increase the sensitivity at FLD.

19.8.1.3.4 Applications

A list of selected applications of HPLC for vitamin K analysis is reported in Table 19.12.

19.8.1.4 Vitamin D

Vitamin D is a group of fat-soluble compounds and the main forms are vitamin D_3 (cholecalciferol) (Figure 19.14), which is the physiological form and the synthetic analogue vitamin D_2 (ergocalciferol). They only differ by the side chain of the sterol skeleton.

Blood levels of vitamin D are influenced both by dietary intake and the amount of daylight exposure to the skin. Indeed, exposure of the skin to ultraviolet light catalyzes the synthesis of vitamin D_3 (cholecalciferol) from 7-dehydrocholesterol; thus vitamin D is more like a hormone and not strictly a vitamin. Furthermore, the UV radiation catalyzes the synthesis of ergocalciferol from ergosterol. This latter compound is found in plants, especially yeast and fungi, but the conversion to ergocalciferol

TABLE 19.12
HPLC Applications for Vitamin K Analysis in Foods

Application	Sample Pre-Treatment	Column	Mobile Phase	Detector	Reference
Determination of vitamin K in food	Extraction with different solvents, according to matrices, re-dissolution in Hex and extraction with MeOH/H$_2$O (9/1 v/v), centrifugation	Gynkotek ODS Hypersil (250 mm × 4.6 mm, 5 μm)	CH$_2$Cl$_2$/MeOH with zinc chloride, NaOAc, HAc (1/9 v/v); flow rate: 1.0 mL/min	FLD (λ_{ex} = 243 nm; λ_{em} = 430 nm), reduction column (20 mm × 4.6 mm) packed with zinc powder	[507]
Determination of vitamin K$_1$ in emulsified nutritional supplements	SPE (Bond Elut C18)	Inertsil ODS-2 (150 mm × 4.6 mm, 5 μm)	MeOH/EtOH (50/50 v/v); flow rate: 0.6 mL/min	FLD (λ_{ex} = 320 nm; λ_{em} = 430 nm), reduction column RC-10 platinum oxide catalyst (30 mm × 4.0 mm)	[518]
Determination of vitamin K$_1$ in baby food	Extraction with CH$_2$Cl$_2$/MeOH (2/1 v/v), filtration through dehydrated Na$_2$SO$_4$, re-dissolved in Hex, extraction with MeOH/H$_2$O (9/1 v/v), centrifugation	Gynkotek ODS Hypersil (250 mm × 4.6 mm, 5 μm)	CH$_2$Cl$_2$/MeOH with zinc chloride, NaOAc, HAc (1/9 v/v); flow rate: 1.0 mL/min	FLD (λ_{ex} = 243 nm; λ_{em} = 430 nm), reduction column (20 mm × 4.6 mm) packed with zinc powder	[443]
Determination of K$_1$ in oil, butter, and margarines	Dilution in Hex, filtration, preparative HPLC: μ-Porasil (300 mm × 3.9 mm, 5 μm)	Vydac 201 TP54 (250 mm × 4.6 mm, 5 μm)	MeOH/0.05 M NaOAc buffer; flow rate: 1.0 mL/min	ED	[508]
Vitamin K in grains, cereals, fast-food breakfast, baked products	Extraction with 2-IP/Hex (3/2 v/v), centrifugation; SPE (500 mg Bond Elut silica)	BDS Hypersil C18 (150 mm × 3, 5 μm)	A: MeOH with 10 mM ZnCl$_2$, 5 mM NaOAc, 5 mM HAc; B: CH$_2$Cl$_2$	FLD (λ_{ex} = 244 nm; λ_{em} = 430 nm), reduction column (50 mm × 2.0 mm) packed with zinc powder	[519]
Determination of phylloquinone in vegetables	Sonication with Hex, SPE with 500 mg silica cartridge	Ultrasphere (45 mm × 4.6 mm, 5 μm)	MeOH with SDS	FLD (λ_{ex} = 342 nm; λ_{em} = 426 nm), post column photoreactor	[513]

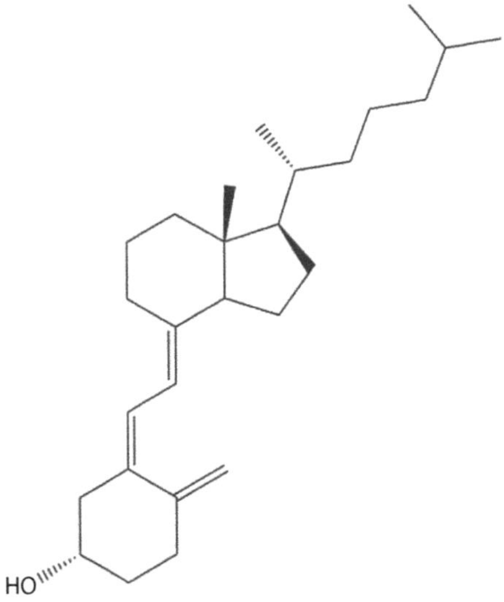

FIGURE 19.14 Main structure of vitamin D (cholecalcipherol).

hardly takes place in nature, thereby plants are a poor source of vitamin D_2. Synthetic vitamin D_2 is the main form added to food or given as a supplement, but also cholecalciferol is allowed in fortified foods [403].

Vitamin D is converted in the liver and kidneys to 1,25-dihydroxyvitamin D, which is the hormone-active compounds. The principal physiological function is to maintain the serum calcium and phosphorus concentrations in a range that support cellular processes, neuromuscular function, and bone ossification [417]. Only a few foods contain vitamin D in quantities that have an impact on the dietary intake: fish liver, fish liver oils, fatty fish, and egg yolks. Thus, some countries practice fortification of certain foods with vitamin D, most often milk, margarine, and/or butter.

Deficiency can lead to a bone-softening disease, like rickets in children and osteomalacia in adults, and it can contribute to osteoporosis. Vitamin D overdose is rare, but it can lead to hypercalcemia and it can develop in anorexia, nausea, and vomiting. Vitamin D is relatively stable in fat solutions, e.g., it is not inactivated by pasteurization or sterilization. It oxidizes in contact with air and in acid solutions.

Vitamin D is present in very low concentration in food, thereby its analysis is challenging, owing to the need to remove substances present in much higher amounts, such as cholesterol, vitamins A and E. When the total amount of vitamin D is determined in animal products, 25-hydroxyvitamin D_3 has to be included due to its high amount.

Very few papers have been published in the last 10 years related to vitamin D determination in food. Thereby, for general consideration about sample preparation the book chapter published by Nollet et al. [407] was used as reference.

19.8.1.4.1 Sample Preparation

In solution, vitamin D (both D_2 and D_3) isomerizes to previtamin D and forms a temperature-dependent equilibrium mixture [520], which leads to quantification problems. Previtamin D is difficult to quantify because of interference from co-eluted contaminants. The reversibility of the isomerization is very slow, therefore the percentage of previtamin can be considered constant during the entire analysis. The quantification of the potential vitamin D can be performed using an external standard that has undergone saponification procedure as the sample [521]. Vitamin D_2 and D_3 can be used as an internal standard to quantify the other one. Indeed, the isomerization rates of vitamins D_2 and D_3 are virtually the same; thereby the previtamin D/vitamin D ratio will be the same for both vitamers at any temperature. The isomerization problem can be resolved by converting previtamin D and vitamin D to isotachysterol by adding to the unsaponificable residue an acidified butanol solution [522].

Saponification is the most widely used method to remove an excess of triglycerides. Saponification can be carried out by heating the solution (resulting in thermal isomerization) or at room temperature for a longer time. The following extraction is usually carried out using petroleum ether/diethyl ether, allowing the coextraction of vitamin D and A. SPE [523,524] is largely applied as a clean-up method, using a silica cartridge, or gel permeation chromatography (GPC) as a preparative step [525,526]. Some authors [527] perform both SPE and preparative HPLC, using a silica column connected to an amino column.

19.8.1.4.2 Chromatography

NP chromatography is unable to separate vitamin D_2 from vitamin D_3. So it is usually used as semipreparative chromatography [527–531]. Instead, RP chromatography is able to resolve vitamin D_2 and D_3, thus it is widely applied as analytical chromatography. Mattila et al. [532] describe a two dimensional LC procedure. The sample is saponified and an NP semipreparative column is used before the quantification in a tandem column set (Zorbax ODS × Vydac 201 TP54 C18).

19.8.1.4.3 Detection

Vitamin D_2 and D_3 exhibit identical UV absorption spectra and they do not possess fluorescence. Electrochemical detection is limited and only few methods are applied in food analysis [530,533]. MS detection has been applied achieving satisfactory detection limit (10^{-6} mol/mL) [534,535].

19.8.1.4.4 Applications

Selected applications of HPLC for vitamin D analysis are listed in Table 19.13.

19.8.1.5 Simultaneous Determination of Fat-Soluble Vitamin

The main challenge in simultaneous determination of fat-soluble vitamins is the different amount of these compounds in foods, especially of vitamin D, which is usually present in very low levels, except when added. SFE has been used for a simultaneous determination of vitamins A and E in cheese and salami [537]. Some authors [538] use the same extraction method (saponification and liquid extraction) and then an aliquot is analyzed with an NP chromatography with UV and FLD to determine vitamins A and E, while another aliquot is treated with semipreparative HPLC, followed by RP chromatography with UV detection. Heudi et al. [426] use NP chromatography-MS to quantify vitamins A, D, and E in one run. Delagado-Zamareno et al. [539] tried to set up a unique method for the determination of vitamins A, D, and E in yogurt using RP chromatography, but vitamin D required some modification of the method and it was quantified only in fortified samples. A liquid extraction method (using hexane/chloroform, 2/1 v/v), followed by an SPE clean-up, was compared with a saponification procedure. The former method gave good recoveries for vitamins A and E, but saponification was found to be necessary for vitamin D.

Qian and Sheng [483] use an acetone/chloroform mixture (30/70 v/v) to extract vitamins A, D, and E from animal feed, obtaining good recoveries for vitamins D and E, and slightly lower for vitamin A (87%). An interesting work on improved HPLC separation of fat-soluble vitamins, using β-cyclodextrins in the mobile phase, was done by Spencer and Purdy [540]. Indeed, vitamins D and E retention factors were reduced; furthermore, the resolutions of vitamin D_2 and D_3, and of α- and δ-tocopherols were increased, but β- and γ-isomers were still coeluted. Using the same kind of additivation in a micellar electrokinetic chromatography (MEKC), it was possible to resolve the entire fat-soluble vitamins considered (vitamins A, D_2, D_3, and the four tocopherol-isomers).

To the best of our knowledge, at present, the simultaneous separation of all four fat-soluble vitamins has been achieved only in biological fluids [541], because in food matrices, the extraction step is fundamental.

TABLE 19.13
HPLC Applications for Vitamin D Analysis in Foods

Application	Sample Pre-Treatment	Column	Mobile Phase	Detector	Reference
Determination of cholecalciferol in fish	Saponification, extraction with petroleum ether/Et_2O (1/1), semipreparative HPLC (μ-Porasil, 3000 mm × 3.9 mm)	Vydac 201 TP 54	MeOH/H_2O (93/7 v/v);	DAD	[528]
Determination of vitamin D_2 in emulsified nutritional supplements	SPE (Bond Elut C18 cartridge)	Precolumn: Hitachigel 3011-0 (100 mm × 4.6 mm, 5 μm), analytical column: Inertsil ODS-5 (150 mm × 4.6 mm, 5 μm)	Precolumn: MeOH; flow rate 0.8 mL/min; column: ACN/MeOH (75/25 v/v); flow rate: 0.6 mL/min	UV (265 nm)	[536]
Determination of vitamin D_3 in raw and cooked pork	Saponification; extraction with petroleum ether/Et_2O (1/1), SPE, preparative HPLC (Luna Silica column)	C18	ACN/MeOH	DAD (220–320 nm); UV (256 nm)	[529]
Determination of vitamin D_3 and 25OHD_3	Saponification;extraction with petroleum ether/Et_2O (1/1), SPE (silica, 2 g), preparative HPLC (Luna Silica column + amino Spereclone)	D_3: Vydac201 TP54 (250 mm × 4.6 mm, 5 μm) 25OHD_3;Luna C18 (250 × 4.6 mm, 5 μm)	D_3: MeOH/ACN (20/80 v/v) 25OHD_3: MeOH/H_2O (90/10 v/v)	DAD (220–320 nm); UV (265 nm)	[527]
Determination of vitamin D and provitamin D in fish	Saponification; extraction using a Chem Elut column, semi-preparative HPLC (Lichrosorb Si 60, 250 mm × 4.0 mm, 5 μm)	Discovery C18 (250 mm × 4.6 mm, 5 μm)	MeOH with 8.4 mM $HClO_4$ and 57 mM $NaClO_4$; flow rate: 1 mL/min	ED	[530]
Determination of D_3 and 25-hydroxyvitamin D_3 (25OHD_3) in beef and lamb	Saponification;extraction with petroleum ether/Et_2O (1/1), semi-preparative HPLC (Luna Silica column, 250 mm × 4.6 mm, 5 μm)	Lunar C18 (250 × 4.6 mm, 5 μm)	MeOH/ACN (25/75 v/v); flow rate 1.0 mL/min	UV	[531]

19.8.2 WATER-SOLUBLE VITAMINS

19.8.2.1 Vitamin C

Vitamin C (3-oxo-L-gulofuranolactone or L-threo-hex-2-enonic acid) is a six-carbon hydroxy-lactone that is structurally related to glucose (Figure 19.15). It is reversely oxidized to L-dehydroascorbic acid (DHAA), in which the unsaturated 2,3-dihydroxy group is replaced by a saturated 2,3-diketone function.

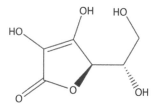

FIGURE 19.15 Structure of vitamin C (ascorbic acid).

Vitamin C is a powerful antioxidant and protects cells against oxidative stress. Therefore, it has been investigated in a variety of clinical conditions, including cancer, vascular disease, and cataracts. Vitamin C deficiency in humans leads to clinical syndromes known as scurvy in adults and Moeller-Barlow disease in children [400]. The primary symptoms in adults include fatigue, weakness, anemia, and aching joints and muscles. Later stages of deficiency are characterized by capillary fragility causing bleeding from the gums and hemorrhages, and delayed wound healing due to impaired collagen synthesis.

The absorption of vitamin C is saturated at high doses, and therefore intakes above 1 g/day would be associated with negligible increased uptake at tissue levels, but they increase the risk of adverse gastrointestinal effects. Indeed, acute gastrointestinal intolerance (e.g., abdominal distension, flatulence, diarrhea, transient colic) has been observed.

The main sources of vitamin C are green vegetables and citrus fruit. Animal tissue contains vitamin C, mainly in the kidneys and liver. The level of vitamin C in food is rapidly reduced during cooking or storage due to oxidation or water dissolution. It is added to food as an antioxidant (with no specified limit on the level of use) or as a supplement (with a maximum recommended daily intake of 3000 mg/day). The forms admitted are L-ascorbic acid (AA), L-ascorbyl 6-palmitate, sodium, calcium, or potassium L-ascorbate [403].

D-Ascorbic acid (IAA) is not present in natural products, but it is present in vitamin C-enriched foods or as an antioxidant. Therefore, it is important to distinguish between AA and IAA [417].

19.8.2.1.1 Sample Preparation

The most important point during sample preparation is to prevent oxidation of ascorbic acid. Indeed, it is easily oxidized by an alkaline pH, heavy metal ions (Cu^{2+} and Fe^{2+}), the presence of halogens compounds, and hydrogen peroxide. The most suitable solvent for this purpose is metaphosphoric acid, which inhibits L-ascorbic oxidase and metal catalysis, and it causes the precipitation of proteins. However, it can cause serious analytical interactions with silica-based column, e.g., C18 or amino bonded-phases [542] and it could co-elute with AA.

19.8.2.1.2 Chromatography

NP chromatography, using highly polar columns and mobile phase consisting of acetonitrile and methanol with diluted phosphate-citric acid-acetic buffers, has been applied, but RP chromatography is widely preferred. Lloyd et al. [543] first introduced a method based on a PS-DVB polymer column (PLRP-S), which is compatible with injection of high concentration of metaphosphoric acid. The composition of the mobile phase may affect the result of the analysis preventing vitamin C from oxidation. The most used solution contains inorganic salts, such as sodium or potassium phosphate buffers, with ethylene–diaminetetraacetic acid disodium dihydrate (EDTA · 2Na · 2H$_2$O). The possibility to using an organic acid, such as monosodium L-glutamate (MSG) and nucleic acids, as guanosine-5′-monophosphate (GMP) has also been tested successfully [544,545].

Recently, Tai, and Gohda [546] have reported a method using a novel technique named hydrophilic interaction chromatography, which use a polar stationary phase with aqueous-organic mobile phase. It enables to quantify accurately AA and related compounds, to run at high flow rate thanks to the low back pressure, and to be easily coupled with an MS.

19.8.2.1.3 Detection

AA and IAA can be easily detected using UV or ED. Compared to AA, DHAA has a weak UV absorption and no response on ED. Therefore, derivatization prior to or after the chromatographic separation is needed to increase sensitivity. Different strategies can be applied: (a) DHAA may be reduced to AA, prior to HPLC by L-cysteine [547] or dithiothreitol [548], or using a postanalytical column for a solid-state reduction [549]; (b) DHAA may be derivatized with o-phenyldiamine to form the fluorophore 3(1,2-dihydroxy-ethyl)furo[3,4-b]quinoxaline-1-one (DFQ) [550]; (c) AA may be oxidized to DHAA by an on-line postcolumn oxidation with Cu^{2+} or Hg^{2+} followed by derivatization with o-phenyldiamine [551,552].

19.8.2.1.4 Applications

A list of selected applications of HPLC for vitamin C analysis in food by is reported in Table 19.14.

19.8.2.2 Folate

Folate is the generic name used to refer to a family of vitamers with related biological activity. Instead, folic acid (pteroylglutamin acid, PGA) (Figure 19.16) refers to the most oxidized, stable, and easily adsorbable synthetic form (monoglutamate). It is commonly used in food supplements and in food fortification because of its stability and becomes biologically active after reduction.

Natural folate, introduced through the diet, are mostly reduced folates, i.e., derivatives of tetrahydrofolate (THF), such as 5-methyl-THF (5-MTHF), 5-formyl-THF, and 5,10-methylene-THF, and exist mainly as pteroylpolyglutamates, with up to nine additional glutamate molecules attached to the pteridine ring.

Folate play an important role in the biosynthesis of DNA bases and in amino acid metabolism. An adeguate intake of folate reduces the risk of abnormalities in early embryonic brain development, specifically the risk of malformations of the embryonic brain/spinal cord. Therefore a proper intake is strictly recommended for pregnant women. Megaloblastic anemia is the ultimate consequence of an inadequate folate intake. No adverse effects have been associated with the consumption of excess folate from foods [417].

The major sources of folate are green vegetables, citrus fruits, legumes, egg yolk, wheat germ, and yeast [417]. This vitamin can be added only in the form of pteroylmonoglutamic acid [402]. The multiplicity of forms, low stability, low concentration, and the complex extraction and detection techniques make the analysis of folate in food a difficult task.

19.8.2.2.1 Sample Preparation

The sample preparation is usually independent of the following method used to quantify folate. The most widespread technique involves homogenization of food in a suitable buffer, such as phosphate-ascorbate buffer, acetate-ascorbate buffer, 2-(cyclohexylamino)ethanesulfonic acid (CHES)/ 4-(2-hydroxyethyl)-1-piperazineethanesulfonic acid (HEPES) buffer, with pH between 4.5 and 7.85, depending on the optimal pH of the enzyme used for subsequent deconjugation [562,563]. A preservative is always added to the extraction buffer to prevent oxidative loss (ascorbic acid, 2-mercaptoethanol or both) [564]. The solution is then heated in order to denaturize folate-protein binding and enzymes that may catalyze the folate degradation. Boiling water [565,566], autoclaving [567,568], or microwave [564] have been applied, but it is difficult to suggest the best condition, as the food matrix largely affects the release and susceptibility of folate. Furthermore, Tamura et al. [567] suggests that the enzyme treatment should be performed before the heating to ensure the complete release of folate from the matrix.

HPLC does not give a good response to long-chain derivatives of folate, therefore, conversion of the natural form of folate, polyglutamates, to mono- or di-glutamate is required. The enzyme used is γ-glutamylcarboxypeptidase, extracted from chicken pancreas, hog kidney, or to a lesser extent from

TABLE 19.14
HPLC Applications for Vitamin C Analysis in Foods

Application	Sample Pre-Treatment	Column	Mobile Phase	Detector	Reference
Determination of AA and IAA in meat	Homogenization with HPO_3, centrifugation, filtration, dilution in mobile phase buffer	$2 \times$ PLRP-S column ($250 \, mm \times 4.6 \, mm$, $5 \, \mu m$) connected in series	$20 \, mM \, NaH_2PO_4 \cdot H_2O$ containing $0.17\% \, HPO_3$	ED	[553]
Determination of AA and DHAA in food	Homogenization with $1\% \, HPO_3$ with 0.5% oxalic acid; centrifugation; filtration	Jupiter C18 ($250 \, mm \times 4.6 \, mm$, $5 \, \mu m$)	$2.3 \, mM$ dodecyltrimethylammonium chloride and $2.3 \, mM \, Na_2EDTA$ in a $66 \, nM$ phosphate–$20 \, mM$ acetate buffer; flow rate $= 1.2 \, mL/min$	AA: UV ($247 \, nm$) or ED DHAA: FLD ($\lambda_{ex} = 350 \, nm$; $\lambda_{em} = 430 \, nm$) post column reagent o-phenyldiamine	[542]
Determination of AA in food supplements	Dilution in $0.2\% \, H_3PO_4$ with $20 \, \mu M$ L-methionine, filtration	Inertsil ODS-3 ($150 \, mm \times 3.0 \, mm$, $5 \, \mu m$)	$2.0\% \, H_3PO_4$ (pH 2.1); flow rate $= 0.4 \, mL/min$	ED	[554]
Determination of AA in cactus pear	Homogenization with potassium phosphate buffer, filtration	Spherisorb ODS	Potassium phosphate buffer; flow rate $= 1.0 \, mL/min$	UV ($264 \, nm$)	[555]
Determination of AA in tropical fruits	Homogenization with metaphosphoric acid containing EDTA; centrifugation; SPE (Sep-Pack C18)	PLRP-S ($250 \, mm \times 2.1 \, mm$, $5 \, \mu m$)	$20 \, mM \, NaH_2PO_4$; flow rate $= 0.25 \, mL/min$	DAD ($254 \, nm$)	[556]
Determination of AA and DHAA in fortified foods	Homogenization with TCEP·HCl solution; addition of TCA; filtration	Lichrospher RP-18 ($250 \, mm \times 4.6 \, mm$, $5 \, \mu m$)	ACN/H_2O ($8/82 \, v/v$) with decylamine and NaOAc; flow rate $= 1.0 \, mL/min$	UV ($265 \, nm$)	[557]
Determination of AA and total vitamin C in milk	Homogenization with metaphosphoric acid; centrifugation; DTT was used to reduce DHAA to AA	Spherisorb ODS-2 C18 ($250 \, mm \times 4.6 \, mm$, $5 \, \mu m$)	$MeOH/H_2O$ with HAc ($5/95 \, v/v$); flow rate $= 0.7 \, mL/min$	UV-Vis ($254 \, nm$)	[558]
Determination of total vitamin C in different foods	Homogenization with metaphosphoric acid; filtration; SPE (C18 Sep-Pak); DTT used to reduce DHAA	Nova Pak C18 ($150 \, mm \times 3.9 \, mm$)	$0.2\% \, H_3PO_4$ in H_2O	UV ($254 \, nm$)	[559]
Determination of AA in pepper	Homogenization with metaphosphoric acid with EDTA; centrifugation; SPE (Sep-Pak C18)	Nucleosil-5 C18 ($250 \, mm \times 4.6 \, mm$)	KH_2PO_4 solution; flow rate: $0.7 \, mL/min$	UV-Vis ($254 \, nm$)	[560]
Determination of vitamin C in fruits	Homogenization with 4.5% metaphosphoric solution; centrifugation; filtration; DTT or DMP used to reduce DHAA to AA	C18 Spherisorb ($250 \, mm \times 4.6 \, mm$)	$0.01\% \, H_2SO_4$; flow rate: $1.0 \, mL/min$	UV ($245 \, nm$)	[561]
Determination of AA in dried fruits and beverages	Sonication with ACN/H_2O ($30/70 \, v/v$); centrifugation	Inertsil Diol ($250 \, mm \times 4.6 \, mm$, $5 \, \mu m$)	ACN/H_2O with NH_4OAc ($85/15 \, v/v$); flow rate: $0.7 \, mL/min$	UV-Vis ($260 \, nm$)	[546]

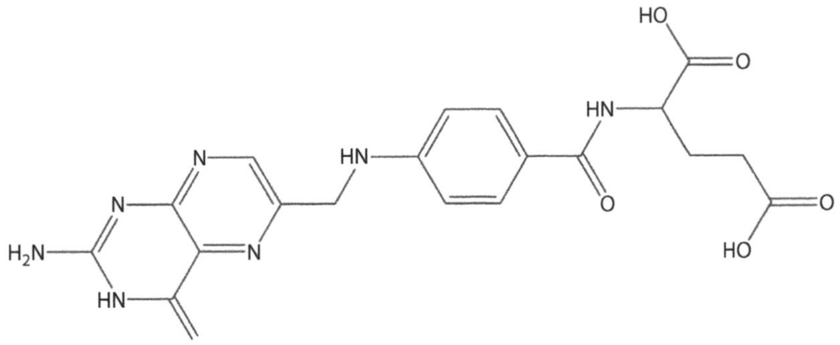

FIGURE 19.16   Sintetic structure of folates (folic acid).

the human or rat plasma, rat pancreas, and rat liver. The optimal conditions (pH, mode of action) depend on the type of conjugase and its origin. It has been known since the 1980s that the use of the conjugase alone is usually not enough to liberate food-bound folate. In 1990, Eitenmiller et al. [569,570] proposed for the first time a so-called "tri-enzyme treatment," which uses in addition a protease and an α-amylase. The conditions and the order of addition of the three enzymes might be different for each type of food.

Separation and purification chromatography has to be applied before the quantitative chromatography. Some purification techniques applied are IEC, such as DEAE-Sephadex A-25 [571], weak anion-exchange column [572,573], strong anion-exchange column (564), cation-exchange column (574), or affinity chromatography columns prepared with immobilized folate-binding protein (FBP) [575]. The latter does not bind 5-formyltetrathydrofolate; therefore it requires preliminary conversion to 10-formyltetrahydrofolate. Affinity chromatography is usually followed by ion-pair liquid chromatography [576]. The subsequent quantification could be carried out by a microbiological assay (assessing the growth of *Lactobacillus rhamnosus*), biospecific procedures (protein binding assay or immunoassay), or HPLC.

19.8.2.2.2   Chromatography

Since the 1970s numerous HPLC methods using IEC, RP and ion-pair chromatography have been proposed. In the last years, RP chromatography has become the most used method, thanks to its simplicity, sensitivity, and compatibility with different detection techniques. The stationary phases usually used are C18 or phenyl-bonded silica-based phases. More recently, alternative stationary phases, such as polar-embedded, polar endcapped, and perfluorinated phases, have been successfully tested for folate analysis [577]. The mobile phase is usually a mixture of phosphate or acetate buffer and acetonitrile or methanol.

19.8.2.2.3   Detection

Some commonly used detectors are UV (at 280 nm), FLD, ED and microbiological assay of collected fractions. UV presents a low sensitivity, but all folate derivatives respond to this detection. FLD is used even if some compounds, like folic acid, do not fluoresce. Therefore, a postcolumn derivatization, involving hypochlorite to cleave folic acid, di- and tetra-hydrofolic acid oxidatively to fluorescence pterins [571], has been introduced. Fewer reports on the use of LC–MS for folate detection are available in the literature [578–580].

19.8.2.2.4   Applications

Selected applications of HPLC for the determination of folate in food matrices are listed in Table 19.15.

TABLE 19.15

HPLC Applications for Folate Analysis in Foods

Application	Sample Pre-Treatment	Column	Mobile Phase	Detector	Reference
Determination of folates in baker's yeast	Homogenization with extraction buffer (phosphate buffer with sodium ascorbate and 2-mercaptoethanol); boiled for 12 min; centrifugation; enzyme digestion; centrifugation; SPE (SAX cartridge)	Aquasil C18 (150 mm × 4.6 mm, 3 μm)	ACN/30 mM phosphate buffer; flow rate: 0.4 mL/min	FLD (λ_{ex} = 290 nm; λ_{em} = 360 nm)	[581]
Determination of folic acid in fortified beverages	SPE (C18 Sep-Pak)	Simmetry C18 (150 mm × 4.6 mm, 5 μm)	MeOH/NH$_4$OAc solution (8mM); flow rate: 1.0 mL/min	DAD (280 nm)	[582]
Determination of folic acid in fortified food	Extraction in ultrasound bath with KOH solution; addition of H$_2$PO$_4$, phosphate buffer and TCA; filtration	Microsorb ODS-2 (150 mm × 4.6 mm, 5 μm)	Acetate buffer/ACN; flow rate: 0.5 mL/min	DAD (290 nm)	[583]
Determination of folates in lyophilized foods, salami and ham	Extraction with CHES-HEPES boiled for 10 min (salami and ham extracted using phosphate buffer; enzyme: hog kidney conjugase (for salami and ham samples also a treatment with protease); purification on FBP column	Hypersil ODS (150 mm × 4.6 mm, 3 μm)	ACN/30 mM phosphate buffer; flow rate: 0.8 mL/min	UV (290 nm); FLD (λ_{ex} = 290 nm; λ_{em} = 356 nm)	[563]
Determination of folate and folic acid in vegetables and fruits	Homogenization with extraction buffer (phosphate buffer with ascorbic acid, sodium azide and 2-mercaptoethanol); enzyme digestion: amylase, protease and rat plasma for conjugase; centrifugation; SPE (SAX cartridge); for LC-MS injection additional purification in Bond Elut Ph column	LC-FLD: Adsorbosphere C18 HS (150 mm × 4.6 mm, 3 μm) LC-MS: Luna C18 (150 mm × 4.6 mm, 5 μm)	LC-FLD: ACN/30 mM potassium phosphate; flow rate: 1.0 mL/min LC-MS: A: 0.1% HCOOH; B: ACN/H$_2$O/ MeOH (26/60/14 v/v/v)	FLD (λ_{ex} = 290 nm; λ_{em} = 355 nm; folic acid: (λ_{ex} = 230 nm; λ_{em} = 440 nm) after photolysis reaction; MS	[579]
Determination of folates in a traditional Indonesian soy-based food	Homogenized with CHES-HEPES buffer and boiled for 10 min; enzyme digestion: protease, conjugase (human plasma) and amylase; boiled; centrifugation	Luna C18 (150 mm × 4.6 mm, 5 μm)	ACN/30mM K$_3$PO$_4$; flow rate: 0.8 mL/min	UV (290 nm)	[584]
Determination of folic acid in vegetables	Addition of the extraction buffer and the deuterated internal standard; Enzymatic digestion with protease and conjugase at 37°C over night; centrifugation; SPE (SAX cartridge)	Aqua C18 (250 mm × 4.6 mm, 5 μm)	Aqueous HCOOH (0.1%)/ACN; flow rate: 0.8 mL/min	MS/MS	[585]
Determination of folic acid in Italian pasta	Extraction with sodium phosphate-sodium citrate/ ascorbate buffer at 100°C for 10 min; enzymatic digestion with papain and amylase	Discovery RP-Amide C16 (150 mm × 4.6 mm, 5 μm)	Aqueous CHO$_2$NH$_4$/ MeOH (95/5 v/v); flow rate: 0.75 mL/min	ESI-MS/MS	[586]

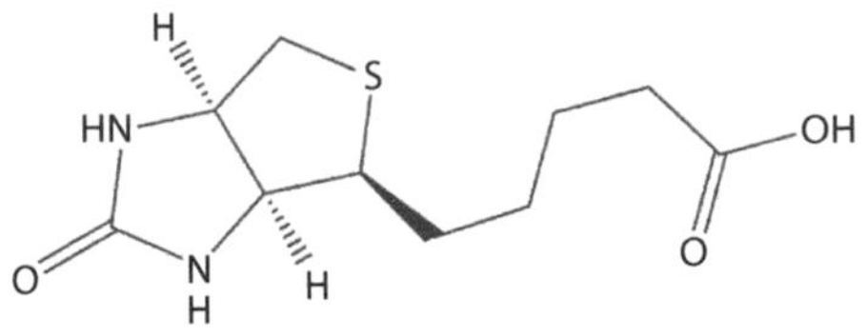

FIGURE 19.17   Structure of biotin.

19.8.2.3 Biotin

Biotin is composed of an imidazolidone ring joined to a tetrahydrothiophene ring (Figure 19.17). The latter possesses a valeric acid side chain. The unique stereoisomer occurring in nature is D(+)-biotin.

Biotin cannot be synthesized by mammals. The contribution of the biotin synthesized by intestinal bacteria to the human requirements is still controversial. Biotin is an essential cofactor for carboxylases involved in production of fatty acids, cell growth, and metabolism of fats and amino acids.

When a normal diet is followed, biotin deficiency is rare. Clinical symptoms of deficiency are alopecia and cutaneous abnormalities such as seborrhoeic dermatitis (especially in patients affected by phenylketonuria), periorificial erythema, and fungal infection. The main sources of biotin are liver, kidney, egg yolk, some vegetables such as soybeans, nuts, spinach, mushrooms, and lentils. In green plants and fruits, biotin occurs in water-extractable forms, whereas in yeast and animal products, it is a firmly bound complex [417]. The variability on the amounts of biotin in foods is due to both natural variation, but also to methodological problems.

Biotin is soluble in water and insoluble in organic solvents. It is stable at pH 5–8 and to heating in strong acid. However it can be oxidized in the sulfur atom and the shortening of the valeric acid side chain results in the loss of vitamin activity.

The basic methods to quantify biotin in food are bioassays, avidin-binding assays or fluorescent derivative assays. Although the avidin-binding assay of biotin and its metabolites after the separation by HPLC is considered as one of the best currently available methods [587], it is still not largely diffused.

19.8.2.3.1 Sample Preparation

Biotin is present in nature predominantly bonded to protein and is relatively stable; therefore it can be extracted under hard conditions, such as autoclaving in sulfuric acid. Enzyme digestion is also applied to break the protein–biotin bond. A following purification is usually performed by adsorption on charcoal or by IEC.

19.8.2.3.2 Chromatography

RP chromatography is the most applied technique to determine biotin in food samples. Instead, only some works report the use of HPLC as preparative chromatography [588]. Water, acetonitrile, methanol, and different buffers are the most widespread mobile phases used. Recently, Höller et al. [589] have reported the use of a hurt-cut HPLC system follow by MS detection to increase the sensitivity of the biotin response.

19.8.2.3.3 Detection

Biotin does not exhibit UV absorbance. It neither shows fluorescence nor electrochemical activity. Therefore, it needs to be derivatized. 4-Bromomethylmethoxiycoumarin (BMMC) [590], 9-anthryldiazomethane (ADAM) [591], and 1-pyrenyldiazomethane (PDAM) [592] have been used as precolumn reagents to convert biotin to fluorescent absorbing derivatives. Instead, to obtain derivatives that are UV detectable, hydrazines are used, such as 2-nitrophenylhydrazine hydrochloride

(2-NPH·HCl) with 1-ethyl-3-(3-dimethylaminopropyl)-carbodiimide hydrochloride (EDC·HCl) as a coupling agent [593,594]. The advantage is that with hydrazines it is possible to carry out the derivatization in milder conditions. An efficient postcolumn derivatization is obtained using avidin or streptavidin. In fact, these two proteins react in a very specific way with biotin, therefore they are bound with a fluorescent marker, such as fluorescein 5-isothiocyanate (FITC) to obtain fluorescent derivatives. Some authors [595] report the use of MS or MS/MS for biotin detection, but this method seems to be less sensitive than FLD.

19.8.2.3.4 Applications

Selected applications of HPLC for the determination of biotin in foods are listed in Table 19.16.

19.8.2.4 Niacin

Niacin is also known as vitamin PP or vitamin B_3. The term niacin describes two related compounds, nicotinic acid and nicotinamide (Figure 19.18), both with biological activity. Niacin is formed from the metabolism of tryptophan, and therefore it is not strictly a vitamin. It is a precursor of two cofactors: nicotinamide adenine dinucleotide (NAD) and nicotinamide adenine dinucleotide phosphate (NADP), which are essential for the functioning of a wide range of enzymes involved in redox reactions.

A deficiency in both tryptophan and the preformed niacin leads to pellagra, which is characterized by spinal pains, "magenta tongue," digestive disturbances, and subsequently erythema with drying and expurgation of the skin. Excessive intake has been observed in patients who are treated with niacin for its therapeutic effects, leading to a lowering of blood cholesterol and blood hyperlipidaemias. The hazards are vasodilator effect, gastrointestinal, hapatotoxicity, and glucose intolerance.

Niacin is present in foods mainly as coenzyme NAD and NADP, which are hydrolyzed in the intestine, and it is adsorbed as nicotinamide or nicotinic acid. The free forms, nicotinamide and nicotinic acid, only allowed to be added in fortified foods [403], occur naturally in limited amounts. Instead, niacin occurs as nicotynil ester bonded to polysaccharides, peptides, and glycopeptides. In general, niacin is widespread in foodstuffs (cereals, seeds, meat, and fish). High concentrations are present in roasted coffee beans as a primarily product of the roasting process [417].

19.8.2.4.1 Sample Preparation

Either acid or alkaline hydrolysis can be applied, converting nicotinamide to nicotinic acid. Alkaline hydrolysis releases also the unavailable vitamers providing the estimation of the total niacin content. Acid hydrolysis, instead, is slower than alkaline hydrolysis; therefore the former is usually coupled with enzymatic digestion by using takadiastase, papain, and clarase. Extraction with water and dilute sulfuric or hydrochloride acid has been applied to release the vitamers from the matrix without degrading nicotinamide [598].

19.8.2.4.2 Chromatography

Microbiological methods are common in niacin determination, but they are time consuming and laborious. Instead, HPLC with RP column enables a better separation and identification of compounds.

19.8.2.4.3 Detection

Niacin can be detected by UV, ED, or FLD. UV is a widespread technique but it needs a longer preparation step and it does not reach high sensitivity. The FLD is more sensitive but it needs a pre or postcolumn derivatization to make niacin fluorescent. Krishnan et al. [599] describe a postcolumn derivatization using cyanogens bromide and p-aminophenol, but this method involves toxic reagents. Mawatari et al. [600], instead, propose a fast, highly specific derivatization procedure, which involves UV irradiation at 300 nm in the presence of hydrogen peroxide and copper(II) ions.

TABLE 19.16
HPLC Applications for Biotin Analysis in Foods

Application	Sample Pre-Treatment	Column	Mobile Phase	Detector	Reference
Determination of biotin in tablets	Homogenization with H_2O; centrifugation; filtration; SPE (Sep-Pak Plus tC18ENV); derivatization with EDC · HCl and NPH · HCl	LC-PDA: L-column ODS (250 mm × 4.6 mm, 5 μm); LC–MS: Capcell Pak C18 UG120 (250 mm × 2.0 mm)	LC-PDA: 20 mM KH_2PO_4/ACN (75/25 v/v); flow rate: 1.0 mL/min; LC–MS: 5 mM NH_4OAc/MeOH (1/1 v/v); flow rate: 0.1 mL/min	DAD (400 nm) ESI-MS	[596]
Determination of biotin in foods	Homogenization with H_2O; addition of HCl and incubation at 100°C for 120 min; centrifugation; filtration	C18 used as preparative column, collected different fractions		Avidin assay for biotin and spectrophotometer at 490 nm	[588]
Determination of biotin in cereals, multivitamin tablet, fruit juice, infant milk, yogurt, meat, and vegetables	Homogenization with H_2O, glutathione, EDTA, citrate buffer, papain and takadiastase (for high starch samples); incubation over night; filtration	Lichrosphere 100 RP 18 endcapped (250 mm × 5.0 mm, 5 μm)	0.1 M phosphate buffer solution/MeOH (81/19 v/v); flow rate: 0.4 mL/min	FLD (λ_{ex} = 490 nm; λ_{em} = 520 nm); post column reagent:avidin-FITC in phosphate buffer solution	[597]
Determination of biotin in fortified foods	Alkaline extraction (for high biotin content foods) with $Na_2S_2O_3 \cdot 9H_2O$ plus NH_4OH solution; filtration; Acid hydrolysis with H_2SO_4 and enzyme digestion (papain) for low biotin content foods	High biotin content: Symmetry Shield C18 (50 mm × 2.1 mm, 3.5 μm); Low biotin content: heart-cut system: Symmetry Shield C18 (50 mm × 2.1 mm, 3.5 μm) × Vydac 218TP5215 (150 mm × 2 mm, 3.5 μm)	H_2O/MeOH with 1% HCOOH; flow rate 1°column: 0.25 mL/min.	MS	[589]

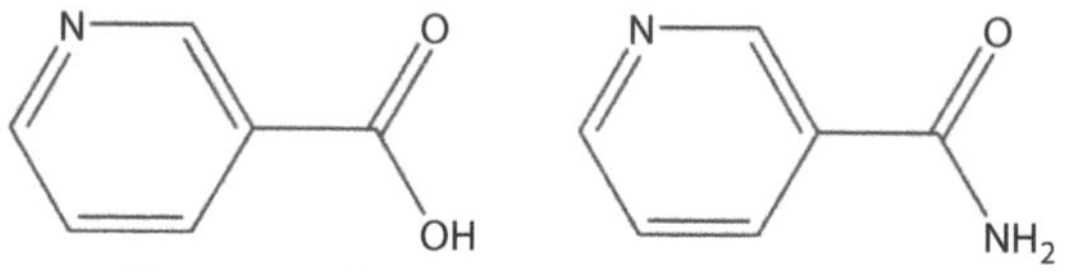

FIGURE 19.18 Structures of niacin.

19.8.2.4.4 Applications

Selected applications of HPLC for the determination of niacin in food matrices are listed in Table 19.17.

19.8.3 B Group Vitamins

The B group vitamins refer mainly to five kinds of vitamins: pantothenic acid or vitamin B_5, cyanocobalamin or vitamin B_{12}, thiamine or vitamin B_1, riboflavin or vitamin B_2, and pyridoxine or vitamin B_6. They have been mainly analyzed in biological fluids or pharmaceutical products rather than in foods. Furthermore, they may be determined together, for instance, in the determination of vitamin B_1 and B_2 by using the same sample preparation [606] or vitamin B_1 B_2 and B_6 [607]. Therefore, due to the low number of applications in food matrices for each vitamin and the frequent determination of more than one vitamin for study, the applications regarding HPLC analysis of these vitamins are reported together in one table at the end of the paragraph concerning B group vitamins (Table 19.18).

19.8.3.1 Pantotenic Acid

Pantothenic acid (Figure 19.19) is known also as vitamin B_5. It is N-(2,4-dihydroxy-3,3-dimethyl-1-oxobutyl)-β-alanine. It occurs in nature only in the D-form. Free pantothenic acid and its sodium salt are chemically unstable, and therefore the usual pharmacological preparation is the calcium salt (calcium pantothenate). The latter is stable in neutral solutions but it is destroyed by heat and acid solution.

The synthetic form is the alcohol, panthenol, which can be oxidized in vivo to pantothenic acid. It is included in the list of substances that may be added in foods and in food supplements [403]. Pantothenic acid is part of the coenzyme A (CoA) molecule; therefore it is involved in acylation reactions, such as in fatty acid and carbohydrate metabolism.

Deficiency in humans is rare because of the widespread availability of pantothenic acid in the usual diet. However, a deficiency of this nutrient can lead to a burning sensation in feet and numb toes.

No data have been reported on pantothenic acid or panthenol toxicity in humans [417]. Pantothenic acid is widely distributed among foods, especially in yeast and organ meat. It is usually present in the bound form (CoA), thereby it requires an enzymatic treatment for the analysis of the total contents.

19.8.3.1.1 Sample Preparation

Pantothenic acid occurs in foods both in the free form and bonded to coenzyme (CoA) or acyl carrier protein (ACP); therefore hydrolysis is needed to extract it totally. Since it is degraded by acid and alkaline hydrolysis, only an enzymatic digestion can be applied. Enzyme hydrolysis with papain, diastase, clarase, takadiastase, intestinal phosphatase, pigeon liver pantetheinase, or combination of them has been used.

19.8.3.1.2 Chromatography

Several authors report the use of GC for pantothenic determination after derivatization [585,608], but mainly RP chromatography is applied in pantothenic acid analysis. Different columns

TABLE 19.17
HPLC Applications for Niacin Analysis in Foods

Application	Sample Pre-Treatment	Column	Mobile Phase	Detector	Reference
Determination of niacin in cereal based products	Extraction with Ca(OH)$_2$ in autoclave at 121°C for 2 h; centrifugation; C18 Sep-Pak column in series with SCX column	C8 Nova-Pak (100 mm × 8 mm)	MeOH/H$_2$O (15/85 v/v); flow rate: 1.5 mL/min	UV (254 nm)	[601]
Determination of nicotinic acid in different foods	Extraction with HCl at 100°C for 1 h; filtration; extraction with NaOH at 120°C for 1 h; filtration	Lichrosphere 100 RP 18 endcapped (250 mm × 5.0 mm, 5 μm)	0.07 M KH$_2$PO$_4$/0.075 M hydrogen peroxide and copper sulfate solution; flow rate: 1.0 mL/min	FLD (λ_{ex} = 322 nm; λ_{em} = 380 nm); photochemical derivatization in PTFE tube at 300–400 nm	[602]
Determination of niacin cereal, vegetables, and meat	Enzyme digestion with NADase at 37°C for 18 h	Uptisphere C18 HDO (150 mm × 4.6 mm, 5 μm)	0.07 M KH$_2$PO$_4$/0.075 M hydrogen peroxide and copper sulfate solution; flow rate: 1.0 mL/min	FLD (λ_{ex} = 322 nm; λ_{em} = 380 nm); photochemical derivatization in PTFE tube at 300–400 nm	[603]
Determination of niacin in milk and cereal based products	Acid extraction with HCl at 100°C for 1 h	Inertsil ODS-2 (250 mm × 4.6 mm, 5 μm)	A: KH$_2$PO$_4$/hydrogen peroxide and copper sulfate solution; B: A plus 10% ACN; flow rate: 1.0 mL/min	FLD (λ_{ex} = 322 nm; λ_{em} = 380 nm); photochemical derivatization in PTFE tube at 300–400 nm	[604]
Determination of niacin in meat and fish	Acid extraction with H$_2$SO$_4$ at 121°C for 2 h, centrifugation; C18 Sep-Pak cartridge in series with SCX column	C8 Nova-Pak (100 mm × 8 mm)	MeOH/H$_2$O (15/85 v/v); flow rate: 1.5 mL/min	UV (254 nm)	[605]

TABLE 19.18
HPLC Applications for Vitamin B_5, Vitamin B_{12}, Vitamin B_1, Vitamin B_2, and Vitamin B_6 Analysis in Foods

Application	Sample Pre-Treatment	Column	Mobile Phase	Detector	Reference
Determination of B_5 in infant formula	Homogenization with H_2O; Addition of HAc; centrifugation; filtration	Luna C8 (250 mm×4.6 mm, 5 μm)	Phosphate buffer/ACN (97/3 v/v)	Multiwavelength UV (200, 205, 240 nm)	[609]
Determination of B_5 in foods	Homogenization with CHES-HEPES buffer; addition of isotope internal standard; filtration	Aqua C18 (250 mm×4.6 mm, 5 μm)	0.1% HCOOH in H_2O/ACN; flow rate: 0.8 mL/min	MS/MS	[622]
Determination of B_5 in foods	Enzymatic digestion with pepsin, alkaline phosphatase, pantetheinase solution; centrifugation; strong anion exchange SPE; strong cation exchange SPE	Lichrosphere 100 RP18 endcapped (250 mm×5.0 mm, 5 μm)	MeOH/Phosphate buffer (90/10 v/v); flow rate: 1.0 mL/min	FLD ($\lambda_{ex}=345$ nm; $\lambda_{em}=455$ nm), post-column derivatization reagent: NaOH, OPA, 3-MPA	[610]
Determination of B_5 in Italian pasta	Homogenization with acetate buffer at 121°C for 5 min; addition of CHO_2NH_4; centrifugation; filtration	Discovery RP-Amide C16 (150 mm×4.6 mm, 5 μm)	Aqueous CHO_2NH_4/MeOH (95/5 v/v); flow rate: 0.75 mL/min	ESI-MS/MS	[586]
Determination of vitamin B_5 in different kind of foods	Homogenization with hot H_2O, autoclaved at 103°C for 20 min; filtration	YMC C18 (150 mm×3.0 mm, 3 μm)	H_2O with 0.025% TFA/ACN; flow rate: 0.5 mL/min	ESI-MS	[623]
Determination of vitamin B_{12} in multivitamin tablets	Dilution in phosphate buffer	μ Bondpak C18 (30×3.9 cm, 10 μm)	MeOH/H_2O (30/70 v/v); flow rate: 0.8 mL/min	FLD ($\lambda_{ex}=275$ nm; $\lambda_{em}=305$ nm)	[624]
Determination of vitamin B_{12} in foods	Homogenization in NaOAc buffer and Na cyanide; addition of enzyme: amylase and pepsin (for total vitamin); incubation at 32°C for 30 min (or 37°C for 3h); immuno-affinity column	Narrow bore C18 Ace 3 AQ (150 mm×3.0 mm)	H_2O with 0.025% of TFA/ACN; flow rate: 250 μL/min	UV (361 nm)	[614]
Determination of B_{12} in foods	Dilution in Na_2SO_4 with 1 mM $Na_2EDTA\cdot H_2O$; filtration; SPE (Bond Elut C18)	Inertsil ODS-2 (150 mm×4.6 mm, 5 μm)	H_2PO_4/ACN (90/10 v/v); flow rate: 1.0 mL/min	UV-Vis (650 nm)	[625]
Determination of total B_1 in dried yeast	Extraction with 10% HCl at 80°C for 30 min; centrifugation; enzymatic digestion with takadiastase; CM-cellulose column	Nuclesil 5 C18 (150 mm×4.5 mm, 5 μm)	Phosphate buffer with 1-octanesulfonate/ACN (4/1 v/v); flow rate: 1.0 mL/min	UV (254 nm)	[626]

Analyte	Sample preparation	Column	Mobile phase	Detection	Ref.
Determination of B_1 in algae and crustacean	Extraction with HCl and sonication; addition of $K_4Fe(CN)_6$ with NaOH and MeOH to oxidase thiamine to thiochrome; filtration	Reprosil-Pur NH2 (250 mm × 4.6 mm, 5 μm)	MeOH/Phosphate buffer (43/57 v/v); flow rate: 1.0 mL/min	FLD ($\lambda_{ex} = 375$ nm; $\lambda_{em} = 450$ nm)	[627]
Determination of B_1, B_2, and B_6 in bakery products	Homogenization with NaOAc solution at 100°C for 10 min; Enzymatic digestion with acid phosphatase, papain, amylase, glucosidase with addition of glyoxylic acid, ferrous sulfate solution and Gh at 37°C for 18h; centrifugation; B_1: Derivatization with $K_4Fe(CN)_6$ with NaOH; SPE (SepPack C18); B_2: filtration; B_6: addition of Na borohydride and of glacial HAc	B_1 and B_2: μBondpak C18 (150 mm × 3.9 mm); B_6: Hypersil C18 (250 mm × 4.6 mm)	B_1 and B_2: NaOAc/MeOH (30/70 v/v); flow rate: 1.0 mL/min B_6: NaH_2PO_4 buffer/ACN (90/10 v/v); flow rate 1.0 mL/min.	B_1: FLD ($\lambda_{ex} = 365$ nm; $\lambda_{em} = 435$ nm); B_2: FLD ($\lambda_{ex} = 422$ nm; $\lambda_{em} = 522$ nm); B_6: post column derivatization with sodium metabisulfite; FLD ($\lambda_{ex} = 325$ nm; $\lambda_{em} = 400$ nm)	[607]
Determination of B_1 and B_2 in foods	Acid hydrolysis with HCl at 121°C for 30 min; Enzymatic digestion with amylase, papain, acid phosphatase with Gh for 1h; filtration	Supelcosil LC-18-DB (250 mm × 4.6 mm, 5 μm)	MeOH/ buffer with Na heptasulfonate (12.7 mM), TEA (0.1%) and KH_2PO_4 (50mM) (35/65 v/v); flow rate: 1.0 mL/min	B_1: FLD ($\lambda_{ex} = 368$ nm; $\lambda_{em} = 420$ nm), post-column derivatization $K_4Fe(CN)_6$; B_2: FLD ($\lambda_{ex} = 468$ nm; $\lambda_{em} = 520$ nm)	[606]
Determination of B_1 and B_2 in foods	Acid hydrolysis with HCl at 100°C for 15 min; Enzyme digestion with acid phosphatase at 37°C for 18h; Addition of TCA; filtration	Inertsil 5 ODS-2 (200 mm × 3.0 mm, 5 μm)	B_1: $MeOH/H_2O$ (80/20 v/v); flow rate: 0.3 mL/min; B_2: $MeOH/H_2O$ (40/60 v/v); flow rate: 0.4 mL/min	B_1: FLD ($\lambda_{ex} = 366$ nm; $\lambda_{em} = 434$ nm), pre-column derivatization and extraction in isobutanol $K_4Fe(CN)_6$; B_2: FLD ($\lambda_{ex} = 450$ nm; $\lambda_{em} = 510$ nm)	[628]
Determination of B_1, B_2, B_6 and niacin in Italian pasta	Acid hydrolysis with HCl at 121°C for 30 min; filtration	Discovery RP-Amide C16 (150 mm × 4.6 mm, 5 μm)	Aqueous CHO_2NH_4/MeOH (95/5 v/v); flow rate: 0.75 mL/min	ESI-MS/MS	[586]
Robotic determination of B_2 vitamers	Homogenization with MeOH, CH_2Cl_2; addition of citrate-phosphate buffer; centrifugation; filtration	Two PLRP-S in series (150 mm × 4.6 mm, + 250 mm × 4.6 mm)	ACN/citrate-phosphate buffer; flow rate 1.2 mL/min	FLD ($\lambda_{ex} = 450$ nm; $\lambda_{em} = 522$ nm)	[629]
Determination of B_2, FAD and FMN in wine, juice, and beer	Dilution with H_2O	Hypersil ODS C18 (200 mm × 2.1 mm, 5 μm)	Na_2PO_4 buffer/ACN (95/5 v/v); flow rate: 0.6 mL/min	FLD ($\lambda_{ex} = 265$ nm; $\lambda_{em} = 525$ nm)	[630]

(continued)

TABLE 19.18 (continued)
HPLC Applications for Vitamin B_5, Vitamin B_{12}, Vitamin B_1, Vitamin B_2, and Vitamin B_6 Analysis in Foods

Application	Sample Pre-Treatment	Column	Mobile Phase	Detector	Reference
Determination of vitamin B_1, B_2, B_6, and niacin in multivitamic tablets	Dilution with mobile phase A	XTerra RP 18 (250 mm × 4.6 mm, 5 μm)	A: Methanolic Hex sulfonic acid Na salt; B: Phosphate buffer with 0.1% Hex sulfonic acid Na salt; flow rate: 1.0 mL/min	UV-Vis (220 nm)	[631]
Determination of B_6 in food bittering extracts	Dilution in MeOH and then in TEA phosphate buffer	Prodigy ODS (250 mm × 3.2 mm, 5 μm)	TEA Phosphate buffer/ACN (88/12 v/v); flow rate: 0.4 mL/min	FLD ($\lambda_{ex} = 330$ nm; $\lambda_{em} = 420$ nm)	[632]
Determination of vitamin B_6 in animal and vegetal products	Acid hydrolysis with HCl at 121°C for 30 min (5 min for vegetables); enzymatic digestion with acid phosphatase (and glucosidase for vegetables); addition of HCl; centrifugation; filtration	Hypersil C18 (150 mm × 4.6 mm, 3 μm)	A: Phosphate buffer with 1-octan sulfonic acid and TEA; B: ACN; flow rate: 1.0 mL/min	FLD ($\lambda_{ex} = 333$ nm; $\lambda_{em} = 375$ nm); post-column derivatization Phosphate buffer	[633]
Determination of B vitamers in beef liver, egg yolk, baby food	Acid hydrolysis with PCA in ultrasonic bath; enzymatic digestion with takadiastase and phosphatase; centrifugation	Discovery RP-Amide C16 (150 mm × 4.6 mm, 5 μm)	ACN/dihydrogenphosphate buffer; flow rate: 1.0 mL/min	FLD ($\lambda_{ex} = 335$ nm; $\lambda_{em} = 389$ nm); post-column derivatization with sodium hydrogensulfite	[634]
Determination of B_6, B_1, B_2, niacin, B_5, folic acid, B_{12}, and vitamin C in supplemented foods	Homogenization with H_2O; addition of a precipitation solution containing zinc acetate, phosphotungstic polyhydrated, and glacial HAc; filtration	Spherisorb ODS-2 C18 (250 mm × 4.6 mm, 3 μm)	Phosphate buffer modified with 1-octanelsulfonic acid sodium salt and TEA/MeOH (99.4/0.6 v/v)	B_6: FLD ($\lambda_{ex} = 290$ nm; $\lambda_{em} = 410$ nm); B_2: FLD ($\lambda_{ex} = 400$ nm; $\lambda_{em} = 520$ nm); B_1: UV (245 nm); B_{12}: UV (370 nm); niacin:UV (261 nm), B_5: UV (195 nm); vitamin C and folic acid: UV (282 nm)	[635]

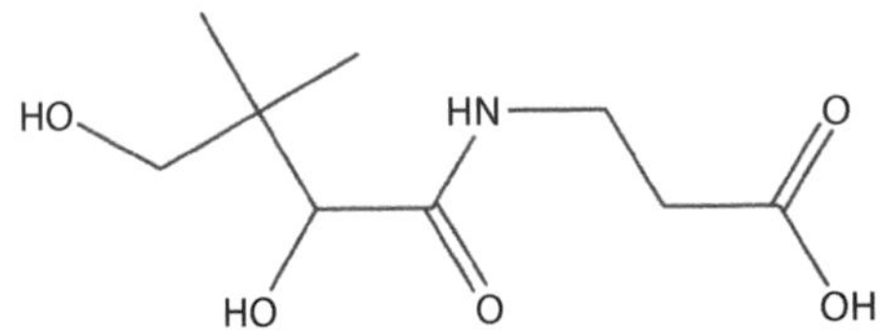

FIGURE 19.19 Structure of pantothenic acid.

(Resolve C18, Nomura Devosil ODS-MG-5, YMC AQ312, Platinum EPS, Luna Phenyl C6 and Luna phenyl C8) were compared with slight modification in the mobile phase gradient and the Luna Phenyl C8 resulted the most selective and robust [609].

19.8.3.1.3 Detection

UV absorption occurs only below 220 nm, thereby it is affected by the interference from mobile phase and from artifacts in complex foods. A multiwavelength UV detection has been experimented successfully for unambiguous evaluation of pantothenic acid [609]. However, UV detection presents a low sensitivity, compared to other techniques, like FLD or MS. FLD is applied by using a post-column derivatization. Pantothenic acid is converted to β-alanine by hot alkaline hydrolysis and a reaction with OPA [610]. Also MS is successfully applied to increase the sensitivity of pantothenic acid analysis.

19.8.3.2 Vitamin B₁₂

The name vitamin B_{12} indicates a group of cobalt-containing corrinoids, also described as cobalamins. Hydroxycobalamin (HOCbl), adenosylcobalamin (AdoCbl), and methylcobalamin (MeCbl) are the natural occurring forms. Instead, cyanocobalamin (Figure 19.20) is the commercially available form used for supplements and food fortification, thanks to its greater relative stability. Occasionally, sulfitocobalmin can occur in processed foods. Vitamin B_{12} functions as a coenzyme and it is linked to human growth, cell development, and is involved in metabolism of certain amino acids. Vitamin B_{12} is present mainly in meat and diary foods, therefore a deficiency can occur in

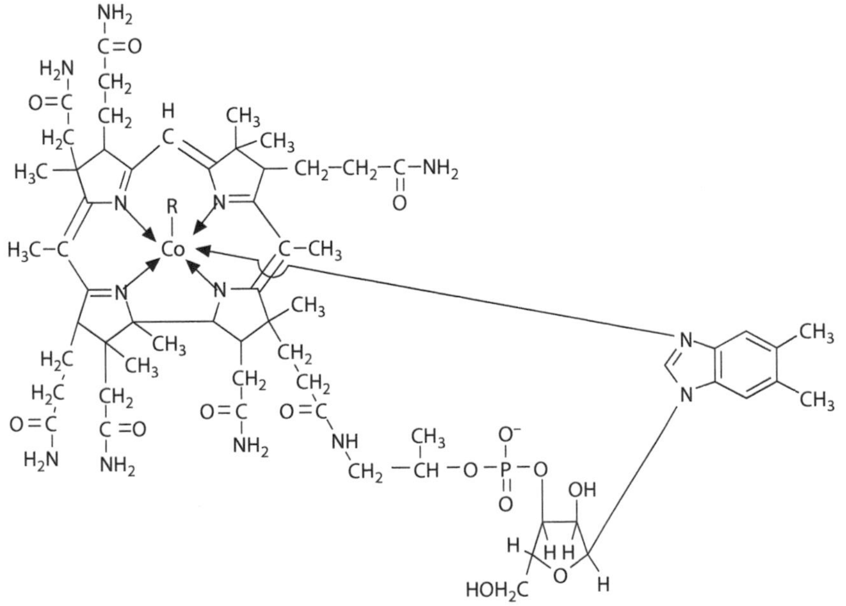

FIGURE 19.20 Structure of cyanocobalamin.

strict vegetarian diet or after diseases affecting cobalamin absorption. The main effects of vitamin deficiency are pernicious anemia, macrocytosis, and neurological problems. A particularity of this vitamin is that it can be stored especially in the liver and kidneys.

No adverse effects have been associated with excess vitamin B12 intake from food or supplements in healthy individuals [417].

Several different techniques have been proposed for vitamin B_{12} determination, such as microbiological assay, spectrophotometry, chemiluminescence, atomic absorption spectrometry, capillary electrophoresis, and HPLC.

19.8.3.2.1   Sample Preparation

Since the main sources of vitamin B_{12} are meat and diary products, the first step is the protein denaturation. To save the vitamers, an enzymatic digestion can be applied using proteinase and extraction into ethanol at 80°C [611] or extraction in dimethylsulfoxide containing ammonium pyrrolidine dithiocarbamate and citric acid [612]. Heating in a cyanide- or sulfate-containing buffer also enables to free the protein-bounded vitamers, but all the vitamers are converted into cyano- or sulfito-cobalamin forms. A succeeding purification is needed and it is performed in anion-exchange chromatography or SPE (C18).

19.8.3.2.2   Chromatography

The most common method is RP-HPLC. Microbore-HPLC [613] and narrow bore columns [614] are applied with good results in increasing sensitivity. A recent comparison between NP and RP microbore columns confirms the better suitability of RP for this purpose [615]. The mobile phase is usually composed of acetonitrile or methanol.

19.8.3.2.3   Detection

These vitamers are UV absorbers, but their detection is complicated by the low level present in foods and the low sensitivity of this detector. Other detectors, like flame atomic absorption spectrometry and inductively coupled plasma (ICP)-MS, may be applied, but without much increase in sensitivity.

19.8.3.3   Thiamine (Vitamin B₁)

Vitamin B_1 or thiamine (Figure 19.21) is 3-(4-amino-2-methylpyrimidin-5-ylmethyl)-5-(2-hydroxyethyl)-4-methylthiazolium. It is isolated, synthesized, and used in food supplements and in food fortifications as a solid thiazolium salt in the form of thiamine hydrochloride or thiamine mononitrate [403].

Vitamin deficiency of B_1 leads to the disease known as Beriberi. However, nowadays in the Western hemisphere, vitamin B_1 deficiency is mainly found as a consequence of extreme alcoholism. In fact, the vitamin absorption by the gut is decreased and its excretion is increased by alcohol. Alcohol also inhibits the activation of vitamin B_1 to its coenzyme form, thiamine pyrophosphate ester (TPP). There is no evidence of adverse effects of oral intake of thiamine [417]. The main food sources of vitamin B_1 are lean pork, legumes, and cereal grains (germ fraction). It is soluble in water and stable at higher temperature and at pH lower than 5.0, but it is destroyed rapidly by boiling at pH 7.0 or above.

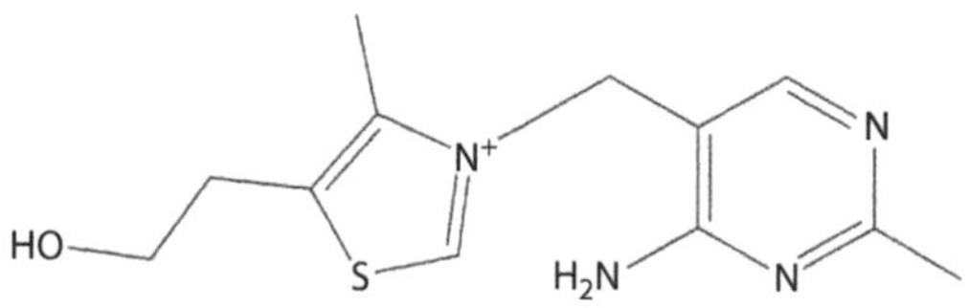

FIGURE 19.21   Structure of vitamin B₁.

19.8.3.3.1 Sample Preparation

Extraction is commonly carried out by hydrolysis in boiling acid such as chloridric acid or sulfuric acid. To release thiamine bonded to phosphate enzyme, hydrolysis with phosphatase, alone or together with claradiastase or takadiastase, is carried out. After the enzymatic digestion, an acid treatment is applied in order to precipitate the protein and denaturate the enzymes. Ndaw et al. [603] proved that for extraction of vitamins B_1, B_2, and B_6, acid hydrolysis is always superfluous if the activity of the enzymes chosen is sufficiently high. SPE or column chromatography may be used in further purification, mainly to remove excess of derivatization reagents used to convert thiamine to a highly fluorescent thiochrome derivatives. IEC may be used in purification step, as well.

19.8.3.3.2 Chromatography

RP chromatography (C18 or amino stationary phase) or ion-pair RP chromatography are used for thiamine analysis. The mobile phases used are mainly methanol or acetonitrile within sodium acetate or phosphate acetate.

19.8.3.3.3 Detection

When the amount of thiamine is relatively high, UV detection at 254 nm can be applied. Fluorometric detection, however, enables to reach higher sensitivity, when coupled with precolumn or postcolumn derivatization. The derivatization is carried out by using potassium ferricyanide under alkaline conditions to convert thiamine to highly fluorescent thiochrome derivatives.

19.8.3.4 Riboflavin (Vitamin B_2)

Riboflavin (vitamin B_2) is chemically specified as a 7,8-dimethyl-10-(1′-D-ribityl) isoalloxazine (Figure 19.22). It is a precursor of certain essential coenzymes, such as flavin mononucleotide (FMN) and flavin-adenine dinucleotide (FAD); in these forms vitamin B_2 is involved in redox reactions, such as hydroxylations, oxidative carboxylations, dioxygenations, and the reduction of oxygen to hydrogen peroxide. It is also involved in the biosynthesis of niacin-containing coenzymes from tryptophan.

In foods vitamin B_2 occurs free or combined both as FAD and FMN and complexed with proteins. Riboflavin is widely distributed in foodstuffs, but there are very few rich sources. Only yeast and liver contain more than 2 mg/100 g. Other good sources are milk, the white of eggs, fish roe, kidney, and leafy vegetables. Since riboflavin is continuously excreted in the urine, deficiency is quite common when dietary intake is insufficient. The symptoms of deficiency are cracked and red lips, inflammation of the lining of the mouth and tongue, mouth ulcers, cracks at the corner of the mouth, and sore throat. Overdose of oral intake present low toxicity, probably explained by the limited capacity of the intestinal absorption mechanism [417].

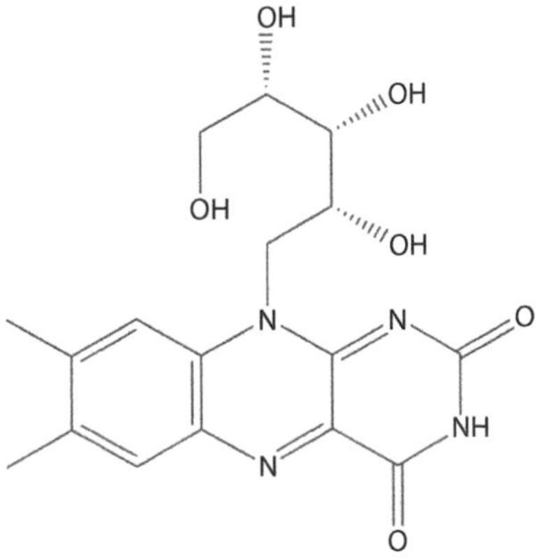

FIGURE 19.22 Structure of vitamin B_2.

19.8.3.4.1 Sample Preparation

The extraction procedure has to be performed in subdued light due to the photosensitivity of vitamin B_2. A selective extraction for free riboflavin can be carried out by using benzyl alcohol. To quantify the total amount of vitamin B_2-active compounds, the release of the coenzyme (FMN and FAD) from the protein to which they are bonded and the following hydrolysis to riboflavin is necessary. For this reason, acid hydrolysis is widely reported [616,617], but, as specified for thiamine, Ndaw et al. [603] proves that an efficient enzymatic digestion is enough. SPE is then used to clean-up the extract before the final quantification. A milder extraction for the coenzyme forms was proposed by performing a first extraction with methanol and a second one with citrate–phosphate buffer [618]. The use of an immobilized enzyme reactor (IMERs) was reported by Yamato et al. [619] to release riboflavin from the phosphate forms in pharmaceutical products. An acid phosphatase was immobilized by covalent coupling with glutaraldehyde to aminpropyl controlled-pore glass and then packed into a column.

19.8.3.4.2 Chromatography

The chromatography conditions applied are usually the same used for thiamine analysis. A RP chromatography is used for analytical determination, such as ODS stationary phase, C18 or amide phases. The mobile phases are methanol, water or acetonitrile with addition of buffer.

19.8.3.4.3 Detection

The riboflavin vitamers have UV absorbance and fluorescence, even if FAD's fluorescence is 10 times lower than riboflavin and FMN.

19.8.3.5 Pyridoxine (Vitamin B_6)

Vitamin B_6 is a mixture of six interrelated forms: pyridoxine (or pyridoxol) (Figure 19.23), pyridoxal, pyridoxamine, and their 5′-phosphates derivatives. Interconversion is possible between all forms. The active form of the vitamin is pyridoxal phosphate, which is a coenzyme correlated with the function of more than 60 enzymes involved in transamination, deamination, decarboxylation, or desulfuration reactions.

Pyridoxine is present in food in the free form and as a glucoside, which may undergo partial hydrolysis in the gut lumen, or may be absorbed intact. Although pyridoxine is associated with the enzyme glycogen phosphorylase in muscles, it is not released in response to a dietary deficiency; therefore it cannot be regarded as a storage form of the vitamin.

Deficiency of this coenzyme can lead to many manifestations. Clinical signs include retarded growth, acrodynia, alopecia, skeletal changes and anemia, while changes in neurotransmitters, such as dopamine, serotonin, norepinephrine (noradrenaline), tryptamine, tyramine, histamine, γ-aminobutyric acid, and taurine, affect the brain function and can lead to seizures and convulsions. An overdose of vitamin B_6 leads to neuronal damage and sensory and motor effects [417].

Vitamin formulations that may be added to foods are pyridoxine hydrochloride, pyridoxine 5-phosphate, and pyridoxine dipalmitate [403].

19.8.3.5.1 Sample Preparation

Vitamin B_6 has been traditionally extracted by using acid hydrolysis (chloridric acid or sulfuric acid) at high temperature followed by enzymatic hydrolysis with takadiastase or phosphatase to release the phosphorilated vitamers. In the 1990s, a rapid method was proposed involving the transformation of all the vitamers into pyridoxol and then analyzed by ion pair chromatography [620]. The following researches anyway focused the attention on separation and determination of all the different forms of vitamin B_6.

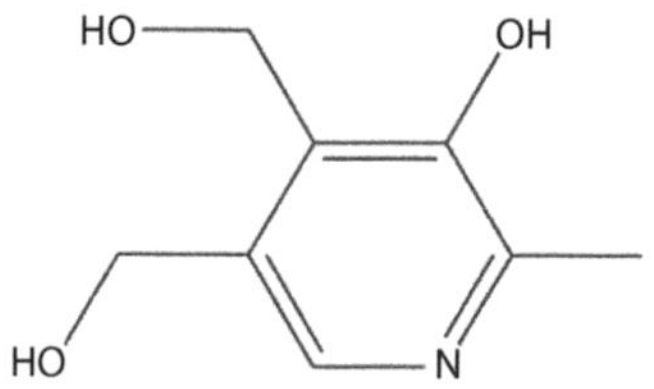

FIGURE 19.23 Main structure of vitamin B_6.

Milder extraction conditions, such as trichloroacetic or perchloric acid, are needed in order not to degrade the vitamers. As for vitamins B_1 and B_2, enzymatic digestion has been proved to be enough if the enzymes show high activity [603].

19.8.3.5.2 Chromatography

RP chromatography is used for analytical determination of vitamin B_6, such as ODS, C18, and amide. The mobile phases used are, like for vitamin B_1 and B_2, acetonitrile, methanol, and water within a percentage of buffer.

19.8.3.5.3 Detection

All the vitamers are UV absorbers and fluorescence, but to increase the sensitivity derivatization is usually applied. K_2HPO_4 buffer containing sodium bisulfite has been used as postcolumn reagent for this purpose. ED may also be applied to quantify pyridoxine [621].

19.8.3.6 Multivitamin Determination

A large number of methods have been developed for analysis of water-soluble vitamins simultaneously in pharmaceutical products (like multivitamin tablet supplements). In fact, for these products no particular sample preparations are required and the high concentrations simplify the detection, enabling the use of UV [636]. The use of MS is also reported [637]. As well, Moreno and Salvadó [638] reports also the use of a unique SPE cartridge (C18) for separating fat-soluble and water-soluble vitamins, which are, then analyzed using different chromatographic systems.

Instead, foodstuffs are a greater challenge due to the complexity of their matrices and the low natural concentration of vitamins in food. The different stability and the numerous vitamers present in foodstuffs required dedicated extraction procedures. The simultaneous extraction of vitamin B_1, B_2 has been performed in a unique step several times. However their detection, if MS is not applied, usually needs to be performed under different conditions, which means carrying out two different chromatographic analyses.

Thiamine, riboflavin, nicotinamide, pyridoxine, and folic acid can be determined together by using DAD, but pantothenic acid and biotin do not have adequate sensitivity for UV detection in complex matrices.

More recently [635], a unique extraction step in supplemented foods, by using hot water and a precipitation solution, following by HPLC-FLD/UV analysis has been performed for the simultaneous determination of pyridoxine, thiamine, riboflavin, niacin, pantothenic acid, folic acid, cyanocobalamin, and ascorbic acid. The mobile phase consisting of phosphate buffer and methanol has been modified in order to perform ion-liquid chromatography by adding 1-octanesulfonic acid sodium salt. Furthermore, triethylamine has been also added to improve peak symmetry.

Recently, the use of a novel stationary phase (p-tert-butyl-calix[8]arene-bonded silica gel, CABS) has been reported for the separation of six water-soluble vitamins (B_1, B_2, B_6, B_5, B_{12}, and C) by using isocratic elution with methanol/acetate buffer [639].

19.9 FOOD CONTAMINANTS

Nowadays HPLC is widely used for analytical determination of a large number of food contaminants. A number of recent works well review HPLC determination of major food contaminants such as mycotoxins [640–643], antimicrobial residues [644–646], residues of growth promoters [647], pesticide residues [648–651], and nitrosamines [652–654].

HPLC determination of food contaminants always involves an extraction step, generally followed by sample clean-up, prior to the analytical determination. When the analytes are present at very low concentrations, sample preparation, performed by means of classical procedures, involves long and tedious extraction and purification steps, high solvent consumption, and can be responsible for losses of the analyte and sample contamination. In recent years, a number of innovative sample preparation techniques, aimed at reducing solvent consumption and saving time have been applied for the analysis of food contaminants.

A thorough insight into analytical possibilities for sample preparation and HPLC determination of one of the most important classes of pollutants that can contaminate foodstuffs (e.g.. polycyclic aromatic hydrocarbons), will be given in the following sections.

19.9.1 POLYCYCLIC AROMATIC HYDROCARBONS

Polycyclic aromatic hydrocarbons (PAHs) represent a heterogeneous class of environmental contaminants formed by incomplete combustion or pyrolysis of organic matter. They comprise compounds with 2–6 fused benzene rings, most of which (4–6 ring compounds) are known human mutagens and carcinogens.

Environmental PAHs arise from two main anthropogenic sources, the combustion of fossil fuels, and the direct release of oil and oil products [655]. The occurrence of PAHs in food is mainly due to environmental contamination (through atmospheric fall-out) and food processing involving thermal treatment at high temperature or direct contact with combustion fumes (grilling, roasting, smoking, etc.). Vegetables are mainly contaminated by deposition of small airborne particulates. Broad-leaved vegetables, grown in areas exposed to industrial emission and/or motor vehicle exhausts, can be heavily contaminated with PAHs [656]. Sources of contamination for marine organism can include oil spills and run-off from land and industrial effluents [657]. Aquatic organisms present a degree of contamination that depends on the degree of contamination of the water where they live and their capability to metabolize these xenobiotics [658]. The practice to dry seeds or olive pomace with systems involving direct contact with combustion fumes, before oil extraction, always lead to high contamination levels, which fortunately can be reduced during oil refining [659,660].

For nonsmokers, the major routes of exposure to PAHs are from food and to some extent from inhaled air [661]. A number of investigations on PAH dietary intake have been carried out in different European countries [662–667]. Most of these studies indicated the group of fats and oils, together with cereals and vegetables, as the major contributors to the total dietary intake of PAHs [668]. Smoked and grilled foods may contribute significantly to PAH intake if such foods are part of the usual diet [663].

The Scientific Committee on Food (SCF) concluded in its opinion of December 4, 2002 [669] that a number of heavy PAHs have both carcinogenic and genotoxic properties and that benzo[a]pyrene (BaP) can be used as a marker for the occurrence of the entire class of carcinogenic and genotoxic PAHs in foods. On the basis of this opinion, Commission Regulation (EC) 208/2005 [670] introduced maximum levels of BaP in different foodstuffs: oils and fats, baby foods, smoked meat, fish, crustaceans, cephalopods, and bivalve molluscs. To gather more data on levels of PAHs in foodstuffs and to clarify whether BaP could be maintained as a marker of the whole group of PAHs, the commission asked the member states to monitor PAHs in different foods (Commission Recommendation (EC) 108/2005) [671]. A very recent report, issued on June 29, 2007 by European Food Safety Authority (EFSA) [661], concluded that it remains dubious that BaP is a good indicator of any PAH contamination. Before 2005, the PAHs usually researched in foods were the 16 Environmental Protection Agency (EPA)-priority PAHs (that comprised 2–6 ring terms). The list of 16 EU-priority PAHs (4–6 ring PAHs), comprising of 15 PAHs highlighted to be carcinogenic by the SCF and the one PAH highlighted by the Joint FAO/WHO Expert Committee on Food Additives (JEFCA), included eight heavy PAH compounds for which no analytical data are available before 2005. The inclusion of these new heavy PAHs in the list of researched compounds complicated analytical determination and demanded for the development of new validated methods for the determination of all the 16 EU-priority PAHs.

Concerning drinking water, the Council Directive 98/83/EC [672] fixed a maximum level for BaP of 0.010 µg/L and for the sum of benzo[b]fluoranthene, benzo[k]fluoranthene, benzo[ghi]perylene, and indeno[1,2,3-cd]pyrene of 0.10 µg/L.

According to the Commission Directive (EC) 10/2005 [673], laying down the sampling methods and the methods of analysis for the official control of the levels of BaP in foodstuffs, laboratories may select any validated method that satisfies the performance criteria indicated in the directive.

19.9.1.1 Sample Preparation

Due to the low amounts in which PAHs are usually found in foodstuffs, their accurate quantitative determination is very difficult. Depending on the complexity of the matrix, sample preparation usually includes an extraction step followed by one or more purification steps. In few cases, extraction and purification can be performed in one single step.

19.9.1.1.1 Water and Alcoholic Beverages

Before the introduction of SPE, PAHs extraction from aqueous samples required long and tedious LLE steps employing high amount of organic solvents [674,675]. Nowadays, most of the methods proposed for PAH determination in water and alcoholic beverages employ SPE on C18 [676–678], C8, and PS-DVB cartridges. Bouzige et al. [679] reported the successful use of "anti-fluorene" sorbents for selective isolation of PAH from water samples. EPA [680] recommended SPE on C18 for PAH analysis in tap water (EPA method 525). The introduction of SPE discs (with a glass fiber matrix) allowed to shorten the extraction time by between 3 and 12 times, in comparison with other SPE systems [681]. To increase recoveries of more hydrophobic PAHs, little amounts of an organic solvent (methanol, acetonitrile, or isopropanol) or tensioactives can be added to the sample before loading [679]. This allows increasing analyte solubility and avoiding analyte adsorption on the surfaces of contact. As demonstrated by Urbe and Ruana [681], the presence of chlorine in tap water can be responsible for low analyte recovery. To increase recovery of BaP from 10% to 80%, it is sufficient to add to the water sample sodium sulfite, able to remove residual chlorine. SPE (on cartridge or disk) has also the advantage to allow on-line coupling with HPLC systems, minimizing sample manipulation, possibility of contamination, and reducing analysis time [675]. Other possible approaches for PAH extraction from aqueous samples before HPLC analysis include the use of SPME [682] and stir bar sorptive extraction (SBSE) [683].

19.9.1.1.2 Meat, Fish, and Vegetables

Two excellent reviews on PAH determination in smoked fish [684] and smoked meat products [685] have been published recently. Traditional extraction procedures for PAH analysis in these complex matrices involve a saponification step, followed by LLE [686] and laborious purification steps on packed columns (silica, florisil, alumina), or, more recently, on disposable SPE cartridges. GPC and TLC can also be used for sample clean-up. According to some authors, saponification is absolutely necessary to obtain quantitative PAH recoveries from these complex matrices, since PAHs are strongly linked to high molecular structures such as proteins and can diffuse inside tissue cells [685,686]. On the other hand, saponification can be responsible for PAH losses due to decomposition during alkaline hydrolysis and formation of emulsions difficult to break up during the LLE step [687], and many researchers disagree with the necessity of a saponification. Different authors recommend direct extraction with organic solvents [688–691]. Soxhlet extraction represents a common method of choice for fatty tissues [692,693], but it is rather time-consuming and requires high solvent consumption. Sonication has been advantageously used to improve PAH extraction efficiency, saving time [694].

In order to accelerate sample preparation, new extraction methodologies such as accelerated solvent extraction (ASE) and MAE, based on the use of elevated temperature and pressure to heat the mixture sample–solvent, have been recently developed and applied for PAH extraction from meat [695] and vegetables [696–698]. García Falcón et al. [699] used microwave treatment with hexane to accelerate PAH extraction from freeze-dried foods. The fat extracted in this way underwent microwave assisted saponification with ethanolic KOH. Hernández-Borges et al. [700] combined microwave-assisted hydrolysis and extraction to isolate organic pollutants from mussels, while

Samsøe-Petersen et al. [701] used focused MAE to extract PAHs with a mixture of hexane and methylene chloride from fruits and vegetables grown in contaminated soils. A further purification step (often performed by SPE) is always needed before HPLC determination.

19.9.1.1.3   Fats and Oils

Classical and innovative sample preparation techniques for PAH analysis in edible oils and lipidic extracts have been exhaustively reviewed by Moret and Conte [659]. In this case, it is possible to realize simultaneous extraction/analyte enrichment and purification. This is also the approach of the AOAC method (Official Method Cd 21-91) for BaP determination in animal and vegetable fats and oils, approved in 1997 and afterward adopted as an ISO method (ISO 15302:1998). Briefly 400 mg of oil (diluted in 2 mL of solvent) are loaded onto a column packed with alumina (22 g). Sample elution is carried out with light petroleum ether or hexane. After discharging the first 20 mL of eluate, the next 60 mL are collected, taken to dryness and the residue, dissolved in tetrahydrofuran, and injected into the HPLC apparatus. Using selective excitation and emission wavelengths (284–406 nm) there is no interference by other PAHs and BaP can be detected with a detection limit of 0.1 μg/kg. This method has the disadvantages of being time-consuming, requiring the consumption of high solvent volumes, and allowing the quantification of BaP only. Recently some researchers of the University of Udine set up a rapid method for the determination of 15 of the 16 EPA priority PAHs in vegetable oils and lipid extract [702]. Briefly, 250 mg of oil in n-hexane are loaded onto a 5 g silica cartridge and PAH fraction is eluted by means of 8 mL of n-hexane/dichloromethane (70/30 v/v). After solvent evaporation, volume was adjusted at 100 μL and injected in a HPLC equipped with a C18 column and spectrofluorometric detector. A similar approach using PS-DVB SPE cartridges has been proposed by other authors [703,704]. In 2005, an interlaboratory survey was organized in Italy to test the capability of these two methods for routine analysis in quality control of vegetable oils. The results of the study were statistically evaluated and both methods satisfied the validation criteria provided by EU.

19.9.1.2   Analytical Determination and Detection

PAHs arising from combustion sources consist predominantly of parent, unalkylated PAHs, while PAH fractions derived from mineral oil contain large amounts of alkylated PAHs. Consequently, chromatograms of the aromatic fractions of samples contaminated by oil and oil products are usually more complex than those of the samples for which combustion is the major source of contamination [655]. Due to the higher resolving power of a capillary column, GC is the best technique to use with samples contaminated with mineral oil residues. Even with the high resolving power of capillary GC, a detailed compositional analysis is far from possible and the approach is toward group-type analysis. NP-HPLC on polar stationary phases, such as silica (for fat preseparation) and amino (for PAH fractioning based on the number of aromatic rings), have been used successfully prior to GC analysis of such complex PAH mixtures [705,706].

When parent nonalkylated PAHs predominate, HPLC is the preferred technique. HPLC is somewhat faster than high-resolution GC and, even though it offers lower resolution efficiency in separating low-molecular PAHs, RP columns can readily separate a number of PAH isomers that are difficult to separate by GC [707,708]. HPLC has some other advantages over GC: it allows determination of high-molecular-mass PAHs, which could not be detected by GC methods because of the thermal decomposition occurring at high temperatures [659,709,710], and it can be used in connection with spectofluorometric detector. As a lot of PAHs exhibit strong fluorescence, HPLC, coupled to a spectrofluorometric detector, represents the most powerful technique concerning sensitivity and selectivity (comparable only to that obtained with MS-(single ion monitoring) SIM detection) [659,709].

Alumina and silica columns, formerly used for PAH separation, were later replaced with chemically bonded phases [685]. RP-HPLC based on the use of C18 columns is nowadays the most popular mode for PAH separation, and specially designed columns from different vendors are commercially available. Columns of 100–250 cm length (3.0–4.6 mm i.d.), packed with 3–5 μm particles, are usually employed. Greater resolution can be obtained by the use of microbore HPLC systems, but at the cost of greatly increased run-times [655]. A typical gradient elution program consists of 50%–60% of acetonitrile (or methanol) in water, maintained isocratic for 2–5 min, and then linearly programmed at 100% acetonitrile (or methanol).

Figure 19.24 shows the HPLC trace obtained in our laboratory for a complex mixture of the 16-EU priority PAHs plus other EPA-priority PAHs, benzo[e]pyrene and benzo[b]chrysene. The column was a Supelcosil LC-PAH (150×4.6 mm i.d., 5 μm) working with an optimized gradient of water and acetonitrile. Two spectrofluorometric detectors (using different wavelength settings) and one UV detector connected in series were used to optimize detection and allow peak purity confirmation. The use of optimized wavelength settings allowed to quantify PAHs that were not chromatographically separated (benzo[j]pyrene and benzo[e]pyrene).

As PAHs absorb in the UV region, variable-wavelengths UV detectors can be used for their detection. To allow positive confirmation of PAH identity DAD can be used. The use of FLD makes it possible to optimize the limit of detection by programmed selection of the excitation and emission wavelengths. Some instruments allow simultaneous PAH detection in several detection channels. Under such conditions, the limit of detection can be decreased to the low picogram range [685], and sensitivity 20–320 times higher than UV detection can be achieved [711]. Some models of detectors allow on-line scanning of fluorescence spectra, which can be very useful to confirm peak

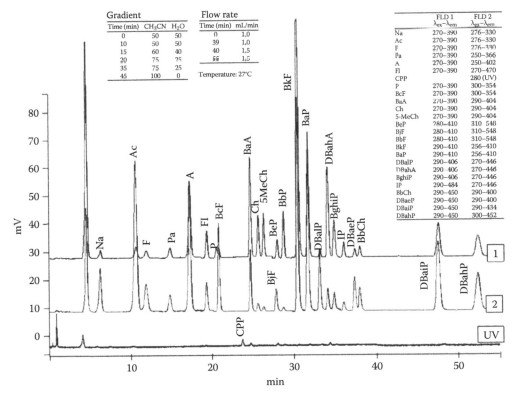

FIGURE 19.24 HPLC separation of a standard mixture of 16 EU-priority PAHs plus EPA-priority PAHs, benzo[e]pyrene and benzo[b]chrysene. Na = naphthalene, Ac = acenaphthylene, F = fluorene, Pa = phenanthrene, A = anthracene, Fl = fluoranthene, P = pyrene, BcF = benzo[c]fluoranthene, CPP = cyclopenta[c,d] pyrene, BaA = benz[a]anthracene, Ch = chrysene, 5-MeCh = 5 methylchrysene, BeP = benzo[e]pyrene, BjF = benzo[j]fluoranthene, BbF = benzo[b]fluoranthene, BkF = benzo[k]fluoranthene, BaP = benzo[a]pyrene, DBahA = dibenz[a,h]anthracene, DBalP = dibenzo[a,l]pyrene, BghiP = benzo[g,h,i]perylene, IP = indeno [1,2,3-cd]pyrene, DBaeP = dibenzo[a,e]pyrene, BbCh = benzo[b]chrisene, DBaiP = dibenzo[a,i]pyrene, DBahP = dibenzo[a,h]pyrene.

identity and degree of pureness. Care must be taken to deoxygenate solvents used in HPLC as the fluorescence of some PAHs (e.g., pyrene) can be dramatically reduced in the presence of oxygen [710]. Amperometric detection is also possible for PAH detection: it allows obtaining 5–10 times better sensitivity than UV detection and is universally applicable (since all PAHs can be determined via electrooxidation that gives rise to radical cations) [711].

HPLC equipped with MS detectors is an effective tool for the characterization of high-molecular, thermally unstable compounds. The analysis of high-molecular-weight PAHs using HPLC–MS with an APCI source has been reported in a range of matrices including zebra mussels [712]. However, because of the high cost of this instrumentation, this technique cannot be applied at present for routine control in food monitoring [711].

19.9.1.3   Coupled Technique for PAH Analysis

Untreated silica column can be advantageously used for HPLC preseparation of PAHs from triglycerides. The capacity of a silica column to retain fat (for columns of the same particle size) depends on the column size, the mobile phase composition, as well as the type and by-products (free acids and polymerized material) of the fat injected [706,713]. Off-line HPLC-HPLC, employing silica column (250 × 4.6 mm i.d., 5 μm of particle size) for sample preparation before RP-HPLC and spectrofluorometric detection, was successfully applied for PAH determination in edible oils [659,691] and fish [714]. After PAH elution, the silica column needs to be backflushed with dichloromethane to remove the fat. The entire sample preparation step can be automated by using a backflush valve and a programmable switching valve box [691].

Off-line methods do not permit a quantitative transfer of the fraction eluted from the first column and are subject to sample contamination. On-line methods allow overcoming these problems and can be easily automated. Van Stijn et al. [715] set up an on-line HPLC-HPLC method involving a clean-up of the sample on a preconcentration donor acceptor complex chromatography column, followed by separation of the PAH fraction on a conventional C18 column. A completely automated HPLC-HPLC system, coupling a first silica column with a second RP column via a solvent evaporator, was developed by Moret et al. [716] for heavy PAH determination in edible oils and lipid extracts. The solvent evaporator (SE) was a prototype [717] consisting of an aluminum block heated at 40°C and containing the vaporizing chamber: a steel capillary tube packed with silica gel (35–70 mesh) and deactivated by mild silylation. The SE works on the principles of concurrent eluent evaporation and overflow. When the PAH fraction starts to be eluted, the eluent is fed into the SE and the analytes are retained from the packed chamber, while solvent vapors are discharged by their expansion. The outlet of the SE is connected to the vacuum to keep the evaporation temperature low, in order to increase the retentive power of the vaporizing chamber. At the end of the transfer, the silica column is backflushed to remove triglycerides, and the PAHs retained in the vaporizing chamber are transferred into the C18 column with a solvent (acetonitrile) compatible with chromatographic separation in the second dimension. The same SE was also employed for on-line HPLC-HPLC-GC determination of alkylated PAHs in samples contaminated with mineral oil residues [705,706]. In this case, the SE connected on-line a first silica column (to retain fat) with a second amino column able to fractionate PAHs according to the number of rings in the molecule.

LIST OF ACRONYMS AND ABBREVIATIONS

2-NPH	2-nitrophenylhydrazine
3,4-DHPEA-EDA	3,4-Dihydroxyphenylethanol-elenolic acid dialdehyde
3-MPA	3-mercapto propionic acid
AA	L-ascorbic acid
AABA	α-amino-n-butanoic acid

AC	affinity chromatography
ACN	acetonitrile
ACP	acyl carrier protein
ADAM	9-anthryldiazomethane
AdoCbl	adenosylcobalamin
AED	amperometric electrochemical detection
APCI	atmospheric pressure chemical ionization
AQC	6-aminoquinolyl-N-hydroxysuccinimidyl carbamate
ASE	accelerated solvent extraction
BaP	benzo[a]pyrene
BHT	butyl hydroxy toluene
BMMC	4-bromomethylmethoxiycoumarin
BTI	bis(thrifluoroacetoxy)iodobenzene
CABS	calix[8]arene-bonded silica gel
CHES	2-(cyclohexylamino)ethanesulfonic acid
CN	carbon number
CoA	coenzyme A
COP	cholesterol oxidation product
DAD	diode array detector
DBB	number of double bonds
DFQ	3(1,2-dihydroxy-ethyl)furo[3,4-b]quinoxaline-1-one
DHAA	L-dehydroascorbic acid
DMP	dimercaptopropanol
DNPH	dinitrophenylhydrazine
Dns-Cl	dansyl chloride
DTT	dithiothreitol
EC	European Commission
ED	electrochemical detector
EDC	1-ethyl-3-(3-dimethylaminopropyl)-carbodiimide
EDTA	ethylene-diaminetetraacetic acid
EFSA	European Food Safety Authority
ELS	evaporative light scattering
EPA	Environmental Protection Agency
ESI	electrospray ionization
Et$_2$O	diethyl ether
EtAc	ethyl acetate
EtOH	ethanol
FAB	fast atom bombardment
FAD	flavin–adenine dinucleotide
FBP	folate-binding protein
FDA	Food and Drug Administration
FDNB	1-fluoro-2,4-dinitrobenzene
FEP	fluorinated ethylene–propylene
FITC	fluorescein 5-isothiocyanate
FLD	fluorometric detector
FMN	flavin mononucleotide
FMOC-Cl	9-fluorenylmethylchloroformate
GC	gas chromatography
Gh	glutathione
GMP	guanosine-5′-monophosphate
GPC	gel permeation chromatography

HAc	acetic acid
HEPES	4-(2-hydroxyethyl)-1-piperazine-ethanesulfonic acid
Hex	*n*-hexane
HIC	hydrophobic interaction chromatography
HMF	hydroxymethylfurfural
HOCbl	hydroxycobalamin
HPLC	high-performance liquid chromatography
IAA	D-ascorbic acid
ICP	inductively coupled plasma
IEC	ion exchange chromatography
IMAC	metal-ion affinity chromatography
IOOC	International Olive Oil Council
IP	isopropanol
ISO	International Standard Organisation
IUPAC	International Union of Pure and Applied Chemistry
JEFCA	Joint FAO/WHO Expert Committee on Food Additives
LC	liquid chromatography
LLE	liquid–liquid extraction
LLL	glycerol trilinoleyl ester (trilinoleine)
LnLnLn	glycerol trilinolenyl ester trilinolenine
LnPP	glycerol dipalmityl linoleyl ester
LOD	limit of determination
LOO	glycerol dioleyl linoleyl ester
LOQ	limit of quantification
MAE	microwave-assisted extraction
MALDI	matrix-assisted laser desorption/ionization
MAO	monoamino oxidase
MDA	malonildialdehyde
MeCbl	methylcobalamin
MeOH	methanol
MS	mass spectrometry
MSG	monosodium L-glutamate
MTBE	methyl *tert*-butyl ether
MW	molecular weight
NAD	nicotinamide adenine dinucleotide
NADase	nicotinamide adenine dinucleotide glycohydrolase
NADP	nicotinamide adenine dinucleotide phosphate
NaOAc	sodium acetate
NH_4OAc	ammonium acetate
NP	normal-phase
NQS	1,2-naphthoquinone-4-sulfonate
ODS	octadecylsilane
OLL	glycerol oleyl inoleil ester
OLLn	glyceryl oleyl linoleyl linolenyl ester
OOLn	glyceryl dioleyl linolenyl esterf
OOO	glyceryl tri oleyl ester (Trioleine)
OPA	*o*-phthalaldehyde
PAD	pulsed amperometric detector
PAGE	polyacrylamide gel electrophoresis
PAH	polycyclic aromatic hydrocarbon
PCA	perchloric acid

PDAM	1-pyrenyldiazomethane
PGA	pteroylglutamin acid
PHL	phospholipid
PITC	phenylisothiocyanate
PLE	pressurized liquid extraction
PLL	glycerol palmityl dilinoleyl ester
PLLn	glycerol palmityl linoelyl linolenyl ester
PLO	glycerol palmityl linoelyl oleyl ester
PLP	glycerol dipalmityl linoelyl ester
PN	partition number
PNBC	p-nitro benzoyl chloride
POA	glycerol palmityl oleyl arachidyl ester
PO-CL	peroxyoxalate chemiluminescence
PoLL	glycerol palmitoleyl di linoleyl ester
POLn	glycerol palmityl oleyl linolenyl ester
POO	glycerol palmityl di oleyl ester
PoOP	glycerol palmitoleyl oleyl palmityl ester
POP	glycerol di palmityl oleyl ester
POS	glycerol palmityl oleyl stearyl ester
PS-DVB	polystyrene-divinylbenzene
PTFE	polytetrafluoroethylene
PUFA	polyunsaturated fatty acids
PVDF	polyvinylidene fluoride
RDA%	relative daily allowance
RI	refractive index
RP	reversed phase
RPC	reversed phase chromatography
RSD	relative standard deviation
SBSE	stir bar sorptive extraction
SCF	Scientific Committee on Food
SDS	sodium dodecyl sulfate
SE	solvent evaporator
SEC	size exclusion chromatography
SFC	supercritical fluid chromatography
SIM	single ion monitoring
SLL	glycerol stearyl dilinoelyl ester
SOL	glycerol stearyl oleyl linoelyl ester
SOO	glycerol stearyl dioleyl ester
SPE	solid phase extraction
SPME	solid phase microextraction
SSS	tristearine
TAG	triacylglycerol
TCA	tetraethyl ammonium
TCA	trichloroacetic acid
TCEP	tris[2-carboxyethyl]phosphine
TEA	triethylamine
TFA	trifluoroacetic acid
TGP	triglyceride polymer
THF	tetrahydrofolate
TLC	thin layer chromatography
TPP	thiamine pyrophosphate ester

ACKNOWLEDGMENT

We would like to thank Dr. Dobrila Braunstein for revising the language in the manuscript.

REFERENCES

1. V. Ruiz-Gutierrez, L. J. R. Barron, *J. Chromatogr. A* 671, 133–168, 1995.
2. International Olive Oil Council, /T.15/NC. N. 3 /Rev.2, 2006.
3. Codex Alimentarius Commissione—ALINORM 07/30/17, July 2007.
4. International Olive Oil Council, method COI/T.20/Doc. no. 20 /Rev. 1, 2001.
5. EEC Regulation 2568/91, annex XVIII, as amended by Reg. (EC) 2472/97, *EEC Off. J. L* 341, 12/12/1997.
6. International Olive Oil Council, method COI/T.26, 2006.
7. H. J. Fiebig, *Fette Seifen Anstrichmittel* 87, 53–57, 1985.
8. W. Moreda, M. C. Perez-Camino, A. Cert, *Grassa y Aceites* 54, 175–179, 2003.
9. D. Ollivier, J. Artaud, C. Pinatel, J. P. Durbec, M. Guérère, *Food Chem.* 97, 382–393, 2006.
10. W. W. Christe, *J. Chromatogr.* 454, 273–276, 1988.
11. W. W. Christi, B. Nikolava-Damayanova, P. Laakso, B. Herslof, *J. Am. Oil Chem. Soc.* 68, 695–701, 1991.
12. W. W. Christie, *J. High Resolut. Chromatogr. Chromatogr. Commun.* 10, 148–152, 1987.
13. L. Mondello, P. Q. Tranchida, V. Stanek, P. Jandera, G. Dugo, P. Dugo, *J. Chromatogr. A* 1086, 91–98, 2005.
14. T. Gomes, *Riv. Ital. Sostanze Grasse* 65, 125–127, 1988.
15. T. Gomes, *J. Am. Oil Chem. Soc.* 69, 1219–1223, 1992.
16. T. Gomes, F. Caponio, *Riv. Ital. Sostanze Grasse* 73, 97–100, 1996.
17. T. Gomes, F. Caponio, *J. Agric. Food Chem.* 46, 1137–1142, 1998.
18. T. Gomes, F. Caponio, *J. Chromatogr. A* 844, 77–86 1999.
19. S. R. Eder, *Fette Seifen Anstrichmitel* 84, 136–141, 1982.
20. C. Gertz, *Eur. J. Lipid Sci. Technol.* 102, 329–336, 2000.
21. C. Gertz, *Eur. J. Lipid Sci. Technol.* 102, 566–572, 2000.
22. C. Gertz, *Eur. J. Lipid Sci. Technol.* 103, 114–116, 2001.
23. C. Gertz, *Eur. J. Lipid Sci. Technol.* 103, 181–184, 2001.
24. German Standard Methods, DGF C-III-3c, Deutsche Gesellschaft für Fettwissenschaft, 2007.
25. German Standard Methods, DGF C-III-3d, Deutsche Gesellschaft für Fettwissenschaft, 2007.
26. P. Rovellini, N. Cortesi, A. Mattei, F. Marotta, *Riv. Ital. Sostanze Grasse* 79, 429–438, 2002.
27. P. Rovellini, *Riv. Ital. Sostanze Grasse* 83, 213–220, 2006.
28. M. F. Caboni, S. Menotta, G. Lercker, *J. Chromatogr.* 683, 59–65, 1994.
29. S. Néron, F. El Amrani, J. Potus, J. Nicolas, *J. Chromatogr. A* 1047, 77–83, 2004.
30. A. Avalli, G. Contarini, *J. Chromatogr. A* 1071, 185–190, 2005.
31. D. Pacetti, H. W. Hulan, M. Schreiner, E. Boselli, N. Frega, *J. Sci. Food Agric.* 85, 1703–1714, 2005.
32. D. Pacetti, E. Boselli, H. W. Hulan, N. G. Frega, *J. Chromatogr. A* 1097, 66–73, 2005.
33. M. F. Caboni, A. Costa, M. T. Rodriguez-Estrada, G. Lercker, *Chromatographia* 46, 151–155, 1997.
34. E. Razzazi-Fazeli, S. Kleineisenb, W. Luf, *J. Chromatogr. A* 896, 321–334, 2000.
35. M. Amelio, R. Rizzo, F. Varazini, *J. Chromatogr.* 606, 179–185, 1992.
36. M. Amelio, R. Rizzo, F. Varazini, *J. Am. Oil Chem. Soc.* 75, 527–530, 1998.
37. ISO/EN 15788-2 European Committee for Standardisation—Brussels, 2004.
38. G. N. Jham, R. Velikova, B. Nikolova-Damyavova, S. C.Rabelo, J. C. Texeira da Silva, K. A. de Paula Souza, V. M. Moreira Valente, P. R. Cecon, *Food Res. Int.* 38, 121–126, 2005.
39. C. Silverstand, C. Haux, *J. Chromatogr. A* 703, 7–14, 1997.
40. S. Vichi, L. Pizzale, L.S. Conte, *Eur. J. Lipid Sci. Technol.* 109, 72–78, 2007.
41. A. Gasparoli, S. Tagliabue, C. Mariani, *Riv. Ital. Sostanze Grasse* 80, 210–211, 2003.
42. N. Cortesi, P. Fusari, *Riv. Ital. Sostanze Grasse* 82, 167–172, 2005.
43. T. Gomes, F. Caponio, M. T. Bilancia, C. Summo, E. Lamparelli, *Riv. Ital. Sostanze Grasse* 81, 161–163, 2004.
44. P. Dugo, O. Favonio, P. Q. Tranchida, G. Dugo, L. Mondello, *J. Chromatogr. A* 1041, 135–142, 2004.
45. S. C. Cunha, M. B. P. P. Oliveira, *Food Chem.* 95, 518–524, 2006.
46. P. Dugo, T. Kumm, M. L. Crupi, A. Cotroneo, L. Mondello, *J. Chromatogr. A* 112, 269–275, 2006.
47. G. Versini, A. Dalla Serra, G. Margheri, *Vignevini* 3, 41–47, 1984.

48. J. J. Mangas, J. Moreno, A. Picinelli, D. Blanco, *J. Agric. Food Chem.* 46, 4174–4178, 1998.
49. EEC Council Directive 2001/110/EC of 20 December 2001, *EEC Off. J. L* 10, 47–52, 12.01.2002
50. E. Marconi, M. F. Caboni, M. C. Messia, G. Panfili, *J. Agric. Food Chem.* 50, 2825–2829, 2002.
51. M. C. Messia, M. F. Caboni, E. Marconi. *J. Agric. Food Chem.* 53, 4440–4443, 2005.
52. E. Mendes, E. Brojo, I. M. P. L. V. O. Ferreira, M. A. Ferreira, *Carbohydr. Polym.* 37, 219–223, 1998.
53. P. Lehtonon, R. Hurme, *J. Inst. Brew* 100, 343–346, 1994.
54. E. Verette, F. Qian, F. Mangani, *J. Chromatogr. A* 705, 195–203, 1995.
55. S. Vendrell, A. I. Castellote, M. C. Lopez, *J. Chromatogr. A* 881, 591–597, 2000.
56. M. D. Rodrıguez, M. J. Villanueva, A. Ridondo, *Food Chem.* 66, 81–85, 1999.
57. E. H. Soufleros, I. Pissa, D. Petridis, M. Lygerakis, K. Mermelas, G. Boukouvalas, E. Tsimitakis, *Food Chem.* 75, 487–500, 2001.
58. Y. Wei, M. Y. Ding, *J. Chromatogr. A* 904, 113–117, 2000.
59. E. F. Lopez, E. F. Gomez, *J. Chromatogr. Sci.* 34, 254–257, 1996.
60. X. Huang, J. J. Pot, W. T. Kok, *Chromatographia* 40, 684–689, 1995.
61. J. Bouzas, C. A. Kantt, F. Bodyfelt, J. A. Torres, *J. Food Sci.* 56, 276–278, 1991.
62. M. Calull, E. Lopez, R. M. Marce, J. C. Olucha, F. Borrull, *J. Chromatogr.* 589, 151–158, 1992.
63. G. Doyon, G. Gaudreau, D. St-Gelais, Y. Bealieu, C. J. Randall, *Can. Inst. Food Sci. Technol.* 24, 87–94, 1991.
64. B. Picinelli, B. Suarez, J. Moreno, R. Rodrıguez, L. M. Caso, J. J. Mangas, *J. Agric. Food Chem.* 8, 3997–4002, 2000.
65. Z. L. Chen, D. B. Hibbert, *J. Chromatogr. A* 766, 27–33, 1997.
66. J. P. Yuan, F. Chen, *Food Chem.* 64, 423–427, 1999.
67. M. Castellari, A. Versari, U. Spinabelli, S. Galassi, A. Amati, *J. Liquid Chromatogr. Relat. Technol.* 23, 2047–2056, 2000.
68. S. Akiyama, K. Nakashima, K. Yamada, *J. Chromatogr. A* 626, 266–270, 1992.
69. K. K. Gey, M. H. Unger, G. Battermann, *Fresenius J. Anal. Chem.* 356, 339–343, 1996.
70. W. J. Mullin, D. B. Emmons, *Food Res. Int.* 30, 147–151, 1997.
71. D. Edelmann, C. Ruzicka, J. Frank, B. Lendl, W. Schrenk, E. Gornik, G. Strasser, *J. Chromatogr. A* 934, 123–128, 2001.
72. R. J. Weston, L. K. Brocklebank, *Food Chem.* 64, 33–37, 1999.
73. J. Wang, P. Sporns, N. H. Low, *J. Agric. Food Chem.* 47, 1549–1557, 1999.
74. R. Andersen, A. Srensen, *J. Chromatogr. A* 897, 195–204, 2000.
75. T. R. I. Castaldi, C. Campa, I. G. Casella, S. A. Bufo, *J. Agric. Food Chem.* 47, 157–163, 1999.
76. C. Corradini, G. Canali, E. Cogliandro, I. Nicoletti, *J. Chromatogr. A* 791, 343–349, 1997.
77. P. Hitos, A. Pons, M. I. M. Hinojosa, J. Lombardero, S. Camacho, P. Dapena, *Bulletin de L'O.I.V.* 775–776, 746–760, 1995.
78. M. Del Alamo, J. L. Bernal, M. J. Del Nozal, C.Gomez-Cordove, *Food Chem.* 71, 189–193, 2000.
79. C. Martinez Montero, M. C. Rodriguez Dodero, D. A. Guillén Sánchez, C. G. Barroso, *Chromatographia* 59, 15–30, 2004.
80. R. Ârez-Jimenez, B. Garcõa-Villanova, E. Guerra-Hernandez, *Food Res. Int.* 33, 833–838, 2000.
81. D. Baunsgaard, C. A. Andresson, A. Arndal, L. Munk, *Food Chem.* 70, 113–121, 2000.
82. M. Herbreteau, M. La Fosse, L. Morin-Allory, M. Dreux, *J. High Resolut. Chromatogr.* 13, 239–243, 2005.
83. M. Miguez Bernardez, J. De la Montana Miguelez, J. Garcia Queijero, *J. Food Compos. Anal.* 17, 63–67, 2004.
84. F. J. Morales, A. Arnoldi, *Food Chem.* 67, 185–191, 1999.
85. J. M. Ames, A. B. Defaye, R. G. Bailey, L. Bate, *Food Chem.* 61, 521–524, 1998.
86. F. Chinnici, U. Spinabelli, C. Riponi, A. Amati, *J. Food Compos. Anal.* 18, 121–130, 2005.
87. E. Forni, A. Soriani, S. Scalise, D. Torreggiani, *Food Res. Int.* 30(2), 87–94, 1997.
88. C. Pompei, A. Spagnolello, *Meat Sci.* 46, 139–146, 1997.
89. M. Zappalà, B. Fallico, E. Arena, A. Verzera, *Food Control* 16, 273–277, 2005.
90. E. Edris, M. Murkovic, B. Siegmund, *Food Chem.* 104, 1310–1314, 2007.
91. L. A. Ameur, G. Trystram, I. Birlouez-Aragon, *Food Chem.* 98, 790–796, 2006.
92. J. Alcázar, J. M. Jurado, F. Pablos, A. G.Gonzalez, M. J. Martın, *Microchem. J.* 82, 22–28, 2006.
93. F. Lo Coco, C. Valentini, V. Novelli, L. Ceccon, *J Chromatogr. A* 749, 95–102, 1996.
94. J. Fiori, G. Serra, A. G. Sabatini, P. Zucchi, R. Barbattini, F. Gazziola, *Ind. Alim.* 39, 463–466, 2000.
95. M. C. Tarczynski, D. N. Byrne, W. B. Miller, *Plant Physiol.* 99, 753–756, 1992.

96. Italian Official Journal, no 185, 11/08/2003, pp. 24–54, 2003.
97. M. Rada-Mendoza, M. Sanz, A. Olano, M. Villamiel, *Food Chem.* 85, 605–609, 2004.
98. T. Barboni, N. Chiaramonti, *Chromatographia*, 63, 445–448, 2006.
99. IDF, 1998, Method no 147B. International Dairy Federation Brussels, Belgium.
100. J. L. Garcıa-Banos, N. Corzo, M. L. Sanz, A. Olano, *Food Chem.* 88, 35–38, 2004.
101. G. Sesta, *Apidologie* 37, 84–90, 2006.
102. D. Lefebvre, V. Gabriel, Y. Vayssier, C. Fontagnè-Faucher, *Lebensm.-Wiss. U.-Technol.* 35, 407–414, 2002.
103. S. Nojiri, N. Taguchi, M. Oishi, S. Suzuki, *J. Chromatogr. A* 893, 195–200, 2000.
104. T. Herraz, *Anal. Chim. Acta* 352, 119–139, 1997.
105. J. Leonil, V. Gagnaire, D. Molle, S. Pezennec, S. Bouhallab, *J. Chromatogr. A* 881, 1–21, 2000.
106. M. V. Moreno-Arribas, E. Pueyo, M. C. Polo, *Anal. Chim. Acta* 458, 63–75, 2002.
107. S. A. Dierckx, A. Huyghebaert, in *Food Analysis by HPLC*, L. M. L. Nollet (Ed.), Marcel Dekker Inc., New York/Basel, Switzerland, 2000, pp. 127–167.
108. C. Luguera, V. Moreno-Arribas, E. Pueyo, M. C. Polo, *J. Agric. Food Chem.* 45, 3766–3770, 1997.
109. C. Luguera, V. Moreno-Arribas, E. Pueyo, B. Baltolomé, M. C. Polo, *Food Chem.* 63, 465–471, 1998.
110. H. P. J. Bennett, *J. Chromatogr.* 359, 383–390, 1986.
111. K. Štulík, V. Pacáková, J. Suchánková, J. H. A. Claessens, *Anal. Chim. Acta* 352, 1–19, 1997.
112. P. Gustavsson, P. Larsson, *J. Chromatogr. A* 734, 231–240, 1996.
113. M. C. García, M. L. Marina, M. Torre, *J. Chromatogr. A* 880, 169–187, 2000.
114. N. B. Afeyan, N. F. Gordon, L. Mazsaroff, L. Varady, S. P. Fulton, Y. B. Yang, F. E. Regnier, *J. Chromatogr.* 519, 1–29, 1990.
115. M. C. Polo, M. Ramos, D. Gonzáles de Llano, in *Food Analysis by HPLC*, L. M. L. Nollet (Ed.), Marcel Dekker Inc., New York, 2000, pp. 99–124.
116. F. Kálmán, T. Cserháti, K. Valkó, K. Neubert, *Anal. Chim. Acta* 268, 247–254, 1992.
117. D. L. Crimmins, *Anal. Chim. Acta* 352, 21–30, 1997.
118. M. C. García, *J. Chromatogr. B* 825, 11–123, 2005.
119. C. T. Mant, R. S. Hodges, in *High-Performance Liquid Chromatography of Peptides and Proteins: Separation, Analysis, and Conformation*, C. T. Mant, R. S. Hodges (Eds.), CRC Press, Boca Raton, FL, 1991, p. 125.
120. B. G. Irvine, *Anal. Chim. Acta* 352, 387–397, 1997.
121. N. G. Hentz, V. Vukasinovic, S. Daunert, *Anal. Chem.* 68, 1550–1555, 1996.
122. K. A. Lerro, R. Orlando, H. Zhang, P. N. R. Usherwood, K. Nakanashi, *Anal. Biochem.* 215, 38–44, 1993.
123. W. E. Collins, *Sep. Purif. Methods* 26, 215–253, 1997.
124. F. E. Regnier, *Nature* 350, 634–635, 1991.
125. S. P. Fulton, A. J. Shahidi, N. F. Gordon, N. B. Afeyan, *Biotechnology* 10, 635–639, 1992.
126. T. J. Ward, *LC-GC* 9, 429–435, 1996.
127. I. Ilsz, R. Berkecz, A. Péter, *J. Sep. Sci.* 29, 1305–1321, 2006.
128. M. Kempe, K. Mosbach, *J. Chromatogr. A* 691, 317–323, 1995.
129. M. Koller, H. Eckert, *Anal. Chim. Acta* 353, 31–59, 1997.
130. J. Dziuba, D. Nałęcz, P. Minkiewicz, *Anal. Chim. Acta* 449, 243–252, 2001.
131. H. F. Alomirah, I. Alli, Y. Konishi, *J. Chromatogr. A* 893, 1–21, 2000.
132. M. Careri, A. Mangia, *J. Chromatogr. A* 1000, 609–635, 2003.
133. M. Careri, F. Bianchi, C. Corradini, *J. Chromatogr. A* 970, 3–64, 2002.
134. E. A. Pastorello, C. Trambaioli, *J. Chromatogr. B* 756, 71–84, 2001.
135. G. Bordin, F. C. Raposo, B. de la Calle, A. R. Rodriguez, *J. Chromatogr. A* 928, 63–76, 2001.
136. A. J. Trujillo, I. Casals, B. Guamis, *J. Chromatogr. A* 870, 371–380, 2000.
137. A.-J. Trujillo, I. Casals, B. Guamis, *J. Dairy Sci.* 83, 11–19, 2000.
138. I. M. P. L. V. O. Ferreira, H. Caçote, *J. Chromatogr. A* 1015, 111–118, 2003.
139. G. Enne, D. Elez, F. Fondrini, I. Bonizzi, M. Feligini, R. Oleandri, *J. Chromatogr. A* 1094, 169–174, 2005.
140. D. M. A. M. Luykx, J. H. G. Cordewener, P. Ferranti, R. Frankhuizen, M. G. E. G. Bremer, H. Hooijerink, A. H. P. America, *J. Chromatogr. A* 1164, 189–197, 2007.
141. S. S. Hong, J. H. Park, S. W. Kwon, *J. Chromatogr. B* 845, 69–73, 2007.
142. J. Hau, L. Bovetto, *J. Chromatogr. A* 926, 105–112, 2001.
143. G. Henry, D. Mollé, F. Morgan, J. Fauquant, *J. Agric. Food Chem.* 50, 185–191, 2002.
144. L. Monaci, A. J. van Hengel, *J. Agric. Food Chem.* 55, 2985–2992, 2007.
145. L. H. L. M. L. M. Santos, I. M. P. L. V. O. Ferreira, *Anal. Biochem.* 362, 293–295, 2007.
146. J. N. Losso, R. R. Bansode, H. A. Bawadi, *J. Agric. Food Chem.* 51, 7122–7126, 2003.
147. F. M. DuPont, R. Chain, R. Lopez, W. H. Vensel, *J. Agric. Food Chem.* 53, 1575–1584, 2005.

148. J. M. Saz, M. L. Marina, *J. Sep. Sci.* 30, 431–451, 2007.
149. J. M. Heras, M. L. Marina, M. C. García, *J. Chromatogr. A* 1153, 97–103, 2007.
150. M. C. García, J. M. Heras, M. L. Marina, *J. Sep. Sci.* 30, 475–482, 2007.
151. M. C. García, M. L. Marina, M. Torre, *Anal. Chem.* 69, 2217–2220, 1997.
152. M. C. García, M. Torre, M. L. Marina, *J. Chromatogr. Sci.* 36, 527–534, 1998.
153. M. C. García, M. L. Marina, M. Torre, *J. Chromatogr. A* 880, 37–46, 2000.
154. M. C. García, M. L. Marina, M. Torre, *J. Liquid Chromatogr. Relat. Technol.* 23, 3165–3174, 2000.
155. M. C. García, M. L. Marina, M. Torre, *J. Chromatogr. A* 822, 225–232, 1998.
156. M. L. Marina, R. Thorogate, N. W. Smith, *J. Sep. Sci.* 29, 979–985, 2006.
157. J. Dziuba, D. Nałęcz, P. Minkiewicz, *Milchwissenschaft* 59, 366–369, 2004
158. J. Dziuba, D. Nałęcz, P. Minkiewicz, B. Dziuba, *Anal. Chim. Acta* 521, 17–24, 2004.
159. M. C. García, M. Torre, M. L. Marina, *J. Chromatogr. A* 881, 47–57, 2000.
160. S. Delgado, M. C. García, M. L. Marina, M. Torre, *J. Sep. Sci.* 26, 1363–1375, 2003.
161. F. Castro-Rubio, R. Rodríguez, M. C. García, M. L. Marina, *J. Agric. Food Chem.* 53, 220–226, 2005.
162. F. Castro-Rubio, M. L. Marina, R. Rodríguez, M. C. García, *Food Addit. Contam.* 22, 1209–1218, 2005.
163. A. Castro-Rubio, F. F. Castro-Rubio, M. C. García, M. L. Marina, *Food Chem.* 100, 948–955, 2007.
164. M. Criado, F. Castro-Rubio, F. García-Ruiz, M. C. García, M. L. Marina, *J. Sep. Sci.* 28, 987–995, 2005.
165. M. C. García, M. Domínguez, F. García-Ruiz, M. L. Marina, *Anal. Chim. Acta* 559, 215–220, 2006.
166. M. Kruså, M. Torre, M. L. Marina, *Anal. Chem.* 72, 1814–1818, 2000.
167. E. Espeja, M. C. García, M. L. Marina, *J. Sep. Sci.* 24, 856–864, 2001.
168. A. Castro-Rubio, F. Castro-Rubio, M. C. García, M. L. Marina, *J. Sep. Sci.* 28, 996–1004, 2005.
169. A. Castro-Rubio, M. C. García, M. L. Marina, *Anal. Chim. Acta* 558, 28–34, 2006.
170. J. M. Rodriguez-Nogales, M. C. García, M. L. Marina, *J. Chromatogr. A* 1104, 91–99, 2006.
171. J. M. Rodriguez-Nogales, M. C. García, M. L. Marina, *J. Agric. Food Chem.* 54, 8702–8709, 2006.
172. I. Alli, M. Okoniewska, B. F. Gibbs, Y. Konishi, *Int. Dairy J.* 8, 643–649, 1998.
173. F. Gaucheron, D. Mollé, V. Briard, J. Léonil, *Int. Dairy J.* 9, 515–521, 1999.
174. U. Bütikofer, E. Baumann, R Sieber, J. O. Bosset, *Lebensm.-Wiss. U.-Technol.* 31, 297–301, 1998.
175. C. Benfeldt, J. Sørensen, *Int. Dairy J.* 11, 445–454, 2004.
176. H. Mallatou, E. C. Pappa, V. A. Boumpa, *Int. Dairy J.* 14, 977–987, 2004.
177. R. A. Verdini, S. E. Zorilla, A. C. Rubiolo, *Int. Dairy J.* 14, 445–454, 2004.
178. R. A. Verdini, S. E. Zorilla, A. C. Rubiolo, *Int. Dairy J.* 15, 363–370, 2005.
179. Y. Ardö, H. Lilbæk, K. R. Kristiansen, M. Zakora, J. Otte, *Int. Dairy J.* 17, 513–524, 2007.
180. S. Kochhar, K. Gartenmann, M. A. Juillerat, *J. Agric. Food Chem.* 48, 5593–5599, 2000.
181. E. Buyukpamukcu, D. M. Goodall, C. E. Hansen, B. J. Keely, S. Kochhar, H. Wille, *J. Agric. Food Chem.* 49, 5822–5827, 2001.
182. M. V. T. Mota, I. M. P. L. V. O. Ferriera, M. B. P. Oliveira, C. Rocha, J. A. Teixera, D. Torres, M. P. Gonçalves, *Food Chem.* 94, 278–286, 2006.
183. Shakeel-Ur-Rehman, N. Y. Farkye, B. Yim, *Int. Dairy J.* 16, 697–700, 2006.
184. L. Pellegrino, A. Tirelli, *Int. Dairy J.* 10, 435–442, 2000.
185. S. Stoeva, C. E. Byrne, A. M. Mullen, D. J. Troy, W. Voelter, *Food Chem.* 69, 365–370, 2000.
186. M. Flores, V. J. Moya, M. C. Aristoy, F. Toldrá, *Food Chem.* 69, 371–377, 2000.
187. V. J. Moya, M. Flores, M. C. Aristoy, F. Toldrá, *Meat Sci.* 58, 197–206, 2001.
188. L. Martín, T. Antequera, J. Ventanas, R. Benítez-Donoso, J. J. Córdoba, *Meat Sci.* 59, 363–368, 2001.
189. M. C. Hughes, J. P. Kerry, E. K. Arendt, P. M. Kenneally, P. L. H. McSweeney, E. E. O'Neil, *Meat Sci.* 62, 205–216, 2002.
190. M. Fountoulakis, H. Lahm, *J. Chromatogr. A* 826, 109–134, 1998.
191. H. Yamanaka, R. Shimada, *Fisheries Sci.* 62, 821–824, 1996.
192. D. I. Sánchez-Machado, J. Lopez-Cervantes, J. Lopez-Hernandez, P. Paseiro-Losada, J. Simal-Lozano, *Chromatographia* 58, 159–163, 2003.
193. S. Fuke, S. Konosu, *Physiol. Behav.* 49, 863–868, 1991.
194. J.-H. Yang, H.-C. Lin, J.-L. Mau, *Food Chem.* 72, 465–471, 2001.
195. A. Harada, S. Yoneyama, S. Doi, M. Aoyama, *Food Chem.* 83, 343–347, 2003.
196. J. T. Kivi, in *Food Analysis by HPLC*, L. M. L. Nollet (Ed.), Marcel Dekker Inc., New York, 2000, pp. 55–98.
197. P. Rozan, Y.-H. Kuo, F. Lambein, *J. Agric. Food. Chem.* 48, 716–723, 2000.
198. P. Rozan, Y.-H. Kuo, F. Lambein, *Phytochemistry* 58, 281–289, 2001.
199. Y. H. Kuo, P. Rozan, F. Lambein, J. Frias, C. Vidal-Valverde, *Food Chem.* 86, 537–545, 2004.
200. S. Blackburn, in *Amino Acid Determination*, S. Blackburn (Ed.), Marcel Dekker, New York, 1978, p. 8.

201. A. J. Darragh, D. J. Garrik, P. J. Moughan, W. H. Hendriks, *Anal. Biochem.* 236, 199–207, 1996.
202. R. W. Zumwalt, J. S. Absheer, F. E. Kaiser, C. W. Gehrke, *J. Assoc. Off. Anal. Chem.* 70, 147–151, 1987.
203. S. H. Chiou, *Biochem. Int.* 17, 981–987, 1988.
204. M. F. Malmer, L. A. Schroeder, *J. Chromatogr.* 514, 227–239, 1990.
205. C. W. Gehrke, L. L. Wall, J. S. Absheer, F. E. Kaiser, R. W. Zumwalt, *J. Assoc. Off. Anal. Chem.* 68, 811–821, 1985.
206. S. T. Chen, S. H. Chiou, Y. H. Chu, K. T. Wang, *Int. J. Peptide Protein Res.* 30, 572–576, 1987.
207. W. G. Engelhart, *Am. Biotech. Lab.* 8, 30–35, 1990.
208. E. Tatar, M. Khalifa, G. Zaray, I. Molnar-Perl, *J. Chromatogr. A* 672, 109–115, 1994.
209. M. Weiss, M. Manneberg, J. F. Juranville, H. W. Lahm, M. Fountoulakis, *J. Chromatogr. A* 795, 263–275, 1998.
210. L. Joergensen, H. N. Thestrup, *J. Chromatogr. A* 706, 421–428, 1995.
211. A. D'aniello, L. Petruccelli, C. Gardner, G. Fisher, *Anal. Biochem.* 213, 290–295, 1993.
212. J. Csapó, Z. Csapó-Kiss, L. Wagner, *Amino Acids* 13, 26–33,1977.
213. C. H. W. Hirs, *Method Enzymol.* 11, 59–62, 1967.
214. M. Manneberg, H. W. Lahm, M. Fountoulakis, *Anal. Biochem.* 231, 349–353, 1995.
215. A. S. Inglis, *Method Enzymol.* 91, 26–36, 1983.
216. J. E. Hale, D. E. Beidler, R. A. Jue, *Anal. Biochem.* 216, 61–66, 1994.
217. Y. H. Tuan, R. D. Phillips, *J. Agric. Food Chem.* 45, 3535–3540, 1997.
218. T. E. Hugli, S. Moore, *J. Biol. Chem.* 247, 2828–2834, 1972.
219. I. Molnar-Perl, *J. Chromatogr. A* 763, 1–10, 1997.
220. F. F. Shih, A. D. Kalmer, *J. Liquid Chromatogr.* 7, 1196–1183, 1984.
221. M. Hauk, *Dtsch. Lebensm. Rundsch.* 86, 12–15, 1990.
222. S. Albala-Hurtado, C. Bover-Cid, M. Izquierdo-Pulido, M. T. Vociano-Nogues, M. C. Vidal-Carou, *J. Chromatogr. A* 778, 235–241, 1997.
223. N. Naulet, P. Tesson, N. Boiler, G. J. Martin, *Analusis* 25, 42–47, 1997.
224. M. J. Gonzáles-Castro, J. López-Hernández, J. Simal-Lozano, M. J. Oruna-Concha, *J. Chromatogr. Sci.* 35, 181–185, 1997.
225. P. Herbert, L. Santos, A. Alves, *J. Food Sci.* 66, 1319–1325, 2001.
226. T. Lida, M. Matsunaga, T. Fukushima, T. Santa, H. Homma, K. Imai, *Anal. Chem.* 69, 4463–4468, 1997.
227. A. J. Smith, *Methods Mol. Biol.* 64, 139–146, 1997.
228. L. Dou, J. Mazzeo, I. S. Krull, *BioChromatography* 5, 74–96, 1990.
229. R. W. Peace, G. S. Gilani, *J. Assoc. Off. Anal. Chem. Int.* 88, 877–885, 2005.
230. J. F. Cotte, H. Casabianca, B. Giroud, M. Albert, J. Lheritier, M. F. Grenier-Loustalot, *Anal. Bioanal. Chem.* 378, 1342–1350, 2004.
231. M. Giraudo, H. Sanchez-Tuero, R. Pavesi, I. Markowski, G. Guirin, J. Montesano, *Alimentaria* 350, 97–102, 2004.
232. L. H. Seron, E. Gerrigós, M. S. Prats, M. L. Martín, V. Bereuguer, N. Grané, *Food Chem.* 61, 455–459, 1998.
233. N. Grané, M. C. Luna, M. S. Prats, V. Berenguer, M. L. Martín, *Food Chem.* 65, 23–28, 1999.
234. R. Loiudice, M. Impembo, B. Laratta, G. Villari, A Lo Voi, P. Siviero, D. Castaldo, *Food Chem.* 53, 81–89, 1995.
235. E. H. Soufleros, E. Bouloumpasi, C. Tsarchopoulos, C. G. Biliaderis, *Food Chem.* 80, 261–273, 2003.
236. M. Cornet, J. Bousset, *Meat Sci.* 51, 215–219, 1999.
237. E. R. Brierley, P. L. R. Bonner, A. H. Cobb, *Plant Sci.* 127,17–24, 1997.
238. P. L. Hurst, G. Boulton, R. E. Lill, *Food Chem.* 61, 381–384, 1998.
239. E. Derbali, J. Makhlouf, L. P. Vezina, *Postharvest Biol. Technol.* 13, 191–204, 1998.
240. S. B. Boggio, J. F. Palatnik, H. W. Heldt, E. M. Valle, *Plant Sci.* 159,125–133, 2000.
241. G. Pratta, R. Zorzoli, S. B. Boggio, L. A. Picardi, E. M. Valle, *Sci. Horticululturae* 100, 341–347, 2004.
242. M. E. Hansen, H. Sørensen, M. Cantwell, *Postharvest Biol. Technol.* 22, 227–237, 2001.
243. C. M. Alfaia, M. F. Castro, V. A. Reis, J. M. Prates, I. T. de Almeida, A. D. Correia, M. A. Dias, *Food Sci. Technol. Int.* 10, 297–304, 2004.
244. C. Ruiz-Capillas, A. Moral, *Food Chem.* 86, 85–91, 2004.
245. C. Ruiz-Capillas, A. Moral, *Eur. Food Res. Technol.* 212, 302–307, 2001.
246. P. K. Sarkar, L. J. Jones, G. S. Craven, S. M. Somerset, C. Palmer, *Food Chem.* 59, 69–75, 1997.
247. E. Rosa, M. H. Gomes, *J. Sci. Food Agric.* 82, 61–64, 2001.
248. Y. Ding, H. Yu, S. Mou, *J. Chromatogr. A* 982, 237–244, 2002.
249. B. Herranz, L. de la Hoz, E. Hierro, M. Fernández, J. A. Ordóñez, *Food Chem.* 91, 673–682, 2005.

250. H.-C. Wu, C.-Y. Shiau, *J. Food Drug Anal.* 10, 170–177, 2002.
251. H.-C. Wu, H.-M. Chen, C.-Y. Shiau, *Food Res. Int.* 36, 949–957, 2003.
252. J. López-Cervantes, D. I. Sánchez-Machado, J. A. Rosas-Rodriguez, *J. Chromatogr. A* 1105, 106–110, 2006.
253. J. M. Izco, A. Irigoyen, P. Torre, Y. Barcina, *J. Chromatogr. A* 881, 69–79, 2000.
254. N. Muñoz, M. Ortigosa, P. Torre, J. M. Izco, *Food Chem.* 83, 329–338, 2003.
255. R. A. Verdini, S. E. Zorilla, A. C. Rubiolo, *J. Food Sci.* 67, 3264–3270, 2002.
256. P. Zunin, F. Evangelisti, *Int. Dairy J.* 9, 653–656, 1999.
257. M. H. Silla Santos, *Int. J. Food Microbiol.* 29, 213–231, 1996.
258. B. Ten Brink, C. Mamink, H. M. L. Joosten, J. H. J. Huisin'l Veld, *Int. J. Food Microbiol.* 11, 73–84, 1990.
259. J. E. Stratton, R. W. Hutkins, S. L. Taylors, *J. Food Prot.* 54, 460–470, 1991.
260. J. Lange, K.Thomas, C. Wittmann, *J. Chromatogr. B*, 779, 229–239, 2002.
261. G. Vinci, M. L. Antonelli, *Food Control* 13, 519–524, 2002.
262. S. A. N. Tailor, K. I. Shulman, S. E. Walker, J. Moss, D. Gardner, *J. Clin. Psychopharmacol.* 14, 5–14, 1994.
263. European Union. 1991. Directive 91/493/EEC: Directive of July 22, 1991 establishing standards to be applied to the production and commercialization of fishery products. *Off. J. L* 268:15–34.
264. S. Angelino, M. C. Gennaro, in *Food Analysis by HPLC*, L. M. L. Nollet (Ed.), Marcel Dekker Inc., New York, 2000, pp. 881–896.
265. S. Wohrl, W. Hemmer, M. Focke, K. Rappersberger, R. Jarisch, *Allergy Asthma Proc.* 25, 305–311, 2004.
266. M. Fernández, B. del Río, D. M. Linares, M. C. Martín and M. A. Alvarez, *J. Dairy Sci.* 89, 3763–3769, 2006.
267. S. T. Edwards, W. E. Sandine, *J. Dairy Sci.* 64, 2431–2438, 1981.
268. A. Halász, A. Baráth, L. Simon-Sarkadi, W. Holzapfel, *Trends Food Sci. Technol.* 5, 42–49, 1994.
269. C. P. Oliveira, M. B. A. Glória, J. F. Barbour, R. A. Scanlan, *J. Agric. Food Chem.* 43, 967–969, 1995.
270. S. Bardócz, *Trend Food Sci. Technol,* 6, 341–346, 1995.
271. K. A. Eliassen, R. Reistad, U. Risøen, H. F. Rønning, *Food Chem.* 78, 273–280, 2002.
272. A. Boucherau, P. Guénot, F. Larher, *J. Chromatogr. B* 747, 49–67, 2000.
273. L. Simon-Sarkadi, W. H. Holzapfel, *Z. Lebensm. Unters. Forsch.* 198, 230–233, 1994.
274. A. R. Shalaby, *Food Res. Int.* 29, 675–690, 1996.
275. A. Okamoto, E. Sugi, Y. Koizumi, F. Yanagida, S. Udaka, *Biosci. Biotechnol. Biochem.* 61, 1582–1584, 1997.
276. P. Kalač, S. Šveková, T. Pelikánová, *Food Chem.* 77, 349–351, 2002.
277. D. Valero, D. Martinez-Romero, M. Serrano, *Trends Food Sci. Technol.* 13, 228–234, 2002.
278. N. Anderton, P. A. Cockrum, D. W. Walker, J. A. Edgar, in *Plant-Associated Toxins: Agricultural, Phytochemical and Ecological Aspects*, S. M. Colgate, P. R. Doring (Eds.), CAB International, Wallingford, U.K., 1994, pp. 269–274.
279. T. D. A. Forbes, I. J. Pemberton, G. R. Smith, C. M. Hensarling, *J. Arid Environ.* 30, 403–415, 1995.
280. S. Moret, D. Smela, T. Populin, L. S. Conte, *Food Chem.* 89, 355–361, 2005.
281. A. Önal, *Food Chem.* 103, 1475–1468, 2007.
282. S. Moret, L. S. Conte, *J. Chromatogr. A* 729, 363–369, 1996.
283. J. H. Mah, H. K. Han, Y. J. Oh, M. G. Kim, H. J. Hwang, *Food Chem.* 79, 239–243, 2002.
284. N. Innocente, M. Biasutti, M. Padovese, S. Moret, *Food Chem.* 101, 1285–1289, 2007.
285. F. Özogul, K. D. A. Taylor, P. Quantick, Y. Özogul, *Int. J. Food Sci. Technol.* 37, 515–522, 2002.
286. M. Vallè, P. Malle, S. Bouquelet, *J. Assoc. Off. Anal. Chem. Int.* 80, 49–56, 1997.
287. G. Duflos, C. Dervin, P. Malle, S. Bouquelet, *J. Assoc. Off. Anal. Chem. Int.* 82, 1097–1101, 1999.
288. L. Yongmei, L. Xin, C. Xiaohong, J. Mei, L. Chao, D. Mingshen, *Food Chem.* 100, 1424–1428, 2007.
289. P. Hernández-Orte, A. Peña-Gallego, M. J. Ibarz, J. Cacho, V. Ferriera, *J. Chromatogr. A* 1129, 160–164, 2006.
290. G. Dugo, F. Vilasi, G. L. La Torre, T. M. Pellicanó, *Food Chem.* 95, 672–676, 2006.
291. R. Romero, M. Sánchez-Viñas, D. Gázquez, M. Gracia Bagur, *J. Agric. Food Chem.* 50, 4713–4717, 2002.
292. D. Torrea, C. Ancín, *J. Agric. Food Chem.* 50, 4895–4899, 2002.
293. G. J. Soleas, M. Carey, D. M. Goldberg, *Food Chem.* 64, 49–58, 1999.
294. M. B. A. Gloría, M. Izquierdo-Pulido, *J. Food Compos. Anal.* 12, 129–136, 1999.
295. N. García-Villard, J. Saurina, S. Hernández-Cassou, *Anal. Chim. Acta* 575, 97–105, 2006.
296. S. Buiatti, O. Boschelle, M. Mozzon, F. Battistutta, *Food Chem.* 52, 199–202, 1995.
297. O. Busto, M. Mestres, J. Gulasch, F. Borrul, *Chromatographia* 40, 404–409, 1995.
298. O. Busto, M. Miracle, J. Guasch, F. Borrull, *J. Chromatogr. A* 757, 311–318, 1997.

299. E. H. Soufleros, E. Bouloumpasi, A. Zotou, Z. Loukou, *Food Chem.* 101, 704–716, 2007.

300. C. Molins-Legua, P. Campins-Falcó, *Anal. Chim. Acta* 546, 206–220, 2005.

301. M. T. Salazar, T. K. Smith, A. Harris, *J. Agric. Food Chem.* 48, 1708–1712, 2000.

302. G. Saccani, E. Tanzi, P. Pastore, S. Cavalli, M. Rey, *J. Chromatogr. A* 1082, 43–50, 2005.

303. J. Karovičová, Z. Kohajdov, *Chem. Pap.* 59, 70–79, 2005.

304. R. M. Linares, J. H. Ayala, A. M. Afonso, V. G. Diaz, *J. Chromatogr.* 808, 87–93, 1998.

305. I. Molnár-Perl, *J. Chromatogr. A* 987, 291–309, 2003.

306. I. Krause, A. Bockhardt, H. Neckermann, T. Henle, H. Klostermeyer, *J. Chromatogr. A* 715, 67–79, 1995.

307. R. Draisci, L. Giannetti, P. Boria, L. Lucentini, L. Palleschi, S. Cavalli, *J. Chromatogr. A* 798, 109–116, 1998.

308. S.-M. Hwang, *J. Assoc. Off. Anal. Chem.* 68, 684–689, 1985.

309. J. Kirschbaum, K. Rebscher, H. Brückner, *J. Chromatogr. A* 881, 517–530, 2000.

310. F. Gosetti, E. Mazzucco, V. Giannotti, S. Polati, M. C. Gennaro, *J. Chromatogr. A* 1149, 151–157, 2007.

311. N. M. Tamim, L. W. Bennett, T. A. Shellem, J. A. Doerr, *J. Agric. Food Chem.* 50, 5012–5015, 2002.

312. A. Askar, H. Treptow, *Ernahrung Nutr.* 13, 425–429, 1989.

313. G. V. Celano, C. Cafarchia, F. Buja, G. Tiecco, *Ind. Alim.* 31, 764–768, 1992.

314. N. Innocente, P. D'Agostin, *J. Food Prot.* 65, 1498–1501, 2002.

315. C. Cantoni, S. Moret, G. Comi, *Ind. Alim.* 32, 842–845, 1993.

316. K. V. Kumudavally, A. Shobha, T. S. Vasundhara, K. Radhakrishna, *Meat Sci.* 59, 411–415, 2001.

317. P. Antila, V. Antila, J. Mattila, H. Hakkarainen, *Milchwissenschaft* 39, 400–404, 1984.

318. M. Marino, M. Maifreni, S. Moret, G. Rondinini, *Lett. Appl. Microbiol.* 31, 169–173, 2000.

319. G. Suzzi, F. Gardini, *Int. J. Food Microbiol.* 88, 41–54, 2003.

320. I. M. Ferreira, O. Pinho, *J. Food Prot.* 69, 2293–303, 2006.

321. J. Gromus, S. Lustig, *Bull. Eur. Brewery Conv.*, 192–217, 1998, Brewing Science Group.

322. J. S. Hough, D. E. Briggs, J. S. Stevens, T. W. Young, in *Malting and Brewing Science*, J. J. Hough, D. E. Briggs, J. S. Stevens, T. W. Young, (Eds.), Chapman & Hall, London, U.K., 1981, pp. 821–828.

323. M. Moll, in *Brewing Science* Vol. 3, J. R. A. Pollock (Ed.), Academic Press, New York, 1987, pp 46–47.

324. E.L. Bradley, P. S. Hughes, *Bull. Eur. Brewery Conv.* 175–191, 1998.

325. F. Garcia-Sanchez, C. Carnero, A. Hereida, *J. Agric. Food Chem.* 36, 80–82, 1988.

326. D. Madigan, I. Mc Murrough, M.R. Smyth, *J. Am. Soc. Brew. Chem.* 52, 152–155, 1994.

327. L. Montanari, G. Perretti, F. Natella, A. Guidi, P. Fantozzi, *Lebensm.-Wiss. U.-Technol.* 12, 535–539, 1999.

328. F. G. Montedoro, M. Servili, M. Baldioli, E. Miniati, *J. Agric. Food Chem.* 40, 1571–1576, 1992.

329. E. D. Coppola, *Food Technol.* 38, 88–91, 1984.

330. G. Shui, L. P. Leong, *J. Chromatogr. A* 977, 89–96, 2002.

331. H. Chen, Y. Zuo, Y. Deng, *J. Chromatogr.* 913, 387–395, 2001.

332. C. Cantarelli, *Riv. Ital. Sostanze Grasse* 38, 69–72, 1961.

333. G. F. Montedoro, C. Cantarelli, *Riv. Ital. Sostanze Grasse* 46, 115–124, 1969.

334. A. Carrasco-Pancorbo, L. Cerretani, A. Bendini, A. Segura-Carretero, T. Gallina-Toschi, A. Fernandez-Gutierrez, *J. Sep. Sci.* 28, 837–858, 2005.

335. F. F. Montedoro, *Sci. Technol. Alim.* 3, 177–186, 1972.

336. A. Vazquez Roncero, *Rev. Fr. Corps Gras.* 25, 21–26, 1978.

337. G. F. Montedoro, M. Servili, M. Baldioli, E. Miniati, *J. Agric. Food Chem.* 40, 1577–1580, 1992.

338. G. F. Montedoro, M. Servili, M. Baldioli, R. Selvaggini, E. Miniati, *J. Agric. Food Chem.* 41, 2228–2234, 1993.

339. R. W. Owen, W. Mier, A. Giacosa, W. E. Hull, B. Spiegel-Halder, H. Bartsch, *Clin. Chem.* 38, 647–659, 2000.

340. E. Perri, A. Raffaelli, G. Sindona, *J. Agric. Food Chem.* 47, 4156–4160, 1999.

341. R. W. Owen, W. Mier, A. Giacosa, W. E. Hull, B. Spiegel-Halder, H. Bartsch, *Clin. Chem.* 46, 976–988, 2000.

342. M. Brenes, F. J. Hidalgo, A. Garcia, J. J. Rios, P. Garcia, R. Zamora, A. Garrido, *J. Am. Oil Chem. Soc.* 7, 715–720, 2000.

343. A. Bianco, F. Coccioli, M. Guiso, C. Marra, *Food Chem.* 77, 405–411, 2001.

344. M. Servili, R. Selvaggini, S. Esposto, A. Taticchi, G. F. Montedoro, G. Morozzi, *J. Chromatogr.* 1054, 113–127, 2004.

345. M. Solinas, A. Cichelli, *Riv. Soc. Ital. Sci. Alim.* 11, 223–226, 1982.

346. N. Cortesi, E. Fedeli, *Riv. Ital. Sostanze Grasse* 60, 341–351, 1983.

347. M. Solinas, *Riv. Ital. Sostanze Grasse* 64, 255–259, 1987.

348. L. Lesage-Meessen, D. Navarro, S. Maunier, J.-C. Sigoillot, J. Lorquin, M. Delatore, J. L. Simon, M. Asher, M. Labat, *Food Chem.* 75, 501–507, 2001.
349. M. Tasioula-Margari, O. Okogeri, *Food Chem.* 74, 377–383, 2001.
350. S. Mannino, M. S. Cosio, M. Bertuccioli, *Ital. J. Food Sci.* 4, 363–366, 1993.
351. F. M. Pirisi, A. Angioni, P. Cabras, V. L. Garau, M. T. Sanjust di Teulada, M. K. dos Santos, G. Bandino, *J. Chromatogr. A* 768, 207–243, 1997.
352. M. Servili, M. Baldioli, R. Selvaggini, E. Miniati, A. Macchioni, G. Montedoro, *J. Agric. Food Chem.* 47, 12–20, 1999.
353. M. Servili, M. Baldioli, R. Selvaggini, E. Miniati, A. Macchioni, G. Montedoro, *J. Am. Oil. Chem. Soc.* 76, 873–882, 1999.
354. F. M. Pirisi, P. Cabras, C. Falqui Cao, M. Migliorini, M. Muggelli, *J. Agric. Food Chem.* 48, 1191–1196, 2000.
355. M. J. Oliveras-Lopez, M. Innocenti, C. Giaccherini, F. Ieri, A. Romani, N. Mulinacci, *Talanta* 73, 726–732, 2007.
356. G. Zgorka, S. Kawka, *J. Pharm. Biomed. Anal.* 24, 1065–1072, 2001.
357. J. S. Amaral, R. M. Seabra, P. Andrade, P. Valentao, J. A. Pereira, F. Ferres, *Food Chem.* 88, 373–379, 2004.
358. A. M. Nuutila, K. Kammiovirta, K.-M. Oksman-Caldentey, *Food Chem.* 76, 519–525, 2002.
359. A. C. Hoefler, P. Coggon, *J. Chromatogr.* 129, 460–463, 1976.
360. L. Yao, Y. Jiang, N. Datta, R. Singanusong, X. Liu, J. Duan, K. Raymont, A. Lisle, Y. Xu, *Food Chem.* 84, 253–263, 2004.
361. M. Bonoli, M. Pelillo, T. Gallina Toschi, G. Lercker, *Food Chem.* 81, 631–638, 2003.
362. M. Pelillo, M. Bonoli, B. Biguzzi, A. Bendini, T. Gallina Toschi, G. Lercker, *Food Chem.* 87, 465–470, 2004.
363. D. Sterbova, D. Matejicek, J. Vlcek, V. Kuban, *Anal. Chim. Acta* 513, 435–444, 2004.
364. D. L. Luthria, S. Mukhopadhyay, D. T. Krizek, *J. Food Compos. Anal.* 19, 771–777, 2006.
365. C. Proestos, D. Sereli, M. Komaitis, *Food Chem.* 95, 44–52, 2006.
366. M. Krizman, D. Baricevic, M. Prosek, *J. Pharm. Biomed. Anal.* 43, 481–485, 2007.
367. S. Tian, K. Nakamura, T. Cui, H. Kayahara, *J. Chromatogr. A* 1063, 121–128, 2005.
368. E. Mataix, M. D. Luque de Castro, *J. Chromatogr. A* 910, 255–263, 2001.
369. M. A. Rodriguez-Delgado, S. Malovana, J. P. Perez, T. Borges, F. J. Garcia Montelongo, *J. Chromatogr. A* 912, 249–257, 2001.
370. A. V. Sakkidi, M. N. Stavrakakis, S. A. Haroutounian, *Lebensm.-Wiss. U.-Technol.* 34, 410–413, 2001.
371. S. Malovana, F. J. Garcia Montelomingo, J. P. Pérez, M.A.Rodrigueéz-Delgado, *Anal. Chim. Acta* 428, 245–253, 2001.
372. E. Celotti, R. Ferrarini, R. Zironi, L. Conte, *J. Chromatogr. A* 730, 47–52, 1996.
373. G. L. La Torre, M. Saitta, F. Vilasi, T. Pellicano, G. Dugo, *Food Chem.* 94, 640–650, 2006.
374. N. G. Frega, E. Boselli, E. Bendia, M. Minardi, A. Benedetti, *Anal. Chim. Acta* 563, 375–381, 2006.
375. M. Tsimidou, M. Lyrridou, D. Boskou, A. Paooa-Lousi, F. Kotsifaki, C. Petrakis, *Grasa Aceites* 47, 151–157, 1996.
376. F. Gutiérrez, M. A. Albi, R. Palma, J. J. Rìos, J. M. Olìas, *J. Food Sci.* 54, 68–70, 1989.
377. M. Litridou, J. Linssen, H. Schols, M. Bergams, M. Posthumus, M. Tsimidou, D. Boskou, *J. Sci. Food Agric.*, 74, 169–174, 1997.
378. P. Rovellini, N. Cortesi, E. Fedeli, *Riv. Ital. Sostanze Grasse* 74, 273–279, 1997.
379. L. Cinquanta, M. Esti, E. La Notte, *J. Am. Oil Chem. Soc.* 74, 1259–1264, 1997.
380. A. Romani, N. Mulinacci, P. Pinelli, F. F. Vincieri, A. Cimato, *J. Agric. Food Chem.* 47, 964–967, 1999.
381. G. Beltran, A. Jiménez, M. P. Aguilera, M. Uceda, *Grasa Aceites* 51, 320–324, 2000.
382. C. Cartoni, F. Coccioli, R. Jasionowska, D. Ramires, *Ital. J. Food Sci.* 12, 163–173, 2000.
383. R. Mateos, J. L. Espartero, M. Trujillo, J. J. Rios, M. Léon-Camacho, F. Alcudia, A. Cert, *J. Agric. Food Chem.* 49, 2185–2192, 2001.
384. S. Gomes-Alonso, M. D. Salvador, G. Fregapane, *J. Agric. Food Chem.* 50, 6812–6817, 2002.
385. N. Troncoso, H. Sierra, L. Carvajal, P. Delpiano, G. Gunther, *J. Chromatogr. A* 1100, 20–25, 2005.
386. E. Nishitani, Y. M. Sagesaka, *J. Food Compos. Anal.* 17, 675–685, 2004.
387. B. Berente, D. De la Calle García, M. Reichenbächer, K. Danzer, *J. Chromatogr. A* 871, 95–103, 2000.
388. T. Nurmi, S. Heinonen, W. Mazur, T. Deyama, S. Nishibe, H. Adlercreutz, *Food Chem.* 83, 303–309, 2003.
389. C. Garcia-Viguera, P. Bridle, *Food Chem.* 54, 349–352, 1995.
390. L. Gambelli, G. P. Santaroni, *J. Food Compos. Anal.* 17, 613–618, 2004.

391. S. Carando, P. L. Teissedre, P. Waffo-Teguo, J. C. Cabanis, G. Deffieux, J. M. Merillon, *J. Chromatogr. A* 849, 617–620, 1999.
392. OIV—Recueil International des Methods d'Analyses—MA-F-AS315-11-ANCYAN, November 11, 2007.
393. M. Lopez, F. Martinez, C. Del Valle, C. Orte, M. Mirò, *J. Chromatogr. A* 922, 359–363, 2001.
394. M. Castellari, E. Sartini, A. Fabiani, G. Arfelli, A. Amati, *J. Chromatog. A* 973, 221–227, 2002.
395. M. Del Alamo, L. Casado, V. Hernandez, J. J. Jimenez, *J. Chromatogr. A* 1049, 97–105, 2004.
396. R. Sladvosky, P. Solch, M. Urbanek, *J. Chromatogr.* 1040, 179–184, 2004.
397. E. Boselli, M. Minardi, A. Giomo, N. G. Frega, *Anal. Chim. Acta* 563, 93–100, 2006.
398. S. Gomez-Alonso, E. Garcia-Romero, I. Hermosin-Gutierez, *J. Food Compos. Anal.* 20, 618–626, 2007.
399. R. J. Kutsky, *Handbook of Vitamins and Hormones*, Van Nostrand Reinhold Co., New York, 1973.
400. SCF, Scientific Committee for Food, Nutrient and energy intakes for the European Community. Reports of the Scientific Committee for Food, 31st Series. European Commission, Luxembourg, 1993.
401. European Union, Dir. 90/496/ECC, *Off. J. Eur.Commun. L* 276/40, October 6, 1990.
402. European Union, Reg. 1924/2006, *Off. J. Eur. Union L* 404, December 30, 2006.
403. European Union, Reg. 1925/2006, *Off. J. Eur. Union L* 404/26, December 30, 2006.
404. D. B. Parrish, *CRC Crit. Rev. Food. Sci. Nutr.* 9, 375–390, 1977.
405. D. B. Parrish, *CRC Crit. Rev. Food. Sci. Nutr.* 13, 161–187, 1979.
406. D. B. Parrish, *CRC Crit. Rev. Food. Sci. Nutr.* 13, 337–352, 1980.
407. G. F. M. Ball, in *Food Analysis by HPLC*, L. M. L. Nollet (Ed.), Marcel Dekker, New York, 2000, p. 321.
408. B. Barua, H. C. Furr, A. J. A. Olson, R. B. van Breemen, Vitamin A and carotenoids. In *Modern Chromatographic Analysis of Vitamins*, 2nd edn., A. P. De Leenheer, W. E. Lambert, and H. J. Nelis (Eds.), Marcel Dekker, New York, 1992, pp. 1–74.
409. G. F. M. Ball, The fat-soluble vitamins. In L. M. L. Nollet (Ed.) *Food Analysis by HPLC*, Marcel Dekker Inc., New York, 2000, pp. 321–402.
410. J. M. Gaziano, J. E. Manson, J. E. Buring, C. H. Hennekens, *Ann. N. Y. Acad. Sci.* 669, 249–259, 1992.
411. IARC, International Agency for Research on Cancer, *Handbooks of Cancer Prevention*, IARC, Lyon, France, 1998.
412. M. Tsuchiya, G. Scita, D. F. T. Thompson, L. Paker, V. E. Kagan, M. A. Livrea, in *Retinoids. Progress in Research and Clinical Application*, M. A. Lizrea and L. Parker (Eds.), Marcel Dekker, New York, 1993, pp. 525–536.
413. S. A. Everett, M. F. Dennis, K. B. Patel, S. Maddix, S. C. Kundu, R. Wilson, *J. Biol. Chem.* 271, 3988–3994, 1996.
414. G. S. Omenn, *Ann. Rev. Public Health* 19, 73–99, 1998.
415. R. Edge, T. G. Truscott, *Nutrition* 13, 992–994, 1997.
416. P. Palozza, *Nutr. Rev.* 56, 257–265, 1998.
417. EFSA, European Food Safety Authority, Tolerable upper intake levels for vitamins and minerals. EFSA, Parma, Italy, February 2006.
418. R. Wyss, *J. Chromatogr. B* 671, 381–425, 1995.
419. L. F. Russel, in *Food Analysis by HPLC*, L. M. L. Nollet (Ed.), Marcel Dekker, New York, 2000, p. 403.
420. A. P. De Leenheer, W. F. Lambert, F. Meyer, in *Retinoids, Progress in Research and Clinical Applications*, M. A. Livrea and L. Packer (Eds.), Marcel Dekker, New York, 1993, p. 551.
421. A. B. Barua, A. J. Olson, *J. Chromatogr. B* 707, 69–79, 1998.
422. H. Iwase, *Anal. Chim. Acta* 463, 21–29, 2002.
423. H. Iwase, *J. Chromatogr. A* 1008, 81–87, 2003.
424. M. Marx, A. Schieber, R. Carle, *Food Chem.* 70, 403–408, 2000.
425. M. A. Schneiderman, A. K. Sharma, D. C. Locke, *J. Chromatogr. A* 765, 215–220, 1997.
426. O. Heudi, M. J. Trisconi, C. J. Blake, *J. Chromatogr. A* 1022, 115–123, 2004.
427. C. H. Azevedo-Meleiro, D. B. Rodriguez-Amaya, *J. Food Compos. Anal.* 17, 385–396, 2004.
428. L. Englberger, J. Schierle, G. C. Marks, M. H. Fitzgerald, *J. Food Compos. Anal.* 16, 3–19, 2003.
429. P. J. M. Hulshof, C. Xu, P. van de Bovenkamp, Muhilal, C.E. West, *J. Agric. Food Chem.* 45, 1174–1179, 1997.
430. L. Ferreira de França, G. Reber, M. A. A. Meireles, N. T. Machado, G. Brunner, *J. Supercrit. Fluids* 14, 247–256, 1999.
431. E. E. Moros, D. Darnoko, M. Cheryan, E. G. Perkins, J. Jerrell, *J. Agric. Food Chem.* 50, 5787–5790, 2002.
432. F. Márkus, H. G. Daood, J. Kapitány, P. A. Biacs, *J. Agric. Food Chem.* 47, 100–107, 1999.

433. M. Murkovic, U. Mülleder, H. Neunteufl, *J. Food Compos. Anal.* 15, 633–638, 2002.
434. A. C. Kurilich, S. J. Britz, B. A. Clevidence, J. A. Novotny, *J. Agric. Food Chem.* 51, 4877–4883, 2003.
435. H. S. Lee, *J. Agric. Food Chem.* 49, 2563–2568, 2001.
436. M. Careri, L. Furlattini, A. Mangia, M. Musci, E. Anklam, A. Theobald, C. von Holst, *J. Chromatogr. A* 912, 61–71, 2001.
437. L. Mathiasson, C. Turner, H. Berg, L. Dahlberg, A. Theobald, E. Anklam, R. Ginn, M. Sharman, F. Ulberth, R. Gabernig, *Food Addit. Contam.* 19, 632–646, 2002.
438. R. Marsili, D. Callahan, *J.Chromatogr. Sci.* 31, 422–428, 1993.
439. L. Ye, W. O. Landen, R. R. Eitenmiller, *J. Agric. Food Chem.* 48, 4003–4008, 2000.
440. A. Rodrìguez-Bernaldo, H. S. Costa, *J. Food Compos. Anal.* 19, 97–111, 2006.
441. S. Casal, B. Macedo, M. B. P. P. Oliveira, *J. Chromatogr. B* 763, 1–8, 2001.
442. J. L. Chávez-Servín, A. I. Castellote, M. C. López-Sabater, *J. Chromatogr. A* 1122, 138–143, 2006.
443. D. Majchrzak, I. Elmadfa, *Food Chem.* 74, 275–280, 2001.
444. R. Gatti, M. G. Gioia, A. M. Di Pietra, M. Cini, *J. Chromatogr. A* 905, 345–350, 2001.
445. T. Lacker, S. Strohschein, K. Albert, *J. Chromatogr. A* 854, 37–44, 1999.
446. C. W. Huck, M. Popp, H. Scherz, G. K. Bonn, *J. Chromatogr. Sci.* 38, 441–449, 2000.
447. D. Giuffrida, L. La Torre, M. Stlitano, T. M. Pellicano, G. Dugo, *Flav. Fragr. J.* 21, 319–323, 2006.
448. C. Turner, L. Mathiasson, *J Chromatogr. A* 874, 275–283, 2000.
449. M. Aké, A. M. Contento, M. D. Blanchin, H. Fabre, *Chromatographia* 47 (11/12), 716–720, 1998.
450. M. W. Davey, J. Keulemans, R. Swennen, *J. Chromatogr. A* 1136, 176–184, 2006.
451. L. Englberger, W. Aalbersberg, M. H. Fitzgerald, G. C. Marks, K. Chand, *J. Food Compos. Anal.* 16, 237–247, 2003.
452. M. Rodríguez-Comesañ, M. S. García-Falcón, J. Simal-Gándara, *Food Chem.* 79, 141–144, 2002.
453. A. Schieber, M. Marx, R. Carle, *Food Chem.* 76, 357–362, 2002.
454. W. J. Blot, J. Y. Li, P. R. Taylor, W. Guo, S. Dawsey, G. Q. Wang, C. S. Yang, S. F. Zheng, M. Gail, G. Y. Li, *J. Natl. Cancer Inst.* 85, 1483–1492, 1993.
455. O. P. Heinonen, D. Albanes, J. Virtamo, P. R. Taylor, J. K. Huttunen, A. M. Hartman, J. Haapakoski et al., *J. Natl. Cancer Inst.* 90, 440–446, 1998.
456. E. R. Greenberg, J. A. Baron, T. D. Tosteson, D. H. Jr. Freeman, G. J. Beck, J. H. Bond, T. A. Colacchio, J. A. Coller, H. D. Frankl, R. W. Haile, *N. Eng. J. Med.* 331, 141–147, 1994.
457. A. Demo, C. Petrakis, P. Kefalas, D. Boskou, *Food Res. Int.* 31, 351–354, 1998.
458. C. J. Hogarty, C. Ang, R. R. Eitenmiller, *J. Food Compos. Anal.* 2, 200–209, 1989.
459. A. P. De Leenheer, V. O. Bevere, M. G. De Ruyter, A. E. Claeys, *J. Chromatogr.* 162, 408–413, 1979.
460. M. M. Delgado-Zamarreño, M. Bustamante-Rangel, A. Sánchez-Pérez, J. Hernández Méndez, *J. Chromatogr. A* 935, 77–86, 2001.
461. G. W. Chase, A. R. Log, *J. AOAC Int.* 81, 582–586, 1998.
462. G. W. Chase, R. R. Eitenmiller, A. R. Long, *J. Liquid Chromatogr.* 20, 3317–3327, 1997.
463. D. B. Gomis, M. P. Fernández, M. D. G. Alvarez, *J. Chromatogr. A* 891, 109–114, 2000.
464. E. J. Rogers, S. M. Rice, R. J. Nicolosi, D. R. Carpenter, C. A. Mc Clelland, L. J. Romanecyk, *J. Am. Oil Chem. Soc.* 70, 301–307, 1993.
465. G. W. Chase, R. R. Eitenmiller, A. R. Long, *J. AOAC Int.* 82, 107–111, 1999.
466. J. F. Bonvehi, F. V. Coll, I. A. Rius, *J. AOAC Int.* 83, 627–634, 2000.
467. M. M. Delgado-Zamarreño, M. Bustamante-Rangel, A. Sánchez-Pérez, R. Carabias-Martínez, *J. Chromatogr. A* 1056, 249–252, 2004.
468. D. J. M. Gómez-Coronado, E. Ibañez, F. Javier Rupérez, C. Barbas, *J. Chromatogr. A* 1054, 227–233, 2004.
469. M. K. Balz, E. Schulta, H. P. Their, *Fat Sci. Technol.* 95, 215–220, 1993.
470. J. K. G. Kramer, L. Blais, R. C. Fouchard, R. A. Melnyk, K. M. R. Kallury, *Lipids* 32, 323–330, 1997.
471. S. L. Richheimer, M. C. Kent, M. W. Bernart, *J. Chromatogr. A* 677, 75–80, 1994.
472. S. L. Abidi, T. L. Mounts, *J. Chromatogr. A* 7882, 25–32, 1997.
473. S. L. Abidi, *J. Chromatogr. A* 844 67–75, 1999.
474. C. Rentel, S. Strohschein, K. Albert, E. Bayer, *Anal. Chem.* 70, 4394–4400, 1998.
475. S. Strohschein, C. Rentel, T. Lacker, E. Bayer, K. Albert, *Anal. Chem.* 71, 1780–1785, 1999.
476. R. R. Eitenmiller, W. O. Landen, in *Vitamin Analysis for the Health and Food Science*, CRC Press, Boca Raton, FL, 1999, pp. 119–192.
477. S. L. Abidi, *J. Chromatogr. A* 881, 197–216, 2000.
478. A. Kamal-Eldin, S. Görgen, J. Pettersson, A. M. Lampi, *J. Chromatogr. A* 881, 217–227, 2000.
479. Y. M. Choo, A. N. Ma, H. Yahaya, Y. Yamauchi, M. Bounoshita, M. Saito, *J. Am. Oil Chem. Soc.* 73, 523–525, 1996.

480. J. W. King, F. Favati, S. L. Taylor, *Sep. Sci. Technol.* 31, 1843–1857, 1996.
481. T. Yarita, A. Nomura, K. Abe, Y. Takeshita, *J. Chromatogr. A* 679, 329–334, 1994.
482. P. Manzi, G. Panfili, L. Pizzoferrato, *Chromatographia* 43, 89–94, 1996.
483. H. Qian, M. Sheng, *J. Chromatogr. A* 825, 127–133, 1998.
484. G. Carlucci, P. Mazzeo, S. Del Governatore, G. Di Giacomo, G. Del Re, *J. Chromatogr. A* 935, 87–91, 2001.
485. J. Torre, M. P. Lorenzo, M. P. Martínez-Alcázar, C. Barbas, *J. Chromatogr. A* 919, 305–311, 2001.
486. M. H. Chen, C. J. Bergman, *J. Food Compos. Anal.* 18, 139–151, 2005.
487. A. López, A. Montaño, A. Garrido, *J. Am. Oil Chem. Soc.* 82 (2), 129–133, 2005.
488. S. C. Cunha, J. S. Amaral, J. O. Fernandes, M. B. P. P. Oliveira, *J. Agric. Food Chem.* 54, 3351–2256, 2006.
489. J. Kalinova, J. Triska, N. Vrchotova, *J. Agric. Food Chem.* 54, 5330–5335, 2005.
490. C. M. López Ortiz, M. S. Moya, V. Berenguer Navarro, *J. Food Compos. Anal.* 19, 141–149, 2006.
491. P. Sookwong, K. Nakagawa, K. Murata, Y. Koijma, T. Miyazawa, *J. Agric. Food Chem.* 55, 461–466, 2007.
492. E. Vági, B. Simándi, K. P. Vásárhelyiné, H. Daood, A. Kéry, F. Doleschall, B. Nagy, *J. Supercrit. Fluids* 40, 218–226, 2007.
493. I. Sundl, M. Murkovic, D. Bandoniene, M. W. Winklhofer-Roob, *Clin. Nutr.* 26, 145–153, 2007.
494. I. K. Cho, J. Rima, C. L. Chang, Q. X. Li, *J. Food Compos. Anal.* 20, 57–62, 2007.
495. S. L. Booth, J. A. T. Pennington, J. A. Sadowski, *J. Am. Diet Assoc.* 96, 149–154, 1996.
496. S. T. Fenton, R. J. Price, C. Bolton-Smith, D. Harrington, M. J. Shearer, *Proc. Nutr. Soc.* 56, 301A, 1997.
497. M. A. Schneiderman, A. K. Sharma, K. R. R. Mahanama, D. C. Locke, *J. Assoc. Off. Anal. Chem.* 71, 815–817, 1988.
498. S. A. Barnett, L. W. Frick, H. M. Baine, *Anal. Chem.* 52, 610–614, 1980.
499. F. Zonta, B. Stancher, *J. Chromatogr.* 329, 257–263, 1985.
500. T. Sakano, T. Nagaoka, A. Morimoto, K. Hirauchi, *Chem. Pharm. Bull.* 34, 4322–4326, 1986.
501. A. Hiraishi, *J. Appl. Bacteriol.* 64, 103–105, 1988.
502. Y. Haron, P. V. Hauschka, *J. Lipid Res.* 24, 481–484, 1983.
503. S. L. Booth, K. W. Davidson, J. A. Sadowski, *J. Agric. Food Chem.* 42, 295–300, 1994.
504. Z. H. Gao, R. G. Ackman, *Food Res. Int.* 28, 61–69, 1995.
505. G. Ferland, J. A. Sadowski, *J. Agric. Food Chem.* 40, 1869–1873, 1992.
506. M. P. Bueno, M. C. Villalobos, *J. Ass. Off. Anal. Chem.* 66, 1063–1066, 1983.
507. E. Jacob, I. Elmadfa, *Food Chem.* 56, 87–91, 1996.
508. V. Piironen, T. Koivu, O. Tammisalo, P. Mattila, *Food Chem.* 59, 473–480, 1997.
509. V. Piironen, T. Koivu, *Food Chem.* 68, 223–226, 2000.
510. T. Koivu, V. Piironen, A. M. Lampi, P. Mattila, *Food Chem.* 64, 411–414, 1999.
511. W. E. Lambert, L. Vanneste, A. P. De Leeheer, *Clin. Chem.* 38, 1743–1748, 1992.
512. H. E. Indyk, *J. Micronutr. Anal.* 4, 61–70, 1988.
513. T. Pérez-Ruiz, C. Martínez-Lozano, M. D. García, J. Martín, *J. Chromatogr. A* 1141, 67–72, 2007.
514. J. P. Langenberg, U. R. Tjaden, E. M. De Vogel, D. I. Langerak, *Acta Alim.* 15, 187–198, 1986.
515. Y. Usui, N. Nishimura, N. Kobayashi, T. Okanoue, K. Kimoto, K. Ozawa, *J. Chromatogr.* 489, 291–301, 1989.
516. S. Ahmed, N. Kishikawa, K. Nakashima, N. Kuroda, *Anal. Chim. Acta* 591, 148–154, 2007.
517. E. Jakob, I. Elmadfa, *Food Chem.* 68, 219–221, 2000.
518. H. Iwase, *J. Chromatogr. A* 881, 261–266, 2000.
519. D. W. Ferriera, D. B. Haytowitz, M. A. Tassinari, J. W. Peterson, S. L. Booth, *J. Food Sci.* 71, S66–S70, 2006.
520. J. A. Keverling Buisman, K. H. Hanewald, F. J. Mulder, J. R. Roborgh, J. Keuning, *J. Pharm. Sci.* 57, 1326–1329, 1968.
521. T. Kobayashi, T. Okano, A. Takeushi, *J. Micronutr. Anal.* 2, 1–24, 1986.
522. V. K. Agarwal, *J. Assoc. Off. Anal. Chem.* 72, 1007–1009, 1989.
523. M. H. Bui, *J. Assoc. Off. Anal. Chem.* 70, 802–805, 1987.
524. H. E. Indyk, D. C. Woollard, *J. Micronutr. Anal.* 1, 121–141, 1985.
525. W. O. Landen, R. R, Eitenmiller, *J. Assoc. Off. Anal. Chem.* 62, 283–289, 1979.
526. W. O. Landen, *J. Assoc. Off. Anal. Chem.* 68, 183–187, 1985.
527. J. Jakobsen, I. Clausen, T. Leth, L. Ovesen, *J. Food Compos. Anal.* 17, 777–787, 2004.
528. P. H. Mattila, V. Piironen, E. Uusi-Rauva, P. Koivistoinen, *J. Food Compos. Anal.* 8, 232–243, 1995.
529. I. Clausen, J. Jakobsen, T. Leth, L. Ovesen, *J. Food Compos. Anal.* 16, 575–585, 2003.

530. U. Ostermeyer, T. Schmidt, *Eur. Food Res. Technol.* 222, 403–413, 2006.
531. R. Purchas, M. Zou, P. Pearce, F. Jackson, *J. Food Compos. Anal.* 20, 90–98, 2007.
532. S. R. Wilson, Q. Lu, M. L. Tulchinsky, Y. Wu, *J. Chem. Soc. Chem. Commun.* 8 664–665, 1993.
533. P. H. Mattila, V. I. Piironen, E. J. Uusi-Rauva, P. E. Koivistoinen, *J. Agric. Food Chem.* 43, 2394–2399, 1995.
534. M. M. D. Zamareno, A. S. Péerez, C. G. Pérez, J. H. Méndez, *J. Chromatogr.* 623, 69–74, 1992.
535. T. Wang, G. Bengtsson, I. Kärnefelt, L. O. Björn, *J. Photochem. Photobiol.* 62, 118–122, 2001.
536. H. Iwase, *J. Chromatogr. A* 881, 189–196, 2000.
537. G. Perretti, O.Marconi, L. Montanari, P. Fantozzi, *Lebensm.-Wiss. U.-Technol.* 37, 87–92, 2004.
538. P. Salo-Väänänen, V. Ollilainen, P. Mattila, K. Lehikoinen, E. Salmela-Mölsä, V. Piironen, *Food Chem.* 71, 535–543, 2000.
539. M. M. Delgado-Zamareno, A. Sanchez-Perez, M. Sanchez-Rodriguez, M. C. Gomez-Perez, J. Hernandez-Mendez, *Talanta* 43, 1555–1563, 1996.
540. B. J. Spencer, W. C. Purdy, *J. Chromatogr. A* 782,227–235, 1997.
541. P. F. Chatzimichalakis, V. F. Samanidou, I. N. Papadoyannis, *J. Chromatogr. B* 805, 289–296, 2004.
542. M. A. Kall, C. Andersen, *J. Chromatogr. B* 730, 101–111, 1999.
543. L. L. Lloyd, F. P. Warner, J. F. Kennedy, C. A. White, *Food Chem.* 28, 257–268, 1988.
544. H. Iwase, *J. Chromatogr. A* 881, 317–326, 2000.
545. H. Iwase, *J. Chromatogr. A* 881, 327–330, 2000.
546. A. Tai, E. Gohda, *J. Chromatogr. B* 853, 214–220, 2007.
547. H. Iwase, I. One, *J. Chromatogr. A* 654, 215–220, 1993.
548. J. Lykkesfeldt, S. Loft, H. E. Poulsen, *Anal. Biochem.* 229, 329–335, 1995.
549. S. Karp, C. S. Helt, N. H. Soujari, *Microchem. J.* 47, 157–162, 1993.
550. M. J. Deutsch, C. E. Weeks, *J. Assoc. Off. Anal. Chem.* 48, 1248–1256, 1965.
551. J. T. Vanderslice, D. J. Higgs, *J. Chromatogr. Sci.* 22, 485–489, 1984.
552. L. S. Liau, D. L. Lee, A. L. New, C. N. Ong, *J. Chromatogr.* 612, 63–70, 1993.
553. N. Hidiroglou, R. Madere, W. Behrens, *J. Food Compos. Anal.* 11, 89–96, 1998.
554. H. Iwase, *Talanta* 60, 1011–1021, 2003.
555. J. O. Kuti, *Food Chem.* 85, 527–533, 2004.
556. M. M. Wall, *J. Food Compos. Anal.* 19, 655–663, 2006.
557. P. Fontannaz, T. Kilinç, O. Heudi, *Food Chem.* 94, 626–631, 2006.
558. M. Romeu-Nadal, S. Morea-Pons, A. I. Castellote, M. C. López-Sabater, *J. Chromatogr. B* 830, 41–46, 2006.
559. S. Kumar, B. Aalbersberg, *J. Food Compos. Anal.* 19, 311–320, 2006.
560. A. Topuz, F. Ozdemir, *J. Food Compos. Anal.* 20, 596–602, 2007.
561. L.Odriozola-Serrano, T. Hernandéz-Jover, O. Martín-Belloso, *Food Chem.* 105, 1151–1158, 2007.
562. G. F. M. Ball, *Water-Soluble Vitamin Assay in Human Nutrition*, Chapman & Hall, Cornwell, U.K., 1994.
563. S. Ruggeri, L. T. Vahteristo, A. Aguzzi, P. Finglas, E. Carnovale, *J. Chromatogr.* 855, 237–245, 1999.
564. L. T. Vahteristo, V. Ollilainen, E. K. Pekka, P. Varo, *J. Agric. Food Chem.* 44, 477–482, 1996.
565. C. M. Pffeifer, L. M. Rogers, J. F. Gregory, *J. Agric. Food Chem.* 45, 407–413,1997.
566. K. Aiso, T. Tamura, *J. Nutr. Sci. Vitamin* 44, 361–370, 1998.
567. T. Tamura, Y. Mizuno, K. E. Johnson, R. A. Jacob, *J. Agric. Food Chem.* 45, 135–139, 1997.
568. J. I. Rader, C. M. Weaver, G. Angyal, *Food Chem.* 62, 451–465,1998.
569. S. De Souza, R. Eitenmiller, *J. Micronutr. Anal.* 7, 37–57, 1990.
570. J. I. Martin, W. I. Landen, A. G. M. Soliman, R. R. Eitenmiller, *J. Assoc. Anal. Chem.* 73, 805–808, 1990.
571. J. F. Gregory, D. B. Sartain, B. P. F. Day, *J. Nutr.* 114, 341–353, 1984.
572. T. Rebello, *Anal. Biochem.* 166, 55–64, 1987.
573. D. M. Goli, J. T. Vanderslice, *Food Chem.* 43, 57–64, 1992.
574. D. S. Duch, S. W. Bowers, C. A. Nochol, *Anal. Biochem.* 130, 385–392, 1983.
575. J. Selhub, O. Ahmad, I. H. Rosenberg, in *Methods in Enzymology*, Vol. 66, D. B. McCormick, L. D. Wright (Eds.), Academic Press, New York, 1980, pp. 686–690.
576. E. Seyoum, J. Selhub, *J. Nutr. Biochem.* 4, 488–494, 1993.
577. M. Johansson, J. Jastrebova, A. Grahn, M. Jägerstad, *Chromatographia* 62, 33–40, 2006.
578. P. Stokes, K. Webb, *J. Chromatogr. A* 864, 59–67, 1999.
579. R. J. Pawlosky, V. P. Flanagan, R. F. Doherty, *J. Agric. Food Chem.* 51(13), 3726–3730, 2003.
580. A. Frieslben, P. Schieberle, M. Rychilik, *Anal. Bioanal. Chem.* 376, 149–156, 2003.

581. J. D. Patring, J. A. Jastrebova, S. B. Hjortmo, T. A. Andlid, I. M. Jägerstad, *J. Agric. Food Chem.* 53, 2406–2411, 2005.
582. S. Pérez Prieto, B. Cancho Grande, S. García Falcón, J. Simal Gándara, *Food Control* 17, 900–904, 2006.
583. R. R. Catharino, J. A. Lima, H. T. Godoy, *Acta Alim.* 36(1), 143–151, 2007.
584. E. Ginting, J. Arcot, *J. Agric. Food Chem.* 52, 7752–7758, 2004.
585. M. Rychlik, A. Freisleben, *J. Food Compos. Anal.* 15, 399–409, 2002.
586. A. Leporati, D. Catellani, M. Suman, R. Andreoli, P. Manini, W. M. A. Niessen, *Anal. Chim. Acta* 531, 87–95, 2005.
587. D. M. Mock, in *Present Knowledge in Nutrition*, 7th edn., Ziegler E. E., Filer L. J. (Eds.), ILSI Press, Washington, DC, 1996, pp. 220–235.
588. C. G. Staggi, W. M. Sealey, B. J. McCabe, A. M. Teague, D. M. Mock, *J. Food Compos. Anal.* 17, 767–776, 2004.
589. U. Höller, F. Wachter, C. Wehrli, C. Fizet, *J. Chromatogr. B* 831, 8–16, 2006.
590. P. L. Desbene, S. Coustal, F. Frappier, *Anal. Biochem.* 128, 359–362, 1983.
591. K. Hayakawa, J. Oizumi, *J. Chromatogr.* 413, 247–250, 1987.
592. T. Yoshida, A. Uetake, C. Nakai, N. Nimura, T. Kinoshita, *J. Chromatogr.* 456, 421–426, 1988.
593. H. Miwa, *J. Chromatogr. A* 881, 365–385, 2000.
594. R. Peters, J. Hellenband, Y. Mengerink, A. Ven der Wal, *J. Chromatogr. A* 1031, 35–50, 2004.
595. R. Wolf, C. Huschka, K. Raith, W. Wohlab, R. Neubert, *Anal. Commun.* 34, 355, 1997.
596. C. Yomota, Y. Ohnishi, *J. Chromatogr. A* 1142, 231–235, 2007.
597. S. Lahély, M. Bergaentzlé, C. Hasselmann, *Food Chem.* 65, 129–133, 1999.
598. P. M. Finglas, R. M. Faulks, *J. Micronutr. Anal.* 3, 251–283, 1987.
599. P. G. Krishnan, I. Mahmud, D. P. Matthees, *Cereal Chem.* 76(4), 515–518, 1999.
600. K. Mawatari, F. Iinuma, M. Watanabe, *Anal. Sci.* 7, 733–736, 1991.
601. S. M. Juraja, V. C.Trenerry, R. G. Millar, P. Scheelings, D. R. Buick, *J. Food Compos. Anal.* 16, 93–106, 2003.
602. S. Lahély, S. Ndaw, F. Arella, C. Hasselmann, *Food Chem.* 65, 253–258, 1999.
603. S. Ndaw, M. Bergaentzlé, D. Aoudé-Werner, C. Hasselmann, *Food Chem.* 71, 129–138, 2000.
604. C. Rose-Sallin, C. J. Blake, D. Genoud, E. R. Tagliaferri, *Food Chem.* 73, 473–480, 2001.
605. K. L. Windhal, V. C. Trenerry, C. M. Ward, *Food Chem.* 65, 263–270, 1998.
606. J. Jakobsen, *Food Chem.* 106, 1209–1217, 2008.
607. F. Batifoulier, M.-A. Verny, E. Chanliaud, C. Rémésy, C. Demigné, *Eur. J. Agron.* 25, 163–169, 2006.
608. M. Rychlik, *J. Mass Spectrom.* 36, 555–562, 2001.
609. D. C. Wollard, H. E. Indyk, S. K. Christiansen, *Food Chem.* 69, 201–208, 2000.
610. C. Pakin, M. Bergaentzlé, V. Hubscher, D. Aoudé-Werner, C. Hasselmann, *J. Chromatogr. A* 1035, 87–95, 2004.
611. P. Gimsing, E. Nexo, E.Hippe, *Anal. Biochem.* 129, 296–304, 1983.
612. T. S. Hudson, S. Subramanian, R. J. Allen, *J. Assoc. Off. Anal. Chem.* 67, 994–998, 1984.
613. H. Chassaigne, R. Lobinski, *Anal. Chim. Acta* 359, 227–235, 1998.
614. O. Heudi, T. Kilinç, P. Fontannaz, E. Marley, *J. Chromatogr. A* 1101, 63–68, 2006.
615. E. G. Yanes, J. N. Miller-Hhli, *Spectrochim. Acta B* 59, 891–899, 2004.
616. P. C. H. Hollmann, J. H. Slangen, P. J. Wagstaffe, U. Faure, D. A. T Southgate, P. M. Finglas, *Analyst* 118, 481–488, 1993.
617. H. Van den Berg, F. van Schaik, P. M. Finglas, I. de Froidmont-Görtz, *Food Chem.* 57, 101–108, 1996.
618. L. F. Russel, J. T. Vanderslice, *Food Chem.* 43, 151–162, 1992.
619. S. Yamato, N. Kawakami, K. Shimada, M. Ono, N. Idei, Y. Itoh, *J. Chromatogr. A* 896, 171–181, 2000.
620. M. Reitzer-Bergaentzlé, E. Marchioni, C. Hasselmann, *Food Chem.* 48, 321–324, 1993.
621. N. Yang, Q. Wan, X. Wang, *Electrochim. Acta* 50, 1275–1280, 2005.
622. M. Rychlik, *Anal. Chim. Acta* 495, 133–141, 2003.
623. R. Mittermayr, A. Kalman, M.-J. Trisconi, O. Heudi, *J. Chromatogr. A* 1032, 1–6, 2004.
624. H.-B. Li, F.Chen, Y. Jiang, *J. Chromatogr. A* 891, 243–247, 2000.
625. H. Iwase, I. Ono, *J. Chromatogr. A* 771, 127–134, 1997.
626. K. Yamanake, M. Matsuoka, K. Banno, *J. Chromatogr. A* 726, 237–240, 1996.
627. E. Pinto, M. Pedersén, P. Snoeijs, L. Van Nieuwerburgh, *Biochem. Biophys. Res. Commun.* 291, 344–348, 2002.
628. X. Tang, D. A. Cronin, N. P. Brunton, *J. Food Compos. Anal.* 19, 831–837, 2006.
629. L. F. Russell, L. Brooks, K. B. McRae, *Food Chem.* 63 (1), 125–131, 1998.
630. C. Andrés-Lacueva, F. Mattivi, D. Tonon, *J. Chromatogr. A* 823, 355–363, 1998.

631. A. El-Gindy, F. El-Yazby, A. Mostafa, M. M. Maher, *J. Pharm. Biomed. Anal.* 35, 703–713, 2004.
632. R. Gatti, M. G. Gioia, V. Cavrini, *Anal. Chim. Acta* 512, 85–91, 2004.
633. M. A. Kall, *Food Chem.* 82, 315–327, 2003.
634. P. Viñas, N. Balsalobre, C. López-Erroz, M. Hernández-Córdoba, *Chromatographia* 59, 381–386, 2004.
635. A. Zafra-Gómez, A. Garballo, J. C. Morale, L. E. Garcia-Ayuso, *J. Agric. Food Chem.* 54, 4531–4536, 2006.
636. O. Heudi, T. Kilinç, P. Fontannaz, *J. Chromatogr. A* 1070, 49–56, 2005.
637. Z. Chen, B. Chen, S. Yao, *Anal. Chim. Acta* 569, 169–175, 2006.
638. P. Moreno, V. Salvadó, *J. Chromatogr. A* 870, 207–215, 2000.
639. L.-S. Li, S.-L. Da, Y.-Q. Feng, M. Liu, *Talanta* 64, 373–379, 2004.
640. M. Miraglia, C. Brera, in *Food Analysis by HPLC*, L. M. L. Nollet (Ed.), Marcel Dekker Inc., New York, 2000, pp. 493–522.
641. G. S. Shephard, N. L. Leggott, *J. Chromatogr. A* 882, 17–22, 2000.
642. J. Jaimez, C. A. Fente, B. I. Vazquez, C. M. Franco, A. Cepeda, G. Mahuzier, P. Prognon, *J. Chromatogr. A* 882, 1–10, 2000.
643. G. S. Shephard, *J. Chromatogr. A* 815, 31–39, 1998.
644. E. Brandšteterová, P. Kubalec, L. Bovanová, in *Food Analysis by HPLC*, L. M. L. Nollet (Ed.), Marcel Dekker Inc., New York, 2000, pp. 621–692.
645. H. Oka, Y. Ito, H. Matsumoto, *J. Chromatogr. A* 882, 109–133, 2000.
646. S. Joshi, *J. Pharm. Biomed. Anal.* 28, 795–809, 2002.
647. C. Van Peteghen, E. Daeseleire, A. Heeremans, in *Food Analysis by HPLC*, L. M. L. Nollet (Ed.), Marcel Dekker Inc., New York, 2000, pp. 965–986.
648. J. L. Tadeo, C. Sánchez-Brunete, R. A. Pérez, M. D. Fernández, *J. Chromatogr. A* 882, 175–191, 2000.
649. J. L. Tadeo, C. Sánchez-Brunete, in *Food Analysis by HPLC*, L. M. L. Nollet (Ed.), Marcel Dekker Inc., New York, 2000, pp. 693–716.
650. Y. Picó, G. Font, J. C. Moltó, J. Manes, in *Food Analysis by HPLC*, L. M. L. Nollet (Ed.), Marcel Dekker Inc., New York, 2000, pp. 717–760.
651. Y. Picó, G. Font, J. C. Moltó, J. Manes, *J. Chromatogr. A* 882, 153–173, 2000.
652. H. Kalaf, J. Steinert, in *Food Analysis by HPLC*, L. M. L. Nollet (Ed.), Marcel Dekker Inc., New York, 2000, pp. 939–964.
653. K. Skog, *Food Chem. Toxicol.* 40, 1197–1203, 2002.
654. M. S. Alaejos, V. Gonzalez, A. M. Afonso, *Food Addit. Contam.* 18, 1–23, 2007.
655. R. J. Law, J. L. Biscava, *Mar. Pollut. Bull.* 29, 235–241, 1994.
656. M. C. R. Camargo, M. C. F. Toledo, *Food Control* 14, 49–53, 2003.
657. D. H. Phillips, *Mutation Res.* 443, 139–147, 1999.
658. J. P. Meador, J. E. Stein, W. L. Reichert, U. Varanesi, *Rev. Environ. Contam. Toxicol.* 143, 79–165, 1995.
659. S. Moret, L. S. Conte, *J. Chromatogr. A* 882, 245–253, 2000.
660. S. Moret, A. Dudine, L. S. Conte, *J. Assoc. Off. Anal. Chem.* 77, 1289–1292, 2000.
661. EFSA, A report from the Unit of data collection and exposure on a request from the European Commission, Findings of the EFSA data collection on polycyclic aromatic hydrocarbons in food, Issued on June 29, 2007.
662. M. Lodovici, P. Dolora, C. Canalini, S. Cappellano, G. Testolin, *Food Addit. Contam.* 12, 703–713, 1995.
663. N. Kazerouni, R. Sinha, C. H. Hsu, A. Greenberg, N. Rothman, *Food Chem. Toxicol.* 39, 423–436, 2001.
664. Food Safety Authority of Ireland (2006), available at http://www.fsai.ie/surveillance/food_safety/chemical/PAH_levels.pdf.
665. Food Standards Agency (2002), available at http://www.food.gov.uk/science/surveillance/fsis-2002/31pah.
666. J. M. Llobet, G. Falcó, A. Bocio, J. L. Domingo, *J. Food Prot.* 69, 2493–2499, 2006.
667. R. Ibáñez, A. Agudo, A. Berenguer, P. Jakszyn, M. J. Tormo, M. J. Sanchéz, J. R. Quirós et al., *J. Chromatogr. A* 882, 1–10, 2000.
668. E. Menichini, B. Bocca, in *Encyclopedia of Food Sciences & Nutrition*, B. Caballero, L. Trugo, L. P. M. Finglas, (Eds.), Academic Press, Amsterdam, the Netherlands, 2003, pp. 4616–4625.
669. SCF, European Commission, Opinion of the on the risks to human health of polycyclic aromatic hydrocarbons in food (expressed on December 4, 2002).
670. European Commission, Commission Regulation 2005/208/EC, Off. J. Eur. Union L34/3, February 4, 2005.

671. European Commission, Commission Recommendation 2005/108/EC, Off. J. Eur. Union L34/43, February 4, 2005.

672. European Commission, Council Directive 98/83/EC, Off. J. Eur. Union L330, November 3, 1998.

673. European Commission, Commission Directive 2005/10/EC, Off. J. Eur. Union L34/15, February 4, 2005.

674. M. L. Lee, M. V. Novotny, K. D. Bartle, in *Analytical Chemistry of Polycyclic Aromatic Compounds*, Academic Press Inc., London, U.K., 1981, pp. 24–25.

675. E. Manoli, C. Samara, *Trends Anal. Chem.* 18, 417–428, 1999.

676. S. Moret, S. Amici, R. Bortolomeazzi, G. Lercker, Z. *Lebensm. Unters. Forsch.* 201, 322–326, 1995.

677. C. Da Porto, S. Moret, S. Soldera, *Anal. Chim. Acta* 563, 396–400, 2006.

678. C. Da Porto, S. Moret, *Food Chem. Toxicol.* 45, 2096–2071, 2007.

679. M. Bouzige, V. Pichon, M. C. Hennion, *J. Chromatogr. A* 823, 197–210, 1998.

680. Environmental Protection Agency, EPA method 525, in Method for the determination of organic compounds in drinking water, EPA/600/4-88/039. Office of Research and Development, Washington, DC, 1988.

681. I. Urbe, J. Ruana, *J. Chromatogr. A* 778, 337–345, 1997.

682. M. R. Negrão, M. F. Alpendurada, *J. Chromatogr. A* 823, 211–218, 1998.

683. M. S. García-Falcón, B. Cancho-Grande, J. Simal-Gándara, *Water Res.* 38, 1679–1684, 2004.

684. A. Stolyhwo, Z. E. Sikorski, *Food Chem.* 91, 303–311, 2005.

685. P. Šimko, *J. Chromatogr. B* 770, 3–18, 2002.

686. G. Grimmer, H. Böhnke, *J. Assoc. Off. Anal. Chem.* 58, 725–733, 1975.

687. K. Takatsuki, S. Suzuki, N.Sato, I. Ushizawa, *J. Assoc. Off. Anal. Chem.* 68, 945–949, 1985.

688. M. M. Khran, L. K. Moore, R. G. Bogar, C. A. Wigren, S. Chan, D. W. Brown, *J. Chromatogr.* 437, 161, 1988.

689. J. F. Uthe, C. J. Musial, *J. Assoc. Off. Anal. Chem.* 71, 363–368, 1988.

690. J. S. Burt, G. F. Ebell, *Mar. Pollut. Bull.* 30, 723–732, 1995.

691. S. Moret, L. S. Conte, *J. High Resolut. Chromatogr.* 21, 253–257, 1998.

692. D. A. Birkholz, R. T. Couts, S. E. Hrudei, *J. Chromatogr.* 449, 251–60, 1988.

693. M. Solé, C. Porte, D. Barcelo, J. Albaiges, *Mar. Pollut. Bull.* 40, 746–753, 2000.

694. M. S. García-Falcón, S. Gonsáles Amigo, M. A. Lage Yusti, M. J. Lopez de Alda Villaizan, J. Simal Lozano, *J. Chromatogr. A* 753, 207–215, 1996.

695. W. Jira, *Eur. Food Res. Technol.* 218, 208–212, 2004.

696. J. Fismes, C. Perrin-Ganier, P. Empereur-Bissonet, J. L. Morel, *J. Environ Qual.* 31, 1649–1656, 2002.

697. A. Hubert, P. Popp, K. D. Wenzel, W. Engewald, G. Schüürmann, *Anal. Bioanal. Chem.* 376, 53–60, 2003.

698. S. Tao, Y. H. Cui, F. L. Xu, B. G. Li, J. Cao, W. X. Liu, G. Schmitt et al., *Sci. Total Environ.* 320, 11–24, 2004.

699. M. S. García-Falcón, J. Simal Gandara, S. T. Carril Gonzales Barros, *Food Addit. Contam.* 17, 957–964, 2000.

700. J. Hernández-Borges, M. A. Rodríguez-Delgado, F. J. Garcia-Montelongo, *Chromatographia* 63, 155–160, 2006.

701. L. Samsøe-Petersen, E. K. Larsen, P. B. Larsen, P. Bruun, *Environ. Sci. Technol.* 36, 3057–3063, 2002.

702. S. Moret, L. Conte, *J. Sep. Sci.* 25, 96–100, 2002.

703. N. Cortesi, *Riv. Ital. Sostanze Grasse* 82, 129–133, 2005.

704. N. Cortesi, P. Rovellini, P. Fusari, *Riv. Ital. Sostanze Grasse* 79, 145–150, 2002.

705. S. Moret, K. Grob, L. S. Conte, *J. Chromatogr. A* 750, 361–368, 1996.

706. S. Moret, K. Grob, L. S. Conte, Z. *Lebensm. Unters. Forsch. A* 204, 241–246, 1997.

707. K. D. Bartle, M. L. Lee, S. A. Wise, *Chem. Soc. Rev.* 10, 113–137, 1981.

708. J. F. Lawrence, B. S. Das, *Int. J. Environ. Anal. Chem.* 24, 113–131, 1986.

709. M. D. Guillen, *Food Addit. Contam.* 11, 669–684, 1994.

710. J. de Boer, R. J. Law, *J. Chromatogr. A* 1000, 223–251, 2003.

711. H. K. Lee, in *Environmental Analysis, Handbook of Analytical Separation*, W. Kleiböhmer (Ed.), Elsevier Science, Amsterdam, the Netherlands, 2001, pp. 69–74.

712. C. H. Marvin, R. W. Smith, D. W. Bryant, B. E. McCarry, *J. Chromatogr. A* 863, 13–24, 1999.

713. K. Grob, I. Kaelin, A. Artho, *J. High Resolut. Chromatogr.* 14, 373–376, 1991.

714. S. Moret, L. S. Conte, D. Dean, *J. Agric. Food Chem.* 47, 1367–1371, 1999.

715. F. Van Stijn, M. A. T. Kerkhoff, B. M. G. Vandeginste, *J. Chromatogr. A* 750, 263–273, F1996.

716. S. Moret, V. Cericco, L. S. Conte, *J. Microcol Sep.* 13, 13–18, 2001.

717. S. Moret, K. Grob, L. S. Conte, *J. High Resolut. Chromatogr.* 19, 434–438, 1996.

20 HPLC in Forensic Sciences

Aldo Polettini

CONTENTS

20.1 INTRODUCTION

Forensic science is defined as the application of science to the processes of law in order to achieve justice. As a consequence of such a broad definition, forensic science encompasses a wide variety of disciplines, from medicine to chemistry, from toxicology to genetics, from bromatology to engineering. In general, however, most of the different activities of a forensic chemist fall within one of the following categories: (a) identification and/or determination of drugs, explosives, soils, paints, fingerprints, fibers, ignitable materials, biological materials; (b) comparison of the formers in order to link evidence from the crime scene to the suspect; and (c) authentication in cases of forgery (e.g., of ancient objects, questioned documents, currency).

Because of the potential legal consequences on the suspect(s) and the subjects involved, forensic analyses require high quality standards and have to be subjected to extensive quality assurance and rigid quality control programs. For example, the enforcement of laws establishing a sanction based on the presence/concentration of a substance in blood (e.g., cocaine in a driver's blood) requires that the substance is correctly identified and reliably determined with acceptable precision and accuracy.

As a consequence, analytical techniques providing the highest selectivity of detection and adequate sensitivity are generally adopted in forensic science. It is likely owing to this reason that the driving force in the spread of different chromatographic techniques in forensic laboratories has been and still is the possibility of a reliable and robust coupling with the most selective available detection technique, mass spectrometry (MS). During the 1980s and early 1990s, gas chromatography (GC) combined with MS was considered as the "golden standard" in the forensic analysis [1]. It was only in the mid-1990s that high-pressure liquid chromatography combined with mass spectrometry (HPLC–MS) began to spread in the forensic chemistry field. This happened thanks to the introduction of atmospheric pressure ionization (API), enabling a robust and rugged hyphenation of the two techniques [2]. After more than 10 years, HPLC–MS has definitely gone beyond the pioneer age and it is currently opened up to forensic scientists, providing higher versatility as compared to GC-MS, particularly in applications requiring the analysis of polar, heavy, and thermally labile molecules [3].

The great majority of the applications of HPLC in the forensic field published in the literature within the last 5 years concern detection, identification, and determination of substances of abuse, pharmaceutical drugs, and other poisons in different substrates in order to carry out a diagnosis of acute, lethal, or chronic intoxication (classical forensic toxicology), to ascertain alteration of individual performance while driving (driving under influence), at work (workplace drug testing), and in sports (doping control), or to monitor abstinence/exposure to substances of abuse. In addition, HPLC has been applied to the analysis of illicit drug preparations as well as of toxic alkaloids and toxins in foods. Other forensic HPLC applications include the analysis of explosives, inks, dyes, and, more recently, DNA analysis for personal identity testing.

20.2 FORENSIC TOXICOLOGY

20.2.1 Substances of Abuse

Analysis of substances of abuse involves different fields of application including diagnosis of acute/lethal intoxication, differentiation between chronic and occasional substance abuse (as it may imply different legal consequences for the substance abuser), enforcement of drug traffic safety (driving under influence), and the identification of the source of origin of illicit drugs.

Among the different chromatographic techniques, HPLC provides two important advantages: (a) the possibility to detect the whole metabolic profile of a substance, including both phase I and II, active and inactive, metabolites, within a single run, which provides an important tool in toxicological interpretation (for example, the metabolic profile of a drug enables to infer on the duration of survival after drug intake and/or on individual metabolic tolerance to the substance); and (b) the efficient separation of a wide variety of compounds including polar, high molecular weight and thermally labile one, thus offering an efficient tool for substance screening in cases of unclear or lacking anamnestic and circumstantial information.

20.2.1.1 Alcohol

In order to assess impairment in driving performance, blood alcohol concentration (BAC) is effectively determined using gas chromatographic separation techniques. However, as ethanol is rapidly eliminated from blood, the diagnosis of recent alcohol intake (e.g. in the case of late sampling after a car crash) as well as of chronic alcohol use (e.g., in cases of withdrawal treatment, workplace testing, driving license reissue/renewal, minors' adoption, divorce proceedings, etc.) requires the use of other markers. Indirect as well as direct markers of ethanol intake have been proposed. Among the first group, carbohydrate-deficient transferrin (CDT) [4–6] is certainly the most used, whereas in the second group, ethyl glucuronide (EtG) and, more recently, ethyl sulfate (EtS) [7–13] are gaining increasing consideration.

CDT is the collective name of a group of minor isoforms of serum transferrin [14] with a low degree of glycosylation, including asialo-, monosialo-, and disialo-Tf. CDT concentrations in serum are elevated during prolonged alcohol overconsumption (≥50–80 g/day for at least 7 days) and decrease, after cessation of drinking, with a half life of about 14 days [15]. CDT is considered to be the most specific indirect marker for alcohol abuse. Immunochemical, capillary zone electrophoretic, and ion exchange HPLC methods are available in order to carry out CDT isoform quantification (asialo- + disialo-Tf/ total serum transferrin, i.e., sum of the peak areas for all isotransferrins from asialo- to pentasialo-transferrin), the latter two providing sufficient selectivity and sensitivity for applications in the forensic environment as well as in other areas requiring high diagnostic reliability and objectivity [4].

The separation of CDT isoforms using anion exchange HPLC was developed by Jeppsson et al. [16] and adopted with minor modifications by different authors. Helander et al. [6] separated transferrin glycoforms using an anion-exchange column by linear salt gradient pH elution (10 mmol/L Bis-Tris) at a flow rate of 1.0 mL/min. After the 40 min run, the column was regenerated and cleaned by washing with 2.0 mol/L NaCl (buffer D) and was finally reequilibrated with the starting buffer. Detection was carried out exploiting the selective absorbance of the iron–transferrin complex at 470 nm. Sample preparation consisted of Tf saturation with iron with ferric nitrilotriacetic acid and precipitation of lipoproteins in serum with a dextran sulfate-$CaCl_2$ solution. Recently, commercial reagents for HPLC analysis (Recipe®) of CDT isoforms have been proposed [4].

EtG and EtS are minor phase II metabolites of alcohol and their presence in a biological sample unequivocally attests exposure to this substance. They can be measured in serum [7] and urine [9,11–13] even after ethanol has been eliminated. Recently, the determination of EtG in hair has been proposed as a sensitive and specific marker of chronic ethanol abuse [8,10]. The determination of EtG in postmortem blood enables the discrimination between ethanol postmortem formation and alcohol intake before death [17].

HPLC is considered as the method of choice for the separation of these two polar metabolites. Both reversed phase [18] separation (e.g., using a 100×3 mm i.d., 3 m particle size C18 column equipped with a RP guard column eluted under isocratic mode with a mobile phase of 0.1% formic acid and acetonitrile, 99:1 v/v, at 0.2 mL/min [9]) or mixed-modal reversed-phase/weak anion exchange phase (50×2 mm i.d. column eluted under isocratic mode with a mixture of acetonitrile/water, 77:23 v/v containing 25 mM acetic acid adjusted at pH 7.4 [11]) have been proposed. Highly sensitive detection is typically carried out in electrospray [8] negative-ion MS-MS selected reaction monitoring (SRM) mode, enabling to reduce sample preparation to dilution and centrifugation in the case of urine, and to protein precipitation with acetonitrile (ACN) followed by dilution and centrifugation in the case of serum. Extraction from hair is effectively achieved by incubating the sample, after decontamination with dichloromethane and methanol and cutting in 1–2 mm segments, in distilled water (see Figure 20.1).

20.2.1.2 Illicit Drugs

20.2.1.2.1   Analysis of Seized Material

The identification and quantification of active ingredient(s) in a seized material is of great forensic relevance as it enables to qualify the material as an "illicit drug" or not. In addition, the profiling of "street" drugs may allow to infer on origin and distribution routes of a confiscated drug.

Dams et al. [19] determined seven different opium alkaloids and derivatives in seized heroin using fast LC–MS analysis. Analytes were separated in 5 min on a monolithic silica column with a gradient elution system and an optimized flow of 5 mL/min. Detection was carried out using a sonic spray ion source [20]; a modified ESI source were ionization is achieved using a nebulizer gas at sonic speed instead of applying an electrical field.

RP-HPLC with UV (230 nm) detection was used by Zelkowicz et al. [21] to reconstruct a simulated heroin distribution chain in order to group sample having the same origin even after adding different amounts of adulterants.

HPLC separation at elevated pressure using 2 mm i.d., 1.7 μm C18 stationary phase columns was found to provide a 12× faster separation and better resolution of different analytes in seized drugs compared to conventional HPLC [22]. Twenty-four solutes of different drug classes, including

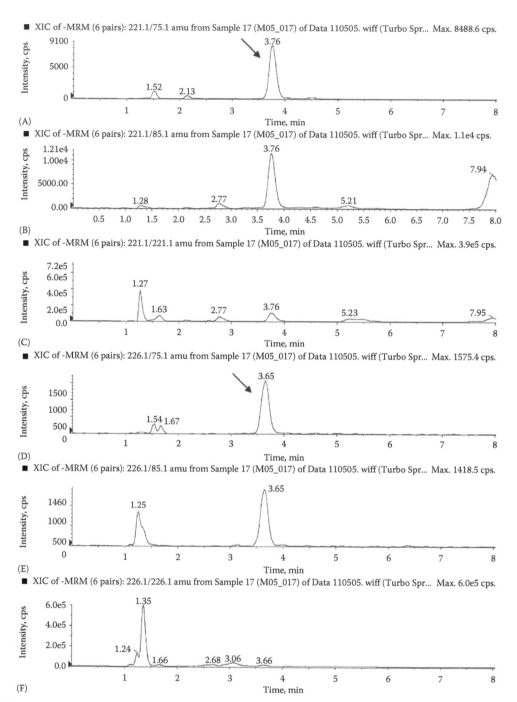

FIGURE 20.1 SRM chromatograms of a real ethyl glucuronide (EtG) positive hair sample collected from a patient under withdrawal treatment from alcohol. Concentration of EtG was 434 pg/mg. Separation: Chrompack Inertsil ODS-3 column (100×3 mm i.d., 3 μm particle size) equipped with a Chrompack (10×2 mm) RP guard column, kept at 25°C; mobile phase: mixture of 0.1% (v/v) formic acid (A) and ACN (B); gradient program: constant flow of 0.2 mL/min at 99% A for 8.0 min, rinsing step at 10% A for 2 min and reequilibration at 99% A up to 21 min; post-column addition of ACN: 0.1 mL/min. Detection: ESI-MS-MS negative ion mode. Reactions monitored for EtG: m/z 221.1 → 75.1 (A), 221.1 → 85.1 (B), 221.1 → 221.1 (C); reactions monitored for internal standard (D$_5$-EtG): m/z 226.1 → 75.1 (D), 226.1 → 85.1 (E), 226.1 → 226.1 (F). (Reprinted from Morini, L. et al., *J. Mass Spectrom.*, 41, 34, 2006. With permission from John Wiley & Sons, Ltd.)

narcotic analgesics, stimulants, depressants, hallucinogens, and anabolic steroids, were fully separated in a 13.5 min gradient.

Gradient RP-HPLC separation with 0.1% formic acid and methanol as mobile phase components (0.25 mL/min) combined either with ESI-time-of-flight [23] MS detection or with chemiluminescence nitrogen detection (CLND) was used to identify and, respectively, quantify illicit drugs in seized material without primary reference standards [24]. The method exploits the accurate mass measurement provided by TOF MS, enabling the unequivocal identification of molecular formula of the unknown analyte, and the CNLD equimolar response to nitrogen.

The hallucinogenic principles of "magic mushrooms" psilocin and psilocybin were selectively detected and quantified in mushrooms samples using HPLC ODS separation coupled with tandem mass spectrometry, reaching detection limits in the picogram range [25].

20.2.1.2.2   Analysis of Biological Samples

Cocaine is one of the oldest and currently one of the most abused drugs in the western world. After intake through nasal or intravenous route, the parent drug undergoes extensive biotransformation leading to the formation of a wide number of metabolites of which benzoylecgonine, methylecgonine, norcocaine, and, in the case of simultaneous intake of alcohol, cocaethylene are the most important. Johansen and Bhatia [26] developed an HPLC–MS-MS method for the determination of cocaine and metabolites in whole blood and urine. After protein precipitation, purification of analytes was obtained by mixed-mode solid phase extraction (SPE). Chromatographic separation was performed at 30°C on a reverse phase C18 column with a gradient system consisting of formic acid, water and acetonitrile. Analysis was performed in ESI positive-ion using a triple quadropole mass spectrometer operating in SRM mode. The lower limit of quantification (LLOQ) of cocaine was 0.008 µg/g. This method proved to be specific and adequately sensitive, offering an excellent alternative to GC-MS that involves derivatization in order to achieve a good chromatographic separation of polar metabolites.

Heroin, the diacetyl derivative of morphine, is the most important illicit drug derived from the opium poppy. Together with heroin and its metabolites, other synthetic (e.g., methadone, tramadol) and semi-synthetic (e.g., buprenorphine) opioids have been quantified in different biological fluid using HPLC.

Fernandez et al. [27] applied HPLC with diode array detection (DAD) to the determination of heroin, methadone, cocaine and metabolites in plasma after mixed-mode SPE. Analytes were separated using a RP8 column (250 mm × 4.6 mm i.d., 5 µm particle size) and acetonitrile-phosphate buffer pH 6.53 as mobile phase with elution in the gradient mode. The method, which provides a LLOQ of 0.1 µg/mL for all compounds, was successfully applied to 21 plasma samples from fatal overdoses.

Musshoff et al. [28] developed an automated and fully validated LC–MS-MS procedure for the simultaneous determination of 11 opioids used in palliative care, including five metabolites. Buprenorphine, codeine, fentanyl, hydromorphone, methadone, morphine, oxycodone, oxymorphone, piritramide, tilidine, and tramadol together with their metabolites bisnortilidine, morphineglucuronides, norfentanyl, and nortilidine were quantified in plasma after automated SPE, and 35 min separation was achieved on a C12 column (4 µm, 150 × 2 mm) using a gradient of ammonium formiate buffer (pH 3.5) and acetonitrile. The method enabled the specific diagnosis of overdose situations as well as monitoring the compliance of patients under palliative care.

Hupka et al. [29] developed a method for the determination of morphine and its phase II metabolites, morphine-3-beta-D-glucuronide and morphine-6-beta-D-glucuronide in the blood of heroin victims. The method is based on immunoaffinity SPE, RP-HPLC isocratic separation (mobile phase: 90% 10 mmol KH_2PO_4, 2 mmol 1-heptanesulfonic acid, adjusted to pH 2.5 with H_3PO_4 and 10% acetonitrile; flow rate: 1.5 mL/min), and laser-induced native fluorescence detection.

Polettini and Huestis [30] determined buprenorphine (BUP) and its metabolites norBUP and BUP glucuronide in plasma after C18 SPE, RP-HPLC and ESI positive ion MS-MS SRM mode, reaching an LLOQ of 0.1 ng/mL for all analytes. Using similar conditions, direct injection of diluted urine was investigated by Kronstrand et al. [31] in order to speed up the analysis. The authors concluded that the direct method for quantitation worked for concentrations above 20 ng/mL, whereas for lower levels the procedure had to be complemented with a concentration by SPE, thus lowering the cutoff from 20 to 1 ng/mL urine.

BUP and nor BUP glucuronides were determined in urine by HPLC–MS-MS after dilution and filtration, reaching LLOQs of 4.6 and 11.8 ng/mL, respectively [32]. Separation was carried out using a C18 column using a 5 mM ammonium acetate/acetonitrile mobile phase in gradient mode. Detection was carried out by monitoring the transitions from the glucuronide to the respective aglycone in positive-ion mode. The method was applied to 1000 samples from law enforcement, prison inmates, probation services, and hospitals and proved to be adequately robust and specific.

Favretto et al. [33] proposed a procedure for the analysis of BUP and nor BUP based on RP-HPLC coupled with ion trap MS using an ESI ion source. The use of the ion trap allowed to obtain intense product ions under MS-MS conditions, thus achieving better selectivity of detection with respect to LC-ESI-MS-MS methods with triple quadrupoles where effective fragmentation in the collision cell was found difficult to obtain. However, this method does not include phase II metabolites of BUP, which can be only indirectly determined after enzymatic hydrolysis.

Racemic methadone (MET) is administered to heroin addicts as a substitution therapy. However, methadone enantiomers possess different pharmacological effects, and the drug has been demonstrated to be enantioselectively metabolized to its two major metabolites, 2-ethylidene-1,5-dimethyl-3,3-diphenylpyrrolidine (EDDP) and 2-ethyl-5-methyl-3,3-diphenyl-1-pyrroline (EMDP). Stereoselective separation of MET, EDDP, and EMDP using an alpha-glycoprotein stationary phase and MS-MS detection was proposed by Kelly et al. [34]. Optimal separation conditions were 20 mM acetic acid: isopropanol (93:7, pH 7.4), with a flow rate of 0.9 mL/min.

Musshoff et al. [35] developed a method for the enantiomeric separation of the synthetic opioid agonist tramadol and its desmethyl metabolite using a Chiralpak AD column containing amylose tris-(3,5-dimethylphenylcarbamate) as chiral selector and a n-hexane/ethanol, 97:3 v/v (5 mM TEA) mobile phase under isocratic conditions (1 mL/min). After atmospheric pressure chemical ionization (APCI), detection was carried out in positive-ion MS-MS SRM mode. The method allowed the confirmation of diagnosis of overdose or intoxication as well as monitoring of patients' compliance.

Fentanyl, a potent synthetic narcotic analgesic, and norfentanyl were determined in postmortem samples using liquid–liquid extraction (LLE) and HPLC-ESI(+)-MS-MS enabling the authors to determine the distribution of parent compound and metabolite in tissues and organs of a suicidal fatality case [36]. However, matrix effects were not investigated.

Amphetamine-type molecules can be effectively determined in biological samples using HPLC. After and effective mixed-mode SPE, six different amphetamine-type analytes were determined in urine using RPLC-UV by Soares et al. [37] reaching LODs in the range 5.3–84 ng which, according to the authors, enable the detection of the analytes after ingestion of fatal and non fatal doses. Recently, Kumihashi et al. [38] developed an automated column switching procedure for the determination of MAMP and AMP in urine with UV detection at 210 nm. HPLC combined with fluorescence detection has been also successfully applied to the determination of methylenedioxy amphetamines in urine [39] and oral fluid [40].

Using HPLC–MS-MS, Wood et al. [41] developed a rapid and sensitive procedure for the determination of amphetamine (AMP), methamphetamine (MAMP), MDMA, MDA, MDEA, and ephedrine in plasma and oral fluid. After 1:5 dilution, protein precipitation and centrifugation at 13,000 rpm, samples were separated into a C18 column operated under isocratic mode (10 mM ammonium acetate and acetonitrile, 75:25, delivered at a flow rate of 0.3 mL/min). Ionization and detection were performed using positive-ion ESI SRM mode. A sample throughput of better than 3 h^{-1} and LODs of 2 ng/mL or better using only 50 μL of sample were achieved. Despite the non selective sample preparation adopted, matrix effects were not investigated, though this was partly compensated by the use of deuterated standards for each of the analytes.

The diagnosis of MAMP and *N,N*-dimethyl-AMP abuse was achieved by Cheng et al. [42] by urinalysis after C18 SPE, isocratic RPLC elution, and ESI (+) MS-MS SRM mode detection of the parent drugs and metabolites. The separation of analytes was achieved in a 5 min chromatographic run. Using similar conditions, AMP, MAMP, MDA, MDMA, MDEA, MBDB, and PMA were determined in urine by solvent extraction and fast separation (4 min) using gradient RPLC-ESI(+)-MS-MS [43]. Hair is increasing in popularity for the diagnosis to chronic exposure to drugs of abuse. Many substances are transported into hair by diffusion from blood to the hair bulb or by transfer from sebum and, once incorporated, remain trapped into the keratinic matrix for weeks to months, thus significantly widening the detection window of drug use. Different HPLC methods for the detection of amphetamine-like compounds [44–46] as well as of other drugs of abuse [33,47,48] in hair have been published in the last 5 years.

Ketamine, a general anesthetic producing hallucinations and exhibiting an addiction potential, and norketamine have been determined in urine using SPE combined with RPLC-ESI(+)-MS-MS [49]. Cheng and Mok [49] developed a fast HPLC–MS-MS method able to separate and detect ketamine in a 2.5 min chromatographic run with a LOD of 5 ng/mL. The method was applied to routine ketamine urine screening to a maximum of 200 samples per day.

LSD is one of the most potent hallucinogenic drugs. It is rapidly and extensively metabolized, so that the parent drug can be detected in urine for not longer than 12–22 h after use [50]. As a consequence, the main metabolite 2-oxo-3-hydroxy-LSD is the target analyte to detect LSD use. Horn et al. [51] developed a procedure for the determination of this metabolite using anion exchange SPE and RP gradient HPLC–MS in selected ion monitoring (SIM) mode reaching a LOQ of 250 pg/mL (matrix effects not investigated). Later, Johansen and Jensen [52] developed an HPLC-ESI(+)-MS-MS method for the determination of LSD, its epimer iso-LSD and 2-oxo-3-hydroxy-LSD in blood and urine after LLE at pH 9.8 with butyl acetate. Separation was carried out using a C18 column eluted under gradient mode. An LLOQ of 10 pg/g was achieved. Using a similar procedure based on ion trap MS-MS detection, Favretto et al. [53] determined LSD and metabolite and iso-LSD in blood, urine, and vitreous humor reaching a LLOQ of 20 pg/mL.

Cannabis derivatives (e.g., marijuana and hashish) are among the most common abused drugs. Apart from their illicit use, natural (delta-9-tetrahydrocannabinol, THC; cannabidiol, CBB; cannabinol CBN) as well as synthetic cannabinoids (nabilone, levonantradol) have been also tested for different therapeutic purposes. The availability of methods for monitoring the recreational use of Cannabis derivatives is therefore of great importance in forensic and clinical toxicology labs. GC-MS has been extensively used in the past, though HPLC–MS methods are increasingly proposed in the literature offering the important advantage of not requiring a derivatization step prior to chromatographic separation. Together with THC, the active metabolite 11-hydroxy-THC (11-OH-THC), and the major urinary metabolite 11-nor-9-carboxy-THC (THC-COOH), which is excreted in urine mainly as the glucuronide, are of forensic relevance. Maralikova and Weinmann [54] developed an HPLC–MS-MS method for the simultaneous analysis of THC, 11-OH-THC, THC-COOH and the glucuronide of the latter in urine. After SPE, separation was carried out using a phenylhexyl column (50×2 mm i.d., 3 μm) and acetonitrile-5 mM ammonium acetate gradient elution. Detection was carried out in ESI(+)-MS-MS SRM mode reaching LLOQs in the range 0.8–4.2 ng/mL. Concheiro et al. [55] determined the parent drug in oral fluid by HPLC–MS in SIM mode (two fragment ions). Separation was carried out using a 2.1 mm × 100 mm × 3.5 μm C18 column eluted isocratically (0.1% formic acid/acetonitrile 15:85, v/v) at a flow rate of 0.25 mL/min. A LOD of 2 ng/mL was reached using 0.2 mL of sample. However, validation did not include the evaluation of matrix effects. The detection of THC in oral fluid is linked to the contamination of the oral cavity during smoking and provides an indication of recent inhalation of hashish/marihuana smoke. According to Laloup et al. [56] a good correlation between detection of THC in oral fluid and plasma—using a validated quantitative HPLC–MS-MS method—exists. When using 0.5 and 1.2 ng/mL as cutoff for plasma and oral fluid, respectively, the sensitivity and specificity of THC detection in the latter were 94.7% and 92.0%, respectively.

Delta-9-tetrahydrocannabinolic acid A (9-THCA-A) is an inactive ingredient of Cannabis, which is converted into THC during smoking. Using HPLC–MS-MS, trace amounts of 9-THCA-A in urine/serum of hashish/marijuana smokers were detected [57]. According to the authors, the determination of 9-THCA-A may prove promising for the interpretation of consumption habits (single vs. frequent intake) or for estimating the time elapsed between intake and blood sampling.

Recently, Quintela et al. [58] have determined THC and the carboxy metabolite in oral fluid using HPLC coupled to a quadrupole-TOF mass spectrometer. Extreme selectivity of detection and LLOQs of 0.1 and 0.5 ng/mL, respectively, were achieved through accurate mass measurement. None of the real positive samples examined was found to contain THC-COOH.

Together with the HPLC methods devoted to target analytes described so far, some multi-target methods for different classes of drugs of abuse have been published in the literature. This multitarget strategy significantly reduces the cost of screening a large number of samples, which makes it competitive in terms of cost-effectiveness with the usual strategy of screening with class-specific immunochemical tests followed by confirmation of positives using hyphenated chromatographic/mass spectroscopic techniques. By modifying a conventional gradient elution HPLC method, Stoll and coworkers [59,60] developed a high-throughput screening of biological samples to detect different drugs of abuse using UV diode array detection. Gradient elution and reconditioning was completed in only 2.80 min.

Dams et al. [18] developed a validated quantitative LC-APCI-MS-MS method for simultaneous determination of multiple illicit drugs and their metabolites in oral fluid. This substrate is being increasingly popular for forensic applications as it provides information on recent use, similarly to blood plasma/serum, although it can be obtained with a simple, noninvasive, collection. Sample pretreatment, though limited to protein precipitation with acetonitrile, was sufficient to avoid matrix effect (see Figure 20.2).

Kronstrand et al. [48] developed an HPLC–MS-MS method able to screen simultaneously for opiates, cocaine, amphetamines, nicotine, and the metabolite cotinine, and selected benzodiazepines in hair.

Isolation of analytes from hair was carried out by incubating the sample overnight in the initial mobile phase (acetonitrile:methanol:20 mM formate buffer, pH 3.0, 10:10:80). The method was applied to autopsy cases adopting a 1 ng/mg cutoff for all analytes.

20.2.1.3 Pharmaceutical Drugs

Methods for the identification and determination of pharmaceutical drugs in the forensic setting are required in order to detect their abuse and misuse (e.g., accidental or suicidal overdosage, homicidal poisoning, illicit performance enhancement). Some pharmaceutical drugs may also enhance the toxic potential of illicit drugs and/or alcohol and their determination is necessary in order to ascertain cases of mixed-drug intoxications.

Among the different classes of pharmaceutical drugs, benzodiazepines (BZD) are probably the most often encountered in forensic toxicology. Being widely prescribed for the therapy of anxiety, sleep disorders, and convulsive attacks, these substances are often abused and misused. As a consequence, they are frequently found in biological fluids collected in fatal cases of drug intoxication, from drivers involved in traffic accidents, from victims of sexual assaults, and, together with opiates, alcohol, and antidepressants, from drug addicts.

Eight BDZs among the most frequently encountered in forensic toxicology (clonazepam, desalkylflurazepam, diazepam, flunitrazepam, lorazepam, midazolam, nordiazepam and oxazepam) were determined in whole blood after solvent extraction with butyl chloride and fast isocratic separation using a C18 (100 × 4.6 mm × 5 μm) column [61]. The mobile phase was composed of phosphate buffer (35 mM, pH 2.1) and acetonitrile (70:30, v/v) and the flow rate was 2 mL/min. Within less than 4 min of analysis time, the analytes could be successfully determined starting from therapeutic concentrations. Using HPLC coupled with APCI-MS-MS, Rivera et al. [62] set up a method for the detection of 18 BDZ and metabolites after butyl chloride extraction at alkaline pH in 0.5 mL

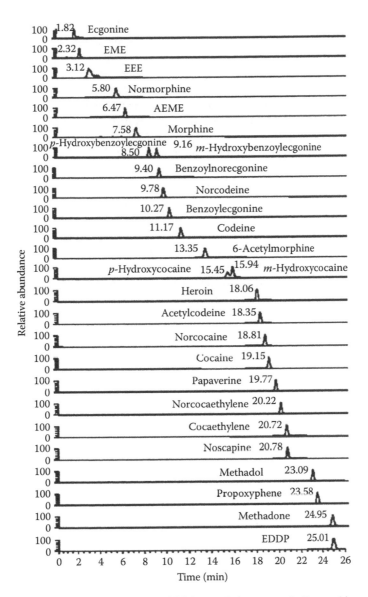

FIGURE 20.2 SRM chromatograms of a group of 27 drugs of abuse, metabolites and impurities simultaneously determined in oral fluid samples. Separation conditions: Synergi Polar RP column (150×2.0 mm, 4 μm), fitted with a guard column with identical packing material (4×2.0 mm) maintained at 25°C; gradient elution with (A) 10 mM ammonium formate in water, 0.001% formic acid (pH = 4.5) and (B) acetonitrile, at a flow rate of 300 mL/min; gradient program: 5% B increasing to 26% B in 13 min, with a final composition of 90% B in 9 min. (Reprinted from Dams, R. et al., *Anal. Chem.*, 75, 798, 2003. Copyright 2003. With permission from American Chemical Society.)

of blood. Kratzsch et al. [63] developed a fully validated method for the screening, library-assisted mass spectral identification, and validated quantification of 23 BDZs, their antagonist flumazenil and the benzodiazepine BZ1 (omega 1) receptor agonists zaleplone, zolpidem, and zopiclone in plasma by HPLC–MS with APCI. After solvent extraction, separation was achieved in RP gradient mode using aqueous ammonium formate and acetonitrile. After detection and identification in full scan mode by library search, quantification was carried out in SIM. The method was successfully used for the diagnosis of overdose situations as well as for the monitoring of patients compliance

(LLOQs in the range 0.5–100 ng/mL). Using SPE followed by RP HPLC combined with ion trap MS(-MS) detection, Smink et al. [64] determined 33 BDZs, metabolites and BDZ-like compounds in whole blood with LODs in the range 0.1–12.6 ng/mL. ElSohly et al. [65] proposed a method for the determination of 22 BDZs in urine collected from alleged drug-facilitated sexual assault victims. After solvent extraction and RP HPLC separation, detection was carried out in full-scan high-resolution MS mode using TOF MS reaching LOQs in the range 2–10 ng/mL. Similar approaches to the multitarget analysis of BDZs have been proposed by different authors in whole blood [66,67], oral fluid [68,69], plasma [69], urine [65,66,70], and hair [66,71,72].

Antidepressant drugs are widely used for the treatment of depression and, owing to their toxic potential, are frequently encountered in emergency toxicology screening and forensic medical examinations. Antidepressants belongs to different classes, namely tricyclic antidepressants (e.g., amitriptyline), selective serotonin reuptake inhibitors (SSRI, e.g., fluoxetine), and monoamine oxidase inhibitors (MAOI, e.g., moclobemide). An HPLC method capable of being used with either DAD or MS detection for the identification and quantitation of 9 SSRI antidepressants and metabolites has been proposed by Goeringer et al. [73]. A careful investigation of the effects of varying buffers and mobile phase pH and of adding modifying agents on resolution and capacity factors was carried out. Using a C18 column (2.1 × 150 mm × 5 μm) operated at 21°C and eluted at a flow rate of 0.25 mL/min the optimum mobile phase was 0.05 M ammonia/methanol/THF (32.5:67.0:0.5) at pH 10.0.

Shinozuka et al. [74] developed an HPLC–MS method with SSI ionization for the simultaneous determination of 20 antidepressant drugs (imipramine, amitriptyline, desipramine, trimipramine, nortriptyline, clomipramine, amoxapine, lofepramine, dosulepin, maprotiline, mianserin, setiptiline, trazodone, fluvoxamine, paroxetine, milnacipran, sulpiride, tandspirone, methylphenidate, and melitracen) in human plasma for forensic toxicology investigations. RP separation was carried out using a C8 column with methanol:10 mM ammonium acetate (pH 5.0):acetonitrile (70:20:10) as mobile phase at 0.10 mL/min at a temperature of 35°C. Solid-phase extraction of these drugs added to the human plasma was performed with an Oasis HLB cartridge column. LODs were between 30 and 630 ng/mL.

A fully automated on-line SPE-HPLC–MS-MS method was developed and validated for the direct analysis of 14 antidepressants and their metabolites in plasma by de Castro et al. [75]. After direct injection of 50 μL of plasma without prior sample pre-treatment, gradient RP separation was completed in 14 min, with a sample throughput of 3 h^{-1}. LOQs were estimated to be at 10 ng/mL. Analytes proved to be stable during sample process with the exception of clomipramine and norclomipramine.

Recently, Titier et al. [76] were able to reach LOQs of 2 ng/mL in blood by applying solvent extraction, RP C18 separation under gradient mode (acetonitrile/ammonium formate buffer 4 mmol/L, pH 3.2), and ESI(+)-MS-MS SRM mode to the determination of different tricyclic and MAOI antidepressants and their metabolites.

Multitarget forensic applications of HPLC for other drug classes are also available in the literature. Josefsson et al. [77] applied HPLC–MS-MS to the determination of 19 neuroleptics and their major metabolites in human tissues and body fluids. Optimal separation was achieved using a cyano column within a 9 min gradient run. Detection was curried out in SRM reaching LODs down to the lower ng/mL level, although more than a 10-fold difference in signal response was observed between analytes. The method was subjected to partial validation only.

β-Blockers (BB) are widely used, in particular in the treatment of angina, hypertension, arrhythmia, and cardiac failure. Their determination in plasma may offer useful information in cases of intoxication. In addition, BB may be abused by athletes, particularly in disciplines where psychomotor coordination is important. Adequate GC repartition of these analytes require derivatization, therefore HPLC appears as a valid alternative for the efficient analysis of BB.

The simultaneous identification and quantification in human plasma of 14 BB (atenolol, sotalol, diacetolol, carteolol, nadolol, pindolol, acebutolol, metoprolol, celiprolol, oxprenolol, labetalol, propranolol, tertatolol, and betaxolol) was achieved by Delamoye et al. [78] using LLE, RPLC C18 (250 mm × 4.6 mm × 5 μm) separation using a gradient of acetonitrile-phosphate buffer pH 3.8 at a flow rate of 1 mL/min. Total analysis time was 26 min per sample, and LODs ranged from 5 to 10 ng/mL. Dupuis et al. [20] validated a method for the determination of the BB atenolol, metoprolol, and propranolol in rabbit postmortem blood, gastric content, vitreous humor, and urine. After LLE using Extrelut columns, C18 separation was performed using a 150 mm × 1 mm × 5 µm column under a gradient of acetonitrile in 2 mM, pH 3 ammonium formate. Ionization and detection were carried out using ESI(+) MS in SIM mode applying in source-collisionally induced dissociation to achieve sufficient fragmentation of the pseudomolecular ions. The determination of selected BB in postmortem fluid and tissue specimens collected from victims of aviation accidents was carried out by Johnson and Lewis [79] using RP gradient elution coupled to ion trap MS. Detection was carried out either in MS-MS or MS^3 to achieve the required selectivity. Sample preparation involved homogenization of tissues, protein precipitation with acetonitrile, and mixed mode SPE. A method for the simultaneous determination of the BB atenolol, sotalol, metoprolol, bisoprolol, propranolol, and carvedilol and, in addition, the calcium-channel antagonists diltiazem, amlodipine, and verapamil, the angiotensin-II antagonists losartan, irbesartan, valsartan, and telmisartan, and the antiarrhythmic drug flecainide, in whole blood samples from forensic autopsies was developed by Kristoffersen et al. [23]. Quantification was performed by RP HPLC with ESI(+) MS detection. The method was optimized using experimental designs including factorial and response surface designs, and allowed to cover low therapeutic concentration levels for all the BB.

Quaternary nitrogen muscle relaxants are sometimes involved in homicide and suicide cases and the availability of reliable methods for their determination in forensic situations is of importance. Because of the presence of a permanent positive charge, and of their thermal instability, HPLC is the method of choice for the separation of these compounds. Cirimele et al. [80] proposed a screening procedure for the identification and the quantification of eight quaternary nitrogen muscle relaxants, including D-tubocurarine, alcuronium, pancuronium, vecuronium, atracurium, mivacurium, rocuronium and mebezonium, in blood samples. The procedure involved ion-pair extraction with methylene chloride at pH 5.4, reversed-phase HPLC separation and ESI MS. LODS of 100 ng/mL were reached. The screening test was applied in two fatal cases, one involving vecuronium (1.2 mg/L in blood) and the other mebezonium (6.5 mg/L). The simultaneous determination of pancuronium, vecuronium, and their metabolites in human serum was performed by Usui et al. [81] using RP HPLC-ESI-MS after SPE with a weak cation exchange cartridge. Full separation was achieved in 6.5 min. Ferrara et al. [82] developed a procedure for the determination of pancuronium and its metabolite laudanosine in postmortem specimens. After LLE and RPLC separation, detection was carried out with ESI(+) ion trap MS under collision-induced dissociation conditions. The method was validated in post-mortem blood, and tissues with LOQs of 1 ng/mL in blood and 5 ng/g in tissues.

Other HPLC methods devoted to the determination of selected pharmaceutical drugs and mainly applied to fatal poisoning cases have been published in the literature within the last 5 years: chloroquine [83], embutramide [84], fenarimol [85], flecainide [86], imidacloprid [87], loperamide [88], metformin [89], and pholedrine [90].

20.2.1.4 Alkaloids

Alkaloids are pharmaceutically active compounds contained in plants. Some of them exhibit also important toxic properties and may be encountered in forensic toxicology as a cause of accidental, suicidal, or homicidal poisonings.

The root of plants of the genus *Aconitum* (*Ranunculaceae*) have been used for thousands of years in Eurasia as powerful poisons since they contain highly toxic diester–diterpene type alkaloids. Owing to the nonvolatile nature of these compounds, HPLC offers significant advantages over GC for their separation. Aconitum alkaloids were determined by Wang et al. in four species of the genus Aconitum [91] as well as in blood and urine samples [92] using offline mixed mode SPE combined with RP gradient HPLC and DAD detection (237 nm). The determination of different *Aconitum* alkaloids and their metabolites (aconitine, mesaconitine, hypaconitine, jesaconitine, benzoylaconine, benzoylmesaconine, benzoylhypaconine, and 14-anisoylaconine) in serum was performed by

Hayashida et al. [93] using a column-switching HPLC–MS procedure. Analytes were trapped in a hydrophilic copolymer gel containing N-vinylacetamide and then separated under RPLC gradient mode. ESI(+) MS SIM was the detection mode. The total analysis time, including system equilibration, was 26 min. LODs in the range 0.2–50 ng/mL were reached. By applying mixed mode SPE, RPLC, and ESI(+) tandem MS detection in SRM mode, Beike et al. [94] developed a fully validated method for aconitine in blood achieving a LOD of 0.1 ng/mL. Using similar conditions, Wang et al. [95] determined yunaconitine, crassicauline A, and foresaconitine in urine specimens reaching a LOD of 0.03–0.05 ng/mL. Recently, HPLC-ESI-TOF MS has been successfully applied to the determination of Aconitum alkaloids in biofluids and root sample [96].

Strychnine is an alkaloid molecule found in the seeds of the plant of the *Strychnos* genus. Despite its high toxicity—strychnine intoxication is characterized by very dramatic and painful symptoms—this compound has some history of use for therapeutic purposes and has been also abused for doping purposes as stimulant. Wang et al. [97] and Duverneuil et al. [98] developed RP HPLC-DAD methods for the determination of this alkaloid in biological samples collected from fatal poisoning cases. SPE with isocration elution and LLE-gradient elution were adopted by the two groups, providing LODs in cadaveric blood of 0.5 and 0.06 ng/mL, respectively. Van Eenoo et al. [99] determined strychnine and the related compound brucine in urine by HPLC-APCI(+)-MS-MS. After LLE with ethyl acetate at pH 9.2, separation was carried out using a cyanopropyl column (100 mm × 4.6 mm × 3 μm eluted with an acetonitrile—1% acetic acid in water (v/v) gradient at a flow rate of 1.0 mL/min. The method (LOD, 0.5 ng/mL) was used to determine the excretion profile of strychnine after the ingestion of an over-the-counter herbal preparation of Strychnos nux-vomica, showing that its use by athletes can lead to a positive doping case.

Atropine and scopolamine, the main toxic alkaloids of *Datura stramonium* and *Datura ferox*, were detected and quantified using HPLC-ESI-MS both in Datura seeds and in the gastric content of a man whose death was ascribed to a fatal heart attack [100].

Ibogaine is a long-acting hallucinogen that occurs naturally in a number of dogbane plants, and particularly in Tabernanthe iboga, otherwise known as iboga. This alkaloid and the desmethyl metabolite were determined in plasma and cadaveric blood collected in a poisoning case involving iboga roots using HPLC-ESI-MS SIM mode by Kontrimaviciute et al. [101].

Pyrrolizidine alkaloids (PAs) and their N-oxides are contained in different plants throughout the world, including species used for the preparation of herbal teas. These compounds exhibit liver and lung toxicity and may cause a number of pathologies including acute liver failure, cirrhosis, and pneumonitis. PAs are also carcinogenic to animals. The quantity of PAs in some herbal teas and dietary supplements is sufficient to be carcinogenic in exposed individuals. HPLC-ESI-MS has been used for the analysis of these compounds in herbal remedies and in comfrey containing products [102,103].

Cardiac glycosides are a combination of an aglycone (genin) with from one to four sugar molecules. These drugs, sharing a powerful action on the myocardium, are found in a number of plants of the genera *Digitalis*, *Strophantus*, and others. Frommherz et al. [104] optimized an LC–MS method for the forensic analysis of the cardio glycosides digoxin and digitoxin in whole blood and tissues. After RP-8 chromatographic separation, detection of the Na-adducts of the analytes and the deuterated internal standard was achieved under ESI SIM mode. SPE clean-up enabled to limit ion suppression to less than 10% and to reach LLOQs of 0.2 and 2 ng/g for digoxin and digitoxin, respectively. The method was fully validated and successfully applied in forensic cases.

20.2.1.5 Toxins

Ricin is a potent cellular protein toxin contained in the beans of the castor been plant (*Ricinus communis*), which is extensively cultivated for oil production and is also a common ornamental garden plant. Ricin is able to inhibit ribosomal protein synthesis eventually causing cell death, and owing to these properties it has been allegedly used in terrorist and criminal activities. After trypsin digestion of castor bean crude extracts, Ostin et al. [105] were able to unambiguously identify diagnostic tryptic fragments of the toxin by LC-ESI-MS-MS. Separation was performed on a $0.3 \times 150\,mm \times 5\,\mu m$ PepMap300 capillary column using a two-step, $7\,\mu L/min$ linear gradient starting with a linear increase from 5% to 10% acetonitrile over 4 min, followed by an increase to 80% acetonitrile over 42 min, with 0.2% formic acid and 0.02% trifluoroacetic acid. An important advantage of the procedure is that digestion does not cleave the disulfide bonds in the protein, thus providing forensic evidence of the presence of intact toxin in the sample.

A number of HPLC methods for the determination of different mycotoxins (aflatoxins, ochratoxin) in biological samples as well as foods are available in the literature. Aflatoxins are produced by the fungi *Aspergillus flavus*, from which their name is derived. They are extremely toxic compounds with additional mutagenic and teratogenic properties. The mycotoxin ochratoxin A (7-carboxy-5-chloro-8-hydroxy-3,4-dihydro-3*R*-methylisocoumarin–β-phenylalanine, OTA) is produced by several species of the fungal genera *Penicillium* and Aspergillus. These mycotoxins have been determined in food and beverages by RP-HPLC with fluorimetric (FL) detection [106–108].

Different aflatoxins (AFB1, AFB2, AFG1, and AFG2) and two phase I oxidative metabolites of AFB1 (the most studied), AFM1, and AFP1, were extracted from urine/blood/serum using automated immunoaffinity SPE and separated on a $2.0 \times 150\,mm \times 3\,\mu m$ phenyl-hexyl column at 50°C under gradient conditions using a mobile phase consisting of H_2O and ACN both containing 0.1% formic acid [109]. Detection was achieved using positive ion ESI-MS–MS after optimizing individual compound specific parameters (i.e., declustering potentials, entrance potentials, and collision cell exit potentials).

A similar procedure has been proposed for the determination of ochratoxin A in foods [110].

20.2.1.6 Pesticides

Organophosphorus pesticides (OP) are a relatively frequent cause of accidental, homicidal and, more often, suicidal poisoning and death. Inoue et al. [111] have recently proposed a method for the rapid determination of different OP (acephate, methidathion, dichlorvos, fenthion, EPN, diazinon, phenthoate, malathion, fenitrothion, and cyanophos) in human serum by HPLC–MS for clinical and forensic purposes. The method involves ACN deproteinization and injection into a C_{18} column. Separation was carried out under gradient conditions using a 10 mM ammonium formate/methanol mobile phase. Detection was performed in APCI-MS SIM mode by monitoring either positive or negative ions, depending on the analyte. The method was validated by establishing linearity, intra- and interassay accuracy and precision, LODs (0.125–1 mg/L) and LOQs (0.25–1.25 mg/L), recoveries, and stabilities. However, matrix effects were not evaluated.

The quaternary ammonium compounds paraquat and diquat are widely used non-selective contact herbicides, which are extremely toxic to humans. Lee et al. [112] established an HPLC–MS-MS procedure for the determination of these herbicides in whole blood and urine using ethyl paraquat as internal standard. After extraction with Sep-Pak C18 cartridges, analytes were separated using ion pair chromatography with heptafluorobutyric acid in 20 mM ammonium acetate and acetonitrile gradient elution. Detection was carried out in ESI MS-MS SRM mode. Using similar separation and detection conditions, paraquat, diquat, difenzoquat, and a number of structurally related quaternary nitrogen muscle relaxants (see Section 20.2.1.3) were determined in whole blood by Ariffin and Anderson [113].

20.2.1.7 Doping Agents

Doping control is a challenging task for both analytical and interpretive reasons: (1) the extreme differentiation of doping substances (from small molecules to proteins); (2) the high sensitivity of detection required for many compounds that, being administered long before competition, are expected to be present in urine at the low microgram-per-liter level at the time of competition; (3) the short term often required to provide results, particularly during high-level sport events; and (4) the discrimination of doping from other possible reasons for positive, such as the use of drugs for recreational purposes or physiological/pathological alterations of endogenous steroids levels. Although gas chromatography–mass spectrometry (GC–MS), mainly owing to its robustness and high level of standardization, is currently still the standard technique in antidoping analysis, liquid chromatography–mass spectrometry (LC–MS) has a number of features that can be effectively exploited in sports drug testing or that offer good perspectives of application in the near future [114].

Doping substances include most of the drugs already discussed in the previous sections (e.g., drugs of abuse like amphetamines, cocaine, opiates; pharmaceutical drugs like diuretics, beta-blockers, etc.) and, in most cases, the described methods may be extended to doping control applications.

Steroids represent another typical class of doping substances. The determination of endogenous steroids (e.g., testosterone), synthetic anabolic steroids, their esters and their conjugates, and corticosteroids by HPLC–MS has been recently object of an exhaustive review [115] and will not be further discussed here.

A number of large molecules are included in the lists of banned substances because of their anabolic effects (human growth hormone, hGH; human corionic gonadotropin, hCG) or because their ability to increase oxygen transportation capacity of blood (e.g., erythropoietin, EPO). However, it must be said that the application of HPLC, in combination with MS, to the detection and identification of peptides is restricted by the very low limits of sensitivity that need to be reached. A typical strategy involves the digestion of the protein followed by separation and identification of specific fragments in the digests [115].

20.2.1.8 General Unknown Screening of Pharmacologically and Toxicologically Relevant Compounds

The methods illustrated in the previous paragraphs, which are aimed at the determination of specific compounds or drug classes, are typically applied when the presence of the drug in the sample is known/strongly suspected based on circumstantial/anamnestic information. As this information is not always available or reliable, forensic toxicology laboratories require also methods for the broad spectrum screening of pharmaco/toxicologically relevant compounds (PTRC). This obviously represents one of the hardest challenges for forensic toxicologists. The extremely wide range of PTRC—in terms of molecular weight, polarity, pK_a, chemical/thermal stability—on one side, and the high number and amount of interfering compounds in the substrate on the other make this goal similar to the search for the needle in the haystack.

In the past, PTRC screening was mainly based on gas chromatography-mass spectrometry (GC-MS) [116]. The choice of GC-MS was based on a number of good reasons (separation power of GC, selectivity of detection offered by MS, inherent simplicity of information contained in a mass spectrum, availability of a well established and standardized ionization technique, electron ionization, which allowed the construction of large databases of reference mass spectra, fast and reliable computer aided identification based on library search) that largely counterbalanced the pitfalls of GC separation, i.e., the need to isolate analytes from the aqueous substrate and to derivatize polar compounds [117].

As stated earlier, during the 1990s the advent of atmospheric pressure interfaces (API) brought about ruggedness in the coupling of HPLC to MS giving new impulse to research in PTRC screening. The much higher versatility of HPLC (suitable for conjugates and other polar analytes, and for thermally labile compounds; no need to hydrolyze or derivatize) indeed appeared to widen the range of compounds identifiable within a screening procedure [2]. Different research groups investigated a library-search-based HPLC–MS approach through the setup of PTRC spectral libraries [118–120]. However, some obstacles were found to hamper the development of broad-range procedures as previously had happened with GC-MS. First of all, API techniques provide soft ionization, accounting for little fragmentation of the pseudomolecular ion and poor structural information [121]. In addition, API full scan have been shown to be scarcely reproducable at interlaboratory level [122]. Although tuning procedures have been proposed [123], it is a fact that this problem has limited thus far the development of spectral databases as large as those used in GC-MS. Currently, HPLC–MS libraries contain spectra of up to 1200 unique compounds [124,125].

Recently, bench-top time-of-flight [23] mass spectrometers have become available enabling acquisition of ions at high mass resolution and providing mass accuracies of few ppm. This feature, combined with isotopic pattern (IP), allows unequivocal identification of the elemental composition of small ions (*m/z* 200–500) [126]. Studies carried out so far with HPLC-TOF have demonstrated the suitability of this approach for screening [127,128], its considerable advantage being that the build-up of mass spectral libraries is not necessary anymore [121,122]. It must be noted that this strategy enables the identification of a specific elemental composition, not of a specific compound, and that different compounds may have the same elemental composition. However, strategies have been recently proposed to tackle this problem through the application of metabolomics and chemometrics [130].

20.3 OTHER FORENSIC APPLICATIONS OF HPLC

20.3.1 EXPLOSIVES

The analysis of explosives has rather obvious forensic implications. HPLC applications have been published concerning both the analysis of postblast residues in the attempt to identify the explosive used, and the detection of explosive traces on the hands and clothing of a suspect in order to provide evidence in court.

Because of the thermal lability of many explosives, together with the high sensitivity requirements, particularly in postexplosion residue analysis, HPLC is considered as the separation method of choice in combination with MS detection [131].

As explosives belong to various chemical groups possessing quite different chemical and physical properties, their analysis by a single method is difficult. Tachon et al. [132] where able to separate 16 different explosives (i.e., nitrate esters, nitramines and nitroaromatics) using a porous graphitic carbon (PGC) Hypercarb column with a binary gradient elution (solvent A: mixture of 1 mM ammonium formate solution/ACN, 70:30 (v/v); solvent B: mixture of ACN and isopropanol, 40:60 v/v). The authors carried out MS acquisition in negative mode comparing ESI and APCI ionization and found the latter providing better sensitivity and allowing easy identification of investigated compounds with limits of detection ranging from 0.04 to 1.06 mg/L (see Figure 20.3).

The introduction of additives in the mobile phase may promote the formation of characteristic adduct ions, thereby improving sensitivity and allowing unambiguous identification of the

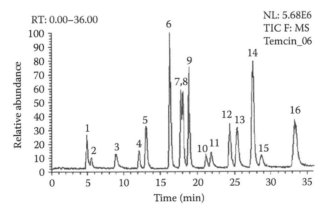

FIGURE 20.3 Total ion chromatogram from the HPLC-APCI-MS analysis of a mixture of mixture of nitrate ester, nitramine and nitroaromatic explosives and by-products (1 = 1,3,5-trinitro-1,3,5-triazacyclohexane; 2 = nitroglycerine; 3 = 1,3,5,7-tetranitro-1,3,5,7-tetraazacyclooctane; 4 = pentaerythritol tetranitrate; 5 = 1,2-dinitrobenzene; 6 = 2,4,6,*N* tetranitro-*N*-methylaniline; 7 = 2,6-dinitrotoluene; 8 = 3,4-dinitrotoluene; 9 = 2,3- dinitrotoluene; 10 = 1,3-dinitrobenzene; 11 = 1,4-dinitrobenzene; 12 = 2,5-dinitrotoluene; 13 = 2,4-dinitrotoluene; 14 = 2,4,6-trinitrotoluene; 15 = 3,5-dinitrotoluene; 16 = 1,3,5-trinitrobenzene). (Reprinted from Tachon, R. et al., *J. Chromatogr. A*, 1154, 174, 2007. Copyright 2007. With permission from Elsevier.)

explosive compounds. Zhao and Yinon obtained an enhanced response of nitramine and nitrate esters by adding ammonium nitrate post-column, thus forming $[M + NO_3]^-$ adduct ions in negative ion mode [133].

Peroxide explosives are potent explosives that can be made starting from common and easy to obtain raw materials. The analysis of triacetone triperoxide (TATP) and hexamethylenetriperoxidediamine (HMTD) was successfully carried out by HPLC-APCI-MS in a powder sample as well as in post-blast extracts originating from a forensic case [134]. After RP separation on a C18 column using a methanol:water (75:25 v/v) mobile phase containing ammonium acetate (2.5 mM) at a 0.4 mL/min flow rate, detection was carried out in positive ion mode. MS-MS analysis of $[TATP + NH_4]^+$ and $[HMTD - H]^+$ as precursor ions was necessary in order to achieve the required sensitivity in the analysis of postblast extracts (LOD 0.8 and 0.08 ng on column, respectively).

In addition to the type of explosive, the analysis may enable in some cases to identify the country of origin and even the manufacturer by examining the type and amount of byproducts, impurities and additives present in the material. Zhao and Yinon set up an HPLC-APCI-MS method for the analysis of the byproducts of industrial 2,4,6-trinitrotoluene (TNT), including isomers of TNT, dinitrotoluene, trinitrobenzene, and dinitrobenzene in order to build a profile for the characterization of TNT samples of different origins [135]. Separation was carried out in reversed-phase mode using a methanol/water mobile phase, and detection was carried out either in single-stage MS or MS-MS (confirmation only) in negative ion mode.

20.3.2 DYES AND INKS

Textile fibers found at a crime scene can be used as evidence linking a suspect to the victim. The typical situation is when a fiber matches with the clothing of the suspect. The mass production of textiles may make this a difficult task and the characterization of textile dyes may be very helpful in providing evidence that a close match does (or does not) exist between the fiber and the suspected source.

Huang et al. [136] applied LC-ESI-MS to the identification of dyes extracted from textile fibers and demonstrated the utility of the method for forensic trace analysis. LC–MS was shown to provide a high degree of chemical structural information, making dye identification highly specific in comparison to optical and/or chromatographic methods of dye analysis. By placing a UV-Vis detector in series before MS, the authors were able to monitor the elution of dyes in the presence of other nondye components extracted from textile fibers, thus permitting dye identification even when the pure standard is unavailable.

Petrick and Wilson [137] recently developed an HPLC method using both UV-Vis and ESI-MS detection able to separate and detect 15 basic dyes and 13 disperse dyes. Separation was carried out in RP-mode using an acidified (formic acid) water-acetonitrile mobile phase in gradient mode. The method enabled the discrimination of fibers with the same apparent colour based on their different chromatographic and mass spectrometric profile.

Chemical comparative analysis of writing inks may help in establishing if different documents have been written with a similar ink and even in dating a sign on a document. Ion-pairing HPLC using tetrabutylammonium bromide as ion-pairing reagent was used by Liu et al. [138] to classify and dating black gel pen ink entries on paper. The authors investigated compositional changes of the dye components in the black gel pen ink entries on paper under different aging conditions and found that the extent of decomposition of the dye components could be related to the aging time.

20.3.3 DNA ANALYSIS

DNA fingerprinting is currently based on the analysis of autosomal short tandem repeats (STRs). STRs are DNA segments that are typically found in noncoding regions of genome and are composed of repeating units of 3–5 nucleotides sequence patterns. Because of their high mutation rate,

STRs possess a high degree of polymorphism in humans, which makes STR typing a powerful approach for identity testing.

STR typing is generally carried out by selective amplification using the polymerase chain reaction technique (PCR) followed by indirect determination of the length of the amplified fragments (with 1 base-pair resolution) by capillary electrophoresis. The determination of the mass and sequence of a nucleotide fragment obviously provides much more information that may be used for identity testing, and LC–MS has proved to be able to unequivocally detect single sequence variations in DNA fragments with lengths up to 250 nucleotides. The first attempts to use ion-pairing reversed-phase HPLC coupled to ESI-MS for DNA fragments identification and typing date back to about 10 years ago [139,140]. Recently, Oberacher et al. [141] proved that HPLC-ESI-MS enables the simultaneous detection of nucleotide variability and measure of fragment length of forensically important STR markers. However, the technology is still far from routine application, mainly because of the difficulties in developing software tools for fast and reliable data interpretation.

20.3.4 Determination of Postmortem Interval

Time of death is almost invariably among the requests that detectives ask to forensic scientists. Nonetheless, this is a very difficult question to answer, at least with an acceptable precision.

Most methods are based on the measurement of physical and chemical changes that occur to a body after death. However, most of these changes are influenced by different variables (e.g., external temperature, physical activity immediately before death, etc.) that make the correlation between a measured variable and postmortem interval (PMI) rather inaccurate.

Among the methods used, the determination of selected of analytes in vitreous humor (and of potassium in particular, based on the observation that its concentration progressively increases in this substrate after death) has been often adopted in the attempt to reduce imprecision of the estimate. Recently, for example, Munoz et al. [142] have developed an HPLC method for the determination of hypoxanthine, another substance whose concentration has been found to increase after death in vitreous humor. Separation was carried out under RP conditions using a mobile phase of KH_2PO_4 0.05 M (pH 3) containing 1% (v/v) methanol at a flow rate of 1.5 mL/min. UV spectra were recorded in the range 200–400 nm. Based on the analysis of samples collected at different PMIs, the authors found that about 53% of the variation in the data is explained by PMI.

20.3.5 Determination of Age

The determination of age in both living and dead unidentified subjects is another application with relevant forensic implications. Besides morphological and radiological methods, the use of age related chemical changes that occur in amino acids and, in particular, racemization have been proposed. Amino acids of proteins that compose the living body are normally present in the L-form. However, with increasing age, a gradual transformation into the D-form (racemization) occurs through a chemical reaction that is influenced by a number of factors such as temperature, pH, humidity, etc.

After Helfman and Bada [143] found that a very good correlation exists between chronological age and racemization ratio of aspartic acid, as compared to the other amino acids, in dentin of teeth, a number of methods, mainly based on GC separation, have been published.

Recently, Yekkala et al. [144,145] have proposed an HPLC method with fluorescence detection. The method involves a rather laborious sample preparation due to the peculiar nature of the substrate involved (teeth), including pulverization, demineralization, hydrolysis of dentin and derivatization with o-phthaldialdehyde and N-acetyl-L-cysteine in order to obtain the enantioseparation of aspartic acid. Using a similar procedure, Benesova et al. [146] found that, in comparison with GC, HPLC provides shorter analysis time and higher sensitivity.

Currently, the determination of racemization of aspartic acid in dentin is considered as one of the most accurate methods for the estimation of age.

20.4 CONCLUSIONS AND PERSPECTIVES

After the introduction into the market of rugged HPLC–MS interfaces the number of HPLC applications in forensic science has progressively increased. The high versatility of HPLC separation, as compared, for example, to capillary GC, and the selectivity of MS detection have certainly made of HPLC–MS a good supplement, if not an effective alternative, to GC-MS. Small and large, stable and unstable, little or highly polar compounds can be efficiently separated and detected using different chromatographic/mass spectrometric instrumental configurations.

The compatibility of the chromatographic system with aqueous biological fluids (urine, serum), and the direct analysis of highly polar compounds with no need for hydrolysis and derivatization allow to reduce sample manipulation and the probability of artifact formation/analyte degradation/contamination. In addition, the possibility of carrying out separate LC and MS experiments accounts for rapid and cost-effective method development. These features are highly considered in the forensic field where often the analyst is requested to deal with the setup of a method for an unusual analyte or substrate.

However, some obstacles still have to be removed before HPLC–MS is fully accepted as a routine technique in the forensic labs. First of all the cost of equipment is still high, despite the reduction trend in the last years, as compared to GC-MS. Second, some problems have to be tackled, i.e., the susceptibility of ion sources (and particularly ESI) to matrix effects on analyte's ionization efficiency (suppression or enhancement) and the scarce reproducibility of MS fragmentation. The third obstacle is actually the other side of the coin of versatility: the wide choice of technical alternatives makes HPLC–MS still far from being a highly standardized, one-button technique as GC-MS.

REFERENCES

1. Maurer, H. H. *Anal Bioanal Chem* **2007**, *388*, 1315–1325.
2. Polettini, A. *Applications of LC-MS in Toxicology*, Pharmaceutical Press: London, U.K., 2006.
3. Wood, M.; Laloup, M.; Samyn, N.; Del Mar Ramirez Fernandez, M.; de Bruijn, E. A.; Maes, R. A.; De Boeck, G. *J Chromatogr A* **2006**, *1130*, 3–15.
4. Bortolotti, F.; De Paoli, G.; Pascali, J. P.; Trevisan, M. T.; Floreani, M.; Tagliaro, F. *Clin Chem* **2005**, *51*, 2368–2371.
5. Alden, A.; Ohlson, S.; Pahlsson, P.; Ryden, I. *Clin Chim Acta* **2005**, *356*, 143–146.
6. Helander, A.; Husa, A.; Jeppsson, J. O. *Clin Chem* **2003**, *49*, 1881–1890.
7. Morini, L.; Politi, L.; Zucchella, A.; Polettini, A. *Clin Chim Acta* **2007**, *376*, 213–219.
8. Morini, L.; Politi, L.; Groppi, A.; Stramesi, C.; Polettini, A. *J Mass Spectrom* **2006**, *41*, 34–42.
9. Politi, L.; Morini, L.; Groppi, A.; Poloni, V.; Pozzi, F.; Polettini, A. *Rapid Commun Mass Spectrom* **2005**, *19*, 1321–1331.
10. Politi, L.; Morini, L.; Leone, F.; Polettini, A. *Addiction* **2006**, *101*, 1408–1412.
11. Bicker, W.; Lammerhofer, M.; Keller, T.; Schuhmacher, R.; Krska, R.; Lindner, W. *Anal Chem* **2006**, *78*, 5884–5892.
12. Dresen, S.; Weinmann, W.; Wurst, F. M. *J Am Soc Mass Spectrom* **2004**, *15*, 1644–1648.
13. Weinmann, W.; Schaefer, P.; Thierauf, A.; Schreiber, A.; Wurst, F. M. *J Am Soc Mass Spectrom* **2004**, *15*, 188–193.
14. Mountfort, K. A.; Bronstein, H.; Archer, N.; Jickells, S. M. *Anal Chem* **2007**, *79*, 2650–2657.
15. Tagliaro, F.; Bortolotti, F.; Pascali, J. P. *Anal Bioanal Chem* **2007**, *388*, 1359–1364.
16. Jeppsson, J. O.; Kristensson, H.; Fimiani, C. *Clin Chem* **1993**, *39*, 2115–2120.
17. Hoiseth, G.; Karinen, R.; Christophersen, A. S.; Olsen, L.; Normann, P. T.; Morland, J. *Forensic Sci Int* **2007**, *165*, 41–45.
18. Dams, R.; Murphy, C. M.; Choo, R. E.; Lambert, W. E.; De Leenheer, A. P.; Huestis, M. A. *Anal Chem* **2003**, *75*, 798–804.

19. Dams, R.; Benijts, T.; Gunther, W.; Lambert, W.; De Leenheer, A. *Anal Chem* **2002**, *74*, 3206–3212.
20. Dupuis, C.; Gaulier, J. M.; Pelissier-Alicot, A. L.; Marquet, P.; Lachatre, G. *J Anal Toxicol* **2004**, *28*, 674–679.
21. Zelkowicz, A.; Magora, A.; Ravreby, M. D.; Levy, R. *J Forensic Sci* **2005**, *50*, 849–852.
22. Lurie, I. S. *J Chromatogr A* **2005**, *1100*, 168–175.
23. Kristoffersen, L.; Oiestad, E. L.; Opdal, M. S.; Krogh, M.; Lundanes, E.; Christophersen, A. S. *J Chromatogr B Analyt Technol Biomed Life Sci* **2007**, *850*, 147–160.
24. Laks, S.; Pelander, A.; Vuori, E.; Ali-Tolppa, E.; Sippola, E.; Ojanpera, I. *Anal Chem* **2004**, *76*, 7375–7379.
25. Kamata, T.; Nishikawa, M.; Katagi, M.; Tsuchihashi, H. *J Forensic Sci* **2005**, *50*, 336–340.
26. Johansen, S. S.; Bhatia, H. M. *J Chromatogr B Analyt Technol Biomed Life Sci* **2007**, *852*, 338–344.
27. Fernandez, P.; Morales, L.; Vazquez, C.; Bermejo, A. M.; Tabernero, M. J. *Forensic Sci Int* **2006**, *161*, 31–35.
28. Musshoff, F.; Trafkowski, J.; Kuepper, U.; Madea, B. *J Mass Spectrom* **2006**, *41*, 633–640.
29. Hupka, Y.; Beike, J.; Roegener, J.; Brinkmann, B.; Blaschke, G.; Kohler, H. *Int J Legal Med* **2005**, *119*, 121–128.
30. Polettini, A.; Huestis, M. A. *J Chromatogr B Biomed Sci Appl* **2001**, *754*, 447–459.
31. Kronstrand, R.; Selden, T. G.; Josefsson, M. *J Anal Toxicol* **2003**, *27*, 464–470.
32. Hegstad, S.; Khiabani, H. Z.; Oiestad, E. L.; Berg, T.; Christophersen, A. S. *J Anal Toxicol* **2007**, *31*, 214–219.
33. Favretto, D.; Frison, G.; Vogliardi, S.; Ferrara, S. D. *Rapid Commun Mass Spectrom* **2006**, *20*, 1257–1265.
34. Kelly, T.; Doble, P.; Dawson, M. *J Chromatogr B Analyt Technol Biomed Life Sci* **2005**, *814*, 315–323.
35. Musshoff, F.; Madea, B.; Stuber, F.; Stamer, U. M. *J Anal Toxicol* **2006**, *30*, 463–467.
36. Coopman, V.; Cordonnier, J.; Pien, K.; Van Varenbergh, D. *Forensic Sci Int* **2007**, *169*, 223–227.
37. Soares, M. E.; Carvalho, M.; Carmo, H.; Remiao, F.; Carvalho, F.; Bastos, M. L. *Biomed Chromatogr* **2004**, *18*, 125–131.
38. Kumihashi, M.; Ameno, K.; Shibayama, T.; Suga, K.; Miyauchi, H.; Jamal, M.; Wang, W.; Uekita, I.; Ijiri, I. *J Chromatogr B Analyt Technol Biomed Life Sci* **2007**, *845*, 180–183.
39. da Costa, J. L.; da Matta Chasin, A. A. *J Chromatogr B Analyt Technol Biomed Life Sci* **2004**, *811*, 41–45.
40. Concheiro, M.; de Castro, A.; Quintela, O.; Lopez-Rivadulla, M.; Cruz, A. *Forensic Sci Int* **2005**, *150*, 221–226.
41. Wood, M.; De Boeck, G.; Samyn, N.; Morris, M.; Cooper, D. P.; Maes, R. A.; De Bruijn, E. A. *J Anal Toxicol* **2003**, *27*, 78–87.
42. Cheng, W. C.; Mok, V. K.; Chan, K. K.; Li, A. F. *Forensic Sci Int* **2007**, *166*, 1–7.
43. Concheiro, M.; Simoes, S. M.; Quintela, O.; de Castro, A.; Dias, M. J.; Cruz, A.; Lopez-Rivadulla, M. *Forensic Sci Int* **2007**, *171*, 44–51.
44. Stanaszek, R.; Piekoszewski, W. *J Anal Toxicol* **2004**, *28*, 77–85.
45. Miki, A.; Katagi, M.; Tsuchihashi, H. *J Anal Toxicol* **2003**, *27*, 95–102.
46. Cheze, M.; Deveaux, M.; Martin, C.; Lhermitte, M.; Pepin, G. *Forensic Sci Int* **2007**, *170*, 100–104.
47. Cheze, M.; Duffort, G.; Deveaux, M.; Pepin, G. *Forensic Sci Int* **2005**, *153*, 3–10.
48. Kronstrand, R.; Nystrom, I.; Strandberg, J.; Druid, H. *Forensic Sci Int* **2004**, *145*, 183–190.
49. Cheng, J. Y.; Mok, V. K. *Forensic Sci Int* **2004**, *142*, 9–15.
50. Nelson, C. C.; Foltz, R. L. *J Chromatogr* **1992**, *580*, 97–109.
51. Horn, C. K.; Klette, K. L.; Stout, P. R. *J Anal Toxicol* **2003**, *27*, 459–463.
52. Johansen, S. S.; Jensen, J. L. *J Chromatogr B Analyt Technol Biomed Life Sci* **2005**, *825*, 21–28.
53. Favretto, D.; Frison, G.; Maietti, S.; Ferrara, S. D. *Int J Legal Med* **2007**, *121*, 259–265.
54. Maralikova, B.; Weinmann, W. *J Mass Spectrom* **2004**, *39*, 526–531.
55. Concheiro, M.; de Castro, A.; Quintela, O.; Cruz, A.; Lopez-Rivadulla, M. *J Chromatogr B Analyt Technol Biomed Life Sci* **2004**, *810*, 319–324.
56. Laloup, M.; Del Mar Ramirez Fernandez, M.; Wood, M.; De Boeck, G.; Maes, V.; Samyn, N. *Forensic Sci Int* **2006**, *161*, 175–179.
57. Jung, J.; Kempf, J.; Mahler, H.; Weinmann, W. *J Mass Spectrom* **2007**, *42*, 354–360.
58. Quintela, O.; Andrenyak, D. M.; Hoggan, A. M.; Crouch, D. J. *J Anal Toxicol* **2007**, *31*, 157–164.
59. Stoll, D. R.; Paek, C.; Carr, P. W. *J Chromatogr A* **2006**, *1137*, 153–162.
60. Porter, S. E.; Stoll, D. R.; Paek, C.; Rutan, S. C.; Carr, P. W. *J Chromatogr A* **2006**, *1137*, 163–172.
61. Bugey, A.; Staub, C. *J Pharm Biomed Anal* **2004**, *35*, 555–562.

62. Rivera, H. M.; Walker, G. S.; Sims, D. N.; Stockham, P. C. *Eur J Mass Spectrom (Chichester, Eng)* **2003**, *9*, 599–607.
63. Kratzsch, C.; Tenberken, O.; Peters, F. T.; Weber, A. A.; Kraemer, T.; Maurer, H. H. *J Mass Spectrom* **2004**, *39*, 856–872.
64. Smink, B. E.; Brandsma, J. E.; Dijkhuizen, A.; Lusthof, K. J.; de Gier, J. J.; Egberts, A. C.; Uges, D. R. *J Chromatogr B Analyt Technol Biomed Life Sci* **2004**, *811*, 13–20.
65. ElSohly, M. A.; Gul, W.; ElSohly, K. M.; Avula, B.; Khan, I. A. *J Anal Toxicol* **2006**, *30*, 524–538.
66. Laloup, M.; Ramirez Fernandez Mdel, M.; De Boeck, G.; Wood, M.; Maes, V.; Samyn, N. *J Anal Toxicol* **2005**, *29*, 616–626.
67. Bugey, A.; Rudaz, S.; Staub, C. *J Chromatogr B Analyt Technol Biomed Life Sci* **2006**, *832*, 249–255.
68. Kintz, P.; Villain, M.; Concheiro, M.; Cirimele, V. *Forensic Sci Int* **2005**, *150*, 213–220.
69. Quintela, O.; Cruz, A.; Castro, A.; Concheiro, M.; Lopez-Rivadulla, M. *J Chromatogr B Analyt Technol Biomed Life Sci* **2005**, *825*, 63–71.
70. Hegstad, S.; Oiestad, E. L.; Johansen, U.; Christophersen, A. S. *J Anal Toxicol* **2006**, *30*, 31–37.
71. Miller, E. I.; Wylie, F. M.; Oliver, J. S. *J Anal Toxicol* **2006**, *30*, 441–448.
72. Quintela, O.; Sauvage, F. L.; Charvier, F.; Gaulier, J. M.; Lachatre, G.; Marquet, P. *Clin Chem* **2006**, *52*, 1346–1355.
73. Goeringer, K. E.; McIntyre, M.; Drummer, O. H. *J Anal Toxicol* **2003**, *27*, 30–35.
74. Shinozuka, T.; Terada, M.; Tanaka, E. *Forensic Sci Int* **2006**, *162*, 108–112.
75. de Castro, A.; Fernandez Mdel, M.; Laloup, M.; Samyn, N.; De Boeck, G.; Wood, M.; Maes, V.; Lopez-Rivadulla, M. *J Chromatogr A* **2007**, *1160*, 3–12.
76. Titier, K.; Castaing, N.; Le-Deodic, M.; Le-Bars, D.; Moore, N.; Molimard, M. *J Anal Toxicol* **2007**, *31*, 200–207.
77. Josefsson, M.; Kronstrand, R.; Andersson, J.; Roman, M. *J Chromatogr B Analyt Technol Biomed Life Sci* **2003**, *789*, 151–167.
78. Delamoye, M.; Duverneuil, C.; Paraire, F.; de Mazancourt, P.; Alvarez, J. C. *Forensic Sci Int* **2004**, *141*, 23–31.
79. Johnson, R. D.; Lewis, R. J. *Forensic Sci Int* **2006**, *156*, 106–117.
80. Cirimele, V.; Villain, M.; Pepin, G.; Ludes, B.; Kintz, P. *J Chromatogr B Analyt Technol Biomed Life Sci* **2003**, *789*, 107–113.
81. Usui, K.; Hishinuma, T.; Yamaguchi, H.; Saga, T.; Wagatsuma, T.; Hoshi, K.; Tachiiri, N.; Miura, K.; Goto, J. *Leg Med (Tokyo)* **2006**, *8*, 166–171.
82. Ferrara, S. D.; Nalesso, A.; Castagna, F.; Montisci, M.; Vogliardi, S.; Favretto, D. *Rapid Commun Mass Spectrom* **2007**, *21*, 2944–2950.
83. Yonemitsu, K.; Koreeda, A.; Kibayashi, K.; Ng'walali, P.; Mbonde, M.; Kitinya, J.; Tsunenari, S. *Leg Med (Tokyo)* **2005**, *7*, 113–116.
84. Abe, E.; Delamoye, M.; Mathieu, B.; Durigon, M.; de Mazancourt, P.; Advenier, C.; Alvarez, J. C. *J Anal Toxicol* **2004**, *28*, 118–121.
85. Proenca, P.; Pinho Marques, E.; Teixeira, H.; Castanheira, F.; Barroso, M.; Avila, S.; Vieira, D. N. *Forensic Sci Int* **2003**, *133*, 95–100.
86. Benijts, T.; Borrey, D.; Lambert, W. E.; De Letter, E. A.; Piette, M. H.; Van Peteghem, C.; De Leenheer, A. P. *J Anal Toxicol* **2003**, *27*, 47–52.
87. Proenca, P.; Teixeira, H.; Castanheira, F.; Pinheiro, J.; Monsanto, P. V.; Marques, E. P.; Vieira, D. N. *Forensic Sci Int* **2005**, *153*, 75–80.
88. Johansen, S. S.; Jensen, J. L. *J Chromatogr B Analyt Technol Biomed Life Sci* **2004**, *811*, 31–36.
89. Moore, K. A.; Levine, B.; Titus, J. M.; Fowler, D. R. *J Anal Toxicol* **2003**, *27*, 592–594.
90. Romhild, W.; Krause, D.; Bartels, H.; Ghanem, A.; Schoning, R.; Wittig, H. *Forensic Sci Int* **2003**, *133*, 101–106.
91. Wang, Z.; Wen, J.; Xing, J.; He, Y. *J Pharm Biomed Anal* **2006**, *40*, 1031–1034.
92. Wang, Z. H.; Guo, D.; He, Y.; Hu, C.; Zhang, J. *Phytochem Anal* **2004**, *15*, 16–20.
93. Hayashida, M.; Hayakawa, H.; Wada, K.; Yamada, T.; Nihira, M.; Ohno, Y. *Leg Med (Tokyo)* **2003**, *5(Suppl 1)*, S101–S104.
94. Beike, J.; Frommherz, L.; Wood, M.; Brinkmann, B.; Kohler, H. *Int J Legal Med* **2004**, *118*, 289–293.
95. Wang, Z. H.; Wang, Z.; Wen, J.; He, Y. *J Pharm Biomed Anal* **2007**.
96. Kaneko, R.; Hattori, S.; Furuta, S.; Hamajima, M.; Hirata, Y.; Watanabe, K.; Seno, H.; Ishii, A. *J Mass Spectrom* **2006**, *41*, 810–814.
97. Wang, Z.; Zhao, J.; Xing, J.; He, Y.; Guo, D. *J Anal Toxicol* **2004**, *28*, 141–144.

98. Duverneuil, C.; de la Grandmaison, G. L.; de Mazancourt, P.; Alvarez, J. C. *Forensic Sci Int* **2004**, *141*, 17–21.
99. Van Eenoo, P.; Deventer, K.; Roels, K.; Delbeke, F. T. *Forensic Sci Int* **2006**, *164*, 159–163.
100. Steenkamp, P. A.; Harding, N. M.; van Heerden, F. R.; van Wyk, B. E. *Forensic Sci Int* **2004**, *145*, 31–39.
101. Kontrimaviciute, V.; Breton, H.; Mathieu, O.; Mathieu-Daude, J. C.; Bressolle, F. M. *J Chromatogr B Analyt Technol Biomed Life Sci* **2006**, *843*, 131–141.
102. Altamirano, J. C.; Gratz, S. R.; Wolnik, K. A. *J AOAC Int* **2005**, *88*, 406–412.
103. Wuilloud, J. C.; Gratze, S. R.; Gamble, B. M.; Wolnik, K. A. *Analyst* **2004**, *129*, 150–156.
104. Frommherz, L.; Kohler, H.; Brinkmann, B.; Lehr, M.; Beike, J. *Int J Legal Med* **2007**.
105. Ostin, A.; Bergstrom, T.; Fredriksson, S. A.; Nilsson, C. *Anal Chem* **2007**.
106. Bognanno, M.; La Fauci, L.; Ritieni, A.; Tafuri, A.; De Lorenzo, A.; Micari, P.; Di Renzo, L.; Ciappellano, S.; Sarullo, V.; Galvano, F. *Mol Nutr Food Res* **2006**, *50*, 300–305.
107. Aresta, A.; Palmisano, F.; Vatinno, R.; Zambonin, C. G. *J Agric Food Chem* **2006**, *54*, 1594–1598.
108. Monaci, L.; Tantillo, G.; Palmisano, F. *Anal Bioanal Chem* **2004**, *378*, 1777–1782.
109. Everley, R. A.; Ciner, F. L.; Zhan, D.; Scholl, P. F.; Groopman, J. D.; Croley, T. R. *J Anal Toxicol* **2007**, *31*, 150–156.
110. Lindenmeier, M.; Schieberle, P.; Rychlik, M. *J Chromatogr A* **2004**, *1023*, 57–66.
111. Inoue, S.; Saito, T.; Mase, H.; Suzuki, Y.; Takazawa, K.; Yamamoto, I.; Inokuchi, S. *J Pharm Biomed Anal* **2007**, *44*, 258–264.
112. Lee, X. P.; Kumazawa, T.; Fujishiro, M.; Hasegawa, C.; Arinobu, T.; Seno, H.; Ishii, A.; Sato, K. *J Mass Spectrom* **2004**, *39*, 1147–1152.
113. Ariffin, M. M.; Anderson, R. A. *J Chromatogr B Analyt Technol Biomed Life Sci* **2006**, *842*, 91–97.
114. Politi, L.; Groppi, A.; Polettini, A. *J Anal Toxicol* **2005**, *29*, 1–14.
115. Thieme, D. In *Applications of LC-MS in Toxicology*; Polettini, A., Ed.; Pharamceutical Press: London, U.K., 2006, pp. 193–215.
116. Polettini, A. *J Chromatogr B Biomed Sci Appl* **1999**, *733*, 47–63.
117. Polettini, A.; Groppi, A.; Vignali, C.; Montagna, M. *J Chromatogr B Biomed Sci Appl* **1998**, *713*, 265–279.
118. Weinmann, W.; Wiedemann, A.; Eppinger, B.; Renz, M.; Svoboda, M. *J Am Soc Mass Spectrom* **1999**, *10*, 1028–1037.
119. Marquet, P. *Ther Drug Monit* **2002**, *24*, 125–133.
120. Gergov, M.; Ojanpera, I.; Vuori, E. *J Chromatogr B Analyt Technol Biomed Life Sci* **2003**, *795*, 41–53.
121. Politi, A., Groppi, A., Polettini, A. In *Applications of LC-MS in Toxicology*; Polettini, A., Ed.; Pharmaceutical Press: London, U.K., 2006, pp. 1–22.
122. Jansen, R.; Lachatre, G.; Marquet, P. *Clin Biochem* **2005**, *38*, 362–372.
123. Weinmann, W.; Stoertzel, M.; Vogt, S.; Wendt, J. *J Chromatogr A* **2001**, *926*, 199–209.
124. Sauvage, F. L.; Saint-Marcoux, F.; Duretz, B.; Deporte, D.; Lachatre, G.; Marquet, P. *Clin Chem* **2006**, *52*, 1735–1742.
125. Dresen, S.; Kempf, J.; Weinmann, W. *Forensic Sci Int* **2006**, *161*, 86–91.
126. Pelzing, M.; Neusüß, C.; Macht, M. *LC-GC Europe* **2004**, *17*, 38–39.
127. Ojanpera, S.; Pelander, A.; Pelzing, M.; Krebs, I.; Vuori, E.; Ojanpera, I. *Rapid Commun Mass Spectrom* **2006**, *20*, 1161–1167.
128. Kolmonen, M.; Leinonen, A.; Pelander, A.; Ojanpera, I. *Anal Chim Acta* **2007**, *585*, 94–102.
129. Polettini, A.; Gottardo, R.; Pascali, J. P.; Tagliaro, F. *Anal. Chem.* 80, 3050–3057, 2008.
130. Lotta, E.; Gottardo, R.; Bertaso, A.; Polettini, A. *J. Mass Spectrom.* 45, 261–271, 2009.
131. Yinon, J. In *LC-MS Applications in Toxicology*; Polettini, A., Ed.; Pharmaceutical Press: London, U.K., 2006, pp. 257–269.
132. Tachon, R.; Pichon, V.; Le Borgne, M. B.; Minet, J. J. *J Chromatogr A* **2007**, *1154*, 174–181.
133. Zhao, X.; Yinon, J. *J Chromatogr A* **2002**, *977*, 59–68.
134. Xu, X.; van de Craats, A. M.; Kok, E. M.; de Bruyn, P. C. *J Forensic Sci* **2004**, *49*, 1230–1236.
135. Zhao, X.; Yinon, J. *J Chromatogr A* **2002**, *946*, 125–132.
136. Huang, M.; Yinon, J.; Sigman, M. E. *J Forensic Sci* **2004**, *49*, 238–249.
137. Petrick, L. M.; Wilson, T. A.; Ronald Fawcett, W. *J Forensic Sci* **2006**, *51*, 771–779.
138. Liu, Y. Z.; Yu, J.; Xie, M. X.; Liu, Y.; Han, J.; Jing, T. T. *J Chromatogr A* **2006**, *1135*, 57–64.
139. Laken, S. J.; Jackson, P. E.; Kinzler, K. W.; Vogelstein, B.; Strickland, P. T.; Groopman, J. D.; Friesen, M. D. *Nat Biotechnol* **1998**, *16*, 1352–1356.

140. Oberacher, H.; Parson, W.; Muhlmann, R.; Huber, C. G. *Anal Chem* **2001**, *73*, 5109–5115.
141. Oberacher, H.; Pitterl, F.; Huber, G.; Niederstatter, H.; Steinlechner, M.; Parson, W. *Hum Mutat* **2007**.
142. Munoz, J. I.; Costas, E.; Rodriguez-Calvo, M. S.; Suarez-Penaranda, J. M.; Lopez-Rivadulla, M.; Concheiro, L. *Hum Exp Toxicol* **2006**, *25*, 279–281.
143. Helfman, P. M.; Bada, J. L. *Nature* **1976**, *262*, 279–281.
144. Yekkala, R.; Meers, C.; Van Schepdael, A.; Hoogmartens, J.; Lambrichts, I.; Willems, G. *Forensic Sci Int* **2006**, *159(Suppl 1)*, S89–S94.
145. Yekkala, R.; Meers, C.; Hoogmartens, J.; Lambrichts, I.; Willems, G.; Van Schepdael, A. *J Sep Sci* **2007**, *30*, 118–121.
146. Benesova, T.; Ales, H.; Pilin, A.; Votruba, J.; Flieger, M. *J Sep Sci* **2004**, *27*, 330–334.

Index

C